Standard Graphs of Functions

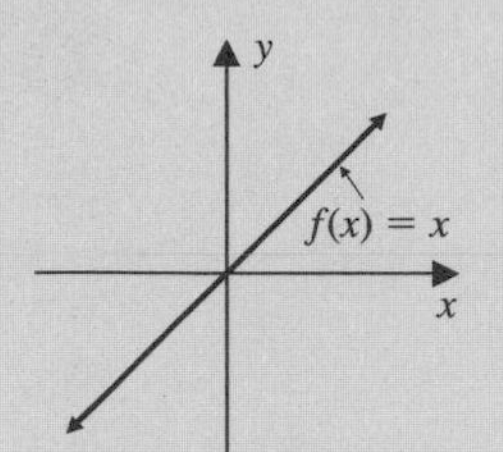

1. **Identity function**

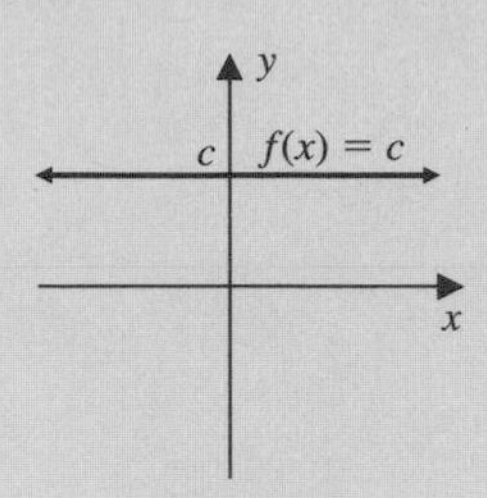

2. **Constant function**
(*c* is a constant)

3. **Squaring function**

4. **Square root function**

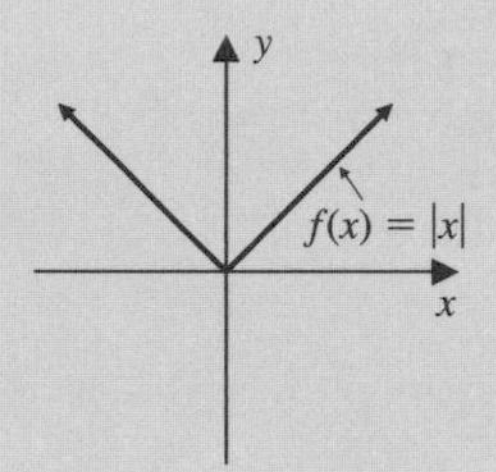

5. **Absolute value function**

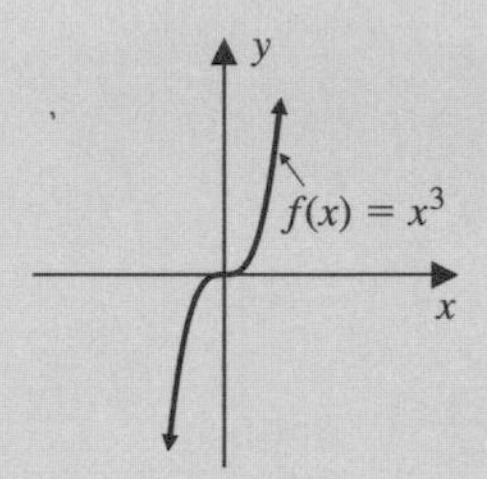

6. **Cubing function**

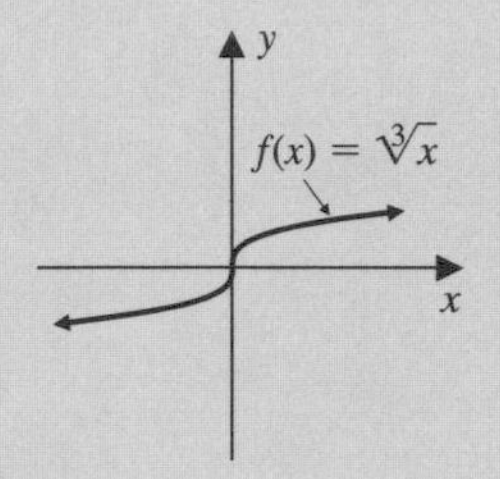

7. **Cube root function**

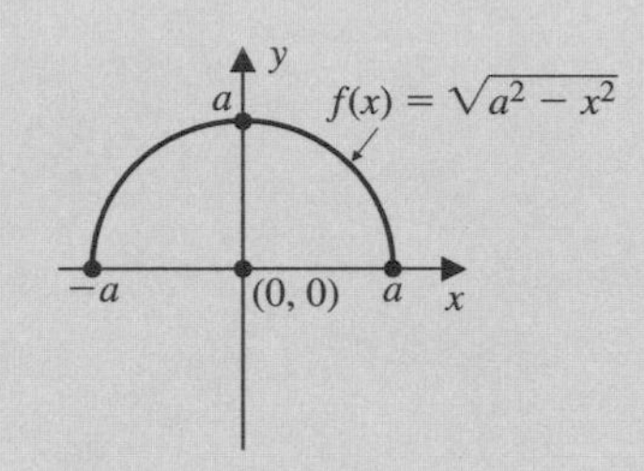

8. **Upper half-circle function**
(center: (0, 0); radius *a*)

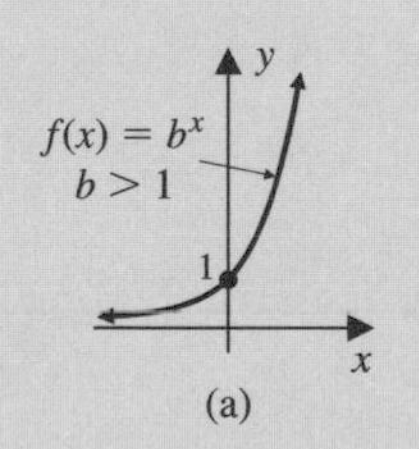

(a)

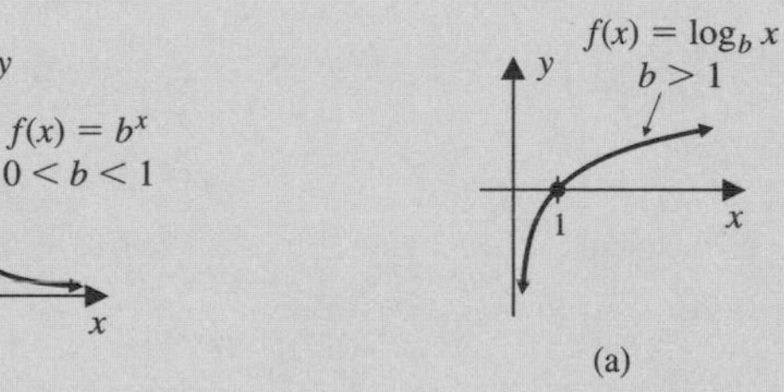

(b)

9. **Exponential functions**

(a)

(b)

10. **Logarithmic functions**

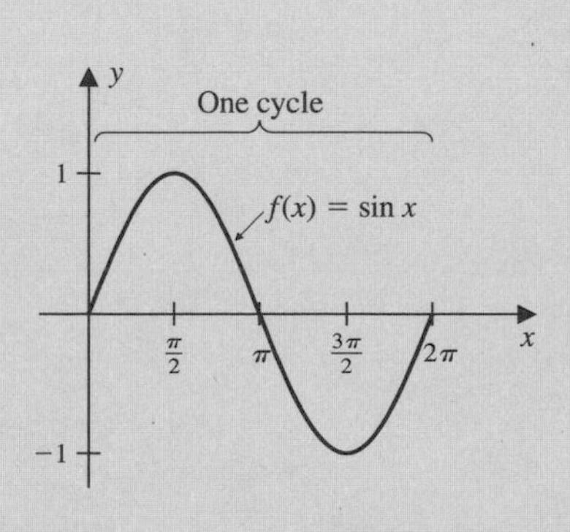

(a)

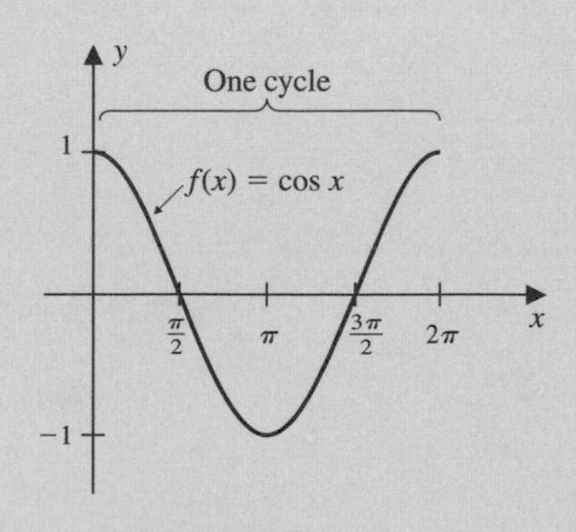

(b)

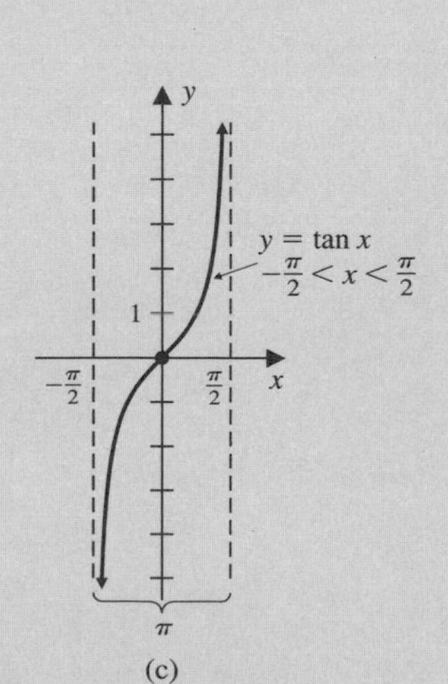

(c)

11. **Trigonometric functions**

Precalculus
FUNCTIONS AND GRAPHS

SIXTH EDITION

Precalculus

FUNCTIONS AND GRAPHS

M. A. Munem

J. P. Yizze

Macomb College

WORTH PUBLISHERS

Precalculus: Functions and Graphs, Sixth Edition

Printed in the United States of America

Library of Congress Catalog Card Number: 96-061159

ISBN: 1–57259–157–9

Printing: 1 2 3 4 5—00 99 98 97

Production: Phyllis Niklas

Design: Janet Bollow

Line art: Scientific Illustrators

Printing and binding: R.R. Donnelley & Sons

Cover: *Bathing at Long Beach, California,* c. 1900. Photo by Henry G. Peabody

Worth Publishers

33 Irving Place

New York, New York 10003

Contents

Preface

Since our last edition, there has been much discussion about reform in the precalculus curriculum and in methods of teaching the subject. In response, the sixth edition is the most extensive revision since the first edition was published twenty-six years ago. To produce a flexible text that contains all the appropriate elements, we have given considerable attention to the development of mathematical models, to the value of collaborative learning, and to the proper integration of modern technology. We also have taken into account market surveys, reviews, and thoughtful comments and suggestions from colleagues and users of previous editions. But most important is that *we have tested our ideas in the classroom.*

In today's world, each teacher of precalculus has a particular approach to help students understand the mathematics and to prepare them for further study. Some believe that more emphasis should be put on the use of technology as a way of accomplishing these goals. Others embrace the traditional approach and view the use of technology with great skepticism. Our philosophy is that technology should be used as an additional tool to enhance the teaching and learning of precalculus. In writing this edition, our goal remains the same—*to establish effective strategies to help students learn mathematics at the precalculus level.*

Features

This book presents precalculus by using graphical, numerical, and analytical points of view. Several features of this edition contribute to its effectiveness.

RELATING IDEAS: Special attention has been paid to pointing out the *connection* between the *algebraic* and *geometric* interpretations of important concepts.

GRAPHS: Sketching graphs by hand-drawing and equation recognition is a fundamental part of learning mathematics at this level. The text uses graphing tools to supplement and extend this process. The ability to generate accurate graphs efficiently increases the use of visualization and thus helps to convey ideas along with word explanations. *Interpreting graphs, exploring new ideas* from graphs, and *extracting information* from graphs are of primary importance in this book.

USE OF TECHNOLOGY: In this text, we use the term *graphing utility* to refer to *graphing calculators and computer software.* We regard graphing utilities as instructional tools, and we integrate their use in the exposition of graphical and numerical viewpoints. This book exploits the capabilities of these tools to help students comprehend the main ideas of a precalculus course. Graphing utilities are used to facilitate graphing equations, to provide a modern approach to introducing new material, to explore mathematical concepts, and to reinforce and demonstrate the validity of analytical results. In addition, we rely on graphing utilities to solve equations and to help interpret real-world situations when analytic methods are impractical or impossible. This text does not assume the availability of any particular graphing utility. Rather, it adopts a language appropriate for the various kinds of software packages and graphing calculators generally used today. For instance, the terms "ZOOM" and "TRACE" on the graph of a function suggest a way of using a graphing utility to explore powerful ideas without the need for listing specific instructions for particular calculators or software. The many examples and problems in this book that require the use of a graphing utility are identified by the symbols GU and GU.

VARIED VIEWPOINTS: In keeping with modern reform, we often require students to manipulate and compare graphical, numerical, and analytical representations of the same concept. This approach is especially effective in helping students to understand topics such as limit and asymptotic behaviors. We have included many examples and problems that promote such *visual thinking.* Such an approach does not always require the use of technology, but its availability supports the students' numerical and geometric intuition.

APPLICATIONS AND MODELS: In this edition, we have placed emphasis on real-world applied problems drawn from a variety of disciplines. They are integrated into the material throughout the book. We believe that we offer *an approach to mathematical modeling that will capture the students' attention.* Based on our experiences in classroom teaching, we have selected certain types of problems. We have organized their solutions through a sequence of questions that help students through the difficult thought processes of developing mathematical models and using them to interpret real-world situations. Students have many opportunities to put function concepts and graphs to use in a variety of lifelike situations. Solving these problems enables students to practice their skills and acquire insights needed in calculus and other fields.

PROBLEM SETS: All problem sets have been reorganized and rewritten. They have been checked and rechecked for accuracy by professors who class-tested this edition of the book.

The problems in each problem set have been grouped into three categories:

1. The first category is designed to provide the practice needed to *master the concepts* in the book. They are generally modeled after the examples and follow the order of presentation of the book.

2. The second category provides a broad range of applications and models that require students to *apply the concepts.* Special emphasis has been put on the design of these problems as indicated above.

3. The third category challenges students to *develop and extend the concepts.* These problems encourage *critical thinking,* and provide an excellent opportunity to stimulate class discussions that foster *collaborative learning* among students.

In any case, the problems in the last two categories offer the opportunity for writing assignments such as a report form for solving an applied problem or a class project.

MARGIN NOTES: Throughout the book, we have included notes to students in the margins. These notes provide *additional insights,* help students *avoid common errors,* and give students an opportunity to be involved in the *development of the material.*

NEW DESIGN: The new design is more open and readable. Each section begins with a list of specific objectives, which provides the organizational breakdown of the section. This makes it easy for students to connect each objective with the material covered in the text. The use of color is intended to highlight the important parts of graphs and important statements. We have ensured the accuracy of the graphical art by using computer-generated graphs. All art has been redrawn for this edition.

ORGANIZATION: A variety of organizational changes have been made. The review material of basic algebra has been placed separately in the beginning of the book. Another change involves the organization of trigonometry. Trigonometry is now covered in three chapters rather than two. As before, we present three perspectives—the trigonometric functions defined on acute angles of right triangles, the trigonometric functions defined on general angles, and the trigonometric functions defined on real numbers. We also show the connection between these latter functions and the unit circle in an approach that we believe is more pedagogically sound than in the previous edition. We also introduce graphing early and make effective use of graphing utilities to demonstrate the validity of identities, to solve equations, and to solve inequalities graphically.

Supplements

The sixth edition is accompanied by an extensive supplements package.

FOR THE INSTRUCTOR

An **Instructor's Solution Manual** contains worked-out solutions to all problems in the textbook.

An **Instructor's Test Manual** includes six different test forms for each chapter of the textbook. These tests have been written at graded levels of difficulty, with three of the tests for each chapter made up of multiple-choice questions and three of standard problem-solving questions. Answers to all test questions are provided.

MICROTEST Test-Generating Software is available in IBM and Macintosh versions. This streamlined program allows instructors to manipulate a large collection of test questions as a basis for building examinations and quizzes.

FOR THE STUDENT

A **Student Solution Guide** offers worked-out solutions to every other odd-numbered problem in the book. Chapter objectives are also included.

A new graphing calculator **videotape** is available. It explores and demonstrates features of the calculator. Demonstrations include: using zoom, math, matrix, and angle menus; creating tables and split screens; and graphing a variety of functions.

Acknowledgments

In preparing this edition, we have drawn from our own experience in teaching from *Precalculus* as well as the feedback provided by our colleagues and students. The suggestions obtained from instructors using the fifth edition were of great help. We wish to thank all of these people and, in particular, to express our gratitude to the following:

Robert Boner, Western Maryland College

Carol Flakus, Lower Columbia College

Gilbert Steiner, Fairleigh Dickinson University

We are grateful to the following individuals who assisted us in planning the revision of this text:

Margaret Dolgas, University of Delaware

Richard Jetton, Arkansas State University

Jane Pinnow, University of Wisconsin, Parkside

Elise Price, Tarrant County Junior College

And we wish to thank the following instructors who reviewed portions of the manuscript for the sixth edition:

Georgianna Aviola, College of William and Mary

Elizabeth Baton, University of North Texas at Denton

Arlene Blasius, SUNY College at Old Westbury

Harold Bowman, University of Nevada, Las Vegas

Harvey Greenwald, California Polytechnic State University

Richard Marshall, Eastern Michigan University

Elmo Moore, Humboldt State University

Stephen Murdock, Tulsa Junior College

Special thanks are due to our many colleagues at Macomb College who taught from previous editions of the book and shared their experiences with us. We especially wish to thank Steve Fassbinder, who checked all the problems in the book and assisted in proofreading, as well as both Paul Williams and William Hamilton for writing the test manual.

Finally, we wish to thank Robert Gebhardt of County College of Morris, who solved all the problems in the book, and who was meticulous in his checking and proofing of the entire text at all stages. We also appreciate the outstanding work of Lisa Durak, who prepared the *Instructor's Solution Manual* and *Student Solution Guide* and checked the accuracy of every answer in the book.

M. A. Munem
J. P. Yizze

OBJECTIVES

1. Identify types of real numbers.
2. Perform operations on polynomials.
3. Factor polynomials.
4. Perform operations on rational expressions.
5. Simplify expressions involving integer exponents.
6. Simplify expressions involving rational exponents and radicals.
7. Perform operations on radical expressions.
8. Perform operations on complex numbers.

Review of Basic Algebra

The concepts and techniques of basic algebra that will be used throughout this text are reviewed in this introductory material. Our review includes coverage of algebraic manipulations needed to master the material in this book.

1 Identifying Types of Real Numbers

Real numbers are encountered in everyday life and can be thought of as numbers that have decimal representations. Table 1 identifies various types of real numbers. The symbol $\mathbb{R}$ is used to represent the set of real numbers.

TABLE 1
Real Number Sets

Name	Description	Examples
Natural numbers, or **positive integers**	Counting numbers	1, 2, 3, 10, 25
Integers	Includes negative integers, zero, and positive integers	$-3, -2, -1, 0, 1, 2, 3$
Rational numbers	All numbers that can be written in the form a/b, where a and b are integers, $b \neq 0$	$-3, 0, -\frac{3}{4}, \frac{5}{7}, 5\frac{1}{3}, 0.37$
Irrational numbers	Real numbers that are not rational numbers	$\sqrt{2}, \sqrt[3]{5}, \pi$
Real numbers	All rational numbers and irrational numbers	$2, -7, \frac{5}{11}, \sqrt{3}, \pi$

The diagram in Figure 1 indicates how the sets of numbers listed in Table 1 are related to each other. Each set includes all numbers in the set(s) connected to and shown below it in Figure 1.

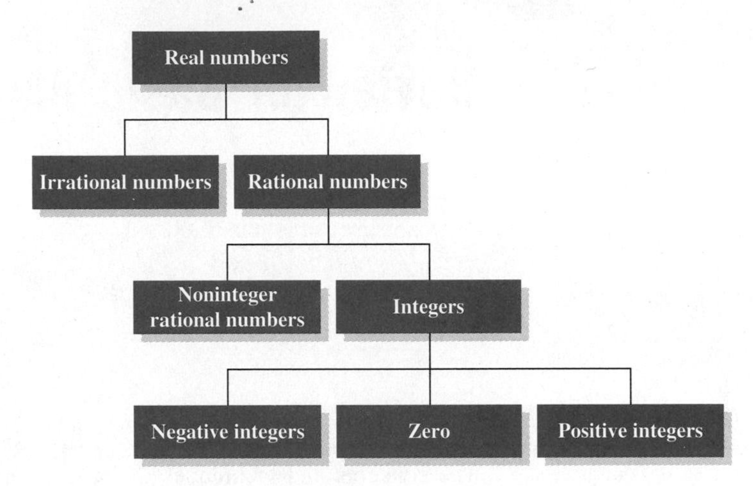

FIGURE 1

The set of real numbers consists of rational and irrational numbers. A decimal representation of a rational number a/b, where a and b are integers and $b \neq 0$, is obtained by dividing the numerator a by the denominator b. The result is either a terminating decimal such as

$$\tfrac{2}{5} = 0.4$$

or a nonterminating decimal with a repeating pattern such as

$$\tfrac{2}{3} = 0.66666\ldots = 0.\overline{6} \qquad \text{and} \qquad \tfrac{15}{11} = 1.363636\ldots = 1.\overline{36}$$

The overbars used in $\overline{6}$ and $1.\overline{36}$ identify the digits that repeat.

The decimal representation of an irrational number will neither terminate nor have a repeating decimal pattern. A calculator is commonly used to find decimal approximations of irrational numbers.

Often the symbol $\approx$ is used to indicate an approximation. For instance, rounded off to two decimal places, $\sqrt{127} \approx 11.27$. However, in this text we normally give rounding instructions and then use the equal sign for such approximations. So we write

$$\sqrt{127} = 11.27 \text{ (approx.)}$$

EXAMPLE 1 Finding decimal representations

Use a calculator to write each number in decimal form rounded off to four decimal places.

(a) $\tfrac{23}{17}$ (b) $(3.45)^5$ (c) $\sqrt[3]{75}$

Solution Using a calculator, we get the following results:

(a) $\frac{23}{17} = 1.3529$ (approx.) (b) $(3.45)^5 = 488.7598$ (approx.)

(c) $\sqrt[3]{75} = 4.2172$ (approx.)

Technically, the number is not the point, nor the point the number. However, it is customary to use the terms *real number* and *point* interchangeably.

A number line establishes a one-to-one correspondence between the real numbers and points on the line. Each real number is associated with a point on the line, and conversely, each point associates with a particular real number called its **coordinate.**

Real numbers can be **located**, or **plotted**, on a number line by using their decimal representations. Figure 2 shows a number line with some plotted points.

When locating points on a number line, we are primarily interested in displaying the relative positions of various points rather than the precise location of each point.

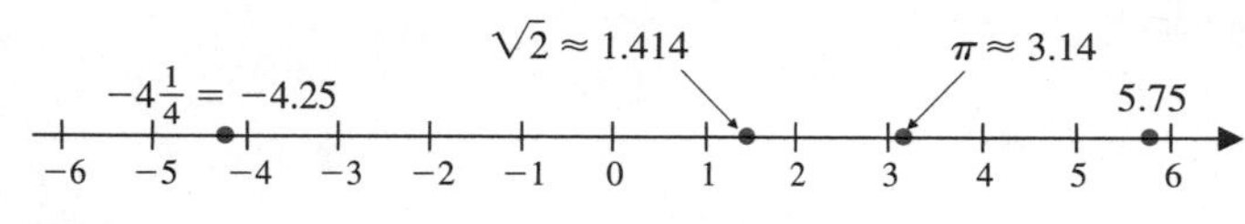

FIGURE 2

PROBLEM SET 1

In problems 1–4, express each rational number as a terminating or repeating decimal.

1. **(a)** $\frac{3}{4}$ **(b)** $-\frac{7}{4}$ **2.** **(a)** $\frac{11}{3}$ **(b)** $-\frac{16}{5}$

3. **(a)** $5\frac{5}{16}$ **(b)** $\frac{0.81}{2.7}$ **4.** **(a)** $-2\frac{5}{33}$ **(b)** $\frac{3.2}{0.037}$

In problems 5–8, convert each decimal to a fraction in lowest terms.

5. **(a)** 0.7 **(b)** 0.55 **6.** **(a)** 2.4 **(b)** 9.05

7. **(a)** 0.4 **(b)** 0.75 **8.** **(a)** 3.5 **(b)** 1.125

In problems 9 and 10, locate the number sets on a number line.

9. **(a)** The set of positive integers less than 5
(b) The set of negative integers greater than −4
(c) The set of rational numbers $-2, -\frac{1}{2}, \frac{2}{3}, 2.3, 3\frac{5}{8}, 4.7$
(d) The set of irrational numbers $\sqrt{3}, \pi, \sqrt[3]{17}$

10. **(a)** The set of all odd positive integers 1, 3, 5, 7, . . .
(b) The set of negative integers −7, −5, −2, −1
(c) The set of noninteger rational numbers $-0.3, 0.5, \frac{2}{3}, 5\frac{1}{4}$
(d) The set of irrational numbers $-\sqrt{2}, \sqrt[3]{16}, \sqrt{7}$

In problems 11–14, use a calculator to write each number in decimal form rounded off to four decimal places.

11. **(a)** $\frac{\sqrt[3]{317}}{11}$ **(b)** $\left(\frac{180}{\pi}\right)^2$

12. **(a)** $\frac{-\sqrt{2}}{13}$ **(b)** $\frac{\sqrt[3]{5}}{17}$

13. **(a)** $\frac{\sqrt{6} - \sqrt{2}}{2}$ **(b)** $\left(\frac{4\pi}{13}\right)^2$

14. **(a)** $-\frac{7}{\pi}$ **(b)** $\left(\frac{\pi}{180}\right)^2$

15. The formula for the volume V of a sphere of radius r is given by

$$V = \frac{4}{3}\pi r^3$$

Find the volumes of spheres of radius 1 inch; 2.7 inches; 11 inches. Round off the answers to two decimal places.

16. The formula for the volume V of a rectangular solid of length ℓ, width w, and height h is given by

$$V = \ell w h$$

Find the volume of a concrete slab in the form of a rectangular solid of length 75.23 feet, width 13.57 feet, and height 1.78 feet. Round off the answer to two decimal places.

17. The formula for the surface area S of a right circular cylinder of radius r and height h is given by

$$S = 2\pi rh + 2\pi r^2$$

Find the surface area of a right circular cylinder of radius 2.3 centimeters and height 7.9 centimeters. Round off the answer to two decimal places.

2 Performing Operations on Polynomials

We begin by listing the algebraic properties of the real number system. They serve as the basis for performing algebraic operations.

ALGEBRAIC PROPERTIES OF THE REAL NUMBER SYSTEM

Assume that a, b, and c are real numbers.

1. The Commutative Properties
(i) *For addition:* $a + b = b + a$
(ii) *For multiplication:* $ab = ba$

2. The Associative Properties
(i) *For addition:* $a + (b + c) = (a + b) + c$
(ii) *For multiplication:* $a(bc) = (ab)c$

3. The Distributive Properties
(i) $a(b + c) = ab + ac$ (ii) $(b + c)a = ba + ca$

4. The Identity Properties
(i) *For addition:* There is a unique real number called the **additive identity**, represented by 0, such that

$$a + 0 = 0 + a = a$$

(ii) *For multiplication:* There is a unique real number called the **multiplicative identity**, represented by 1, such that

$$a \cdot 1 = 1 \cdot a = a$$

5. The Inverse Properties
(i) *For addition:* For each real number a, there is a unique number $-a$ called the **additive inverse** such that

$$a + (-a) = (-a) + a = 0$$

(ii) *For multiplication:* For each real number $a \neq 0$, there is a unique number $1/a$ called the **multiplicative inverse** or **reciprocal** such that

$$a \cdot \frac{1}{a} = \frac{1}{a} \cdot a = 1$$

6. The Closure Properties
(i) *For addition:* The sum $a + b$ is a real number.
(ii) *For multiplication:* The product ab is a real number.

These properties apply not only to real numbers but also to polynomial expressions. Note that in the expression $a + b$, a and b are called the **terms**. For example, the terms in the expression $3x + 5y + 6t$ are $3x$, $5y$, and $6t$.

The building blocks of mathematical expressions in algebra are variables and constants. When a letter is used to represent any one of the members of a set of real numbers, it is called a **variable**. Letters used to designate fixed but unspecified numbers are called **constants**; their values remain unchanged throughout the discussion. Specific numbers are also considered to be constants. Thus, for the expression $5x + 3$, where x is a positive integer, we refer to x as a variable and 5 and 3 as constants.

Exponential notation is used to represent repeated multiplication.

EXPONENTIAL NOTATION

$$x^n = \overbrace{x \cdot x \cdot x \cdot \cdots \cdot x}^{n \text{ factors}} \qquad \text{where } n \text{ is a positive integer}$$

The expression x^n is read as the ***n*th power of *x***; x is called the **base** and n is called the **exponent**.

For example, $5^4 = 5 \cdot 5 \cdot 5 \cdot 5$; 5 is the base and 4 is the exponent. Similarly, $(x + y)^2 = (x + y) \cdot (x + y)$; $(x + y)$ is the base and 2 is the exponent. Also,

$$x^1 = x$$

Expressions formed by any combination of addition, subtraction, multiplication, division, exponentiation, and/or computation of roots of numbers and variables are called **algebraic expressions**. For example,

$$5x + 7, \quad y^2 + 5y - 3, \quad \frac{6x^3y^2 + 4}{y^3 - x}, \quad \text{and} \quad \sqrt{x^2 - y^2}$$

are algebraic expressions, where x and y are variables.

If specific numbers are substituted for the variables in an algebraic expression, the resulting real number is called the **value** of the expression for these numbers. For example, the value of the expression

$$\frac{6x^3y^2 + 4}{y^3 - x}$$

Evaluations of algebraic expressions are facilitated by the use of a calculator.

for $x = -1$ and $y = 2$ is obtained by replacing x by -1 and y by 2 in the expression to obtain

$$\frac{6(-1)^3(2)^2 + 4}{(2)^3 - (-1)} = \frac{-24 + 4}{8 + 1} = \frac{-20}{9}$$

An algebraic expression that results from applying only the operations of addition, subtraction, and multiplication on a set of numbers and variables is called a *polynomial*. Numbers are also considered to be polynomials. The expressions

$5x + 2$, $3x^2 - 2x + 7$, and $4xy^4 - 3xy + 5x + 6y^2$ are examples of polynomials. However,

$$\frac{6x^3 + 4}{2x^2 + 5} \quad \text{and} \quad 5y^2 + \sqrt{y} - 7$$

are not polynomials, since the first expression involves division by a variable expression and the second expression contains a root of a variable.

Much emphasis in algebra is given to polynomials that contain one variable.

DEFINITION
POLYNOMIAL IN x

A **polynomial** in x is an algebraic expression of the form

$$a_n x^n + a_{n-1} x^{n-1} + \cdots + a_1 x + a_0$$

where n is a nonnegative integer and $a_n, a_{n-1}, \ldots, a_1$, and a_0 are real numbers called **numerical coefficients**. Also, a_n is called the **leading coefficient**, and a_0 is called the **constant term**. The polynomial is said to have **degree *n*** if $a_n \neq 0$. A nonzero constant is called a **constant polynomial of degree 0**.

The number 0 is considered to be a zero polynomial with no degree assigned to it.

A polynomial containing one term is called a **monomial**; a polynomial of two terms is called a **binomial**; a polynomial of three terms is called a **trinomial**. Examples of polynomials are shown in Table 1.

TABLE 1
Classifications of Polynomials

Polynomial Type	Terms	Example
Monomial	One	$3x$
Binomial	Two	$4x + 5$
Trinomial	Three	$2x^2 + 5x - 9$

Terms of a polynomial, such as $5x^2$ and $-7x^2$, that differ only in their numerical coefficients, are called **like terms**. Polynomials are added by using the algebraic properties to arrange, group, and combine like terms as illustrated in the following example. Note that subtraction of polynomials can be rewritten as addition by using the fact that

$$a - b = a + (-b)$$

EXAMPLE 1 Adding and subtracting polynomials

Perform each operation.

(a) $(4x^3 + 7x - 13) + (-2x^3 + 5x + 17)$
(b) $(2x^3 + 3x^2 - 5x + 11) - (4x^3 - 5x^2 + 9)$

Solution

(a) $(4x^3 + 7x - 13) + (-2x^3 + 5x + 17)$

$= (4x^3 - 2x^3) + (7x + 5x) + (-13 + 17)$ Rearrange and regroup

$= 2x^3 + 12x + 4$ Combine like terms

(b) $(2x^3 + 3x^2 - 5x + 11) - (4x^3 - 5x^2 + 9)$

$= (2x^3 + 3x^2 - 5x + 11) + (-4x^3 + 5x^2 - 9)$	$a - b = a + (-b)$
$= (2x^3 - 4x^3) + (3x^2 + 5x^2) - 5x + (11 - 9)$	Rearrange and regroup
$= -2x^3 + 8x^2 - 5x + 2$	Combine like terms

To multiply monomials we use the algebraic properties along with the following property of positive integer exponents:

PRODUCT PROPERTY OF EXPONENTS

Suppose x is a real number. If m and n are positive integers, then

$$x^m \cdot x^n = x^{m+n}$$

For example, we multiply monomials $-2x^2$ and $7x^3$ as follows:

$(-2x^2) \cdot (7x^3) = (-2 \cdot 7)(x^2 \cdot x^3)$	Rearrange and regroup
$= -14x^5$	Product property

To multiply polynomials with more than one term, we use the distributive properties to multiply every term in the first polynomial by every term in the second.

EXAMPLE 2 **Multiplying polynomials**

Perform each multiplication.

(a) $(3x - 5)(4x + 1)$ (b) $(x^2 + 2x - 1)(x^2 + x - 2)$

Solution

(a) To multiply $(3x - 5)(4x + 1)$ we apply the distributive properties as follows:

$(3x - 5)(4x + 1)$

$= 3x(4x + 1) - 5(4x + 1)$	Distribute $4x + 1$
$= (3x \cdot 4x) + (3x \cdot 1) + (-5 \cdot 4x) + (-5 \cdot 1)$	Distribute $3x$ and -5
$= 12x^2 + 3x - 20x - 5$	Multiply monomials
$= 12x^2 - 17x - 5$	Combine like terms

Observe that each term in the first binomial is multiplied by each term in the second binomial. Another way of multiplying these binomials is to use a technique called the **FOIL method**, which is illustrated as follows:

$(3x - 5)(4x + 1) = 12x^2 \quad -17x \quad -5$

First term: $(3x)(4x)$

Middle term: $(3x)(1) + (-5)(4x) = -17x$
Product of **O**uter terms + Product of **I**nner terms

Last term: $(-5)(1)$

(b) To simplify the work of multiplying $(x^2 + 2x - 1)(x^2 + x - 2)$, we use the following vertical arrangement:

$$\begin{array}{rrrrrl}
x^2 & + 2x & - 1 & & & \\
x^2 & + \ x & - 2 & & & \\
\hline
x^4 + 2x^3 & - \ x^2 & & & & \text{Distribute } x^2 \\
& x^3 + 2x^2 & - \ x & & & \text{Distribute } x \text{ and align like terms} \\
& & - 2x^2 - 4x + 2 & & & \text{Distribute } -2 \text{ and align like terms} \\
\hline
x^4 + 3x^3 & - \ x^2 & - 5x + 2 & & & \text{Combine like terms}
\end{array}$$

PROBLEM SET 2

In problems 1 and 2, identify the algebraic property of real numbers that justifies the given statement.

1. **(a)** $t + (3 + t) = t + (t + 3)$
(b) $-4(5 - x) = -20 + 4x$
(c) $-(4x) + (4x) = 0$ **(d)** $(0.25) \cdot 4 = 1$

2. **(a)** $(x + 3) - 1 = x + (3 - 1)$
(b) $x + (y + 5) = (x + y) + 5$
(c) $1 \cdot (x + 2) = x + 2$
(d) $(a + 7) + c = (7 + a) + c$

In problems 3 and 4, evaluate the algebraic expression for the given values. Round off each answer to two decimal places.

3. $17 - \sqrt{x^2 + y} + \dfrac{y^3}{x}$; $x = 2$ and $y = -1$

4. $\dfrac{\sqrt{3x^3 - y^2}}{5xy + y}$; $x = 5.1$ and $y = -4.3$

In problems 5 and 6, identify which of the algebraic expressions are polynomials. For each polynomial determine the degree and the nonzero numerical coefficients, and specify whether the polynomial is a monomial, binomial, or trinomial.

5. **(a)** $4x^2 - x$
(b) $3x^5 - x^3 + \dfrac{1}{x}$

6. **(a)** $\frac{4}{5}x^2 - x^4 + \sqrt{2x}$ **(b)** $-\frac{1}{2}x$

In problems 7–16, simplify by performing the specified operations.

7. **(a)** $4xy^2 - 3x^2y^2 - 2xy^2 + xy^2$
(b) $(2x^3 - 5x^2 - 7x - 1) + (x^3 - 2x^2 + 7x + 3)$

8. **(a)** $-2t^2x^3 + 3t^3x^3 + 4t^2x^3 + tx^2$
(b) $(-7x^3 + 4x + 3) + (3x^3 + 4x^2 - 7x - 8)$

9. $(3x^4 - 4x^3 + 6x^2 + x - 1) - (4 - x + 3x^2 - 3x^3 - x^4)$

10. $(4x^2 - 7x - 8) - (-2x^2 + 4x - 3)$

11. **(a)** $(2x^4)(-x^4)(7x)$ **(b)** $(2x^2y^3)(3xy^5)$

12. **(a)** $(-5x^4y^2)(3xy + x - 1)$
(b) $(11x^4)(2x^2 + 5x + 2)$

13. **(a)** $(2x + 3)(-5x - 2)$
(b) $(y + 7)(y^2 + 3y + 1)$

14. **(a)** $(3x - 5)(x + 2)$
(b) $(x + 3)(x^2 - 4x + 5)$

15. **(a)** $(3x - 4)(9x^2 + 12x + 16)$
(b) $(2x^2 + x + 1)(3x^2 - x + 5)$

16. **(a)** $(2x^2 - x + 1)(x^2 + x + 3)$
(b) $(3x + 1)(3x - 2)(-x + 4)$

3 Factoring Polynomials

To **factor** a polynomial expression means to write it as a product of other polynomials. For instance, since

$$x(x - 2) = x^2 - 2x$$

we say that x and $x - 2$ are **factors** of $x^2 - 2x$. A polynomial with integer coefficients is said to be **prime** if it cannot be factored into polynomials (other than 1 or -1). When a polynomial is expressed as a product of prime factors, we say that it is **factored completely**.

If there is a common factor in every term of an expression, then we use the distributive property to factor the expression by removing the *greatest common factor.*

EXAMPLE 1 **Removing the greatest common factor**

Factor $3x^2y^5 - 5xy^3$ by removing the greatest common factor.

Solution The greatest common factor is xy^3. So, by the distributive property, we get

$$3x^2y^5 - 5xy^3 = xy^3(3xy^2) + xy^3(-5) = xy^3(3xy^2 - 5)$$

Skill in factoring relates directly to skill in multiplying. Factoring proficiency can be increased by recognizing multiplication patterns such as those shown in the special products listed in Table 1.

TABLE 1
Special Products

Name	
1. Square of a binomial	$(a + b)^2 = a^2 + 2ab + b^2$
2. Square of a binomial	$(a - b)^2 = a^2 - 2ab + b^2$
3. Difference of squares	$(a + b)(a - b) = a^2 - b^2$
4. Sum of cubes	$(a + b)(a^2 - ab + b^2) = a^3 + b^3$
5. Difference of cubes	$(a - b)(a^2 + ab + b^2) = a^3 - b^3$

EXAMPLE 2 **Using special products**

Factor each of the given polynomials by using one of the special products.

(a) $16x^2 - 49y^2$ (b) $8x^3 + 27y^3$

Solution

(a) We apply special product 3 to get

$$16x^2 - 49y^2 = (4x)^2 - (7y)^2 = (4x + 7y)(4x - 7y)$$

(b) By using special product 4, we obtain

$$\begin{aligned} 8x^3 + 27y^3 &= (2x)^3 + (3y)^3 \\ &= (2x + 3y)[(2x)^2 - (2x)(3y) + (3y)^2] \\ &= (2x + 3y)(4x^2 - 6xy + 9y^2) \end{aligned}$$

When none of the multiplication patterns of the special products apply, we use trial and error to factor trinomials of the form $ax^2 + bx + c$ (where a, b, and c are integer coefficients) into a product of two binomials, if such a factorization is possible. In a sense, we are reversing the FOIL method used in multiplication.

EXAMPLE 3 **Factoring a trinomial**
Factor the expression $3x^2 + x - 2$.

Solution After trying different combinations, we discover that

$$3x^2 + x - 2 = (3x - 2)(x + 1)$$

Sometimes a factorization is mixed in the sense that it may involve more than one technique.

EXAMPLE 4 **Factoring by using two techniques**
Factor $5x^3 - 45x$ completely.

Solution We factor this expression as follows:

$$\begin{aligned} 5x^3 - 45x &= 5x(x^2 - 9) && \text{5x is the greatest common factor} \\ &= 5x(x + 3)(x - 3) && \text{Special product 3} \end{aligned}$$

PROBLEM SET 3

In problems 1–4, factor out the greatest common factor from each polynomial.

1. (a) $5y^4 + 2y^3$ (b) $8x^2 + 12x^4$
2. (a) $-x^5 - 2x^4 + x^3$ (b) $-2y^2 - 6y + 8$
3. (a) $17x^3y^2 - 34x^2y$ (b) $81x^3y + 12x^2y^6$
4. (a) $-4x^3y^3 - 12x^2y^2 - 18xy$
   (b) $-9x^6y^4 - 15x^3y^2 + 18x^2y$

In problems 5–10, use the special products to factor each polynomial completely.

5. (a) $9y^2 - 25$ (b) $4x^2 - 49y^2$
6. (a) $x^4 - 16y^4$ (b) $(3x + 2y)^2 - 25z^2$
7. (a) $x^3 - 8$ (b) $8x^3 - y^3$
8. (a) $u^3 + 27v^3$ (b) $27a^3 + 8b^3$
9. (a) $x^2 + 4x + 4$ (b) $x^2 - 6x + 9$
10. (a) $x^2 - 2xy + y^2$ (b) $4x^2 + 4xy + y^2$

In problems 11–16, factor each polynomial completely.

11. (a) $x^2 - 16x + 63$
    (b) $x^2 - 7xy + 10y^2$
12. (a) $5x^2 + 14x - 24$
    (b) $3x^2 - 13xy - 10y^2$
13. (a) $-2x^3 - 5x^2 + 12x$
    (b) $24x^3 + 34x^2 - 10x$
14. (a) $6x^4 + 20x^3 + 6x^2$
    (b) $6x^2y - 21xy + 9y$
15. (a) $y^3 - 3y^2 - 25y$
    (b) $2t^3 + 2t^2 - 50t$
16. (a) $x^5 - x^3 - x^2 + 1$
    (b) $u^3 - 5u - 2u^2 + 10$

4 Performing Operations on Rational Expressions

If the numerator and denominator of a fraction are polynomials, then it is called a **rational expression**. Examples of rational expressions are

$$\frac{2}{3}, \quad \frac{1}{x}, \quad \frac{x+2}{x-5}, \quad \frac{7}{t^2-4}, \quad \text{and} \quad \frac{4y^3+1}{8y^4+13y^2-8}$$

Since division by 0 is *not* defined, it is always understood that the *denominator of a rational expression cannot represent 0*. Thus, for

$$\frac{5x}{x^2-3x} = \frac{5x}{x(x-3)} \qquad x \neq 0 \text{ and } x \neq 3$$

Whenever we use a rational expression throughout this book, we assume that the variables involved are restricted to numerical values that will give a nonzero denominator.

Rational expressions, as with rational numbers, can be written in **lowest terms**, where the numerator and the denominator have no common factor other than 1 or -1. Thus, to **simplify**, or **reduce**, a rational expression, we factor both the numerator and the denominator into prime factors and then **cancel** common factors. This procedure is based on the *fundamental principle of fractions*.

FUNDAMENTAL PRINCIPLE OF FRACTIONS

$$\frac{PK}{QK} = \frac{P}{Q} \qquad \text{where } Q \neq 0 \text{ and } K \neq 0$$

EXAMPLE 1 Reducing a rational expression

Reduce the rational expression: $\dfrac{2x^2+5x-3}{10x^2+9x-7}$

Solution

$$\frac{2x^2+5x-3}{10x^2+9x-7} = \frac{(2x-1)(x+3)}{(2x-1)(5x+7)} \qquad \text{Factor}$$

$$= \frac{x+3}{5x+7} \qquad \text{Reduce}$$

Multiplication and division of rational expressions is accomplished by using the following rules:

$$\text{(i)}\ \frac{P}{Q}\cdot\frac{R}{S} = \frac{PR}{QS} \qquad \text{(ii)}\ \frac{P}{Q}\div\frac{R}{S} = \frac{P}{Q}\cdot\frac{S}{R} = \frac{PS}{QR} \qquad \text{provided } \frac{R}{S} \neq 0$$

The key to writing the results of these operations in simplified form is factoring and reducing.

EXAMPLE 2 Multiplying and dividing rational expressions

Perform each operation and simplify the result. Express the answer in factored form.

(a) $\dfrac{x^2 + 4x + 4}{2x^2 + 2x - 4} \cdot \dfrac{2x - 2}{x^2 + 2x}$ (b) $\dfrac{x^2 - 10x + 25}{x^2 - 100} \div \dfrac{x^2 - 7x + 10}{x^2 + 12x + 20}$

Solution

(a)
$$\begin{aligned}\frac{x^2 + 4x + 4}{2x^2 + 2x - 4} \cdot \frac{2x - 2}{x^2 + 2x} &= \frac{(x^2 + 4x + 4)(2x - 2)}{(2x^2 + 2x - 4)(x^2 + 2x)} && \text{Multiplication rule}\\ &= \frac{(x + 2)^2(2)(x - 1)}{(2)(x + 2)(x - 1)(x)(x + 2)} && \text{Factor}\\ &= \frac{1}{x} && \text{Reduce}\end{aligned}$$

(b)
$$\begin{aligned}&\frac{x^2 - 10x + 25}{x^2 - 100} \div \frac{x^2 - 7x + 10}{x^2 + 12x + 20}\\ &= \frac{x^2 - 10x + 25}{x^2 - 100} \cdot \frac{x^2 + 12x + 20}{x^2 - 7x + 10} && \text{Division rule}\\ &= \frac{(x^2 - 10x + 25)(x^2 + 12x + 20)}{(x^2 - 100)(x^2 - 7x + 10)} && \text{Multiplication rule}\\ &= \frac{(x - 5)^2(x + 2)(x + 10)}{(x - 10)(x + 10)(x - 2)(x - 5)} && \text{Factor}\\ &= \frac{(x - 5)(x + 2)}{(x - 10)(x - 2)} && \text{Reduce}\end{aligned}$$

■

To add or subtract two rational expressions with the same denominator, we use the following rules:

(i) $\dfrac{P}{Q} + \dfrac{R}{Q} = \dfrac{P + R}{Q}$ (ii) $\dfrac{P}{Q} - \dfrac{R}{Q} = \dfrac{P - R}{Q}$

EXAMPLE 3 Subtracting rational expressions with the same denominators

Perform the subtraction and simplify: $\dfrac{x}{4 - x^2} - \dfrac{2}{4 - x^2}$

Solution
$$\begin{aligned}\frac{x}{4 - x^2} - \frac{2}{4 - x^2} &= \frac{x - 2}{4 - x^2} && \text{Subtraction rule}\\ &= \frac{\overset{-1}{\cancel{x - 2}}}{(2 + x)\cancel{(2 - x)}} && \text{Factor and reduce}\\ &= \frac{-1}{2 + x}\end{aligned}$$

■

To add or subtract rational expressions with different denominators, we must first find a common denominator. One common denominator for P/Q and R/S is the product QS of the two denominators because both rational expressions can be expressed with this denominator as follows:

$$\frac{P}{Q} = \frac{PS}{QS} \quad \text{and} \quad \frac{R}{S} = \frac{QR}{QS}$$

So

$$\frac{P}{Q} + \frac{R}{S} = \frac{PS}{QS} + \frac{QR}{QS} = \frac{PS + QR}{QS}$$

Subtraction can be done similarly. Normally, it is more efficient to find the *least common denominator (LCD)* of both rational expressions to make the denominators the same, as illustrated in the next example.

EXAMPLE 4 Adding rational expressions with different denominators

Perform the addition and simplify: $\dfrac{2x + 10}{x^2 + 6x + 5} + \dfrac{3}{x^2 + 2x + 1}$

Solution First we factor each denominator completely. The LCD is the product of the factors with the largest exponent each factor has in any factored denominator. Thus, we obtain the result given in the table.

Denominators	Prime Factors
$x^2 + 6x + 5$	$(x + 1)(x + 5)$
$x^2 + 2x + 1$	$(x + 1)^2$
LCD	$(x + 1)^2(x + 5)$

Next we express each rational expression in terms of the LCD in the denominator by multiplying the numerator and denominator of each expression by the appropriate factors. We get

$$\begin{aligned}
\frac{2x + 10}{x^2 + 6x + 5} + \frac{3}{x^2 + 2x + 1} &= \frac{2x + 10}{(x + 1)(x + 5)} + \frac{3}{(x + 1)^2} && \text{Factor each denominator} \\
&= \frac{(2x + 10)(x + 1)}{(x + 1)^2(x + 5)} + \frac{3(x + 5)}{(x + 1)^2(x + 5)} && \text{Convert to the LCD} \\
&= \frac{2x^2 + 12x + 10 + 3x + 15}{(x + 1)^2(x + 5)} && \text{Multiply and add} \\
&= \frac{2x^2 + 15x + 25}{(x + 1)^2(x + 5)} && \text{Combine like terms} \\
&= \frac{(2x + 5)(x + 5)}{(x + 1)^2(x + 5)} && \text{Factor the numerator} \\
&= \frac{2x + 5}{(x + 1)^2} && \text{Reduce}
\end{aligned}$$

Sometimes the numerator or denominator (or both) of a fraction contains one or more fractions. In such a case, the fraction is called a **complex fraction**.

Examples of complex fractions are

$$\frac{\frac{1}{2}}{3}, \quad \frac{\frac{1}{x}}{\frac{3}{x+1}+7}, \quad \text{and} \quad \frac{x-\frac{1}{x}}{\frac{2}{1+x}-\frac{x}{1-x}}$$

To *simplify* a complex fraction means to write the fraction in a reduced form without any fractions in the numerator or denominator. The next example illustrates one procedure for simplifying complex fractions by using the LCD.

EXAMPLE 5 Simplifying a complex fraction

Simplify: $\dfrac{\frac{x}{2}-\frac{2}{x}}{\frac{1}{x}+\frac{1}{2}}$

Solution First we recognize that the LCD for $x/2$, $2/x$, $1/x$, and $1/2$ is $2x$. Then we multiply both the numerator and denominator of the given expression by the LCD, $2x$, to get

$$\begin{aligned}
\frac{\left(\frac{x}{2}-\frac{2}{x}\right)}{\left(\frac{1}{x}+\frac{1}{2}\right)}\cdot\frac{2x}{2x} &= \frac{\left(\frac{x}{2}\right)(2x)-\left(\frac{2}{x}\right)(2x)}{\left(\frac{1}{x}\right)(2x)+\left(\frac{1}{2}\right)(2x)} && \text{Distribute}\\
&= \frac{x^2-4}{2+x} && \text{Multiply}\\
&= \frac{(x+2)(x-2)}{2+x} && \text{Factor}\\
&= x-2 && \text{Reduce}
\end{aligned}$$

PROBLEM SET 4

In problems 1 and 2, determine the values of x for which the expression is undefined.

1. **(a)** $\dfrac{4}{x+7}$ **(b)** $\dfrac{x+2}{(x+1)(x-3)}$

2. **(a)** $\dfrac{x-1}{x+8}$ **(b)** $\dfrac{7}{x(x-7)}$

In problems 3–6, reduce each rational expression.

3. **(a)** $\dfrac{3x^3y^7a}{15x^4y^2a^3}$ **(b)** $\dfrac{2-x}{x^2-4}$

4. **(a)** $\dfrac{4x^2-9}{9x-6x^2}$ **(b)** $\dfrac{1-x^2}{x^3-1}$

5. **(a)** $\dfrac{x^3+4x^2}{x^3+6x^2+8x}$ **(b)** $\dfrac{7x^2-5xy}{49x^3-25xy^2}$

6. **(a)** $\dfrac{6x^2-7x-3}{3x^2-5x-2}$ **(b)** $\dfrac{x^2y^4-x^4y^2}{x^2y^4+2x^3y^3+x^4y^2}$

In problems 7–16, perform the indicated operation and simplify the result. Express the answer in factored form. Specify any restrictions on the variables.

7. $\dfrac{3x+6}{5x+5}\cdot\dfrac{x+1}{x^2+5x+6}$

8. $\dfrac{a^2-1}{a+1}\cdot\dfrac{7a^2-5a-2}{a^2-2a+1}$

9. $\dfrac{x^2-11x+10}{9x^2-25}\div\dfrac{x^2-8x-20}{12x^2+20x}$

10. $\dfrac{6y^2 + 11y - 10}{3y^2 - 5y - 12} \div \dfrac{2y^2 + 9y + 10}{3y^2 + 10y + 8}$

11. $\dfrac{x^2}{x + 3} - \dfrac{9}{x + 3}$ **12.** $\dfrac{x}{x - 1} - \dfrac{1}{x - 1}$

13. $\dfrac{x}{x^2 - 25} + \dfrac{1}{x + 5}$ **14.** $\dfrac{x}{x^2 + 5x - 6} + \dfrac{3}{x + 6}$

15. $\dfrac{x}{x^2 - 9} - \dfrac{x - 1}{x^2 - 5x + 6}$

16. $\dfrac{5b - 1}{3b^2 - 2b - 8} - \dfrac{3b + 2}{2b^2 - 3b - 2}$

In problems 17 and 18, simplify the complex fractions.

17. **(a)** $\dfrac{\frac{3}{4x}}{7}$ **(b)** $\dfrac{x + \frac{3}{x}}{1 + \frac{3}{x^2}}$ **(c)** $\dfrac{\frac{x - 1}{x + 1} - 3}{\frac{x - 1}{x + 1} + 4}$

18. **(a)** $\dfrac{5x}{\frac{1}{3}}$ **(b)** $\dfrac{\frac{x}{y} - \frac{y}{x}}{\frac{x}{y} + \frac{y}{x}}$ **(c)** $\dfrac{x + \frac{y}{y - x}}{y + \frac{x}{x - y}}$

5 Simplifying Expressions Involving Integer Exponents

So far we have worked only with positive integer exponents. Now we extend the notion of exponents to include zero and negative integer exponents.

DEFINITION

ZERO AND NEGATIVE INTEGER EXPONENTS

If $x \neq 0$ and n is a positive integer, then

$$x^0 = 1 \quad \text{and} \quad x^{-n} = \frac{1}{x^n}$$

For instance,

$$7^0 = 1, \quad 2^{-3} = \frac{1}{2^3} = \frac{1}{8}, \quad \text{and} \quad x^{-2} = \frac{1}{x^2}$$

Note that 0^0 is not defined.

Using the definitions of positive, negative, and zero integer exponents, we have the following properties:

PROPERTIES OF EXPONENTS

Suppose x and y are real numbers. If m and n are integers, then

(i) $x^m \cdot x^n = x^{m+n}$ (ii) $(x^m)^n = x^{mn}$ (iii) $(xy)^n = x^n y^n$

(iv) $\left(\dfrac{x}{y}\right)^n = \dfrac{x^n}{y^n}$, where $y \neq 0$ (v) $\dfrac{x^m}{x^n} = x^{m-n}$, where $x \neq 0$

These properties are used to simplify expressions involving exponents.

EXAMPLE 1 Simplifying exponent expressions

Rewrite each expression without negative exponents and simplify the result.

(a) $(2x^{-3}y^2)^{-4}$ (b) $\left(\dfrac{x^{-4}}{y^3}\right)^{-5}$

(c) $\dfrac{(x+h)^{-1} - x^{-1}}{h}$

Solution

(a)
$$\begin{aligned}(2x^{-3}y^2)^{-4} &= (2)^{-4}(x^{-3})^{-4}(y^2)^{-4} && \text{Property (iii)}\\ &= \frac{1}{2^4}x^{12}y^{-8} && 2^{-4} = \frac{1}{2^4} \text{ and Property (ii)}\\ &= \frac{x^{12}}{16y^8} && 2^4 = 16 \text{ and } y^{-8} = \frac{1}{y^8}\end{aligned}$$

(b)
$$\begin{aligned}\left(\frac{x^{-4}}{y^3}\right)^{-5} &= \frac{(x^{-4})^{-5}}{(y^3)^{-5}} && \text{Property (iv)}\\ &= \frac{x^{20}}{y^{-15}} = \frac{x^{20}}{\left(\frac{1}{y^{15}}\right)} && \text{Property (ii) and } y^{-15} = \frac{1}{y^{15}}\\ &= \frac{x^{20}\cdot y^{15}}{\left(\frac{1}{y^{15}}\right)\cdot y^{15}} = x^{20}y^{15} && \text{Multiply numerator and denominator by LCD, } y^{15}\end{aligned}$$

(c)
$$\begin{aligned}&\frac{(x+h)^{-1} - x^{-1}}{h}\\ &= \frac{\frac{1}{x+h} - \frac{1}{x}}{h} && (x+h)^{-1} = \frac{1}{x+h};\ x^{-1} = \frac{1}{x}\\ &= \frac{\left(\frac{1}{x+h} - \frac{1}{x}\right)\cdot x(x+h)}{h\cdot x(x+h)} && \text{Multiply numerator and denominator by LCD, } x(x+h)\\ &= \frac{\frac{1}{x+h}\cdot x(x+h) - \frac{1}{x}\cdot x(x+h)}{hx(x+h)} && \text{Distribute}\\ &= \frac{x-(x+h)}{hx(x+h)} = \frac{x-x-h}{hx(x+h)} = \frac{-h}{hx(x+h)} && \text{Simplify the numerator}\\ &= \frac{-1}{x(x+h)} && \text{Reduce}\end{aligned}$$

PROBLEM SET 5

In problems 1–12, rewrite each expression without negative exponents and simplify the result.

1. **(a)** $(1 + 7y^2)^0$ **(b)** 5^{-2} **(c)** x^4y^{-7}
(d) $5^{-3} \cdot 5^{-2}$ **(e)** $\dfrac{x^m}{x^{-m}}$

2. **(a)** $(3 + 5x^4)^0$ **(b)** $x^{-3}y^5$ **(c)** $7a^{-4}b^2$
(d) $\left(\dfrac{3}{p^{-1}}\right)^{-2}$ **(e)** $\dfrac{x^{-n}}{x^n}$

3. **(a)** $\dfrac{3 \cdot 2^{-1} \cdot 4^{-1}}{2^2}$ **(b)** $\dfrac{2^{-3} \cdot 5^{-2}}{10^{-1} \cdot 16}$
(c) $x^{-2n} \cdot x^{2n+5}$

4. **(a)** $\dfrac{2^3 \cdot 2^4 \cdot 6^{-2}}{6^2}$ **(b)** $\dfrac{3^{-5} \cdot 4^{-2} \cdot 2^{-1}}{2^3 \cdot 3^2}$
(c) $\dfrac{(x + 3y)^{-12}}{(x + 3y)^{10}}$

5. **(a)** $16^{-2}[(2^{-1})(2)(2^5)]^4$ **(b)** $[(5^{-1})(5^2 \cdot 5^{-3})]^{-1}$

6. **(a)** $\left(\dfrac{3^2 \cdot 3^4 \cdot 9^{-1}}{4^3 \cdot 3^{-2} \cdot 5^0}\right)^{-1}$ **(b)** $\left(\dfrac{3^0 \cdot 2^{-6}}{2^{-2} \cdot 4 \cdot 7^0}\right)^{-2}$

7. $(-3x^{-5}y^4z^3)^{-4}$

8. $\left(\dfrac{3x^{-2}y^5}{9x^{-4}y^0}\right)^{-3}$

9. $\dfrac{x^{-4}y^2z^{-4}}{(xy)^{-2}(yz)^{-4}}$

10. $\left[\dfrac{(ab)^2(bc)^{-3}}{(ab)^3(cd)^{-2}}\right]^{-2}$

11. $\dfrac{x^2 \cdot x^3 \cdot x^{-1} \cdot (x^{-2})^3}{x^{-3}}$

12. $\left[\dfrac{x^{-4}y^2z^{-3}}{x^3(yz)^{-2}}\right]^{-4}$

In problems 13–18, rewrite each expression as a simplified fraction without negative exponents.

13. **(a)** $\dfrac{5^{-1} - 3^{-1}}{(45)^{-1}}$ **(b)** $\dfrac{1}{4^{-2} - 3^{-2}}$

14. **(a)** $\dfrac{4 - 3^{-2}}{2 + 3^{-1}}$ **(b)** $\dfrac{1}{(4 + 3)^{-2}}$

15. **(a)** $\dfrac{a^{-1} - b^{-1}}{a - b}$ **(b)** $a^{-1}b + ab^{-1}$

16. **(a)** $\dfrac{a^{-1} + b^{-1}}{a^{-1} - b^{-1}}$ **(b)** $\dfrac{x^{-2} - y^{-2}}{x^{-1} - y^{-1}}$

17. $\dfrac{(x + h)^{-2} - x^{-2}}{h}$

18. $\dfrac{(x + h)^{-3} - x^{-3}}{h}$

6 Simplifying Expressions Involving Rational Exponents and Radicals

To raise a number to a power is the reverse process of finding the root of a number. So we define the nth root of a number in terms of the nth power.

DEFINITION
*n*th ROOT of *x*

> Suppose x is a real number and $r^n = x$, where n is an integer greater than 1. Then r is called an ***n*th root** of x. If $r^2 = x$, then r is a **square root** of x. If $r^3 = x$, then r is a **cube root** of x.

Examples of roots are listed in Table 1.

TABLE 1
Examples of nth Roots

Given Number	nth Root(s)
9	-3 and 3 are square roots of 9 since $3^2 = (-3)^2 = 9$
-8	-2 is the only real cube root of -8 since only $(-2)^3 = -8$
625	-5 and 5 are fourth roots of 625 since $(-5)^4 = 5^4 = 625$

Note that if n is an even positive integer and x is a positive real number, then there are *two* real number nth roots of x, one positive and one negative. If n is an odd positive integer, then there is only one real number nth root of x, regardless of whether x is a negative or a positive real number or zero.

Radical notation, $\sqrt[n]{x}$, is used to denote the *principal nth root of x*.

DEFINITION

$\sqrt[n]{x}$, PRINCIPAL nTH ROOT OF x

Suppose x is a real number and n is a positive integer greater than 1. Then the **principal *n*th root of *x***, denoted by $\sqrt[n]{x}$, is defined as follows:

(i) If n is even and x is positive, then $\sqrt[n]{x}$ is the *positive* nth root of x.
(ii) If n is odd, then $\sqrt[n]{x}$ is the nth root of x.
(iii) If n is even and x is negative, then $\sqrt[n]{x}$ is not a real number.
(iv) $\sqrt[n]{0} = 0$.

Note that $(\sqrt[n]{x})^n = x$ whenever $\sqrt[n]{x}$ is a real number. In the expression $\sqrt[n]{x}$, the symbol $\sqrt[n]{}$ is called the **radical**. The positive integer n is called the **index**, and the real number x under the radical is called the **radicand**. If the index is not written, it is understood to be 2.

For example,

$$\sqrt{25} = 5 \text{ (not } -5) \quad \text{because } 5^2 = 25, \text{ and 5 is positive.}$$
$$\sqrt[3]{-8} = -2 \quad \text{because } (-2)^3 = -8.$$
$$\sqrt{-4} \text{ is not a real number.}$$
$$\sqrt{x^2} = x \quad \text{if } x > 0.$$
$$\sqrt[3]{x^3} = x$$

EXAMPLE 1 Simplifying radical expressions

Simplify each radical expression.

(a) $\sqrt[3]{64x^3}$ (b) $\sqrt[5]{x^{10}}$ (c) $\sqrt{x^6}$ where $x > 0$

Solution

(a) $\sqrt[3]{64x^3} = 4x$ because $(4x)^3 = 64x^3$.
(b) $\sqrt[5]{x^{10}} = x^2$ because $(x^2)^5 = x^{10}$.
(c) If $x > 0$, $\sqrt{x^6} = x^3$ because $(x^3)^2 = x^6$.

Consider the equation $x = 27^{1/3}$. Let us assume that the properties of exponents hold in this case. Then

$$x^3 = (27^{1/3})^3 = 27^{(1/3)(3)}$$

So $$x^3 = 27$$

This latter equation indicates that x is a cube root of 27; that is, $x = \sqrt[3]{27}$. Since we started with $x = 27^{1/3}$, we conclude that

$$27^{1/3} = \sqrt[3]{27} = 3$$

In general, if x is a real number and n is a positive integer for which $\sqrt[n]{x}$ is a real number, then

$$x^{1/n} = \sqrt[n]{x}$$

For instance, $49^{1/2} = \sqrt{49} = 7$, and $(-125)^{1/3} = \sqrt[3]{-125} = -5$.

Now we can define rational number exponents.

DEFINITION

RATIONAL NUMBER EXPONENTS

Suppose m/n is a rational number reduced to lowest terms and n is a positive integer. Then, if each of the roots exists,

$$x^{m/n} = (x^{1/n})^m = \left(\sqrt[n]{x}\right)^m = \sqrt[n]{x^m}$$

For example,

$$32^{2/5} = \left(\sqrt[5]{32}\right)^2 = 2^2 = 4$$

$$(81x^4)^{(-3/4)} = \left(\sqrt[4]{81x^4}\right)^{-3} = (3x)^{-3} = \frac{1}{(3x)^3} = \frac{1}{27x^3} \qquad x > 0$$

The properties of integer exponents listed on page 15 are also true for rational exponents.

EXAMPLE 2 Simplifying expressions involving rational exponents

Simplify each expression and write the answer so that it contains only positive exponents. Assume that the variables are restricted to values for which all expressions are defined.

(a) $x^{(-1/2)} \cdot x^{3/2}$ (b) $(x^{(-5/7)})^{-14}$ (c) $\left(\frac{x^{-9}}{y^{-6}}\right)^{(-2/3)}$

Solution

(a) $x^{(-1/2)} \cdot x^{3/2} = x^{(-1/2)+(3/2)} = x^{2/2} = x$ (b) $(x^{(-5/7)})^{-14} = x^{(-5/7)(-14)} = x^{10}$

(c) $\left(\frac{x^{-9}}{y^{-6}}\right)^{(-2/3)} = \frac{(x^{-9})^{(-2/3)}}{(y^{-6})^{(-2/3)}} = \frac{x^{(-9)(-2/3)}}{y^{(-6)(-2/3)}} = \frac{x^6}{y^4}$

We are now able to derive properties of radicals directly from known properties of exponents. Assume that x and y are real numbers, and m and n are positive integers. Then, provided all expressions are defined, we have the following results:

PROPERTIES OF RADICALS

Exponent Property	Equivalent Radical Property
(i) $(xy)^{1/n} = x^{1/n}y^{1/n}$	$\sqrt[n]{xy} = \sqrt[n]{x}\sqrt[n]{y}$
(ii) $\left(\frac{x}{y}\right)^{1/n} = \frac{x^{1/n}}{y^{1/n}}$	$\sqrt[n]{\frac{x}{y}} = \frac{\sqrt[n]{x}}{\sqrt[n]{y}}$

To simplify a radical expression, we use the properties of radicals and write the expression in a form that satisfies the following conditions:

1. The power of any term under the radical is less than the index of the radical; that is, in $\sqrt[n]{x^m}$, $m < n$.
2. The exponent of any term under the radical and the index of the radical have no common factor other than 1 or -1; that is, in $\sqrt[n]{x^m}$, m and n have no common factors.
3. The radicand contains no fractions.
4. There are no radicals in the denominator.

EXAMPLE 3 Simplifying radicals

Use the properties of radicals to simplify each expression. Assume that all variables are positive.

(a) $\sqrt{63}$ (b) $\sqrt{125x^2}$ (c) $\sqrt[4]{\frac{32x^9}{81y^4}}$

Solution

(a) $\sqrt{63} = \sqrt{9 \cdot 7} = \sqrt{9}\sqrt{7} = 3\sqrt{7}$

(b) $\sqrt{125x^2} = \sqrt{25x^2 \cdot 5} = \sqrt{25x^2}\sqrt{5} = 5x\sqrt{5}$

(c) $\sqrt[4]{\frac{32x^9}{81y^4}} = \frac{\sqrt[4]{32x^9}}{\sqrt[4]{81y^4}} = \frac{\sqrt[4]{16x^8 \cdot 2x}}{3y} = \frac{2x^2\sqrt[4]{2x}}{3y}$

PROBLEM SET 6

In problems 1–4, evaluate each expression.

1. (a) $\sqrt[3]{-64}$ (b) $\sqrt[5]{32}$
   (c) $-\sqrt{\frac{25}{144}}$ (d) $\sqrt[4]{-16}$
2. (a) $-\sqrt[5]{-32}$ (b) $-\sqrt[3]{-125}$
   (c) $\sqrt[3]{3^{-9}}$ (d) $\sqrt[4]{(-3)^4}$
3. (a) $\sqrt[3]{x^6}$ (b) $\sqrt{x^4}$
   (c) $\sqrt[4]{(x+1)^8}$ (d) $\sqrt[4]{x^8}$
4. (a) $\sqrt[4]{x^{16}}$ (b) $\sqrt{(x^2+1)^2}$
   (c) $-\sqrt{(1+2x)^4}$ (d) $\sqrt[5]{x^{15}}$

In problems 5 and 6, evaluate each expression.

5. (a) $8^{2/3}$ (b) $(-32)^{4/5}$
6. (a) $(-64)^{2/3}$ (b) $-64^{(-2/3)}$

In problems 7–10, perform each operation and express the result in simplified form with positive exponents.

7. (a) $7^{(-1/2)} \cdot 7^{5/2}$ (b) $(x^{(-3/4)})^{(-8/2)}$
8. (a) $\left(\frac{32}{x^{-5}}\right)^{-2/5}$ (b) $\frac{x^{2/3}}{x^{(-4/5)}}$

9. (a) $x^{(-2/3)} \cdot x^{5/3}$ (b) $(8^{(-2/3)})^{-6}$

10. (a) $(32u^{-5})^{(-3/5)}$ (b) $\left(\frac{125}{y^3}\right)^{(-1/3)}$

In problems 11–18, simplify each expression. Assume all variables are restricted to values for which the radical expressions are defined.

11. (a) $\sqrt[5]{-64}$ (b) $5\sqrt[3]{-16}$

12. (a) $\sqrt{25x^3}$ (b) $\sqrt[3]{(x^3 + y^3)^6}$

13. (a) $2xy^2\sqrt[4]{x^7y^5}$ (b) $\sqrt[3]{32x^5}$

14. (a) $\sqrt[3]{250x^4y^7}$ (b) $\sqrt[5]{-x^3y^4t^5z^{27}}$

15. $\dfrac{\sqrt[3]{200uv^7}}{\sqrt[3]{25u^4v}}$

16. $\dfrac{\sqrt{9r} \cdot \sqrt{3r^4}}{\sqrt{3r^3}}$

17. $\dfrac{\sqrt[3]{9t^5} \cdot \sqrt[3]{27t^2}}{\sqrt[3]{216t^4}}$

18. $\dfrac{\sqrt[3]{p^2q^3} \cdot \sqrt[3]{125p^3q^2}}{\sqrt[3]{8p^3q^4}}$

7 Performing Operations on Radical Expressions

To add or subtract radicals, we simplify each of the radicals and then combine like terms, as we did with polynomials.

EXAMPLE 1 Adding and subtracting radicals

Find: $4\sqrt[3]{3x} + \sqrt[3]{24x} - \sqrt[3]{81x}$

Solution

$$\begin{aligned} 4\sqrt[3]{3x} + \sqrt[3]{24x} - \sqrt[3]{81x} &= 4\sqrt[3]{3x} + \sqrt[3]{8}\cdot\sqrt[3]{3x} - \sqrt[3]{27}\cdot\sqrt[3]{3x} \\ &= 4\sqrt[3]{3x} + 2\sqrt[3]{3x} - 3\sqrt[3]{3x} \\ &= 3\sqrt[3]{3x} \end{aligned}$$

Simplify radicals

Combine like terms

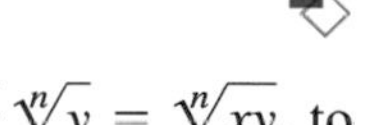

We use the distributive property, along with the property $\sqrt[n]{x}\sqrt[n]{y} = \sqrt[n]{xy}$, to multiply expressions containing radicals. Note that if $x > 0$, then

$$\sqrt{x}\cdot\sqrt{x} = \sqrt{x^2} = x \qquad \text{that is,} \qquad \left(\sqrt{x}\right)^2 = x.$$

EXAMPLE 2 Multiplying radicals

Perform each multiplication and simplify the result.

(a) $2\sqrt{3}\left(\sqrt{6} - \sqrt{5}\right)$ (b) $\left(3\sqrt{x} + 5\sqrt{y}\right)\left(3\sqrt{x} - 5\sqrt{y}\right)$

(c) $\left(3\sqrt{2} - 2\sqrt{3}\right)^2$

Solution

$$\begin{aligned} \text{(a) } 2\sqrt{3}\left(\sqrt{6} - \sqrt{5}\right) &= 2\sqrt{3}\cdot\sqrt{6} - 2\sqrt{3}\cdot\sqrt{5} && \text{Distribute} \\ &= 2\sqrt{18} - 2\sqrt{15} && \sqrt{x}\cdot\sqrt{y} = \sqrt{xy} \\ &= 2\sqrt{9\cdot 2} - 2\sqrt{15} && \text{Simplify} \\ &= 2\cdot 3\sqrt{2} - 2\sqrt{15} && \\ &= 6\sqrt{2} - 2\sqrt{15} && \text{Multiply} \end{aligned}$$

(b) $\left(3\sqrt{x} + 5\sqrt{y}\right)\left(3\sqrt{x} - 5\sqrt{y}\right) = \left(3\sqrt{x}\right)^2 - \left(5\sqrt{y}\right)^2$ — Special product 3

$= 9x - 25y$ — $\left(\sqrt{a}\right)^2 = a$

(c) $\left(3\sqrt{2} - 2\sqrt{3}\right)^2 = \left(3\sqrt{2}\right)^2 - 2\left(3\sqrt{2}\right)\left(2\sqrt{3}\right) + \left(2\sqrt{3}\right)^2$ — Special product 2

$= 18 - 12\sqrt{6} + 12$ — Multiply

$= 30 - 12\sqrt{6}$ — Combine like terms

In Example 2b, the product of the two radical expressions contains no radical. In such a case, we say that the two expressions are **rationalizing factors** for each other. We often rewrite a fraction so that there are no radicals in the denominator by multiplying the numerator and denominator by a rationalizing factor for the denominator. This process is called **rationalizing the denominator**.

EXAMPLE 3 Rationalizing the denominator

Rationalize the denominator of each fraction.

(a) $\dfrac{6}{\sqrt{3}}$ (b) $\dfrac{4}{2\sqrt{x} - 3\sqrt{y}}$

Solution

(a) The rationalizing factor is $\sqrt{3}$ since $\sqrt{3} \cdot \sqrt{3} = 3$. So

$$\frac{6}{\sqrt{3}} = \frac{6}{\sqrt{3}} \cdot \frac{\sqrt{3}}{\sqrt{3}} = \frac{6\sqrt{3}}{3} = 2\sqrt{3}$$

(b) Here the rationalizing factor is $2\sqrt{x} + 3\sqrt{y}$. So

$$\frac{4}{2\sqrt{x} - 3\sqrt{y}} = \frac{4}{2\sqrt{x} - 3\sqrt{y}} \cdot \frac{2\sqrt{x} + 3\sqrt{y}}{2\sqrt{x} + 3\sqrt{y}}$$

$$= \frac{4\left(2\sqrt{x} + 3\sqrt{y}\right)}{\left(2\sqrt{x}\right)^2 - \left(3\sqrt{y}\right)^2} = \frac{8\sqrt{x} + 12\sqrt{y}}{4x - 9y}$$

PROBLEM SET 7

In problems 1–16, perform the indicated operations and simplify the result. Assume each variable is a positive real number.

1. (a) $\sqrt{27} - \sqrt{12}$ (b) $\sqrt{45} + \sqrt{80}$
2. (a) $5\sqrt{32} + 7\sqrt{72}$ (b) $10\sqrt{20} - 3\sqrt{45}$
3. (a) $\sqrt{72x} - \sqrt{50x}$ (b) $\sqrt{80x} + \sqrt{20x}$
4. (a) $\sqrt[4]{162} - \sqrt[4]{32}$ (b) $\sqrt[3]{250} + \sqrt[3]{54}$
5. $-3\sqrt{32} + \sqrt{8} - \sqrt{2}$
6. $5\sqrt[3]{54} - 2\sqrt[3]{16}$
7. $9\sqrt{27x^2} - 5x\sqrt{3} - 3\sqrt{12x^2}$
8. $\sqrt{p^3} + \sqrt{25p^3} + \sqrt{9p}$

9. $\sqrt[3]{8x^4} + 2\sqrt[3]{-125x^4} - x\sqrt[3]{x}$

10. $a\sqrt{2a^3b} + a^2\sqrt{32ab} - \sqrt{162a^5b}$

11. $(4\sqrt{x} - 9\sqrt{y})(4\sqrt{x} + 9\sqrt{y})$

12. $(\sqrt{3} - 5\sqrt{2})^2$

13. $(\sqrt{5} + \sqrt{10})^2$

14. $(\sqrt{10x} + \sqrt{2y})(\sqrt{10x} - \sqrt{2y})$

15. $\sqrt{2}(\sqrt{6} - 2\sqrt{14} + \sqrt{2})$

16. $(5\sqrt{2} - \sqrt{6})(-2\sqrt{2} + \sqrt{6})$

In problems 17–20, rationalize the denominator and simplify the result.

17. (a) $\dfrac{2}{\sqrt{2}}$ (b) $\dfrac{-6}{\sqrt{14}}$

18. (a) $\dfrac{10x}{3\sqrt{5}}$ (b) $\dfrac{5}{\sqrt{3} - 1}$

19. (a) $\dfrac{1 - \sqrt{x}}{1 + \sqrt{x}}$ (b) $\dfrac{\sqrt{x} + \sqrt{y}}{\sqrt{x} - \sqrt{y}}$

20. (a) $\dfrac{\sqrt{y+3} + \sqrt{y}}{\sqrt{y+3} - \sqrt{y}}$ (b) $\dfrac{1}{\sqrt{t} + \sqrt{t-1}}$

In problems 21 and 22, rewrite each expression by rationalizing the numerator.

21. $\dfrac{\sqrt{x+h} - \sqrt{x}}{h}$

22. $\dfrac{\sqrt{x+3} - \sqrt{x}}{3}$

8 Performing Operations on Complex Numbers

Earlier we saw that symbols such as $(-9)^{1/2}$, $\sqrt{-4}$, and $\sqrt{-x}$ (where x is positive) do not represent real numbers because there is no real number whose square is negative. In order to define square roots of negative numbers, we have a set of numbers that contains the set of real numbers and also square roots of negative numbers. We call this set of numbers the set of *complex numbers* and denote it by C. In general, a number of the form

$$a + bi \qquad \text{where } a \text{ and } b \text{ are real numbers}$$

is called a **complex number**. The symbol i denotes a number whose square is -1; that is, we define i by the equation

$$\boxed{i^2 = -1}$$

and we write $\sqrt{-1} = i$.

Using i, we can define the principal square root of -4 as follows:

$$\sqrt{-4} = \sqrt{4(-1)} = \sqrt{4} \cdot \sqrt{-1} = 2i$$

and, if x is nonnegative,

$$\sqrt{-25x^2} = \sqrt{25x^2(-1)} = \sqrt{25x^2} \cdot \sqrt{-1} = 5xi$$

Also, if i is raised to a positive integer power, we obtain results such as

$$i^3 = i^2 \cdot i = -i$$
$$i^4 = i^2 \cdot i^2 = (-1)(-1) = 1$$
$$i^5 = i^4 \cdot i = 1 \cdot i = i$$

TABLE 1

Examples of Complex Numbers

Number	Real Part	Imaginary Part
$3 + 5i$	3	5
$\sqrt{5} - 2i$	$\sqrt{5}$	-2
$-1 - 7i$	-1	-7
$8i = 0 + 8i$	0	8
$5 = 5 + 0i$	5	0

For the complex number $z = a + bi$, a is called the **real part** of z and b is called the **imaginary part** of z. Some examples of complex numbers are given in Table 1.

Since any real number a can be expressed in the complex number form $a + 0i$, we conclude that the complex number set includes all real numbers. Equality, addition, subtraction, and multiplication of the two complex numbers $a + bi$ and $c + di$ are defined as follows:

ALGEBRA OF COMPLEX NUMBERS

(i) **Equality:** $a + bi = c + di$ if and only if $a = c$ and $b = d$
(ii) **Addition:** $(a + bi) + (c + di) = (a + c) + (b + d)i$
(iii) **Subtraction:** $(a + bi) - (c + di) = (a - c) + (b - d)i$
(iv) **Multiplication:** $(a + bi)(c + di) = (ac - bd) + (ad + bc)i$

It is unnecessary to memorize the above definitions. When working with complex numbers, we may treat them the same as polynomials, where i is treated like a variable with the added property that $i^2 = -1$.

EXAMPLE 1 Combining complex numbers

Perform each operation and simplify.

(a) $(3 + 2i) + (5 - 8i)$ (b) $(4 - 7i) - (1 + 2i)$ (c) $(3 - 5i)(2 + 3i)$

Solution

(a) $(3 + 2i) + (5 - 8i) = (3 + 5) + (2 - 8)i = 8 - 6i$

(b) $(4 - 7i) - (1 + 2i) = (4 - 1) + (-7 - 2)i = 3 - 9i$

(c) $(3 - 5i)(2 + 3i) = 6 - 10i + 9i - 15i^2$
$= 6 - i - 15(-1)$
$= 6 + 15 - i = 21 - i$

We divide complex numbers by applying a process similar to the one we used to rationalize the denominators of radical expressions. In this case, the factor that plays the role of the rationalizing factor is called the *complex conjugate*. If $z = a + bi$, then the **complex conjugate** is denoted by $\bar{z} = \overline{a + bi}$, and it is defined as

$$\bar{z} = \overline{a + bi} = a - bi$$

For example:

If $z = 3 + 2i$, then its complex conjugate is $\bar{z} = \overline{3 + 2i} = 3 - 2i$.
If $z = -5 - 7i$, then $\bar{z} = -5 + 7i$.

Note that, if $z = a + bi$, then

$$z\bar{z} = (a + bi)(a - bi) = a^2 - b^2i^2 = a^2 - b^2(-1) = a^2 + b^2$$

which is a real number.

The *division* of two complex numbers is accomplished by multiplying the numerator and the denominator by the conjugate of the denominator. This results in a fraction that contains a real number in the denominator, as the next example shows.

EXAMPLE 2 Dividing complex numbers

Perform each division and express the answer in the form $a + bi$.

(a) $\dfrac{1}{5 + 3i}$ (b) $\dfrac{5 - 10i}{3 - 4i}$

Solution By multiplying the numerator and denominator of each expression by the conjugate of the denominator, we have:

(a) $$\frac{1}{5 + 3i} = \frac{1}{5 + 3i} \cdot \frac{5 - 3i}{5 - 3i}$$
$$= \frac{5 - 3i}{25 + 9} = \frac{5 - 3i}{34} = \frac{5}{34} - \frac{3}{34}i = \frac{5}{34} + \left(\frac{-3}{34}\right)i$$

(b) $$\frac{5 - 10i}{3 - 4i} = \frac{5 - 10i}{3 - 4i} \cdot \frac{3 + 4i}{3 + 4i} = \frac{55 - 10i}{9 + 16} = \frac{55 - 10i}{25}$$
$$= \frac{5(11 - 2i)}{25} = \frac{11 - 2i}{5} = \frac{11}{5} - \frac{2}{5}i = \frac{11}{5} + \left(\frac{-2}{5}\right)i$$

PROBLEM SET 8

In problems 1 and 2, write each expression in the form $a + bi$.

1. (a) $\sqrt{-16}$ (b) $3 + \sqrt{-8}$
   (c) $7 - 3\sqrt{-9}$ (d) $\dfrac{2 - \sqrt{-4}}{6}$
   (e) $\sqrt{-x^2}$, where $x > 0$
   (f) $i^6 + 5$
2. (a) $2 - \sqrt{-4x^4}$ (b) $6 - \sqrt{-16y^4}$
   (c) $17 + 5\sqrt{-17}$ (d) $\dfrac{9 + \sqrt{-27}}{3}$
   (e) $-\sqrt{-x^2}$, where $x < 0$
   (f) $i^7 - 2$

In problems 3–10, perform each operation and simplify.

3. (a) $(3 - 2i) + (7 - 3i)$ (b) $(4 - i) - (7 - 3i)$
4. (a) $(-3 + 6i) - (3 + 7i)$
   (b) $(7 + 24i) + (-3 - 4i)$
5. (a) $(3 - 6i)(2 + i)$ (b) $(-5 + 2i)(-3 + 2i)$
6. (a) $(3 + 2i)(1 + i)$ (b) $(2 + i)(2 - i)$
7. (a) $(2 + i)^2$ (b) $(3 - 4i)^2$
8. (a) $\dfrac{4 + 3i}{2 - 4i}$ (b) $\dfrac{4 - 3i}{-1 - 2i}$
9. (a) $\dfrac{6 + 4i}{1 + 5i}$ (b) $\dfrac{7 + 6i}{-5 + 2i}$
10. (a) $\dfrac{4 - \sqrt{3}\,i}{2 + \sqrt{3}\,i}$ (b) $\dfrac{3 - 5i}{\sqrt{5} + 4i}$

CUMULATIVE PROBLEM SET

1. Express each rational number as a terminating or a repeating decimal.
(a) $-\frac{5}{3}$ **(b)** $\frac{0.41}{2.7}$ **(c)** $-\frac{9}{5}$ **(d)** $\frac{4.2}{0.037}$

2. Convert each decimal to a fraction in lowest terms.
(a) 0.11 **(b)** 0.22 **(c)** 8.05 **(d)** −2.125

3. Locate the numbers 0, −12, $\frac{3}{5}$, 2.41, and $\sqrt{29}$ on a number line.

4. State the property that justifies each statement
(a) $x + (y + 7) = x + (7 + y)$
(b) $2x(y + z) = 2xy + 2xz$
(c) $x + (-x) = 0$ **(d)** $2(xy) = (2x)y$

5. The volume V and the surface area S of a sphere of radius r are given by the formulas

$$V = \frac{4\pi r^3}{3} \quad \text{and} \quad S = 4\pi r^2$$

Find V and S for:
(a) $r = 7.53$ inches **(b)** $r = 0.17$ inch
Round off the answers to two decimal places.

6. Identify each polynomial as a monomial, binomial, or trinomial. Give the degree and nonzero numerical coefficients.
(a) $8x - 5$ **(b)** $-7x^2$
(c) $-2y^3 - 4y^2 + 7y + 3$

7. Evaluate the expression $5x^2 + 2xy + 7y^2$ for $x = -2.7$ and $y = 3.5$.

8. Perform each operation and simplify the result.
(a) $(2x^2 - 4) + (-5x^2 + 3)$
(b) $(3x^2 + 7x + 8) - (2x^2 - 3x + 7)$
(c) $(-x)^5(2x)^4$
(d) $(-y)^2(-y)^3(-y)^7$
(e) $-4x^3(-2x^2 - 3x + 1)$
(f) $(2x - 3)(x + 7)$
(g) $(2x - 1)^2$
(h) $(2x - y)(4x^2 + 2xy + y^2)$

9. Factor each polynomial completely.
(a) $3x^2y^2 + 6x^2z^2 - 9x^2$
(b) $x^2 + 7x - 18$ **(c)** $x^4 - 16y^4$
(d) $27x^3 + y^3$

10. Reduce the fraction to lowest terms:

$$\frac{6x^2 - 7x - 3}{4x^2 - 8x + 3}$$

11. Perform the indicated operation and simplify the result. Express the answer in factored form.
(a) $\frac{c}{c - 1} \cdot \frac{c^2 - 1}{c^3}$
(b) $\frac{x^2 + 5x + 6}{x^2 - 4} \div \frac{x^2 + 4x + 4}{x^2 - 4x + 4}$
(c) $\frac{6}{x^2 - 2x - 8} + \frac{5}{x^2 + 2x}$
(d) $\frac{5}{x - 3} - \frac{5}{3 - x}$ **(e)** $\frac{x - \frac{1}{y}}{1 - \frac{x}{y}}$

12. Rewrite each expression so that it contains only positive exponents.
(a) $(3x)^{-3}$ **(b)** $\left(\frac{p^4}{2}\right)^0$
(c) $3^{-2} - 5^{-2}$ **(d)** $\frac{1}{2^{-2} + 3^{-2}}$
(e) $(125)^{(-4/3)}$ **(f)** $\left(\frac{32}{x^{-5}}\right)^{(-2/5)}$
(g) $(2x^{1/3}y^{3/2})^6$

13. Simplify each expression, where x is positive.
(a) $\sqrt{(x^2 + 7)^2}$ **(b)** $-\sqrt[4]{x^8}$
(c) $\sqrt[6]{(x + 1)^6}$

14. Use the properties of radicals to simplify each expression.
(a) $\sqrt{27}$ **(b)** $\sqrt[3]{-54}$
(c) $\sqrt[7]{\frac{-2}{x^{14}}}$ **(d)** $\sqrt{4x^5}$

15. Perform each operation and simplify.
(a) $5\sqrt{8} - \sqrt{32}$ **(b)** $2\sqrt[3]{54} + 2\sqrt[3]{16}$
(c) $(\sqrt{2} + 4\sqrt{3})^2$
(d) $(\sqrt{3} - \sqrt{2})(2\sqrt{3} + \sqrt{2})$
(e) $(2\sqrt{10} - 5)(2\sqrt{10} + 5)$

16. Rationalize the denominator and simplify.
(a) $\frac{5}{\sqrt{6}}$ **(b)** $\frac{5}{\sqrt{3} - 1}$

17. Perform each operation and simplify.
(a) $(5 - 7i) - (5 - 13i)$ **(b)** $(6 - 8i) + (5 + 3i)$
(c) $(-3 - 2i)(2 + 5i)$ **(d)** $\frac{-3}{3 + 4i}$

1

Coordinate Geometry and Models

CHAPTER CONTENTS

In this chapter we focus on topics that are important to the study of functions and their graphs. We also introduce mathematical modeling and graphing technology. Throughout this text we take advantage of graphing technology to discover and explore new concepts, to extend our investigation and understanding of topics, to demonstrate results, and to solve problems.

1.1 Inequalities

OBJECTIVES

1. Order real numbers and define intervals.
2. Solve linear inequalities.
3. Solve absolute value inequalities.
4. Solve applied problems.

Ordering Real Numbers and Defining Intervals

A number line enables us to establish *order* relationships between real numbers. If point a lies to the left of point b on a number line, we say that a is *less than* b (or equivalently, b is *greater than* a), and we write

$$a < b \qquad (\text{or } b > a) \quad (\text{Figure 1})$$

Number line — a, b; $a < b\ (b > a)$

FIGURE 1

For example, $5 > 3$ and $-8 < -5$.

The order relationship is defined more formally as follows:

DEFINITION
ORDER

> Assume that a and b are real numbers. We say that ***a* is less than *b***, or equivalently, ***b* is greater than *a***, if $b - a$ is a positive number. This **order** is denoted by the **inequality** $a < b$ (or $b > a$).

For instance, -5 is less than 2, since $2 - (-5) = 7$ (which is positive); so we write $-5 < 2$ or $2 > -5$. The inequality symbol $\leq$ means **less than or equal to**, while the symbol $\geq$ means **greater than or equal to**. Also, $a > 0$ means that a is positive, and $a < 0$ means that a is negative.

Certain sets of numbers, defined in terms of the order relation, can be expressed in a special notation called **interval notation**. If a and b are real numbers, with

$a < b$, there are four types of **bounded intervals** consisting of real numbers between a and b, as listed in Table 1.

TABLE 1
Bounded Interval Notation

Terminology for Bounded Intervals	Interval Notation	Inequality Notation	Number Line Representation
Open interval	(a, b)	$a < x < b$	a b
Closed interval	$[a, b]$	$a \leq x \leq b$	a b
Half-open interval	$[a, b)$	$a \leq x < b$	a b
Half-open interval	$(a, b]$	$a < x \leq b$	a b

The numbers a and b are called the **end points** of the closed interval $[a, b]$ from a to b. The brackets indicate the *inclusion* of a and b in the interval. The parentheses for the open interval (a, b) from a to b indicate the *exclusion* of the end points a and b.

EXAMPLE 1 Illustrating bounded intervals

Illustrate each interval on a number line and describe it using inequality notation.

(a) $[-4, 3]$ (b) $(-1, 4)$ (c) $[-2, 1)$

Solution Figure 2 shows each interval and its inequality description.

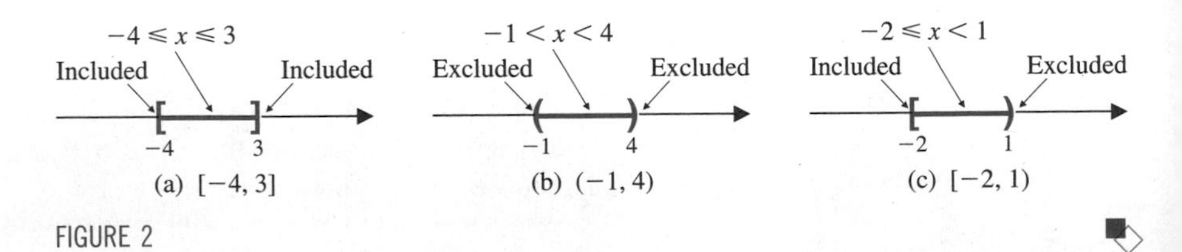

FIGURE 2

If a is a real number, there are five types of **infinite**, or **unbounded**, intervals consisting of real numbers as listed in Table 2.

Note that ∞ ("infinity") and $-\infty$ are convenient symbols; they are *not* real numbers. For instance, the interval $(0, \infty)$ consists of the positive real numbers, and the interval $(-\infty, 0)$ consists of the negative real numbers. Note that the interval $(-\infty, \infty)$ may also be represented by $\mathbb{R}$.

TABLE 2

Unbounded Interval Notation

Interval Notation	Inequality Notation	Number Line Representation
(a, ∞)	$a < x$	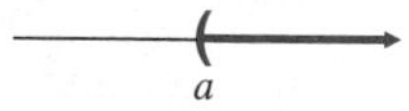
$[a, \infty)$	$a \leq x$	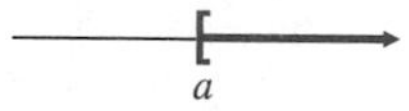
$(-\infty, a)$	$x < a$	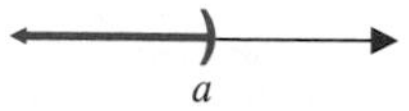
$(-\infty, a]$	$x \leq a$	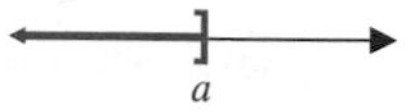
$(-\infty, \infty)$	$-\infty < x < \infty$	

EXAMPLE 2 Illustrating unbounded intervals

Illustrate each interval on a number line and describe it with inequality notation.

(a) $[2, \infty)$ (b) $(-\infty, 3)$

Solution Figure 3 shows each interval and its inequality description.

(a) $[2, \infty)$ (b) $(-\infty, 3)$

FIGURE 3

Solving Linear Inequalities

Inequalities such as

$$3x + 2 \leq 5 \qquad \text{and} \qquad 5x - 2 \geq 7x + 11$$

are called *linear inequalities* in one variable. To solve a linear inequality, we use the following properties:

PROPERTIES OF INEQUALITIES

Assume a, b, and c are real numbers.

The Addition and Subtraction Properties:

(i) If $a < b$, then $a + c < b + c$ and $a - c < b - c$.

The Multiplication and Division Properties:

(ii) If $a < b$ and $c > 0$, then $ac < bc$ and $\frac{a}{c} < \frac{b}{c}$.

(iii) If $a < b$ and $c < 0$, then $ac > bc$ and $\frac{a}{c} > \frac{b}{c}$.

To **graph** the solution of an inequality on a number line is to locate the set of all real numbers that satisfy the inequality.

EXAMPLE 3 Solving linear inequalities

Solve each inequality, express the solution in interval notation, and graph it on a number line.

(a) $2x - 5 > 3$ (b) $2 - 3x \leq 11$ (c) $-5 \leq 3x + 1 < 7$

Solution

(a) $2x - 5 > 3$ Given

$2x > 8$ Add 5 to each side; Property (i)

$x > 4$ Divide each side by 2; Property (ii)

FIGURE 4a

So the solution includes all real numbers greater than 4. The interval notation of the solution is $(4, \infty)$, and the graph is shown in Figure 4a.

(b) $2 - 3x \leq 11$ Given

$-3x \leq 9$ Subtract 2 from each side; Property (i)

$x \geq -3$ Divide each side by -3 and reverse the inequality sign; Property (iii)

FIGURE 4b

So the solution consists of all real numbers greater than or equal to -3. The interval notation of this solution is $[-3, \infty)$. Its graph is shown in Figure 4b.

(c) $-5 \leq 3x + 1 < 7$ Given

$-6 \leq 3x < 6$ Subtract 1 from all three parts; Property (i)

$-2 \leq x < 2$ Divide each term by 3; Property (ii)

FIGURE 4c

The solution consists of all real numbers between -2 and 2, including -2. In interval notation, the solution is $[-2, 2)$, and its graph is shown in Figure 4c.

Solving Absolute Value Inequalities

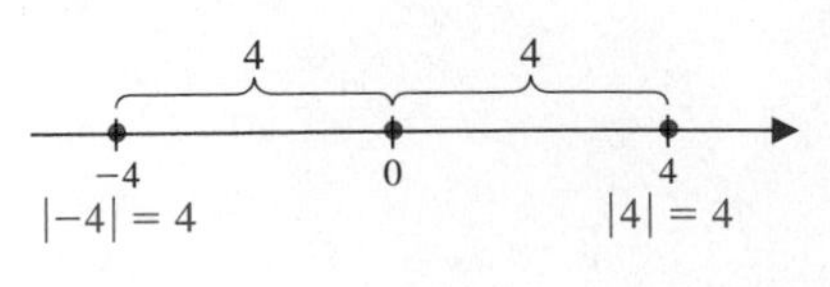

FIGURE 5

The number line provides us with a way of introducing the concept of *absolute value*. If a is a real number, then $|a|$, called the *absolute value* of a, represents the distance on a number line between point a and the 0 point. If $a \neq 0$, then $-a$ and a are the two numbers whose distance to 0 is $|a|$ units. For instance, -4 and 4 are the two numbers whose distance to the 0 point is 4 units (Figure 5), and we write

$$|-4| = |4| = 4$$

From an algebraic point of view, we have the following definition:

DEFINITION

ABSOLUTE VALUE

If a is a real number, then the **absolute value** of a, denoted by $|a|$, is defined as follows:

$$|a| = \begin{cases} a & \text{if } a \text{ is positive or zero; that is, if } a \geq 0 \\ -a & \text{if } a \text{ is negative; that is, if } a < 0 \end{cases}$$

For instance, $|7| = 7$ because 7 is positive and $|-7| = -(-7) = 7$ because -7 is negative; $|\sqrt{3} - 3| = -(\sqrt{3} - 3) = 3 - \sqrt{3}$ because $\sqrt{3} - 3$ is negative.

Sometimes in calculus it is necessary to rewrite an absolute value expression in a form that contains no absolute value.

EXAMPLE 4 Rewriting an absolute value expression

Rewrite $|w - 2|$ in a form without absolute value.

Solution According to the definition of absolute value,

$$|w - 2| = \begin{cases} w - 2 & \text{if } w - 2 \geq 0\text{; that is, if } w \geq 2 \\ -(w - 2) = 2 - w & \text{if } w - 2 < 0\text{; that is, if } w < 2 \end{cases}$$

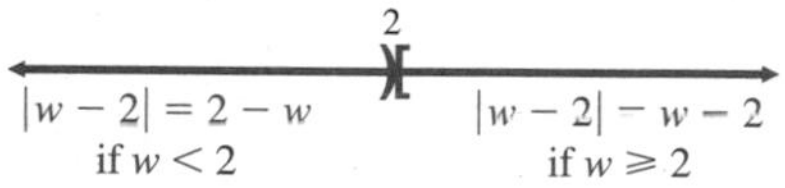

2

$|w - 2| = 2 - w$ if $w < 2$

$|w - 2| = w - 2$ if $w \geq 2$

FIGURE 6

In other words, $|w - 2|$ can be replaced with $w - 2$ if $w \geq 2$, and $|w - 2|$ can be replaced with $2 - w$ if $w < 2$ (Figure 6).

The following properties of absolute value should be noted:

(i) $|a| \geq 0$ for any real number a.
(ii) $|0| = 0$.
(iii) $|a| = |-a|$ for any real number a.

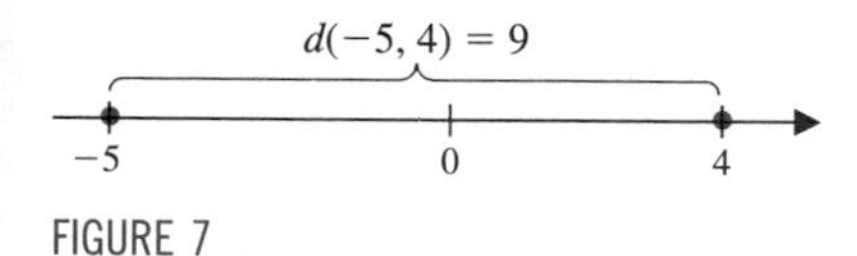

FIGURE 7

We can use absolute value to find the distance between numbers on a number line. Consider the number line in Figure 7, which displays the locations of -5 and 4. The distance between them, which we denote as $d(-5, 4)$, can be found in either of the following ways:

$$d(-5, 4) = |4 - (-5)| = |9| = 9$$

or

$$d(-5, 4) = |(-5) - 4| = |-9| = 9$$

This observation leads to the following general property:

DISTANCE BETWEEN TWO POINTS

The **distance** between two points on a number line whose coordinates are a and b, denoted by $d(a, b)$, is given by

$$d(a, b) = |a - b| \qquad \text{or by} \qquad d(a, b) = |b - a|$$

EXAMPLE 5 **Finding the distance on a number line**

Find the distance on a number line between each pair of points with the given coordinates. Display the result on a number line.

(a) -5 and 9 (b) -8 and -1 (c) a and 3, where $a < 3$

Solution

	Given Numbers	Distance Between Points	Graphical Display
(a)	-5 and 9	$d(-5, 9) = \lvert 9 - (-5) \rvert = \lvert 14 \rvert = 14$ or $d(-5, 9) = \lvert (-5) - 9 \rvert = \lvert -14 \rvert = 14$	14 -5 9
(b)	-8 and -1	$d(-8, -1) = \lvert -1 - (-8) \rvert = \lvert 7 \rvert = 7$ or $d(-8, -1) = \lvert -8 - (-1) \rvert = \lvert -7 \rvert = 7$	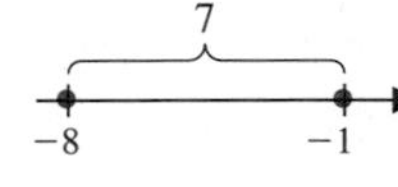
(c)	a and 3, where $a < 3$	$d(a, 3) = \lvert 3 - a \rvert = 3 - a$ or $d(a, 3) = \lvert a - 3 \rvert = -(a - 3) = 3 - a$	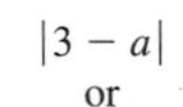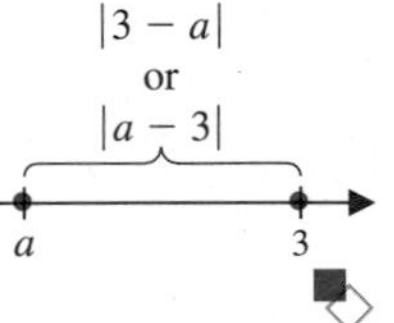

To solve inequalities involving absolute values we use the following properties:

PROPERTIES OF INEQUALITIES INVOLVING ABSOLUTE VALUES

If $p > 0$, then:

(i) $|u| < p$ if and only if $-p < u < p$ (Figure 8a)
(ii) $|u| > p$ if and only if $u < -p$ or $u > p$ (Figure 8b)

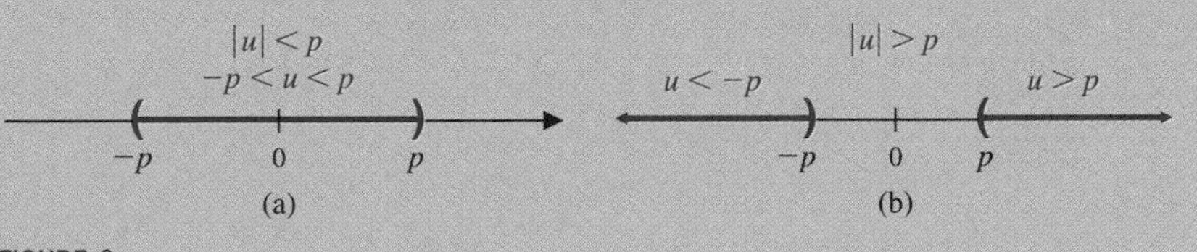

FIGURE 8

These properties also apply if the signs $<$ and $>$ are replaced by $\leq$ and $\geq$, respectively.

EXAMPLE 6 **Solving absolute value inequalities**

Solve each absolute value inequality, write the solution in interval notation, and graph it on a number line.

(a) $|x - 1| \leq 2$ (b) $|2x - 3| > 5$

Solution

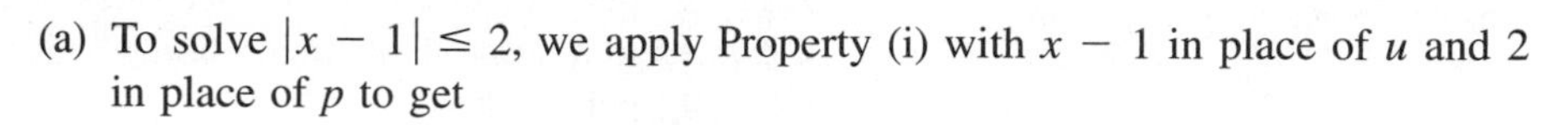

(a) To solve $|x - 1| \leq 2$, we apply Property (i) with $x - 1$ in place of u and 2 in place of p to get

FIGURE 9a

$$-2 \leq x - 1 \leq 2$$
$$-1 \leq x \leq 3 \qquad \text{Add 1 to each part}$$

Thus, the solution consists of all numbers in the interval $[-1, 3]$ (Figure 9a).

(b) We apply Property (ii) with $2x - 3$ in place of u and 5 in place of p to get the following inequalities:

$$2x - 3 < -5 \quad \text{or} \quad 2x - 3 > 5$$
$$2x < -2 \quad \Big| \quad 2x > 8$$
$$x < -1 \quad \Big| \quad x > 4$$

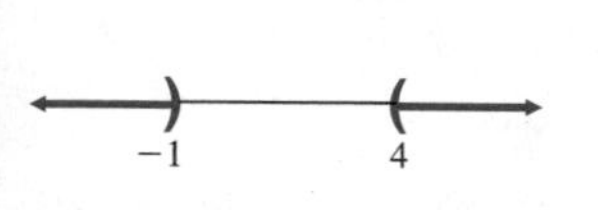

FIGURE 9b

Therefore, the solution consists of all numbers in *either* of the intervals $(-\infty, -1)$ or $(4, \infty)$ (Figure 9b).

Solving Applied Problems

The following examples illustrate how to represent real-life situations with inequality statements.

EXAMPLE 7 Representing exercise restrictions

A doctor advises a heart patient to exercise at least 20 minutes but no more than 30 minutes per day.

(a) Use an inequality to express the range of minutes of exercise each day.
(b) Use an inequality to express the range of hours of exercise for 10 days.

Solution

The phrase "at least 20" means 20 or more and the phrase "no more than 30" indicates 30 or less.

(a) If M represents the number of minutes of daily exercise recommended by the doctor, then

$$20 \leq M \leq 30$$

(b) Since 20 minutes is the same as $\frac{20}{60} = \frac{1}{3}$ hour, and since 30 minutes is the same as $\frac{30}{60} = \frac{1}{2}$ hour, then the daily range of h hours is given by $\frac{1}{3} \leq h \leq \frac{1}{2}$. Over a 10 day period, the range D of hours of exercise is given by

$$10\left(\tfrac{1}{3}\right) \leq D \leq 10\left(\tfrac{1}{2}\right)$$

or

$$\tfrac{10}{3} \leq D \leq 5$$

That is, the patient should have done at least $3\frac{1}{3}$ hours (or 3 hours and 20 minutes) of exercise, but no more than 5 hours, during a 10 day period.

EXAMPLE 8 Finding the number of newspaper home deliveries

A newspaper carrier earns \$12 per week plus 60¢ per week for each newspaper delivered to a home. How many home deliveries must the carrier make to earn at least \$42 per week?

Solution Let x represent the number of home deliveries. Since the carrier earns \$12 per week plus 60¢ for each delivery, the weekly earnings (in dollars) are represented by

$$12 + 0.60x \text{ dollars}$$

In order to earn at least \$42 per week, x must satisfy the following inequality: $12 + 0.60x \geq 42$. So

$$\begin{aligned} 12 + 0.60x &\geq 42 \\ 0.60x &\geq 30 \quad \text{Subtract 12 from each side} \\ x &\geq 50 \quad \text{Divide each side by 0.60} \end{aligned}$$

Therefore, the carrier must deliver at least 50 newspapers to earn at least \$42 per week.

PROBLEM SET 1.1

Mastering the Concepts

In problems 1–4, express each statement using inequality notations.

1. (a) -2 is less than or equal to 3
   (b) -5 is greater than -7
2. (a) x is at most 4
   (b) t is at least -8
3. (a) x lies to the left of a and to the right of 6
   (b) x is nonnegative
4. (a) x is less than 4, and y is at least 4
   (b) x is greater than -5, and y is at most -5

In problems 5 and 6, illustrate each interval on a number line and describe it using inequality notation.

5. (a) $(0, 6)$ (b) $[-1, 4)$ (c) $(1, \infty)$ (d) $(-\infty, 5]$
6. (a) $[1, 7)$ (b) $(-2, 3]$ (c) $(-\infty, 0)$ (d) $[4, \infty)$

In problems 7–16, solve each inequality and express the solution in interval notation. Also graph the solution on a number line.

7. (a) $2x \leq -4$ (b) $-2x \leq 4$
8. (a) $2x > -6$ (b) $-2x > 6$
9. (a) $3x < 9$ (b) $4x + 3 \geq 12$
10. (a) $5x \geq 15$ (b) $3x - 2 > 7$
11. (a) $-21w \leq -63$ (b) $-8t - 4 \leq -16 - 2t$
12. (a) $-4x > -12$ (b) $5 + x < -x + 3$
13. $2 < 16 + 4x < 10$
14. $-5 \leq 3x + 2 < 4$
15. $\frac{x}{3} + 2 \leq \frac{x}{4} - 2x$
16. $\frac{x + 6}{5} \leq 4 - \frac{3x}{7}$

In problems 17–22, rewrite each expression without absolute value notation.

17. (a) $|\pi - 5|$ (b) $|\sqrt{7} - 2|$
18. (a) $|\sqrt{5} - 3|$ (b) $|2 - \sqrt{11}|$
19. (a) $|x - 5|$ (b) $|x + 3|$
20. (a) $|y + 0.1|$ (b) $|t - \frac{5}{2}|$
21. $|2x + 1|$
22. $|3x - 2|$

In problems 23–28, find the distance between each pair of numbers.

23. 2 and -6
24. -7 and -3
25. -4 and 4
26. -3 and 3
27. $-4t$ and $6t$
28. $-8t$ and $-6t$

In problems 29–38, solve each absolute value inequality. Express the solution in interval notation and graph it on a number line.

29. **(a)** $|x| \le 3$ **(b)** $|x| > 2$

30. **(a)** $|x| < 1$ **(b)** $|x| > 4$

31. **(a)** $|x - 2| < 3$ **(b)** $|x - 2| \ge 3$

32. **(a)** $|x + 1| \le 2$ **(b)** $|x + 1| > 2$

33. **(a)** $|x + 2| \ge 5$ **(b)** $|x + 2| < 5$

34. **(a)** $|x + 1| > 7$ **(b)** $|x + 1| \le 7$

35. $|2x + 3| < 1$

36. $|3x - 6| \le 4$

37. $|5x - 3| \ge 2$

38. $|5x - 3| < 2$

Applying the Concepts

39. Temperature Range: The average daily range of temperatures T for a city during a summer month varies from at least 72°F to no more than 98°F.

(a) Use an inequality to describe the temperature range for one day.

(b) If the average temperature decreases 1°F per day, use an inequality to describe the temperature at the end of one week.

40. Electrical Energy Consumption: The average American home uses at least 90 but no more than 120 kilowatt-hours of electricity per month.

(a) Use an inequality to express the average range of kilowatt-hours per day, assuming that one month is 30 days.

(b) Use an inequality to express the average range of kilowatt-hours per hour; per week; per year.

41. Plumber's Wages: A plumber charges $18 per hour plus $30 per service call for home repair work. How long must the plumber work to earn at least $84 for a service call?

42. Investment: Suppose that $5600 is invested in a mutual fund that pays simple interest for 1 year. If an investor wishes to earn at least $280 in interest for the year, what interest rate is required?

43. Diet Program: A clinic advertises that a person can reduce his or her weight by at least 1.5 pounds per week by exercising and special dieting. At the beginning of the diet program, a person weighs 195 pounds. What is the maximum number of weeks before this person's weight will be reduced to 170 pounds?

44. Business Profit: A tire manufacturing company shows a profit; that is, the revenue obtained from the sales of its tires is greater than the cost of producing them. If each tire costs $30 in materials, and if it costs $7500 a week in labor and overhead to produce and sell the tires, how many tires must be sold each week at $85 each in order for the company to show a profit?

45. Stock-Brokerage: A stockbroker uses a computer program to be alerted on the price changes of stocks. Suppose that a certain stock is selling for $18.25 per share, and the price of the stock changes by at least $1.25. The prices x that would set the alert are given by the inequality

$$|18.25 - x| \ge 1.25$$

Solve this inequality and interpret the solution.

46. Body Temperature: During and after surgery, a patient's temperature T (in degrees Fahrenheit) is given by

$$|T - 98.6| \le 2.3$$

Determine the interval over which the patient's temperature varies.

Developing and Extending the Concepts

In problems 47 and 48, two given sets of numbers are combined to form a third set by using the *set operations* of *union* and *intersection*. If A and B are sets of numbers, then the set consisting of all numbers that belong to A or to B (or to both) is called the **union of A and B** and is written as $A \cup B$. The set of all numbers that are in both A and B is called the **intersection of A and B** and is written as $A \cap B$. For example, if we are given intervals $(-\infty, 2]$ and $(1, 9)$, then

$$(-\infty, 2] \cup (1, 9) = (-\infty, 9) \quad \text{(Figure 10a)}$$
$$(-\infty, 2] \cap (1, 9) = (1, 2] \quad \text{(Figure 10b)}$$

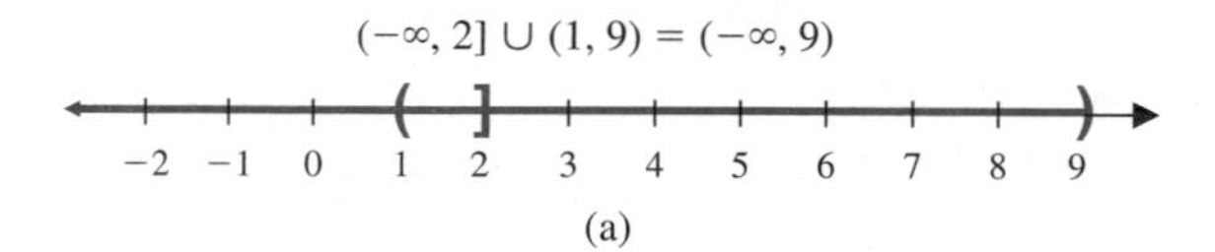

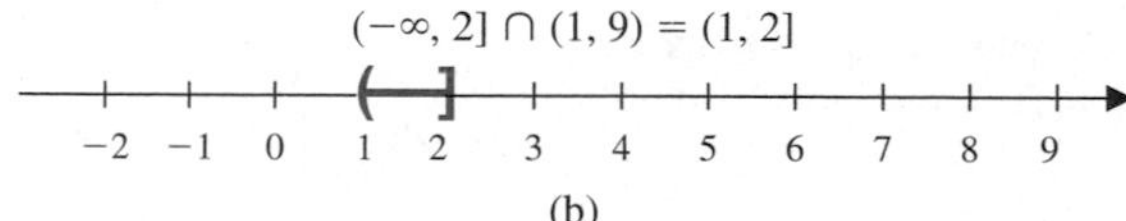

FIGURE 10

Rewrite each set using inequality notation and graph the set on a number line.

47. **(a)** $[0, \infty) \cap [-1, 2)$ **(b)** $(-2, 4) \cup (3, 8]$

48. **(a)** $(-\infty, 2) \cup (-\infty, 5)$ **(b)** $(-\infty, 2) \cap (-\infty, 5)$

49. Try different values of x and y to illustrate the validity of each equation. Then explain why these equations are true for all values of x and y, except $y \neq 0$ in part (b).

(a) $|xy| = |x||y|$ **(b)** $\left|\frac{x}{y}\right| = \frac{|x|}{|y|}$

(c) $|x - y| = |y - x|$

50. Insert $=$ or $<$ between the expressions

$$|x + y| \quad \text{and} \quad |x| + |y|$$

so that the result is true for the given conditions. Give examples to support each assertion.

(a) $x > 0, y > 0$ **(b)** $x < 0, y < 0$

(c) $x < 0, y > 0$ **(d)** $x > 0, y < 0$

51. A student tried to solve the inequality

$$|3x - 4| \geq 5$$

Instead of writing $3x - 4 \geq 5$ or $3x - 4 \leq -5$, which should be done, a mistake was made. In error, the student wrote $5 \leq 3x - 4 \leq -5$. What is wrong with this approach?

52. The **triangle inequality** asserts that $|a + b| \leq |a| + |b|$ for any real numbers a and b. Use the triangle inequality to verify each of the following statements:

(a) If $|x| < 3$ and $|y| < 1$, then $|x + y| < 4$.

(b) If $|x - y| < \frac{1}{10}$ and $|y - t| < \frac{1}{10}$, then $|x - t| < \frac{1}{5}$.

OBJECTIVES

1. Plot points in the plane.
2. Find the distance and the midpoint between two points.
3. Recognize and graph equations of circles.
4. Use mathematical modeling.

1.2 The Cartesian Coordinate System and Modeling

A number line associates points on the line with real numbers. We now extend this concept by introducing the *Cartesian coordinate system* (also called the *rectangular coordinate system*), which enables us to associate points in the plane with *ordered pairs* of real numbers.

Plotting Points in the Plane

Figure 1 displays the **Cartesian**, or **rectangular**, **coordinate system**. It consists of two perpendicular number lines, one horizontal (positive portion to the right and negative portion to the left) and one vertical (positive portion above and negative portion below), called the **coordinate axes**. These axes intersect at a point called the **origin**, which is the zero point on each number line. The horizontal axis is often labeled as the ***x* axis** and the vertical as the ***y* axis**. A plane endowed with a Cartesian coordinate system is called a **Cartesian plane** or the ***xy* plane**.

In applications, the axis may have different labels. For instance, to explore the relation between time t and the distance d traveled by a car, we might label the horizontal axis as the t axis and the vertical axis as the d axis.

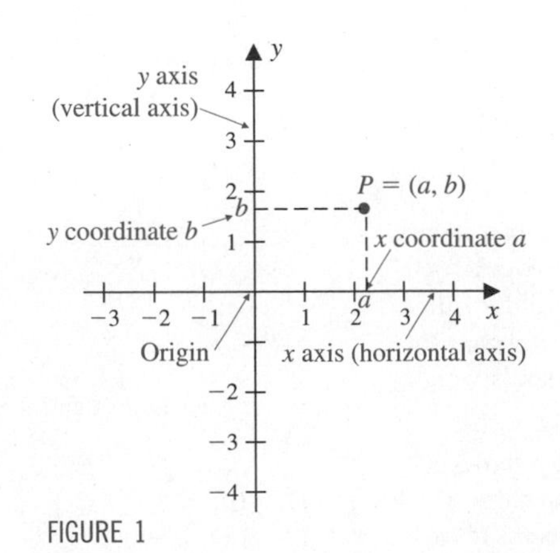

FIGURE 1

Any **ordered pair** of real numbers (a, b) can be represented as a point P in the coordinate system. The first number, a, which is called the ***x* coordinate** (or *abscissa*) of P, indicates the location of P to the right or left of the y axis. The second number, b, which is called the ***y* coordinate** (or *ordinate*) of P, indicates the location of P above or below the x axis. The point P with coordinates (a, b) is called the **graph** of the ordered pair (a, b). We plot, or locate, the position of (a, b) in the plane by placing a dot at the point P (Figure 1). The association between point P and ordered pair (a, b) seems so natural that in practice we write

$$P = (a, b)$$

The coordinate axes divide the plane into four regions called **quadrants**, which are described and displayed in Figure 2.

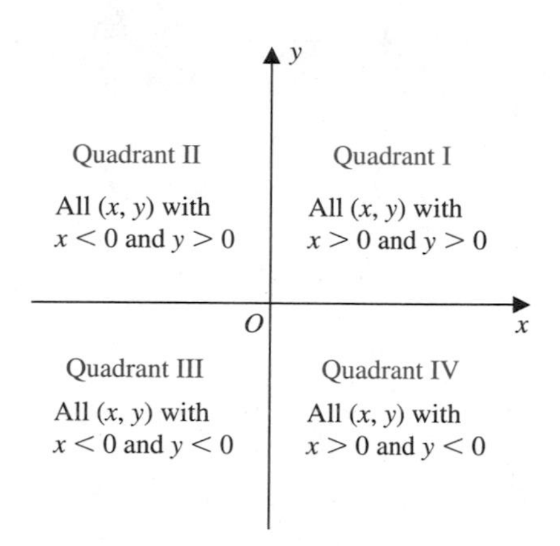

FIGURE 2

EXAMPLE 1 Plotting points

Plot the points (1, 2), (−3, 4), (−2, −6), (4, 0), (0, 0), (0, −5), and (3, −5), and specify their quadrant locations.

Solution Figure 3 shows that the points (1, 2), (−3, 4), (−2, −6), and (3, −5) lie, respectively, in quadrants I, II, III, and IV. The points (4, 0), (0, 0), and (0, −5) lie on the coordinate axes.

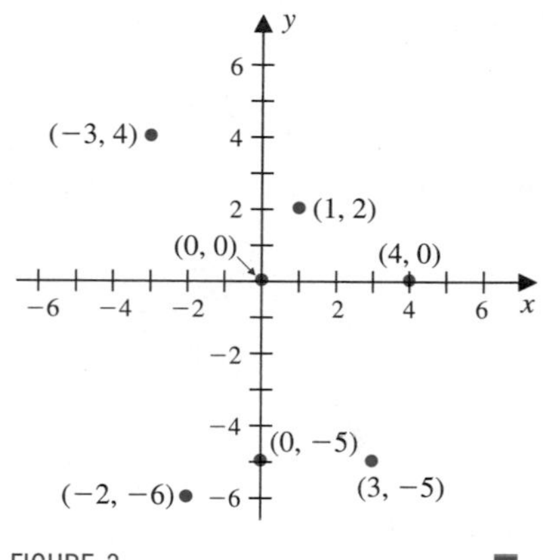

FIGURE 3

Finding the Distance and the Midpoint Between Two Points

Subscripts are used as a reminder that x_1 and y_1 are the coordinates of point P_1, and that x_2 and y_2 are the coordinates of point P_2.

The distance between two points $P_1 = (x_1, y_1)$ and $P_2 = (x_2, y_2)$ can be found in terms of their coordinates by using the following formula:

THE DISTANCE FORMULA

If $P_1 = (x_1, y_1)$ and $P_2 = (x_2, y_2)$ are two points in a Cartesian plane, then the distance d between P_1 and P_2 is given by

$$d = d(P_1, P_2) = \sqrt{(x_2 - x_1)^2 + (y_2 - y_1)^2}$$

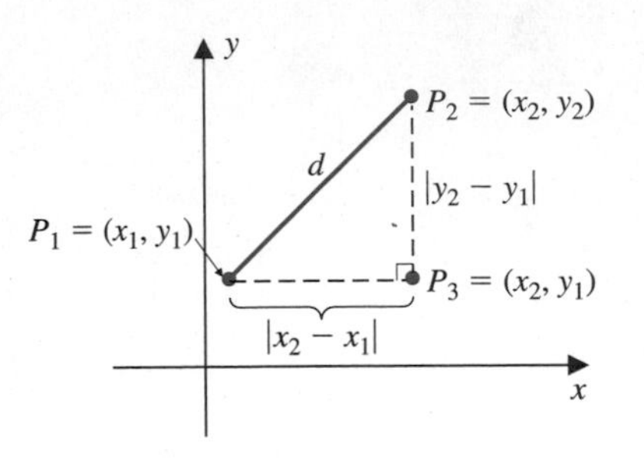

FIGURE 4

The Pythagorean theorem states that, in a right triangle, the square of the length of the hypotenuse is equal to the sum of the squares of the lengths of the other two sides.

Proof To establish this formula, we construct a right triangle $P_1P_2P_3$ by connecting P_1 and P_2 and drawing line segments from P_1 and P_2 parallel to the x and y axis, respectively, as shown in Figure 4. We see that the horizontal distance between P_1 and P_3 is given by

$$|\overline{P_1P_3}| = |x_2 - x_1|$$

and the vertical distance between P_2 and P_3 is given by

$$|\overline{P_2P_3}| = |y_2 - y_1|$$

By applying the Pythagorean theorem, the length d of the hypotenuse is

$$\begin{aligned} d = d(P_1, P_2) &= \sqrt{|\overline{P_1P_3}|^2 + |\overline{P_2P_3}|^2} \\ &= \sqrt{|x_2 - x_1|^2 + |y_2 - y_1|^2} \\ &= \sqrt{(x_2 - x_1)^2 + (y_2 - y_1)^2} \end{aligned}$$

Note that the formula is true even if the two given points line up vertically or horizontally.

EXAMPLE 2 Finding the distance between two points

Find the distance between the two given points. Round off the answer to two decimal places.

(a) $P_1 = (-1, -2)$ and $P_2 = (3, -4)$
(b) $P = (71.37, -27.04)$ and $Q = (14.86, 11.73)$

Solution

The distance formula can also be written as

$$\begin{aligned} d &= d(P_1, P_2) \\ &= \sqrt{(x_1 - x_2)^2 + (y_1 - y_2)^2} \end{aligned}$$

(a) The distance between P_1 and P_2 is given by

$$\begin{aligned} d(P_1, P_2) &= \sqrt{[3 - (-1)]^2 + [-4 - (-2)]^2} \\ &= \sqrt{4^2 + (-2)^2} = \sqrt{20} = 2\sqrt{5} \quad \text{or approx. } 4.47 \end{aligned}$$

(b) The distance between P and Q is given by

$$d(P, Q) = \sqrt{(71.37 - 14.86)^2 + (-27.04 - 11.73)^2} \quad \text{or approx. } 68.53$$

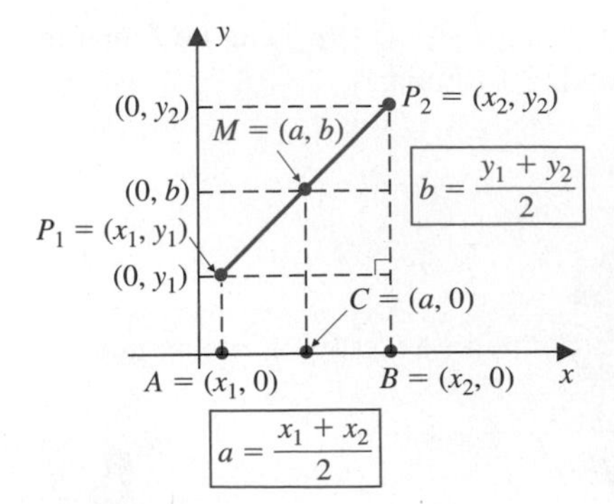

FIGURE 5

At times, it is necessary to determine the midpoint of a line segment. We can determine the coordinates of the midpoint of a line segment in a Cartesian plane by using the coordinates of the end points. Figure 5 shows a line segment with end points $P_1 = (x_1, y_1)$ and $P_2 = (x_2, y_2)$. Let $M = (a, b)$ be the midpoint of $\overline{P_1P_2}$. If $A = (x_1, 0)$, $B = (x_2, 0)$, and C is the midpoint of the line segment $\overline{AB}$, then $|\overline{AC}| = |\overline{BC}|$. The midpoint $C = (a, 0)$ is found by taking the average of the x coordinates of A and B. That is,

$$a = \frac{x_1 + x_2}{2}$$

Similarly, it can be shown that

$$b = \frac{y_1 + y_2}{2}$$

In general, we have the following formula:

THE MIDPOINT FORMULA

The midpoint M of the line segment with end points $P_1 = (x_1, y_1)$ and $P_2 = (x_2, y_2)$ is given by

$$M = \left(\frac{x_1 + x_2}{2}, \frac{y_1 + y_2}{2}\right)$$

EXAMPLE 3 **Calculating a midpoint**

Find the midpoint of the line segment with end points $(-4, 1)$ and $(3, 5)$.

Solution Using the midpoint formula, the coordinates of the midpoint M of the line segment are given by

$$M = \left(\frac{x_1 + x_2}{2}, \frac{y_1 + y_2}{2}\right) = \left(\frac{-4 + 3}{2}, \frac{1 + 5}{2}\right) = \left(-\frac{1}{2}, 3\right)$$

Recognizing and Graphing Equations of Circles

The Cartesian coordinate system enables us to visually represent equations involving two variables. Such representations are called *graphs*.

DEFINITION

GRAPH OF AN EQUATION

The **graph** of an equation in two variables x and y consists of all points (x, y) in a Cartesian plane whose coordinates x and y satisfy that equation.

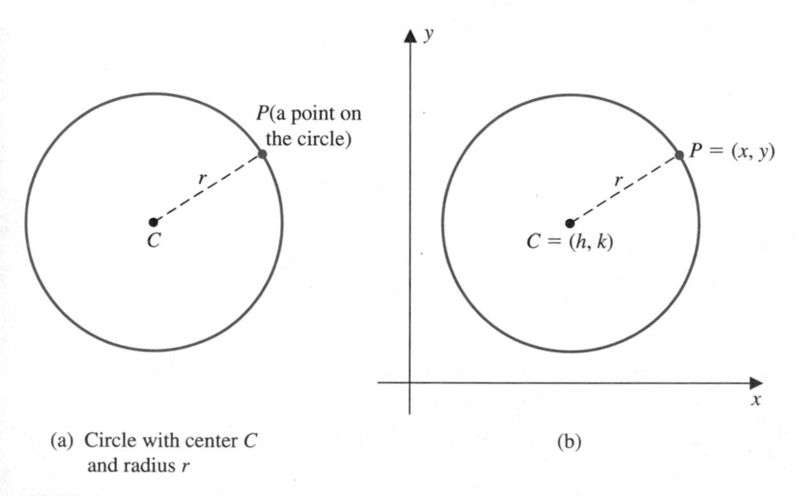

(a) Circle with center C and radius r

(b)

FIGURE 6

It is possible to recognize and graph equations of circles with relative ease. We know from geometry that a circle consists of all points that are at a fixed distance r, its *radius*, from a fixed point C, its *center* (Figure 6a). Suppose that a circle is located in a Cartesian plane so that its center C is (h, k) and its radius is r. If a point $P = (x, y)$ is on the circle, its distance from $C = (h, k)$ has to be r units (Figure 6b). By the distance formula, x and y satisfy the equation

$$\sqrt{(x - h)^2 + (y - k)^2} = r$$

or

$$(x - h)^2 + (y - k)^2 = r^2$$

Conversely, any ordered pair (x, y) that satisfies this last equation defines a point $P = (x, y)$ that lies on the circle with center $C = (h, k)$ and radius r. Thus, we have the following general result:

STANDARD EQUATION OF A CIRCLE

Any point (x, y) on the circle with center (h, k) and radius r must satisfy the equation

$$(x - h)^2 + (y - k)^2 = r^2$$

This equation is called the **standard equation of the circle.**

Table 1 lists some examples of standard equations of circles along with their graphs.

TABLE 1

Examples of Graphs of Circles

Equation of a Circle	Center (h, k)	Radius	Graph
$(x - 1)^2 + (y - 2)^2 = 9$	$(1, 2)$	3	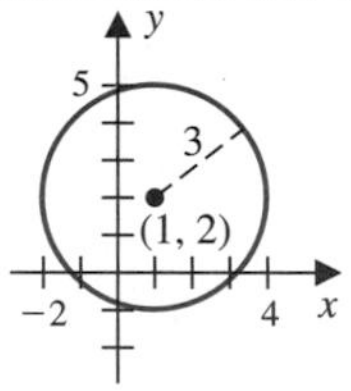
$x^2 + (y + 3)^2 = 25$	$(0, -3)$	5	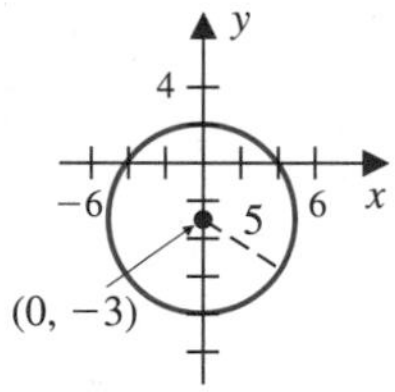
$x^2 + y^2 = 4$	$(0, 0)$	2	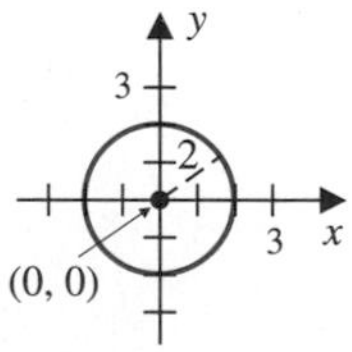

A circle whose center is at $(0, 0)$ and has a radius of 1 unit has the standard equation $x^2 + y^2 = 1$. This is called the **unit circle**.

Given the radius and the center of a circle we can determine its standard equation.

EXAMPLE 4 **Finding the standard equation of a circle**

Find the standard equation for the circle of radius 4 with center at the point $(4, -1)$ and sketch its graph.

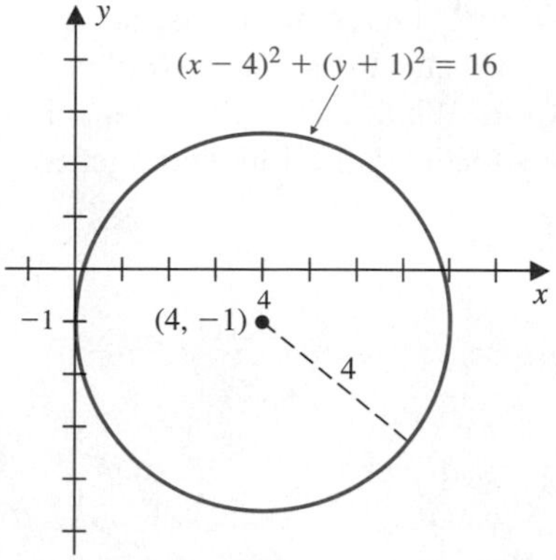

Solution Substituting 4 for r and $(4, -1)$ for (h, k) in the equation

$$(x - h)^2 + (y - k)^2 = r^2$$

we get

$$(x - 4)^2 + [y - (-1)]^2 = 4^2$$
$$(x - 4)^2 + (y + 1)^2 = 16$$

The graph is shown in Figure 7.

FIGURE 7

At times, an alternate algebraic form of an equation of a circle is given, such as $x^2 + y^2 + 6x - 2y - 15 = 0$. The graph of this equation can be determined by converting it to its standard equation. To perform such a conversion, we use the fact that a polynomial of the form $x^2 + bx$ can be changed into a perfect square trinomial by adding $(b/2)^2$:

$$x^2 + bx + \left(\frac{b}{2}\right)^2 = \left(x + \frac{b}{2}\right)^2$$

Adding $(b/2)^2$ to the expression $x^2 + bx$ is called *completing the square*. For instance,

$$x^2 + 14x + 49 = (x + 7)^2$$
$$\left(\tfrac{14}{2}\right)^2 = 7^2$$

That is, adding 49 to $x^2 + 4x$ completes the square.

EXAMPLE 5 **Finding the center and radius of a circle**

Show that the graph of the equation

$$x^2 + y^2 + 6x - 2y - 15 = 0$$

is a circle by converting it to its standard equation. Determine the center and radius and sketch the circle.

Solution We proceed as follows:

$$(x^2 + 6x \qquad) + (y^2 - 2y \qquad) = 15$$
Isolate variable terms on one side by grouping the x terms and the y terms

$$(x^2 + 6x + 9) + (y^2 - 2y + 1) = 15 + 9 + 1$$
Complete the square in each pair of parentheses

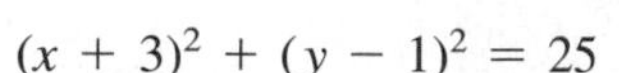

$$(x + 3)^2 + (y - 1)^2 = 25$$
Factor and simplify

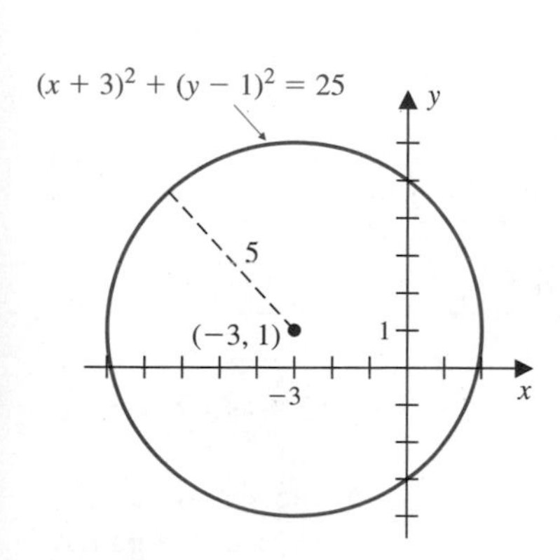

FIGURE 8

We recognize the form of this last equation. Its graph is a circle of radius $r = 5$ with center $(h, k) = (-3, 1)$ (Figure 8).

Using Mathematical Modeling

The Cartesian plane plays an important role in *mathematical modeling*—a process of extracting and interpreting data, graphs, geometric figures, formulas, or equations from specific information describing real-world situations. Real-world problems often are presented in written or verbal form. While there are no set rules for constructing mathematical models for such situations, the following general guidelines may help.

1. **Understand the Problem:** Study the problem carefully to determine what information is given and what information is being sought.
2. **Organize the Information:** First, represent each quantity with an appropriate mathematical expression, taking into account restrictions or limitations. Then express each relationship in the problem by a table, a graph, a geometric figure, an equation, or a formula.
3. **Analyze the Information:** Perform the necessary mathematical procedures for the model such as solving equations, or finding minimum or maximum values.
4. **Interpret the Information:** Draw conclusions about the real-world situation being modeled and determine whether the model provides a good tool for understanding and interpreting the situation.

The next example illustrates the use of graphs in mathematical modeling. Here we plot given data to get a clearer view of the trend in the average annual 30 year fixed mortgage rates.

EXAMPLE 6 Modeling by using graphs

The average percent annual 30 year fixed mortgage rate A for each year n from 1985 through 1994 is listed in Table 2.

TABLE 2

Year (n)	1985	1986	1987	1988	1989	1990	1991	1992	1993	1994
Average Percent 30 Year Fixed Mortgage Rate (A)	11.12	9.82	8.94	8.81	9.76	9.68	9.02	7.98	7.02	7.13

Organizing and graphing data are important tools in *data analysis*.

(a) Plot the data in the table, using n as the horizontal axis and A as the vertical axis.

(b) Connect each consecutive pair of points in part (a) with line segments. The resulting graph is referred to as a **line graph** of the data. How does the pattern of the line graph relate to the data in Table 2?

Solution

(a) Figure 9a shows the points plotted from the data in Table 2. The plot of the data is called a **scattergram**, or **scatter diagram**.

(b) By connecting the points plotted in Figure 9a with line segments, we obtain the graph in Figure 9b. The geometric pattern of the line graph clearly shows the trend in the average 30 year fixed mortgage rates over the time period from 1985 through 1994. The rates have generally been dropping.

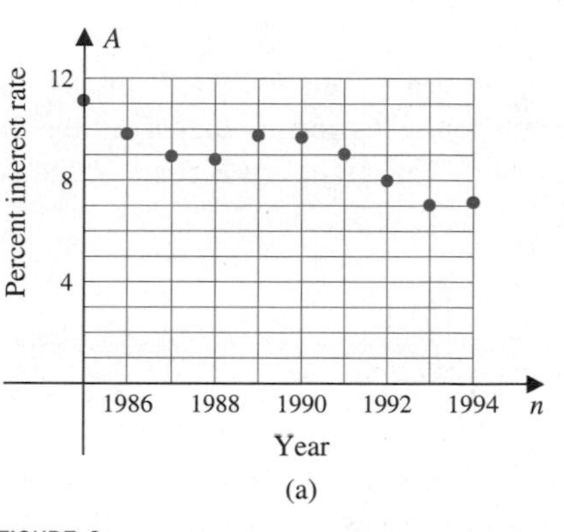

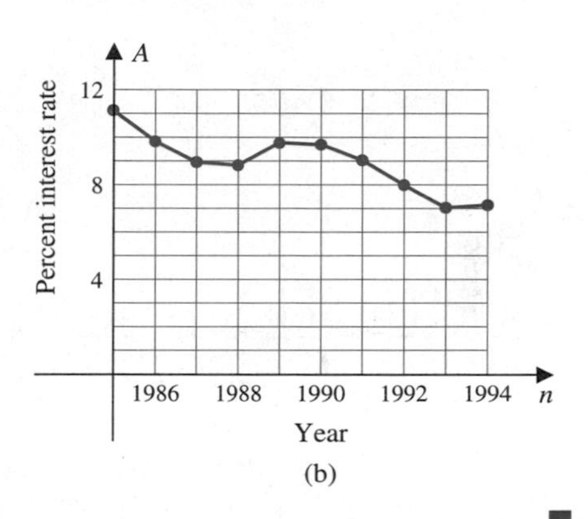

FIGURE 9

The spacing between successive units on the horizontal or vertical axis is called the **scale** of the axis. Note that the scales on the two coordinate axes in Figure 9a are different from each other.

Next, we examine a mathematical model based on the notion of **direct variation**. If there is a constant k such that

$$y = kx$$

holds for all values of x, we say that y **is directly proportional to** x, or that y **varies directly as** x. The constant k is called the **constant of variation**, or **constant of proportionality**. For instance, the formula for the area A of a circle of radius r, $A = \pi r^2$, indicates that A varies directly as r^2.

EXAMPLE 7 Modeling by direct variation

The number of British pounds B is directly proportional to the number of U.S. dollars D. On June 20, 1995, every 100 British pounds bought 159 U.S. dollars.

(a) Write the equation that expresses B in terms of D. Find the constant of variation k_1 to four decimal places and interpret it.

(b) Write an equation that expresses D in terms of B to show that D varies directly as B. Interpret the constant of variation k_2.

Solution

(a) Because B is directly proportional to D, we have

$$B = k_1 D \qquad \text{where } k_1 \text{ is the constant of variation}$$

Since $B = 100$ when $D = 159$, we have

$$100 = k_1 \cdot 159 \qquad \text{or} \qquad k_1 = \tfrac{100}{159}$$

So

$$B = \left(\frac{100}{159}\right)D$$

is the equation that expresses B in terms of D. From above, $k_1 = \frac{100}{159} = 0.6289$, rounded to four decimal places. This means that k_1 is the exchange rate of converting dollars to pounds. That is, every U.S. dollar is worth approximately 0.6289 British pound.

(b) From part (a), we have $B = \left(\frac{100}{159}\right)D$. Solving this equation for D, we get

$$D = \left(\frac{159}{100}\right)B \quad \text{or} \quad D = 1.59B$$

So, by definition, D varies directly as B and $k_2 = 1.59$ is the constant of variation. This means that k_2 is the exchange rate for converting pounds to dollars. That is, every British pound is worth 1.59 U.S. dollars.

PROBLEM SET 1.2

Mastering the Concepts

In problems 1 and 2, plot the points on the same coordinate system and indicate which quadrant or coordinate axis contains each point.

1. $(3, 4), (-2, 4), (\sqrt{5}, -1), (0, \sqrt{7})$
2. $(3, -\sqrt{2}), (0, -\frac{8}{3}), (-2, -\sqrt{11}), (-\pi, \sqrt[3]{71})$

In problems 3–8:
(i) Use the distance formula to find $d(A, B)$, the distance between the points A and B. Round off the numerical problems to two decimal places.
(ii) Use the midpoint formula to find the midpoint of the line segment $\overline{AB}$.

3. (a) $A = (1, 1), B = (-3, 2)$
   (b) $A = (-2, 5), B = (3, -1)$
4. (a) $A = (1, -2), B = (7, 10)$
   (b) $A = (2, 3), B = (-\frac{1}{2}, 1)$
5. (a) $A = (3, -4), B = (-5, -7)$
   (b) $A = (-\frac{2}{5}, \frac{1}{5}), B = (\frac{1}{5}, \frac{3}{5})$
6. (a) $A = (5, -t), B = (7, 5)$
   (b) $A = (t, 8), B = (t, 7)$
7. (a) $A = (t, u + 1), B = (t + 1, u)$
   (b) $A = (-3.65, 8.22), B = (4.73, 5.32)$
8. (a) $A = (w, w/4), B = (3w, w)$
   (b) $A = (-6.13, 1.87), B = (5.25, -4.28)$

In problems 9–12, find the standard equation of the circle with the given characteristics and graph it.

9. (a) Center: $(3, -2)$; radius: 5
   (b) Center: $(-1, 3)$; radius: 2
10. (a) Center: $(3, 4)$; contains the point $(0, 0)$
    (b) Center: $(6, 1)$; contains the point $(-2, 1)$
11. (a) End points of a diameter: $(-3, 7), (7, 1)$
    (b) End points of a diameter: $(2, -3), (-4, 7)$
12. (a) Center: $(4, 7)$; tangent to the y axis
    (b) Center: $(h, 4)$ in quadrant I; radius: 4; tangent to the x axis

In problems 13 and 14, write the standard equation of each of the graphed circles.

13. (a)

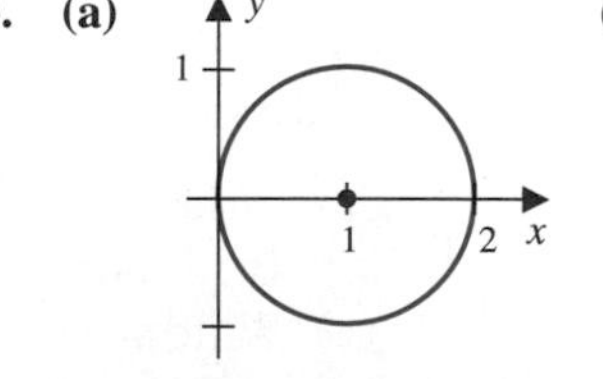

(b)

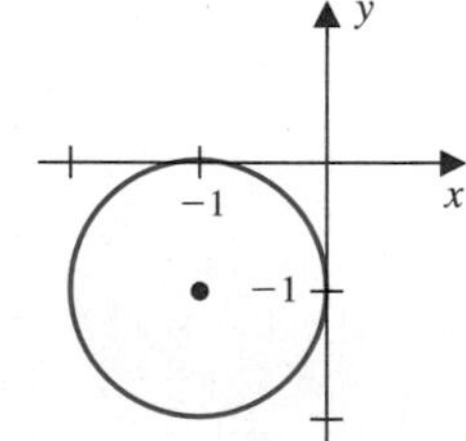

14. (a)

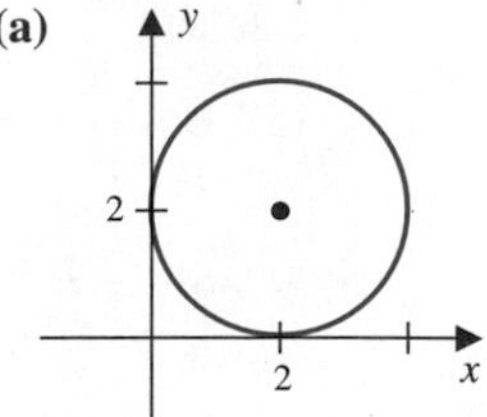

(b)

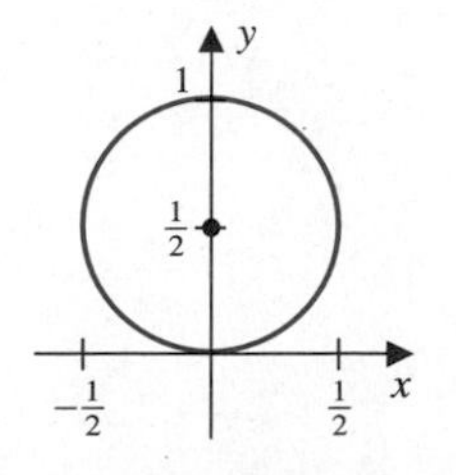

In problems 15–24, find the center and radius of the circle with the given equation. Then sketch its graph.

15. $(x - 1)^2 + (y + 2)^2 = 25$
16. $(x + 2)^2 + (y - 1)^2 = 4$
17. $(x + 2)^2 + (y - 1)^2 = 16$
18. $(x + 3)^2 + y^2 = 9$
19. $x^2 + y^2 - 3x + 4y + 4 = 0$
20. $x^2 + y^2 + 8x - 6y = 15$
21. $x^2 + y^2 - 4x - 6y = 0$
22. $x^2 + y^2 + 10x = 75$
23. $2x^2 + 2y^2 + 50x + 20y = -24$
24. $3x^2 + 3y^2 + 12x - 18y = -12$

In problems 25–28, data are given for a specific situation.

(a) Construct a scattergram for the data. The variable in the top row of each table represents the horizontal axis, and the variable in the second row represents the vertical axis.
(b) Also draw a line graph for the data (see Example 6). Use the line graph to describe the trend indicated in each table.

25.

Temperature, T *Degrees Fahrenheit*	28	31	34	35	37	39	41	43
Humidity, H *Percent*	42	44	48	50	54	57	58	61

26.

Pressure, P *Pounds per square inch*	20	25	30	35	40	45	50	55
Volume, V *Cubic inches*	75	60	50	42	37	33	30	25

27.

Miles Driven by a Taxi, M	1	2	3	5	8	9
Fare, F *Dollars*	1.70	2.90	4.20	6.50	10.10	11.30

28.

Long Distance Time, T *Minutes*	1	3	6	10	15	30
Charges, C *Dollars*	0.37	0.60	0.94	1.40	1.95	3.50

29. Suppose that y is directly proportional to x and $y = 16$ when $x = 4$. Find y when $x = 8$.

30. Suppose that W is directly proportional to S^3 and $W = 27$ when $S = 2$. Find W when $S = 2.4$.

31. Suppose that y is directly proportional to x^2.
(a) Express y in terms of x^2 if $y = 4$ when $x = 2$.
(b) What happens to y if x is tripled?
(c) What happens to y if x is multiplied by a factor of 4?

32. Suppose that w is directly proportional to the product of x^2 and y^3.
(a) Express w in terms of x and y if $w = 16$ when $x = 3$ and $y = -2$.
(b) What happens to w if x is doubled and y is tripled?
(c) What happens to w if x is tripled and y is halved?

Applying the Concepts

33. Insurance Rate: Insurance companies expect younger people who are the principal drivers of motor vehicles to pay more for coverage than older drivers. Suppose the following data are obtained from a study conducted by an insurance company for people between the ages of 16 and 25 years old:

Age A of Person *Years*	16	17	18	19	20	21	22	23	24	25
Annual Insurance Cost, C *Dollars*	1500	1400	1300	1200	1100	1000	900	800	700	500

(a) Draw a line graph that represents the age A as the horizontal axis and the cost C as the vertical axis.
(b) Use the line graph to discuss the trend of the model.

34. Wind-Chill Index: When the wind is blowing on a cold day, the air feels colder than it really is. The following data show that if the actual temperature on a given day is 30°F, then the human skin is exposed to a cooling equivalent temperature, called the **wind-chill index**, according to the wind speed.

Wind Speed, W *Miles per hour*	5	10	15	20	25	30
Equivalent Temperature, T *Degrees Fahrenheit*	27	16	9	3	0	−2

(a) Draw a line graph that represents the wind speed W as the horizontal axis, and the equivalent temperature T as the vertical axis.
(b) Use the line graph to discuss the trend of this model.

35. Engineering—Automobile Skid Marks: The speed V at which an automobile stops when its brakes are applied varies directly as the square root of the length d of the skid marks. Suppose that skid marks from a car involved in an accident measured 88 feet when the car had been going 40 miles per hour.
(a) Find a mathematical model that expresses V in terms of d.
(b) Estimate the speed of a car if its skid marks are 173 feet. Was this car exceeding the speed limit of 55 miles per hour?

36. Astronomy—Planetary Motion: Kepler's third law of planetary motion states that the square of time t (in days) required for a planet to make one revolution around the Sun varies directly as the cube of its average distance d (in miles) from the Sun. Construct a mathematical model that expresses t in terms of d if $t = 365$ days when $d = 93$ million miles.

37. Pollution: The amount of pollution A (in tons) entering the atmosphere varies directly as the number of people N living in a certain area. A population of 50,000 people generates 35,000 tons of pollutants entering the atmosphere.
(a) Write a mathematical model that expresses A in terms of N.
(b) Predict how many tons of pollutants enter the atmosphere in a city of 600,000 people.

38. Manufacturing: A manufacturer of graphing calculators determines that the daily cost C (in dollars) varies directly as the number of units n produced, where $3000 \le n \le 6000$.
(a) Find a mathematical model that expresses C in terms of n if the cost is \$236,500 when 4200 units are produced.
(b) What will the cost be if the production is increased to 5600 units?

Developing and Extending the Concepts

39. Let $a > 0$ and $b < 0$. State the quadrant in which each point lies.
(a) (a, b) (b) $(a, -b)$
(c) $(b, -a)$ (d) $(-a, -b)$

40. The center of a rectangle is at the origin. If the four sides are parallel to the coordinate axes and if the coordinates of two vertices are $(-4, 3)$ and $(4, 3)$, find the coordinates of the other two vertices.

41. (a) Which of the three points $P = (1, 5)$, $Q = (2, 4)$, and $R = (3, 3)$ is the farthest from the origin?
(b) Suppose that we interchange the x and y coordinates of each point in part (a). That is, suppose that $P_1 = (5, 1)$, $Q_1 = (4, 2)$, and $R_1 = (3, 3)$. Is the answer the same as the answer in part (a)? Explain.

42. Use the distance formula to determine whether the triangle ABC with vertices $A = (-5, 1)$, $B = (-6, 5)$, and $C = (-2, 4)$ is an isosceles triangle (a triangle with two sides of equal length).

43. Suppose that P_1, P_2, and P_3 are points in the plane such that P_2 lies on the line segment $\overline{P_1P_3}$. Then P_1, P_2, and P_3 are said to be **collinear**. This condition exists if and only if

$$d(P_1, P_3) = d(P_1, P_2) + d(P_2, P_3)$$

(a) Draw a diagram to illustrate whether the points $P_1 = (-3, -2)$, $P_2 = (1, 2)$, and $P_3 = (3, 4)$ are collinear.
(b) Use the distance formula to confirm the observation of part (a).

44. Use the distance formula and the Pythagorean theorem to determine whether $P = (-4, 1)$, $Q = (6, -1)$, and $R = (-3, 6)$ are the vertices of a right triangle.

45. Describe the graph of $x^2 + y^2 = r^2$ if
(a) $r = 0$ (b) $r < 0$ (c) $r > 0$

46. Which of the following is an equation of a circle? Give the reason in each case.
(a) $2x^2 + 2y^2 + 4x - 7y + 2 = 0$
(b) $2x^2 + 2y^2 + 4x - 7y + 10 = 0$

47. Find the standard equation of the circle with center at (6, 1) and radius 8. Then use the equation to determine whether point P is inside, outside, or on the circle.
(a) $P = (1, 8)$ (b) $P = (10, 7)$ (c) $P = (6, 9)$

48. Given the equation of a circle $(x - h)^2 + (y - k)^2 = r^2$, indicate whether point $P = (x_1, y_1)$ is inside, outside, or on the circle if:
(a) $(x_1 - h)^2 + (y_1 - k)^2 < r^2$
(b) $(x_1 - h)^2 + (y_1 - k)^2 > r^2$
(c) $(x_1 - h)^2 + (y_1 - k)^2 = r^2$
(d) Examine the results in parts (a)–(c) and give specific examples to illustrate your assertions.

49. Find an equation for (i) the upper half, (ii) the lower half, (iii) the right half, and (iv) the left half of each circle. Then graph each equation.

(a) $x^2 + y^2 = 16$

(b) $(x - 2)^2 + y^2 = 9$

50. Consider the equation of the circle

$$(x + 3)^2 + (y - 4)^2 = 36$$

Determine the midpoint of *all* chords that contain the center of the circle.

OBJECTIVES

1. Find and interpret the slope of a line.
2. Graph linear equations and determine the intercepts.
3. Find equations of lines.
4. Determine parallel and perpendicular lines.
5. Solve applied problems.

1.3 Linear Graphs

Our goals in this section are to recognize equations whose graphs are straight lines, and to determine equations for lines with given characteristics. Such equations are called *linear equations*, and they occur in many applications. A key geometric feature of a straight line is how steeply it rises or falls as we move horizontally from left to right. To measure this steepness, we use an important characteristic called the *slope* of a line.

Finding and Interpreting the Slope of a Line

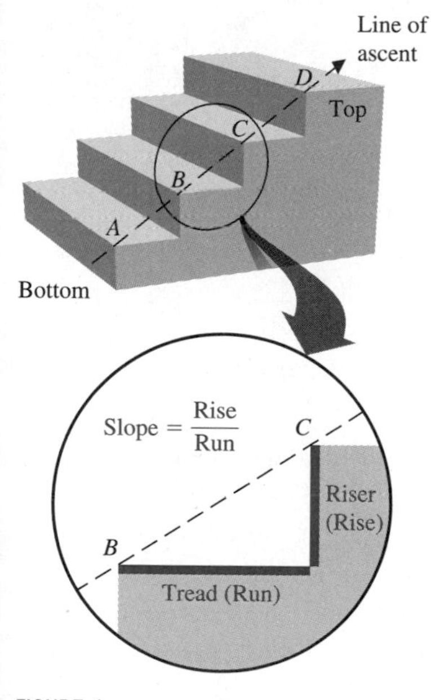

FIGURE 1

The idea of the slope of a line is nicely illustrated by examining a straight flight of stairs. Each step of the stairway consists of a **riser** and a tread (or **run**) (Figure 1).

The **slope** of the line of ascent is given by the ratio

$$\text{Slope} = \frac{\text{Length of riser}}{\text{Length of tread}}$$

or simply

$$\boxed{\text{Slope} = \frac{\text{Rise}}{\text{Run}}}$$

Thus, if a stairway has risers of length 7 inches and treads of length 11 inches (see Figure 2, at the top of the next page), the slope of the line of ascent is given by

$$\frac{\text{Rise}}{\text{Run}} = \frac{7}{11}$$

If two steps were used (Figure 2), the slope would be given by

$$\frac{\text{Rise}}{\text{Run}} = \frac{14}{22} = \frac{7}{11}$$

the same value as obtained from one step.

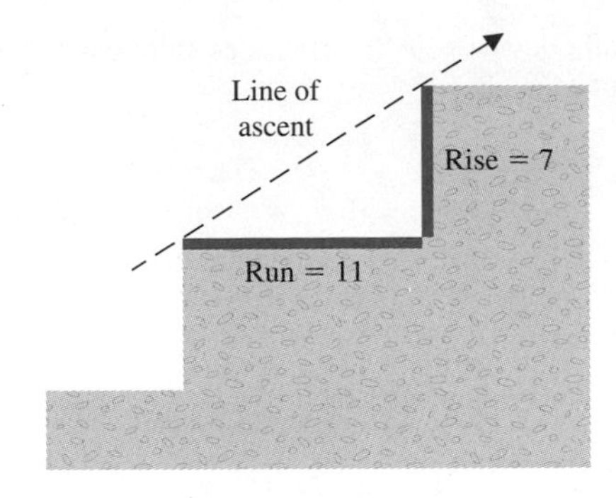

FIGURE 2

This illustration leads us to the general definition of the slope of a line.

DEFINITION

SLOPE OF A LINE

If $P_1 = (x_1, y_1)$ and $P_2 = (x_2, y_2)$ are two different points on a nonvertical line, the **slope *m*** of the line is given by

$$m = \frac{\text{Change in } y}{\text{Change in } x} = \frac{\Delta y}{\Delta x} = \frac{y_2 - y_1}{x_2 - x_1} \quad \text{(Figure 3)}$$

where the symbol Δy is read as "delta *y*" and Δx is read as "delta *x*."

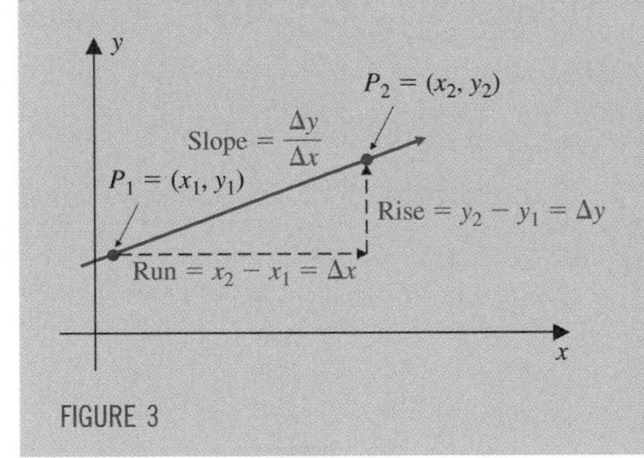

FIGURE 3

When finding the slope of a line containing the points P_1 and P_2, it is not important which point is called $P_1 = (x_1, y_1)$ and which point is called $P_2 = (x_2, y_2)$, since

$$\frac{y_2 - y_1}{x_2 - x_1} = \frac{-(-y_2 + y_1)}{-(-x_2 + x_1)} = \frac{y_1 - y_2}{x_1 - x_2}$$

The slope of a line is the same no matter which two distinct points are selected to compute its value (Problem 65).

EXAMPLE 1 Finding and interpreting a slope

Sketch the line containing points P_1 and P_2, find the slope, and interpret the slope in terms of the direction of the line.

(a) $P_1 = (-2, -1)$ and $P_2 = (2, 5)$ (b) $P_1 = (-1, 3)$ and $P_2 = (4, 1)$
(c) $P_1 = (-2, 4)$ and $P_2 = (3, 4)$ (d) $P_1 = (4, -2)$, and $P_2 = (4, 3)$

Solution The graphs and slopes, along with a description of the direction of each line, are given below.

Given Points	Sketch of Line	Slope of Line, $m = \Delta y/\Delta x$	Direction of Line
(a) $P_1 = (-2, -1)$; $P_2 = (2, 5)$	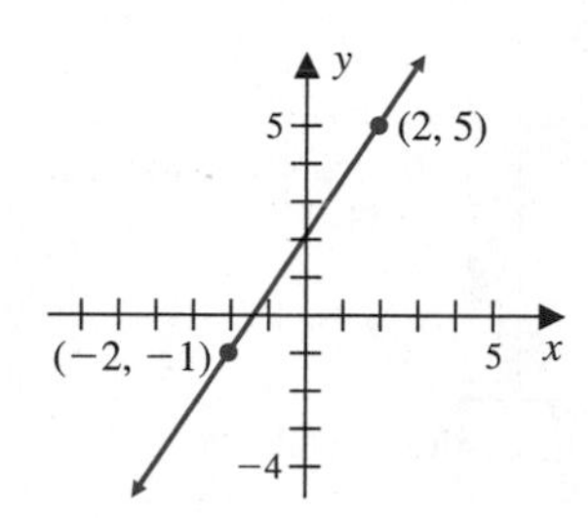	$\dfrac{5-(-1)}{2-(-2)} = \dfrac{5+1}{2+2} = \dfrac{6}{4} = \dfrac{3}{2}$	Rises (to the right)
(b) $P_1 = (-1, 3)$; $P_2 = (4, 1)$	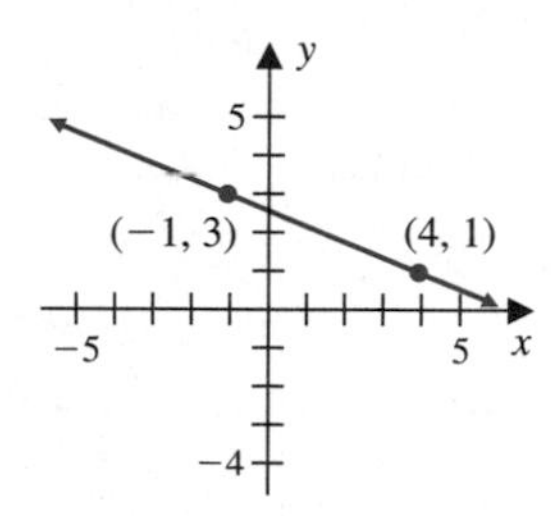	$\dfrac{1-3}{4-(-1)} = \dfrac{-2}{5}$	Falls (to the right)
(c) $P_1 = (-2, 4)$; $P_2 = (3, 4)$	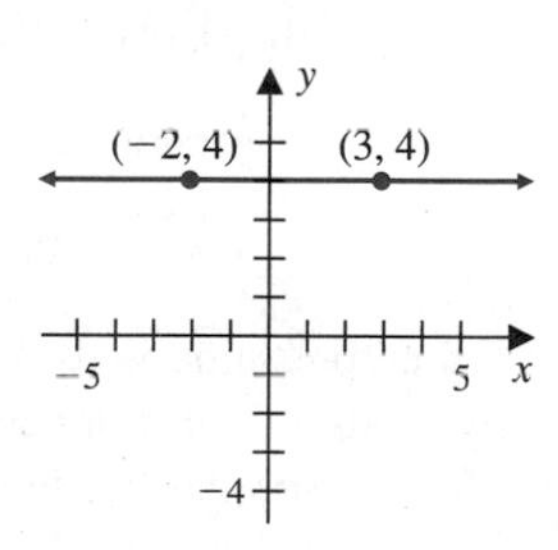	$\dfrac{4-4}{3-(-2)} = \dfrac{0}{5} = 0$	Horizontal
(d) $P_1 = (4, -2)$; $P_2 = (4, 3)$	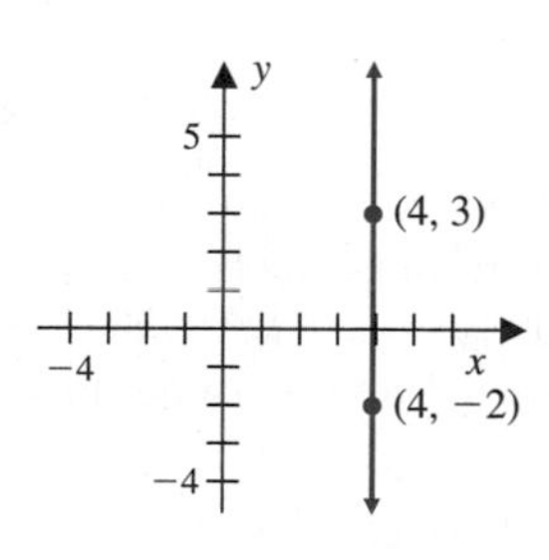	$\dfrac{3-(-2)}{4-4} = \dfrac{5}{0}$ Undefined	Vertical

In general, the connections between slopes and directions of lines are summarized in Table 1 on the following page.

TABLE 1
Slope and Direction of a Line

Slope Value (m)	Line Direction	Sample Graph
Positive	Rises from left to right	$m > 0$
Negative	Falls from left to right	$m < 0$
Zero	Horizontal	$m = 0$
Not defined	Vertical	m undefined

Graphing Linear Equations and Determining the Intercepts

The graph of the first-degree equation

$$Ax + By = C$$

where A and B are constants, with A and B not both equal to 0, is a straight line. Conversely, any straight line in a Cartesian coordinate system has an equation of this form. For this reason, such equations are called **linear equations**, and x and y are said to be **linearly related**. For instance,

$$y = 2x, \quad y = 3, \quad x = 2, \quad \text{and} \quad 3x - 2y = 6$$

are examples of linear equations.

We use the terms *straight line* and *line* interchangeably.

We know from geometry that a straight line is defined by any two of its points. Consequently, we can sketch the graph of a linear equation by plotting any two points and then drawing a line through them.

EXAMPLE 2 Graphing a linear equation

Graph the linear equation $3x - y = 6$.

Solution If $x = 1$, then $3(1) - y = 6$, or $y = -3$, so $(1, -3)$ is a point on the line. If $x = 0$, then $-y = 6$, or $y = -6$, so $(0, -6)$ is another point on the line.

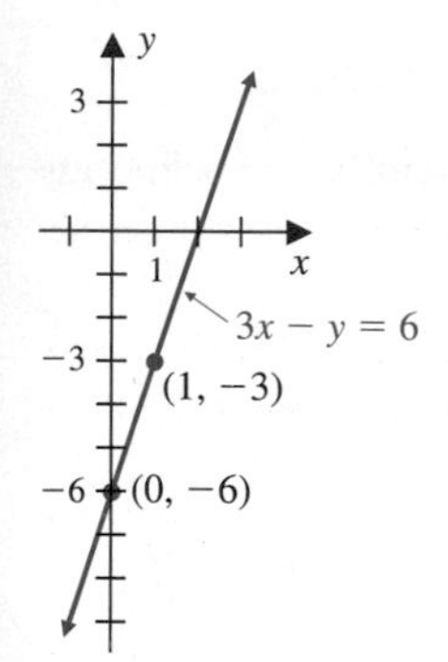

FIGURE 4

To graph the equation, we plot $(1, -3)$ and $(0, -6)$ and then draw a line through them (Figure 4). ■

When graphing an equation, we often need to know the points where the graph intersects the coordinate axes. The x and y coordinates of these points, which are called **intercepts**, are described, illustrated, and determined in Table 2.

TABLE 2

Intercepts of the Graph of an Equation That Relates x and y

Terminology	Description	Illustration	Determination of Intercepts
x intercepts	The x coordinates of points where the graph intersects the x axis	y; x intercept; x	Set $y = 0$ and solve for x
y intercepts	The y coordinates of points where the graph intersects the y axis	y; y intercept; x	Set $x = 0$ and solve for y

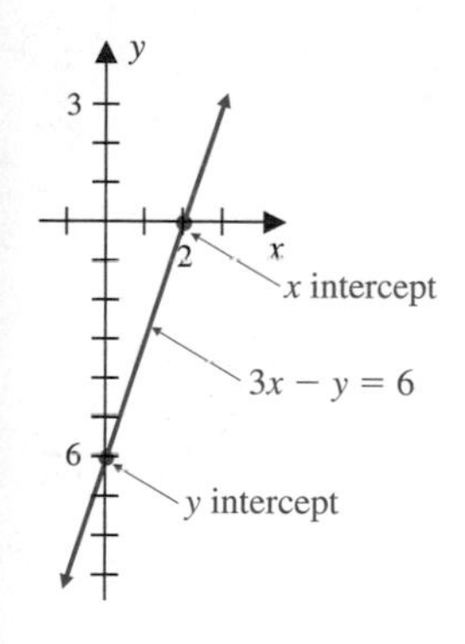

FIGURE 5

Thus, for $3x - y = 6$, the x intercept is obtained by setting $y = 0$ to get $3x = 6$, or $x = 2$. Similarly, we set $x = 0$ and solve $-y = 6$ to get the y intercept, $y = -6$ (Figure 5).

A line parallel to the x axis intersects the y axis at some point, say $(0, b)$. This line consists precisely of those points whose y coordinates are equal to b; its equation is $y = b$ (Figure 6a). Similarly, a line parallel to the y axis intersects the x axis at some point, say $(a, 0)$. This line consists precisely of those points whose x coordinates are equal to a, and its equation is $x = a$ (Figure 6b).

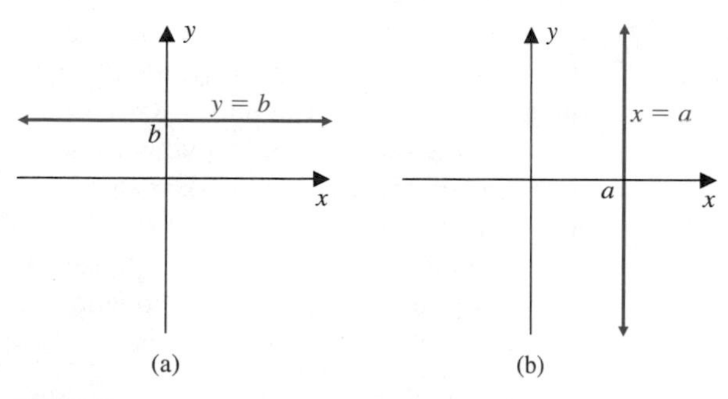

FIGURE 6

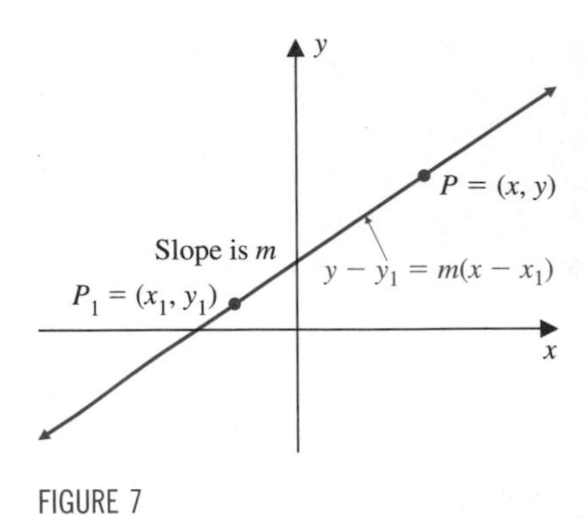

FIGURE 7

Finding Equations of Lines

Given sufficient information about a line in the Cartesian system, we can find an equation whose graph is that line. The slope of the line plays an important role in this process.

Suppose that we are given the slope m and a point $P_1 = (x_1, y_1)$ on a nonvertical line (Figure 7). If $P = (x, y)$ represents any other point on the line, then the slope of the line containing P_1 and P is given by

$$\frac{y - y_1}{x - x_1} = m$$

so

$$y - y_1 = m(x - x_1)$$

Any point $P = (x, y)$ whose coordinates satisfy the above equation lies on the line. Conversely, any point that lies on the line has coordinates that satisfy the equation. Therefore, we have the following:

POINT–SLOPE EQUATION OF A LINE

An equation for the line that contains the point $P_1 = (x_1, y_1)$ and has the slope m is given by

$$y - y_1 = m(x - x_1)$$

This equation is referred to as a **point–slope equation of the line**.

EXAMPLE 3 Finding a point–slope equation of a line

Graph the line that contains point (2, 3) and has slope 5, and then find a point–slope equation for the line.

Solution In order to graph the line we need to find another point in addition to the given point (2, 3). Since the slope $m = 5 = \frac{5}{1}$, we can locate a second point by starting at the point (2, 3) and then moving 1 unit to the right (run) and 5 units up (rise) to get point (3, 8). Then the graph is obtained by drawing a line through points (2, 3) and (3, 8), as shown in Figure 8.

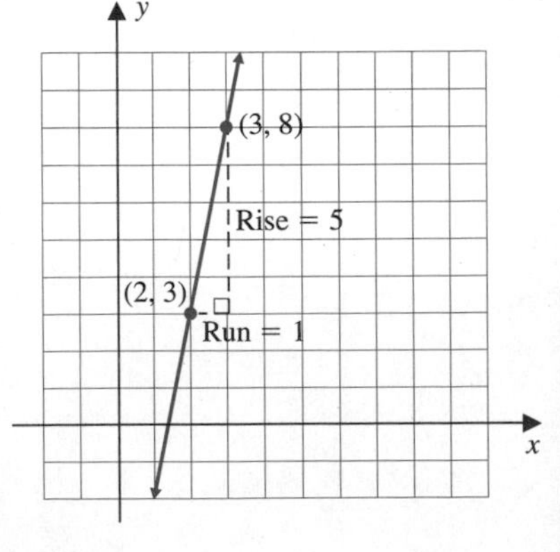

FIGURE 8

Substituting $x_1 = 2$, $y_1 = 3$, and $m = 5$ into the point–slope equation $y - y_1 = m(x - x_1)$ we get

$$y - 3 = 5(x - 2)$$

A point–slope equation of a line is not *unique*. If we used (3, 8) instead of (2, 3) to get the equation, the result would be $y - 8 = 5(x - 3)$.

Suppose that a nonvertical line has slope m. Since the line is not parallel to the y axis, it must intersect this axis at some point $(0, b)$ (Figure 9). In other words, b is the y intercept of the line. Applying the point–slope equation with $(x_1, y_1) = (0, b)$, we get

$$y - b = m(x - 0)$$

so

$$y = mx + b$$

This latter equation is called the *slope–intercept equation of the line*, and we have the following result:

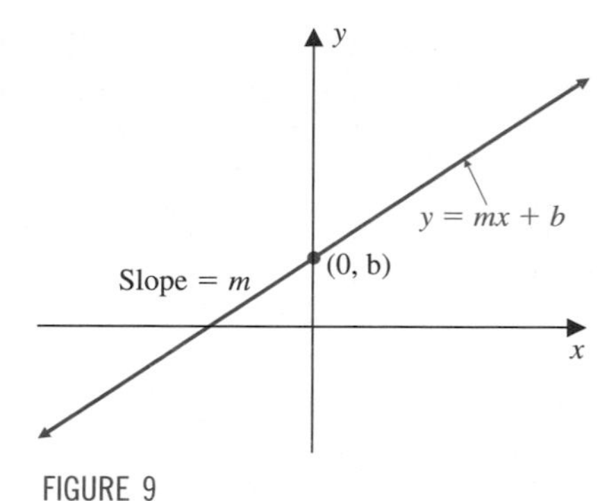

FIGURE 9

SLOPE–INTERCEPT EQUATION OF A LINE

The graph of the equation $y = mx + b$, which is called the **slope–intercept equation**, is a line with slope m and y intercept b.

A nonvertical line can have only one slope–intercept equation, since a line can have only one slope m and one y intercept b.

When an equation of a line is written in slope–intercept form the slope and y intercept can be determined by inspection. Table 3 shows some examples.

TABLE 3
Examples of Slope–Intercept Forms of Linear Equations

Equation	Slope	y Intercept
$y = 3x + 5$	$m = 3$	$b = 5$
$y = -2x + 7$	$m = -2$	$b = 7$
$y = x - 4 = (1)x + (-4)$	$m = 1$	$b = -4$
$y = 3 = (0)x + 3$	$m = 0$	$b = 3$

Thus, by rewriting a linear equation in slope–intercept form, we can easily find its slope and y intercept. For instance, suppose we are given the linear equation $3x + 5y = 20$. We solve for y to get the slope–intercept equation $y = -\frac{3}{5}x + 4$, so the line has slope $-\frac{3}{5}$ and y intercept 4.

The next example shows how we can find an equation of a line in a Cartesian system when we are given the location of two of its points.

EXAMPLE 4 Finding the slope–intercept equation given two points

Find the slope–intercept equation of a line that contains the points $(-2, 5)$ and $(3, -4)$.

Solution The slope of the line is given by

$$m = \frac{5 - (-4)}{-2 - 3} = -\frac{9}{5}$$

Since the point $(-2, 5)$ belongs to the line, we can use $(x_1, y_1) = (-2, 5)$ in a point–slope equation to get

$$y - 5 = -\tfrac{9}{5}[x - (-2)] \quad \text{or} \quad y - 5 = -\tfrac{9}{5}(x + 2)$$

Solving the latter equation for y, we obtain

$$y = -\tfrac{9}{5}x - \tfrac{18}{5} + 5 \quad \text{Distribute } -\tfrac{9}{5} \text{ and add 5 to each side}$$

$$y = -\tfrac{9}{5}x + \tfrac{7}{5} \quad \text{Combine like terms}$$

The slope is $-\frac{9}{5}$ and the y intercept is $\frac{7}{5}$.

Determining Parallel and Perpendicular Lines

Since parallel lines have the same direction, they have the same slope. Two lines that intersect at a right angle are said to be *perpendicular*, and their slopes satisfy the condition stated in the following properties:

PROPERTIES

SLOPES OF PARALLEL AND PERPENDICULAR LINES

Suppose L_1 and L_2 are two distinct nonvertical lines with slopes m_1 and m_2, respectively.

1. L_1 is **parallel** to L_2 if and only if $m_1 = m_2$.
2. L_1 is **perpendicular** to L_2 if and only if $m_1 = -1/m_2$, or $m_1 m_2 = -1$ (Problem 66).

EXAMPLE 5 Finding equations of parallel or perpendicular lines

Let L be a line whose equation is given by $2x - y + 6 = 0$. Find the slope–intercept equation of the line that contains the point $(3, 2)$ and is:

(a) Parallel to L (b) Perpendicular to L

In both situations, graph the two lines in the same coordinate system.

Solution The slope–intercept equation for L is obtained by solving for y in terms of x to get

$$y = 2x + 6$$

So the slope is $m_1 = 2$ and the y intercept is $b = 6$.

(a) Since parallel lines have the same slope, the required line has slope 2. Thus, a point–slope equation of the line parallel to L and containing the point $(3, 2)$ is given by

$$y - 2 = 2(x - 3)$$

We solve this equation for y to get the slope–intercept equation

$$y = 2x - 4$$

Figure 10a shows the graphs of the two lines.

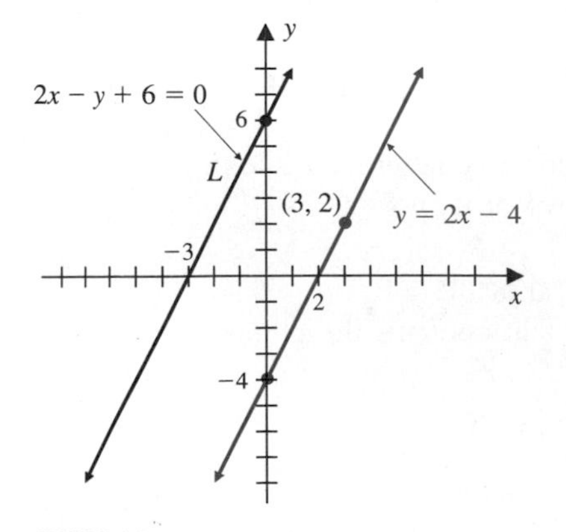

FIGURE 10a

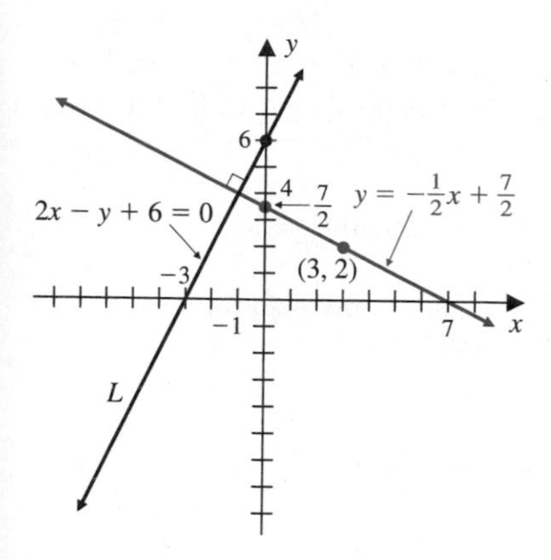

FIGURE 10b

(b) The slope m_2 of a line perpendicular to L, which has slope $m_1 = 2$, must satisfy the condition $m_1 m_2 = -1$ or $m_2 = -1/m_1$, so $m_2 = -1/2$. Hence, a point–slope equation of the line perpendicular to L and containing the point (3, 2) is given by

$$y - 2 = -\tfrac{1}{2}(x - 3)$$

Solving for y, we get the slope–intercept equation

$$y = -\tfrac{1}{2}x + \tfrac{7}{2}$$

Figure 10b shows the graphs of the two lines.

Solving Applied Problems

Many practical situations are modeled by linear equations and their graphs.

EXAMPLE 6 Modeling the resale value of an automobile

The resale value R (in dollars) of a certain automobile is modeled by the linear equation

$$R = P(1 - 0.07t)$$

where P is the original price (in dollars) of the automobile and t is the number of years after it was purchased.

(a) Express the resale value R as an equation in terms of t if the original price was \$16,000.
(b) Find the resale value of the car after 1 year, 3 years, and 5 years.
(c) Graph the equation from part (a) by representing t on the horizontal axis and R on the vertical axis.
(d) Find and interpret the intercepts.
(e) Find and interpret the slope of the line.
(f) Does this model make sense if the car is sold after 15 years? Explain.

Solution

(a) Since the original price of the automobile was \$16,000, we substitute $P = 16{,}000$ into the given equation to get

$$R = 16{,}000(1 - 0.07t)$$
$$R = -1120t + 16{,}000$$

The latter equation is in slope–intercept form.

(b) After 1 year, $t = 1$, so $R = -1120(1) + 16{,}000 = 14{,}880$

After 3 years, $t = 3$, so $R = -1120(3) + 16{,}000 = 12{,}640$

After 5 years, $t = 5$, so $R = -1120(5) + 16{,}000 = 10{,}400$

Thus, the resale values after 1 year, 3 years, and 5 years are, respectively, \$14,880; \$12,640; \$10,400.

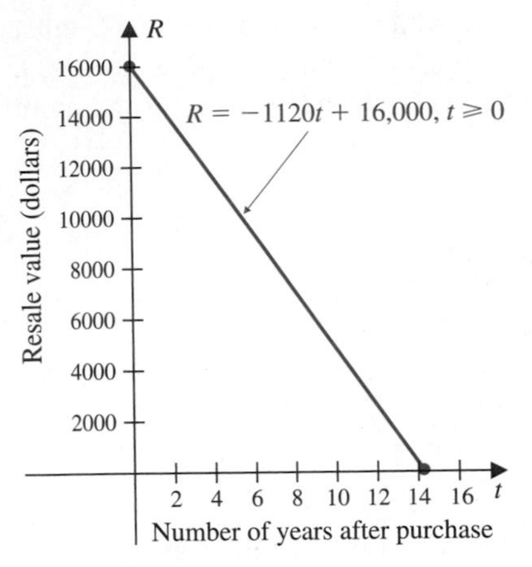

FIGURE 11

Because of the way we labeled the vertical and horizontal axes, we refer to the y intercept as the R intercept and the x intercept as the t intercept.

(c) Since t represents elapsed time, it follows that $t \geq 0$. The graph of the linear equation from part (a) with the restriction $t \geq 0$ is displayed in Figure 11.

(d) The R intercept is found by setting $t = 0$ to get

$$R = -1120(0) + 16{,}000 = 16{,}000$$

This means that when no time has elapsed, the automobile is new, so its resale value is the same as the original price, \$16,000. The t intercept is found by setting $R = 0$ to get

$$\begin{aligned} 0 &= -1120t + 16{,}000 \\ 1120t &= 16{,}000 \\ t &= \frac{16{,}000}{1120} = 14.29 \text{ (approx.)} \end{aligned}$$

This means that after about 14.29 years the automobile will have no resale value, according to this model.

(e) The slope of the line can be read directly from the slope–intercept equation

$$R = -1120t + 16{,}000$$

to be -1120. This means that for each elapsed year (1 unit change in t) there is a reduction of \$1120 in value.

(f) If the automobile is more than 15 years old, then according to the model, we have

$$\begin{aligned} t &\geq 15 && \\ -1120t &\leq -1120(15) && \text{Multiply each side by } -1120 \\ -1120t + 16{,}000 &\leq -1120(15) + 16{,}000 && \text{Add 16,000 to each side} \\ R &\leq -800 && \text{Substitute and simplify} \end{aligned}$$

If the automobile is still operational after 15 years, it is unlikely that it would have a negative resale value. So the model does not make sense after 14.29 years.

EXAMPLE 7 Modeling distance traveled

Suppose that a minivan travels 42 miles on 1.5 gallons of gasoline, and 98 miles on 3.5 gallons. Assume a linear relationship between d, the distance traveled in miles, and n, the number of gallons of gasoline consumed.

The slope of a line is often described as the average rate of change of one quantity with respect to another quantity.

(a) Find the rate of gasoline consumption in miles per gallon by computing a slope.

(b) Find an equation that expresses d in terms of n and graph it.

(c) How many miles can the minivan travel without a refill on 12 gallons of gasoline? On 14 gallons of gasoline?

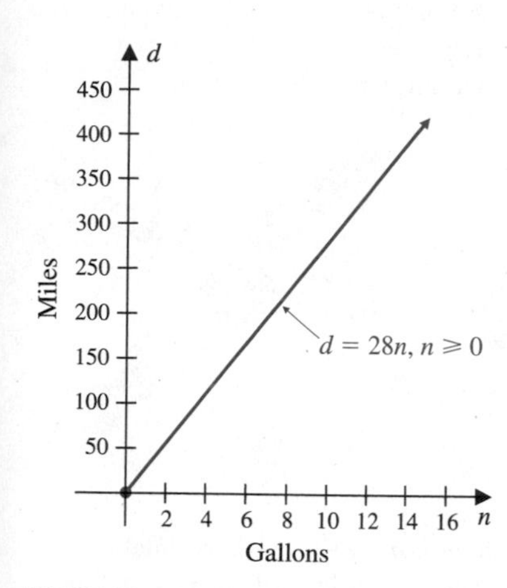

FIGURE 12

Solution

(a) If the distance traveled, d, is represented on the y axis and n, the number of gallons of gasoline consumed, is shown on the x axis, then the slope is given by

$$m = \frac{y_1 - y_2}{x_1 - x_2} = \frac{d_1 - d_2}{n_1 - n_2} = \frac{42 - 98}{1.5 - 3.5} = \frac{-56}{-2} = 28$$

So the rate of gasoline consumption is 28 miles per gallon.

(b) The relationship between d and n is linear, the slope is 28, and when $n = 1.5$, $d = 42$, so the point–slope equation gives us

$$d - 42 = 28(n - 1.5)$$
$$d = 28n$$

Since n represents the number of gallons, $n \geq 0$. The graph of the equation $d = 28n$, where $n \geq 0$, is displayed in Figure 12.

(c) If $n = 12$, then $d = 28(12) = 336$. When $n = 14$, $d = 28(14) = 392$. So the minivan can travel 336 miles on 12 gallons of gasoline, and 392 miles on 14 gallons.

PROBLEM SET 1.3

Mastering the Concepts

In problems 1–14, sketch the line containing points P_1 and P_2, find the slope, and interpret the slope in terms of the direction of the line.

1. $P_1 = (2, 3), P_2 = (-1, -2)$
2. $P_1 = (4, 1), P_2 = (3, -1)$
3. $P_1 = (-1, 7), P_2 = (3, 7)$
4. $P_1 = (0, 1), P_2 = (1, -3)$
5. $P_1 = (-3, -1), P_2 = (-12, 11)$
6. $P_1 = (0, 1), P_2 = (-1, 3)$
7. $P_1 = (-2, -6), P_2 = (4, -4)$
8. $P_1 = (-3, -4), P_2 = (-1, 6)$
9. $P_1 = \left(-\frac{2}{3}, 1\right), P_2 = \left(-1, \frac{5}{3}\right)$
10. $P_1 = \left(\frac{5}{7}, -\frac{2}{7}\right), P_2 = \left(\frac{5}{7}, \frac{2}{7}\right)$
11. $P_1 = (-2, 5), P_2 = (3, 5)$
12. $P_1 = (-2, -5), P_2 = (3, -5)$
13. $P_1 = (4, 2), P_2 = (4, 0)$
14. $P_1 = (-1, 1), P_2 = (-1, -5)$

In problems 15–30, graph each linear equation, and locate the x and y intercepts.

15. $2x + y = 2$
16. $-x + 2y = 4$
17. $y + 7 = -2(x - 5)$
18. $y - 8 = 4(x - 2)$
19. $3x - 2y = 6$
20. $2x - 3y = -6$
21. $y = -2x + 2$
22. $y = -2x - 1$
23. $y = \frac{1}{2}x + 3$
24. $y = -\frac{1}{3}x + 2$
25. $2x - 6 = 0$
26. $3x + 6 = 0$
27. $3y + 9 = 0$
28. $2y - 8 = 0$
29. $\frac{x}{3} + \frac{y}{4} = 1$
30. $\frac{y}{2} - \frac{3x}{5} = 1$

In problems 31 and 32, sketch the line that contains the point P and has the slope m and then find a point–slope equation for the line.

31. **(a)** $P = (4, -3), m = 3$ **(b)** $P = (-2, 3), m = 0$
32. **(a)** $P = (-3, 1), m = -\frac{2}{3}$
(b) $P = (-2, 0), m = -\frac{1}{2}$

In problems 33 and 34, find the slope–intercept equation of the line whose graph is given.

33. **(a)**

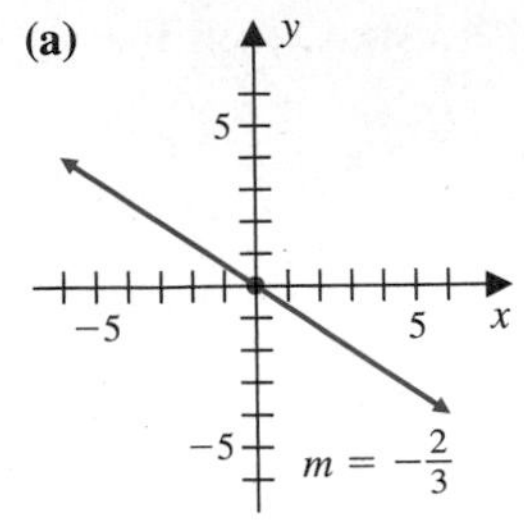

(b)

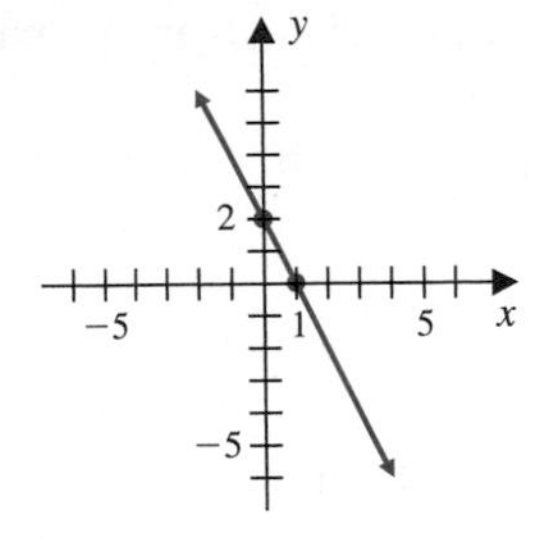

34. **(a)**

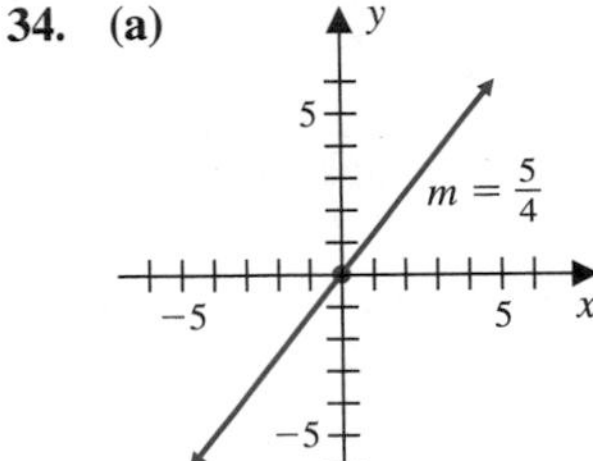

(b)

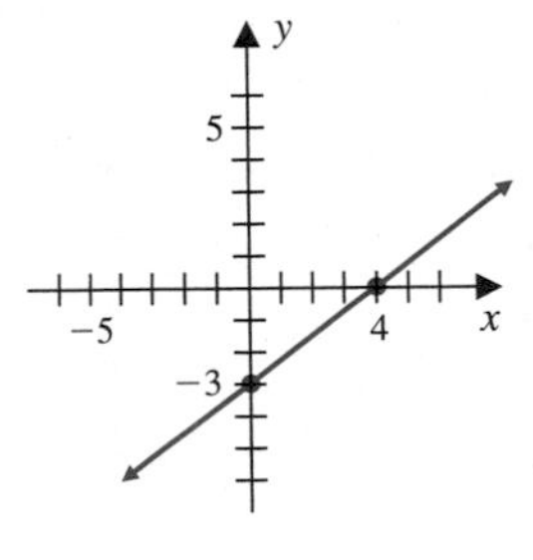

In problems 35–44, if possible, find the slope–intercept equation of the line that satisfies the given conditions. Then graph the line.

35. Slope: 3; y intercept: 4

36. Slope: -3; contains point $P = (-2, 0)$

37. Contains points $P_1 = (-2, 5)$ and $P_2 = (2, -3)$

38. Contains points $P_1 = (3, 3)$ and $P_2 = (1, -2)$

39. Contains point $P = (2, 1)$; parallel to the line with equation $x - 3y = 6$

40. Contains point $P = (-1, 4)$; parallel to the line with equation $3x + y = 5$

41. Contains point $P = (-3, 2)$; perpendicular to the line with equation $4x - 2y = 7$

42. Contains point $P = (-3, -5)$; perpendicular to the line with equation $3x + 2y = -2$

43. Contains point $P = (3, -1)$; parallel to the y axis

44. Contains point $P = (-2, 5)$; perpendicular to the x axis

Applying the Concepts

45. Grade of a Ramp: The *grade of a ramp* is the slope of the ramp. It is equal to the vertical distance the road rises divided by the length of the horizontal over which the road rises. At an exit ramp off an expressway, cars ascend the ramp to an overpass that is 38 feet above the expressway. The horizontal distance from the beginning of the exit ramp to the point on the expressway below the overpass is 1640 feet (Figure 13). Find the grade of the exit ramp to the nearest hundredth percent.

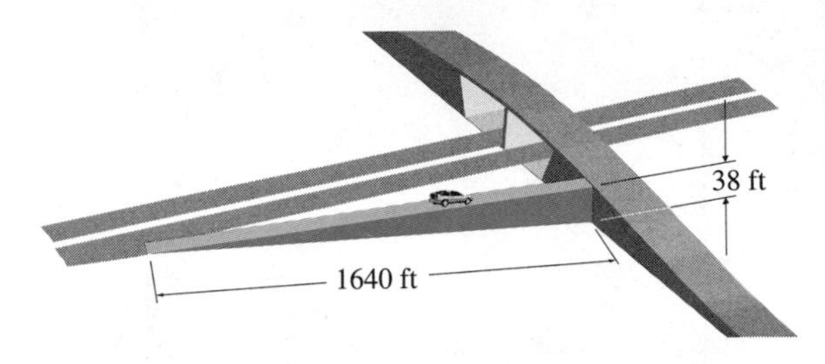

FIGURE 13

46. Pitch of a Roof: The *span of a roof* is the horizontal distance between the outside extensions of the roof, while the *rise* is the vertical distance from the top of the rafters to the center of the span (Figure 14). The *pitch* is the slope of the rafters with respect to the span. Determine the rise for a house whose span is 30 feet with a pitch of $\frac{1}{4}$, $\frac{1}{3}$, $\frac{1}{2}$, or $\frac{7}{12}$.

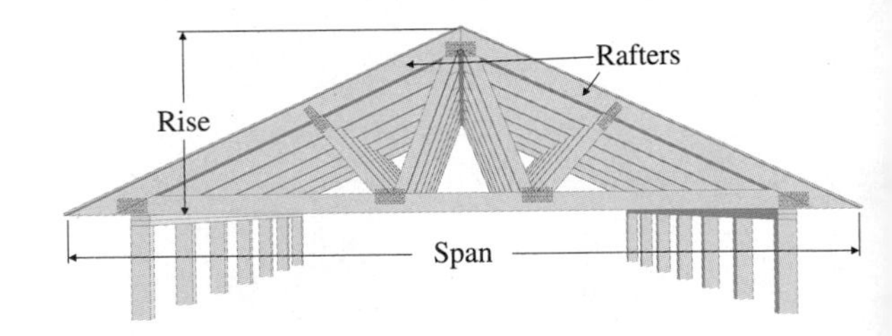

FIGURE 14

47. Temperature Conversion: The relationship between temperature in degrees Celsius, C, and degrees Fahrenheit, F, is given by the equation

$$C = \tfrac{5}{9}(F - 32)$$

(a) Graph this equation using F as the horizontal axis and C as the vertical axis, and interpret it.
(b) Find and interpret the intercepts.
(c) Find and interpret the slope of the line.
(d) Is there any temperature that is the same on both scales?

48. Weight vs. Age: A study conducted by a weight clinic determines that the average weight W (in pounds) of a male is related to his age A (in years) by the mathematical model

$$W = \tfrac{9}{5}A + 114 \qquad 20 \le A \le 45$$

(a) Predict the average weight of a 35-year-old man.
(b) Using this model, predict the age of a male whose weight is 186 pounds.
(c) Graph this equation using A as the horizontal axis and W as the vertical axis, and interpret it.
(d) Find the slope of the line and interpret it.
(e) Do you think this model is applicable to males over 45 years old? How about under 20 years old?

49. Manufacturing: In planning to manufacture a laser printer for a computer, a manufacturer determines that the number n of printers produced and the total cost C (in dollars) of producing these printers are related by a linear equation. Projections are that the total cost of producing 50 printers is \$9000 and that 125 printers will cost \$15,000 to produce.
(a) Find an equation that expresses C in terms of n.
(b) Graph the equation and interpret it.
(c) Give an interpretation of the slope of this equation.
(d) According to this model, what is the total cost of producing 200 laser printers?
(e) In economics, the difference between the cost of producing n items and $n + 1$ items is called the **marginal cost**. Compute the marginal cost for this situation and interpret it.

50. Baseball Statistics: Suppose that a major league baseball player has hit 7 home runs in the first 29 games of the season, and assume this pace continues throughout the 162 game season.
(a) If the relationship is linear, express the number of home runs, N, in terms of the number t of games played.
(b) Give an interpretation of the slope of the equation.
(c) Use the model from part (a) to predict approximately how many home runs the player will hit for the entire season.

51. Physiology: A jogger's heart rate of N beats per minute is related to the jogger's speed V (in feet per second) by a linear mathematical model. Assume a jogger's heart rate is 75 beats per minute at a speed of 12 feet per second and 80 beats per minute at a speed of 15 feet per second.
(a) Find an equation that expresses N in terms of V.
(b) Graph the equation and interpret it.
(c) Find the slope and interpret it.
(d) Find the speed V if the heart rate N is 85 beats per minute.

52. Physics: Hooke's law for a perfectly elastic spring states that the distance d (in meters) a spring is stretched is directly proportional to the stretching force F (in newtons). Suppose that a force of 3 newtons stretches the spring a distance of 0.2 meter, and a force of 9 newtons stretches the spring a distance of 0.6 meter.
(a) Find a linear equation that expresses F in terms of d.
(b) Give an interpretation of the slope of this equation.
(c) What is the length of the spring if a force of 6 newtons is applied?
(d) How much force is needed to stretch the spring a distance of 0.5 meter?

53. Real Estate: The management of an apartment complex charges \$475 per month for each of its 90 apartments, and the complex is filled. Because of rising maintenance costs, the management needs to raise the rent to \$500 for each apartment. As a result, the number of rented apartments decreases to 65. Assume that this trend continues and the relationship between the rent R and the number n of apartments rented is linear, where $n \geq 20$.
(a) Find an equation that expresses R in terms of n.
(b) Graph the equation and interpret the slope.
(c) What is the monthly rent per unit if 75 apartments are rented?
(d) How many apartments are rented if the monthly rent per apartment is \$525?

54. State Income Tax: Michigan's 1994 individual income tax is 4.4% on all adjusted gross income over \$1500.
(a) Assuming the relationship between an individual's income tax T and the adjusted gross income I is linear, find an equation that expresses T in terms of I for someone earning more than \$1500.
(b) Graph the equation and interpret the slope.
(c) Use the model to find an individual's adjusted gross income for 1 year if the income tax owed is \$1460.

55. Straight-Line Depreciation: A straight-line depreciation is one way a business can find the value of an item for tax purposes. An athletic club buys weight-lifting equipment for \$6400 and determines that the equipment is expected to last for 6 years. Suppose that the equipment will have a salvage value of \$500 at the end of the 6 year period, and assume that the decline in value is the same each year.
(a) Find an equation that expresses the salvage value V in terms of t years of elapsed time.
(b) What are the restrictions on t?
(c) Find and interpret the slope of the line.
(d) Determine by how many dollars the equipment depreciates each year.

56. Recycling: Suppose that a recycling center recycled 495 tons of newspapers in the month of January and 525 tons of newspapers in the month of April. Suppose the increase was at a uniform rate. Let T denote the number of tons of newspaper recycled in the nth month (where $n = 1$ represents January). Assume the relationship between T and n is linear, where $1 \leq n \leq 12$.
(a) Find an equation that expresses T in terms of n.
(b) Find and interpret the slope of the line.
(c) In how many months is the amount of paper 565 tons?
(d) Does this model apply for more than 1 year?

57. Air Pollution: A study shows that the relationship of the concentration C (in parts per million) of carbon monoxide in the atmosphere and the number of years t after 1989 is linear. A fuel power plant emitted 13.8 parts per million of carbon monoxide in the atmosphere in 1990 ($t = 1$), and 16.9 parts per million in 1993 ($t = 4$).
(a) Find an equation that expresses C in terms of t.
(b) Find and interpret the slope of the line.
(c) If this trend continues to hold, determine how many parts per million of carbon monoxide will be emitted in the atmosphere in the year 2000.

58. Communications Satellite: A communications satellite handled 1.8 million messages during its first year of operation and 11.8 million messages during its ninth year of operation. Assume the relationship between the number N (in millions) of messages handled and the year t of operation is linear.
(a) Find an equation that expresses N in terms of t.
(b) Find and interpret the slope.
(c) Predict the number of messages that will be handled during the twelfth year of operation.

Developing and Extending the Concepts

59. Suppose that A, B, C, and D are nonzero real numbers. Determine whether the two lines with equations

$$Ax + By + C = 0 \quad \text{and} \quad Ax + By + D = 0$$

are parallel, perpendicular, or neither.

60. Use slopes to determine whether $P = (9, 6)$, $Q = (-1, 2)$, and $R = (1, -3)$ are vertices of a right triangle. If they are, which vertex is the right angle?

61. Suppose that $P = (3, 4)$ is a point on the circle whose equation is $x^2 + y^2 = 25$. From plane geometry, the tangent line to the circle at point P is perpendicular to the radius $\overline{OP}$, where O is the center of the circle. Find the equation of the tangent line at point P.

62. Let L be a line that is neither horizontal nor vertical, and assume the x and y intercepts are the nonzero values a and b, respectively. Then (x, y) is on L provided that

$$\frac{x}{a} + \frac{y}{b} = 1$$

This equation is called the **intercept equation** of a line. Write this equation in the slope–intercept form, and show that the slope is $-b/a$. Also, find the intercept form of the equation of a line for each of the following situations:
(a) x intercept 4; and y intercept -6
(b) x intercept 1; and y intercept $\frac{2}{5}$
(c) Line contains points $(1, -2)$ and $(3, 5)$

63. Suppose that the product of the slopes of two lines containing the origin is positive, what can be said about the directions of the two lines?

64. (a) Figure 15a shows four lines L_1, L_2, L_3, and L_4 with slopes m_1, m_2, m_3, and m_4, respectively. Explain why the slopes are positive. How do the values of m_1, m_2, m_3, and m_4 compare to each other? Generalize your conclusion.
(b) Figure 15b shows four lines L_1, L_2, L_3, and L_4 with slopes m_1, m_2, m_3, and m_4, respectively. Explain why the slopes are negative. How do the values of m_1, m_2, m_3, and m_4 compare to each other? Generalize your conclusion.

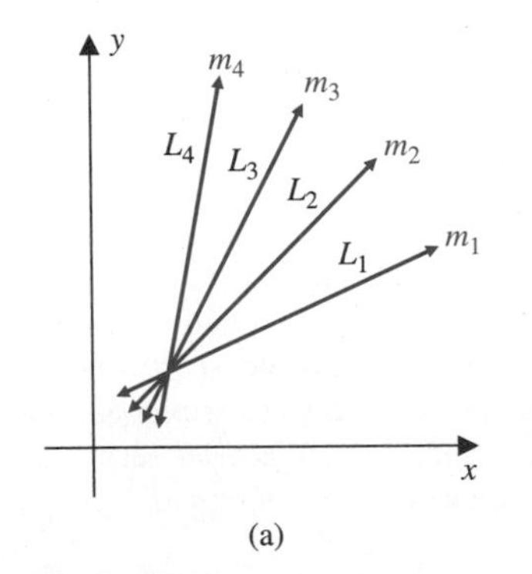

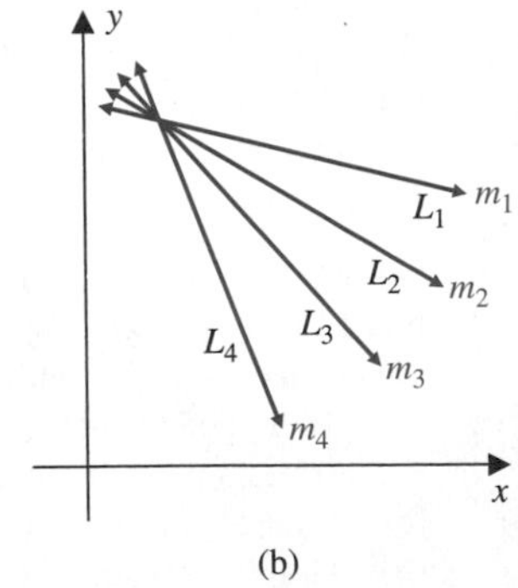

FIGURE 15

65. Suppose that we choose the two points $P_1 = (x_1, y_1)$ and $P_2 = (x_2, y_2)$ on a nonvertical line L, and then choose two other points $P_1' = (x_1', y_1')$ and $P_2' = (x_2', y_2')$ on L (Figure 16).

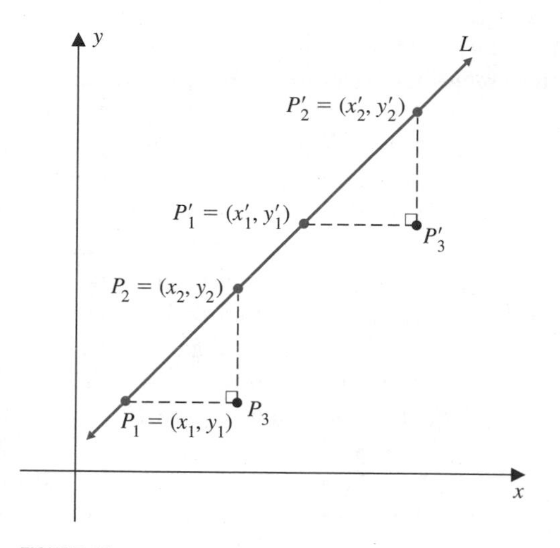

FIGURE 16

(a) Prove that right triangles $P_1P_2P_3$ and $P'_1P'_2P'_3$ are similar.

(b) Use the result in part (a) to fill in the following equal ratios of corresponding sides.

$$\frac{|\overline{P_2P_3}|}{|\overline{P_1P_3}|} = \underline{\quad\quad}$$

Then rewrite this equation in terms of the coordinates of the points.

(c) What can be concluded if P_1 and P_2 or P'_1 and P'_2 are the two points selected to compute the slope of line L?

66. Figure 17 shows the slopes of perpendicular lines L_1 and L_2 to be m_1 and m_2 respectively. Suppose $P_1 = (1, m_1)$ is a point on L_1 and $P_2 = (1, m_2)$ is a point on L_2. Use the distance formula, together with the Pythagorean theorem, to show that $m_1m_2 = -1$. [*Hint:* Since triangle P_1OP_2 is a right triangle, then we have $[d(O, P_1)]^2 + [d(O, P_2)]^2 = [d(P_1, P_2)]^2$.]

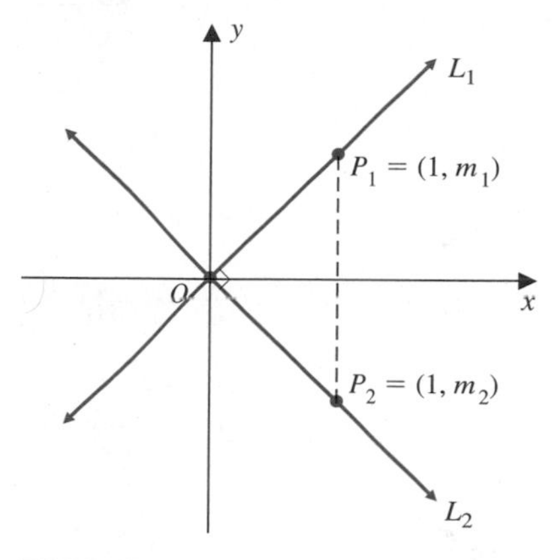

FIGURE 17

OBJECTIVES

1. Graph equations by point-plotting.
2. Determine the symmetry of a graph.
3. Graph equations by using graphing utilities.
4. Solve applied problems.

1.4 Graphs and Graphing Technology

So far we have used equation recognition to sketch graphs of equations for circles and lines. Another technique used to sketch the graph of an equation is called the *point-plotting method.*

Graphing Equations by Point-Plotting

We indicated in Section 1.2 that the graph of an equation in two variables is the set of all points whose coordinates satisfy the equation. The point-plotting method of graphing equations is outlined below.

GRAPHING BY POINT-PLOTTING

Step 1. Determine several ordered pairs of numbers that satisfy the given equation. To do so, it is helpful to express one of the variables in terms of the other, and list the ordered pairs in a table.
Step 2. Plot these ordered pairs in a coordinate system.
Step 3. Connect the points to form a curve.

EXAMPLE 1 **Graphing an equation by point-plotting**

Use point-plotting to sketch the graph of $y = x^2 - 9$. Also, locate the x and y intercepts.

TABLE 1

$y = x^2 - 9$

x	y
-4	7
-3	0
-2	-5
-1	-8
0	-9
1	-8
2	-5
3	0
4	7

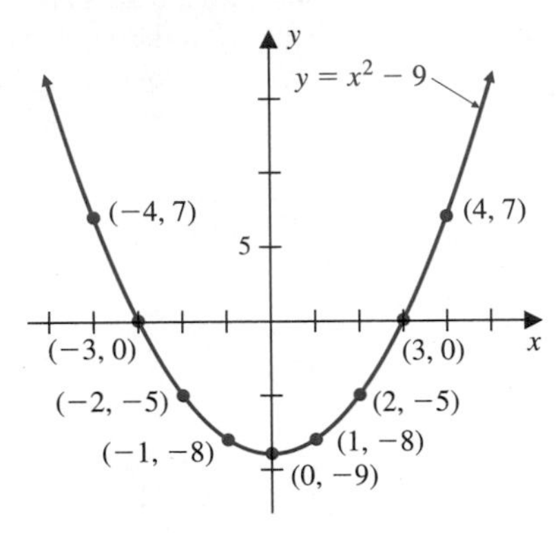

FIGURE 1

Solution Countless ordered pairs of numbers satisfy this equation. A sample of these pairs is determined by substituting arbitrary x values into the equation and calculating the corresponding y values (Table 1). By plotting these ordered pairs as points in a coordinate system and then connecting them with a smooth curve, we obtain the graph (Figure 1).

Upon examining Table 1 we see that the x intercepts are -3 and 3, and the y intercept is -9. These intercepts are plotted in Figure 1.

Determining the Symmetry of a Graph

Recognizing symmetry is helpful when graphing equations. Graphs that are symmetric with respect to the y axis, the x axis, or the origin are defined as follows:

DEFINITION

SYMMETRIC GRAPHS

Symmetry with Respect To:	Whenever Point (x, y) Is on the Graph, Then Its Reflection:	Example
y axis	$(-x, y)$ is on the graph	$y = \frac{1}{x^2 + 1}$; $(-x, y)$, (x, y)
x axis	$(x, -y)$ is on the graph	$y^2 = x$; (x, y), $(x, -y)$
Origin	$(-x, -y)$ is on the graph	$y = x^3 - x$; (x, y), $(-x, -y)$

Point-plotting should be used to verify the accuracy of the graphs of

$y = \frac{1}{x^2 + 1}$, $y^2 = x$, and $y = x^3 - x$

shown in the definition.

Such symmetries can be detected by examining an equation of a graph before graphing, as explained in Table 2.

TABLE 2

Equation Tests for Symmetry

Symmetry with Respect To:	If an Equivalent Equation Is Obtained When:	Example *Each pair of equations is equivalent*
y axis	x is replaced by $-x$	$y = \frac{1}{x^2 + 1}$ and $y = \frac{1}{(-x)^2 + 1}$
x axis	y is replaced by $-y$	$y^2 = x$ and $(-y)^2 = x$
Origin	x and y are replaced by $-x$ and $-y$, respectively	$y = x^3 - x$ and $(-y) = (-x)^3 - (-x)$ or $-y = -x^3 + x$

EXAMPLE 2 Using symmetry as an aid to graphing

Use symmetry to sketch the graph of each equation.

(a) $y = x^2$ (b) $y = x^3$

Solution

(a) Substituting $-x$ for x, we obtain $y = (-x)^2$ or $y = x^2$, which is equivalent to the original equation, so the graph is symmetric with respect to the y axis. To obtain the graph, we first use point-plotting to sketch the part of the graph in the first quadrant (Figure 2a). Finally, we reflect the result about the y axis to complete the graph (Figure 2b).

x	$y = x^2$
0	0
1	1
2	4
3	9

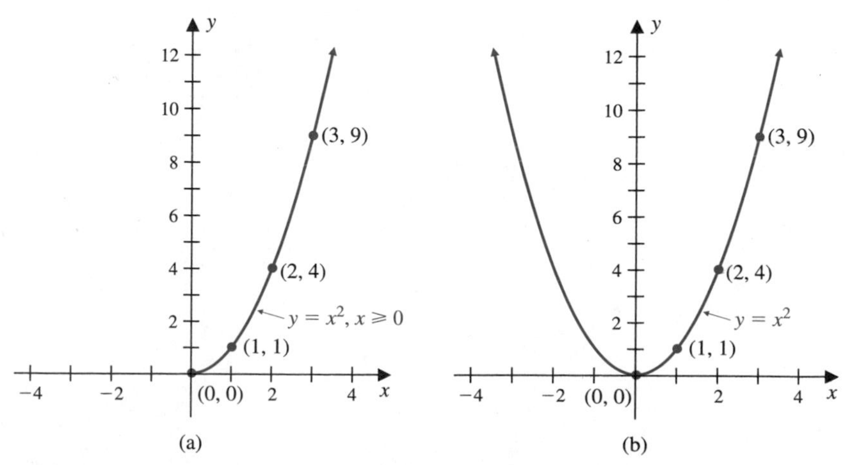

FIGURE 2

(b) The substitution of $-x$ for x and $-y$ for y produces the equation $-y = (-x)^3$ or $-y = -x^3$. Since this equation is equivalent to the equation $y = x^3$, the graph is symmetric with respect to the origin. Thus, we first sketch the part of the graph in the first quadrant (Figure 3a). Then we reflect the result about the origin to get the complete graph (Figure 3b).

x	y
0	0
$\frac{1}{2}$	$\frac{1}{8}$
1	1
2	8

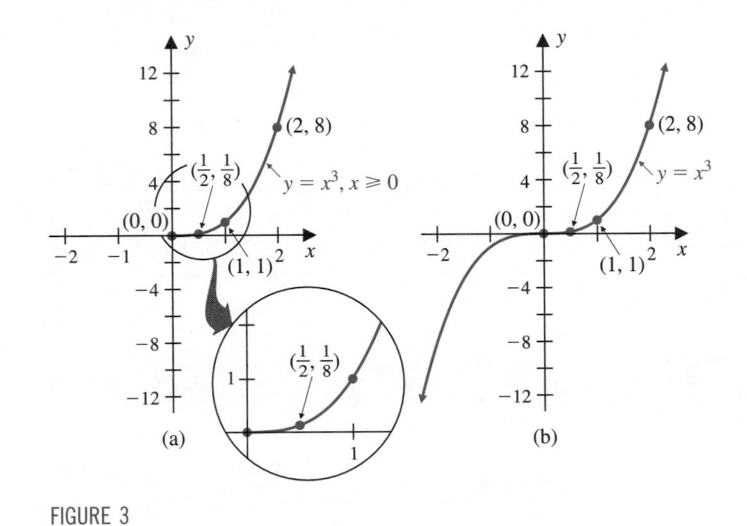

FIGURE 3

Graphing Equations by Using Graphing Utilities

Graphing equations by the point-plotting method can present misleading results if not enough points are used. For instance, suppose that we are to graph the equation

$$y = x^4 - 8x^2 - 9$$

by using the ordered pairs listed in Table 3. Figure 4a shows a graph obtained by connecting some points from Table 3 with a smooth curve.

TABLE 3

x	$y = x^4 - 8x^2 - 9$
−4	119
−3	0
0	−9
3	0
4	119

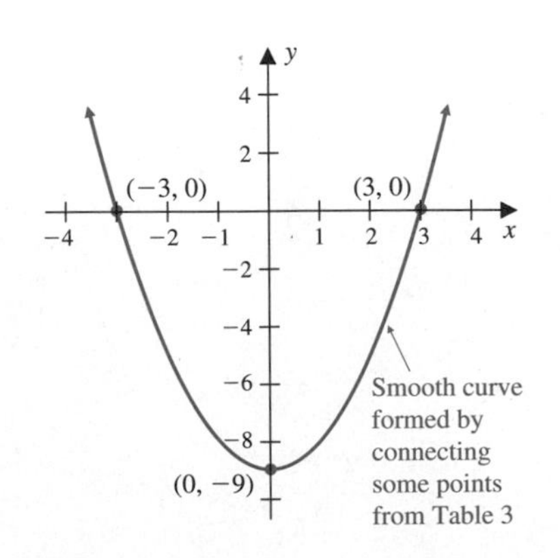

FIGURE 4a

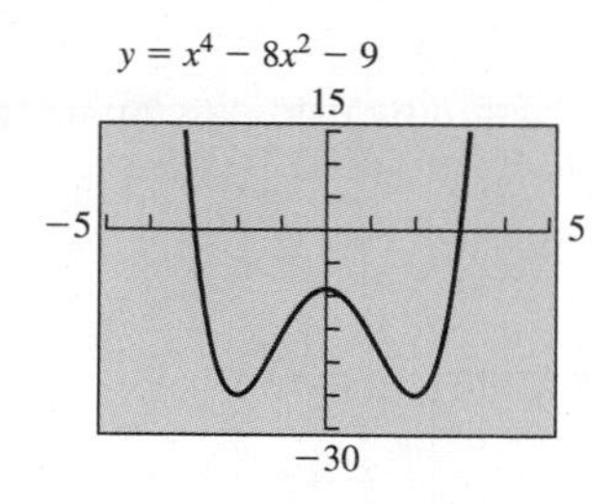

FIGURE 4b

The more points we plot, the more accurate our graph becomes; however, the task also becomes more tedious and laborious. Graphing utilities make the process of obtaining accurate graphs much easier. The term **graphing utilities** refers to both **graphing calculators** and **graphing software**. These tools generate graphs by plotting several points in close proximity to each other with speed and accuracy. In fact, if we use a graphing calculator to graph the equation $y = x^4 - 8x^2 - 9$, we obtain the graph in Figure 4b. Upon comparing Figure 4a to Figure 4b, it is apparent that Figure 4a is not an accurate representation of the graph.

The portion of the graph that is displayed at any one time by a graphing utility is called a **viewing rectangle** or **viewing window**. Although graphing calculators differ in features (your instruction manual should be consulted), most have a RANGE or WINDOW menu that allows us to choose the minimum and maximum values of both x and y for the viewing window. Figure 5 illustrates a viewing window selection.

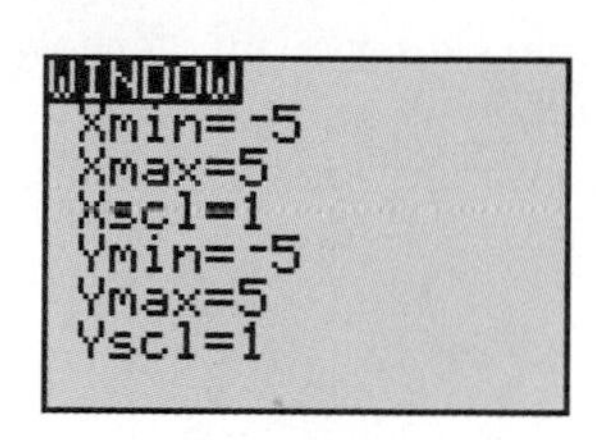

(a) Settings on menu
Note: "scl" stands for "scale"

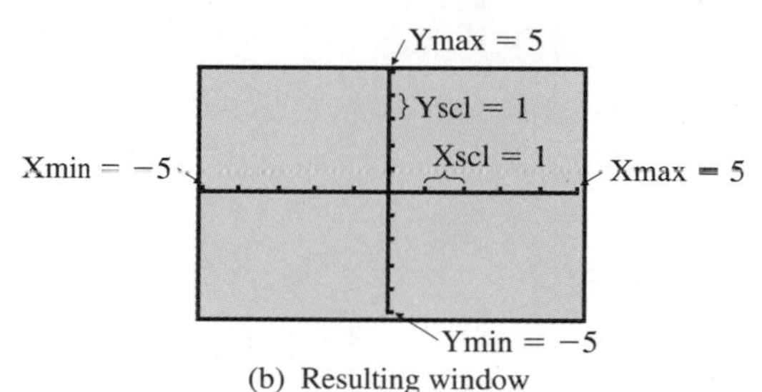

(b) Resulting window

FIGURE 5

Throughout this text, all graphs are either hand-drawn or outputs of graphing utilities. Those graphs produced by graphing utilities are enclosed by viewing rectangles with displayed WINDOW selections as in Figure 6a. Those representing hand-drawn graphs are not enclosed by rectangles, as in Figure 6b. Examples and problems for which a graphing utility is used are marked with the symbol GU.

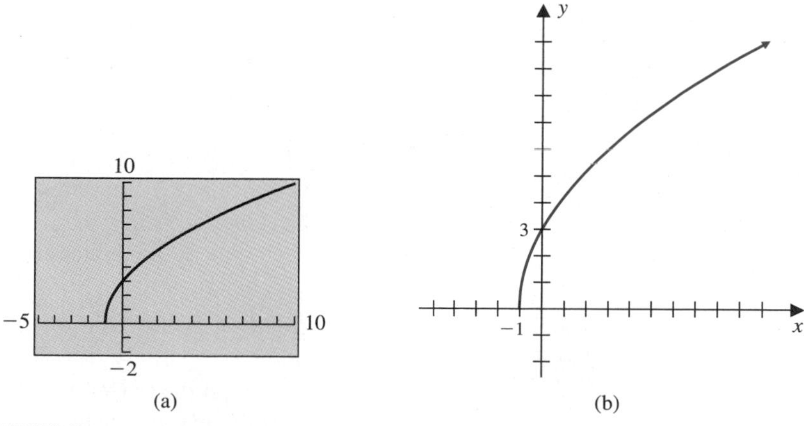

(a) (b)

FIGURE 6

EXAMPLE 3 **Graphing equations using a graphing utility**

GU

Use the specified WINDOW selections to graph each equation by using a graphing utility.

(a) $y = x^2 + 2x - 3$ (b) $y = x^3 - 4x$

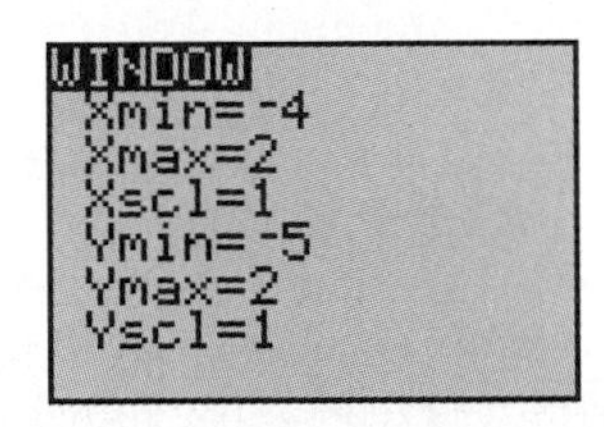

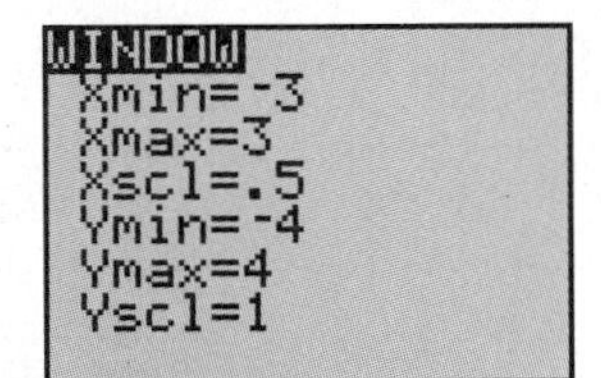

Solution

(a) Figure 7a shows the viewing window of the graph of $y = x^2 + 2x - 3$ for the given WINDOW selections.

(b) Figure 7b shows the viewing window of the graph of $y = x^3 - 4x$ for the given WINDOW selections.

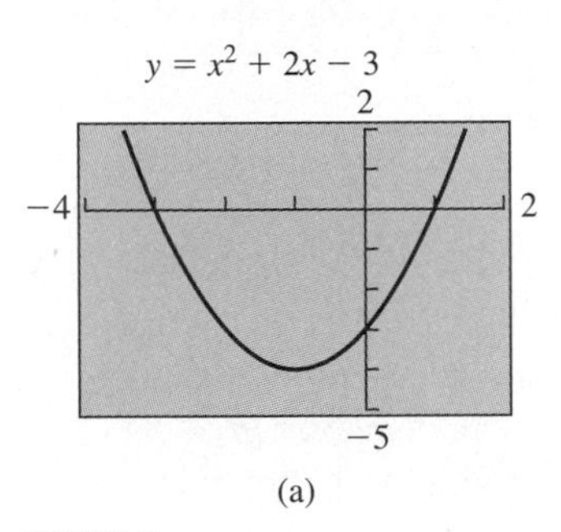

(a)

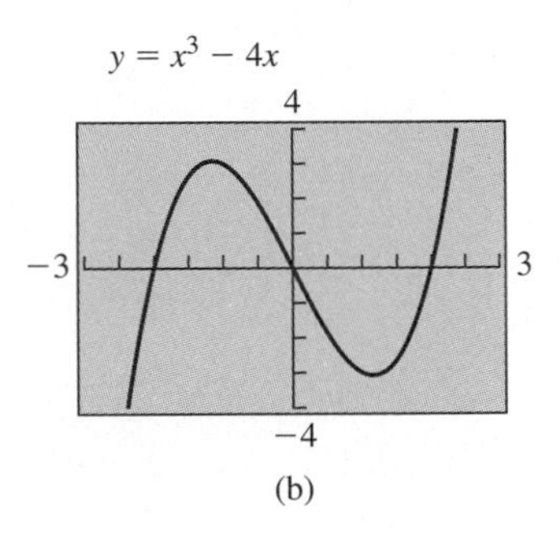

(b)

FIGURE 7

When graphing equations in x and y on a graphing utility, we first "isolate y" by solving the given equation for y in terms of x. Sometimes the process is straightforward. For instance, to graph $2x^2 - y + 3 = 0$, we first solve for y to get $y = 2x^2 + 3$; also, we graph $x^4 - 2x^3 + 5x + 3y = 0$ by solving for y to get $y = -\frac{1}{3}x^4 + \frac{2}{3}x^3 - \frac{5}{3}x$. At times, solving for y may result in two different equations. If this occurs we superimpose two graphs, one for each of the two equations, in the same viewing window in order to get the graph of the original equation. This technique is illustrated in the next example.

Most graphing utilities can graph more than one equation in the same viewing window.

EXAMPLE 4 **Graphing two equations in the same viewing window**

Given the equation $x^2 - 4y^2 - 4 = 0$, solve for y in terms of x. Then use a graphing utility to graph the two resulting equations in the same viewing window to get the graph of the given equation.

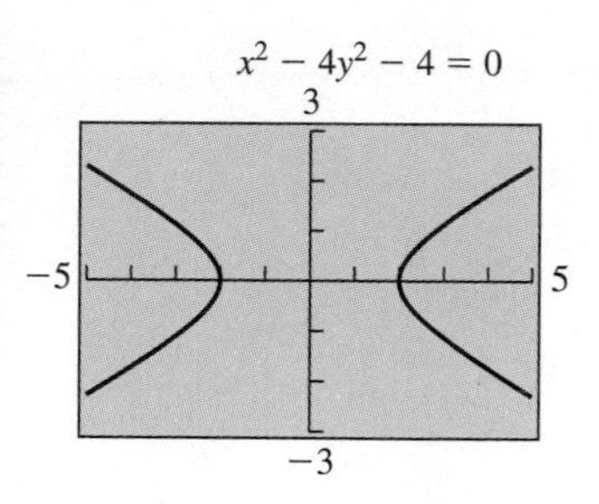

FIGURE 8

Solution First we solve for y in terms of x:

$$x^2 - 4y^2 - 4 = 0$$
$$-4y^2 = -x^2 + 4$$
$$y^2 = \frac{x^2}{4} - 1$$

so

$$y = \sqrt{\frac{x^2}{4} - 1} \quad \text{or} \quad y = -\sqrt{\frac{x^2}{4} - 1}$$

Figure 8 displays a viewing window of the graphs of $y_1 = \sqrt{(x^2/4) - 1}$ (above the x axis) and $y_2 = -\sqrt{(x^2/4) - 1}$ (below the x axis) superimposed in the same viewing window. These two graphs taken together represent the graph of $x^2 - 4y^2 - 4 = 0$.

Solving Applied Problems

If two quantities x and y are related by the equation $y = k/x$, where k is a constant, we say that y is **inversely proportional** to x, or y **varies inversely** as x. In the next example, we use a formula called *Boyle's law* to model the inverse relationship between the volume and pressure of a gas.

EXAMPLE 5 Using an inverse variation model

According to Boyle's law, under constant temperature, the volume V of a gas is inversely proportional to the pressure P. Hence,

$$V = \frac{k}{P} \qquad \text{where } k \text{ is a constant}$$

(a) Suppose that for a certain gas, $k = 3975$, V is expressed in cubic inches, and P represents pounds per square inch. Express V in terms of P.
(b) What is the restriction on P?
(c) Use point-plotting to graph the equation in part (a) with the restriction in part (b), where P is scaled on the horizontal axis and V is scaled on the vertical axis.
(d) Use the result in part (c) to examine the trend of the volume as the pressure increases.

Solution

(a) Since $V = k/P$ and $k = 3975$, it follows that $V = 3975/P$.
(b) Since P represents pressure, then $P > 0$.
(c) By plotting a few points and then connecting them with a smooth curve, we get the graph displayed in Figure 9 (at the top of the next page).

P	V
10	397.5
20	198.75
40	99.38
100	39.75
500	7.95
1000	3.98

Could the pressure equal 0 in this model? Explain. Could the volume ever equal 0? Explain.

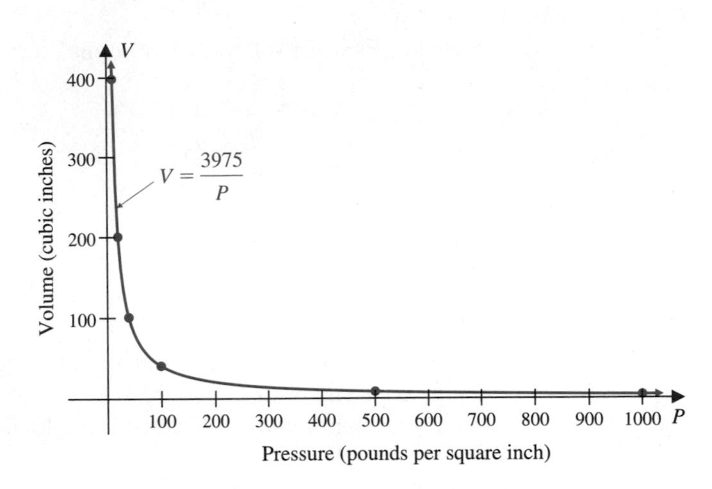

FIGURE 9

(d) Upon examining the graph in Figure 9 we see that as the pressure increases, the volume correspondingly decreases.

The following example uses formulas from geometry to construct a model for interpreting cost.

EXAMPLE 6 **Modeling the cost of a steam boiler**
A company manufactures cylindrical-shaped steam boilers in different sizes. The cost of each boiler is \$75 per square foot of its total surface area. Suppose that a boiler is 24.5 feet long and has a radius of r feet (Figure 10).

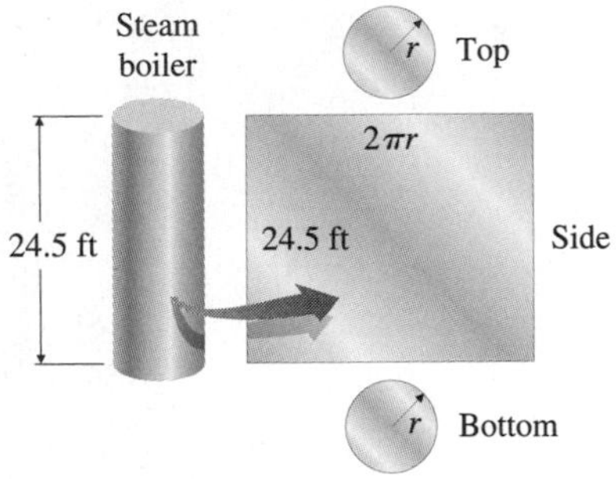

FIGURE 10

(a) Construct an equation that expresses the cost C (in dollars) of the boiler in terms of the radius r.

(b) What restriction is there on r?

(c) Use a graphing utility to graph the equation in part (a) with the restriction in part (b) and interpret the graph.

(d) Adjust the equation in part (a) to express the new cost K in terms of the radius if the price per square foot of the side of the boiler is reduced by 20%.

(e) Graph the equation in part (d) and compare this graph to the graph in part (c).

Solution

(a) We derive the equation that expresses the cost C in terms of the radius r as follows:

$C = 75\pi(49r + 2r^2)$

250,000

0 0 15

(a)

$$\begin{aligned} C &= 75[(\text{Area of the side}) + (\text{Area of top}) + (\text{Area of bottom})] \\ &= 75[(2\pi r \cdot 24.5) + \pi r^2 + \pi r^2] \\ &= 75(49\pi r + 2\pi r^2) \\ &= 75\pi(49r + 2r^2) \end{aligned}$$

(b) Since r represents the radius, $r > 0$.

(c) A viewing window of the graph of the equation is displayed in Figure 11a. The pattern of the graph visually conveys the fact that the bigger the radius, the more costly the boiler.

(d) If the side is discounted by 20%, then the new cost K becomes

$$\begin{aligned} K &= 75\pi(49r + 2r^2) - 0.20(75 \cdot 49\pi r) \\ &= 75\pi(49r + 2r^2) - 735\pi r \\ &= 3675\pi r - 735\pi r + 150\pi r^2 \\ &= 2940\pi r + 150\pi r^2 \end{aligned}$$

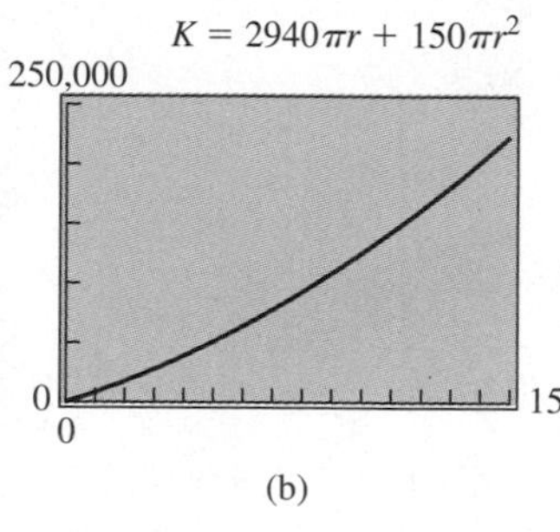

(b)

FIGURE 11

(e) Figure 11b displays a viewing window of the graph of this equation. Upon comparing the two graphs, we see that as the radius increases, the cost with the discount increases at a slower rate than the cost without the discount. As a result, the cost with the discount is less than the cost without the discount.

PROBLEM SET 1.4

Mastering the Concepts

In problems 1–8, use the point-plotting method to sketch the graph of each equation. Also, locate the x and y intercepts.

1. $y = -2x^2$ **2.** $y = \frac{1}{3}x^2$

3. $y = 3x^2 + 1$ **4.** $y = -\frac{1}{2}x^2 + 4$

5. $y = \sqrt{x - 3}$ **6.** $y = \sqrt{-x} + 3$

7. $y = -x^3 - 1$ **8.** $y = 4x^3 - 4$

In problems 9–12, first solve for y in terms of x and then sketch the graph by using point-plotting. Also, locate the x and y intercepts.

9. $7x - 2y = 5$ **10.** $x^2 - 2y = 6$

11. $5y^2 - x = 6$ **12.** $y^2 + x = 6$

In problems 13–16, plot the point P in the Cartesian plane and then locate and give the coordinates of the following points:

(a) Q, symmetric to P with respect to the x axis

(b) R, symmetric to P with respect to the y axis

(c) S, symmetric to P with respect to the origin

13. $P = (1, 4)$ **14.** $P = (-4, -2)$

15. $P = (-5, 2)$ **16.** $P = (3, -1)$

In problems 17–22, part of the graph of an equation is displayed. Sketch the complete graph assuming the graph is symmetric to the

(a) x axis **(b)** y axis **(c)** Origin

17.

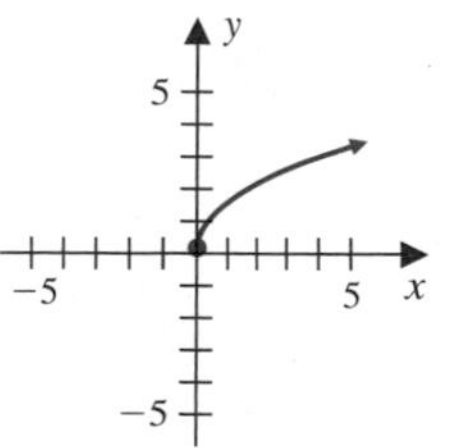

18.

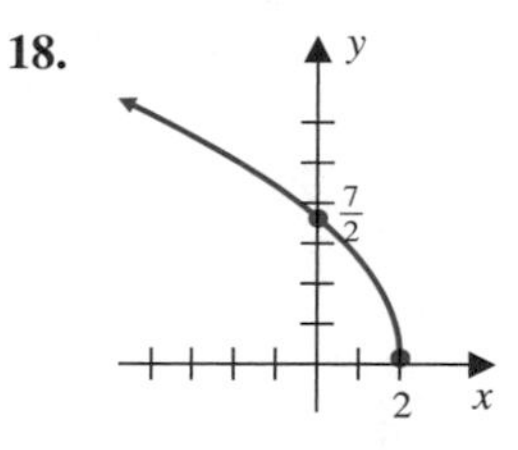

19.

y 5 −4 5 x −5

20.

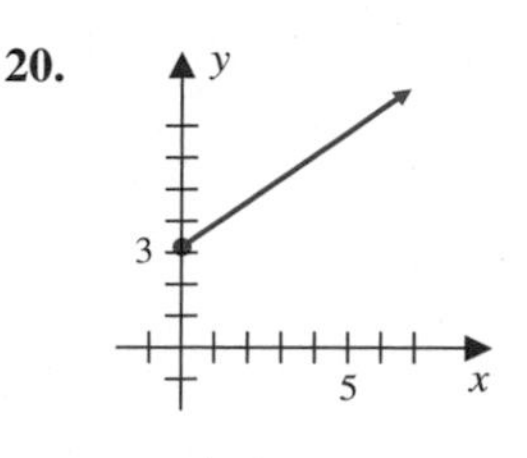

21.

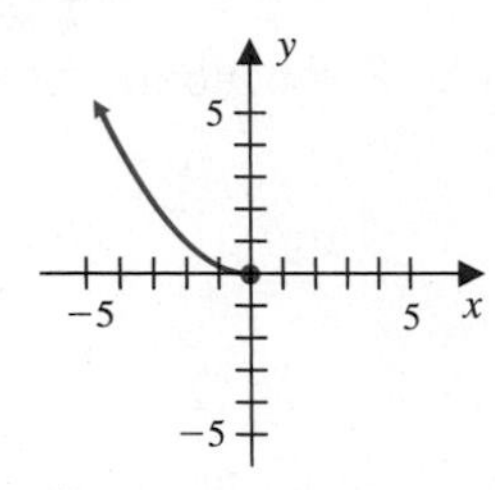

22.

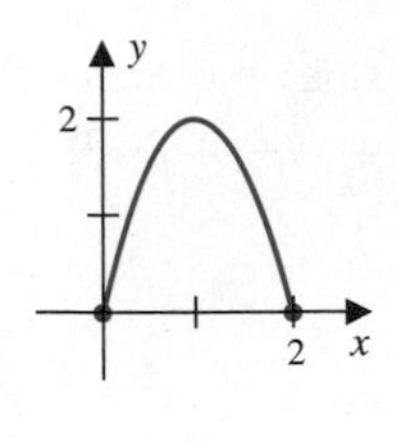

In problems 23–30, use the equation tests in Table 2 to determine symmetry with respect to the x axis, y axis, or origin. Then use the symmetry, if it exists, to sketch the graph of each equation.

23. $y = |x| + 2$ **24.** $y = 3|x| - 1$

25. $x = y^2 + 2$ **26.** $x + 1 = 2y^2$

27. $y = \sqrt{9 - x^2}$ **28.** $y = -\sqrt{9 - x^2}$

29. $y = \sqrt[3]{x}$ **30.** $y = 2x^3 + x$

In problems 31–36, match each equation with its graph.

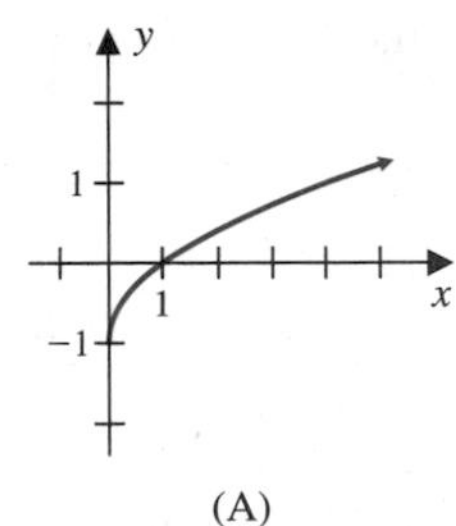

(A)

(B)

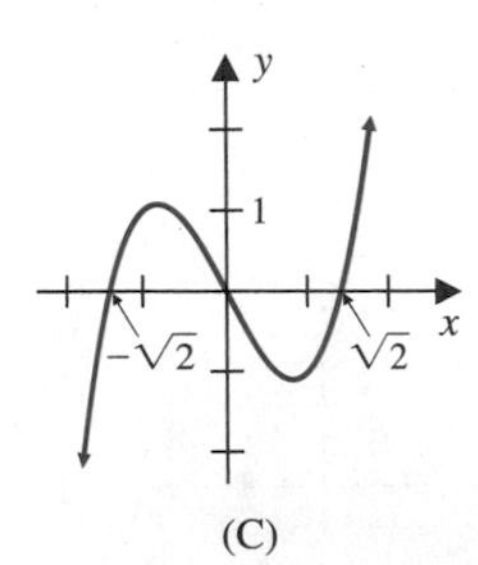

(C)

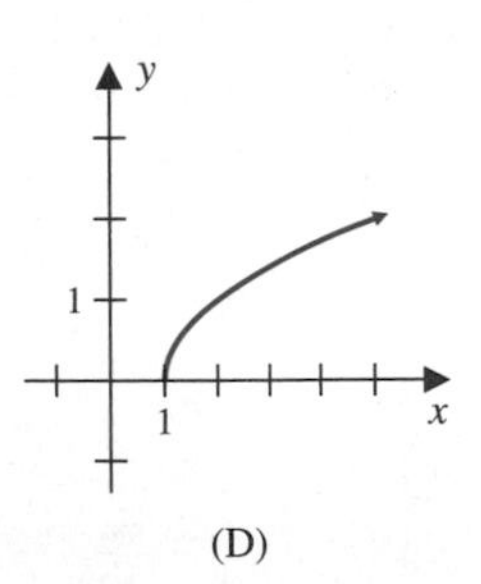

(D)

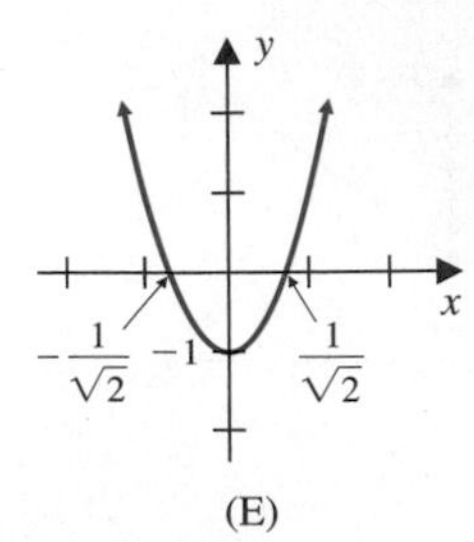

(E)

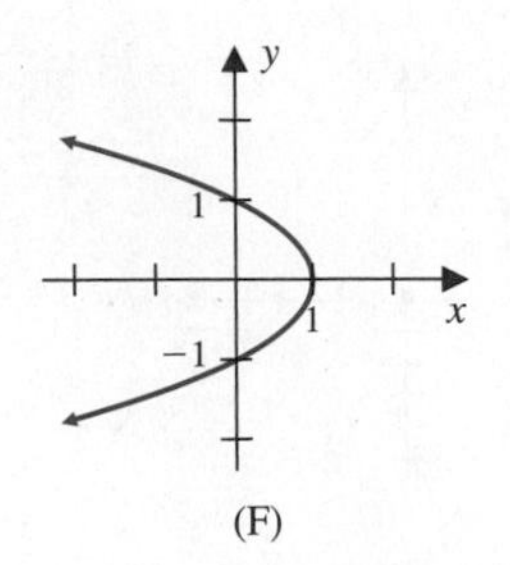

(F)

31. $y = 2x^2 - 1$ **32.** $y = \sqrt{x} - 1$

33. $y = \sqrt{x - 1}$ **34.** $x = -y^2 + 1$

35. $y = -x^3 + x$ **36.** $y = x^3 - 2x$

GU In problems 37–40, use a graphing utility to sketch the graph of the given equation in each of the specified viewing windows. Describe the differences between the resulting graphs in parts (a) and (b). Which of the two graphs appear to be more accurate? Explain.

37. $y = x^2 - 5x + 6$

(a) Xmin: −2; Xmax: 8; Xscl: 1;
Ymin: −2; Ymax: 8; Yscl: 1

(b) Xmin: −2; Xmax: 2; Xscl: 1;
Ymin: −2; Ymax: 8; Yscl: 1

38. $y = -x^2 - 3x + 4$

(a) Xmin: −5; Xmax: 5; Xscl: 1;
Ymin: −5; Ymax: 5; Yscl: 1

(b) Xmin: −6; Xmax: 4; Xscl: 1;
Ymin: −2; Ymax: 8; Yscl: 1

39. $y = x^3 - 3x^2 + 1$

(a) Xmin: −2; Xmax: 2; Xscl: 1;
Ymin: −2; Ymax: 8; Yscl: 1

(b) Xmin: −5; Xmax: 5; Xscl: 1;
Ymin: −5; Ymax: 5; Yscl: 1

40. $y = -x^3 + 7x + 10$

(a) Xmin: −5; Xmax: 5; Xscl: 1;
Ymin: −20; Ymax: 10; Yscl: 1

(b) Xmin: −5; Xmax: 5; Xscl: 1;
Ymin: −5; Ymax: 20; Yscl: 1

GU In problems 41–46, use a graphing utility to sketch the graph of each equation. Use the graph to decide whether the resulting curve is symmetric with respect to the x axis, y axis, or origin. Verify the answer by using the equation tests for symmetry given in Table 2.

41. $y = \dfrac{1}{x^3}$ **42.** $y = x^3 + x$

43. $y = 5x - x^2$ **44.** $y = x^2 + 3x$

45. $y = 3x^2 + \dfrac{1}{x^2}$ **46.** $y = x^2 - \dfrac{1}{x^2}$

GU In problems 47–54, solve for y in terms of x and then use a graphing utility to graph the given equation. Superimpose two graphs on the same viewing window, if necessary.

47. $5x + 8y = 7$ **48.** $3x - 2y = 1$

49. $x^2 - 4x - y = 0$ **50.** $2x^2 + 3x + y = 3$

51. $4x - 16y^2 = 16$ **52.** $3x^2 + y^2 = 9$

53. $100x^2 + y^2 = 25$ **54.** $9y^2 - 3x = 27$

55. Suppose that w is inversely proportional to t.
(a) Express w in terms of t if $w = 3$ when $t = 2$.
(b) What happens to w if t is doubled?
(c) What happens to w if t is multiplied by a factor of 4?

56. Suppose that x is inversely proportional to y and y is inversely proportional to z. How are x and z related?

Applying the Concepts

57. Physics: Ohm's law states that the current I (in amperes), voltage E (in volts), and resistance R (in ohms) in a simple electric circuit are related by the formula

$$I = \frac{E}{R}$$

(a) What restriction should be imposed on R?
(b) If the voltage is 110, use point-plotting to graph the resulting equation with the restriction from part (a).
(c) What is the value of the current if the resistance is 8 ohms? 10 ohms? 12 ohms?
(d) If the trend continues, discuss the relationship between the current and the resistance.

58. Advertising: An advertising agency determines that its revenue R (in thousands of dollars) is given by the model

$$R = 50x^2 - 200x + 800$$

where x represents the number of clients.
(a) What restriction should be imposed on x?
GU **(b)** Use a graphing utility to sketch the graph of the equation that represents this model using the restriction from part (a).
(c) How much revenue will be generated if the agency has 150 clients? 200 clients? 500 clients?
(d) If the trend continues, discuss the relationship between the revenue and number of clients.

59. Ice Cream Cone: An ice cream machine dispenses soft ice cream to fill a cone-shaped shell with a height h equal to twice the radius r (in inches) of its base (Figure 12).

FIGURE 12

(a) Express the volume V of the ice cream cone in terms of r.
(b) What restriction is there on r?
(c) Determine the volume if the radius is 1 inch; 1.5 inches; 2 inches. Round off each answer to two decimal places.
GU **(d)** Use a graphing utility to graph the equation in part (a) with the restriction in part (b).
(e) Use the graph to discuss the trend of the volume as the radius increases.

60. Cost of a Walkway: The length and width of a backyard are 50 feet and 40 feet, respectively. A homeowner wishes to construct a concrete walk of uniform width x feet that borders the yard and to sod the interior with grass. Suppose that the cost of laying concrete is \$5 per square foot and the cost of sodding grass is \$3 per square foot.
(a) Draw a sketch of the backyard and concrete border. Use the sketch to determine the restrictions on x.
(b) Construct an equation that expresses the total cost C (in dollars) of the concrete and sod in terms of x.
GU **(c)** Use a graphing utility to sketch the graph of the equation in part (b) with the restrictions in part (a).
(d) Adjust the equation in part (b) to express the cost K in terms of x if the price per square foot of the concrete is reduced by 10%.
GU **(e)** Use a graphing utility to graph the equation in part (d) and compare it to the graph in part (c). Use the graphs to compare the trends of the cost as the width x increases.

Developing and Extending the Concepts

61. GU Use a graphing utility to graph the following four equations on the same coordinate system.

$$y_1 = \sqrt{x} \quad y_2 = \sqrt{-x} \quad y_3 = -\sqrt{x} \quad y_4 = -\sqrt{-x}$$

Describe the symmetric relationships among the graphs.

62. Given the equation $ax^2 + by^2 = 1$, where a and b are positive constants:
(a) In what quadrants will the graph lie?
(b) Explain why it is sufficient to graph the equation in quadrant I and then use symmetry to get the complete graph.
(c) Select specific values for a and b to illustrate the situation.

63. Given the equation $y^2 = x^2$:
(a) Test the equation for symmetry.
(b) Graph the equation by using point-plotting.
GU **(c)** How can the graph be obtained by using a graphing utility?

64. **(a)** Explain why the graph of $y^2 = 1 - x$ is different from the graph of $y = \sqrt{1 - x}$.
(b) Explain why the graph of $y^3 = 1 - x$ is the same as the graph of $y = \sqrt[3]{1 - x}$.
(c) Use the results from parts (a) and (b) to compare the graph of $y^n = 1 - x$ to the graph of $y = \sqrt[n]{1 - x}$ according to whether n is a positive even integer or a positive odd integer.

OBJECTIVES

1. Solve linear equations.
2. Solve equations involving rational expressions.
3. Solve equations involving absolute value.
4. Solve quadratic equations.
5. Solve equations involving radicals.
6. Solve applied problems.

1.5 Solutions of Equations and Graphs

This section deals with equations in one variable. Examples of such equations are

$$-2x + 7 = 13, \quad |2x - 1| = 3, \quad x^2 + 5x - 6 = 0, \quad \text{and} \quad \sqrt{5y - 7} = 11$$

To **solve** an equation in one variable, we need to find all values which, when substituted for the variable, make the statement of equality a true statement. These values are called **solutions**, or **roots**, of the equation. For example, -3 is the solution for $-2x + 7 = 13$ because $-2(-3) + 7 = 13$ is a true statement, and no other value for x satisfies this equation. Here we review techniques learned in algebra for solving equations, and we also relate graphs to solutions of equations.

Solving Linear Equations

First-degree, or *linear*, *equations in one variable*, such as $4x - 7 = 1$, $2y - 7 = y + 1$, and $2x - 5x = 4 - 3x$, involve only the first power of the variable and constants. To solve this type of equation we transform the equation into an equivalent one (one with the same solution) in which the variable is isolated on one side of the equal sign. This technique relies on the following properties:

PROPERTIES THAT GENERATE EQUIVALENT EQUATIONS

Assume a, b, and c represent numbers.

(i) **Addition and Subtraction Properties:**
If $a = b$, then $a + c = b + c$ and $a - c = b - c$.

(ii) **Multiplication and Division Properties:**
If $a = b$ and $c \neq 0$, then $ac = bc$ and $\dfrac{a}{c} = \dfrac{b}{c}$.

The next example shows how to relate the solution of an equation to a graph.

EXAMPLE 1 Solving a linear equation

(a) Solve the linear equation $4(x - 3) = 2(3x + 1) + 5x$ algebraically.
(b) Demonstrate the solution graphically.

Solution

(a) To solve this equation algebraically, we first simplify each side and then apply the above properties as follows:

$$\left.\begin{aligned} 4(x-3) &= 2(3x+1)+5x \\ 4x-12 &= 6x+2+5x \\ 4x-12 &= 11x+2 \end{aligned}\right\} \quad \text{Simplify each side}$$

$$\begin{aligned} 4x &= 11x+14 && \text{Add 12 to each side} \\ -7x &= 14 && \text{Subtract } 11x \text{ on each side} \\ x &= -2 && \text{Divide each side by } -7 \end{aligned}$$

So the solution of the equation is -2.

(b) We can relate the solution of the given equation to the graph of an associated equation in two variables as follows. Every equation in one variable is equivalent to an equation that has 0 on one side. The equation

$$4(x - 3) = 2(3x + 1) + 5x$$

can be rewritten as

$$4(x - 3) - 2(3x + 1) - 5x = 0$$

or simply,

$$-7x - 14 = 0$$

If a real number is a solution of this equation, then it is also the x intercept of the graph of the equation

$$y = -7x - 14$$

and vice versa. Figure 1 shows that the x intercept of the graph of the equation $y = -7x - 14$ is -2. This demonstrates graphically that the solution of the original equation is -2.

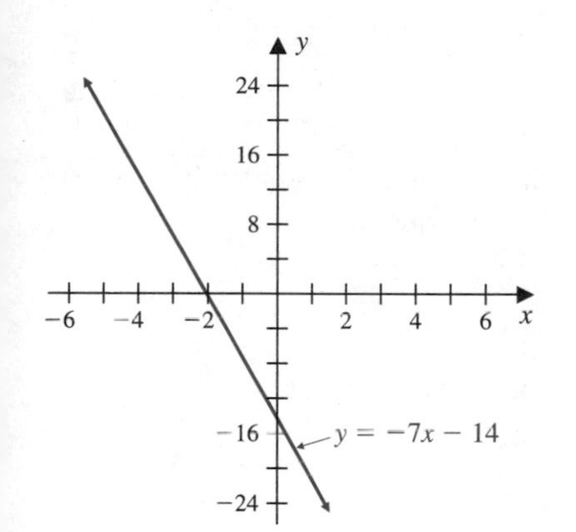

FIGURE 1

Solving Equations Involving Rational Expressions

If an equation involves rational expressions, we can convert it to a simpler equivalent form by multiplying both sides of the equation by a common denominator (we normally use the least common denominator, LCD) of all the fractions. Then we solve the resulting equation. The proposed solution should always be checked if variables are in the denominator of the original equation, since it may be an *extraneous root*—that is, a misleading value that does not satisfy the original equation.

EXAMPLE 2 **Solving a rational equation**

Solve the equation: $\frac{3}{x} - \frac{1}{2x} = \frac{2}{x-1}$

Solution The solution of the equation proceeds as follows:

$$\frac{3}{x} - \frac{1}{2x} = \frac{2}{x-1} \qquad \text{Given}$$

$$2x(x-1)\left(\frac{3}{x} - \frac{1}{2x}\right) = 2x(x-1)\left(\frac{2}{x-1}\right) \qquad \text{Multiply both sides by the LCD}$$

$$2(x-1)(3) - (x-1) = 2x(2) \qquad \text{Distribute}$$

$$\left.\begin{aligned} 6(x-1) - (x-1) &= 4x \\ 6x - 6 - x + 1 &= 4x \\ 5x - 5 &= 4x \end{aligned}\right\} \qquad \text{Simplify}$$

$$x = 5 \qquad \text{Add 5 and subtract } 4x \text{ on each side}$$

Next, we check the proposed solution by substituting $x = 5$ into the given equation:

$$\frac{3}{5} - \frac{1}{10} = \frac{2}{4} \quad \text{or} \quad \frac{6-1}{10} = \frac{2}{4} \quad \text{or} \quad \frac{5}{10} = \frac{2}{4}$$

which is true. Therefore, the solution is 5.

To examine the solution of the equation in Example 2 graphically, we first observe that the given equation is equivalent to the equation

$$\frac{3}{x} - \frac{1}{2x} - \frac{2}{x-1} = 0$$

Thus, the solution is the same as the x intercept of the graph of

$$y = \frac{3}{x} - \frac{1}{2x} - \frac{2}{x-1}$$

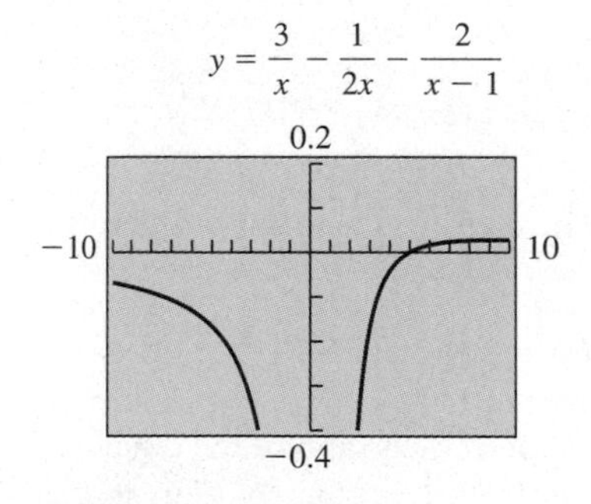

FIGURE 2

Figure 2 shows a viewing window of the graph of the latter equation. The graph suggests that there is one x intercept at 5, thus displaying the solution found algebraically.

Solving Equations Involving Absolute Value

To solve equations involving absolute value, we use the following properties, which follow from the definition of absolute value given in Section 1.1:

ABSOLUTE VALUE PROPERTIES

Let p be a real number.

(i) If $p > 0$, then $|u| = p$ if and only if $u = -p$ or $u = p$.
(ii) If $p < 0$, then $|u| = p$ has no solution.
(iii) If $p = 0$, then $|u| = p$ if and only if $u = 0$.

EXAMPLE 3 Solving an absolute value equation

(a) Solve the equation $|2x - 3| = 5$ algebraically.
(b) Display the solution graphically.

Solution

(a) By using Property (i) and replacing u by $2x - 3$ and p by 5, we have

$$\begin{array}{rcl|rcl} 2x - 3 &=& -5 \quad \text{or} & 2x - 3 &=& 5 \\ 2x &=& -2 & 2x &=& 8 \\ x &=& -1 & x &=& 4 \end{array}$$

Thus, the solution includes -1 and 4.

(b) We use a graphing utility to graph the associated equation

$$y = |2x - 3| - 5$$

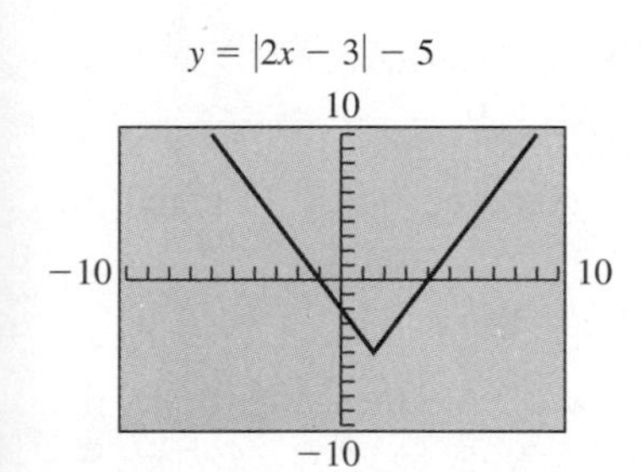

FIGURE 3

From the graph (Figure 3) we observe that -1 and 4 are the x intercepts, the same values as the solutions of the original equation.

Solving Quadratic Equations

Any equation that can be written in the form

$$ax^2 + bx + c = 0$$

where a, b, and c are constants, $a \neq 0$, is called a **second-degree**, or **quadratic, equation in one variable**. This form is referred to as the **standard form** of a quadratic equation. For instance, the equation $2x^2 - 3x = -1$ has standard form $2x^2 + (-3)x + 1 = 0$. We'll review three methods for solving quadratic equations with real number coefficients: the *factor method*, the *completing-the-square method*, and the *quadratic formula*.

1. The Factor Method

At times, we solve quadratic equations in standard form by factoring the quadratic expression as a product of two first-degree polynomials. The equation can then be solved by setting each factor equal to 0 and solving the resulting first-degree equations. This procedure is based on the following property:

ZERO FACTOR PROPERTY

$a \cdot b = 0$ if and only if $a = 0$ or $b = 0$, or both a and b equal 0.

EXAMPLE 4 Solving an equation by factoring

(a) Solve the equation $x^2 - x = 6$ by the factor method.

(b) Display the solution graphically.

Solution

(a) We solve the equation by the factor method as follows:

$$x^2 - x = 6$$

$$x^2 - x - 6 = 0 \quad \text{Subtract 6 on each side}$$

$$(x - 3)(x + 2) = 0 \quad \text{Factor}$$

$$x - 3 = 0 \quad \text{or} \quad x + 2 = 0 \quad \text{Set each factor equal to 0}$$

$$x = 3 \quad | \quad x = -2 \quad \text{Solve each equation}$$

So the solution of the original equation includes -2 and 3.

(b) To display the solutions graphically, we use a graphing utility to graph the associated equation

$$y = x^2 - x - 6$$

From the graph (Figure 4) we observe that -2 and 3 are the x intercepts, the same values as the solution of the equation.

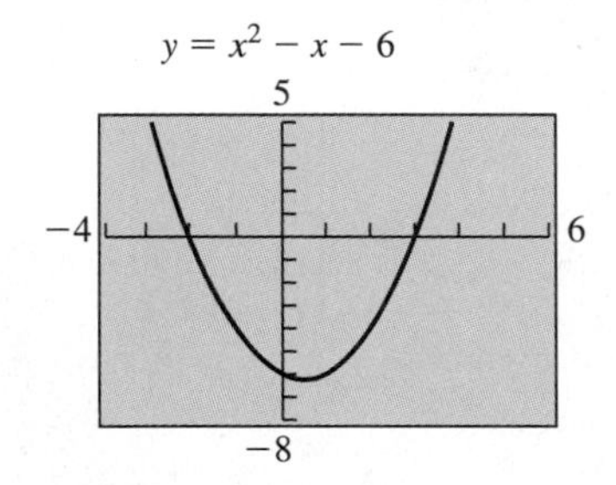

FIGURE 4

2. The Completing-the-Square Method

The equation $x^2 = 9$ can be solved by **extracting the roots** of each side to get

$$x = -\sqrt{9} \quad \text{or} \quad x = \sqrt{9}$$

$$x = -3 \quad | \quad x = 3$$

So the solutions of $x^2 = 9$ are -3 and 3. The same argument works for any positive real number k; that is:

The solutions of $x^2 = k$ are $x = -\sqrt{k}$ and $x = \sqrt{k}$, where $k > 0$.

We now use this idea to develop another method for solving a quadratic equation, based on the technique of completing the square that was introduced in Section 1.2.

EXAMPLE 5 Solving a quadratic equation by completing the square

Solve the equation $2x^2 - 2x - 1 = 0$ by using the completing-the-square method.

Solution To solve the equation, we proceed as follows:

$$2x^2 - 2x - 1 = 0 \quad \text{Given}$$

$$2x^2 - 2x = 1 \quad \text{Add 1 to each side}$$

$$x^2 - x = \frac{1}{2} \quad \text{Divide each side by 2 to make the coefficient of } x^2 \text{ equal to 1}$$

$$x^2 - x + \frac{1}{4} = \frac{1}{2} + \frac{1}{4} \quad \text{Complete the square of the left-hand side by adding } \left[\tfrac{1}{2}(-1)\right]^2 = \tfrac{1}{4} \text{ to each side}$$

$$\left(x - \frac{1}{2}\right)^2 = \frac{3}{4} \qquad \text{Factor and simplify}$$

$$x - \frac{1}{2} = -\frac{\sqrt{3}}{2} \quad \text{or} \quad x - \frac{1}{2} = \frac{\sqrt{3}}{2} \qquad \text{Extract the roots of each side}$$

$$x = \frac{1}{2} - \frac{\sqrt{3}}{2} \quad \Big| \quad x = \frac{1}{2} + \frac{\sqrt{3}}{2} \qquad \text{Solve for } x$$

So the solution includes $(1 - \sqrt{3})/2$ (approx. -0.37) and $(1 + \sqrt{3})/2$ (approx. 1.37).

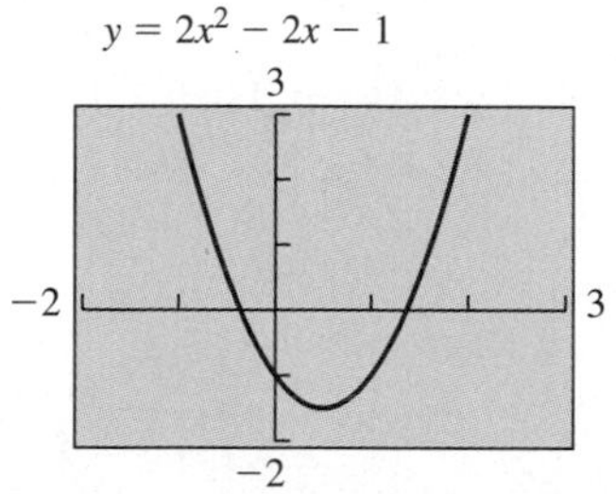

$y = 2x^2 - 2x - 1$

FIGURE 5

A viewing window of the graph of the associated equation for Example 5, $y = 2x^2 - 2x - 1$, is shown in Figure 5. It shows that there are two x intercepts, one between -1 and 0 and another between 1 and 2, consistent with the values of the solution of the equation.

Later, in Chapter 3, we will investigate a way of using graphs to approximate the x intercepts.

3. The Quadratic Formula

We can generalize the method of completing the square to arrive at a formula that enables us to solve any quadratic equation (Problem 60).

QUADRATIC FORMULA

If $ax^2 + bx + c = 0$, with a, b, and c real numbers and $a \neq 0$, then the solutions of the equation can be determined by the formula

$$x = \frac{-b \pm \sqrt{b^2 - 4ac}}{2a}$$

where $b^2 - 4ac$ is called the **discriminant** of the quadratic equation.

EXAMPLE 6 **Using the quadratic formula**

Solve each equation by using the quadratic formula.

(a) $2x^2 - 8x + 3 = 0$ (b) $5x^2 + 2x + 1 = 0$

Solution

(a) Here, $a = 2$, $b = -8$, and $c = 3$, so by using the quadratic formula we have

$$x = \frac{-b \pm \sqrt{b^2 - 4ac}}{2a}$$

$$= \frac{-(-8) \pm \sqrt{64 - 4(2)(3)}}{2(2)} = \frac{8 \pm \sqrt{64 - 24}}{4}$$

$$= \frac{8 \pm \sqrt{40}}{4} = \frac{8 \pm 2\sqrt{10}}{4} = \frac{2(4 \pm \sqrt{10})}{4} = \frac{4 \pm \sqrt{10}}{2}$$

Therefore, the solutions are

$$2 + \frac{\sqrt{10}}{2} \text{ (approx. 3.58)} \qquad \text{and} \qquad 2 - \frac{\sqrt{10}}{2} \text{ (approx. 0.42)}$$

(b) Using the quadratic formula with $a = 5$, $b = 2$, and $c = 1$, we have

$$x = \frac{-2 \pm \sqrt{4 - 4(5)(1)}}{10}$$
$$= \frac{-2 \pm \sqrt{-16}}{10} = \frac{-2 \pm 4i}{10}$$
$$= \frac{2(-1 \pm 2i)}{10} = \frac{-1 \pm 2i}{5}$$

Notice that the solutions are complex conjugates of each other. Later, in Chapter 3, we will discuss complex roots of polynomial equations in more detail.

So, the solutions are the complex numbers

$$\frac{-1}{5} + \frac{2}{5}i \qquad \text{and} \qquad \frac{-1}{5} - \frac{2}{5}i$$

The viewing window of the graph of $y = 2x^2 - 8x + 3$ (Example 6a) shown in Figure 6a indicates that there are two intercepts, one between 0 and 1 and another between 3 and 4. This is consistent with the solutions found in the example. In contrast, the viewing window of the graph of $y = 5x^2 + 2x + 1$ is shown in Figure 6b. Example 6b shows that there are no x intercepts, thus indicating that the equation $5x^2 + 2x + 1 = 0$ has no real number solutions.

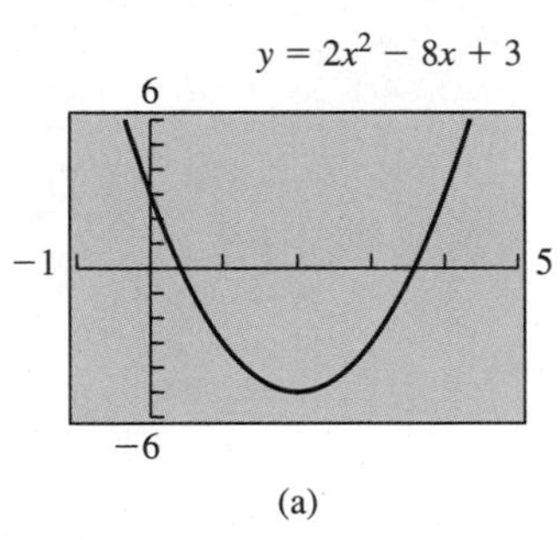

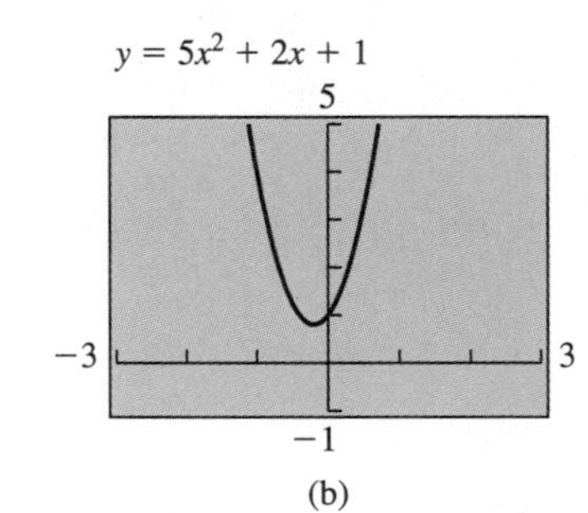

FIGURE 6

Solving Equations Involving Radicals

Next, we consider real number solutions of algebraic equations that contain *radicals* or *rational exponents*. Examples of these equations are

$$\sqrt{x} = 5, \quad \sqrt{5y + 1} = 4, \quad \sqrt{x} = \sqrt{x + 16} - 4 \quad \text{and} \quad (3u + 1)^{1/3} = u + 1$$

A method of solving equations of this type is to isolate one of the radicals and then raise each side of the equation to the same positive integer power that eliminates the isolated radical. It may be necessary to repeat this step until all the radicals are eliminated. Then the resulting equation is simplified and solved. It is important to realize that this process can introduce one or more extraneous roots when dealing with even-numbered roots. Therefore, it is necessary to check all the proposed solutions in the original equation whenever a radical with an even index is involved, to determine whether each solution should be accepted or rejected.

EXAMPLE 7 Solving a radical equation

(a) Solve the equation $\sqrt{x+4} - 2 = \sqrt{x}$ algebraically.
(b) Display the solution graphically.

Solution

(a) We proceed as follows:

$$\left(\sqrt{x+4} - 2\right)^2 = \left(\sqrt{x}\right)^2 \quad \text{Square each side of the given equation}$$
$$x + 4 - 4\sqrt{x+4} + 4 = x \quad \text{Multiply}$$
$$x - 4\sqrt{x+4} + 8 = x \quad \text{Collect terms on the left side}$$
$$-4\sqrt{x+4} = -8 \quad \text{Isolate the radical expression on one side}$$
$$\sqrt{x+4} = 2 \quad \text{Divide each side by } -4$$

We repeat the process by squaring each side of the latter equation to obtain

$$\left(\sqrt{x+4}\right)^2 = 2^2$$
$$x + 4 = 4 \quad \text{Multiply}$$
$$x = 0 \quad \text{Solve for } x$$

Since even roots are involved, we check $x = 0$ to determine whether it is extraneous:

$$\sqrt{0+4} - 2 = \sqrt{0}$$
$$2 - 2 = 0$$

Therefore, the solution is $x = 0$.

(b) To display the solution graphically, we use a graphing utility to graph the associated equation

$$y = \sqrt{x+4} - 2 - \sqrt{x}$$

From the graph (Figure 7), we observe that 0 is the x intercept, which is the same value as the solution of the equation.

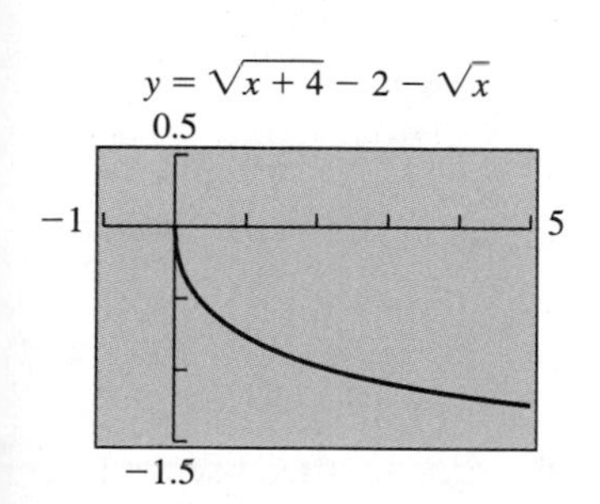

FIGURE 7

Solving Applied Problems

In the next example, we solve an applied problem by setting up an equation containing one unknown to model the situation. Then we solve the equation and interpret the solution.

EXAMPLE 8 **Solving an investment problem**

A person invests part of $5000 in a certificate of deposit (CD) that yields 5.5% simple annual interest and the rest in a mutual fund (MF) that yields 6% simple annual interest. At the end of the year, the combined interest from the two investments is $282.50. How much was invested in the CD and how much in the MF?

Solution Here, we let x represent the money invested in the CD, so $(5000 - x)$ represents the remaining amount invested in the MF. The quantities involved in the problem are listed in the following table:

Investment Type	Principal *Amount invested*	Interest Rate	Simple Interest *Earned in 1 year*
CD	x dollars	0.055	$0.055x$
MF	$(5000 - x)$ dollars	0.06	$0.06(5000 - x)$

Since the combined interest is $282.50, we have

$$\begin{pmatrix}\text{Interest}\\ \text{from CD}\end{pmatrix} + \begin{pmatrix}\text{Interest}\\ \text{from MF}\end{pmatrix} = \begin{pmatrix}\text{Total}\\ \text{interest}\end{pmatrix}$$

$$0.055x + 0.06(5000 - x) = 282.50$$

$$0.055x + 300 - 0.06x = 282.50$$

$$-0.005x = -17.5$$

$$x = 3500$$

Therefore, $3500 was invested in the CD. Since $5000 - x = 5000 - 3500 = 1500$, it follows that $1500 was invested in the MF.

In the following two applications, we use a graph to help interpret the situation.

EXAMPLE 9 **Solving a physics model**

A ball is thrown vertically upward with an initial speed of 48 feet per second. Neglecting air resistance, its height (in feet) above the ground after t seconds is given by the equation

$$h = 48t - 16t^2$$

(a) How long will it take the ball to reach a height of 32 feet on the way up?

(b) How long will it take the ball to reach the ground?

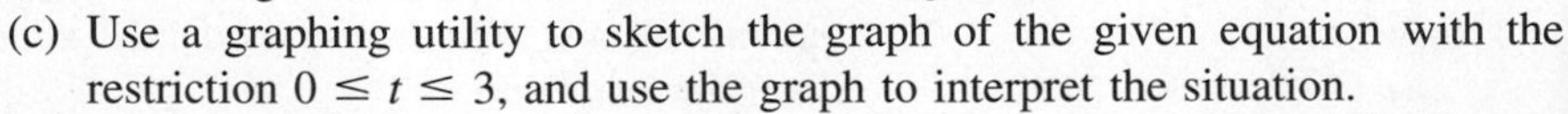

(c) Use a graphing utility to sketch the graph of the given equation with the restriction $0 \leq t \leq 3$, and use the graph to interpret the situation.

Solution

(a) We want to find the time t for the ball to reach a height of 32 feet on the way up. So we substitute 32 for h to get

$$32 = 48t - 16t^2$$

Writing the equation in standard form, we get

$$\begin{aligned} 16t^2 - 48t + 32 &= 0 \\ t^2 - 3t + 2 &= 0 \quad \text{Divide each side by 16} \\ (t-1)(t-2) &= 0 \quad \text{Factor} \\ t - 1 = 0 \quad \text{or} \quad t - 2 &= 0 \quad \text{Solve} \\ t = 1 \quad \Big| \quad t &= 2 \end{aligned}$$

Since we want the time t for the ball to reach a height of 32 feet on the way up, the desired solution is 1 second. The ball begins to fall and returns to a height of 32 feet after 2 seconds.

(b) Since the height $h = 0$ when the ball hits the ground, we substitute 0 for h and solve for t:

$$\begin{aligned} 0 &= 48t - 16t^2 \\ 0 &= 16t(3 - t) \quad \text{Factor} \\ 16t = 0 \quad &\text{or} \quad 3 - t = 0 \quad \text{Solve} \\ t = 0 \quad &\Big| \quad t = 3 \end{aligned}$$

The solution $t = 0$ represents the starting point on the ground, so it takes the ball 3 seconds to return to the ground.

(c) Figure 8 shows a viewing window of the graph of $h = 48t - 16t^2$, where $0 \le t \le 3$. It shows that the ball rises to a maximum height and then falls. The graph also shows that it takes 3 seconds for the ball to return to the ground.

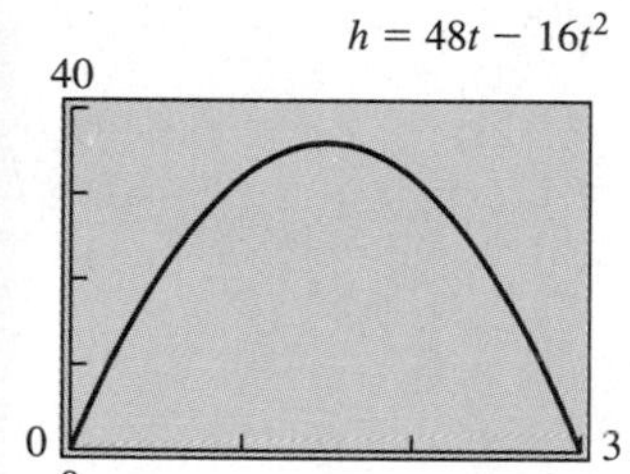

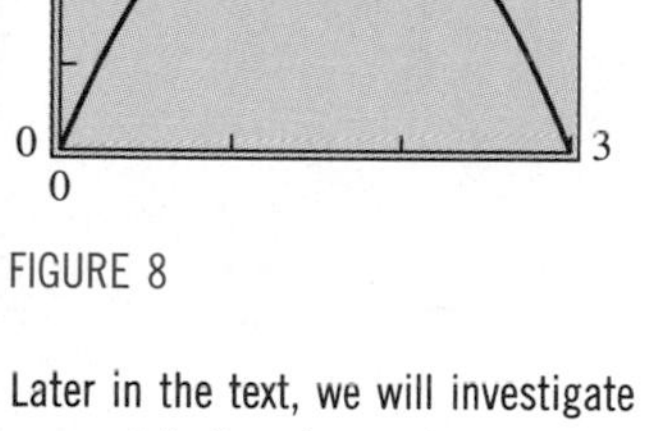

FIGURE 8

Later in the text, we will investigate ways of finding the maximum height and when it occurs.

EXAMPLE 10 **Modeling a volume problem**

A rectangular sheet of aluminum whose length is 2 inches more than its width is to be made into a tray by cutting 2 inch squares from each corner and bending up the flaps.

(a) Express the volume V in terms of x, where x (in inches) is the original width of the sheet of aluminum.

(b) Find the width x when the volume is 336 cubic inches.

GU

(c) Use a graphing utility to sketch the graph of the equation associated with the equation used to do part (b) and interpret it in terms of the results found in part (b).

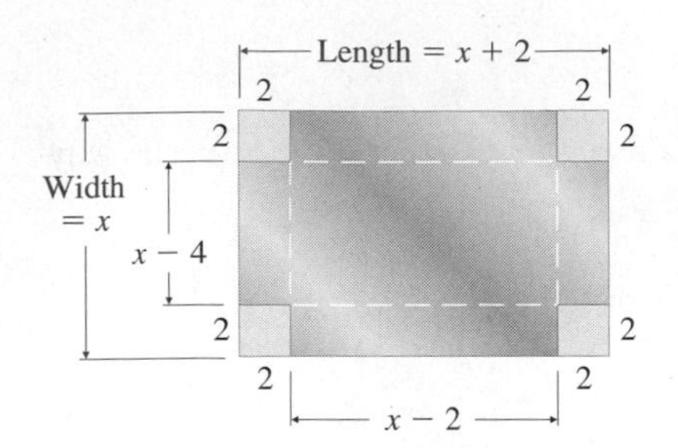

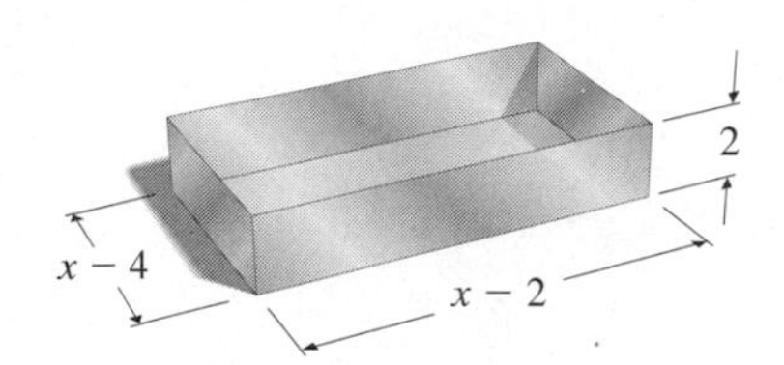

FIGURE 9

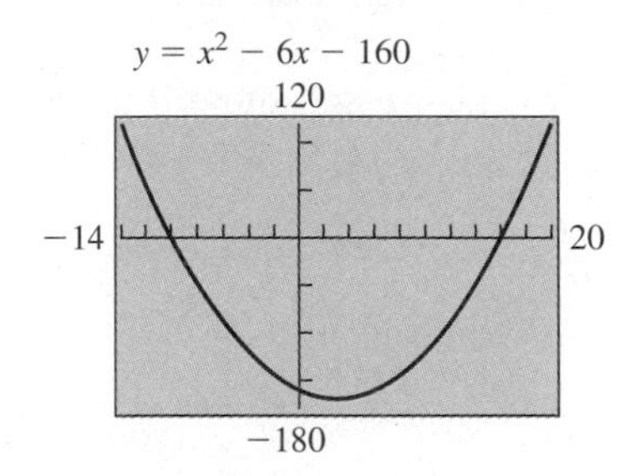

FIGURE 10

Solution

(a) A diagram of the situation is shown in Figure 9. From the figure we can see that the width of the tray is $x - 4$ and the length is $x - 2$. Since V represents the volume, it follows that

$$\begin{aligned}\text{Volume} &= (\text{Length}) \times (\text{Width}) \times (\text{Height})\\ &= (x-2)(x-4)(2)\\ &= 2x^2 - 12x + 16\end{aligned}$$

(b) To find x when $V = 336$, we replace V by 336 in the equation to get

$$\begin{aligned}336 &= 2x^2 - 12x + 16\\ 2x^2 - 12x - 320 &= 0 && \text{Write in standard form}\\ x^2 - 6x - 160 &= 0 && \text{Divide each side by 2}\\ (x-16)(x+10) &= 0 && \text{Factor}\end{aligned}$$

$$x - 16 = 0 \quad \text{or} \quad x + 10 = 0 \qquad \text{Solve}$$

$$x = 16 \quad \Big| \quad x = -10$$

Here we use only the positive solution since the dimensions of the tray must be positive numbers. So a width of 16 inches yields a tray with volume 336 cubic inches.

(c) Figure 10 shows a viewing window of the graph of $y = x^2 - 6x - 160$. The graph indicates that the x intercepts are -10 and 16. So when $x = 16$, $y = 0$. Thus, the volume is 336 cubic inches when $x = 16$.

PROBLEM SET 1.5

Mastering the Concepts

GU In problems 1–6, solve the equation algebraically and display the result graphically by using a graphing utility.

1. **(a)** $34 - 3x = 4(3 - 2x) + 27$
(b) $8(5x - 1) + 36 = -3(x + 5)$

2. **(a)** $3y - 2(y + 1) = 2(y - 1)$
(b) $7(t - 3) = 4(t + 5) - 47$

3. **(a)** $5 + 8(u + 2) = 23 - 2(2u - 5)$
(b) $11 - 7(1 - 2p) = 9(p + 1)$

4. **(a)** $11 = 3v - 8(7 - v) + 23$
(b) $6(y - 10) + 3(2y - 7) = -45$

5. $\dfrac{3x-2}{3} + \dfrac{x-3}{2} = \dfrac{5}{6}$

6. $\dfrac{x-14}{5} + 4 = \dfrac{x+16}{10}$

In problems 7–10, solve each equation.

7. **(a)** $\dfrac{1}{y} + \dfrac{2}{y} = 3 - \dfrac{3}{y}$
(b) $\dfrac{2}{3}u + \dfrac{1}{6}u = \dfrac{1}{4}$

8. **(a)** $\dfrac{2t}{t-2} = \dfrac{4}{t-2} + 1$
(b) $\dfrac{9}{5y-3} = \dfrac{5}{3y+7}$

9. $\dfrac{2}{y-3} - \dfrac{1}{y+5} = \dfrac{1}{y^2+2y-15}$

10. $\dfrac{1}{x-2} - \dfrac{3}{x+2} = \dfrac{2}{x^2-4}$

GU In problems 11–16, solve each equation algebraically and display the result graphically by using a graphing utility.

11. **(a)** $|x - 2| = 4$ **(b)** $|2t + 6| = 18$
12. **(a)** $|3x + 9| = 12$ **(b)** $|14 - 7y| = |-21|$
13. **(a)** $\left|1 - \frac{2}{3}x\right| = 5$ **(b)** $\left|3 - \frac{2x}{5}\right| - 2 = 5$
14. **(a)** $|6 - 3y| - 5 = 7$ **(b)** $|7 - 2t| + 4 = |-9|$
15. **(a)** $|x - 2(x - 2)| = |-3|$
(b) $|t - 2(t + 5)| = |-2|$
16. **(a)** $|y - 9| = |3y + 1|$
(b) $|y - 9| = |2y - 3|$

GU In problems 17–20, solve each equation by factoring and display the result graphically by using a graphing utility.

17. **(a)** $x^2 - x - 12 = 0$ **(b)** $x^2 - x - 20 = 0$
18. **(a)** $6x^2 + 6x - 36 = 0$
(b) $2y^2 + 10y - 28 = 0$
19. **(a)** $3t^2 + 6t - 24 = 0$ **(b)** $42 - 7x - 7x^2 = 0$
20. **(a)** $40 - 2x - 2x^2 = 0$
(b) $66 - 27x - 3x^2 = 0$

GU In problems 21–24, use the method of completing the square to solve each equation. Round off each answer to two decimal places. Use the graph of the associated equation to display the result graphically by using a graphing utility.

21. **(a)** $x^2 - 2x - 2 = 0$ **(b)** $x^2 + 10x + 3 = 0$
22. **(a)** $7y^2 - 4y - 1 = 0$ **(b)** $2p^2 + 6p - 7 = 0$
23. **(a)** $9t^2 - 30t = -21$ **(b)** $5m^2 - 8m = 17$
24. **(a)** $2r^2 + 3r = 2$ **(b)** $4y^2 + 7y = 5$

In problems 25–28, solve each equation by the quadratic formula. Round off each answer to two decimal places.

25. **(a)** $2x^2 + x - 1 = 0$ **(b)** $3x^2 - 5x + 1 = 0$
26. **(a)** $2y^2 - 6y + 3 = 0$ **(b)** $3u^2 + 4u - 4 = 0$
27. $1.5x^2 - 7.5x - 4.3 = 0$
28. $5.13x^2 - 7.14x - 3.57 = 0$

GU In problems 29–36, solve each equation algebraically and display the result graphically by using a graphing utility.

29. **(a)** $\sqrt{x - 5} = 2$ **(b)** $\sqrt{3x + 1} = 4$
30. **(a)** $\sqrt[3]{6x - 4} = 2$ **(b)** $\sqrt[3]{3x + 4} = 3$
31. **(a)** $4 + \sqrt{y - 5} = 7$ **(b)** $6 + \sqrt{x - 1} = 7$
32. **(a)** $x - 5 = \sqrt{x + 7}$ **(b)** $\sqrt{5x + 9} = x - 1$
33. $\sqrt{x - 2} = \sqrt{2x} - 2$ **34.** $\sqrt{y} = \sqrt{y + 16} - 4$
35. $\sqrt{m + 3} = \sqrt{m - 2} + 1$
36. $\sqrt{r^2 + 5r - 10} = r$

GU In problems 37–42, solve each equation algebraically. Round off each answer to two decimal places. Use the graph of the associated equation to display the result graphically by using a graphing utility.

37. $\sqrt{u^2 + 3u} = u + 1$ **38.** $\sqrt{\sqrt{c + 4} + c} = 4$
39. $4 - (2t + 1)^{3/2} = 5$ **40.** $(4x + 5)^{(-1/4)} = 2$
41. $(t^2 + 6t)^{1/3} = 3$ **42.** $(5x^2 + 7x - 3)^{3/2} = 1$

Applying the Concepts

In problems 43–54, set up an appropriate equation to model the situation and then solve each problem.

43. Investment: An investor wishes to borrow \$10,000. There is an annual simple interest rate charge of 7% on part of the loan, and an annual simple interest rate charge of 6.25% on the other part. If the annual interest on both parts totals \$656.50, how much money is borrowed at each rate?

44. Investment: A family invests a total of \$85,000 in two tax-free municipal bonds to reduce its income tax. One bond pays 5% tax-free simple annual interest, and the other bond pays 5.5% simple annual interest. The total nontaxable income from both investments at the end of 1 year is \$4475. How much did the family invest in each bond?

45. Investment: A retiree deposited \$6200 in two different savings accounts. The annual interest rate on one account was 5.5% and the annual interest rate on the other account was 5.8%. At the end of the first year, the interest earned on the first account was \$92.40 more than the interest earned on the second. Determine the amount deposited in each account.

46. Budgeting: A small business allocated four times as much for utility expenses as for interest expenses. The annual total amount budgeted for these expenses was \$82,000. How much was budgeted for utilities?

47. Mixture: A medicine contains 25% alcohol. How much water should be added to 120 cubic centimeters of the medicine if the final mixture is to contain 20% alcohol?

48. Mixture: A bottle of mineral water contains 16 ounces of liquid with a 40% calcium content. How many ounces of this liquid should be poured out and replaced with a 60% calcium solution to obtain a 45% calcium solution?

49. Cycling and Jogging Rates: A jogger and a bicycle rider head for the same destination at the same time and from the same point (Figure 11). The average speed of the bicycle rider is 3 times the speed of the jogger. At the end of 2 hours, the bicycle rider is 12 miles ahead of the jogger. How fast is the bicycle rider traveling, assuming that the speeds are constant?

FIGURE 11

50. Price of a Graphing Calculator: When the price of a graphing calculator is \$75 per unit, an electronics store sells 95 units per week. Each time the price is reduced by \$10, the sales increase by 30 units per week. What selling price will result in weekly revenues of \$8525?

51. Time Required To Do a Job: An electronic computer can be used to perform a large stock portfolio analysis in 6 minutes. With the help of a new computer, the analysis can be run in 2 minutes. How long would it take the new computer to run the analysis alone?

52. Filling a Swimming Pool: Two hoses are being used to fill a swimming pool. The larger hose supplies twice as much water as the smaller hose. If both hoses can fill the swimming pool in 8 hours, how long would it take each hose alone to fill the pool?

53. Designing an Advertising Flier: A 20 by 30 centimeter brochure is to be used for an advertising flier. The margins at the sides, top, and bottom all have the same width. Find the width of the margins if the printed area is 416 square centimeters.

54. Enclosing a Region: Suppose a livestock rancher wants to enclose a rectangular feedlot. One side of the lot is centered against the back of the barn, which is 50 feet in length, and no fencing is needed along the back of the barn (Figure 12). The rancher needs an area of 2800 square feet to accommodate the cattle. Determine the dimensions necessary to enclose the required area of the lot if 170 feet of fencing are used.

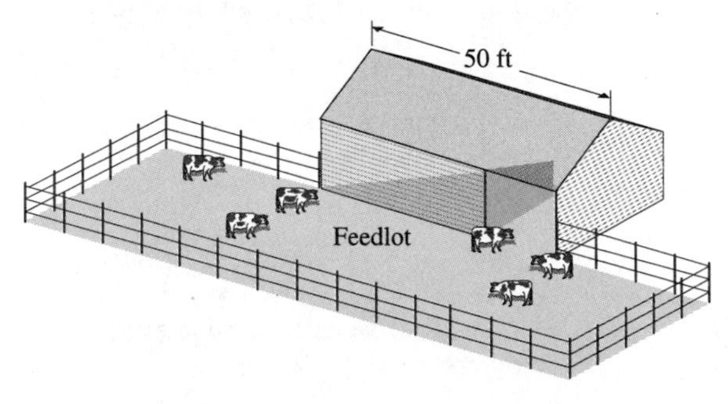

FIGURE 12

55. Height of a Toy Rocket: The height h (in feet) reached by a model rocket t seconds after it is launched is given by

$$h = -16t^2 + 96t$$

(a) How long will it take the rocket to reach a height of 10 feet above ground?
(b) How long will it take the rocket to hit the ground?
GU **(c)** Use a graphing utility to graph the given equation, and then use the graph to interpret the situation.

56. Physics: A ball is thrown vertically upward from the top of a building 80 feet high. After t seconds, the height h (in feet) is given by

$$h = 80 + 32t - 16t^2$$

(a) How long will it take the ball to reach a height of 70 feet above the ground?
(b) How long will it take the ball to reach the ground?
GU **(c)** Use a graphing utility to graph the given equation, and then use the graph to interpret the situation.

57. Geometry: The length of a rectangle is 1 inch more than its width.
(a) Express the area A of the rectangle in terms of its width x.
(b) Find the dimensions of the original rectangle if the area will increase by 30 square inches when the length is doubled (the width does not change).

58. Geometry: An open box is formed from a square piece of cardboard with side length of x inches by cutting 2 inch squares from each corner and folding up the sides (Figure 13).

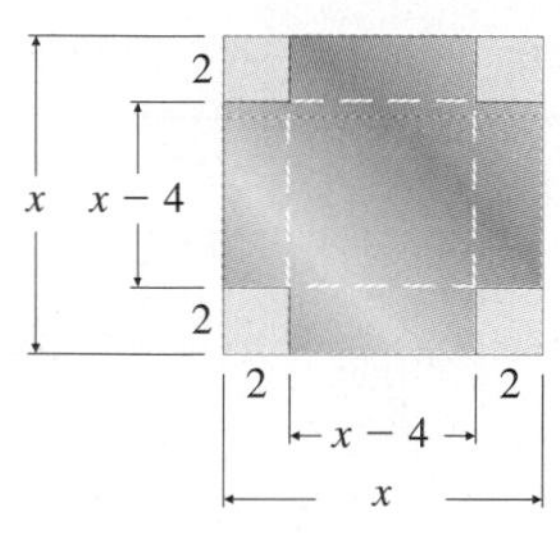

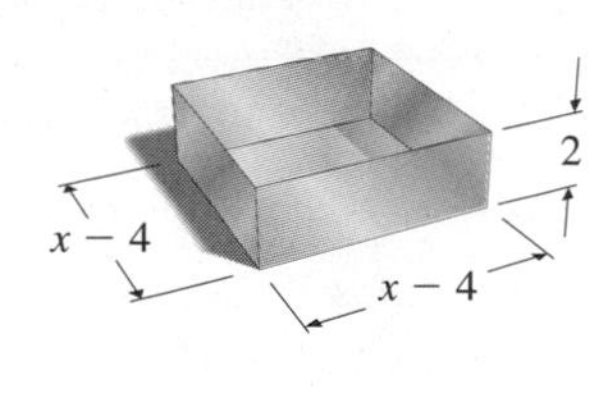

FIGURE 13

(a) Express the volume V of the box in terms of x.
(b) If the volume of the box is 72 cubic inches, what are the dimensions of the original cardboard?

Developing and Extending the Concepts

59. Use the discriminant to determine the number and kind of roots for each quadratic equation. Use a graph to demonstrate the results.
GU
(a) $2x^2 - 4x + 1 = 0$ (b) $x^2 - 6x + 9 = 0$
(c) $2x^2 + 3x + 2 = 0$

60. (a) Show that if $a \neq 0$, then $ax^2 + bx + c = 0$ can be written in the form

$$x^2 + \frac{b}{a}x = -\frac{c}{a}$$

(b) Complete the square to obtain

$$\left(x + \frac{b}{2a}\right)^2 = \frac{b^2 - 4ac}{4a^2}$$

(c) Solve the equation in part (b) for x.

61. Given a quadratic equation with real coefficients

$$ax^2 + c = 0$$

determine the type of solutions of the equation if:
(a) $ac = 0$ (b) $ac < 0$

62. The number of diagonals d of a polygon with n sides is given by the formula

$$d = \frac{n(n-3)}{2}$$

Determine the number of sides of a polygon with 27 diagonals.

In problems 63–66, use the given substitution to transform each equation to an equivalent equation that is quadratic in form. Solve the resulting equation and then use the result to solve the original equation.

63. (a) $y^{-2} + y^{-1} = 2$; use $u = y^{-1}$
(b) $x^{1/2} - 3x^{1/4} + 2 = 0$; use $u = x^{1/4}$

64. (a) $2t^{-2} - 5t^{-1} - 3 = 0$; use $u = t^{-1}$
(b) $x^{2/3} - 3x^{1/3} - 10 = 0$; use $u = x^{1/3}$

65. $(t^2 - t)^2 - 4(t^2 - t) = 12$; use $u = t^2 - t$

66. $\left(3x - \frac{2}{x}\right)^2 + 6\left(3x - \frac{2}{x}\right) + 5 = 0$; use $u = 3x - \frac{2}{x}$

67. Without solving, use a graph to explain why each equation has no real number solution.
GU
(a) $\sqrt{x+3} = -5$ (b) $|x^2 - 4| = -3$

CHAPTER 1 REVIEW PROBLEM SET

1. Fill in the blanks with the <, >, or = symbol.
(a) If $x < 0$, then $-x$ ____ 0.
(b) If $x > 0$, then $-x$ ____ 0.
(c) If $x > 0$, then $|x|$ ____ $-x$.
(d) If $x < 0$, then $|x|$ ____ $-x$.

2. For what values of x is each of the given statements true?

(a) $|x| = |-x|$ (b) $\frac{|-x|}{-x} = -1$

(c) $|1 - x| = |x|$

In problems 3 and 4, illustrate each interval on a number line and describe it using inequality notation.

3. (a) $[-4, 5]$
(b) $(-\infty, -2]$
(c) $(1, \infty) \cap (-1, 3)$

4. (a) $(-\infty, 3]$
(b) $[-2, 5] \cup [4, 7]$
(c) $[7, \infty)$

In problems 5 and 6, solve each inequality and express the solution in interval notation. Also, graph the solution on a number line.

5. **(a)** $5x - 9 > 2x + 3$ **(b)** $\frac{2x}{3} + \frac{1}{5} > \frac{7}{15} + \frac{4x}{5}$

(c) $\frac{5}{x} < \frac{10}{3x}$

6. **(a)** $3(x - 5) \le 6 - 3x$

(b) $\frac{x}{2} + \frac{5}{3} < \frac{5x}{6} + 1$ **(c)** $\frac{5x - 1}{-5} > \frac{3}{10}$

7. Write each expression without absolute value notation.

(a) $|3 - \sqrt{7}|$ **(b)** $|5x - 3|$ **(c)** $|1 - 7x|$

8. Find the distance between each pair of numbers.

(a) -3 and 7 **(b)** $-7t$ and $-4t$ **(c)** $-5x$ and $3x$

In problems 9 and 10, solve each absolute value inequality. Express the solution in interval notation and graph it on a number line.

9. **(a)** $|3t - 7| < 1$ **(b)** $|3x + 2| \ge 8$

(c) $|x + 4| \le |x - 1|$

10. **(a)** $|3 - 2x| \le 17$ **(b)** $|1 - 3x| \ge 7$

(c) $|x + 6| < |x - 4|$

In problems 11 and 12:

(i) Use the distance formula to find $d(P, Q)$, the distance between P and Q.

(ii) Use the midpoint formula to find the midpoint of the line segment $\overline{PQ}$.

11. **(a)** $P = (2, 1)$; $Q = (4, -5)$

(b) $P = (-3, 2)$; $Q = (6, -1)$

12. **(a)** $P = (-6, -3)$; $Q = (2, 1)$

(b) $P = (5.82, -3.71)$; $Q = (4.73, 6.28)$ (Round off to two decimal places.)

In problems 13 and 14, find the center and radius of the circle with the given equation. Sketch the graph.

13. **(a)** $(x + 1)^2 + (y + 3)^2 = 16$

(b) $x^2 + y^2 - 6x + 2y - 26 = 0$

14. **(a)** $\left(x - \frac{3}{2}\right)^2 + (y + 4)^2 = 25$

(b) $x^2 + y^2 - 6x + 4y - 12 = 0$

In problems 15 and 16, find the constant of variation for each case.

15. **(a)** y is directly proportional to $\sqrt[3]{x}$, and $y = 12$ when $x = 8$.

(b) p is inversely proportional to q^2, and $p = 3$ when $q = 4$.

16. **(a)** r is directly proportional to s, and $r = 150$ when $s = 2$.

(b) y is inversely proportional to $\sqrt{x}$, and $y = 7$ when $x = 4$.

In problems 17 and 18, determine the slope of the line containing the given points.

17. **(a)** $P_1 = (-1, -2)$; $P_2 = (3, 4)$

(b) $P_1 = (-2, 3)$; $P_2 = (3, 1)$

18. **(a)** $P_1 = (2, -2)$; $P_2 = (-1, 5)$

(b) $P_1 = (-3, 1)$; $P_2 = (2, 5)$

In problems 19 and 20, graph each linear equation and locate the x and y intercepts. Also, find the slope of each line.

19. **(a)** $x - 2y + 3 = 0$ **(b)** $2x + 3y - 5 = 0$

20. **(a)** $4x + y - 9 = 0$ **(b)** $-3x - 5y + 10 = 0$

In problems 21 and 22, determine the value of k so that the given condition is satisfied for the given line.

21. **(a)** $kx + 7y = 9$; the slope of the line is -3.

(b) $3x - ky = 2$; the line is parallel to $3x + 2y = 11$.

22. **(a)** $5kx - 3y = 4$; the line is perpendicular to $2x + y = 0$.

(b) $kx - 3y = 10$; the line is parallel to the x axis.

In problems 23 and 24, find an equation of each line in slope–intercept form and sketch the graph.

23. **(a)** $m = 3$ and contains $P_1 = (4, 5)$

(b) Contains $P_1 = (3, -2)$ and $P_2 = (5, 6)$

24. **(a)** $m = 0$ and contains $P_1 = (-2, 3)$

(b) Perpendicular to the line $8x + 7y + 3 = 0$ and contains $P_1 = (2, -4)$

25. Find the value k so that the line $x + 2y = 6k$ has the same y intercept as $y = -\frac{1}{2}(1 - x)$.

26. Suppose that $a > 0$ and $a + k > 0$. Show that the line containing the points $(a, \sqrt{a})$ and $(a + k, \sqrt{a + k})$ has slope

$$\frac{1}{\sqrt{a + k} + \sqrt{a}}$$

In problems 27 and 28, use the point-plotting method to sketch the graph of each equation. Also, find the x and y intercepts. Discuss the symmetry of each graph.

27. **(a)** $y = |x + 5|$ **(b)** $y = x^2 - |x|$

28. **(a)** $y = \sqrt{4 - x^2}$ **(b)** $y = x^2 - 4x$

GU In problems 29 and 30, solve for y in terms of x to obtain two equations. Then use a graphing utility to superimpose the graphs of the two equations on the same viewing window to get the graph of the original equation.

29. **(a)** $x + 4y^2 = 4$ **(b)** $x^2 + x + y^2 = 1$

30. **(a)** $25x^2 + y^2 = 9$ **(b)** $4y^2 - x = 9$

GU In problems 31–34, solve each equation algebraically and display the solution graphically by using a graphing utility.

31. **(a)** $2(2x - 1) = 3(x + 1)$

(b) $\dfrac{2x + 3}{5} - \dfrac{2x - 1}{3} = \dfrac{8}{15}$

(c) $|3x - 2| = 4$ **(d)** $2x^2 - x - 1 = 0$

32. **(a)** $12(x - 2) + 8 = 5(x - 1) + 2x$

(b) $\dfrac{10 - x}{x} + \dfrac{3x + 3}{3x} = 3$

(c) $\sqrt[3]{2x + 5} = 3$ **(d)** $|x - 1| = 2x$

33. **(a)** $|2 - 3x| = 5$

(b) $\sqrt{1 - x} = 2 - \sqrt{1 - 5x}$

(c) $x(x - 2) = 15$ **(d)** $3x^2 + 6x - 7 = 0$

34. **(a)** $|7x - 1| = -3$ **(b)** $(5x - 1)^{3/2} = 8$

(c) $3x^2 - 2x + 12 = 0$

(d) $|x - 2| = |2 - x|$

35. Track Record: The following data are obtained from a study conducted on women's 400 meter track records for the international Olympic Games from 1920 to 1980:

Year	1920	1930	1940	1950	1960	1970	1980
Time *Seconds*	65	59	57	56	53	51	48

(a) Construct a scattergram for these data, representing the year as the horizontal axis and the time as the vertical axis.

(b) Draw a line graph that represents the track records for the given years and then use it to discuss the trend of this model.

36. Bowling Average: The following data show the relationship between the bowling averages of six people and their ages.

Age *Years*	25	37	42	50	54	60
Bowling Average	195	180	170	165	160	100

(a) Construct a scattergram to display the information. Consider the age to be the horizontal axis and the average as the vertical axis.

(b) Draw a line graph that describes the situation and use it to discuss the trend for this model.

37. Temperature Conversion: A temperature of 10° on a Celsius scale corresponds to a temperature of 50° on a Fahrenheit scale. Also, 40°C corresponds to 104°F. Assume the relationship between F and C is linear.

(a) Find the equation that relates F to C.

(b) Find the value of F that corresponds to 52°C.

(c) Sketch the graph of the equation in part (a).

(d) Interpret the slope of the line.

38. Chemistry: The concentration C of carbon monoxide in the atmosphere is related to the number of years t since 1986 by a linear equation. Assume there were 9.9 parts per million (ppm) of carbon monoxide in the atmosphere in 1986, and there were 6.8 ppm in the atmosphere in 1995.

(a) Find the equation that relates C to t.

(b) Use the equation to predict C in 1998.

(c) Graph the equation in part (a).

(d) Interpret the slope of the line.

39. Investment: A small company had \$24,000 to invest. It invested some of the money in a bank that paid 5% annual simple interest. The rest of the money was invested in stocks that paid dividends equivalent to 8% annual simple interest. At the end of 1 year, the combined income from these investments was \$1620. How much money was originally invested in stocks?

40. Investment: A businesswoman had \$18,000 to invest. She invested some of it in a bank certificate that paid 6% annual simple interest, and the rest in another certificate that paid 5% annual simple interest. At the end of 1 year, the combined income from these investments was \$1030. How much money did she originally invest at 6% interest?

41. Parking Fees: The manager of a parking garage agreed to pay a troop of Boy Scouts a fixed amount for washing some cars—enough to give each boy in the troop \$10. When 25 boys failed to show up, those already present agreed to do the work, which meant that each boy who worked got \$15. How many boys are in the troop?

42. Landscaping: A landscaper wants to put a cement walk of uniform width around a rectangular garden that measures 20 by 60 feet. The landscaper has enough cement to cover 516 square feet. How wide should the walk be in order to use all the cement?

43. Projectile: An arrow is shot upward from the ground with an initial speed of 112 feet per second. Its height h (in feet) after t seconds is given by the model

$$h = 112t - 16t^2$$

(a) What restriction should be imposed on t?
GU (b) Use a graphing utility to sketch the graph of the equation that represents this model.
(c) What is the height of the arrow after 4 seconds?
(d) When does the arrow hit the ground?
(e) Find the time it takes for the arrow to first reach a height of 144 feet.

44. Manufacturing: Suppose that a company's profit P (in dollars) for a given week when producing n items in that week is given by the model

$$P = -n^2 + 100n - 1000$$

(a) What restriction should be imposed on n?
GU (b) Use a graphing utility to sketch the graph of the equation that represents this model.
(c) What is the company's profit in 1 week if 40 items were produced and sold?
(d) How many units should be produced in 1 week if the company's profit for that week is \$1500?

CHAPTER 1 TEST

1. (a) Graph each interval on a number line, and describe the interval using inequality notation.
(i) $(-2, 5]$ (ii) $(-\infty, 4]$
(b) Write each number set in interval notation and graph it on a number line.
(i) $-2 \le x < 4$ (ii) $x \ge 3$ or $2 \le x \le 5$

2. Solve each inequality. Express the solution in interval notation and graph the solution on a number line.
(a) $5x - 2 < -8$ (b) $|x + 5| \le 7$
(c) $|3x - 2| > 4$

3. Find the radius and center of the circle with equation $x^2 + y^2 + 6x - 2y - 15 = 0$. Sketch the graph.

4. Let $P_1 = (-3, 4)$ and $P_2 = (-5, 1)$.
(a) Find $d(P_1, P_2)$.
(b) Find the coordinates of the midpoint M of the line segment $\overline{P_1P_2}$.
(c) Find the slope of the line containing points P_1 and P_2.
(d) Find an equation of the line containing P_1 and P_2. Express the answer in slope–intercept form.
(e) Sketch the graph of the line and locate its intercepts.
(f) Find an equation of the line that contains P_1 and is:
(i) Parallel to the line $3x + y = 1$
(ii) Perpendicular to the line $3x + y = 1$

5. Use the point-plotting method to sketch the graph of each equation. Find the intercepts of the graph, and discuss the symmetry of each graph.
(a) $y = |x - 2|$ (b) $y = -3x^2$
(c) $y = x^3 + x$

6. Solve each equation algebraically, and then display the solution graphically.
(a) $x + 3(3x - 1) = 4(x + 2) + 4$
(b) $(x + 6)(x - 2) = -7$
(c) $|3x + 2| = 5$
(d) $\sqrt{x - 2} = x - 4$

7. A manufacturer finds that the cost C (in dollars) of producing a new product line and the number of items, x, produced are related by a linear equation. Suppose that the cost of producing 50 items will be \$3750 and the cost of producing 400 items will be \$12,500. Use this information to predict the cost of producing 470 items.

8. If the temperature of a gas is constant, the volume V occupied by that gas varies inversely as the pressure P to which the gas is subjected. If the volume of a gas is 8 cubic feet when the pressure is 12 pounds per square inch, find the volume of the gas if the pressure is 16 pounds per square inch.

9. An investment of \$8000 is split into a savings account paying 4% simple annual interest and a mutual fund paying 7% simple annual interest. If the total annual interest from the two accounts is \$485, how much is invested at each rate?

2

Functions and Graphs

CHAPTER CONTENTS

The study of functions begins in this chapter. Functions are used to represent relationships between quantities. Such representations are especially important in modeling real-life situations. As we shall see, functions help us understand the dependence of one quantity on another. Graphs of functions are an integral part of our investigation because they give us additional insight into the properties and behavior of functions.

OBJECTIVES

1. Define functions.
2. Use function notation and terminology.
3. Graph functions.
4. Identify graphs that represent functions.
5. Determine the domain and range.
6. Solve applied problems.

2.1 The Function Concept

A *function* establishes a relationship or correspondence between two sets of elements. For each element in a first set there corresponds one and only one element in a second set. To help understand the concept of a function, let us consider some real-life illustrations:

1. For each videocassette in a store, there is one corresponding price.
2. For each car in a state, there is one corresponding license number.
3. For each person, there is one corresponding weight at any given time.

There are three characteristics common to all functions—an input set, an output set, and a rule that determines the association between the members of the two sets. The following table lists these characteristics for the above three examples:

In the Set of Inputs	In the Set of Outputs	Rule (Established By)
1. Videocassettes	Prices	Store pricing
2. Cars	License numbers	Licensing procedure
3. Persons	Weights	Scale reading

Functions provide us with an important tool used in mathematical modeling.

Defining Functions

The diagrams in Figures 1a and 1b (on the next page) depict correspondences from one set of numbers, labeled D, to another set of numbers, labeled R. Figure 1a lists corresponding values for the squares of some numbers, while Figure 1b lists corresponding values whose squares equal each given number.

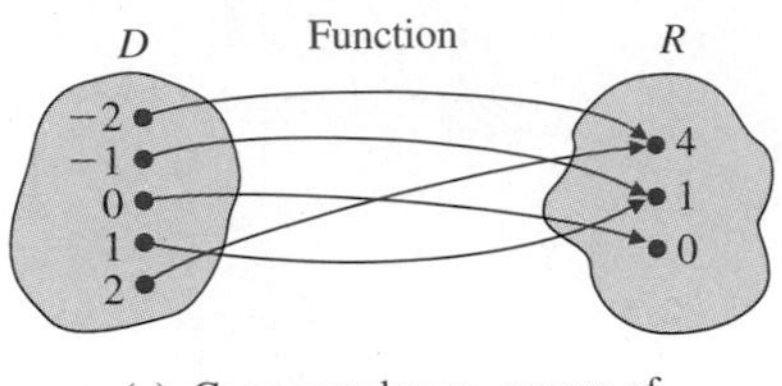

(a) Correspondence: square of a number

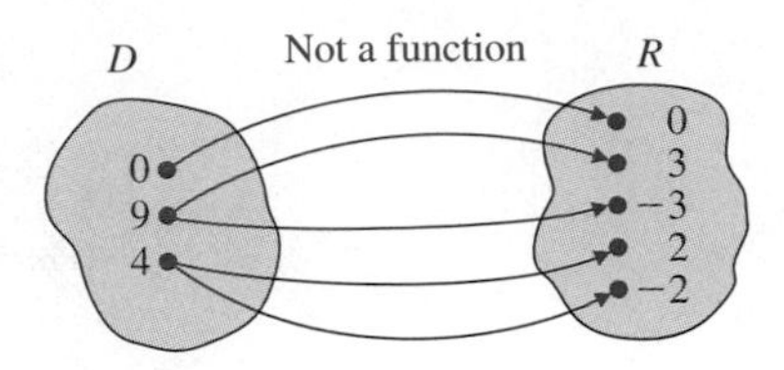

(b) Correspondence: squares of corresponding numbers equal the given number

FIGURE 1

The correspondence in Figure 1a defines a function, since each value in set D has one and only one corresponding value in set R. However, the correspondence in Figure 1b does not define a function, because at least one value in set D has more than one corresponding value in set R. For instance, 9 has two corresponding values, 3 and -3. The common features—input, rule, and output—are incorporated in the definition of a function.

DEFINITION

FUNCTION

A **function** from a set D to a set R is a rule of correspondence that assigns to each element in set D one and only one element in set R. The members of sets D and R, respectively, are referred to as the *inputs* and *outputs* of the function.

In this definition, the set D of all possible inputs is also called the **domain** of the function and the set R of the corresponding outputs is called the **range**. Throughout this book, domains and ranges will usually be sets of real numbers.

EXAMPLE 1

Identifying functions

Table 1 summarizes the results of a 5 point quiz given to a class of 21 students.

TABLE 1

Set S *Possible scores*	Set N *Number of students who earned the score*
5	3
4	1
3	8
2	6
1	3
0	0

(a) Does the correspondence in Table 1 define a function from set S to set N?
(b) Does the correspondence in Table 1 define a function from set N to set S?

If the correspondence is a function, identify the domain and range.

Solution

(a) Figure 2a is a diagram that depicts the correspondence from set S to set N. Each number (input) in set S has one and only one corresponding number (output) in set N. Consequently, the correspondence in Table 1 defines a function from set S to set N. The domain is S, the set of possible scores, and the range is N, the set of the numbers of students who earned the scores.
(b) Figure 2b depicts the correspondence from set N to set S. It shows that the number 3 has two corresponding values, 1 and 5, so the correspondence does not define a function from set N to set S.

Even though the domain numbers 5 and 1 have the same corresponding range number, 3, this does not violate the definition of a function.

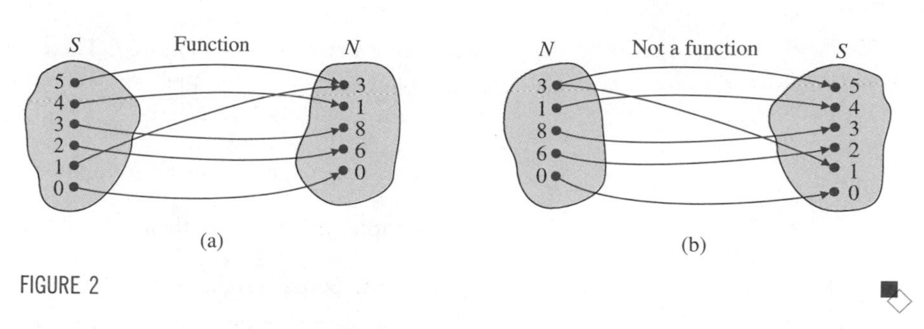

FIGURE 2

Functions may be defined by a table of data, by a graph, or by an equation. For example, Table 2 gives the average annual inflation rate from 1984 through 1994. We can interpret the data in Table 2 as a function if we consider each year as a member of the domain and each inflation rate as a member of the range.

TABLE 2

History of Inflation Rate

Year	1984	1985	1986	1987	1988	1989	1990	1991	1992	1993	1994
Average Annual Inflation Rate *Percent*	4.3	3.6	2.0	3.6	4.1	4.8	5.4	4.2	3.0	3.0	2.6

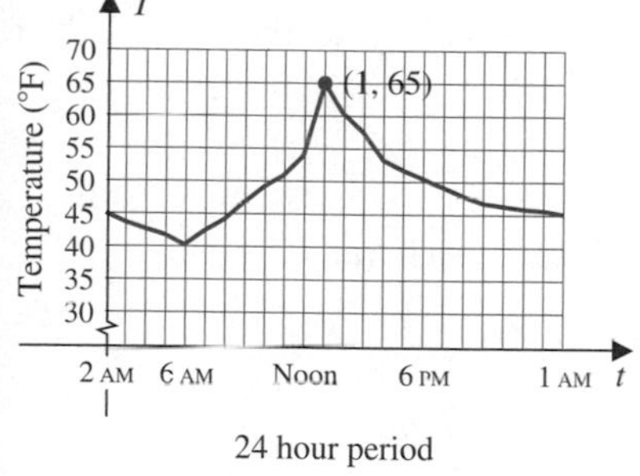

FIGURE 3

For each year, there is one and only one average annual inflation rate. An examination of the data indicates the trend of these rates from 1984 through 1994.

Functions also arise naturally as graphs. For instance, Figure 3 shows the temperature (in degrees Fahrenheit) in a midwestern city over a 24 hour period in May. This graph defines a function in which the times t constitute the domain and the temperatures T make up the range. The location of each point on the graph, such as (1 PM, 65°F) provides us with an ordered pair of numbers that enables us to identify each time (domain member) and its corresponding temperature (range member).

Real-world situations are often modeled by functions defined by tables or charts. However, much of the emphasis in mathematics is given to functions defined by equations. For example, if we represent the radius of a circle by r, then its area A is given by the equation

$$A = \pi r^2$$

For each value of the radius r there corresponds one value for the area A, and we say that "A is a function of r." The domain of this function, which is the set of all possible radii r, includes all positive real numbers, and the range is the set of the associated values of the areas A, which consists of all positive real numbers.

Using Function Notation and Terminology

Function notation provides us with a convenient shorthand for writing functions. Suppose a function is given. We can denote this function by the letter f, and let

x be a number in the domain of f. Then $f(x)$, read "f of x" or "the value of f at x," denotes the number in the range determined by the number x according to the rule of the function f. That is:

$f(x)$ is the output produced by input x.

For example, if $f(x) = x^2$, then:

$f(-1)$ is the output produced by input -1, and we write $f(-1) = (-1)^2 = 1$.
$f(0)$ is the output produced by input 0, and we write $f(0) = 0^2 = 0$.
$f(2)$ is the output produced by input 2, and we write $f(2) = 2^2 = 4$.

Any choice of letters of the alphabet such as f, g, and F can be used to designate functions. Sometimes, instead of writing $f(x) = x^2$, we might write $y = x^2$ where it is understood that the value of y is calculated after choosing a value of x. That is, the value of y *depends* on the choice of the value of x. Because of this dependence of the values of y on the values of x we refer to x as the **independent variable** and to y as the **dependent variable**.

In dealing with a function defined by $y = f(x)$, it is important to understand the following notation:

1. The *function* f is a rule of correspondence.
2. The *output* value y or $f(x)$ is a number depending on the *input* value of x.
3. The *equation* $y = f(x)$ relates the dependent variable y to the independent variable x.

EXAMPLE 2 Finding output values of a function

Let f be the function defined by $f(x) = 7x + 2$. Find each value.

(a) $f(1)$ (b) $f(-2)$ (c) $[f(4)]^2$ (d) $\sqrt{f(2)}$

Solution Here, f is defined by the equation $f(x) = 7x + 2$.

(a) We replace x by 1 to get $f(1) = 7(1) + 2 = 7 + 2 = 9$.
(b) Replacing x by -2, we get $f(-2) = 7(-2) + 2 = -14 + 2 = -12$.
(c) Similarly, $[f(4)]^2 = [7(4) + 2]^2 = (28 + 2)^2 = 30^2 = 900$.
(d) Finally, $\sqrt{f(2)} = \sqrt{7(2) + 2} = \sqrt{14 + 2} = \sqrt{16} = 4$.

Algebraic expressions can be substituted for the independent variable of a function. Thus, to compute $f(a + 3)$ for $f(x) = 7x + 2$, the output corresponding to input $a + 3$ is found by replacing x with $a + 3$ to get

$$\begin{aligned} f(a+3) &= 7(a+3) + 2 \\ &= 7a + 21 + 2 \\ &= 7a + 23 \end{aligned}$$

EXAMPLE 3 Replacing the independent variable with an algebraic expression
Let $g(x) = 3x^2 - 2$. Find and simplify each expression.

(a) $g(2t)$ (b) $g(k + 1)$ (c) $g(k) + 1$

Solution

(a) $g(2t) = 3(2t)^2 - 2 = 3(4t^2) - 2 = 12t^2 - 2$
(b) $g(k + 1) = 3(k + 1)^2 - 2 = 3(k^2 + 2k + 1) - 2$
$= 3k^2 + 6k + 3 - 2 = 3k^2 + 6k + 1$
(c) $g(k) + 1 = (3k^2 - 2) + 1 = 3k^2 - 2 + 1 = 3k^2 - 1$

Note that $g(k + 1) \neq g(k) + 1$.

In calculus it is necessary to evaluate expressions of the form

$$\frac{f(t + h) - f(t)}{h} \qquad h \neq 0$$

This is called the **difference quotient** of a function f.

EXAMPLE 4 Finding difference quotients
Find the difference quotient for each function.

(a) $f(x) = 3x - 5$ (b) $f(x) = 2x^2 + 1$

Solution

(a) If $f(x) = 3x - 5$, then

$$\frac{f(t + h) - f(t)}{h} = \frac{[3(t + h) - 5] - [3t - 5]}{h}$$
$$= \frac{3t + 3h - 5 - 3t + 5}{h}$$
$$= \frac{3h}{h} = 3$$

The difference quotient, 3, for $f(x) = 3x - 5$ is the same as the slope of the line $y = 3x - 5$ (see Problem 72).

(b) For $f(x) = 2x^2 + 1$, we have

$$\frac{f(t + h) - f(t)}{h} = \frac{[2(t + h)^2 + 1] - [2t^2 + 1]}{h}$$
$$= \frac{[2(t^2 + 2th + h^2) + 1] - [2t^2 + 1]}{h}$$
$$= \frac{2t^2 + 4th + 2h^2 + 1 - 2t^2 - 1}{h}$$
$$= \frac{4th + 2h^2}{h} = \frac{\not{h}(4t + 2h)}{\not{h}} = 4t + 2h$$

Graphing Functions

The graph of a function consists of all points (x, y) in the Cartesian plane such that x is in the domain and y is the corresponding range number.

EXAMPLE 5 **Graphing a function defined by a table**
Graph the function defined by the data in Table 2 on page 91. Use t to represent each year and I to represent the corresponding inflation rate.

Solution The graph of the function is obtained by plotting the points with the coordinates given in Table 2. The result is the scattergram shown in Figure 4.

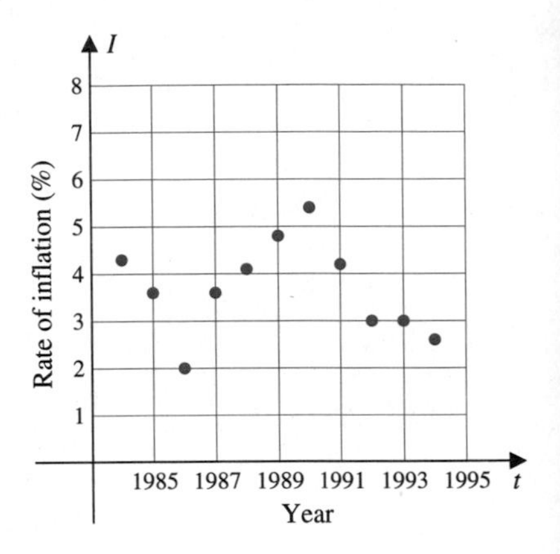

FIGURE 4

If a function is defined by the equation $y = f(x)$, the graph of f includes all points $(x, f(x))$ that satisfy the equation, where x is any number in the domain of f.

EXAMPLE 6 **Graphing a function defined by an equation**
Sketch the graph of $f(x) = 2x - 3$.

Solution The graph of the function is the same as the graph of the equation $y = 2x - 3$, a line with slope $m = 2$ and y intercept -3 (Figure 5).

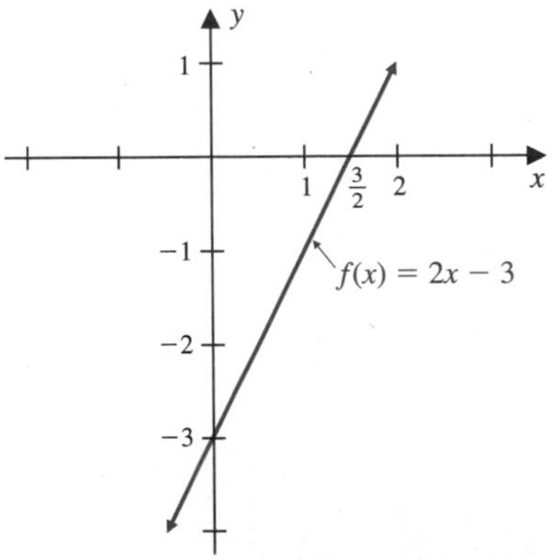

FIGURE 5

Generalizing Example 6, the graph of any function f defined by the linear equation

$$y = f(x) = mx + b$$

is a line with slope m and y intercept b. For this reason, a function of the form

$$f(x) = mx + b$$

is called a **linear function**.

Identifying Graphs That Represent Functions

Graphs of functions can be easily distinguished from graphs of other equations. To see why, consider the graphs of $y = \sqrt{9 - x^2}$ and $x^2 + y^2 = 9$ in Figures 6a and 6b, respectively.

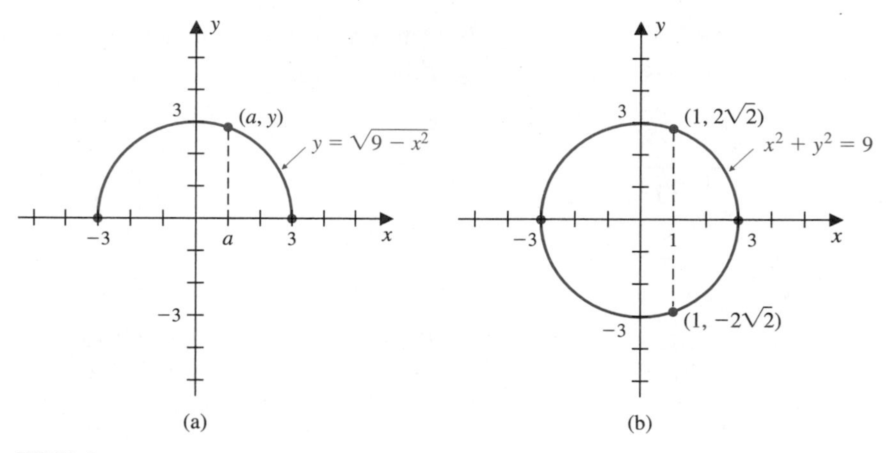

FIGURE 6

In Figure 6a, each input value a, where $-3 \leq a \leq 3$, corresponds to exactly one output value y, so the equation $y = \sqrt{9 - x^2}$ represents a function of x. Notice that the vertical line through any value of $x = a$ intersects the graph at exactly one point. In contrast, Figure 6b shows that the input value 1 has two different corresponding output values, $-2\sqrt{2}$ and $2\sqrt{2}$, so that the equation $x^2 + y^2 = 9$ does not represent a function of x. In Figure 6b, the vertical line through 1 intersects the graph at two points. These observations lead to the following generalization:

VERTICAL-LINE TEST

A set of points in a Cartesian plane is the graph of a function if and only if no vertical line intersects the graph more than oncc.

EXAMPLE 7

Using graphs to determine whether equations represent functions

Use a graphing utility to graph each equation and then use the graph to determine whether the equation represents a function.

(a) $y = 2x^2 - 4x - 3$ (b) $\dfrac{x^2}{4} + y^2 = 1$

Solution Our strategy here is to graph each equation and then use the vertical-line test to see if the equation represents a function. The graphing is expedited by using a graphing utility.

(a) Figure 7a displays a viewing window of the graph of the equation $y = 2x^2 - 4x - 3$. Since no vertical line intersects the graph more than once, the equation represents a function.

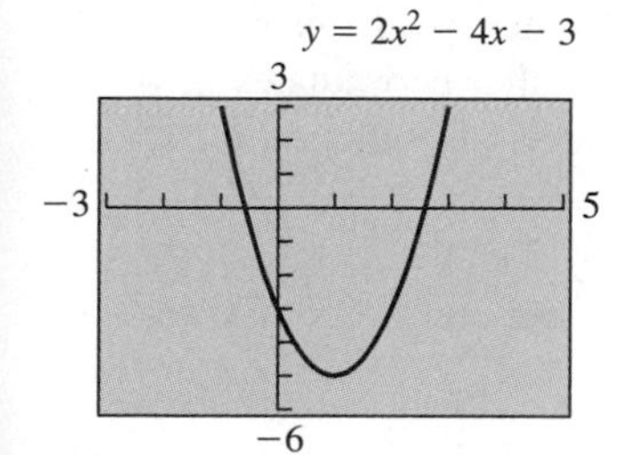

FIGURE 7a

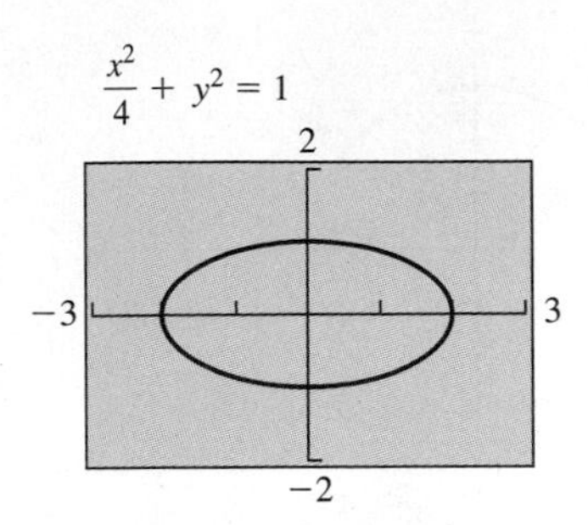

FIGURE 7b

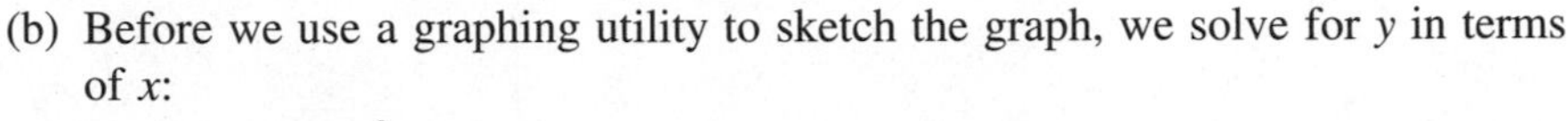

(b) Before we use a graphing utility to sketch the graph, we solve for y in terms of x:

$$\frac{x^2}{4} + y^2 = 1$$

$$y^2 = 1 - \frac{x^2}{4}$$

$$y = \sqrt{1 - \frac{x^2}{4}} \quad \text{or} \quad y = -\sqrt{1 - \frac{x^2}{4}}$$

Then we graph both equations, $y = \sqrt{1 - (x^2/4)}$ and $y = -\sqrt{1 - (x^2/4)}$, on the same coordinate system to get the graph of the given equation (Figure 7b). Here, we see that it is possible to draw a vertical line, such as $x = 1$, that intersects the graph more than once, so the equation does not represent a function of x.

Determining the Domain and Range

Unless otherwise specified, the domain of a function defined by an equation consists of all real number values of the input variable that result in real number output values. The domain is often determined by examining the equation that defines the function. Table 3 lists examples of some functions and their domains.

TABLE 3

Examples of Functions and Their Domains

Function	Domain	Reason
1. $f(x) = x^2 + 3$	$\mathbb{R}$	$x^2 + 3$ is a real number no matter what real value is assigned to x.
2. $g(x) = \sqrt{x}$	All real numbers x such that $x \geq 0$	x must be nonnegative in order for $\sqrt{x}$ to be a real number.
3. $h(x) = \frac{1}{x + 2}$	All real numbers such that $x \neq -2$	Division by 0 is not defined.

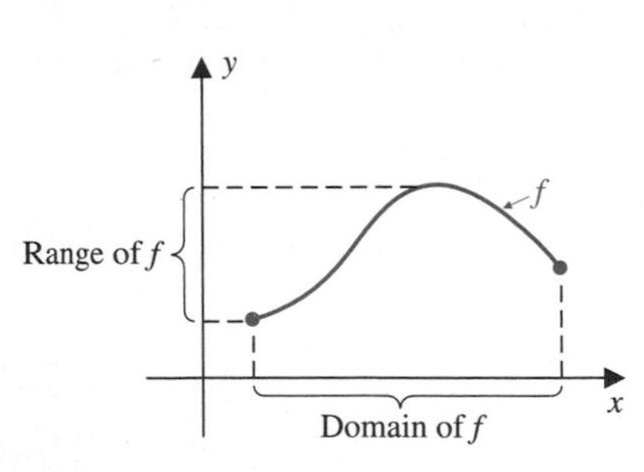

FIGURE 8

It is not always easy to find the range of a function from its equation. However, the graph of a function helps us to recognize its domain and range. Figure 8 illustrates that the domain of a function f is the set of all x coordinates of points on its graph, and the range is the set of all y coordinates of points on its graph.

EXAMPLE 8 Using graphs to determine the domain and range of a function

Use the graph of each function to determine the domain and range. Express the domain and range in interval notation.

(a) $f(x) = 2x^2 - 4x - 3$ (Figure 9a) (b) $g(x) = \sqrt{1 - \frac{x^2}{4}}$ (Figure 9b)

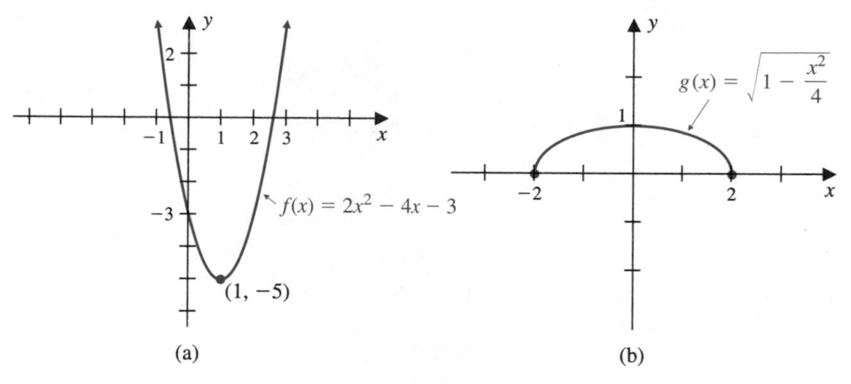

FIGURE 9

Solution

(a) Figure 9a suggests that the domain of f contains all real numbers and the range includes all real numbers y such that $y \geq -5$. In interval notation, the domain is $(-\infty, \infty)$ and the range is $[-5, \infty)$.

(b) Figure 9b suggests that the domain of g includes all real numbers x such that $-2 \leq x \leq 2$ and the range consists of all real numbers y such that $0 \leq y \leq 1$. So, in interval notation, the domain is $[-2, 2]$ and the range is $[0, 1]$.

Solving Applied Problems

Functions are used in modeling real-life situations as illustrated in the next example.

EXAMPLE 9 Modeling the cost of laying cable

A cable television company has its master antenna located at point A on the bank of a straight river 0.6 mile wide (Figure 10). It is going to run a cable from A to a point P on the opposite bank of the river and then straight along the bank to a town T situated 1.8 miles downstream from Q, a point on the shore directly opposite point A so that $\overline{AQ}$ is perpendicular to $\overline{QP}$. It costs \$5 per foot to run the cable underwater and \$3 per foot to run the cable along the bank.

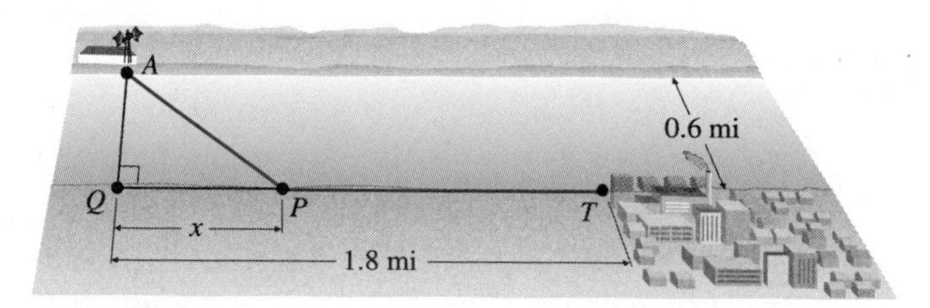

FIGURE 10

(a) Express the cost C (in dollars) as a function of the distance x (in miles) between point P and Q.
(b) What are the restrictions on x?
GU (c) Use a graphing utility to graph the function from part (a) with the restrictions from part (b).
(d) What is the cost if the cable is run directly underwater from A to Q and then along the bank to T?
(e) What is the cost if the cable is run totally underwater?
(f) Generally, what happens to the cost as x increases from 0 to 1.8 miles?

Solution First, we denote the lengths of line segments $\overline{AQ}$, $\overline{AP}$, $\overline{QP}$, and $\overline{PT}$ as $|\overline{AQ}|$, $|\overline{AP}|$, $|\overline{QP}|$, and $|\overline{PT}|$, respectively. Notice that $|\overline{QP}| = x$, and $|\overline{AQ}| = 0.6$.

(a) The cost C can be found as follows:

$$\begin{aligned} C &= (\text{Cost underwater from } A \text{ to } P) + (\text{Cost along bank from } P \text{ to } T) \\ &= [(\$5 \text{ per foot}) \times |\overline{AP}| \text{ (in feet)}] + [(\$3 \text{ per foot}) \times |\overline{PT}| \text{ (in feet)}] \\ &= \left[5\sqrt{(0.6)^2 + x^2} \cdot 5280\right] + [3(1.8 - x) \cdot 5280] \quad \text{Pythagorean theorem; 5280 feet per mile} \end{aligned}$$

Thus,

$$C = 5280\left[5\sqrt{0.36 + x^2} + 5.4 - 3x\right]$$

This equation expresses C as a function of x.

Selections for Ymin and Ymax were made so that a smooth connected graph is shown for $0 \le x \le 1.8$. Generally, appropriate selections for Ymin and Ymax on a graphing utility require practice and experience.

(b) Upon examining Figure 10, we see that x is restricted to values such that $0 \le x \le 1.8$.
(c) In view of the restrictions in part (b), we select Xmin as 0 and Xmax as 1.8. A viewing window of the graph of the function is displayed in Figure 11.

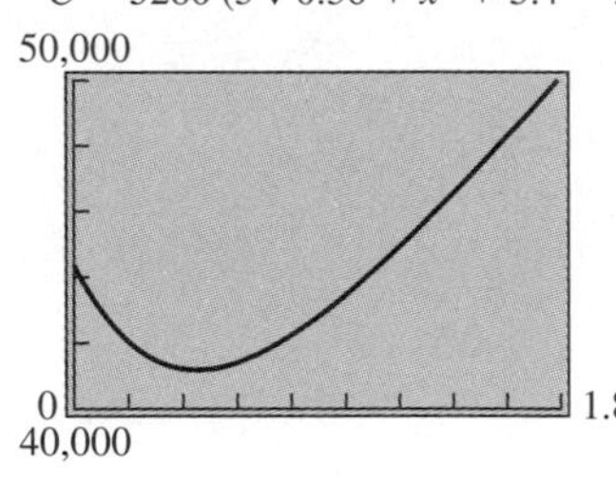

FIGURE 11

(d) If the cable is run from A to Q and then to T, $x = 0$. So

$$\begin{aligned} C &= 5280\left[5\sqrt{0.36} + 5.4\right] \\ &= 44{,}352 \end{aligned}$$

Thus, this option would cost \$44,352.
(e) If the cable is run entirely underwater from A to T, then $x = 1.8$. So

$$\begin{aligned} C &= 5280\left[5\sqrt{0.36 + (1.8)^2} + 5.4 - 3(1.8)\right] \\ &= 50{,}090.48 \quad \text{(approx.)} \end{aligned}$$

Thus, it would cost \$50,090.48 if the cable is run totally underwater.

Later in Chapter 3, we will discuss ways of finding the lowest point on a graph. For this example, the location of the lowest point would enable us to find the least expensive route for the cable.

(f) The graph in Figure 11 suggests that as the distance x increases from 0 to 1.8 miles, the cost appears to decrease at first and then increase to reach its highest value. This highest value occurs when $x = 1.8$, that is, when the cable is run totally underwater. ■

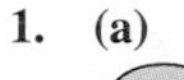

PROBLEM SET 2.1

Mastering the Concepts

In problems 1 and 2, indicate which of the correspondences depicted in the diagrams are functions. If the correspondence is not a function, explain why not.

1. **(a)**

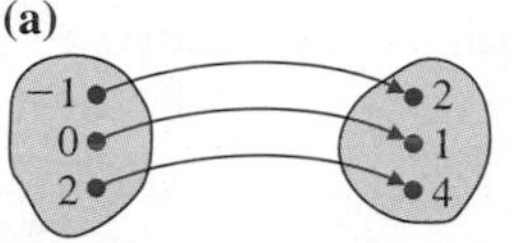

(b)

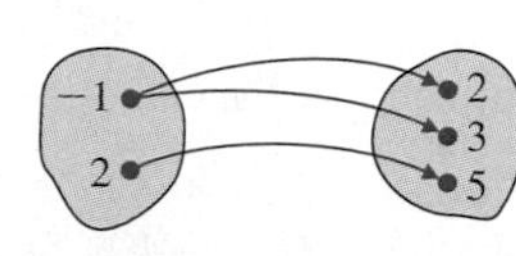

2. **(a)**

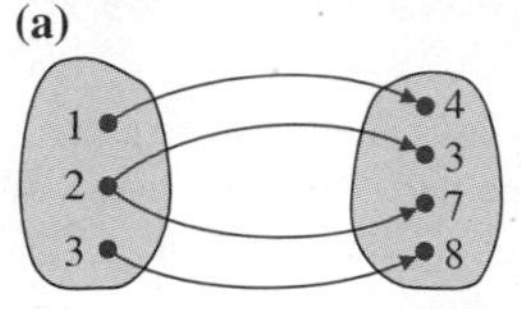

(b)

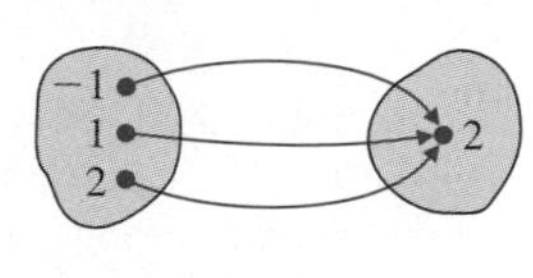

3. Table 4 lists the top five ticket sales for a high school play.

TABLE 4

Set T *Number of tickets sold*	20	15	14	10	8
Set N *Number of students who sold that many tickets*	1	3	1	2	5

(a) Does Table 4 define a function from set T to set N?
(b) Does Table 4 define a function from set N to set T?

4. Table 5 lists the number of employees of a company between the ages of 55 and 60.

TABLE 5

Set A *Age*	55	56	57	58	59	60
Set N *Number of employees of that age*	5	3	1	0	2	4

(a) Does Table 5 define a function from set A to set N?
(b) Does Table 5 define a function from set N to set A?

In problems 5–14, find each value.

5. $f(x) = -5x + 3$
(a) $f(2)$ **(b)** $f(-2)$
(c) $f\left(\frac{3}{10}\right)$ **(d)** $f\left(-\frac{3}{10}\right)$

6. $g(x) = 3x - 7$
(a) $g(-5)$ **(b)** $g(5)$
(c) $g\left(-\frac{2}{3}\right)$ **(d)** $g\left(\frac{2}{3}\right)$

7. $F(x) = x^2 - 4x + 2$
(a) $F(-3)$ **(b)** $F(3)$
(c) $F\left(\frac{2}{3}\right)$ **(d)** $F\left(-\frac{2}{3}\right)$

8. $G(x) = -x^2 + 3x - 5$
(a) $G(-5)$ **(b)** $G(5)$
(c) $G\left(\frac{3}{5}\right)$ **(d)** $G\left(-\frac{3}{5}\right)$

9. $V(t) = \dfrac{2t + 1}{t - 2}$
(a) $V(1)$ **(b)** $V(-1)$
(c) $V\left(\frac{1}{2}\right)$ **(d)** $V\left(-\frac{1}{2}\right)$

10. $W(r) = \dfrac{-r + 3}{5 - 2r}$
(a) $W(2)$ **(b)** $W(-2)$
(c) $W\left(-\frac{1}{4}\right)$ **(d)** $W\left(\frac{1}{4}\right)$

11. $h(w) = \sqrt{w - 5} + 3$ [Round off to three decimal places in parts (c) and (d).]
(a) $h(6)$ **(b)** $h(9)$
(c) $h(7.256)$ **(d)** $h(2\pi)$

12. $f(y) = 6 - \sqrt{4y + 1}$ [Round off to three decimal places in parts (c) and (d).]
(a) $f(0)$ **(b)** $f\left(\frac{1}{4}\right)$
(c) $f(2.875)$ **(d)** $f(\pi^2)$

13. $g(x) = \begin{cases} x + 2 & \text{if } x \leq 3 \\ x + 4 & \text{if } x > 3 \end{cases}$
(a) $g(-1)$ **(b)** $g(3)$
(c) $g(0)$ **(d)** $g(4)$

14. $f(x) = \begin{cases} 2x^2 + 5 & \text{if } x \leq -1 \\ 3 - x^2 & \text{if } x > -1 \end{cases}$
(a) $f(-2)$ **(b)** $f(0)$
(c) $f(7)$ **(d)** $f(-1)$

In problems 15–22, find and simplify each expression.

15. $h(t) = -2t - 5$
(a) $h(a + 3)$ **(b)** $h(a) + h(3)$
(c) $h\left(\dfrac{4}{a + 3}\right)$ **(d)** $\dfrac{h(4)}{h(a + 3)}$

16. $g(w) = \frac{1}{3}w - 2$
(a) $g(3y + 6)$ **(b)** $g(3y) + g(6)$
(c) $3g(y) + 6$ **(d)** $3g(y) + g(6)$

17. $f(x) = 2x^2 - 5x + 1$

(a) $f(a + 1)$ (b) $f(a) + f(1)$
(c) $2f(a) - f(2a)$ (d) $f(a^2)$

18. $G(r) = -r^2 + 3r - 6$

(a) $G(\sqrt{t})$ (b) $\sqrt{G(t)}$
(c) $G(t + h) - G(t)$ (d) $G(1/h)$

19. $H(x) = 3 - 2|x|$

(a) $H(7a)$ (b) $7H(a)$
(c) $H(a^2)$ (d) $H(-a^2)$

20. $V(t) = |t| - |t + 3|$

(a) $V(x - 3)$ (b) $V(x) - V(3)$
(c) $V(5x) + 3$ (d) $V(5x + 3)$

21. $Q(y) = \dfrac{1}{y + 3}$

(a) $Q\left(\dfrac{1}{y + 3}\right)$ (b) $\dfrac{Q(1)}{Q(y + 3)}$
(c) $\dfrac{1}{Q(y) + Q(3)}$ (d) $\dfrac{Q(1)}{Q(y) + Q(3)}$

22. $F(t) = \dfrac{4}{t - 3}$

(a) $F\left(\dfrac{3y + 4}{y}\right)$ (b) $\dfrac{F(3y + 4)}{F(y)}$
(c) $\dfrac{3F(y) + F(3y)}{F(y)}$ (d) $\dfrac{3y + 4}{F(y)}$

In problems 23–26, find the difference quotient $\dfrac{f(t + h) - f(t)}{h}$, $h \neq 0$.

23. (a) $f(x) = 4x + 1$ (b) $f(x) = x^2 - 3$

24. (a) $f(x) = -2x + 3$ (b) $f(x) = -4x^2 + 7$

25. (a) $f(x) = \dfrac{1}{x}$ (b) $f(x) = \dfrac{3}{x - 1}$

26. (a) $f(x) = -2x^2 + 3x - 1$

(b) $f(x) = x^2 - \dfrac{1}{x}$

In problems 27 and 28, graph each function defined by the given table. Find the domain and range.

27. (a)

x	-2	-1	0	1	2	3
y	-3	-1	1	3	5	7

(b)

t	-2	-1	0	1	2	3
w	-3	0	1	0	-3	-8

28. (a)

r	1	2	3	4	5
A	1	$\sqrt{2}$	$\sqrt{3}$	2	$\sqrt{5}$

(b)

x	$\frac{1}{2}$	1	2	3	4
y	2	1	$\frac{1}{2}$	$\frac{1}{9}$	$\frac{1}{16}$

In problems 29 and 30, graph each function by using equation recognition. Find the domain and range of each function.

29. (a) $h(x) = -\frac{1}{2}x + 3$ (b) $f(x) = \sqrt{49 - x^2}$

30. (a) $G(t) = 4 - 3t$ (b) $f(x) = -\sqrt{49 - x^2}$

GU In problems 31–36, use a graphing utility to sketch the graph of each function in the specified viewing window. What is the domain and range suggested by the graph of each function?

31. $y = x^3 + 2$
Xmin: -10; Xmax: 10; Xscl: 1;
Ymin: -10; Ymax: 10; Yscl: 1

32. $y = -2x^3 + 1$
Xmin: -10; Xmax: 10; Xscl: 1;
Ymin: -20; Ymax: 10; Yscl: 1

33. $y = \sqrt{x - 5}$
Xmin: -2; Xmax: 30; Xscl: 2;
Ymin: -1; Ymax: 10; Yscl: 1

34. $y = \sqrt{5 - x}$
Xmin: -20; Xmax: 10; Xscl: 1;
Ymin: -1; Ymax: 10; Yscl: 1

35. $y = |x - 2| - 5$
Xmin: -5; Xmax: 10; Xscl: 1;
Ymin: -7; Ymax: 5; Yscl: 1

36. $y = -|x - 2| + 3$
Xmin: -5; Xmax: 8; Xscl: 1;
Ymin: -5; Ymax: 10; Yscl: 1

In problems 37 and 38, indicate whether the graph represents a function.

37. (a)

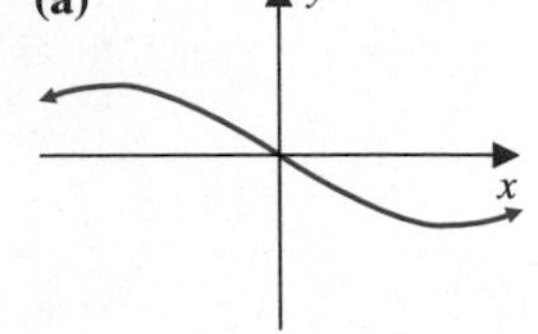

(b)

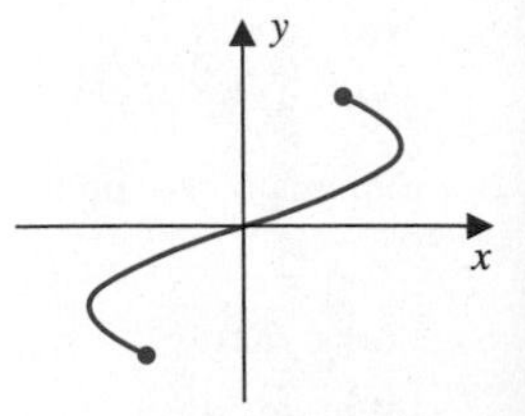

38. **(a)**

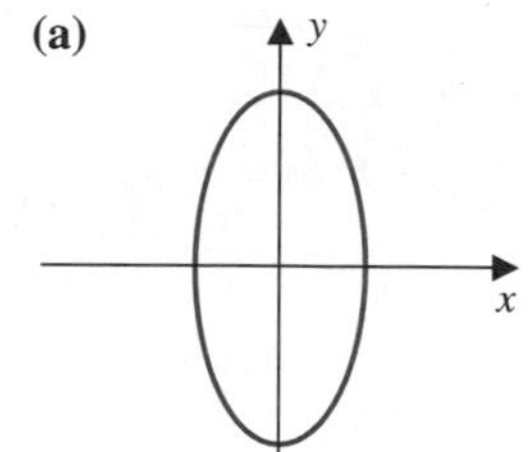

(b)

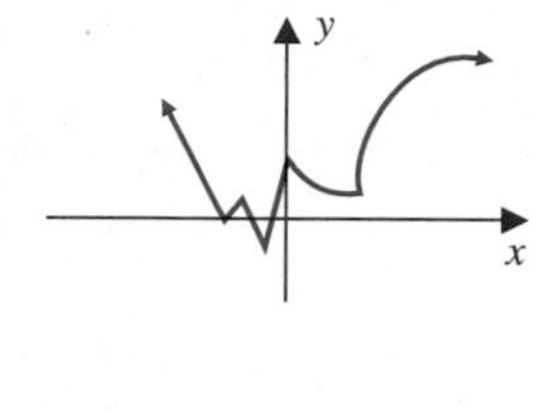

In problems 39–42, sketch the graph of each equation, and use the graph to identify those equations that define y as a function of x.

39. **(a)** $x^2 + y^2 = 25$ **(b)** $y = 2x^2 - 1$

40. **(a)** $-2x^2 + y = 4$ **(b)** $(x - 1)^2 + y^2 = 4$

41. **(a)** $y = 3$ **(b)** $y = \sqrt{4 - x}$

42. **(a)** $x = 2$ **(b)** $y = \sqrt{x + 5}$

In problems 43–52, find the domain of each function without graphing.

43. $f(x) = 5 - 3x$

44. $g(x) = 5x - 7$

45. $h(x) = \dfrac{7}{x + 4}$

46. $H(x) = \dfrac{-3}{-x - 2}$

47. $F(x) = \dfrac{5}{x^2 - 9}$

48. $f(x) = \dfrac{x^2 + 3}{x^2 + 2x - 8}$

49. $f(x) = \sqrt{3x - 2}$

50. $F(x) = \sqrt{3 - 5x}$

51. $g(x) = \dfrac{x}{x^2 - x}$

52. $h(x) = \dfrac{x - 2}{x^2 - 4}$

In problems 53 and 54, use the graph to determine the domain and range of each function. Express the domain and range in interval notation.

53. **(a)**

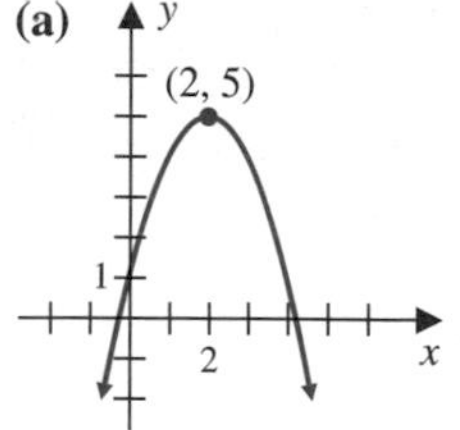

(b)

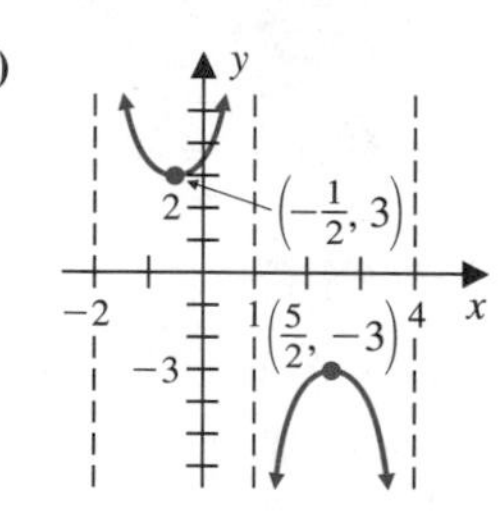

54. **(a)**

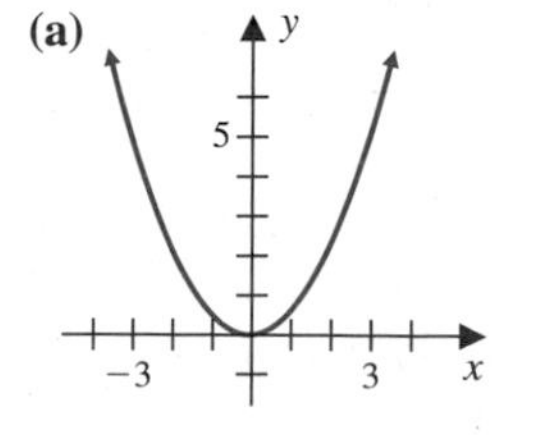

(b)

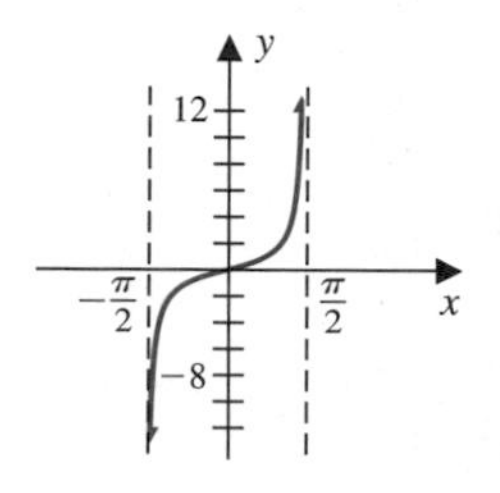

Applying the Concepts

55. Fuel Economy: Figure 12 displays a graph that relates the fuel economy F (in miles per gallon) of a car to different average speeds V (in miles per hour) of the car between 40 and 80 miles per hour.

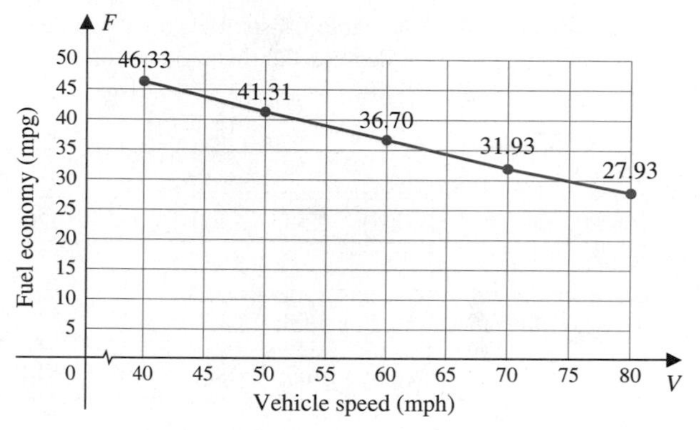

FIGURE 12

(a) Explain why the graph defines a function of V.

(b) Determine the domain and range.

(c) Estimate the fuel economy when the car has an average speed of 40 miles per hour; 60 miles per hour; 75 miles per hour. Round off to two decimal places.

(d) Estimate the average speed of the car when the fuel economy is 41.31 miles per gallon; 35 miles per gallon; 30 miles per gallon. Round off to two decimal places.

56. Biology: Figure 13 shows the relationship between the weight W (in ounces) and the length L (in inches, up to 16 inches) of a certain type of fish.

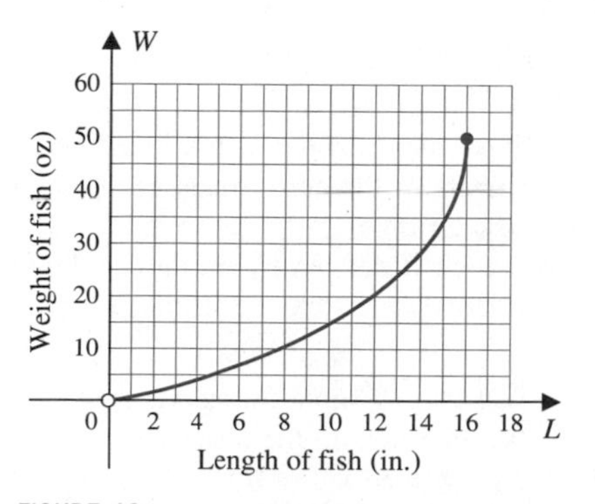

FIGURE 13

(a) Explain why the graph defines a function of L.
(b) Determine the domain and range.
(c) Estimate the weight (in ounces) if the length (in inches) is 10; 13; 16.
(d) Estimate the length (in inches) if the weight (in ounces) is 15; 30; 40.

57. Global Temperature: Table 6 lists the annual average global temperatures (in degrees Fahrenheit) for 10 year periods from 1880 to 1990.

TABLE 6

Year (t)	1880	1890	1900	1910	1920	1930
Average Temperature (T) *Degrees Fahrenheit*	58.2	58.5	58.7	58.3	58.8	59.2

Year (t)	1940	1950	1960	1970	1980	1990
Average Temperature (T) *Degrees Fahrenheit*	59.1	59.0	59.15	58.4	59.8	60.1

(a) Explain why Table 6 defines a function, where t represents members of the domain and T represents members of the range.
(b) Graph the function defined by the data in Table 6.
(c) Discuss the trend in the change of global temperature.

58. Stock Market: Table 7 shows the Dow Jones industrial average for various times during one day.

TABLE 7

Time of Day (t)	10 AM	11 AM	12 noon	1 PM	2 PM	3 PM	4 PM
Dow Jones Average (I)	4562	4564	4562	4561	4563	4552	4547

(a) Explain why Table 7 defines a function, where t represents members of the domain and I represents members of the range.
(b) Graph the function defined by the data in Table 7.
(c) Discuss the trend in the change of the Dow Jones average.

59. Dimensions of a Play Area: A home owner wants to enclose a rectangular play area next to the house using 40 meters of fencing. The house wall is to form one side of the play area (Figure 14).

FIGURE 14

(a) Express the length ℓ of the play area as a function of its width w.
(b) Express the area A of the play area as a function of its width w.
(c) What are the restrictions on w?
GU (d) Use a graphing utility to sketch the graph of the function from part (b) using the restrictions in part (c), and interpret the graph.

60. Outdoor Track: An outdoor track is to be constructed in the shape shown in Figure 15 and is to have a perimeter of P meters.

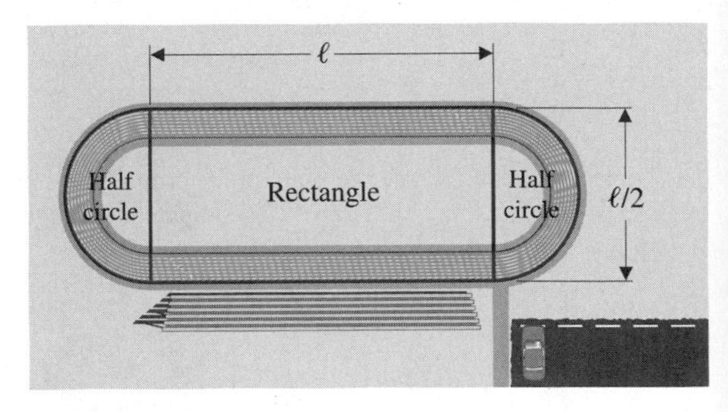

FIGURE 15

(a) Express the perimeter P as a function of ℓ.
(b) Use the result from part (a) to express ℓ as a function of P.
(c) Express the total area A enclosed by the track as a function of ℓ.
(d) Use the results from parts (b) and (c) to express the area A as a function of P.
GU (e) Use a graphing utility to sketch the graph of the function in part (d) and interpret it.

61. Offshore Oil Well: An offshore oil well is located at point A, which is 13 kilometers from the nearest point Q on a straight shoreline. The oil is to be piped from A to a point P on the shoreline and then straight along the shoreline to a terminal at point T, which is 10 kilometers from point Q (Figure 16). Suppose that it costs $90,000 per kilometer to lay pipe underwater, and it costs $60,000 per kilometer to lay the pipe along the shoreline.

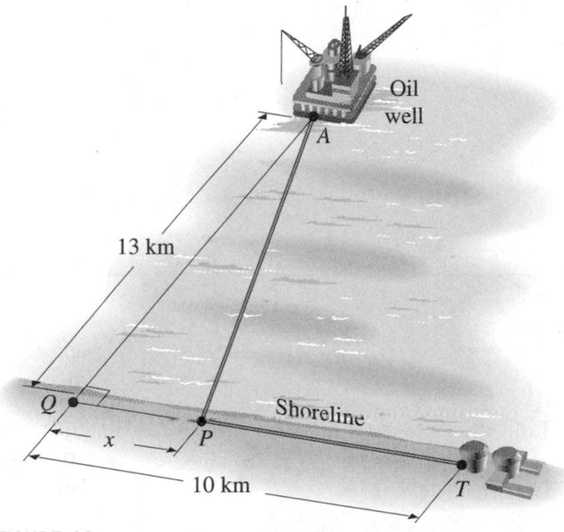

FIGURE 16

(a) Express the cost C (in dollars) as a function of x, where x is the distance between Q and P.
(b) What restrictions are there on x?
GU (c) Use a graphing utility to graph the function in part (a) with the restrictions in part (b).
(d) What is the cost if the pipe is laid from A to Q and then from Q to T?
(e) What is the cost if the pipe is laid totally underwater from A to T? Round off to two decimal places.
(f) Use the graph to describe what happens to the cost as x increases from 0 to 10 kilometers.

62. Landscaping: A landscape designer is planning to plant a rectangular tulip bed of dimensions $2x$ and $2y$ on a circular plot of land of radius 5 yards, as displayed in Figure 17. It is determined that the cost of planting the tulip bed is $20 per square yard. Let C (in dollars) be the total cost of planting the tulip bed.
(a) Draw a circle of radius 5 whose center is at the origin of a Cartesian coordinate system, and find its equation.
(b) Use the result from part (a) to express the total cost C of planting the tulip bed as a function of x.
(c) What are the restrictions on x?

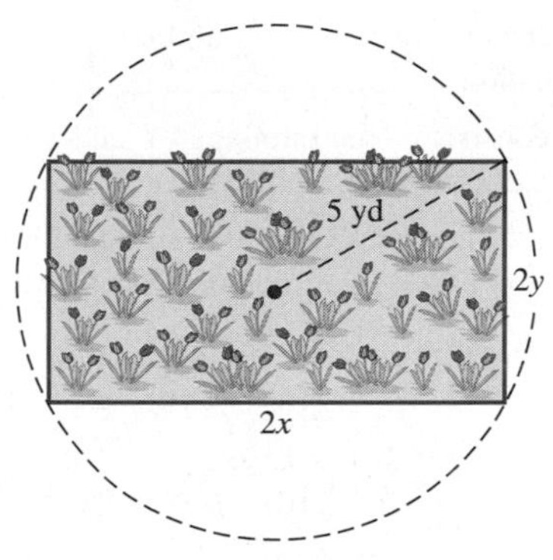

FIGURE 17

GU (d) Use a graphing utility to graph the function in part (b) with the restrictions in part (c).
(e) Use the graph to describe the cost trend as x increases.

Developing and Extending the Concepts

In problems 63 and 64, indicate which of the given situations always defines a function from set A to set B. Explain.

63. (a) For a 50 item multiple-choice final examination in Calculus I, the following correspondence is defined:

Set A consists of the possible scores (number of correct answers) on the examination.

Set B consists of the numbers of students who had the same score.

Each possible score is associated with the number of students who had that score.

(b) Set A consists of all possible gross taxable incomes reported to the Internal Revenue Service in a given year.

Set B consists of the amounts paid in Federal income tax for the same year.

Each gross taxable income is associated with the corresponding amount of taxes paid for that income.

64. (a) Set A consists of the heights of a group of people.

Set B consists of the weights of the same group of people.

Each height is associated with the weights of people of that height.

(b) On a 50 item multiple-choice examination, the following correspondence is defined:

Set A consists of the item numbers on the examination.

Set B consists of the numbers of students who answered each item incorrectly.

Each test item number is associated with the number of students who answered that item incorrectly.

65. Determine whether $f(a + b) = f(a) + f(b)$:
(a) $f(x) = 3x$ **(b)** $f(x) = 2x + 5$

66. Determine whether $f(ka) = kf(a)$:
(a) $f(x) = 3x$ **(b)** $f(x) = 2x + 5$

67. In calculus, the notation $g(x)\Big|_a^b$ is defined by

$$g(x)\Big|_a^b = g(b) - g(a)$$

Evaluate each of the following expressions:

(a) $(-3x + 5)\Big|_{-3}^{1}$ **(b)** $5\left(\frac{x^{-2}}{-2} + x^3\right)\Big|_{-1}^{4}$

68. **(a)** In a Cartesian system, is the y axis the graph of a function? What is its equation?
(b) In a Cartesian system, is the x axis the graph of a function? What is its equation?

69. Birthdate: Suppose that more than 366 people are in a room. For each calendar date we associate every person in the room who has that birthdate.
(a) Explain why this association does not define a function with domain all calendar dates and range all the people.
(b) Suppose we "reverse" the association; that is, we consider the people to be the domain and the birthdays to be the range. Would this reverse association define a function? Explain.

70. The area A of a circle of radius r is given by the formula $A = \pi r^2$. Explain why the graph in Figure 18 is *not* the graph of this function.

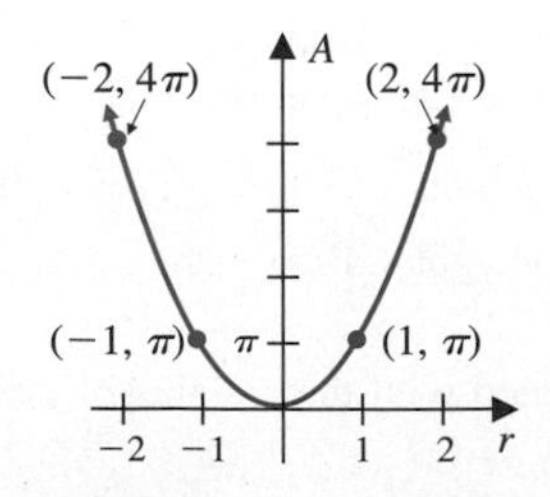

FIGURE 18

71. Use the slope formula to find the slope of the straight line L containing points $(t, f(t))$ and $(t + h, f(t + h))$ on the graph of $y = f(x)$ (Figure 19). Compare this slope to the difference quotient of f.

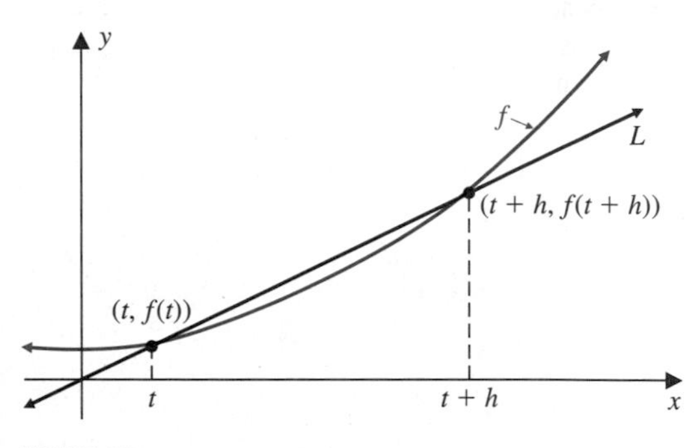

FIGURE 19

72. Find the difference quotient of a linear function $f(x) = mx + b$ and compare it to the slope of the line with equation $y = mx + b$.

73. Linear Approximation: The running times (in seconds) for the winners in the Olympic Games women's 200 meter dash (from 1968 to 1992) are given in the following table:

Year (N)	1968	1972	1976	1980	1984	1988	1992
Time (T) *Seconds*	22.50	22.40	22.37	22.03	21.81	21.34	21.81

(a) Construct a scattergram for the data. The year N represents the horizontal axis and the time T represents the vertical axis. For ease of plotting, use $N = 0$ for 1968, 4 for 1972, and so on.
(b) The linear function $T = -0.04241N + 22.546$ best approximates the data in the scattergram. Sketch the graph of this function on the same coordinate system used in part (a).
(c) Compare the data in the table to the results given by the equation in part (b).
(d) Use the equation in part (b) to predict the running time for the winner in this Olympic event in the year 1996 and in the year 2000. Round off the answers to two decimal places.

OBJECTIVES

1. Determine intervals where a function is increasing, decreasing, or constant.
2. Identify even and odd functions.
3. Recognize standard graphs.
4. Graph piecewise-defined functions.
5. Solve applied problems.

2.2 Graph Features

In this section we explore geometric features of graphs of functions. Also, we expand our recognition of standard graphs.

Determining Intervals Where a Function Is Increasing, Decreasing, or Constant

Figure 1 shows a graph of the immigration into the United States during the twentieth century. Notice that until 1930 the number of immigrants consistently *decreases*, and correspondingly, the graph *falls* from left to right. Then, beginning in the 1930s, the number of immigrants *increases*, and correspondingly, the graph *rises* from left to right. In other words, a falling and then rising pattern in the graph conveys a decrease followed by an increase in the number of immigrants.

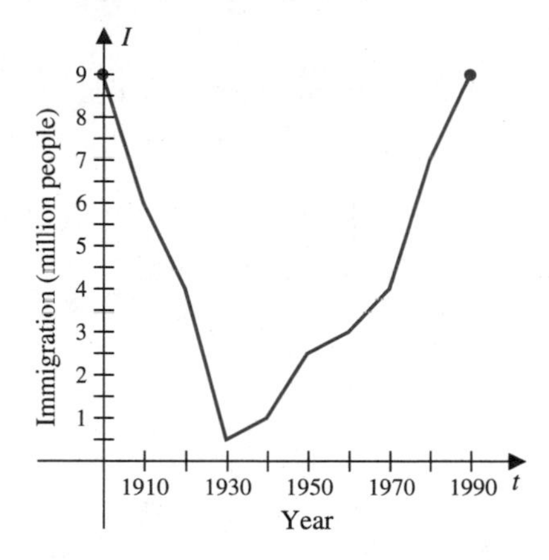

FIGURE 1
Twentieth-century immigration: Total immigration in the 1990's is expected to match the 1901–1910 high of 8.8 million newcomers. Some analysts predict that 9.5 million immigrants will arrive between 1991 and 2000.
Source: U.S. Census Bureau

This example leads us to the notions of increasing, decreasing, and constant function behavior, which are described in the box on the following page:

PROPERTIES

INCREASING, DECREASING, AND CONSTANT FUNCTION BEHAVIOR

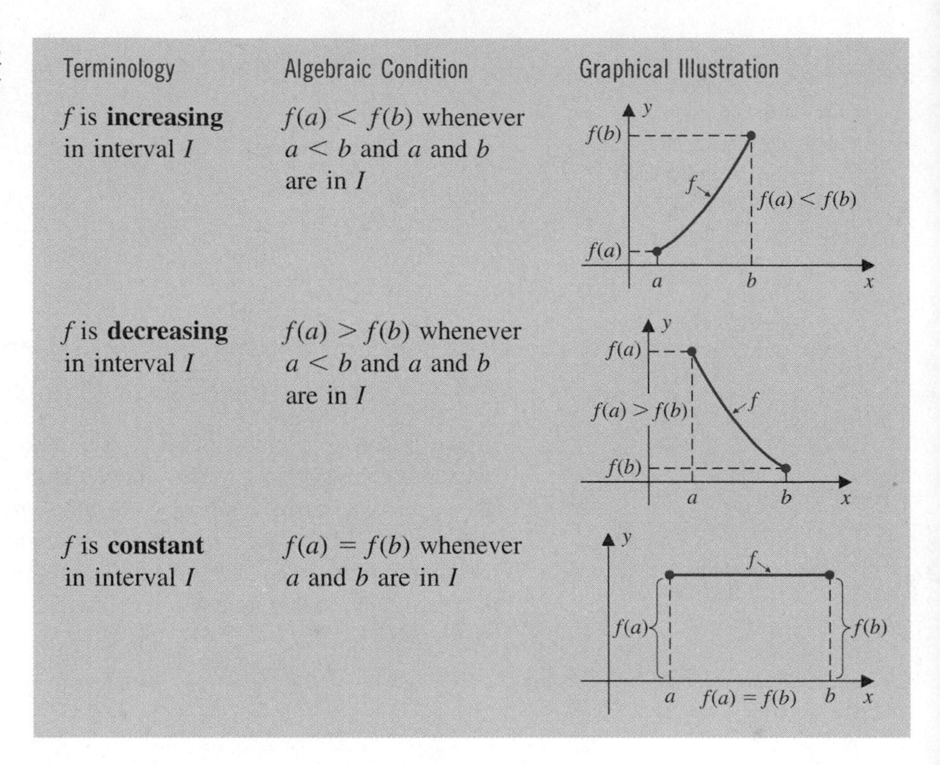

Terminology	Algebraic Condition	Graphical Illustration
f is **increasing** in interval I	$f(a) < f(b)$ whenever $a < b$ and a and b are in I	
f is **decreasing** in interval I	$f(a) > f(b)$ whenever $a < b$ and a and b are in I	
f is **constant** in interval I	$f(a) = f(b)$ whenever a and b are in I	

We describe where increasing, decreasing, or constant behavior occurs in terms of the x values.

A function's domain can be divided into subintervals over which the function is increasing, decreasing, or constant.

EXAMPLE 1 Determining intervals where a function is increasing, decreasing, or constant

Figure 2 shows the graph of a function f. Find the intervals where f is increasing, decreasing, or constant.

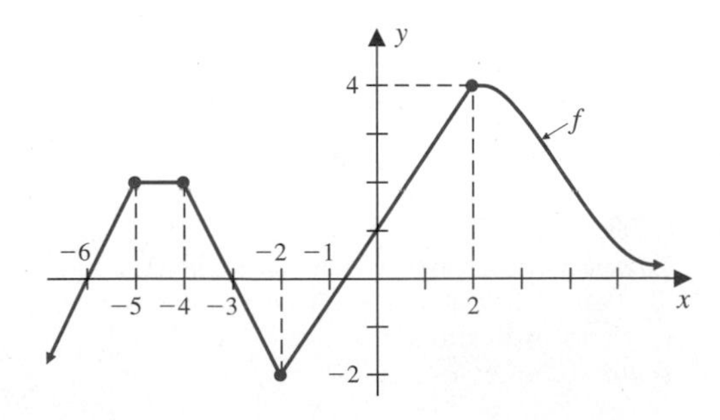

FIGURE 2

Solution The function is increasing in the intervals $(-\infty, -5]$ and $[-2, 2]$, decreasing in the intervals $[-4, -2]$ and $[2, \infty)$, and constant in the interval $[-5, -4]$.

Identifying Even and Odd Functions

In Section 1.4 we introduced the notion of symmetry of the graph of an equation. Now we connect that discussion to function behavior. A function whose graph is symmetric with respect to the y axis is called an *even function*; a function whose graph is symmetric with respect to the origin is called an *odd function*. Even and odd functions can also be described by using function notation as explained in the following property:

PROPERTY

EVEN AND ODD FUNCTIONS

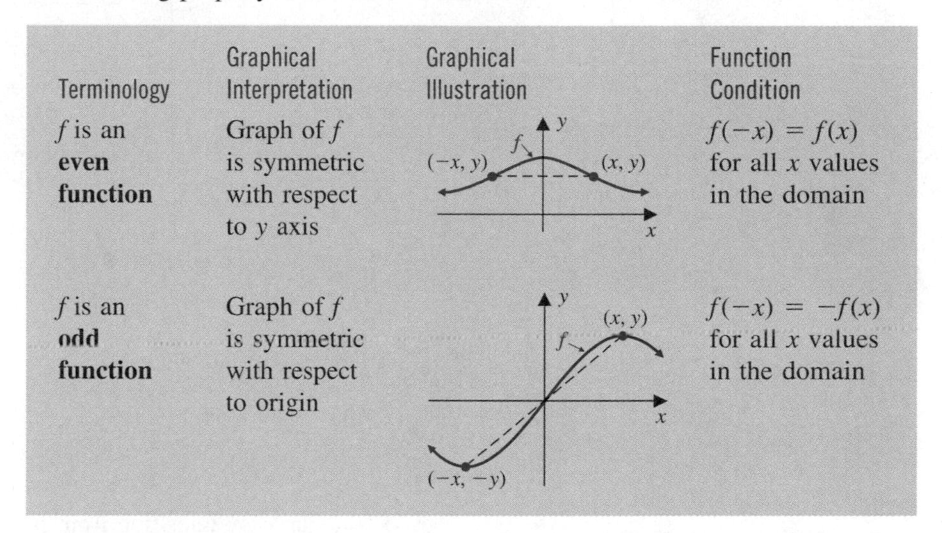

Terminology	Graphical Interpretation	Graphical Illustration	Function Condition
f is an **even function**	Graph of f is symmetric with respect to y axis		$f(-x) = f(x)$ for all x values in the domain
f is an **odd function**	Graph of f is symmetric with respect to origin		$f(-x) = -f(x)$ for all x values in the domain

EXAMPLE 2 Identifying even and odd functions

Use function notation to determine whether each function is even, odd, or neither. Then graph the function on a graphing utility to demonstrate the conclusion.

(a) $f(x) = 3x^4$ (b) $g(x) = x^3 - 2x$ (c) $h(x) = x^2 + x$

Solution

(a) Since $f(-x) = 3(-x)^4 = 3x^4 = f(x)$, it follows that f is an even function. As expected, the graph of f is symmetric with respect to the y axis (Figure 3a, page 108).

(b) Here,

$$\begin{aligned} g(-x) &= (-x)^3 - 2(-x) \\ &= -x^3 + 2x \\ &= -(x^3 - 2x) \\ &= -g(x) \end{aligned}$$

so g is an odd function and its graph is symmetric with respect to the origin (Figure 3b, page 108).

(c)

$$\begin{aligned} h(-x) &= (-x)^2 + (-x) \\ &= x^2 - x \neq x^2 + x = h(x) \end{aligned}$$

Also,

$$-h(x) = -(x^2 + x) = -x^2 - x \neq x^2 - x = h(-x)$$

Thus, $h(-x) \neq h(x)$ and $-h(x) \neq h(-x)$, so h is neither even nor odd. Its graph is shown in Figure 3c.

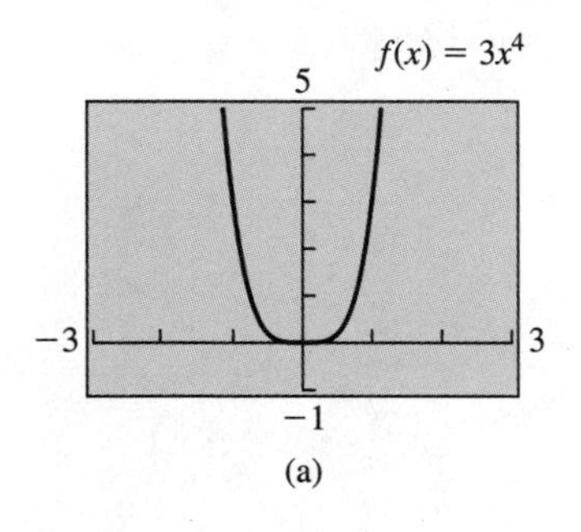

(a)

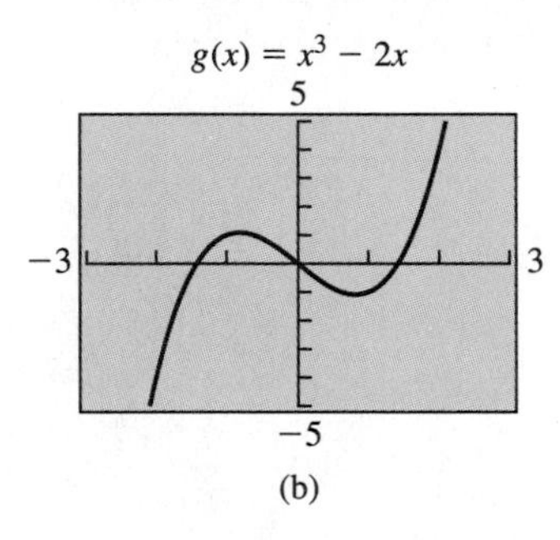

(b)

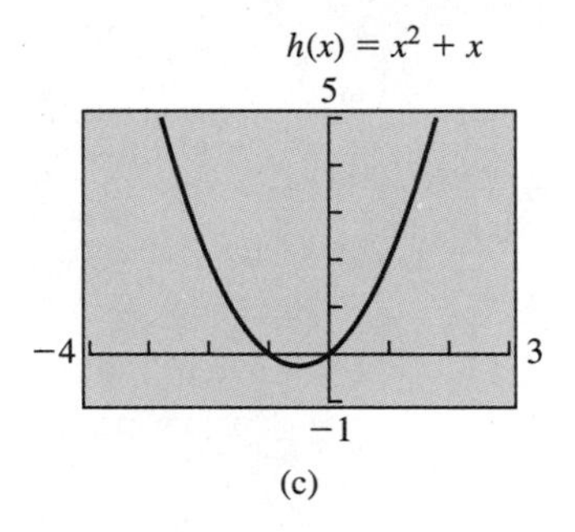

(c)

FIGURE 3

Recognizing Standard Graphs

Certain functions and their graphs play a special role in the study of functions. The recognition of these special functions and their graphs is important for two reasons:

1. They provide an easy transition from familiar ideas to new concepts.
2. Knowing their graphs, which we shall refer to as *standard graphs*, can greatly simplify sketching graphs of other functions with equations that are variations of their equations.

These special functions and their graphs are displayed in Figure 4. Their properties are listed in Table 1.

TABLE 1

Standard Graphs—Properties of Functions

Function	Domain	Range	Increases in Interval	Decreases in Interval	Symmetry with Respect To	Even or Odd
1. $f(x) = x$	$\mathbb{R}$	$\mathbb{R}$	$(-\infty, \infty)$	—	Origin	Odd
2. $f(x) = c$, a constant	$\mathbb{R}$	One number, c	Constant	Constant	y axis	Even
3. $f(x) = x^2$	$\mathbb{R}$	$[0, \infty)$	$[0, \infty)$	$(-\infty, 0]$	y axis	Even
4. $f(x) = \sqrt{x}$	$[0, \infty)$	$[0, \infty)$	$[0, \infty)$	—	—	—
5. $f(x) = \lvert x \rvert$	$\mathbb{R}$	$[0, \infty)$	$[0, \infty)$	$(-\infty, 0]$	y axis	Even
6. $f(x) = x^3$	$\mathbb{R}$	$\mathbb{R}$	$(-\infty, \infty)$	—	Origin	Odd
7. $f(x) = \sqrt[3]{x}$	$\mathbb{R}$	$\mathbb{R}$	$(-\infty, \infty)$	—	Origin	Odd
8. $f(x) = \sqrt{a^2 - x^2}$	Interval $[-a, a]$	Interval $[0, a]$	$[-a, 0]$	$[0, a]$	y axis	Even

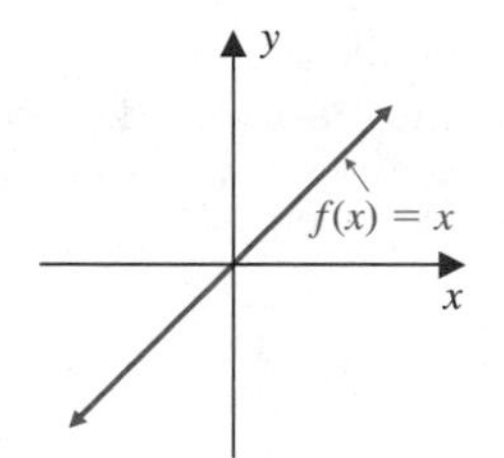

1. **Identity function**

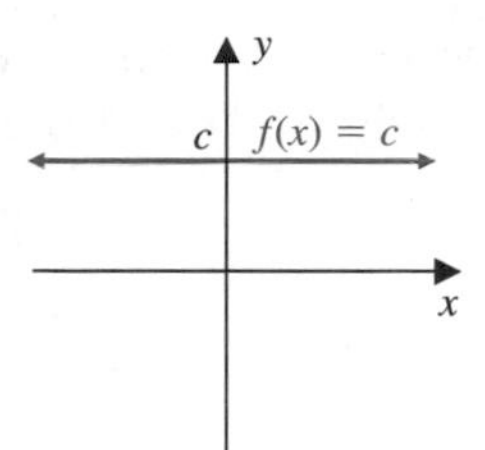

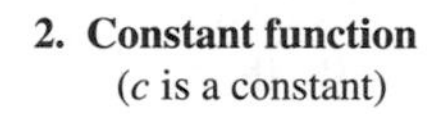

2. **Constant function**
(c is a constant)

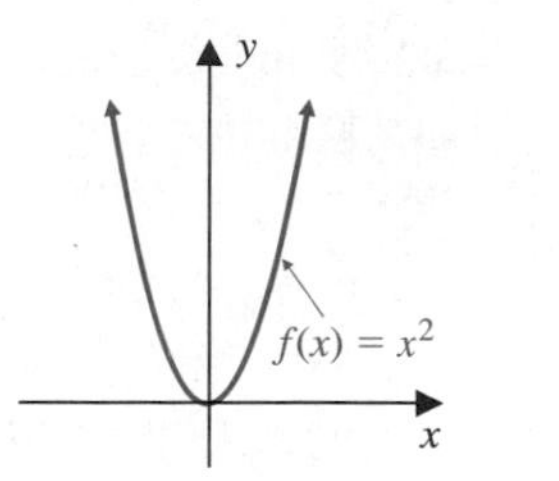

3. **Squaring function**

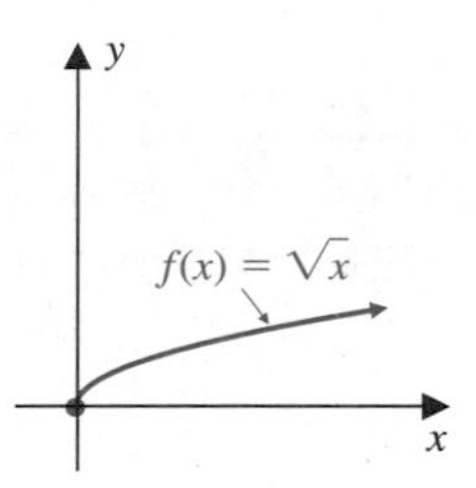

4. **Square root function**

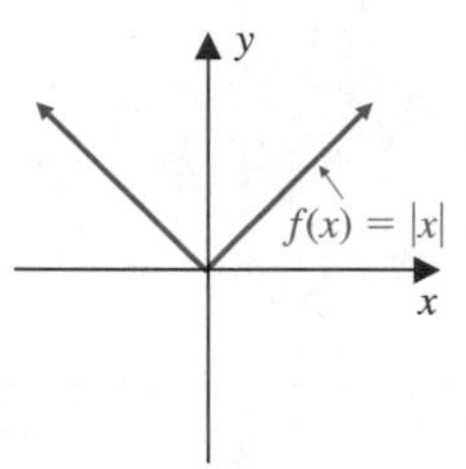

5. **Absolute value function**

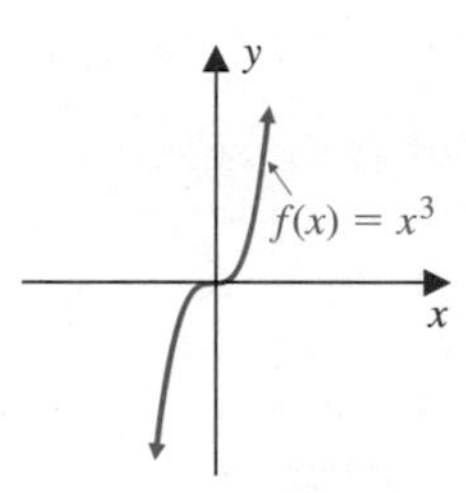

6. **Cubing function**

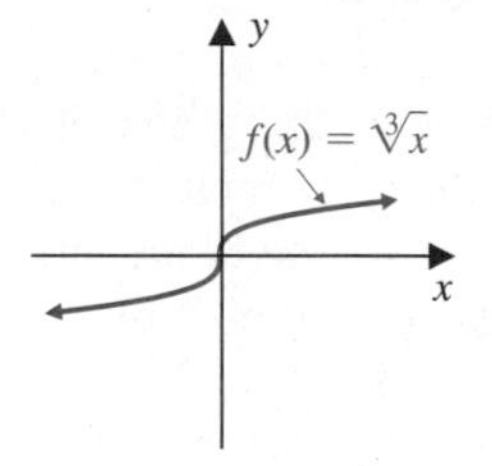

7. **Cube root function**

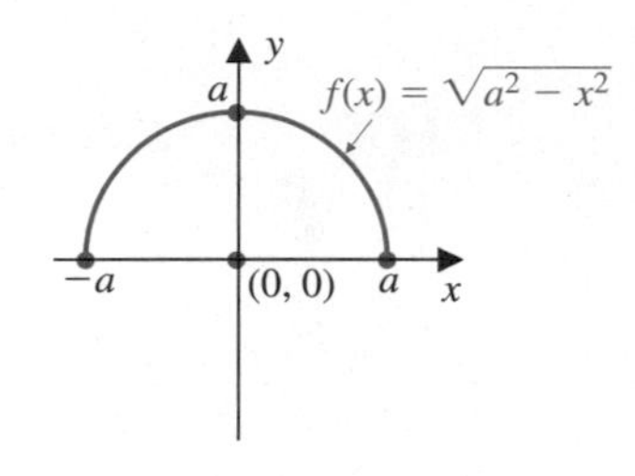

8. **Upper half-circle function**
(center: (0, 0); radius a)

FIGURE 4
Standard graphs of functions

EXAMPLE 3 Comparing standard graphs

Graph $f(x) = x^2$ and $g(x) = x^3$ on the same coordinate system. Then use the fact that the graphs intersect at points (0, 0) and (1, 1) to describe where the graph of g is below or above the graph of f.

Solution Figure 5 displays the graphs of $f(x) = x^2$ and $g(x) = x^3$. Upon examining the graphs we observe that the graph of $g(x) = x^3$ is below the graph of $f(x) = x^2$ over the intervals $(-\infty, 0)$ and $(0, 1)$, and is above the graph of f over the interval $(1, \infty)$. ■

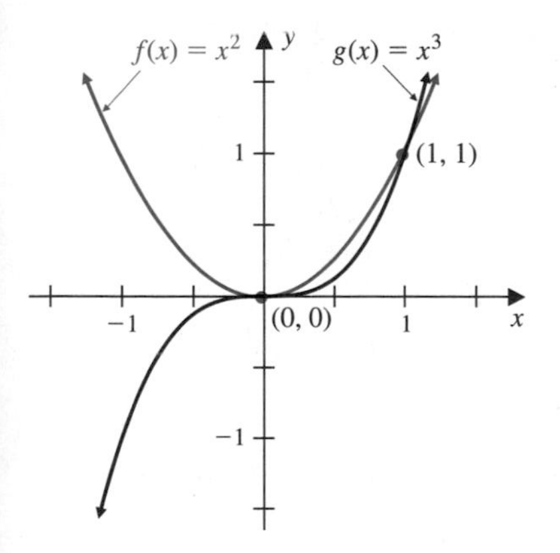

FIGURE 5

Graphing Piecewise-Defined Functions

Just as we have different rules governing our behavior at different times, a function is not always defined by only one rule. When a function is defined by using different equations on different intervals, we call it a **piecewise-defined function**. The graph of such a function is obtained by sketching the graphs of the equations with their restrictions on the same coordinate system.

Some graphing utilities have special features for graphing piecewise-defined functions.

EXAMPLE 4 **Using standard graphs to graph a function defined by two equations**

Use the appropriate standard graphs to sketch the graph of the piecewise-defined function f given by

$$f(x) = \begin{cases} x^2 & \text{if } x < 1 \\ x & \text{if } x \geq 1 \end{cases}$$

Also, identify the domain and range.

Solution If $x < 1$, then $f(x) = x^2$. This portion of the graph coincides with the standard graph of the squaring function $y = x^2$ for x in the interval $(-\infty, 1)$. For $x \geq 1$, the graph of f coincides with the standard graph of the identity function $y = x$ restricted to the interval $[1, \infty)$. Figure 6 displays the graph of the given function f. The domain of f is the set of real numbers, and the range consists of all numbers in the interval $[0, \infty)$.

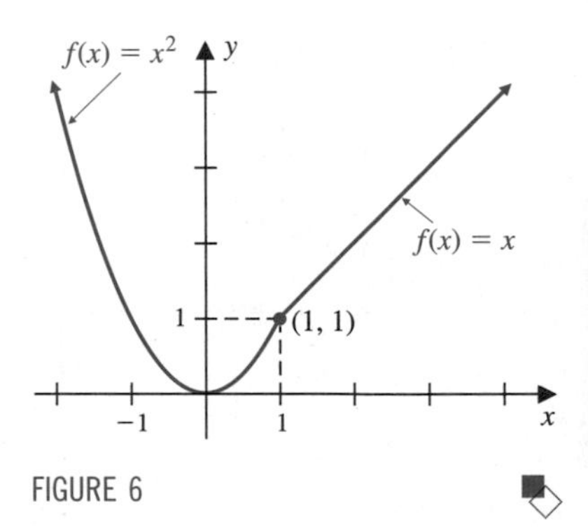

FIGURE 6

EXAMPLE 5 **Graphing a function defined by three equations**

Sketch the graph of the piecewise-defined function g given by

$$g(x) = \begin{cases} 3x + 1 & \text{if } x \leq 0 \\ 1 & \text{if } 0 < x < 9 \\ \sqrt{x} & \text{if } x \geq 9 \end{cases}$$

Determine the domain and range.

Solution The graph of g, which consists of parts of three graphs with equations $y = 3x + 1$, $y = 1$, and $y = \sqrt{x}$, is displayed in Figure 7.

From the graph we see that the domain includes all real numbers, and the range includes all values of y such that $y \leq 1$ or $y \geq 3$.

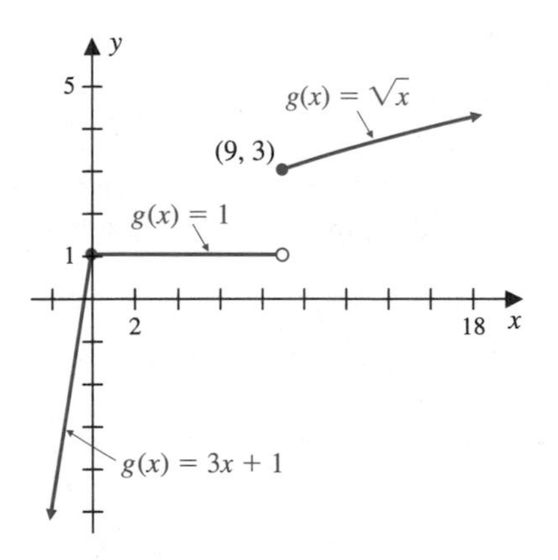

FIGURE 7

The graph in Figure 6 has no breaks or gaps in it. In this case, we say that the function f is **continuous** for all values of x. On the other hand, Figure 7 shows a graph that is "disconnected" at $x = 9$, so we say that the function is **discontinuous** at $x = 9$. Another example of a discontinuous function is the **greatest-integer function**, which is defined by the equation

$$f(x) = [\![x]\!]$$

where $[\![x]\!]$ is the greatest integer less than or equal to x (Problem 41). For instance:

$[\![4]\!] = 4$, since 4 is the greatest integer less than or equal to 4.

$[\![5\tfrac{1}{2}]\!] = 5$, since 5 is the greatest integer less than or equal to $5\tfrac{1}{2}$.

$[\![-\tfrac{5}{4}]\!] = [\![-1.25]\!] = -2$; $[\![0.7]\!] = 0$; and $[\![0]\!] = 0$.

Solving Applied Problems

Functions are used to model real-life situations, as illustrated in the following examples. The first example uses a function defined by a graph.

EXAMPLE 6 Using a graph as a model

The number n of deer in a forest t years after the beginning of a 10 year conservation study is shown in Figure 8.

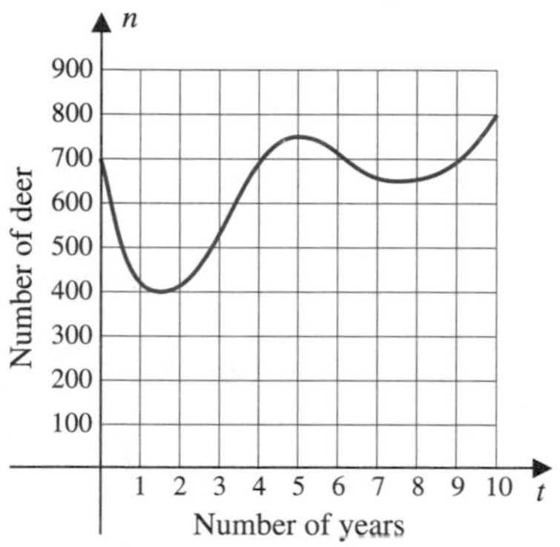

FIGURE 8

(a) Approximately what was the deer population after 1.5 years? After 5 years? After 9 years?
(b) Determine the time intervals over which the deer population decreased and increased.
(c) Estimate the lowest and highest populations and when they occurred.

Solution

(a) The graph contains the points (1.5, 400), (5, 750), and (9, 700), so after 1.5 years there were about 400 deer; after 5 years 750 deer; and after 9 years 700 deer.
(b) Upon examining the graph, we see that the function decreases over the intervals [0, 1.5] and [5, 7.5]. So the deer population declined over the first 1.5 years of the study, and then again between 5 and 7.5 years later. The function increases over the intervals [1.5, 5] and [7.5, 10]. This means that the deer population increased between 1.5 and 5 years after the study began and again later, between 7.5 and 10 years after the beginning of the study.
(c) The lowest population (smallest value of n) is 400 and it occurred when $t = 1.5$, that is, after 1.5 years. The highest population (largest value of n) is 800, and it occurred when $t = 10$, that is, after 10 years.

In the next example, we use a piecewise-defined function to model a situation.

EXAMPLE 7 Using a piecewise-defined function as a model

A youth soccer club decides to organize a 1 hour fund-raising ice skating party at the local rink. The rink charges a nonrefundable rental fee of \$150 per hour. The fee includes one attendant for the first 40 skaters. For each additional group of skaters up to and including 40, another attendant is required at a cost of \$30 per attendant. The rink also limits the number of skaters to 155 people, and the soccer club charges \$5 per ticket.

(a) Construct a piecewise-defined function that expresses the profit P (in dollars) as a function of the number of sold tickets t.
(b) What are the restrictions on t?
(c) Graph the function with the restrictions from part (b).
(d) Find the range and interpret it. Find the worst possible loss and the maximum possible profit.

Solution Table 2 summarizes the situation.

TABLE 2

Number of Sold Tickets (t)	Income *Dollars*	Cost *Dollars*	Profit (P) = Income − Cost
$0 \le t \le 40$	$5t$	150	$5t - 150$
$40 < t \le 80$	$5t$	$150 + 30 = 180$	$5t - 180$
$80 < t \le 120$	$5t$	$150 + 30 + 30 = 210$	$5t - 210$
$120 < t \le 155$	$5t$	$150 + 30 + 30 + 30 = 240$	$5t - 240$

(a) By using the information in Table 2, we are able to write P as a piecewise-defined function of t as follows:

$$P = \begin{cases} 5t - 150 & \text{if } 0 \le t \le 40 \\ 5t - 180 & \text{if } 40 < t \le 80 \\ 5t - 210 & \text{if } 80 < t \le 120 \\ 5t - 240 & \text{if } 120 < t \le 155 \end{cases} \qquad \text{where } t \text{ is an integer}$$

(b) Since the number of skaters is limited to 155, it follows that t is an integer such that $0 \le t \le 155$.

(c) Even though t technically represents integers, we'll graph the function in a Cartesian plane so that t represents all real numbers in the interval $[0, 155]$ (Figure 9).

(d) The range includes values of P such that $-150 \le P \le 535$. There is a loss if P is negative. If no tickets are sold ($t = 0$), the club would lose \$150, its worst possible loss. The maximum profit, or largest value of P, is \$535, which occurs if 155 tickets are sold.

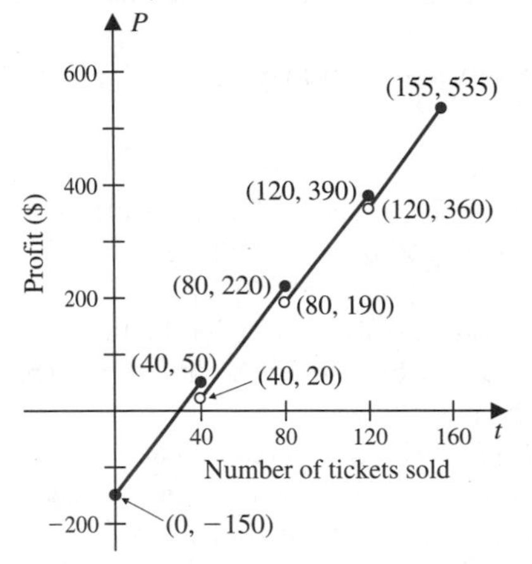

FIGURE 9

PROBLEM SET 2.2

Mastering the Concepts

In problems 1–6, use the given graph to find the intervals where the function is increasing, decreasing, or constant.

1.

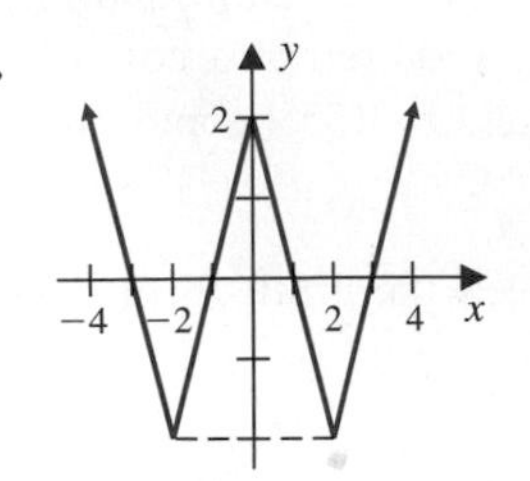

2.

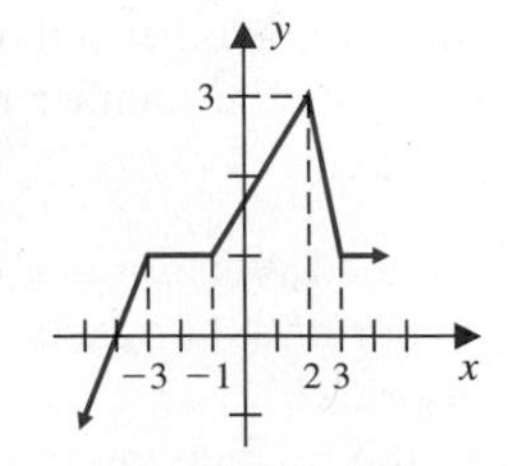

3.

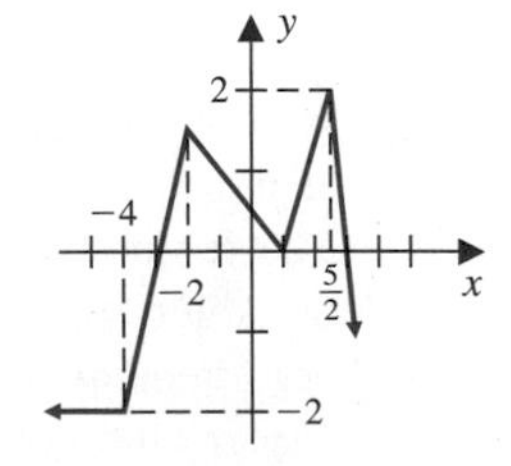

4.

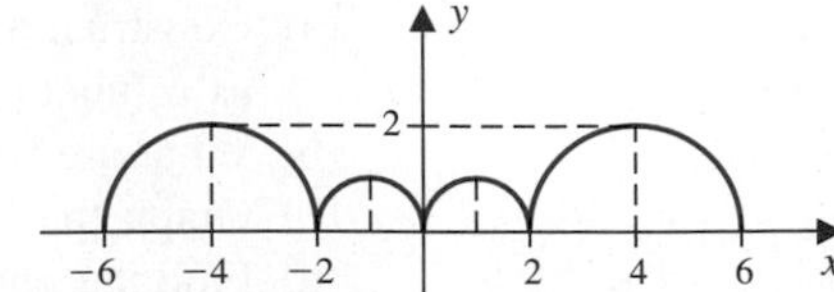

5.

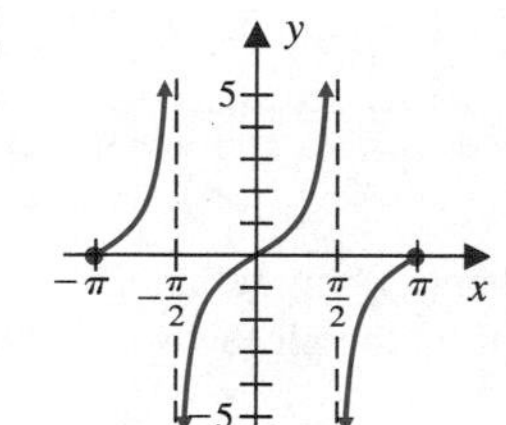

6.

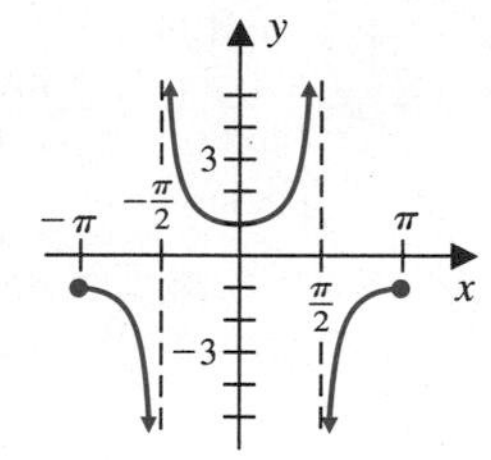

In problems 7–16:

(a) Determine without graphing whether each function is even, odd, or neither by using function notation.

GU **(b)** Demonstrate the validity of the answer in part (a) by using a graphing utility to graph the function.

7. $f(x) = 8x^2 + 3$

8. $g(x) = -4x^3 + x$

9. $h(x) = -x^4 + 3$

10. $F(x) = \sqrt{7x^4 + 3}$

11. $f(x) = 5x^3 - 2x$

12. $f(x) = 5x^2 + x^3$

13. $f(x) = -x^2 - |x|$

14. $G(x) = \dfrac{\sqrt{x^2 + 4}}{|x|}$

15. $f(x) = \dfrac{1}{x} + x^2$

16. $h(x) = \dfrac{5}{x^2 + 4}$

In problems 17–26, sketch the graph of each function. Use the graph to determine whether the function is even or odd, and to indicate the intervals where it is increasing, decreasing, or constant.

17. $f(x) = 3x$

18. $g(x) = -2x$

19. $h(x) = -4x + 5$

20. $f(x) = 3x + 2$

21. $F(x) = -5x^2 + 10$

22. $G(x) = -7x^2 + 1$

23. $F(x) = \sqrt{4 - x^2}$

24. $H(x) = -\sqrt{4 - x^2}$

25. $g(x) = -x^2$

26. $H(x) = x^2 - 1, x \geq 0$

In problems 27–30, graph both functions on the same coordinate system. Then use the points of intersection to describe where the graph of g is below or above the graph of f.

27. $f(x) = x^2$ and $g(x) = \sqrt{x}$ intersect at points (0, 0) and (1, 1).

28. $f(x) = x^3$ and $g(x) = \sqrt[3]{x}$ intersect at points (−1, −1), (0, 0), and (1, 1).

29. $f(x) = \sqrt[3]{x}$ and $g(x) = \sqrt{x}$ intersect at points (0, 0) and (1, 1).

30. $f(x) = x^2$ and $g(x) = |x|$ intersect at points (−1, 1), (0, 0), and (1, 1).

In problems 31–34, use standard graphs to sketch the graph of each piecewise-defined function. (Refer to Figure 4.)

31. $g(x) = \begin{cases} 2 & \text{if } x < 4 \\ \sqrt{x} & \text{if } x \geq 4 \end{cases}$

32. $h(x) = \begin{cases} x & \text{if } x \leq 0 \\ \sqrt{x} & \text{if } x > 0 \end{cases}$

33. $F(x) = \begin{cases} x^3 & \text{if } x \leq 0 \\ \sqrt{1 - x^2} & \text{if } 0 < x \leq 1 \end{cases}$

34. $f(x) = \begin{cases} \sqrt[3]{x} & \text{if } x < -1 \\ |x| & \text{if } x \geq -1 \end{cases}$

In problems 35–40, sketch the graph of each piecewise-defined function. Determine the domain and range. Identify each point of discontinuity.

35. $f(x) = \begin{cases} x + 3 & \text{if } x \leq 1 \\ 4x^2 & \text{if } x > 1 \end{cases}$

36. $g(x) = \begin{cases} 2x^2 + 1 & \text{if } x < 3 \\ 10 - x & \text{if } x \geq 3 \end{cases}$

37. $h(x) = \begin{cases} 3x + 1 & \text{if } x < -2 \\ 0 & \text{if } -2 \leq x \leq 2 \\ -2x + 3 & \text{if } x > 2 \end{cases}$

38. $f(x) = \begin{cases} -x & \text{if } x < 0 \\ 4 - 2x & \text{if } 0 \leq x < 3 \\ 2x + 1 & \text{if } x \geq 3 \end{cases}$

39. $f(x) = \begin{cases} -2 & \text{if } x < 1 \\ 3x & \text{if } 1 \leq x < 3 \\ 2 - x^2 & \text{if } x \geq 3 \end{cases}$

40. $h(x) = \begin{cases} 1 - x & \text{if } 0 < x \leq 1 \\ 3 + x & \text{if } 1 < x \leq 2 \\ -1 & \text{if } 2 < x \leq 3 \end{cases}$

In problems 41–44, graph the function and determine its range.

41. $f(x) = [\![x]\!]$

42. $f(x) = [\![3x]\!]$

43. $f(x) = [\![x/2]\!]$

44. $f(x) = [\![x/3]\!]$

Applying the Concepts

45. Cost Trend: The function defined by the graph in Figure 10 (at the top of the next page) indicates the average cost C (in thousands of dollars) for a manufacturer to produce x (in hundreds) items of a product up to 1000 items.

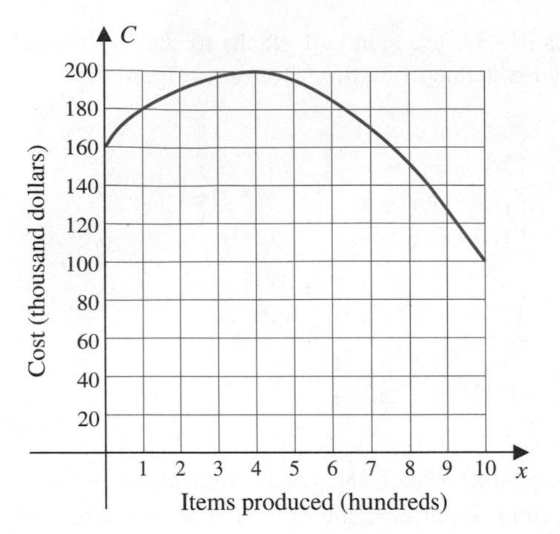

FIGURE 10

(a) Find the average cost if the number of items produced is 100, 400, 700, 800, 1000.

(b) Find the number of items that would yield an average cost of $140,000.

(c) Find the intervals where the function is increasing and decreasing, and interpret this behavior in terms of average cost and items produced.

(d) Find the highest and lowest average costs and when they occur.

46. Water Level: Figure 11 indicates the height h (in feet) of the tide above mean sea level in a harbor at time t over a 24 hour period.

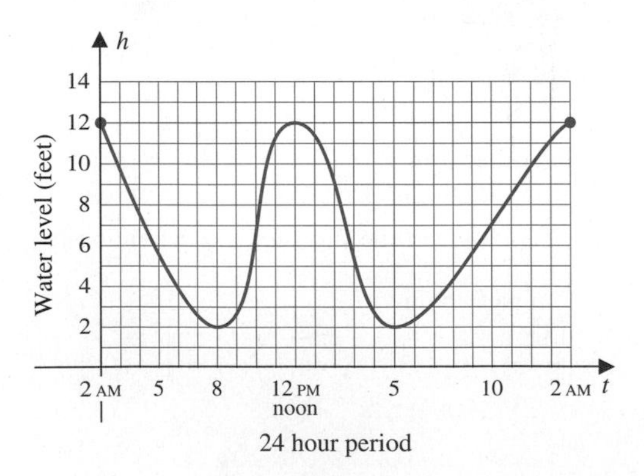

FIGURE 11

(a) Find the height at 5 AM, noon, 5 PM, 2 AM.

(b) Approximate the times when the water level above mean sea level will be 10 feet.

(c) Determine the time intervals when the water level is rising and falling.

(d) Find the highest and lowest levels and when they occur.

47. Water Charges: The monthly charge C (in dollars) for water in a small town is given by the piecewise-defined function

$$C(x) = \begin{cases} 18 & \text{if } 0 \le x \le 20 \\ 18 + 0.1(x - 20) & \text{if } x > 20 \end{cases}$$

where x is in hundreds of gallons.

(a) Sketch the graph of C.

(b) What is the monthly fixed charge?

(c) Suppose a proposed rate change would fix the monthly charge at $12 for the first 1500 gallons and charge an additional 25¢ for each hundred gallons beyond 1500. Write a piecewise-defined function that expresses the monthly charge K (in dollars) in terms of x (in hundreds of gallons of water) for the proposed rate change.

(d) Graph the function from part (c).

(e) Compare the charges under each rate structure.

48. Telephone Charges: Suppose the cost of a telephone call between two cities is $0.83 for the first minute and $0.75 for each additional minute (or portion of a minute). So the cost C (in dollars) of a call that lasts t minutes is given by the equation

$$C = 0.83 + 0.75[\![1 - t]\!]$$

(a) Sketch the graph of this function.

(b) What is the fixed charge?

(c) A proposed rate change would charge $0.65 up to 1 minute and $0.85 for each additional minute (or portion of a minute). Under this proposal, use the greatest-integer function to express the cost K (in dollars) as a function of time t, the number of minutes the call lasts.

(d) Graph the function from part (c).

(e) Compare the costs of a call under each rate structure.

49. Car Rental: A car rental agency offers special rates for a 4 day convention. It charges C dollars for renting a car for N days as specified in Table 3.

TABLE 3

Number of days (N)	Charge (C) *Dollars*
1 day, or part of a day	20
2 days, or 1 day and part of a day	38
3 days, or 2 days and part of a day	56
4 days, or 3 days and part of a day	74

(a) Determine a piecewise-defined function that expresses C in terms of N.
(b) What are the restrictions on N?
(c) Graph the function in part (a) with the restrictions from part (b).
(d) Suppose that a competing car agency offers a flat rate of \$19.50 per day or any part of a day. Express the cost K (in dollars) as a piecewise-defined function of N, where $0 < N \leq 4$, and graph it.
(e) Compare the competing rate structures.

50. Stockbroker's Fee: Suppose that the fee C (in dollars) charged by a discount stockbroker for executing a trade of N shares is given in Table 4.

TABLE 4

Number of Shares (N)	Fee (C) *Dollars*
$0 < N < 100$	$25 + 0.10$ per share traded
$100 \leq N < 500$	$30 + 0.05$ per share traded
$N \geq 500$	$45 + 0.02$ per share traded

(a) Find the fee if the broker executes a trade of 50 shares, 100 shares, 275 shares, 923 shares.
(b) Determine a piecewise-defined function that expresses C in terms of N.
(c) What are the restrictions on N?
(d) Graph the function in part (b) with the restrictions from part (c).
(e) What would the graph look like if the fee structure in Table 4 were changed so that a flat fee of \$55 is charged for trades involving at least 100 shares? Compare this graph to the graph from part (d). Compare the two fee structures.

51. Post Office Rates: The 1995 U.S. postal rates R (in dollars) for mailing a first class item weighing up to and including 6 ounces are summarized in Table 5.

TABLE 5

Weight w *Ounces*	Rate R *Dollars*
$0 < w \leq 1$	0.32
$1 < w \leq 2$	0.55
$2 < w \leq 3$	0.78
$3 < w \leq 4$	1.01
$4 < w \leq 5$	1.24
$5 < w \leq 6$	1.47

(a) Use the greatest-integer function to determine a function that expresses the rate R in terms of w.
(b) What are the restrictions on w?
(c) Graph the function in part (a) with the restrictions from part (b).
(d) If this rate structure applies to items that weigh up to 11 ounces, then what would be the rate for a letter weighing 7 ounces? 9.5 ounces? 10 ounces?

52. Parking Charges: A public parking structure with an 8 hour limit charges its customers \$2.50 for the first hour (or part of the hour) and \$1.50 for each additional hour (or part of the hour) up to 8 hours.
(a) Use the greatest-integer function to determine a function that expresses the cost C (in dollars) of parking as a function of t, the number of hours parked.
(b) Sketch the graph of the function with the appropriate restrictions on t.
(c) A rate change is made that charges the customer \$3.25 for the first hour and \$1.25 for each additional hour or part of an hour up to 8 hours. Under this change, use the greatest-integer function to determine a function that expresses the cost K (in dollars) in terms of t, the number of hours parked, where $0 < t \leq 8$.
(d) Graph the function in part (c).
(e) Compare the parking charges under each rate structure.

Developing and Extending the Concepts

53. **(a)** If f is an odd function, does the graph of f necessarily contain the origin? Support your assertion.

(b) Use a graphing utility to sketch the graph of the function $f(x) = 1/x$.
(c) Is your assertion in part (a) consistent with the graph in part (b)?

54. **(a)** If g is an even function, does the graph of g necessarily intersect the y axis? Support your assertion.

GU **(b)** Use a graphing utility to sketch the graph of the function $g(x) = 1/x^2$.

(c) Is your assertion in part (a) consistent with the graph in part (b)?

In problems 55–60, for each given pair of equations, the second equation has been formed from the first equation by switching the x and y variables.

(a) Graph both equations on the same coordinate system.

(b) Does the second equation represent a function, where x is the domain variable? If so, compare its domain and range to the domain and range of the first equation, which is a function of x.

55. $y = 2;\ x = 2$

56. $y = x^2;\ x = y^2$

57. $y = \sqrt{x};\ x = \sqrt{y}$

58. $y = x^3;\ x = y^3$

59. $y = |x|;\ x = |y|$

60. $y = \sqrt{9 - x^2};\ x = \sqrt{9 - y^2}$

61. Given the piecewise-defined equation

$$y = \begin{cases} x & \text{if } x \le 5 \\ x - 1 & \text{if } x > k \end{cases}$$

(a) Does the equation represent a function if $k = 1$? Explain.

(b) Does the equation represent a function if $k = 6$? Explain.

(c) What conditions must be placed on the value of k in order to have a function?

62. **(a)** Assume the graph of $f(x) = x^n + x^2$ is symmetric with respect to the y axis. What condition must n, a positive integer, satisfy? Explain.

(b) Is it possible for the graph of $g(x) = x^n + x^3$, n a positive integer, to be symmetric with respect to the y axis? Explain.

63. **(a)** Is it possible to have a function with a graph that is symmetric with respect to the x axis? Explain.

(b) Is it possible for a function to be both even and odd? Explain.

64. Rewrite each function as a piecewise-defined function that does not involve absolute value; then sketch the graph. Find any points of discontinuity.

(a) $f(x) = \dfrac{|x|}{x}$ **(b)** $g(x) = 4|x| - 1$

(c) $h(x) = 2x + \dfrac{|x - 3|}{x - 3}$

GU In problems 65 and 66, use a graphing utility to graph each piecewise-defined function. (Some graphing utilities have built-in programs to generate these types of graphs.)

65. **(a)** $f(x) = \begin{cases} 2x + 7 & \text{if } x \ge -2 \\ 1 - x & \text{if } x < -2 \end{cases}$

(b) $g(x) = \begin{cases} x^2 + 1 & \text{if } x < 1 \\ x + 1 & \text{if } x \ge 1 \end{cases}$

66. **(a)** $f(x) = \begin{cases} -x^2 + 1 & \text{if } x \le 1 \\ x - 2 & \text{if } x > 1 \end{cases}$

(b) $g(x) = \begin{cases} x^3 & \text{if } x \le 1 \\ 3 - x & \text{if } x > 1 \end{cases}$

OBJECTIVES

1. Graph by vertical shifting.
2. Graph by horizontal shifting.
3. Graph by reflecting.
4. Graph by vertical scaling.
5. Graph by using more than one transformation.

2.3 Transformations of Graphs

In this section we present graphing techniques, called **transformations**, where we geometrically transform standard graphs to obtain graphs of other functions. Graphing displays produced by a graphing utility are used to help us discover the concepts presented here.

Graphing by Vertical Shifting

We begin by using a graphing utility to graph $f(x) = x^2$, $g(x) = x^2 + 2$, and $h(x) = x^2 - 1$ on the same coordinate system. The results are displayed in Figure 1. By examining the geometric features of the three graphs, we make the following observations:

$g(x) = x^2 + 2$
$h(x) = x^2 - 1$
$f(x) = x^2$
$(a, a^2 + 2)$
(a, a^2)
$(a, a^2 - 1)$
8
−5
5
−3

FIGURE 1

1. The three graphs have the same shape but different locations.

2. 

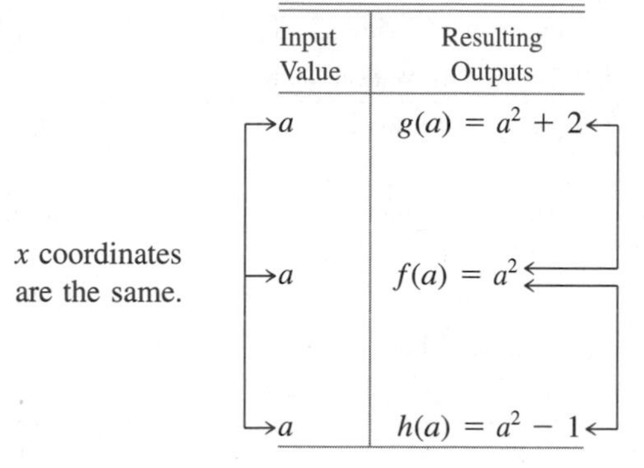

Thus, we conclude that the graph of $g(x) = x^2 + 2$ can be obtained by *vertically shifting* or *translating* the graph of $f(x) = x^2$ upward 2 units. Similarly, the graph of $h(x) = x^2 - 1$ can be obtained by vertically shifting the graph of $f(x) = x^2$ downward 1 unit.

EXAMPLE 1 **Vertically shifting a graph**

Use the graph of $f(x) = |x|$ to sketch the graph of each function.

(a) $F(x) = |x| + 2$ (b) $G(x) = |x| - 3$

Solution

(a) For each value of x, the value of $F(x) = |x| + 2$ is exactly 2 units more than $f(x) = |x|$. So the graph of F can be obtained by shifting the graph of f upward 2 units (Figure 2a).

(b) Since for every value of x, $G(x) = |x| - 3$ is exactly 3 units less than $f(x) = |x|$, the graph of G can be obtained by shifting the graph of f downward 3 units (Figure 2b).

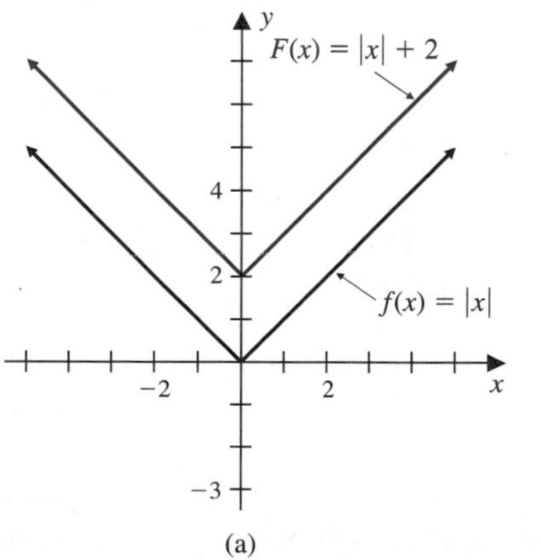

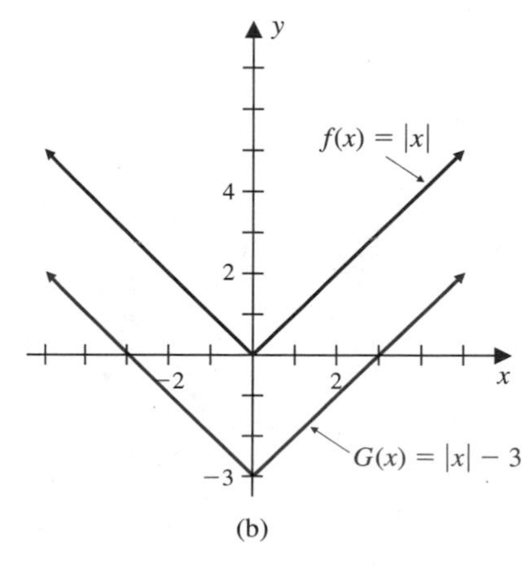

FIGURE 2

Our illustrations can be generalized to obtain the results given in Table 1.

TABLE 1
Graphing by Vertical Shifting

Operation on $y = f(x)$	New Function Equation	Geometric Effect on Graph of $y = f(x)$	Illustration
Add a positive constant k to $f(x)$.	$y = f(x) + k$	Shifts the graph of f vertically k units upward.	$(a, b + k)$ $y = f(x) + k$ y k (a, b) a $y = f(x)$ x
Subtract a positive constant k from $f(x)$.	$y = f(x) - k$	Shifts the graph of f vertically k units downward.	y $y = f(x)$ (a, b) a k x $(a, b - k)$ $y = f(x) - k$

Graphing by Horizontal Shifting

We can also sketch graphs by shifting or translating known graphs horizontally. We begin by comparing the graphs of $f(x) = x^2$ and $g(x) = (x - 1)^2$. For convenience we obtain these graphs by using a graphing utility (Figure 3). Once again, the graphs lead us to important observations:

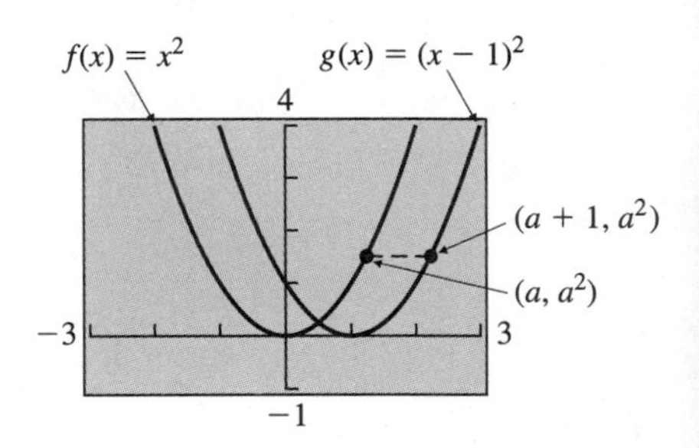

FIGURE 3

1. The shapes of the graphs are the same, but their locations are different.

2. For each point on the graph of f, there is a point on the graph of g 1 unit to the right.

Input Value	Resulting Output
a	$f(a) = a^2$
$a + 1$	$g(a + 1) = [(a + 1) - 1]^2 = a^2$

Same y coordinates

Thus, the graph of $g(x) = (x - 1)^2$ can be obtained by *horizontally shifting* or *translating* the graph of $f(x) = x^2$ to the right 1 unit. This illustration leads to the general results given in Table 2.

TABLE 2
Graphing by Horizontal Shifting

Operation on $y = f(x)$	New Function Equation	Geometric Effect on Graph of $y = f(x)$	Illustration
Add a positive constant h to x.	$y = f(x + h)$	Shifts the graph of f horizontally h units to the left.	
Subtract a positive constant h from x.	$y = f(x - h)$	Shifts the graph of f horizontally h units to the right.	

EXAMPLE 2 Horizontally shifting a graph

Use the graph of $y = \sqrt{x}$ to sketch the graph of each function.

(a) $f(x) = \sqrt{x + 1}$ (b) $g(x) = \sqrt{x - 2}$

Solution

(a) The graph of $f(x) = \sqrt{x + 1}$ can be obtained by shifting the graph of $y = \sqrt{x}$ to the left 1 unit (Figure 4a).

(b) The graph of $g(x) = \sqrt{x - 2}$ can be obtained by shifting the graph of $y = \sqrt{x}$ to the right 2 units (Figure 4b).

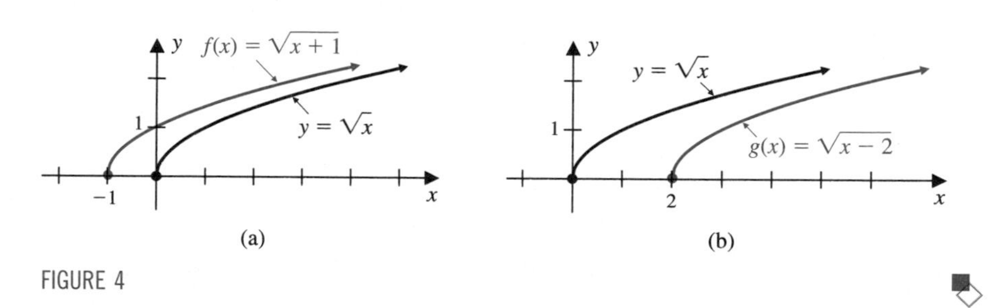

FIGURE 4

Graphing by Reflecting

Another type of transformation is called a *reflection*. Let us consider the graphs of $f(x) = \sqrt{x}$ and $g(x) = -\sqrt{x}$ (Figure 5, page 120). The geometric features of these graphs suggest that they are symmetric about the x axis. We can formally prove this symmetry as follows:

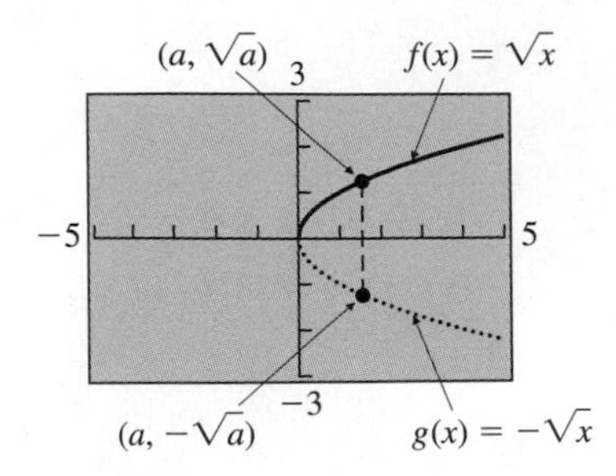

FIGURE 5

Input Value	Resulting Output
a	$f(a) = \sqrt{a}$
a	$g(a) = -\sqrt{a}$

Same x coordinates

y coordinates have opposite signs, so the points on the graph are symmetric about the x axis.

Thus, the graph of $g(x) = -\sqrt{x}$ can be obtained by reflecting the graph of $f(x) = \sqrt{x}$ about the x axis. Similarly, the graphs of $f(x) = \sqrt{x}$ and $h(x) = \sqrt{-x}$ shown in Figure 6 suggest that they are symmetric about the y axis. We prove this symmetry as follows:

FIGURE 6

Input Value	Resulting Output
a	$f(a) = \sqrt{a}$
$-a$	$h(-a) = \sqrt{-(-a)} = \sqrt{a}$

x coordinates have opposite signs.

Same y coordinates

Thus, the graph of $h(x) = \sqrt{-x}$ can be obtained by reflecting the graph of $f(x) = \sqrt{x}$ about the y axis. These observations lead to the general results in Table 3.

TABLE 3
Graphing by Reflecting

Operation on $y = f(x)$	New Function Equation	Geometric Effect on Graph of $y = f(x)$	Illustration
Multiply $f(x)$ by -1.	$y = -f(x)$	Reflects the graph of f about the x axis.	y, $y = f(x)$, (a, b), b, a, x, $-b$, $(a, -b)$, $y = -f(x)$
Replace x by $-x$.	$y = f(-x)$	Reflects the graph of f about the y axis.	$y = f(-x)$, y, $y = f(x)$, $(-a, b)$, b, (a, b), $-a$, a, x

EXAMPLE 3 **Reflecting a graph about the x axis**
Use the graph of $y = x^2$ to sketch the graph of $f(x) = -x^2$.

Solution The graph of $f(x) = -x^2$ is obtained by reflecting the graph of $y = x^2$ about the x axis (Figure 7).

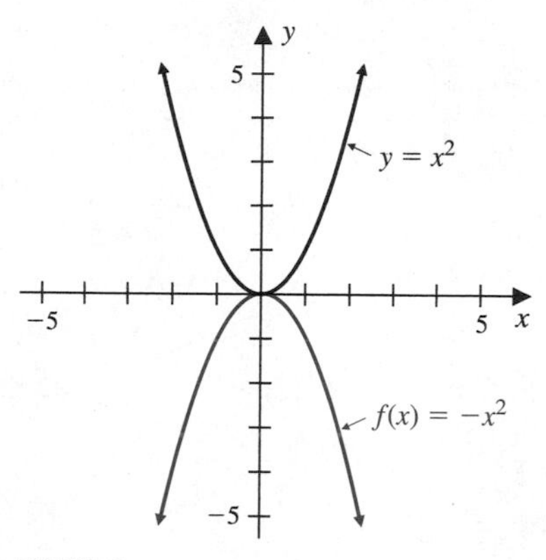

FIGURE 7

Graphing by Vertical Scaling

Translations and reflections change only the position of the graph, not its shape. Now we consider a transformation, called *vertical scaling*, that actually changes the shape of a graph. We begin by using a graphing utility to graph $f(x) = x^2$, $g(x) = 2x^2$, and $h(x) = \frac{1}{2}x^2$ in the same viewing window (Figure 8). Upon comparing the graphs we make the following observations:

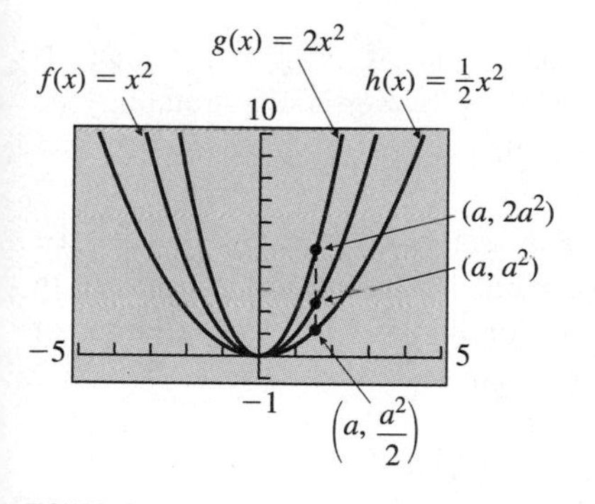

FIGURE 8

1. In comparison to the graph of f, the graph of g appears to be vertically stretched whereas the graph of h appears to be vertically compressed or flattened.

2.

Input Value	Resulting Outputs
a	$g(a) = 2a^2$
a	$f(a) = a^2$
a	$h(a) = \frac{1}{2}a^2$

Same x coordinates

y coordinates of graph of g are double those of graph of f.

y coordinates of graph of h are half those of graph of f.

Thus, we conclude that the graph of $g(x) = 2x^2$ can be obtained by vertically stretching the graph of $f(x) = x^2$ by a factor of 2. Similarly, the graph of $h(x) = \frac{1}{2}x^2$ can be obtained by vertically compressing or flattening the graph of f by a factor of $\frac{1}{2}$.

Table 4, at the top of the next page, generalizes these results.

TABLE 4

Graphing by Vertical Stretching or Compressing

Operation on $y = f(x)$	New Function Equation	Geometric Effect on Graph of $y = f(x)$	Illustration
Multiply $f(x)$ by a, where $a > 1$.	$y = af(x)$	Vertically stretches the graph of f by a factor of a.	y; $y = af(x)$; (c, ad); $y = f(x)$; (c, d); c; x
Multiply $f(x)$ by a, where $0 < a < 1$.	$y = af(x)$	Vertically compresses the graph of f by a factor of a.	y; $y = f(x)$; (c, d); $y = af(x)$; (c, ad); c; x

EXAMPLE 4 Vertically scaling a graph

Use the graph of $y = x^3$ to sketch the graph of each function.

(a) $g(x) = 5x^3$ (b) $h(x) = \frac{1}{5}x^3$

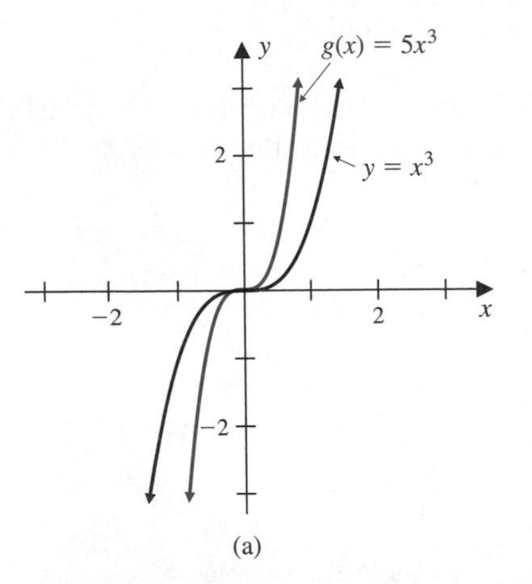

(a)

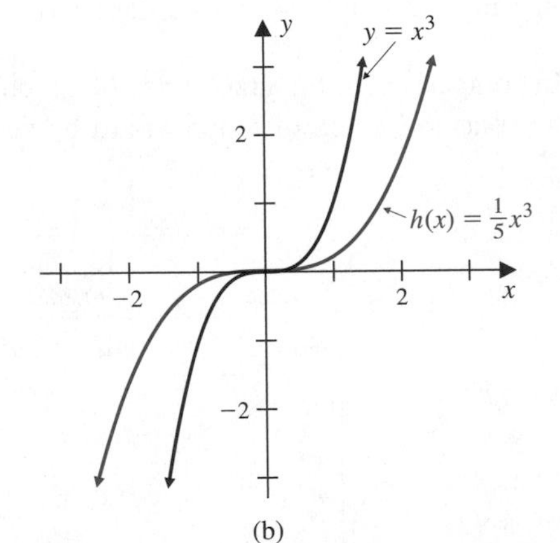

(b)

FIGURE 9

Solution

(a) The graph of $g(x) = 5x^3$ can be obtained by vertically stretching the graph of $y = x^3$ by a factor of 5 (Figure 9a).

(b) The graph of $h(x) = \frac{1}{5}x^3$ can be obtained by vertically compressing the graph of $y = x^3$ by a factor of $\frac{1}{5}$ (Figure 9b).

Graphing by Using More Than One Transformation

When combining more than one transformation in the same graph, the order in which these transformations are performed is important. Generally, the graph of the function

$$F(x) = af(x - h) + k$$

can be sketched from the graph of $y = f(x)$ by following the sequence of transformations in the order specified in Table 5.

TABLE 5

Sketching the Graph of $F(x) = af(x - h) + k$ from the Graph of $y = f(x)$

Order of Transformations	Equation Change	Geometric Change
Step 1. Vertically stretch or compress by a factor of $\lvert a \rvert$.	$y = af(x)$	Stretch if $\lvert a \rvert > 1$; compress if $0 < \lvert a \rvert < 1$.
Step 2. Reflect about x axis if $a < 0$.	$y = af(x)$	Reflect about x axis.
Step 3. Horizontally shift.	$y = af(x - h)$	Shift to the right if $h > 0$, to the left if $h < 0$.
Step 4. Vertically shift.	$y = af(x - h) + k$	Shift up if $k > 0$, down if $k < 0$.

EXAMPLE 5 **Using transformations to sketch a graph**

Use transformations of the graph $y = \lvert x \rvert$ to sketch the graph of $f(x) = -2\lvert x + 1 \rvert + 3$.

Solution We begin with the graph of $y = \lvert x \rvert$ and then proceed by following the order of transformations given in Table 5.

Step 1. Vertically stretch the graph of $y = \lvert x \rvert$ by a factor of 2 to obtain the graph of $y_1 = 2\lvert x \rvert$ (Figure 10a).

Step 2. Reflect the graph of $y_1 = 2\lvert x \rvert$ about the x axis to graph $y_2 = -2\lvert x \rvert$ (Figure 10b).

Step 3. Horizontally shift the graph of $y_2 = -2\lvert x \rvert$ to the left 1 unit to get the graph of $y_3 = -2\lvert x + 1 \rvert$ (Figure 10c).

Step 4. Vertically shift the graph of $y_3 = -2\lvert x + 1 \rvert$ up 3 units to obtain the desired graph of $f(x) = -2\lvert x + 1 \rvert + 3$ (Figure 10d).

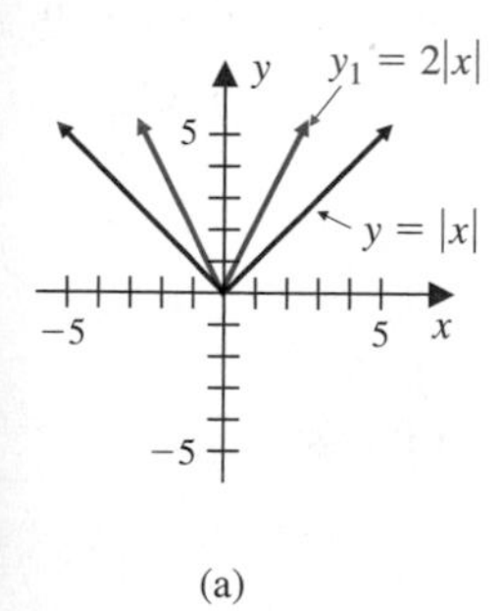

(a)

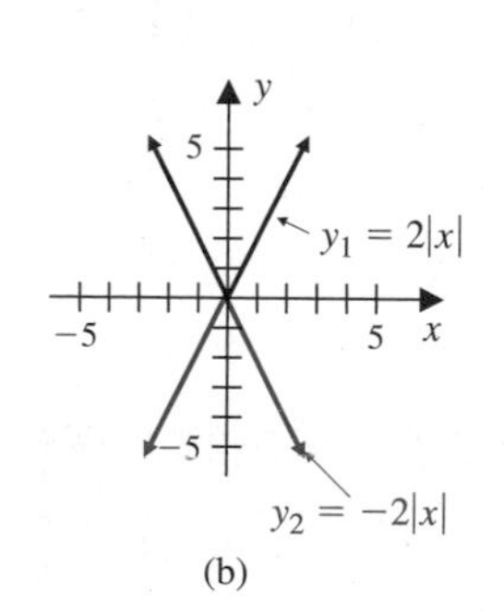

(b)

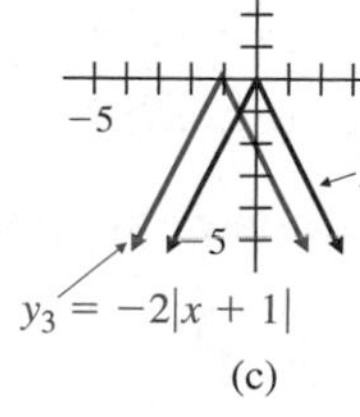

(c)

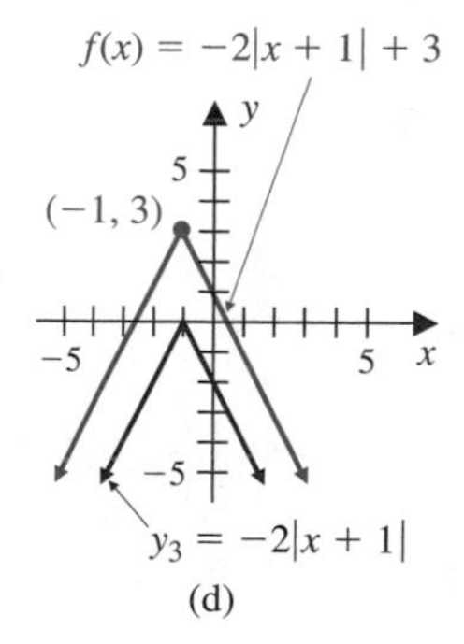

(d)

FIGURE 10

The graph of $y = f(-x)$ can be obtained by reflecting the graph of $y = f(x)$ about the y axis. This reflection is carried out in place of or in addition to step 2 in Table 5.

EXAMPLE 6 **Reflecting a graph about the y axis**

Use transformations of the graph of $y = \sqrt{x}$ to sketch a graph of $f(x) = \sqrt{2 - x}$.

Solution First, we rewrite the function as $f(x) = \sqrt{-(x - 2)}$. Next, we follow the order of transformations given in Table 5.

If we follow a different order of transformations, we may get an erroneous result (see Problem 57).

Step 1. There is no vertical stretching or compressing.

Step 2. Reflect the graph of $y = \sqrt{x}$ about the y axis to graph $y_1 = \sqrt{-x}$ (Figure 11a).

Step 3. Shift the graph of $y_1 = \sqrt{-x}$ to the right 2 units to obtain the desired graph of $f(x) = \sqrt{-(x - 2)}$ (Figure 11b).

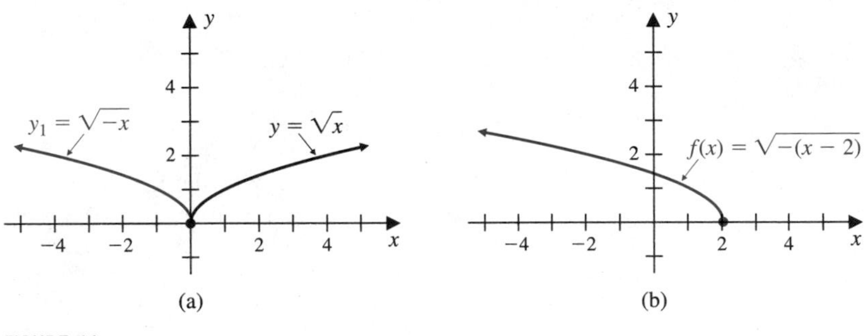

FIGURE 11

The next example illustrates how to start with a standard graph and follow a given sequence of transformations to construct another graph and determine its equation.

EXAMPLE 7 **Constructing the function equation from given transformations**

Determine the function equation for F if the graph of F is obtained from the graph of $f(x) = x^3$ as follows:

First, the graph of f is reflected about the x axis.

The resulting graph is shifted 4 units to the right.

Finally, the latter graph is shifted 3 units down.

Solution Construction of the equation for F is developed as follows:

Start

$$f(x) = x^3 \xrightarrow{\text{Reflect about } x \text{ axis}} y_1 = -x^3 \xrightarrow{\text{Shift 4 units right}} y_2 = -(x - 4)^3$$

$$\xrightarrow{\text{Shift 3 units down}} F(x) = -(x - 4)^3 - 3$$

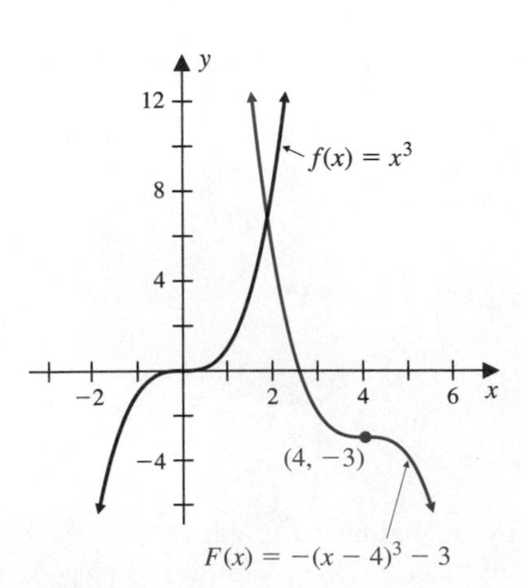

FIGURE 12

The graphs of f and F are shown in Figure 12.

PROBLEM SET 2.3

Mastering the Concepts

In problems 1–6, use transformations to sketch the graphs of f for the given values of each constant on the same coordinate system.

1. $f(x) = \sqrt{x} + k$, $k = -2, 1, 2$
2. $f(x) = \sqrt{1 - x^2} + k$, $k = -2, 1, 2$
3. $f(x) = (x + h)^3$, $h = -2, 1, 2$
4. $f(x) = \sqrt[3]{x + h}$, $h = -2, 1, 2$
5. $f(x) = a|x|$, $a = -3, -1, -\frac{1}{2}, \frac{1}{2}, 3$
6. $f(x) = ax^3$, $a = -3, -1, -\frac{1}{2}, \frac{1}{2}, 3$

In problems 7–12, use transformations to explain how the graph of f is related to the given function, and sketch the graph of f.

7. $y = x^2$; $f(x) = 2x^2 + 3$
8. $y = \sqrt{x}$; $f(x) = 3\sqrt{x} - 2$
9. $y = |x|$; $f(x) = 3|x - 1| - 2$
10. $y = x^3$; $f(x) = -(x + 1)^3 + 2$
11. $y = \sqrt{x}$; $f(x) = -2\sqrt{x - 1} + 3$
12. $y = \sqrt[3]{x}$; $f(x) = 2\sqrt[3]{x - 1} - 3$

In problems 13 and 14, use the given graph of f along with transformations to graph each of the following functions:

(a) $y = 4f(x)$
(b) $y = f(x) - 2$
(c) $y = f(x - 3)$
(d) $y = -\frac{1}{2}f(x) + 3$
(e) $y = 2f(x + 1) - 3$

13.

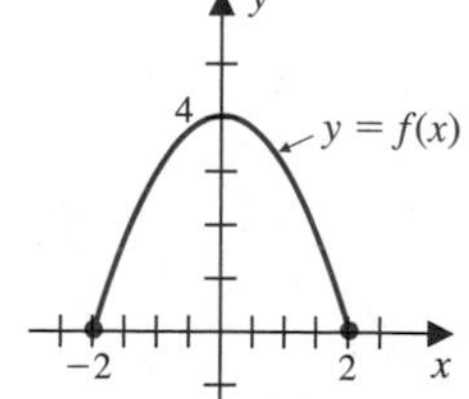

14.

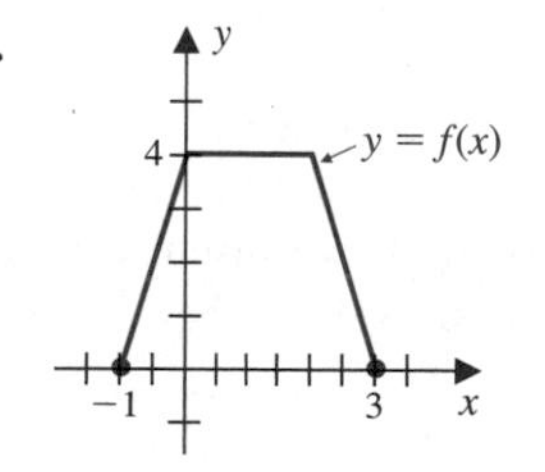

In problems 15 and 16, the graph of a function f is given along with its equation. The graphs of G, T, and S, obtained by transformations of the graph of f, are shown on the same coordinate system. Find equations for G, T, and S.

15.

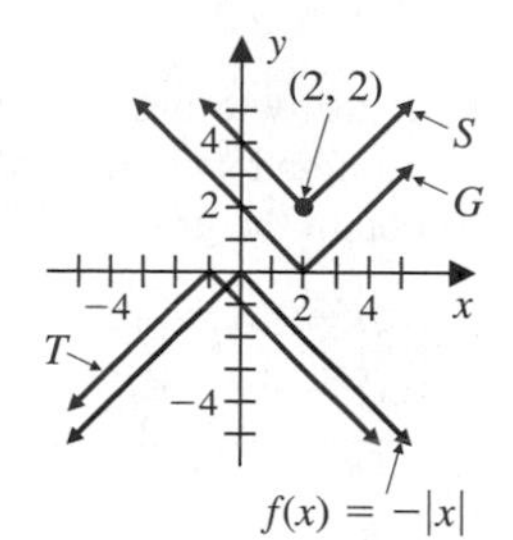

16.

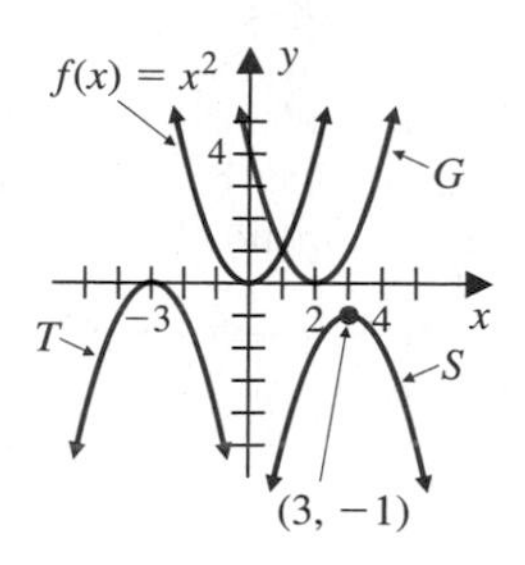

In problems 17–24, the graph of function G is obtained from the graph of the given function f as specified. Determine the function equation for G. Graph f and G on the same coordinate system.

17. Shift the graph of $f(x) = x$ up 3 units.
18. Shift the graph of $f(x) = |x|$ down 3 units.
19. Shift the graph of $f(x) = x^2$ to the right 2 units.
20. Shift the graph of $f(x) = x^3$ to the left 2 units.
21. Reflect the graph of $f(x) = \sqrt{x}$ across the x axis, and then shift it to the right 5 units.
22. Reflect the graph of $f(x) = \sqrt{x}$ across the x axis, and then shift it $\frac{1}{2}$ unit to the right.
23. Shift the graph of $f(x) = x^2$ up 1 unit and to the left 2 units, and then vertically scale it by a multiple of $\frac{1}{3}$.
24. Reflect the graph of $f(x) = x^2$ across the x axis, and then vertically scale it by a multiple of 2.

In problems 25–44, use a standard graph along with transformations to graph each function. Identify the equation of the standard graph. Also, find the domain and range of the function.

25. $f(x) = x + 2$
26. $g(x) = \sqrt{x} - 2$
27. $G(x) = 1 - x^3$
28. $h(x) = 1 + \sqrt[3]{x}$
29. $f(x) = \sqrt{x + 4}$
30. $g(x) = |x + 3|$
31. $F(x) = 2|x - 1|$
32. $h(x) = -2|x + 1|$
33. $H(x) = -5\sqrt{x + 1} + 1$
34. $f(x) = 7 + |x - 2|$
35. $f(x) = -\frac{1}{3}(x + 1)^3$
36. $g(x) = \dfrac{(x - 3)^2}{2}$
37. $g(x) = 4x^2 - 1$
38. $f(x) = 2(x - 1)^2 + 1$
39. $h(x) = 2(x - 3)^2 + 4$
40. $h(x) = (x + 1)^3 - 2$
41. $G(x) = 3\left(x - \frac{1}{2}\right)^3 - 1$
42. $H(x) = -2\sqrt{x + 1} + 3$
43. $g(x) = \frac{1}{2}\sqrt{1 - x^2}$
44. $F(x) = 3\sqrt{4 - x^2} + 1$

In problems 45 and 46, use transformations of an appropriate standard graph to sketch the graph of each given function.

45. $f(x) = (5 - x)^3$ **46.** $g(x) = \sqrt[3]{5 - x}$

GU In problems 47–50, use a graphing utility to graph each function. Then begin with a standard graph and state the order of transformations that can be used to obtain the graph of f.

47. $f(x) = -2(x + 1)^2 + 3$ **48.** $f(x) = 3(x - 1)^2 - 2$

49. $f(x) = 3\sqrt{x - 2} - 4$ **50.** $f(x) = -\frac{1}{2}\sqrt{x + 2} + 3$

Developing and Extending the Concepts

51. Suppose that the point (a, b) lies on the graph of $y = f(x)$. What are the coordinates of the resulting point location in terms of a and b after (a, b) is transformed according to the graph of each equation?
(a) $y = f(x - 2)$ **(b)** $y = f(x + 4)$
(c) $y = -f(-x) + 1$

52. Suppose that the function f is decreasing on the interval $(-1, 1)$ and increasing on the intervals $(-\infty, -1)$ and $(1, \infty)$. On what interval(s) is the function defined by $y = -f(-x)$ decreasing?

In problems 53–56, suppose that h, k, and a, $a \neq 0$, are constants. Let:
(a) $g(x) = f(x) + k$ **(b)** $g(x) = f(x + h)$
(c) $g(x) = af(x)$

53. Suppose that f is increasing. Which of the given g functions is increasing?

54. Suppose that f is decreasing. Which of the given g functions is decreasing?

55. Suppose that f is even. Which of the given g functions is even?

56. Suppose that f is odd. Which of the given g functions is odd?

57. In Example 6, the graph of $f(x) = \sqrt{2 - x}$ was obtained by reflecting the graph of $y = \sqrt{x}$ about the y axis, followed by shifting 2 units to the right.
(a) Reverse the order of transformations; that is, shift the graph of $y = \sqrt{x}$ to the right 2 units and then reflect the result about the y axis.
(b) The graph obtained in part (a) is not the same as the graph in Example 6. Explain the differences and why they occurred.
(c) How can such erroneous results be avoided?

58. Suppose that the graph of $y = f(x)$ is symmetric about the y axis. What kind of symmetry, if any, does the graph of $y = f(x - 3)$ have?

GU In problems 59–62, a function f is given.
(a) Determine the function equation for G.
(b) Use a graphing utility to sketch graphs of f and G on the same coordinate system.
(c) Describe how the graph of G can be obtained from the graph of f if transformations are used.

59. $f(x) = \sqrt[3]{x}(x + 2)$; $G(x) = f(x - 1)$

60. $f(x) = \sqrt[4]{x}(x - 3)$; $G(x) = f(x + 1)$

61. $f(x) = \dfrac{x^2}{x + 1}$; $G(x) = f(x - 1) + 2$

62. $f(x) = \dfrac{\sqrt[3]{x}}{x^2 + 1}$; $G(x) = f(x - 1) + 2$

2.4 Combinations of Functions

OBJECTIVES

1. Determine the sum, difference, product, and quotient functions.
2. Form composite functions.
3. Express a function as a composition of other functions.
4. Solve applied problems.

Just as two numbers can be added, subtracted, multiplied, and divided to produce other numbers, so two functions can be combined using these operations to produce other functions. In this section, we discuss these operations along with another, called *composition*, which has no analogy in the arithmetic of numbers.

Determining the Sum, Difference, Product, and Quotient Functions

If $f(x) = x^2 - 1$ and $g(x) = 2x + 1$, then the sum, difference, product, and quotient of f and g, respectively, define the following new functions:

1. Sum: $(f + g)(x) = f(x) + g(x)$
$= (x^2 - 1) + (2x + 1) = x^2 + 2x$
2. Difference: $(f - g)(x) = f(x) - g(x)$
$= (x^2 - 1) - (2x + 1) = x^2 - 2x - 2$
3. Product: $(f \cdot g)(x) = f(x) \cdot g(x)$
$= (x^2 - 1)(2x + 1) = 2x^3 + x^2 - 2x - 1$
4. Quotient: $\left(\frac{f}{g}\right)(x) = \frac{f(x)}{g(x)} = \frac{x^2 - 1}{2x + 1}, x \neq -\frac{1}{2}$

This illustration leads us to the following definitions:

DEFINITIONS

SUM, DIFFERENCE, PRODUCT, AND QUOTIENT FUNCTIONS

Let f and g be any two functions. We define the functions $f + g, f - g, f \cdot g$, and f/g as follows:

(i) **Sum function**: $(f + g)(x) = f(x) + g(x)$
(ii) **Difference function**: $(f - g)(x) = f(x) - g(x)$
(iii) **Product function**: $(f \cdot g)(x) = f(x) \cdot g(x)$
(iv) **Quotient function**: $\left(\frac{f}{g}\right)(x) = \frac{f(x)}{g(x)}$, for $g(x) \neq 0$

The domains of the sum, difference, product, and quotient functions consist of all values of x *common* to the domains of f and g, except for the quotient function, in which case the values of x for which the denominator is 0 are also excluded.

EXAMPLE 1 Finding the sum, difference, and product functions

Let $f(x) = 3x^3 + 7$ and $g(x) = x^2 - 3x - 4$. Find each of the following functions:

(a) $(f + g)(x)$ (b) $(f - g)(x)$ (c) $(f \cdot g)(x)$

Solution

(a) $(f + g)(x) = f(x) + g(x) = (3x^3 + 7) + (x^2 - 3x - 4)$
$= 3x^3 + x^2 - 3x + 3$
(b) $(f - g)(x) = f(x) - g(x) = (3x^3 + 7) - (x^2 - 3x - 4)$
$= 3x^3 - x^2 + 3x + 11$
(c) $(f \cdot g)(x) = f(x) \cdot g(x) = (3x^3 + 7) \cdot (x^2 - 3x - 4)$
$= 3x^5 - 9x^4 - 12x^3 + 7x^2 - 21x - 28$

EXAMPLE 2 Finding the quotient function and its domain

Find the quotient function f/g and its domain for each pair of functions.

(a) $f(x) = 3x^3 + 7$ and $g(x) = x^2 - 4$
(b) $f(x) = \sqrt{9 - x^2}$ and $g(x) = \sqrt{x - 1}$

Solution

(a) $$\left(\frac{f}{g}\right)(x) = \frac{f(x)}{g(x)} = \frac{3x^3 + 7}{x^2 - 4}$$

The domain for both f and g consists of all real numbers. However, the expression $(3x^3 + 7)/(x^2 - 4)$ represents a real number for all real values of x except those for which the denominator $x^2 - 4 = (x - 2)(x + 2) = 0$. Thus, the domain of f/g includes all real numbers except 2 and -2.

(b) $$\left(\frac{f}{g}\right)(x) = \frac{f(x)}{g(x)} = \frac{\sqrt{9 - x^2}}{\sqrt{x - 1}}$$

The domain of f is the interval $[-3, 3]$ and the domain of g is the interval $[1, \infty)$. The values of x common to the domains of f and g consist of all numbers in the interval $[1, 3]$. However, we must exclude 1 from the domain of f/g because $g(1) = 0$. Thus, the domain of f/g includes all numbers in the interval $(1, 3]$.

Forming Composite Functions

The *composition* of two functions provides us with another way of combining two functions to form a third function. The idea is to apply the functions in a specific order. Suppose we are given two functions $y = f(x)$ and $y = g(x)$. We define a new function h as follows:

First, put an input number, say x, into the g function.

Second, take the output of the g function, $g(x)$, and use it as input into the f function.

The resulting output, $f[g(x)]$, is the output of the function $h(x) = f[g(x)]$; the function h is referred to as a *composite function.*

For instance, if $f(x) = x^3$ and $g(x) = 2x + 3$, then

$$h(1) = f[g(1)] = f[2(1) + 3] = f(5) = 5^3 = 125$$
$$h(2) = f[g(2)] = f[2(2) + 3] = f(7) = 7^3 = 343$$
$$h(-1) = f[g(-1)] = f[2(-1) + 3] = f(1) = 1^3 = 1$$

In a sense, the function h is obtained by a "chain reaction" in which we first apply the g function followed by the f function. The composite function h is sometimes written as

$$h = f \circ g \quad \text{Read "}f\text{ circle }g\text{."}$$

For $f(x) = x^3$ and $g(x) = 2x + 3$, the equation for the composite function h can be determined by the following substitutions:

$$
\begin{aligned}
h(x) = (f \circ g)(x) &= f[g(x)] && \text{Input } x \\
&= f(2x + 3) && \text{Output of } g \text{ function is input of } f \text{ function} \\
&= (2x + 3)^3 && \text{Output of } f \text{ function} \\
(f \circ g)(x) &= (2x + 3)^3 && \text{Output of composite function } h
\end{aligned}
$$

DEFINITION

COMPOSITION OF FUNCTIONS

Let f and g be two functions. The **composition** of g *followed by* f, which is denoted by $f \circ g$, is the function defined by

$$(f \circ g)(x) = f[g(x)]$$

$f \circ g$ is called the **composite function** whose domain consists of all x in the domain of g for which $g(x)$ is in the domain of f.

Although the symbol $f \circ g$ may suggest some kind of product, it should not be confused with the actual product $f \cdot g$ of f and g.

Figure 1 illustrates how the composite function $f \circ g$ is formed by applying the functions g and f in sequence: First apply g to x; then apply f to $g(x)$.

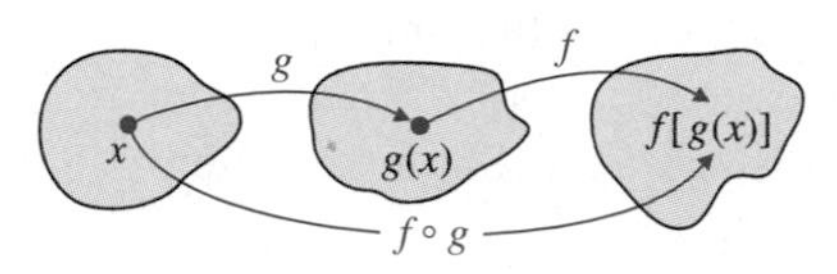

FIGURE 1

EXAMPLE 3 Finding the composition of two functions

Let $f(x) = 5x^2$ and $g(x) = 2x - 3$. Find each of the following results:

(a) $(f \circ g)(4)$ (b) $(g \circ f)(4)$ (c) $(f \circ g)(x)$
(d) $(g \circ f)(x)$ (e) $(f \circ f)(x)$ (f) $(g \circ g)(x)$

Solution

(a) $(f \circ g)(4) = f[g(4)] = f(2 \cdot 4 - 3) = f(5) = 5 \cdot 5^2 = 125$

(b) $(g \circ f)(4) = g[f(4)] = g(5 \cdot 4^2) = g(80) = 2 \cdot 80 - 3 = 157$

(c) $(f \circ g)(x) = f[g(x)] = f(2x - 3) = 5(2x - 3)^2$
$= 5(4x^2 - 12x + 9) = 20x^2 - 60x + 45$

(d) $(g \circ f)(x) = g[f(x)] = g(5x^2) = 2 \cdot 5x^2 - 3 = 10x^2 - 3$

(e) $(f \circ f)(x) = f[f(x)] = f(5x^2) = 5(5x^2)^2 = 125x^4$

(f) $(g \circ g)(x) = g[g(x)] = g(2x - 3) = 2(2x - 3) - 3$
$= 4x - 6 - 3 = 4x - 9$

Notice that the definition of the function $f \circ g$ specifies that the input x must be in the domain of g and must also result in an output $g(x)$ that is in the domain of f.

EXAMPLE 4 **Finding the domains of composite functions**

Let $f(x) = 2x + 3$ and $g(x) = \sqrt{x}$.

(a) Find $(f \circ g)(x)$ and the domain of $f \circ g$.
(b) Find $(g \circ f)(x)$ and the domain of $g \circ f$.

Solution

(a)
$$(f \circ g)(x) = f[g(x)] = f\left(\sqrt{x}\right) = 2\sqrt{x} + 3$$

The domain and range of g are both $[0, \infty)$. Since the domain of f consists of all real numbers $\mathbb{R}$, all output values from the g function are acceptable input for the f function. Thus, the domain of $f \circ g$ consists of all values x in the interval $[0, \infty)$.

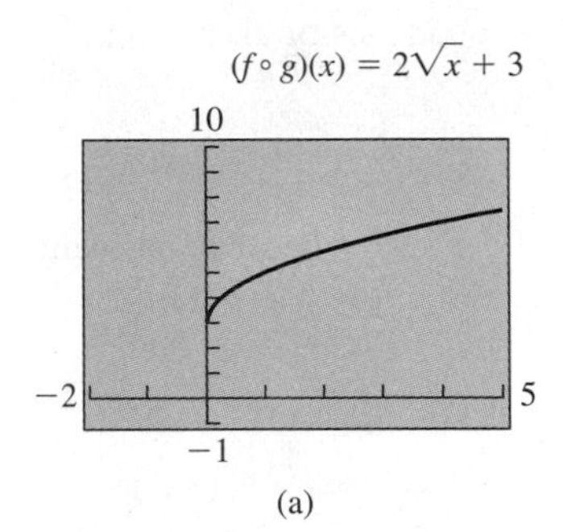

(a)

(b)
$$(g \circ f)(x) = g[f(x)] = g(2x + 3) = \sqrt{2x + 3}$$

The domain and range of f consist of all real numbers $\mathbb{R}$. Since the domain of g is $[0, \infty)$, the only acceptable input values for the g function must be nonnegative. Output values from the f function must satisfy $2x + 3 \geq 0$, that is, $x \geq -\frac{3}{2}$. So the domain of $g \circ f$ is the interval $\left[-\frac{3}{2}, \infty\right)$.

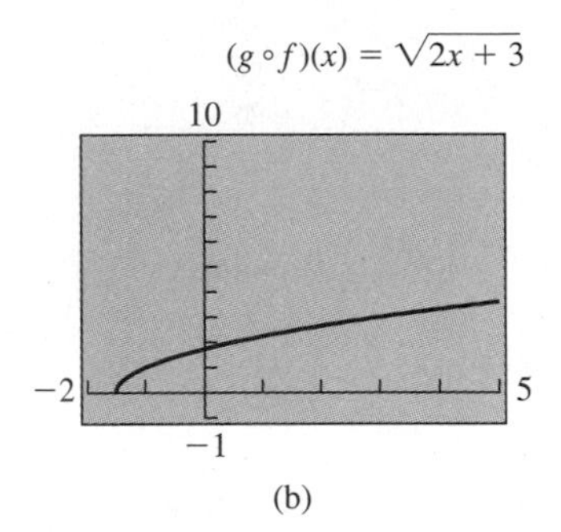

(b)

FIGURE 2

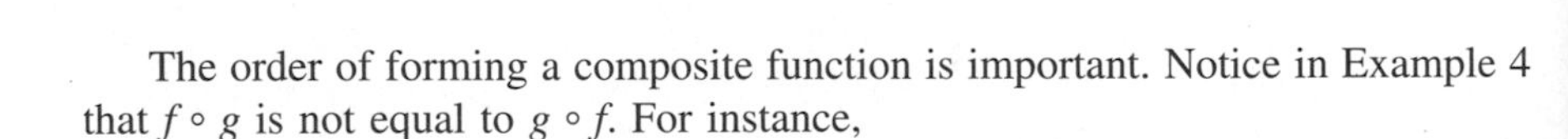

The order of forming a composite function is important. Notice in Example 4 that $f \circ g$ is not equal to $g \circ f$. For instance,

$$(f \circ g)(1) = 2\sqrt{1} + 3 = 5 \qquad \text{whereas} \qquad (g \circ f)(1) = \sqrt{2 \cdot 1 + 3} = \sqrt{5}$$

This fact can also be demonstrated graphically. Figure 2a shows a viewing window of the graph of $f \circ g$, and Figure 2b shows the graph of $g \circ f$. Notice that the two graphs are not the same. In general, $f \circ g \neq g \circ f$.

Care must be taken to make sure that the correct domain is used when graphing a composite function.

EXAMPLE 5 **Graphing a composite function**

Let $f(x) = \sqrt{x^2 + 2}$ and $g(x) = \sqrt{x - 1}$.

(a) Find the equation $y = (f \circ g)(x) = f[g(x)]$.
(b) Graph the equation determined in part (a).
(c) Find the domain of $f \circ g$.
(d) What is the domain suggested by the graph found in part (b), and how does it compare to the answer in part (c)?
(e) What adjustment has to be made to the sketch in part (b) to obtain the correct graph of $f \circ g$?

Some graphing utilities provide for the efficient graphing of the composite function $f \circ g$ by allowing the function expression $g(x)$ to be used as input to the f function.

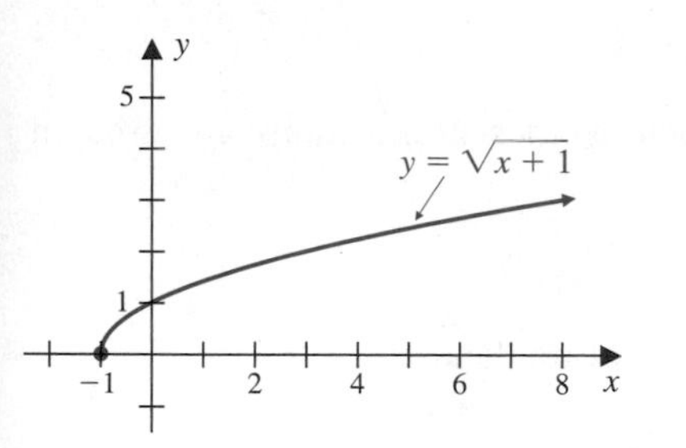

(a)

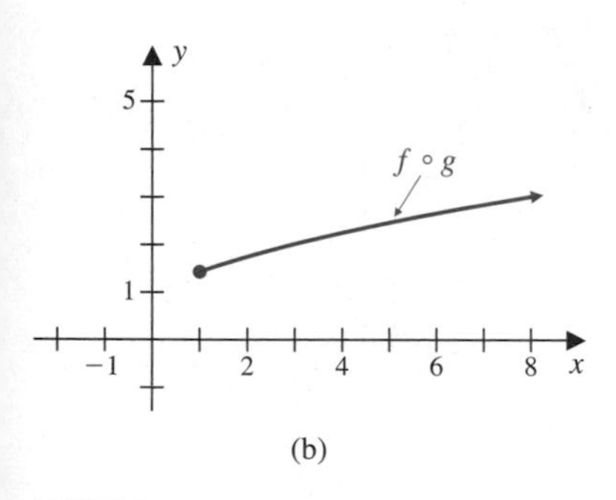

(b)

FIGURE 3

Solution For descriptive purposes, let us agree to call g the *inner function* and f the *outer function* because of the positions they occupy in the expression $f[g(x)]$.

(a) For $f(x) = \sqrt{x^2 + 2}$ and $g(x) = \sqrt{x - 1}$, we have

$$y = f[g(x)] = f\left(\sqrt{x-1}\right) = \sqrt{\left(\sqrt{x-1}\right)^2 + 2} = \sqrt{x - 1 + 2} = \sqrt{x+1}$$

that is, $y = \sqrt{x + 1}$.

(b) Figure 3a shows the graph of $y = \sqrt{x + 1}$.

(c) Since the inner function, $g(x) = \sqrt{x - 1}$, is defined for $x \geq 1$ and the outer function f is defined for all real numbers, it follows that the domain of $f \circ g$ includes all real numbers x such that $x \geq 1$ or $[1, \infty)$.

(d) The graph in Figure 3a suggests that the domain includes all x such that $x \geq -1$; however, the domain of $f \circ g$ includes all x such that $x \geq 1$ or $[1, \infty)$.

(e) Restricting x so that $x \geq 1$ gives us the graph of $f \circ g$, which is displayed in Figure 3b.

Expressing a Function as a Composition of Other Functions

In calculus we often need to interpret a given function as a composition of two functions. In a sense we "unravel" the equation used to define the given function. Example 6 illustrates the process.

EXAMPLE 6 **Finding a composite form of a function**

Express $h(x) = (5x - 1)^2$ as a composition $h = f \circ g$; that is, find f and g so that $h(x) = f[g(x)]$.

Solution In order to see that h can be obtained as a composition $h = f \circ g$, we must be able to recognize the inner function g and the outer function f in the equation that defines h. Of course, g is the first function and f is the second function applied in the composition $f \circ g$. The diagram in Figure 4 helps us to define f and g so that $h(x) = f[g(x)]$. We can see that $h = f \circ g$, where $g(x) = 5x - 1$ and $f(x) = x^2$.

This assignment of f and g is not the only possibility. Another is, say, $g(x) = 5x$ and $f(x) = (x - 1)^2$.

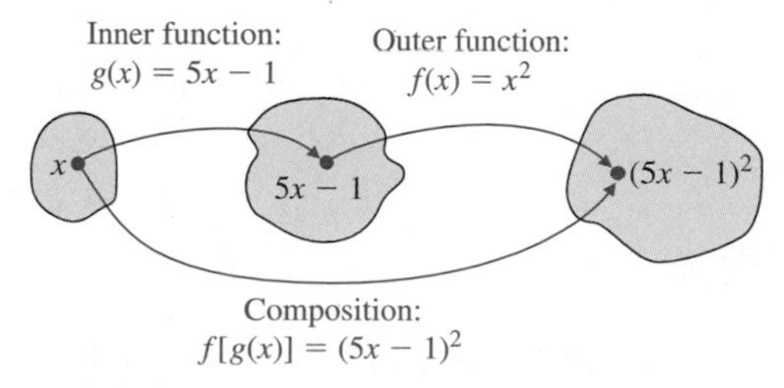

FIGURE 4

Solving Applied Problems

The following examples illustrate how combinations of functions are used in modeling.

EXAMPLE 7 **Modeling currency exchanges**

Table 1 lists the foreign currency exchange rates for January 12, 1996.

TABLE 1

Dollar's Worth, *Wall Street Journal* January 12, 1996

	Foreign Currency in U.S. Dollars	One U.S. Dollar in Foreign Currency
Britain (pound)	1.5445	0.6475
Canada (dollar)	0.7338	1.3627
France (franc)	0.2022	4.9453
Germany (mark)	0.6942	1.4405
Japan (yen)	0.009542	104.80
Switzerland (franc)	0.8604	1.1622

(a) Find a function f that expresses the number of Japanese yen $f(d)$ in terms of the number of U.S. dollars d.

(b) Find a function g that expresses the number of U.S. dollars $g(m)$ in terms of the number of German marks m.

(c) Form the composite function $f \circ g$ and interpret it.

Solution

(a) According to the exchange rate in Table 1, 1 U.S. dollar is worth 104.8 Japanese yen. Consequently, d dollars are worth $104.8d$ yen. So, $f(d) = 104.8d$ is a function that expresses the number of yen $f(d)$ in terms of the number of U.S. dollars d.

(b) Table 1 indicates that 1 German mark is worth 0.6942 U.S. dollar. Thus, m marks are worth $0.6942m$ U.S. dollars. So $g(m) = 0.6942m$ expresses the number of U.S. dollars $g(m)$ in terms of the number of marks m.

(c) The composite function $f \circ g$ is formed as follows:

$$\begin{aligned}(f \circ g)(m) &= f[g(m)] \\ &= f(0.6942m) \\ &= (104.8)(0.6942m) \\ &= 72.75m \text{ (approx.)}\end{aligned}$$

Figure 5 depicts how the composite function $f \circ g$ is formed and helps us to interpret it.

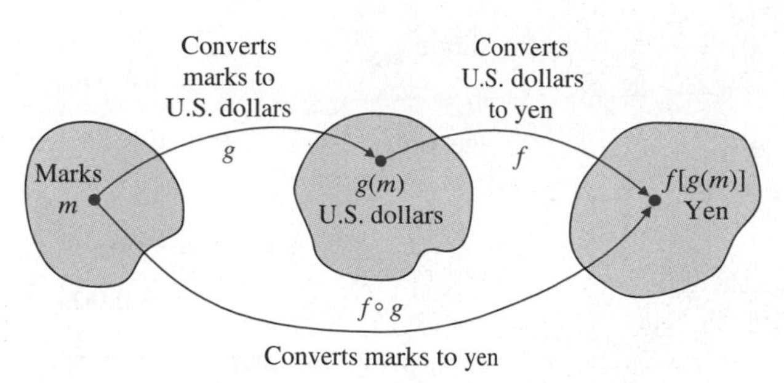

FIGURE 5

Thus, the function $f \circ g$ expresses the number of yen in terms of the number of marks, so it can be used to convert from marks to yen. For instance, 150 German marks corresponds to the number of Japanese yen given by

$$(f \circ g)(150) = (72.75)(150) = 10{,}912.5 \text{ (approx.)}$$

EXAMPLE 8 Modeling a manufacturer's profit

A pharmaceutical company finds that the cost function C for producing a certain antibiotic is $C(x) = 0.003x^2 + 80x + 500{,}000$, where $C(x)$ is the cost (in dollars), x is the number of units of the antibiotic produced, and $0 \le x \le 30{,}000$. The firm sets the selling price at \$200 per unit.

(a) Express the total revenue R as a function of x.
(b) Use a graphing utility to graph the cost function C and the revenue function R from part (a) in the same viewing window. Interpret the intersection of the two graphs.
(c) Write an expression for the total profit P as a function of x and use a graphing utility to graph it. Interpret the graph in terms of profit and loss.

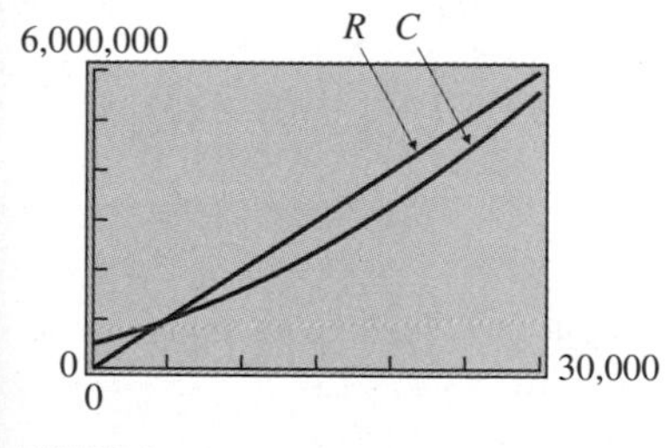

FIGURE 6a

In Chapter 3 we will investigate another way of finding x by using special features of graphing utilities.

Solution

(a) If x units are sold at \$200 per unit, then the revenue $R(x)$ (in dollars) is given by $R(x) = 200x$.
(b) Figure 6a displays a viewing window showing the graphs of the cost function C and the revenue function R, taking into account that $0 \le x \le 30{,}000$. We observe that the two graphs intersect where $C(x) = R(x)$, that is, when $0.003x^2 + 80x + 500{,}000 = 200x$. Solving this quadratic equation for x yields two results, which are approximately 4725 and 35,275. Because of the restriction on x, the value 4725 is the only feasible solution. So, when about 4725 units are sold, both revenue and cost are equal. If fewer units are sold, $R(x) < C(x)$, indicating a loss. If more than 4725 units are sold, $R(x) > C(x)$, indicating a profit. We call the value $x = 4725$ the *break-even point.*

(c) Since

$$\text{Profit} = \text{Revenue} - \text{Cost}$$

the profit function P in this situation is given by the difference function,

$$\begin{aligned} P(x) &= R(x) - C(x) \\ &= 200x - (0.003x^2 + 80x + 500{,}000) \\ &= -0.003x^2 + 120x - 500{,}000 \end{aligned}$$

In Chapter 3 we will investigate techniques that enable us to find this highest profit value and the value of x when it occurs. For this particular situation, a maximum profit of \$700,000 occurs when 20,000 units are sold.

A viewing window of the graph of the profit function P is shown in Figure 6b. We see that there is a loss if $P < 0$ (when $x < 4725$) and a profit if $P > 0$ (when $x > 4725$). Also, as more units are sold beyond 4725, the profit increases to a "peak" level and then begins to decrease.

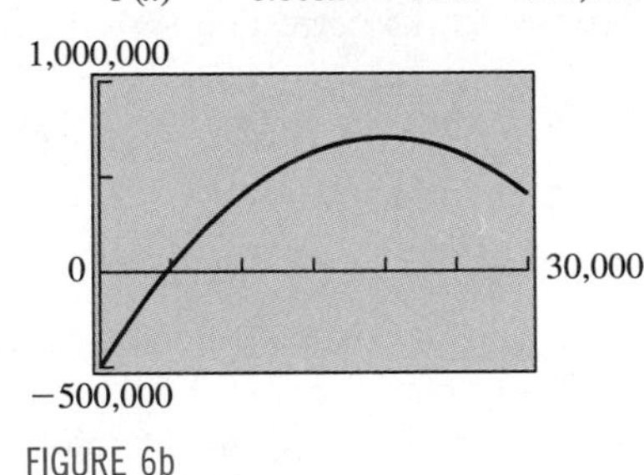

FIGURE 6b

PROBLEM SET 2.4

Mastering the Concepts

In problems 1–8, find each of the following functions:
(a) $(f + g)(x)$ **(b)** $(f - g)(x)$ **(c)** $(f \cdot g)(x)$

1. $f(x) = 3x + 1$ and $g(x) = -2x - 7$
2. $f(x) = -4x - 5$ and $g(x) = x + 6$
3. $f(x) = x^2 - 3x + 2$ and $g(x) = 4x^2 + 1$
4. $f(x) = 7x^2 - 1$ and $g(x) = -3x^2 + 8x - 5$
5. $f(x) = x^3 + 5x$ and $g(x) = 2x^3 - 3x + 4$
6. $f(x) = x^2 + 5$ and $g(x) = -x^3 + x^2 - 7x + 2$
7. $f(x) = \dfrac{2}{x - 5}$ and $g(x) = \dfrac{x}{3 - 4x}$
8. $f(x) = \dfrac{2x + 3}{x - 5}$ and $g(x) = \dfrac{2 - 7x}{3x + 1}$

In problems 9–14, find each quotient function and its domain.
(a) $\dfrac{f}{g}$ **(b)** $\dfrac{g}{f}$

9. $f(x) = 3x + 1$ and $g(x) = -2x - 7$
10. $f(x) = -4x - 5$ and $g(x) = 6 - x$
11. $f(x) = x^2 + 4x + 3$ and $g(x) = x^3 + 1$
12. $f(x) = 2x^2 - 8x$ and $g(x) = x^3 + 1$
13. $f(x) = \sqrt{2 - x}$ and $g(x) = \dfrac{x}{x - 1}$
14. $f(x) = \sqrt{x + 3}$ and $g(x) = \dfrac{2x - 3}{x}$

In problems 15 and 16, find each of the following values:
(a) $(f \circ g)(2)$ **(b)** $(g \circ f)(2)$
(c) $(f \circ f)(2)$ **(d)** $(g \circ g)(2)$
(e) $g[f(4)]$ **(f)** $f[g(3)]$
(g) $f[g(5)]$ **(h)** $g[f(5)]$
(i) $f[f(-1)]$ **(j)** $g[g(-1)]$

15. $f(x) = 2x^2 + 6$ and $g(x) = 7x + 2$
16. $f(x) = 5 - 3x^2$ and $g(x) = -3x + 1$

In problems 17 and 18, use the given graphs of f and g to compute each of the following values:

(a) $(g \circ f)(2)$ **(b)** $(g \circ g)(3)$
(c) $(f \circ g)(3)$ **(d)** $(g \circ f)(7)$
(e) $(f \circ f)(2)$ **(f)** $(g \circ g)(7)$

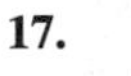
17.

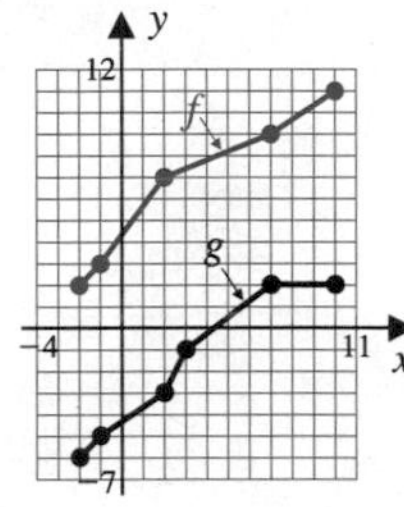

18.

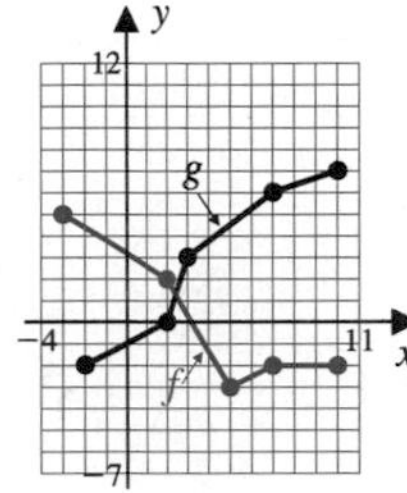

19. Table 2 lists values of two functions f and g. Find each of the following values:
(a) $(g \circ f)(2)$ **(b)** $(f \circ f)(0)$
(c) $(f \circ g)(3)$ **(d)** $(g \circ f)(-3)$

TABLE 2

x	−3	−2	−1	0	1	2	3
$f(x)$	−5	−3	−1	1	3	5	7

x	−5	0	1	3	5
$g(x)$	−15	0	−9	−1	15

20. Table 3 lists values of two functions f and g. Find each of the following values:
(a) $(g \circ f)(-2)$ **(b)** $(f \circ g)(1)$
(c) $(f \circ f)(-1)$ **(d)** $(g \circ g)(-3)$

TABLE 3

x	−2	−1	0	3	4	7	10
$f(x)$	1	3	5	11	5	3	−4

x	−3	−2	−1	1	2	3
$g(x)$	−2	1	0	10	9	5

GU In problems 21–30, find each of the following functions and their domains. Then use a graphing utility to graph the function with the appropriate domain.
(a) $y = f[g(x)]$ **(b)** $y = g[f(x)]$
(c) Compare the graphs of $f \circ g$ and $g \circ f$ to decide if $f \circ g$ and $g \circ f$ are always equal.

21. $f(x) = 2x$ and $g(x) = 5x - 3$
22. $f(x) = -3x$ and $g(x) = 3x$
23. $f(x) = 2x^2 + 5$ and $g(x) = 7\sqrt{x}$
24. $f(x) = \sqrt{2x - 1}$ and $g(x) = x^2 + 9$
25. $f(x) = 11x + 2$ and $g(x) = \dfrac{x}{11} - \dfrac{2}{11}$
26. $f(x) = x^3 + 1$ and $g(x) = \sqrt[3]{x - 1}$
27. $f(x) = \sqrt{x - 1}$ and $g(x) = x + 5$
28. $f(x) = 2x - 2$ and $g(x) = \sqrt{x + 2}$
29. $f(x) = \dfrac{1}{3x + 2}$ and $g(x) = \dfrac{3}{2x - 5}$
30. $f(x) = \dfrac{x}{3x - 5}$ and $g(x) = \dfrac{1 - x}{3 + 2x}$

In problems 31–36, find two functions f and g so that the given function h can be expressed as a composition $h = f \circ g$.

31. $h(x) = (5x - 2)^3$ **32.** $h(x) = (x^2 + 2x - 1)^4$
33. $h(t) = (t^2 - 2)^{-2}$ **34.** $h(w) = \left(\dfrac{w + 1}{w - 1}\right)^3$
35. $h(x) = \sqrt[3]{x + x^{-1}}$ **36.** $h(t) = (t^2 + 3)^{4/5}$

37. Let $D(t) = 3t + 1$ and $R(x) = -5x + 2$
(a) Find $(D \circ R)(x)$
(b) For what value of x does $D[R(x)] = 2$?

38. Let $S(r) = r^3 + 2$ and $D(t) = \sqrt[3]{t + 7}$
(a) Find $(S \circ D)(t)$
(b) For what value of t does $S[D(t)] = 13$?

In problems 39–44:
(a) Find the equation $y = f[g(x)]$.
(b) Sketch the graph of the equation determined in part (a).
(c) Find the domain of $f \circ g$.
(d) What adjustment, if any, has to be made to the sketch in part (b) in order to obtain the correct graph of $f \circ g$?

39. $f(x) = x^2$ and $g(x) = \sqrt{x}$
40. $f(x) = \sqrt{x}$ and $g(x) = x^2$
41. $f(x) = \dfrac{1}{x}$ and $g(x) = \dfrac{1}{x}$
42. $f(x) = \dfrac{1}{x - 1}$ and $g(x) = \dfrac{1}{x}$
43. $f(x) = \sqrt{4 - x^2}$ and $g(x) = \sqrt{2 - x}$
44. $f(x) = \sqrt{25 - x^2}$ and $g(x) = \sqrt{x - 5}$

Applying the Concepts

45. Currency Exchange: Use the currency exchange rates given in Table 1 (page 132) to find a composite function that converts from:
(a) British pounds to French francs
(b) Swiss francs to Canadian dollars

46. Automobile Rebate: A car dealer offers a \$1000 rebate and a 15% discount off the sticker price x (in dollars) of a new car at the end of a model year.

(a) Express the cost R of the car as a function of x if only the rebate is given.

(b) Express the cost D of the car as a function of x if only the discount is given.

(c) Find $(D \circ R)(x)$ by calculating the rebate first, then the discount. Graph $D \circ R$.

(d) Find $(R \circ D)(x)$ by calculating the discount first, then the rebate. Graph $R \circ D$.

(e) Which of the options in parts (c) and (d) is a better deal? Interpret the results if the sticker price of the car is \$17,500.

47. Meteorology: A spherical weather balloon is being inflated at a constant rate. The radius r (in inches) of the balloon is increasing with time t (in minutes) according to the function $r = g(t) = 1.85t$. The volume V of the balloon is given by the formula $V = f(r) = \frac{4}{3}\pi r^3$.

(a) Find a composite function that expresses the volume V of the balloon as a function of time t, and interpret its meaning.

(b) The surface area S of the balloon is given by $S = h(r) = 4\pi r^2$. Find a composite function that expresses the surface area S as a function of time t, and interpret its meaning.

48. Ecology: An offshore oil well begins to leak, and the oil slick starts to spread on the surface of the water in a circular pattern in such a way that the radius r is increasing at the rate of 0.5 kilometer per hour.

(a) Express the radius r as a function f of the number of elapsed hours t.

(b) The area of a circle of radius r is given by $A(r) = \pi r^2$. Form the composite function $A \circ f$, and interpret the meaning of the function.

(c) Find the area of the oil slick 5 hours after the beginning of the leak.

49. Baseball: A baseball diamond is a square 90 feet on each side. Suppose that a ball is hit directly down the third base line at the rate of 55 feet per second. Let x denote the distance (in feet) the ball travels, y denote the distance (in feet) of the ball from first base, and t denote the elapsed time (in seconds) since the ball was hit (Figure 7). Here, y is a function of x, say $y = f(x)$, and x is a function of t, say $x = g(t)$.

(a) Find equations that define $f(x)$ and $g(t)$.

(b) Find the expression for $(f \circ g)(t)$ and interpret $f \circ g$.

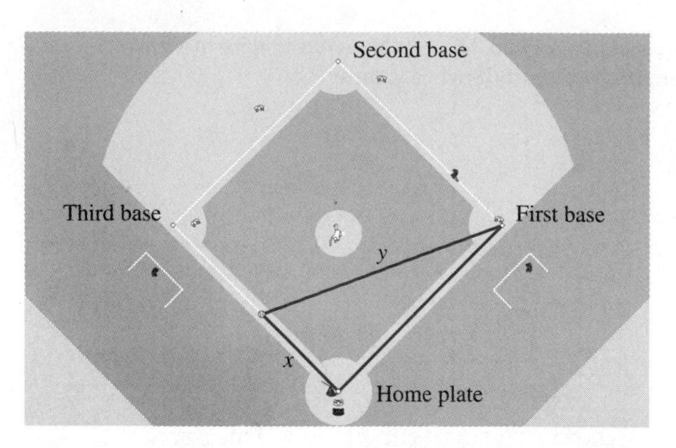

FIGURE 7

50. Pollution: A study by air quality engineers in a certain city shows that the concentration C (in parts per million) of carbon monoxide in the center of the city is given by the function

$$C(x) = 10^{-9}x^2 + 10^{-5}x + 1$$

where x is the average number of cars driven in the center of the city each day. Traffic engineers forecast that after t years, the average number x of cars that will be driven in the center of the city each day is given by the function

$$x = g(t) = 10{,}000t + 100{,}000$$

with $t = 0$ representing 1995 when there are 100,000 cars present.

(a) Write a composite function $C \circ g$ that gives the carbon monoxide concentration after t years.

(b) According to this model, what is the carbon monoxide concentration initially? After 1 year? After 2 years? After 5 years?

GU **(c)** Use a graphing utility to sketch a graph of the function $C \circ g$ found in part (a).

(d) Assume that the initial number of cars is 50,000 instead of 100,000. Construct a composite function as described in part (a). Use a graphing utility to graph this function and compare it to the graph from part (c).

51. GU Manufacturing: A company's cost C (in dollars) for producing a certain product is given by

$$C(x) = x^2 + 30x + 500 \qquad 0 \le x \le 45$$

where x is the number of units manufactured and sold. The selling price is set at \$90 per unit.

(a) Determine the revenue R as a function of x. Then use a graphing utility to sketch graphs of the cost and revenue functions in the same coordinate system. Interpret the result.

(b) Determine the profit P as a function of x and use a graphing utility to graph it. Interpret the graph in terms of profit and loss.

52. Marketing: The cost C (in dollars) of buying x bushels of corn is given by

GU

$$C(x) = 5x + 1000 \qquad \text{where } 0 \le x \le 2500$$

The revenue R (in dollars) from selling x bushels of corn is given by

$$R(x) = 8x - \frac{x^2}{1000}$$

(a) Use a graphing utility to sketch graphs of the cost and revenue functions in the same coordinate system. Interpret the result.

(b) Determine the profit function P and use a graphing utility to graph it. Interpret the graph in terms of profit and loss.

Developing and Extending the Concepts

53. Assume that f and g are increasing functions for all real numbers.

(a) Is $f + g$ necessarily an increasing function?

(b) Give an example of two increasing functions so that $f - g$ is a decreasing function for all real numbers.

54. Assume that f and g are decreasing functions for all real numbers.

(a) Is $f + g$ necessarily a decreasing function? Explain.

(b) Give an example of two decreasing functions so that $f - g$ is an increasing function for all real numbers.

55. Assume f and g are even functions. Are $f + g$, $f - g$, $f \cdot g$, and f/g all necessarily even functions? Explain and give examples to support your assertions.

56. Assume f and g are odd functions. Are $f + g$, $f - g$, $f \cdot g$, and f/g all necessarily odd functions? Explain and give examples to support your assertions.

57. **(a)** If g is an even function, is $f \circ g$ necessarily even?

(b) If f and g are both odd functions, is $f \circ g$ necessarily odd? Explain.

58. Composing a function f with itself is called *function iteration*. The successive functions

$$f \circ f, \quad f \circ f \circ f, \quad f \circ f \circ f \circ f, \quad \text{and so on}$$

are called the *iterates* of f. If $n \ge 2$ is an integer, we use the notation

$$f^{[n]} = f \circ f \circ f \circ f \circ f \circ f \circ f \circ \cdots \circ f \ (n \text{ times})$$

for the *nth iterate* of f. For instance, $f^{[2]} = f \circ f$, $f^{[3]} = f \circ f \circ f$, and so on.

(a) Find $f^{[2]}(x)$ if $f(x) = 2x + 1$.

(b) Find $f^{[3]}(x)$ if $f(x) = -5x + 2$.

(c) Find $f^{[10]}(x)$ if $f(x) = 1/x$.

59. The composition $f \circ g \circ h$ of three functions f, g, and h is defined by $(f \circ g \circ h)(x) = f[g[h(x)]]$. Find $(f \circ g \circ h)(x)$ for the following functions:

(a) $f(x) = x - 3$, $g(x) = x^3$, $h(x) = \sqrt[5]{x}$

(b) $f(x) = x^2 + 4$, $g(x) = 3x + 2$, $h(x) = 2x - 5$

60. **(a)** Give an example of two functions f and g so that $(f \circ g)(x) = (f \cdot g)(x)$.

(b) Give an example of two functions f and g so that $(f \circ g)(x) = (g \circ f)(x)$.

2.5 Inverse Functions

OBJECTIVES

1. Define inverse functions.
2. Recognize the symmetry of the graphs of a function and its inverse.
3. Determine graphically whether a function has an inverse.
4. Determine whether a function is one-to-one.
5. Find the inverse of a function.
6. Solve applied problems.

Our goal in this section is to develop the notion of *inverse functions*. This concept will be especially important later in this text when we study exponential and trigonometric functions.

Defining Inverse Functions

Let us begin by examining the composition of two functions $f(x) = 5x$ and $g(x) = x/5$.

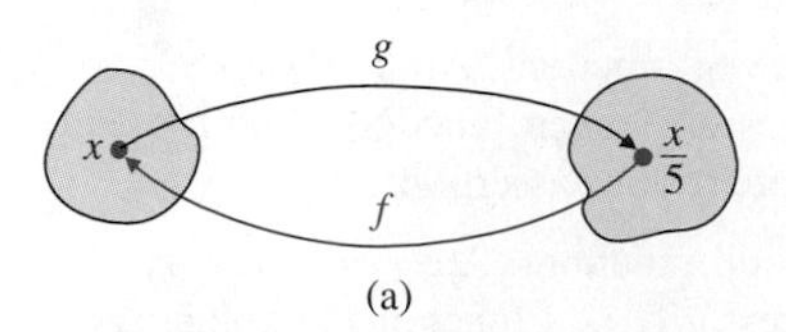

(a)

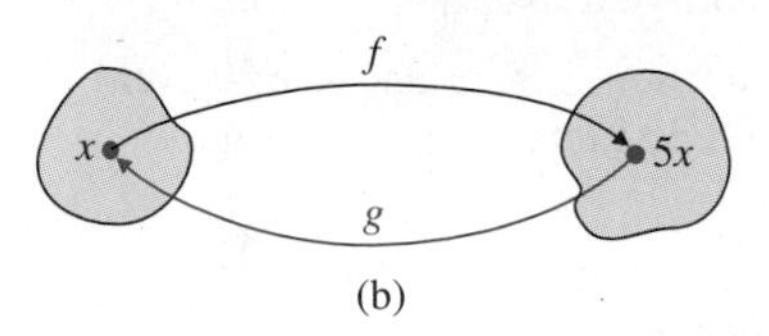

(b)

FIGURE 1

$$(f \circ g)(x) = f[g(x)] = f\left(\frac{x}{5}\right) = 5 \cdot \frac{x}{5} = x$$

$$(g \circ f)(x) = g[f(x)] = g(5x) = \frac{5x}{5} = x$$

Two observations are in order:

1. For $f \circ g$, f "undoes" the output of function g, $x/5$, and produces a final output of x. The final output is the same as the original input (Figure 1a).
2. Similarly, for $g \circ f$, g "undoes" the output of function f and returns x as the final output (Figure 1b).

Two functions related in such a way are said to be *inverses* of each other.

DEFINITION

INVERSE FUNCTIONS

Two functions f and g are **inverses** of each other if and only if

$$(f \circ g)(x) = f[g(x)] = x$$

for every value of x in the domain of g, and

$$(g \circ f)(x) = g[f(x)] = x$$

for every value of x in the domain of f.

A function f is said to be *invertible* if such a function g exists.

EXAMPLE 1 Verifying that two functions are inverses of each other

Show that the functions $f(x) = 3x + 2$ and $g(x) = \frac{1}{3}x - \frac{2}{3}$ are inverses of each other.

Solution From the definition, we must verify that $f[g(x)] = x$ and $g[f(x)] = x$. We have

$$f[g(x)] = f\left(\tfrac{1}{3}x - \tfrac{2}{3}\right) = 3\left(\tfrac{1}{3}x - \tfrac{2}{3}\right) + 2 = (x - 2) + 2 = x$$
$$g[f(x)] = g(3x + 2) = \tfrac{1}{3}(3x + 2) - \tfrac{2}{3} = \left(x + \tfrac{2}{3}\right) - \tfrac{2}{3} = x$$

for all x. Thus, we conclude that f and g are indeed inverses of each other.

When f and g are inverses of each other, we refer to g as the *inverse function of* f, and vice versa, and we write

$$g = f^{-1} \qquad \text{or} \qquad g^{-1} = f$$

Care must be taken not to confuse $y = f^{-1}(x)$, the inverse function of f, with $[f(x)]^{-1} = 1/f(x)$, the reciprocal of $f(x)$.

Thus, in Example 1, we write either

$$f(x) = 3x + 2 \qquad \text{and} \qquad f^{-1}(x) = \tfrac{1}{3}x - \tfrac{2}{3}$$

or

$$g(x) = \tfrac{1}{3}x - \tfrac{2}{3} \qquad \text{and} \qquad g^{-1}(x) = 3x + 2$$

The conditions stated in the definition of an inverse function can be restated as follows:

$$f^{-1}[f(x)] = x \qquad \text{for every } x \text{ in the domain of } f$$
$$f[f^{-1}(x)] = x \qquad \text{for every } x \text{ in the domain of } f^{-1}$$

Recognizing the Symmetry of the Graphs of a Function and Its Inverse

A certain symmetry exists between the graph of a function and the graph of its inverse. First, we observe that the points $M = (a, b)$ and $N = (b, a)$ in Figure 2a are *symmetric about the line* $y = x$ because this line is the perpendicular bisector of the line segment $\overline{MN}$. Now suppose that (a, b) is any point on the graph of f and that function f has an inverse function f^{-1}. Then

$b = f(a)$	The values $x = a$ and $y = b$ satisfy the function equation
$f^{-1}(b) = f^{-1}[f(a)]$	Apply the function f^{-1} to each side
$f^{-1}(b) = a$	$f^{-1}[f(x)] = x$, by definition

So (b, a) is on the graph of f^{-1}. Thus, whenever (a, b) is a point on the graph of f, the point (b, a) is on the graph of f^{-1} (Figure 2b). We therefore have the following property:

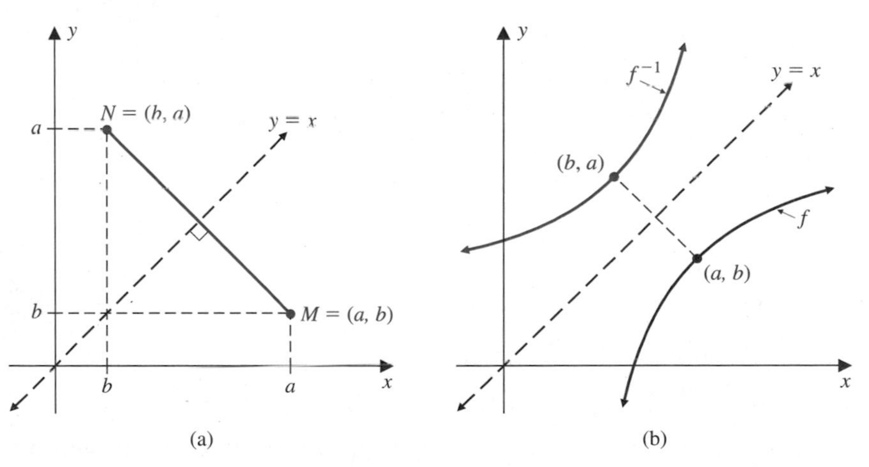

FIGURE 2

PROPERTY

GRAPHS OF f AND f^{-1}

The graphs of f and f^{-1} are reflections of each other about the line $y = x$.

EXAMPLE 2 **Illustrating the symmetry of the graphs of f and f^{-1}**

GU

Use a graphing utility to graph $f(x) = 3x + 2$, $f^{-1}(x) = \frac{1}{3}x - \frac{2}{3}$, and $y = x$ in the same viewing window to display that the graphs of f and f^{-1} are symmetric about the line $y = x$.

Solution Figure 3 demonstrates the symmetry of the graphs of f and f^{-1} about the line $y = x$.

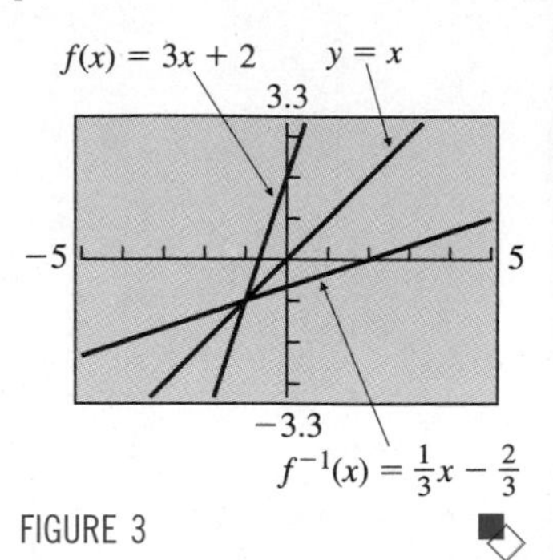

FIGURE 3

EXAMPLE 3 **Graphing f^{-1} from the graph of f**

Use the graph of f given in Figure 4a to sketch the graph of f^{-1}. Identify the domain and range of f and f^{-1}.

Solution Because the graphs of f and f^{-1} are symmetric about the line $y = x$, the graph of f^{-1} is obtained by reflecting the graph of f about the line $y = x$ (Figure 4b). From the graphs, we observe that the domain of f is the interval $[0, \infty)$ and its range is the interval $[1, \infty)$. Also, we see that the domain of f^{-1} is the interval $[1, \infty)$ and its range is the interval $[0, \infty)$. In other words, the domain and range of f are, respectively, the range and domain of f^{-1}.

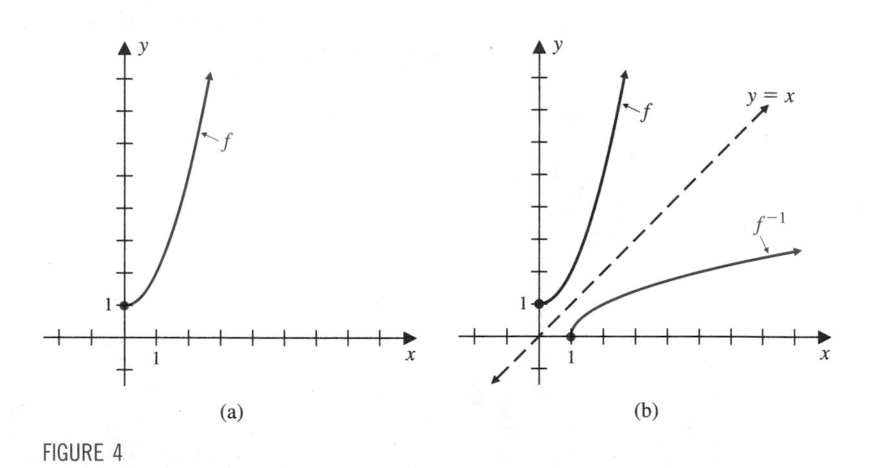

FIGURE 4

Example 3 illustrates the following property:

PROPERTY

DOMAIN AND RANGE OF f AND f^{-1}

The domain of f is the same as the range of f^{-1}, and the range of f is the same as the domain of f^{-1}.

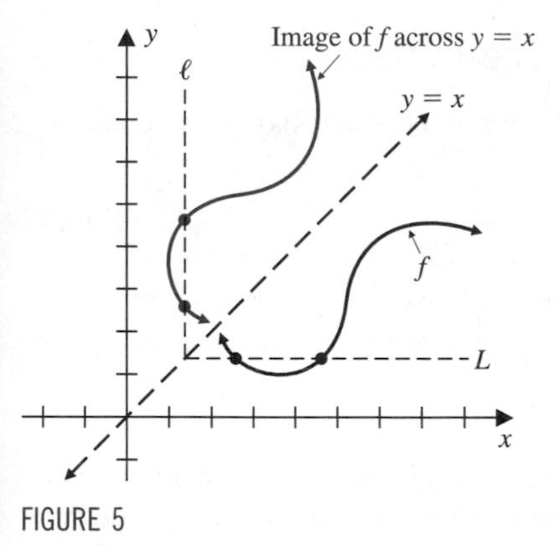

FIGURE 5

Determining Graphically Whether a Function Has an Inverse

We can use the graph of a function to determine whether it has an inverse. Consider the graph of the function f in Figure 5. The reflected image of the graph of f about the line $y = x$ is not the graph of a function because we can draw a vertical line ℓ that intersects the graph more than once. Thus, f cannot have an inverse. Notice that the horizontal line L, obtained by reflecting ℓ across the line $y = x$, intersects the graph of f more than once.

This observation provides the basis for using the graph of a function to determine whether it has an inverse.

PROPERTY

HORIZONTAL-LINE TEST

A function f has an inverse if and only if no horizontal straight line intersects its graph more than once.

EXAMPLE 4 Determining whether a function has an inverse

Use the horizontal-line test to determine whether each function has an inverse.

(a) $f(x) = \sqrt{x}$ (b) $g(x) = x^2$ GU (c) $h(x) = \dfrac{\sqrt{x^4 + 1}}{x + 1}$

Solution Functions f and g have standard graphs, which are shown in Figures 6a and 6b, respectively. The graph of h was obtained by using a graphing utility (Figure 6c).

(a) No horizontal line intersects the graph of f more than once (Figure 6a), so f has an inverse.
(b) Any horizontal line drawn above the x axis will intersect the graph of g twice (Figure 6b), so g does not have an inverse.
(c) A horizontal line can be drawn that will intersect the graph twice (Figure 6c). Thus, h does not have an inverse.

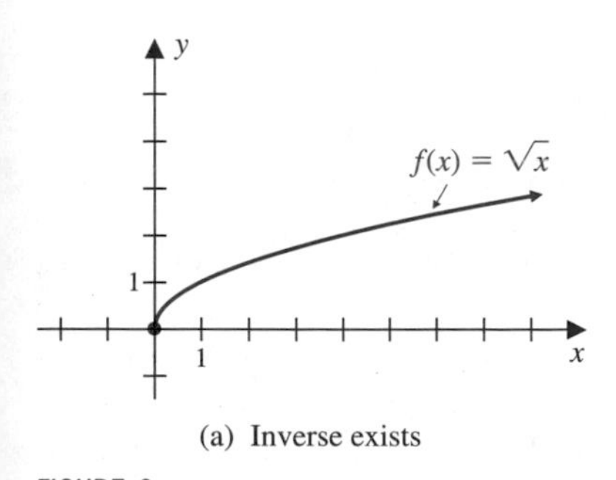

(a) Inverse exists

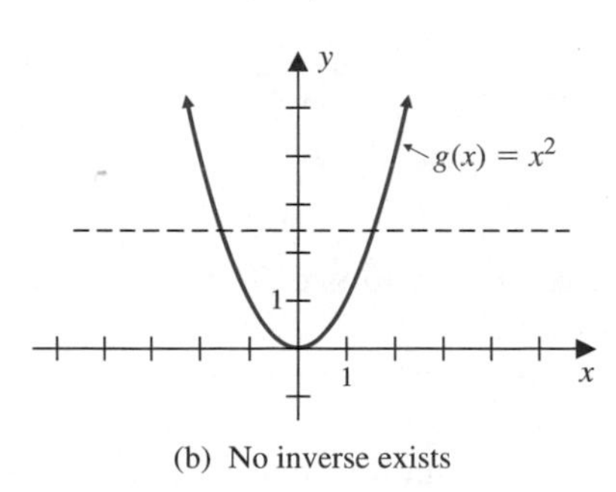

(b) No inverse exists

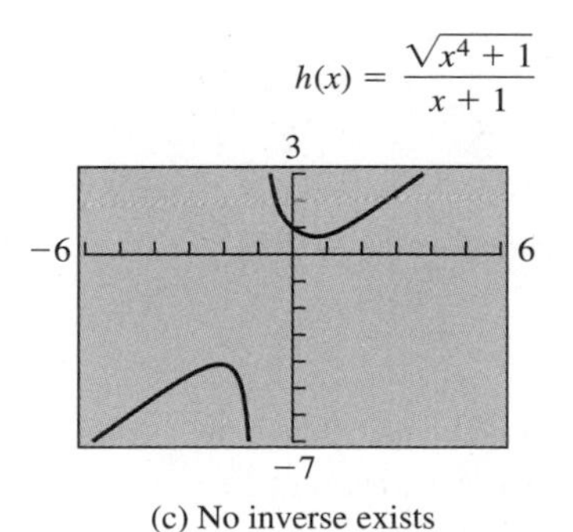

(c) No inverse exists

FIGURE 6

Determining Whether a Function Is One-to-One

Let us review two characteristics of a function $y = f(x)$ that has an inverse:

1. The graph of the function satisfies the vertical-line test. Thus, for every input value x there is one and only one output value for y.
2. Because the function has an inverse, the graph must also satisfy the horizontal-line test. This means that every output value y has one and only one corresponding input value for x.

It follows that a function has an inverse if and only if each distinct input value always results in a distinct output value. A function that has this latter property is called a *one-to-one function.*

DEFINITION

ONE-TO-ONE FUNCTION

A function f is said to be **one-to-one** if, whenever a and b are in the domain of f and $f(a) = f(b)$, it follows that $a = b$.

EXAMPLE 5 **Determining one-to-one functions**

Use the definition of a one-to-one function to determine whether each function is one-to-one. Graph each function and use the graph to demonstrate the validity of the result.

(a) $f(x) = 3x - 2$ (b) $g(x) = 4x^2$

Solution

(a) We assume that $f(a) = f(b)$ for some numbers a and b in the domain of f. According to the definition, we must show that $a = b$. We proceed as follows:

$$
\begin{aligned}
f(a) &= f(b) && \text{Assumption} \\
3a - 2 &= 3b - 2 && \text{Evaluate } f(a) \text{ and } f(b). \\
3a &= 3b && \text{Add 2 to each side.} \\
a &= b && \text{Divide each side by 3.}
\end{aligned}
$$

Hence, f is one-to-one. The horizontal-line test confirms that f is one-to-one (Figure 7a).

(b) Here we select $x = -1$ and $x = 1$, and observe that $g(-1) = 4(-1)^2 = 4$ and $g(1) = 4 \cdot 1^2 = 4$. Hence, $g(-1) = g(1)$ but $-1 \neq 1$, so g does not satisfy the definition and g is not one-to-one. The horizontal-line test confirms that g is not one-to-one (Figure 7b).

The selection of specific numbers to show that a general property does *not* hold, as in Example 5b, illustrates a method of proof called *proof by counterexample.*

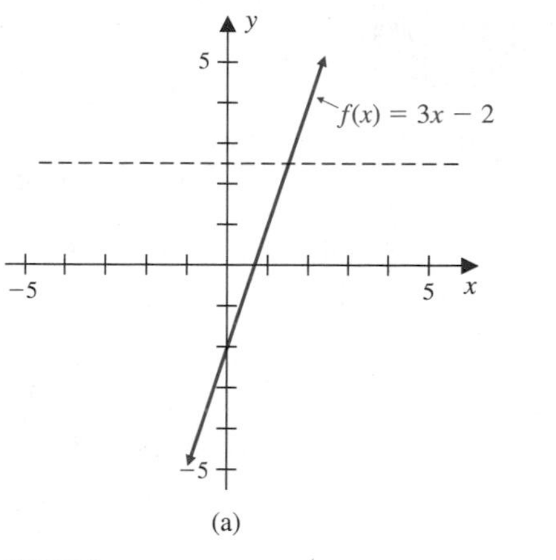

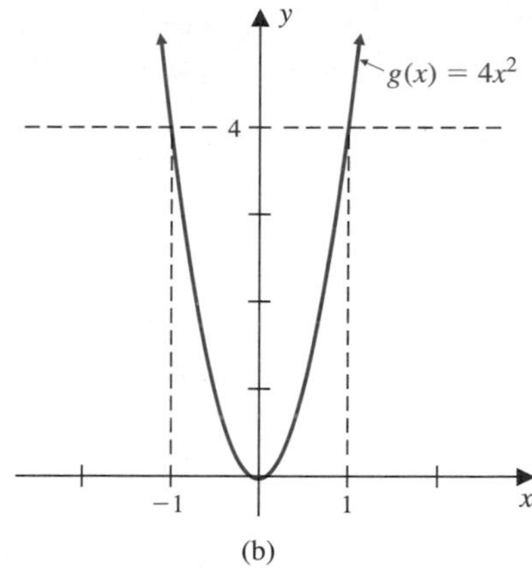

FIGURE 7

Note that functions that are either increasing or decreasing throughout their domains are one-to-one functions, and so they have inverses.

Finding the Inverse of a Function

We now turn our attention to answering the question:

If a function f has an inverse, how do we find the inverse function f^{-1}?

Earlier, we discovered that whenever (a, b) is on the graph of f, then (b, a) is on the graph of f^{-1}. This means that f^{-1} is the function that can be formed by interchanging the roles of the domain and range values of f. For instance, suppose that f is a function defined by Table 1a, then f^{-1} is formed by switching the roles of the domain and range values as shown in Table 1b.

TABLE 1
Example of f and f^{-1}
(a) Function f

Domain Value	−3	1	5	7	23
Corresponding Range Value	0	−4	2	11	−6

(b) Function f^{-1}

Domain Value	0	−4	2	11	−6
Corresponding Range Value	−3	1	5	7	23

If a function f has an inverse f^{-1} and f is defined by the equation $y = f(x)$, then we have:

$$y = f(x) \qquad y \text{ is expressed in terms of } x$$
$$f^{-1}(y) = f^{-1}[f(x)] \qquad \text{Apply } f^{-1} \text{ to each side}$$
$$f^{-1}(y) = x \qquad x \text{ is expressed in terms of } y$$

This observation provides the basis for the following procedure, which enables us to find the equation for f^{-1} if f is defined by an equation.

PROCEDURE

ALGEBRAIC METHOD FOR FINDING f^{-1}

Step 1. Write the equation $y = f(x)$ that defines the one-to-one function f.
Step 2. Solve the equation $y = f(x)$ for x in terms of y, obtaining an equation for the inverse function $x = f^{-1}(y)$.
Step 3. Verify that the domain and range of f are, respectively, the same as the range and domain of f^{-1}.

EXAMPLE 6 **Finding the inverse of a function**
Let $f(x) = 2x - 3$.

(a) Use the above method to find $f^{-1}(x)$.
(b) Graph f and f^{-1} on the same coordinate system.

Solution The function $f(x) = 2x - 3$ is a linear function. Since the slope $m = 2$, the graph of f is increasing for all real numbers, so f is one-to-one and thus f^{-1} exists.

(a) Step 1. Let $y = f(x) = 2x - 3$.
Step 2. Solve the equation for x in terms of y as follows:

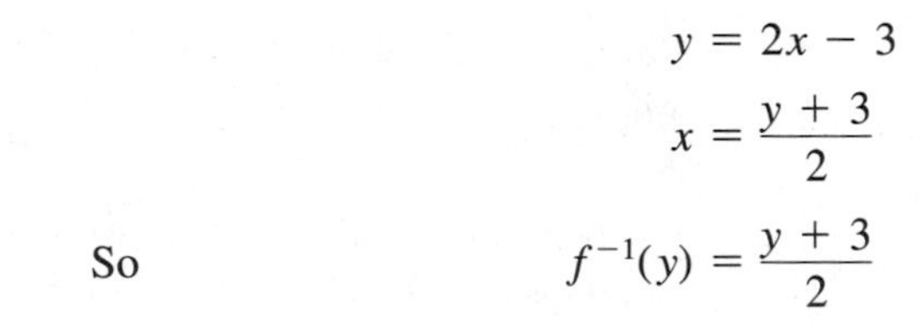

$$y = 2x - 3$$

$$x = \frac{y + 3}{2}$$

So

$$f^{-1}(y) = \frac{y + 3}{2}$$

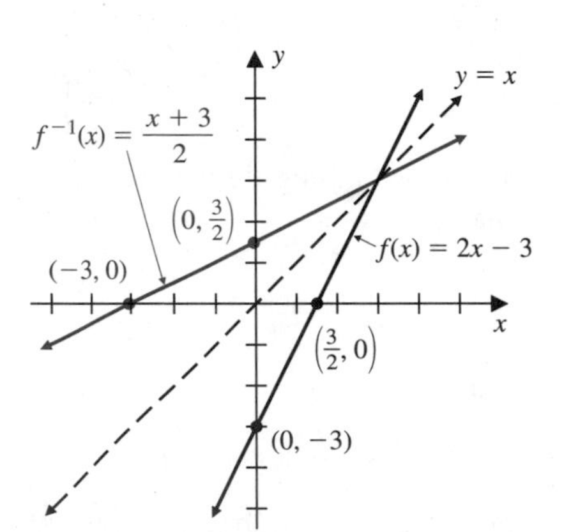

FIGURE 8

Some graphing utilities are programmed to graph f and f^{-1} on the same coordinate system by giving only the equation that defines f.

Following the convention of using x to represent the independent variable, we rewrite the latter equation as

$$f^{-1}(x) = \frac{x + 3}{2}$$

Step 3. The domain and range of f consist of all real numbers, which is the same for f^{-1}.

(b) The graphs of f and f^{-1} are shown in Figure 8. As expected, they are symmetric about the line $y = x$.

Even though two functions have the same defining equation, one may have an inverse and the other may not.

EXAMPLE 7 **Comparing two functions with the same defining equation**
Let $f(x) = x^2 + 1$ and $g(x) = x^2 + 1$, where $x \geq 0$.

(a) Use the graphs of f and g to explain why f does not have an inverse and g does have an inverse.
(b) Find $g^{-1}(x)$.

Solution

(a) By inspecting the graph of f in Figure 9a, we see that it does not satisfy the horizontal-line test, so f does not have an inverse. On the other hand, the graph of g (Figure 9b) does satisfy the horizontal-line test, so g has an inverse function. Notice that both functions have the same defining equation, but g has a restricted domain that allows it to have an inverse.

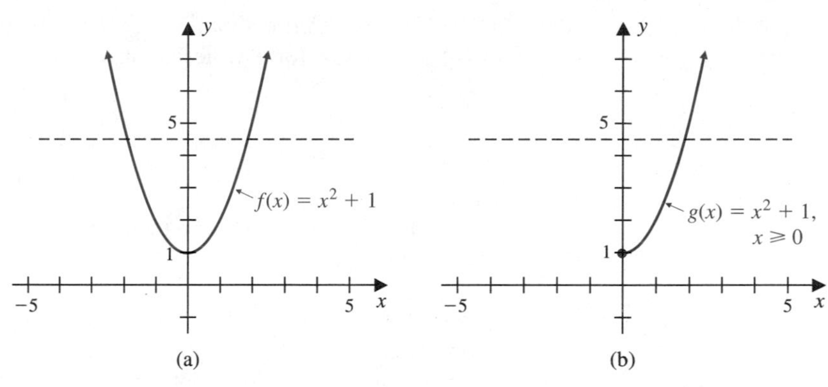

FIGURE 9

(b) We obtain an equation for the inverse function g^{-1} as follows:

Step 1. Let $y = x^2 + 1$.
Step 2. Solve for x:

$$x^2 = y - 1$$
$$x = \sqrt{y - 1} \quad x \geq 0 \text{ is a given restriction}$$

Thus, $$g^{-1}(y) = \sqrt{y - 1}$$

By changing the notation, we get $g^{-1}(x) = \sqrt{x - 1}$.

Step 3. The domain of g and the range of g^{-1} are the same, the interval $[0, \infty)$. Also, the range of g is the same as the domain of g^{-1}, the interval $[1, \infty)$.

Solving Applied Problems

The next example shows an application of inverse functions.

EXAMPLE 8 Using the inverse function

Insurance companies reward students who have good grades and good driving records by discounting their insurance premiums. Suppose that a student who qualifies is awarded a 20% discount in premiums.

(a) Find a function f that expresses the reduced premium R in terms of the original premium P.
(b) Find the inverse function f^{-1} of part (a) and interpret it in terms of the reduced and original premiums.

Solution

(a) A 20% reduction in the original premium means the new premium is 80% of the original one. Therefore, the function f that expresses the reduced premium R in terms of the original premium P is given by

$$R = f(P) = 0.80P$$

(b) The inverse function f^{-1} of $f(P) = 0.80P$ is obtained by solving the equation $R = 0.80P$ for P in terms of R:

$$P = \frac{1}{0.80}R = 1.25R$$

So,
$$P = f^{-1}(R) = 1.25R$$

This means that a 25% increase in the reduced premium is needed to obtain the original premium.

PROBLEM SET 2.5

Mastering the Concepts

In problems 1–6, show that the functions f and g are inverses of each other by verifying that $f[g(x)] = x$ and $g[f(x)] = x$. Sketch the graphs of f and g on the same coordinate system to show that their graphs are symmetric about the line $y = x$.

1. $f(x) = 7x - 2$; $g(x) = \frac{x}{7} + \frac{2}{7}$

2. $f(x) = 1 - 5x$; $g(x) = \frac{1}{5} - \frac{x}{5}$

3. $f(x) = x^4$, where $x \geq 0$; $g(x) = \sqrt[4]{x}$

4. $f(x) = x^3$; $g(x) = \sqrt[3]{x}$

5. $f(x) = \frac{1}{x - 2}$; $g(x) = \frac{1}{x} + 2$

6. $f(x) = \sqrt{x + 3}$; $g(x) = x^2 - 3$, where $x \geq 0$

In problems 7 and 8, use the horizontal-line test to determine whether each of the functions whose graph is given has an inverse.

7. (a)

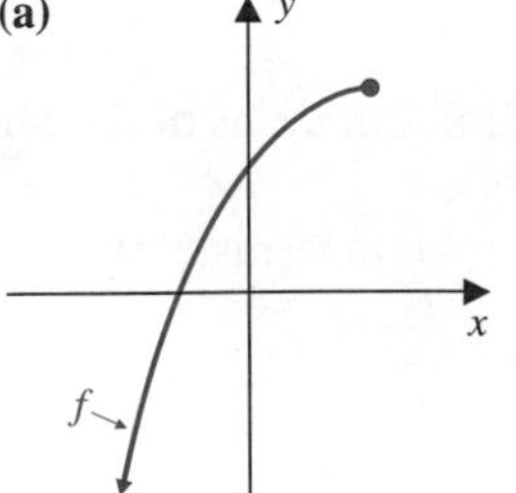

(b)

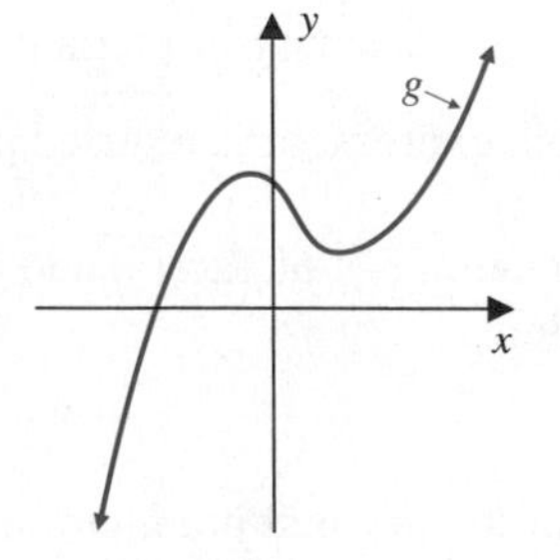

8. (a)

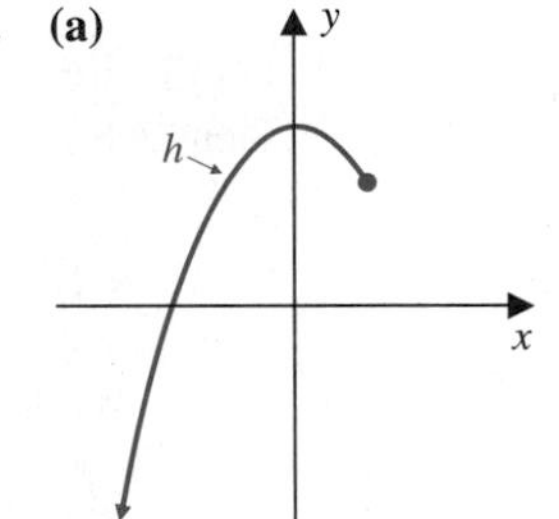

(b)

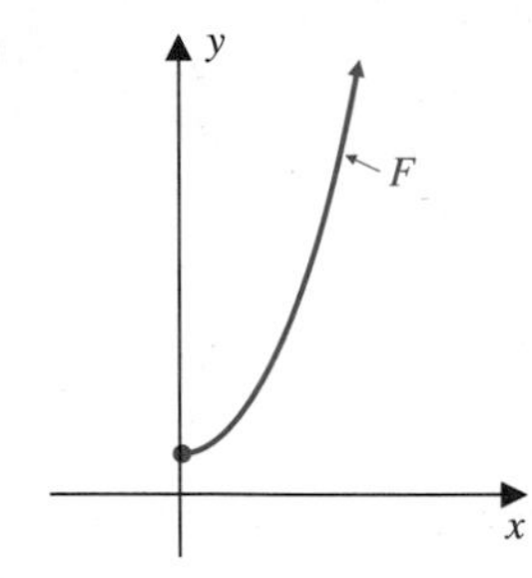

In problems 9 and 10, use the given graph of the one-to-one function f to do the following:

(i) Sketch the graph of f^{-1} by reflecting the graph of f about the line $y = x$.

(ii) Determine and compare the domains and ranges of f and f^{-1}.

9. (a)

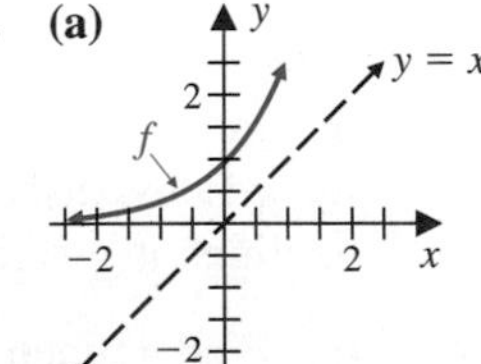

(b)

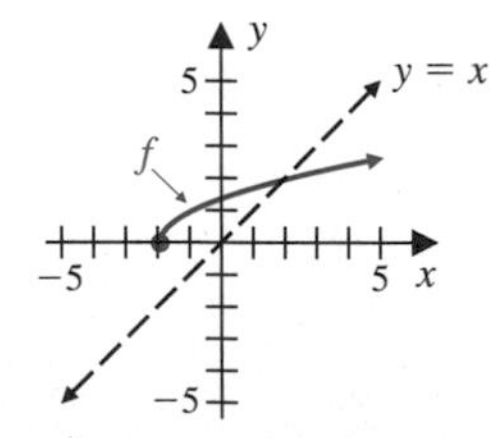

10. (a)

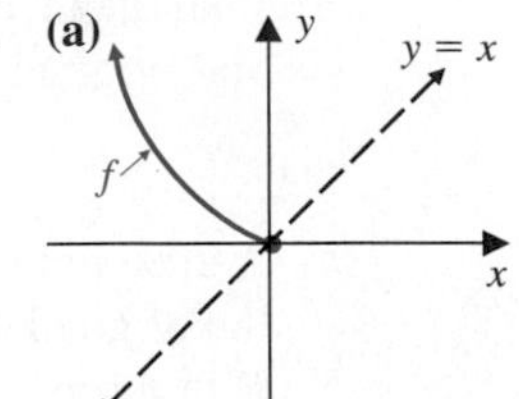

(b)

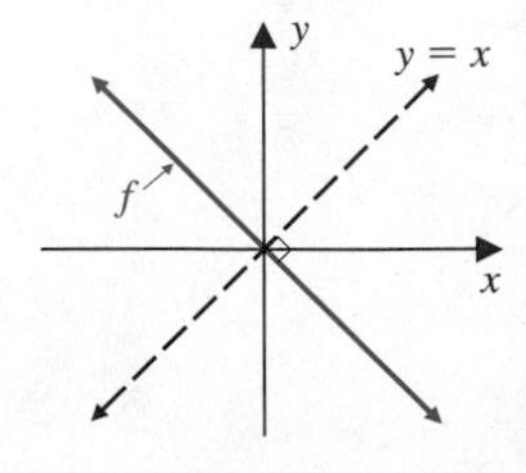

11. Write a table of values for f^{-1}, where f is given in Table 2. What are the domain and range of f? What are the domain and range of f^{-1}?

TABLE 2

x	1	2	3	4	5	6	7	8
$f(x)$	2	−3	3	5	7	100	−4	1

12. Determine whether each function has an inverse.
(a) $f(x)$ is the volume (in liters) of x kilograms of water.
(b) $f(n)$ is the number of students in a precalculus class whose birthday is on the nth day of the year.
(c) $f(t)$ is the number of customers in a department store at t minutes past 6:00 PM.
(d) $f(w)$ is the cost (in dollars) of express mailing a package that weighs w ounces.

In problems 13–18, use the definition of a one-to-one function to determine whether f is one-to-one. Graph f to demonstrate the validity of the result.

13. $f(x) = -2x + 7$
14. $f(x) = -\frac{2}{3}x + 5$
15. $f(x) = 2x^2 + 1$
16. $f(x) = -|x|$
17. $f(x) = \sqrt{4 - x^2}$
18. $f(x) = \sqrt{5 + 3x}$

In problems 19–32, sketch the graph of each function and use the graph to determine whether f^{-1} exists. If f^{-1} does not exist, explain why. If f^{-1} exists, use the procedure on page 143 to find $f^{-1}(x)$. Sketch the graphs of f and f^{-1} on the same coordinate system.

19. $f(x) = 7x + 5$
20. $f(x) = 1 - 3x$
21. $f(x) = \dfrac{3}{x - 1}$
22. $f(x) = \dfrac{2}{x + 3}$
23. $f(x) = \sqrt{x - 3}$
24. $f(x) = \sqrt{3 - 2x}$
25. $f(x) = x^3 - 8$
26. $f(x) = x^3 + 1$
27. $f(x) = |x + 4|$
28. $f(x) = -2x^2 + 5$
29. $f(x) = 2 - \sqrt[3]{x}$
30. $f(x) = \sqrt[3]{x + 1} - 2$
31. $f(x) = -x^2, x \geq 0$
32. $f(x) = \sqrt{1 - x^2}, 0 \leq x \leq 1$

In problems 33–36, the functions f and g have the same defining equation.
(a) Use the graphs of f and g to explain why f does not have an inverse but g does.
(b) Find $g^{-1}(x)$.

33. $f(x) = |x|$; $g(x) = |x|, x \leq 0$
34. $f(x) = -x^2 + 4$; $g(x) = -x^2 + 4, x \geq 0$
35. $f(x) = \sqrt{1 - x^2}$; $g(x) = \sqrt{1 - x^2}, 0 \leq x \leq 1$
36. $f(x) = \sqrt{4 - x^2}$; $g(x) = \sqrt{4 - x^2}, -2 \leq x \leq 0$

GU In problems 37–40, use a graphing utility to sketch the graph of each function. Use the horizontal-line test to determine whether the function has an inverse.

37. $f(x) = \dfrac{x + 2}{\sqrt[3]{x} + 1}$
Xmin: −3; Xmax: 1.5; Xscl: 1;
Ymin: −3; Ymax: 7; Yscl: 1

38. $f(x) = \dfrac{3x}{1 - \sqrt{x}}$
Xmin: 0; Xmax: 5; Xscl: 1;
Ymin: −20; Ymax: 20; Yscl: 5

39. $f(x) = \dfrac{x^3 - |x|}{x}$
Xmin: −5; Xmax: 5; Xscl: 1;
Ymin: −4; Ymax: 4; Yscl: 1

40. $f(x) = \dfrac{x^3 - 4x}{x^3 - 4}$
Xmin: −5; Xmax: 5; Xscl: 1;
Ymin: −4; Ymax: 4; Yscl: 1

41. GU Match each function f with its inverse g. Then use a graphing utility to graph the function and its inverse on the same coordinate system.

(a) $f(x) = 5x^3 + 10$ **(A)** $g(x) = \dfrac{\sqrt[3]{x} - 10}{5}$

(b) $f(x) = (5x + 10)^3$ **(B)** $g(x) = \sqrt[3]{\dfrac{x - 10}{5}}$

(c) $f(x) = 5(x + 10)^3$ **(C)** $g(x) = \sqrt[3]{\dfrac{x}{5}} - 10$

(d) $f(x) = (5x)^3 + 10$ **(D)** $g(x) = \dfrac{\sqrt[3]{x - 10}}{5}$

42. Figure 10 shows the graph of f.
(a) Explain why f has an inverse.
(b) What is $f^{-1}(2)$? $f^{-1}(3)$? $f^{-1}(-5)$?
(c) Sketch the graph of f^{-1}.

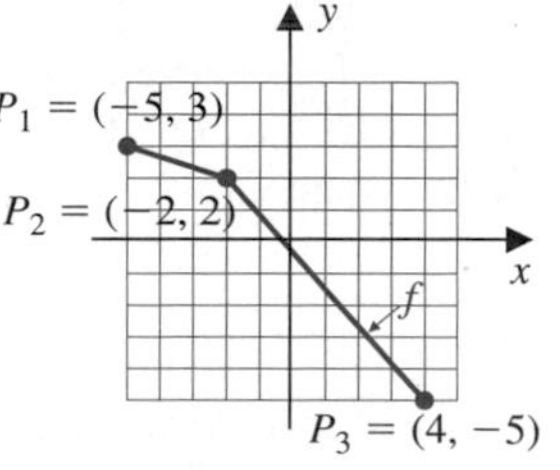

FIGURE 10

43. Figure 11 shows the graph of a function f. Use symmetry and transformations to sketch the graph of each function.

(a) $y = f^{-1}(x)$
(b) $y = f^{-1}(x) + 1$
(c) $y = f^{-1}(x - 1)$
(d) $y = f^{-1}(-x)$

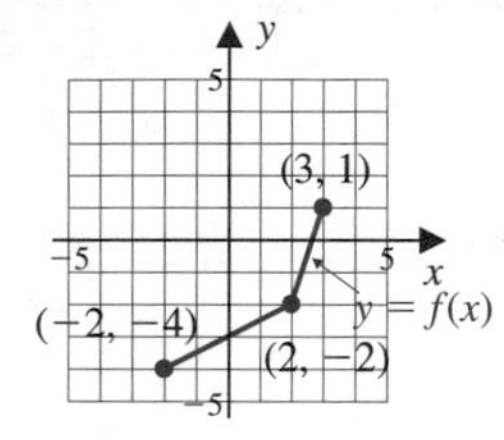

FIGURE 11

44. Figure 12 shows the graph of a function f. Use symmetry and transformations to sketch the graph of each function.

(a) $y = f^{-1}(x)$
(b) $y = f^{-1}(x) - 2$
(c) $y = f^{-1}(x - 2)$
(d) $y = -f^{-1}(x)$

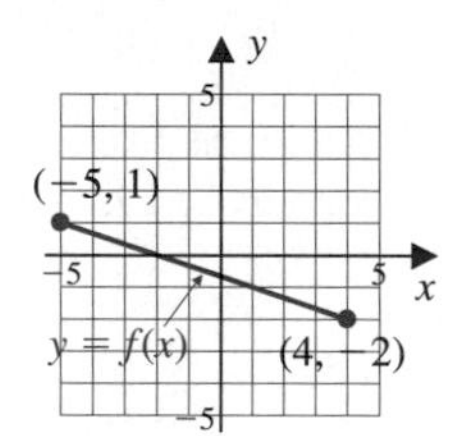

FIGURE 12

Applying the Concepts

45. Wages: To avoid layoffs, workers at a certain company accept a 15% reduction in wages.

(a) Find a function f that expresses the reduced wages W in terms of the original wages P.
(b) Find the inverse function of part (a) and interpret it in terms of the reduced and original wages.

46. Currency Exchange: On January 12, 1996, \$100 U.S. was worth 116 Swiss francs:

(a) Write a function f that expresses the number of dollars $f(c)$ in terms of the number of Swiss francs c.
(b) Write a function g that expresses the number of Swiss francs $g(d)$ in terms of the number of U.S. dollars d.
(c) What is the relationship between f and g? Explain.

47. Physics—Temperature Conversion: The function that converts degrees Celsius (C) to degrees Fahrenheit (F) is defined by the equation

$$F = f(C) = \frac{9}{5}C + 32$$

(a) Find f^{-1} and interpret the results in terms of Fahrenheit and Celsius readings.
(b) Graph f and f^{-1} on the same coordinate system.
(c) Find the Celsius measurements corresponding to 85°F and 113°F.

48. Business—Cost Function: A lumber yard will deliver wood for \$6 per board foot plus a fixed delivery charge of \$30.

(a) Write the equation that expresses the cost function C (in dollars) in terms of x board feet of lumber delivered; that is, find $C = f(x)$.
(b) Find f^{-1} and interpret the result in terms of cost and board feet.
(c) Graph f and f^{-1} on the same coordinate system.
(d) Find the number of board feet delivered if the cost is \$270.

49. Sales Commission: A salesperson makes a commission of 40% of total sales plus a salary of \$25,000 per year.

(a) Determine the function that describes the total annual income y (in dollars) in terms of the total sales x (in dollars).
(b) Find the inverse of this function and graph it.
(c) Interpret the inverse function in terms of sales and income.

50. Dress Sizes: A size 4 dress in the United States corresponds to a European size 32 dress, and a size 12 dress in the United States corresponds to a European size 48. Assume that the relationship between the sizes in the United States and Europe is linear.

(a) Express the European dress size D as a function f of the dress size x in the United States.
(b) Use the result in part (a) to find the European dress sizes that correspond to sizes 6, 8, and 10 in the United States.
(c) Explain why f has an inverse. Find f^{-1}, graph it, and interpret it in terms of dress sizes.
(d) Use the inverse function to find dress sizes in the United States to correspond to sizes 40, 52, and 60 in Europe.

Developing and Extending the Concepts

51. Use the horizontal-line test to determine whether each function is one-to-one. If the function is one-to-one, find its inverse. Determine the domains and ranges of f and f^{-1}.

(a) $f(x) = \begin{cases} \frac{1}{2}x - 4 & \text{if } x < 0 \\ x - 4 & \text{if } x \geq 0 \end{cases}$

(b) $f(x) = \begin{cases} x^3 & \text{if } x < 0 \\ -\sqrt{x} & \text{if } x \geq 0 \end{cases}$

52. GU Use a graphing utility to graph

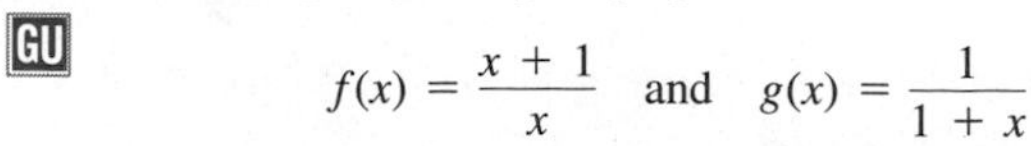

$$f(x) = \frac{x + 1}{x} \quad \text{and} \quad g(x) = \frac{1}{1 + x}$$

Are these functions inverses of each other? Confirm your conclusion algebraically.

53. Explain why $(f^{-1})^{-1} = f$.

54. Give examples of functions f and g to confirm that the inverse of the quotient function $(f/g)^{-1}$ does not necessarily equal the quotient function g/f. That is, give examples of functions f and g to show that $(f/g)^{-1} \neq g/f$.

55. Suppose $f(x) = 2x + 4$ and $g(x) = 3x - 1$.

(a) Determine f^{-1}, g^{-1}, and $(f \circ g)^{-1}$.

(b) Show that $(f \circ g)^{-1} \neq f^{-1} \circ g^{-1}$.

(c) Show that $(f \circ g)^{-1} = g^{-1} \circ f^{-1}$.

(d) Generalize your results to express $(g \circ f)^{-1}$ in terms of g^{-1} and f^{-1} for any two invertible functions f and g. Give examples to support your assertion.

In problems 56 and 57, graph f and then use symmetry with respect to the line $y = x$ to graph f^{-1} without first finding the equation for f^{-1}. Then verify the result by finding the equation for f^{-1} and graphing it.

56. **(a)** $f(x) = 3x + 2$

(b) $f(x) = 2x^3 - 1$

57. **(a)** $f(x) = 1 - 7x$

(b) $f(x) = \sqrt[3]{x + 1}$

CHAPTER 2 REVIEW PROBLEM SET

1. Indicate whether each correspondence depicted in the diagrams is a function. Explain your reasoning.

(a)

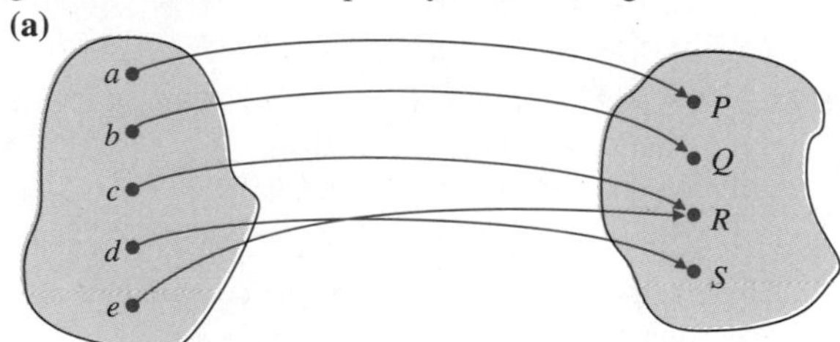

(b)

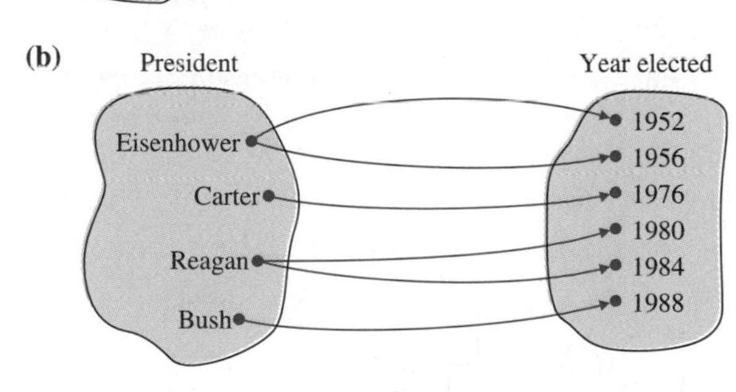

2. Rewrite each equation in the function form $y = f(x)$.

(a) $3x - 7y + 3 = 0$ **(b)** $17x^2 + 3y - 20 = 0$

3. Let $f(x) = x^3$, $g(x) = 2x + 3$, and $h(x) = x^2 - 5x$. Find each value:

(a) $f(-1)$ **(b)** $g(-1)$

(c) $h(-5)$ **(d)** $f(\sqrt[3]{2})$

(e) $h(\sqrt{2})$ **(f)** $g(a + b) - g(a)$

(g) $\dfrac{g(a + b) - g(a)}{b}$

4. Let $f(x) = 2x - 5$, $g(x) = 3x^2 + 4x$, and $h(x) = \sqrt{4x - 3}$. Find each value:

(a) $f(-2)$

(b) $g(-3)$

(c) $h(3)$

(d) $f(t + 2)$

(e) $g(b) - g(1)$

(f) $g(t + k)$

(g) $[f(a)]^2$

5. Let $f(x) = \begin{cases} x^3 & \text{if } x < 1 \\ \sqrt{x} & \text{if } x \geq 1 \end{cases}$

(a) Find $f(-2)$, $f(0)$, $f(1)$, and $f(4)$.

(b) Find the domain and range of f.

(c) Sketch the graph of f.

6. Figure 1 shows the graph of a function f.

(a) Find $f(-5)$, $f(-4)$, $f(1)$, and $f(5)$.

(b) In what intervals is f increasing?

(c) In what intervals is f decreasing?

(d) Use the graph to solve the equation $f(x) = 3$.

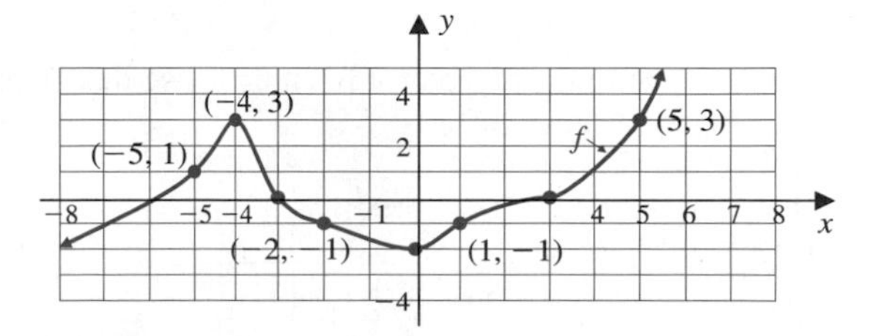

FIGURE 1

7. Determine which of the graphs shown represent a function.

(a)

(b)

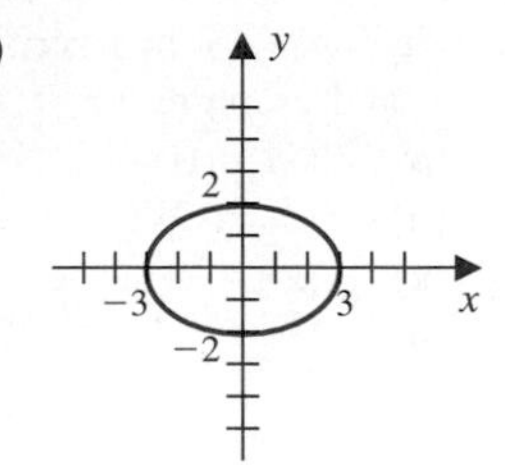

(c)

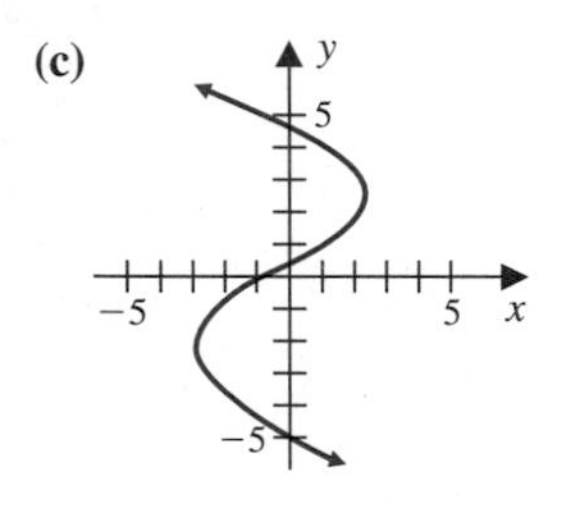

(d)

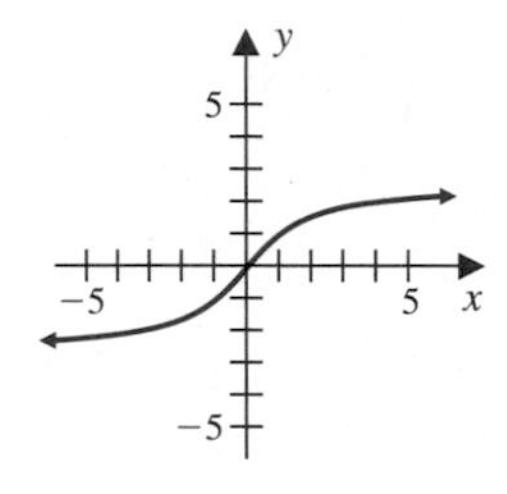

8. GU Use a graphing utility to graph each of the following functions:
(i) $f(x) = x^3 - 3x$ **(ii)** $g(x) = -x^4 + 3x^2 + 1$
Then:
(a) Discuss the symmetry of each graph.
(b) Use symmetry to determine whether each function is even or odd.
(c) Verify the results in part (b) algebraically.

9. Sketch the graph of each equation and determine whether each graph represents a function.
(a) $|y| = 4x^2$ **(b)** $|y| = 2$

10. GU **(a)** Use a graphing utility to graph $g(x) = x^4 + 4x^2$.
(b) Use the graph to determine which of the following assertions hold for all real numbers:
(i) $g(-x) = g(x)$ **(ii)** $g(-x) = -g(x)$

11. Describe a sequence of transformations that will transform the graph of f into the graph of g.
(a) $f(x) = x^2 + 1$; $g(x) = (x + 2)^2 + 4$
(b) $f(x) = x^2 - x$; $g(x) = (x + 1)^2 - (x + 1) + 3$
(c) $f(x) = x^2 - 5$; $g(x) = 5 - x^2$
(d) $f(x) = \sqrt{x + 4}$; $g(x) = -2\sqrt{x + 4} + 2$

12. Figure 2 shows the graph of a function f. Use this graph to sketch each graph of g:
(a) $g(x) = f(x) - 3$
(b) $g(x) = f(x - 3)$
(c) $g(x) = -3f(x)$

FIGURE 2

13. The graph of g is obtained from the graph of the given function f as specified. Determine the function equation for g. Graph f and g on the same coordinate system.
(a) Shift the graph of $f(x) = -x^2 + 2$ horizontally 3 units to the right and then vertically downward 4 units.
(b) Shift the graph of $f(x) = \sqrt[3]{x}$ horizontally 5 units to the left, vertically stretch it by a factor of 2, and then vertically shift the result upward 3 units.
(c) Shift the graph of $f(x) = \sqrt{-x}$ horizontally 4 units to the left and then reflect it across the x axis.

14. Graph each of the following functions and find the x intercepts.
(a) $f(x) = |x - 6| - 2$
(b) $f(x) = -2(x - 1)^2 - 4$

15. For the given functions f and g, find an expression for:
(i) $(f + g)(x)$ **(ii)** $(f - g)(x)$ **(iii)** $(f \cdot g)(x)$
(iv) $(f \circ g)(x)$ **(v)** $(g \circ f)(x)$ **(vi)** $(f/g)(x)$
(a) $f(x) = x + 2$; $g(x) = x - 1$
(b) $f(x) = x^3$; $g(x) = -x$
(c) $f(x) = x^2$; $g(x) = 2x + 1$
(d) $f(x) = 2x^2 - 1$; $g(x) = 2x^2 + 1$

16. Let $f(x) = x^3$ and $g(x) = \sqrt[3]{x}$.
(a) Find $(f \circ g)(x)$ and its domain.
(b) Find $(g \circ f)(x)$ and its domain.
(c) Are $(f \circ g)(x)$ and $(g \circ f)(x)$ the same? Explain.

17. Express the given function h as the composition of two other functions f and g so that $h = f \circ g$.
(a) $h(x) = (7x + 2)^5$ **(b)** $h(t) = \sqrt{t^2 + 17}$

18. **(a)** Let $f(x) = 7 - 5x$. Find $g(x)$ so that $(f \circ g)(x) = x$.
(b) Let $f(x) = 2x + 1$. Find $g(x)$ so that $(f \circ g)(x) = 3x - 1$.

19. Show that g is the inverse of f.
(a) $f(x) = 5x + 4$; $g(x) = \dfrac{x - 4}{5}$
(b) $f(x) = \sqrt[5]{x} + 3$; $g(x) = (x - 3)^5$
(c) $f(x) = \dfrac{1}{x + 1}$; $g(x) = \dfrac{1 - x}{x}$

20. GU Use a graphing utility and the horizontal-line test to determine whether f has an inverse.
(a) $f(x) = \sqrt{3x - 2} + 5$
(b) $f(x) = x^3 + x$ **(c)** $f(x) = x^{3/5} + 1$

21. Determine the inverse of each function. Verify that

$$f[f^{-1}(x)] = f^{-1}[f(x)] = x$$

Graph f and f^{-1} on the same coordinate system.
(a) $f(x) = 7 - 13x$ **(b)** $f(x) = \sqrt[5]{x}$

22. Figure 3 shows the graph of a function f. Sketch the graph of each function defined by the given equation.

(a) $y = f^{-1}(x)$
(b) $y = f^{-1}(x) - 2$
(c) $y = f^{-1}(x - 2)$
(d) $y = f^{-1}(-x)$
(e) $y = -f^{-1}(x)$

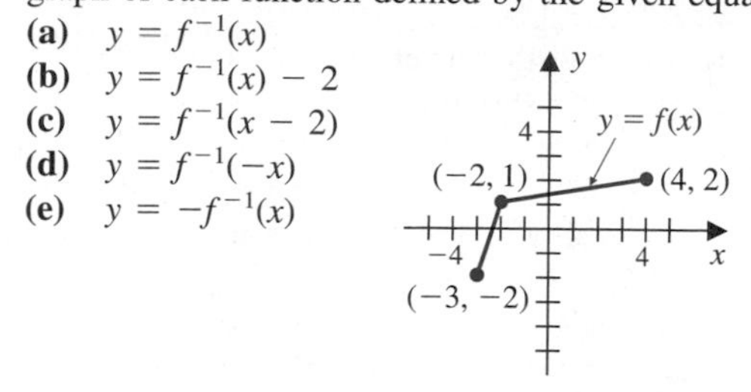

FIGURE 3

23. Sketch the graph of the inverse function f^{-1} for each of the following:

(a) $f(x) = \sqrt{x + 2}$
(b) $f(x) = \begin{cases} x^2 - 1 & \text{if } x \le 0 \\ -0.5x - 1 & \text{if } x > 0 \end{cases}$

24. Each of the given functions has no inverse. State one way to restrict the domain of each function so that the restricted function has an inverse.

(a) $f(x) = |x - 2|$ **(b)** $f(x) = 2x^2$

25. Marketing: A produce distributor prices oranges at \$4 per bag when the stores it supplies can sell 10,000 bags. When the demand for oranges rises to 14,000 bags, the price is reduced to \$3 per bag. Assume that the price P (in dollars) per bag is a linear function of the quantity x bags, where $10{,}000 \le x \le 20{,}000$.

(a) Find an equation that expresses P as a function x.
(b) Sketch the graph of this function.
(c) Discuss the trend. What is the price of oranges when the demand is 16,000 bags?
(d) Suppose the price per bag is \$2, how many bags of oranges are needed?

26. Coal Reserve: The amount A of coal (in tons) available from a supplier t weeks after opening a new yard is given by the following piecewise-defined function:

$$A(t) = \begin{cases} -\frac{275}{3}t + 300 & \text{if } 0 \le t < 3 \\ -\frac{125}{3}t + 375 & \text{if } 3 \le t < 6 \\ -\frac{350}{3}t + 1100 & \text{if } 6 \le t < 9 \end{cases}$$

(a) Sketch the graph of A.
(b) How many tons were available initially? After 2 weeks? After 5 weeks? After 8 weeks?
(c) What is the domain of this function?
(d) Determine both times when 65 tons of coal were available.

27. Advertising: Figure 4 displays a graph that relates the profit (or loss) P (in thousands of dollars) to the amount of money A (in thousands of dollars) spent on advertising, up to \$45,000, for a certain product.

(a) Explain why the graph defines a function.
(b) From the graph, approximately determine the domain and range.
(c) Estimate the profit when the advertising expenditure is \$2500; \$10,000; \$25,000; \$40,000.
(d) What expenditure on advertising will achieve maximum profit?
(e) Discuss the profit trend. For instance, what will happen to the profit if more than \$40,000 is spent on advertising?

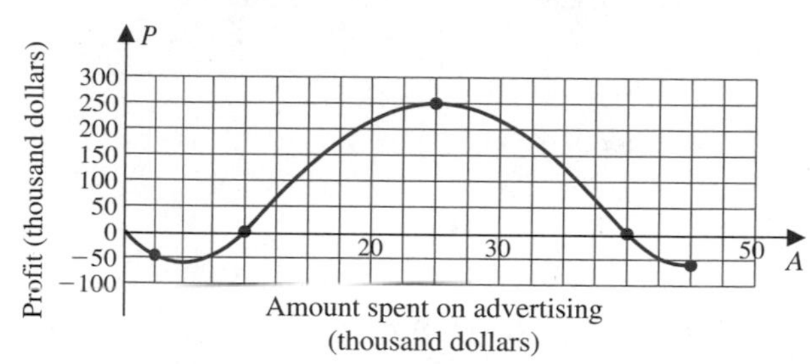

FIGURE 4

28. Modeling Profit: A manufacturer makes and sells x home television satellite systems per day. The daily cost function C (in dollars) is given by

$$C(x) = 225x + 450$$

and the daily revenue function R from selling x systems is

$$R(x) = -0.03x^2 + 675x$$

(a) Express the profit P from making and selling x systems as a function of x.
GU **(b)** Use a graphing utility to graph the profit function P.
(c) Determine the profit on a day when 200 systems are made and sold.

29. Manufacturing; A radio manufacturer finds that its production cost C (in dollars) for producing x radios is given by the function

$$C(x) = 50{,}000 + 10{,}000\sqrt[3]{x + 1}$$

It sells the radios to distributors for \$10 each, and the demand is so high that all the manufactured radios are sold.

(a) Express the revenue R (in dollars) as a function of x if x radios are produced and sold.
(b) Express the profit P (in dollars) as a function of x.
GU **(c)** Use a graphing utility to graph the function P, and then use the graph to discuss the profit trend.

30. Geometry: The area A of an equilateral triangle is given by the function

$$A = f(x) = \frac{\sqrt{3}}{4}x^2$$

where x is the side length of the triangle, and x is given by $x = g(P) = \frac{1}{3}P$, where P is the perimeter of the triangle.

(a) Find $(f \circ g)(P)$.
(b) Interpret the resulting expression in part (a).

31. Business: A group of senior citizens determines that the cost C per person (in dollars) of chartering an airplane to a nearby casino is given by the function

$$C(x) = 205 + \frac{100}{x} \qquad 1 \le x \le 200$$

where x is the number of people.

(a) Determine $C^{-1}(x)$.
(b) Interpret C^{-1}.

32. Biology: A biologist discovers that the number N of bacteria (in thousands) in a culture at a temperature T (in degrees Celsius) is given by the function $N(T) = -90(T + 1)^{-1} + 20$. The temperature T at time t (in hours) is given by $T(t) = 3t + 4$, where $0 \le t \le 12$.

(a) Find $(N \circ T)(t)$.
(b) What does the function $N \circ T$ represent?

(c) Use a graphing utility to graph $N \circ T$.
(d) Discuss what happens to the number of bacteria as the time elapses.

CHAPTER 2 TEST

1. Determine whether the correspondence described by each given diagram is a function.

(a)

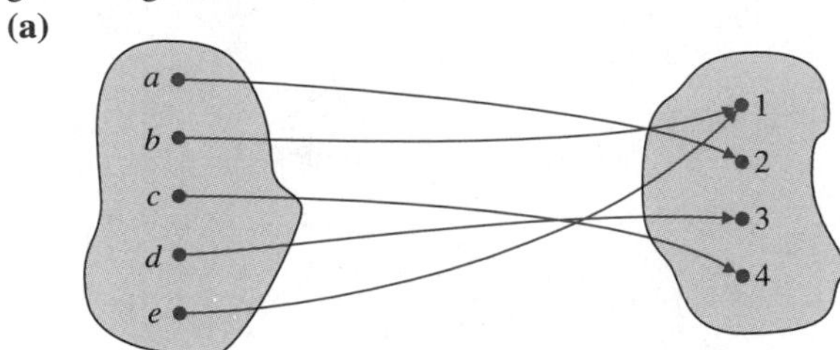

(b)

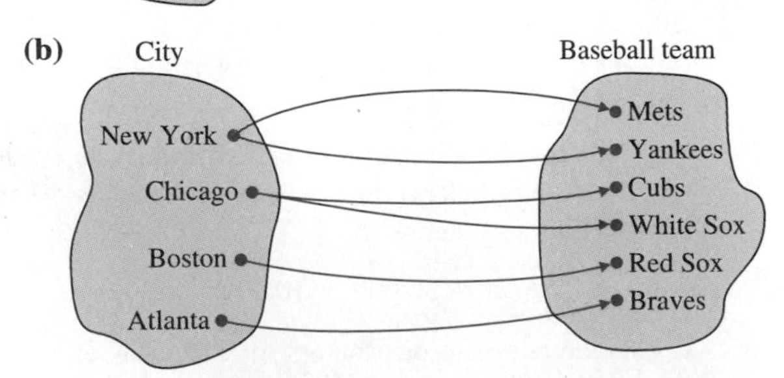

2. **(a)** Plot the following points on a Cartesian system: $(-2, 3)$, $(-1, 4)$, $(0, 0)$, $(3, 2)$.
(b) Suppose the points plotted in part (a) lie on the graph of an odd function f. Plot each point: $(2, f(2))$, $(1, f(1))$, $(0, f(0))$, $(-3, f(-3))$.

3. Let $f(x) = 4x^2 + 3$. Complete the table:

x	-2	-1	0	1	2	$-a$	a	$3t$
$f(x)$								

4. Let $f(x) = |3 - x|\sqrt{x}$. Find $f(4)$, $f(9)$, and $f\left(\frac{1}{16}\right)$.

5. Use a graphing utility to graph each function. Find the domain of each function.
(a) $g(x) = -\sqrt{4 - x}$
(b) $g(x) = \sqrt{x - 1} + \sqrt{x - 1}$

6. Sketch the graph of

$$f(x) = \begin{cases} -x & \text{if } x \le -1 \\ x^2 & \text{if } -1 < x < 1 \\ \sqrt{x} & \text{if } x \ge 1 \end{cases}$$

Find the domain and range of f.

7. Let $f(x) = 2x - 8$ and $g(x) = 1 - x^2$. Find:
(a) $(f + g)(x)$
(b) $(f \cdot g)(x)$
(c) $(f/g)(x)$
(d) $(f \circ g)(x)$
(e) $(g \circ f)(3)$
(f) $\dfrac{f(t + 3) - f(t)}{3}$

8. Figure 1 shows the graph of f.

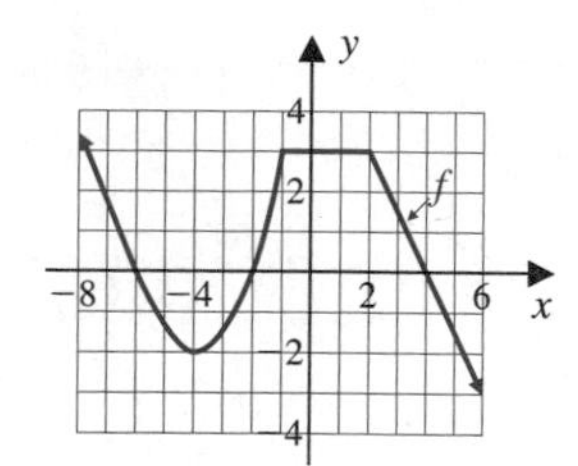

FIGURE 1

(a) Use the graph to find $f(-6)$, $f(-4)$, $f(1)$, $f(2)$, and $f(4)$.
(b) On what intervals is f increasing?
(c) On what intervals is f decreasing?
(d) On what interval is f constant?
(e) What is the domain of f?
(f) What is the range of f?

9. Use the given tables for $f(x)$ and $g(x)$ to fill in the blanks in the tables for $f[g(x)]$ and $g[f(x)]$.

x	1	2	3	4
$f(x)$	3	4	2	1

x	1	2	3	4
$g(x)$	4	1	3	2

x	1	2	3	4
$f[g(x)]$				

x	1	2	3	4
$g[f(x)]$				

10. Let $f(x) = 4x - 1$.
(a) Find $f^{-1}(x)$.
(b) Verify that $f[f^{-1}(x)] = x$.
(c) Sketch the graphs of f and f^{-1} on the same coordinate system.

11. Describe a sequence of transformations that will change the graph of $f(x) = |x|$ into the graph of $g(x) = -2|x + 1| + 3$.

12. Suppose that a right triangle has a hypotenuse of length c, a leg of length a, and a leg of length 2 inches.
(a) Express c as a function of a.
(b) Express a as a function of c.
(c) Show that the functions in parts (a) and (b) are inverses of each other.

13. Medical researchers determine that the blood sugar of a person with diabetes is regulated by a mixture of time-released insulin. Figure 2 shows the amount of sugar A (in milligrams per deciliter) t hours after the insulin was taken.
(a) What is the domain of this function?
(b) When is the blood sugar 150 milligrams per deciliter?
(c) When is the blood sugar lowest?
(d) What is the lowest blood sugar?

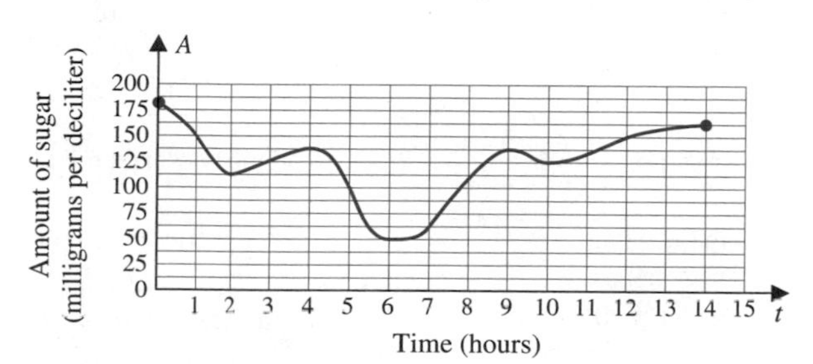

FIGURE 2

14. A magazine publisher has a profit of $95,000 per year when it distributes the magazine to 31,000 subscribers. When its distribution is increased by 4000 subscribers, its profit is increased by $20,000. Assume the profit P is a linear function of the number of subscribers S.
(a) Express P as a function of S.
(b) What will the profit be if the publisher distributes the magazine to 40,000 subscribers?
(c) Graph the function P.
(d) How many subscribers receive the magazine if no profit is obtained for that year?

3

Polynomial and Rational Functions

CHAPTER CONTENTS

In Chapter 2, we explored the general concepts associated with functions and their graphs. In this chapter, we consider *polynomial functions*—those functions whose rules are given by polynomials. In addition, we study *rational functions*—that is, functions formed by quotients of polynomials. We will continue to emphasize graphs and applications, including mathematical modeling of real-world phenomena.

OBJECTIVES

1. Graph quadratic functions by using transformations.
2. Solve quadratic inequalities graphically.
3. Determine extreme values.
4. Solve applied problems.

3.1 Quadratic Functions

So far we have discussed constant functions [functions whose rules are given by $f(x) = b$, $b \neq 0$] and linear functions [functions of the form $g(x) = mx + b$]. These are special cases of polynomial functions of zero degree and first degree, respectively. We begin this chapter with a study of second-degree polynomial functions, referred to as *quadratic functions*.

DEFINITION
QUADRATIC FUNCTION

A function f of the form

$$f(x) = ax^2 + bx + c$$

where a, b, and c are constants and $a \neq 0$, is called a **quadratic function**.

Examples of quadratic functions are

$$h(x) = x^2, \quad f(x) = 3x^2 - 1, \quad \text{and} \quad g(x) = -0.5x^2 - x + 3$$

Graphing Quadratic Functions by Using Transformations

Figures 1a and 1b show the graphs of the quadratic functions $f(x) = x^2$ and $g(x) = -x^2$, respectively.

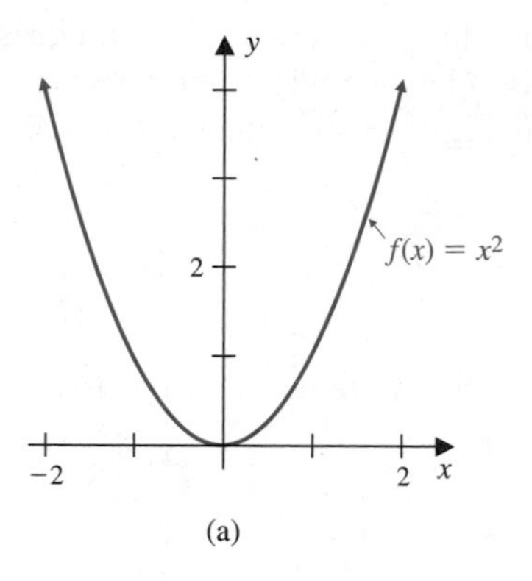

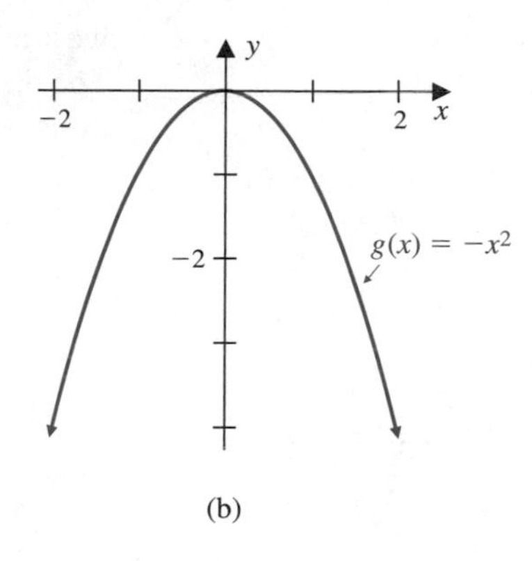

FIGURE 1

The first successful manned U.S. space flight—flown by Alan Shepard, Jr. in 1961—followed a suborbital path that was parabolic.

These graphs are examples of curves called *parabolas*. In fact, the graph of any quadratic function is a **parabola**. All such graphs have the same basic shape, and they have the following geometric characteristics, as shown in Figure 2:

1. Each graph is symmetric with respect to a vertical line called the **axis of symmetry**.
2. Each graph **opens upward** (when the coefficient of x^2 is *positive*) or **downward** (when this coefficient is *negative*).
3. The **vertex** of a parabola is either the lowest point on the graph (if it opens upward) or the highest point (if it opens downward).

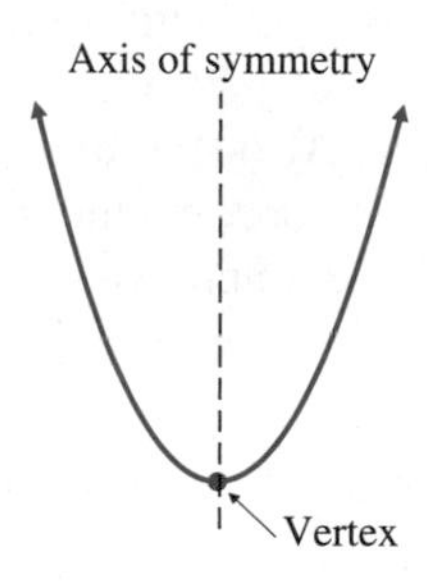

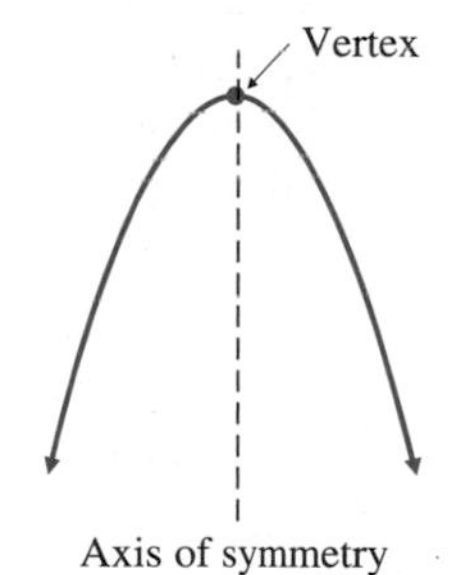

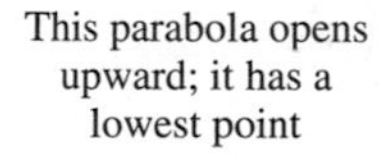

This parabola opens upward; it has a lowest point

This parabola opens downward; it has a highest point

FIGURE 2
Characteristics of parabolas

When a quadratic function $f(x) = ax^2 + bx + c$ is rewritten in the special algebraic form we can use the transformation techniques introduced in Section 2.3 to graph it. The diagram below reviews the effects of the values of a, h, and k on transformations, and Figure 3 illustrates a transformation from the graph of $y = x^2$ to the graph of f.

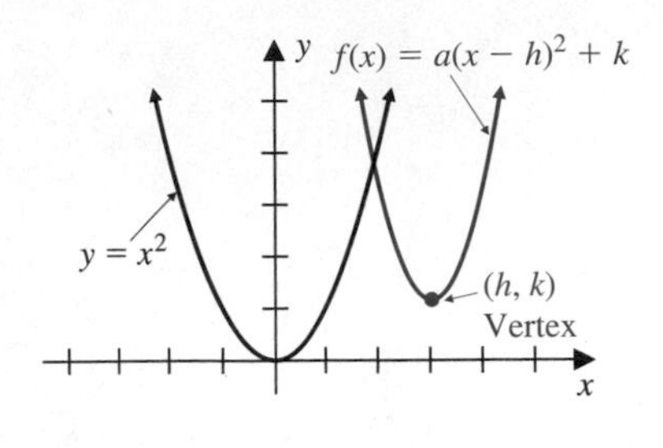

FIGURE 3

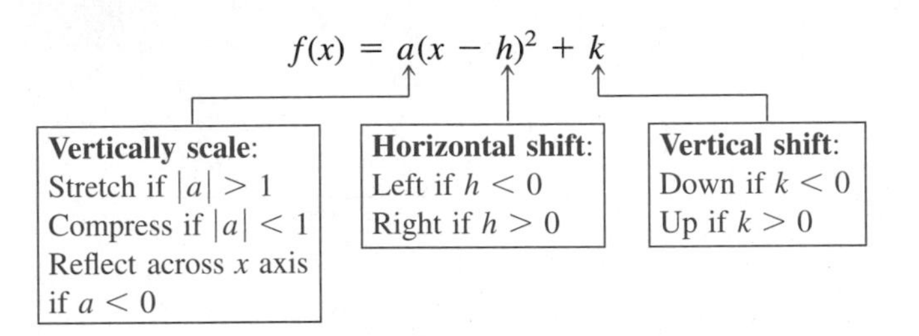

EXAMPLE 1 **Sketching the graph of a parabola by using transformations**

Use the graph of $y = x^2$ along with transformations to graph the equation $g(x) = -2(x - 1)^2 + 3$. Find the vertex and axis of symmetry of the parabola. Also determine the domain and range.

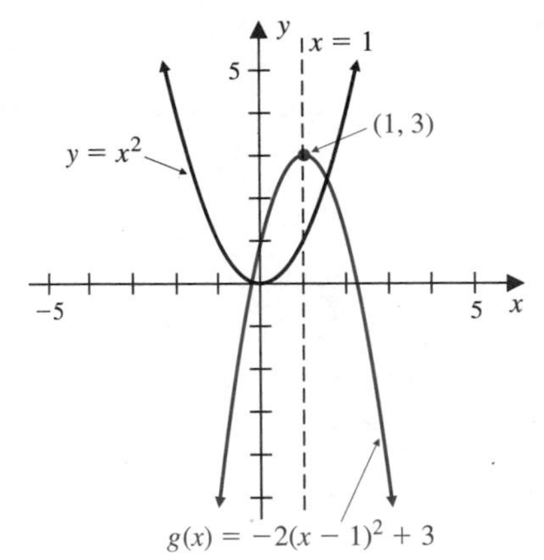

FIGURE 4

Solution The graph of g is obtained by transforming the graph of $y = x^2$ as follows:

(i) Vertically stretch the graph of $y = x^2$ by a factor of 2.
(ii) Reflect the resulting graph across the x axis.
(iii) Shift the latter graph to the right by 1 unit.
(iv) Finally, shift the graph obtained in step (iii) upward by 3 units to get the graph of $g(x) = -2(x - 1)^2 + 3$ (Figure 4).

Because of the 1 unit horizontal shift to the right and the 3 unit shift upward, the vertex (0, 0) of the graph of $y = x^2$ is shifted to the point (1, 3) to become the vertex of the graph of g. Also, the axis of symmetry of $y = x^2$ (the y axis) is shifted 1 unit to the right to become the line $x = 1$, which is the axis of symmetry of the graph of g. From the graph, we see that the domain of g is $\mathbb{R}$ and the range includes all numbers in the interval $(-\infty, 3]$.

A quadratic function given in the *standard form*

$$f(x) = ax^2 + bx + c \qquad a \neq 0$$

can be converted to the above special algebraic form by using the technique of completing the square.

EXAMPLE 2 **Converting a quadratic function to the special algebraic form**

Rewrite $f(x) = 3x^2 + 6x + 5$ in the form $f(x) = a(x - h)^2 + k$. Find the vertex and axis of symmetry, and sketch the graph of f. Also determine the range of f.

Solution To rewrite the quadratic function f in the special algebraic form, we proceed as follows:

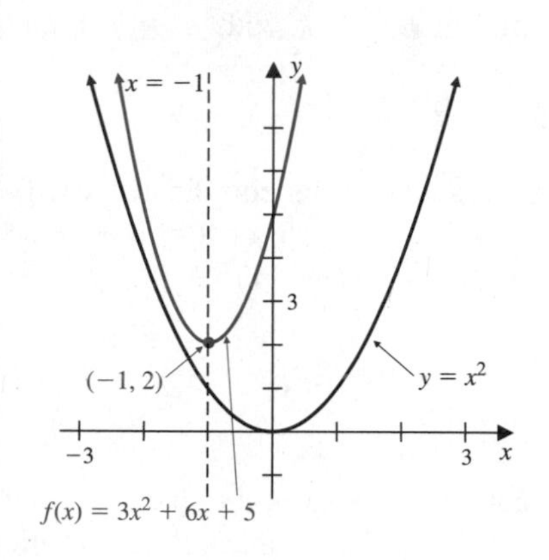

FIGURE 5

$f(x) = 3x^2 + 6x + 5$	Given
$= 3(x^2 + 2x) + 5$	Group the x terms and factor out 3
$= 3(x^2 + 2x + 1 - 1) + 5$	Complete the square for $x^2 + 2x$ by adding and subtracting 1 inside the parentheses
$= 3(x^2 + 2x + 1) - 3 + 5$ $= 3(x + 1)^2 + 2$	Regroup, factor, and simplify
$= 3[x - (-1)]^2 + 2$	Special form

The graph of f can be obtained by first vertically stretching the graph of $y = x^2$ by a factor of 3 ($a = 3$), then shifting the resulting curve 1 unit to the left ($h = -1$), and finally shifting this latter curve 2 units upward ($k = 2$) (Figure 5). The graph of f is a parabola that opens upward with vertex at $(-1, 2)$; its axis of symmetry is the line $x = -1$. The range includes all numbers in the interval $[2, \infty)$.

Table 1 summarizes the information about the graph of a quadratic function that can be read from its special algebraic form.

TABLE 1

Summary: Characteristics of the Graph of $f(x) = a(x - h)^2 + k$

If:	Parabola Opens:	Vertex	Axis of Symmetry	Illustration
$a > 0$	Upward	(h, k)	$x = h$	
$a < 0$	Downward	(h, k)	$x = h$	

Up to now we have sketched graphs of quadratic functions defined by given equations. Sometimes we can reverse the process and use data from a parabolic graph to derive an equation that defines the graph.

EXAMPLE 3 **Determining an equation of a parabola**

Find an equation of the parabola, expressed in the form $f(x) = a(x - h)^2 + k$, that fits the data in Figure 6 (at the top of page 158).

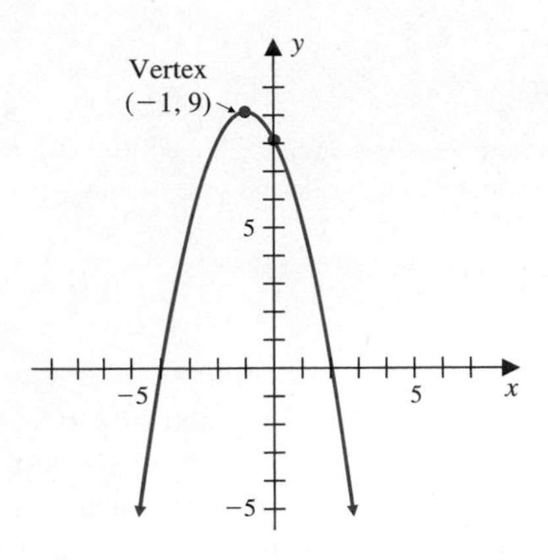

FIGURE 6

Solution Figure 6 shows the vertex at the point $(-1, 9)$, and the parabola opens downward. Using the special form $f(x) = a(x - h)^2 + k$ with $h = -1$ and $k = 9$, we get

$$f(x) = a(x + 1)^2 + 9$$

To find a, notice that $(0, 8)$ is a point on the parabola, so its coordinates satisfy the equation $f(x) = a(x + 1)^2 + 9$. Thus, $8 = a(0 + 1)^2 + 9$, or $8 = a + 9$, so $a = -1$. Hence, an equation of the parabola is $f(x) = -(x + 1)^2 + 9$.

Solving Quadratic Inequalities Graphically

The real number solutions of the quadratic equation $ax^2 + bx + c = 0$ correspond to the x intercepts of the graph of $f(x) = ax^2 + bx + c$. For instance, if a quadratic equation has two distinct real solutions, then the graph of its associated function has two x intercepts (Figure 7a). If the equation has one real solution, then the graph has one x intercept (Figure 7b), and if the equation has imaginary solutions, then the graph has no x intercepts (Figure 7c).

We can use the locations of the x intercepts along with the graph of a quadratic function to *solve a quadratic inequality* by determining where the graph is above or below the x axis.

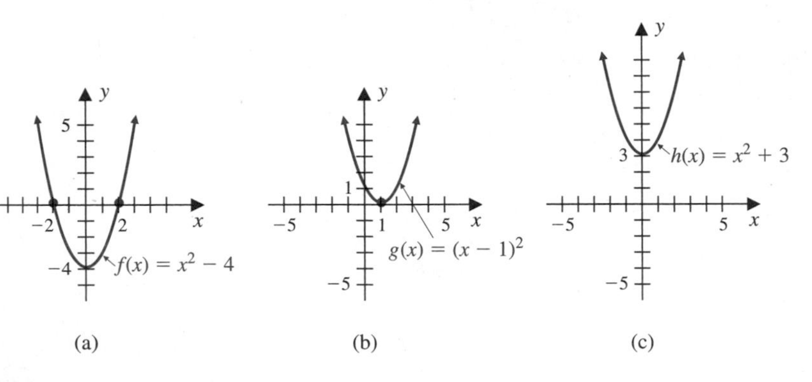

FIGURE 7

EXAMPLE 4 Solving quadratic inequalities graphically

Use the graph of $f(x) = 3x^2 - 5x + 2$ to solve each inequality.

(a) $3x^2 - 5x + 2 \geq 0$ (b) $3x^2 - 5x + 2 < 0$

Solution We begin by locating the x intercepts of the graph of f. To find them, we set $f(x) = 0$ and then solve the resulting equation. In this case, we have

$$3x^2 - 5x + 2 = (3x - 2)(x - 1) = 0 \qquad \text{to get} \qquad x = \tfrac{2}{3} \quad \text{or} \quad x = 1$$

So the x intercepts are $\frac{2}{3}$ and 1. Figure 8 shows the graph of f and the locations of the x intercepts.

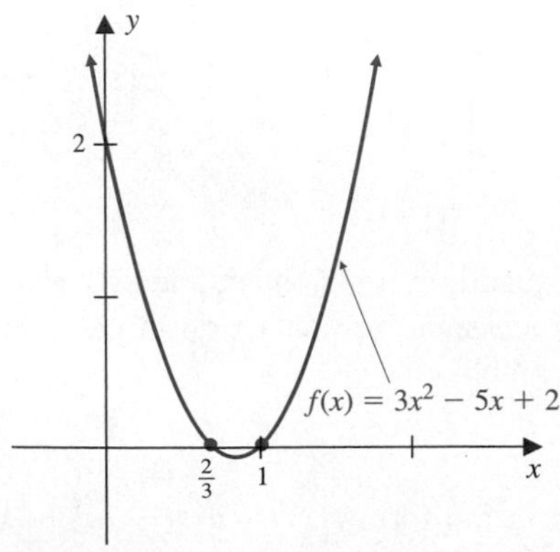

FIGURE 8

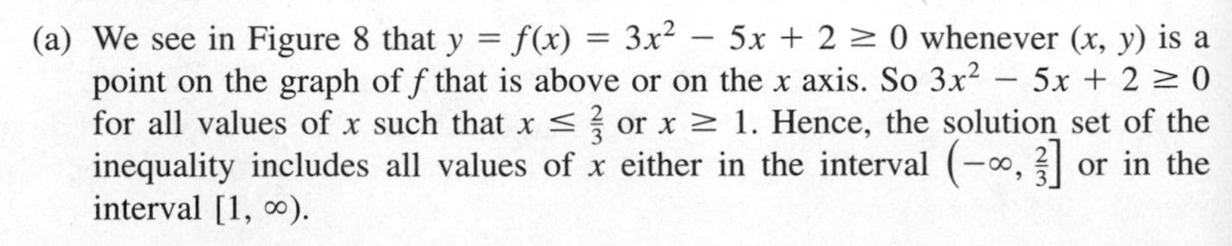

(a) We see in Figure 8 that $y = f(x) = 3x^2 - 5x + 2 \geq 0$ whenever (x, y) is a point on the graph of f that is above or on the x axis. So $3x^2 - 5x + 2 \geq 0$ for all values of x such that $x \leq \frac{2}{3}$ or $x \geq 1$. Hence, the solution set of the inequality includes all values of x either in the interval $\left(-\infty, \frac{2}{3}\right]$ or in the interval $[1, \infty)$.

(b) Again, we observe in Figure 8 that $y = f(x) = 3x^2 - 5x + 2 < 0$ when the graph of f is below the x axis. Thus, the solution set of $3x^2 - 5x + 2 < 0$ includes all values of x such that $\frac{2}{3} < x < 1$, that is, all numbers in the interval $\left(\frac{2}{3}, 1\right)$.

Determining Extreme Values

We know that the graph of a quadratic function is a parabola that opens either upward or downward. If the graph opens upward, the vertex, which occurs at the lowest point on the graph, is called the **minimum point** of the function (see Figure 8). Analogously, if the graph opens downward, the vertex, or highest point on the graph, is called the **maximum point** (see Figure 6). The minimum or maximum point is called the **extreme point** of the graph, and its y coordinate is called the **extreme value** (either the minimum or the maximum value) of the function.

By employing the technique of completing the square, we can rewrite $f(x) = ax^2 + bx + c$ as

$$f(x) = a\left(x + \frac{b}{2a}\right)^2 + \left(c - \frac{b^2}{4a}\right) \quad \text{(Problem 56)}$$

Comparing the latter form with the special form

$$f(x) = a(x - h)^2 + k$$

we see that $h = -b/(2a)$, so the extreme point is

$$(h, k) = \left(-\frac{b}{2a}, f\left(-\frac{b}{2a}\right)\right)$$

This result provides us with an easy way of finding the extreme value of a quadratic function (Table 2).

TABLE 2

Extreme Value of $f(x) = ax^2 + bx + c$

If:	Parabola Opens:	Extreme Point, or Vertex	Extreme Value	Illustration
$a > 0$	Upward	$\left(-\frac{b}{2a}, f\left(-\frac{b}{2a}\right)\right)$	$f\left(-\frac{b}{2a}\right)$ (Minimum)	y, x; Extreme point (Minimum) $\left(-\frac{b}{2a}, f\left(-\frac{b}{2a}\right)\right)$
$a < 0$	Downward	$\left(-\frac{b}{2a}, f\left(-\frac{b}{2a}\right)\right)$	$f\left(-\frac{b}{2a}\right)$ (Maximum)	y, x; Extreme point (Maximum) $\left(-\frac{b}{2a}, f\left(-\frac{b}{2a}\right)\right)$

EXAMPLE 5 **Finding the extreme value algebraically**

Find the extreme value of $f(x) = -2x^2 + 12x - 16$, and determine whether the value is a maximum or minimum.

Solution The function $f(x) = -2x^2 + 12x - 16$ is in the standard form $f(x) = ax^2 + bx + c$, where $a = -2$ and $b = 12$. Since a is negative, the graph opens downward and there is a maximum value when

$$x = -\frac{b}{2a} = -\frac{12}{-4} = 3$$

The maximum value of f is given by

$$f\left(-\frac{b}{2a}\right) = f(3) = -2(3)^2 + 12(3) - 16 = 2$$

Consult your manual for the proper terminology and keystroke instructions.

Graphing utilities have special features that enable us to find the locations of extreme points with a high degree of accuracy. For instance, suppose we want to use a graphing utility to approximate the location of the extreme point of the graph of $f(x) = 2.4x^2 - 3.7x + 1.7$ to four decimal places. We begin by graphing the function $f(x) = 2.4x^2 - 3.7x + 1.7$ with a graphing utility (Figure 9a). Next, we use a feature of the graphing utility usually called the ZOOM feature to fill the screen with a magnified viewing window of a smaller portion of the curve that contains the extreme point. Repeating this use of the ZOOM feature gives us closer and closer views of the extreme point (Figure 9b). Then, by using the TRACE feature, we move the cursor along the curve so we are able to monitor the coordinates of its location, which are shown at the bottom of the screen. Since this graph has a minimum point, we obviously are looking for the location that displays the smallest possible y value. One result of this technique is shown in Figure 9c, which indicates that the approximate location of the minimum point, rounded to four decimal places, is (0.7708, 0.2740).

Some graphing utilities are programmed to locate extreme points on the original graph without using the ZOOM and TRACE features.

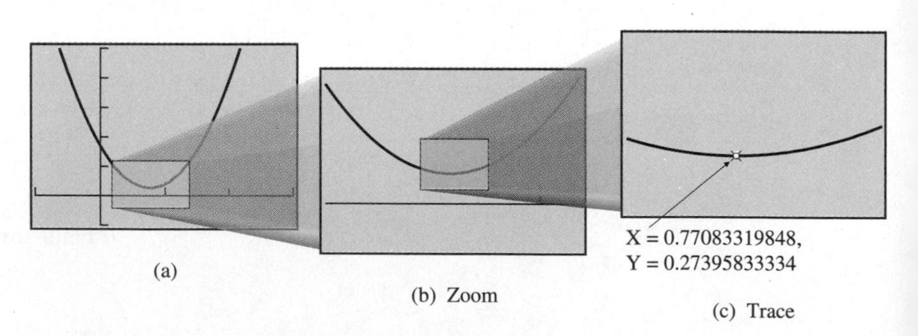

FIGURE 9

This result can be confirmed algebraically. From Table 2, we know that the extreme point of $f(x) = 2.4x^2 - 3.7x + 1.7$ occurs when

$$x = \frac{-b}{2a} = \frac{-(-3.7)}{2(2.4)} = \frac{3.7}{4.8} = \frac{37}{48}$$

So the minimum value of $f(x)$ is given by

$$f\left(\tfrac{37}{48}\right) = 2.4\left(\tfrac{37}{48}\right)^2 - 3.7\left(\tfrac{37}{48}\right) + 1.7 = \tfrac{263}{960}$$

Since $\frac{37}{48}$ is approximately 0.7708 and $\frac{263}{960}$ is approximately 0.2740, the result is confirmed.

Solving Applied Problems

The solutions of many applied problems depend on finding the extreme value of a quadratic function. In Example 6, we analyze a model defined by a quadratic function; and in Example 7, we create a quadratic function to model a situation.

EXAMPLE 6 **Examining profit**

(See Example 8 in Section 2.4.) Suppose that the profit $P(x)$ (in dollars) for selling x units of a product, where $0 \le x \le 30{,}000$, is given by

$$P(x) = -0.003x^2 + 120x - 500{,}000$$

(a) Find the profit if the company sells 15,000 units.
(b) Find the maximum profit and when it occurs.

(c) Use a graphing utility to graph P with the given restrictions, and then use the graph to discuss the profit–loss pattern.

Solution

(a) Since $P(15{,}000) = -0.003(15{,}000)^2 + 120(15{,}000) - 500{,}000 = 625{,}000$, it follows that if 15,000 units are sold, the profit is \$625,000.

(b) The quadratic function P has the form

$$P(x) = ax^2 + bx + c = -0.003x^2 + 120x - 500{,}000$$

where $a = -0.003$, $b = 120$, and $c = -500{,}000$, so we can use the properties in Table 2. Since $a < 0$, the function P has a maximum value that occurs when

$$x = -\frac{b}{2a} = \frac{-120}{2(-0.003)} = 20{,}000$$

So the maximum value of P is given by

$$P(20{,}000) = -0.003(20{,}000)^2 + 120(20{,}000) - 500{,}000 = 700{,}000$$

Thus, when 20,000 units are sold, a maximum profit of \$700,000 is earned.

(c) Figure 10 shows a viewing window of the graph of the function P. From the graph, we see that the company initially experiences a loss (P is negative). Then as sales increase, there is an increasing profit (P is positive) up to a maximum of \$700,000 when 20,000 units are sold, as found in part (b). Finally, the profit decreases as sales increase from 20,000 to 30,000. Perhaps the decreasing profits (as sales increase from 20,000 to 30,000) occur because of an increase in manufacturing costs or a price reduction.

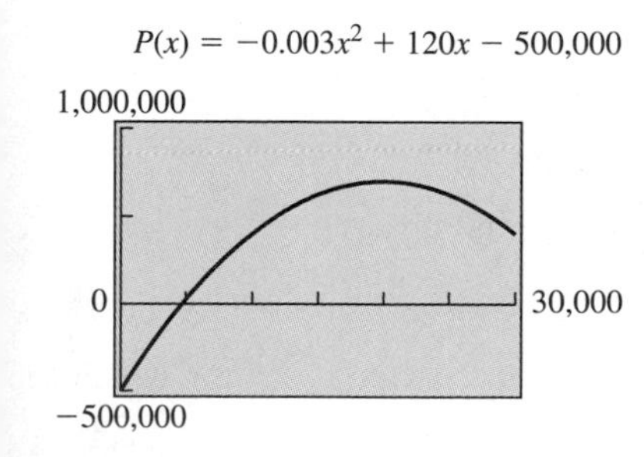

FIGURE 10

EXAMPLE 7 **Constructing a rectangular corral of maximum area**

A rancher wishes to construct a rectangular corral. To save fencing costs, a river is used along one side of the corral and the other three sides are formed by 3000 feet of fence. Assume that x represents the width (in feet) of two of the fenced sides, y the length (in feet) of the third fenced side, and A the area of the corral (Figure 11).

FIGURE 11

(a) Express the area A as a function of x.
(b) What are the restrictions on x?
(c) Find the dimensions of the corral that encloses the maximum area. What is this maximum area?

(d) Use a graphing utility to graph the function defined in part (a) with the restrictions from part (b). Discuss what happens as the width increases from 0 to 1500 feet.

Solution

(a) Since the rancher plans to use 3000 feet of fence for three sides, the sketch in Figure 11 indicates that

$$2x + y = 3000$$

so

$$y = 3000 - 2x$$

Thus, the area A of the corral is given by

$$\begin{aligned} A &= xy && \text{Area} = (\text{Width}) \cdot (\text{Length}) \\ &= x(3000 - 2x) && \text{Substitute for } y \\ &= 3000x - 2x^2 && \text{Multiply} \\ &= -2x^2 + 3000x && \text{Write in standard form} \end{aligned}$$

So the area A is a quadratic function of the width x.

(b) Since the sum of the widths, $2x$, cannot be longer than the total amount of available fencing, it follows that

$$\begin{aligned} 0 &< 2x < 3000 \\ 0 &< x < 1500 && \text{Divide each part by 2} \end{aligned}$$

(c) In order to find the maximum value of A, we first observe that

$$A = -2x^2 + 3000x = ax^2 + bx + c \qquad \text{with } a = -2 \text{ and } b = 3000$$

So by using the results in Table 2, we find that the value of x for which the maximum value of A occurs is given by

$$x = -\frac{b}{2a} = -\frac{3000}{2(-2)} = 750$$

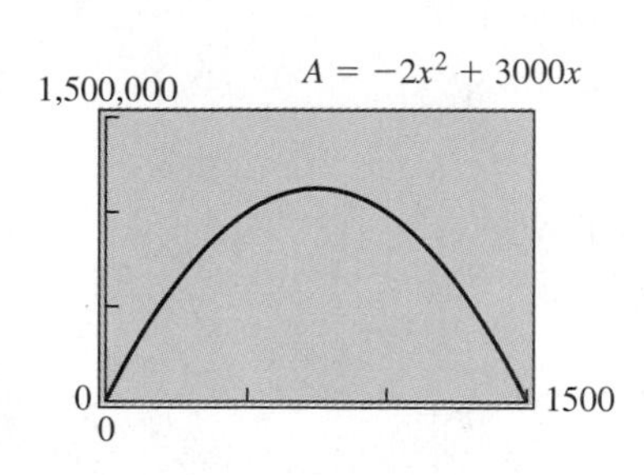

FIGURE 12

The corresponding value of y is given by

$$y = 3000 - 2x = 3000 - 2(750) = 1500$$

Therefore, the maximum value of A is given by

$$A = (750)(1500) = 1,125,000$$

So a maximum area of 1,125,000 square feet is obtained when the corral is 750 feet wide and 1500 feet long.

(d) Next, we use a graphing utility to sketch a graph of the quadratic function derived in part (a), taking into account the restrictions on x determined in part (b). The resulting graph (Figure 12) reveals that as the width x increases from 0 to 1500, the corresponding area A first increases to attain a maximum value and then decreases to a minimum value.

PROBLEM SET 3.1

Mastering the Concepts

1. Match each quadratic function with its graph in Figure 13.

(a) $f(x) = -(x - 2)^2$ **(b)** $g(x) = (x - 1)^2 - 2$
(c) $h(x) = -(x + 2)^2 - 3$
(d) $F(x) = (x + 1)^2 - 1$

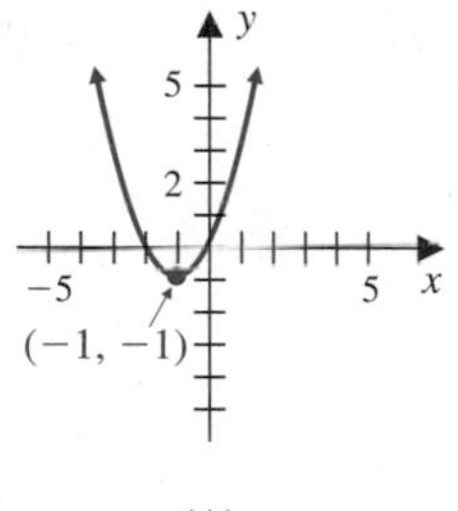

(A)

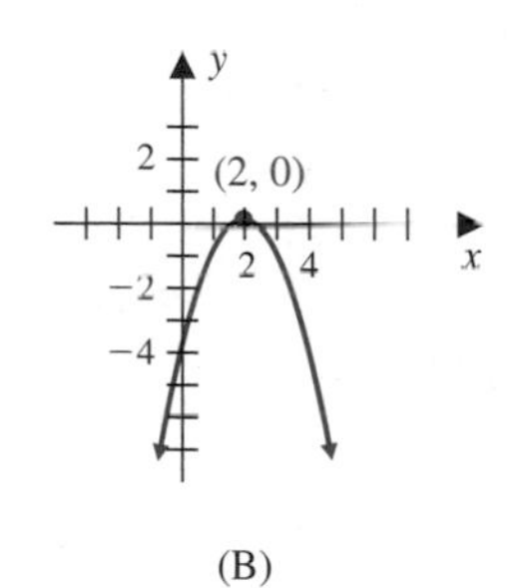

(B)

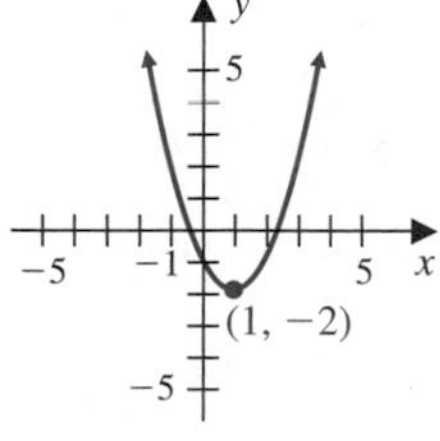

(C)

(D)

FIGURE 13

2. Describe the transformations that can be used to change the graph of $f(x) = x^2$ into the graph of g. Sketch the graph of g.

(a) $g(x) = 3x^2 + 4$
(b) $g(x) = (x + 1)^2 - 1$
(c) $g(x) = -(x + 1)^2 + 2$

In problems 3–8, use the graph of $y = x^2$ along with transformations to graph each function. Find the vertex and axis of symmetry of each parabola. Also determine the domain and range.

3. $f(x) = (x - 3)^2 + 2$
4. $g(x) = -(x + 1)^2 + 1$
5. $h(x) = -2(x + 2)^2 - 3$
6. $F(x) = 3(x - 1)^2 + 2$
7. $F(x) = \frac{1}{2}(x + 3)^2 + 1$
8. $G(x) = -\frac{1}{3}(x + 2)^2 - 1$

In problems 9–18, rewrite each function f in the form

$$f(x) = a(x - h)^2 + k$$

Find the vertex and axis of symmetry, and sketch the graph of each function. Also determine the range.

9. $f(x) = x^2 - 4x - 5$
10. $f(x) = x^2 - 10x + 20$
11. $f(x) = -x^2 - 6x - 8$
12. $f(x) = -x^2 + 5x + 11$
13. $f(x) = 3x^2 - 5x - 21$
14. $f(x) = 2x^2 - 11x + 5$
15. $f(x) = -5x^2 + 3x + 4$
16. $f(x) = -3x^2 - 6x + 5$
17. $f(x) = -\frac{1}{2}x^2 - 2x + 1$
18. $f(x) = \frac{1}{3}x^2 + \frac{1}{5}x - \frac{3}{7}$

In problems 19 and 20, find equations for the given parabolas.

19. **(a)**

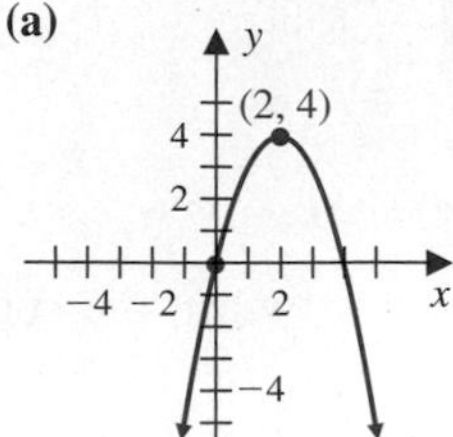

(b)

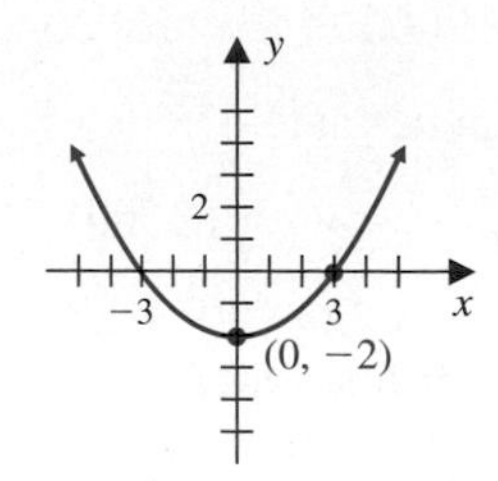

20. **(a)**

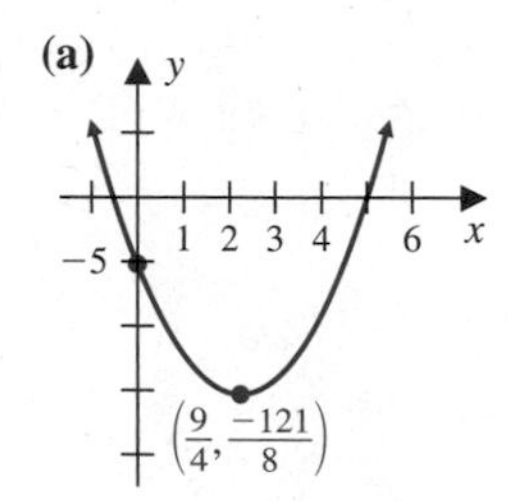

(b)

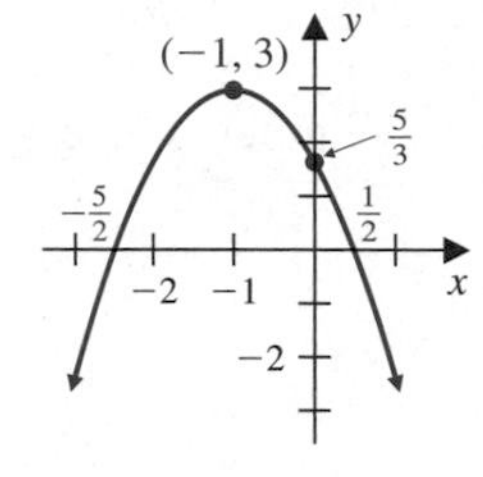

In problems 21–24, find the equation in the form $f(x) = a(x - h)^2 + k$ for the quadratic function f whose graph satisfies the given conditions. Also sketch the graph of f.

21. Vertex at $(2, -2)$ and contains the point $(0, 0)$
22. Vertex at $(-1, -9)$ and contains the point $(-4, 0)$
23. Vertex at $(0, 1)$ and contains the point $(2, -3)$
24. Vertex at $(0, 3)$ and contains the point $(-3, 0)$

In problems 25–30, use the locations of the x intercepts and the graph of the given function f to solve the associated inequality. Express the solution set in interval notation.

25. $f(x) = x^2 + 5x - 14;\ x^2 + 5x - 14 > 0$
26. $f(x) = -x^2 + x + 20;\ -x^2 + x + 20 \geq 0$
27. $f(x) = -2x^2 + 5x - 3;\ -2x^2 + 5x \leq 3$
28. $f(x) = 2x^2 - x - 1;\ 2x^2 - x \geq 1$
29. $f(x) = 6x^2 + 7x - 20;\ 6x^2 + 7x - 20 \geq 0$
30. $f(x) = 2x^2 - 9x - 5;\ 2x^2 - 9x - 5 < 0$

In problems 31–36, find the extreme value algebraically. Determine whether the value is a maximum or a minimum.

31. $f(x) = x^2 + 8x + 12$
32. $g(x) = 10x^2 + x - 2$
33. $h(x) = -x^2 + 4x - 5$
34. $h(x) = -9x^2 - 6x + 8$
35. $g(x) = 2x^2 + x - 1$
36. $h(x) = -2x^2 - 9x + 5$

GU In problems 37 and 38, use a graphing utility to approximate the location of the extreme point to four decimal places. Determine whether the extreme point is a maximum or a minimum. Confirm the result algebraically.

37. **(a)** $f(x) = x^2 - 7.58x + 1.45$
(b) $g(x) = -1.8x^2 + 2.56x - 4$

38. **(a)** $f(x) = -0.56x^2 - x - 5.87$
(b) $g(x) = 2.01x^2 - 3.41x - 1.12$

Applying the Concepts

39. Minimum Product: The difference of two real numbers is 6.
(a) Express the product P as a function of one of the numbers.
(b) Find the numbers that produce a minimum value of P. What is the minimum value?

40. Geometry: The sum of the height y (in centimeters) and the base x (in centimeters) of a triangle is 24 centimeters.
(a) Express the area A of the triangle as a function of x.
(b) Find the base that produces the maximum area of the triangle. What is the maximum area?

41. Maximum Profit: Suppose that the profit P (in thousands of dollars) of a manufacturing company is related to the number x (in hundreds) of full-time employees by the quadratic function

$$P = -1.25x^2 + 156.25x \qquad 0 \leq x \leq 100$$

(a) What is the profit if the number of employees is 500? 1000? 1500?
(b) Find the maximum profit and the corresponding number of employees.
GU **(c)** Use a graphing utility to graph the function. Discuss the profit trend as the number of employees increases.

42. Minimum Cost: The cost C (in dollars) to produce x tons of an alloy used for engine blocks is modeled by the function

$$C = 2.97x^2 - 603x + 62{,}836 \qquad 0 \leq x \leq 110$$

(a) How much does it cost if the company produces 50 tons? 100 tons? 110 tons?
(b) Find the minimum cost and the corresponding number of tons.
GU **(c)** Use a graphing utility to graph the function. Discuss the cost as the number of tons increases.

In problems 43–52, round off the answers to two decimal places when necessary.

43. Maximum Height of a Baseball: The trajectory of a baseball hit by a player at home plate is approximated by the parabola whose function is given by

$$h(x) = -0.004x^2 + x + 3.5$$

In this model, home plate is located at the origin, $h(x)$ (in feet) is the height of the ball above ground, and x is the horizontal distance (in feet) the ball has traveled from home plate.

(a) How high is the ball after it has traveled horizontally 50 feet? 100 feet? 200 feet?

GU **(b)** Use a graphing utility to graph the function. From the graph, approximate the maximum height and how far the ball travels horizontally to reach this height.

(c) Confirm the results in part (b) algebraically.

(d) If the ball is not touched in flight, how far will it have traveled horizontally when it hits the ground?

44. Maximum Height of a Rocket: A model rocket is launched and then it accelerates until the propellant burns out, after which it coasts upward to its highest point. The height h of the rocket (in meters) above ground level t seconds after the burnout is modeled by the function

$$h = -4.9t^2 + 343t + 2010$$

(a) Find the height of the rocket above ground level 10 seconds, 20 seconds, and 30 seconds after the burnout.

GU **(b)** Use a graphing utility to graph the function and approximate the maximum height attained by the rocket. How many seconds after the burnout does it take to reach the maximum height?

(c) Confirm the results in part (b) algebraically.

(d) How long after the burnout will it take for the rocket to reach ground level?

45. Construction: A lookout tower is to be constructed on the peak of a hill. A portion of a cross section of the crown of the hill is approximately parabolic in shape. If a coordinate system is superimposed on this cross section as shown in Figure 14, then the parabolic portion is modeled by the function

$$y = -0.0000333x^2 + 0.03x + 120$$

where $0 \le x \le 1200$, and x and y are measured in feet above sea level. (The parabolic shape is barely visible in Figure 14 due to the scale of the drawing.)

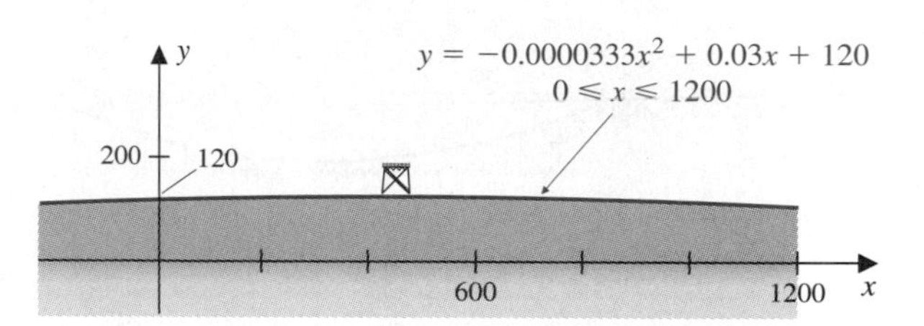

FIGURE 14

(a) Find the height of the hill above sea level when x is 200 feet, 500 feet, and 1000 feet.

GU **(b)** Use a graphing utility to graph the function and approximate the coordinates of the peak of the hill.

(c) Confirm the results in part (b) algebraically.

(d) If the observation booth of the tower is to be 150 feet above sea level, how tall should the tower be?

46. Maximum Current of a Circuit: In an electrical circuit, the power P (in watts) delivered to the load is modeled by the function

$$P = EI - RI^2$$

where E is the voltage (in volts), R is the resistance (in ohms), and I is the current (in amperes). For a circuit of 120 volts and a resistance of 12 ohms, find the current that produces the maximum power, and find the maximum power.

47. Constructing a Playground: A recreation department plans to build a rectangular playground enclosed by 1000 feet of fencing. Assume that A represents the area of the enclosure and that x and y, respectively, represent the width and length (in feet).

(a) Express the area A as a function of the width x.

(b) What are the restrictions on x?

(c) Find the dimensions of the playground that encloses the maximum area. What is this area?

GU **(d)** Use a graphing utility to graph the function defined in part (a). Discuss what happens to the area as the width increases.

48. Shape of a Suspension Bridge: A suspension bridge has a main span of 1200 meters between the towers that support the parabolic main cables, and these towers extend 160 meters above the road surface (Figure 15, page 166).

(a) Assuming that the main cable is tangent to the road at the center of the bridge and that the road is horizontal, find an equation of the main parabolic cable with respect to an xy coordinate system (with a vertical y axis, the x axis running along the road surface, and the origin at the center of the bridge).

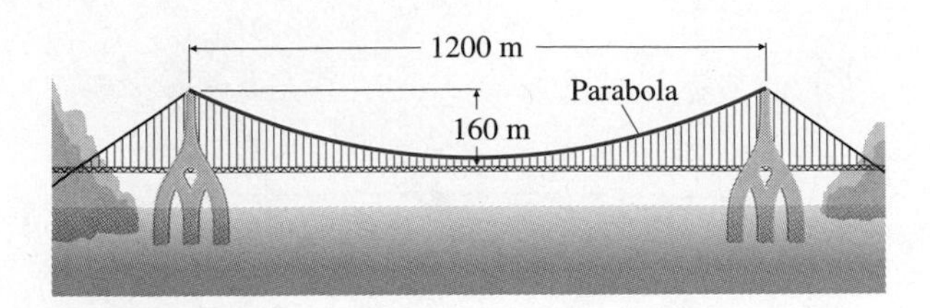

FIGURE 15

(b) Graph the equation in part (a).
(c) Find the distance between the roadway and the cable at a point 300 meters from the center of the bridge.

49. Construction: A rancher has 450 feet of fencing to enclose three sides of a rectangular pen. A barn will be used to form the fourth side of the pen. What dimensions of the pen would enclose the maximum area? What is the maximum area?

50. Construction: A breeder of horses wants to fence in two grazing areas along a river, including one interior fence separation perpendicular to the river as shown in Figure 16. The river will serve as one side, and 600 yards of fencing are available. What is the largest area that can be enclosed? What are the exterior dimensions that yield the maximum area?

FIGURE 16

51. Architecture: A Norman style window has the shape of a rectangle surmounted by a semicircular arch (Figure 17). Find the height y of the rectangle and the radius x of the semicircle of the Norman window with maximum area whose perimeter is $8 + 2\pi$ feet. (Use $\pi = 3.14$.)

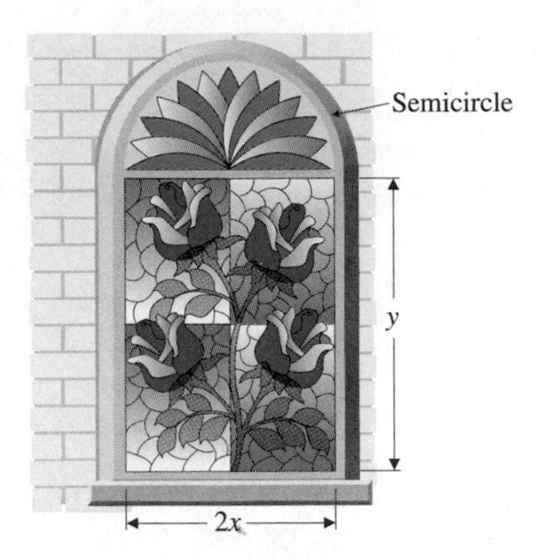

FIGURE 17

52. Agriculture: An orange grower has determined that if 120 orange trees are planted per acre, each will yield approximately 60 small boxes of oranges per tree over the growing season. The grower plans to add or delete trees in the orchard. The state agricultural service advises that because of crowding, each additional tree will reduce the average yield per tree by about 2 boxes over the growing season. How many trees per acre should be planted to maximize the total yield of oranges, and what is the maximum yield of the crop?

Developing and Extending the Concepts

53. Decide how many x intercepts the graph of the function $f(x) = ax^2 + bx + c$ has if the *discriminant*, defined as $D = b^2 - 4ac$ of the corresponding quadratic equation $ax^2 + bx + c = 0$, satisfies the given condition. Give examples of functions and their graphs to support your assertions.

(a) $D > 0$ **(b)** $D = 0$ **(c)** $D < 0$

54. GU Use the results from Problem 53 to determine how many x intercepts the graph of the function f has. Then use a graphing utility to graph the function and demonstrate the result.

(a) $f(x) = -x^2 - 6x + 7$
(b) $f(x) = 4x^2 + 12x + 9$
(c) $f(x) = 2x^2 + 3x + 2$
(d) $f(x) = -4x^2 + 28x - 49$

55. Suppose that x and y are interchanged in a quadratic function (equation) $y = a(x - h)^2 + k$. How does the shape of the resulting curve compare with the shape and location of the original curve? Give illustrations of the situation when $a < 0$ and when $a > 0$.

56. **(a)** Use the method of completing the square to rewrite the quadratic function $f(x) = ax^2 + bx + c$ in the form $f(x) = a(x - h)^2 + k$. Express h in terms of a and b.

(b) Use the result from part (a) to show that the extreme point of the graph of f is given by

$$(h, k) = \left(-\frac{b}{2a}, f\left(-\frac{b}{2a}\right)\right)$$

57. The following table shows the results of a biological experiment that relates the temperature T in degrees Celsius to the number N of insects found in a sample of water.

T	0	10	20	30	40	50
N	20	620	950	920	670	35

(a) Construct a scattergram that exhibits the data.

(b) Assuming that this relationship is approximately quadratic, determine how closely the function $N = -1.5T^2 + 75T + 20$ fits the data.

(c) Sketch the graph of the function in part (b).

(d) Assuming the function in part (b) provides an accurate approximation of the relationship between N and T for this experiment, find the value of T that predicts the maximum number of insects.

OBJECTIVES

1. Graph functions of the form $f(x) = x^n$.
2. Determine limit behavior.
3. Graph general polynomial functions.
4. Solve polynomial inequalities graphically.
5. Solve applied problems.

3.2 Polynomial Functions of Higher Degree

Up to this point we have analyzed polynomial functions of degree less than 3. Now we turn our attention to polynomial functions of higher degree.

Graphing Functions of the Form $f(x) = x^n$

Polynomial functions of the form

$$f(x) = x^n$$

are called **power functions.**

If $n \geq 4$ is an even integer, then the graph of $f(x) = x^n$ resembles the graph of $y = x^2$. For instance, Figure 1 displays the graphs of $f(x) = x^4$ and $y = x^2$. The graph of f has the following properties:

1. **Symmetry**: Since $f(-x) = (-x)^4 = x^4 = f(x)$, f is an even function, so the graph of f is symmetric with respect to the y axis.

2. **Shape of Curve**: The graph of f forms a U-shaped curve that contains the points $(-1, 1)$, $(0, 0)$, and $(1, 1)$. Figure 1 shows that the graph of f is flatter and lies below the parabola $y = x^2$ over the intervals $(-1, 0)$ and $(0, 1)$. It lies above the parabola over the intervals $(-\infty, -1)$ and $(1, \infty)$. The graph of f falls more rapidly than the graph of $y = x^2$ over the interval $(-\infty, -1)$ and rises more sharply over the interval $(1, \infty)$.

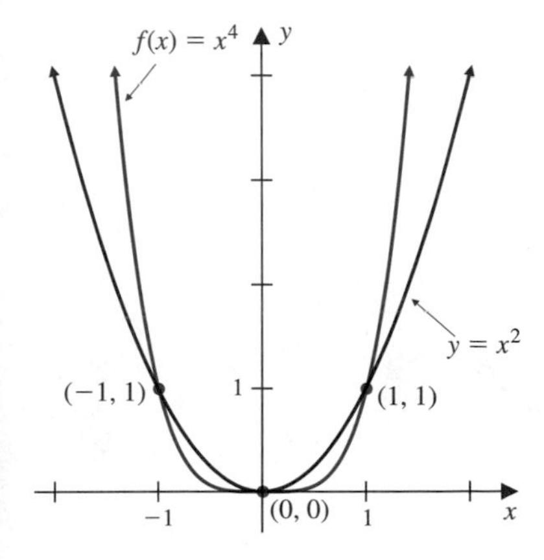

FIGURE 1

If $n \geq 5$ is an odd integer, then the graph of $f(x) = x^n$ resembles the graph of $y = x^3$. Figure 2 shows the graphs of $y = x^3$ and $f(x) = x^5$.

The graph of f has the following properties:

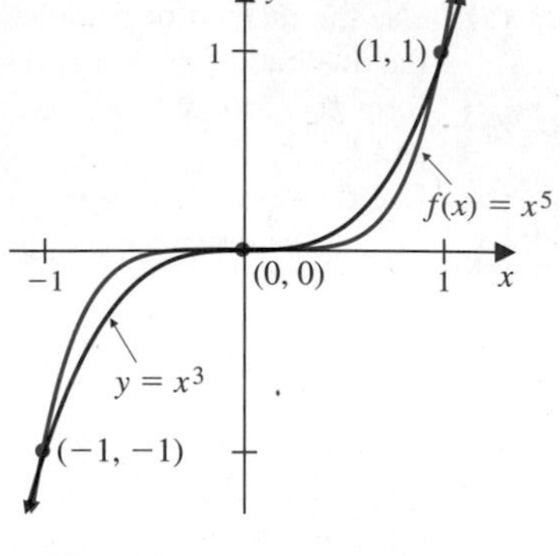

FIGURE 2

1. **Symmetry**: Since $f(-x) = (-x)^5 = -x^5 = -f(x)$, f is an odd function, so the graph of f is symmetric with respect to the origin.
2. **Shape of Curve**: The graph of f contains the origin and the points $(-1, -1)$ and $(1, 1)$. Comparing the graph of $f(x) = x^5$ to the graph of $y = x^3$, we notice that the graph of f is flatter over the interval $(-1, 1)$, and rises more sharply to the left of $x = -1$ and to the right of $x = 1$.

EXAMPLE 1 Describing graphs of $f(x) = x^n$

First indicate the symmetry and general shape of the graph of each function. Then confirm the description by sketching its graph using a graphing utility.

(a) $g(x) = x^6$ (b) $h(x) = x^7$

Solution

(a) Since the power of $g(x) = x^6$ is even, the graph is symmetric with respect to the y axis and it is U-shaped, resembling the graph of $y = x^2$. The graphs of g and $y = x^2$ are shown in the viewing window in Figure 3a.

(b) For $h(x) = x^7$, the power is odd, so the graph is symmetric with respect to the origin. The shape of the graph resembles the graph of $y = x^3$. The graphs of h and $y = x^3$ are displayed in the viewing window in Figure 3b.

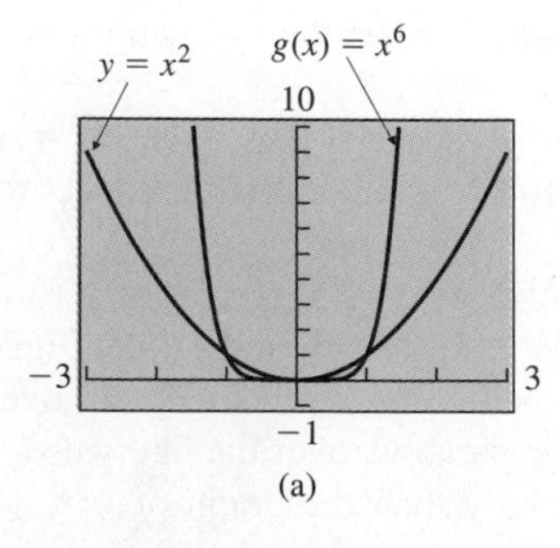

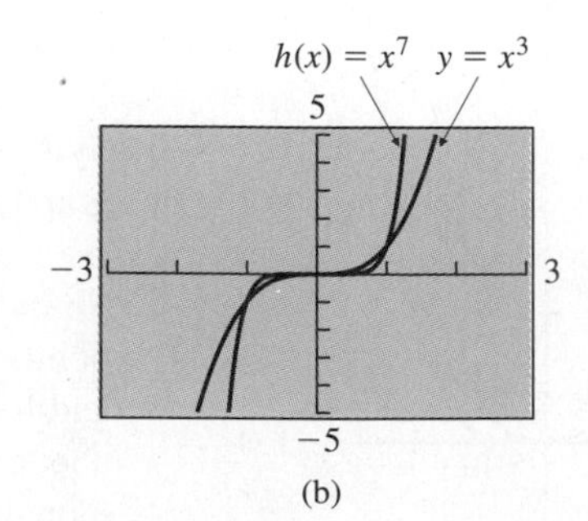

FIGURE 3

Transformations of the graphs of $y = x^n$ can be used to graph functions of the form

$$f(x) = a(x - h)^n + k$$

where $n \geq 3$ is an integer.

EXAMPLE 2 **Using transformations of a power function**

Use transformations of the graph of $y = x^4$ to sketch the graphs of $f(x) = 3x^4$, $g(x) = -3x^4$, and $h(x) = -3x^4 + 1$ on the same coordinate system.

Solution As shown in Figure 4, we vertically stretch the graph of $y = x^4$ by a factor of 3 to obtain the graph of $f(x) = 3x^4$. Reflecting the graph of $f(x) = 3x^4$ across the x axis, we get the graph of $g(x) = -3x^4$. Finally, we shift the graph of $g(x) = -3x^4$ up 1 unit to obtain the graph of $h(x) = -3x^4 + 1$. ■

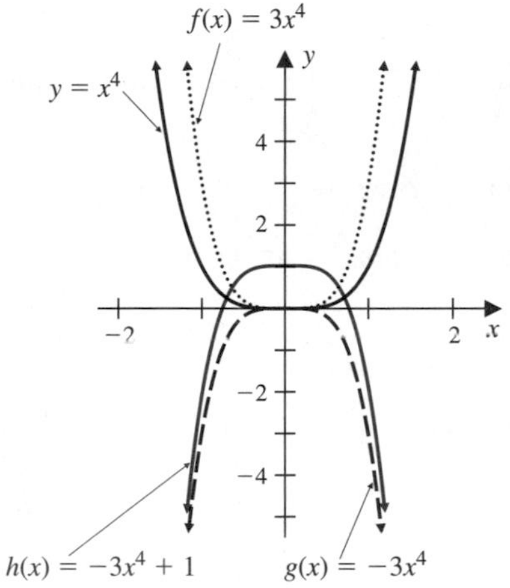

FIGURE 4

Determining Limit Behavior

Since a polynomial can be evaluated for any real number, it follows that the domain of any polynomial function consists of all real numbers. Let's examine what happens to the values (and graph) of a polynomial function $y = f(x)$ as $|x|$ gets very large for positive or negative values of x. Such behavior is referred to as the **limit behavior** of f. For instance, the data in Table 1 and the graph of $f(x) = x^4$ in Figure 1 reveal the limit behavior of $f(x) = x^4$.

TABLE 1

Limit Behavior Pattern of $f(x) = x^4$

x	−100	−10	−5	−1	0	1	5	10	100
$f(x) = x^4$	100,000,000	10,000	625	1	0	1	625	10,000	100,000,000

We see that as x takes on larger and larger positive values (increases), $f(x)$ **increases without bound**. We say that **$f(x)$ approaches positive infinity as x approaches positive infinity**, and we describe this limit behavior symbolically by writing

$$f(x) \to +\infty \quad \text{as} \quad x \to +\infty$$

Similarly, we observe that $f(x)$ increases without bound as x takes on smaller and smaller negative values (decreases). So we say that **$f(x)$ approaches positive infinity as x approaches negative infinity**. Accordingly, we write

$$f(x) \to +\infty \quad \text{as} \quad x \to -\infty$$

By examining the graph of $y = x^3$ in Figure 2, we see that $g(x) = x^3$ increases without bound as positive values of x become larger and larger (increase). Also,

$g(x)$ decreases without bound as negative values of x become smaller and smaller (decrease). So we describe the limit behavior of g by writing, respectively,

$$g(x) \to +\infty \quad \text{as} \quad x \to +\infty \qquad \text{and} \qquad g(x) \to -\infty \quad \text{as} \quad x \to -\infty$$

EXAMPLE 3 **Determining Limit Behavior**

Use the graph of each function to determine its limit behavior.

(a) The function F graphed in Figure 5a.
(b) The function G graphed in Figure 5b.

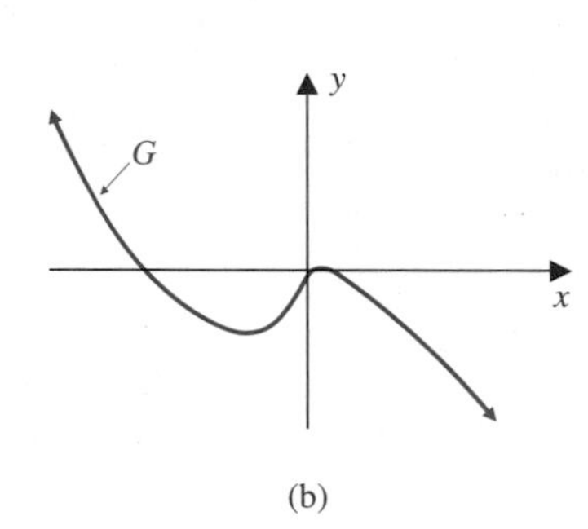

FIGURE 5

Solution

(a) Figure 5a indicates that the values of $F(x)$ increase without bound as the values of x become larger and as the values of x become smaller. Thus, we write

$$F(x) \to +\infty \quad \text{as} \quad x \to +\infty \qquad \text{and} \qquad F(x) \to +\infty \quad \text{as} \quad x \to -\infty$$

(b) As shown in Figure 5b, the values of $G(x)$ decrease without bound as x takes on larger positive values. So we write

$$G(x) \to -\infty \quad \text{as} \quad x \to +\infty$$

As the values of x become smaller (negative) numbers, $G(x)$ increases without bound. Thus, we write

$$G(x) \to +\infty \quad \text{as} \quad x \to -\infty$$

Graphing General Polynomial Functions

Accurately graphing a higher-degree polynomial function that is not expressed in the form

$$f(x) = a(x - h)^n + k \qquad \text{where } n \geq 3 \text{ is an integer}$$

often requires advanced techniques usually studied in calculus. However, computer software and graphing calculators are regularly used to sketch the graphs of such functions. Whether we are graphing such a function by hand or by using a graphing utility, it helps to know the following three properties.

Properties of Polynomial Functions

PROPERTY 1
CONTINUITY

A polynomial function is **continuous** for all real numbers.

This means that the graph of a polynomial function is an unbroken curve, with *no jumps, gaps,* or *holes*. In addition, graphs of polynomial functions have *no sharp corners*.

PROPERTY 2

LIMIT BEHAVIOR

The **limit behavior** of a polynomial function

$$f(x) = a_nx^n + a_{n-1}x^{n-1} + \cdots + a_1x + a_0 \qquad a_n \neq 0$$

is the same limit behavior of the function defined by its leading term,

$$g(x) = a_nx^n$$

This means that the shape of the graph of a polynomial function at the *far left* and *far right* closely resembles the graph of $g(x) = a_nx^n$. We refer to the function g as the *limit behavior model of f.*

Table 2 gives two examples of Property 2.

TABLE 2

Examples of Limit Behavior Models

Given Function	Limit Behavior Model	Limit Behavior Comparison	Graphical Display
$f(x) = 2x^3 - 3x^2 - 3x + 2$	$g(x) = 2x^3$	$g(x) \to +\infty$ as $x \to +\infty$ $g(x) \to -\infty$ as $x \to -\infty$ and so $f(x) \to +\infty$ as $x \to +\infty$ $f(x) \to -\infty$ as $x \to -\infty$	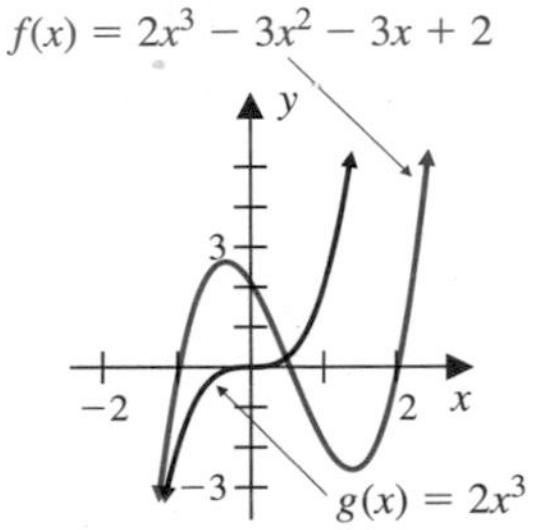 $f(x) = 2x^3 - 3x^2 - 3x + 2$ y, 3, −2, 2, x, −3 $g(x) = 2x^3$
$h(x) = -2x^4 + 2x^2$	$g(x) = -2x^4$	$g(x) \to -\infty$ as $x \to +\infty$ $g(x) \to -\infty$ as $x \to -\infty$ and so $h(x) \to -\infty$ as $x \to +\infty$ $h(x) \to -\infty$ as $x \to -\infty$	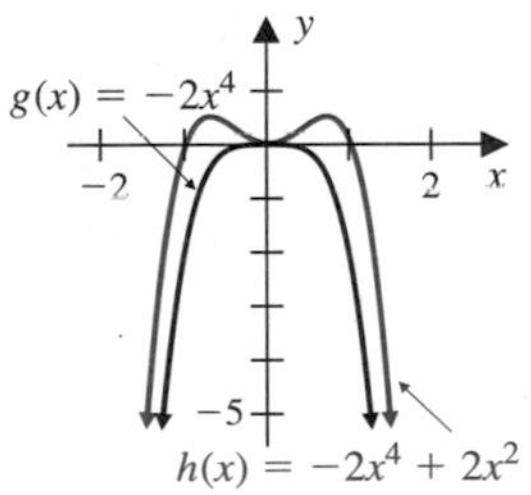 y $g(x) = -2x^4$ −2, 2, x, −5 $h(x) = -2x^4 + 2x^2$

A point on the graph of a polynomial function where the graph changes direction from increasing to decreasing, or vice versa, is called a **turning point** of the graph.

PROPERTY 3

TURNING POINTS

The graph of a polynomial function of degree n has at most $n - 1$ turning points.

The graph of a second-degree (quadratic) function has one turning point—its vertex.

This means that the total number of "peaks" and "valleys" on the graph of a polynomial function of degree n is at most $n - 1$. Figure 6 displays a graph that

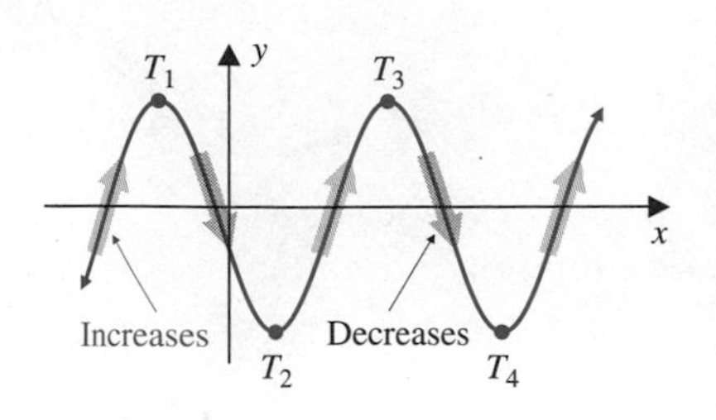

FIGURE 6

has four turning points labeled T_1, T_2, T_3, and T_4. This figure also shows that the graph has peaks at T_1 and T_3 and valleys at T_2 and T_4.

Graphing can be done more efficiently if we use the above properties to anticipate the general behavior of a function. For instance, if a graphing utility were used to graph

$$f(x) = x^3 - 4x^2 + x + 6$$

using the WINDOW (specified in Figure 7a) and no effort was made to anticipate the behavior of the function, then it would appear that the graph of f is as shown in Figure 7b.

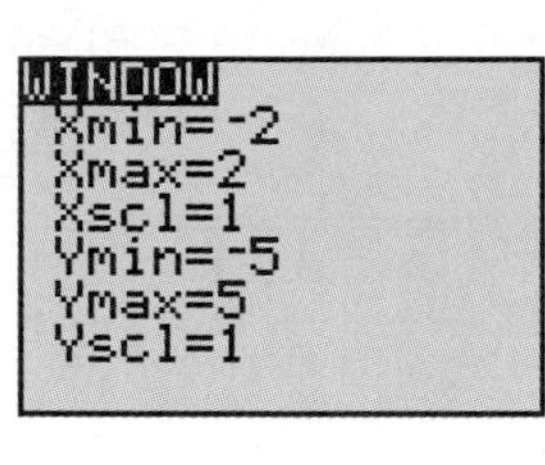

(a)

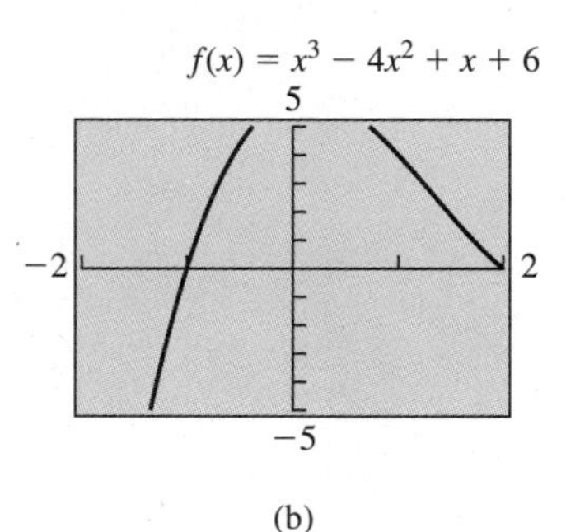

(b)

FIGURE 7

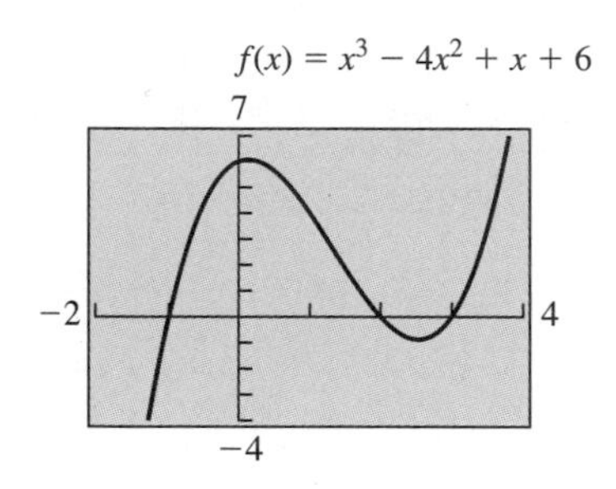

FIGURE 8

Since the graph of f is supposed to be continuous, the Ymax choice needs to be modified. Also, the limit behavior model for f is $g(x) = x^3$ (Figure 2). Consequently, the limit behavior for f should be

$$f(x) \to +\infty \quad \text{as} \quad x \to +\infty \qquad \text{and} \qquad f(x) \to -\infty \quad \text{as} \quad x \to -\infty$$

Since Figure 7b does not include the appropriate limit behavior of f for large positive values of x, we need to change the Xmax choice. Figure 8 displays one possible WINDOW selection that yields a more accurate graph of f, showing both the continuity and limit behavior. More confidence in the accuracy of the graph is gained by observing that this third-degree polynomial function has the maximum number of turning points, $3 - 1 = 2$.

EXAMPLE 4 **Using a limit behavior model to sketch a graph**

Determine the limit behavior model of the graph of $f(x) = x^4 + 2x^3 - 1$. Then select an appropriate viewing window to graph f. Indicate the number and type of turning points.

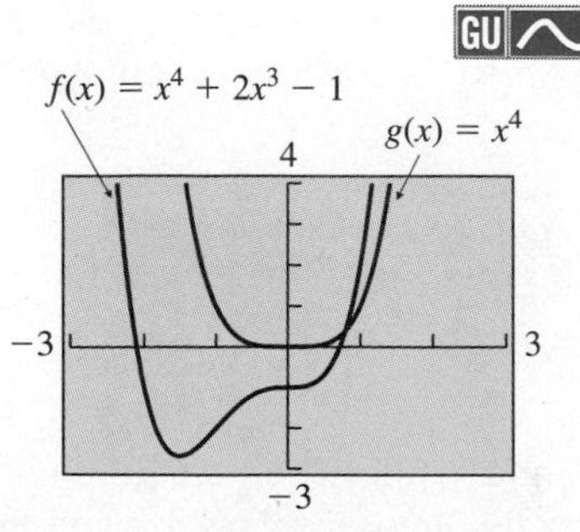

FIGURE 9

Solution The function $g(x) = x^4$ is the limit behavior model of the function $f(x) = x^4 + 2x^3 - 1$. Consequently,

$$f(x) \to +\infty \quad \text{as} \quad x \to +\infty \qquad \text{and} \qquad f(x) \to +\infty \quad \text{as} \quad x \to -\infty$$

Figure 9 shows one possible viewing window for the graph of f that displays the appropriate limit behavior. The graph of $g(x) = x^4$ is also included for comparison purposes. Note that the graph of f is continuous and has one turning point that is a valley.

Solving Polynomial Inequalities Graphically

Inequalities involving power functions can be solved graphically.

EXAMPLE 5 **Solving inequalities graphically**

Graph $f(x) = x^2$ and $g(x) = x^3$ on the same coordinate system, and then use the graphs to solve each inequality.

(a) $x^2 \leq x^3$ (b) $x^2 > x^3$

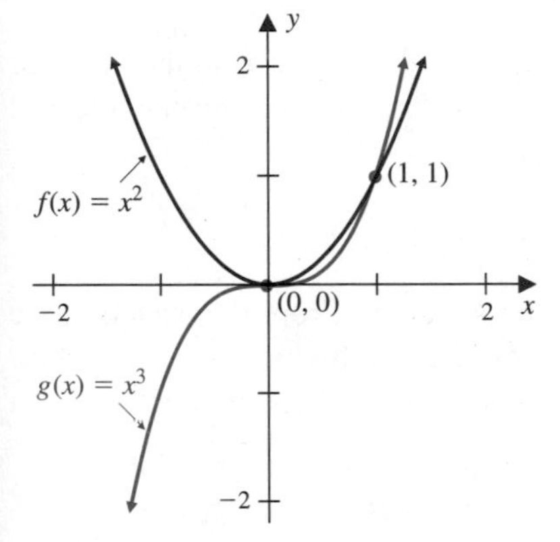

FIGURE 10

Solution We begin by graphing f and g on the same coordinate system (Figure 10). Notice that the two graphs intersect at (0, 0) and (1, 1).

(a) The inequality $x^2 \leq x^3$ is true for those values of x where the graphs intersect [where $f(x) = g(x)$] or where the graph of f is below the graph of g [where $f(x) < g(x)$]. Thus, the solution includes $x = 0$ and all values of x in the interval $[1, \infty)$.

(b) The inequality $x^2 > x^3$ is true for those values of x where the graph of f is above the graph of g, that is, where $f(x) > g(x)$. Thus, the solution includes all values of x such that $x < 0$ or $0 < x < 1$, that is, all numbers in the interval $(-\infty, 0)$ or the interval $(0, 1)$.

The techniques discussed in Section 3.1 for solving quadratic inequalities by using graphs, along with the locations of x intercepts, also can be used to solve polynomial inequalities of higher degrees. For now, we limit our efforts to solving inequalities where the polynomials are expressed in complete factored form.

EXAMPLE 6 **Solving a polynomial inequality graphically**

GU

Use a graphing utility to graph

$$f(x) = x^3 - 4x^2 + x + 6 = (x + 1)(x - 2)(x - 3)$$

Find the exact locations of the x intercepts of the graph, and then use the graph and the intercepts to solve the inequality

$$x^3 - 4x^2 + x + 6 \leq 0$$

The factorization made it easy to find the x intercepts. Later, we will learn how to locate x intercepts for polynomial functions that are difficult to factor.

Solution A viewing window of the graph of f is displayed in Figure 8 (page 172). By setting each of the factors $x + 1$, $x - 2$, and $x - 3$ equal to 0 and solving for x, we find that the x intercepts are -1, 2, and 3. The portions of the graph on or below the x axis indicate the values of x that satisfy the inequality. Thus, the solution set of the inequality consists of all values of x such that $x \leq -1$ or $2 \leq x \leq 3$, that is, all values of x in the intervals $(-\infty, -1]$ or $[2, 3]$.

Solving Applied Problems

The next example illustrates the use of a higher-degree polynomial function in mathematical modeling.

EXAMPLE 7 **Maximizing volume**

The liner for an outside window flower box is to be made by cutting equal squares from the corners of a rectangular sheet of perforated aluminum and turning up the sides. Assume the aluminum sheet is 40 inches by 16 inches and the length of each square is x inches.

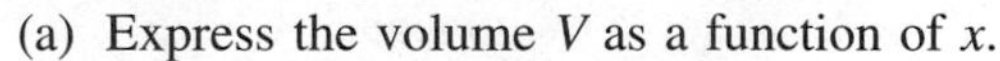

(a) Express the volume V as a function of x.

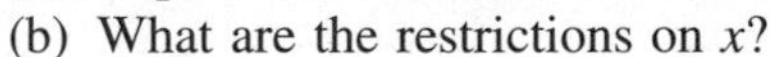

(b) What are the restrictions on x?

(c) Use a graphing utility to sketch the graph of the function in part (a) incorporating the restrictions from part (b).

(d) Use the graph and the ZOOM and TRACE features to estimate the value of x that gives the maximum volume. Round off the answers to one decimal place.

Solution

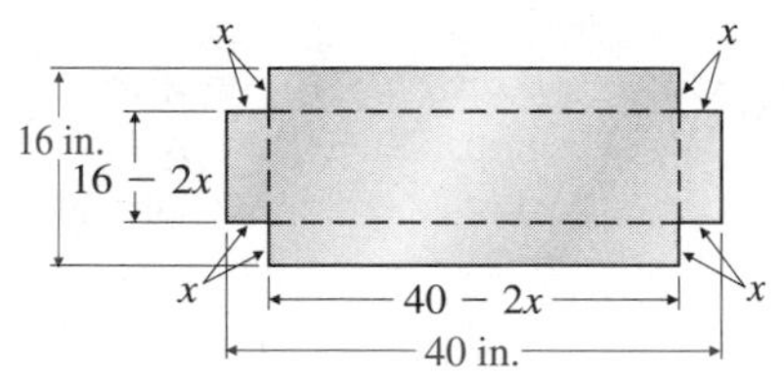

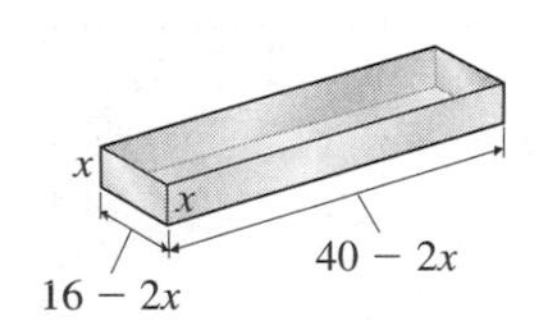

FIGURE 11

(a) Figure 11 displays a physical model of the situation. From the sketches, we see that the height h of the liner is x inches, the width w is $16 - 2x$ inches, and the length ℓ is $40 - 2x$ inches. Thus, the volume V of the liner is given by

$$\begin{aligned} V &= \ell wh && \text{Formula from geometry} \\ &= (40 - 2x)(16 - 2x)x && \text{Substitute} \\ &= 4x(20 - x)(8 - x) && \text{Factor} \end{aligned}$$

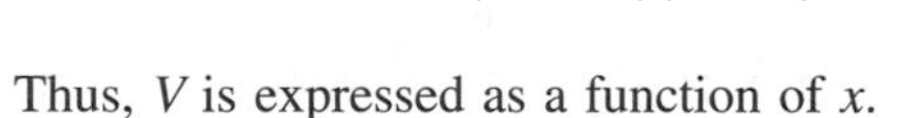

Thus, V is expressed as a function of x.

(b) Each physical dimension must be positive; that is, $x > 0$; $40 - 2x > 0$, so $20 - x > 0$ or $x < 20$; and $16 - 2x > 0$, so $8 - x > 0$ or $x < 8$. It follows that $0 < x < 8$.

(c) Figure 12a shows a viewing window of the graph for $0 < x < 8$.

(d) After using the ZOOM and TRACE features repeatedly, we see from Figure 12b that the maximum volume of about 1039.5 cubic inches is obtained when x is about 3.5 inches.

In this example we relied on the features of a graphing utility to locate the maximum value of V. Advanced techniques for finding this value analytically (without reliance on a graphing utility) are studied in calculus.

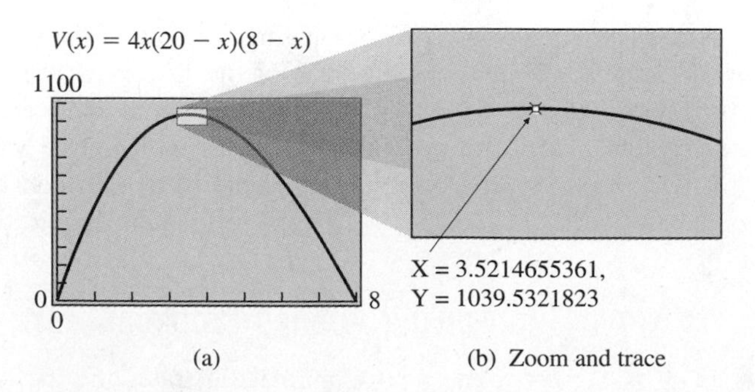

FIGURE 12

PROBLEM SET 3.2

Mastering the Concepts

GU In problems 1 and 2, first indicate the symmetry and general shape of the graph of each function. Then confirm the description by using a graphing utility to sketch the graph.

1. **(a)** $f(x) = x^8$ **(b)** $h(x) = x^9$

2. **(a)** $f(x) = x^{10}$ **(b)** $h(x) = x^{11}$

In problems 3–14, use transformations of the graph of an appropriate power function to sketch the graph of the given function.

3. $g(x) = \frac{1}{3}x^4$

4. $h(x) = -5x^4$

5. $f(x) = -2x^5$

6. $g(x) = \frac{1}{2}x^5$

7. $h(x) = x^5 + 3$

8. $f(x) = x^6 - 7$

9. $f(x) = -\frac{1}{5}(x + 1)^4$

10. $h(x) = 4(x - 4)^5$

11. $g(x) = 2(x - 1)^4 + 3$

12. $f(x) = \frac{1}{3}(x + 2)^5 + 2$

13. $h(x) = -4(x - 2)^3 + 1$

14. $g(x) = -\frac{1}{2}(x + 3)^4 - 2$

In problems 15–20, use the graph of f to determine the limit behavior of $f(x)$ as:

(a) $x \to +\infty$ **(b)** $x \to -\infty$

Indicate the number of turning points on each graph, and determine whether the graph has peaks or valleys at these points.

15.

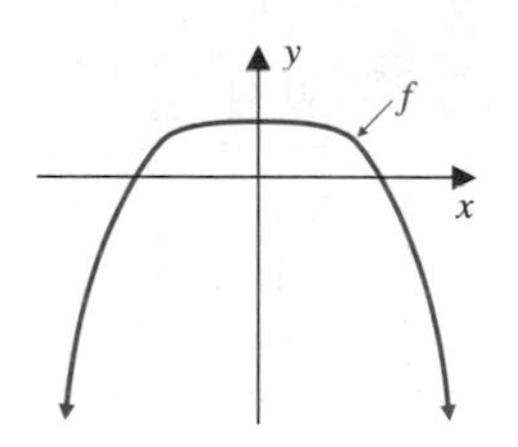

16.

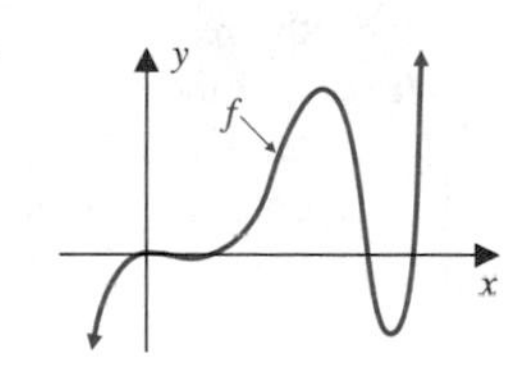

17.

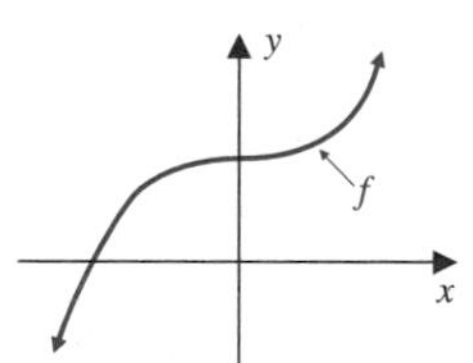

18.

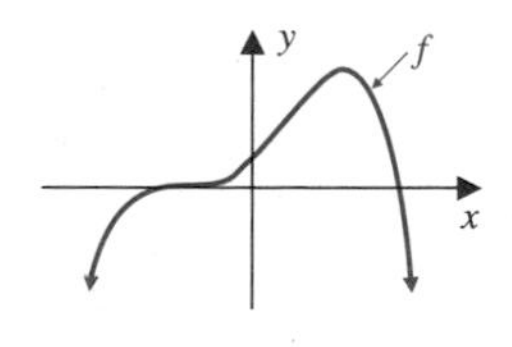

19.

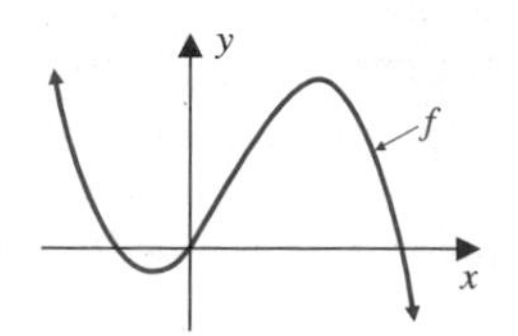

20.

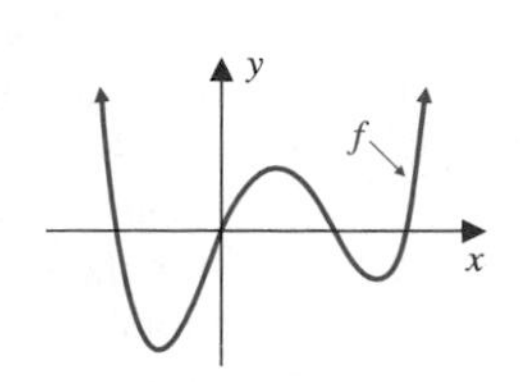

In problems 21 and 22, graph the limit behavior model function and then use the result to match the given polynomial function with its graph.

21. **(a)** $f(x) = -x^3 + 2x - 5$

(b) $g(x) = x^4 - 8x^3 + 20x^2 - 16x + 8$

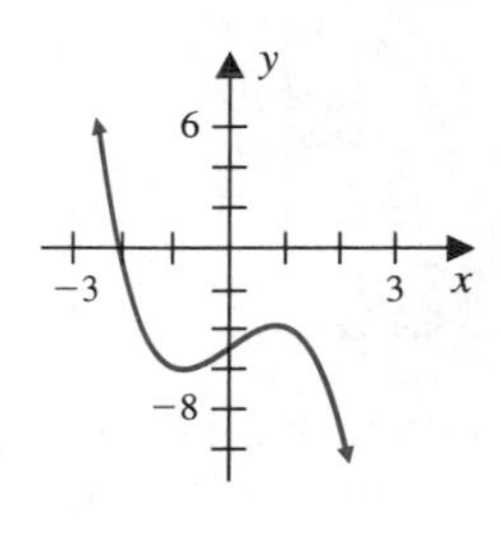

(A)

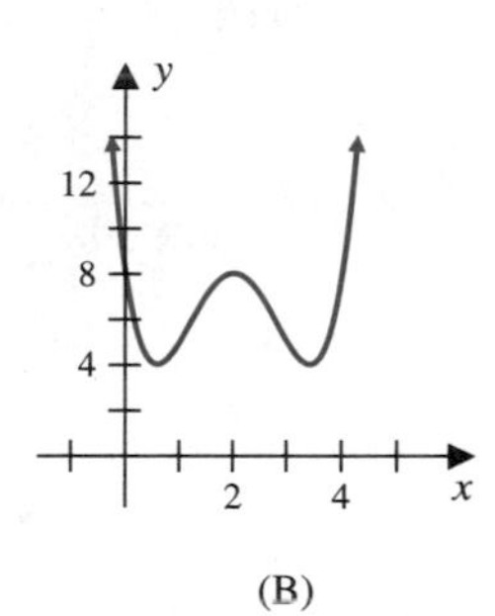

(B)

22. **(a)** $h(x) = x^3 + 3x^2 - 2x - 5$

(b) $F(x) = -2x^4 - 8x^3 - 12x^2 - 8x + 1$

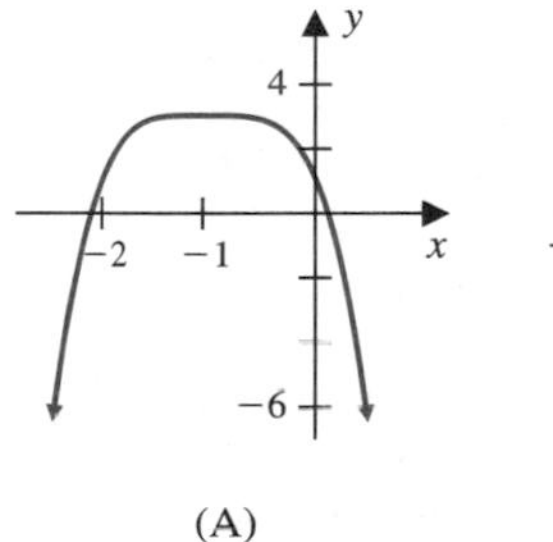

(A)

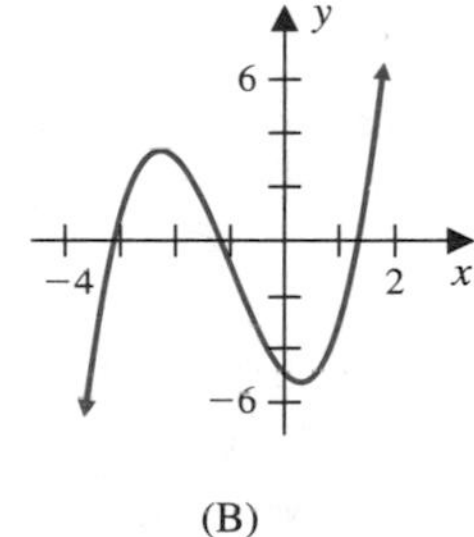

(B)

GU In problems 23 and 24, graph the given function for the specified WINDOW selections. Explain why the selection is inappropriate.

23. $f(x) = x^3 - 6x^2 + 11x - 6$

(a)

(b)

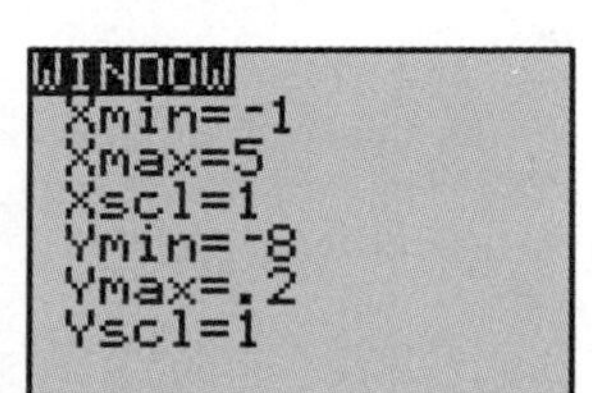

24. $h(x) = -x^4 + 2x + 3$

(a)

WINDOW
Xmin=-3
Xmax=3
Xscl=1
Ymin=-3
Ymax=3
Yscl=1

(b)

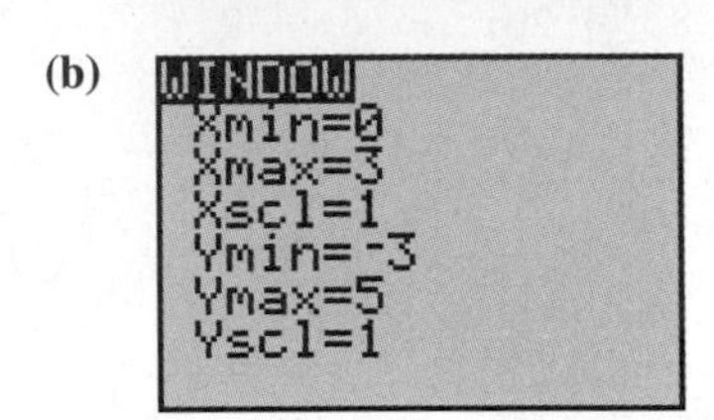

GU In problems 25–34, use the limit behavior model of the given polynomial function f to determine the limit behavior of $f(x)$:
(a) As $x \to +\infty$ **(b)** As $x \to -\infty$
Then use a graphing utility to sketch a graph of f by selecting an appropriate viewing window. Indicate the number and types of turning points.

25. $f(x) = x^4 - x^2 - 20$
26. $f(x) = x^4 - 4x^2 - 3x + 12$
27. $f(x) = x^3 - 8x^2 + 16x$ **28.** $f(x) = -x^3 - 6x^2$
29. $f(x) = -2x^4 + 2x^3 - x + 3$
30. $f(x) = x(x + 1)(x - 2)$
31. $f(x) = -4x^3 + 2x^2$ **32.** $f(x) = -5x^2(x - 1)^3$
33. $f(x) = 3x(x + 2)(x - 5)$
34. $f(x) = 2x^3 - 100x^2 + 15x$

In problems 35–38, graph f and g on the same coordinate system and then use the graphs to solve each inequality.

35. $f(x) = x^2$ and $g(x) = x^5$
(a) $x^2 \le x^5$ **(b)** $x^2 > x^5$
36. $f(x) = x^3$ and $g(x) = x^4$
(a) $x^3 \le x^4$ **(b)** $x^3 > x^4$
37. $f(x) = x^2$ and $g(x) = x^6$
(a) $x^2 \le x^6$ **(b)** $x^2 > x^6$
38. $f(x) = x^3$ and $g(x) = x^5$
(a) $x^3 \le x^5$ **(b)** $x^3 > x^5$

GU In problems 39–44, use a graphing utility to graph each given function. Find the exact locations of the x intercepts of the graph, and then use the graph and the intercepts to solve the associated inequality.

39. $f(x) = x(x - 1)(x + 2)$; $x(x - 1)(x + 2) > 0$
40. $h(x) = x^3 - x^2 - 2x = x(x - 2)(x + 1)$; $x^3 - x^2 - 2x \le 0$
41. $f(x) = x^3 - x^2 - 12x = x(x - 4)(x + 3)$; $x^3 - x^2 - 12x < 0$
42. $g(x) = -2x^4 + 2x^2 = -2x^2(x + 1)(x - 1)$; $-2x^4 + 2x^2 > 0$
43. $f(x) = (2x - 5)(x - 1)^2$; $(2x - 5)(x - 1)^2 \ge 0$
44. $f(x) = 2x^4 - 3x^3 - 12x^2 + 7x + 6 = (2x + 1)(x - 1)(x - 3)(x + 2)$; $2x^4 - 3x^3 - 12x^2 + 7x + 6 > 0$

Applying the Concepts

In problems 45–48, round off the answers to two decimal places.

45. Unemployment Rate: A study conducted by a state unemployment commission determined the percentage of the work force expected to be unemployed in the state over a 4 year period beginning in January 1995. The unemployment R (in percent) over that period is approximated by the model

$$R = 0.0002t^3 - 0.0147t^2 + 0.216t + 5.9$$

where t is the number of months, with $t = 0$ representing the beginning of the 4 year period.
(a) What are the restrictions on t?
GU **(b)** Use a graphing utility to graph this function over the restricted interval.
(c) Use the graph and the ZOOM and TRACE features to determine when the unemployment is expected to be the lowest over this period. What is the lowest expected rate of unemployment?
(d) What is the highest unemployment rate expected over this period of time? When is it expected to occur?

46. Drug Dosage: Data collected from a study conducted by medical researchers show the amount A of glucose (measured in tenths of a percent) in a patient's bloodstream at time t (in minutes) during a stress test. The amount A is approximated by the model

$$A = 0.1t^3 - 2.1t^2 + 12t + 2 \qquad 1.5 < t \le 11$$

GU **(a)** Use a graphing utility to graph this function.
(b) Use the graph and the ZOOM and TRACE features to determine the maximum amount of glucose and when it occurs.
(c) What is the minimum amount of glucose, and when does it occur?

47. Package Design: An open-top box is to be constructed by removing equal squares, each of length x inches, from the corners of a piece of cardboard that measures 20 inches by 20 inches and then folding up the sides.

(a) Draw a sketch of the situation and then express the volume V of the box as a function of x.

(b) What are the restrictions on x?

GU **(c)** Use a graphing utility to sketch the graph of the function from part (a) incorporating the restrictions from part (b).

(d) Use the graph and the ZOOM and TRACE features to determine the value of x for which the volume is a maximum. Also find the maximum volume.

48. U.S. Postal Package Design: A closed box with a square end x inches on a side is to be mailed. The U.S. Postal Service will accept the box for domestic shipment only if the length L of the box plus its girth $(4x)$ is at most 108 inches (Figure 13). Suppose that the length plus girth of a box is exactly 108 inches.

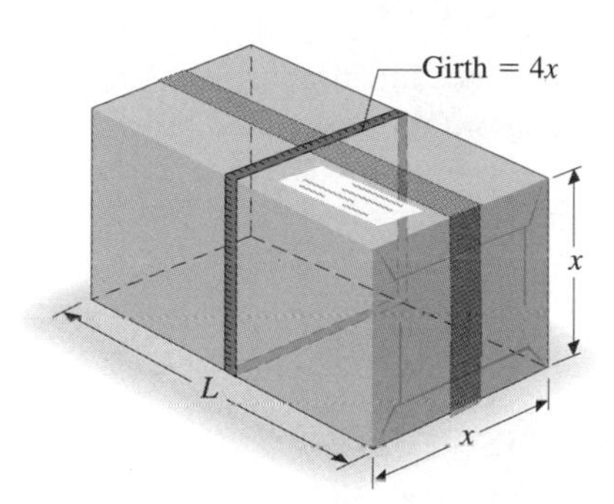

FIGURE 13

(a) Express the volume V of the box as a function of x.

(b) What are the restrictions on x?

GU **(c)** Use a graphing utility to sketch the graph of the function from part (a) incorporating the restrictions from part (b).

(d) Use the graph and the ZOOM and TRACE features to determine the value of x for which the volume is a maximum. Also find the maximum volume.

Developing and Extending the Concepts

49. Compare the limit behaviors of the functions $f(x) = ax^4$ and $g(x) = bx^4$ if:

(a) $ab > 0$ **(b)** $ab < 0$

50. Compare the limit behaviors of the functions $f(x) = ax^3$ and $g(x) = bx^3$ if:

(a) $ab > 0$ **(b)** $ab < 0$

51. Examine the graph of $y = x^3$. For $x < 0$ the shape of the curve is **concave down** (cupped downward), and for $x > 0$ the shape is **concave up** (cupped upward). The point on the graph where a curve changes concavity is said to be a **point of inflection**. For $y = x^3$, the point of inflection is (0, 0). With this in mind, graph each given function. For what values of x is the curve concave down? For what values of x is the curve concave up? Where is the point of inflection?

(a) $f(x) = 2x^3 + 1$ **(b)** $g(x) = -(x - 4)^3 + 1$

52. Graph $f(x) = x^3$ and $g(x) = |x^3|$ on the same coordinate system. How can the graph of g be obtained from the graph of f? In general, explain how the graph of $y = |f(x)|$ can be geometrically obtained from the graph of the polynomial function $y = f(x)$. When does $|f(x)| = f(x)$?

In problems 53–56, display a general sketch of the limit behavior of the graph of the polynomial function

$$f(x) = a(x - h)^n + k$$

under the given conditions.

53. $a > 0$ and $n \geq 1$ is an odd integer.

54. $a < 0$ and $n \geq 1$ is an odd integer.

55. $a > 0$ and $n \geq 2$ is an even integer.

56. $a < 0$ and $n \geq 2$ is an even integer.

In problems 57 and 58, complete the following table:

$x =$	-10^3	-10^2	-10	-1	1	10	10^2	10^3
$f(x)$								
$g(x)$								
$R(x)$								

What happens to the values of $R(x)$ as $x \to -\infty$ and as $x \to +\infty$? How do the data in the table show that g is the limit behavior model for f?

GU Demonstrate the result by using a graphing utility to graph f and g on the same coordinate system.

57. $f(x) = x^3 + 2x^2 - 5x + 1$

$$= x^3\left(1 + \frac{2}{x} - \frac{5}{x^2} + \frac{1}{x^3}\right)$$

$$= g(x) \cdot R(x)$$

58. $f(x) = 2x^4 - x^2 + x - 7$

$$= 2x^4\left(1 - \frac{1}{2x^2} + \frac{1}{2x^3} - \frac{7}{2x^4}\right)$$

$$= g(x) \cdot R(x)$$

59. What can we conclude about the degree of a polynomial function if its graph has exactly three turning points?

60. Give an example of a polynomial function of degree 3 that has:
(a) No turning point
(b) Two turning points
Is it possible for such a function to have only one turning point? Explain and illustrate.

OBJECTIVES

1. Perform long division of polynomials.
2. Evaluate polynomial functions using the remainder theorem.
3. Perform synthetic division.
4. Factor polynomials using synthetic division.
5. Relate zeros, factors, and x intercepts.

3.3 Polynomials and Their Division Properties

In this section we develop results regarding evaluation and factoring of polynomials. We also explore the connection between solutions of polynomial equations and the x intercepts of graphs of the associated functions. We introduce synthetic division as a tool in developing our ideas.

Performing Long Division of Polynomials

An *algorithm* is a step-by-step systematic procedure used to perform a computation. For instance, we use the long division algorithm to divide 372 by 5 to obtain the *quotient* and the *remainder*, as shown below:

```
                 74  ← Quotient     Dividend
Divisor  →  5 )372  ←────────────────┘
         ①   35                 ① Multiply 7 · 5
         ②    22                ② Subtract 372 − 350
         ③    20                ③ Multiply 4 · 5
         ④     2                ④ Subtract 22 − 20
Remainder ─────↑
```

The result of the division can be expressed either in the *multiplicative form* as

$$372 = (5)(74) + 2 \qquad \text{Dividend} = (\text{Divisor})(\text{Quotient}) + \text{Remainder}$$

or in the *fractional form* as

$$\frac{372}{5} = 74 + \frac{2}{5} \qquad \frac{\text{Dividend}}{\text{Divisor}} = \text{Quotient} + \frac{\text{Remainder}}{\text{Divisor}}$$

The long division of polynomials follows the same pattern. For example, to divide the polynomial $f(x) = 2x^3 - 3x^2 + 5x + 2$ by $D(x) = x + 3$, we note that both the divisor $x + 3$ and the dividend $2x^3 - 3x^2 + 5x + 2$ are arranged in descending powers of x. (If they weren't, we would rewrite them so that they were before beginning the process.) Then we proceed as shown at the top of the next page:

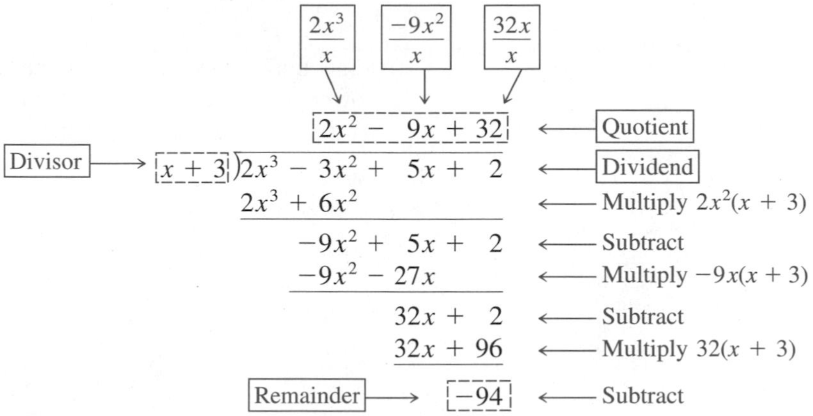

$\frac{2x^3}{x}$ $\quad \frac{-9x^2}{x}$ $\quad \frac{32x}{x}$

	$2x^2 - 9x + 32$	← Quotient
Divisor →	$x + 3\overline{)2x^3 - 3x^2 + 5x + 2}$	← Dividend
	$2x^3 + 6x^2$	← Multiply $2x^2(x + 3)$
	$-9x^2 + 5x + 2$	← Subtract
	$-9x^2 - 27x$	← Multiply $-9x(x + 3)$
	$32x + 2$	← Subtract
	$32x + 96$	← Multiply $32(x + 3)$
Remainder →	-94	← Subtract

As in the above numerical example, we can write the result either in multiplicative form:

$$2x^3 - 3x^2 + 5x + 2 = (x + 3)(2x^2 - 9x + 32) + (-94)$$

or in fractional form:

$$\frac{2x^3 - 3x^2 + 5x + 2}{x + 3} = 2x^2 - 9x + 32 + \frac{(-94)}{x + 3}$$

The preceding example illustrates the following property:

THE DIVISION ALGORITHM RESULT

> If $f(x)$ and $D(x)$ are nonconstant polynomials such that the degree of $f(x)$ is greater than or equal to the degree of $D(x)$, then there exist unique polynomials $Q(x)$ and $R(x)$ such that
>
> $$f(x) = D(x) \cdot Q(x) + R(x)$$
>
> where the degree of $R(x)$ is less than the degree of $D(x)$ [$R(x)$ may be 0]. The expression $D(x)$ is called the **divisor**, $f(x)$ is the **dividend**, $Q(x)$ is the **quotient**, and $R(x)$ is the **remainder**.

The result of the division algorithm can also be expressed in the fractional form

$$\boxed{\frac{f(x)}{D(x)} = Q(x) + \frac{R(x)}{D(x)}}$$

which carries implicit restrictions on the values of x since $D(x)$ cannot equal 0.

EXAMPLE 1 Using long division to divide polynomials

(a) Use long division to divide $f(x) = 3x^3 - 2x + 1$ by $D(x) = x - 2$. Identify the quotient and the remainder. Then express the result in multiplicative form, and verify the multiplicative form algebraically.

GU (b) Use a graphing utility to demonstrate the validity of the result graphically.

Solution

(a) First, notice that the divisor and the dividend are arranged in descending powers of x. Because there is a *missing* x^2 term in the polynomial, we insert the term $0x^2$ so that the process is easier to organize:

$$\begin{array}{rrl}
 & 3x^2 + 6x + 10 & \longleftarrow \boxed{\text{Quotient}} \\
\boxed{\text{Divisor}} \longrightarrow x - 2 & \overline{)\,3x^3 + 0x^2 - 2x + 1} & \longleftarrow \boxed{\text{Dividend}} \\
 & \underline{3x^3 - 6x^2} & \longleftarrow 3x^2(x-2) \\
 & 6x^2 - 2x + 1 & \longleftarrow \text{Subtract} \\
 & \underline{6x^2 - 12x} & \longleftarrow 6x(x-2) \\
 & 10x + 1 & \longleftarrow \text{Subtract} \\
 & 10x - 20 & \longleftarrow 10(x-2) \\
\boxed{\text{Remainder}} \longrightarrow & 21 & \longleftarrow \text{Subtract}
\end{array}$$

We conclude that the quotient is $Q(x) = 3x^2 + 6x + 10$ and the remainder is $R(x) = 21$. So the multiplicative form is

$$\begin{aligned}
f(x) &= 3x^3 - 2x + 1 \\
&= D(x) \cdot Q(x) + R(x) \\
&= (x - 2)(3x^2 + 6x + 10) + 21
\end{aligned}$$

To verify the result algebraically, we expand the right side of the latter equation to get

$$\begin{aligned}
D(x) \cdot Q(x) + R(x) &= (x - 2)(3x^2 + 6x + 10) + 21 \\
&= (3x^3 - 2x - 20) + 21 \\
&= 3x^3 - 2x + 1 \\
&= f(x)
\end{aligned}$$

Thus, the expressions are equal.

(b) To demonstrate the validity of the result graphically, we proceed as follows. First, let $f(x) = 3x^3 - 2x + 1$ and $g(x) = (x - 2)(3x^2 + 6x + 10) + 21$. Next, we use a graphing utility to get the graphs of f (Figure 1a) and g (Figure 1b) for the same viewing window settings. The two graphs are identical, demonstrating the validity of the result. ■

$f(x) = 3x^3 - 2x + 1$

5 / −3 / 3 / −5

(a)

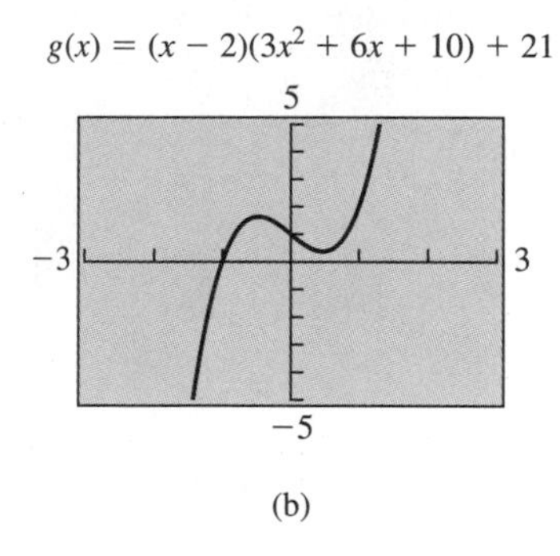

(b)

FIGURE 1

Evaluating Polynomial Functions Using the Remainder Theorem

Further examination of the results in Example 1 enables us to discover a connection between the division of polynomials and the evaluation of polynomial functions. Consider the polynomial function defined by

$$f(x) = 3x^3 - 2x + 1$$

If $x = 2$, then $f(2) = 3(2)^3 - 2(2) + 1 = 21$. From the results in Example 1, we know that

$$f(x) = (x - 2)(3x^2 + 6x + 10) + 21$$

So, if we substitute 2 for x in this expression, we get

$$\begin{aligned} f(2) &= (2-2)[3(2)^2 + 6(2) + 10] + 21 \\ &= 0 + 21 \\ &= 21 \end{aligned}$$

Thus, we observe that $f(2) = 21$, the same value as the remainder when $f(x)$ is divided by $x - 2$.

When the remainder is a constant, we'll denote it as R from here on.

In general, if $f(x)$ is a polynomial of degree $n \geq 1$ and it is divided by $x - c$, then according to the division algorithm we get a result of the form

$$f(x) = (x - c) \cdot Q(x) + R$$

where $Q(x)$ is the quotient and R is the remainder. Consequently, if we let $x = c$, it follows that

$$\begin{aligned} f(c) &= (c - c) \cdot Q(c) + R \\ &= 0 \cdot Q(c) + R \\ &= R \end{aligned}$$

So we have established the following general result:

THE REMAINDER THEOREM

> If a polynomial $f(x)$ of degree $n \geq 1$ is divided by $x - c$, then the remainder $R = f(c)$.

EXAMPLE 2 Evaluating a polynomial by using the remainder theorem

Use the remainder theorem to find $f(2)$ if $f(x) = 3x^3 - x^2 + 2x - 18$.

Solution First, we divide $3x^3 - x^2 + 2x - 18$ by $x - 2$:

$$\begin{array}{r} 3x^2 + 5x + 12 \\ x - 2 \overline{)\, 3x^3 - x^2 + 2x - 18} \\ \underline{3x^3 - 6x^2 } \\ 5x^2 + 2x \\ \underline{5x^2 - 10x } \\ 12x - 18 \\ \underline{12x - 24} \\ 6 \end{array}$$

Thus, the quotient $Q(x) = 3x^2 + 5x + 12$ and the remainder $R = 6$. Therefore, by the remainder theorem, $R = f(2) = 6$.

The result in Example 2 can be verified by a direct calculation of $f(2)$ to get

$$\begin{aligned} f(2) &= 3(2)^3 - (2)^2 + 2(2) - 18 \\ &= 3(8) - 4 + 4 - 18 = 6 \end{aligned}$$

Performing Synthetic Division

The synthetic division method can be used only when the divisor is of the form $x - c$.

Applications of the remainder theorem involve the division of a polynomial $f(x)$ of degree $n \geq 1$ by a *first-degree* polynomial of the form $D(x) = x - c$. To simplify the task of performing such a division, we use an algorithm called **synthetic division**. In Example 2 we used long division to divide the polynomial $f(x) = 3x^3 - x^2 + 2x - 18$ by $D(x) = x - 2$. Now, we perform this same division by using the synthetic division method as follows:

	Explanation of Process	Result
Step 1.	Arrange the terms of the polynomial $f(x)$ in descending powers of x, and use the 0 coefficient for any missing powers of x (as in Example 1).	$f(x) = 3x^3 - x^2 + 2x - 18$
Step 2.	Consider the divisor in the form $x - c$. Write down the value of c (here, $c = 2$), draw a vertical line, and then list the coefficients of $f(x)$.	2 | 3 −1 2 −18
Step 3.	Leave space below the row of coefficients, draw a horizontal line, and copy the leading coefficient below the line.	2 | 3 −1 2 −18 / 3
Step 4.	Multiply 3, the leading coefficient, by 2, the value for c, and place the product, 6, above the horizontal line under the second coefficient, −1. Then add −1 to this product and place the result, 5, below the horizontal line.	2 | 3 −1 2 −18 / 6 / 3 5

Now multiply 5 by 2 and place the product, 10, above the horizontal line under the third coefficient, 2. Then add 2 to this product and place the result, 12, below the horizontal line.

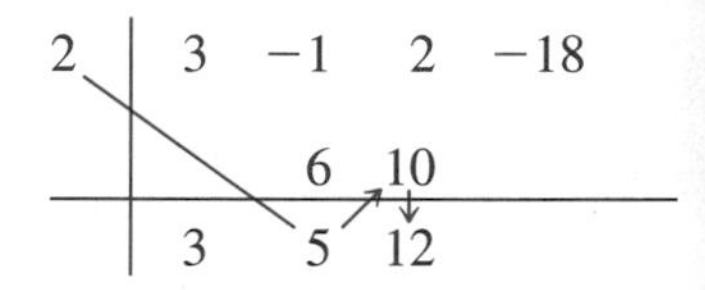

Continue in this way, multiplying 12 by 2 and placing the product, 24, above the horizontal line under the next coefficient, −18. Then add −18 and 24 to obtain 6, placing 6 below the horizontal line. Isolate the very last sum, 6, by drawing a short vertical line.

Step 5. In the last row, the numbers 3, 5, and 12 are the coefficients of the quotient $Q(x)$ and the last number, 6, is the remainder R.

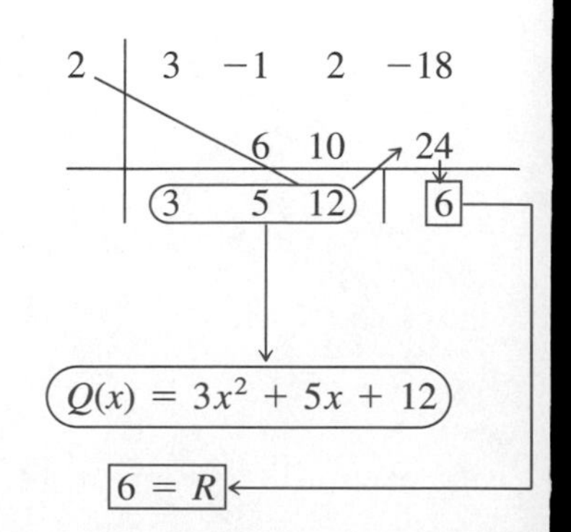

Notice how the long division in Example 2 compares with the synthetic division performed above:

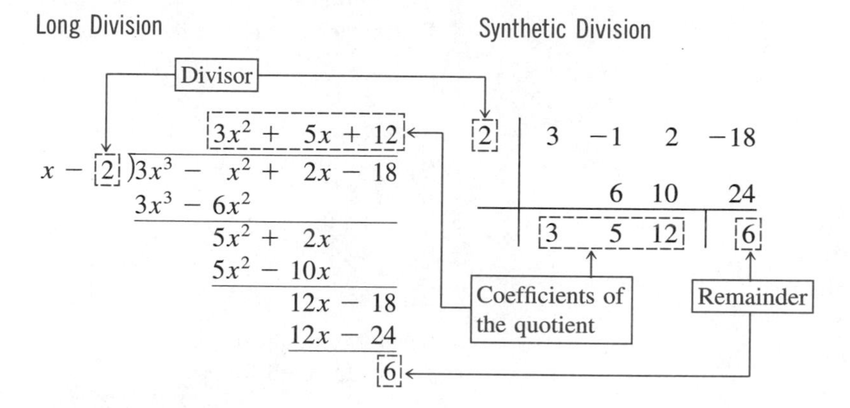

EXAMPLE 3 **Using synthetic division to find the quotient and the remainder**

Use synthetic division to obtain the quotient and the remainder if the polynomial $f(x) = 2x^4 - 3x^2 + 5x + 7$ is divided by $D(x) = x + 3$.

Solution The terms of $f(x)$ are given in descending powers of x, but we need to use 0 to denote the coefficient of the missing x^3 term. So the coefficients of $f(x)$ are listed as 2, 0, -3, 5, and 7.

The divisor has the form $D(x) = x + 3 = x - (-3) = x - c$, so $c = -3$. By synthetic division, we get

$$\begin{array}{r|rrrrr} -3 & 2 & 0 & -3 & 5 & 7 \\ & & -6 & 18 & -45 & 120 \\ \hline & 2 & -6 & 15 & -40 & \vert\ 127 \end{array}$$

Hence, the quotient is $Q(x) = 2x^3 - 6x^2 + 15x - 40$ and the remainder is $R = 127$.

EXAMPLE 4 **Using synthetic division to evaluate a polynomial function**

Synthetic division is sometimes called *synthetic substitution.*

Use synthetic division and the remainder theorem to evaluate $f(4)$ if $f(x) = 3x^5 - 45x^3 + 1$.

Solution We use synthetic division to divide the polynomial $3x^5 + 0x^4 - 45x^3 + 0x^2 + 0x + 1$ by $x - c$ with $c = 4$:

$$\begin{array}{r|rrrrrr} 4 & 3 & 0 & -45 & 0 & 0 & 1 \\ & & 12 & 48 & 12 & 48 & 192 \\ \hline & 3 & 12 & 3 & 12 & 48 & \vert\ 193 \end{array}$$

The remainder is 193, which is the last number in the last row. So, by the remainder theorem, $f(4) = 193$. This substitution can be carried out more efficiently by using a calculator or a computer.

Factoring Polynomials Using Synthetic Division

The synthetic division process provides an effective tool for factoring polynomials. For instance, consider the polynomial function

$$f(x) = x^4 + 3x^3 + 2x^2 - 7x - 39$$

We can determine whether $x + 3$ is a factor of the polynomial by using synthetic division to divide $f(x)$ by $x + 3$, to obtain

$$\begin{array}{r|rrrr|r} -3 & 1 & 3 & 2 & -7 & -39 \\ & & -3 & 0 & -6 & 39 \\ \hline & 1 & 0 & 2 & -13 & 0 \end{array}$$

Thus, the quotient is $x^3 + 2x - 13$ and the remainder is 0. So we can express $f(x)$ in factored form as

$$f(x) = x^4 + 3x^3 + 2x^2 - 7x - 39 = (x + 3)(x^3 + 2x - 13) + 0$$

or simply,

$$f(x) = (x + 3)(x^3 + 2x - 13)$$

and we conclude that $x + 3$ is a factor of $f(x)$.

In general, suppose that we examine a polynomial function written as $f(x) = (x - c) \cdot Q(x) + R$, where $Q(x)$ is the quotient and $R = f(c)$ is the remainder when $f(x)$ is divided by $x - c$. If $f(c) = 0$, then $f(x) = (x - c) \cdot Q(x)$, and $x - c$ is a factor of $f(x)$. Conversely, if $x - c$ is a factor of $f(x)$, then $f(x) = (x - c) \cdot Q(x)$ and by the remainder theorem $f(c) = 0$.

Thus, we have the following theorem:

THE FACTOR THEOREM

> If $f(x)$ is a polynomial of degree $n \geq 1$ and $f(c) = 0$, then $x - c$ is a factor of the polynomial $f(x)$; conversely, if $x - c$ is a factor of $f(x)$, then $f(c) = 0$.

The factor theorem, along with synthetic division, enables us to determine whether an expression of the form $x - c$ is a factor of a polynomial, and at the same time, we find the other factor. This idea is illustrated in the next example.

EXAMPLE 5 **Using the factor theorem to determine factors of a polynomial**

Let $f(x) = x^3 + 5x^2 + 5x - 2$. Use the factor theorem to determine whether $D(x)$ is a factor of $f(x)$. If it is, use synthetic division to find the factorization.

(a) $D(x) = x + 2$ (b) $D(x) = x - 1$

Solution

(a) We have $x + 2 = x - (-2)$, which has the form $x - c$ with $c = -2$. Using synthetic division to divide $f(x)$ by $x + 2$, we get

$$\begin{array}{r|rrrr} -2 & 1 & 5 & 5 & -2 \\ & & -2 & -6 & 2 \\ \hline & 1 & 3 & -1 & 0 \end{array}$$

It follows that $f(-2) = 0$, so by the factor theorem, we know $x + 2$ is a factor of $f(x)$. The synthetic division indicates that the factorization is

$$x^3 + 5x^2 + 5x - 2 = (x + 2)(x^2 + 3x - 1)$$

(b) Here, $x - 1$ has the form $x - c$ with $c = 1$. Using synthetic division to divide $f(x)$ by $x - 1$, we get

$$\begin{array}{r|rrrr} 1 & 1 & 5 & 5 & -2 \\ & & 1 & 6 & 11 \\ \hline & 1 & 6 & 11 & 9 \end{array}$$

Since $f(1) = 9 \neq 0$, it follows from the factor theorem that $x - 1$ is not a factor of $f(x)$.

Relating Zeros, Factors, and x Intercepts

In the context of functions, we refer to a solution of the equation $f(x) = 0$ as a *zero* of the function defined by $y = f(x)$. The zeros of linear and quadratic functions are relatively easy to determine because the techniques for solving linear and quadratic equations are well-established. For example, the only zero of $f(x) = x - 2$ is 2. The zeros of $g(x) = x^2 - 4x - 5$ are found by solving the equation $x^2 - 4x - 5 = 0$ to obtain $x = 5$ and $x = -1$. Although the zeros of a function may include nonreal complex numbers, for now we focus our attention on the real (number) zeros of polynomial functions.

The real zeros of the polynomial function $y = f(x)$ are the same as the x intercepts of the graph of f. For example, the graph of

$$f(x) = x^3 + 2x^2 - 5x - 6$$

in Figure 2 shows that the x intercepts of f are -3, -1, and 2. Since these same numbers are the solutions of the equation $f(x) = x^3 + 2x^2 - 5x - 6 = 0$, it follows that -3, -1, and 2 are also the real zeros of f. In addition, the factor theorem tells us that $x + 3$, $x + 1$, and $x - 2$ are factors of $f(x)$, so the factored form of f is given by

$$\begin{aligned} f(x) &= x^3 + 2x^2 - 5x - 6 \\ &= (x + 3)(x + 1)(x - 2) \end{aligned}$$

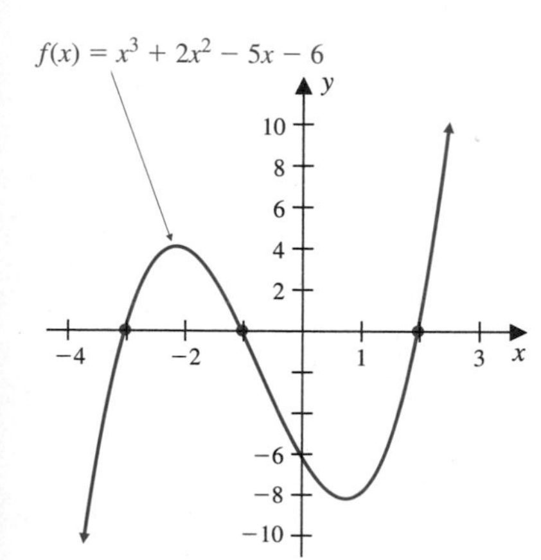

FIGURE 2

This illustration leads us to the following generalization:

EQUIVALENT STATEMENTS

If f is a polynomial function and c is a real number, then the following statements are equivalent:

1. The number c is a zero of f.
2. The number c is a solution or root of the equation $f(x) = 0$.
3. The polynomial $x - c$ is a factor of $f(x)$.
4. The number c is an x intercept of the graph of f.

Thus, each zero c of a polynomial function f corresponds to a first-degree factor $(x - c)$ of $f(x)$. When $f(x)$ is factored, the same factor may occur more than once, in which case c is called a *repeated* or *multiple zero* of f.

DEFINITION

MULTIPLICITY OF A ZERO

If $f(x) = (x - c)^s \cdot Q(x)$, $Q(c) \neq 0$, and s is a positive integer, then we say that c is a **zero of multiplicity** s of the function f.

For instance, $f(x) = (x - 1)^2(x + 3)$ has 1 as a zero of multiplicity 2 and -3 as a zero of multiplicity 1.

EXAMPLE 6 Finding multiplicities of zeros

Find the zeros of $f(x) = (x + 2)^3(x - 3)^4(x - 5)^2$. State the multiplicity of each.

Solution The factored form shows that $f(x)$ has three different zeros, -2, 3, and 5, with the following multiplicities:

Zero	Multiplicity
-2	3
3	4
5	2

PROBLEM SET 3.3

Mastering the Concepts

In problems 1–8:

(a) Use long division to divide $f(x)$ by $D(x)$. Identify the quotient $Q(x)$ and the remainder $R(x)$. Also express each result in the multiplicative form

$$f(x) = D(x) \cdot Q(x) + R(x)$$

GU (b) Verify the multiplicative form algebraically and use a graphing utility to demonstrate the validity of the result graphically.

1. $f(x) = 4x^3 - 2x^2 + 7x - 1$; $D(x) = x + 1$
2. $f(x) = 8x^3 - x^2 - x + 5$; $D(x) = x - 2$
3. $f(x) = 3x^4 - 2x^3 + 4x^2 + 3x + 7$; $D(x) = x - 3$
4. $f(x) = 2x^5 + 3x^4 - x^2 + x - 4$; $D(x) = x + 2$
5. $f(x) = 6x^4 + 10x^2 + 7$; $D(x) = 2x + 3$
6. $f(x) = 15x^4 + 5x^3 + 6x^2 - 2$; $D(x) = 5x - 3$
7. $f(x) = 4x^4 - 2x^3 - 19x^2 + 7x + 1$; $D(x) = 2x^2 + 1$
8. $f(x) = 8x^4 + 11x^3 - 61x + 30$; $D(x) = x^2 - 2x + 3$

9. Use the remainder theorem to find the value of each function.

(a) $f(-1)$ for $f(x)$ in Problem 1
(b) $f(2)$ for $f(x)$ in Problem 2
(c) $f(3)$ for $f(x)$ in Problem 3

10. Use the remainder theorem to find the value of each function.
(a) $f(-2)$ for $f(x)$ in Problem 4
(b) $f\left(-\frac{3}{2}\right)$ for $f(x)$ in Problem 5
(c) $f\left(\frac{3}{5}\right)$ for $f(x)$ in Problem 6

In problems 11–18, use synthetic division to find the quotient $Q(x)$ and the remainder R if:

	Dividend $f(x)$ Is:	Divisor $D(x)$ Is:
11.	$2x^3 + 17x^2 - 11x - 21$	$x - 3$
12.	$x^3 - 2x^2 - 1$	$x + 2$
13.	$4x^4 - 3x^3 + 2x^2 + 5$	$x - 1$
14.	$10x^4 - 30x^3 + 63$	$x - 6$
15.	$-5x^4$	$x + 2$
16.	$3x^6 - 7$	$x + 3$
17.	$3 + x^2 - 5x^5$	$x + 2$
18.	$8 - 2x^3 + 3x^4 - 2x^5$	$x + 3$

In problems 19–22, use synthetic division and the remainder theorem to evaluate each polynomial function as indicated.

19. $f(x) = 3x^3 + 6x^2 - 10x + 7; f(2)$
20. $f(x) = 3x^3 + 4x^2 - 7x + 16; f(-1)$
21. $f(x) = 2x^3 - 5x^2 + 5x + 11; f\left(\frac{1}{2}\right)$
22. $f(x) = 4x^3 + 3x^2 - x - 3; f\left(\frac{3}{2}\right)$

23. Use the factor theorem and synthetic division to determine whether each binomial is a factor of the polynomial $f(x) = x^3 - 28x - 48$. If it is, find the factorization.
(a) $x + 2$ (b) $x + 4$
(c) $x + 6$ (d) $x - 6$

24. Use the factor theorem and synthetic division to determine whether each binomial is a factor of the polynomial $f(x) = 2x^4 - 9x^3 - 34x^2 - 9x + 14$. If it is, use synthetic division to find the factorization.
(a) $x - 7$ (b) $x + 2$
(c) $x - \frac{1}{2}$ (d) $x - 1$

In problems 25–28:
(a) Use synthetic division to show that the binomial is a factor of $f(x)$, and then factor $f(x)$.
GU (b) Verify the result algebraically and use a graphing utility to demonstrate the validity of the result graphically.

25. $f(x) = x^3 - 6x^2 + 11x - 6; x - 1$
26. $f(x) = x^3 + 6x^2 - 11x - 16; x + 1$
27. $f(x) = 2x^4 + 3x - 26; x + 2$
28. $f(x) = 6x^3 - 25x^2 - 29x + 20; x - 5$

In problems 29–32, use synthetic division to show that c is a zero of $f(x)$.

29. $f(x) = 4x^3 - 9x^2 - 8x - 3; c = 3$
30. $f(x) = x^3 + x^2 - 12x + 12; c = 2$
31. $f(x) = x^4 - 8x; c = 2$
32. $f(x) = x^5 - 5x^4 - 5x^3 + 45x^2 - 108; c = 3$

In problems 33–36, find all zeros of $f(x)$, and indicate the multiplicity of each zero.

33. (a) $f(x) = (x - 1)(x^2 - 3x + 2)^2$
(b) $f(x) = (x - 1)^3(x - 2)(x - 5)$
34. (a) $f(x) = (x + 1)(x - 5)^3(x^2 - 25)$
(b) $f(x) = (x^4 - x^3)(x + 3)^2$
35. (a) $f(x) = (x^3 - 9x)(x - 3)^2$
(b) $f(x) = [(x - 1)^2 - 9]^2$
36. (a) $f(x) = (2x - 5)(3x + 1)^2(x^2 - 5x)$
(b) $f(x) = (2x - 5)^3(x^2 - 1)^2$

Developing and Extending the Concepts

37. Find k so that $x - 2$ is a factor of $f(x) = 3x^3 + 4x^2 + kx - 20$.

38. Find k so that $x - 10$ is a factor of $f(x) = kx^3 - 25x^2 + 47x + 30$.

39. Find k so that $f(3) = -2$ for $f(x) = 2x^3 - 6x^2 + kx - 20$.

40. Are there any real numbers k for which $x - k$ is a factor of $x^4 + 3x^2 + 1$? Explain.

41. (a) Explain how the factor theorem can be used to determine whether $x - 1$ is a factor of $x^n - 1$, where n is a positive integer.
(b) What restrictions must be imposed on n (a positive integer) so that $x + 1$ is a factor of $x^n - 1$?
(c) Is $x - 2$ a factor of $x^{40} - 2^{40}$? Is $x + 2$ a factor of $x^{40} + 2^{40}$? Explain.

42. Find all possible positive integer values of m and n so that the polynomial function

$$f(x) = (x - 1)^m(x + 1)(x + 2)^n$$

is fifth degree. For each possible situation identify the zeros and their multiplicities.

43. Give an example of a quadratic function f that has the given property, and graph the function.
(a) The function f has no real zeros.
(b) The function f has one real zero of multiplicity 2.
(c) The function f has two different real zeros.
(d) What can you conclude regarding the relationship between the graph of a quadratic function and the number of real zeros?

44. Give an example of a quadratic function to show that a complex number zero of a polynomial function is not necessarily an x intercept of the graph of the function.

45. Compare the multiplicative form

$$f(x) = D(x) \cdot Q(x) + R(x)$$

of the division algorithm to the fractional form

$$\frac{f(x)}{D(x)} = Q(x) + \frac{R(x)}{D(x)}$$

Are the two forms equivalent for all values of x? Explain. Give an example to support your assertion.

46. Suppose that c_1, c_2, and c_3 are the zeros of f and that

$$\begin{aligned} f(x) &= (x - c_1)(x - c_2)(x - c_3) \\ &= x^3 + a_2x^2 + a_3x + a_4 \end{aligned}$$

Express a_2, a_3, and a_4 in terms of c_1, c_2, and c_3.

47. Suppose that a polynomial function f has zeros 1, 4, and 5, and no others. What can be said about the degree of $f(x)$?

48. Find a polynomial function g of degree 4 whose zeros are -1, 1, 2, and 3, where $g(0) = 12$.

OBJECTIVES

1. Determine the number and types of zeros.
2. Find rational zeros.
3. Solve applied problems.

3.4 Rational Zeros

So far we know how to find the zeros of linear and quadratic functions. In addition, if a higher-degree polynomial function is factored as a product of linear polynomials, then the zeros are easy to determine. For example, the zeros of $f(x) = x(x - 1)(x + 2)$ are 0, 1, and -2.

Our work in the next two sections will focus on finding the zeros of higher-degree polynomial functions that are not in factored form. In this section, we learn a technique for finding rational numbers that are zeros of polynomial functions with integer coefficients. We begin by establishing some general properties regarding the zeros of polynomial functions.

Determining the Number and Types of Zeros

According to the factor theorem, if c_1 is a zero of a polynomial function f of degree n, $n \geq 1$, then $f(x)$ can be factored as

$$f(x) = a(x - c_1) \cdot q_1(x)$$

where a is the leading coefficient of $f(x)$.

If c_2 is another zero of f, then $f(x)$ can be factored as

$$f(x) = a(x - c_1)(x - c_2) \cdot q_2(x)$$

Continuing this process, we observe that the factored form of $f(x)$ cannot have more than n linear factors. (If there were more than n factors, then multiplying them together would result in a polynomial whose degree would exceed n, the known degree of f.) Thus, we are led to the following property:

PROPERTY 1

MAXIMUM NUMBER OF ZEROS

> If f is a polynomial function of degree n, then f cannot have more than n zeros, not necessarily all different.

Since each real zero corresponds to an x intercept, we are led to the following property:

PROPERTY 2

MAXIMUM NUMBER OF x INTERCEPTS

> If f is a polynomial function of degree n, then the graph of f cannot have more than n x intercepts.

For instance, all the functions $f(x) = x^3 + x$, $g(x) = x^3 - 4x^2 + 4x$, and $h(x) = x^3 - 7x - 2$ have degree 3. After using a graphing utility to sketch their graphs (Figures 1a, 1b, and 1c, respectively), we see that the graph of f has one x intercept, the graph of g has two x intercepts, and the graph of h has three x intercepts.

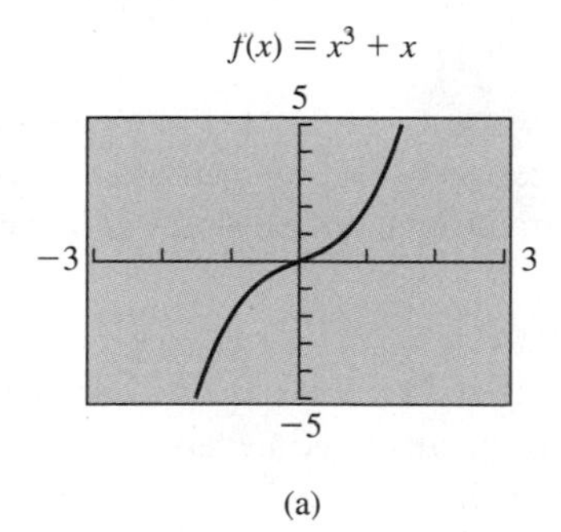

(a)

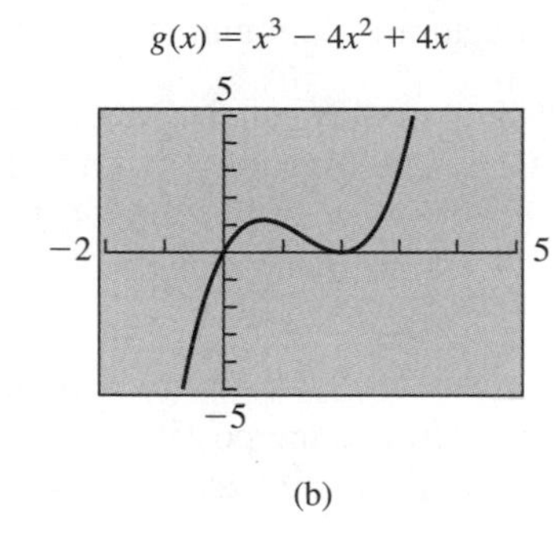

(b)

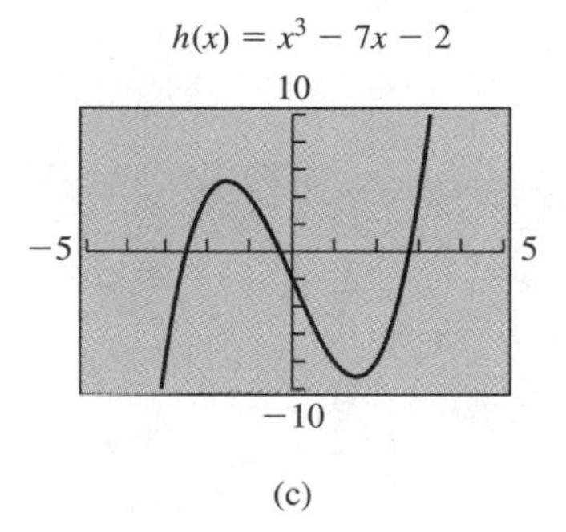

(c)

FIGURE 1

It should be noted that some of the zeros of a polynomial function f may be multiple zeros, and some may be *imaginary numbers* (nonreal complex numbers). For instance,

$$f(x) = x(x - 3)^2(x^2 + 4)$$

has four distinct zeros. Two of the zeros are the real numbers 0 and 3, which has multiplicity 2. The other two zeros are the imaginary numbers $2i$ and $-2i$, since these are the solutions of the equation $x^2 + 4 = 0$. If the zeros that are imaginary numbers are counted and if zeros of multiplicity m are counted as m zeros, we can state the following property:

PROPERTY 3

NUMBER OF ZEROS OF A POLYNOMIAL FUNCTION

> If f is a polynomial function of degree $n > 0$ and if a zero of multiplicity m is counted m times, then f has precisely n zeros.

Because of Property 3, a polynomial function of degree n that has k real zeros must have $n - k$ zeros that are imaginary numbers.

Graphs of polynomial functions help us to determine the number and types of zeros.

EXAMPLE 1 **Determining the number of imaginary zeros**

Suppose that f is a polynomial function whose graph is displayed in Figure 2. Determine the number of imaginary zeros of f if f has degree 3.

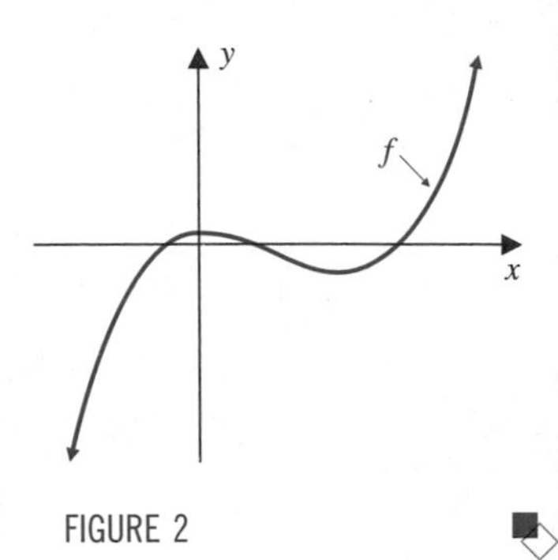

FIGURE 2

Solution From the graph of f we see that there are three x intercepts. So f has three real zeros. Since f has degree 3, it follows from Property 3 that f has $3 - 3 = 0$ imaginary zeros.

EXAMPLE 2 **Determining the types of zeros from graphs**

Use a graphing utility to graph $f(x) = 2x^4 - 4x^3 - 20x^2 + 28x + 32$, and then use the graph to determine the number and types of zeros.

$f(x) = 2x^4 - 4x^3 - 20x^2 + 28x + 32$

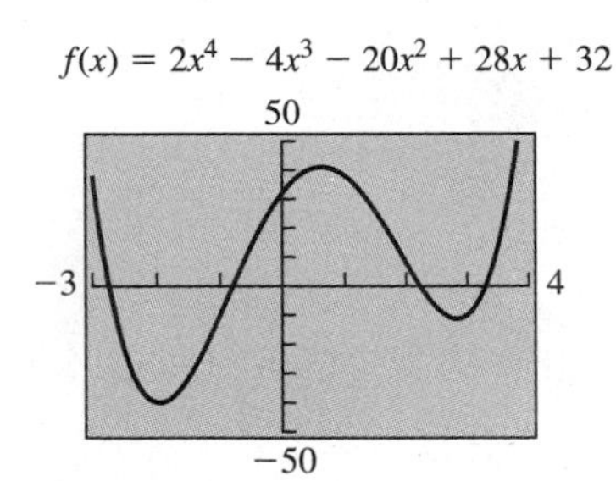

FIGURE 3

Solution A viewing window of the graph of f is shown in Figure 3. There are four x intercepts. So the function has four real zeros; two are negative and two are positive. Since the degree of $f(x)$ is 4, the graph includes all possible zeros, and f has $4 - 4 = 0$ imaginary zeros.

The French mathematician and philosopher René Descartes is credited with finding a way to determine the possible number of positive and negative zeros of a polynomial function with real coefficients. His method is based on the *variation of signs* of a polynomial.

Consider the polynomial

$$f(x) = 3x^5 - 2x^4 + 4x^2 + 6x - 7$$

This polynomial is arranged in descending powers of x. Notice that there are three variations in the signs of the coefficients, reading from left to right:

$$\underbrace{+3x^5 -}_{①} \underbrace{2x^4 +}_{②} \underbrace{4x^2 + 6x -}_{③} 7$$

If we replace x by $-x$ in $f(x)$, we obtain

$$\begin{aligned} f(-x) &= 3(-x)^5 - 2(-x)^4 + 4(-x)^2 + 6(-x) - 7 \\ &= \underbrace{-3x^5 - 2x^4 +}_{①} \underbrace{4x^2 -}_{②} 6x - 7 \end{aligned}$$

Here, we have two variations of signs.

In general, if the terms of a polynomial are arranged in order of descending powers, we say that a variation of signs occurs when two successive terms have opposite signs. Missing terms (with 0 coefficients) are ignored when counting the total number of variations of signs. Now we can state Descartes' rule.

DESCARTES' RULE OF SIGNS

Suppose that $f(x)$ is a polynomial with real coefficients arranged in descending powers of x.

(i) The **number of positive zeros** of f is either equal to the number of variations of signs for $f(x)$, or it is fewer than that number by an even positive integer.

(ii) The **number of negative zeros** of f is either equal to the number of variations of signs for $f(-x)$, or it is fewer than that number by an even positive integer.

When using Descartes' rule, a zero of multiplicity k is counted k times.

EXAMPLE 3 Using Descartes' rule of signs

Use Descartes' rule of signs to determine the possible number of positive real zeros, negative real zeros, and imaginary zeros of

$$f(x) = 3x^5 - 2x^4 + 3x^3 + 2x^2 + 7x - 3$$

Solution There are three variations of signs in $f(x)$. Therefore, $f(x)$ has either three positive zeros or one positive zero. Since

$$\begin{aligned} f(-x) &= 3(-x)^5 - 2(-x)^4 + 3(-x)^3 + 2(-x)^2 + 7(-x) - 3 \\ &= -3x^5 - 2x^4 - 3x^3 + 2x^2 - 7x - 3 \end{aligned}$$

$f(-x)$ has two variations in signs. Consequently, $f(x)$ has either two negative zeros or no negative zero. Since f has precisely five zeros (with a zero of multiplicity k counted as k zeros), the actual number and types of zeros of f will be one of the four possibilities listed in Table 1.

TABLE 1

$f(x) = 3x^5 - 2x^4 + 3x^3 + 2x^2 + 7x - 3$

	Possible Number of Positive Zeros	Possible Number of Negative Zeros	Possible Number of Imaginary Zeros
1.	3	2	0
2.	3	0	2
3.	1	2	2
4.	1	0	4

We conclude from Table 1 that the function has at least one positive real zero (or one positive x intercept).

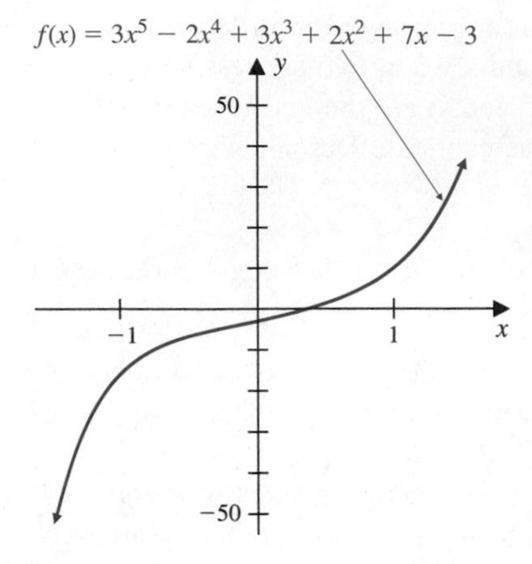

FIGURE 4

Further analysis is necessary in order to determine precisely which of the four situations listed in Table 1 of Example 3 is true. From the graph of $f(x) = 3x^5 - 2x^4 + 3x^3 + 2x^2 + 7x - 3$ (Figure 4) it appears that f has one positive real zero and no negative real zeros, so there are $5 - 1 = 4$ imaginary zeros.

Finding Rational Zeros

Let us now focus our attention on finding rational number zeros, which we refer to simply as *rational zeros*. Consider

$$
\begin{aligned}
f(x) &= 24x^3 - 22x^2 - 5x + 6 \\
&= (4x - 3)(2x + 1)(3x - 2)
\end{aligned}
$$

The zeros of f are found by solving

$$f(x) = (4x - 3)(2x + 1)(3x - 2) = 0$$

to obtain the rational zeros $\frac{3}{4}$, $-\frac{1}{2}$, and $\frac{2}{3}$.

We observe the following relationship between the rational zeros and the coefficients of $f(x)$:

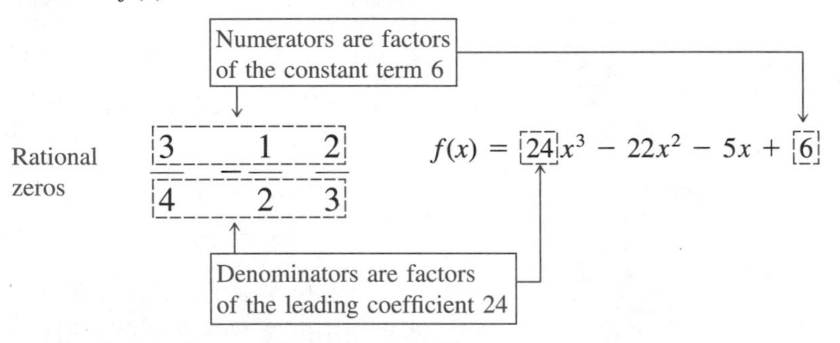

This observation leads us to the next theorem.

RATIONAL ZERO THEOREM

Assume all the coefficients in the polynomial function

$$f(x) = a_n x^n + a_{n-1}x^{n-1} + \cdots + a_1 x + a_0$$

are integers, where $a_n \neq 0$ and $a_0 \neq 0$. If f has a rational zero p/q (where p and q are integers) in lowest terms, then

$$p \text{ must be a factor of } a_0 \qquad \text{and} \qquad q \text{ must be a factor of } a_n$$

It is important to understand the following points about the rational zero theorem:

1. The polynomial must have integer coefficients.
2. Acceptable factors for a_0 and a_n include -1 and 1.

3. The theorem does not guarantee the existence of rational zeros; indeed, there may be none at all.

The rational zero theorem leads to the following strategy for finding all possible rational zeros of a polynomial function with integer coefficients.

STRATEGY FOR FINDING RATIONAL ZEROS OF A POLYNOMIAL FUNCTION f

Step 1. List all factors of the constant term a_0 in the polynomial. (This list provides us with all possible values of p.) List all factors of the coefficient a_n of the term of the highest degree in the polynomial. (This list provides us with all possible values of q.)

Step 2. Use the lists from step 1 to determine all possible rational numbers p/q in reduced form. These rational numbers are the only possible rational zeros of f.

Step 3. Test the values produced in step 2. If none of them is a root of $f(x) = 0$, we conclude that f has no rational zeros.

If p/q is a rational zero of f, then by the factor theorem, $x - (p/q)$ is a factor of $f(x)$.

EXAMPLE 4 Finding rational zeros

Use the rational zero theorem to find the rational zeros of

$$f(x) = x^3 - 2x^2 - x + 2$$

and then use the results to factor $f(x)$.

Solution Assume that p/q is a rational zero of f and p/q is reduced to lowest terms. We apply the above strategy as follows:

Step 1. The factors of $a_0 = 2$ provide the possibilities for p, which are

$$p: \quad -1, 1, -2, 2$$

The possible values for q are the factors of $a_3 = 1$, which are

$$q: \quad -1, 1$$

Step 2. The possible values for p/q are

$$p/q: \quad -1, 1, -2, 2$$

Step 3. Test p/q values. For instance, $f(-1) = 0$, so -1 is a rational zero of f. Since -1 is a zero of f, $x - (-1) = x + 1$ is a factor of $f(x)$. By using synthetic division,

$$\begin{array}{r|rrrr} -1 & 1 & -2 & -1 & 2 \\ & & -1 & 3 & -2 \\ \hline & 1 & -3 & 2 & 0 \end{array}$$

we obtain the factorization

$$x^3 - 2x^2 - x + 2 = (x + 1)(x^2 - 3x + 2)$$

So the remaining zeros of f are found by solving

$$x^2 - 3x + 2 = 0$$
$$(x - 2)(x - 1) = 0$$
$$x = 2 \quad \text{or} \quad x = 1$$

If we had first tested 1 or 2 in step 3, the final result would turn out to be the same.

Hence, the rational zeros of f are -1, 1, and 2. By the factor theorem,

$$f(x) = (x + 1)(x - 1)(x - 2)$$

If the rational zero theorem produces several possibilities for rational zeros (step 2 in the above procedure), then testing each possibility (step 3) can become rather tedious. The next example shows how the testing process can be shortened by using a graph and the fact that the zeros of a polynomial function are the x intercepts of its graph.

EXAMPLE 5 GU

Using a graph to search for rational zeros

Use the rational zero theorem, along with the graph of f, to find the rational zeros of

$$f(x) = 4x^3 + 16x^2 + 9x - 9$$

Solution We begin by assuming that p/q, in reduced form, is a rational zero of f. Next, we apply the three-step strategy.

Step 1. The possible values of p and q are

$$p: \quad -1, 1, -3, 3, -9, 9$$
$$q: \quad -1, 1, -2, 2, -4, 4$$

Step 2. The possible values of p/q are

$$\frac{p}{q}: \quad -1, 1, -3, 3, -9, 9, -\tfrac{1}{2}, \tfrac{1}{2}, -\tfrac{1}{4}, \tfrac{1}{4}, -\tfrac{3}{2}, \tfrac{3}{2}, -\tfrac{3}{4}, \tfrac{3}{4}, -\tfrac{9}{2}, \tfrac{9}{2}, -\tfrac{9}{4}, \tfrac{9}{4}$$

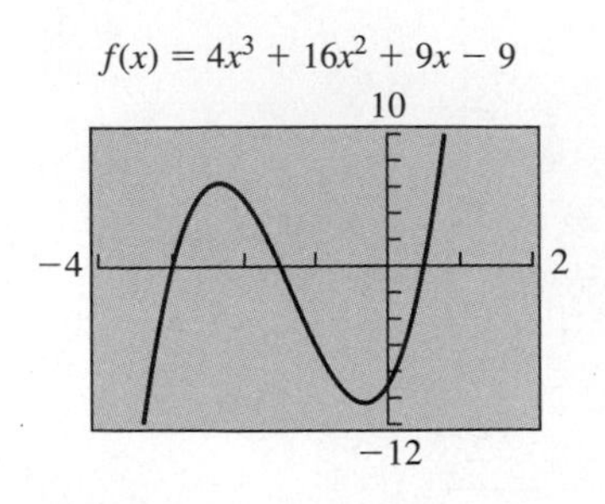

FIGURE 5

Instead of testing all these possibilities, we use the graph of f to eliminate some of the numbers in step 2. Figure 5 shows a viewing window of the graph of f. The general locations of the x intercepts suggest that the list of possibilities for p/q can be narrowed down to -3, $-\frac{3}{2}$, $\frac{1}{4}$, and $\frac{1}{2}$.

Step 3. We use synthetic division to test one of these values, say $\frac{1}{2}$, as follows:

$$\begin{array}{r|rrrr} \frac{1}{2} & 4 & 16 & 9 & -9 \\ & & 2 & 9 & 9 \\ \hline & 4 & 18 & 18 & 0 \end{array}$$

Thus, $\frac{1}{2}$ is a rational zero and we get the factorization

$$\begin{aligned} 4x^3 + 16x^2 + 9x - 9 &= \left(x - \tfrac{1}{2}\right)(4x^2 + 18x + 18) \\ &= \left(x - \tfrac{1}{2}\right)(2)(2x^2 + 9x + 9) \\ &= 2\left(x - \tfrac{1}{2}\right)(x + 3)(2x + 3) \end{aligned}$$

If we had first tested -3 or $-\frac{3}{2}$ in step 3, the final result would turn out to be the same.

It follows that the rational zeros of f are $\frac{1}{2}$, -3, and $-\frac{3}{2}$.

If the polynomial equation $f(x) = 0$ has noninteger rational coefficients, then we first multiply both sides of the equation by the least common denominator (LCD) of the coefficients to convert the equation to an equivalent form containing only integer coefficients. Then the rational zero theorem can be applied. This technique is illustrated in the next example.

EXAMPLE 6 **Using an equivalent form to find rational zeros**

Find a rational zero and use the result to find the other real zeros of

$$f(x) = \tfrac{1}{2}x^3 - 3x^2 + 5x - \tfrac{3}{2}$$

Solution The zeros of f are the same as the roots of

$$\tfrac{1}{2}x^3 - 3x^2 + 5x - \tfrac{3}{2} = 0$$

We multiply both sides of the equation by the LCD of the coefficients, 2, to obtain the equivalent equation

$$x^3 - 6x^2 + 10x - 3 = 0$$

in which all the coefficients are integers. So the zeros of f are the same as the zeros of

$$g(x) = x^3 - 6x^2 + 10x - 3$$

Now we can apply the rational zero theorem to the function g. Assume p/q, in reduced form, is a rational zero of g.

Step 1. Possibilities for p and q are

$$\begin{aligned} p&: \ -1, 1, -3, 3 \\ q&: \ -1, 1 \end{aligned}$$

Step 2. Possibilities for p/q are

$$\frac{p}{q}: \ -1, 1, -3, 3$$

Step 3. After testing these values, we find that $g(3) = 0$, so 3 is a zero of g. In fact,

$$g(x) = x^3 - 6x^2 + 10x - 3 = (x - 3)(x^2 - 3x + 1)$$

The remaining zeros are found by solving

$$x^2 - 3x + 1 = 0$$

Here, we use the quadratic formula with $a = 1$, $b = -3$, and $c = 1$ to get

$$x = \frac{3 + \sqrt{5}}{2} \quad \text{or} \quad x = \frac{3 - \sqrt{5}}{2}$$

Hence, the real zeros of g and f are

$$3, \quad \frac{3 + \sqrt{5}}{2} \text{ (approx. 2.618),} \quad \text{and} \quad \frac{3 - \sqrt{5}}{2} \text{ (approx. 0.382)}$$

Solving Applied Problems

Many applications of mathematics involve solving polynomial equations. Finding the solution of an equation is equivalent to finding the zeros of a function. The next example illustrates an application of the use of this equivalency between solutions and zeros. It also stresses the importance of carefully interpreting the results of a mathematical model.

EXAMPLE 7 Modeling the dimensions of a container

A large container for filling bags of low-nitrogen fertilizer is to be constructed by attaching an inverted cone to the bottom of a right circular cylinder of radius r. The total depth of the container is 12 feet and the radius is the same as the height of the cone, as shown in Figure 6.

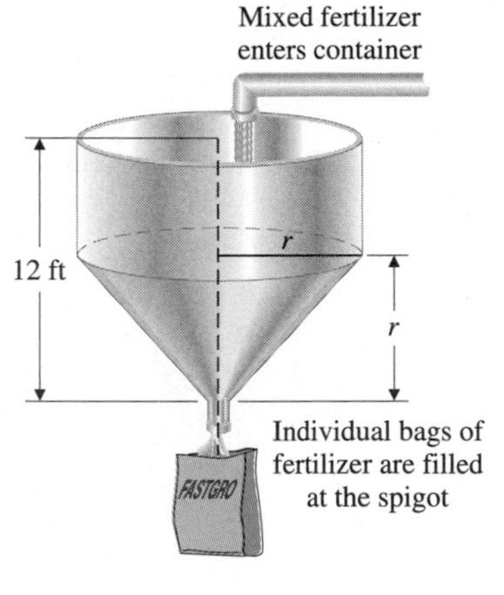

FIGURE 6

(a) Express the volume V of the container as a function of r.

(b) Taking into account appropriate restrictions on r, use a graphing utility to graph the function. Then use the graph to determine the maximum volume under the given conditions.

(c) Suppose that the container is to have a capacity of 90π cubic feet. What is the radius?

Solution

(a) The volume V of the container is the sum of the volumes of the cylinder and the cone. Since the height of the cone is r feet, then the height of the cylinder is $12 - r$ feet. So the volume V is given by

$$\begin{aligned} V &= (\text{Volume of cylinder}) + (\text{Volume of cone}) \\ &= \pi r^2(12 - r) + \tfrac{1}{3}\pi r^3 \qquad \text{Volume of cylinder} = \pi r^2 h; \ \text{Volume of cone} = \tfrac{1}{3}\pi r^2 h \\ &= 12\pi r^2 - \pi r^3 + \tfrac{1}{3}\pi r^3 \\ &= -\tfrac{2}{3}\pi r^3 + 12\pi r^2 \end{aligned}$$

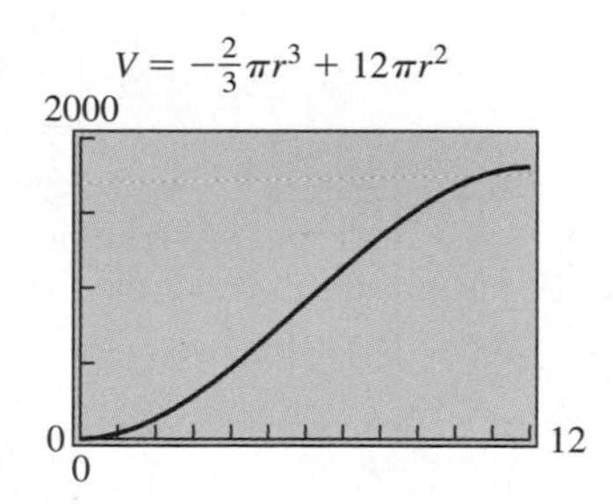

FIGURE 7

Thus, the equation $V = -\frac{2}{3}\pi r^3 + 12\pi r^2$ defines V as a function of r.

(b) Since r and V represent physical properties, $r > 0$ and $V > 0$. However, since the total depth of the container is 12 feet and r is the same as the height of the cone, then r is at most 12 feet, that is, $r \leq 12$. So we can conclude that $0 < r \leq 12$. A viewing window of the graph of $V = -\frac{2}{3}\pi r^3 + 12\pi r^2$, where $0 < r \leq 12$, is displayed in Figure 7. It clearly shows that the volume increases as r increases from 0 to 12. Thus, the maximum volume is attained when $r = 12$; its value is

$$V = -\tfrac{2}{3}\pi(12)^3 + 12\pi(12)^2 = 576\pi$$

So the maximum volume is 576π cubic feet, or approximately 1810 cubic feet, and it is attained when the radius is 12 feet. This situation occurs when the container is a cone 12 feet high (there is no cylindrical part).

(c) Using the function found in part (a), we let $V = 90\pi$, so we need to solve the equation

$$90\pi = -\tfrac{2}{3}\pi r^3 + 12\pi r^2$$

This equation is equivalent to

$$r^3 - 18r^2 + 135 = 0$$

Solving the latter equation is equivalent to finding the zeros of the function

$$f(r) = r^3 - 18r^2 + 135$$

The rational zero theorem leads to the possible rational zeros:

$$\frac{p}{q}\text{:} \quad \pm 1, \pm 3, \pm 5, \pm 9, \pm 15, \pm 27, \pm 45, \pm 135$$

However, we established in part (b) that $0 < r \leq 12$, so the only feasible potential zeros from this list are 1, 3, 5, and 9. After testing these possibilities, we get $f(3) = 0$, so 3 is a zero of f and $r - 3$ is a factor of $f(r)$. In fact,

$$f(r) = (r - 3)(r^2 - 15r - 45)$$

Solving $r^2 - 15r - 45 = 0$ yields two additional zeros of f:

$$\frac{15 - 9\sqrt{5}}{2} \quad \text{and} \quad \frac{15 + 9\sqrt{5}}{2}$$

Hence, the real zeros of f are 3, $(15 - 9\sqrt{5})/2$, and $(15 + 9\sqrt{5})/2$. However, r cannot equal $(15 - 9\sqrt{5})/2$ because this number is negative (approx. -2.56). Furthermore, r cannot equal $(15 + 9\sqrt{5})/2$ (approx. 17.56) since we have already established that $r \leq 12$. Consequently, the radius r must be 3 feet.

PROBLEM SET 3.4

Mastering the Concepts

In problems 1 and 2, suppose that a polynomial function f with the given graph has zeros of multiplicity 1. Determine the number of positive real zeros, negative real zeros, and imaginary zeros of f.

1. **(a)** f has degree 4
(b) f has degree 6

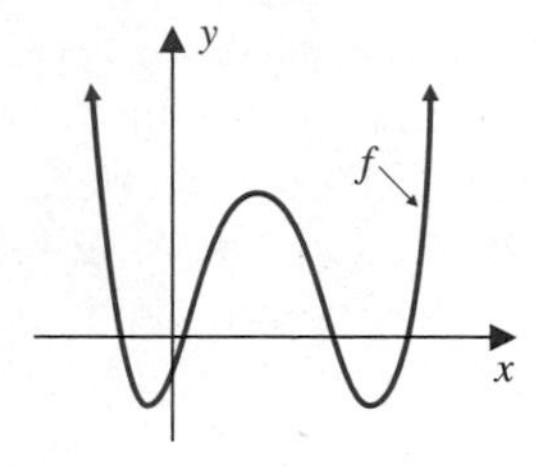

2. **(a)** f has degree 3
(b) f has degree 5

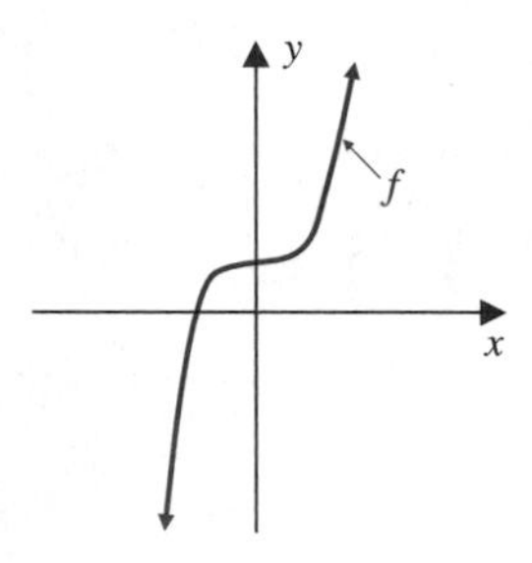

GU In problems 3–8, use a graphing utility to sketch the graph of the polynomial function. Then use the graph to determine the possible number and types of zeros.

3. $f(x) = x^4 - 10$
4. $g(x) = x^5 + 1$
5. $g(x) = 3x^3 - 4x + 1$
6. $h(x) = x^3 + x^2 + x + 1$
7. $h(x) = x^4 - 2x^3 + 6x^2 - 8$
8. $g(x) = 7x^4 - x^3 + 11x - 1$

In problems 9–16, use Descartes' rule of signs to determine the possible number of positive real zeros, negative real zeros, and imaginary zeros of each given function.

9. $f(x) = x^3 - 8x - 2$
10. $g(x) = x^3 + x^2 + x + 1$
11. $g(x) = 3x^3 - 4x + 1$
12. $h(x) = -x^3 + x^2 - 2x + 7$
13. $h(x) = 7x^4 - x^3 + 11x - 1$
14. $g(x) = 2x^4 - 4x^3 - 6x^2 - 8$
15. $f(x) = x^5 - 6x - 5$
16. $h(x) = x^5 + 4x^4 - 3x^2 + x$

In problems 17–26, use the rational zero theorem to find the rational zeros of the function. Use the results to factor the function expression.

17. $f(x) = x^3 - x^2 - 4x + 4$
18. $g(x) = x^3 + 2x - 12$
19. $h(x) = 5x^3 - 12x^2 + 17x - 10$
20. $f(x) = x^3 - x^2 - 14x + 24$
21. $f(x) = x^5 - 5x^4 + 7x^3 + x^2 - 8x + 4$
22. $h(x) = 4x^5 - 23x^3 - 33x^2 - 17x - 3$
23. $f(x) = x^4 - x^3 - \frac{25}{4}x^2 + \frac{x}{4} + \frac{3}{2}$
24. $g(x) = \frac{2}{5}x^4 - x^3 - \frac{8}{5}x^2 + 5x - 2$
25. $g(x) = \frac{2}{3}x^3 + \frac{x^2}{15} - \frac{7}{15}x + \frac{2}{15}$
26. $f(x) = \frac{2}{3}x^4 - \frac{5}{2}x^2 + \frac{5}{6}x + 1$

GU In problems 27–34, use the rational zero theorem, along with the graph of the given function, to find the rational zeros. Use the results to factor the function expression.

27. $h(x) = 4x^4 - 4x^3 - 7x^2 + 4x + 3$
28. $g(x) = x^4 - 9x^2 + 20$
29. $f(x) = x^4 - 3x^3 - 12x - 16$
30. $h(x) = 2x^3 + x^2 + x$
31. $f(x) = x^3 - \frac{7}{2}x^2 + 3x + \frac{5}{2}$
32. $g(x) = x^3 - \frac{x^2}{2} - 4x + 2$
33. $g(x) = 2x^3 - \frac{5}{2}x^2 - \frac{23}{2}x + 3$
34. $f(x) = \frac{1}{4}x^3 - \frac{3}{8}x^2 + x$

In problems 35–40, find a rational zero and use the result to find the other real zeros.

35. $f(x) = x^3 + 2x^2 - 3x - 6$
36. $h(x) = x^3 + 3x^2 - 6x - 18$
37. $h(x) = x^3 + x^2 - 3x + 1$
38. $g(x) = x^3 - 5x^2 - 25x - 3$
39. $g(x) = x^3 + \frac{x^2}{2} - 12x - 6$
40. $f(x) = 2x^3 - 3x^2 - \frac{16}{5}x + \frac{12}{5}$

Applying the Concepts

41. Radar Enclosure: A *radome* is a geometric solid that is formed by removing a section from a sphere so that it has a flat base, as shown in Figure 8. Such structures are used to enclose radar antennas for protection from rain, wind, and snow. The volume V of a radome of radius r and height x is given by the formula

$$V = \pi r x^2 - \frac{\pi}{3}x^3$$

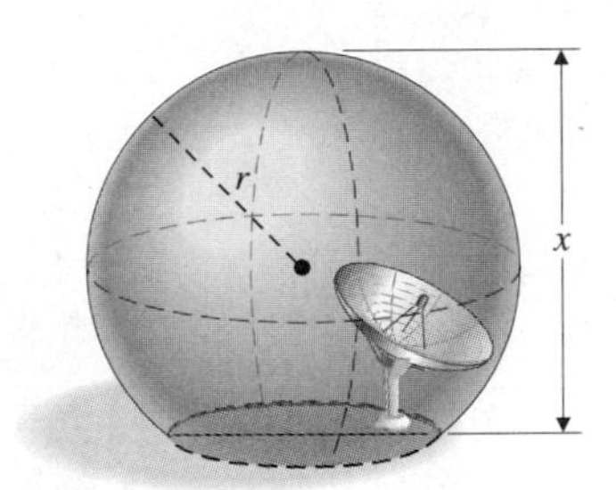

FIGURE 8

(a) Suppose that a radome has a radius of 21 feet and a height of at most 40 feet. Express the volume V as a function of the height x.

(b) What are the restrictions on x?

GU **(c)** Use a graphing utility to graph the function and determine the maximum volume for the restrictions.

(d) Find the height x of the radome if it encloses a volume of $11{,}664\pi$ cubic feet.

42. Percent of Income: Suppose that the relationship between the percent of population x (expressed as a decimal) and the percent of the total income y (expressed as a decimal) of a certain community is modeled by the function

$$y = 2.7x^3 - 2.4x^2 + 0.7x$$

(a) What are the restrictions on x?

GU **(b)** Use a graphing utility to graph the function using the restrictions.

(c) What percent of the total income does 40% of the population in the community account for?

(d) What percent of the population accounts for 20% of the total income of the community?

43. Storage: A cistern to be fabricated in the shape of a right circular cylinder with a hemisphere at the bottom is to have a depth of 30 feet (Figure 9).

(a) Express the volume V of the cistern as a function of r, the radius of the cylinder.

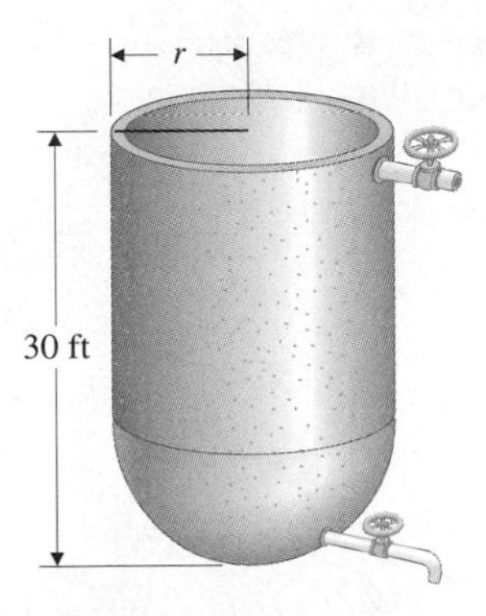

FIGURE 9

(b) Determine the restrictions on r.

GU **(c)** Use a graphing utility to graph the function in part (a) with the restrictions on r from part (b). Determine the maximum volume under the given conditions.

(d) What is the radius if the volume is 1008π cubic feet?

44. Construction: A rectangular toy chest has a square base that is 2 feet by 2 feet and a height of 1 foot. Suppose that each dimension of the chest is increased by x feet, where x is at most 2 feet.

(a) Express the volume V of the new chest as a function of x.

(b) Determine the restrictions on x.

GU **(c)** Use a graphing utility to graph the function in part (a). Determine the maximum volume of the chest under the given restrictions.

(d) What is the value of x if the volume of the new chest is 4.5 times the volume of the old one?

Developing and Extending the Concepts

In problems 45–49, indicate whether the statement is true or false. If the statement is false, give an example to disprove it; if it is true, justify it with an explanation.

45. If the graph of a polynomial function f has three x intercepts, then the degree of $f(x)$ is 3.

46. The numbers -1, 0, 2, and 3 cannot all be zeros of a polynomial function of degree 3.

47. If $\frac{2}{3}$ is a rational zero of the polynomial function

$$g(x) = a_n x^n + a_{n-1}x^{n-1} + \cdots + a_1 x + a_0$$

and the coefficients are all integers, then a_0 is an even integer.

48. If a polynomial function f has three rational zeros, then the graph of f has exactly three x intercepts.

49. If the graph of a polynomial function of degree 5 has one x intercept, then the function has four imaginary zeros.

OBJECTIVES

1. Find bounds of real zeros.
2. Approximate zeros by using a graphing utility.
3. Relate complex zeros to factoring.
4. Solve applied problems.

3.5 Irrational and Complex Zeros

The techniques we have used so far for finding real zeros have limitations. For instance, consider the function

$$f(x) = x^4 - 12x^2 + 35$$

A viewing window of the graph of f is shown in Figure 1. The graph clearly shows that f has four real zeros. However, if we use the rational zero theorem, it turns out that none of the zeros is a rational number. So they must be irrational numbers. In this section, we study a way to approximate such irrational zeros. Also, we investigate the connection between complex zeros and factors of polynomials.

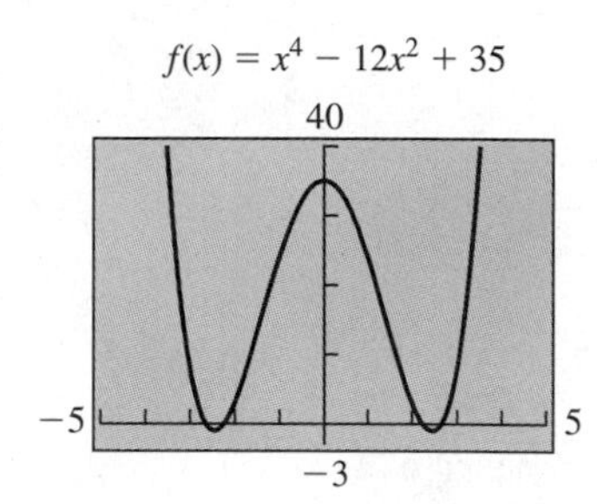

FIGURE 1

Finding Bounds of Real Zeros

Polynomial functions have two properties that help us find the values of real zeros.

PROPERTY 1

INTERMEDIATE-VALUE PROPERTY

Suppose that f is a polynomial function with real coefficients. If the values of $f(a)$ and $f(b)$ have opposite algebraic signs, where $a < b$, then there is at least one number c in the interval (a, b) such that $f(c) = 0$.

This property is illustrated in Figure 2, which shows part of the graph of a polynomial function f that changes sign on an interval $[a, b]$. Because the graph is a continuous or unbroken curve, it must cross the x axis at least at one point, say $(c, 0)$, between the points $(a, f(a))$ and $(b, f(b))$. In other words, there must be a real zero in the interval (a, b).

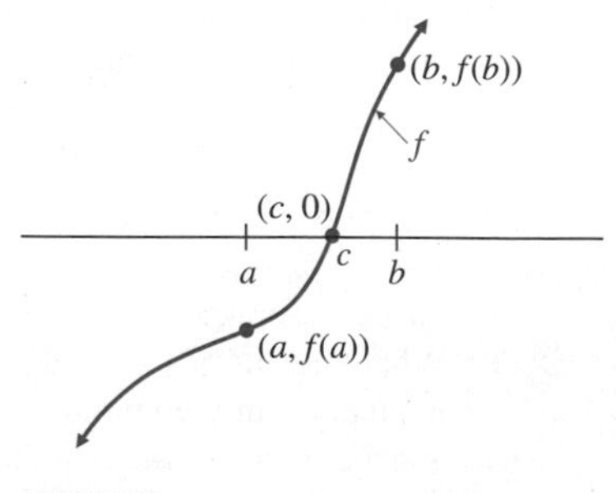

FIGURE 2

EXAMPLE 1 Using the intermediate-value property

(a) Use the intermediate-value property to show that

$$f(x) = x^3 + 3x^2 - 2x - 5$$

has a real zero in the interval $(1, 2)$.

(b) Demonstrate the result by using a graphing utility to sketch the graph of f.

Solution

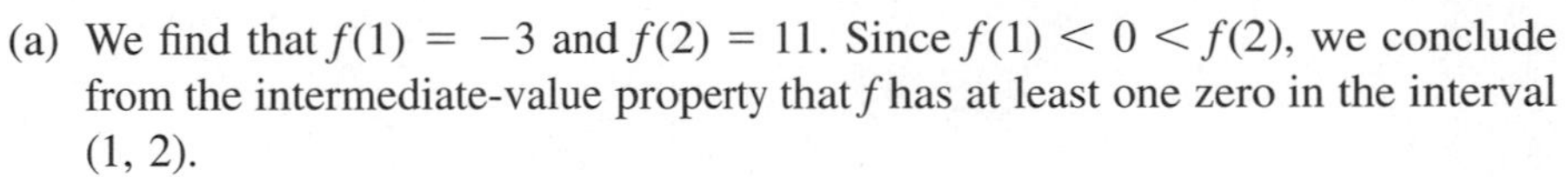

$f(x) = x^3 + 3x^2 - 2x - 5$

FIGURE 3

(a) We find that $f(1) = -3$ and $f(2) = 11$. Since $f(1) < 0 < f(2)$, we conclude from the intermediate-value property that f has at least one zero in the interval $(1, 2)$.

(b) The viewing window of the graph of $f(x) = x^3 + 3x^2 - 2x - 5$ in Figure 3 shows that f has exactly one zero in the interval $(1, 2)$.

Each x intercept on the graph of a polynomial function locates a real zero. But how do we know whether a graph shows all the x intercepts? For instance, consider the function

$$f(x) = 2x^5 - 19x^4 + 47x^3 - x^2 - 49x + 20$$

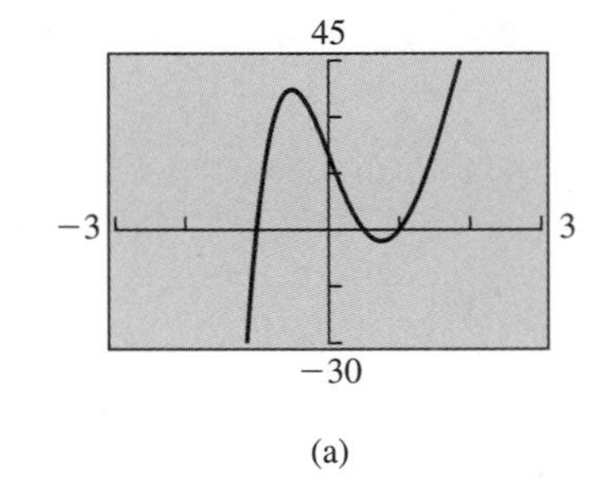

(a)

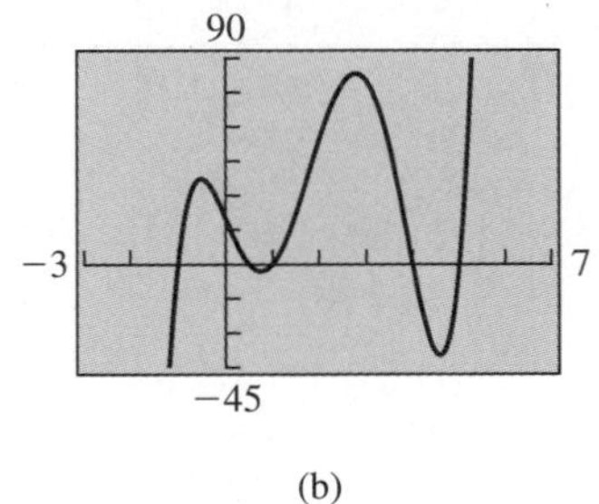

(b)

FIGURE 4

The viewing window in Figure 4a seems to display an accurate graph of f. It is continuous and its limit behavior (which is the same as $y = 2x^5$) appears to be correct. We might deduce from this graph that f has 3 real zeros and 2 imaginary zeros, but this conclusion is wrong. The viewing window in Figure 4b reveals that f actually has 5 real zeros.

This example leads us to the following question:

What interval on the x axis should be selected in order to ensure the inclusion of all the x intercepts for a given polynomial function?

The next property helps us answer this question.

PROPERTY 2

UPPER AND LOWER BOUND PROPERTY

Suppose that $y = f(x)$ is a polynomial function with real coefficients and a positive leading coefficient, and that $f(x)$ is divided by $x - c$ using synthetic division.

(i) If $c > 0$ and the numbers in the last row of the synthetic division process are nonnegative (positive or 0), then c is an **upper bound** for the real zeros of f; that is, any zero of f is less than or equal to c.

(ii) If $c < 0$ and the numbers in the last row of the synthetic division process are alternately nonnegative (positive or 0) and nonpositive (negative or 0), then c is a **lower bound** for the real zeros of f; that is, any zero of f is greater than or equal to c.

EXAMPLE 2 **Finding upper and lower bounds for zeros**

(a) Use Property 2 to find upper and lower integer bounds for the real zeros of

$$f(x) = 10x^4 - 3x^3 - 45x^2 - 3x - 55$$

(b) Use the result in part (a) to graph f with a graphing utility so that all real zeros are displayed.

Solution

(a) Our strategy is to check for an integer upper bound by beginning with 1 and working upward to 2, then 3, and so forth. The results of the synthetic divisions by $x - 1$, $x - 2$, and $x - 3$ are as follows:

1	10	−3	−45	−3	−55
		10	7	−38	−41
	10	7	−38	−41	−96

2	10	−3	−45	−3	−55
		20	34	−22	−50
	10	17	−11	−25	−105

3	10	−3	−45	−3	−55
		30	81	108	315
	10	27	36	105	260

Since all numbers in the last row of the synthetic division by $x - 3$ are positive, it follows that 3 is an upper bound for the real zeros of f; that is, if $\bar{c}$ is a real zero of f, then $\bar{c} \leq 3$. Similarly, to check for a lower bound, we begin with -1, then -2, then -3, to get the following result:

−3	10	−3	−45	−3	−55
		−30	99	−162	495
	10	−33	54	−165	440

Any integer greater than 3 is also an integer upper bound for the real zeros of f, and any integer less than -3 is also an integer lower bound.

Since the signs of the numbers in the last row alternate, we conclude that -3 is a lower bound; that is, if $\bar{c}$ is a real zero of f, then $\bar{c} \geq -3$. Thus, any real zero of f is a number between -3 and 3. So -3 and 3 are, respectively, lower and upper integer bounds for real zeros of f.

(b) Our selection of an appropriate viewing window for the graph of f is influenced by four factors:

(i) The graph is continuous.
(ii) The graph has the limit behavior of $y = 10x^4$.
(iii) The graph has at most three turning points.
(iv) The graph includes all real zeros.

According to the result in part (a), condition (iv) is guaranteed as long as we select an interval on the x axis that includes the interval $[-3, 3]$. Figure 5 shows a viewing window of the graph that displays the four characteristics mentioned above. The graph shows conclusively that f has two distinct real zeros, one between -2 and -3 and one between 2 and 3. ■

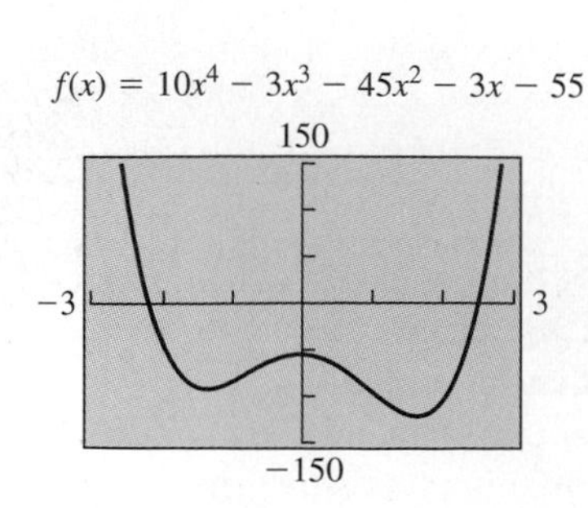

FIGURE 5

Once upper and lower bounds for real zeros are determined, we can apply the rational zero theorem to find the possible rational number zeros that fall between the bounds. Finally, if there are irrational zeros remaining, we can use a graph of the function sketched on the interval defined by the upper and lower bounds, along with an approximating method, to estimate these values.

Approximating Zeros by Using a Graphing Utility

Some graphing utilities are programmed so that we only need to enter an equation and the solutions are displayed. Our goal here is to use properties of graphs to find solutions.

Many techniques such as the *bisection method* (Problems 57–62) and *Newton's method* (studied in calculus) have been developed to approximate the real zeros of a function. Here, we'll use the ZOOM and TRACE features available on graphing utilities to approximate values of real zeros.

It is common to express the desired accuracy of the approximation of a zero in terms of an absolute value inequality. For instance, if we are asked to find an approximation c for a real zero such that $|f(c)| \leq 0.0005$, then we continue the approximating process until we find a value c such that $-0.0005 \leq f(c) \leq 0.0005$.

EXAMPLE 3

Using a graphing utility to approximate a zero

Use a graphing utility to sketch the graph of $f(x) = x^3 - 3x + 1$. Then use the ZOOM and TRACE features to approximate the value c of the largest zero so that $|f(c)| \leq 0.0005$. Round off the answer to four decimal places.

Some graphing utilities are programmed to locate the x intercept without having to use the ZOOM and TRACE features.

Solution Figure 6a shows a viewing window of the graph of the function $f(x) = x^3 - 3x + 1$ that displays all real zeros of f. From this graph, we see that the largest zero of f is between 1 and 2. To approximate the value of this zero, we use successively more restricted viewing windows that ZOOM in on the zero and then use the TRACE feature as we did before for finding minimum and maximum points.

For the selections made here, we get an approximation of $x = 1.5320888862$ (Figure 6b). Since $f(1.5320888862) = -0.0000000001533$, the approximation satisfies the condition $|f(c)| \leq 0.0005$. So the approximate value of this zero is 1.5321 (rounded to four decimal places).

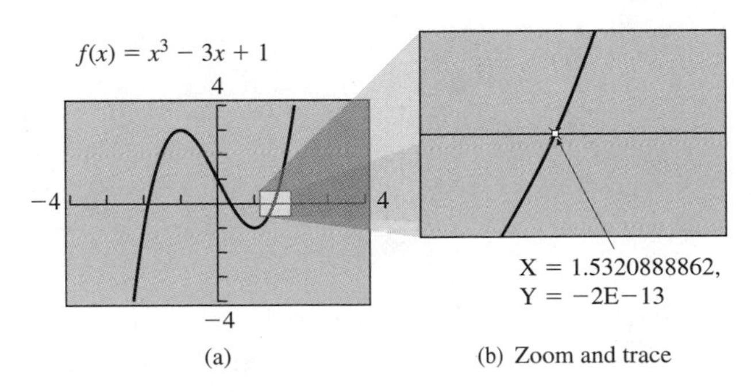

FIGURE 6

In Section 3.2 we solved polynomial inequalities where the polynomial was given in factored form. The next example shows how to approximate the solutions of polynomial inequalities when the polynomial is not easily factored.

EXAMPLE 4 **Solving an inequality graphically**

GU

Use a graphing utility to solve the inequality $-x^4 - 2x^3 + 4x^2 + 6x + 3 \geq 0$. Round off the answers to two decimal places.

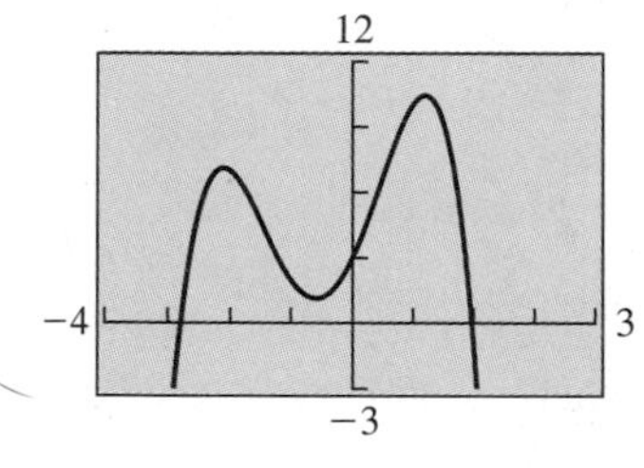

FIGURE 7

Solution We begin by using a graphing utility to sketch the graph of $f(x) = -x^4 - 2x^3 + 4x^2 + 6x + 3$ so that all the real zeros are displayed (Figure 7).

The solution of the inequality includes all values of x for which $f(x) \geq 0$; that is, all values of x for which the graph is above or on the x axis. In order to find these values, we need to determine the x intercepts, or zeros, of f. By using the ZOOM and TRACE features we find that the intercepts are approximately -2.80 and 1.97, so the solution set of the inequality includes approximately all values of x such that $-2.80 \leq x \leq 1.97$, that is, all values of x in the interval $[-2.80, 1.97]$.

Relating Complex Zeros to Factoring

In Section 3.4, we saw that a polynomial function of degree n has exactly n zeros, possibly including repeated zeros. This result follows from the *Fundamental Theorem of Algebra.*

THE FUNDAMENTAL THEOREM OF ALGEBRA

Every polynomial function of degree $n \geq 1$ has at least one complex zero.

Because the complex numbers have the same basic properties as the real numbers, many of our previous results, such as the factor theorem, extend to complex zeros. By combining the Fundamental Theorem of Algebra with the factor theorem, we obtain the following property:

THE COMPLETE LINEAR FACTORIZATION PROPERTY

If $f(x)$ is a polynomial of degree $n \geq 1$, then there are n complex numbers $c_1, c_2, \ldots, c_n$ that form the linear factorization

$$f(x) = a(x - c_1)(x - c_2) \cdot \cdots \cdot (x - c_n)$$

where a is the leading coefficient of $f(x)$.

The numbers $c_1, c_2, \ldots, c_n$ are the n complex zeros of f.

EXAMPLE 5 **Finding linear factors**

Find the complete linear factorization for each polynomial function.

(a) $f(x) = 2x^2 - 5x - 3$ (b) $f(x) = x^3 - 3x^2 + 2$

Solution

(a) Here, we have

$$f(x) = 2x^2 - 5x - 3 = (2x + 1)(x - 3) = 2\left(x + \tfrac{1}{2}\right)(x - 3)$$

(b) After using the rational zero theorem and synthetic division, we find that 1 is a zero of f and that

$$f(x) = x^3 - 3x^2 + 2 = (x - 1)(x^2 - 2x - 2)$$

Next, we use the quadratic formula to solve $x^2 - 2x - 2 = 0$ to get $x = 1 + \sqrt{3}$ or $x = 1 - \sqrt{3}$. So, by the factor theorem,

$$\begin{aligned} f(x) &= (x - 1)(x^2 - 2x - 2) \\ &= (x - 1)\left[x - \left(1 + \sqrt{3}\right)\right]\left[x - \left(1 - \sqrt{3}\right)\right] \end{aligned}$$

Consider the polynomial function

$$f(x) = x^2 - 6x + 25$$

We can find the zeros of f by using the quadratic formula to solve the equation $x^2 - 6x + 25 = 0$ to get

$$3 + 4i \quad \text{and} \quad 3 - 4i$$

Notice that these zeros of f are complex conjugates of each other. In general, we have the following theorem:

CONJUGATE ZEROS THEOREM

Let f be a polynomial function with real coefficients. If $a + bi$ is a zero of f, then the complex conjugate $a - bi$ is also a zero of f.

EXAMPLE 6 Finding complex zeros

Find the zeros of

$$f(x) = x^4 - 4x^3 + 5x^2 - 4x + 4$$

given that i is a zero of f. Also, factor $f(x)$ and determine the multiplicity of each zero.

Solution The conjugate zero theorem tells us that since i is a zero of f, then the conjugate $-i$ is also a zero of f. By the factor theorem, we have

$$\begin{aligned} f(x) &= x^4 - 4x^3 + 5x^2 - 4x + 4 \\ &= (x - i)(x + i)Q(x) \\ &= (x^2 + 1)Q(x) \end{aligned}$$

After dividing $f(x)$ by $x^2 + 1$ using long division, we get the factorization

$$\begin{aligned} f(x) &= (x^2 + 1)(x^2 - 4x + 4) \\ &= (x - i)(x + i)(x - 2)^2 \end{aligned}$$

So i and $-i$ are each zeros of f of multiplicity 1, and 2 is a zero of f of multiplicity 2.

EXAMPLE 7 **Using zeros to form a polynomial function**

Form a polynomial function with real coefficients, a leading coefficient of 1, the smallest possible degree, and the zeros -2, 3, and $1 - 3i$.

Solution Since we want a polynomial of the smallest possible degree, we assume that each zero has multiplicity 1. Because we want the coefficients to be real numbers, we must include $1 + 3i$, the complex conjugate of $1 - 3i$, as a zero. By the factorization theorem we obtain

$$\begin{aligned} f(x) &= [x - (-2)](x - 3)[x - (1 - 3i)][x - (1 + 3i)] \\ &= (x^2 - x - 6)(x^2 - 2x + 10) \\ &= x^4 - 3x^3 + 6x^2 + 2x - 60 \end{aligned}$$

It is important to understand what is meant by a *complete factorization* of a polynomial with real coefficients of degree $n > 1$. In the complex number system it is possible to write such a polynomial completely as a product of *linear* factors. However, in the real number system, we have the following property:

PROPERTY

COMPLETE FACTORIZATION IN THE REAL NUMBER SYSTEM

In the *real number system*, any polynomial function of degree $n > 1$ with real coefficients can be written as the product of linear and/or quadratic factors with real coefficients, where the quadratic factors have no real zeros.

The examples given in Table 1 compare complete factorizations in the complex number system to those in the real number system.

TABLE 1

Examples of Complete Factorizations of Polynomials

Polynomial Function with Real Coefficients	Complete Factorization	
	Real number system	*Complex number system*
$x^2 + 4$	$x^2 + 4$	$(x + 2i)(x - 2i)$
$x^3 - 1$	$(x - 1)(x^2 + x + 1)$	$(x - 1)\left[x - \left(\frac{-1 + \sqrt{3}i}{2}\right)\right]\left[x - \left(\frac{-1 - \sqrt{3}i}{2}\right)\right]$
$x^4 - 14x^3 + 61x^2 - 54x - 130$	$(x + 1)(x - 5)(x^2 - 10x + 26)$	$(x + 1)(x - 5)[x - (5 - i)][x - (5 + i)]$

Solving Applied Problems

Mathematical models of real-world situations often involve approximating real zeros.

EXAMPLE 8 Modeling a suspension bridge

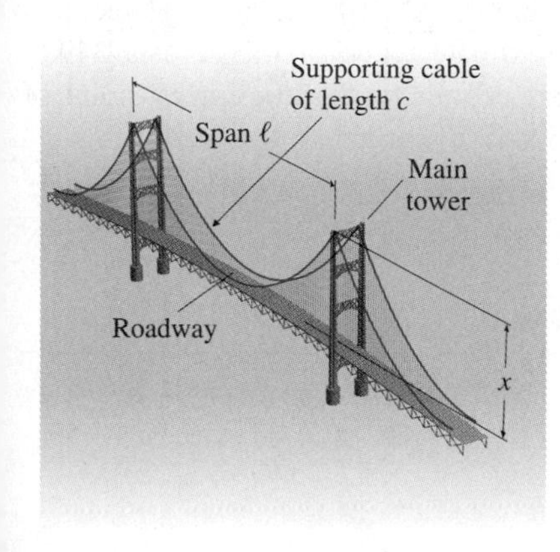

FIGURE 8

Figure 8 shows a sketch of the main span of a suspension bridge. Assume the relationship between the length c of the supporting cable between two vertical towers and the *sag* x, the vertical distance from the top of either tower to the lowest point of the supporting cable at or above the roadway, is given by the formula

$$c = \ell + \frac{8x^2}{3\ell} - \frac{32x^4}{5\ell^3}$$

where ℓ is the horizontal distance of the *span* between the towers.

The Mackinac Bridge, one of the world's longest suspension bridges, connects the upper and lower peninsulas of Michigan. The distance between its main towers is 3800 feet, the top of each main tower is 552 feet above water level, and the roadway is 199 feet above water, as shown in Figure 9.

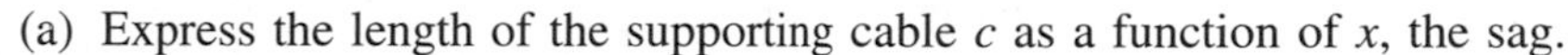

(a) Express the length of the supporting cable c as a function of x, the sag.
(b) Determine the restrictions on x.

(c) How high above the roadway is the lowest point of the cable if the supporting cable is 3884 feet long? Round off the answer to one decimal place.

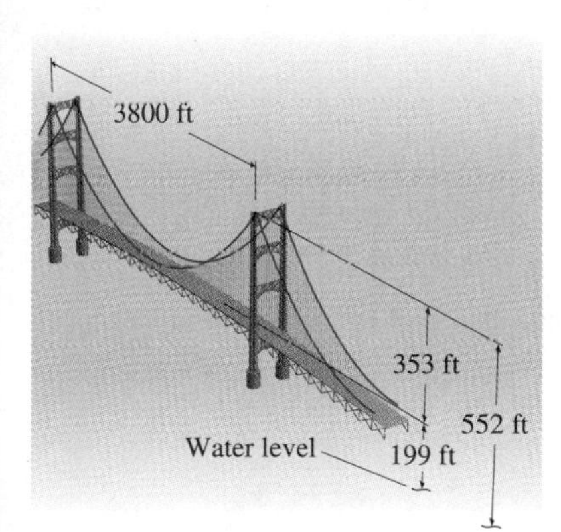

FIGURE 9

Solution

(a) Since ℓ is given to be 3800, we substitute this value into the above formula to get

$$c = 3800 + \frac{8x^2}{3(3800)} - \frac{32x^4}{5(3800)^3}$$
$$= 3800 + \frac{x^2}{1425} - \frac{x^4}{857{,}375 \cdot 10^4}$$

This latter equation expresses c as a function of x.

(b) Since the lowest point of the supporting cable is at or above the roadway, then the sag x must be less than or equal to the tower's height above the roadway. Upon examining the given data, which are shown in Figure 9, it follows that $0 < x \leq 353$.

(c) We begin by finding the sag x, given that the cable is 3884 feet long. To determine x, we substitute $c = 3884$ into the function in part (a) to get

$$3884 = 3800 + \frac{x^2}{1425} - \frac{x^4}{857{,}375 \cdot 10^4}$$
$$0 = -84 + \frac{x^2}{1425} - \frac{x^4}{857{,}375 \cdot 10^4}$$

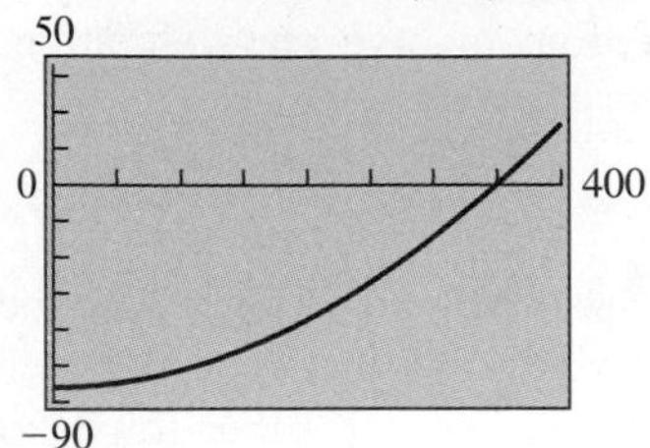

FIGURE 10

The solutions of this latter equation are the same as the zeros of the function f defined by

$$f(x) = -84 + \frac{x^2}{1425} - \frac{x^4}{857{,}375 \cdot 10^4} \qquad \text{where } 0 < x \leq 400$$

The viewing window of the graph of f shown in Figure 10 indicates that there is one zero in the interval (0, 400). By using the ZOOM and TRACE features on the graph of f we find that an approximate value of the zero is 349.5. Consequently, the sag is about 349.5 feet. It follows that the lowest point of the cable is approximately $353 - 349.5 = 3.5$ feet above the roadway.

PROBLEM SET 3.5

Mastering the Concepts

In problems 1–6:

(a) Use the intermediate-value property to show that the function f has a zero in the given interval.

GU **(b)** Demonstrate the result by using a graphing utility to sketch the graph of f.

1. $f(x) = x^3 - 3x^2 + 4x - 5$; [1, 3]
2. $f(x) = x^3 - 4x^2 + 3x + 1$; [2, 4]
3. $f(x) = 3x^3 - 10x + 9$; [−3, −2]
4. $f(x) = 2x^3 + 6x^2 - 8x + 2$; [−5, −4]
5. $f(x) = x^4 - 4x^3 + 10$; [0.9, 2.1]
6. $f(x) = x^4 + 6x^3 - 18x^2$; [2.1, 2.2]

In problems 7–12, for each given function:

(a) Use the upper and lower bound property to find upper and lower integer bounds for the real zeros.

GU **(b)** Use the result from part (a) to graph the function with a graphing utility so that all real zeros are displayed.

7. $f(x) = x^3 - 3x + 1$
8. $g(x) = x^4 + 2x^3 + 2x^2 - 4x - 8$
9. $h(x) = x^3 - 6x^2 + 3x + 13$
10. $h(x) = -x^3 + 3x + 1$
11. $h(x) = 23x - x^3$
12. $g(x) = 2x^5 - 4x^4 + x^2 - 10$

GU In problems 13–18, use a graphing utility to sketch the graph of each function, and then use the ZOOM and TRACE features to approximate the value c of each real zero so that $|f(c)| \leq 0.0005$. Round off the answers to four decimal places, if necessary.

13. $f(x) = x^3 - 3x - 2$
14. $f(x) = x^3 - 3x^2 + 5$
15. $f(x) = x^4 - 3x^2 + 2$
16. $f(x) = x^3 + 4x^2 - 3x - 10$
17. $f(x) = 2x^5 - 4x^4 + x^2 - 10$
18. $f(x) = x^4 - 2x^3 - 3x^2 + 4x + 2$

GU In problems 19–24, use the graph of an appropriate function along with the values of the zeros to approximate the solutions of each inequality. Round off to two decimal places.

19. $x^3 - 3x > 2$ (see Problem 13)
20. $x^3 + 5 \leq 3x^2$ (see Problem 14)
21. $x^4 + 2 \leq 3x^2$ (see Problem 15)
22. $x^3 + 4x^2 \leq 3x + 10$ (see Problem 16)
23. $2x^5 + x^2 \geq 4x^4 + 10$ (see Problem 17)
24. $x^4 + 4x + 2 > 2x^3 + 3x^2$ (see Problem 18)

In problems 25–30, use the given zero of f to find all other complex zeros and express $f(x)$ in complete linear factored form. Also determine the multiplicity of each zero.

25. $f(x) = x^4 + 13x^2 + 36$; $2i$
26. $f(x) = 2x^5 - 7x^4 + 12x^3 - 8x^2 + 4$; $1 + i$ (multiplicity 2)
27. $f(x) = x^4 + 2x^2 + 1$; i (multiplicity 2)
28. $f(x) = x^5 - 2x^4 + 2x^3 + 8x^2 - 16x + 16$; $1 + i$
29. $f(x) = x^4 + 2x^3 - 4x - 4$; $-1 + i$
30. $f(x) = 2x^6 + x^5 + 2x^3 - 6x^2 + x - 4$; i (multiplicity 2)

In problems 31–36, form a polynomial function of the specified degree, with real coefficients, a leading coefficient of 1, and the given numbers as zeros.

31. $1 + \sqrt{2}$, $1 - \sqrt{2}$, 3; degree 3

32. $\sqrt{3}, -\sqrt{3}$, 4 (multiplicity 2); degree 4
33. $1 + i$; degree 2
34. 3, -3, $3i$; degree 4
35. $1 - 3i$ (multiplicity 2); degree 4
36. $1 + 5i$, 2 (multiplicity 2); degree 4

In problems 37–44, factor $f(x)$ completely:
(a) In the real number system
(b) In the complex number system

37. $f(x) = x^2 - x + 1$
38. $f(x) = 2x^2 + 4x + 6$
39. $f(x) = 2x^3 + 4x$
40. $f(x) = 4x^3 + 2x$
41. $f(x) = x^4 - 81$
42. $f(x) = x^4 + 6x^2 + 5$
43. $f(x) = 3x^3 - x^2 + 3x - 1$
44. $f(x) = x^3 + x^2 + x - 3$

Applying the Concepts

In problems 45–48, round off the answers to two decimal places.

45. Rate of Inflation: Suppose that a typical family in the United States will save approximately s percent of its income per year if the rate of inflation is x percent, where s is modeled by the function

GU

$$s = -0.001x^3 + 0.06x^2 - 1.2x + 12 \qquad 0 \le x \le 10$$

Use a graphing utility to approximate what the rate of inflation would be if the average saving percent were 10 percent, 8 percent, and 5 percent.

46. Hot-Air Balloon: Atmospheric pressure P (in pounds per square inch) is approximated by the function

GU

$$P = 15 - 6h + 1.2h^2 - 0.16h^3 \qquad 0 \le h \le 3.5$$

where height h is in thousands of feet above sea level. Use a graphing utility to approximate the height of a hot-air balloon if the atmospheric pressure at that height is 10 pounds per square inch, 5 pounds per square inch, and 3.4 pounds per square inch.

47. The Golden Gate Bridge Cable: The Golden Gate Bridge in San Francisco has twin supporting main towers extending 525 feet above the road surface. The two main towers are 4200 feet apart.
(a) Draw a sketch of the situation.
(b) Use the formula in Example 8 to express the length of the supporting cable c between the two towers as a function of the sag x.
(c) Determine the restrictions on x.
GU **(d)** Use a graphing utility to find how high the lowest point of the cable is above the roadway if the length of the supporting cable is 4363 feet.

48. Radius of a Propane Tank: A steel propane gas storage tank is to be constructed in the shape of a right circular cylinder with a hemisphere attached at each end. The total length of the tank is 12 feet (Figure 11).
(a) Express the volume V of the tank as a function of the radius x (in feet) of the cylinder.
(b) What are the restrictions on x?
GU **(c)** Use a graphing utility to determine the radius so that the resulting volume is 260 cubic feet.

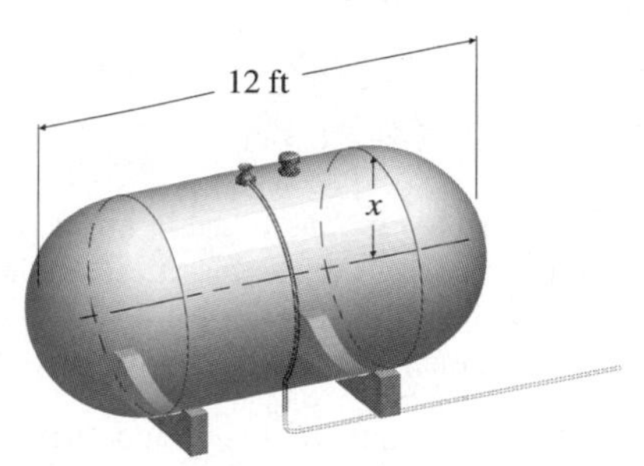

FIGURE 11

Developing and Extending the Concepts

In problems 49 and 50, give an example to show that the given statement is false.

49. Suppose that f is a polynomial function with real coefficients. If the values of $f(a)$ and $f(b)$ are both positive, where $a < b$, then f has no zero between a and b.

50. Suppose that f is a polynomial function with real coefficients. If the values of $f(a)$ and $f(b)$ have opposite algebraic signs, where $a < b$, then there is exactly one zero between a and b.

51. Let $f(z) = z^3 - iz^2 + 2iz + 2$.
(a) Determine whether i and its conjugate $-i$ are zeros of f.
(b) If both are not zeros of f, does this contradict the conclusion of the conjugate zero theorem? Explain.

52. Use a graphing utility to determine the number of imaginary zeros of

GU

$$f(z) = z^5 + 1.1z^4 - 3.21z^3 - 2.84z^2 + 2.70z + 1.62$$

assuming there are no zeros of multiplicity greater than 1.

53. Give examples of two different polynomial functions such that both functions have zeros 1, 2, and 3.

54. Give examples of two different polynomial functions of degree 4 with real coefficients such that both functions have zeros 1, 2, and i.

GU In problems 55 and 56, use a graphing utility to graph f and to approximate the solution of the equation in part (a); then confirm the solution algebraically. Finally, use the graph of f and the solution of the equation to solve the inequality in part (b).

55. $f(x) = \sqrt{2x + 1} - \sqrt{5x - 11}$
  (a) $\sqrt{2x + 1} - \sqrt{5x - 11} = 0$
  (b) $\sqrt{2x + 1} - \sqrt{5x - 11} < 0$

56. $f(x) = \sqrt{x + 12} - \sqrt{x} - 2$
  (a) $\sqrt{x + 12} - \sqrt{x} - 2 = 0$
  (b) $\sqrt{x + 12} - \sqrt{x} - 2 > 0$

In problems 57–62, refer to the graphs found in problems 7–12 for the given functions and use the **bisection method** outlined below to approximate the value of the zero. Round off the answers to four decimal places.

Suppose f is continuous on the interval $[a, b]$, and $f(a)$ and $f(b)$ have opposite signs, so that the function f has a zero at some number, say c, in the interval $[a, b]$. For specificity, assume $f(a) > 0$ and $f(b) < 0$ (Figure 12a). If $f(m) > 0$, where $m = (a + b)/2$ is the midpoint of the interval $[a, b]$, then the zero c is in the interval $[m, b]$ (Figure 12b). If m_1 is the midpoint of interval $[m, b]$ and $f(m_1) < 0$, then the interval $[m, m_1]$ contains the zero c. This interval is one-half the length of interval $[m, b]$ (Figure 12c). We continue this bisection process until we get a value for a midpoint, say M, so that $|f(M)|$ is less than or equal to some preassigned value such as 0.005. This value M approximates the zero c.

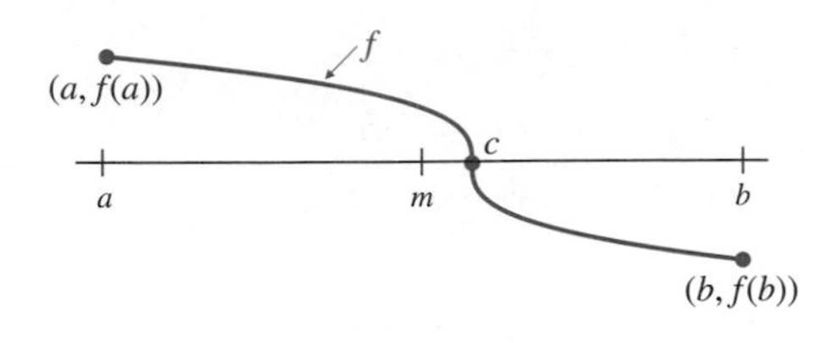

(a)

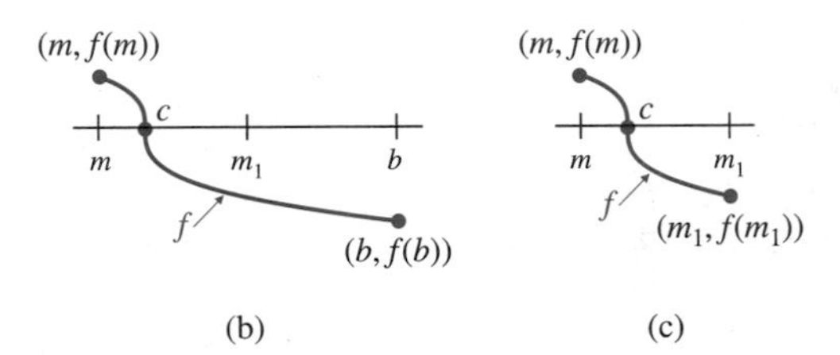

(b) (c)

FIGURE 12

57. $f(x) = x^3 - 3x + 1$ (see Problem 7); $0 < c < 1$

58. $g(x) = x^4 + 2x^3 + 2x^2 - 4x - 8$ (see Problem 8); $1 < c < 2$

59. $h(x) = x^3 - 6x^2 + 3x + 13$ (see Problem 9); smallest zero c

60. $h(x) = -x^3 + 3x + 1$ (see Problem 10); largest zero c

61. $h(x) = 23x - x^3$ (see Problem 11); largest zero c

62. $g(x) = 2x^5 - 4x^4 + x^2 - 10$ (see Problem 12); $2 < c < 3$

OBJECTIVES

1. Graph functions of the form $R(x) = 1/x^n$.
2. Locate asymptotes.
3. Graph rational functions.
4. Solve applied problems.

3.6 Rational Functions

Just as rational numbers are defined in terms of quotients of integers, *rational functions* are defined in terms of quotients of polynomials.

DEFINITION

RATIONAL FUNCTION

A function R of the form

$$R(x) = \frac{p(x)}{q(x)}$$

where $p(x)$ and $q(x)$ are polynomials and $q(x) \neq 0$, is called a **rational function**.

Examples of rational functions are

$$f(x) = \frac{3}{x}, \quad g(x) = \frac{2x}{x^2 - 4}, \quad \text{and} \quad h(x) = \frac{3x - 1}{x^2 + 4}$$

The **domain** of a rational function $R(x) = p(x)/q(x)$ consists of all real numbers x for which $q(x) \neq 0$ since division by 0 is not defined. For instance, the domain of $f(x) = 3/x$ is the set of all real numbers except 0. Also, the domain of $g(x) = (2x)/(x^2 - 4)$ is the set of all real numbers except -2 and 2, because $x^2 - 4 = 0$ if $x = -2$ or $x = 2$.

Graphing Functions of the Form $R(x) = \dfrac{1}{x^n}$

The zeros of $q(x) = x^n$ are not in the domain of R, but understanding the behavior of $R(x)$ for x near these zeros is particularly useful in sketching the graph of R. To illustrate this behavior, consider the rational function

$$f(x) = \frac{1}{x}$$

The domain of f includes all real numbers except $x = 0$. So the graph has no y intercept. The data in Table 1 suggest that as x takes on positive values closer and closer to 0 (that is, as x approaches 0 from the right), the positive function values increase without bound. We describe this behavior symbolically by writing

$$f(x) \to +\infty \quad \text{as} \quad x \to 0^+$$

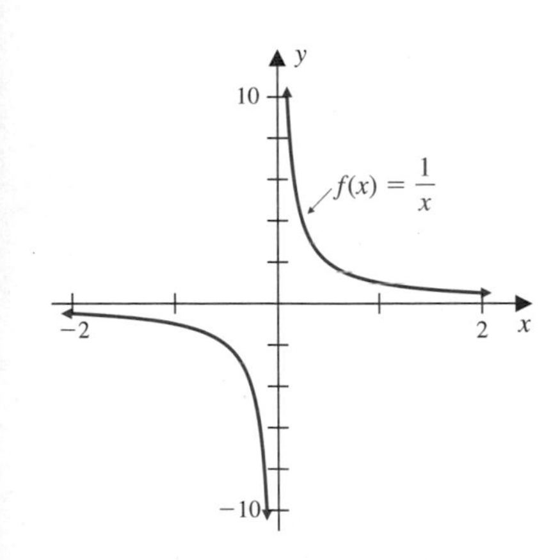

FIGURE 1

TABLE 1
As $x \to 0^+$, $f(x) \to +\infty$

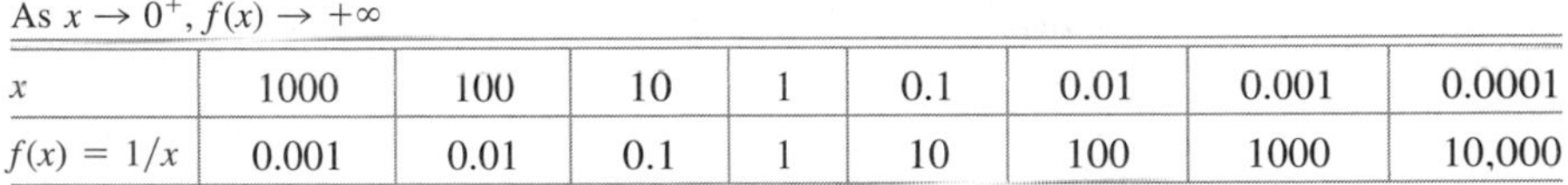

x	1000	100	10	1	0.1	0.01	0.001	0.0001
$f(x) = 1/x$	0.001	0.01	0.1	1	10	100	1000	10,000

Correspondingly, the graph of $f(x) = 1/x$ (Figure 1) gets higher and higher; and approaches the y axis, but never touches it, as x approaches 0 from the right.

Similarly, as Table 2 suggests, the negative function values $f(x)$ decrease without bound as negative x values get closer and closer to 0 (that is, as x approaches 0 from the left), and we write

$$f(x) \to -\infty \quad \text{as} \quad x \to 0^-$$

TABLE 2
As $x \to 0^-$, $f(x) \to -\infty$

x	−1000	−100	−10	−1	−0.1	−0.01	−0.001	−0.0001
$f(x) = 1/x$	−0.001	−0.01	−0.1	−1	−10	−100	−1000	−10,000

This means that as x approaches 0 from the left, the graph of $f(x) = 1/x$ gets lower and lower; and approaches the y axis, but never touches it, as shown in Figure 1.

Because of this behavior as x approaches 0 from either the right or the left, we say that the graph of $f(x) = 1/x$ approaches the y axis *asymptotically* and the y axis ($x = 0$) is referred to as a *vertical asymptote*. More formally, we have the following definition:

DEFINITION

VERTICAL ASYMPTOTE

The line $x = k$, where k is constant, is called a **vertical asymptote** of the graph of a function f if

$$f(x) \to -\infty \qquad \text{or} \qquad f(x) \to +\infty$$

either as $x \to k^-$ (from the left) or as $x \to k^+$ (from the right) (Figure 2).

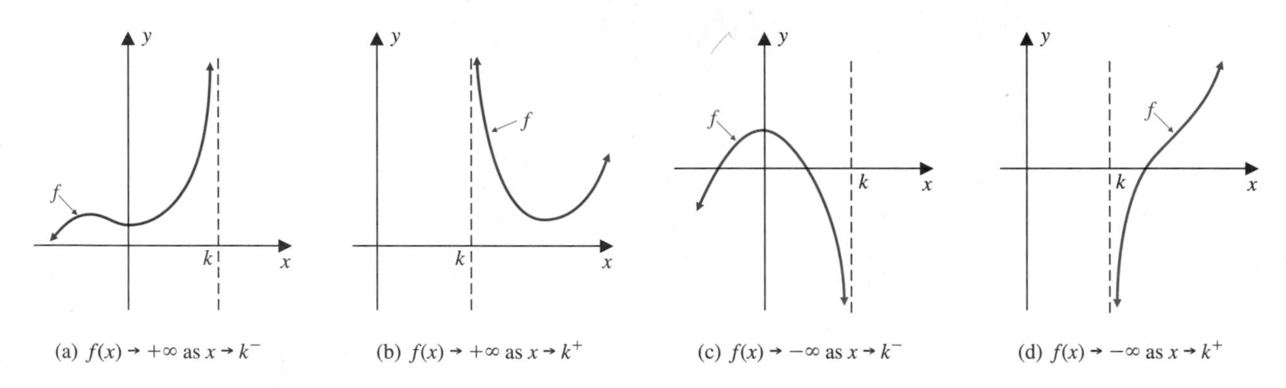

FIGURE 2
Sample graphs; $x = k$ is a vertical asymptote

Also, the data in Tables 1 and 2, in conjunction with the graph in Figure 1, indicate that the function $f(x) = 1/x$ has the following limit behavior:

$$f(x) \to 0 \text{ (through positive values)} \quad \text{as} \quad x \to +\infty$$
$$f(x) \to 0 \text{ (through negative values)} \quad \text{as} \quad x \to -\infty$$

In other words, the graph gets closer and closer to the x axis either as $x \to +\infty$ or as $x \to -\infty$. Because of this limit behavior, the x axis (which has the equation $y = 0$) is referred to as a *horizontal asymptote* of the graph of f. In general, we have the following definition:

DEFINITION

HORIZONTAL ASYMPTOTE

The line $y = k$, where k is a constant, is called a **horizontal asymptote** for the graph of a function f if

$$f(x) \to k \quad \text{as} \quad x \to -\infty \qquad \text{or} \qquad f(x) \to k \quad \text{as} \quad x \to +\infty \quad \text{(Figure 3)}$$

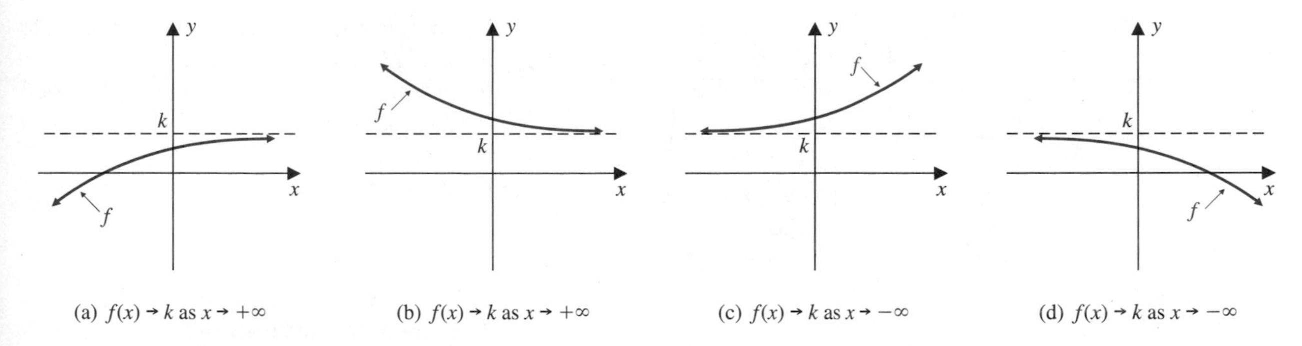

FIGURE 3
Sample graphs; $y = k$ is a horizontal asymptote

EXAMPLE 1 Graphing $g(x) = 1/x^2$
Examine the behavior of

$$g(x) = \frac{1}{x^2}$$

as $x \to 0^+$. Also, describe the limit behavior as $x \to +\infty$. Then use this information to sketch the graph of g.

Solution The domain of g includes all real numbers except $x = 0$, so the graph has no y intercept. The data in Table 3 suggest that

$$g(x) \to +\infty \quad \text{as} \quad x \to 0^+$$

so the y axis is a vertical asymptote. Also, the data in Table 3 help us to conclude that g has the following limit behavior:

$$g(x) \to 0 \text{ (through positive values)} \quad \text{as} \quad x \to +\infty$$

TABLE 3
As $x \to 0^+$, $g(x) \to +\infty$

x	1000	100	10	1	0.1	0.01	0.001
$g(x) = 1/x^2$	0.000001	0.0001	0.01	1	100	10,000	1,000,000

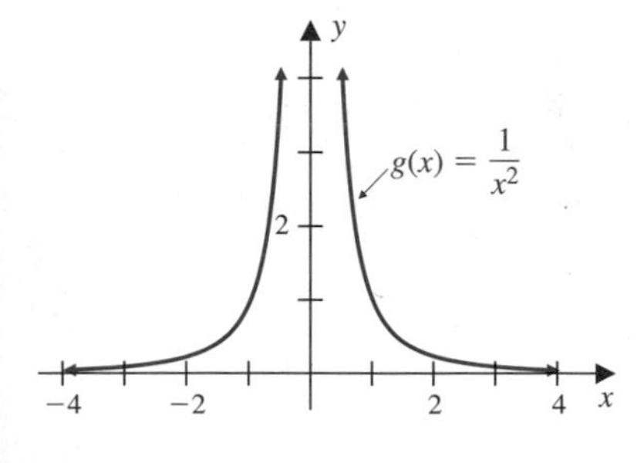

FIGURE 4

Consequently, the x axis is a horizontal asymptote of the graph of the function g. Since $g(-x) = 1/(-x)^2 = 1/x^2 = g(x)$, it follows that g is an even function and its graph is symmetric with respect to the y axis. By using the asymptotic behavior and symmetry, we get the graph of $g(x) = 1/x^2$ (Figure 4). Note that there is no x intercept since $1/x^2 = 0$ has no solution.

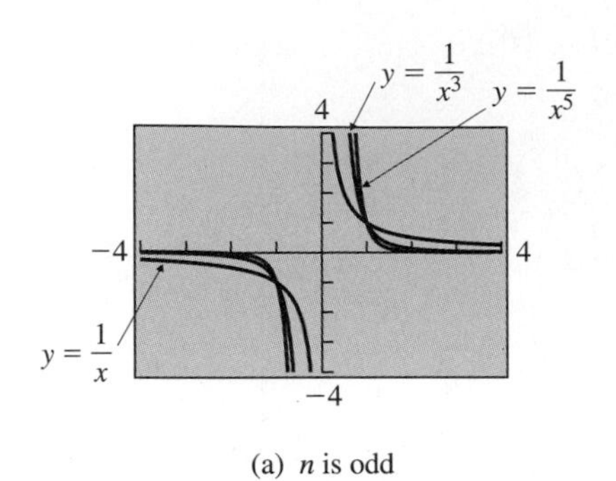

(a) n is odd

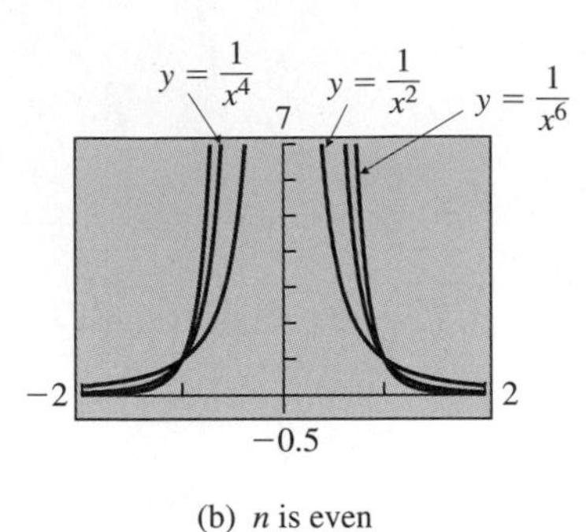

(b) n is even

FIGURE 5

Sample graphs of $R(x) = \frac{1}{x^n}$

In general, the graphs of the functions $R(x) = 1/x^n$, where n is a positive integer, resemble the graph of $y = 1/x$, if n is odd (Figure 5a), and the graph of $y = 1/x^2$, if n is even (Figure 5b). In either case, the graphs of $R(x) = 1/x^n$ have the following characteristics:

1. The y axis is a vertical asymptote.
2. The x axis is a horizontal asymptote.
3. There is no y intercept and no x intercept.
4. The graphs are symmetric with respect to the origin if n is odd; they are symmetric with respect to the y axis if n is even.

From our discussion in Section 2.3, transformations of graphs of $R(x) = 1/x^n$ can be used to obtain graphs of other rational functions.

EXAMPLE 2 **Using transformations to graph rational functions**

Use transformations of the graph of either $y = 1/x$ or $y = 1/x^2$ to graph each function. Identify the asymptotes.

(a) $f(x) = \dfrac{1}{x-2}$ (b) $g(x) = \dfrac{1}{x^2} + 2$

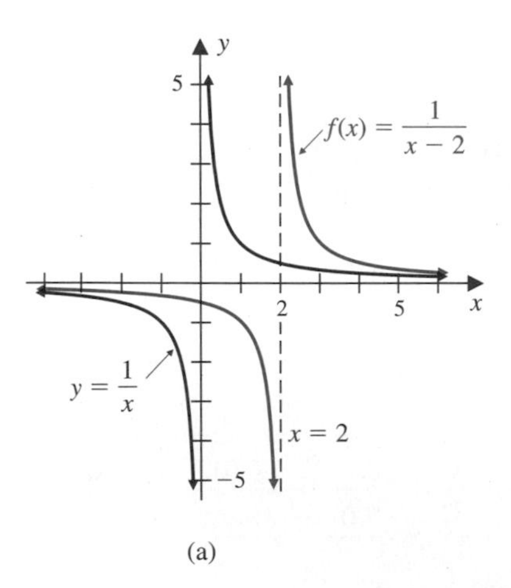

(a)

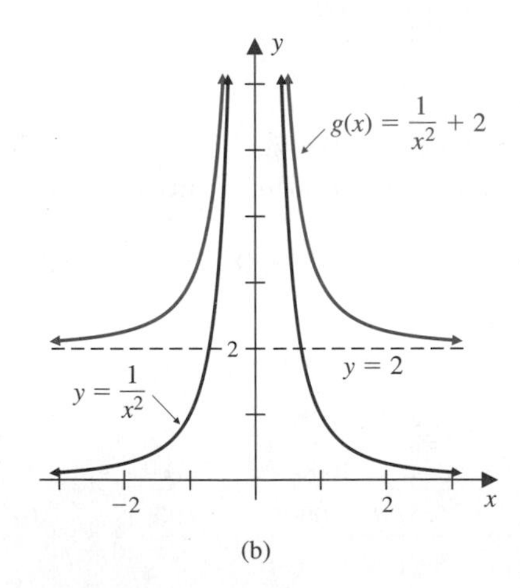

(b)

FIGURE 6

Solution

(a) The graph of f can be obtained by shifting the graph of $y = 1/x$ horizontally 2 units to the right (Figure 6a). Note that the vertical asymptote is shifted from the y axis to the line $x = 2$, but the horizontal asymptote remains the x axis.
(b) The graph of g can be obtained by shifting the graph of $y = 1/x^2$ vertically upward 2 units (Figure 6b). Note that the vertical asymptote is still the y axis, but the horizontal asymptote is shifted 2 units up from the x axis to become the line $y = 2$.

Locating Asymptotes

In order to get an accurate graph of a rational function we need to locate its asymptotes first. Our investigation of functions of the form $R(x) = 1/x^n$ illustrates the following behavior: For any fraction $p(x)/q(x)$, as the values of the denominator

$q(x)$ get closer to 0 and the values of the numerator $p(x)$ approach a value different from 0, the cumulative effect is that $p(x)/q(x)$ will either increase or decrease without bound. This behavior provides the basis for determining the vertical asymptotes of a rational function from its equation.

PROPERTY 1

LOCATING VERTICAL ASYMPTOTES

A rational function

$$R(x) = \frac{p(x)}{q(x)}$$

has a vertical asymptote at the line $x = c$ if c is a zero of q and not a zero of p.

For instance, $R(x) = 1/(x - 2)$ has a vertical asymptote at the line $x = 2$, since 2 is a zero of $q(x) = x - 2$ but not of $p(x) = 1$ (see Figure 6a). Also,

$$H(x) = \frac{8x}{x^2 - 9}$$

has two vertical asymptotes; one at the line $x = 3$ and another at the line $x = -3$, because 3 and -3 are zeros of $q(x) = x^2 - 9$ but not of $p(x) = 8x$.

In general, a rational function may have many vertical asymptotes, one for each zero of the denominator, provided these zeros are not also zeros of the numerator.

Care must be taken when using Property 1. If the number c is a zero of both functions p and q for the rational function $R(x) = p(x)/q(x)$, then there are two possibilities for the line $x = c$. One possibility is that it may *not* be a vertical asymptote for the graph of R. For instance, the line $x = 2$ is not a vertical asymptote for the graph of $g(x) = (x^2 - 4)/(x - 2)$. By writing the numerator of g in factored form, we obtain

$$g(x) = \frac{x^2 - 4}{x - 2} = \frac{(x - 2)(x + 2)}{x - 2} = x + 2 \qquad \text{for } x \neq 2$$

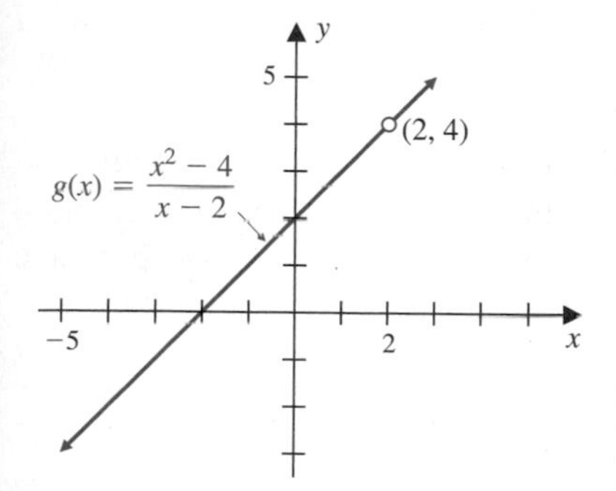

FIGURE 7

If a graphing utility is used to graph a function such as $g(x) = (x^2 - 4)/(x - 2)$, the resulting display may not show the hole.

Note that the function $y = x + 2$ is *not* the same as the function g, because they have different domains. The graph of $y = x + 2$ is a straight line that includes the point (2, 4), while the graph of g is the same straight line except that it has a *hole* at the point (2, 4) (Figure 7).

The other possibility is that the graph of R may have a vertical asymptote. For example, the line $x = 2$ is a vertical asymptote for the graph of $h(x) = (x^2 - 2x)/(x^2 - 4x + 4)$, even though 2 is a zero of both the numerator and denominator of h. By writing both the numerator and denominator in factored form, we have

$$h(x) = \frac{x^2 - 2x}{x^2 - 4x + 4} = \frac{x(x - 2)}{(x - 2)(x - 2)} = \frac{x}{x - 2}$$

This situation does not contradict Property 1 (see Problem 46).

A rational function can have at most one horizontal asymptote. A technique for locating such an asymptote is illustrated in the next example.

EXAMPLE 3 Locating a horizontal asymptote

Locate the horizontal asymptote of

$$g(x) = \frac{3x}{2x + 1}$$

and display this asymptotic behavior graphically.

Solution As $x \to +\infty$, both the numerator and the denominator of the ratio $(3x)/(2x + 1)$ are increasing in value. To determine the limit behavior of this ratio, we begin by dividing both the numerator and the denominator by x (the largest power of x) to get

$$g(x) = \frac{3x}{2x + 1} = \frac{\frac{3x}{x}}{\frac{2x}{x} + \frac{1}{x}} = \frac{3}{2 + \frac{1}{x}}$$

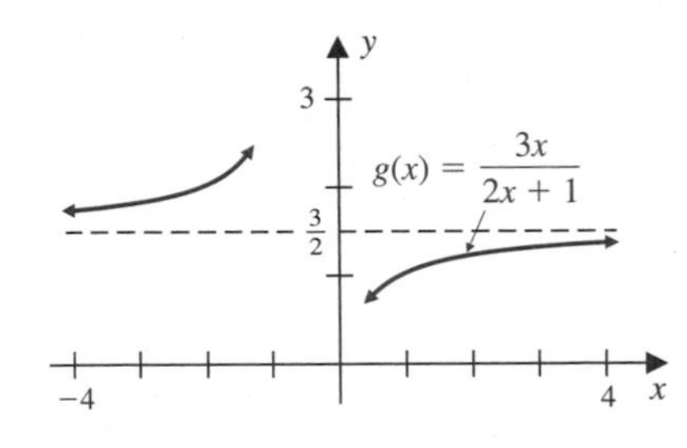

FIGURE 8

When $|x|$ gets very large, the value of $1/x$ approaches 0; that is, as $|x| \to +\infty$, $1/x \to 0$. So,

$$g(x) \to \frac{3}{2 + 0} = \frac{3}{2}$$

That is,

$$g(x) \to \tfrac{3}{2} \quad \text{as} \quad x \to +\infty \qquad \text{and} \qquad g(x) \to \tfrac{3}{2} \quad \text{as} \quad x \to -\infty$$

Thus, the line $y = \frac{3}{2}$ is the horizontal asymptote of the graph of g, as displayed in Figure 8.

In general, to determine the limit behavior of the graph of a rational function when $|x| \to \infty$, we first divide both the numerator and the denominator by x raised to its highest power. Then we use arguments similar to the one used in Example 3 to help locate any horizontal asymptote. The result depends on the degrees of the polynomials in the numerator and the denominator, as indicated in the following property:

PROPERTY 2

LOCATING HORIZONTAL ASYMPTOTES

Assume R is a rational function (in reduced form) defined by

$$R(x) = \frac{p(x)}{q(x)} = \frac{a_m x^m + a_{m-1}x^{m-1} + \cdots + a_1 x + a_0}{b_n x^n + b_{n-1}x^{n-1} + \cdots + b_1 x + b_0}$$

where $m \geq 0$ and $n > 0$ are integers, and $a_m \neq 0$ and $b_n \neq 0$.

1. If $m < n$, then the line $y = 0$ (the x axis) is the horizontal asymptote.
2. If $m = n$, then the line $y = a_m/b_n$ is the horizontal asymptote.
3. If $m > n$, then there is no horizontal asymptote.

For instance, the line $y = 3$ is the horizontal asymptote of the graph of $R(x) = (3x^2)/(x^2 + 2)$, because $m = n = 2$. The function $g(x) = (8x + 7)/(x^2 + 1)$ has the line $y = 0$ (the x axis) as its horizontal asymptote, since $m = 1$ is less than $n = 2$. Also, Property 2 indicates that the rational function

$$h(x) = \frac{5x^2 - 2}{x}$$

does not have a horizontal asymptote, since $m = 2$ is greater than $n = 1$. Figure 9 shows a viewing window of the graph of h. We observe that as $x \to +\infty$ or $x \to -\infty$, that is, as $|x| \to +\infty$, the graph moves away from the x axis. To understand the behavior of the graph of h as $|x|$ gets larger, we divide the numerator $5x^2 - 2$ by the denominator x to obtain

$$h(x) = \frac{5x^2 - 2}{x} = \frac{5x^2}{x} - \frac{2}{x} = 5x - \frac{2}{x}$$

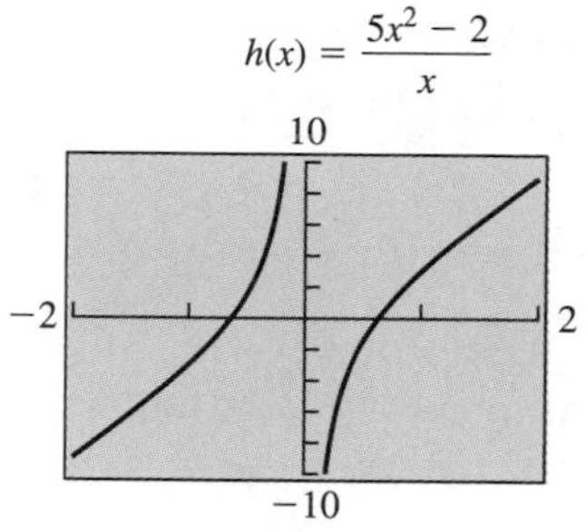

$h(x) = \frac{5x^2 - 2}{x}$

10

−2 2

−10

FIGURE 9

Now, when $|x|$ is large, $2/x$ is close to 0. Correspondingly, $h(x)$ is close to $5x$, so the graph of h gets closer and closer to the line $y = 5x$. We refer to the line $y = 5x$ as an **oblique asymptote** for the graph of h.

Graphing Rational Functions

The following strategy provides a guideline for graphing a rational function:

GUIDELINE FOR GRAPHING A RATIONAL FUNCTION $R(x) = p(x)/q(x)$

Step 1. **Locate Vertical Asymptotes and Holes**:
- (a) Factor the numerator $p(x)$ and the denominator $q(x)$ and reduce the fraction.
- (b) Use Property 1 to help find vertical asymptotes.
- (c) If $p(c) = 0$ and $q(c) = 0$, then there may be a hole in the graph of R at $x = c$.

Step 2. **Find the Intercepts**:
- (a) If $p(c) = 0$ and $q(c) \neq 0$, then there is an x intercept at $x = c$.
- (b) If R is defined at $x = 0$, then $R(0)$ is the y intercept of the graph.

Step 3. **Locate any Horizontal Asymptote**: Use Property 2 to find any horizontal asymptote.

Step 4. **Get Other Information**:
- (a) Determine whether the graph has symmetry with respect to the y axis or origin.
- (b) Use the intercepts and vertical asymptotes and plot a few points.

Step 5. **Sketch the Graph**: Combine the above results to get a sketch of the graph.

EXAMPLE 4 Sketching the graph of a rational function

Use the guideline above to sketch the graph of

$$f(x) = \frac{3x + 17}{x + 5}$$

Solution We apply the guideline as follows:

Step 1. **Locate Vertical Asymptotes and Holes**: The fraction is in reduced form. Since -5 is a zero of the denominator but not the numerator, it follows that the line $x = -5$ is a vertical asymptote. By examining some values of $f(x)$ near -5 we discover that

$$f(x) \to -\infty \quad \text{as} \quad x \to -5^- \qquad \text{and} \qquad f(x) \to +\infty \quad \text{as} \quad x \to -5^+$$

Step 2. **Find the Intercepts**: To find the x intercept, we solve $3x + 17 = 0$ to get $x = -\frac{17}{3}$. Since $x + 5$ does not equal 0 for $x = -\frac{17}{3}$, it follows that $-\frac{17}{3}$ is the x intercept. The y intercept is given by $f(0) = \frac{17}{5}$.

Step 3. **Locate any Horizontal Asymptote**: Since the degree of the numerator is the same as the degree of the denominator for f, it follows from Property 2 that the line $y = \frac{3}{1} = 3$ is a horizontal asymptote. By calculating $f(x)$ for some larger and larger positive, and smaller and smaller negative values of x we find that

$$f(x) \to 3 \text{ (from below)} \quad \text{as} \quad x \to -\infty$$

and

$$f(x) \to 3 \text{ (from above)} \quad \text{as} \quad x \to +\infty$$

Step 4. **Get Other Information**: Since the function is neither even nor odd, there is no symmetry with respect to the y axis or origin. Next, we select a few points, such as those listed in Table 4, and plot them.

Step 5. **Sketch the Graph**: By using the above results we get the graph of f (Figure 10).

TABLE 4

Selected x Value	$f(x)$	Point on Graph
-6	1	$(-6, 1)$
$-\frac{26}{5}$	-7	$\left(-\frac{26}{5}, -7\right)$
-1	$\frac{7}{2}$	$\left(-1, \frac{7}{2}\right)$
4	$\frac{29}{9}$	$\left(4, \frac{29}{9}\right)$

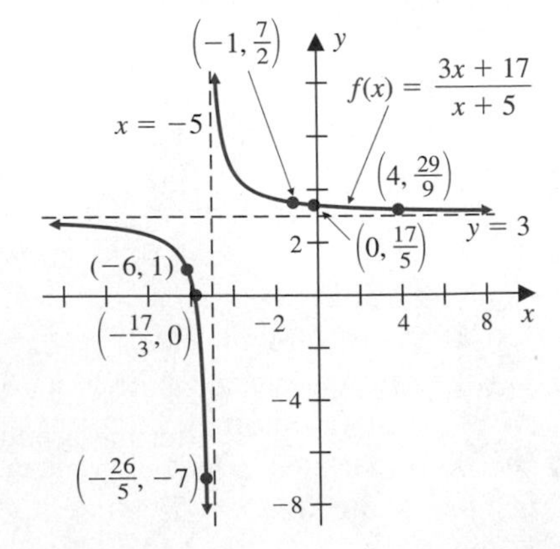

FIGURE 10

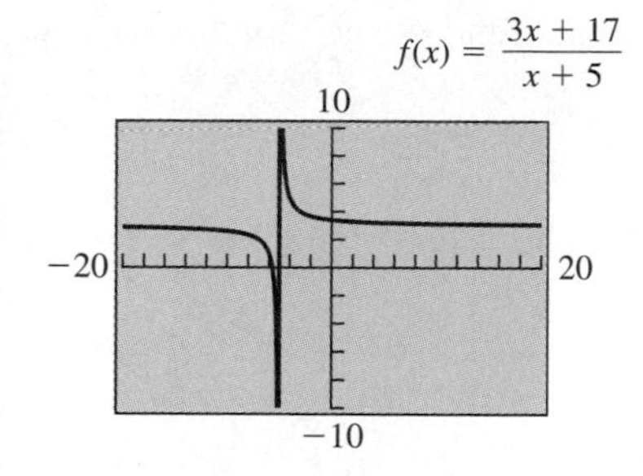

FIGURE 11

Some graphing utilities have a special feature that generates graphs without connecting separate branches. The instruction manual should be consulted.

When a graphing utility is used to sketch the graph of a rational function the result may not correctly represent the behavior of the function near a vertical asymptote. The utility may erroneously connect separate branches of the graph across a vertical asymptote. For instance, the viewing window of the graph of $f(x) = (3x + 17)/(x + 5)$ (see Example 4) shown in Figure 11 incorrectly displays a graph that includes a curve that appears to be the vertical asymptote. Actually, the utility is erroneously connecting the two separate branches of the graph. Inaccuracies such as this one may occur near vertical asymptotes, depending on the selected viewing window.

Finally, we note that the graph of a rational function will never intersect any of its vertical asymptotes. However, as Figure 12 shows, the graph of $f(x) = (2x^2 + 1)/(2x^2 - 3x)$ does intersect its horizontal asymptote $y = 1$ at the point $\left(-\frac{1}{3}, 1\right)$.

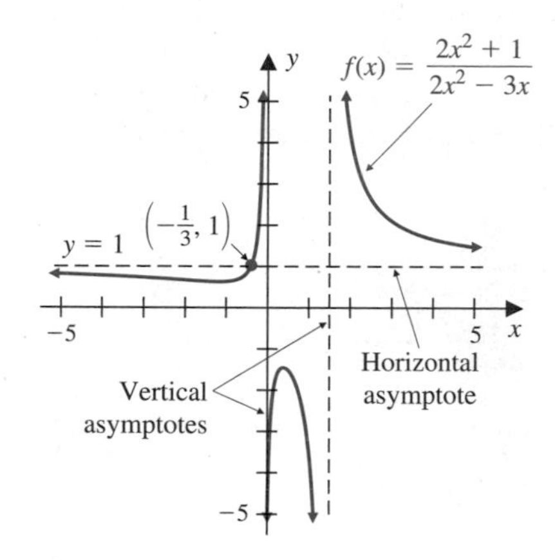

FIGURE 12

Solving Applied Problems

Trends that occur in real-world situations are sometimes modeled by rational functions.

EXAMPLE 5 **Examining the trend of monthly sales**

The model

$$s(t) = \frac{2000(1 + t)}{2 + t}$$

relates the monthly sales s (in dollars) of a certain new product to the time t (in months), beginning 1 month after the product has been introduced.

(a) What are the restrictions on t?

(b) Use a graphing utility to sketch the graph of the function, taking into account the restrictions found in part (a). Also, identify the asymptotes.

(c) Use the horizontal asymptote to determine the trend of the monthly sales as time elapses.

Solution

(a) We use $t \geq 1$, because the model is applicable at the beginning of the first month.

(b) Figure 13 shows a viewing window of the graph of s, where $t \geq 1$. Since -2 is not in the domain of this function, no vertical asymptote shows in the graph. However, since

$$s(t) = \frac{2000(1+t)}{2+t} = \frac{2000t + 2000}{t+2} = \frac{2000 + \frac{2000}{t}}{1 + \frac{2}{t}}$$

it follows that as $t \to +\infty$, $s(t) \to 2000$; that is, the line $y = 2000$ is a horizontal asymptote.

(c) As time elapses, the monthly sales trend will approach but not exceed \$2000. We refer to \$2000 as the *limiting value* of the monthly sales.

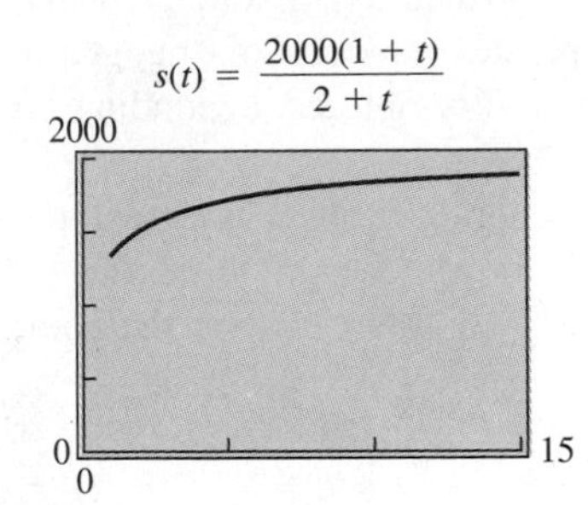

FIGURE 13

PROBLEM SET 3.6

Mastering the Concepts

GU In problems 1 and 2, use a graphing utility to graph each function. Identify the asymptotes and demonstrate the results numerically with a table of values.

1. (a) $F(x) = \frac{1}{x^6}$ (b) $H(x) = \frac{1}{x^5}$

2. (a) $G(x) = \frac{1}{x^8}$ (b) $h(x) = \frac{1}{x^7}$

In problems 3–6, use the given graph of each function to identify the vertical and horizontal asymptotes. Describe the asymptotic behaviors of the function.

3.

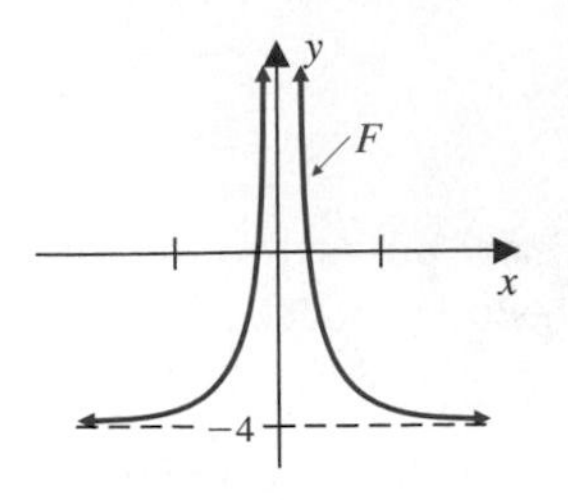

4.

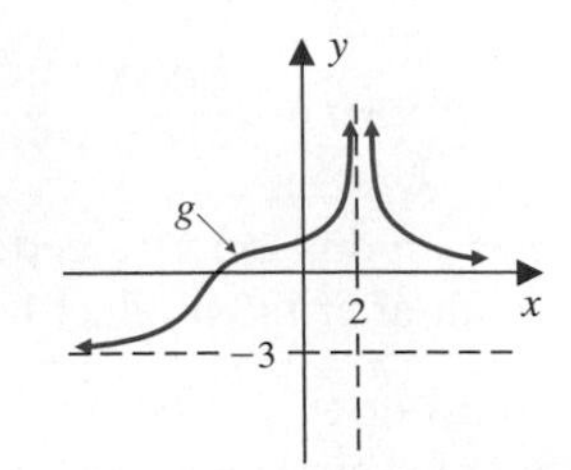

5.

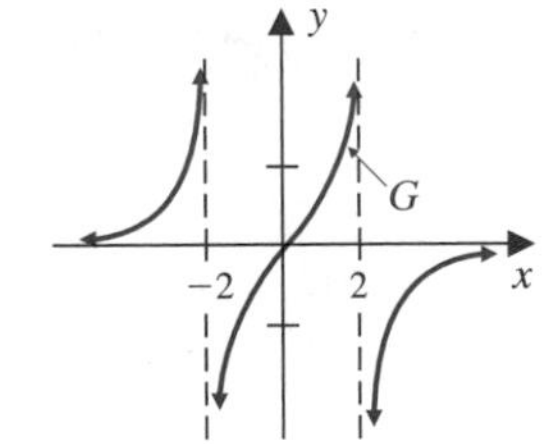

6.

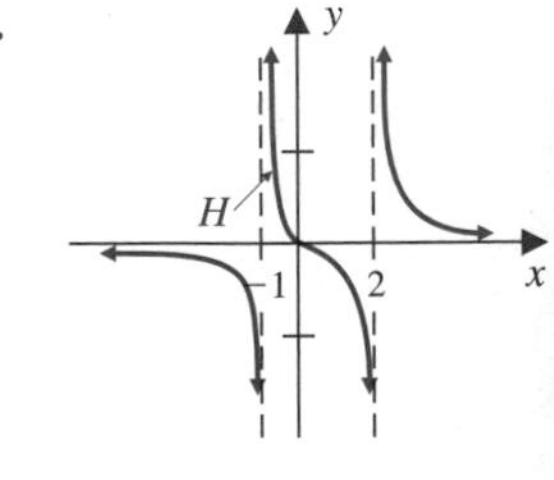

In problems 7–12, describe a sequence of transformations that will transform the graph of either $y = 1/x$ or $y = 1/x^2$ into the graph of each function. Sketch the graph of each function, and identify the asymptotes in each case.

7. (a) $f(x) = \frac{4}{x+3}$ (b) $g(x) = \frac{1}{x} + 3$

8. (a) $g(x) = \frac{1}{(x-3)^2}$ (b) $h(x) = \frac{4}{x^2} - 1$

9. (a) $f(x) = \frac{-3}{(x+1)^2}$ (b) $g(x) = \frac{2}{x-1} + 3$

10. (a) $h(x) = \frac{3x+2}{x}$ (b) $f(x) = \frac{-2}{(x+1)^2} - 1$

11. **(a)** $f(x) = \dfrac{2}{(x+3)^2} - 1$

(b) $g(x) = \dfrac{2x^2 + 16x + 33}{(x+4)^2}$

[*Hint:* Use long division first.]

12. **(a)** $g(x) = \dfrac{1}{x-2} + 3$

(b) $h(x) = \dfrac{3x+7}{x+2}$ [*Hint:* Use long division first.]

In problems 13 and 14, find any vertical asymptotes. Graph each function and locate any holes in the graph.

13. **(a)** $f(x) = \dfrac{x^2 - 1}{x - 1}$ **(b)** $g(x) = \dfrac{x^2 - 2x - 3}{x - 3}$

14. **(a)** $f(x) = \dfrac{x - 4}{x^2 - 16}$ **(b)** $g(x) = \dfrac{x^3 - 1}{x - 1}$

In problems 15 and 16, locate the horizontal asymptote of the graph of each function, and then display this asymptotic behavior graphically.

15. **(a)** $f(x) = \dfrac{-2x}{x + 1}$ **(b)** $g(x) = \dfrac{1 + 3x}{2x - 1}$

16. **(a)** $f(x) = \dfrac{x^2 + 1}{x^2 - 1}$ **(b)** $g(x) = \dfrac{2x^2 + x}{x^2 - 2x + 3}$

GU In problems 17 and 18, find the oblique asymptote of the graph of each function, and then use a graphing utility to graph the function.

17. **(a)** $f(x) = \dfrac{-4x^2 + 1}{x + 2}$ **(b)** $g(x) = \dfrac{x^2 - 4x + 7}{x - 1}$

18. **(a)** $f(x) = \dfrac{3x^2 - 12x + 7}{x + 1}$

(b) $g(x) = \dfrac{x^3 - 2x^2}{x^2 - 1}$

In problems 19–22:

GU **(a)** Use a graphing utility to sketch the graph of each function, and then match the function to its hand-drawn graph in Figure 14.

(b) Locate all asymptotes.

19. $f(x) = \dfrac{x - 1}{x^2 + x}$ **20.** $g(x) = \dfrac{x^2}{x^2 - 4}$

21. $h(x) = \dfrac{2x}{x + 1}$ **22.** $f(x) = \dfrac{x^2 - 3x + 4}{x + 2}$

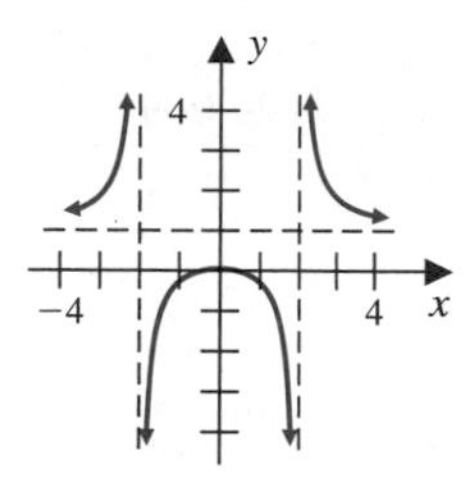

(A)

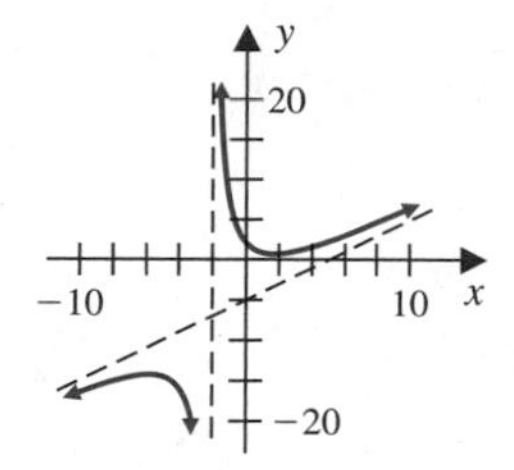

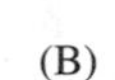
(B)

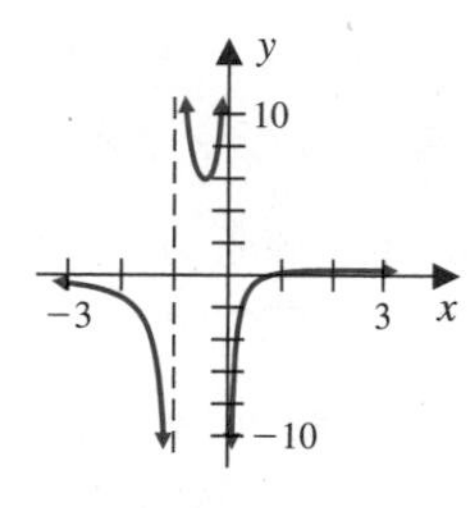

(C)

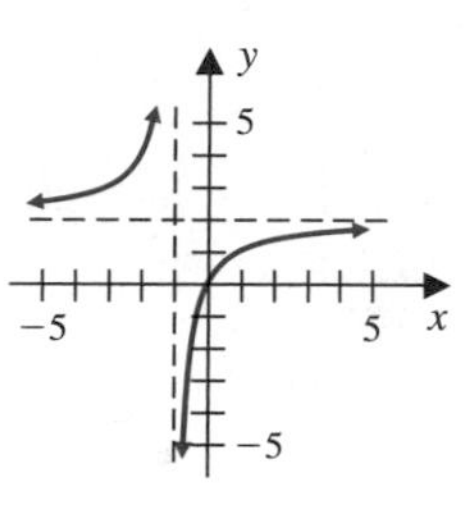

(D)

FIGURE 14

In problems 23 and 24, use the given graph of the rational function to find the equation of each asymptote.

23.

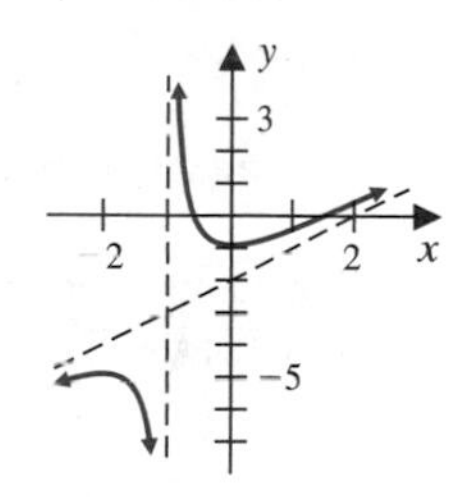

24.

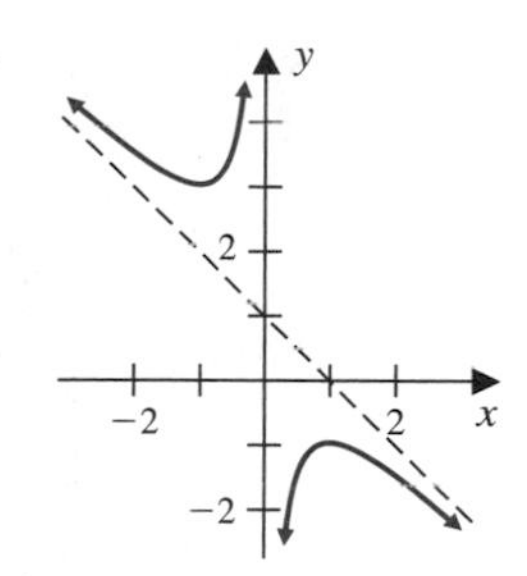

For each function in problems 25–34:

(a) Use the guideline on page 217 to sketch its graph.

GU **(b)** Use a graphing utility to sketch its graph.

(c) Compare the result in part (a) to the result in part (b).

25. $f(x) = \dfrac{2x}{x + 1}$ **26.** $g(x) = \dfrac{-3x + 1}{x + 4}$

27. $f(x) = \dfrac{3}{x^2 - 2x - 3}$ **28.** $h(x) = \dfrac{2}{x^2 - x - 2}$

29. $g(x) = \dfrac{-x}{x^2 - x - 6}$ **30.** $f(x) = \dfrac{-2x}{x^2 - 3x - 4}$

31. $f(x) = \dfrac{x^2 + 2}{x^2 - 4x + 3}$ **32.** $g(x) = \dfrac{x^2 - 4}{x^2 + 1}$

33. $g(x) = \dfrac{x^2 + 2}{x - 1}$ **34.** $f(x) = \dfrac{x^2 + 1}{x}$

Applying the Concepts

35. Business Sales Trend: Suppose that the sale of S units of a certain product is related to the daily advertising expenditure x (in dollars) by the model

$$S(x) = \frac{5000x}{x + 100}$$

(a) What are the restrictions on x?

GU (b) Use a graphing utility to sketch the graph of S using the restrictions from part (a). Also, locate the asymptotes.

(c) What is the sales trend as the advertising expenditures increase?

36. Government Securities: Each weekday, national newspapers such as the *New York Times* and the *Wall Street Journal* publish a curve called a *yield curve* for U.S. government securities. Suppose that the percentage yield p on government securities that mature after t years is modeled by the function

$$p = \frac{t + 0.1}{0.1t + 0.03} \qquad 0.25 \leq t \leq 30$$

GU (a) Use a graphing utility to graph this function and locate the asymptotes.

(b) Explain the implication of the horizontal asymptote on the graph of p. That is, analyze the trend of the yield as a function of maturity.

37. Air Pollution: Suppose that the cost C (in dollars) of removing x percent of air pollutants caused by automobile emissions in a certain town is given by the model

$$C = \frac{100{,}000x}{100 - 1.67x} \qquad 0 < x < 60$$

GU (a) Use a graphing utility to graph the function C.

(b) Use the graph to analyze the trend of the cost of pollution control by examining what happens as the air pollutants approach 60%.

38. Pain Relief: Researchers have determined that the percentage p of pain relief from x grams of aspirin is given by the model

$$p = \frac{100x^2}{x^2 + 0.02} \qquad 0 < x \leq 1$$

GU (a) Use a graphing utility to graph p.

(b) Use the graph to approximate the value of x when the pain relief is very close to its maximum value of 100%.

(c) Explain the implication of the horizontal asymptote of the graph of p.

Developing and Extending the Concepts

39. Let $f(x) = 1/x$. Compute and simplify the expression

$$q = \frac{f(x) - f(2)}{x - 2}$$

Use a graphing utility to sketch the graph of q.

40. Let $f(x) = 1/x^2$. Compute and simplify the expression

$$q = \frac{f(x) - f(1)}{x - 1}$$

Use a graphing utility to sketch the graph of q.

41. GU Use a graphing utility to sketch the graph of each function. Identify the intercepts and the asymptotes of each graph.

(a) $f(x) = \left|\dfrac{1}{x}\right| - 3$ (b) $g(x) = \left|\dfrac{1}{x} - 3\right|$

42. The function

$$f(x) = \frac{x^3 + x^2 - 1}{x + 1}$$

can be written as

$$f(x) = \frac{x^3 + x^2 - 1}{x + 1} = x^2 - \frac{1}{x + 1}$$

GU (a) Use a graphing utility to sketch graphs of f and $g(x) = x^2$ in the same viewing window.

(b) Use the graphs to compare the limit behaviors of f and g.

(c) Describe the asymptotic behavior of f with respect to the graph of g.

In problems 43 and 44:

GU (a) Use a graphing utility to graph each given function, and find its x intercepts.

(b) Use the graph from part (a) to solve the given equation and inequality.

(c) Confirm the results in part (b) algebraically.

43. $f(x) = \frac{7}{3x - 9} - \frac{2}{x - 3} - \frac{4}{9}$; $\frac{7}{3x - 9} - \frac{2}{x - 3} = \frac{4}{9}$; $\frac{7}{3x - 9} - \frac{2}{x - 3} < \frac{4}{9}$

44. $f(x) = \frac{5}{2x - 6} - \frac{1}{x - 1} - \frac{1}{x + 3}$; $\frac{5}{2x - 6} - \frac{1}{x - 1} = \frac{1}{x + 3}$; $\frac{5}{2x - 6} - \frac{1}{x - 1} > \frac{1}{x + 3}$

45. **(a)** Explain why the following statement is true: If the line with equation $x = k$ is a vertical asymptote of the graph of the rational function $y = f(x)$, then k is not in the domain of f.

(b) Explain why the following statement is false: If the line with equation $y = k$ is a horizontal asymptote of the graph of the rational function $y = f(x)$, then k is not in the range of f.

46. Given the rational function $h(x) = \frac{x^2 - 2x}{x^2 - 4x + 4}$:

(a) Show that 2 is a zero of both $p(x) = x^2 - 2x$ and $q(x) = x^2 - 4x + 4$.

GU **(b)** Use a graphing utility to graph h and locate the vertical asymptotes of the graph.

(c) Does this function contradict Property 1 on page 215? Explain.

CHAPTER 3 REVIEW PROBLEM SET

In problems 1 and 2, use transformations of the graph of $y = x^2$ to graph the given function. Find the vertex, axis of symmetry, domain, and range.

1. $f(x) = -\frac{1}{2}(x + 5)^2 + 3$ **2.** $g(x) = 3(x + 1)^2 - 2$

In problems 3 and 4, rewrite the function f in standard form $f(x) = a(x - h)^2 + k$. Sketch the graph, and locate the vertex and axis of symmetry.

3. $f(x) = x^2 + 4x - 7$ **4.** $f(x) = -2x^2 - 5x - 1$

In problems 5 and 6, find the standard equation $f(x) = a(x - h)^2 + k$ of the quadratic function satisfying the given conditions.

5. y intercept: -8; vertex at $(-1, -9)$

6. x intercept: 1; vertex at $\left(\frac{3}{2}, \frac{1}{2}\right)$

In problems 7 and 8, use the graph of the given quadratic function and the location of the x intercepts to solve the corresponding inequality.

7. $f(x) = 2x^2 - x - 1$; $2x^2 - x - 1 < 0$

8. $g(x) = x^2 + 5x - 6$; $x^2 + 5x \geq 6$

In problems 9 and 10, use transformations of an appropriate power function to graph the given function.

9. $h(x) = 2(x - 1)^3 - 3$ **10.** $g(x) = -\frac{1}{3}(x - 2)^4 + 5$

In problems 11 and 12, use the graph of f to determine the limit behavior of $f(x)$ as:

(a) $x \to -\infty$ **(b)** $x \to +\infty$

11.

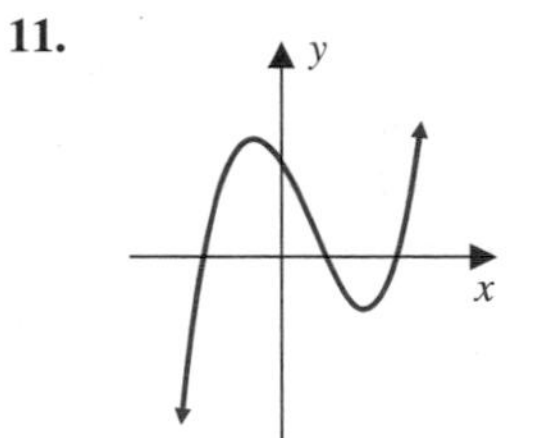

12.

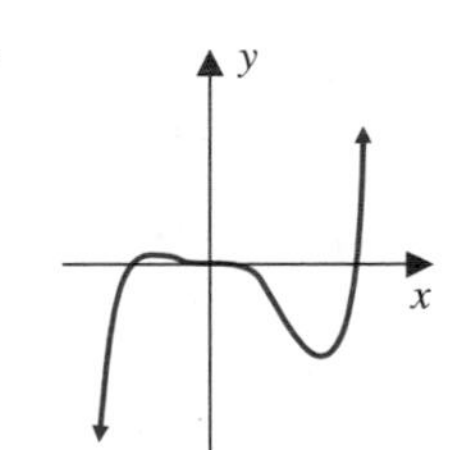

GU In problems 13 and 14, use a graphing utility to graph the function. Determine the x intercepts of the given function, and then use the graph to discuss the limit behavior of the function and to solve the associated inequality.

13. $h(x) = x^3 - x$; $x^3 - x < 0$

14. $g(x) = x^4 - 2x^2 + 1$; $x^4 - 2x^2 + 1 \geq 0$

In problems 15 and 16, use long division to divide $f(x)$ by $D(x)$. Identify the quotient $Q(x)$ and remainder R. Also, express each result in the multiplicative form of the division algorithm.

15. $f(x) = 3x^4 + 5x^3 - 7x^2 - x + 1$; $D(x) = 2x - 1$

16. $f(x) = 5x^3 + 7x - 6$; $D(x) = 3x + 2$

In problems 17 and 18, use the remainder theorem to find the value of each function if $x = c$.

17. $f(x) = 4x^3 + 2x^2 - 6x - 2$; $c = -2$
18. $g(x) = -3x^3 - 17x + 11$; $c = 4$

19. Use synthetic division to evaluate

$$f(x) = x^5 + x^4 - 3x^3 - 5x^2 + x + 7$$

for the given value of x.
(a) $x = -3$ **(b)** $x = 5$

20. Use synthetic division to find the quotient $Q(x)$ and remainder R if

$$f(x) = 5x^4 - 2x^3 + 11x^2 + 5x + 36$$

is divided by $D(x) = x + 1$.

In problems 21 and 22, verify that $f(c) = 0$. Then use the factor theorem and synthetic division to factor $f(x)$ completely.

21. $f(x) = x^3 + 6x^2 - 11x - 10$; $c = 2$
22. $f(x) = x^4 + 5x^3 - 19x^2 - 65x + 150$; $c = -5$

In problems 23 and 24, use the given graph of the function to determine the number of positive and negative zeros. Also, find lower and upper integer bounds of the real zeros. For each real zero c, find consecutive integers k and $k + 1$ such that $k \leq c < k + 1$.

23. $f(x) = 6x^4 - 19x^3 + 13x^2 + 4x - 4$

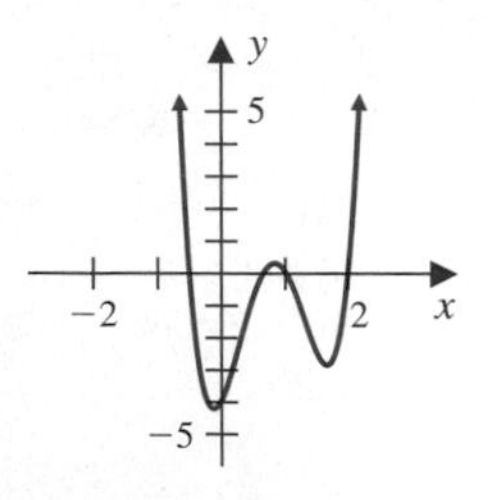

24. $g(x) = -4x^3 - 16x^2 - 9x + 9$

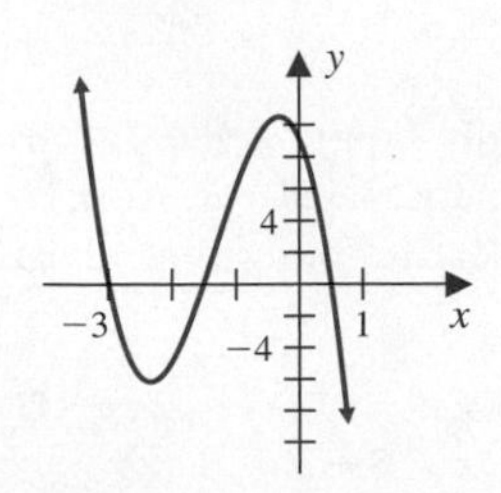

In problems 25 and 26, use the rational zero theorem to find all rational zeros of f. Use the results to factor $f(x)$ and to find other real zeros.

25. $f(x) = x^3 - 4x^2 + x + 6$
26. $f(x) = \frac{x^4}{2} - \frac{7}{2}x^3 + 9x^2 - 10x + 4$

In problems 27 and 28, use the intermediate-value property to show that the function f has a zero in the given interval.

27. $f(x) = x^3 - x - 5$; $[a, b] = [0, 2]$
28. $f(x) = x^4 - x - 1$; $[a, b] = [-1, 1]$

GU In problems 29 and 30, use a graphing utility to sketch the graph of each function and to approximate all zeros c so that $|f(c)| \leq 0.005$. Round off the answers to four decimal places.

29. $f(x) = x^4 - 9x^2 + 14$ **30.** $g(x) = -x^4 - 4x + 1$

In problems 31 and 32, find the complex zeros and their multiplicities.

31. $h(x) = (x^4 - 1)(x^2 + 8x + 16)$
32. $g(x) = x^3(x^2 + 9)(x^2 - 4x + 7)$

In problems 33 and 34, find the other complex zeros of f.

33. $f(x) = x^4 - 10x^3 + 37x^2 - 110x + 286$; $5 + i$ is a zero of f
34. $f(x) = x^4 - 12x^3 + 43x^2 - 22x - 78$; $5 - i$ is a zero of f

In problems 35 and 36, form a polynomial function with leading coefficient 1 that has the smallest possible degree and real coefficients with the given zeros.

35. $2, 2 - i$; degree 3 **36.** $i, 1 + i$; degree 4

In problems 37 and 38, use Descartes' rule of signs to determine the number of positive zeros, negative zeros, and imaginary zeros.

37. $f(x) = x^3 + 2x^2 - 3x - 11$
38. $h(x) = 3x^4 + 3x^3 - 13x - 6$

In problems 39 and 40, factor $f(x)$ as far as possible:
(a) In the real number system
(b) In the complex number system

39. $f(x) = x^4 - 8x^3 + 16x^2 + 8x - 17$
40. $f(x) = x^4 - 3x^2 + 6x + 8$

In problems 41 and 42, find the domain of each function.

41. $g(x) = \dfrac{3x - 1}{x^2(x^2 - 9)}$

42. $h(x) = \dfrac{x^3}{x^3 - 3x}$

In problems 43–46, graph the rational function, locate the vertical and horizontal asymptotes, and find the intercepts.

43. $f(x) = \dfrac{3}{(x - 1)^2} - 5$

44. $g(x) = -\dfrac{1}{2(x + 1)} + 2$

45. $h(x) = \dfrac{4x + 2}{x - 7}$

46. $f(x) = \dfrac{3x^2}{x^2 - 6x + 9}$

47. Baseball Path: A baseball is hit so that the height h (in feet) of the ball t seconds later is given by the model

$$h(t) = -16t^2 + 64t + 5$$

- **(a)** Use this model to find $h(0)$ and $h(1)$. What do these results mean in this situation?
- **(b)** What restriction should be imposed on t in this model?
- **(c)** Graph the path of the baseball and use the graph to describe the change in the height as the time increases.
- **(d)** When does the ball reach its maximum height?
- **(e)** What is the maximum height?

48. Revenue: The revenue R (in dollars) from a grove of oranges is given by the equation

$$R(x) = (800 - x)x$$

where x is the price (in dollars) of each bin of oranges sold.

- **(a)** For what price of a bin of oranges is the revenue \$57,600?
- **(b)** What price of a bin of oranges will maximize the revenue?
- **(c)** What is the maximum revenue?

49. Population Growth: A marine biologist discovers that the number N of insects found in a sample of water is given by the model

$$N(T) = -1.5T^2 + 75T + 20$$

where T is the temperature in degrees Celsius.

- **(a)** Calculate $N(0)$, $N(10)$, and $N(25)$. Explain the meaning of each value.
- **(b)** Find the temperatures when the number of insects is 800. Round off the answers to one decimal place.
- GU **(c)** Use a graphing utility to sketch the graph of the function, and then use the graph to determine the temperature at which the population will be a maximum.
- **(d)** Find the size of the maximum population. Round off the answer to the nearest insect.

50. Population Growth: A biologist finds that the number N of bacteria in a culture t hours after the start of an experiment is given by the mathematical model

$$N(t) = t^3 - 6t^2 + 9t + 100$$

- **(a)** Calculate $N(0)$, $N(10)$, and $N(25)$. Explain the meaning of each value.
- GU **(b)** Use a graphing utility to find the time when the number of bacteria is 1000. Round off the answer to one decimal place.

51. Ranching: A rancher wants to enclose a rectangular pen with an area of 100 square feet by fencing the entire perimeter. Suppose that the length of the pen is x feet.

- **(a)** What restriction should be placed on x?
- **(b)** Express the perimeter p as a function of x.
- **(c)** If the fencing costs \$6.60 per foot, express the total cost C (in dollars) as a function of x.
- GU **(d)** Use a graphing utility to sketch the graph of C and to determine the value of x for which the cost is minimum. What is the minimum cost?

52. Chemistry: By adding x milliliters of a 10% acid solution to 10 milliliters of a 30% acid solution, a chemist obtains a mixture with an acid concentration of C%, where C (expressed in decimal form) is given by the function

$$C(x) = \frac{3 + 0.1x}{10 + x}$$

- **(a)** What restriction should be imposed on x in this situation?
- GU **(b)** Use a graphing utility to sketch the graph of C.
- **(c)** Discuss the asymptotic behavior of the graph of C and interpret it.
- **(d)** Use the graph to determine what value of x results in a 25% acid solution. Round off the answer to one decimal place.

CHAPTER 3 TEST

1. Match each function with its graph:
   (a) $f(x) = x^3 - 2x^2$ (b) $g(x) = -x^5 + 1$
   (c) $h(x) = \dfrac{-2x + 1}{x}$

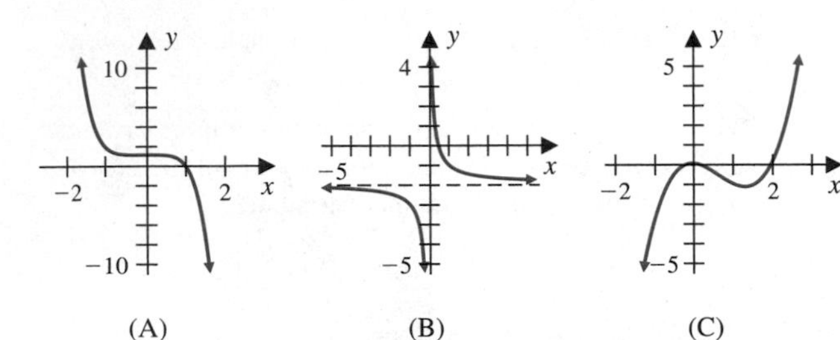

2. Let $f(x) = -2x^2 - 5x + 3$.
   (a) Rewrite $f(x)$ in the form of $f(x) = a(x - h)^2 + k$. Use transformations of the graph of $y = x^2$ to sketch the graph.
   (b) Find the vertex and axis of symmetry.
   (c) Determine whether the graph opens upward or downward, and indicate if the vertex is a maximum or a minimum point.
   (d) Find the domain and range of f.
   (e) Solve the inequality $-2x^2 - 5x + 3 \leq 0$ graphically. Express the solution in interval notation.

3. Describe a sequence of transformations that will transform the graph of $y = x^4$ to the graph of $f(x) = -2(x - 1)^4 + 1$.
   (a) Sketch the graph of f.
   (b) Use the graph to determine the limit behavior of $f(x)$ as $x \to -\infty$ and as $x \to +\infty$.

4. Find the zeros of $f(x) = (x^5 - x^4)(x + 1)^2$, and classify the zeros by multiplicity.

5. Let $f(x) = x^5 - 6x^4 + 11x^3 - 2x^2 - 12x + 8$.
   (a) Use synthetic division to obtain the quotient $Q(x)$ and the remainder R if the divisor is $D(x) = x + 2$ and the dividend is $f(x)$.
   (b) Use the remainder theorem to find $f(-2)$.
   (c) Use the factor theorem to determine whether $x + 1$ is a factor of $f(x)$.

6. Let $f(x) = 9x^3 + 6x^2 - 5x - 2$.
   (a) Use a graphing utility to sketch the graph of f.
   (b) Use the graph from part (a) to determine the number and types of zeros.
   (c) Use the rational zero theorem along with the graph of f to determine the rational zeros of f.
   (d) Use the result from part (c) to express $f(x)$ in factored form.
   (e) Use the graph to find the intervals for which $9x^3 + 6x^2 - 5x - 2 < 0$.

7. Form a polynomial function of degree 4 with real coefficients that has 2, 3, and $1 + i$ as its zeros and 1 as the leading coefficient.

8. Consider the function f defined by $f(x) = k/x$, where $k > 0$.
   (a) What happens to the graph of f if we assume the values of x are small negative values?
   (b) What happens when the input value of x is 0?
   (c) What happens when the input values of x assume large positive values?
   (d) What happens when we change the function by choosing negative values of k?

9. Let $f(x) = \dfrac{4x - 5}{x - 2}$.
   (a) Find the asymptotes of the graph of f.
   (b) Find the intercepts of the graph of f.
   (c) Graph f.
   (d) Use the graph to specify the domain and range of f.

10. An experimental rocket is launched from a hill 206 feet high. The height h (in feet) of the rocket from the ground t seconds after it is launched is given by the function
$$h(t) = -16t^2 + 468t + 206$$
    (a) Calculate $h(0)$, $h(2)$, and $h(4)$. Explain the meaning of each value.
    (b) Graph h with an appropriate restriction on t.
    (c) For what value of t does the object reach its maximum height?
    (d) What is the maximum height of the object?

11. A psychologist estimates that the percent p of the information remembered by a student in an experiment t months after the student learned the material is given by the function
$$p(t) = \frac{100}{1 + t}$$
    (a) What is the restriction on t?
    (b) Sketch the graph of p.
    (c) Use the graph to determine what happens to $p(t)$ as $t \to +\infty$. Explain what this behavior means in terms of the experiment.

4

Exponential and Logarithmic Functions

CHAPTER CONTENTS

Up to now we have studied algebraic functions formed by sums, differences, products, and quotients of polynomial functions. Nonalgebraic functions are called *transcendental functions*. In this chapter, we will introduce two types of transcendental functions that are closely related—*exponential* and *logarithmic* functions. We will examine their properties, their graphs, and their use in a variety of applications.

4.1 Exponential Functions

OBJECTIVES

1. Define exponential functions.
2. Graph exponential functions.
3. Solve applied problems.

As we will see, exponential functions are especially useful in modeling the growth or decline of such varied phenomena as population, radioactivity, and investment yields.

Defining Exponential Functions

In Section 6 (page 19), we defined b^x, where b is a positive constant, for all rational values x. For instance,

$$3^2 = 9, \quad 4^{3/2} = \left(\sqrt{4}\right)^3 = 2^3 = 8, \quad \text{and} \quad 7^{1/3} = \sqrt[3]{7}$$

It is possible to extend the notion of exponents to include all irrational numbers. However, a thorough explanation depends on concepts studied in calculus. For now, it is enough to know that if r is a rational number approximately equal to an irrational number x, then b^r is approximately equal to b^x. The better that r approximates x, the better that b^r approximates b^x. For instance, Table 1 (which was formed by using the $\boxed{y^x}$ key on a calculator) shows a pattern of improving approximations of 4^π rounded to four decimal places. Thus, as the value of the exponent gets closer and closer to π, the value of 4^π gets closer and closer to a real number whose approximate decimal value is 77.8802.

With this background we are ready to define exponential functions.

TABLE 1
Approximations of 4^π

	Approximation of π	Approximation of 4^π *Calculator value*	
Improved accuracy ↓	3.14	77.7085	Improved accuracy ↓
	3.142	77.9242	
	3.1416	77.8810	
	3.14159	77.8799	
	⋮	⋮	

DEFINITION

EXPONENTIAL FUNCTION—BASE b

A function of the form

$$f(x) = b^x \qquad \text{where } b > 0 \text{ and } b \neq 1$$

is called an **exponential function of base *b***. The independent variable x represents any real number.

In the definition, $b \neq 1$ because $f(x) = 1^x = 1$, which is considered to be a constant function rather than an exponential function.

Thus, $f(x) = 3^x$ is an exponential function with base 3 and $g(x) = 5^{-x} = \left(\frac{1}{5}\right)^x$ is an exponential function with base $\frac{1}{5}$.

EXAMPLE 1 Evaluating an exponential function

Find each function value for $f(x) = 3^x$. Round off the answers to two decimal places.

(a) $f(-2)$ (b) $f(-2.7)$ (c) $f(\sqrt{2})$ (d) $f\left(\frac{3}{2}\right)$

Solution

(a) $f(-2) = 3^{-2} = \frac{1}{9} = 0.11$ (approx.)
The remaining parts are found by using a calculator.
(b) $f(-2.7) = 3^{-2.7} = 0.05$ (approx.) (c) $f(\sqrt{2}) = 3^{\sqrt{2}} = 4.73$ (approx.)
(d) $f\left(\frac{3}{2}\right) = 3^{3/2} = 5.20$ (approx.)

Graphing Exponential Functions

To examine the common features of exponential functions, we use the point-plotting method to graph two such functions—one with a base greater than 1 and another with a base less than 1.

EXAMPLE 2 Graphing exponential functions

Sketch the graph of each exponential function. Determine the domain, range, and any asymptotes.

(a) $f(x) = 2^x$ (b) $g(x) = \left(\frac{1}{2}\right)^x$

Solution

(a) Table 2 lists some ordered pairs of numbers that satisfy $f(x) = 2^x$. After plotting these points and then connecting them with a smooth continuous curve, we obtain the graph (Figure 1a). Since 2^x is defined for any real number x, the domain of f consists of all real numbers $\mathbb{R}$. From the graph, we see that the range consists of all positive real numbers. Because of the limit behavior

$$f(x) \to 0 \qquad \text{as} \qquad x \to -\infty$$

the x axis is a horizontal asymptote.

Since $2^x = 0$ has no solution, there is no x intercept; that is, the graph does not intersect the x axis.

(b) Again, Table 2 lists some ordered pairs of numbers that satisfy the equation $g(x) = \left(\frac{1}{2}\right)^x$. After plotting these points and then connecting them with a smooth continuous curve, we obtain the graph (Figure 1b). From the graph, we see that the domain consists of all real numbers $\mathbb{R}$ and the range consists of all positive real numbers. Since $g(x) \to 0$ as $x \to +\infty$, the x axis is a horizontal asymptote.

TABLE 2

x	$f(x) = 2^x$	$g(x) = \left(\frac{1}{2}\right)^x$
-3	$\frac{1}{8}$	8
-2	$\frac{1}{4}$	4
-1	$\frac{1}{2}$	2
0	1	1
1	2	$\frac{1}{2}$
2	4	$\frac{1}{4}$
3	8	$\frac{1}{8}$

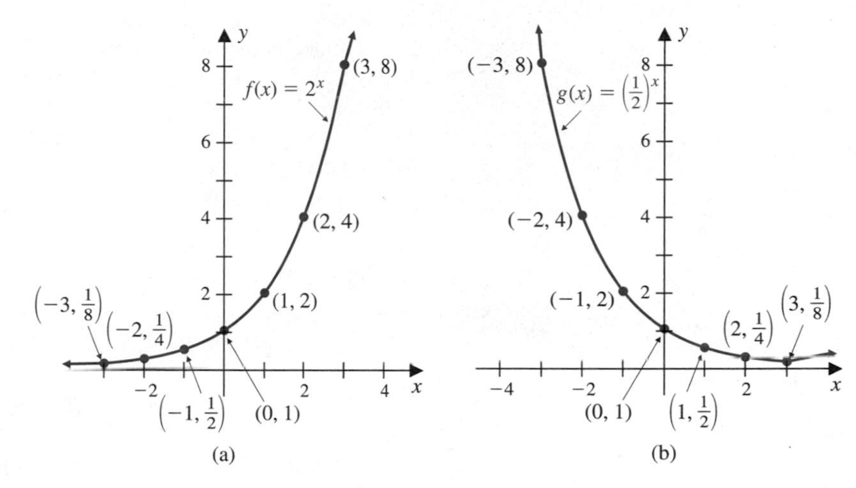

FIGURE 1

As illustrated in Figure 2a, the graphs of the functions $f(x) = b^x$, where $b > 1$, resemble the graph of $y = 2^x$. Similarly, if $0 < b < 1$, the graphs of $f(x) = b^x$ have the same general shape and characteristics as the graph of $y = \left(\frac{1}{2}\right)^x$ (Figure 2b).

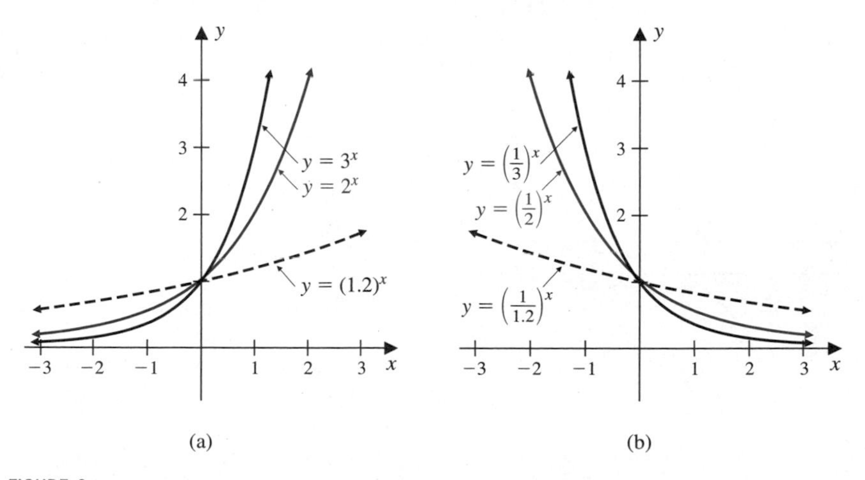

FIGURE 2

These observations lead us to the general results given in Table 3.

TABLE 3

General Graphs and Properties of $f(x) = b^x$, $b > 0$, $b \neq 1$

Properties	General Graphs
1. The domain includes all real numbers. **2.** The range is the interval $(0, \infty)$. **3.** The function increases everywhere if $b > 1$ and decreases everywhere if $0 < b < 1$. **4.** The x axis is a horizontal asymptote. **5.** The function is one-to-one. **6.** The graph contains the point $(0, 1)$.	$f(x) = b^x$, $b > 1$; $f(x) = b^x$, $0 < b < 1$

EXAMPLE 3 Comparing exponential functions

Use the graphs of $f(x) = 3^x$ and $g(x) = 5^x$ to solve the inequality $3^x < 5^x$.

Solution For convenience, we use a graphing utility to graph the two functions in the same viewing window (Figure 3). The graph of the function $f(x) = 3^x$ is below the graph of $g(x) = 5^x$ when $x > 0$. So the solution of $3^x < 5^x$ consists of all values of x in the interval $(0, \infty)$.

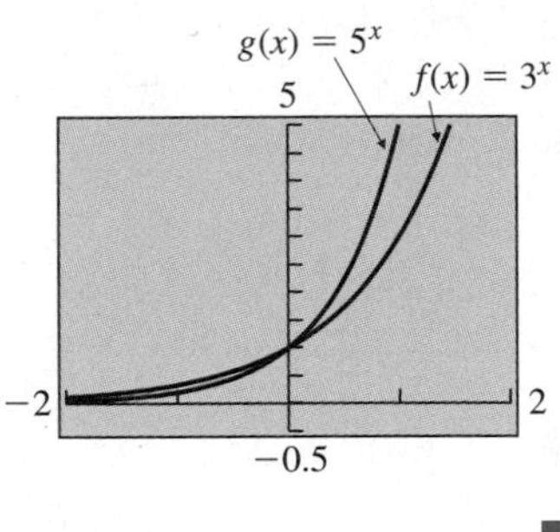

FIGURE 3

The transformation techniques of graph sketching presented in Section 2.3 and illustrated throughout Chapter 3 can be applied to exponential functions.

EXAMPLE 4 Using transformations to graph

Describe a sequence of transformations that will transform the graph of $y = \left(\frac{1}{3}\right)^x$ into the graph of $G(x) = -\left(\frac{1}{3}\right)^x + 1$. Sketch the graph of G. Determine the domain and range of G, and identify any horizontal asymptote.

Solution As displayed in Figure 4, the graph of G can be obtained by performing the following sequence of transformations:

(i) Reflect the graph of $y = \left(\frac{1}{3}\right)^x$ across the x axis to get the graph of $g(x) = -\left(\frac{1}{3}\right)^x$.

(ii) Vertically shift the graph of g up 1 unit to get the graph of G.

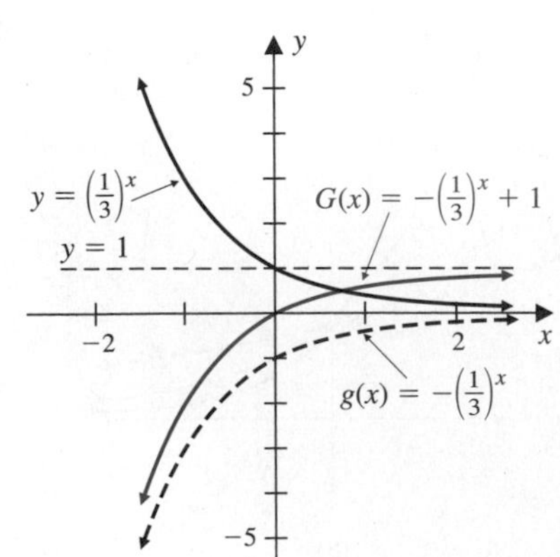

FIGURE 4

From the graph, we see that the domain of G consists of all real numbers $\mathbb{R}$. Because of the 1 unit vertical shift upward, the horizontal asymptote of the graph of g (the x axis) moves 1 unit above the x axis to become the line $y = 1$, which is the horizontal asymptote of the graph of G. So $G(x) \to 1$ as $x \to +\infty$, and the range of G is the interval $(-\infty, 1)$.

Functions containing exponential expressions can be graphed most efficiently by using a graphing utility.

EXAMPLE 5 **Using a graphing utility to graph**

Use a graphing utility to graph the function $f(x) = 3^x - x^2$, and determine the domain and range. Identify any horizontal asymptote.

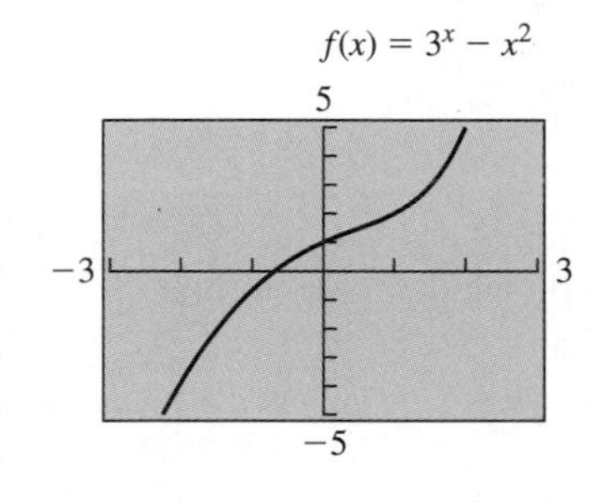

FIGURE 5

Solution A viewing window of the graph of $f(x) = 3^x - x^2$ is displayed in Figure 5. The domain and the range of f appear to include all real numbers $\mathbb{R}$. The graph suggests that the limit behavior is given by

$$f(x) \to +\infty \quad \text{as} \quad x \to +\infty \qquad \text{and} \qquad f(x) \to -\infty \quad \text{as} \quad x \to -\infty$$

So it seems that the graph has no horizontal asymptote.

Solving Applied Problems

Exponential functions are used to model certain phenomena such as *exponential growth* and *compound interest*. We examine these types of models in the following examples.

EXAMPLE 6 **Modeling population growth**

The population of Australia in 1995 was about 18.35 million with an annual growth rate of approximately 1.5%.

(a) Determine an exponential function that models the population (in millions) as a function of t, the number of years elapsed since 1995.
(b) Use the model to predict the population (in millions) in the year 2005. Round off the answer to two decimal places.
(c) Use a graphing utility to predict when the population will reach 20 million. Round off the answer to the nearest year.

Solution

(a) We begin by letting $P(t)$ represent the population (in millions) t years after 1995. Under the assumption that the annual growth rate of 1.5% continues into the future, we are led to the pattern of population growth given in Table 4, where the results are rounded off to two decimal places.

TABLE 4

Year	t	$P(t)$, Annual Growth Rate of 1.5%
1995	0	$P(0) = 18.35 = 18.35(1.015)^0$
1996	1	$P(1) = 18.35(1.015)^1 = 18.63$
1997	2	$P(2) = [18.35(1.015)](1.015) = 18.35(1.015)^2 = 18.90$
1998	3	$P(3) = [18.35(1.015)^2](1.015) = 18.35(1.015)^3 = 19.19$
⋮	⋮	⋮

Even though the data used to derive the function restricted the values of t to nonnegative *integers*, we follow the convention of letting t represent all *real numbers* such that $t \geq 0$.

This pattern suggests that the function defined by the equation

$$P(t) = 18.35(1.015)^t \qquad \text{where } t \geq 0$$

will serve as a model for the population growth of Australia starting from 1995.

(b) In the year 2005, $t = 10$, so according to the model, the approximate population is given by

$$\begin{aligned} P(10) &= 18.35(1.015)^{10} \\ &= 21.30 \end{aligned}$$

Thus, the population in 2005 is predicted to be about 21.30 million.

(c) If the population is 20 million, we must find a number t such that $P(t) = 20$; that is, we must solve the equation

$$\begin{aligned} 18.35(1.015)^t &= 20 \\ 18.35(1.015)^t - 20 &= 0 \end{aligned}$$

This latter equation can be solved graphically by finding the t intercept of the graph of the function

$$f(t) = 18.35(1.015)^t - 20 \quad \text{(Figure 6)}$$

$f(t) = 18.35(1.015)^t - 20$

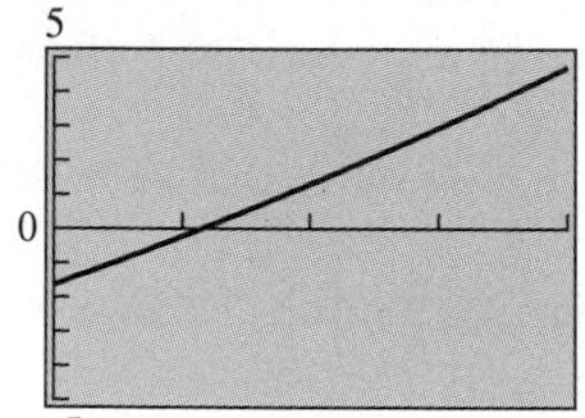

5

0 20

−5

FIGURE 6

By using the ZOOM and TRACE features, we find the intercept for f on the t axis to be approximately 5.78. So the population will reach 20 million after about 6 years.

Exponential functions also enable us to analyze financial transactions involving *compound interest.* Consider what happens when a sum of money, called the **principal**, is put in an investment that earns an **annual interest rate *r*** credited at regular intervals of time, n times per year. **Compound interest** is the interest paid on interest previously earned as well as on the initial deposit. The **interest rate per compounding period** is given by r/n, and the **number of compounding periods after *t* years** is nt.

For example, suppose that \$3000 (the principal) is invested in a savings account that pays 3.6% annual interest compounded semiannually. The interest rate per compounding period is given by $r/n = 0.036/2$, and after t years, the number of compounding periods is $2t$. The pattern of growth of the investment is shown in Table 5.

From the table, we see that the accumulated amount S after t years, which includes both the initial investment and interest earned, is given by

$$S = 3000\left(1 + \frac{0.036}{2}\right)^{2t}$$

TABLE 5

Number of Compounding Periods (nt)	Principal (P) *Dollars*	Interest Earned (I) *Dollars*	Accumulated Amount ($P + I$), or S *Dollars*
End of 6 months or 0.5 year: $2(0.5) = 1$	3000	$3000\left(\frac{0.036}{2}\right) = 54$	$3000 + 3000\left(\frac{0.036}{2}\right)$ $= 3000\left(1 + \frac{0.036}{2}\right) = 3054$
End of 1 year: $2(1) = 2$	$3000\left(1 + \frac{0.036}{2}\right)^1$	$3000\left(1 + \frac{0.036}{2}\right)^1\left(\frac{0.036}{2}\right)$ $= 54.97$	$3000\left(1 + \frac{0.036}{2}\right)^2 = 3108.97$
End of 18 months or 1.5 years: $2(1.5) = 3$	$3000\left(1 + \frac{0.036}{2}\right)^2$	$3000\left(1 + \frac{0.036}{2}\right)^2\left(\frac{0.036}{2}\right)$ $= 55.96$	$3000\left(1 + \frac{0.036}{2}\right)^3 = 3164.93$
⋮	⋮	⋮	⋮
End of t years: $2t$	$3000\left(1 + \frac{0.036}{2}\right)^{2t-1}$	$3000\left(1 + \frac{0.036}{2}\right)^{2t-1}\left(\frac{0.036}{2}\right)$	$3000\left(1 + \frac{0.036}{2}\right)^{2t}$

This illustration leads us to the following general formula for compound interest:

COMPOUND INTEREST FORMULA

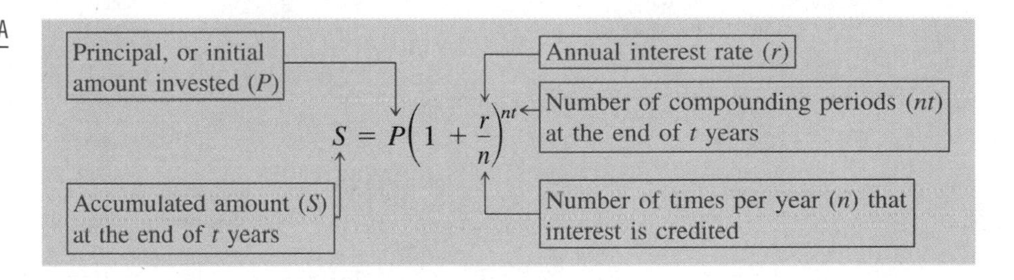

A graphing utility with the TRACE feature enables us to determine the trend of the accumulated amount for an investment that pays compound interest.

EXAMPLE 7 Examining compound interest

Suppose that \$5250 is invested in a credit union paying an annual interest rate of 5%, compounded quarterly.

(a) Express the accumulated amount S as a function of the elapsed number of years t.

(b) Use a graphing utility to graph the function in part (a), and then use the TRACE feature to approximate the number of years (to the nearest tenth) it takes for the accumulated amount to exceed \$6000, \$8000, and \$10,000.

(c) How much will accumulate after 5 years?

(d) A competing bank offers to pay an annual interest rate of 5% compounded daily. A service fee of 50¢ per month is charged to handle the account. After 5 years, how does the accumulated amount of a \$5250 investment made in this bank compare to the \$5250 investment made in the credit union?

Solution

(a) In this case, we substitute $P = 5250$, $r = 0.05$, and $n = 4$ into the compound interest formula to get

$$S = 5250\left(1 + \frac{0.05}{4}\right)^{4t}$$

or

$$S = 5250(1.0125)^{4t} \qquad t \geq 0$$

This equation expresses S as a function of t.

(b) Figure 7 shows a viewing window of the graph of $S = 5250(1.0125)^{4t}$. We use the TRACE feature to move the cursor along the graph. By monitoring the changing x values (corresponding to t) and y values (corresponding to S) in the viewing window, we obtain results such as those listed in Table 6.

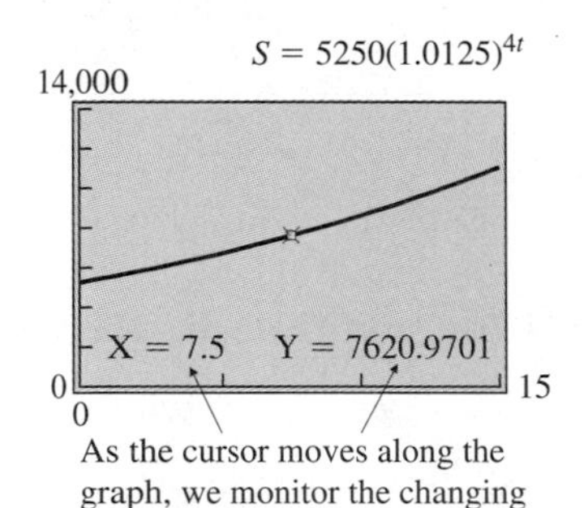

FIGURE 7

TABLE 6

When t (number of years) changes from:	S (accumulated amount) changes from:	Conclusion
2.6190476191 to 2.7380952381	5979.6902102 to 6015.1678124	The accumulated amount exceeds \$6000 after about 2.7 years.
8.4523809524 to 8.5714285714	7990.2940573 to 8037.7006058	The accumulated amount exceeds \$8000 after about 8.5 years.
12.857142857 to 12.976190476	9945.3045341 to 10004.310193	The accumulated amount exceeds \$10,000 in about 13 years.

(c) After 5 years, $t = 5$, so

$$S = 5250(1.0125)^{20} = 6730.70 \text{ (approx.)}$$

Thus, the accumulated amount is \$6730.70.

(d) If \$5250 is invested in the bank where the money is compounded daily, then after 5 years the accumulated amount is given by

$$5250\left(1 + \frac{0.05}{365}\right)^{(365)(5)} = 6741.02 \text{ (approx.)}$$

Thus, the accumulated amount is \$6741.02. However, the 50¢ monthly fee adds up to \$30 over 5 years, and this reduces the accumulated amount to \$6711.02. So the bank investment actually pays \$6730.70 − \$6711.02 = \$19.68 less than the credit union investment after 5 years.

PROBLEM SET 4.1

Mastering the Concepts

In problems 1 and 2, use a calculator to approximate each expression to four decimal places.

1. **(a)** 2^{π} **(b)** $2^{-\sqrt{2}}$
(c) $\sqrt{3}^{\sqrt{3}}$ **(d)** $1000(5^{-\pi})$
(e) $100(1.02)^{-8}$

2. **(a)** π^{3} **(b)** $5^{-\sqrt{7}}$
(c) $2^{\sqrt[3]{5}}$ **(d)** $\sqrt[5]{5^{\sqrt{5}}}$
(e) $10^{\pi}(3.14)^{1/3}$

3. Suppose that $f(x) = 4^{-x/2}$. Find each value. Round off the answers to two decimal places.
(a) $f(-1)$ **(b)** $f(-2)$
(c) $f(2)$ **(d)** $f(0.3)$
(e) $f(-1.2)$ **(f)** $f(\sqrt{3})$

4. Suppose that $g(x) = \left(\frac{2}{3}\right)^x$. Find each value. Round off the answers to two decimal places.
(a) $g(-1)$ **(b)** $g(-2)$
(c) $g(3)$ **(d)** $g(1.7)$
(e) $g(-3.4)$ **(f)** $g(-\sqrt{7})$

In problems 5–8, sketch the graphs of each pair of functions on the same coordinate system, and then use the graphs to solve the inequality. Express each answer in interval notation.

5. $f(x) = 3^x$ and $g(x) = 2^x$; $3^x > 2^x$
6. $f(x) = 10^x$ and $g(x) = 7^x$; $7^x < 10^x$
7. $f(x) = \left(\frac{1}{4}\right)^x$ and $g(x) = \left(\frac{1}{5}\right)^x$; $\left(\frac{1}{4}\right)^x < \left(\frac{1}{5}\right)^x$
8. $f(x) = 4^{-x}$ and $g(x) = 3^{-x}$; $4^{-x} > 3^{-x}$

9. Each graph in Figure 8 is a transformation of the graph of the function $y = 2^x$. Match each given function with its graph in Figure 8.
(a) $f(x) = 2^x + 1$
(b) $g(x) = 2^{x+1}$
(c) $h(x) = -2^x$
(d) $F(x) = 2^x - 1$

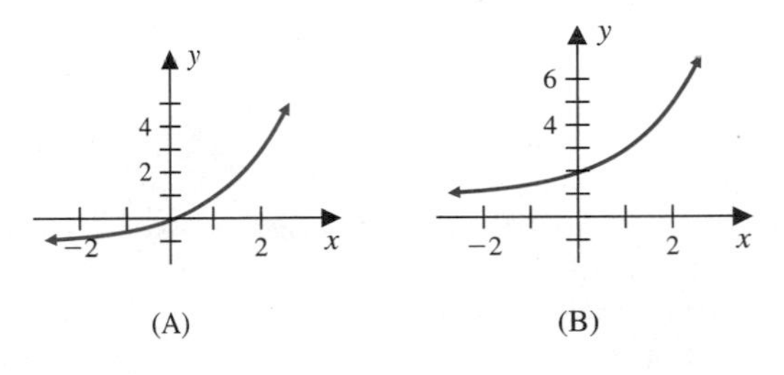

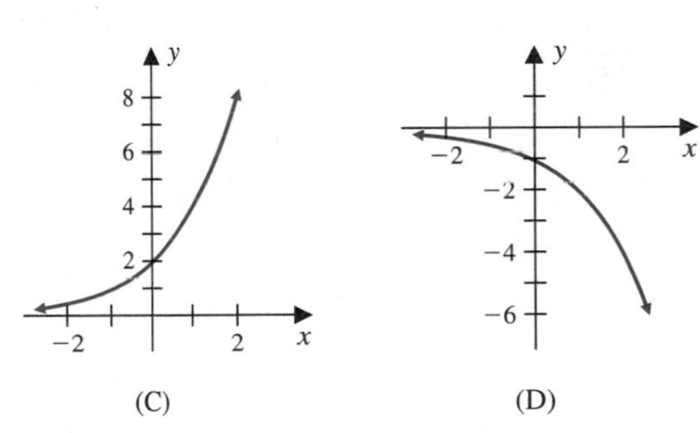

FIGURE 8

10. Each graph in Figure 9 (at the top of page 236) is a transformation of the graph of the function $y = (0.4)^x$. Match each given function with its graph in Figure 9.
(a) $g(x) = (0.4)^x - 1$
(b) $f(x) = (0.4)^{x-1}$
(c) $h(x) = -(0.4)^x$
(d) $G(x) = (0.4)^{x+1}$

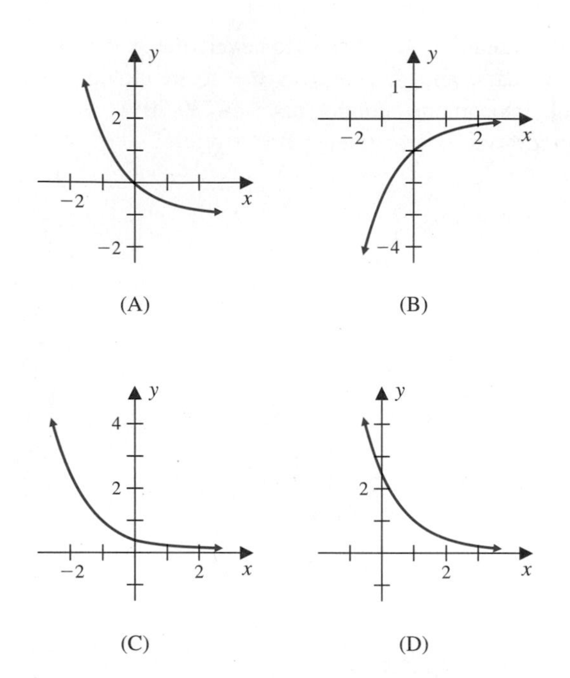

FIGURE 9

In problems 11–20, describe a sequence of transformations that will transform the graph of a function of the form $y = b^x$ to the graph of the given function. Sketch the graph in each case. Determine the domain and range, and indicate whether the function increases or decreases. Also identify any horizontal asymptote.

11. $f(x) = 5^{x+1}$

12. $h(x) = 4^{x-2}$

13. $f(x) = 2^x + 3$

14. $g(x) = 4^x - 1$

15. $h(x) = 4 \cdot 5^x$

16. $f(x) = 3 \cdot 2^{x+2}$

17. $f(x) = 2^{-x} - 3$

18. $g(x) = -2 - 5^{x+1}$

19. $f(x) = -2 \cdot 3^{x+1} - 1$

20. $g(x) = -3 \cdot 2^{x+2} + 4$

21. Each curve in Figure 10 is the graph of an exponential function $y = b^x$ that contains the given point. Find the value of b in each case.

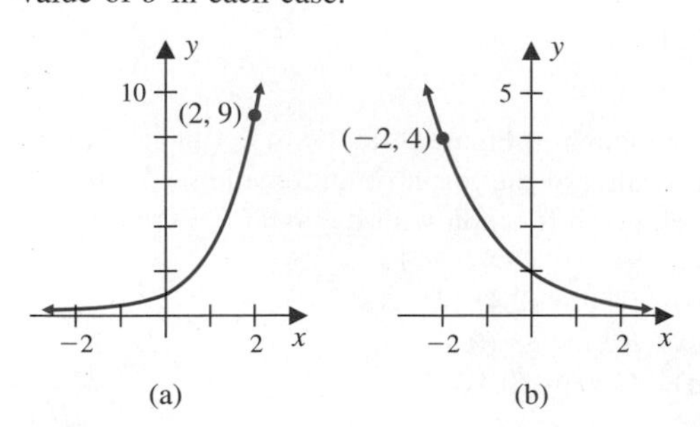

FIGURE 10

22. Each curve in Figure 11 is the graph of an exponential function $y = -b^x$ that contains the given point. Find the value of b in each case.

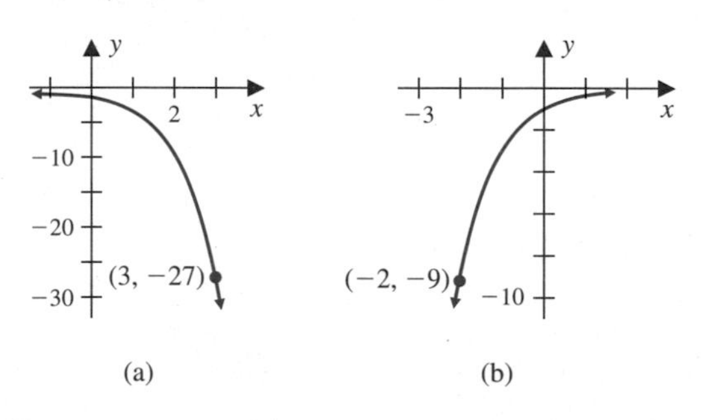

FIGURE 11

23. Graph $f(x) = 5^x$ and $g(x) = 5^{-x}$ on the same coordinate system. Explain how the graph of g can be obtained from the graph of f.

24. Explain why the graph of $g(x) = b^{-x}$ is a reflection of the graph of $f(x) = b^x$ about the y axis, where $b > 0$ and $b \neq 1$.

GU In problems 25–30, use a graphing utility to graph each function. Use the graph to help determine the domain and range. Find any horizontal asymptote. Round off answers to four decimal places.

25. $g(x) = 3^{-x} + x^3$

26. $h(x) = 2^{1-x^2}$

27. $h(x) = 2^{-|x|}$

28. $f(x) = \dfrac{2^x + 2^{-x}}{2}$

29. $f(x) = 2^{x^2+3x-4}$

30. $g(x) = 2^{\sqrt{x}}$

Applying the Concepts

31. Population Growth: In 1991 the population of India was about 866 million with an annual growth rate of 1.9%.

(a) Assuming that this growth rate continues, determine a function P that models the population growth; that is, find a function that expresses the population (in millions) as a function of t, the number of years elapsed since 1991.

(b) Use the model to predict the population (in millions) in the year 2000. Round off the answer to the nearest million.

GU **(c)** Use a graphing utility to predict when the population will reach 1 billion. Round off the answer to the nearest year.

32. Bacteria Growth: A mathematical model that approximates the number N of bacteria grown in a colony under ideal laboratory conditions after t days is given by

$$N = 1500 \cdot (2)^{t/4}$$

(a) According to this model, what is the daily growth rate?
(b) Find the number of bacteria initially present.
(c) Find the number of bacteria present after 13 days.
GU **(d)** Use a graphing utility to predict when the number of bacteria in the colony will be 5000. Round off the answer to the nearest hour.

33. Compound Interest: If \$5000 is invested in a certificate of deposit for 5 years at an annual interest rate of 5.75%, how much money will be in the account if the interest rate is compounded:
(a) Semiannually?
(b) Quarterly?
(c) Monthly?
(d) Weekly?

34. Compound Interest: If \$4000 is invested at an annual rate of 6.15% per year compounded monthly, how much money will be in the account after:
(a) 6 months?
(b) 2 years?

35. Compound Interest: Suppose that \$10,450 is invested in a money market account paying an annual interest rate of 4.25% compounded monthly.
(a) Express the accumulated amount S as a function of elapsed time t.
GU **(b)** Graph the function in part (a) with a graphing utility. Then use the TRACE feature to find out how long it takes for the accumulated amount to exceed \$12,000; \$15,000; \$20,000. Round off the answers to the nearest month.
(c) After 3 years, how will the earned interest in this investment compare to the earned interest if the money is invested in a bank that pays an annual interest rate of 4.015% compounded daily?

36. Compound Interest: The accumulated amount S for an investment of \$8500 in which the interest is compounded daily for t years is given by

$$S = 8500(1.00015)^{365t}$$

GU **(a)** Use a graphing utility to graph S as a function of t and interpret the result.
(b) What annual interest rate is being paid?
(c) How much will accumulate after 3 years?
(d) How much less is earned if the interest is compounded quarterly instead of daily for 3 years at the same annual rate?

37. College Fund: Two grandparents want to invest in a college fund that will pay \$25,000 to their newly born grandchild 18 years from now. The fund pays an annual rate of 7% compounded quarterly. What principal needs to be invested now to meet the goal? Round off the answer to the nearest dollar.

38. Saving for a Car: A college professor wishes to have \$18,000 cash for a purchase of a new car 4 years from now. How much should be placed in an account now if the account pays 6.75% annual interest compounded monthly? Round off the answer to the nearest dollar.

39. Car Trade-in Value: Suppose that the *trade-in value* V (in dollars) of an automobile that is t years old is given by the formula

$$V = P(1 - r)^t$$

where r is the annual percentage of the depreciation and P is the original price of the car (in dollars). A certain make of car is purchased for \$19,545, and it depreciates in value by 20% each year.
(a) Write an equation that expresses the trade-in value V as a function of t.
(b) Find the trade-in value after 1 year, after 3 years, and after 5 years. Round off answers to the nearest dollar.
GU **(c)** Use a graphing utility to sketch the graph of the equation in part (a) for $0 \le t \le 5$ and use the TRACE feature to determine how many years it will take for the value to depreciate to under \$10,000. Round off the answer to the nearest tenth of a year.

40. Real Estate Appreciation: Because of inflation, the value of a home often increases with time. State and local agencies use the formula

$$S = C(1 + r)^t$$

to assess the value S (in dollars) of a home for property tax purposes, where C dollars is the cost of the home when new, t is the age of the home (in years), and r is the annual rate of inflation. Assume that the inflation rate is 3.1% and that a home originally cost \$135,000.
(a) Write an equation that expresses the assessed value S as a function of t.
(b) If the inflation rate remains constant at 3.1%, what is the assessed value after 1 year? 5 years? 10 years? 15 years?

GU (c) Use a graphing utility to graph the equation in part (a) for $0 \leq t \leq 20$ and use the TRACE feature to find out how long it takes for the appreciated value to exceed \$150,000, \$175,000, and \$200,000. Round off the answers to the nearest tenth of a year.

41. Recycling: In the state of Michigan, stores charge an additional 10¢ deposit for each bottle of beverage sold. Customers return these bottles for refunds, and then the stores return them to the beverage companies for recycling. Suppose that 65% of all bottles distributed will be recycled every year. If a company distributed 2.5 million bottles in 1 year, the number N (in millions) of recycled containers still in use after t years is given by the model

$$N = 2.5(0.65)^t$$

(a) How many recycled bottles (in millions) are still in use after 1 year? 3 years? 5 years? Round off the answers to two decimal places.

GU **(b)** Use a graphing utility to graph N as a function of t. Then use the TRACE feature to determine the number of years it takes for the number of recycled containers to fall below 1 million, $\frac{1}{2}$ million, and $\frac{1}{4}$ million. Round off the answers to the nearest tenth of a year.

42. Oil Demand: Suppose that the annual demand D (in millions of barrels) for oil from an oil-producing country is given by the model

$$D = 200(2.8)^{-0.25p}$$

where p is the price (in dollars) of a barrel of oil and $12 \leq p \leq 25$.

GU **(a)** Use a graphing utility to sketch the graph of D and use it to discuss the relationship of demand and price. What happens to demand as price increases?

(b) The revenue R (in millions of dollars) received from selling D million barrels at p dollars per barrel is given by

$$R = D \cdot p$$

Express the revenue R as a function of p.

(c) What is the expected revenue if the price of a barrel of oil is \$14.00? \$14.75? \$15.00? \$20.00?

GU **(d)** Use a graphing utility to graph the function found in part (b), where $12 \leq p \leq 25$. Then use the TRACE feature to find the price per barrel at which the revenue is expected to fall below 60 million dollars, 40 million dollars, and 20 million dollars.

Developing and Extending the Concepts

43. Explain why the graph of the exponential function $f(x) = b^x$ contains the points (0, 1) and (1, b).

44. **(a)** Show algebraically that

$$(3^x + 3^{-x})^2 - (3^x - 3^{-x})^2 = 4$$

GU **(b)** Use a graphing utility to illustrate the validity of the equation in part (a) by comparing the graphs of $f(x) = (3^x + 3^{-x})^2 - (3^x - 3^{-x})^2$ and $g(x) = 4$.

45. **(a)** Find the zeros of $f(x) = 2x^2(3^x) - 7x(3^x) - 4(3^x)$ algebraically.

GU **(b)** Use a graphing utility to sketch the graph of f and find the zeros of f from the graph. Compare the results to part (a).

46. **(a)** Solve the equation below for x algebraically.

$$\frac{(x + 4)(3^x)}{x - 3} = 2x(3^x)$$

(b) Explain why the solutions of the equation in part (a) are the same as the x intercepts of the graph of

$$f(x) = \frac{(x + 4)(3^x)}{x - 3} - 2x(3^x)$$

GU **(c)** Use a graphing utility to graph f and find the x intercepts. Compare the results to the solution in part (a).

47. Let $f(x) = 1 + 2^x$ and $g(x) = \dfrac{1}{1 + 2^x} + \dfrac{1}{1 + 2^{-x}}$

(a) Show that: $\dfrac{1}{f(x)} + \dfrac{1}{f(-x)} = g(x)$

GU **(b)** Use a graphing utility to sketch the graph of g and describe it.

(c) Simplify the expression algebraically:
$$\frac{1}{1 + 2^x} + \frac{1}{1 + 2^{-x}}$$

(d) Discuss the relationship between the graph in part (b) and the conclusion in part (c).

48. GU Which two functions are equal? Justify your assertion algebraically and illustrate the validity of your conclusion graphically.
(i) $f(x) = 2^{-(x-4)}$ **(ii)** $g(x) = (0.5)^{x-4}$
(iii) $h(x) = (0.25)^{2x-8}$

49. Let $f(x) = 2^x$ and $g(x) = |x|$.

GU **(a)** Find $(f \circ g)(x)$ and use a graphing utility to sketch the graph of $f \circ g$.

(b) Find $(g \circ f)(x)$ and use a graphing utility to sketch the graph of $g \circ f$.

(c) Does $f \circ g$ equal $g \circ f$? Explain from both an algebraic and graphical point of view.

50. Inflation: The following data represent the amount A that \$1 will be worth t years from now, assuming that inflation runs steadily at 4% per year.

Year (t)	0	1	2	3	4	5	6	7	8
Amount (A)	\$1	\$0.96	\$0.92	\$0.88	\$0.85	\$0.82	\$0.78	\$0.75	\$0.72

(a) Construct a scattergram that exhibits the given data.
(b) Determine how closely the exponential function $A = (0.96)^t$ predicts the data.
(c) Sketch the graph of the function A in part (b).
(d) Assuming the function A fits the data exhibited in part (a), predict how many years it will be before today's dollar is worth only 20 cents.

OBJECTIVES

1. Define the natural base—base e.
2. Graph exponential functions with base e.
3. Solve applied problems.

4.2 The Natural Base—Base e

In this section we focus on a special base used for exponential functions called the *natural base*. Population growth, continuous compound interest, bacteria counts, drug concentration, and radioactive decay are modeled by exponential functions with the natural base. The importance of this base was first recognized by the Swiss mathematician Leonhard Euler (pronounced "oiler"), 1707–1783, and in his honor it is denoted by the symbol e.

Defining the Natural Base—Base e

In Section 4.1 we introduced the compound interest formula

$$S = P\left(1 + \frac{r}{n}\right)^{nt}$$

If we let $P = 1$, $r = 1$, and $t = 1$, then the formula becomes

$$S = \left(1 + \frac{1}{n}\right)^n$$

which is the value of \$1 invested at 100% interest compounded n times a year for a period of 1 year. Now we ask the question: What happens to the value of S in the latter equation as $n \to +\infty$?

To help answer this question, we graph the function

$$f(x) = \left(1 + \frac{1}{x}\right)^x \qquad x > 0$$

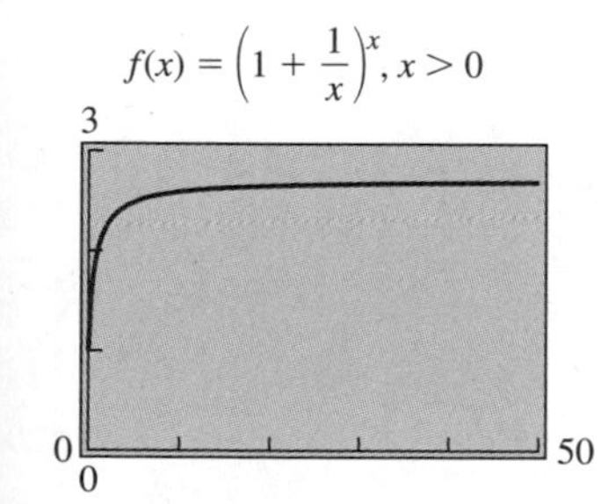

FIGURE 1

with the aid of a graphing utility. The viewing window of the graph shown in Figure 1 suggests that as $n \to +\infty$,

$$\left(1 + \frac{1}{n}\right)^n$$

is asymptotically approaching a fixed value less than 3. To examine this behavior numerically, we use a calculator to list some values of $[1 + (1/n)]^n$ for increasing values of n (Table 1, at the top of page 240).

TABLE 1

Approximate Values of $\left(1 + \frac{1}{n}\right)^n$

n	1	10	100	1000	10,000	100,000	1,000,000
$\left(1 + \frac{1}{n}\right)^n$	2.00000	2.59374	2.70481	2.71692	2.71815	2.71827	2.71828

According to the data in Table 1, the values of $[1 + (1/n)]^n$ appear to be getting closer and closer to a number whose first few digits are 2.7182. . . . In calculus, it can be shown that as $n \to +\infty$ the values of $[1 + (1/n)]^n$ actually approach a fixed irrational number denoted by e. To eleven decimal places,

$$e = 2.71828182846$$

Approximate values of e^x for any real number x can be found easily by using a calculator.

EXAMPLE 1 Approximating values of e^x

Approximate each expression to four decimal places.

(a) $e^{2.3}$ (b) $e^{-\sqrt{3}}$

Solution By using a calculator, we obtain the following approximations to four decimal places.

(a) $e^{2.3} = 9.9742$ (b) $e^{-\sqrt{3}} = 0.1769$

Graphing Exponential Functions with Base e

The exponential function with base e is referred to as the *natural exponential function.* This function plays an important role in calculus and in the application of mathematics to many fields. It is defined as follows:

DEFINITION

THE NATURAL EXPONENTIAL FUNCTION

The **natural exponential function**, or **exponential function with base e**, is defined by the equation

$$f(x) = e^x$$

where x is any real number.

The expression e^x is also denoted by EXP(x).

The properties shown in Table 3 in Section 4.1 (page 230) enable us to sketch the graphs of $f(x) = e^x$ and $g(x) = e^{-x}$ quickly. Since $e > 1$, the function $f(x) = e^x$

is increasing and its graph resembles the model graph of $y = b^x$ for $b > 1$ (Figure 2a). Since $e^{-x} = (1/e)^x$ and $0 < 1/e < 1$, the graph of $g(x) = e^{-x}$ is similar to the graph of the exponential function $y = b^x$ for $0 < b < 1$ (Figure 2b). The x axis is a horizontal asymptote in both graphs.

The graphs of $f(x) = e^x$ and $g(x) = e^{-x}$ are reflections of each other about the y axis.

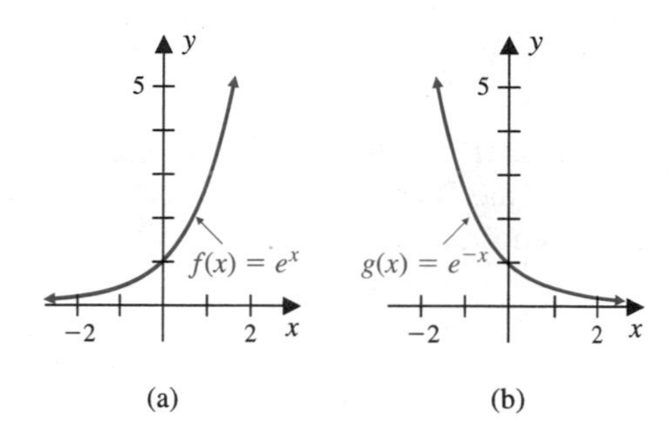

FIGURE 2

Transformation techniques of graph sketching can also be applied to exponential functions with base e.

EXAMPLE 2 Graphing by using transformations

Sketch the graph of $f(x) = -e^{x+3}$ by using a sequence of transformations of the graph of $y = e^x$ and find the horizontal asymptote.

Solution To graph $f(x) = -e^{x+3}$, we first reflect the graph of $y = e^x$ about the x axis to get the graph of $h(x) = -e^x$. Then the graph of $f(x) = -e^{x+3} = h(x + 3)$ is obtained by shifting the graph of h to the left 3 units (Figure 3). The horizontal asymptote of $y = e^x$ is not affected by these transformations; it is the x axis for f as well.

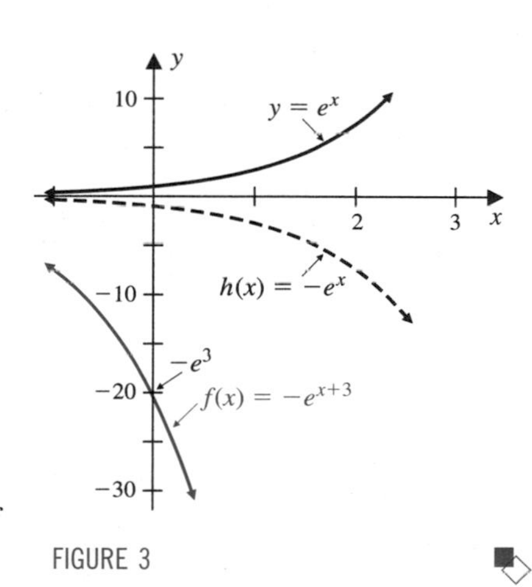

FIGURE 3

Graphing utilities can generate graphs of exponential functions with base e with relative ease because of the built-in e^x function key.

EXAMPLE 3 Graphing by using a graphing utility

Use a graphing utility to graph $f(x) = e^{-x^2}$, where $-2 \leq x \leq 2$. Find any horizontal asymptote.

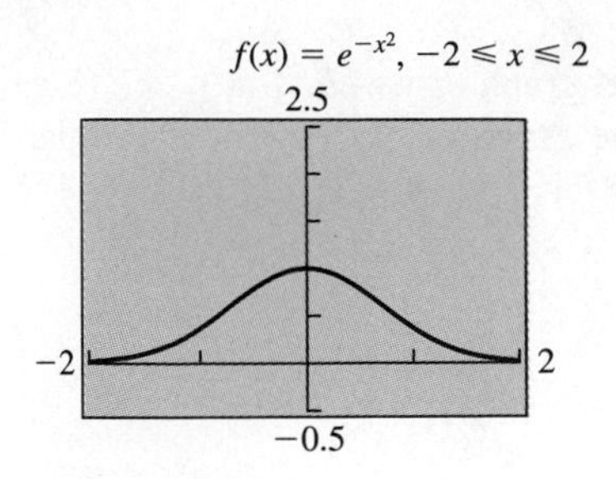

FIGURE 4

Solution Figure 4 shows a viewing window for the graph of $f(x) = e^{-x^2}$, where $-2 \leq x \leq 2$. The graph appears to be symmetric with respect to the y axis, which is confirmed by observing that $f(-x) = e^{-(-x)^2} = e^{-x^2} = f(x)$. The x axis is the horizontal asymptote of the graph of the function, since $f(x) \to 0$ as $x \to -\infty$ and as $x \to +\infty$.

Solving Applied Problems

Exponential functions with base e arise in a variety of applications to real-world phenomena. These applications include *continuous compound interest*, *growth and decay*, and *radiocarbon dating*.

Continuous Compound Interest

In the work on compound interest in Section 4.1, we used compounding periods such as months and quarter years, but compounding periods can be of any length. Many savings institutions compound the interest daily. We could go further and compound interest every minute, every second, every half-second, and so on.

Let us examine the compound interest formula to see what happens when interest is compounded *continuously.* What we must do is to let n become larger and larger in the compound interest formula. That is, we want to determine what happens to the formula as $n \to +\infty$:

$$
\begin{aligned}
S &= P\left(1 + \frac{r}{n}\right)^{nt} && \text{Compound interest formula} \\
&= P\left(1 + \frac{1}{n/r}\right)^{(n/r)rt} && \text{Rewrite algebraically} \\
&= P\left[\left(1 + \frac{1}{m}\right)^{m}\right]^{rt} && \text{Let } m = n/r
\end{aligned}
$$

Now, if r is constant and $n \to +\infty$, it follows that $m = n/r \to +\infty$. We know that as $m \to +\infty$,

$$\left(1 + \frac{1}{m}\right)^{m} \to e$$

Therefore as $n \to +\infty$,

$$P\left[\left(1 + \frac{1}{m}\right)^{m}\right]^{rt} \to Pe^{rt}$$

The resulting formula,

$$\boxed{S = Pe^{rt}}$$

is called the **continuous compound interest formula**, where P is the initial investment (or principal), r is the annual interest rate, t is the number of elapsed years, and S is the accumulated amount.

EXAMPLE 4 **Examining continuously compounding interest**

Assume that \$20,000 is invested in a savings account that pays an annual interest rate of 3.5% compounded continuously.

(a) Express the accumulated amount S as a function of the number of elapsed years t.

(b) Use a graphing utility to graph the function in part (a). Interpret the graph.

(c) What is the accumulated amount after 5 years?

(d) Suppose that at the same time, \$20,000 is invested in a credit union account that pays an annual interest rate of 3.5% compounded monthly. Compare the two accounts after 5 years.

Solution

(a) Using the continuous compound interest formula above with $P = 20{,}000$ and $r = 0.035$, we get the equation

$$S = 20{,}000e^{0.035t}$$

This equation expresses S as a function of t.

(b) Since t represents time, $t \geq 0$. The graph of $S = 20{,}000e^{0.035t}$, where $t \geq 0$, is displayed in the viewing window in Figure 5. The graph shows that the accumulated amount S increases exponentially over time.

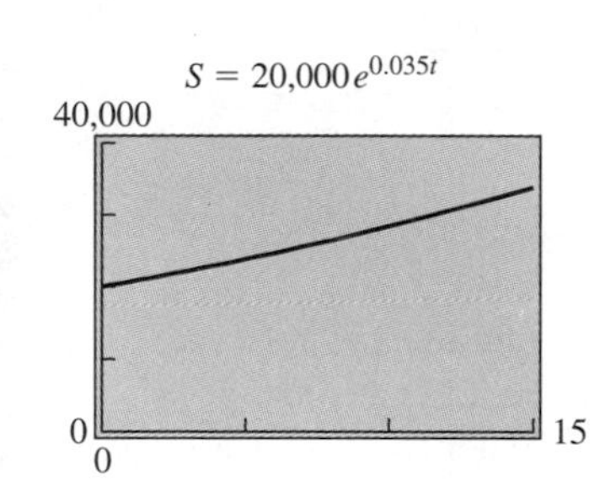

FIGURE 5

(c) After 5 years, $t = 5$. Rounding to two decimal places, we get

$$\begin{aligned} S &= 20{,}000e^{(0.035)(5)} \\ &= 23{,}824.92 \end{aligned}$$

So the accumulated amount after 5 years is \$23,824.92.

(d) Since $P = 20{,}000$, $r = 0.035$, $n = 12$, and $t = 5$, we use the compound interest formula:

$$S = P\left(1 + \frac{r}{n}\right)^{nt}$$

$$S = 20{,}000\left(1 + \frac{0.035}{12}\right)^{(12)(5)} = 23{,}818.86 \text{ (approx.)}$$

So the accumulated amount in the credit union account is \$23,818.86. The credit union investment pays \$23,824.92 − \$23,818.86 = \$6.06 less than the savings account over 5 years.

Growth and Decay

Base e is also used in mathematical modeling of phenomena that exhibit continuous (uninterrupted) growth or decay as follows:

MODELING CONTINUOUS GROWTH OR DECAY

An exponential function of the form

$$P = P_0e^{kt} \qquad \text{where } k > 0$$

is used to model continuous **growth**. Its general graph is shown in Figure 6a.

An exponential function of the form

$$P = P_0e^{-kt} \qquad \text{where } k > 0$$

is used to model continuous **decay**. Its general graph is shown in Figure 6b.

In either case, P_0 is the initial quantity, t is the time, and k is the rate of growth or decay.

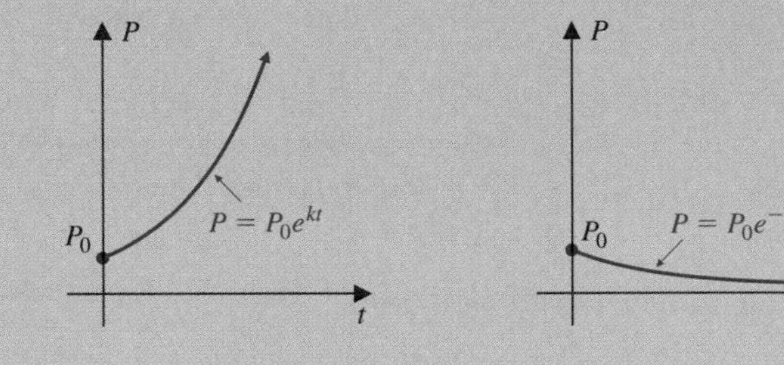

(a) Exponential continuous growth (b) Exponential continuous decay

FIGURE 6

We say that P is growing or decaying *exponentially* at a *continuous* rate of k.

For example, suppose that the initial population of a certain city is 1.013 million, and the population P (in millions) grows exponentially at a continuous rate of about 0.58% per year. The growth model for this population t years later is given by the function

$$P(t) = 1.013e^{0.0058t}$$

If we replace t by 1, 2, 3, and so on, we observe that the corresponding output values increase exponentially as the input values increase.

For the decay situation, the outputs decrease as the input values increase, as illustrated in the next example.

Radiocarbon Dating

Radioactive carbon-14 (C-14) is found in all living things. Once an organism dies, the C-14 begins to decay radioactively. The less C-14 found in a specimen, the older the specimen. Archaeologists are able to date objects such as relics and fossils by determining the remaining amounts of C-14. The process is called *radiocarbon dating*.

EXAMPLE 5 Using radiocarbon dating

The percentage P (in decimals) of C-14 found in a specimen t years after it begins to decay is modeled by the exponential function

$$P = e^{-0.000121t}$$

(a) Use the model to determine the percentage P of C-14 still present in an organism t years after it begins to decay if $t = 0$; 1000; 2000; 5000; 6000; 10,000. Round off the answers to two decimal places.

(b) Suppose that archaeologists unearth a fossil and determine that it currently contains half of its original C-14. Use a graphing utility to date the fossil's age; that is, find out how long ago the fossil began to decay. Round off the answer to the nearest 10 years.

Solution

(a) We substitute each value of t into the function $P = e^{-0.000121t}$ and then use a calculator to evaluate. The results are summarized in Table 2.

TABLE 2

Percentage of C-14 Present After t Years

t Years	P	Percentage of C-14 Present
0	1.00	100
1,000	0.89	89
2,000	0.79	79
5,000	0.55	55
6,000	0.48	48
10,000	0.30	30

(b) Since the fossil has half of its C-14 remaining, it follows that $P = 0.50$. So the solution of the equation

$$0.50 = e^{-0.000121t}$$

or, equivalently,

$$0 = e^{-0.000121t} - 0.50$$

will give us the number of years of decay it took to reach this state. However, the solution of this equation is the same as the x intercept of the graph of the function

$$f(x) = e^{-0.000121x} - 0.50$$

By using a graphing utility to graph f (Figure 7), along with the ZOOM and TRACE features, we find that x is approximately 5730 (to the nearest 10 years). Thus, it takes about 5730 years to lose half the C-14 content.

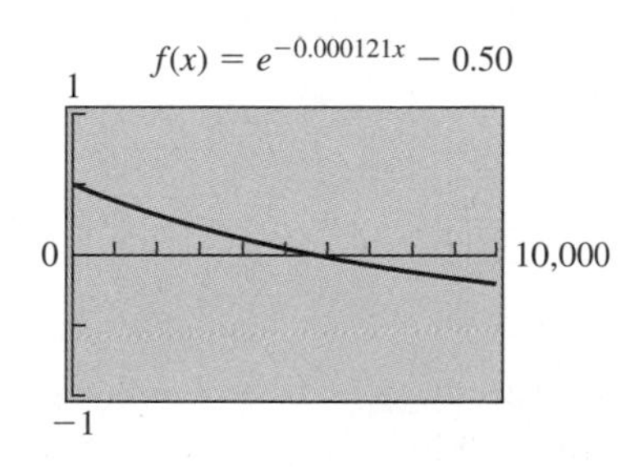

FIGURE 7

The number of years that it takes for half the amount of a radioactive substance to decay is called its **half-life**. So in Example 5b we found that the half-life of C-14 is about 5730 years.

PROBLEM SET 4.2

Mastering the Concepts

In problems 1 and 2, use a calculator to approximate each expression to four decimal places.

1. **(a)** $e^{3.4}$ **(b)** $e^{-\pi}$ **(c)** $e^{\sqrt{2}}$
   (d) $5e^{(-0.5)(7)}$ **(e)** $1 - 4e^{-0.53}$
2. **(a)** $e^{-2.5}$ **(b)** $e^{3/7}$ **(c)** $e^{-\sqrt{3}}$
   (d) $e^{\sqrt{0.5}}$ **(e)** $\dfrac{3}{1 + 2e^{0.4}}$

In problems 3–8, sketch the graphs of each pair of functions on the same coordinate system and then use the graphs to solve the inequality. Express each solution in interval notation.

3. $f(x) = e^x$ and $g(x) = 2^x$; $e^x < 2^x$
4. $f(x) = e^{-x}$ and $g(x) = 3^{-x}$; $e^{-x} > 3^{-x}$
5. $f(x) = e^{0.2x}$ and $g(x) = e^{-0.2x}$; $e^{0.2x} > e^{-0.2x}$
6. $f(x) = e^{2x}$ and $g(x) = e^{-2x}$; $e^{2x} < e^{-2x}$
7. $f(x) = 3e^x$ and $g(x) = -3e^x$; $3e^x > -3e^x$
8. $f(x) = 2e^{-x}$ and $g(x) = -2e^{-x}$; $2e^{-x} < -2e^{-x}$

GU In problems 9–16, use transformations of the graph of $y = e^x$ to sketch the graph of each function. Use a graphing utility to demonstrate the validity of the result. Identify any horizontal asymptote.

9. $f(x) = -2e^x$
10. $g(x) = e^{1-x}$
11. $h(x) = 1 + e^x$
12. $f(x) = -e^x + 2$
13. $f(x) = e^{-x} - 2$
14. $g(x) = e^{x-1} - 1$
15. $f(x) = 2 - 3e^{-x}$
16. $g(x) = 2 + 3e^{-x}$

17. Each graph in Figure 8 is a transformation of the graph of the function $y = e^x$. Match each function with its graph.
    (a) $f(x) = e^x + 1$
    (b) $f(x) = e^{x+1}$
    (c) $f(x) = -e^x$
    (d) $f(x) = e^x - 1$

18. Each graph in Figure 9 is a transformation of the graph of the function $y = e^{-x}$. Match each function with its graph.
    (a) $g(x) = e^{-x} - 1$
    (b) $g(x) = e^{-x-1}$
    (c) $g(x) = -e^{-x}$
    (d) $g(x) = e^{-x} + 1$

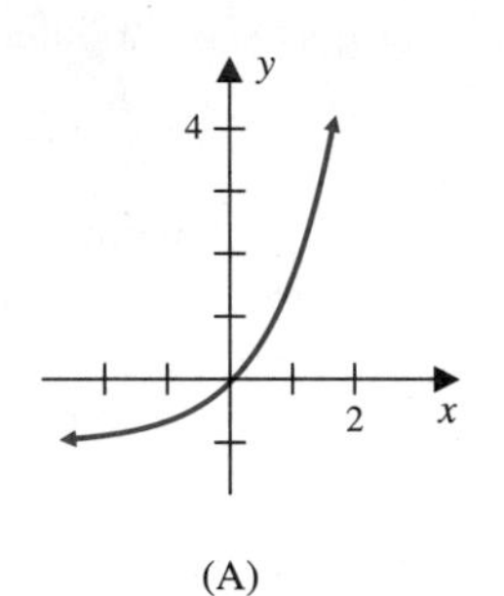

(A)

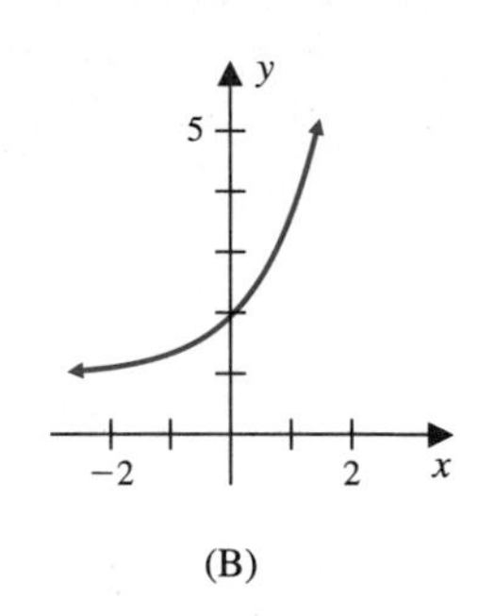

(B)

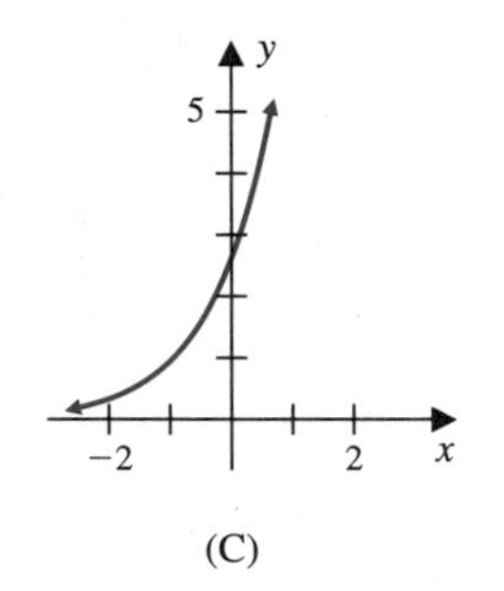

(C)

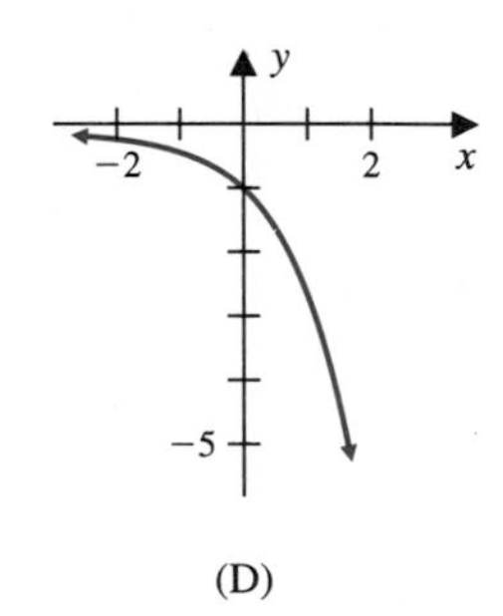

(D)

FIGURE 8

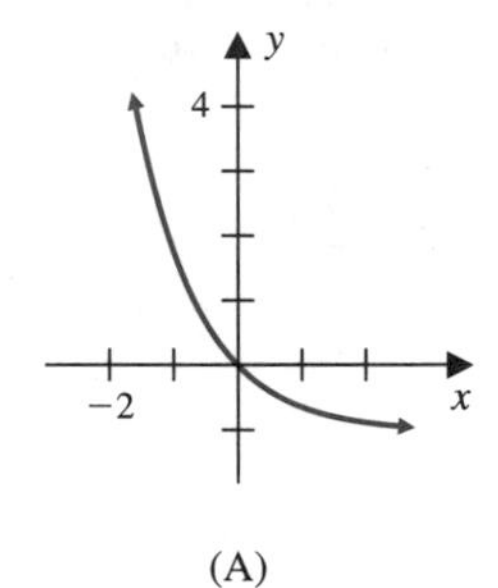

(A)

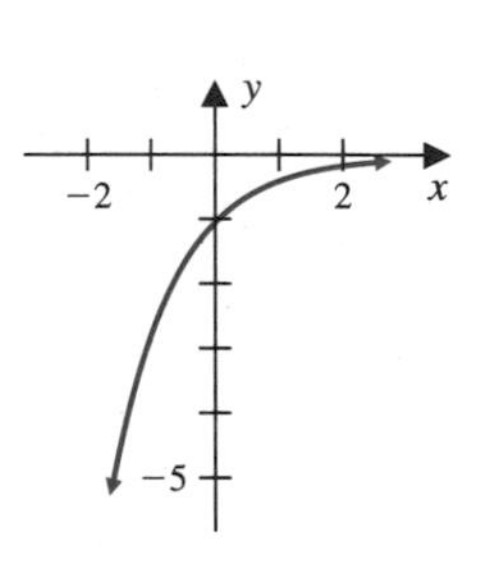

(B)

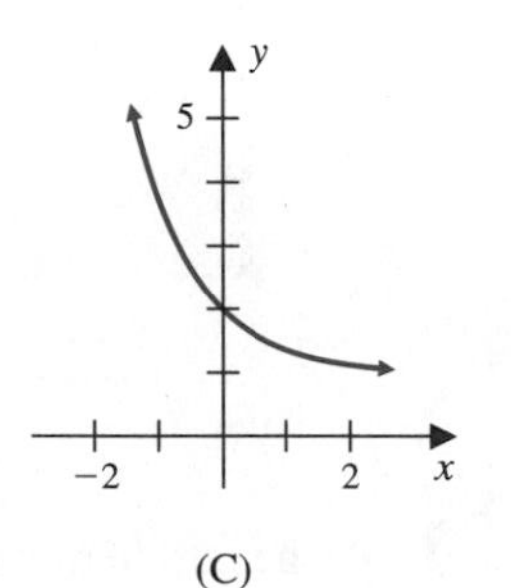

(C)

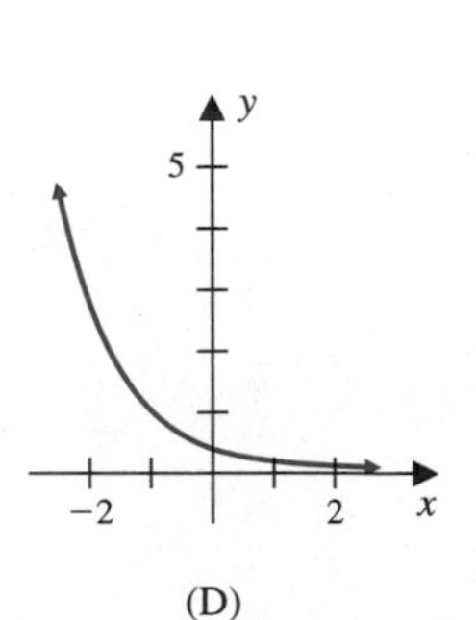

(D)

FIGURE 9

GU In problems 19–24, use a graphing utility to sketch the graph of each function. Find any horizontal asymptote.

19. $f(x) = e^{-|x|}$

20. $f(x) = \dfrac{e^{(-x/2)} + e^{(x/2)}}{2}$

21. $f(x) = e^{x^2}$

22. $f(x) = \dfrac{1}{1 + e^{-x}}$

23. $f(x) = x^2e^{-2x}$

24. $f(x) = e^{3x} - e^{2x-1}$

Applying the Concepts

In problems 25–40, round off the answers to two decimal places.

25. Money Growth: Suppose that $7500 is invested in a certificate of deposit (CD) paying 4.95% annual interest compounded continuously. How much money will be in the account at the end of:
(a) 5.5 years? **(b)** 10 years?

26. Money Growth: Suppose that $5000 is invested at an annual rate of 7% compounded continuously. What is the interest earned after:
(a) 3 years? **(b)** 5 years?

27. Cumulative Earnings: Suppose that $5000 is invested in a money market account paying an annual interest rate of 5.75% compounded continuously.
(a) Express the accumulated amount S as a function of the number of elapsed years t.
GU **(b)** Use a graphing utility to graph the function in part (a). Interpret the graph.
(c) What is the accumulated amount after 4 years?
(d) Will this investment earn more or less than another investment of $5000 that pays an annual interest rate of 6% compounded semiannually? Explain the difference.

28. Cumulative Interest Trend: The accumulated amount S of an investment of $10,000 paying continuously compounded interest over t years is given by

$$S = 10{,}000e^{0.0625t}$$

(a) What is the annual interest rate being paid on this investment?
GU **(b)** Use a graphing utility to graph S as a function of t. Then use the TRACE feature to approximate how many years it will take for the accumulated amount to exceed $15,000, $20,000, and $25,000.

29. Present Value: A savings certificate will pay $10,000 at maturity 5 years from now. How much money is to be invested now if the annual interest rate on the certificate is 6.5% compounded continuously?

30. Present Value: An account is now worth $10,560. It has been accumulating interest at an annual rate of 6% compounded continuously for 5 years. What was the initial deposit?

31. Population Growth: The world population at the beginning of 1991 was about 5.4 billion people, and it was growing at a continuous rate of about 1.7% per year.
(a) Find the exponential growth model for the world population P (in billions) t years after 1991.
(b) Use the model to predict the world population in the year 2000.
GU **(c)** Use a graphing utility to graph P as a function of t. Then use the TRACE feature to approximate the year when the world population will exceed 6.5 billion, 7 billion, and 10 billion.

32. Population Growth: According to the U.S. Census Bureau, the population of the United States in 1995 was approximately 262 million people, and it was growing at a continuous rate of about 0.73% per year.
(a) Find the exponential growth model that expresses the population P (in millions) t years after 1995.
(b) Predict the population in the year 2050.
GU **(c)** Use a graphing utility to determine the year when the population will reach 400 million.

33. Bacteria Growth: A biologist studying bacteria in a swimming pool determines that the initial population of bacteria is 25,000 per cubic inch, and it is growing at a continuous rate given by the model

$$P = 25{,}000e^{0.1t}$$

where P represents the number present after t days.
(a) What is the continuous growth rate of the bacteria?
GU **(b)** Use a graphing utility to predict how long it will take for the number of bacteria to triple.

34. GU Bacteria Growth: The growth rate of bacteria in foods is used to determine the safe shelf-life of various products. Once the bacteria count reaches a certain level, these products are no longer safe to eat. If the daily continuous growth rate of bacteria in a certain brand of cheese is 5% of its population each day, how many times the current count level will the bacteria count be after 140 days?

35. Radioactive Decay: Certain radioactive elements decay exponentially. The decay model for a specific radioactive element is

$$A = A_0 e^{-0.04463t}$$

where A is the amount present after t days and A_0 is the amount present initially. Assume there is a block of 50 grams of the element at the start.

- **(a)** How much of the element remains after 10 days? 15 days? 30 days?
- GU **(b)** Use a graphing utility to graph A as a function of t. Interpret the graph.
- **(c)** Use a graphing utility to determine how many days it takes for the material to decay to 10 grams.
- **(d)** According to this model, will the substance ever decay to 0 gram? Explain.

36. Electric Circuit: The electric current I (in amperes) flowing in a series circuit having inductance L henrys, resistance R ohms, and electromotive force E volts (Figure 10) is given by the model

$$I = \frac{E}{R}(1 - e^{-Rt/L})$$

where t is the time (in seconds) after the current begins to flow.

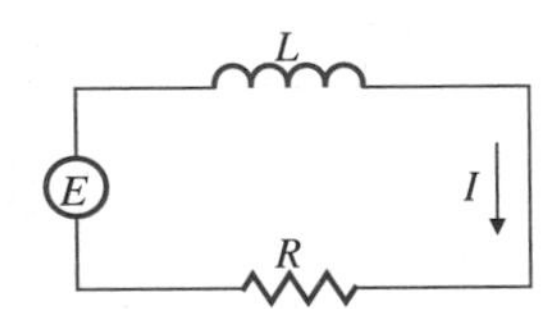

FIGURE 10

- **(a)** If $E = 7.5$ volts, $R = 3$ ohms, and $L = 0.83$ henry, express I as a function of t.
- **(b)** Find the amount of current that flows in the circuit after 0.3 second, 0.5 second, and 1 second.
- GU **(c)** Use a graphing utility to find out how long it takes for the current to reach 1.3 amperes.
- **(d)** What is happening to the current as $t \to +\infty$?

37. Environmental Science: According to the Bouguer–Lambert law, the percentage P of light that penetrates ordinary seawater to a depth of d feet is given by the function

$$P = e^{-0.044d}$$

- **(a)** What is the percentage of light penetration in seawater to a depth of 4 feet? 6 feet?
- GU **(b)** Use a graphing utility to graph P as a function of d and interpret it. What is the range of this function?
- **(c)** Use the graph to predict the depth at which the light penetration is 50%.

38. Physics—Law of Cooling: Suppose that a heated object is placed in cooler surroundings. *Newton's law of cooling* relates the temperature T of the heated object to the temperature T_0 of the cooler surroundings. It states that after t units of time,

$$T = T_0 + Ce^{-kt}$$

where k and C are positive constants associated with the cooling object. Suppose that a baked cake is removed from an oven and cooled to a room temperature of 70°F. The temperature T of the cake after t minutes is given by

$$T = 70 + 280e^{-0.14t}$$

- **(a)** What was the temperature of the cake when it was first removed from the oven?
- **(b)** What is the temperature of the cake after 5 minutes? 10 minutes? 20 minutes?
- GU **(c)** Use a graphing utility to predict how long it will take for the temperature of the cake to drop to 75°F.

39. Civil Engineering: When a flexible cord or cable hangs freely from its ends, it forms an arc called a *catenary*. Its equation has the form

$$y = \frac{k}{2}(e^{cx} + e^{-cx})$$

for appropriate choices of the constants c and k. The famous Gateway Arch in St. Louis, Missouri, has the shape of an *inverted catenary* (Figure 11). Suppose the arch is placed in a coordinate system so that the x axis is ground level and the y axis is the axis of symmetry of the arch. Assume its equation is given by

$$y = -63.85(e^{(x/127.7)} + e^{(-x/127.7)}) + 757.70$$

where x is in feet.

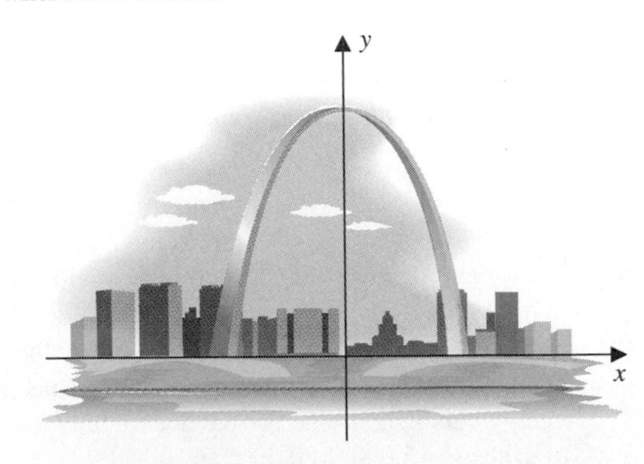

FIGURE 11

GU (a) Use a graphing utility to sketch the graph of this equation, and compare the shape of the curve to Figure 11.
(b) Locate the highest point (*apex*) of the arch.
(c) Find the distance between the bases at ground level.

40. Behavioral Science—Learning Curve: The *learning curve model*

$$N = C(1 - e^{-kt})$$

relates the number of tasks mastered (N) during the tth unit of time, where C and k are positive constants associated with the task. Suppose that learning vocabulary words in a foreign language follows the learning curve expressed by

$$N = 40(1 - e^{-0.2t})$$

where N predicts the number of words a student can learn during the tth day of study.
(a) According to this model, how many words is a foreign language student expected to learn on the 5th day? 15th day?
GU (b) Use a graphing utility to graph N as a function of t and interpret the graph as $t \to +\infty$.
(c) Use the graph from part (b) to predict the day when 35 words will be mastered.

41. Social Science—Spread of a Rumor: Suppose that sociologists estimate that if a person starts a rumor in a small town with a population of 20,000, the rumor will spread according to the function

$$N = 20{,}000[1 + 1.4e^{(-t+3)}]^{-1} - 16{,}700$$

where N is the number of people who have heard the rumor t hours later $(t \geq 5)$.
(a) According to this function, how many people will have heard the rumor after 5 hours? 8 hours? 10 hours?
GU (b) Use a graphing utility to graph the function. Interpret the graph, and then use the TRACE feature to determine how many hours it will take for at least 2000 people to hear the rumor.

42. Animal Population Growth: Suppose that the initial deer population in a wildlife refuge is 160. It is predicted that this population will increase according to the *logistic growth model* given by

$$N = \frac{800}{1 + 4e^{-0.12t}}$$

where N is the number of deer expected in the refuge after t years.
(a) According to this model, how many deer will there be after 1 year? 3 years? 5 years?
GU (b) Use a graphing utility to graph N as a function of t. Then use the graph to estimate the largest deer population the refuge can hold; that is, find N as $t \to +\infty$.

Developing and Extending the Concepts

43. GU Use a graphing utility to sketch the graphs of $f(x) = e^x$ and

$$g(x) = 1 + x + \frac{x^2}{2} + \frac{x^3}{6}$$

on the same coordinate axes for $-3 \leq x \leq 3$. Compare the two graphs. What do you conclude?

44. Use a calculator to determine which is larger, $(e^e)^e$ or e^{e^e}.

45. The *hyperbolic sine function*, denoted by *sinh*, and the *hyperbolic cosine function*, denoted by *cosh*, are defined as follows:

$$\sinh x = \frac{e^x - e^{-x}}{2} \quad \text{and} \quad \cosh x = \frac{e^x + e^{-x}}{2}$$

These functions are used in engineering.
(a) Show algebraically that $(\cosh x)^2 - (\sinh x)^2 = 1$ for every number x.
GU (b) Use a graphing utility to graph the function $f(x) = (\cosh x)^2 - (\sinh x)^2$ to demonstrate the validity of the result in part (a) graphically.

46. Let $f(x) = e^x$.
(a) Show that

$$\frac{f(x+h) - f(x)}{h} = e^x\left(\frac{e^h - 1}{h}\right)$$

is the slope of the line containing the points $(x, f(x))$ and $(x + h, f(x + h))$.
(b) Use this result to express the slope of the line containing $(0, 1)$ and (h, e^h) as a function of h.
GU (c) Use a graphing utility to graph the function found in part (b), and then use the graph to determine what value the slope approaches as $h \to 0^+$.

47. (a) Simplify the expression below algebraically.

$$\frac{(e^x + e^{-x})^2 - (e^x - e^{-x})^2}{(e^x + e^{-x})^2}$$

GU (b) Graph

$$f(x) = \frac{(e^x + e^{-x})^2 - (e^x - e^{-x})^2}{(e^x + e^{-x})^2}$$

and the function defined by the simplified form from part (a) to display graphically that both expressions are equal.

48. Population Growth: The following data represent the population growth P of bacteria in a colony t hours after the start of an experiment.

Hour (t)	0	1	2	3	4	5
Population (P)	37	100	262	747	2065	5500

(a) Construct a scattergram that exhibits the given data.
(b) Determine how well the function $P = 37.19e^{1.017t}$ predicts the data.
(c) Sketch the graph of the function P in part (b).
(d) Assuming the function P fits the data exhibited in part (a), predict the population of bacteria 8 hours after the start of the experiment.

OBJECTIVES

1. Define logarithms.
2. Evaluate logarithms—base 10 and base *e*.
3. Graph logarithmic functions.
4. Solve applied problems.

4.3 Logarithmic Functions

In Section 4.1 we observed that the exponential function $f(x) = b^x$, for $b > 0$ and $b \neq 1$, is one-to-one, and so it has an inverse. Its inverse is called a *logarithmic function*. In this section we investigate the graphs and properties of logarithmic functions.

Defining Logarithms

A logarithm is actually an exponent. For instance, consider the expression $49 = 7^2$. We refer to 2 as the *logarithm* of 49 with base 7, and we write $\log_7 49 = 2$. The logarithm 2 is the exponent to which 7 is raised to get 49. In general, we have the following definition:

DEFINITION

LOGARITHM

Let $b > 0$ and $b \neq 1$. The **logarithm of x with base b**, which is represented by y, is defined by

$$y = \log_b x \quad \text{if and only if} \quad x = b^y$$

for every $x > 0$ and for every real number y.

Thus, the logarithm y is the exponent to which b is raised to get x.

The two equations in the above definition are equivalent and as such can be used interchangeably. The first equation is in *logarithmic form* and the second is in *exponential form*. The following diagram is helpful when changing from one form to the other:

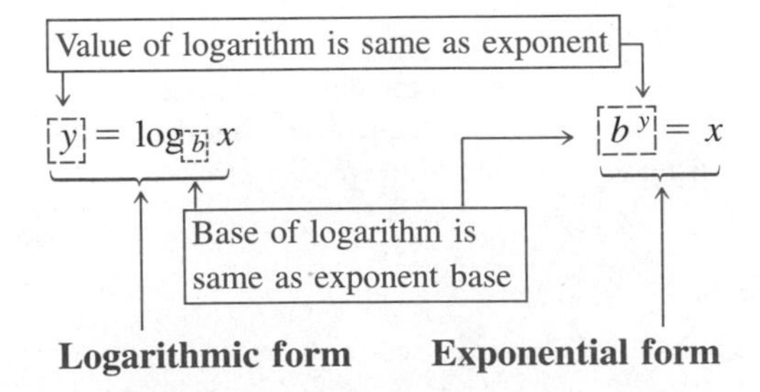

TABLE 1

Examples—Equivalent Logarithmic and Exponential Forms

Logarithmic Form	Exponential Form
$\log_2 16 = 4$	$2^4 = 16$
$\log_9 3 = \frac{1}{2}$	$9^{1/2} = 3$
$\log_5 \frac{1}{25} = -2$	$5^{-2} = \frac{1}{25}$
$\log_{10} 17 = t$	$10^t = 17$
$\log_3(2x - 5) = y$	$3^y = 2x - 5$

Table 1 shows examples of equivalent forms.

Since exponential functions are one-to-one, it follows that if $b > 0$ and $b \neq 1$, then we have the following property:

ONE-TO-ONE PROPERTY FOR EXPONENTS

$$b^u = b^v \quad \text{if and only if} \quad u = v$$

For instance, if $3^x = 3^{2x-1}$, then $x = 2x - 1$, so $x = 1$. At times, we can find the numerical value of a logarithm easily by converting to exponential form and then using the one-to-one property, as the next example shows.

EXAMPLE 1 **Evaluating logarithms**

Evaluate each logarithm.

(a) $\log_4 64$ (b) $\log_3 \frac{1}{9}$ (c) $\log_e e^4$ (d) $\log_{1/4} 16$

Solution

(a) We can represent the value of $\log_4 64$ by u; that is, we let $u = \log_4 64$. Then $4^u = 64$. Since $64 = 4^3$, it follows that $u = 3$. So $\log_4 64 = 3$.
(b) Let $u = \log_3 \frac{1}{9}$. Then $3^u = \frac{1}{9} = 3^{-2}$, so $u = -2$ and $\log_3 \frac{1}{9} = -2$.
(c) Let $u = \log_e e^4$. Since $e^u = e^4$, $u = 4$ and $\log_e e^4 = 4$.
(d) Let $u = \log_{1/4} 16$. Then

$$\left(\frac{1}{4}\right)^u = 16 \quad \text{or} \quad \left(\frac{1}{4}\right)^u = (4^{-1})^u = 4^{-u} = 4^2$$

So $-u = 2$ or $u = -2$ and $\log_{1/4} 16 = -2$.

Since $b^0 = 1$ for $b \neq 0$ and $b^1 = b$, it follows from the definition of a logarithm that

$$\log_b 1 = 0 \quad \text{and} \quad \log_b b = 1$$

EXAMPLE 2 **Solving logarithmic equations**

Solve each logarithmic equation for x by using its exponential form.

(a) $\log_3 x = 4$ (b) $\log_5 \frac{1}{125} = 3x + 2$
(c) $\log_4(2x - 1) = \frac{1}{2}$ (d) $y = 5 + 8 \log_e(3x)$

Solution We rewrite the given equation in exponential form and then solve the resulting equation.

(a) $\log_3 x = 4$ is equivalent to $3^4 = x$, so $x = 81$.
(b) $\log_5 \frac{1}{125} = 3x + 2$ is equivalent to $5^{3x+2} = \frac{1}{125}$, so we have

$$\begin{aligned} 5^{3x+2} &= 5^{-3} && \text{Rewrite } \tfrac{1}{125} \text{ as } 5^{-3} \\ 3x + 2 &= -3 && \text{One-to-one property} \\ x &= -\tfrac{5}{3} && \text{Solve for } x \end{aligned}$$

(c) $\log_4(2x - 1) = \frac{1}{2}$ is equivalent to $4^{1/2} = 2x - 1$, that is, $2 = 2x - 1$, so $x = \frac{3}{2}$.
(d) Since $y = 5 + 8 \log_e(3x)$, then

$$\frac{y - 5}{8} = \log_e(3x)$$

This latter equation is equivalent to $3x = e^{(y-5)/8}$, so $x = \frac{1}{3}e^{(y-5)/8}$.

Evaluating Logarithms—Base 10 and Base *e*

The base of a logarithmic function can be any positive number except 1. However, the two bases that are most widely used are 10 and *e*. A logarithm with base 10 is called a **common logarithm**. Its value at x is denoted by $\mathbf{log}\ \boldsymbol{x}$, that is,

$$\log x = \log_{10} x$$

A logarithm with base *e* is called a **natural logarithm**, and its value at x is denoted by $\mathbf{ln}\ \boldsymbol{x}$, that is,

$$\ln x = \log_e x$$

Since $\log_b b = 1$, it follows that

$$\log 10 = 1 \quad \text{and} \quad \ln e = 1$$

Sometimes we can use the definition of logarithms to evaluate common and natural logarithms easily. For instance:

$$\begin{aligned} \log 1000 &= 3 && \text{Since } 10^3 = 1000 \\ \log 0.0001 &= -4 && \text{Since } 10^{-4} = 0.0001 \\ \ln e^5 &= 5 && \text{Since } e^5 = e^5 \\ \ln \frac{1}{e^2} &= -2 && \text{Since } e^{-2} = \frac{1}{e^2} \end{aligned}$$

For other situations, we use a calculator with a log key for base 10 and an "el-en" ln key for base *e* to approximate values of logarithms. Table 2 lists some calculator approximations of logarithms (rounded to four decimal places) along with the equivalent exponential forms.

TABLE 2

Examples—Calculator Values of Logarithms

Approximate Logarithm Value	Exponential Form
$\log 4819 = 3.6830$	$10^{3.6830} = 4819$
$\log 0.897 = -0.0472$	$10^{-0.0472} = 0.897$
$\ln 45 = 3.8067$	$e^{3.8067} = 45$
$\ln 0.057 = -2.8647$	$e^{-2.8647} = 0.057$

Next, we consider the problem of finding the value of x when the value of $\log x$ or $\ln x$ is given.

EXAMPLE 3 Solving logarithmic equations

Use a calculator to approximate the value of x to four decimal places.

(a) $\log x = 0.7235$ (b) $\ln x = 1.4674$ (c) $\log(x - 1) = -1.7$

Solution Using a calculator and rounding the results to four decimal places, we get:

(a) $\log x = 0.7235$ is equivalent to $x = 10^{0.7235}$, so $x = 5.2905$.
(b) $\ln x = 1.4674$ is equivalent to $x = e^{1.4674} = 4.3379$.
(c) $\log(x - 1) = -1.7$ is equivalent to $x - 1 = 10^{-1.7}$, so $x = 1 + 10^{-1.7}$ or $x = 1.0200$.

Graphing Logarithmic Functions

We can use the notion of inverse functions to graph logarithmic functions. Consider the exponential function $f(x) = 2^x$. Since f is one-to-one, it has an inverse f^{-1}, which we can determine as follows:

First, we write the equation that defines f as

$$y = 2^x$$

Next, we express x in terms of y. In this situation, we use the equivalent logarithmic form $x = \log_2 y$, to get

$$f^{-1}(y) = \log_2 y$$

Finally, we use x to represent the independent variable and write

$$f^{-1}(x) = \log_2 x$$

This latter equation describes a logarithmic function to the base 2.

The argument used above can be applied to any exponential function $f(x) = b^x$, where $b > 0$ and $b \neq 1$. In general, we have the following property:

The inverse of the exponential function $f(x) = b^x$ is the logarithmic function defined by $g(x) = \log_b x$, and vice versa.

Table 3 lists examples of exponential functions along with their inverses.

TABLE 3

Examples—Inverses of Exponential Functions

Exponential Function	Inverse Function
$f(x) = 4^x$	$f^{-1}(x) = \log_4 x$
$f(x) = \left(\frac{1}{3}\right)^x$	$f^{-1}(x) = \log_{1/3} x$
$f(x) = 10^x$	$f^{-1}(x) = \log x$
$f(x) = e^x$	$f^{-1}(x) = \ln x$

Use a calculator to attempt to evaluate log(−4) and ln 0. Describe what happens. Consult your manual.

Since $g(x) = \log_b x$ is the inverse of $f(x) = b^x$, it follows that the domain of f, which includes all real numbers, is the range of g. Also, the range of f, the set of positive real numbers, is the domain of g. Thus, in the real number system, we can only evaluate the logarithm of a positive number. For example, the domain of the function

$$f(x) = \log(2x - 1)$$

in the real number system includes all values of x such that $2x - 1 > 0$, or $x > \frac{1}{2}$. That is, the domain includes all numbers in the interval $\left(\frac{1}{2}, \infty\right)$.

There are different methods for graphing logarithmic functions. One way is to use symmetry. Recall that in Section 2.5 we learned that the graphs of a function and its inverse are reflections of each other across the line $y = x$. We can use this property to graph a logarithmic function $y = \log_b x$. For example, let's consider the function $F(x) = \log_2 x$. We start by graphing its inverse $g(x) = 2^x$. Then we reflect this graph across the line $y = x$ to obtain the graph of $F(x) = \log_2 x$ (Figure 1).

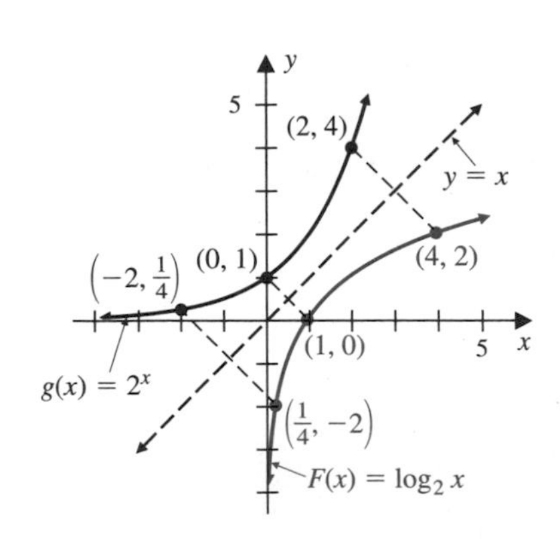

FIGURE 1

Another approach to graphing a logarithmic function $y = \log_b x$ is to first convert its equation to the equivalent exponential form $x = b^y$. Then use the point-plotting method to sketch the graph of the equation $x = b^y$.

EXAMPLE 4 **Graphing logarithmic functions using exponential forms**

Graph each function by using its exponential form. Also, determine the domain and range.

(a) $f(x) = \log_4 x$
(b) $g(x) = \log_{1/4} x$

Solution

(a) To graph f we first convert $y = f(x) = \log_4 x$ to its equivalent form, $x = 4^y$. Next, we form a table of values for $x = 4^y$ by substituting values for y and then finding the corresponding x values (Table 4). Finally, we plot the points and connect them with a smooth curve to get the graph of f (Figure 2a). The domain is the interval $(0, \infty)$ and the range includes all real numbers $\mathbb{R}$.

(b) Similarly, $y = g(x) = \log_{1/4} x$ is converted to the equivalent form $x = \left(\frac{1}{4}\right)^y$. Then this equation is graphed by point-plotting as in part (a) (Table 4, Figure 2b). The domain consists of all values of x in the interval $(0, \infty)$ and the range includes all real numbers $\mathbb{R}$.

TABLE 4

y	$x = 4^y$	$x = \left(\frac{1}{4}\right)^y$
0	1	1
1	4	$\frac{1}{4}$
2	16	$\frac{1}{16}$

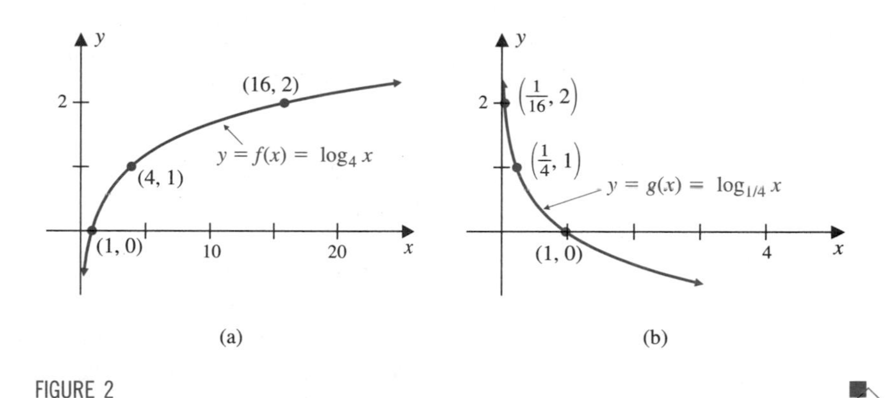

FIGURE 2

The general graphs and properties of the logarithmic functions $f(x) = \log_b x$ have similar general characteristics depending on whether $b > 1$ or $0 < b < 1$ (Table 5).

TABLE 5

General Graphs and Properties of $f(x) = \log_b x$

Properties	General Graphs
1. The domain is the interval $(0, \infty)$. **2.** The range includes all real numbers. **3.** The function increases everywhere if $b > 1$ and decreases everywhere if $0 < b < 1$. **4.** The y axis is a vertical asymptote. **5.** The function is one-to-one. **6.** The graph contains the point $(1, 0)$.	$f(x) = \log_b x$, $b > 1$ (y, x, 1) $f(x) = \log_b x$, $0 < b < 1$ (y, x, 1)

Examples of specific graphs for $b > 1$ and $0 < b < 1$ are shown in Figures 3a and 3b, respectively (at the top of page 256).

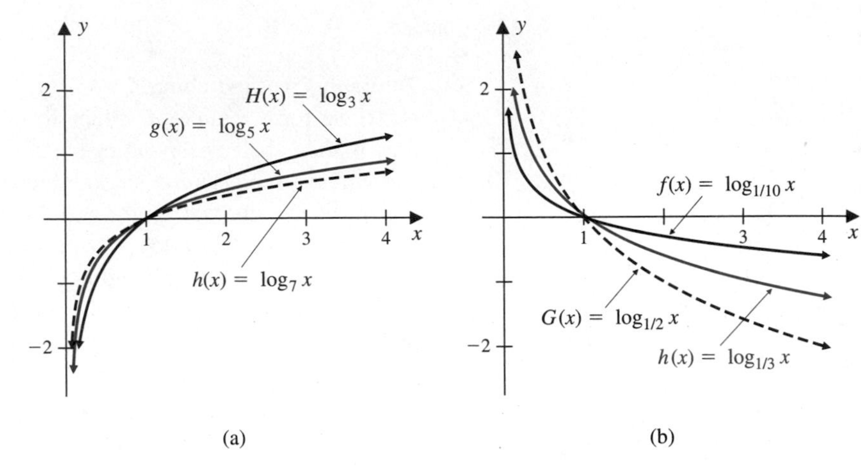

FIGURE 3

We can use transformations of graphs of the functions $y = \log_b x$ to graph other logarithmic functions.

EXAMPLE 5 **Using transformations to graph**

Use a transformation to sketch the graph of $G(x) = \log_4(x - 1)$. Determine the domain and range of G and find the vertical asymptote of its graph.

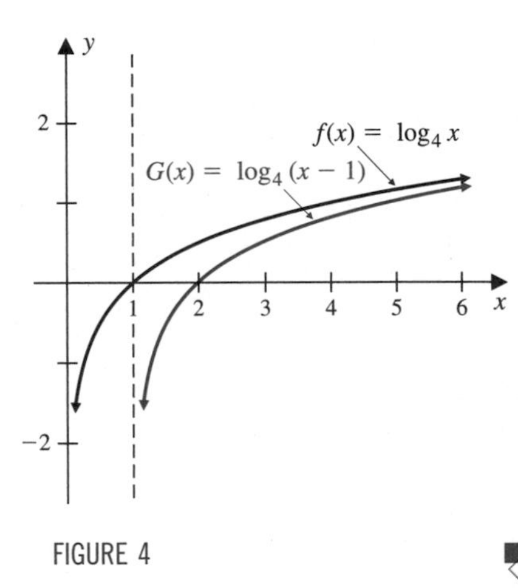

FIGURE 4

Solution The graph of G can be obtained by shifting the graph of $f(x) = \log_4 x$ to the right 1 unit (Figure 4). Since the domain of f is $(0, \infty)$, it follows that the domain of G is $(1, \infty)$, and the line $x = 1$ is a vertical asymptote. The graph indicates that the range of G consists of all real numbers $\mathbb{R}$.

In the next section we will use a graphing utility to graph logarithmic functions with bases different from 10 or e.

Functions involving common and natural logarithmic expressions can be graphed with a graphing utility.

EXAMPLE 6 **Graphing by using a graphing utility**

Given the function $f(x) = \ln(x^2 - x - 6)$:

(a) Use a graphing utility to sketch the graph of f.
(b) Use the graph to determine the domain and any vertical asymptotes.
(c) Confirm algebraically the domain found in part (b).

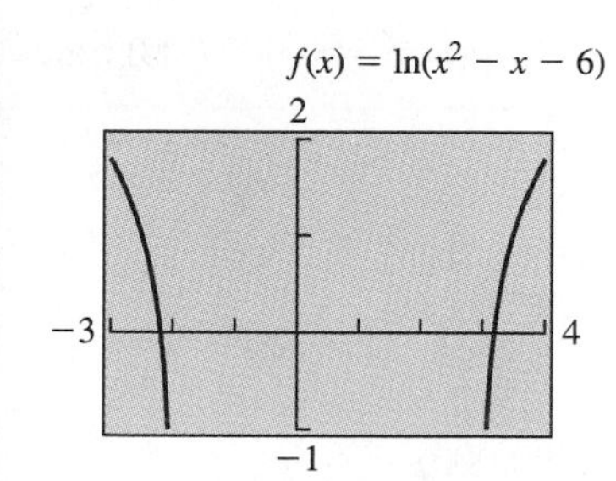

FIGURE 5

Solution

(a) Figure 5 shows a viewing window for the graph of f.
(b) The graph suggests that the domain includes all values of x such that $x < -2$ or $x > 3$. Also, the lines $x = -2$ and $x = 3$ appear to be vertical asymptotes.
(c) To find the domain of f algebraically, we use the fact that $\ln(x^2 - x - 6)$ is defined only if $x^2 - x - 6 = (x - 3)(x + 2) > 0$. Solving this inequality, we get $x < -2$ or $x > 3$. Thus, the domain of f consists of all real numbers in the intervals $(-\infty, -2)$ or $(3, \infty)$. This conclusion confirms the result in part (b).

Solving Applied Problems

Applications as diverse as measuring the strength of an earthquake, determining the acidity of a liquid, and modeling the profit from a business make use of logarithmic functions.

EXAMPLE 7 Measuring the strength of an earthquake

The **Richter scale** provides us with a way of grading the strength of an earthquake on a scale from 1 to 10—the larger the number, the more severe the earthquake. For a given earthquake, suppose that the largest seismic wave recorded on a seismograph is I and the smallest seismic wave recorded for the area is I_0. Then the **magnitude** M of the earthquake is given by

$$M = \log \frac{I}{I_0}$$

where I is called the **amplitude** of the earthquake, I_0 is called the **zero-level amplitude**, and the ratio I/I_0 is called the **intensity** of the earthquake.

(a) The 1964 Alaska earthquake had a magnitude of 8.6. Find the intensity of the earthquake and express its amplitude in terms of the zero-level amplitude.
(b) The 1992 earthquake in Cairo, Egypt, had a magnitude of 7.3. Compare the intensity of this earthquake to the intensity of the 1964 Alaska earthquake.

Solution

(a) Substituting $M = 8.6$ into the formula

$$M = \log \frac{I}{I_0}$$

we get

$$8.6 = \log \frac{I}{I_0}$$

Converting to exponential form, we find that the intensity is given by

$$\frac{I}{I_0} = 10^{8.6} \qquad \text{so that} \qquad I = 10^{8.6}\, I_0$$

That is, the amplitude of the earthquake was $10^{8.6}$ (approx. 400,000,000) times the zero-level amplitude.

(b) For the Cairo earthquake, we have

$$7.3 = \log \frac{I}{I_0}$$

so the intensity is

$$\frac{I}{I_0} = 10^{7.3} \text{ (approx. 20,000,000)}$$

Since $10^{8.6}/10^{7.3} = 10^{1.3}$, it follows that $10^{8.6} = (10^{7.3})(10^{1.3})$, so the intensity of the Alaska earthquake was $10^{1.3}$ or approximately 20 times the intensity of the Cairo earthquake.

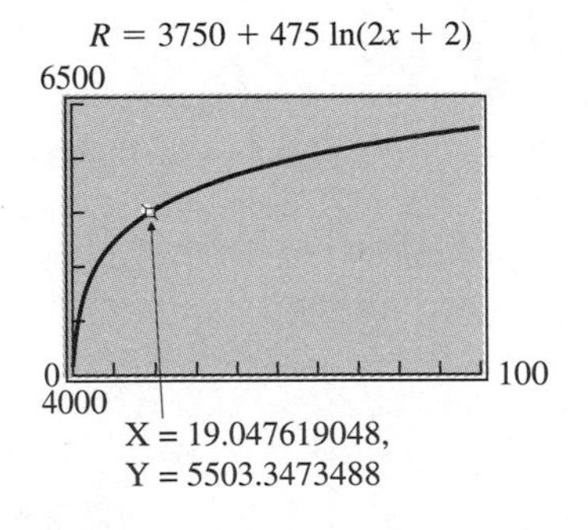

FIGURE 6

There are applications in business where logarithmic functions are used to model the situation. For instance, suppose a computer software company discovers that its annual revenue R (in thousands of dollars) is related to its research and development expenditures x (in tens of thousands of dollars) by the model

$$R = 3750 + 475 \ln(2x + 2)$$

A viewing window of the graph of the function is shown in Figure 6. Initially (when $x = 0$), R is approximately 4079, so the revenue exceeds \$4 million even if no money is expended for research and development. By using the TRACE feature and monitoring the changing x and y values, we discover that the revenue R exceeds \$5.5 million when the research and development expenditure x exceeds about \$190,000. After this, the curve begins to level off, indicating that revenue increases at a slower pace with greater expenditures.

PROBLEM SET 4.3

Mastering the Concepts

In problems 1 and 2, write each exponential equation in logarithmic form.

1. **(a)** $5^3 = 125$ **(b)** $32^{1/5} = 2$
(c) $17 = e^t$ **(d)** $b^x = 13z + 1$

2. **(a)** $\left(\frac{1}{2}\right)^{-3} = 8$ **(b)** $9^{3/2} = 27$
(c) $10^w = x - 3$ **(d)** $a^8 = y - 5$

In problems 3 and 4, write each logarithmic equation in exponential form.

3. **(a)** $\log_9 81 = 2$ **(b)** $\log 0.0001 = -4$
(c) $\log_c 9 = w$ **(d)** $\ln \frac{1}{2} = -1 - 3x$

4. **(a)** $\log_{36} 216 = \frac{3}{2}$ **(b)** $\ln s = t$
(c) $\log x = \sqrt{2}$ **(d)** $\log_b 7 = -3 + 8x$

In problems 5 and 6, solve each equation by expressing both sides in terms of the same base.

5. **(a)** $3^{3x} = 27$ **(b)** $4^{3x-1} = 8^{x-1}$
(c) $2^{-100x} = (0.5)^{x-4}$ **(d)** $3^{x^2+x} = 1$

6. **(a)** $6^{2x-1} = 216$ **(b)** $16^{3x} = 8^{2x-1}$
(c) $6^{7-x} = 6^{2x+1}$ **(d)** $2^{15x^2+14x} = 256$

In problems 7 and 8, find the exact value of each logarithm.

7. **(a)** $\log_2 \frac{1}{8}$ **(b)** $\ln e^4$
(c) $\log \sqrt[3]{10}$ **(d)** $\log_8 4$

8. **(a)** $\log_7 7$ **(b)** $\log_5 \frac{1}{625}$
(c) $\log \sqrt{10}$ **(d)** $\ln \sqrt[3]{e}$

In problems 9–14, solve each logarithmic equation for x by using its exponential form.

9. **(a)** $\log_6 x = 2$ **(b)** $\log_x \frac{1}{4} = -\frac{1}{2}$

10. **(a)** $\log_{27} x = \frac{1}{3}$ **(b)** $\log_x 4 = \frac{2}{3}$

11. **(a)** $\log_2 16 = 3x - 5$ **(b)** $\log_3(x + 1) = 2$

12. **(a)** $\log_5(5x - 1) = -2$
(b) $\log_5(x^2 - 4x) = 1$

13. **(a)** $y = 3 - \log x$ **(b)** $y = 8 + \ln(2x + 1)$

14. **(a)** $y = \ln(x - 1)$ **(b)** $y = 7 + \log(4 - x)$

In problems 15 and 16, use a calculator to approximate the value of each logarithm to four decimal places. Express the result in the equivalent exponential form.

15. **(a)** $\log 39.2$ **(b)** $\ln 961$
(c) $\log \frac{3}{4}$ **(d)** $\ln 0.23$
(e) $\ln 10$

16. **(a)** $\ln 39.2$ **(b)** $\log 961$
(c) $\ln \frac{3}{4}$ **(d)** $\log 0.23$
(e) $\log e$

In problems 17 and 18, use a calculator to approximate the value of x to four decimal places.

17. **(a)** $\ln x = 2.37$ **(b)** $\log x = 0.4137$
(c) $\ln x = -2.5$ **(d)** $\ln(2x - 1) = -3.8$

18. **(a)** $\log x = 2.37$ **(b)** $\ln x = 0.4137$
(c) $\log x = -2.5$ **(d)** $\log(2x - 1) = -3.8$

In problems 19 and 20, find the domain of each function.

19. **(a)** $f(x) = \ln(5x + 1)$ **(b)** $g(x) = \log(x^2 - x)$
(c) $h(x) = \log_3\left(\frac{x - 1}{x + 1}\right)$
(d) $f(x) = \log\left(\frac{1}{x^2 - 1}\right)$

20. **(a)** $g(x) = \log(1 - 3x)$
(b) $h(x) = \ln(x^2 - 2x + 1)$
(c) $h(x) = \ln(2 - x^2)$ **(d)** $f(x) = \log_{1/5}\left(\frac{1}{x^2 - 9}\right)$

In problems 21–24, find the inverse of the given function, and then use symmetry to graph both functions on the same coordinate system. Also find the domain and range of each inverse function.

21. $f(x) = 5^x$ **22.** $g(x) = \left(\frac{1}{7}\right)^x$

23. $h(x) = \log_{1/10} x$ **24.** $F(x) = \log x$

25. Each curve in Figure 7 is the graph of a function of the form $y = \log_b x$. Find the base if its graph contains the given point.

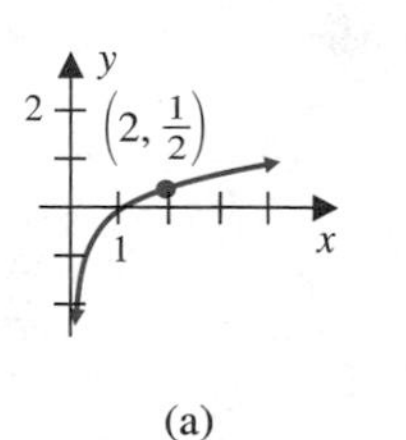

(a)

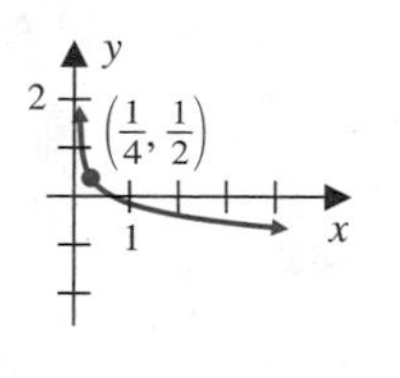

(b)

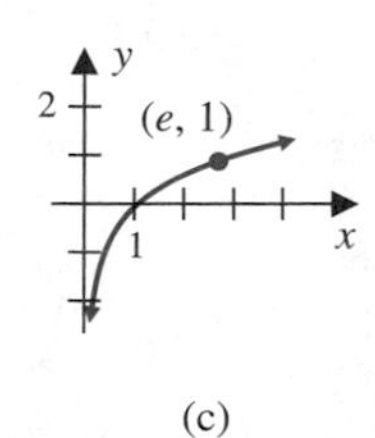

(c)

FIGURE 7

26. Find the inverse of each function.
(a) $f(x) = 3^{2x-5}$ **(b)** $g(x) = 2\ln(5x + 3)$

In problems 27 and 28, graph all three given logarithmic functions on the same coordinate system. Use the graphs to describe the relative steepness and limit behavior of the functions in terms of the different bases.

27. $f(x) = \log_2 x$; $g(x) = \log_{2.5} x$; $h(x) = \log_3 x$

28. $f(x) = \log_{1/3} x$; $g(x) = \log_{1/4} x$; $h(x) = \log_{1/5} x$

In problems 29–36, graph the given function by using transformations of the graph of $y = \log_b x$ with the same base. Determine the domain and vertical asymptote.

29. $f(x) = 1 + \ln x$ **30.** $g(x) = 3\log x$

31. $g(x) = \frac{1}{3}\log_5 x - 2$ **32.** $h(x) = 1 - \log_7 x$

33. $h(x) = \log(x + 3)$ **34.** $f(x) = \frac{1}{2}\ln(x - 1)$

35. $f(x) = -2\ln(x - e)$ **36.** $g(x) = \log(x - 10) - 1$

GU In problems 37–42, use a graphing utility to sketch the graph of each function. Use the graph to find the domain and any vertical asymptotes. Confirm the domain algebraically.

37. $f(x) = x\ln(x - 2)$ **38.** $h(x) = 3[\log(3 - 2x)]^2$

39. $g(x) = \log(x^2 - 5x + 4)$

40. $f(x) = \ln(x^2 + 3x + 2)$

41. $h(x) = \ln(x^2 + 1)$ **42.** $g(x) = \log\left(\frac{1 - x}{1 + x}\right)$

Applying the Concepts

43. Magnitude of an Earthquake: The amplitude of the earthquake in Japan in May 1983 was $10^{7.7}$ times the zero-level amplitude. Find the magnitude of the earthquake on the Richter scale.

44. Earthquake Comparisons: The 1988 Armenian earthquake had a magnitude of 6.8 on the Richter scale, and the 1985 earthquake in Mexico City had a magnitude of 7.8. How many times more powerful was the intensity of the Mexico City earthquake than the Armenian earthquake?

In problems 45 and 46, the concentration of hydrogen ions in a substance is denoted by $[H^+]$, measured in moles per liter. The **pH** of a substance is defined by the logarithmic function $\text{pH} = -\log [H^+]$. This function is used to measure the *acidity* of the substance. The pH of neutral distilled water is 7. A substance with a pH of less than 7 is known as an *acid*, whereas a substance with a pH of more than 7 is called a *base*.

45. pH Scale: Find the pH of each substance and determine whether it is an acid or a base. Round off the answers to one decimal place.
(a) Beer: $[H^+] = 3.16 \times 10^{-3}$ mole per liter
(b) Eggs: $[H^+] = 1.6 \times 10^{-8}$ mole per liter
(c) Milk: $[H^+] = 4 \times 10^{-7}$ mole per liter

46. pH Scale:
(a) Express the hydrogen ion concentration $[H^+]$ as a function of pH.
(b) Use the function in part (a) to find the hydrogen ion concentration (in moles per liter) of: calcium hydroxide, $\text{pH} = 13.2$; vinegar, $\text{pH} = 3.1$; and tomatoes, $\text{pH} = 4.2$. Round off the answers to three decimal places.
GU **(c)** Use a graphing utility to graph the function in part (a) and interpret what happens as $\text{pH} \to 0^+$.

In problems 47 and 48, the loudness L of a sound is measured by

$$L = 10 \log \frac{I}{I_0}$$

Here, L is the number of *decibels,* I is the intensity of the sound (in watts per square meter), and I_0 is the smallest intensity that can be heard. Suppose the intensity I_0 is 10^{-12} watt per square meter.

47. Sound Intensity:
(a) Find the number of decibels of the noise of a truck passing a pedestrian at the side of the road if the sound intensity of the truck is 10^{-3} watt per square meter.
(b) Find the number of decibels of sound from a jet plane with sound intensity 8.3×10^2 watts per square meter. Round off the answer to two decimal places.

48. Sound Intensity:
(a) The threshold of pain from loud sound is considered to be about 120 decibels. Find the intensity of such a sound in watts per square meter.
(b) Find the intensity of a whisper in watts per square meter if the whisper is 25 decibels.

49. Advertising: A company has determined that when its monthly advertising expenditure is x (in thousands of dollars), then the total monthly sales S (in thousands of dollars) is given by the model

$$S = \frac{2000 \ln(x + 5)}{\ln 10}$$

(a) What are the expected monthly sales (to the nearest thousand dollars) if the monthly expenditure for advertising is $10,000? $50,000? $100,000?
GU **(b)** Use a graphing utility to graph S as a function of x. Then use the TRACE feature to determine the minimum amount (to the nearest thousand dollars) that should be spent on advertising if the monthly sales are to exceed $2 million, $3 million, $4 million.

50. Manufacturing: A small auto parts supply company's weekly profit P (in dollars) is given by the function

$$P = 1828 + 914 \log x$$

where x is the number of parts produced each minute.
(a) Find the weekly profit (to the nearest dollar) if the number of parts produced each minute is 10, 50, 100, 500, and 1000.
GU **(b)** Graph P as a function of x and use the TRACE feature to determine how many parts per minute must be produced in order to have the weekly profit exceed $3500, $4000, and $4500.

Developing and Extending the Concepts

51. GU Suppose that $0 < a < b$. Use a graphing utility to sketch the graph of $f(x) = 10x \log x$. Use the graph to determine whether $10a \log a < 10b \log b$.

52. GU Use a graphing utility to sketch the graphs of $y = \ln(x + 1)$ and

$$y = x - \frac{x^2}{2} + \frac{x^3}{3} - \frac{x^4}{4}$$

for $|x| < 1$ on the same coordinate system. Compare the two graphs. What do you conclude?

53. If $P_1 = -\log A$, $P_2 = -\log B$, and $0 < A < B$, then which is larger, P_1 or P_2? Explain.

54. Graph $f(x) = 2$, $g(x) = \log_2 x$, and $h(x) = 3$ on the same coordinate axes. Use the graphs to solve $2 \leq \log_2 x \leq 3$.

55. GU Use a graphing utility to graph $f(x) = \log x - \ln x$. Then use the graph to solve:
(a) $\log x = \ln x$ **(b)** $\log x > \ln x$
(c) $\log x < \ln x$

OBJECTIVES

1. Use the properties of logarithms.
2. Solve equations containing logarithms.
3. Use the change-of-base formula.
4. Solve applied problems.

4.4 Properties of Logarithms

In this section, we'll use the properties of logarithms to simplify expressions and solve equations.

Using the Properties of Logarithms

From the definition of logarithms, we know that for $x > 0$ the equation $y = \log_b x$ is equivalent to the equation $b^y = x$. Substituting $\log_b x$ for y in the equation $b^y = x$, we get the property

$$b^{\log_b x} = x \qquad x > 0$$

For base 10 and base e, this property takes on the forms

$$10^{\log x} = x \quad \text{and} \quad e^{\ln x} = x \qquad x > 0$$

For instance, $e^{\ln(3x+5)} = 3x + 5$.

Since logarithms are exponents, we can use the properties of exponents to derive some basic properties of logarithms.

BASIC PROPERTIES OF LOGARITHMS

Suppose that M, N, and b are positive real numbers, where $b \neq 1$, and r is any real number. Then:

(i) $\log_b MN = \log_b M + \log_b N$ (ii) $\log_b \dfrac{M}{N} = \log_b M - \log_b N$

(iii) $\log_b N^r = r \log_b N$

We verify these properties as follows:

(i) Since $M = b^{\log_b M}$ and $N = b^{\log_b N}$,

$$MN = b^{\log_b M} \cdot b^{\log_b N} = b^{\log_b M + \log_b N}$$

By converting the equation to logarithmic form, we get

$$\log_b MN = \log_b M + \log_b N$$

(ii) Since $M = (M/N)N$, we have

$$\log_b M = \log_b\left[\left(\frac{M}{N}\right)N\right]$$

$$\log_b M = \log_b \frac{M}{N} + \log_b N \quad \text{Property (i)}$$

$$\log_b \frac{M}{N} = \log_b M - \log_b N \quad \text{Isolate } \log_b \frac{M}{N} \text{ on one side}$$

(iii) Since $N = b^{\log_b N}$, then

$$N^r = (b^{\log_b N})^r = b^{r \log_b N} \quad (a^m)^n = a^{mn}$$

After converting to logarithmic form, we get

$$\log_b N^r = r \log_b N$$

The basic properties of logarithms for the special cases of the common and natural logarithms are stated in Table 1.

TABLE 1

Basic Properties of Common and Natural Logarithms

Common Logarithms *Base 10*	Natural Logarithms *Base e*
(i) $\log MN = \log M + \log N$	(i) $\ln MN = \ln M + \ln N$
(ii) $\log \frac{M}{N} = \log M - \log N$	(ii) $\ln \frac{M}{N} = \ln M - \ln N$
(iii) $\log N^r = r \log N$	(iii) $\ln N^r = r \ln N$

In Property (iii), if the value of N is the same as the base, then we get $\log_b b^r = r \log_b b = r \cdot 1 = r$, so

$$\boxed{\log_b b^r = r}$$

Specifically for $b = 10$ or $b = e$, we have

$$\boxed{\log 10^r = r \quad \text{and} \quad \ln e^r = r}$$

For instance, $\ln e^{x^2-x} = x^2 - x$.

The next two examples illustrate how the properties of logarithms are used to manipulate logarithmic expressions algebraically.

EXAMPLE 1 Using the properties of logarithms

Write each expression as a sum or difference of multiples of logarithms.

(a) $\log_8 \dfrac{x}{5}$ (b) $\log(x^7y^{11})$ (c) $\ln\left[\dfrac{(y+7)^3}{\sqrt{y}}\right]$

Solution We assume that all numbers whose logarithms are taken are positive.

(a) $\log_8 \dfrac{x}{5} = \log_8 x - \log_8 5$ Property (ii)

(b) $\log(x^7y^{11}) = \log x^7 + \log y^{11}$ Property (i)

$= 7 \log x + 11 \log y$ Property (iii)

(c) $\ln\left[\dfrac{(y+7)^3}{\sqrt{y}}\right] = \ln(y+7)^3 - \ln\sqrt{y}$ Property (ii)

$= \ln(y+7)^3 - \ln y^{1/2}$ $\sqrt{y} = y^{1/2}$

$= 3\ln(y+7) - \frac{1}{2}\ln y$ Property (iii)

EXAMPLE 2 Combining logarithmic expressions

Write each expression as a single logarithm.

(a) $2\ln x - 3\ln y$, where $x > 0$ and $y > 0$.
(b) $\log(c^2 - cd) - \log(2c - 2d)$, where $c^2 - cd > 0$ and $2c - 2d > 0$

Solution

(a) $2\ln x - 3\ln y = \ln x^2 - \ln y^3$ Property (iii)

$= \ln\left(\dfrac{x^2}{y^3}\right)$ Property (ii)

(b) $\log(c^2 - cd) - \log(2c - 2d) = \log\left[\dfrac{c^2 - cd}{2c - 2d}\right]$ Property (ii)

$= \log\left[\dfrac{c(c-d)}{2(c-d)}\right] = \log\left(\dfrac{c}{2}\right)$ Provided $c \neq d$

In calculus, we often encounter expressions involving exponential and logarithmic functions with base e that have to be simplified by using the properties of logarithms.

EXAMPLE 3 Simplifying expressions involving base e

Simplify each expression.

(a) $e^{3\ln(2x+1)}$ (b) $e^{4-2\ln x}$

Solution

(a) $e^{3\ln(2x+1)} = e^{\ln(2x+1)^3}$ Property (iii)

$= (2x+1)^3$ $e^{\ln u} = u$

(b) $e^{4-2\ln x} = e^4 \cdot e^{-2\ln x}$ $e^{u+v} = e^u \cdot e^v$

$= e^4 \cdot e^{\ln x^{-2}}$ Property (iii)

$= e^4 \cdot x^{-2}$ $e^{\ln u} = u$

$= \dfrac{e^4}{x^2}$

Solving Equations Containing Logarithms

To solve an equation involving logarithms we often use the following strategy:

Step 1. Use the properties of logarithms to rewrite the equation so that it contains one logarithm on one side.
Step 2. Convert the equation from step 1 to exponential form.
Step 3. Solve the resulting equation.

This procedure may introduce extraneous roots, so we need to check the solution found in step 3 in the original equation.

EXAMPLE 4 Solving a logarithmic equation
Solve the equation and check for extraneous roots.

$$\log_3(x+1) + \log_3(x+3) = 1$$

Solution We first note that the domain of the logarithmic function $f(x) = \log_3(x+1) + \log_3(x+3)$ includes all values of x such that $x > -1$ because we can only evaluate logarithms of positive real numbers. Next, we solve the equation as follows:

Step 1. $\log_3(x+1) + \log_3(x+3) = 1$ Given

$\log_3[(x+1)(x+3)] = 1$ Property (i)

Step 2. $(x+1)(x+3) = 3^1$ Exponential form

Step 3. $x^2 + 4x + 3 = 3$, $x^2 + 4x = 0$ Simplify

$x(x+4) = 0$ Factor

$x = 0$ or $x = -4$ Solve for x

Check for extraneous roots Since the domain is the set of all values $x > -1$, the number -4 is eliminated. We check $x = 0$:

$$\log_3(x+1) + \log_3(x+3) = \log_3 1 + \log_3 3 = 0 + 1 = 1$$

Thus, 0 is the only solution of the equation.

Using the Change-of-Base Formula

Most graphing utilities have logarithm function keys for base 10 and base e. In order to determine logarithms with other bases, we use the following *change-of-base formula*:

CHANGE-OF-BASE FORMULA

$$\log_b x = \frac{\log_a x}{\log_a b}$$

where a and b are positive real numbers different from 1, and x is positive.

We can verify the formula as follows:

$y = \log_b x$	Given
$b^y = x$	Exponential form
$\log_a b^y = \log_a x$	Take the logarithm of each side to base a
$y \log_a b = \log_a x$	Property (iii)
$y = \dfrac{\log_a x}{\log_a b}$	Solve for y
$\log_b x = \dfrac{\log_a x}{\log_a b}$	Since $y = \log_b x$

For instance, $\log_3 7$ can be approximated by using

$$\log_3 7 = \frac{\log 7}{\log 3} = 1.771 \qquad \text{or} \qquad \log_3 7 = \frac{\ln 7}{\ln 3} = 1.771$$

In general, $\dfrac{\log x}{\log b} = \dfrac{\ln x}{\ln b}$ where $b > 0$, $b \neq 1$, and $x > 0$.

We can use a graphing utility to graph a logarithmic function with a base different from 10 or e by rewriting the function in terms of base 10 or base e.

EXAMPLE 5 GU

Using a change of base to graph

Use the change-of-base formula along with a graphing utility to sketch the graph of $f(x) = \log_8 x$. Also determine the domain and any vertical asymptotes.

Solution First, we rewrite the function in terms of base 10 or base e. We'll select base 10. Thus,

$$f(x) = \log_8 x = \frac{\log x}{\log 8}$$

Next, we use a graphing utility to graph $f(x) = (\log x)/(\log 8)$. A viewing window of the graph is shown in Figure 1. Since we can evaluate logarithms only for positive real numbers, it follows that the domain includes all values x in the interval $(0, \infty)$. The y axis is a vertical asymptote.

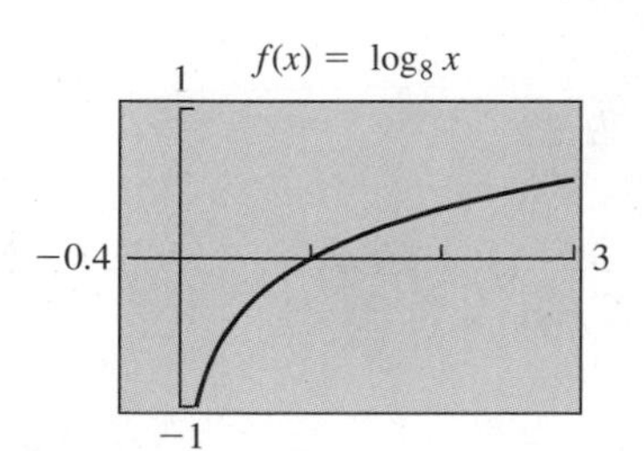

FIGURE 1

Solving Applied Problems

The properties of logarithms are used to rewrite formulas encountered in a variety of applications.

EXAMPLE 6 Determining weight loss

The weight W of a person (in pounds) after t days of dieting is related to the person's initial weight W_0 (in pounds) by the following mathematical model, called the *weight equation*:

$$\ln(W - W_1) = \ln(W_0 - W_1) - 0.005t$$

Here, W_1 is the number of calories consumed per day divided by the constant 17.5, the number of calories needed per pound per day.

(a) Solve this equation for W in terms of W_0, W_1, and t.

(b) Suppose that the initial weight of a person is 210 pounds, and that the person is put on a diet of 1800 calories per day. Express the person's weight W as a function of the number of days t after the dieting begins. How much weight (to the nearest pound) is expected to be lost after 60 days?

Solution

(a) We solve for W in terms of W_0, W_1, and t as follows:

$$\ln(W - W_1) = \ln(W_0 - W_1) - 0.005t \qquad \text{Given}$$

$$\ln(W - W_1) - \ln(W_0 - W_1) = -0.005t \qquad \text{Isolate } t \text{ expression}$$

$$\ln\left(\frac{W - W_1}{W_0 - W_1}\right) = -0.005t \qquad \text{Property (ii)}$$

$$\frac{W - W_1}{W_0 - W_1} = e^{-0.005t} \qquad \text{Exponential form}$$

$$\left.\begin{aligned} W - W_1 &= (W_0 - W_1)e^{-0.005t} \\ W &= (W_0 - W_1)e^{-0.005t} + W_1 \end{aligned}\right\} \qquad \text{Solve for } W$$

(b) The given conditions indicate that $W_0 = 210$ and $W_1 = 1800/17.5$. So, substituting into the equation found in part (a), we get

$$W = \left(210 - \frac{1800}{17.5}\right)e^{-0.005t} + \frac{1800}{17.5}$$

$$W = \left(\frac{750}{7}\right)e^{-0.005t} + \frac{720}{7}$$

Thus, we have expressed W as a function of t. After 60 days, $t = 60$, so the weight W (rounded to the nearest pound) is given by

$$W = \left(\frac{750}{7}\right)e^{-0.005(60)} + \frac{720}{7} = 182$$

indicating a weight loss of about $210 - 182 = 28$ pounds.

PROBLEM SET 4.4

Mastering the Concepts

In problems 1–14, write each expression as a sum or difference of multiples of logarithms.

1. $\log_3 x(x + 1)$

2. $\log_3 \dfrac{18}{x + 2}$

3. $\log_b \dfrac{x^9}{y^7}$

4. $\log_b x^2y^3$

5. $\log_b(x + 3)^4$

6. $\log x\sqrt{2x + 1}$

7. $\log \sqrt[6]{\dfrac{x^5}{y^2}}$

8. $\ln \dfrac{x^5y^2}{\sqrt[5]{z}}$

9. $\ln[y(3x + 1)^{2/3}]$

10. $\ln(x^2 + 7x)$

11. $\log_5 \dfrac{x - 2}{x^2 - 4}$

12. $\log \dfrac{x^3}{\sqrt{x^2 - 10}}$

13. $\log \dfrac{5x(x^2 + 1)^2}{(x + 1)\sqrt{7x + 3}}$

14. $\ln \dfrac{4x^3}{(x + 1)^{2/3}(x + 2)^5}$

In problems 15–22, use the properties of logarithms to write each expression as a single logarithm.

15. $\log_5 \frac{5}{7} + \log_5 \frac{40}{25}$

16. $\log_2 \frac{32}{11} + \log_2 \frac{121}{16} - \log_2 \frac{4}{5}$

17. $\log(a^2 - ab) - \log(7a - 7b)$

18. $\log\left(a + \dfrac{a}{b}\right) - \log\left(c + \dfrac{c}{b}\right)$

19. $\ln(x^2 - 9) - \ln(x^2 - 6x + 9)$

20. $\ln(3x^2 + 7x + 4) - \ln(3x^2 - 5x - 12)$

21. $\log\left(\dfrac{1}{4} - \dfrac{1}{x^2}\right) - \log\left(\dfrac{1}{2} - \dfrac{1}{x}\right)$

22. $3 \ln(x + 2) - \frac{1}{3} \ln x - \frac{1}{3} \ln(1 - 2x)$

23. Simplify each expression.

(a) $e^{\ln 5}$ **(b)** $e^{-2 \ln 3}$ **(c)** $e^{3+4 \ln x}$
(d) $e^{\ln(1/x)}$ **(e)** $e^{-7-\ln x}$ **(f)** $\ln e^{x^2-4}$
(g) $e^{(\ln x^2)-1}$ **(h)** $e^{\ln x-3 \ln y}$

24. Simplify each expression.

(a) $\dfrac{\ln e^{m^2-2m-24}}{m^2 - 36}$ **(b)** $e^{\ln[(x+y)/2]} \cdot e^{-\ln[1/(x+y)]}$

In problems 25–34, solve each equation and check for extraneous roots.

25. $\log 2x = \log 3 + \log(x - 1)$

26. $\log_5 y + \log_5(y - 4) = 1$

27. $\ln x + \ln(x - 2) = \ln(x + 4)$

28. $2 \ln(t + 1) - \ln(t + 4) = \ln(t - 1)$

29. $\log_2(w^2 - 9) - \log_2(w + 3) = 2$

30. $\log(v + 1) - \log v = 1$

31. $\log_4 x + \log_4(6x + 10) = 1$

32. $\log_3 t + \log_3(t - 6) = \log_3 7$

33. $\log(x^2 - 144) - \log(x + 12) = 1$

34. $\log x^2 = 2 \log x$

In problems 35 and 36, compute the value to three decimal places after using the change-of-base formula to write each logarithm as a quotient of:

(i) Common logarithms **(ii)** Natural logarithms

35. **(a)** $\log_2 5$ **(b)** $\log_7 2.89$
(c) $\log_4 \frac{3}{23}$ **(d)** $\log_3 e$

36. **(a)** $\log_8 13$ **(b)** $\log_4 \frac{1}{17}$
(c) $\log_3 0.46$ **(d)** $\log_{11} 7$

GU In problems 37–44, use the change-of-base formula along with a graphing utility to sketch the graph of each function. Also determine the domain and any vertical asymptotes.

37. $g(x) = \log_3 x$

38. $h(x) = \log_{1/3} x$

39. $g(x) = \log_4(5x + 1)$

40. $f(x) = \log_2(3 - 4x)$

41. $f(x) = \log_7(4x^2 - 1)$

42. $h(x) = \log_5(x^2 + 1)$

43. $h(x) = 2 \log_6 \sqrt{x^2 - 1}$

44. $g(x) = \frac{1}{2} \log_9 \sqrt[3]{x^2 - 1}$

Applying the Concepts

45. Installment Loan Payment: An equation that relates a loan of P dollars to be paid off in n equal monthly installment payments of A dollars with an interest rate of r percent per period is given by

$$\ln A - \ln P = \ln r + n \ln(1 + r) - \ln[(1 + r)^n - 1]$$

where $r =$ (Annual loan rate)/12.

(a) Solve this equation for A in terms of P, r, and n.

(b) Express A as a function of n if \$18,000 is borrowed to purchase a boat at an annual rate of 10%. What are the monthly payments if the loan is to be paid off in 3 years? 4 years? 5 years?

46. Depreciation: Used-car dealers determine the value of a car by applying the formula

$$\log(1 - r) = \frac{1}{t}(\log w - \log p)$$

where p (in dollars) is the purchase price of a car when it was new, and w (in dollars) is its value t years later at an annual rate of depreciation r.

(a) Solve this equation for r in terms of t, w, and p.
(b) Express the annual rate of depreciation r as a function of t if a new car was purchased for \$18,300 and sold t years later for \$7500. What is the rate of depreciation if the car was sold after 3 years? 3 years and 6 months? 4 years? Round off the answers to two decimal places.

47. Population Growth—Fruit Flies: The fruit fly *Drosophila melanogaster* is used in some genetic studies in laboratories. The number N of fruit flies in a colony after t days of breeding is given by the equation

$$\ln(230 - N) - \ln N = \ln 6.931 - 0.1702t$$

(a) Express N as a function of t.
(b) How many fruit flies are in the colony initially? After 10 days? After 16 days? After 30 days?
GU **(c)** Use a graphing utility to graph the function in part (a). Then use the TRACE feature to determine approximately how many days it will take for the number in the colony to exceed 150, 200, and 300.

48. Physics—Speed of a Rocket: Suppose a rocket with mass M_0 (in kilograms) is moving in free space at a speed of V_0 (in kilometers per second). When the rocket is fired, its speed V is given by

$$V = V_0 + V_1(\ln M_0 - \ln M)$$

where M is the mass of the propellant that has been burned off and V_1 is the speed of the exhaust gasses produced by the burn.
(a) Solve the equation for M in terms of V, V_0, V_1, and M_0.
GU **(b)** The Saturn rocket used in the Apollo moon landing had an initial mass M_0 of 2.85×10^6 kilograms, an exhaust speed V_1 of 2.46 kilometers per second, and a free-space speed V_0 of 4.10 kilometers per second. Use a graphing utility to graph M as a function of V under these conditions.
(c) Use the graph to approximate V (to two decimal places) if $M = 7.7 \times 10^4$ kilograms.
(d) Confirm the result in part (c) by substituting the values into the given model.

Developing and Extending the Concepts

49. Use $M = 10{,}000$, $N = 10$, $b = 10$, and $p = 3$ to show that each statement is false.
(a) $\log_b \dfrac{M}{N} = \dfrac{\log_b M}{\log_b N}$
(b) $\dfrac{\log_b M}{\log_b N} = \log_b M - \log_b N$
(c) $\log_b M \cdot \log_b N = \log_b M + \log_b N$
(d) $\log_b MN = \log_b M \cdot \log_b N$
(e) $\log_b M^p = (\log_b M)^p$
(f) $(\log_b M)^p = p \log_b M$

50. Show that: $\ln\left(\dfrac{\sqrt{3} + \sqrt{2}}{\sqrt{3} - \sqrt{2}}\right) = 2 \ln(\sqrt{3} + \sqrt{2})$

In problems 51 and 52, find values for a and b that satisfy the given equation.

51. $\log(a + b) = \log a + \log b$
52. $\log(a - b) = \log a - \log b$

In problems 53 and 54, use the change-of-base formula to prove each equation.

53. $\log_b a = \dfrac{1}{\log_a b}$

54. $\dfrac{\log_b x}{\log_{ab} x} = 1 + \log_b a$

55. GU Use graphs to criticize the following statement: Since $\ln x^2 = 2 \ln x$, the graph of $f(x) = \ln x^2$ is the same as the graph of $g(x) = 2 \ln x$.

56. GU Use graphs to criticize the following statement: Since $\ln(4x^2 - 1) = \ln[(2x + 1)(2x - 1)] = \ln(2x + 1) + \ln(2x - 1)$, the graphs of $f(x) = \ln(4x^2 - 1)$ and $g(x) = \ln(2x + 1) + \ln(2x - 1)$ are the same.

OBJECTIVES

1. Solve exponential equations algebraically.
2. Solve equations and inequalities graphically.
3. Solve applied problems.

4.5 Additional Equations and Inequalities

Now we'll use the properties of logarithms and graphing features to develop additional procedures for solving equations and inequalities involving exponential and logarithmic expressions.

Solving Exponential Equations Algebraically

In Section 4.3, we solved exponential equations in which both sides can be written as powers of the same base. Exponential equations involving different bases, such as $4^{2x-1} = 3$ and $e^{-3t} = 0.5$, can be solved algebraically by using logarithms. The technique is illustrated in the next example.

EXAMPLE 1 Solving exponential equations algebraically

Use logarithms to solve each equation. Round off the answer to four decimal places.

(a) $4^{2x-1} = 3$ (b) $e^{-3t} = 0.5$

Solution

(a) Taking the common logarithm of each side of the equation, we have

$$\begin{aligned}
\log 4^{2x-1} &= \log 3 \\
(2x - 1)\log 4 &= \log 3 && \log N^r = r \log N \\
2x - 1 &= \frac{\log 3}{\log 4} \\
2x &= 1 + \frac{\log 3}{\log 4} && \text{Solve for } x \\
x &= \frac{1}{2}\left(1 + \frac{\log 3}{\log 4}\right) \\
x &= 0.8962 \text{ (approx.)} && \text{Use a calculator}
\end{aligned}$$

(b) Taking the natural logarithm of each side of the equation $e^{-3t} = 0.5$, we have

$$\begin{aligned}
\ln e^{-3t} &= \ln 0.5 \\
-3t \ln e &= \ln 0.5 && \ln N^r = r \ln N \\
t &= \frac{-\ln 0.5}{3} && \ln e = 1 \\
t &= 0.2310 \text{ (approx.)} && \text{Use a calculator}
\end{aligned}$$

Solving Equations and Inequalities Graphically

Not all exponential and logarithmic equations and inequalities can be solved using the algebraic methods studied so far. Such problems can sometimes be solved by graphical means.

EXAMPLE 2 Solving an equation graphically

Solve the equation $\ln(2x + 1) = -3x + 2$ graphically. Round off the answer to three decimal places.

Solution We'll use a graphing utility to develop two methods for approximating the solutions of this equation. The first method is based on the fact that the equation

$$\ln(2x + 1) = -3x + 2$$

is equivalent to the equation

$$\ln(2x + 1) + 3x - 2 = 0$$

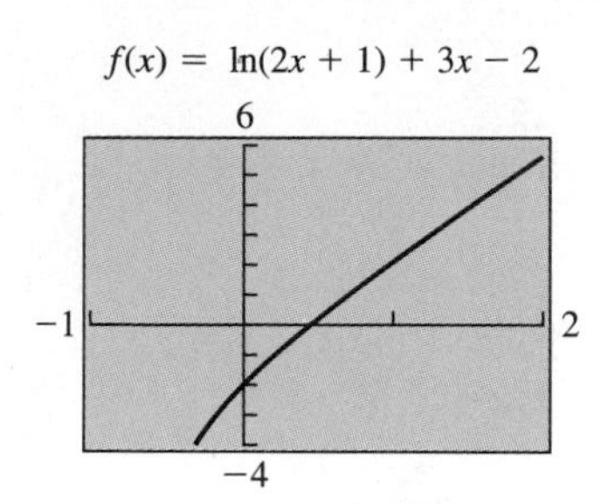

FIGURE 1

The solutions of this latter equation are the same as the x intercepts of the graph of the function

$$f(x) = \ln(2x + 1) + 3x - 2$$

Figure 1 shows a viewing window of the graph of f. By repeatedly using the ZOOM and TRACE features, we find that the x intercept of the graph of f is approximately 0.452, so the solution of the given equation is 0.452. In this approach, the accuracy of the solution can be seen by observing how close the y value is to 0 each time we repeat the process.

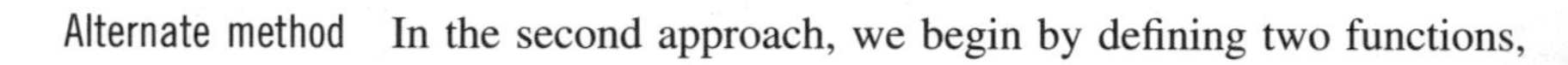

Alternate method In the second approach, we begin by defining two functions,

$$y_1 = \ln(2x + 1) \qquad \text{and} \qquad y_2 = -3x + 2$$

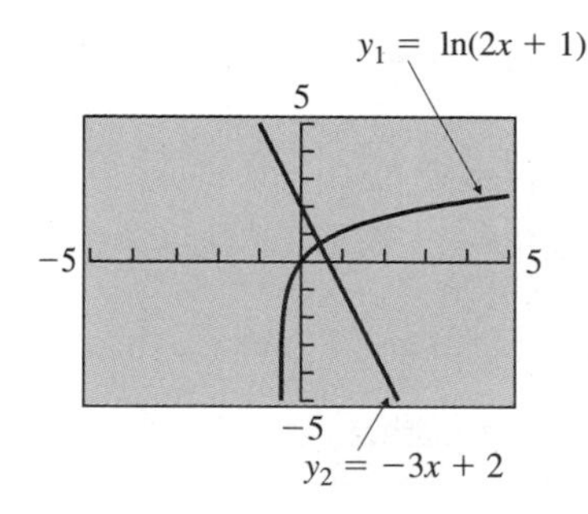

FIGURE 2

Then we use a graphing utility to graph both equations on the same coordinate system (Figure 2). In the viewing window, we see that the graphs intersect at one point. By using the ZOOM and TRACE features, we find that the coordinates of the point of intersection are approximately (0.452, 0.644). The x coordinate of this point is the solution of the original equation, because when we substitute 0.452 in each equation, the y value is approximately the same. That is, $y_1 = \ln[2(0.452) + 1] = 0.644$ and $y_2 = -3(0.452) + 2 = 0.644$.

EXAMPLE 3 GU

Solving an inequality graphically

Solve the inequality $x^2 > 2^x$ graphically. Round off the answers to two decimal places.

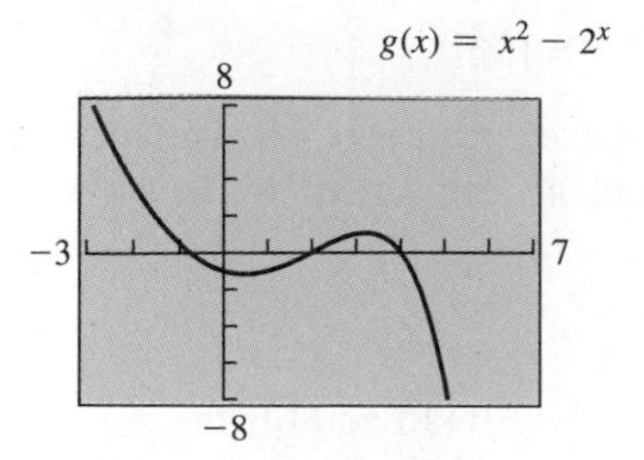

FIGURE 3

Solution The inequality $x^2 > 2^x$ is equivalent to $x^2 - 2^x > 0$. First, we graph $g(x) = x^2 - 2^x$ (Figure 3). Then we use the ZOOM and TRACE features to find that the approximate locations of the three x intercepts are -0.77, 2, and 4. Upon examining the graph, we see that $x^2 - 2^x > 0$ where the graph of g is above the x axis—that is, whenever x is in the interval $(-\infty, -0.77)$ or in the interval $(2, 4)$.

Alternately, we can solve this inequality by first graphing $f(x) = x^2$ and $g(x) = 2^x$ on the same coordinate system, and then locating the values of x where the graph of f is above the graph of g.

Solving Applied Problems

Up to now we have used graphical means to solve applied problems involving exponential and logarithmic expressions. Now we will use algebraic methods to solve such problems.

EXAMPLE 4 Using the compound interest formula

Determine the number of years it will take a $5000 certificate of deposit to double in value if it is invested at a 6% annual interest rate compounded quarterly. Round off the answer to two decimal places.

Solution The final value of S (in dollars) for this investment after x years is given by the compound interest equation

$$S = 5000\left(1 + \frac{0.06}{4}\right)^{4x}$$

If t is the time required to double the money, then

$$10{,}000 = 5000(1.015)^{4t} \qquad \text{or} \qquad 2 = (1.015)^{4t}$$

We solve the latter equation as follows:

$$\begin{aligned} \log 2 &= \log(1.015)^{4t} && \text{Take the logarithm of each side} \\ \log 2 &= 4t \log 1.015 && \log N^r = r \log N \\ t &= \frac{\log 2}{4 \log 1.015} && \text{Solve for } t \\ &= 11.64 \quad \text{(approx.)} && \text{Use a calculator} \end{aligned}$$

So, the investment will double in approximately 11 years, 8 months.

A radioactive substance decomposes according to the decay model

$$A = A_0 e^{-kt} \qquad \text{where } k > 0$$

and t represents time.

EXAMPLE 5 Using the decay model of an element

Polonium-210 is a radioactive element that decays according to the model

$$A = A_0 e^{-0.005t}$$

where A_0 is the initial weight of the sample. How long will it take for a sample of this material to decay by 30% if t represents elapsed time (in days)?

Solution If the sample weighs A_0 grams initially, then after decaying by 30%, its weight A is given by

$$A = A_0 - 0.30A_0 = 0.70A_0$$

Substituting into the decay model, we get

$$\begin{aligned} 0.70A_0 &= A_0 e^{-0.005t} \\ 0.70 &= e^{-0.005t} \end{aligned}$$

Next, we solve for t as follows:

$$\ln 0.70 = \ln e^{-0.005t} \quad \text{Take the natural logarithm of each side}$$
$$\ln 0.70 = -0.005t \ln e \quad \ln N^r = r \ln N$$
$$\ln 0.70 = -0.005t \quad \ln e = 1$$
$$t = \frac{\ln 0.70}{-0.005}$$
$$t = 71.33 \text{ (approx.)} \quad \text{Use a calculator}$$

Thus, it takes a little over 71 days to decay by 30%.

PROBLEM SET 4.5

Mastering the Concepts

In problems 1–14, use logarithms to solve each equation. Round off the answers to four decimal places.

1. $6^x = 12$ **2.** $3^{5x} = 2$
3. $e^{-5x} = 7$ **4.** $e^{4x} = 3$
5. $7^{3x-1} = 5$ **6.** $10^{x+1} = 4$
7. $e^{x+1} = 10$ **8.** $e^{2-3x} = 8$
9. $e^{2x-1} = 5^x$ **10.** $3^{x+1} = 17^{2x}$
11. $e^{2x-1} = 10^x$ **12.** $e^x = 10^{x+1}$
13. $(1.08)^{2x+3} = (1.7)^x$ **14.** $(0.97)^{3x-1} = (3.1)^{2x}$

GU In problems 15–24, use a graphing utility to solve each equation. Round off the answers to three decimal places.

15. $2^x - 1 = 3x$ **16.** $3e^x - 7 = 5x$
17. $e^{-x^2} + x - 1 = 0$ **18.** $3^x - 4 = 4x$
19. $e^{x^2} - 3x - 5 = 0$ **20.** $3^{2x} = x^2$
21. $\ln x + x^2 = 2$ **22.** $\ln(x + 1) + x = 3$
23. $e^{-x} = \ln x$ **24.** $\log(x^2 + 2x) - x = 1$

GU In problems 25–32, use a graphing utility to solve each inequality. Round off the answers to two decimal places.

25. $3^x < x^2$ **26.** $e^{2x} > 3 - x$
27. $\ln x > \log 2x$ **28.** $\ln(x - 3) < \log x$
29. $e^{-x} \le 3 + \ln x$ **30.** $e^{3x} - \ln x \le 5$
31. $\log(x - 1) < 3x$ **32.** $3 \ln x - x^3 > 4$

33. GU Recall from Section 4.2 (Problem 45) that the hyperbolic sine function is denoted by $\sinh x$ and defined by

$$f(x) = \sinh x = \frac{e^x - e^{-x}}{2}$$

Use a graphing utility to graph f. Then use the graph to solve the equation $\sinh x = 1$. Round off the answer to two decimal places.

34. GU Use a graphing utility to sketch the graph of

$$f(x) = \frac{2^x + 2^{-x}}{2}$$

Then approximate x to two decimal places if $f(x) = 3$.

Applying the Concepts

In problems 35–44, round off each answer to two decimal places.

35. Compound Interest: Suppose that \$10,000 is invested in a certificate of deposit at a 4.75% annual interest rate. How many years will it take for the money to double if the interest is compounded:
(a) Semiannually? **(b)** Quarterly?
(c) Monthly? **(d)** Continuously?

36. Compound Interest: Suppose that \$1500 is invested at a 4.65% annual interest rate. How many years will it take for the money to triple if the interest is compounded:
(a) Semiannually? **(b)** Quarterly?
(c) Monthly? **(d)** Continuously?

37. Investment Growth Model: Suppose that P dollars are invested in a money market fund that pays a 4.85% annual interest rate compounded quarterly, and that the accumulated amount S (in dollars) after t years is a multiple of P, that is, $S = kP$.

(a) Use common logarithms to express t as a function of k.
(b) Use the function in part (a) to find out how long it takes for the investment to double, to triple, and to quadruple.
GU (c) Use a graphing utility to graph the function in part (a). Then use the graph to demonstrate the validity of the results in part (b).

38. Investment Growth Model: Suppose that P dollars are invested in a credit union paying a 5.03% annual interest rate compounded continuously, and that the accumulated amount S (in dollars) after t years is a multiple of P, that is, $S = kP$.
(a) Use natural logarithms to express t as a function of k.
(b) Use the function in part (a) to find out how long it takes for the investment to double, to triple, and to quadruple.
GU (c) Use a graphing utility to graph the function in part (a). Then use the graph to demonstrate the validity of the results in part (b).

39. Productivity: In an experiment conducted by a company, it was determined that the number N of days of training needed for a factory worker to produce x automobile parts per day is given by the model

$$N = 100 - 25 \ln(60 - x)$$

(a) How many parts per day is a factory worker able to produce if 25 training days are needed?
GU (b) Use a graphing utility to graph the function. Then use the graph to demonstrate the validity of the result in part (a).

40. Advertising: A company predicts that sales will increase during a 30 day promotional television campaign according to the logistic model

$$N = \frac{400}{1 + 300e^{-0.5t}}$$

where N is the number of daily sales t days after the campaign begins.
(a) How many days of advertising will result in daily sales of 300 units?
GU (b) Use a graphing utility to graph the function. Then use the graph to demonstrate the validity of the result in part (a).

41. Deer Population: Suppose that the population P of a herd of deer newly introduced into a game preserve grows according to the model $P = 300 + 100 \ln(t + 1)$, where t is the time in years.
(a) How many deer are present in the preserve initially?
(b) In how many years will the deer population be 500?
GU (c) Use a graphing utility to graph the function. Then use the graph to demonstrate the validity of the result in part (b).

42. Radioactive Decay: Radioactive potassium-42, which is used by cardiologists as a tracer, has a half-life of about 12.5 hours.
(a) Find the decay model for this element.
(b) How long will it take for 80% of the original sample to decay?

43. Heat-Treating: A steel panel is tested for stress by heating to 385°F and then immediately cooling it in a cooling chamber where the temperature is held at 32°F. The panel cools according to Newton's law of cooling:

$$T = 32 + 353e^{-kt}$$

where T is the temperature in degrees Fahrenheit after t minutes and k is a constant.
(a) Find k if the temperature of the panel is 198°F after 5 minutes. Round off to two decimal places.
(b) How long after the steel panel is placed in the cooling chamber will its temperature be 85°F?
GU (c) Will the temperature of the panel drop to 32°F during the test? Interpret the result by using the graph of the function with the value of k found in part (a).

44. Drug Dosage Model: The concentration C (in milligrams) of a certain drug after t hours in a patient's bloodstream is given by

$$C = 100e^{-0.442t}$$

(a) How long does it take for the drug concentration to reduce by 50%?
GU (b) Use a graphing utility to graph the function. Then use the graph to demonstrate the validity of the result in part (a).

Developing and Extending the Concepts

GU In problems 45–47, solve each equation algebraically. Then use a graphing utility to demonstrate the validity of the solution.

45. $e^x - 3e^{-x} = 2$

46. $\dfrac{\ln(x + 2)}{\ln(x + 1)} = 2$

47. $\ln x = \log x$

48. Solve the equation $x^2 + 5x + 6 = 0$. Substitute e^t for x in the equation to get $e^{2t} + 5e^t + 6 = 0$. Explain why the latter equation has no solution. Use a graphing utility to graph $f(x) = x^2 + 5x + 6$ and $g(t) = e^{2t} + 5e^t + 6$ to support the explanation.

49. Is it possible to solve the equation $\sqrt[x]{x} = 1.2$ algebraically? If so, solve it. If not, use a graphing utility to approximate the solution for $1 \le x \le 15$. Round off to two decimal places.

CHAPTER 4 REVIEW PROBLEM SET

1. Let $f(x) = 2.5^{-x}$. Find each value rounded off to three decimal places.
(a) $f(-1)$ **(b)** $f(2)$ **(c)** $f(0.3)$
(d) $f(-\sqrt{3})$ **(e)** $f(1.7)$

2. Use a graphing utility to sketch the graphs of both given functions on the same coordinate system. Then use the graphs to solve each inequality.
(a) $f(x) = 5^x$ and $g(x) = 3^x$; $5^x > 3^x$
(b) $f(x) = \left(\frac{1}{4}\right)^x$ and $g(x) = \left(\frac{2}{5}\right)^x$; $\left(\frac{1}{4}\right)^x < \left(\frac{2}{5}\right)^x$
(c) $f(x) = e^x$ and $g(x) = 3^x$; $e^x < 3^x$

3. Describe a sequence of transformations to change the graph of $y = 2^x$ into the graph of the given function. Graph the function. Indicate the domain, range, and whether the function is increasing or decreasing. Also, determine the horizontal asymptote.
(a) $f(x) = 2^x + 1$ **(b)** $g(x) = -2^x$
(c) $F(x) = 3(2^x)$ **(d)** $G(x) = 2^x - 3$
(e) $f(x) = 2^{x+1} - 3$ **(f)** $G(x) = -3(2^{x+1}) + 4$

4. Use a graphing utility to graph each function. Use the graph to find the domain and range. Find any horizontal asymptote.
(a) $f(x) = e^{-x^3}$ **(b)** $g(x) = e^{x^2}$
(c) $h(x) = -e^{-|x|}$ **(d)** $G(x) = \dfrac{e^x - e^{-x}}{2}$
(e) $f(x) = xe^{-x}$

5. Let $f(x) = e^x$.
(a) Find $f(3.2)$, $f(\pi)$, $f(-\sqrt{2})$, and $f(\sqrt{0.7})$ to three decimal places.
(b) Find: $\dfrac{f(2 + h) - f(2)}{h}$
(c) Show that:

$$[1 + f(a - b)]^{-1} + [1 + f(b - a)]^{-1} = 1$$

(d) Describe a sequence of transformations to change the graph of $f(x) = e^x$ into the graph of $G(x) = 2 - 3e^x$.

6. Solve the equations by expressing each side in terms of the same base.
(a) $2^x = 8^{x-1}$ **(b)** $4^{-x} = 8^{x+2}$ **(c)** $3^{x^2+x} = 9$

7. Write each exponential equation in logarithmic form.
(a) $e^{3.1} = a$ **(b)** $e^{x^2} = a + b$ **(c)** $5^{-2.7} = b$
(d) $e^{a+b} = c$ **(e)** $10^{a-b} = t$

8. Find the exact value of each logarithm without using a calculator.
(a) $\log_3 9$ **(b)** $\log_4 8$ **(c)** $\log_6 1$
(d) $\ln \sqrt[5]{e}$ **(e)** $\ln e^{-3}$ **(f)** $\log_5 0.04$
(g) $\log_{100} 0.001$

9. Solve each equation for x.
(a) $\log_4(x + 3) = -1$ **(b)** $\log(x^2 - 6) = 1$
(c) $\ln 2x = \ln 8 - \ln 2$
(d) $\log_5(2x - 1) + \log_5(2x + 1) = 1$
(e) $\ln(x + 8) - \ln x = 1$
(f) $\ln \sqrt{x^2 + 1} = 0$
(g) $3 \ln x + t = b$ **(h)** $e^{\ln(2x+y)} = u$

10. Let

$$f(x) = \frac{5^x - 5^{-x}}{5^x + 5^{-x}}$$

Determine whether

$$f(u + v) = \frac{f(u) + f(v)}{1 + f(u)f(v)}$$

11. Solve the equation

$$\frac{e^x + e^{-x}}{e^x - e^{-x}} = u$$

for x in terms of u.

12. Write each expression as a single logarithm and simplify.
(a) $\ln(x^2 - 16) - \ln(x + 4)$
(b) $\ln 3x - 2[\ln x - \ln(3 + x)]$
(c) $\ln\left(\dfrac{e}{\sqrt[3]{x}}\right) - \ln \sqrt[3]{ex}$
(d) $2 \log_4 x^3 + \log_4 \dfrac{2}{x} - \log_4 \dfrac{2}{x^4}$

13. Determine whether each statement is true or false. Use a graphing utility to help support your conclusion. Assume $x > 0$.
GU
(a) $\ln \sqrt[5]{x} = \sqrt[5]{\ln x}$ **(b)** $\ln x^2 = (\ln x)^2$
(c) $\ln\left(\frac{1}{x}\right) = \frac{1}{\ln x}$ **(d)** $\ln|x| = |\ln x|$

14. Describe a sequence of transformations that will change the graph of either $y = \ln x$ or $y = \log x$ to the graph of the given function. Graph the function. Find the domain, range, and vertical asymptotes.
(a) $f(x) = \ln(x - 3)$ **(b)** $g(x) = \ln(3 - x)$
(c) $h(x) = -\log x$

15. Use a graphing utility to graph each function. Then find the domain of the function and the vertical asymptotes.
GU
(a) $f(x) = \ln(-x)$
(b) $g(x) = \log(x^2 - x - 12)$
(c) $h(x) = \ln(-x - 2)$ **(d)** $h(x) = \log \frac{x}{x - 1}$

16. Use logarithms to solve each equation. Round off each answer to three decimal places.
(a) $3^{x+2} = 5$ **(b)** $e^{-0.5t} = 17$
(c) $5^{x+1} = 13^{2x}$

17. Use a graphing utility to solve each inequality. Round off the answers to two decimal places.
GU
(a) $e^{3x} < 4$ **(b)** $\ln x - e^x < -3$
(c) $2^x \leq 3^{x+3}$

18. Bacteria Growth: Suppose that a carton of milk contains 5000 bacteria per cubic inch at the time it was bought, and that the number doubles every day. The number N of bacteria in the carton t days after the milk was bought is given by the model

$$N(t) = 5000 \cdot 2^t$$

(a) Find $N(5)$ and $N(3.5)$. Explain what each value means in this situation.
GU **(b)** Use a graphing utility to sketch the graph of the function. Interpret the graph. Suppose that it is not safe to drink the milk when the bacteria count is 3,000,000. For how many days after the carton was bought can we safely drink the milk?

19. Compound Interest: If \$10,000 is deposited in an account that pays 5% annual interest compounded annually, then the accumulated balance A in the account at the end of t years is given by the function

$$A(t) = 10{,}000(1.05)^t$$

(a) Find $A(3)$, $A(7)$, and $A(10)$. Explain what each value means in this situation.
GU **(b)** Use a graphing utility to determine when the balance will reach \$35,000. Round off the answer to two decimal places.

20. Investment: Suppose that the interest rate allows an investment to double every 10 years. Then the accumulated value A from an initial investment of \$1000 is given by

$$A(t) = 1000 \cdot 2^t$$

where t represents the number of 10 year periods.
(a) Find the accumulated value of a \$1000 investment after 25 years.
GU **(b)** Use a graphing utility to determine when the investment will be worth \$500,000. Round off to two decimal places.

21. Depreciation: Banks depreciate the value of a car according to its age. Suppose that a car is bought for \$15,000 and depreciates about 25% of its value every year. The value V (in dollars) of the car t years later can be predicted by the model

$$V(t) = 15{,}000(0.75)^t$$

(a) Find $V(2)$ and $V(3)$. Explain what each value means in this situation.
GU **(b)** Use a graphing utility to predict when the value of the car will be \$10,500. Round off the answer to two decimal places.

22. Compound Interest: Suppose that \$1000 is put into a savings plan that yields a nominal interest rate of 5%. How much money will be accumulated in the account after 8 years if the interest is compounded:
(a) Annually? **(b)** Semiannually?
(c) Quarterly? **(d)** Monthly?
(e) Continuously?

23. Radioactive Decay: A radioactive substance decays according to the model

$$A = A_0 e^{-0.001t}$$

Thus, A grams of the substance remain after t years, where A_0 is the initial amount present.
(a) If 100 grams of the substance are present initially, how much will be present at the end of 50 years? Round off the answer to two decimal places.
GU **(b)** Use a graphing utility to determine when the substance decays by 70%. Round off the answer to the nearest year.

24. Present Value: Determine how much money must be invested now in order to have \$500,000 in 5 years, if the investment during this period earns a nominal interest rate of 6% compounded annually.

25. Compound Interest: Suppose that \$10,000 is invested at a 4.5% nominal interest rate. How many years will it take for the money to double if the interest is compounded:
(a) Quarterly? **(b)** Continuously?

26. Doubling Time: Banks use the formula

$$T = \frac{\log 2}{\log(1 + r)}$$

to predict the number of years T required to double an investment at an annual interest rate of r percent.
(a) Find the time it takes to double an investment of \$1000 at an annual interest rate of 4%, 5%, and 9%.
(b) According to the popular "rule of 70," the time T it takes for an investment to double is approximately $T = 70/r$, where r is the annual interest rate. Compare the results in part (a) to the results given by the rule of 70.

27. Chemistry: The pH of a solutuion is given by pH $= -\log\,[H^+]$. Find the pH of each substance to two decimal places.
(a) Vinegar: $[H^+] = 1.58 \times 10^{-3}$ mole per liter
(b) Milk of magnesia: $[H^+] = 3.16 \times 10^{-11}$ mole per liter

28. Manufacturing: A manufacturer determines that the daily cost C (in dollars) to manufacture x VCR components is given by the function

$$C(x) = 365[1 + \ln(x^2 - 225)] \qquad \text{where } x > 15$$

(a) Find $C(20)$ and $C(40)$. Explain what each value means in this situation.
(b) If the daily cost to manufacture VCR components is \$3640, how many components are produced each day?

29. Advertising: A health club determines that the number of memberships N sold in a year is related to the number of dollars x spent on advertising in the year by the model

$$N(x) = 50 + 100 \ln\left(\frac{x}{100} + 2\right)$$

(a) How many memberships will be sold in 1 year if \$1000 is spent on advertising? If \$1500 is spent?
(b) If the number of memberships for a year is 400, is it worthwhile to spend \$1000 on advertising? Explain.

30. Annuity: If P dollars is deposited at the end of each period for n periods in an annuity that earns interest at a rate of r percent per period, the *future value* A of the annuity is given by the equation

$$n \log(1 + r) + \log P = \log(Ar + P)$$

(a) Solve this equation for A in terms of P, n, and r.
(b) Suppose that a college professor plans to invest \$210 per month for retirement in a 403b tax-deferred annuity. The annuity pays an annual interest rate of 7.2%. Express the future value A as a function of n, the number of monthly deposits.
(c) What will be the earnings of the investment after 6 months? 1 year? 5 years?
GU **(d)** Use a graphing utility to graph the function in part (a). Then use the TRACE feature to determine how long it will take (to the nearest month) an investment to become worth over \$3000; over \$5000; over \$20,000.

CHAPTER 4 TEST

1. Let $f(x) = 3^{x^2-x}$. Find each value. Round off the answers to three decimal places.
(a) $f(-1)$ **(b)** $f(2)$
(c) $f(0.3)$ **(d)** $f(\sqrt{5})$

2. Let $f(x) = \left(\frac{1}{5}\right)^x$.
(a) Sketch the graph of f.
(b) Specify the domain and range of f.
(c) Indicate whether f is increasing or decreasing.
(d) Identify any asymptotes for the graph of f.
(e) Describe a sequence of transformations to change the graph of f to the graph of $g(x) = 2\left(\frac{1}{5}\right)^x - 3$. Then sketch the graph of g, and identify the asymptotes of the graph.

3. **(a)** Convert $6^u = 7$ to logarithmic form.
(b) Convert $\log_3 x = b$ to exponential form.

4. Without using a calculator, simplify each expression.

(a) $\ln \frac{1}{e^3}$ **(b)** $\ln \sqrt[7]{e}$

(c) $e^{\ln\sqrt{x+3}}$ **(d)** $e^{\ln x^2}$

5. Let $f(x) = \log x$.

(a) Find each value: $f(0.1)$, $f(1)$, $f(3)$, and $f(7)$. Round off each answer to four decimal places.

(b) Sketch the graph of f, and specify its domain and range.

(c) Does f have an inverse? Explain. If it does, find f^{-1} and sketch the graphs of f and f^{-1} on the same coordinate system.

(d) Describe a sequence of transformations to change the graph of f to the graph of $g(x) = -\log(x + 7)$. Sketch the graph of g and find all its asymptotes.

6. Suppose that $f(x) = A \ln x + B$, where A and B are constants. If $f(1) = 5$ and $f(e^3) = 8$, find the values of A and B.

7. Indicate whether each statement is true or false. Assume all variables are positive.

(a) $\ln \frac{u}{7} = \ln u - \ln 7$ **(b)** $\frac{\ln a}{\ln b} = \ln \frac{a}{b}$

(c) $\frac{\log_a u}{t} = \log_a u^{1/t}$ **(d)** $\log(xy)^t = t \log_y x + t$

8. Solve each equation for x.

(a) $\log_3(2x - 1) = 2$ **(b)** $x^5 e^{-4 \ln x} = 3$

(c) $\log_2 x + \log_2(x + 1) = 1$

(d) $3^{x+2} = 7$ Round off to two decimal places.

9. Sketch the graphs of $f(x) = \ln x + \ln 3$ and $g(x) = \ln(x + 3)$ on the same coordinate system. Use the graphs to determine the value of x for which $f(x) = g(x)$.

10. Suppose that $4000 is invested in an account paying an annual interest rate of 5%.

(a) Find the amount in the account at the end of 2 years if the interest is compounded daily.

(b) How long will it take for this investment to triple if interest is compounded continuously?

11. Suppose that the number N of bacteria in a certain culture is approximated by the function

$$N(t) = N_0 \cdot 4^{0.05t}$$

where N_0 represents the initial number and t is the number of elapsed hours.

(a) If 6000 bacteria are present initially, how many will be present after 15 hours?

(b) Sketch the graph of N if $N_0 = 6000$, and interpret the graph.

(c) How many hours after the initial observation will there be 14,000 bacteria if $N_0 = 6000$? Round off the answer to two decimal places.

5 Trigonometric Functions

CHAPTER CONTENTS

The word *trigonometry* stems from the Greek words for triangle measurement because the subject was originally developed to solve geometric problems involving triangles. Today, the study of trigonometry is viewed from two perspectives—trigonometric functions of angles and those of real numbers. Trigonometric functions of angles enable us to solve applied problems dealing with angles and lengths, and they are used in fields such as surveying and navigation. On the other hand, trigonometric functions of real numbers are used to model repeating phenomena such as sound waves, tidal waves, alternating electric current, and business cycles.

5.1 Angles and Arcs

OBJECTIVES

1. Describe and measure angles.
2. Convert angle measurements.
3. Describe angles in standard position.
4. Find the length of an arc.
5. Solve applied problems.

Before introducing the trigonometric functions of angles, we need to review angles and their measurements (see Appendix I for review topics).

Describing and Measuring Angles

In geometry, an angle in the plane consists of two **rays** (half-lines), called the **sides**, and their common end point, the **vertex**. Angles are usually denoted by lowercase Greek letters, such as α (alpha), β (beta), γ (gamma), and θ (theta), or by angle notation, such as $\angle ABC$, where the middle letter B represents the vertex, A is on one ray, and C is on the other (Figure 1).

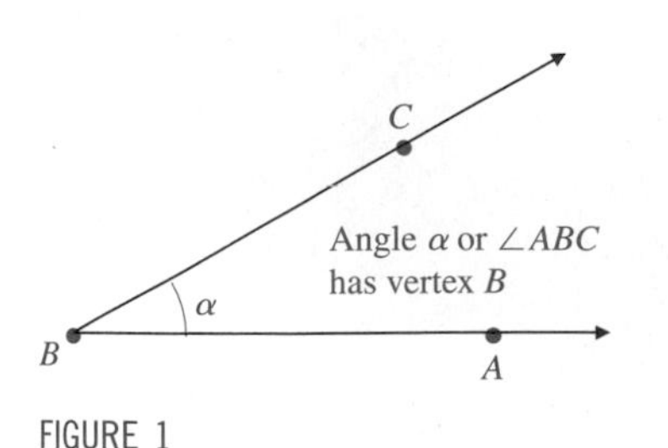

FIGURE 1

In trigonometry, an angle is often viewed as being formed by rotating a ray about its end point. The original position of the ray is called the **initial side** of the angle, and the final position is called the **terminal side**. Angles formed by a *counterclockwise* rotation are considered to have **positive** measure (Figure 2a), whereas those formed by a *clockwise* rotation are considered to have **negative** measure (Figure 2b).

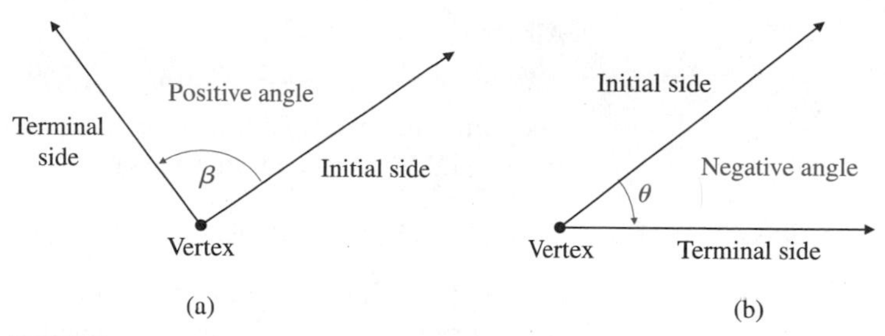

FIGURE 2

By convention, an angle label denotes both the angle and its measure. Thus, $\alpha = 45°$ indicates that angle α has degree measure 45°.

The two most commonly used units for angle measure are *degrees* and *radians*. **One degree (1°)** is the measure of an angle formed by $\frac{1}{360}$ of a complete counterclockwise rotation. Negative angles are measured by negative numbers (for example, $-30°$), while positive angles are measured by positive numbers (for example, 85°). Table 1 lists some terms commonly used to describe angles.

TABLE 1

Angle Terminology

Term	Degree Measure	Illustration
Straight angle	$180° = \frac{1}{2}(360°)$	180°
Right angle	$90° = \frac{1}{4}(360°)$	90°
Acute angle γ	$0° < \gamma < 90°$	γ
Obtuse angle θ	$90° < \theta < 180°$	θ
Complementary angles α and β	$\alpha + \beta = 90°$, where $0 < \alpha < 90°$ and $0 < \beta < 90°$	β α
Supplementary angles θ and γ	$\theta + \gamma = 180°$, where $0 < \theta < 180°$ and $0 < \gamma < 180°$	γ θ

There are two ways of representing a fractional part of a degree. One way is the *decimal form*, such as 37.82°, which means 37° plus 82 hundredths of a degree. The other form is known as *DMS* (*Degrees, Minutes, Seconds*). In this form, a degree is divided into 60 equal parts called **minutes**, denoted by ′. A minute is divided into 60 equal parts called **seconds**, denoted by ″. Thus

$$1° = 60' \qquad 1' = 60'' \qquad 1° = 3600''$$

Also,

$$1' = \left(\tfrac{1}{60}\right)^\circ \quad \text{and} \quad 1'' = \left(\tfrac{1}{3600}\right)^\circ$$

The following example illustrates the process of converting angles from decimal form to DMS form and vice versa.

EXAMPLE 1 Converting degree measures

(a) Express 37.45° in terms of degrees, minutes, and seconds.
(b) Express 23°17′37″ in decimal degree measure to two decimal places.

Many calculators have keys that automatically convert decimal degree measures to degrees, minutes, and seconds, and vice versa.

Solution

(a) $37.45^\circ = 37^\circ$ plus 45 hundredths of a degree, but $0.45^\circ = (0.45)(60') = 27'$. So, $37.45^\circ = 37^\circ 27' 0''$.
(b) $23^\circ 17' 37'' = \left(23 + \frac{17}{60} + \frac{37}{3600}\right)^\circ$, or approximately 23.29°.

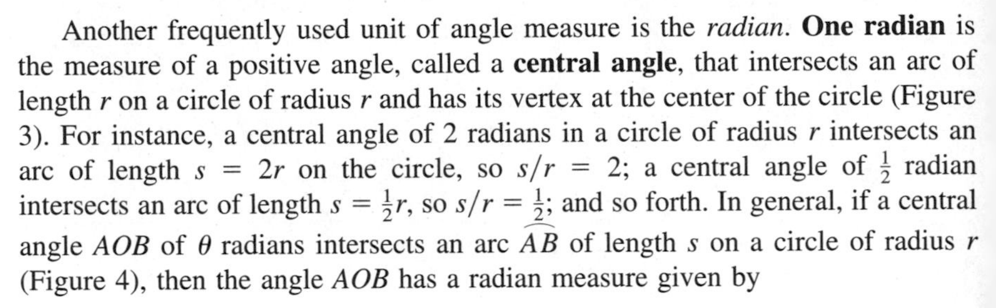

Another frequently used unit of angle measure is the *radian*. **One radian** is the measure of a positive angle, called a **central angle**, that intersects an arc of length r on a circle of radius r and has its vertex at the center of the circle (Figure 3). For instance, a central angle of 2 radians in a circle of radius r intersects an arc of length $s = 2r$ on the circle, so $s/r = 2$; a central angle of $\frac{1}{2}$ radian intersects an arc of length $s = \frac{1}{2}r$, so $s/r = \frac{1}{2}$; and so forth. In general, if a central angle AOB of θ radians intersects an arc $\widehat{AB}$ of length s on a circle of radius r (Figure 4), then the angle AOB has a radian measure given by

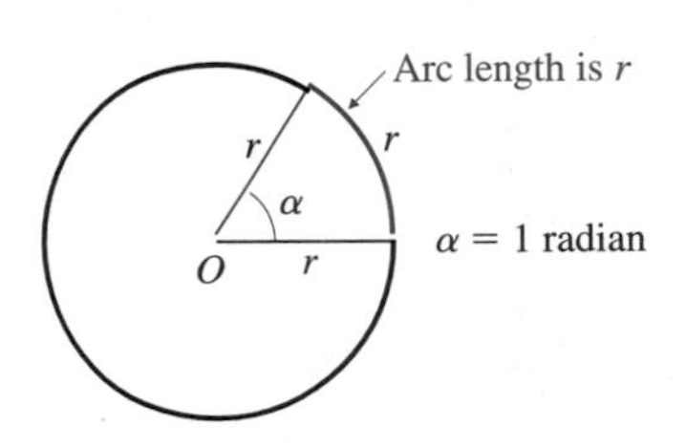

FIGURE 3

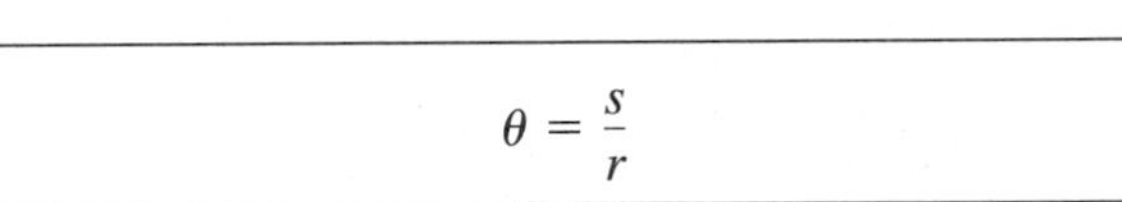

$$\theta = \frac{s}{r}$$

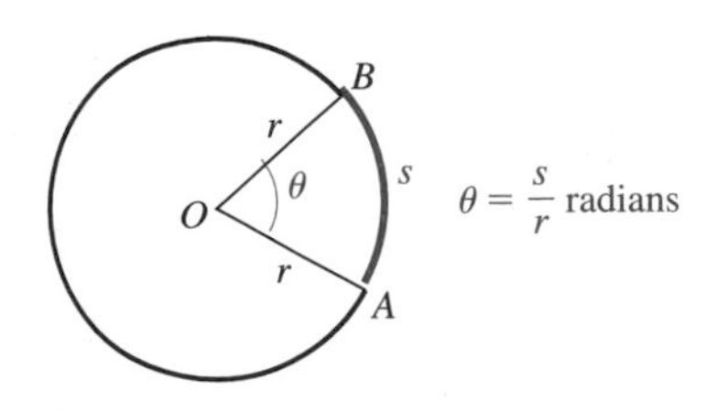

FIGURE 4

Thus,

If $r = 1$ and the angle is positive, then the radian measure is given by $\theta = s/1$; that is, the radian measure is the same as the length of the subtended arc on a circle of radius 1.

If the angle is negative, then $\theta = -s$. Since the circumference C of a circle of radius 1 is given by $C = 2\pi r = 2\pi \cdot 1 = 2\pi$, it follows that an angle formed by one complete counterclockwise rotation has radian measure 2π (Figure 5a); an angle formed by $\frac{1}{2}$ of a clockwise rotation has radian measure $-\pi$ (Figure 5b).

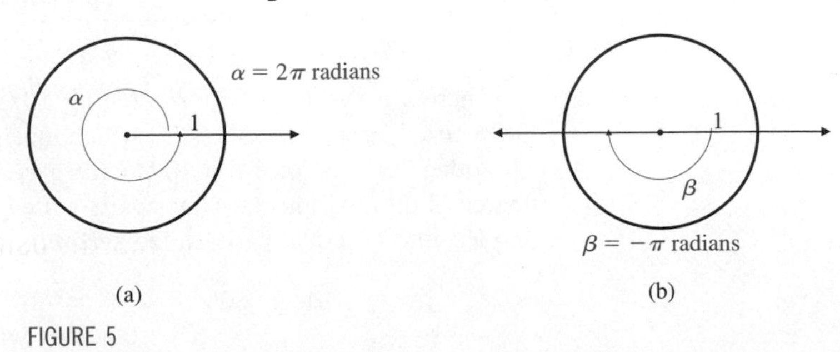

FIGURE 5

Converting Angle Measurements

Let's examine the relationship between the radian and degree measurements of angles. We know that a positive angle formed by one complete revolution has degree measure 360° and radian measure 2π radians. So we write

$$360° = 2\pi \text{ radians} \qquad \text{or} \qquad 180° = \pi \text{ radians}$$

Thus,

$$1° = \frac{\pi}{180} \text{ radian (approx. 0.0175 radian)}$$

and

$$1 \text{ radian} = \left(\frac{180}{\pi}\right)^{\circ} \text{ (approx. 57.30°)}$$

Therefore, we have the following conversion rules:

1. An angle of D degrees has radian measure R given by

$$R = \frac{\pi}{180}D$$

2. An angle of R radians has degree measure D given by

$$D = \frac{180}{\pi}R$$

EXAMPLE 2 Converting angle measures

(a) Convert 26.85° to radian measure (to two decimal places).
(b) Convert -0.3152 radian to degree measure (to two decimal places).

Solution

(a) Since $R = (\pi/180)D$, it follows that 26.85° corresponds to $(\pi/180)(26.85)$ radian, or about 0.47 radian.
(b) Since $D = (180/\pi)R$, it follows that -0.3152 radian corresponds to $\left(\frac{180}{\pi}\right)(-0.3152)$ degrees, or approximately $-18.06°$. ■

It is customary to omit the term *radian* when dealing with radian measure. Thus, we write $\pi/3$ to correspond to 60°, without using the word *radians* after $\pi/3$.

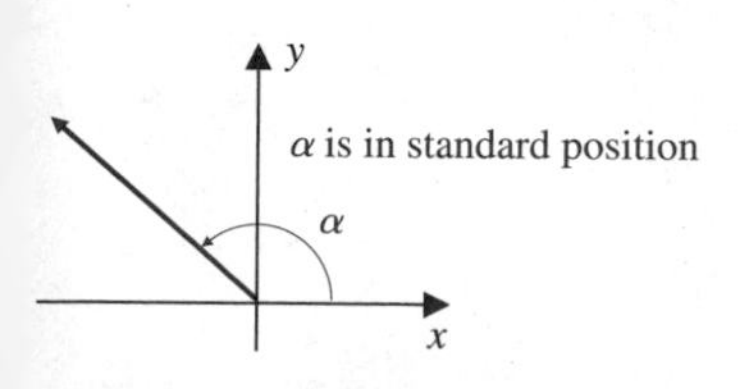

FIGURE 6

Describing Angles in Standard Position

In a Cartesian system, an angle is said to be in **standard position** if its vertex is at the origin and its initial side coincides with the positive x axis (Figure 6).

An angle is said to be in a certain quadrant if the terminal side lies in that quadrant when the angle is in standard position. For instance, a 75° angle is in *quadrant I* (Figure 7a) and a 130° angle is in *quadrant II* (Figure 7b).

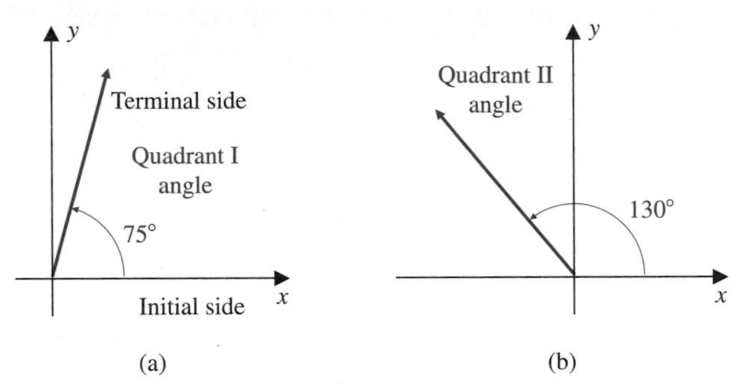

FIGURE 7

If the terminal side of an angle in standard position lies along either the x axis or the y axis, then the angle is called **quadrantal**. For example, $-270°$, $-180°$, $-90°$, $0°$, $90°$, and $180°$ are quadrantal angles. Angles that have the same initial sides and the same terminal sides are called **coterminal angles**.

EXAMPLE 3 Finding the measures of angles in standard position

Suppose that the angles 50°, 120°, and 30° are in standard position. Find the measure of each of the following angles and sketch the angles.

(a) The complement of the 50° angle.
(b) The supplement of the 120° angle.
(c) Two angles that are coterminal with the 30° angle.

Solution

(a) If α is the complement of 50°, then $\alpha + 50° = 90°$. So $\alpha = 40°$ (Figure 8a).
(b) If β is the supplement of 120°, $\beta + 120° = 180°$. So $\beta = 60°$ (Figure 8b).
(c) Figure 8c displays two angles that are coterminal with the 30°: a positive one, 390°, and a negative one, $-330°$.

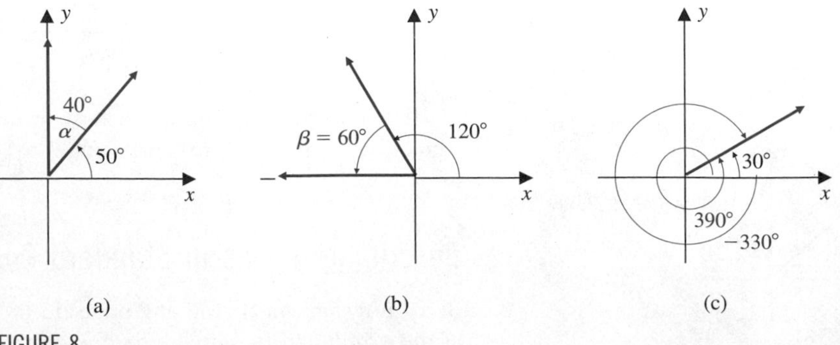

FIGURE 8

Finding the Length of an Arc

Earlier we established that if a central angle of a circle of radius r subtends an arc of length s, then the radian measure of θ is given by the formula $\theta = s/r$. By solving the formula for s, we obtain the following formula for **arc length**:

The formula $s = r\theta$ is applicable *only* when θ is expressed in radian measure.

$$s = r\theta$$

EXAMPLE 4 Finding an arc length

Find the length s of the arc intercepted on a circle of radius 6 centimeters by a central angle θ of 240°.

Solution We first convert 240° to radian measure to get

$$\theta = \frac{\pi}{180}D = \left(\frac{\pi}{180}\right)(240) = \frac{4\pi}{3} \text{ radians}$$

So,

$$s = r\theta = 6\left(\frac{4\pi}{3}\right) = 8\pi$$

Therefore, the arc length is 8π centimeters, or about 25.13 centimeters.

Solving Applied Problems

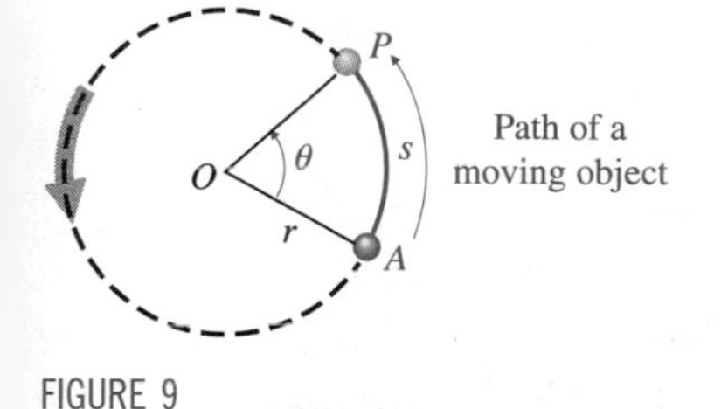

FIGURE 9

Angles measured in radians allow us to study the motion of an object moving at a constant speed around a circle of radius r. Suppose that a moving object starts at point A and after T units of time reaches point P (Figure 9). If the arc $\overparen{AP}$ has length s, then the object has moved s units of distance in T units of time. Since *distance divided by time equals speed*, we represent the **linear speed v** of the moving object as

$$v = \frac{s}{T}$$

The ratio that measures the rate of change of the angle θ, where θ is measured in radians, with respect to T units of time is called the **angular speed ω** (Greek letter omega) and is given by

$$\omega = \frac{\theta}{T}$$

Since the arc length $s = r\theta$, it follows that

$$v \text{ (linear speed)} = \frac{s}{T} = \frac{r\theta}{T} = r\left(\frac{\theta}{T}\right) = r\omega$$

So the relationship between the linear speed v and angular speed ω is given by

$$v = r\omega$$

In other words, the linear speed of an object moving along a circular path of radius r is the product of the angular speed and the radius.

EXAMPLE 5 **Finding the angular speed of a flywheel**

A belt passes over the rim of a flywheel with radius 18 centimeters (Figure 10).

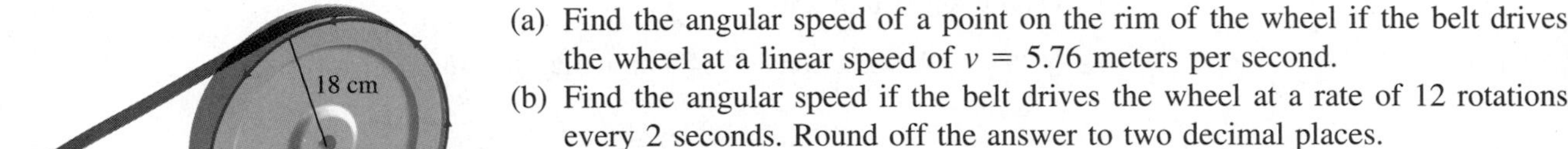

(a) Find the angular speed of a point on the rim of the wheel if the belt drives the wheel at a linear speed of $v = 5.76$ meters per second.

(b) Find the angular speed if the belt drives the wheel at a rate of 12 rotations every 2 seconds. Round off the answer to two decimal places.

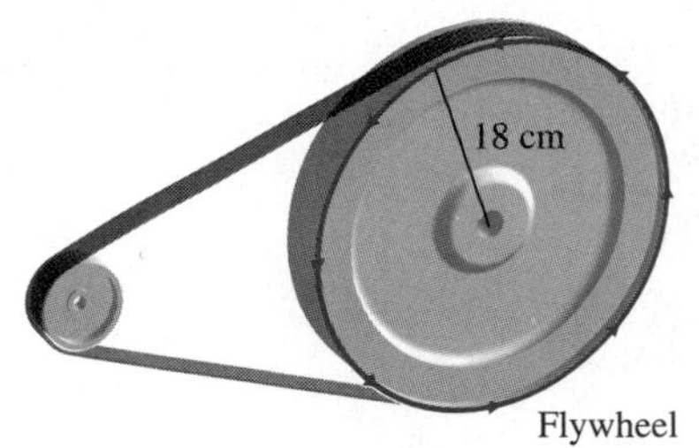

FIGURE 10

Solution

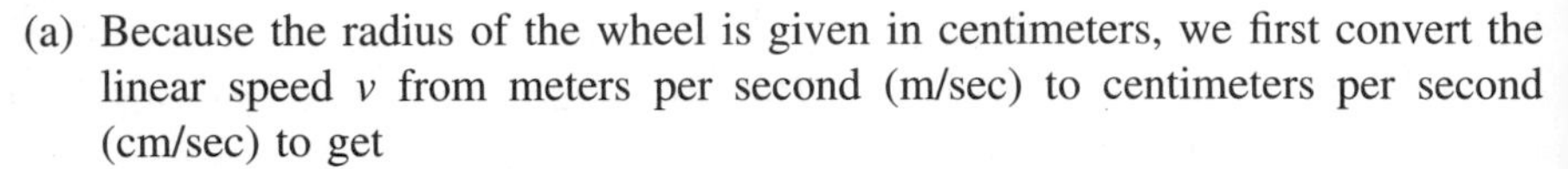

(a) Because the radius of the wheel is given in centimeters, we first convert the linear speed v from meters per second (m/sec) to centimeters per second (cm/sec) to get

$$v = \left(5.76\ \frac{\text{m}}{\text{sec}}\right)\left(100\ \frac{\text{cm}}{\text{m}}\right) = 576 \text{ centimeters per second}$$

Since $v = r\omega$, the angular speed ω is given by

$$\omega = \frac{v}{r} = \frac{576}{18} = 32 \text{ radians per second}$$

(b) Since the belt drives a point on the rim of the wheel 12 times around the circle in 2 seconds, a point on the rim of the wheel rotates $12(2\pi) = 24\pi$ radians in 2 seconds. Therefore, the angular speed of the belt is

$$\omega = \frac{\theta}{T} = \frac{24\pi}{2} = 12\pi, \text{ or approx. } 37.70 \text{ radians per second}$$

EXAMPLE 6 **Finding the speed of a Ferris wheel**

A Ferris wheel with radius 49 feet makes one complete rotation every 24 seconds. Determine the angular and linear speed of the circular motion of a rider. Round off the answers to two decimal places.

Solution In this situation, the angle θ makes one complete rotation, that is, $\theta = 2\pi$. So the angular speed is given by

$$\omega = \frac{\theta}{T} = \frac{2\pi}{24} = \frac{\pi}{12}, \text{ or approx. } 0.26 \text{ radian per second}$$

The linear speed v is found by using the radius $r = 49$ and the angular speed $\omega = \pi/12$ to get

$$v = r\omega = (49)\left(\frac{\pi}{12}\right) = \frac{49\pi}{12}$$

That is, the rider travels at about 12.83 feet per second.

PROBLEM SET 5.1

Mastering the Concepts

In problems 1–4, express each angle measure in terms of degrees, minutes, and seconds.

1. **(a)** $16.31°$ **(b)** $-87.81°$
2. **(a)** $64.14°$ **(b)** $-12.16°$
3. **(a)** $-156.63°$ **(b)** $89.74°$
4. **(a)** $-213.68°$ **(b)** $463.09°$

In problems 5–8, express each angle in terms of decimal degrees. Round off each answer to two decimal places.

5. **(a)** $38°18'$ **(b)** $65°11'23''$
6. **(a)** $35°41'$ **(b)** $-48°15'25''$
7. **(a)** $-141°28'15''$ **(b)** $244°46'15''$
8. **(a)** $1°1'1''$ **(b)** $-10°10'10''$

In problems 9–14, convert each angle measure to radians. Express the answer in terms of π and also as a decimal rounded off to two decimal places.

9. **(a)** $75°$ **(b)** $-135°$
10. **(a)** $240°$ **(b)** $-7.5°$
11. **(a)** $-95°$ **(b)** $444°$
12. **(a)** $-220°$ **(b)** $330°$
13. **(a)** $67.5°$ **(b)** $30°30'36''$
14. **(a)** $17.45°$ **(b)** $115°13'44''$

In problems 15–20, convert each radian measure to degrees. Round off each answer to two decimal places.

15. **(a)** $\frac{2\pi}{3}$ **(b)** $\frac{43\pi}{6}$
16. **(a)** $\frac{11\pi}{6}$ **(b)** $\frac{7\pi}{18}$
17. **(a)** $\frac{7\pi}{12}$ **(b)** $-\frac{4\pi}{9}$
18. **(a)** $-\frac{3\pi}{8}$ **(b)** $-\frac{\pi}{14}$
19. **(a)** 5 **(b)** -2.3
20. **(a)** -1 **(b)** 4.6

In problems 21–24, determine the measure of the angle that is:
(i) Complementary to θ **(ii)** Supplementary to θ

21. **(a)** $\theta = 36°$ **(b)** $\theta = 78°$
22. **(a)** $\theta = 57°$ **(b)** $\theta = 87°$
23. **(a)** $\theta = \frac{\pi}{7}$ **(b)** $\theta = \frac{4\pi}{9}$
24. **(a)** $\theta = \frac{\pi}{5}$ **(b)** $\theta = \frac{3\pi}{7}$

In problems 25–30, s is the arc length corresponding to a central angle θ on a circle of radius r. Find the missing quantity to two decimal places.

25. $r = 7$ inches, $\theta = \frac{3\pi}{14}$, $s = ?$
26. $s = 6$ feet, $\theta = \frac{\pi}{7}$, $r = ?$
27. $r = 1.8$ meters, $\theta = 217°$, $s = ?$
28. $r = 11$ centimeters, $\theta = \frac{7\pi}{5}$, $s = ?$
29. $s = 5$ yards, $\theta = \frac{5\pi}{3}$, $r = ?$
30. $r = 7$ feet, $\theta = 110°$, $s = ?$

In problems 31–36, locate each angle in standard position. Specify its quadrant location, and sketch two angles coterminal with the given angle.

31. **(a)** $\frac{3\pi}{4}$ **(b)** $-\frac{\pi}{6}$
32. **(a)** $-\frac{5\pi}{12}$ **(b)** $\frac{7\pi}{6}$

33. **(a)** $-45°$ **(b)** $120°$

34. **(a)** $210°$ **(b)** $105°$

35. **(a)** $\frac{17\pi}{3}$ **(b)** $-700°$

36. **(a)** $-\frac{11\pi}{3}$ **(b)** $538°$

Applying the Concepts

In problems 37–48, round off the results to two decimal places.

37. Nautical Mile: A nautical mile is the arc length intersected on the surface of the Earth by a central angle of 1 minute (Figure 11). Assume that the radius of the Earth is 3960 miles.

(a) How many miles are there in 1 nautical mile?

(b) How many feet are there in 1 nautical mile?

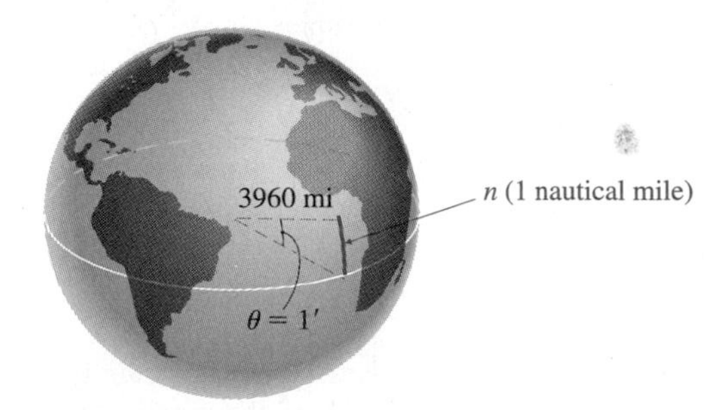

FIGURE 11

38. Pendulum Swing: Find the length of a pendulum if the tip of the pendulum traces an arc of 6 feet and the measure of the angle that this arc subtends is $\pi/5$ radian.

39. Child's Swing: The length of each chain supporting the seat of a child's swing is 8 feet. When the swing moves from its extreme forward position to its extreme backward position, the radian measure of the angle swept out by the chains is $(5\pi)/6$. How far does the seat travel in one trip between these extreme points; that is, what is the length of the arc generated by the seat through one swing between the two extreme points?

40. Revolving Door: A revolving door rotates through an angle of 2.64 radians before it is stopped. If one half of the door is 4 feet wide, through what distance does the edge of the door move?

41. Windshield Wiper: A windshield wiper of a car is 48 centimeters long, and it rotates at an angular speed of 12 degrees per second. Find the linear speed of the tip of the wiper.

42. Central Angle of a Pulley: A pulley of radius 25 centimeters uses a belt to drive another pulley of radius 15 centimeters. Find the radian measure of the angle θ_2 through which the smaller pulley turns as the larger pulley makes an angle θ_1 of 140° (Figure 12).

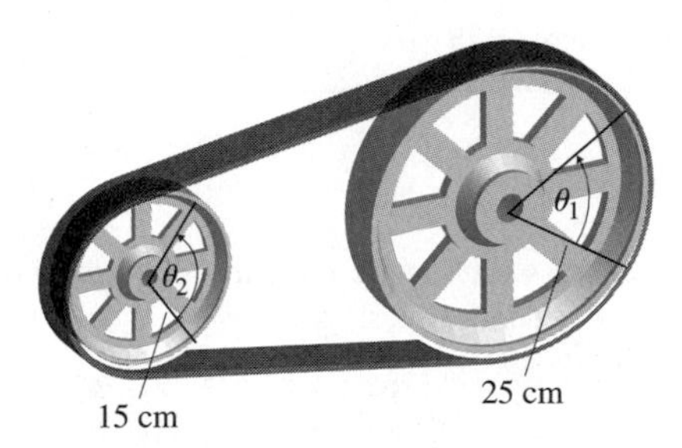

FIGURE 12

43. Tire Rotations: A car with tires that are 22 inches in diameter travels at 55 miles per hour.

(a) Find the angular speed of each tire in radians per minute.

(b) How many rotations per minute does each tire make?

44. Satellite Orbit: A satellite is orbiting the Earth in a circular orbit with a radius of 7,680 kilometers. If it makes $\frac{3}{4}$ of a revolution every hour, find:

(a) Its angular speed **(b)** Its linear speed

45. Bicycle Sprocket Wheels: A bicycle's small sprocket wheel of radius 2.75 inches is connected by a chain to a large sprocket wheel of radius 4.50 inches (Figure 13). The small sprocket wheel turns at 50 rotations per minute.

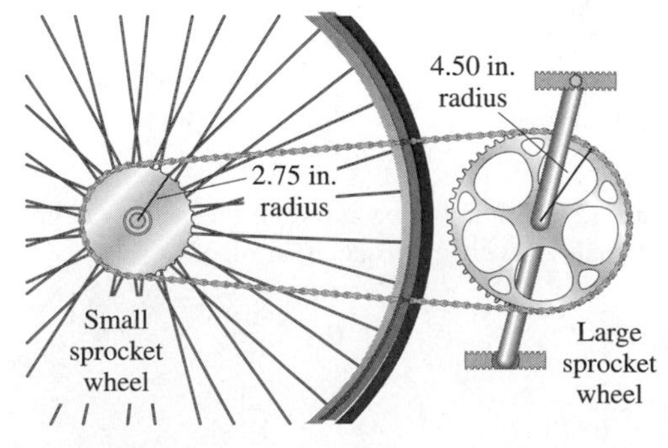

FIGURE 13

Find:

(a) The linear speed (in inches per minute) of the chain connecting the sprocket wheels.

(b) The number of rotations per minute of the large sprocket wheel.

46. The Earth's Rotation: The Earth, which is approximately 93,000,000 miles from the sun, revolves about the sun in a nearly circular orbit in approximately 365 days. Find the linear speed (in miles per day) of the Earth in its orbit.

47. Waterwheel: A point on the rim of a circular waterwheel of radius 8 feet makes one complete rotation every 10 seconds. Find the angular and linear speeds of the circular motion of the moving point.

48. Steamboat: As the paddlewheel with radius 9 feet of a steamboat turns, a point on the end of the paddle blade makes one complete rotation every 8 seconds. Determine the angular and linear speeds of the circular motion of the point on the paddle blade.

Developing and Extending the Concepts

49. Figure 14 shows three circles with the same center, a common central angle θ, and radii r_1, r_2, and r_3, where $r_1 < r_2 < r_3$. A student makes the following assertion: Since the radian measure of a central angle of a circle depends on the radius of the circle, it follows that the circle with a larger radius has a bigger central angle. Explain why this assertion is false.

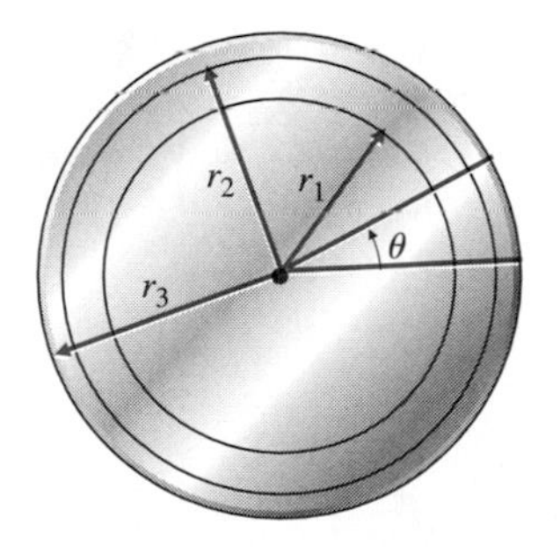

FIGURE 14

50. A sector POQ of a circle with center O is the region inside the circle bounded by the arc $\overset{\frown}{PQ}$ and the radial segments $\overline{OP}$ and $\overline{OQ}$ (Figure 15). If the central angle θ is measured in radians, and r is the radius of the circle, show that the area A of the sector is given by

$$A = \frac{1}{2}r^2\theta \quad \text{or} \quad A = \frac{1}{2}rs$$

where s is the length of arc $\overset{\frown}{PQ}$.

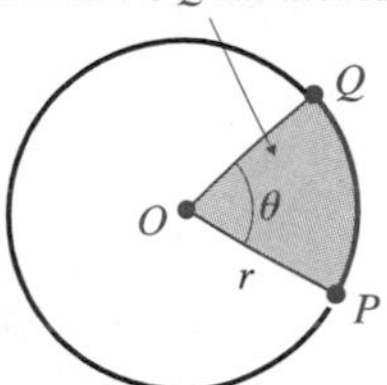

FIGURE 15

In problems 51 and 52, use the result from problem 50 to find the area of the sector determined by central angle θ. Round off the answers to two decimal places.

51. **(a)** $r = 7$ inches and $\theta = \dfrac{3\pi}{14}$

(b) $r = 6$ meters and $\theta = 72°$

52. **(a)** $r = 8$ centimeters and $\theta = 225°$

(b) $r = 6$ feet and $\theta = 4.81$

53. The following assertion is made: Suppose that angle θ_1 has radian measure t and angle θ_2 has radian measure $t + 2\pi$. When both angles are placed in standard position, they have the same terminal side. It follows that $\theta_1 = \theta_2$. Explain why this assertion is wrong.

54. Suppose that you ride on a merry-go-round at a carnival. You realize that you go faster when you sit near the outside than when you sit near the inside, even though the merry-go-round travels at a constant angular speed (Figure 16). Explain this phenomenon.

FIGURE 16

OBJECTIVES

1. Define trigonometric functions of acute angles.
2. Define trigonometric functions of general angles.
3. Determine the signs of trigonometric function values.
4. Use reference triangles to evaluate trigonometric functions.
5. Approximate trigonometric function values.

5.2 Trigonometric Functions of Angles

In this section we begin by defining the trigonometric functions of acute angles in terms of ratios of the lengths of the sides of right triangles. Then we'll extend the definitions to general angles.

Defining Trigonometric Functions of Acute Angles

Consider a right triangle ABC (Figure 1a) where the right angle (90°) is at vertex C. The other two vertex angles at A and B are acute angles. The side $\overline{AB}$ opposite to the right angle is called the **hypotenuse**. For convenience, we denote one of the two acute angles by θ, the length of the **side opposite** θ by **opp**, the **side adjacent** to θ to **adj**, and the **hypotenuse** by **hyp** (Figure 1b).

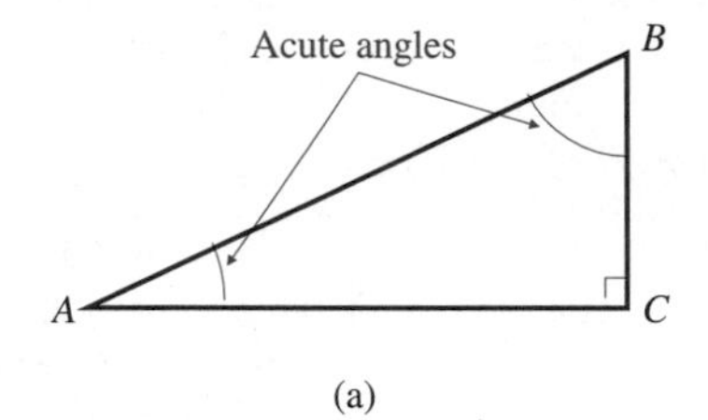

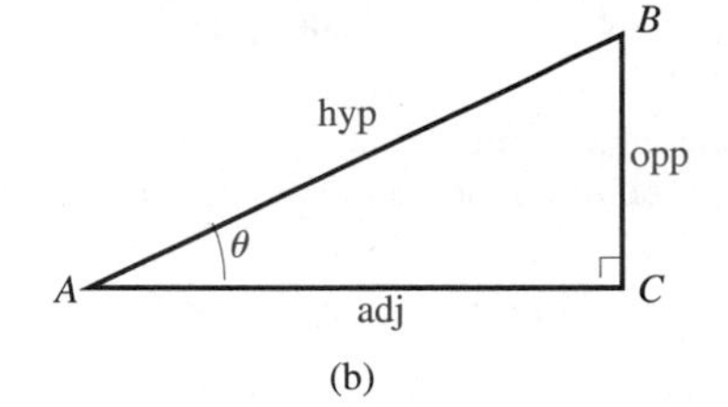

FIGURE 1

We define the **six trigonometric functions of the acute angle θ** as follows:

Name of Function	Abbreviation	Value of Function at θ
sine	sin	$\sin\theta = \dfrac{\text{opp}}{\text{hyp}}$
cosine	cos	$\cos\theta = \dfrac{\text{adj}}{\text{hyp}}$
tangent	tan	$\tan\theta = \dfrac{\text{opp}}{\text{adj}}$
cosecant	csc	$\csc\theta = \dfrac{\text{hyp}}{\text{opp}}$
secant	sec	$\sec\theta = \dfrac{\text{hyp}}{\text{adj}}$
cotangent	cot	$\cot\theta = \dfrac{\text{adj}}{\text{opp}}$

EXAMPLE 1 Evaluating trigonometric functions

Figure 2 shows two right triangles. The first one has sides of lengths 3, 4, and 5 (Figure 2a). The second is similar to the first; each side is 6 times longer than the corresponding side in the first (Figure 2b). Find the exact values of the six trigonometric functions of θ for both triangles and compare these values.

The angle θ is the same in both triangles because the triangles are *similar* (the sides are proportional; see Appendix I).

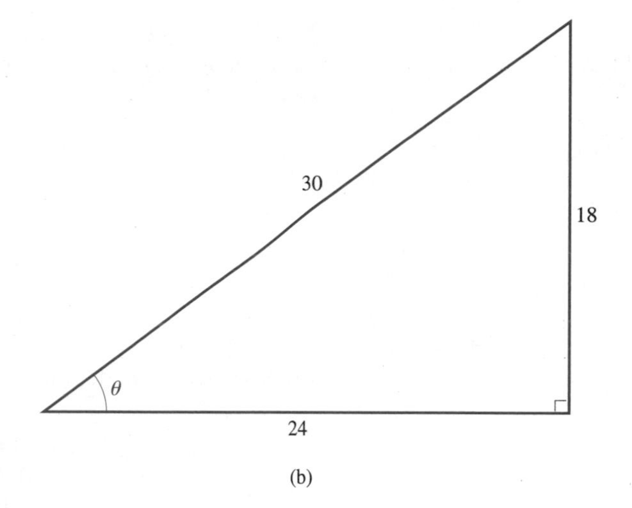

FIGURE 2

Solution From the definitions of the trigonometric functions we get the following results:

Smaller Triangle

$$\sin\theta = \frac{\text{opp}}{\text{hyp}} = \frac{3}{5}$$
$$\cos\theta = \frac{\text{adj}}{\text{hyp}} = \frac{4}{5}$$
$$\tan\theta = \frac{\text{opp}}{\text{adj}} = \frac{3}{4}$$
$$\csc\theta = \frac{\text{hyp}}{\text{opp}} = \frac{5}{3}$$
$$\sec\theta = \frac{\text{hyp}}{\text{adj}} = \frac{5}{4}$$
$$\cot\theta = \frac{\text{adj}}{\text{opp}} = \frac{4}{3}$$

Larger Triangle

$$\sin\theta = \frac{18}{30} = \frac{3}{5}$$
$$\cos\theta = \frac{24}{30} = \frac{4}{5}$$
$$\tan\theta = \frac{18}{24} = \frac{3}{4}$$
$$\csc\theta = \frac{30}{18} = \frac{5}{3}$$
$$\sec\theta = \frac{30}{24} = \frac{5}{4}$$
$$\cot\theta = \frac{24}{18} = \frac{4}{3}$$

Notice that the values of the trigonometric functions of θ depend on the *ratios* of the lengths of the sides of the right triangle, not on the sizes of the triangles that contain θ.

EXAMPLE 2 Using a right triangle to evaluate trigonometric functions

Use the right triangle in Figure 3 (at the top of page 290) to find the values of the six trigonometric functions of acute angle θ.

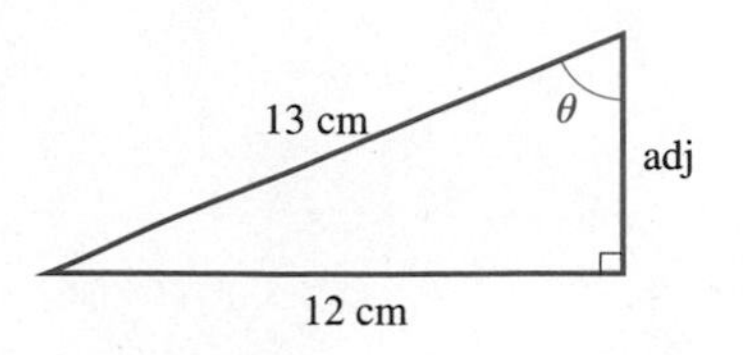

FIGURE 3

Solution For angle θ, we have opp = 12 centimeters and hyp = 13 centimeters, but the length of the adjacent side isn't given. Using the Pythagorean theorem, we have

$$\begin{aligned}(\text{adj})^2 + (\text{opp})^2 &= (\text{hyp})^2\\ (\text{adj})^2 &= (\text{hyp})^2 - (\text{opp})^2\\ &= 13^2 - 12^2 = 25\end{aligned}$$

So,

$$\text{adj} = \sqrt{25} = 5 \text{ centimeters}$$

It follows from the definitions that

$$\sin\theta = \frac{\text{opp}}{\text{hyp}} = \frac{12}{13} \qquad \csc\theta = \frac{\text{hyp}}{\text{opp}} = \frac{13}{12}$$

$$\cos\theta = \frac{\text{adj}}{\text{hyp}} = \frac{5}{13} \qquad \sec\theta = \frac{\text{hyp}}{\text{adj}} = \frac{13}{5}$$

$$\tan\theta = \frac{\text{opp}}{\text{adj}} = \frac{12}{5} \qquad \cot\theta = \frac{\text{adj}}{\text{opp}} = \frac{5}{12}$$

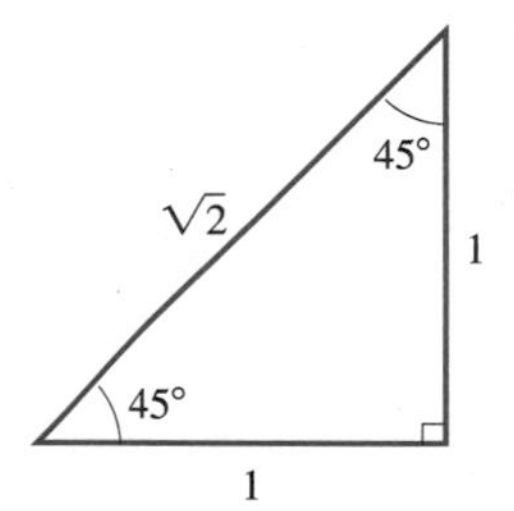

FIGURE 4
45°–45° right triangle

Our knowledge of right triangles provides us with a way to determine the exact values of trigonometric functions for the *special angles* 45°, 30°, and 60° (or equivalently, $\pi/4$, $\pi/6$, and $\pi/3$ radians). For a 45° angle, we first construct an *isosceles right triangle* with two equal sides of length 1 unit (Figure 4). This triangle is referred to as a *45°–45° right triangle*. We know from plane geometry that both of the acute angles in this right triangle measure 45°. Using the Pythagorean theorem, we have $1^2 + 1^2 = (\text{hyp})^2$, or hyp $= \sqrt{2}$. Thus, the trigonometric values of 45° are given by

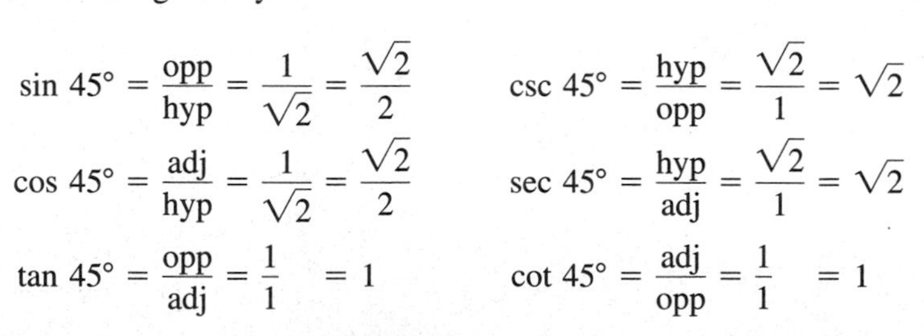

$$\sin 45° = \frac{\text{opp}}{\text{hyp}} = \frac{1}{\sqrt{2}} = \frac{\sqrt{2}}{2} \qquad \csc 45° = \frac{\text{hyp}}{\text{opp}} = \frac{\sqrt{2}}{1} = \sqrt{2}$$

$$\cos 45° = \frac{\text{adj}}{\text{hyp}} = \frac{1}{\sqrt{2}} = \frac{\sqrt{2}}{2} \qquad \sec 45° = \frac{\text{hyp}}{\text{adj}} = \frac{\sqrt{2}}{1} = \sqrt{2}$$

$$\tan 45° = \frac{\text{opp}}{\text{adj}} = \frac{1}{1} = 1 \qquad \cot 45° = \frac{\text{adj}}{\text{opp}} = \frac{1}{1} = 1$$

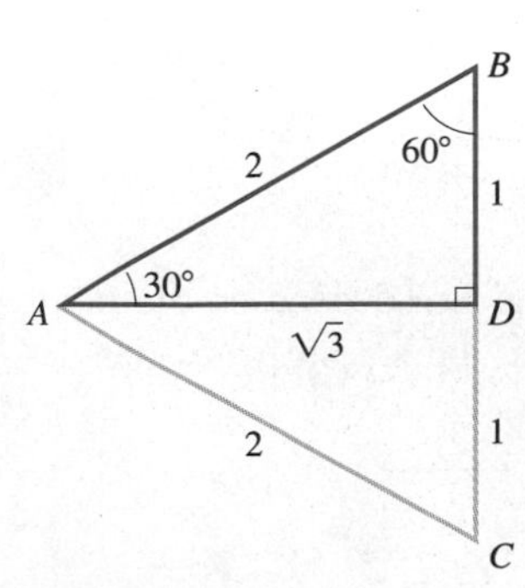

FIGURE 5
A 30°–60° right triangle

For a 30° angle and a 60° angle, we begin by constructing an equilateral triangle, with sides of length 2 units (Figure 5). Recall from plane geometry that the three angles of an equilateral triangle each measure 60°. Assume that $\overline{AD}$ is the perpendicular bisector of $\overline{BC}$. Triangle ADB is referred to as a *30°–60° right triangle*. By using the Pythagorean theorem, we have $|\overline{AD}|^2 = 2^2 - 1^2 = 3$, or $|\overline{AD}| = \sqrt{3}$. Then, using triangle ADB and the definitions of the trigonometric functions, we get

$$\sin 30° = \frac{\text{opp}}{\text{hyp}} = \frac{1}{2} \qquad \csc 30° = \frac{\text{hyp}}{\text{opp}} = \frac{2}{1} = 2$$

$$\cos 30° = \frac{\text{adj}}{\text{hyp}} = \frac{\sqrt{3}}{2} \qquad \sec 30° = \frac{\text{hyp}}{\text{adj}} = \frac{2}{\sqrt{3}} = \frac{2\sqrt{3}}{3}$$

$$\tan 30° = \frac{\text{opp}}{\text{adj}} = \frac{1}{\sqrt{3}} = \frac{\sqrt{3}}{3} \qquad \cot 30° = \frac{\text{adj}}{\text{opp}} = \frac{\sqrt{3}}{1} = \sqrt{3}$$

EXAMPLE 3 **Using a 30°–60° right triangle**

Use a 30°–60° right triangle to determine the trigonometric values of 60°.

Solution By using the 30°–60° right triangle *ADB* in Figure 5, we get

$$\sin 60° = \frac{\text{opp}}{\text{hyp}} = \frac{\sqrt{3}}{2} \qquad \csc 60° = \frac{\text{hyp}}{\text{opp}} = \frac{2}{\sqrt{3}} = \frac{2\sqrt{3}}{3}$$

$$\cos 60° = \frac{\text{adj}}{\text{hyp}} = \frac{1}{2} \qquad \sec 60° = \frac{\text{hyp}}{\text{adj}} = \frac{2}{1} = 2$$

$$\tan 60° = \frac{\text{opp}}{\text{adj}} = \frac{\sqrt{3}}{1} = \sqrt{3} \qquad \cot 60° = \frac{\text{adj}}{\text{opp}} = \frac{1}{\sqrt{3}} = \frac{\sqrt{3}}{3}$$

The exact trigonometric values for the special angles 45°, 30°, and 60° (or equivalently, $\pi/4$, $\pi/6$, and $\pi/3$ radians) are listed in Table 1. Since we work with these angles so often, their trigonometric values should be learned. When trying to recall these values, it is helpful to construct a 45°–45° right triangle or a 30°–60° right triangle, and then use the definitions of the trigonometric functions as we did in Example 3.

TABLE 1

Trigonometric Values for Special Angles

Angle θ							
Degree Measure	Radian Measure	$\sin\theta$	$\cos\theta$	$\tan\theta$	$\csc\theta$	$\sec\theta$	$\cot\theta$
30°	$\frac{\pi}{6}$	$\frac{1}{2}$	$\frac{\sqrt{3}}{2}$	$\frac{\sqrt{3}}{3}$	2	$\frac{2\sqrt{3}}{3}$	$\sqrt{3}$
45°	$\frac{\pi}{4}$	$\frac{\sqrt{2}}{2}$	$\frac{\sqrt{2}}{2}$	1	$\sqrt{2}$	$\sqrt{2}$	1
60°	$\frac{\pi}{3}$	$\frac{\sqrt{3}}{2}$	$\frac{1}{2}$	$\sqrt{3}$	$\frac{2\sqrt{3}}{3}$	2	$\frac{\sqrt{3}}{3}$

Defining Trigonometric Functions of General Angles

The definitions of the trigonometric functions are extended from acute angles to general angles in the box at the top of page 292.

DEFINITION

TRIGONOMETRIC FUNCTIONS OF AN ANGLE

Suppose that θ is an angle in standard position. Assume $(x, y) \neq (0, 0)$ is a point on the terminal side of θ at a distance $r = \sqrt{x^2 + y^2}$ from the origin (Figure 6). Then the trigonometric functions of θ are defined as follows:

$$\sin\theta = \frac{y}{r} \qquad \csc\theta = \frac{r}{y},\quad y \neq 0$$

$$\cos\theta = \frac{x}{r} \qquad \sec\theta = \frac{r}{x},\quad x \neq 0$$

$$\tan\theta = \frac{y}{x},\quad x \neq 0 \qquad \cot\theta = \frac{x}{y},\quad y \neq 0$$

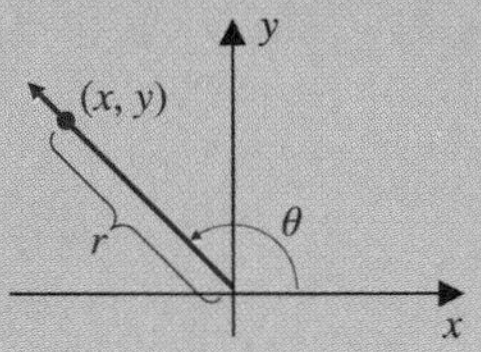

FIGURE 6

There are two important observations to be made about this definition:

1. The angle θ can be measured in degrees or radians and can be positive, negative, or zero.
2. The values of the six trigonometric functions depend on the location of the terminal side of the angle θ and not the choice of any particular point (x, y) on the terminal side of θ (Problem 54).

EXAMPLE 4 Evaluating trigonometric functions of angles

Find the six trigonometric function values of each angle.

(a) α is an angle in standard position that contains the point $(5, -12)$ on its terminal side.

(b) $\beta = 180°$

Solution

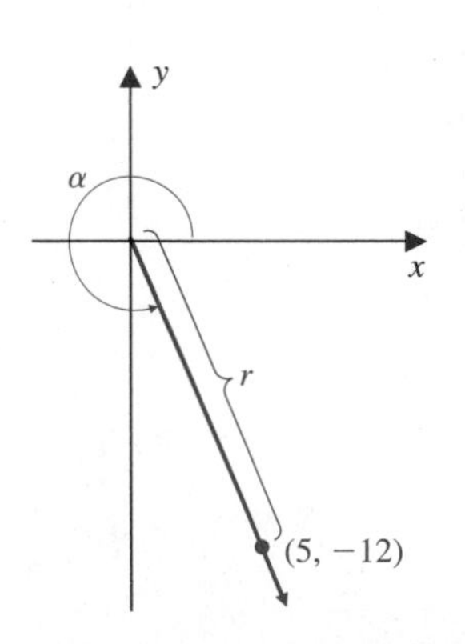

FIGURE 7a

(a) Figure 7a shows one possibility for an angle α in standard position whose terminal side contains the point $(5, -12)$. Since $x = 5$ and $y = -12$, the distance r from the origin to $(5, -12)$ is given by

$$r = \sqrt{x^2 + y^2} = \sqrt{5^2 + (-12)^2} = \sqrt{25 + 144} = \sqrt{169} = 13$$

So, by the definitions,

$$\sin\alpha = \frac{y}{r} = \frac{-12}{13} \qquad \csc\alpha = \frac{r}{y} = \frac{13}{-12} = -\frac{13}{12}$$

$$\cos\alpha = \frac{x}{r} = \frac{5}{13} \qquad \sec\alpha = \frac{r}{x} = \frac{13}{5}$$

$$\tan\alpha = \frac{y}{x} = \frac{-12}{5} \qquad \cot\alpha = \frac{x}{y} = \frac{5}{-12} = -\frac{5}{12}$$

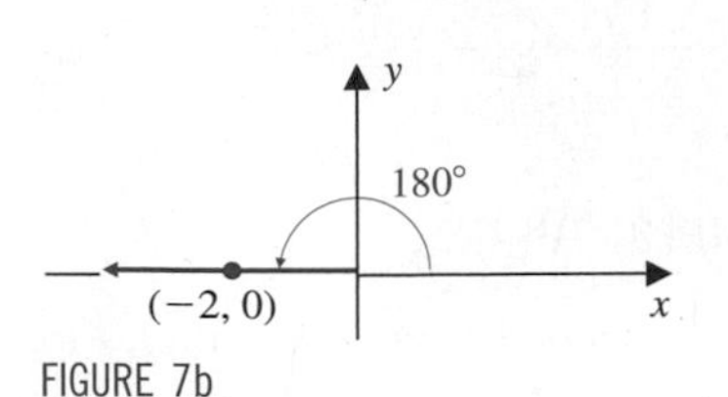

FIGURE 7b

(b) First, we display $\beta = 180°$ in standard position and then select a point on its terminal side. We choose the point $(-2, 0)$ (Figure 7b). For this selection, $x = -2$ and $y = 0$, and its distance from the origin is $r = 2$. By definition, we have

$$\sin 180° = \frac{y}{r} = \frac{0}{2} = 0 \qquad \csc 180° = \frac{r}{y} = \frac{2}{0} \quad \text{(undefined)}$$

$$\cos 180° = \frac{x}{r} = \frac{-2}{2} = -1 \qquad \sec 180° = \frac{r}{x} = \frac{2}{-2} = -1$$

$$\tan 180° = \frac{y}{x} = \frac{0}{-2} = 0 \qquad \cot 180° = \frac{x}{y} = \frac{-2}{0} \quad \text{(undefined)}$$

Determining the Signs of Trigonometric Function Values

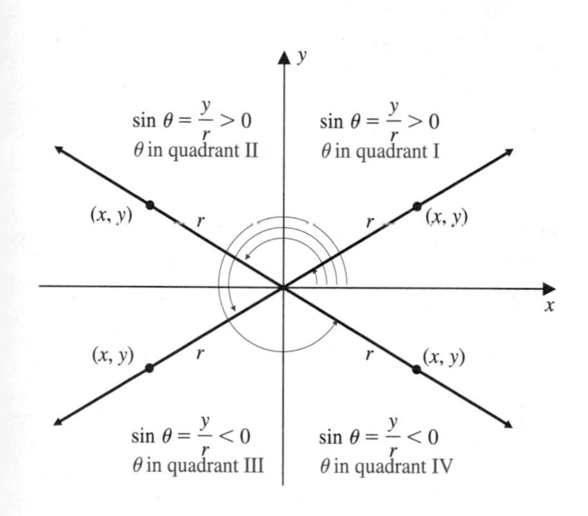

FIGURE 8

The algebraic sign of a trigonometric function value depends on the quadrant in which the angle in standard position lies. For instance, as Figure 8 shows, $\sin\theta$ is positive for θ in quadrants I and II, and $\sin\theta$ is negative in quadrants III and IV. Similarly, we can determine the algebraic signs of the other five trigonometric function values in the various quadrants. The results are listed in Table 2.

TABLE 2

Signs of the Values of the Trigonometric Functions of θ (θ in Standard Position)

Terminal Side of Angle θ in Quadrant:	Positive Values	Negative Values
I	All	None
II	$\sin\theta$, $\csc\theta$	$\cos\theta$, $\sec\theta$, $\tan\theta$, $\cot\theta$
III	$\tan\theta$, $\cot\theta$	$\sin\theta$, $\cos\theta$, $\csc\theta$, $\sec\theta$
IV	$\cos\theta$, $\sec\theta$	$\sin\theta$, $\csc\theta$, $\tan\theta$, $\cot\theta$

EXAMPLE 5 Finding the quadrant location

Find the quadrant in which θ lies if θ is in standard position, and $\cot\theta > 0$ and $\cos\theta < 0$.

Solution Using Table 2, we find that $\cot\theta > 0$ if θ is in quadrant I or III, and $\cos\theta < 0$ if θ is in quadrant II or III. For both conditions to be satisfied, θ must be in quadrant III.

Using Reference Triangles to Evaluate Trigonometric Functions

We will often find it convenient to evaluate trigonometric functions of certain nonacute angles that are integer multiples of the special angles 30°, 45°, or 60°, by making use of the concept of *reference triangles*. This procedure is described as follows:

Step 1. Sketch the given angle θ in standard position.
Step 2. Superimpose a reference triangle; that is, draw either a 45°–45° right triangle or a 30°–60° right triangle in whichever quadrant will allow the hypotenuse to lie on the terminal side of θ and one side of the triangle to lie on the x axis.
Step 3. Use the measurements of that particular right triangle along with the quadrant location to obtain a point (x, y) on the terminal side of θ.
Step 4. Use the definitions of the trigonometric functions of angles.

EXAMPLE 6 Using reference triangles to evaluate trigonometric functions

Use the above four-step procedure to evaluate the six trigonometric functions of $\theta = 135°$.

Solution

Refer to Figure 9.

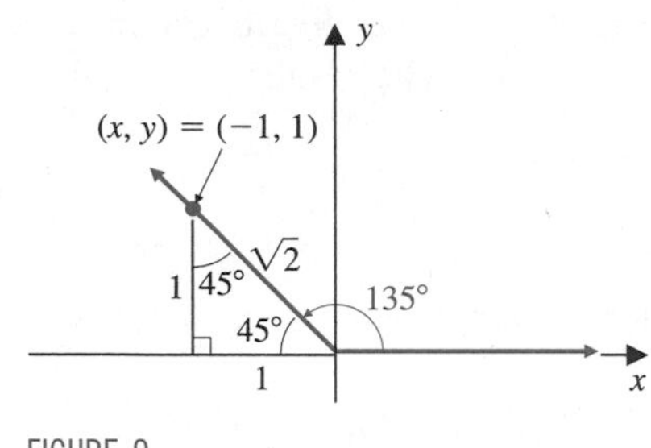

FIGURE 9

Step 1. We sketch $\theta = 135°$ in standard position.
Step 2. Then we draw a 45°–45° right triangle as a reference triangle in quadrant II so that the hypotenuse lies on the terminal side of θ and one leg lies on the negative x axis.
Step 3. By using the measurements of the 45°–45° right triangle, we get the point $(x, y) = (-1, 1)$ on the terminal side of θ. It follows that $r = \sqrt{(-1)^2 + 1^2} = \sqrt{2}$.
Step 4. By definition, we get

$$\sin 135° = \frac{y}{r} = \frac{1}{\sqrt{2}} = \frac{\sqrt{2}}{2} \qquad \csc 135° = \frac{r}{y} = \frac{\sqrt{2}}{1} = \sqrt{2}$$

$$\cos 135° = \frac{x}{r} = \frac{-1}{\sqrt{2}} = -\frac{\sqrt{2}}{2} \qquad \sec 135° = \frac{r}{x} = \frac{\sqrt{2}}{-1} = -\sqrt{2}$$

$$\tan 135° = \frac{y}{x} = \frac{1}{-1} = -1 \qquad \cot 135° = \frac{x}{y} = \frac{-1}{1} = -1$$

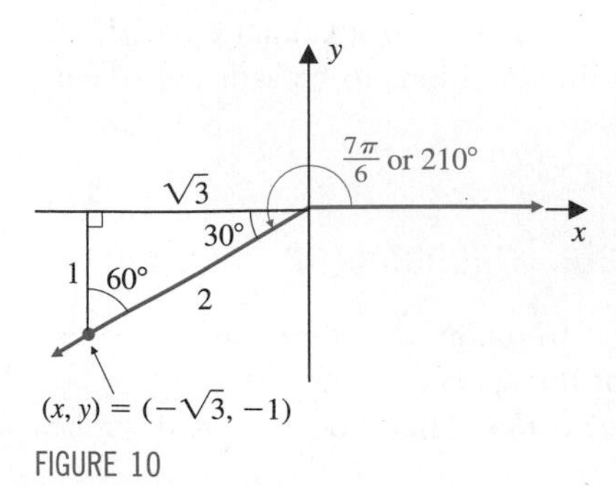

FIGURE 10

To evaluate the trigonometric functions of an angle given in radian measure, we can convert it to degree measure and then proceed as we did in Example 6. For instance, assume $\theta = (7\pi)/6$. Since $(7\pi)/6$ radians correspond to 210°, we sketch $\theta = 210°$ in standard position along with its reference triangle as shown in Figure 10. After locating the point $(x, y) = \left(-\sqrt{3}, -1\right)$ on the terminal side of θ, we use the definitions to find the trigonometric values of $(7\pi)/6$. For example,

$$\sin \frac{7\pi}{6} = \sin 210° = \frac{y}{r} = \frac{-1}{2} \qquad \cos \frac{7\pi}{6} = \cos 210° = \frac{x}{r} = \frac{-\sqrt{3}}{2}$$

Approximating Trigonometric Function Values

If an angle is not a special angle or a multiple of a special angle, then we can find approximate values of the trigonometric functions of the angle, whether it is given in degree or radian measure, by using a calculator. Most calculators have keys for only the sine, cosine, and tangent functions. To evaluate the remaining three trigonometric functions on such calculators, we make use of the following reciprocal relationships:

RECIPROCAL RELATIONSHIPS

For any angle θ, where these functions are defined, the following relationships hold:

$$\csc\theta = \frac{1}{\sin\theta} \qquad \sec\theta = \frac{1}{\cos\theta} \qquad \cot\theta = \frac{1}{\tan\theta}$$

These relationships follow directly from the definitions of the trigonometric functions. For instance,

$$\csc\theta = \frac{r}{y} = \frac{1}{(y/r)} = \frac{1}{\sin\theta}$$

The other two are proved similarly. Thus, the values of the cosecant, secant, and cotangent functions are found by using the reciprocal key 1/x or x^{-1} along with the function keys for the sine, cosine, and tangent, respectively. Before using a calculator to find function values of an angle given in radian measure, we must be sure that the calculator is in radian mode. For function values of an angle given in degree measure, we select degree mode.

Not all calculators follow the same sequence of keystrokes when computing trigonometric values of angles. The manual should be consulted for directions.

EXAMPLE 7 Using a calculator to approximate values of trigonometric functions

Use a calculator to approximate each value to four decimal places.

(a) $\sin 53.4°$ (b) $\cos 233°21'17''$ (c) $\csc(-4.6481)$ (d) $\cot \frac{\pi}{8}$

Solution

(a) After setting the calculator in degree mode, we get

$$\sin 53.4° = 0.8028 \text{ (approx.)}$$

(b) Here again, we use the degree mode to get

$$\cos 233°21'17'' = -0.5969 \text{ (approx.)}$$

(c) First we set the calculator in radian mode. Then we use a reciprocal identity as follows:

$$\csc(-4.6481) = \frac{1}{\sin(-4.6481)} = 1.0021 \text{ (approx.)}$$

(d) After setting the calculator in radian mode, we use the reciprocal identity to get

$$\cot \frac{\pi}{8} = \frac{1}{\tan(\pi/8)} = 2.4142 \text{ (approx.)}$$

PROBLEM SET 5.2

Mastering the Concepts

1. Figure 11 shows two right triangles. The first right triangle has sides of lengths 3, 4, and 5. In the second triangle, which is similar to the first, each side is four times longer than the corresponding side in the first. Find the exact values of the six trigonometric functions of θ for both triangles and compare these values.

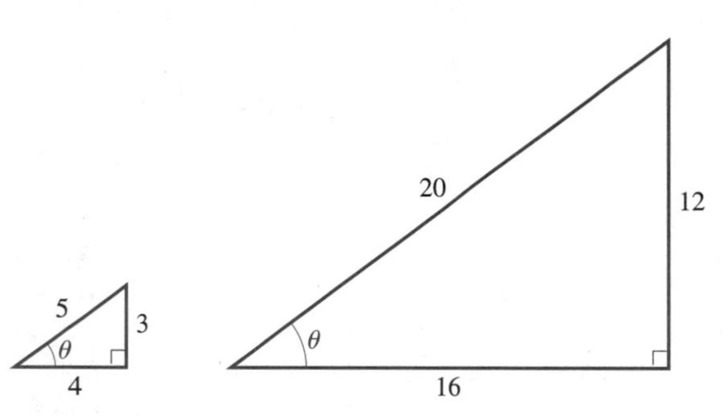

FIGURE 11

2. Figure 12 shows two right triangles. The first has sides of lengths 5, 12, and 13. In the second triangle, which is similar to the first, each side is half the length of the corresponding side in the first. Find the exact values of the six trigonometric functions of θ for both triangles and compare these values.

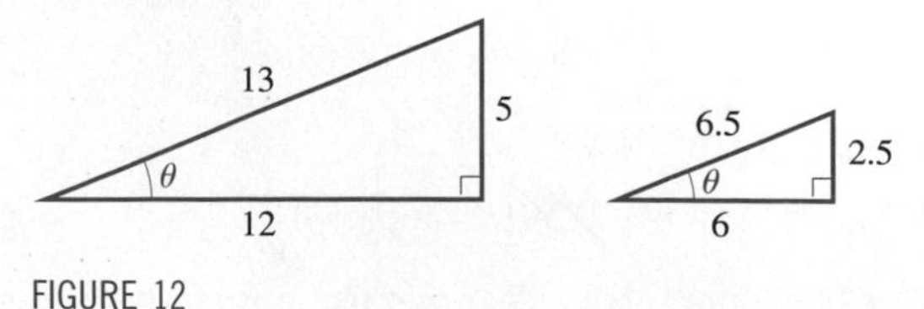

FIGURE 12

In problems 3–8, find the values of the six trigonometric functions of the acute angle θ for each right triangle. Round off the answers to three decimal places.

3. 8 cm, 5 cm, θ

4.

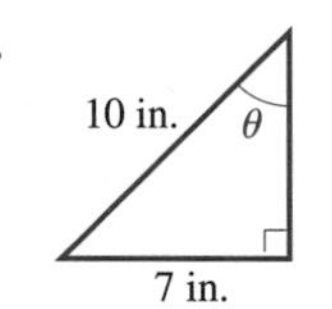

5. θ, 7.3 ft, 2.5 ft

6.

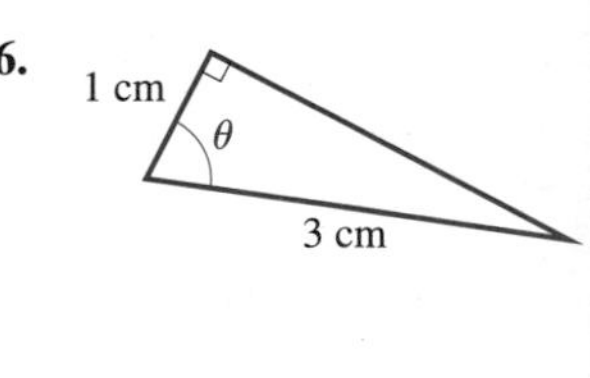

7. 8 m, θ, 17 m

8.

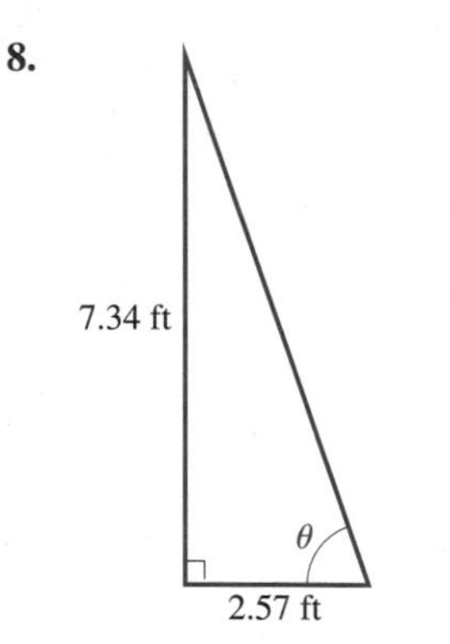

9. Let θ be an acute angle of a right triangle and $\sin \theta = \frac{4}{5}$. Draw such a triangle, and then find the exact values of the other five trigonometric functions of θ.

10. Let θ be an acute angle of a right triangle and $\cos \theta = \frac{12}{13}$. Draw such a triangle, and then find the exact values of the other five trigonometric functions of θ.

In problems 11–16, find the exact value of each expression.

11. $\sin 30° \cos 60° + \sin 60° \cos 30°$
12. $\cos 60° \cos 30° - \sin 60° \sin 30°$

13. $\sin\frac{\pi}{2}\cos\frac{\pi}{4} - \sin\frac{\pi}{4}\cos\frac{\pi}{2}$

14. $\sin\frac{\pi}{3}\cot\frac{\pi}{3} + 4\sec\frac{\pi}{3}$ **15.** $3\tan 45° - 2\cot 45°$

16. $\sec 60°\cos 60° + \csc 30°\sin 30°$

In problems 17–30, angle θ is in standard position. Find the values of the six trigonometric functions of angle θ if the terminal side of θ contains the given point (x, y). Sketch two possible angles, one positive and one negative, that satisfy the conditions on θ. Round off the answers to three decimal places in problems 20–24.

17. $(x, y) = (4, 3)$ **18.** $(x, y) = (-5, 12)$
19. $(x, y) = (-3, -4)$ **20.** $(x, y) = (7, -10)$
21. $(x, y) = (4, -5)$ **22.** $(x, y) = \left(-\sqrt{3}, -\sqrt{2}\right)$
23. $(x, y) = (1.35, -2.76)$ **24.** $(x, y) = (-7.89, 9.39)$
25. $(x, y) = (8, b);\ r = 17,\ b < 0$
26. $(x, y) = \left(a, -\frac{1}{2}\right);\ r = 1,\ a < 0$
27. $(x, y) = (a, a);\ a < 0$ **28.** $(x, y) = (-a, a);\ a > 0$
29. $(x, y) = \left(\pi, \frac{1}{\pi}\right)$ **30.** $(x, y) = \left(e, \frac{1}{e}\right)$

In problems 31 and 32, find the exact values of the six trigonometric functions of each angle.

31. **(a)** $90°$ **(b)** $-\frac{\pi}{2}$ **32.** **(a)** π **(b)** $-270°$

In problems 33–36, indicate the quadrant in which θ lies.

33. **(a)** $\cos\theta > 0$ and $\sin\theta < 0$
(b) $\sec\theta > 0$ and $\cot\theta < 0$
34. **(a)** $\sin\theta > 0$ and $\cos\theta < 0$
(b) $\sin\theta > 0$ and $\sec\theta < 0$
35. **(a)** $\tan\theta < 0$ and $\cos\theta < 0$
(b) $\tan\theta < 0$ and $\cos\theta > 0$
36. **(a)** $\cos\theta < 0$ and $\cot\theta < 0$
(b) $\sec\theta < 0$ and $\cot\theta > 0$

In problems 37–44, use either a 45°–45° or a 30°–60° right triangle as a reference triangle to find the exact value of each expression.

37. **(a)** $\sin 150°$ **(b)** $\cos 225°$
38. **(a)** $\cos\left(-\frac{7\pi}{6}\right)$ **(b)** $\cot 330°$
39. **(a)** $\tan 240°$ **(b)** $\cot\frac{11\pi}{6}$
40. **(a)** $\sec\left(-\frac{11\pi}{6}\right)$ **(b)** $\csc\left(-\frac{7\pi}{6}\right)$
41. **(a)** $\sec 315°$ **(b)** $\csc 300°$
42. **(a)** $\csc(-330°)$ **(b)** $\cot\left(-\frac{4\pi}{3}\right)$
43. **(a)** $\tan\left(-\frac{\pi}{4}\right)$ **(b)** $\cot 210°$
44. **(a)** $\sin\left(-\frac{7\pi}{4}\right)$ **(b)** $\sin\frac{7\pi}{6}$

In problems 45–52, use a calculator to determine the value of each function. Round off the answer to four decimal places.

45. **(a)** $\sin 43°$ **(b)** $\csc(-61.57°)$
46. **(a)** $\cos\left(-\frac{3\pi}{7}\right)$ **(b)** $\sec\frac{3\pi}{8}$
47. **(a)** $\cot 88°41'$ **(b)** $\sin(-103°8'12'')$
48. **(a)** $\sec 1.588$ **(b)** $\csc 1.234$
49. **(a)** $\tan\frac{\pi}{5}$ **(b)** $\cos 71°46'32''$
50. **(a)** $\csc 53.8°$ **(b)** $\sin(-0.453)$
51. **(a)** $\cos(-1.233)$ **(b)** $\cot 37.8°$
52. **(a)** $\csc 72°12'25''$ **(b)** $\tan 80.3°$

Developing and Extending the Concepts

53. A student makes the following assertion: Assume $\tan\theta = \frac{2}{3}$. Since

$$\tan\theta = \frac{\sin\theta}{\cos\theta}$$

it follows that $\sin\theta = 2$ and $\cos\theta = 3$. Consequently, the sine and cosine of an angle can have values larger than 1. Explain why this assertion is wrong.

54. **(a)** Suppose that (x, y) and (x_1, y_1) are two different points in quadrant I on the terminal side of θ, as shown in Figure 13.

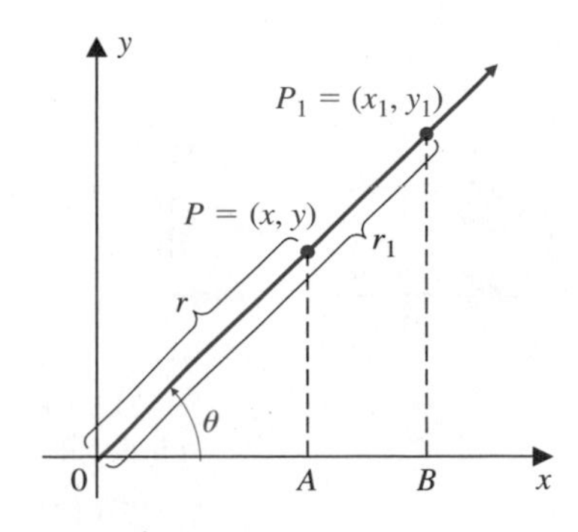

FIGURE 13

Use similar triangles to show that

$$\frac{y_1}{r_1} = \frac{y}{r} \quad \frac{x_1}{r_1} = \frac{x}{r} \quad \frac{y_1}{x_1} = \frac{y}{x}$$

$$\frac{r_1}{y_1} = \frac{r}{y} \quad \frac{r_1}{x_1} = \frac{r}{x} \quad \frac{x_1}{y_1} = \frac{x}{y}$$

(b) Show that the ratios in part (a) also hold if angle θ is in quadrant II.

(c) Show that the ratios in part (a) also hold if angle θ is in quadrant III.

(d) Show that the ratios in part (a) also hold if angle θ is in quadrant IV.

55. Use a right triangle to prove that if α and β are complementary angles, then the following equations are true. Give numerical examples to illustrate this result.

$$\sin \alpha = \cos \beta \qquad \tan \alpha = \cot \beta \qquad \csc \alpha = \sec \beta$$

OBJECTIVES

1. Associate real numbers with points on the unit circle.
2. Define the trigonometric functions of real numbers.
3. Evaluate trigonometric functions.

5.3 Trigonometric Functions of Real Numbers

Until now, we have considered trigonometric functions of angles, which we measured either in degrees or radians. In this section, we consider a second approach to the study of trigonometric functions, where the domains are real numbers. We use the unit circle to define these functions.

Associating Real Numbers with Points on the Unit Circle

Recall that the **unit circle** has a radius 1 with center at (0, 0) and standard equation

$$x^2 + y^2 = 1$$

EXAMPLE 1 Finding coordinates of points on the unit circle

Assume that the point $P = \left(-\frac{3}{5}, b\right)$ is in quadrant II and is on the unit circle. Find b.

Solution Since $P = \left(-\frac{3}{5}, b\right)$ is on the unit circle, its coordinates satisfy the equation $x^2 + y^2 = 1$. So,

$$\left(-\tfrac{3}{5}\right)^2 + b^2 = 1$$

$$\tfrac{9}{25} + b^2 = 1$$

$$b^2 = \tfrac{16}{25}$$

Since the point P is in quadrant II, $b > 0$, so $b = \sqrt{\frac{16}{25}} = \frac{4}{5}$. The coordinates of the point P are given by $\left(-\frac{3}{5}, \frac{4}{5}\right)$.

The circumference of a unit circle is 2π. Thus, the arc length of a quarter-circle is $(2\pi)/4 = \pi/2$, and the arc length of a half-circle is $(2\pi)/2 = \pi$. Table 1 displays some arc lengths on the unit circle, where one end of the arc is the point (1, 0) and the other end is located by moving counterclockwise along the circle.

TABLE 1
Examples of Arc Lengths on the Unit Circle

Arc Length	Decimal Approximation *Rounded off to two decimal places*	Graphical Display
$\frac{\pi}{6}$	0.52	y; $\frac{\pi}{6}$; (1, 0) x
$\frac{\pi}{4}$	0.79	y; $\frac{\pi}{4}$; (1, 0) x
$\frac{\pi}{3}$	1.05	y; $\frac{\pi}{3}$; (1, 0) x
$\frac{\pi}{2}$	1.57	y; $\frac{\pi}{2}$; (1, 0) x
$\frac{2\pi}{3}$	2.09	y; $\frac{2\pi}{3}$; (1, 0) x
π	3.14	y; π; (1, 0) x
$\frac{3\pi}{2}$	4.71	y; $\frac{3\pi}{2}$; (1, 0) x
2π	6.28	y; 2π; (1, 0) x

EXAMPLE 2 Identifying arc lengths on the unit circle

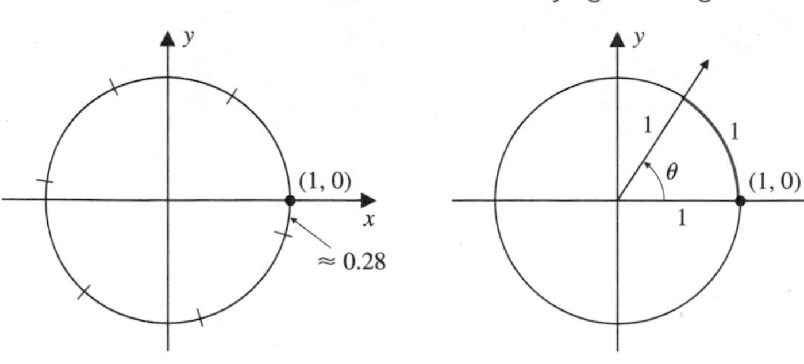

(a) Six arcs of length 1 plus an arc of length approx. 0.28

(b) θ is 1 radian

FIGURE 1

Suppose that the circumference of the unit circle is marked off by arcs of length 1, starting at (1, 0) and moving counterclockwise. Indicate the number of such arcs, and display the result on the unit circle.

Solution Since the circumference is 2π, or approximately 6.28, we conclude that there are 6 arcs of length 1 on the unit circle with an additional arc approximately 0.28 unit long (Figure 1a). As shown in Figure 1b, the central angle θ that intersects an arc of length 1 on the unit circle has a radian measure of 1.

We use arc lengths to associate each real number t with a point on the unit circle as follows:

1. If $t = 0$, associate it with the point (1, 0) (Figure 2a).
2. If $t > 0$, start at the point (1, 0) and determine an arc of length t units by moving *counterclockwise* (Figure 2b). Associate t with the resulting end point of the arc, labeled (x, y).
3. If $t < 0$, start at the point (1, 0) and determine an arc of length $|t|$ units by moving *clockwise* (Figure 2c). Associate t with the resulting end point of the arc, labeled (x, y).

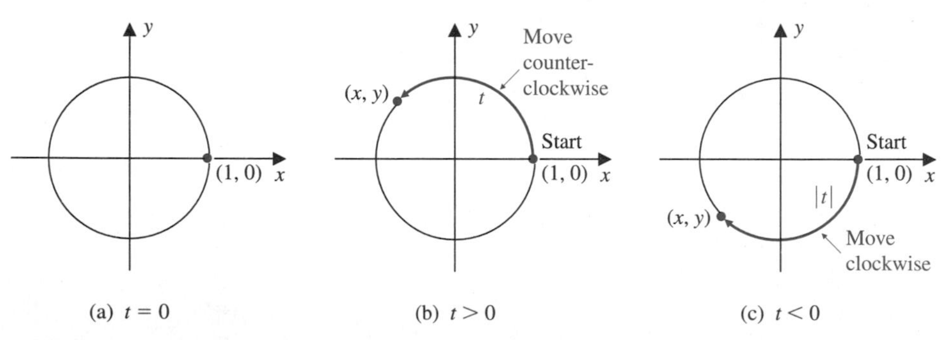

(a) $t = 0$

(b) $t > 0$

(c) $t < 0$

FIGURE 2

For instance, if $t = \pi/2$, we move counterclockwise $\pi/2$ units along the unit circle starting at (1, 0) to the associated point (0, 1) (Figure 3a). If $t = -\pi$, we move clockwise $|-\pi| = \pi$ units from (1, 0) to the associated point $(-1, 0)$ (Figure 3b).

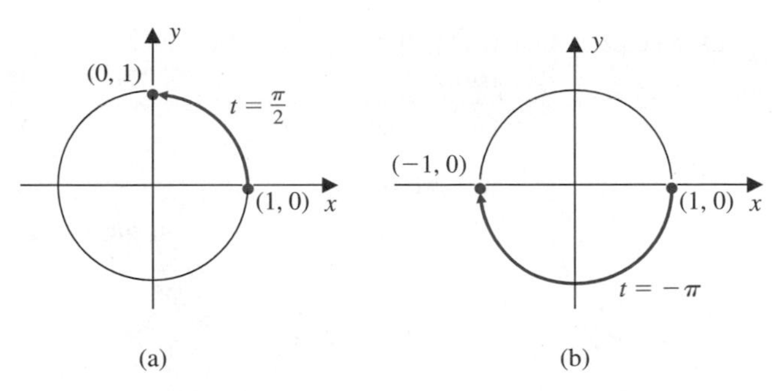

FIGURE 3

We denote this association between t and the coordinates (x, y) by using the function notation

$$P(t) = (x, y)$$

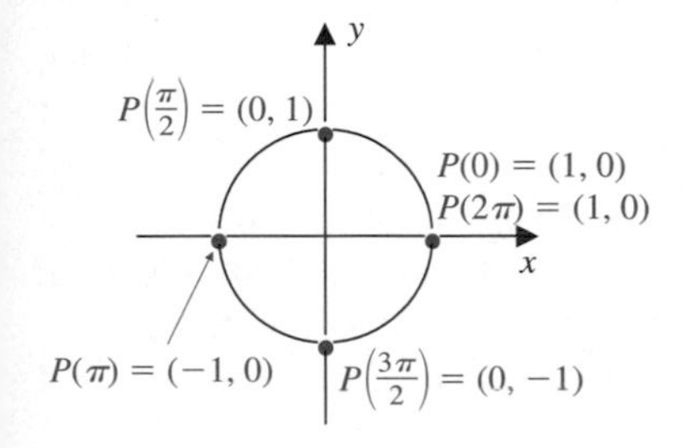

FIGURE 4

Table 2 lists some examples of the coordinates of $P(t)$ and Figure 4 displays their locations on the unit circle.

TABLE 2
Coordinates of $P(t)$

$P(t)$	$P(0)$	$P\left(\frac{\pi}{2}\right)$	$P(\pi)$	$P\left(\frac{3\pi}{2}\right)$	$P(2\pi)$
(x, y)	$(1, 0)$	$(0, 1)$	$(-1, 0)$	$(0, -1)$	$(1, 0)$

Defining the Trigonometric Functions of Real Numbers

The coordinates (x, y) of a point $P(t)$ on the *unit* circle are used to define the six *trigonometric functions of real number t*. These functions are also called the *circular functions of t*.

DEFINITION

TRIGONOMETRIC FUNCTIONS OF A REAL NUMBER

Let t be any real number and assume $P(t) = (x, y)$ (Figure 5) is the associated point on the unit circle. Then the **trigonometric functions (or circular functions) of t** are defined as follows:

$$\sin t = y \qquad \csc t = \frac{1}{y}, \quad y \neq 0$$

$$\cos t = x \qquad \sec t = \frac{1}{x}, \quad x \neq 0$$

$$\tan t = \frac{y}{x}, \quad x \neq 0 \qquad \cot t = \frac{x}{y}, \quad y \neq 0$$

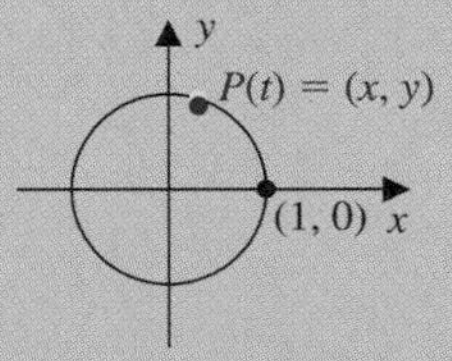

y
$P(t) = (x, y)$
(1, 0) x

FIGURE 5

These definitions hold for any location of $P(t)$.

EXAMPLE 3 **Evaluating the trigonometric functions**

Determine the exact values of the six trigonometric functions of t if the point $P(t) = \left(-\frac{4}{5}, \frac{3}{5}\right)$.

Solution Notice that $x = -\frac{4}{5}$, $y = \frac{3}{5}$, and $x^2 + y^2 = \left(-\frac{4}{5}\right)^2 + \left(\frac{3}{5}\right)^2 = 1$, so the point $P(t) = \left(-\frac{4}{5}, \frac{3}{5}\right)$ lies on the unit circle. It follows from the definitions that

$$\sin t = y = \frac{3}{5} \qquad \csc t = \frac{1}{y} = \frac{1}{\frac{3}{5}} = \frac{5}{3}$$

$$\cos t = x = -\frac{4}{5} \qquad \sec t = \frac{1}{x} = \frac{1}{-\frac{4}{5}} = -\frac{5}{4}$$

$$\tan t = \frac{y}{x} = \frac{\frac{3}{5}}{-\frac{4}{5}} = -\frac{3}{4} \qquad \cot t = \frac{x}{y} = \frac{-\frac{4}{5}}{\frac{3}{5}} = -\frac{4}{3}$$

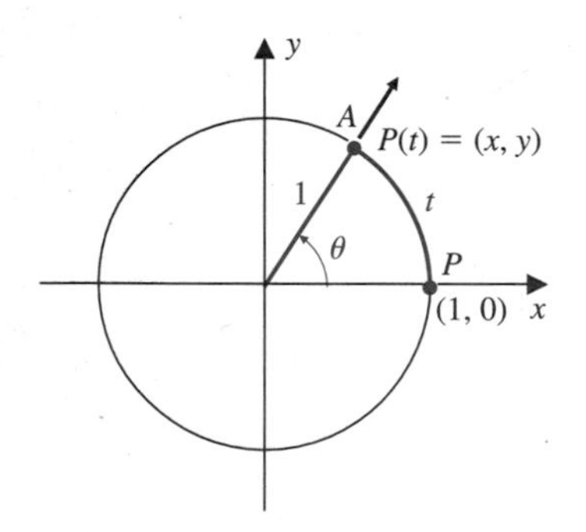

FIGURE 6
θ has radian measure t

Now that we have defined the trigonometric functions of angles and of real numbers, we will explore the relationships between these two kinds of functions. Upon examining $P(t) = (x, y)$ more closely, we observe that each real number t defines an arc $\widehat{AP}$ that subtends a central angle θ in the unit circle, as illustrated in Figure 6. We know from Section 5.1 that the radian measure of θ is t. Thus, we may think of t as a real number *or* as the measure of a directed arc length on the unit circle *or* as the radian measure of a central angle θ. This observation enables us to establish the connection between the trigonometric functions of real numbers and those of angles.

Suppose that t is a real number. Then the definitions of the trigonometric functions of *real numbers* give us

$$\sin t = y \quad \text{and} \quad \cos t = x \quad \text{(Figure 7)}$$

If we consider θ to be the angle in standard position with radian measure t, the definitions of the trigonometric functions of *angles* specify that

$$\sin \theta = \frac{y}{r} = \frac{y}{1} = y \quad \text{and} \quad \cos \theta = \frac{x}{r} = \frac{x}{1} = x$$

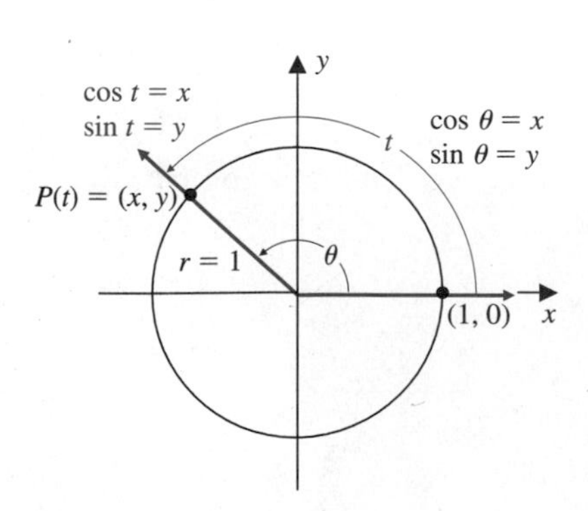

FIGURE 7

Thus, we conclude that

$$\sin t = \sin \theta \quad \text{and} \quad \cos t = \cos \theta$$

We see that the values of the sine and cosine functions of the real number t are the same as the values of the sine and cosine functions of an angle θ with radian measure t, respectively. Similar relationships hold for the other four trigonometric functions. In general, we have the following property:

PROPERTY

> The values of the trigonometric functions of angle θ are the same as the values of the trigonometric functions of real number t, the radian measure of θ.

Thus, the value of the sine of the real number $\pi/5$ is the same as the value of the sine of the angle $\pi/5$ radian, which, in turn, is the same as the value of the sine of 36° ($\pi/5 = 36°$). With the above property in mind, we can now use the term *trigonometric functions* regardless of whether *angles* or *real numbers* are employed. Keep in mind that the reciprocal relationships are also true for real numbers. Furthermore, the signs of values (on page 293) are true for real numbers.

Evaluating Trigonometric Functions

Using the above property, we can find the exact values of trigonometric functions for multiples of real numbers $\pi/6$, $\pi/4$, and $\pi/3$ by using Table 1 on page 291 and reference triangles.

EXAMPLE 4 Evaluating trigonometric functions of real numbers

Find the exact value of each expression.

(a) $\sin \frac{\pi}{4}$ (b) $\cos \frac{5\pi}{6}$

Solution

(a) $\sin \frac{\pi}{4} = \sin \frac{\pi}{4} = \sin 45° = \frac{\sqrt{2}}{2}$

$\sin \frac{\pi}{4}$ =	$\sin \frac{\pi}{4}$ =	$\sin 45° = \frac{\sqrt{2}}{2}$
↑ Real number	↑ Radian measure of angle	↑ Degree measure of angle

(b) $\cos \frac{5\pi}{6} = \cos \frac{5\pi}{6} = \cos 150°$

$\cos \frac{5\pi}{6}$ =	$\cos \frac{5\pi}{6}$ =	$\cos 150°$
↑ Real number	↑ Radian measure of angle	↑ Degree measure of angle

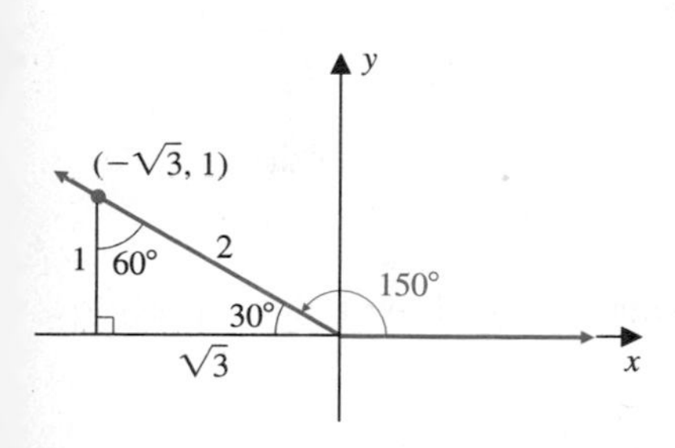

FIGURE 8

By using a 30°–60° reference triangle (Figure 8), we find that

$$\cos 150° = \frac{x}{r} = -\frac{\sqrt{3}}{2}$$

So

$$\cos \frac{5\pi}{6} = -\frac{\sqrt{3}}{2}$$

From the definitions of the trigonometric functions we know that if the point $P(t) = (x, y)$, then $\cos t = x$ and $\sin t = y$. It follows that

$$\boxed{P(t) = (\cos t, \sin t)}$$

Thus,

$$P\left(\frac{\pi}{2}\right) = \left(\cos\frac{\pi}{2}, \sin\frac{\pi}{2}\right) = (0, 1) \quad \text{and} \quad P(\pi) = (\cos\pi, \sin\pi) = (-1, 0)$$

In addition to the points $P(t)$ on the coordinate axes, such as $t = \pi/2, \pi, 0$, or $-\pi$, there are other special points on the unit circle for which we can identify the coordinates of $P(t)$ in exact form. To do so, it is helpful to first recall the 45°–45° or 30°–60° right triangles. For instance,

$$P\left(\frac{\pi}{6}\right) = \left(\cos\frac{\pi}{6}, \sin\frac{\pi}{6}\right) = \left(\frac{\sqrt{3}}{2}, \frac{1}{2}\right)$$

and

$$P\left(\frac{5\pi}{4}\right) = \left(\cos\frac{5\pi}{4}, \sin\frac{5\pi}{4}\right) = \left(-\frac{\sqrt{2}}{2}, -\frac{\sqrt{2}}{2}\right)$$

Since the value of each trigonometric function at a real number t is the same as its value at an angle of t radians, we can approximate values of these functions by using a calculator set in radian mode.

EXAMPLE 5 **Approximating trigonometric function values**

Use a calculator to approximate each value to four decimal places.

(a) $\cos(-1.73)$ (b) $\cot\frac{7\pi}{30}$

Solution First, we set the calculator in radian mode. Then we perform the appropriate sequence of keystrokes to obtain the following approximations:

(a) $\cos(-1.73) = -0.1585$ (b) $\cot\frac{7\pi}{30} = \dfrac{1}{\tan\frac{7\pi}{30}} = 1.1106$

We can also use a calculator to approximate the coordinates of $P(t)$.

EXAMPLE 6 **Locating $P(t)$ by using calculator values**

Find the coordinates of each point. Round off the answers to four decimal places.

(a) $P(2)$ (b) $P(-0.725)$

Solution We set the calculator in radian mode to obtain the following approximations:

(a) $P(2) = (\cos 2, \sin 2) = (-0.4161, 0.9093)$
(b) $P(-0.725) = (\cos(-0.725), \sin(-0.725)) = (0.7485, -0.6631)$

PROBLEM SET 5.3

Mastering the Concepts

In problems 1–6, find the missing coordinate of each point on the unit circle.

1. **(a)** $\left(\frac{8}{17}, b\right)$, in quadrant I
(b) $\left(a, -\frac{15}{17}\right)$, in quadrant IV

2. **(a)** $\left(a, -\frac{1}{3}\right)$, in quadrant III
(b) $\left(-\frac{2}{5}, b\right)$, in quadrant II

3. **(a)** $(a, -2a)$, in quadrant IV
(b) $(3b, 4b)$, in quadrant I

4. $(-0.4561, b)$, in quadrant II (Round off to four decimal places.)

5. $(a, -0.8913)$, in quadrant III (Round off to four decimal places.)

6. $(0.75a, -0.23)$, in quadrant IV (Round off to four decimal places.)

In problems 7–12, mark off arcs of the given length on the unit circle starting at (1, 0) and moving in the specified direction. Indicate the number of such arcs (use $\pi = 3.14$ in problems 9–12).

7. Arcs of length $\pi/5$, moving counterclockwise
8. Arcs of length $\pi/8$, moving clockwise
9. Arcs of length 2, moving clockwise
10. Arcs of length 3, moving counterclockwise
11. Arcs of length 0.5, moving counterclockwise
12. Arcs of length 0.75, moving clockwise

In problems 13–20, display each point on the unit circle and find the quadrant (if any) containing each point.

13. **(a)** $P\left(\frac{\pi}{7}\right)$ **(b)** $P\left(-\frac{3\pi}{5}\right)$

14. **(a)** $P\left(\frac{-2\pi}{5}\right)$ **(b)** $P\left(\frac{17\pi}{8}\right)$

15. **(a)** $P\left(\frac{17\pi}{6}\right)$ **(b)** $P\left(\frac{-8\pi}{7}\right)$

16. **(a)** $P\left(\frac{31\pi}{8}\right)$ **(b)** $P\left(-\frac{9\pi}{5}\right)$

17. **(a)** $P(6)$ **(b)** $P(1.4)$

18. **(a)** $P(3.6)$ **(b)** $P(-5.7)$

19. **(a)** $P(-11.2)$ **(b)** $P(23.5)$

20. **(a)** $P(-3.1)$ **(b)** $P(-13.4)$

In problems 21–26, show that each point $P(t)$ lies on the unit circle, and then find the values of the trigonometric functions of t.

21. $P(t) = \left(-\frac{\sqrt{3}}{2}, \frac{1}{2}\right)$ **22.** $P(t) = \left(-\frac{4}{5}, -\frac{3}{5}\right)$

23. $P(t) = \left(\frac{5}{13}, -\frac{12}{13}\right)$ **24.** $P(t) = \left(\frac{1}{\sqrt{10}}, \frac{3}{\sqrt{10}}\right)$

25. $P(t) = \left(\frac{-3}{\sqrt{13}}, \frac{2}{\sqrt{13}}\right)$ **26.** $P(t) = \left(\frac{-4}{\sqrt{17}}, \frac{-1}{\sqrt{17}}\right)$

In problems 27–30, determine the coordinates of $P(t)$. Then find the exact values of the trigonometric functions of t.

27. **(a)** $t = -\frac{\pi}{2}$ **(b)** $t = \frac{3\pi}{2}$

28. **(a)** $t = \pi$ **(b)** $t = \frac{5\pi}{2}$

29. **(a)** $t = -2\pi$ **(b)** $t = -\frac{5\pi}{2}$

30. **(a)** $t = \frac{-7\pi}{2}$ **(b)** $t = -\frac{3\pi}{2}$

In problems 31–38, find the exact value of each expression.

31. **(a)** $\sin\frac{3\pi}{4}$ **(b)** $\cos\frac{4\pi}{3}$

32. **(a)** $\sin\left(-\frac{3\pi}{4}\right)$ **(b)** $\cos\frac{7\pi}{4}$

33. **(a)** $\sin\frac{11\pi}{6}$ **(b)** $\cos\left(-\frac{5\pi}{6}\right)$

34. **(a)** $\sin\frac{5\pi}{3}$ **(b)** $\cos\left(-\frac{7\pi}{4}\right)$

35. **(a)** $\sin\left(-\frac{\pi}{4}\right)$ **(b)** $\cos\frac{5\pi}{4}$

36. **(a)** $\sin\left(-\frac{7\pi}{4}\right)$ **(b)** $\cos\left(-\frac{5\pi}{4}\right)$

37. **(a)** $\sin\left(-\frac{11\pi}{6}\right)$ **(b)** $\cos\left(-\frac{3\pi}{4}\right)$

38. **(a)** $\sin\left(-\frac{5\pi}{3}\right)$ **(b)** $\cos\left(-\frac{4\pi}{3}\right)$

In problems 39–44, use a calculator to find the approximate value of each expression rounded off to four decimal places.

39. **(a)** $\sin 0.743$ **(b)** $\cos(-2.31)$

40. **(a)** $\cos 1.4$ **(b)** $\sin(-3.42)$

41. **(a)** $\tan 0.572$ **(b)** $\cot 0.537$

42. **(a)** tan 2.178 **(b)** cot 8.213
43. **(a)** sec 0.839 **(b)** csc(−1.093)
44. **(a)** csc 0.536 **(b)** sec(−5.22)

In problems 45–48, use $P(t) = (x, y) = (\cos t, \sin t)$ to find the coordinates of each point. Round off the answers to four decimal places.

45. **(a)** $P(2.7)$ **(b)** $P(-4.18)$
46. **(a)** $P(-3.58)$ **(b)** $P(5.73)$
47. **(a)** $P(-0.432)$ **(b)** $P(5.129)$
48. **(a)** $P(0.572)$ **(b)** $P(-0.636)$

Developing and Extending the Concepts

49. Which trigonometric functions of t are not defined when $P(t)$ is on the y axis?

50. Which trigonometric functions of t are not defined when $P(t)$ is on the x axis?

In problems 51 and 52, find the values of the trigonometric functions of t under the given conditions.

51. $P(t)$ is in quadrant III and lies on the line $y = 3x$.

52. $P(t)$ is in quadrant IV and lies on the line $3x + 2y = 0$.

53. The function

$$f(t) = \frac{\sin t}{t}$$

is used in calculus. For parts (a)–(d), find each value and round off each answer to four decimal places.
(a) $f(0.1)$ **(b)** $f(0.01)$
(c) $f(0.001)$ **(d)** $f(0.0001)$
(e) Discuss the behavior of f as $t \to 0^+$.

54. The function

$$g(t) = \frac{1 - \cos t}{t}$$

is also used in calculus. Apply the instructions of problem 53 to the function g.

55. In calculus it is shown that cos t can be approximated by the polynomial function

$$f(t) = 1 - \frac{t^2}{2} + \frac{t^4}{24} \qquad \text{where } -1 \le t \le 1$$

Compare cos 0.1 and $f(0.1)$, cos 0.2 and $f(0.2)$, cos(−0.3) and $f(-0.3)$, and cos 1 and $f(1)$. Round off the values to four decimal places.

56. In calculus it is shown that sin t can be approximated by the polynomial function

$$g(t) = t - \frac{t^3}{6} + \frac{t^5}{120} \qquad \text{where } -1 \le t \le 1$$

Compare sin 0.1 and $g(0.1)$, sin 0.2 and $g(0.2)$, sin(−0.3) and $g(-0.3)$, and sin(−1) and $g(-1)$. Round off the values to four decimal places.

OBJECTIVES

1. Find the periodicity of trigonometric functions.
2. Establish fundamental identities.
3. Determine even and odd trigonometric functions.
4. Solve applied problems.

5.4 Properties of Trigonometric Functions

In this section we'll use the definitions of the trigonometric functions to develop some of their properties.

Finding the Periodicity of Trigonometric Functions

We'll now consider a basic property of the trigonometric functions that simplifies the work of finding their values and graphs. This property characterizes the repetitive nature of these functions and is referred to as *periodicity.*

Suppose that an angle with radian measure t is placed in standard position and intersects the unit circle at the point P on its terminal side (Figure 1a). Since the circumference of the unit circle is 2π, if we start at the point (1, 0) and go a distance of $t + 2\pi$ along the arc of this circle, we will arrive at the same point P on the circle (Figure 1b).

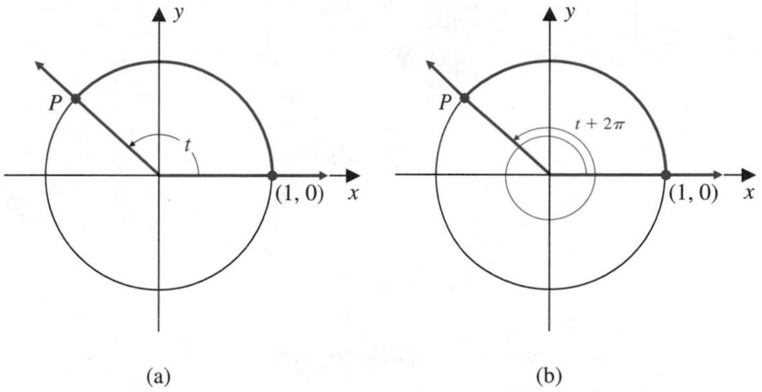

(a) (b)

FIGURE 1

From the definitions of the trigonometric functions, we observe that the coordinates of P are either

$$(\cos t, \sin t) \qquad \text{or} \qquad (\cos(t + 2\pi), \sin(t + 2\pi))$$

Because this is the same point, these coordinates must be equal. That is,

$$\cos t = \cos(t + 2\pi) \qquad \text{and} \qquad \sin t = \sin(t + 2\pi)$$

Similar results are obtained for repeated rotations (positive or negative) of length 2π around the unit circle. Thus, we have the following general property:

$$\sin(t + 2\pi k) = \sin t \qquad \text{and} \qquad \cos(t + 2\pi k) = \cos t$$

for any integer k and real number t. Because of this property, the sine and cosine functions are said to be *periodic* in the sense that their values repeat themselves whenever t increases or decreases by an integer multiple of 2π. In terms of degree measures, we have

$$\sin(\theta + 360° \cdot k) = \sin \theta \qquad \text{and} \qquad \cos(\theta + 360° \cdot k) = \cos \theta$$

EXAMPLE 1 **Using periodicity to evaluate trigonometric functions**

Use periodicity to find the exact value.

(a) $\sin \dfrac{13\pi}{6}$ (b) $\cos(-945°)$

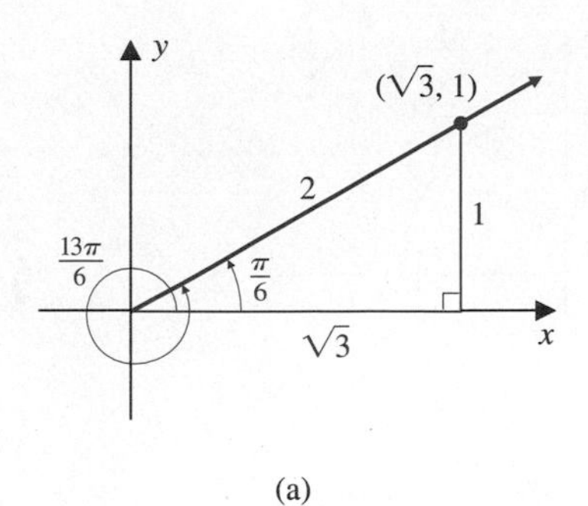

(a)

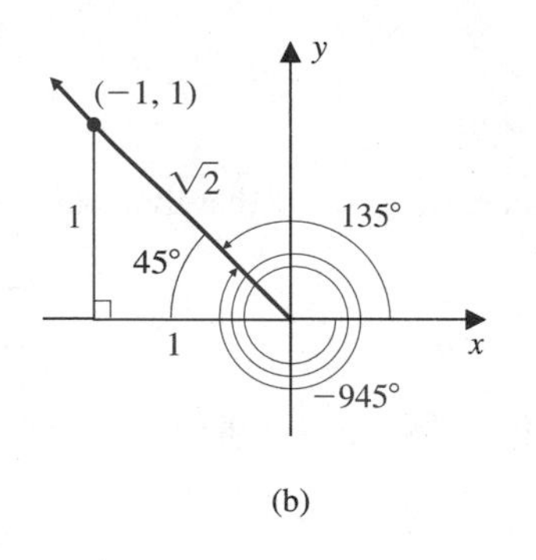

(b)

FIGURE 2

Solution

(a) Since

$$\frac{13\pi}{6} = \frac{\pi}{6} + \frac{12\pi}{6} = \frac{\pi}{6} + 2\pi$$

it follows that

$$\sin\frac{13\pi}{6} = \sin\left(\frac{\pi}{6} + 2\pi\right) = \sin\frac{\pi}{6} = \frac{1}{2} \quad \text{(Figure 2a)}$$

(b) Because

$$-945° = 135° - 1080° = 135° + (-3)\cdot(360°)$$

it follows that

$$\cos(-945°) = \cos[135° + (-3)\cdot 360°] = \cos 135° = -\frac{1}{\sqrt{2}} \quad \text{(Figure 2b)}$$

The notion of periodic, or cyclic, phenomena is encountered in everyday life. For instance, the days of the week repeat every 7 days and the minute hand on a clock repeats its rotation every 60 minutes. In mathematics, a function f is said to be **periodic** if there exists a positive real number p for which

$$f(t + p) = f(t)$$

for all t in the domain of f. The smallest number p is called the **fundamental period** of f. It follows from this definition, for instance, that

$$f(t - p) = f[(t - p) + p] = f(t) \qquad \text{and} \qquad f(t) = f(t + 2p) = f(t - 2p)$$

Hence, the values of a periodic function of period p are repeated whenever t increases or decreases by a multiple of p. Thus, we have established that the sine and the cosine functions are periodic with period 2π. We can use this property to find periods of other functions, as illustrated in the next example.

EXAMPLE 2 Finding the period of a trigonometric function

Use the fact that the sine function has period 2π to find the period of $f(t) = \sin 2t$.

Solution

$f(t) = \sin 2t$	Given
$= \sin(2t + 2\pi)$	The sine has period 2π
$= \sin[2(t + \pi)]$	Factor out 2
$= f(t + \pi)$	Function notation

Thus, f is periodic with period π. This means that the values of f repeat every π units.

Since

$$\sec t = \frac{1}{\cos t} \qquad \text{and} \qquad \csc t = \frac{1}{\sin t}$$

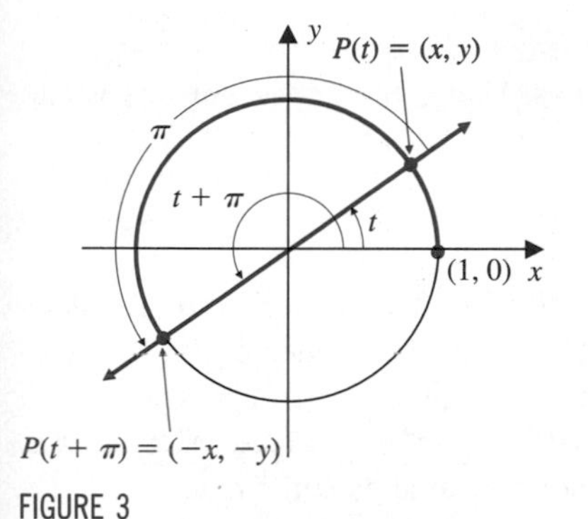

FIGURE 3

each of these two functions has a period of 2π (Problem 45). Each of the tangent and cotangent functions also has a period of 2π; however, each of these functions has a *fundamental* period of π. As illustrated in Figure 3, $P(t) = (x, y)$ lies on the terminal side of an angle with radian measure t, and $P(t + \pi)$ lies on the terminal side of an angle with radian measure $t + \pi$. Since $P(t)$ and $P(t + \pi)$ are symmetric points with respect to the origin, it follows that their coordinates are (x, y) and $(-x, -y)$, respectively.

By definition, $\tan t = y/x$ and $\tan(t + \pi) = (-y)/(-x) = y/x$, so

$$\boxed{\tan(t + \pi) = \tan t}$$

For example,

$$\tan \frac{10\pi}{3} = \tan\left(\frac{\pi}{3} + \frac{9\pi}{3}\right) = \tan\left(\frac{\pi}{3} + 3\pi\right) = \tan \frac{\pi}{3} = \sqrt{3}$$

Similarly, it can be shown that the cotangent also has a period of π (Problem 47); that is,

$$\boxed{\cot(t + \pi) = \cot t}$$

Establishing Fundamental Identities

The trigonometric functions have special relationships usually expressed as equations called *identities*. An **identity** is an equation that is true for all values of the variable for which both sides of the equation are defined. In Section 5.2, we derived the *reciprocal identities*. Here, we'll establish two additional types of **fundamental identities**—the *quotient* and the *Pythagorean identities*. To derive these identities, we consider an angle in standard position with radian measure t so that $P(t) = (x, y)$ (Figure 4).

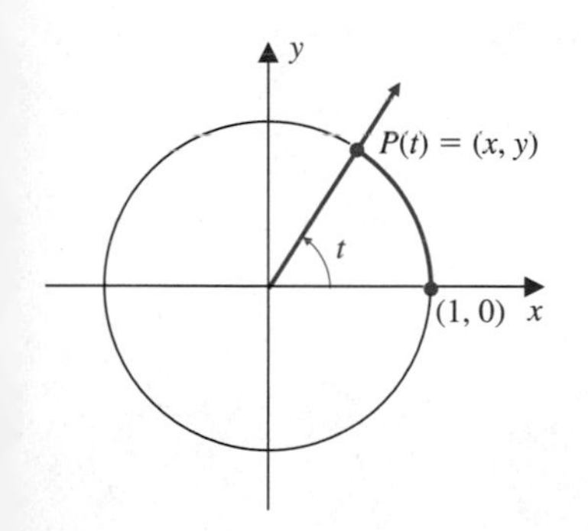

FIGURE 4

It follows from the definitions of the trigonometric functions of t that

$$\tan t = \frac{y}{x} = \frac{\sin t}{\cos t} \qquad \text{and} \qquad \cot t = \frac{x}{y} = \frac{\cos t}{\sin t}$$

Thus, we have the following results:

THE QUOTIENT IDENTITIES

(i) $\tan t = \dfrac{\sin t}{\cos t}$ (ii) $\cot t = \dfrac{\cos t}{\sin t}$

Another important identity is derived by using the equation $x^2 + y^2 = 1$ and the definitions of the cosine and sine functions. Using $\cos t = x$, $\sin t = y$, and $x^2 + y^2 = 1$, it follows that

$$(\cos t)^2 + (\sin t)^2 = 1$$

This relationship is called a Pythagorean identity, because its derivation involves the fact that $x^2 + y^2 = 1$, which can be considered a consequence of the Pythagorean theorem. It is customary to write powers of trigonometric expressions such as $(\sin t)^n$ and $(\cos t)^n$, respectively, as $(\sin t)^n = \sin^n t$ and $(\cos t)^n = \cos^n t$. However, as we shall see later, $(\sin t)^{-1}$ is not the same as $\sin^{-1} t$.

So $$\cos^2 t = (\cos t)^2 \quad \text{and} \quad \sin^2 t = (\sin t)^2$$

Thus, the identity

$$(\cos t)^2 + (\sin t)^2 = 1 \quad \text{is written as} \quad \cos^2 t + \sin^2 t = 1$$

We can use this latter identity to prove two other identities. For instance, if we divide both sides of the identity by $\cos^2 t$, we get

$$\frac{\cos^2 t}{\cos^2 t} + \frac{\sin^2 t}{\cos^2 t} = \frac{1}{\cos^2 t}$$

$$1 + \left(\frac{\sin t}{\cos t}\right)^2 = \left(\frac{1}{\cos t}\right)^2$$

$$1 + \tan^2 t = \sec^2 t$$

Similarly, if we divide both sides of the identity $\cos^2 t + \sin^2 t = 1$ by $\sin^2 t$ and then simplify the result, we get

$$\cot^2 t + 1 = \csc^2 t$$

Thus, we are led to the following results:

PYTHAGOREAN IDENTITIES

(i) $\cos^2 t + \sin^2 t = 1$ (ii) $1 + \tan^2 t = \sec^2 t$
(iii) $1 + \cot^2 t = \csc^2 t$

The fundamental identities can be used to compute values of the trigonometric functions and simplify trigonometric expressions.

EXAMPLE 3 **Evaluating functions by using identities**

Use identities to find the exact values of the other five trigonometric functions of t if $\sin t = \frac{8}{17}$ and $\cos t < 0$.

Solution Since $\sin t$ is positive and $\cos t$ is negative, it follows that $P(t)$ is in quadrant II. Using $\cos^2 t + \sin^2 t = 1$, we have $\cos^2 t = 1 - \sin^2 t$, so $\cos t = \pm\sqrt{1 - \sin^2 t}$. We know that $\cos t$ is negative, so

$$\cos t = -\sqrt{1 - \left(\tfrac{8}{17}\right)^2} = -\sqrt{1 - \tfrac{64}{289}} = -\sqrt{\tfrac{225}{289}} = -\tfrac{15}{17}$$

Next, we use the quotient and reciprocal identities to obtain

$$\tan t = \frac{\sin t}{\cos t} = \frac{\frac{8}{17}}{-\frac{15}{17}} = \frac{-8}{15} \qquad \cot t = \frac{\cos t}{\sin t} = \frac{-\frac{15}{17}}{\frac{8}{17}} = \frac{-15}{8}$$

$$\sec t = \frac{1}{\cos t} = \frac{1}{-\frac{15}{17}} = \frac{-17}{15} \qquad \csc t = \frac{1}{\sin t} = \frac{1}{\frac{8}{17}} = \frac{17}{8}$$

EXAMPLE 4 **Simplifying an expression**

Use identities to simplify the expression: $\cot^2 t \sin^2 t + \sin^2 t$

Solution

$$\cot^2 t \sin^2 t + \sin^2 t = \frac{\cos^2 t}{\cancel{\sin^2 t}} \cdot \cancel{\sin^2 t} + \sin^2 t \quad \text{Quotient identity}$$

$$= \cos^2 t + \sin^2 t = 1 \quad \text{Pythagorean identity}$$

Determining Even and Odd Trigonometric Functions

Now we consider the effects of replacing t by $-t$ in the trigonometric functions. Consider the situation illustrated in Figure 5. By using the definitions of the cosine and sine functions, we get

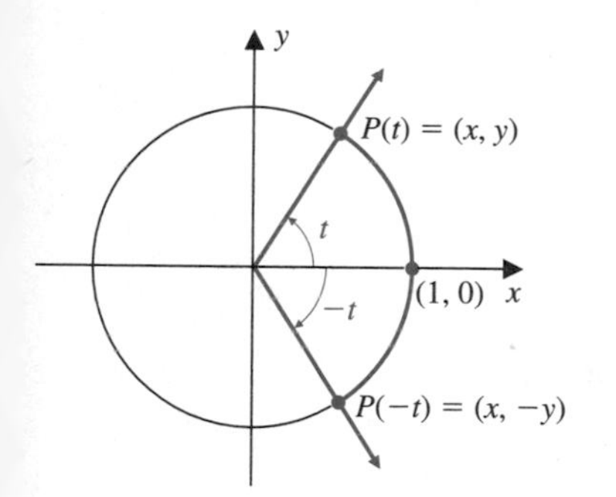

FIGURE 5

$$\cos t = x \quad \text{and} \quad \cos(-t) = x, \quad \text{so} \quad \cos(-t) = \cos t$$

Similarly,

$$\sin t = y \quad \text{and} \quad \sin(-t) = -y, \quad \text{so} \quad \sin(-t) = -\sin t$$

Also,

$$\tan(-t) = \frac{\sin(-t)}{\cos(-t)} = \frac{-\sin t}{\cos t} = -\tan t$$

that is, $\tan(-t) = -\tan t$.

Thus, we obtain the following identities:

EVEN–ODD TRIGONOMETRIC IDENTITIES

(i) $\sin(-t) = -\sin t$ (ii) $\cos(-t) = \cos t$ (iii) $\tan(-t) = -\tan t$

These identities can be used to derive even–odd identities for the cotangent, secant, and cosecant functions (Problem 48).

EXAMPLE 5 **Using the even–odd identities**

Suppose that $\sin t = \frac{3}{5}$ and $\cos t = \frac{4}{5}$. Find:

(a) $\tan(-t)$ (b) $\sec(-t)$

Solution

(a) $\tan(-t) = \dfrac{\sin(-t)}{\cos(-t)} = \dfrac{-\sin t}{\cos t} = \dfrac{-\frac{3}{5}}{\frac{4}{5}} = -\dfrac{3}{4}$

(b) $\sec(-t) = \dfrac{1}{\cos(-t)} = \dfrac{1}{\cos t} = \dfrac{1}{\frac{4}{5}} = \dfrac{5}{4}$

The even–odd identities are used at times to simplify expressions, as the next example shows.

EXAMPLE 6 **Simplifying an expression**

Use identities to simplify the expression: $\dfrac{\cot\theta\cos(-\theta)}{\csc^2\theta - 1}$

Solution

$$\frac{\cot\theta\cos(-\theta)}{\csc^2\theta - 1} = \frac{\cot\theta\cos\theta}{\csc^2\theta - 1} \qquad \cos(-\theta) = \cos\theta$$

$$= \frac{\frac{\cos\theta}{\sin\theta}\cos\theta}{\cot^2\theta} = \frac{\frac{\cos^2\theta}{\sin\theta}}{\frac{\cos^2\theta}{\sin^2\theta}} \qquad \text{Quotient and Pythagorean identities}$$

$$= \frac{\cos^2\theta}{\sin\theta}\cdot\frac{\sin^2\theta}{\cos^2\theta} = \sin\theta \qquad \text{Simplify}$$

Solving Applied Problems

Some periodic phenomena that exhibit repeating behavior can be modeled by the sine and cosine functions. We'll illustrate such modeling in the next example.

EXAMPLE 7 **Modeling a Ferris wheel**

Suppose that a Ferris wheel with a boarding ramp above ground level has a radius of 49 feet (Figure 6). A model for determining the height above ground $h(t)$ (in feet) of a rider after t seconds of travel is given by

$$h(t) = 54 + 49\sin\left(\frac{\pi}{12}t - \frac{\pi}{2}\right)$$

(a) Find the rider's height above ground level at the start, after 1 second, and after 12 seconds. Round off the answers to two decimal places.
(b) Find the period of h and interpret this value.
(c) A ride on the Ferris wheel includes six complete rotations. How long does a ride take?

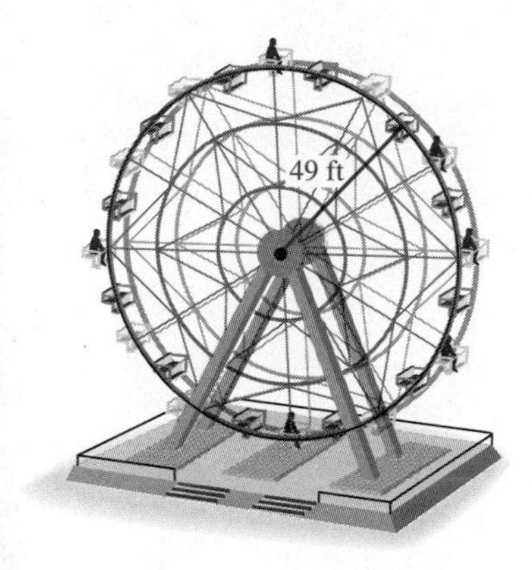

FIGURE 6

Solution

(a) At the start, $t = 0$, so the rider's height is given by

$$\begin{aligned} h(0) &= 54 + 49 \sin\left(\frac{\pi}{12} \cdot 0 - \frac{\pi}{2}\right) \\ &= 54 + 49 \sin\left(-\frac{\pi}{2}\right) \\ &= 54 + 49(-1) = 5 \quad \text{Since } \sin(-\pi/2) = -\sin(\pi/2) = -1 \end{aligned}$$

Thus, the rider is 5 feet above ground level on the boarding ramp when the ride begins. After 1 second,

$$\begin{aligned} h(1) &= 54 + 49 \sin\left(\frac{\pi}{12} - \frac{\pi}{2}\right) \\ &= 54 + 49 \sin\left(-\frac{5\pi}{12}\right) = 6.67 \text{ (approx.)} \end{aligned}$$

So the rider's height above ground level is about 6.67 feet after 1 second. When $t = 12$,

$$\begin{aligned} h(12) &= 54 + 49 \sin\left(\frac{\pi}{\cancel{12}} \cdot \cancel{12} - \frac{\pi}{2}\right) \\ &= 54 + 49 \sin\left(\frac{\pi}{2}\right) = 54 + 49 = 103 \quad \text{Since } \sin(\pi/2) = 1 \end{aligned}$$

So the rider's height is 103 feet after 12 seconds.

(b) We know the sine function has period 2π, so we have

$$\begin{aligned} h(t) &= 54 + 49 \sin\left(\frac{\pi}{12}t - \frac{\pi}{2}\right) && \text{Given} \\ &= 54 + 49 \sin\left[\left(\frac{\pi}{12}t - \frac{\pi}{2}\right) + 2\pi\right] && \text{Sine has period } 2\pi \\ &= 54 + 49 \sin\left(\frac{\pi}{12}t + \frac{24\pi}{12} - \frac{\pi}{2}\right) && 2\pi = \frac{24\pi}{12} \\ &= 54 + 49 \sin\left[\frac{\pi}{12}(t + 24) - \frac{\pi}{2}\right] && \text{Factor out } \frac{\pi}{12} \\ &= h(t + 24) && \text{Function notation} \end{aligned}$$

Thus, the period of h is 24; that is, the values of h repeat every 24 seconds. Consequently, the Ferris wheel makes a complete rotation every 24 seconds. (See Example 6 in Section 5.1.)

(c) It takes 24 seconds for one complete rotation. So a ride, which includes six rotations, requires $(6)(24) = 144$ seconds, or 2 minutes and 24 seconds.

PROBLEM SET 5.4

Mastering the Concepts

In problems 1–10, use periodicity to find the exact value.

1. **(a)** $\cos\left(\frac{5\pi}{6} + 2\pi\right)$ **(b)** $\sin\left(\frac{2\pi}{3} + 4\pi\right)$

2. **(a)** $\sin(315° + 360°)$ **(b)** $\tan(45° + 720°)$

3. **(a)** $\sin 945°$ **(b)** $\cot \frac{41\pi}{4}$

4. **(a)** $\sin(-675°)$ **(b)** $\cos\left(\frac{-9\pi}{4}\right)$

5. **(a)** $\csc\left(\frac{-43\pi}{4}\right)$ **(b)** $\tan 1020°$

6. **(a)** $\sin 7\pi$ **(b)** $\cos 750°$

7. **(a)** $\tan(-1845°)$ **(b)** $\cos\left(\frac{-61\pi}{6}\right)$

8. **(a)** $\csc\left(\frac{-8\pi}{3}\right)$ **(b)** $\cot\left(\frac{-61\pi}{6}\right)$

9. **(a)** $\cos 900°$ **(b)** $\tan\left(\frac{-65\pi}{6}\right)$

10. **(a)** $\cos(-405°)$ **(b)** $\sin\left(\frac{-89\pi}{6}\right)$

In problems 11–16, use the fact that each of the sine and the cosine functions has a period of 2π to find the period of the given function.

11. **(a)** $f(t) = \sin 3t$ **(b)** $g(t) = \cos 4t$

12. **(a)** $f(t) = \cos \frac{3t}{2}$ **(b)** $g(t) = \sin \frac{3t}{4}$

13. **(a)** $f(t) = \cos \frac{t}{3}$ **(b)** $g(t) = \sin \frac{2\pi t}{3}$

14. **(a)** $f(t) = \cos \frac{2t}{3}$ **(b)** $g(t) = \sin \frac{\pi t}{2}$

15. **(a)** $f(t) = \sin \frac{\pi t}{4}$ **(b)** $g(t) = \cos(-6t)$

16. **(a)** $f(t) = \cos 3\pi t$ **(b)** $g(t) = \sin 1.5t$

In problems 17–24, use the fundamental identities to find the exact values of the other five trigonometric functions.

17. $\sin t = \frac{15}{17}$; t is in quadrant I

18. $\cos \theta = \frac{-4}{5}$; θ is in quadrant II

19. $\cos t = \frac{5}{13}$; $\tan t > 0$

20. $\sin t = \frac{-7}{25}$; $\cot t < 0$

21. $\csc \theta = \frac{17}{8}$; θ is in quadrant II

22. $\sec \theta = -\frac{5}{3}$; $\sin \theta < 0$

23. $\tan t = -3$; $\cos t > 0$

24. $\cot \theta = \frac{3}{4}$; $\sin \theta < 0$

In problems 25–28, use the fundamental identities to find the approximate value of the other five trigonometric functions. Round off each answer to two decimal places.

25. $\sin t = 0.57$; $\tan t < 0$

26. $\cos t = -0.83$; $\cot t > 0$

27. $\tan t = -1.53$; $\sin t > 0$

28. $\cot \theta = 3.41$; $\csc \theta > 0$

In problems 29 and 30, write each expression in terms of $\sin t$.

29. **(a)** $\tan t \cos t$ **(b)** $3 + 5\cos^2 t$
(c) $\sin t \csc t - \cos^2 t$

30. **(a)** $(\cos t)(\tan t + \sec t)$
(b) $\sin^3 t + \sin t \cos^2 t$
(c) $\sin t \cos t \cot t$

In problems 31 and 32, write each expression in terms of $\cos t$.

31. **(a)** $\sin t \cot t$ **(b)** $(\sin t)(\cot t + \csc t)$
(c) $\cos^2 t - \sin^2 t$

32. **(a)** $\tan t \sin t$ **(b)** $\cos^3 t + \cos t \sin^2 t$
(c) $(\sin t)(\csc t - \sin t)$

In problems 33–40, carry out each operation and simplify.

33. **(a)** $\cos^4 \theta + \cos^2 \theta - \cos^2 \theta \sin^2 \theta$
(b) $\sec(-\theta)\cot(-\theta)$

34. **(a)** $\frac{4\sin^2 \theta - 1}{2\sin \theta + 1}$ **(b)** $\frac{1 - \tan^4 t}{\sec^2 t}$

35. **(a)** $(\csc t - \cot t)(\sec t + 1)$
(b) $(\sec \theta - \tan \theta)(\sin \theta + 1)$

36. **(a)** $\csc^4 t - \cot^4 t$ **(b)** $\cot^2 t - \csc^2 t$

37. **(a)** $\sin t + \frac{\cos^2 t}{\sin t}$ **(b)** $\cos \theta + \frac{\sin^2 \theta}{\cos \theta}$

38. **(a)** $\frac{1 - (\sin \theta - \cos \theta)^2}{\sin \theta}$
(b) $\frac{1 + \cos t}{\sin t} + \frac{\sin t}{1 + \cos t}$

39. **(a)** $\cos \theta[\tan \theta - \cot(-\theta)]$
(b) $[\sin(-t)\csc(-t) + \cos t \sec t]^2$

40. **(a)** $\frac{1 - \cos^2(-t)}{\sin t}$ **(b)** $\frac{\sec^2(-\theta) - 1}{\sec^2 \theta}$

Applying the Concepts

41. Seasonal Business Profit: The profit P (in thousands of dollars) per month from the sales of scuba equipment in the midwestern United States for 2 years beginning on January 1, 1994, is given by the function

$$P(t) = 700\left(1 - \cos \frac{\pi t}{6}\right)$$

where t is time elapsed (in months) and $t = 0$ corresponds to January 1, 1994.

(a) Compare the profits on the following dates. Round off the answers to two decimal places.
(i) April 1, 1994 and April 1, 1995
(ii) July 1, 1994 and July 1, 1995
(iii) October 1, 1994 and October 1, 1995
(iv) January 1, 1994 and January 1, 1995

(b) From the results in part (a), does the function P appear to be periodic? If so, what is the period?

(c) Explain this model in terms of the product being sold, the region of the country, and the seasonal profit pattern.

42. Electrical Current: An alternating current (ac) generator produces a current I (in amperes) at time t (in seconds). The model relating I and t is given by the function

$$I(t) = 35 \sin 120\pi t$$

(a) Find the current after 0.01 second, 0.03 second, 0.05 second, and 1 second. Round off the answers to two decimal places.

(b) Find the period of the function and interpret it.

43. Electrical Voltage: Power companies provide an alternating voltage at wall outlets that can be modeled by the function

$$V(t) = 170 \cos 377t$$

where V (in volts) is the voltage at time t (in seconds).

(a) Find the voltage after 0.003 second, 0.005 second, 0.007 second, and 1 second. Round off the answers to two decimal places.

(b) Find the period of the function and interpret it.

(c) Approximately how many periods occur over 1 second of elapsed time?

44. Daily Temperature Change: Air temperature T generally varies in a periodic manner, with highs during the day and lows during the night. Assume that the weather pattern is the same for one week, and suppose that T (in degrees Fahrenheit) at a particular time of the day is given by the function

$$T(t) = 15 \sin\left(\frac{\pi t}{12} - \frac{2\pi}{3}\right) + 60$$

where t is the time (in hours), with $t = 0$ at midnight.

(a) Find the temperature at 9 AM, at noon, at 6 PM, and at 10 PM. Round off the answers to two decimal places.

(b) Find the period of the function and interpret it.

Developing and Extending the Concepts

45. Prove that the secant and cosecant functions are periodic functions with a period of 2π.

46. Suppose that f is periodic with period 3 and $f(1) = 7$. Find the value of $f(28)$.

47. Show that $f(t) = \cot t$ is periodic with period π. That is, prove that $\cot(t + \pi) = \cot t$.

48. Show that the following even–odd identities are true:
(a) $\cot(-t) = -\cot t$ **(b)** $\sec(-t) = \sec t$
(c) $\csc(-t) = -\csc t$

In problems 49 and 50, verify that the equation is true for the given value of x by rounding off the values to four decimal places.

49. **(a)** $f(-x) = -f(x)$, where $f(x) = \sin x$ and $x = 3.752$
(b) $f(-x) = f(x)$, where $f(x) = \cos x$ and $x = 5.384$

50. **(a)** $f(-x) = -f(x)$, where $f(x) = \tan x$ and $x = 4.138$
(b) $f(-x) = f(x)$, where $f(x) = \sec x$ and $x = 2.874$

OBJECTIVES

1. Graph the sine and cosine functions.
2. Graph the functions $f(x) = a \sin x + c$ and $g(x) = a \cos x + c$.
3. Graph the functions $f(x) = a \sin(kx + b)$ and $g(x) = a \cos(kx + b)$.
4. Solve applied problems.

5.5 Graphs of Sine and Cosine Functions

This section is devoted to the examination of the graphs of the sine and cosine functions, with particular emphasis on what these graphs can tell us about their properties.

Graphing the Sine and Cosine Functions

To graph the sine function, we recall two important facts:

1. If t is a real number and $P(t) = (x, y)$ is the associated point on the unit circle, then $\sin t = y$ (Figure 1a).
2. As $P(t)$ moves one complete revolution counterclockwise around the unit circle, starting at $P(0) = (1, 0)$ and ending at $P(2\pi) = (1, 0)$, correspondingly, t increases from 0 to 2π (Figure 1b).

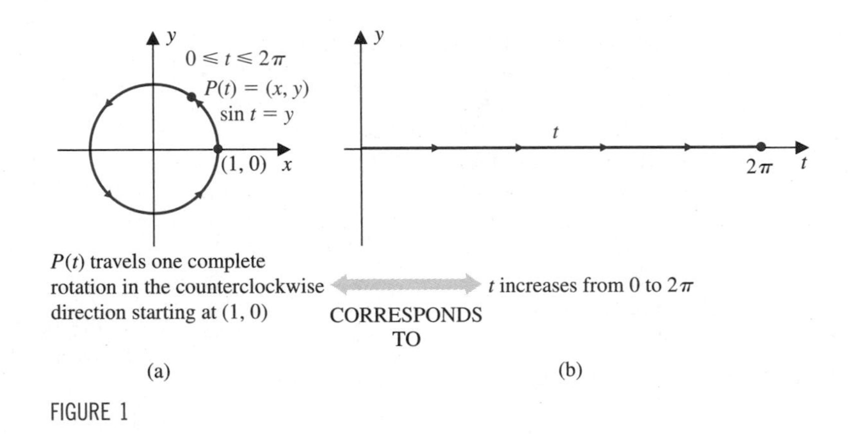

FIGURE 1

Figure 2a shows the locations of several specific points for $P(t) = (x, y)$. Since $y = \sin t$, we can obtain a rough sketch of the graph of $f(t) = \sin t$ by plotting the points (t, y) in a Cartesian coordinate system and then connecting the points with a smooth graph (Figure 2b). The graph that results when t increases from 0 to 2π, corresponding to $P(t)$ moving one counterclockwise rotation, is called **one cycle** of the sine function.

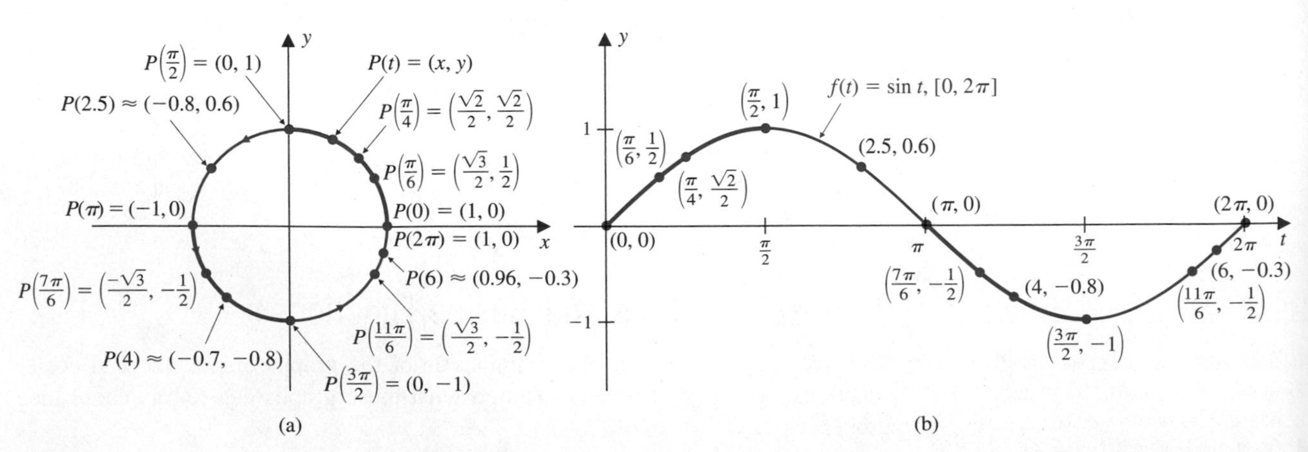

FIGURE 2

The periodicity of the sine function is reaffirmed by noting that if t increases by 2π units beyond 2π, $P(t)$ traverses the unit circle again, and, correspondingly, another sine cycle is generated. *Thus, using the convention of denoting the independent variable of a function by x instead of t, we obtain a complete graph of $y = \sin x$ by repeating the cycle to the right and to the left every 2π units* (Figure 3). Upon examining the sine graph we are led to the properties listed in Table 1.

TABLE 1

Properties of the Sine Function

Property	$y = \sin x$
Domain	All real numbers
Range	Interval $[-1, 1]$; that is, $-1 \le \sin x \le 1$
Period	2π
Symmetry	Symmetric with respect to the origin; that is, the function is *odd*, so $\sin(-x) = -\sin x$

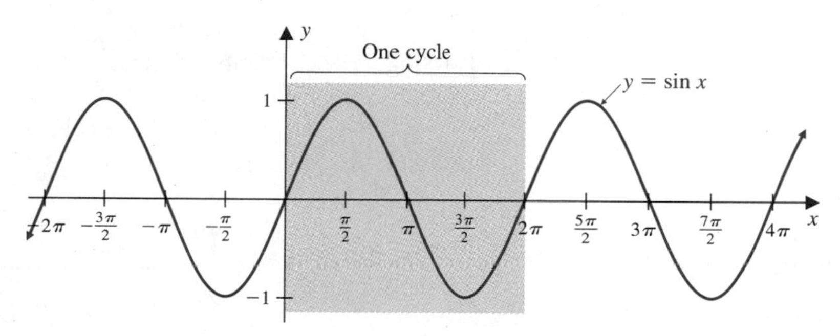

FIGURE 3

EXAMPLE 1 **Locating extreme points and intercepts**

Use the graph of $y = \sin x$ to locate:

(a) The maximum and minimum points (b) The x intercepts

Solution Using the graph of $y = \sin x$ (Figure 3), we are led to the following results:

(a) There is a maximum point at $(\pi/2, 1)$. Because of the periodicity, it follows that there are infinitely many other maximum points. They are located every 2π units apart at the points $\left(\frac{1}{2}\pi + 2\pi k, 1\right)$, where k is any integer. A minimum point occurs at $((3\pi)/2, -1)$. By using periodicity, we get infinitely many other minimum points at $\left(\frac{3}{2}\pi + 2\pi k, -1\right)$, were k is any integer.

(b) The curve has infinitely many x intercepts every π units apart at $x = k\pi$, where k is any integer.

As with the sine function, we sketch one cycle of the graph of the cosine function $y = \cos x$ in the interval $[0, 2\pi]$ by first plotting some points (Table 2) and then connecting them with a smooth graph (Figure 4), as shown on page 318.

TABLE 2

x	$\cos x$
0	1
$\pi/3$	$\frac{1}{2}$
$\pi/2$	0
π	-1
$(4\pi)/3$	$-\frac{1}{2}$
$(3\pi)/2$	0
2π	1

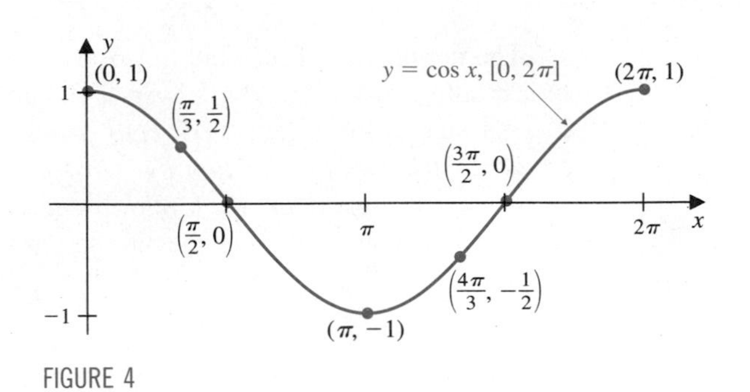

FIGURE 4

Using the periodicity of the cosine function, we repeat the cycle shown in Figure 4 to the right and to the left every 2π units, to get the complete graph of $y = \cos x$ (Figure 5). From the graph, we observe the properties of the cosine function listed in Table 3.

TABLE 3
Properties of the Cosine Function

Property	$y = \cos x$
Domain	All real numbers
Range	Interval $[-1, 1]$; that is, $-1 \leq \cos x \leq 1$
Period	2π
Symmetry	Symmetric with respect to the y axis; that is, the function is *even*, so $\cos(-x) = \cos x$

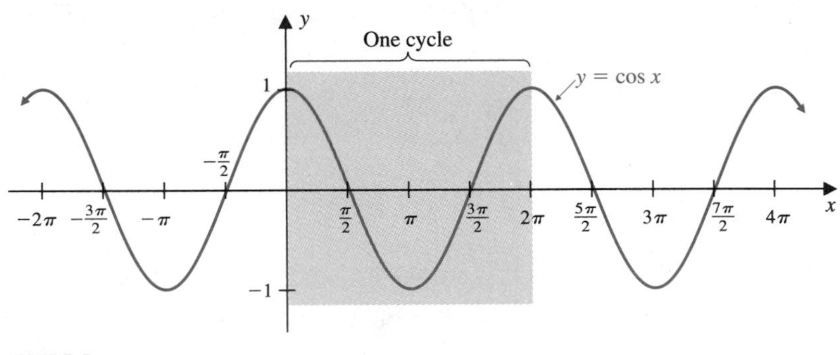

FIGURE 5

The basic characteristics of the graphs of the sine and cosine functions should be learned so their graphs can be sketched quickly. In particular, note the period, the locations of maximum and minimum points, and the locations of the intercepts.

Graphing the Functions $f(x) = a \sin x + c$ and $g(x) = a \cos x + c$

To graph the functions

$$f(x) = a \sin x + c \qquad \text{and} \qquad g(x) = a \cos x + c$$

we use the transformation techniques presented in Section 2.3.

EXAMPLE 2 Using transformations to graph

Use transformations of the sine and cosine graphs to sketch the graph of each function. Also determine the range and period.

(a) $F(x) = 5 \sin x$ (b) $G(x) = \frac{1}{5} \cos x + 1$

Solution

(a) First we sketch the graph of $y = \sin x$; then we multiply each ordinate by 5. Geometrically, this means that the graph of $y = \sin x$ is vertically stretched by a factor of 5 to obtain the graph of $F(x) = 5 \sin x$ (Figure 6a). From the graph, we see that the range of F is the interval $[-5, 5]$ and the period is 2π.

(b) We start by graphing $y = \cos x$. Then the graph of $g(x) = \frac{1}{5} \cos x$ is obtained by vertically shrinking the graph of $y = \cos x$ by a factor of $\frac{1}{5}$ (Figure 6b). Finally, we get the graph of $G(x) = \frac{1}{5} \cos x + 1$ by shifting the graph of g up 1 unit (Figure 6b). The range of G is the interval $\left[-\frac{1}{5} + 1, \frac{1}{5} + 1\right]$, that is, $\left[\frac{4}{5}, \frac{6}{5}\right]$, and the period is 2π.

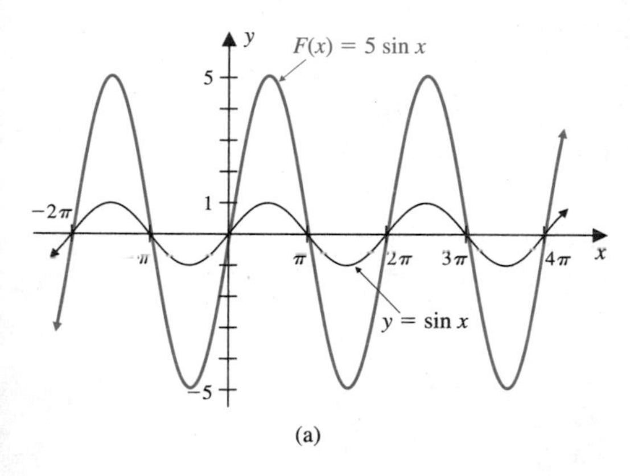

(a)

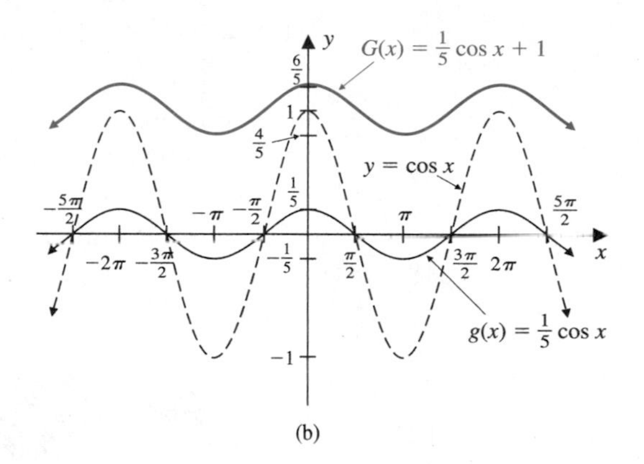

(b)

FIGURE 6

From the graphs in Example 2, we make the following generalization: For any number a, the graphs of $y = a \sin x$ and $y = a \cos x$ can be obtained from the graphs of $y = \sin x$ and $y = \cos x$, respectively, by simply stretching or compressing vertically by a factor of $|a|$. If $a < 0$, we also have to reflect the resulting graph across the x axis. In any case, the number $|a|$ given by

$$|a| = \tfrac{1}{2}(M - m)$$

where M is the *maximum value* and m is the *minimum value* of $y = a \sin x$ or $y = a \cos x$, is called the **amplitude** of each function.

By using the graphs in Figure 6, we get the amplitudes listed in Table 4.

TABLE 4

Function	Maximum Value (M)	Minimum Value (m)	Amplitude
$F(x) = 5 \sin x$	5	-5	$\frac{1}{2}[5 - (-5)] = \frac{10}{2} = 5$
$G(x) = \frac{1}{5} \cos x + 1$	$\frac{6}{5}$	$\frac{4}{5}$	$\frac{1}{2}\left(\frac{6}{5} - \frac{4}{5}\right) = \frac{1}{2}\left(\frac{2}{5}\right) = \frac{1}{5}$

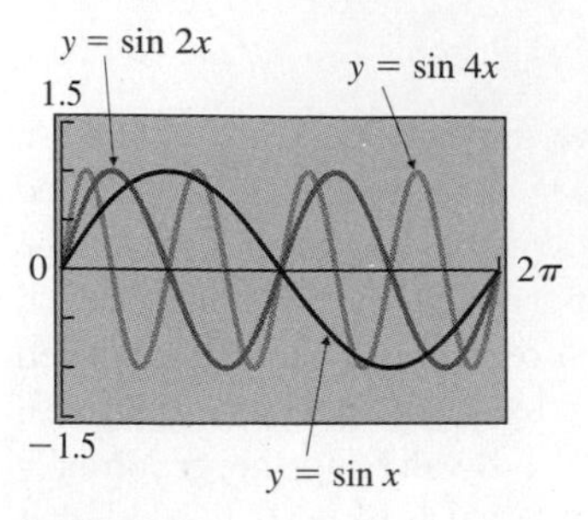

FIGURE 7

Graphing the Functions $f(x) = a \sin(kx + b)$ and $g(x) = a \cos(kx + b)$

To discover what effect the values of k have on the graph of $y = \sin kx$, we compare the graphs of $y = \sin x$, $y = \sin 2x$, and $y = \sin 4x$ on the interval $[0, 2\pi]$ with the aid of a graphing utility (Figure 7). By examining these graphs, we see that on the interval $[0, 2\pi]$:

(i) The graph of $y = \sin x$ generates one sine cycle.
(ii) The graph of $y = \sin 2x$ generates two sine cycles.
(iii) The graph of $y = \sin 4x$ generates four sine cycles.

This graphical exploration indicates that:

(i) The period of $y = \sin x$ is 2π.
(ii) The period of $y = \sin 2x$ is $(2\pi)/2 = \pi$.
(iii) The period of $y = \sin 4x$ is $(2\pi)/4 = \pi/2$.

A similar comparison could be made for the cosine function.

These observations lead us to the following general property:

PROPERTY

PERIODICITY

The **period** of $f(x) = a \sin kx$ or $g(x) = a \cos kx$, is $(2\pi)/k$, where $k > 0$. Thus, one cycle of the graph occurs in the interval $[0, (2\pi)/k)$.

We can prove this property analytically for the sine function as follows:

$f(x) = a \sin kx$	Given
$= a \sin(kx + 2\pi)$	The sine has period 2π
$= a \sin\left[k\left(x + \dfrac{2\pi}{k}\right)\right]$	Factor out k
$= f\left(x + \dfrac{2\pi}{k}\right)$	Function notation

So f repeats its value every $(2\pi)/k$ units.

A similar argument can be developed for $g(x) = a \cos kx$.

EXAMPLE 3 **Finding the period**

GU

Find the period of $f(x) = 5 \sin(x/2)$, and then use a graphing utility to demonstrate the validity of the result by showing one cycle graphically.

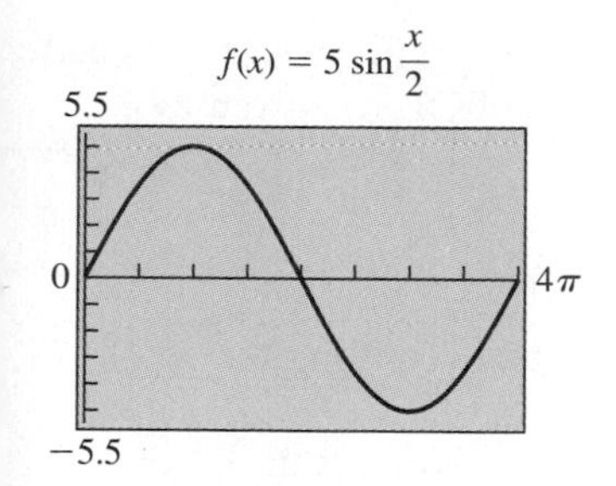

FIGURE 8

Solution The function $f(x) = 5 \sin(x/2)$ is of the form $y = a \sin kx$, where $a = 5$ and $k = \frac{1}{2}$. So the amplitude of f is 5, and its period is given by

$$\frac{2\pi}{k} = \frac{2\pi}{\frac{1}{2}} = 4\pi$$

A viewing window of the graph of f (Figure 8) shows one cycle on the interval $[0, 4\pi]$, thus demonstrating that the period is 4π.

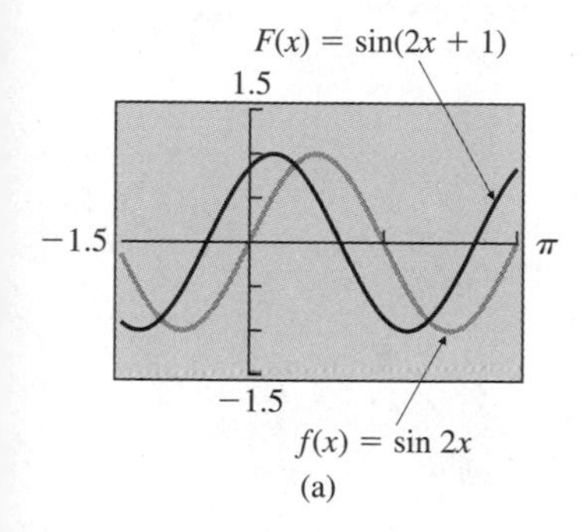

(a)

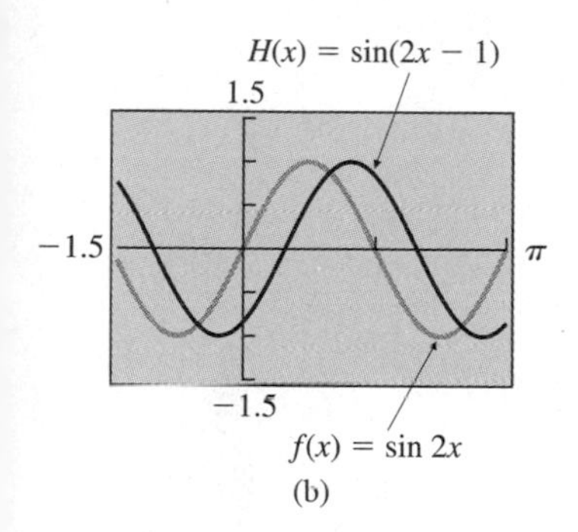

(b)

FIGURE 9

Note that there is no loss in generality by having k positive in the above property. If k is negative, we can rewrite the function using the even–odd properties in such a way that the coefficient of x turns out to be positive. For instance, if $g(x) = \cos(-3x)$, we can rewrite it as $g(x) = \cos 3x$, since the cosine function is even. So, in this case, the period is $(2\pi)/3$.

We can use the graphs of $f(x) = a \sin kx$ and $g(x) = a \cos kx$ to graph the functions $F(x) = a \sin(kx + b)$ and $G(x) = a \cos(kx + b)$. To discover what effect the value of b has on the graph, let's compare the graph of $F(x) = \sin(2x + 1)$ to that of $f(x) = \sin 2x$ with the aid of a graphing utility (Figure 9a). From the figure, it appears that the graph of F is a horizontal shift of the graph of f. Similarly, if we compare the graph of $H(x) = \sin(2x - 1)$ to the graph of $f(x) = \sin 2x$ (Figure 9b), we get the impression that the graph of H is also a horizontal shift of the graph of f.

To confirm analytically what these two figures suggest, we proceed as follows. In the first case, we have

$$\begin{aligned} F(x) &= \sin(2x + 1) && \text{Given} \\ &= \sin\left[2\left(x + \tfrac{1}{2}\right)\right] && \text{Factor out 2} \\ &= f\left(x + \tfrac{1}{2}\right) && \text{Function notation} \\ &= f\left(x - \left(-\tfrac{1}{2}\right)\right) && \text{Rewrite} \end{aligned}$$

By using the notation of horizontal shifting, which was discussed in Section 2.3, we conclude that the graph of F can be obtained by shifting the graph of f horizontally $\frac{1}{2}$ unit to the left. A similar argument can be used to show that the graph of H is obtained by shifting the graph of f horizontally $\frac{1}{2}$ unit to the right.

These observations lead us to the following property:

PROPERTY

PHASE SHIFT

The graph

$F(x) = a \sin(kx + b)$ [or $G(x) = a \cos(kx + b)$] where $k > 0$

can be obtained by horizontally shifting the graph of $f(x) = a \sin kx$ [or $g(x) = a \cos kx$] by $|-b/k|$ units to the left if $-b/k < 0$ and to the right if $-b/k > 0$. The number $-b/k$ is called the **phase shift** of the function.

The phase shift can be found by setting $kx + b = 0$ and solving for x to get $x = -b/k$.

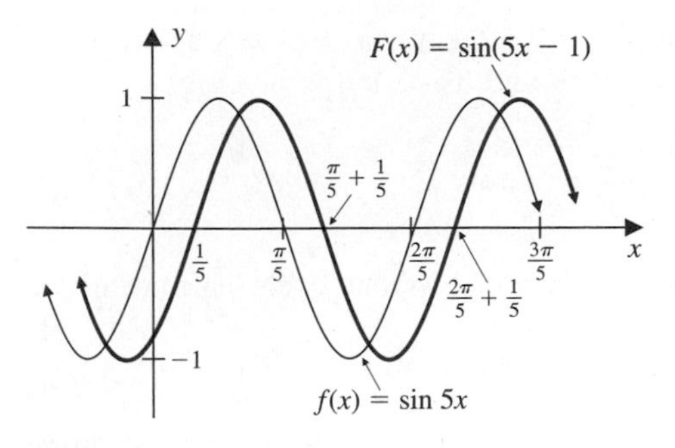

FIGURE 10
The graph of f can be shifted $\frac{1}{5}$ unit to the right to get the graph of F

For example, $F(x) = \sin(5x - 1) = \sin[5x + (-1)]$. Thus, $k = 5$ and $b = -1$ in the above property, so the phase shift of F is given by

$$\frac{-b}{k} = \frac{-(-1)}{5} = \frac{1}{5}$$

It follows that the graph of F can be obtained by shifting the graph of $f(x) = \sin 5x$ by $\frac{1}{5}$ unit to the right (Figure 10).

In a sense, the phase shift locates a starting position for a cycle on the x axis.

Let's summarize what we have learned so far about the connection between sine and cosine function expressions and their graphs.

SUMMARY

If $a \neq 0$ and $k > 0$, then each of the functions

$$F(x) = a\sin(kx + b) \qquad \text{and} \qquad G(x) = a\cos(kx + b)$$

has

$$\text{Amplitude} = |a| \qquad \text{Period} = \frac{2\pi}{k} \qquad \text{Phase shift} = \frac{-b}{k}$$

These properties suggest the following strategy for sketching one cycle of the graphs of sine and cosine functions:

STRATEGY FOR GRAPHING THE FUNCTIONS $y = a\sin(kx + b)$ and $y = a\cos(kx + b)$, $k > 0$

If k, the coefficient of x, is negative, then we use the even–odd properties to rewrite the function so that we can work with a positive value for k.

Step 1. Draw a coordinate system and locate the phase shift $-b/k$ on the x axis.

Step 2. Using the point from step 1 as the left end point, mark off on the x axis an interval equal in length to the period $(2\pi)/k$ to get the right end point of the cycle,

$$\frac{-b}{k} + \frac{2\pi}{k}$$

Step 3. Use the amplitude $|a|$ and the shape of a sine (or cosine) cycle along with horizontal scaling to locate minimum points, maximum points, and x intercepts. Reflect the graph across the x axis if a is negative.

EXAMPLE 4 **Using properties to sketch a cosine graph**

Identify and use the phase shift, period, and amplitude to graph one cycle of the function $f(x) = 3\cos(2x + 1)$.

Solution For $f(x) = 3\cos(2x + 1)$, $|a| = 3$, $k = 2$, and $b = 1$. To graph this function, we proceed as follows:

Step 1. Locate the phase shift $-b/k = -1/2$ on the x axis.

Step 2. Since the period is $(2\pi)/k = (2\pi)/2 = \pi$, start at $-\frac{1}{2}$ and then mark off an interval of length π on the x axis, ending at $-\frac{1}{2} + \pi$ (approx. 2.64).

Step 3. Sketch a cosine cycle in the interval $\left[-\frac{1}{2}, -\frac{1}{2} + \pi\right]$, and use the amplitude $a = 3$ and horizontal scaling to locate the minimum point at

$$\left(-\frac{1}{2} + \frac{\pi}{2}, -3\right)$$

and maximum points at $\left(-\frac{1}{2}, 3\right)$ and $\left(-\frac{1}{2} + \pi, 3\right)$. The x intercepts are

$$-\frac{1}{2} + \frac{\pi}{4} \qquad \text{and} \qquad -\frac{1}{2} + \frac{3\pi}{4}$$

(Figure 11).

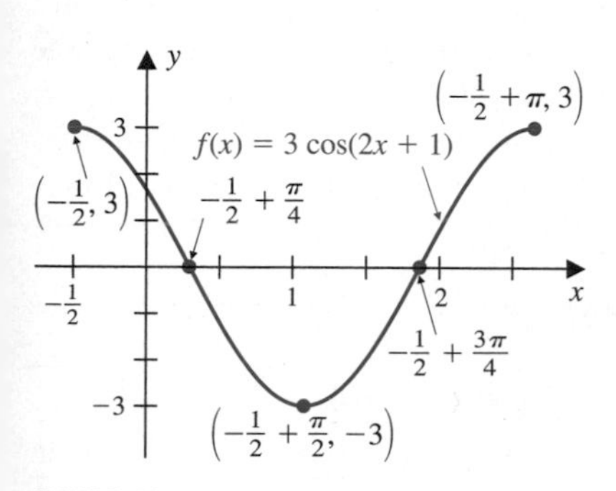

FIGURE 11

The graphs of sine and cosine functions are called **sine waves** or **sinusoidal curves**. The characteristics of the graph of such a curve can be used to find its equation.

EXAMPLE 5 Finding an equation for a sine wave

Given the graph of one cycle of the function $y = a\sin(kx + b)$ in Figure 12, find values for $a > 0$, $k > 0$, and b. Also find the amplitude, period, and phase shift.

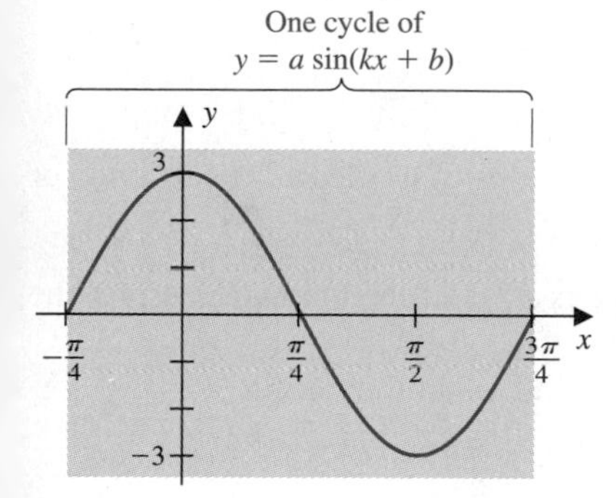

FIGURE 12

Solution The graph shows that the amplitude is 3; that is, $a = 3$. The period, which is the length of the interval containing the cycle, is given by

$$\frac{3\pi}{4} - \left(\frac{-\pi}{4}\right) = \pi$$

so $(2\pi)/k = \pi$, or $k = 2$. Since the phase shift is shown to be $-\pi/4$, it follows that $-b/k = -b/2 = -\pi/4$, or $b = \pi/2$. Thus, the equation is

$$y = 3\sin\left(2x + \frac{\pi}{2}\right)$$

Graphing utilities can be used to graph and analyze the behavior of trigonometric functions.

EXAMPLE 6 Interpreting the behavior of a function graphically

(a) Use a graphing utility to graph

$$f(x) = \frac{1 - \cos x}{x} \qquad \text{where } -4 \le x \le 4$$

Then examine the graph to determine what happens to the values of $f(x)$ as $x \to 0$.

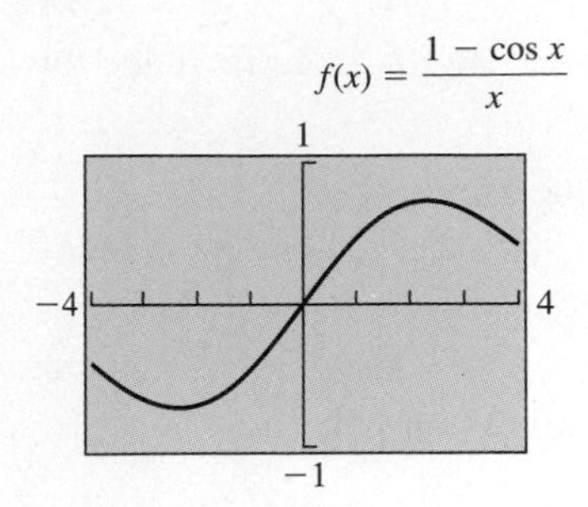

FIGURE 13

(b) Support the result in part (a) numerically by determining $f(x)$ for some values of x as $x \to 0^+$ and as $x \to 0^-$.

Solution

(a) Figure 13 shows a viewing window of the graph of f. It appears from the graph that $f(0) = 0$. Actually, f is not defined at $x = 0$ because its denominator is 0, so the graph should have a hole or gap at (0, 0). On the one hand, the graph generated by the graphing utility is misleading in suggesting that $f(0) = 0$, but on the other hand, it is useful in indicating that as $x \to 0$, $f(x) \to 0$.

(b) Table 5 provides approximate numerical data that support this observation.

TABLE 5

	$x \to 0^-$					$x \to 0^+$			
x	−1	−0.5	−0.01	−0.001	0	0.001	0.01	0.5	1
$f(x)$	−0.46	−0.24	−0.005	−0.0005	—	0.0005	0.005	0.24	0.46

Solving Applied Problems

Graphs help us to interpret periodic phenomena.

EXAMPLE 7 **Graphing the height of a rider on a Ferris wheel as a function of time**

Recall the Ferris wheel model (Section 5.4, Example 7) given by

$$h(t) = 54 + 49 \sin\left(\frac{\pi}{12}t - \frac{\pi}{2}\right)$$

where $h(t)$ (in feet) is the height above ground level of a rider after t seconds of elapsed time.

(a) Use a graphing utility to graph three cycles of the function and interpret the graph.

(b) Find the minimum and maximum heights and when they occur.

Solution

(a) We found in Section 5.4, Example 7, that the period of the function is 24 seconds. We can confirm the period by noting that

$$\frac{2\pi}{k} = \frac{2\pi}{(\pi/12)} = 24$$

So, in order to graph three cycles, we must graph the function from $t = 0$ to $t = 3 \cdot 24 = 72$ (seconds). A viewing window of this graph is shown in Figure 14. Each cycle of the graph represents one rotation of the rider on the Ferris wheel. As the rider makes a rotation, the point on the graph rises to a maximum and then falls to a minimum.

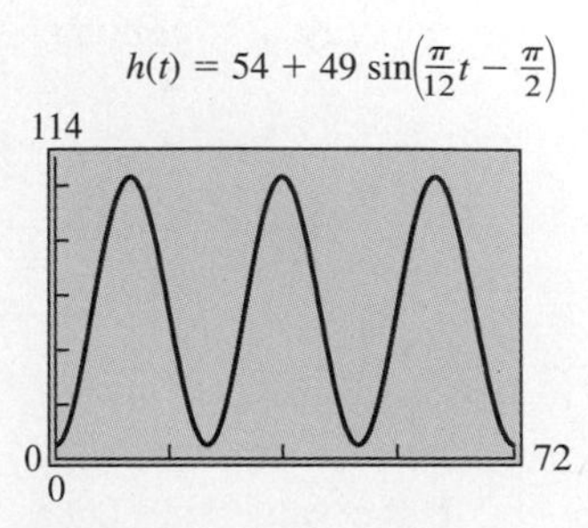

FIGURE 14

(b) Since the minimum value of a sine function is -1, the minimum value of $h(t)$ is given by $54 + 49(-1) = 5$. Therefore, the minimum height is 5 feet. Because the periodicity is 24, the minimum is obtained every 24 seconds from the start. Similarly, because the maximum value of the sine function is 1, the maximum value of $h(t)$ is given by $54 + 49(1) = 54 + 49 = 103$. So the maximum height is 103 feet above ground level, and it occurs at 12 seconds, 36 seconds, and 60 seconds, since the period is 24.

PROBLEM SET 5.5

Mastering the Concepts

1. Use the graph of $y = \cos x$ to locate the minimum points, maximum points, and x intercepts.

2. Use the graph of $y = \sin x$ on the interval $[0, 2\pi]$ to describe its increasing and decreasing behavior.

In problems 3–18, use the graph of the sine or cosine function and transformations to sketch the graph of each function. Indicate the range, amplitude, and period of each function.

3. (a) $y = 3 \cos x$ (b) $y = -3 \cos x$

4. (a) $y = -5 \sin x$ (b) $y = \frac{4}{3} \sin x$

5. (a) $y = \frac{2}{3} \cos(-x)$ (b) $y = \frac{1}{2} \cos x$

6. (a) $y = -2 \cos x$ (b) $y = 2 \cos(-x)$

7. (a) $y = \cos 2x$ (b) $y = \cos \frac{x}{2}$

8. (a) $y = \sin 3x$ (b) $y = \sin \frac{x}{3}$

9. (a) $y = \cos \frac{\pi x}{3}$ (b) $y = \cos(-2\pi x)$

10. (a) $y = \sin \frac{\pi x}{3}$ (b) $y = -\sin \frac{\pi x}{2}$

11. (a) $y = \frac{5}{3} \sin(-6x)$ (b) $y = 5 \sin 10x$

12. (a) $y = 3 \cos\left(\frac{-\pi x}{6}\right)$ (b) $y = -3 \cos 5x$

13. (a) $y = -3 \cos \frac{\pi x}{2}$ (b) $y = \pi \cos\left(\frac{-5x}{2}\right)$

14. (a) $y = -5 \sin \frac{2\pi x}{3}$ (b) $y = -\frac{1}{2} \sin \frac{3x}{2}$

15. (a) $y = 1 + \sin x$ (b) $y = 3 - \sin x$

16. (a) $y = 2 + \cos x$ (b) $y = -\frac{3}{4} + 2 \cos x$

17. (a) $y = 2 + 3 \cos \frac{\pi x}{4}$ (b) $y = -2 - 3 \cos 2\pi x$

18. (a) $y = \frac{1}{2} - \frac{1}{2} \sin \frac{\pi x}{2}$ (b) $y = -2 - \sin \frac{\pi x}{6}$

19. Match each function with one of the graphs shown in Figure 15: $f(x) = 0.5 \sin x$; $g(x) = \sin x$; $h(x) = 2 \sin x$

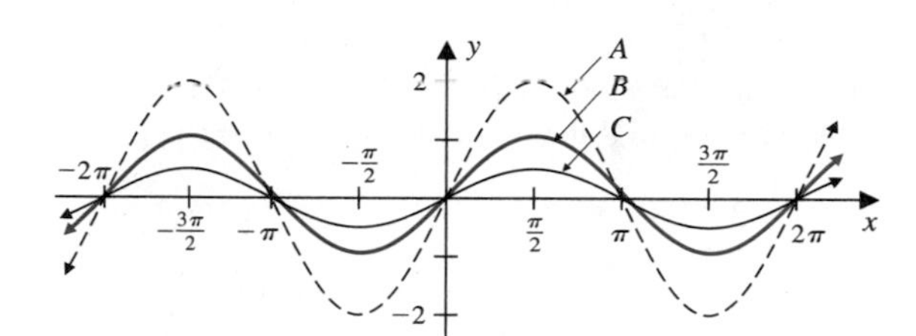

FIGURE 15

20. Match each function with one of the graphs shown in Figure 16: $f(x) = 2 \sin x + 1$; $g(x) = \sin x + 3$; $h(x) = 0.5 \sin x - 2$

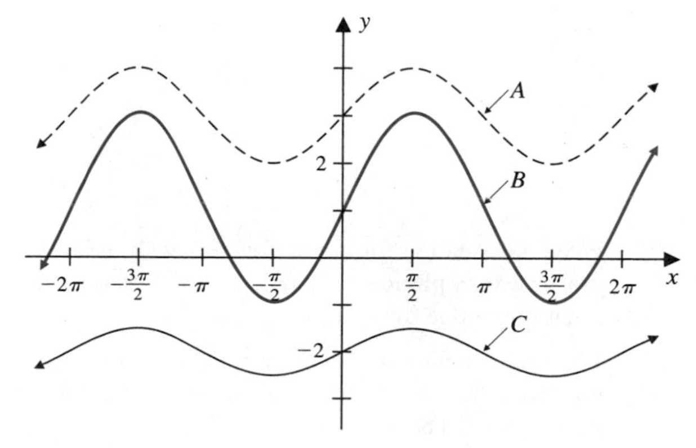

FIGURE 16

In problems 21–36, identify and use the phase shift, period, and amplitude to graph one cycle of each function.

21. $y = 2\sin\left(x - \frac{\pi}{4}\right)$ **22.** $y = -3\sin\left(x + \frac{\pi}{3}\right)$

23. $y = 4\cos\left(x + \frac{\pi}{6}\right)$ **24.** $y = 2\cos\left(x - \frac{\pi}{3}\right)$

25. $y = -2\sin(2x - \pi)$ **26.** $y = -2\cos(3x - \pi)$

27. $y = 2\sin(\pi x - \pi)$ **28.** $y = 3\cos(\pi x - \pi)$

29. $y = -3\cos(3x + 2)$ **30.** $y = -3\sin(4x - 1)$

31. $y = 2\cos\left(\frac{\pi x}{2} - \frac{1}{4}\right)$ **32.** $y = -4\sin\left(\frac{\pi x}{3} + \frac{1}{3}\right)$

33. $y = -2\sin(5 - 3x)$ **34.** $y = -2\cos(3 - 2x)$

35. $y = 2 + 3\cos(2x - 2\pi)$

36. $y = -2 - 3\sin(3x + \pi)$

In problems 37 and 38, each figure shows the graph of one cycle of the function

$$y = a\sin(kx + b)$$

Find values for $a > 0$, $k > 0$, and b, and write the resulting equation. Also find the amplitude, period, and phase shift.

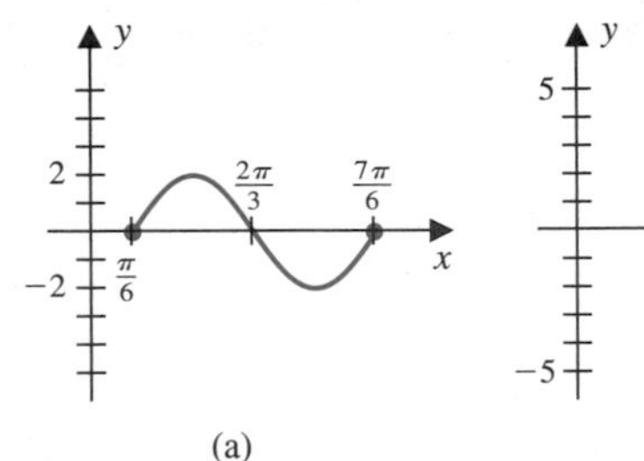

(a) (b)

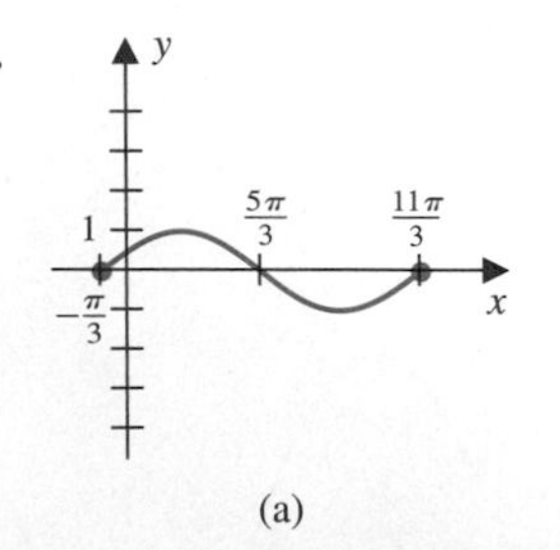

(a)

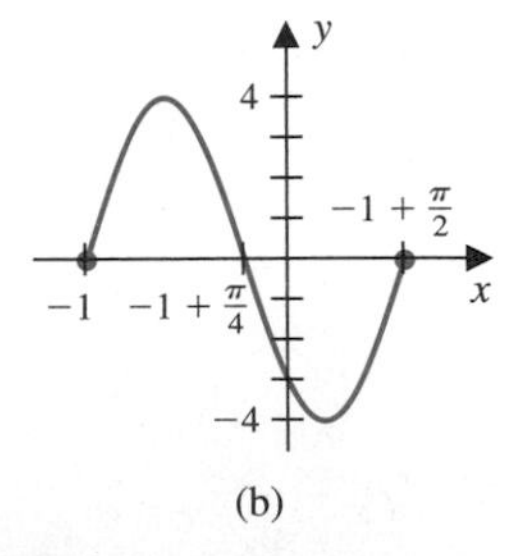

(b)

39. Write an equation for a sine function with the following properties: amplitude = 3, period = $\pi/3$, phase shift = 0. Then graph one cycle.

40. Write an equation for a cosine function with the following properties: amplitude = 2, period = $\pi/6$, phase shift = $\pi/2$. Then graph one cycle.

41. GU Use a graphing utility to sketch the graph of the function $f(x) = \sin^2 x + \cos^2 x$. Interpret the result.

42. GU Assume

$$f(x) = \sin\left(\frac{\pi}{2} - x\right) \quad \text{and} \quad g(x) = \cos x$$

where $-\pi \le x \le \pi$

Compare the graphs. Are they the same? Explain.

43. $f(x) = \frac{\sin x}{x}$

(a) Is $f(0)$ defined?

GU **(b)** Use a graphing utility to graph f, and then use the graph to help determine the behavior of $f(x)$ as $x \to 0$.

(c) Complete the following table to support the conclusion in part (b):

	$x \to 0^-$					$x \to 0^+$			
x	−1	−0.5	−0.01	−0.001	0	0.001	0.01	0.5	1
$f(x) = \frac{\sin x}{x}$					—				

44. GU Assume $y = \cos|x|$, where $-2\pi \le x \le 2\pi$. Compare this graph to the graph of $f(x) = \cos x$. Are the two graphs the same or different? Explain.

Applying the Concepts

GU In problems 45–52, use a graphing utility to graph the functions. Round off the answers to two decimal places.

45. Blood Pressure: The blood pressure of a person is given by the model

$$y = 110 + 30\cos\frac{12\pi t}{5}$$

where t is in seconds and y is in millimeters of mercury (mm Hg).

(a) Sketch the graph of this function for $0 \le t \le 2.5$.

(b) Determine the period of the function.

(c) Use the graph to determine the maximum and minimum blood pressure of the person for the given interval.

46. Electrical Circuit: The electricity supplied to homes is called alternating current (ac), and the current I (in amperes) is related to time t (in seconds) by the equation

$$I = 5\cos 120\pi t$$

The voltage V (in volts) that causes this current to flow is related to time t (in seconds) by the equation

$$V = 170\cos[120\pi(t + 0.031)]$$

(a) Sketch the graphs of the current and voltage functions for $0 \le t \le 0.004$.
(b) Use the graphs to estimate the voltage at the time the current is a maximum.
(c) Use the graphs to estimate the current at the time the voltage is a maximum.
(d) Predict the first time when the voltage reaches 160 volts.

47. Height of a Point on a Waterwheel: Figure 17 shows a rotating waterwheel. At time t (in seconds) the height h (in feet) of a point P on the rim of the wheel above the surface of the water is given by the model

$$h(t) = 7 + 8\cos\left(\frac{\pi t}{5} - \frac{2}{5}\right)$$

FIGURE 17

(a) Find the height of the point P on the wheel after 2 seconds, after 5 seconds, and after 10 seconds.
(b) How long does it take for the point P on the wheel to go around once?
(c) Sketch the graph of h for two cycles.
(d) Find the minimum and maximum heights of P on the wheel.

48. Sound Wave: The difference D (in dynes per square centimeter) between the atmospheric pressure and the air pressure at the eardrum produced by a sound wave at t seconds is given by the function

$$D(t) = 4\sin\left(100\pi t + \frac{\pi}{4}\right)$$

(a) Sketch the graph of this function for $0 \le t \le 0.18$.
(b) Find the period and amplitude.
(c) Find the minimum and maximum values of $D(t)$.

49. Seasonal Business Profit: The profit P (in thousands of dollars) per month from the sales of scuba equipment for 2 years beginning in January 1, 1994, is given by the function

$$P(t) = 900 - 700\cos\frac{\pi t}{6}$$

where t is the elapsed time (in months) and $t = 0$ corresponds to January 1, 1994.
(a) Graph the function for $0 \le t \le 24$.
(b) Find the amplitude and period, and interpret them.
(c) Determine the profit at 3 months, 9 months, 1 year, 18 months, and 2 years.
(d) What is the maximum profit, and when does it occur?

50. Tides: During a 24 hour day, the mathematical model

$$d = 3\cos 0.52t + 9$$

is used to predict the tide. That is, it predicts how much above mean sea level the tide will raise or lower the water depth d (in feet) in a certain harbor, t hours after midnight.
(a) Sketch the graph of this function over a 24 hour day.
(b) Find the amplitude and period, and interpret them.
(c) Find the minimum and maximum depth of water in the harbor, and determine when they occur.
(d) Interpret your answers for part (c) in terms of this tidal wave.

51. Hours of Daylight: Suppose that the hours $h(t)$ of daylight for a certain city are related to the day of the year by the model

$$h(t) = 2.3\sin\left[\frac{2\pi}{365}(t - 80)\right] + 12$$

where t is the day of the year, and $t = 1$ corresponds to January 1.
(a) Sketch the graph of this function for $1 \le t \le 365$.
(b) Find the amplitude and the period.
(c) Determine the maximum and minimum number of hours of daylight and when they occur.

52. Business Profit: A company's profit fluctuates during a 12 month period according to the model

$$P(t) = 30{,}000 - 70{,}000(\sin t + 0.5\cos 0.5t)$$

where $P(t)$ represents the profit (in dollars) after t months.
(a) Find the profit (or loss) at the beginning of the 12 month period, after 5 months, after 8 months, and after 1 year.
(b) Sketch the graph of P for $0 \le t \le 12$.

(c) Interpret the graph to determine when there is a profit.

(d) What are the maximum and minimum profits over the 12 month period, and when do they occur?

Developing and Extending the Concepts

53. Sketch the graph of each function without using a graphing utility. Explain the effect of the absolute value on the graph.

(a) $y = |-3 \cos 2x|$ **(b)** $y = -\left|2 \sin \frac{\pi x}{2}\right|$

54. Sketch the graphs of

$$f(x) = \sin^2 x \quad \text{and} \quad g(x) = \cos^2\left(\frac{\pi}{2} - x\right)$$

on the interval $[0, 2\pi]$. How do the graphs compare?

55. Use a graphing utility to graph both

$$f(x) = \sin x \quad \text{and} \quad g(x) = x - \frac{x^3}{6} + \frac{x^5}{120}$$

for the interval $[-2.75, 2.75]$. Compare the values of $f(x)$ and $g(x)$ in this interval.

56. Use a graphing utility to graph

GU

$$f(x) = \cos x \quad \text{and} \quad g(x) = 1 - \frac{x^2}{2} + \frac{x^4}{24}$$

for the interval $[-2.75, 2.75]$. Compare the values of $f(x)$ and $g(x)$ in this interval.

GU In problems 57–60, use a graphing utility to sketch the graph of each function, where $-2\pi \le x \le 2\pi$. If the function is periodic, find its period.

57. $y = \sin x + 2 \cos x$

58. $y = 4 \sin x + 3 \cos x$

59. $y = \cos x - 3 \sin 2x$

60. $y = \cos 2x + \cos 3x$

OBJECTIVES

1. Graph the tangent function.
2. Graph the cotangent function.
3. Graph the secant and cosecant functions.

5.6 Graphs of Other Trigonometric Functions

In Section 5.5, we graphed the sine and cosine functions. In this section we graph the other four trigonometric functions. Unlike the sine and cosine functions, the other four functions have graphs with vertical asymptotes.

Graphing the Tangent Function

The graphs of $f(x) = \sin x$ and $g(x) = \cos x$, which are shown in Figure 1, can be used to gain information about the graph of $y = \tan x$.

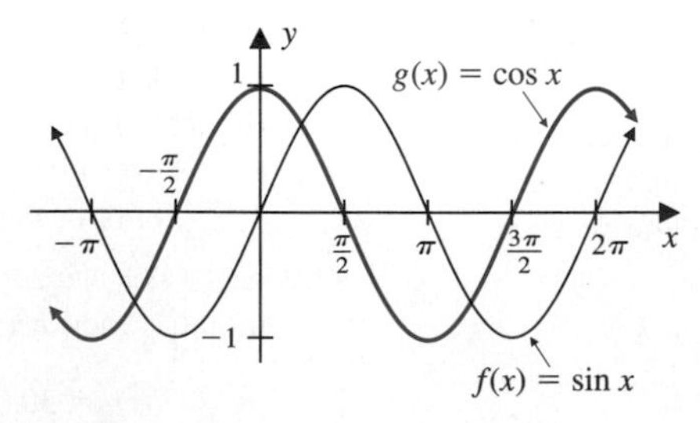

FIGURE 1

Since

$$\tan x = \frac{\sin x}{\cos x}$$

it follows that $y = \tan x$ is undefined for those values of x where $\cos x = 0$. These values, which are the same as the x intercepts of the graph of $y = \cos x$, are $x = \pm\pi/2, \pm(3\pi)/2, \pm(5\pi)/2$, and so on. Consequently, the domain of $y = \tan x$ consists of all values of x for which $x \neq (\pi/2) + n\pi$, where n is an integer.

To help find out what happens to the values of $\tan x$ as x approaches $\pi/2$, we examine the entries in Table 1. They suggest that as x approaches $\pi/2$ from the left, $\tan x$ increases without bound; that is,

$$\tan x \to +\infty \qquad \text{as} \qquad x \to \left(\frac{\pi}{2}\right)^-$$

Thus, the line $x = \pi/2$ is a vertical asymptote of the graph.

TABLE 1

	$x \to (\pi/2)^-$							
x	0	0.5	1.0	1.5	1.56	1.569	1.57	$\pi/2$
$y = \tan x$	0	0.546	1.557	14.101	92.620	556.691	1255.766	Undefined

Similarly, it can be shown that

$$\tan x \to -\infty \qquad \text{as} \qquad x \to -\left(\frac{\pi}{2}\right)^+$$

so $x = -\pi/2$ is also a vertical asymptote. The x intercepts of $y = \tan x$, that is, the values of x for which $\tan x = 0$, are the same as the x intercepts of $f(x) = \sin x$ (Figure 1). They are $x = 0, \pm\pi, \pm 2\pi$, and so on. By using this information and plotting a few points, we obtain a sketch of the graph of $y = \tan x$ in the interval $(-\pi/2, \pi/2)$ (Figure 2a). Since the period of the tangent function is π, we repeat the graph in Figure 2a to the right and left every π units to get a complete graph of the tangent function (Figure 2b).

Graph $y = \tan x$ using a graphing utility and compare the result to Figure 2b.

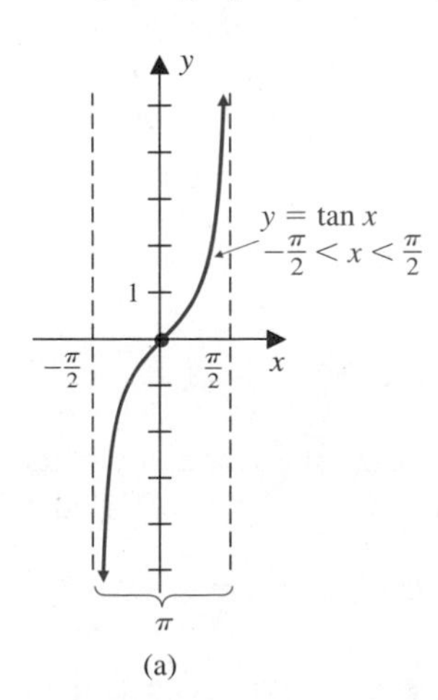

(a)

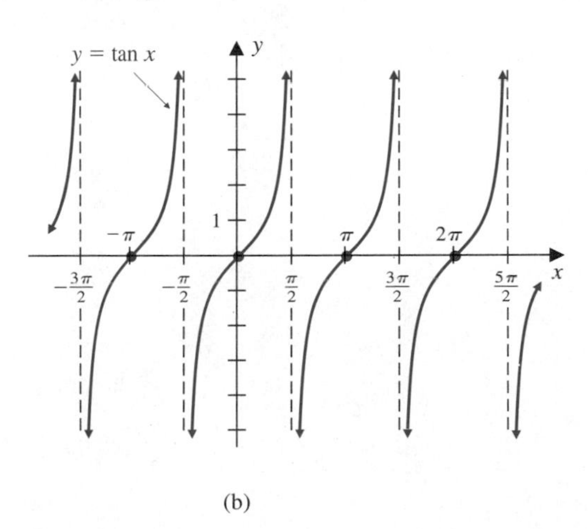

(b)

FIGURE 2

Table 2 lists properties of the tangent function.

TABLE 2
Properties of the Tangent Function

Property	$y = \tan x$
Domain	All real numbers except $(\pi/2) + n\pi$, where n is an integer
Range	All real numbers
Period	π
Symmetry	Symmetric with respect to the origin; that is, the function is *odd*, so $\tan(-x) = -\tan x$
Asymptotes	$x = \pm\pi/2, \pm(3\pi)/2, \pm(5\pi)/2, \ldots, \pm(\pi/2) + n\pi$, where n is an integer
x intercepts	$0, \pm\pi, \pm2\pi, \ldots \pm n\pi$, where n is an integer
Increasing or decreasing behavior	Increases between consecutive asymptotes

In general, graphs of the tangent functions $f(x) = a \tan(kx + b)$ have the following characteristics:

The period of $f(x) = a \tan(kx + b)$ is π/k, and its phase shift is $-b/k$ for $k > 0$ (Problem 46).

If k is negative, we can use the even–odd properties to rewrite the function with a positive coefficient of x, as we did for the sine and cosine functions. For instance,

$$f(x) = \tan(-3x + 1) = \tan[-(3x - 1)] = -\tan(3x - 1)$$

EXAMPLE 1 **Graphing a tangent function**

Sketch the graph of

$$f(x) = \tan\left(x - \frac{\pi}{6}\right)$$

for one period by using the graph of $y = \tan x$ in the interval $(-\pi/2, \pi/2)$. Specify the period, phase shift, and vertical asymptotes.

Solution When $f(x)$ is rewritten in the form $a \tan(kx + b)$, we get

$$f(x) = 1 \tan\left[1x + \left(-\frac{\pi}{6}\right)\right]$$

So $a = 1$, $k = 1$, and $b = -\pi/6$. Thus, the period is $\pi/k = \pi/1 = \pi$, which is the same as the period for $y = \tan x$, and the phase shift is given by

$$\frac{-b}{k} = \frac{-(-\pi/6)}{1} = \frac{\pi}{6}$$

Thus, we can shift the graph of $y = \tan x$ (Figure 3a) $\pi/6$ unit to the right to get the graph of f (Figure 3b). Note that the left asymptote shifts from $x = -\pi/2$ to

$$x = -\frac{\pi}{2} + \frac{\pi}{6} = -\frac{\pi}{3}$$

and the right asymptote shifts from $x = \pi/2$ to

$$x = \frac{\pi}{2} + \frac{\pi}{6} = \frac{2\pi}{3}$$

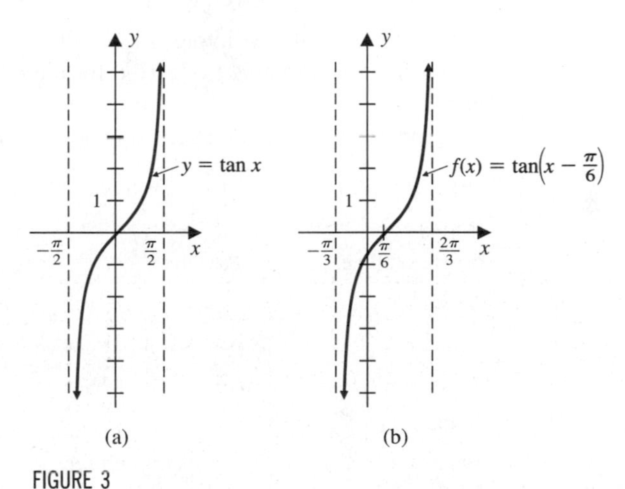

FIGURE 3

Graphing the Cotangent Function

Since

$$\cot x = \frac{\cos x}{\sin x}$$

it follows that the cotangent function $y = \cot x$ is not defined whenever $\sin x = 0$; that is, when $x = 0, \pm\pi, \pm 2\pi$, and so on (see Figure 1). Thus, the domain of $y = \cot x$ includes all values of x for which $x \neq n\pi$, where n is an integer. Since $\cot x = 1/(\tan x)$, by taking the reciprocals of the y values in the graph of $y = \tan x$, we can sketch the graph of the cotangent function (Figure 4a, page 332). Note that the x intercepts and the vertical asymptotes of the tangent and cotangent functions are interchanged; that is, if $x = n\pi$, where n is an integer, then $\tan x = 0$, and $\cot x$ is undefined and has a vertical asymptote. On the other hand, if $x = (\pi/2) + n\pi$, where n is an integer, then $\cot x = 0$, and $\tan x$ is undefined

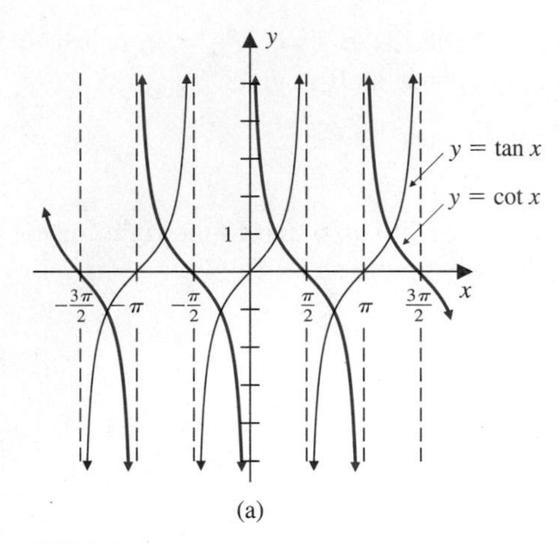

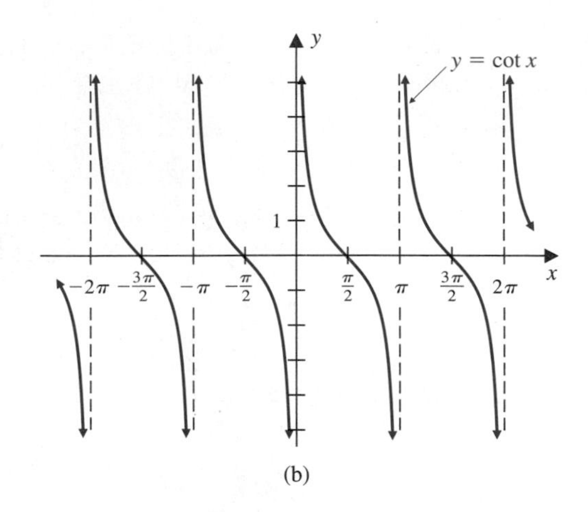

FIGURE 4

and has a vertical asymptote. Figure 4b shows the graph of $y = \cot x$ by itself. As for the tangent, π is the period for the cotangent so its graph repeats every π units.

Properties of the cotangent function are listed in Table 3.

TABLE 3

Properties of the Cotangent Function

Property	$y = \cot x$
Domain	All real numbers except $n\pi$, where n is an integer
Range	All real numbers
Period	π
Symmetry	Symmetric with respect to the origin; that is, the function is *odd*, so $\cot(-x) = -\cot x$
Asymptotes	$x = 0, \pm\pi, \pm 2\pi, \ldots, \pm n\pi$, where n is an integer
x intercepts	$\pm\pi/2, \pm(3\pi)/2, \pm(5\pi)/2, \ldots, \pm(\pi/2) + n\pi$, where n is an integer
Increasing or decreasing behavior	Decreases between consecutive asymptotes

As is the case for the tangent function, graphs of cotangent functions have the following characteristics:

> The period for the function $y = a \cot(kx + b)$ is given by π/k, and its phase shift is $-b/k$ for $k > 0$ (Problem 46).

EXAMPLE 2 **Graphing a cotangent function**
Sketch the graph of one period of the function

$$g(x) = \cot\left(2x - \frac{\pi}{2}\right)$$

by using the graph of $y = \cot x$ in the interval $(0, \pi)$. Find the period and phase shift, and determine the asymptotes.

Solution We begin by graphing $f(x) = \cot 2x$. The graph of one period of $f(x) = \cot 2x$ has the same general shape as the graph of $y = \cot x$ in the interval $(0, \pi)$ (Figure 5a). However, it is shrunk horizontally by a factor of $\frac{1}{2}$, because the period of f is given by $\pi/k = \pi/2$ instead of π (Figure 5b). Correspondingly, the vertical asymptotes for f occur when $2x = 0$ or π, that is, when $x = 0$ or $\pi/2$.

Note that the phase shift is 0, so the graph of f exhibits no horizontal shift from the graph of $y = \cot x$.

The period of the given function g, which is $\pi/2$, is the same as the period of the function f because in both cases $k = 2$. Since the phase shift of

$$g(x) = \cot\left(2x - \frac{\pi}{2}\right) = \cot\left[2x + \left(-\frac{\pi}{2}\right)\right] \quad \text{is} \quad \frac{-b}{k} = \frac{-(-\pi/2)}{2} = \frac{\pi}{4}$$

it follows that the graph of g can be obtained by shifting the graph of $f(x) = \cot 2x$ to the right $\pi/4$ unit (Figure 5c). Note that the vertical asymptotes of the graph of f also shift to the right $\pi/4$ unit to become the vertical asymptotes of the graph of g. The asymptotes are given by

$$x = 0 + \frac{\pi}{4} = \frac{\pi}{4} \quad \text{and} \quad x = \frac{\pi}{2} + \frac{\pi}{4} = \frac{3\pi}{4}$$

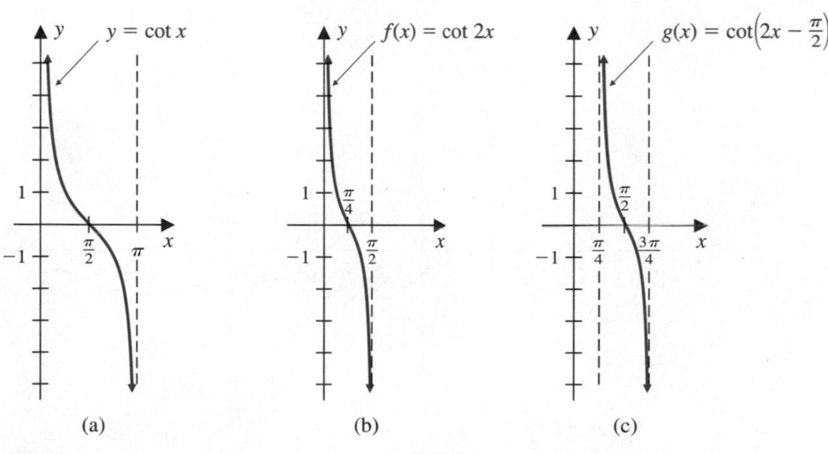

FIGURE 5

Graphing the Secant and Cosecant Functions

The graph of $y = \sec x$ can be sketched with the aid of the reciprocal identity

$$\sec x = \frac{1}{\cos x}$$

The vertical asymptotes occur when $\cos x = 0$, that is, when $x = (\pi/2) + n\pi$, for any integer n. To sketch the graph of $y = \sec x$, we graph the cosine function, locate the vertical asymptotes, and then take the reciprocals of the ordinates (Figure 6). Notice that because of their reciprocal relationship the secant increases where the cosine decreases, and it decreases where the cosine increases. The graph shows that the range of the secant function includes all numbers y, where $y \leq -1$ or $y \geq 1$. In addition, the graph shows that it repeats itself every 2π units. It follows that the function $y = \sec x$ is periodic with period 2π.

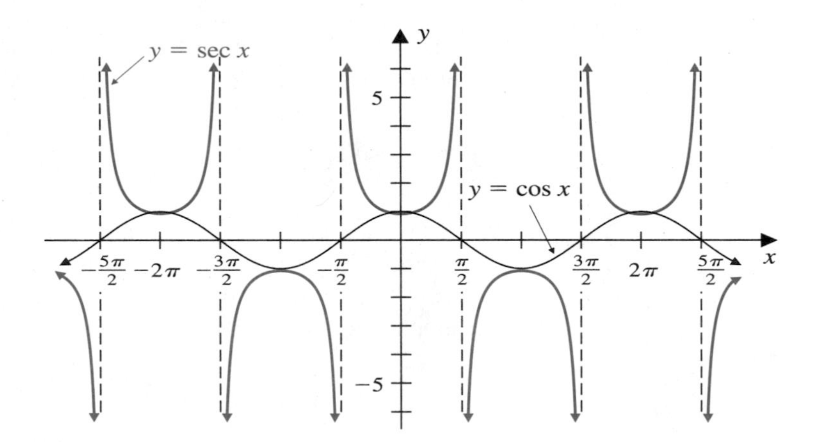

FIGURE 6

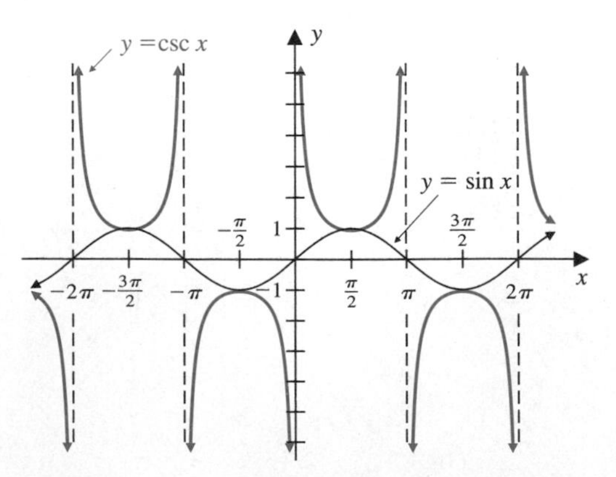

FIGURE 7

Similarly, the graph of the cosecant function $y = \csc x$ can be sketched by using

$$\csc x = \frac{1}{\sin x}$$

The vertical asymptotes occur when $\sin x = 0$, or when $x = n\pi$, for any integer n (Figure 7). The graph reveals that the range of the cosecant function includes all numbers y, where $y \leq -1$ or $y \geq 1$. The graph repeats every 2π units, so the function $y = \csc x$ is periodic with period 2π.

Table 4 summarizes properties of the secant and cosecant functions.

TABLE 4
Properties of the Secant and Cosecant Functions

Property	$y = \sec x$	$y = \csc x$
Domain	All real numbers except $(\pi/2) + n\pi$, where n is an integer	All real numbers except $n\pi$, where n is an integer
Range	All real numbers y such that $y \leq -1$ or $y \geq 1$	All real numbers y such that $y \leq -1$ or $y \geq 1$
Period	2π	2π
Symmetry	Symmetric with respect to the y axis; that is, the function is *even*, so $\sec(-x) = \sec x$	Symmetric with respect to the origin; that is, the function is *odd*, so $\csc(-x) = -\csc x$
Asymptotes	$x = (\pi/2) + n\pi$, where n is an integer	$x = n\pi$, where n is an integer

As with the sine and cosine functions, we have the following properties:

> The functions $f(x) = a \sec(kx + b)$ and $g(x) = a \csc(kx + b)$ have period $(2\pi)/k$ and phase shift $-b/k$ for $k > 0$ (Problem 47).

EXAMPLE 3 **Graphing secant and cosecant functions**

Use transformations to sketch the graph of one period of each function. Find the period, phase shift, and vertical asymptotes.

(a) $f(x) = 3 \csc \dfrac{x}{2}$ (b) $g(x) = \sec(3x + \pi)$

Solution

(a) The graph of $f(x) = 3 \csc (x/2)$ has the same general shape as the graph of $y = \csc x$. However, since $k = \frac{1}{2}$ in the form $y = a \csc(kx + b)$, the period of f is $(2\pi)/k = (2\pi)/\left(\frac{1}{2}\right) = 4\pi$, instead of 2π. So the graph of $y = \csc x$, which has a period of 2π, is stretched horizontally by a factor of 2. Three vertical asymptotes occur when $x/2 = -\pi$, 0, or π, that is, when $x = -2\pi$, 0, or 2π. Since $b = 0$, the phase shift is 0. Because of the multiple of 3, we also need to stretch the graph vertically to get the graph of f (Figure 8a). Note that the vertical stretch by a multiple of 3 causes the range to become all values of y such that $y \leq -3$ or $y \geq 3$.

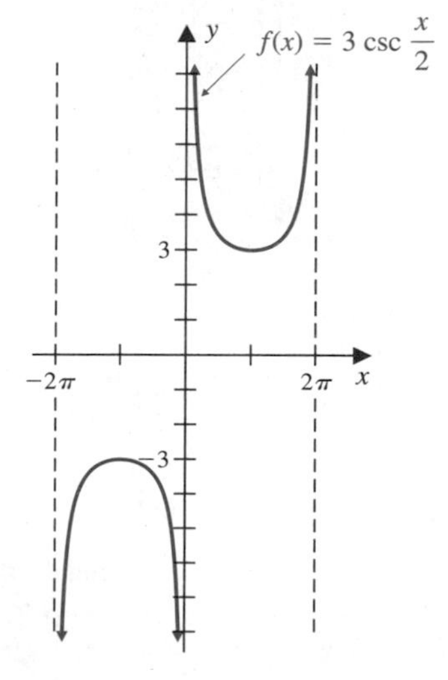

FIGURE 8a

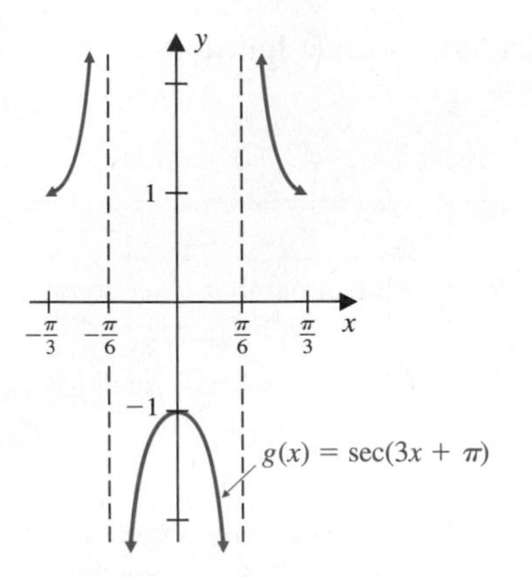

FIGURE 8b

(b) We can obtain the graph of $g(x) = \sec(3x + \pi)$ from the graph of $y = \sec x$. First, note that the period is $(2\pi)/k = (2\pi)/3$. So the graph of $y = \sec x$, which has period 2π, must be horizontally shrunk by a multiple of $\frac{1}{3}$. Finally, we shift the graph to the left by $\pi/3$ units because the phase shift is $-b/k = -\pi/3$. Two vertical asymptotes occur when $3x + \pi = \pi/2$ or $(3\pi)/2$, that is, when $x = -\pi/6$ or $\pi/6$. Figure 8b shows the graph of one period of g.

Most graphing utilities do not have function keys for the cotangent, secant, and cosecant functions. To graph these functions, we use the TAN, COS, and SIN keys, respectively, and the reciprocal key 1/x or x^{-1}, as we did when we evaluated these functions. For instance, the graph of $y = \cot(4x - 3)$ is obtained by graphing

$$y = \frac{1}{\tan(4x - 3)}$$

We graph $f(x) = 5 \sec 3x$ by graphing

$$f(x) = \frac{5}{\cos 3x}$$

and so forth.

EXAMPLE 4 Interpreting a function graphically

Use a graphing utility to sketch the graph of $y = \sec^2 x - \tan^2 x$. Interpret the result.

Solution We enter the equation as

$$y = \left(\frac{1}{\cos x}\right)^2 - (\tan x)^2$$

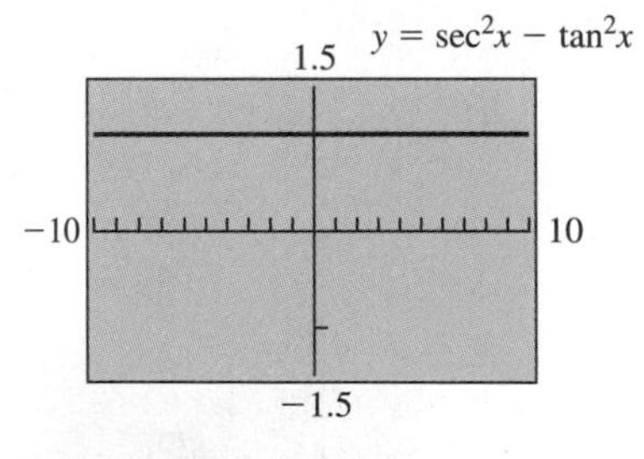

FIGURE 9

A viewing window of the graph of $y = \sec^2 x - \tan^2 x$ is displayed in Figure 9. Upon examining the graph, it appears that it is a horizontal line with a y value of 1. So the graph suggests that $\sec^2 x - \tan^2 x = 1$. Actually, there should be gaps on the graph whenever $x = (\pi/2) + n\pi$, where n is an integer, since these values of x are not in the domains of the tangent and secant functions. The graph demonstrates the validity of the Pythagorean identity $\sec^2 x = 1 + \tan^2 x$.

PROBLEM SET 5.6

Mastering the Concepts

In problems 1–24, use transformations of the graphs of the basic trigonometric functions to sketch the graph of one period for each function. Specify the period and phase shift.

1. $y = 3 \tan x$

2. $y = -3 \tan x$

3. $y = -\frac{3}{2} \cot x$

4. $y = -2 \sec x$

5. $y = 5 \csc x$

6. $y = 2 \sec \frac{x}{4}$

7. $y = 1 + 2 \csc x$

8. $y = 2 - \sec x$

9. $y = -2 \cot 4x$

10. $y = 2 \tan \dfrac{\pi x}{4}$

11. $y = -2 \sec \dfrac{2x}{3}$

12. $y = -3 \csc \dfrac{\pi x}{6}$

13. $y = 2 \csc 4x$

14. $y = -3 \sec \pi x$

15. $y = 2 \csc\left(x - \dfrac{\pi}{6}\right)$

16. $y = 3 \csc\left(x + \dfrac{\pi}{3}\right)$

17. $y = \tan\left(x - \dfrac{\pi}{4}\right)$

18. $y = \tan\left(x + \dfrac{2\pi}{3}\right)$

19. $y = 1.5 \cot\left(x - \dfrac{\pi}{2}\right)$

20. $y = 3 \cot\left(x + \dfrac{5\pi}{6}\right)$

21. $y = 3 \sec\left(\dfrac{x}{4} + \dfrac{\pi}{4}\right)$

22. $y = 2 \csc\left(3x - \dfrac{\pi}{2}\right)$

23. $y = -2 \cot\left(\dfrac{\pi x}{4} - \dfrac{\pi}{4}\right)$

24. $y = -2 \sec\left(\dfrac{\pi x}{6} + \dfrac{\pi}{6}\right)$

25. Use the graphs of the functions to indicate where each function listed in the table is increasing or decreasing.

	As x increases from:			
	0 to $\pi/2$	$\pi/2$ to π	π to $(3\pi)/2$	$(3\pi)/2$ to 2π
(a) $y = \tan x$				
(b) $y = \cot x$				
(c) $y = \sec x$				
(d) $y = \csc x$				

26. **(a)** Explain why $y = \tan x$ is not one-to-one.
(b) Is $f(x) = \tan x$, where $-\pi/2 < x < \pi/2$, one-to-one? Use a graph to support the assertion.
(c) Is $f(x) = \tan x$, where $0 < x < 2\pi$, one-to-one? Use a graph to support the assertion.

In problems 27–30, use the graph of the indicated function to complete each statement.

27. **(a)** As $x \to \left(\dfrac{\pi^+}{4}\right)$, $\tan x \to$ ____

(b) As $x \to \left(\dfrac{\pi^+}{2}\right)$, $\tan x \to$ ____

28. **(a)** As $x \to 0^+$, $\cot x \to$ ____

(b) As $x \to \left(\dfrac{\pi^+}{2}\right)$, $\sec x \to$ ____

29. **(a)** As $x \to \left(\dfrac{\pi^-}{2}\right)$, $\sec x \to$ ____

(b) As $x \to \left(\dfrac{\pi^-}{4}\right)$, $\cot x \to$ ____

30. **(a)** As $x \to \left(\dfrac{\pi^-}{6}\right)$, $\tan x \to$ ____

(b) As $x \to \pi^+$, $\cot x \to$ ____

In problems 31–34, one period of a graph is shown. Find an equation of the given form for the graph.

31. $y = \tan kx$, $k > 0$

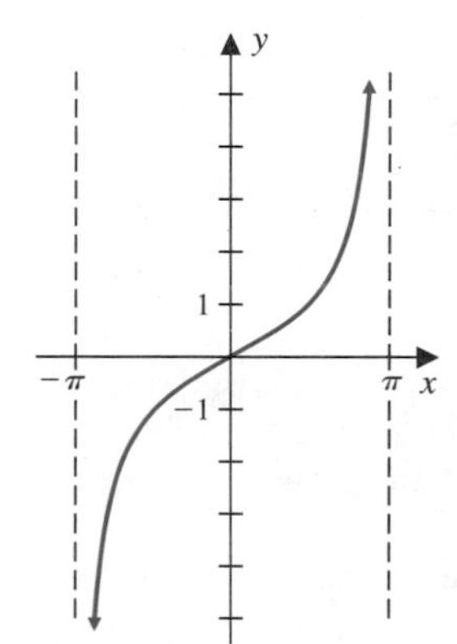

32. $y = a \tan kx$, $a < 0$ and $k > 0$

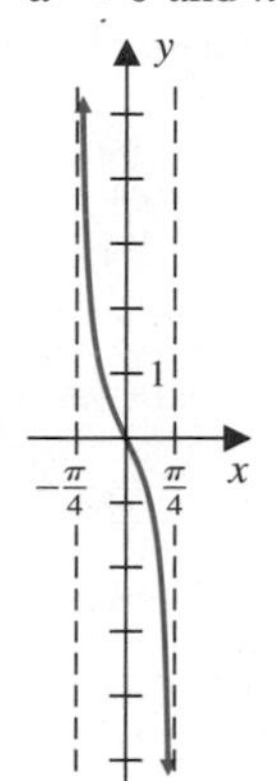

33. $y = a \csc(kx + b)$, $a > 0$, $b > 0$, $k > 0$

34. $y = a \sec kx$, $a > 0$ and $k > 0$

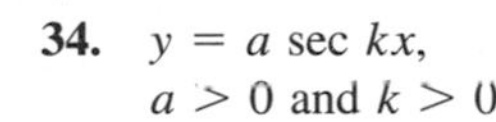

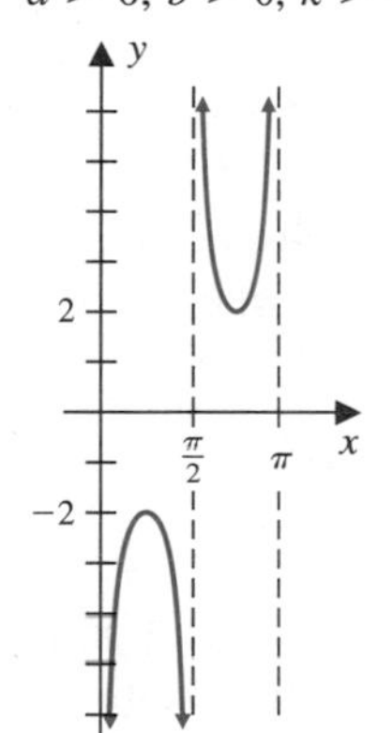

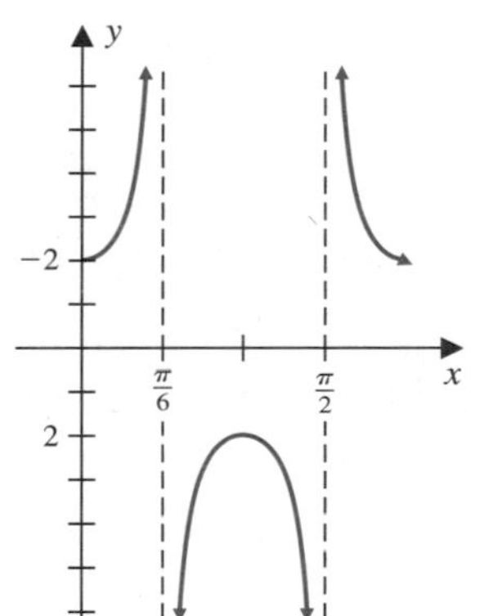

35. GU Use a graphing utility to sketch the graph of the function $f(x) = \csc^2 x - \cot^2 x$. Interpret the result in terms of a Pythagorean identity.

36. What is the distance from one vertical asymptote to the next for each function?

(a) $y = \sec\left(x - \dfrac{\pi}{2}\right)$ **(b)** $y = 2 \tan 3x$

(c) $y = -2 \cot \dfrac{x}{6}$

Developing and Extending the Concepts

37. GU Use a graphing utility to graph $f(x) = (\tan x)(\cos x)$ and $g(x) = \sin x$, where $-2\pi \le x \le 2\pi$. Are the graphs the same? Are the functions equal? Explain.

38. Compare the graphs of $f(x) = \sec x$ and $g(x) = \sec(-x)$. Are they the same or different? Explain.

39. Compare the graphs of $f(x) = \csc x$ and $g(x) = \csc(-x)$. Are they the same or different? Explain.

40. Sketch the graphs of

$$f(x) = \tan x \quad \text{and} \quad g(x) = \cot\left(\frac{\pi}{2} - x\right)$$

on the same coordinate system. How do the graphs compare?

41. Sketch the graphs of

$$f(x) = \sec x \quad \text{and} \quad g(x) = \csc\left(\frac{\pi}{2} - x\right)$$

on the same coordinate system. How do the graphs compare?

42. Explain why the graphs of $y = \tan x$, $f(x) = \tan(x + \pi)$, $g(x) = \tan(x - \pi)$, and $h(x) = \tan(x + 2\pi)$ are the same.

43. Use symmetry to sketch the graphs of both $y = \cot x$ and $x = \cot y$, where $-\pi < x < \pi$, on the same coordinate system.

In problems 44 and 45, write an equation of the form $y = \tan kx$ or $y = \csc kx$ that satisfies the given condition.

44. **(a)** Tangent function of period $\pi/6$
(b) Cosecant function of period $(5\pi)/6$

45. **(a)** Tangent function of period 2
(b) Cosecant function of period 1.5

46. Show that the functions $f(x) = a \tan(kx + b)$ and $g(x) = a \cot(kx + b)$ both have period π/k and phase shift $-b/k$ for $k > 0$.

47. Show that both functions $f(x) = a \sec(kx + b)$ and $g(x) = a \csc(kx + b)$ have period $(2\pi)/k$ and phase shift $-b/k$ for $k > 0$.

OBJECTIVES

1. Define the inverse sine, cosine, and tangent functions.
2. Simplify expressions involving inverse functions.
3. Define the other inverse trigonometric functions.

5.7 Inverse Trigonometric Functions

By examining the graphs of the six basic trigonometric functions, we observe (using the horizontal-line test) that none of them is one-to-one. So none of them has an inverse. However, by restricting domains, we can get trigonometric functions that do have inverses.

Defining the Inverse Sine, Cosine, and Tangent Functions

Figure 1 shows the graphs of $f(x) = \sin x$, $g(x) = \cos x$, and $h(x) = \tan x$, with the domain of f restricted to the interval $[-\pi/2, \pi/2]$ (Figure 1a); the domain of g restricted to the interval $[0, \pi]$ (Figure 1b); and the domain of h restricted to the interval $(-\pi/2, \pi/2)$ (Figure 1c). From the graphs in Figure 1, we see that each of these functions is one-to-one, and therefore has an inverse.

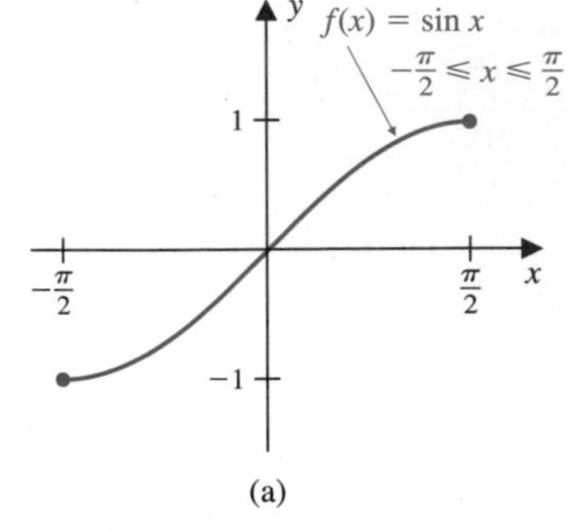

(a)

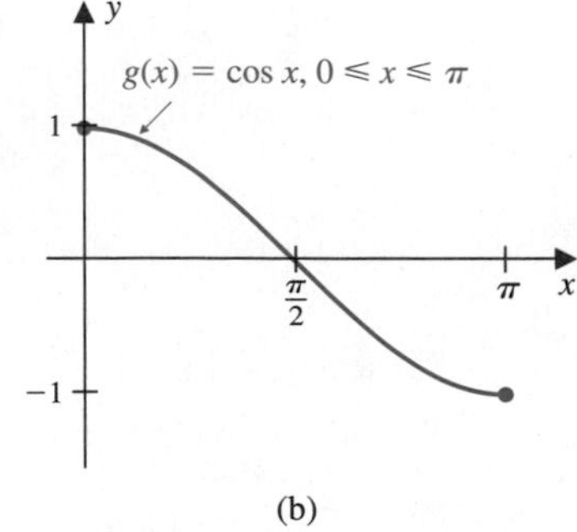

(b)

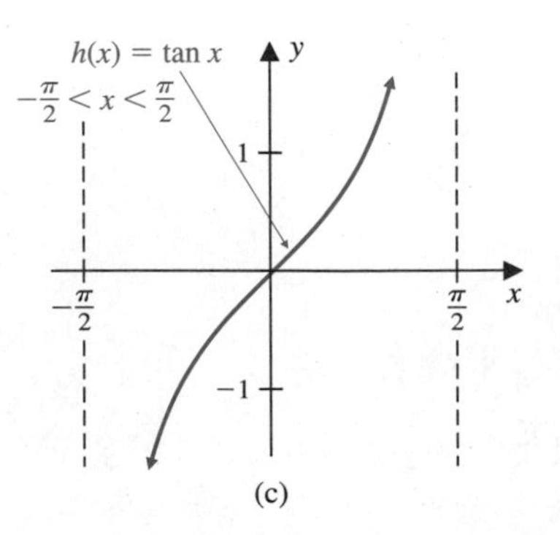

(c)

FIGURE 1

When we refer to the *inverse sine function*, we mean the function that is the inverse of $f(x) = \sin x$, with the restriction $-\pi/2 \le x \le \pi/2$ (Figure 1a). The inverse sine function is denoted by $y = \sin^{-1} x$ (read "the inverse sine of x") or $y = \arcsin x$ (read "the arcsine of x").

DEFINITION

THE INVERSE SINE FUNCTION

The **inverse sine function,**

$$y = \sin^{-1} x \qquad \text{or} \qquad y = \arcsin x$$

is defined by

$$\sin y = x \qquad -\pi/2 \le y \le \pi/2 \quad \text{and} \quad -1 \le x \le 1$$

Note that even though we write $\sin^2 x$ to represent $(\sin x)^2$, we never use $\sin^{-1} x$ to represent $(\sin x)^{-1}$, since

$$(\sin x)^{-1} = \frac{1}{\sin x}$$

which is not the same as the inverse function of $\sin x$. This confusion arises because we use the "exponent" -1 in two different ways: to denote reciprocals and to denote inverse functions.

We can think of $\arcsin x$ or $\sin^{-1} x$ as the number (or the angle) between $-\pi/2$ and $\pi/2$ inclusive, whose sine is x. The special values learned earlier enable us to find exact inverse function values for certain numbers.

EXAMPLE 1 Finding exact values of the inverse sine

Find the exact value of each expression.

(a) $\sin^{-1} \dfrac{1}{2}$ (b) $\sin^{-1}\left(-\dfrac{\sqrt{2}}{2}\right)$

Solution By using the special values along with the restrictions on the inverse sine, we get the solutions.

(a) Suppose that $y = \sin^{-1} \frac{1}{2}$. This equation is equivalent to $\sin y = \frac{1}{2}$, where $-\pi/2 \le y \le \pi/2$. From the special values, we know that $\sin(\pi/6) = \frac{1}{2}$. Since $-\pi/2 \le \pi/6 \le \pi/2$, it follows that $y = \pi/6$. That is, we have $\sin^{-1} \frac{1}{2} = \pi/6$.

(b) Similarly, $y = \sin^{-1}\left(-\sqrt{2}/2\right)$ is equivalent to $\sin y = -\sqrt{2}/2$, where $-\pi/2 \le y \le \pi/2$. The value $-\pi/4$ satisfies the conditions for y, so $\sin^{-1}\left(-\sqrt{2}/2\right) = -\pi/4$.

The restrictions on the inverse functions are very important to keep in mind. For instance, even though $\sin[(7\pi)/4] = -\sqrt{2}/2$, $\sin^{-1}\left(\sqrt{2}/2\right) \ne (7\pi)/4$ because $(7\pi)/4$ does not fall in the interval $[-\pi/2, \pi/2]$.

We now define the inverse of $g(x) = \cos x$, with x restricted to the interval $0 \leq x \leq \pi$ (Figure 1b). The inverse cosine function is denoted by $y = \cos^{-1} x$ or $y = \arccos x$.

DEFINITION

THE INVERSE COSINE FUNCTION

The **inverse cosine function,**

$$y = \cos^{-1} x \qquad \text{or} \qquad y = \arccos x$$

is defined by

$$\cos y = x \qquad 0 \leq y \leq \pi \quad \text{and} \quad -1 \leq x \leq 1$$

We can think of $\arccos x$ or $\cos^{-1} x$ as the number (or the angle) between 0 and π inclusive, whose cosine is x.

EXAMPLE 2 Finding exact values of the inverse cosine

Find the exact value of each expression.

(a) $\cos^{-1}(-1)$ (b) $\arccos\left(-\frac{\sqrt{2}}{2}\right)$

Solution The restrictions on the inverse cosine lead us to the following results:

(a) For $y = \cos^{-1}(-1)$, we know that $\cos \pi = -1$ and $0 \leq \pi \leq \pi$, so $y = \pi$.

(b) For $y = \arccos\left(-\frac{\sqrt{2}}{2}\right)$, we know that $\cos \frac{3\pi}{4} = -\frac{\sqrt{2}}{2}$ and $0 \leq \frac{3\pi}{4} \leq \pi$, so $y = \frac{3\pi}{4}$.

The function $y = \tan^{-1} x$ or $y = \arctan x$ denotes the inverse of the function $h(x) = \tan x$, where $-\pi/2 < x < \pi/2$ (Figure 1c).

DEFINITION

THE INVERSE TANGENT FUNCTION

The values $-\pi/2$ and $\pi/2$ are excluded for the tangent function, since $\tan(-\pi/2)$ and $\tan(\pi/2)$ are not defined.

The **inverse tangent function**

$$y = \tan^{-1} x \qquad \text{or} \qquad y = \arctan x$$

is defined by

$$\tan y = x \qquad -\frac{\pi}{2} < y < \frac{\pi}{2} \text{ and } x \text{ a real number}$$

We can think of $\arctan x$ or $\tan^{-1} x$ as the number (or angle) between $-\pi/2$ and $\pi/2$ whose tangent is x. For instance, $\tan^{-1}(-1) = -\pi/4$ since we know that $\tan(-\pi/4) = -1$ and $-\pi/2 < -\pi/4 < \pi/2$.

Figure 2 shows viewing windows for the graphs of the inverse sine and inverse cosine functions, respectively. Each graph of the inverse trigonometric function is the reflection of the graph of the restricted function across the line $y = x$.

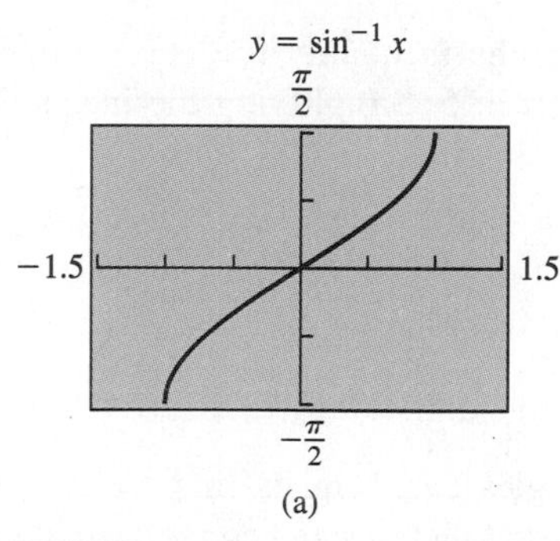

(a)

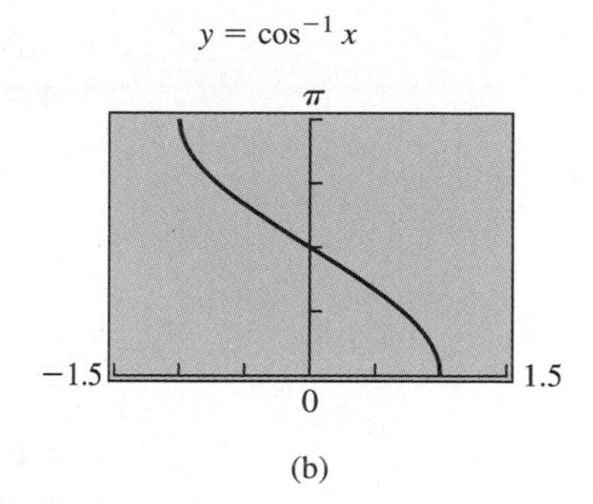

(b)

FIGURE 2

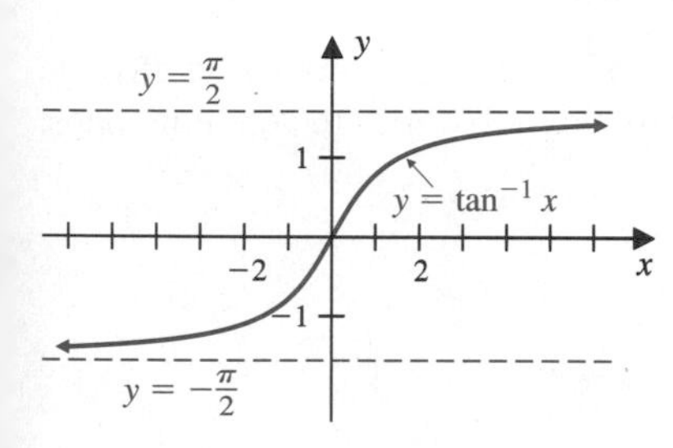

FIGURE 3

Since the tangent function $f(x) = \tan x$ has vertical asymptotes at $x = -\pi/2$ and $x = \pi/2$, it follows that there are horizontal asymptotes for the graph of $y = \tan^{-1} x$ at $y = -\pi/2$ and $y = \pi/2$ (Figure 3).

When a calculator is used to approximate the values of the inverse functions, we use the radian mode.

EXAMPLE 3 **Approximating values of inverse functions**

Use a calculator to approximate each value to two decimal places.

(a) $\sin^{-1} 0.8016$ (b) $\cos^{-1}(-0.3281)$ (c) $\arctan 0.6235$

Solution First we set the calculator in radian mode, and then we use the appropriate inverse function key to get the following approximate values:

(a) $\sin^{-1} 0.8016 = 0.93$ (b) $\cos^{-1}(-0.3281) = 1.91$
(c) $\arctan 0.6235 = 0.56$

When we rewrite an equation containing an inverse trigonometric function as an equation in terms of a trigonometric function, care must be taken to keep track of the restrictions on the variables.

EXAMPLE 4 **Rewriting an inverse function**

If $y = 5 + 3 \cos^{-1} 2x$, express x in terms of y. Determine the restrictions on x and y.

Solution The given equation, $y = 5 + 3 \cos^{-1} 2x$, can be written as

$$\frac{y - 5}{3} = \cos^{-1} 2x$$

This latter equation is equivalent to

$$\cos\left(\frac{y - 5}{3}\right) = 2x \qquad \text{where } 0 \le \frac{y - 5}{3} \le \pi \text{ and } -1 \le 2x \le 1$$

Simplifying the inequality $0 \le (y - 5)/3 \le \pi$ gives us $0 \le y - 5 \le 3\pi$, or $5 \le y \le 3\pi + 5$. Simplifying the inequality $-1 \le 2x \le 1$ yields $-\frac{1}{2} \le x \le \frac{1}{2}$. Thus,

$$x = \frac{1}{2}\cos\left(\frac{y-5}{3}\right) \qquad \text{where } 5 \le y \le 3\pi + 5 \text{ and } -\frac{1}{2} \le x \le \frac{1}{2}$$

Simplifying Expressions Involving Inverse Functions

Right triangles can help us find exact values of expressions involving inverse trigonometric functions and convert trigonometric expressions into algebraic ones.

EXAMPLE 5 Simplifying inverse trigonometric expressions

Use right triangles to simplify each expression.

(a) $\tan\left(\cos^{-1}\frac{3}{4}\right)$ (b) $\sin^{-1} x + \cos^{-1} x$, where $0 < x < 1$

Solution

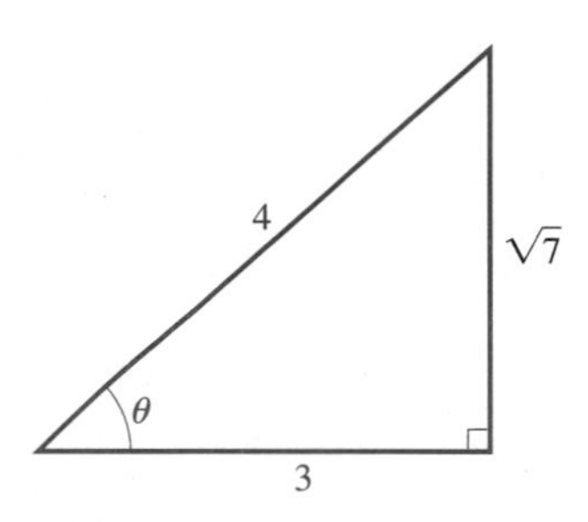

FIGURE 4a

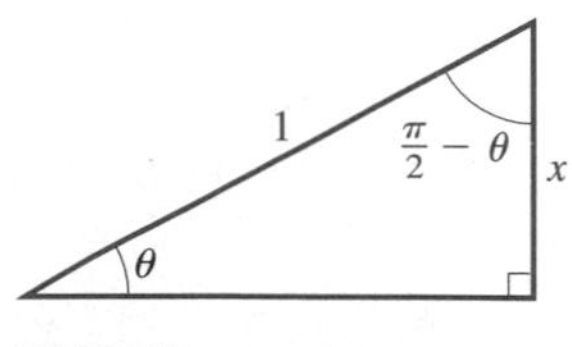

FIGURE 4b

(a) Assume that $\theta = \cos^{-1}\frac{3}{4}$. By definition, $\cos\theta = \frac{3}{4}$, where $0 < \theta < \pi/2$, since the cosine is positive. Next, we construct a right triangle with one side of length 3 adjacent to angle θ and the hypotenuse of length 4 so that $\cos\theta = \frac{3}{4}$ (Figure 4a). By the Pythagorean theorem, the length of the side opposite angle θ is $\sqrt{4^2 - 3^2} = \sqrt{7}$. Using right triangle trigonometry, we obtain

$$\tan\left(\cos^{-1}\frac{3}{4}\right) = \tan\theta = \frac{\text{opp}}{\text{adj}} = \frac{\sqrt{7}}{3}$$

(b) Assume that $\theta = \sin^{-1} x$, so $\sin\theta = x$. Since $0 < x < 1$, it follows that $0 < \theta \le \pi/2$. We can think of θ as an acute angle in a right triangle with sides that satisfy

$$\sin\theta = \frac{\text{opp}}{\text{hyp}} = \frac{x}{1} = x \quad \text{(Figure 4b)}$$

Using right triangle trigonometry, we get

$$\cos\left(\frac{\pi}{2} - \theta\right) = \frac{\text{adj}}{\text{hyp}} = \frac{x}{1} = x \qquad \text{where } 0 < \frac{\pi}{2} - \theta < \frac{\pi}{2}$$

So

$$\cos^{-1} x = \frac{\pi}{2} - \theta$$

It follows that

$$\sin^{-1} x + \cos^{-1} x = \theta + \left(\frac{\pi}{2} - \theta\right) = \frac{\pi}{2}$$

that is, $\sin^{-1} x + \cos^{-1} x = \pi/2$.

Defining the Other Inverse Trigonometric Functions

By restricting the domains of the cosecant, secant, and cotangent functions, we get one-to-one functions that enable us to define the other inverse trigonometric functions.

DEFINITION

OTHER INVERSE TRIGONOMETRIC FUNCTIONS

Function	Defined by
$y = \csc^{-1} x$	$\csc y = x$, where $x \leq -1$ or $x \geq 1$ and $-\pi/2 \leq y \leq \pi/2$, $y \neq 0$
$y = \sec^{-1} x$	$\sec y = x$, where $x \leq -1$ or $x \geq 1$ and $0 \leq y \leq \pi$, $y \neq \pi/2$
$y = \cot^{-1} x$	$\cot y = x$, where x is any real number and $0 < y < \pi$

The notations arccsc x, arcsec x, and arccot x, respectively, are also used for these functions.

The graphs and properties of these inverse functions are shown in Table 1. Note that the choices of the ranges for $g(x) = \sec^{-1} x$ and $h(x) = \csc^{-1} x$ are not universally accepted and may differ in some other textbooks.

TABLE 1

Other Inverse Trigonometric Functions

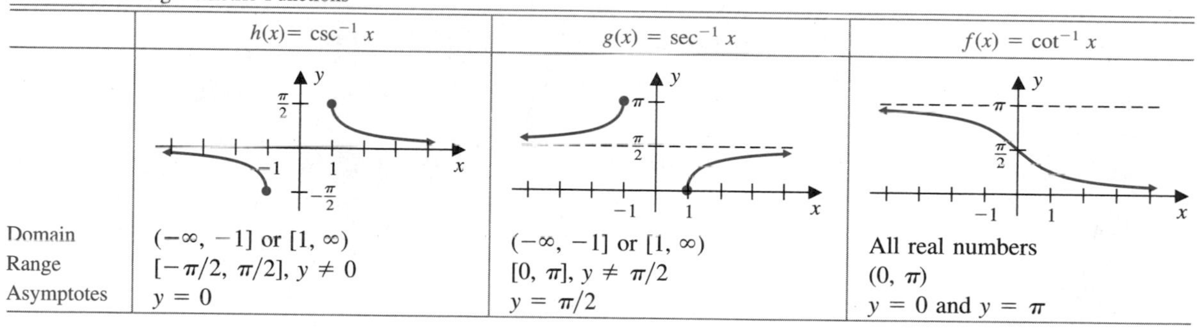

	$h(x) = \csc^{-1} x$	$g(x) = \sec^{-1} x$	$f(x) = \cot^{-1} x$
Domain	$(-\infty, -1]$ or $[1, \infty)$	$(-\infty, -1]$ or $[1, \infty)$	All real numbers
Range	$[-\pi/2, \pi/2]$, $y \neq 0$	$[0, \pi]$, $y \neq \pi/2$	$(0, \pi)$
Asymptotes	$y = 0$	$y = \pi/2$	$y = 0$ and $y = \pi$

EXAMPLE 6 Evaluating inverse functions

Find the exact value of each function.

(a) $\cot^{-1} \sqrt{3}$ (b) $\csc^{-1}(-2)$

Solution

(a) $\cot^{-1} \sqrt{3}$ is the number y in the interval $(0, \pi)$ that satisfies $\cot y = \sqrt{3}$. Therefore, $y = \pi/6$.

(b) $\csc^{-1}(-2)$ is the number y in the interval $[-\pi/2, \pi/2]$, $y \neq 0$, that satisfies $\csc y = -2$. Therefore, $y = -\pi/6$.

Most calculators do not have keys for the inverse cotangent, inverse secant, or inverse cosecant functions. The following identities are used to find the approximate values of these functions on a calculator:

1. If $|x| \geq 1$, then:

(a) $\csc^{-1} x = \sin^{-1} \frac{1}{x}$ (b) $\sec^{-1} x = \cos^{-1} \frac{1}{x}$

2. If x is a real number, then:

$$\cot^{-1} x = \begin{cases} \tan^{-1} \frac{1}{x} & \text{if } x > 0 \\ \tan^{-1} \frac{1}{x} + \pi & \text{if } x < 0 \\ \frac{\pi}{2} & \text{if } x = 0 \end{cases}$$

EXAMPLE 7 Approximating values of inverse trigonometric functions

Use a calculator to approximate each value to three decimal places.

(a) $\cot^{-1} 12.3$ (b) $\sec^{-1} 1.84$ (c) $\csc^{-1}(-1.52)$

Solution First, we set the calculator in radian mode. Then we use the above identities to get the following approximate values:

(a) $\cot^{-1} 12.3 = \tan^{-1}\left(\frac{1}{12.3}\right) = 0.081$

(b) $\sec^{-1} 1.84 = \cos^{-1}\left(\frac{1}{1.84}\right) = 0.996$

(c) $\csc^{-1}(-1.52) = \sin^{-1}\left(\frac{1}{-1.52}\right) = -0.718$

PROBLEM SET 5.7

Mastering the Concepts

In problems 1–10, find the exact value of each expression.

1. (a) $\sin^{-1} \frac{\sqrt{2}}{2}$ (b) $\sin^{-1}\left(-\frac{1}{2}\right)$

2. (a) $\cos^{-1} \frac{\sqrt{3}}{2}$ (b) $\cos^{-1}\left(-\frac{1}{2}\right)$

3. (a) $\tan^{-1} \frac{\sqrt{3}}{3}$ (b) $\tan^{-1}\left(-\frac{\sqrt{3}}{3}\right)$

4. (a) $\arctan(-1)$ (b) $\arccos 1$

5. (a) $\arcsin\left(-\frac{\sqrt{3}}{2}\right)$ (b) $\arccos\left(-\frac{\sqrt{3}}{2}\right)$

6. (a) $\arcsin 1$ (b) $\arctan \sqrt{3}$

7. (a) $\cos^{-1}\left(\cos \frac{\pi}{6}\right)$ (b) $\tan^{-1}\left(\tan \frac{\pi}{4}\right)$

8. (a) $\sin\left(\cos^{-1} \frac{1}{2}\right)$ (b) $\sin^{-1}\left[\sin\left(-\frac{3\pi}{2}\right)\right]$

9. (a) $\sin^{-1}\left(\sin \frac{5\pi}{4}\right)$ (b) $\tan[\tan^{-1}(-1)]$

10. (a) $\sec\left(\sin^{-1} \frac{\sqrt{3}}{2}\right)$ (b) $\csc(\tan^{-1} 1)$

In problems 11–16, find the approximate values of each expression to two decimal places.

11. **(a)** $\sin^{-1} 0.2182$ **(b)** $\arcsin(-0.7771)$
12. **(a)** $\cos^{-1} 0.8628$ **(b)** $\arccos(-0.8473)$
13. **(a)** $\tan^{-1} 1.072$ **(b)** $\arctan(-41.03)$
14. **(a)** $\arccos 0.4037$ **(b)** $\arcsin(2/\pi)$
15. **(a)** $\arccos(-0.7112)$ **(b)** $\tan^{-1} 100$
16. **(a)** $\sin^{-1}(1 - \sqrt{2})$ **(b)** $\sin^{-1}(\sqrt{3} - \sqrt{2})$

In problems 17–24, express x in terms of y. Determine the restrictions on x and y.

17. $y = \sin^{-1}(x + 1)$ **18.** $y = \cos^{-1}(x - 2)$
19. $y = \cos^{-1}(2x - 8)$ **20.** $y = 2\sin^{-1}(3x + 7)$
21. $y = -3\tan^{-1}(3x - 2)$ **22.** $y = -2\cos^{-1}(2x - 1)$
23. $y = \frac{1}{3}\cos^{-1}(2x - 4)$ **24.** $y = 4 - 2\sin^{-1} 4x$

In problems 25–30, use a right triangle whenever possible to find the exact value of each expression.

25. **(a)** $\cos\left(\sin^{-1}\frac{4}{5}\right)$ **(b)** $\sin\left(\cos^{-1}\frac{5}{13}\right)$
26. **(a)** $\sin\left(\tan^{-1}\frac{4}{3}\right)$ **(b)** $\cos^{-1}\left(\cos\frac{3\pi}{2}\right)$
27. **(a)** $\tan\left(\sin^{-1}\frac{2\sqrt{5}}{5}\right)$ **(b)** $\cos\left[\cos^{-1}\left(-\frac{2\sqrt{5}}{5}\right)\right]$
28. **(a)** $\sin[\tan^{-1}(-2)]$ **(b)** $\sec\left(\sin^{-1}\frac{4}{5}\right)$
29. **(a)** $\csc\left(\cos^{-1}\frac{1}{4}\right)$ **(b)** $\sec^{-1}\left(\sec\frac{\pi}{3}\right)$
30. **(a)** $\cot\left[\tan^{-1}\left(-\frac{5}{12}\right)\right]$ **(b)** $\csc\left(\sin^{-1}\frac{12}{13}\right)$

In problems 31 and 32, use a right triangle to rewrite each expression as an algebraic expression in terms of x that does not involve an inverse trigonometric function (where $x > 0$).

31. **(a)** $\sin(\cos^{-1} x)$ **(b)** $\tan(\cos^{-1} x)$
(c) $\cot(\tan^{-1} x)$
32. **(a)** $\cot(\sin^{-1} x)$ **(b)** $\cot(\cos^{-1} x)$
(c) $\csc\left(\tan^{-1}\frac{x}{\sqrt{2}}\right)$

In problems 33–36, find the exact value of each expression.

33. **(a)** $\cot^{-1} 1$ **(b)** $\csc^{-1}(-\sqrt{2})$
34. **(a)** $\cot^{-1}(-1)$ **(b)** $\sec^{-1} 2$
35. **(a)** $\sec^{-1}\sqrt{2}$ **(b)** $\csc^{-1}\left(-\frac{2}{\sqrt{3}}\right)$
36. **(a)** $\sec^{-1}(-\sqrt{2})$ **(b)** $\csc^{-1}\left(\frac{2}{\sqrt{3}}\right)$

In problems 37–40, find the approximate value rounded to two decimal places.

37. **(a)** $\cot^{-1} 0.9713$ **(b)** $\sec^{-1} 3.4182$
38. **(a)** $\csc^{-1} 1.0152$ **(b)** $\cot^{-1} 1.0487$
39. **(a)** $\sec^{-1}(-2.4182)$ **(b)** $\csc^{-1}(-8.8952)$
40. **(a)** $\sec^{-1} 8.4513$ **(b)** $\cot^{-1}(-0.8752)$

Developing and Extending the Concepts

41. GU Use a graphing utility to graph $y = \sin^{-1} x + \cos^{-1} x$, where $0 \le x \le 1$. Explain how the graph demonstrates the result in Example 5b.

42. GU Use a graphing utility to graph $y = \tan^{-1} x^2$. Then use the graph to describe the limit behavior of the function; that is, determine what happens to the values of $\tan^{-1} x^2$ as $x \to +\infty$ and as $x \to -\infty$.

GU In problems 43–50, use transformations of the graphs in Figures 2 and 3 to sketch the graph of each function. Demonstrate the validity of the result by graphing the function with a graphing utility.

43. $y = \frac{1}{2}\sin^{-1} x$ **44.** $y = -3\cos^{-1} x$
45. $y = 1 + \sin^{-1} x$ **46.** $y = \cos^{-1} x - 2$
47. $y = 3 + 2\tan^{-1} x$ **48.** $y = \tan^{-1} x + \frac{\pi}{2}$
49. $y = 2[\sin^{-1}(x - 1)] - \pi$
50. $y = \frac{1}{3}[\cos^{-1}(x + 1)] + \frac{\pi}{2}$

51. Show that $\sin^{-1}(-x) = -\sin^{-1} x$

52. Is it true that $\tan^{-1} x = \dfrac{\sin^{-1} x}{\cos^{-1} x}$? Justify your answer.

53. GU Use a graphing utility to graph

$$f(x) = (\sin x)^{-1} = \frac{1}{\sin x} \quad \text{and} \quad g(x) = \sin^{-1} x$$

for $-1 \le x \le 1$. Are the graphs the same? Explain the use of the superscript $^{-1}$ in both equations.

54.  Use a graphing utility to graph

$$f(x) = (\cos x)^{-1} = \frac{1}{\cos x} \quad \text{and} \quad g(x) = \cos^{-1} x$$

for $-1 \le x \le 1$. Are the graphs the same? Explain the use of the superscript $^{-1}$ in both equations.

55. GU Use a graphing utility to sketch the graphs of

$$f(x) = \sin^{-1} x \quad \text{and} \quad g(x) = x + \frac{x^3}{3} + \frac{x^5}{5}$$

for the interval $[-1, 1]$. Compare the values of $f(x)$ and $g(x)$ in this interval.

56. GU Use a graphing utility to sketch the graphs of

$$f(x) = \tan^{-1} x \quad \text{and} \quad g(x) = x - \frac{x^3}{3} + \frac{x^5}{5}$$

for the interval $[-1, 1]$. Compare the values of $f(x)$ and $g(x)$ in this interval.

57. GU Use a graphing utility to sketch the graph of the function $y = \sin(\sin^{-1} x)$. What is the domain of this function? Explain your result.

OBJECTIVES

1. Solve right triangles.
2. Solve applied problems.
3. Model simple harmonic motion.

5.8 Trigonometric Applications and Models

In this section, we'll investigate right triangle applications and the use of trigonometric functions to model the periodic phenomenon known as *simple harmonic motion.*

Solving Right Triangles

Under certain conditions, it is possible to find the measurements of unknown sides and angles of a right triangle from known sides and angles by using the trigonometric functions, as defined earlier for acute angles. This process is called *solving a right triangle.*

Consider the right triangle ACB shown in Figure 1. We'll adopt the convention of labeling the angles at the vertices A, B, and C by α, β, and γ, respectively, and the lengths of the sides opposite these angles by a, b, and c, respectively. Note that α and β are complementary angles and $\gamma = 90°$.

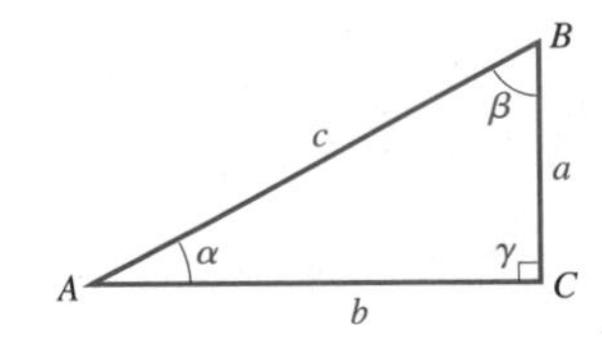

FIGURE 1

When solving a right triangle, we usually draw a sketch of the triangle approximately to scale to gain some insight into the problem. Such a sketch helps us to notice an error if the computed values seem unreasonable.

EXAMPLE 1 Solving a right triangle given a side and an angle

Solve the right triangle ACB if $a = 9$ and $\beta = 36.87°$. Round off the answers to two decimal places.

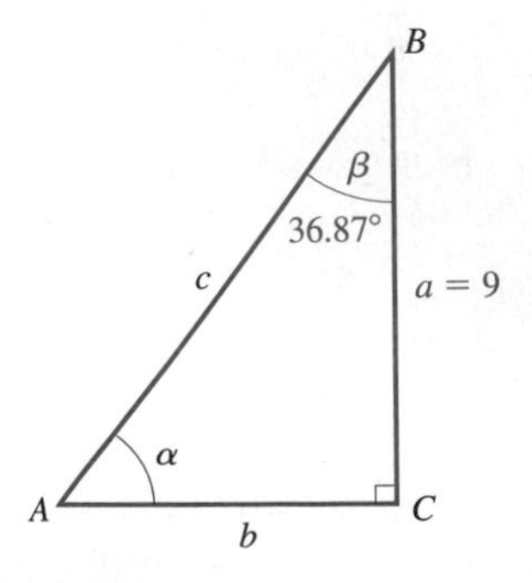

FIGURE 2

Solution Figure 2 shows a sketch of the right triangle. Here, we have to find α b, and c. Because α and β are complementary,

$$\alpha = 90° - \beta = 90° - 36.87° = 53.13°$$

To find b, we use the relationship $\tan \beta = \text{opp/adj} = b/a$ and a calculator to get

$$b = a \tan \beta = 9 \tan 36.87° = 6.75 \text{ (approx.)}$$

To find c, we use the relationship $\cos \beta = \text{adj/hyp} = a/c$ to get

$$c = \frac{a}{\cos \beta} = \frac{9}{\cos 36.87°} = 11.25 \text{ (approx.)}$$

In the next example, we use a calculator and the inverse trigonometric functions to find the degree measures of angles.

EXAMPLE 2 **Solving a right triangle given two sides**
Solve the right triangle ACB if $a = 31.42$ and $b = 26.74$. Round off the answers to two decimal places.

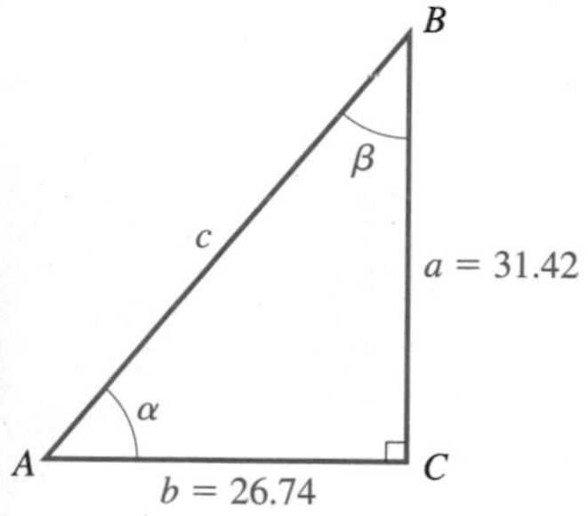

Solution Figure 3 shows a sketch of the right triangle. Here, we have to find α, β, and c. We'll start by finding α. From right triangle trigonometry, we know that

$$\tan \alpha = \frac{\text{opp}}{\text{adj}} = \frac{a}{b} = \frac{31.42}{26.74}$$

FIGURE 3

To find α means to find the acute angle whose tangent is $31.42/26.74$, so we use the inverse tangent function. The inverse trigonometric function keys on a calculator can give us *degree outputs* even though the inverse trigonometric functions technically were defined only for real numbers (radian measures). If we set a calculator in *degree mode*, we get the corresponding degree measure of these angles. In this situation, we get

$$\alpha = \tan^{-1}\left(\frac{31.42}{26.74}\right) = 49.60° \text{ (approx.)}$$

Because α and β are complementary,

$$\beta = 90° - 49.60° = 40.40° \text{ (approx.)}$$

Finally, to find c, we use the Pythagorean theorem to get

$$c = \sqrt{a^2 + b^2} = \sqrt{31.42^2 + 26.74^2} = 41.26 \text{ (approx.)}$$

Solving Applied Problems

One of the major reasons for the development of trigonometry was to solve applications involving right triangles.

EXAMPLE 3 **Finding the height reached by a ladder**
If a 24 foot extension ladder leaning against a house makes a 63° angle with the ground, then how far up the side of the house does the ladder reach? Round off the answer to two decimal places.

Solution The situation is shown in Figure 4. We need to determine the length $|\overline{CB}|$. By using right triangle trigonometry, we obtain

$$\sin 63° = \frac{\text{opp}}{\text{hyp}} = \frac{|\overline{CB}|}{24}$$

or $\quad |\overline{CB}| = 24 \sin 63° = 21.38 \text{ (approx.)}$

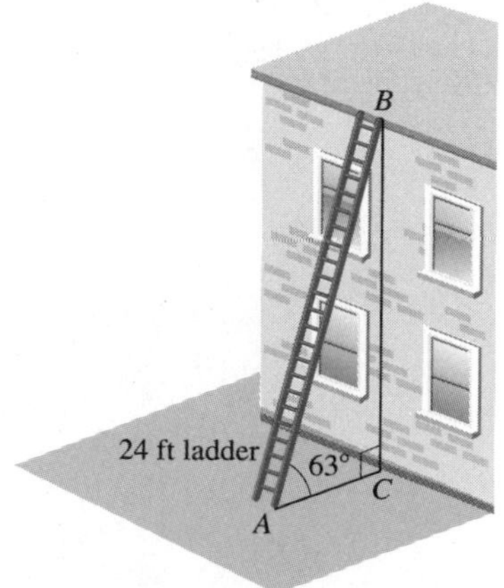

FIGURE 4

Thus, the ladder reaches about 21.38 feet up the side of the house (about 21 feet and 5 inches).

A *transit* is an instrument used by surveyors to determine angles.

At times, right triangle applications involve the measure of an acute angle formed by an observer's direct line of sight to an object and a horizontal line. If the object is above the horizontal line, this angle is called an **angle of elevation** (Figure 5a). If the object is below the horizontal line, the angle is called an **angle of depression** (Figure 5b).

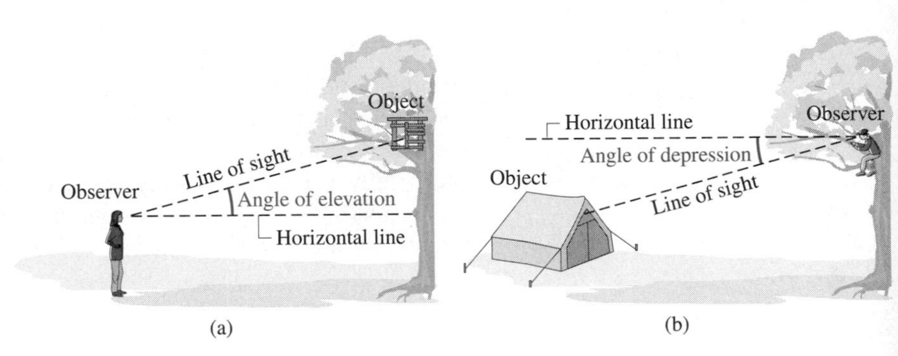

FIGURE 5

EXAMPLE 4 **Using an angle of elevation**

The Empire State Building in New York City is about 1250 feet high. A surveyor's transit is set 5 feet above ground level. The angle of elevation from the transit to a point on top of the Empire State Building is found to be 25.33°. How far is the surveyor from the building? Round off the answer to the nearest 10 feet.

FIGURE 6

Solution The situation is modeled by the right triangle in Figure 6, where d represents the desired distance. Here, we have

$$\tan 25.33° = \frac{\text{opp}}{\text{adj}} = \frac{1250 - 5}{d}$$

or

$$d = \frac{1245}{\tan 25.33°} = 2630 \text{ (approx.)}$$

Consequently, the surveyor is about 2630 feet from the building.

EXAMPLE 5 **Using angles of depression**

An air traffic controller stands in a control tower that is 150 feet high and observes two commercial jets waiting in the same runway for a takeoff. The angle of depression of the first jet is 15.3°, and the angle of depression of the second is 8.2°. Assuming that the tower and the two jets lie on the same plane, find the distance between the jets. Round off the answer to the nearest 10 feet.

Solution In Figure 7, A represents the location of the controller, C the location of the first jet, and D the location of the second jet. We wish to find $|\overline{CD}|$, which is equal to $|\overline{BD}| - |\overline{BC}|$.

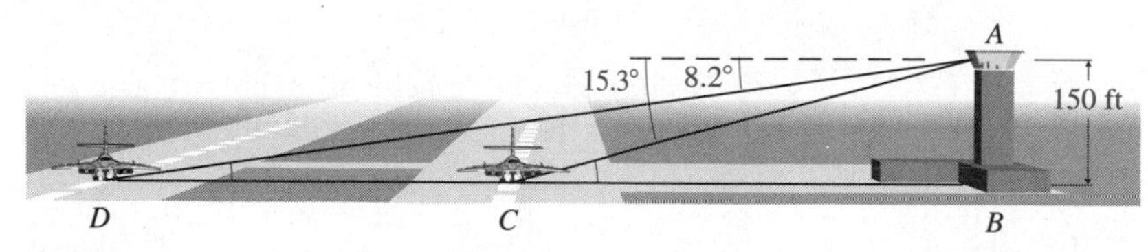

FIGURE 7

From geometry (Appendix I), we know that alternate interior angles are equal, so angle $BCA = 15.3°$ and angle $BDA = 8.2°$. In the right triangle ABC, we have

$$\tan 15.3° = \frac{|\overline{AB}|}{|\overline{BC}|} = \frac{150}{|\overline{BC}|}$$

or

$$|\overline{BC}| = \frac{150}{\tan 15.3°} = 550 \text{ (approx.)}$$

In the right triangle ABD,

$$\tan 8.2° = \frac{|\overline{AB}|}{|\overline{BD}|} = \frac{150}{|\overline{BD}|}$$

or

$$|\overline{BD}| = \frac{150}{\tan 8.2°} = 1040 \text{ (approx.)}$$

Hence, the approximate distance between the two jets is given by

$$|\overline{CD}| = |\overline{BD}| - |\overline{BC}| = 1040 - 550 = 490$$

That is, the jets are approximately 490 feet apart.

EXAMPLE 6 Modeling a viewing angle

A museum plans to hang a painting for public viewing. The painting, which is 4 feet high, is mounted on a wall in such a way that its lower edge is 1 foot 4 inches above the level of the average viewer's eye. The viewer studies the painting x feet from the wall on which the painting is mounted (Figure 8).

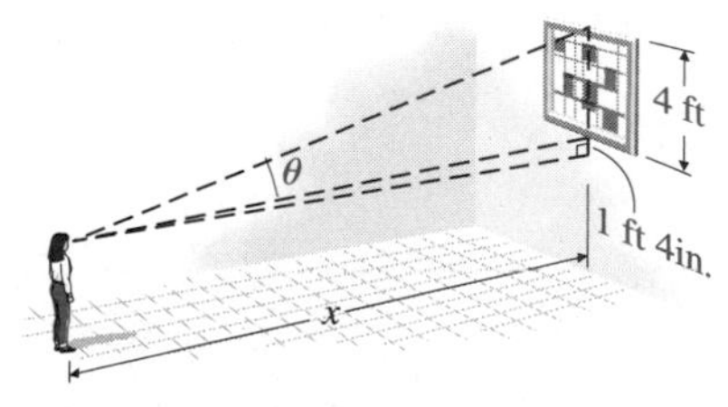

FIGURE 8

(a) Use right triangle trigonometry to express the angle θ formed by the viewer's eye and the top and bottom of the painting as a function of x.

(b) Use a graphing utility to sketch the graph of the function found in part (a). Then use the ZOOM and TRACE features to approximate the maximum value of θ and when it occurs. What is the degree measure of the maximum angle? Round off the answer to the nearest degree.

(c) Does the maximum angle give the best view of the painting? Explain.

Solution

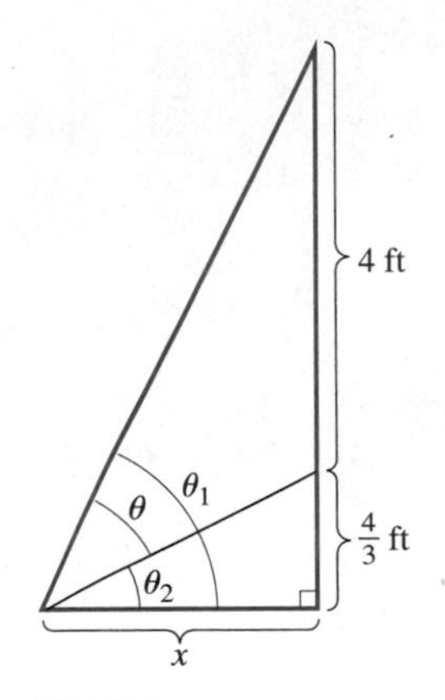

FIGURE 9a

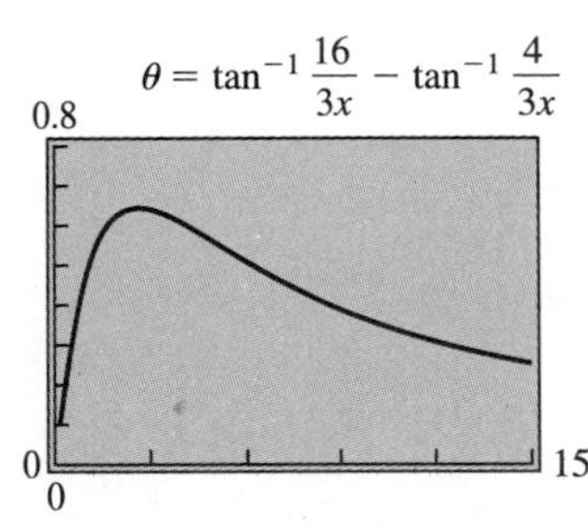

FIGURE 9b

(a) Figure 9a helps us express θ as a function of x. (Note that 1 foot 4 inches is $\frac{4}{3}$ feet.) We see in the figure that

$$\theta = \theta_1 - \theta_2$$

$$\theta = \tan^{-1}\left(\frac{4 + \frac{4}{3}}{x}\right) - \tan^{-1}\left(\frac{\frac{4}{3}}{x}\right)$$

$$\theta = \tan^{-1}\frac{16}{3x} - \tan^{-1}\frac{4}{3x}$$

This latter equation expresses θ as a function of x.

(b) Since x represents distance, x is positive. Figure 9b shows a viewing window of the graph of

$$\theta = \tan^{-1}\frac{16}{3x} - \tan^{-1}\frac{4}{3x}$$

where $x > 0$ and θ is considered to be in radian (real number) measure. By using the ZOOM and TRACE features, we find the maximum point to be approximately (2.6667, 0.6435) rounded to four decimal places. This means that when x is approximately 2.6667, θ attains its maximum value of about 0.6435. The real value 0.6435 can be thought of equivalently as the radian measure of the angle θ. Converting $\theta = 0.6435$ radian to degree measure yields a degree measure of about 37°. Thus, when the viewer stands about 2.6667 feet, or 2 feet 8 inches, from the wall, the maximum angle measure of about 37° is attained.

(c) Standing 2 feet and 8 inches from a painting 4 feet high may be too close to appreciate the artistic qualities of the painting. The maximum viewing angle does not necessarily give the best location for appreciating the art.

Modeling Simple Harmonic Motion

In the real world there are many phenomena that involve oscillation or vibration in a uniform manner, repeating periodically in definite intervals of time. These phenomena can be modeled by *sinusoidal curves*, also called *simple harmonic curves*, which are the graphs of sine and cosine functions. Examples include alternating electrical current, sound waves, light waves, pendulums, and mass–spring systems. The mathematical model for a quantity y that is oscillating in a simple harmonic manner is given by one of the following equations:

$$y = a \cos \omega t \qquad \text{or} \qquad y = a \sin \omega t$$

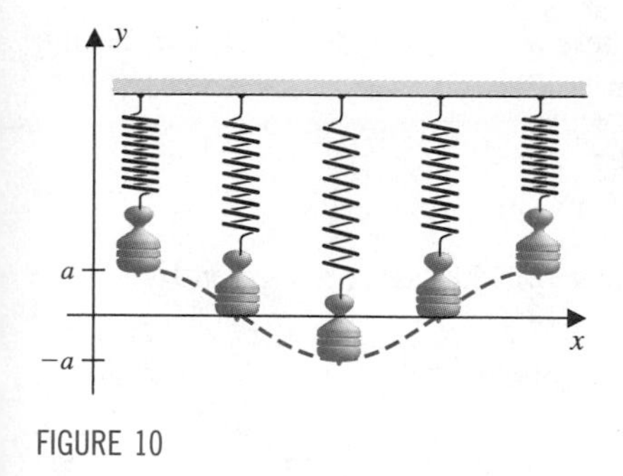

FIGURE 10

For instance, if a weight suspended from a spring is pulled down a distance $|a|$ and then released, it oscillates vertically in periodic motion. To model the situation, we let y (say, in centimeters) represent the *directed distance* of the weight from its rest position after time t (say, in seconds). We consider y to be a function of t. If the values of y are plotted for specific values of t, and if friction is neglected, then the resulting graph (Figure 10) has an equation of one of the following forms:

$$y = a \cos \sqrt{\frac{k}{m}}\, t \quad \text{or} \quad y = a \sin \sqrt{\frac{k}{m}}\, t$$

Here, k, the **stiffness coefficient**, is a constant associated with the particular spring, and m is the **mass** of the weight.

The variable y is also referred to as the **displacement**. A positive value of y indicates that it is above its rest position, and a negative value of y indicates that it is below its rest position. The farther the weight is pulled down before it is released, the greater the amplitude of motion will be. Furthermore, the stiffer the spring, the more rapidly the weight will oscillate, and thus the smaller the time period of repeating motion will be.

The **period** $\boldsymbol{T}$ for either $y = a \cos \omega t$ or $y = a \sin \omega t$ is given by

$$T = \frac{2\pi}{\omega}$$

where $\omega > 0$ is called the **angular frequency**.

Thus, for the spring–weight situation, it takes $(2\pi)/\omega$ units of time for the spring to go through one complete oscillation, where $\omega = \sqrt{k/m}$. The **frequency** f of a simple harmonic motion is the number of oscillations per unit of time, so

$$f = \frac{1}{T} = \frac{\omega}{2\pi}$$

EXAMPLE 7 Finding the frequency of a simple harmonic model

Suppose that alternating electric current is described by the simple harmonic model given by

$$y = 14.1 \sin 120\pi t$$

where y (in amperes) is the current at time t (in seconds). Find the amplitude, period, and frequency of the current.

Solution In this situation, $y = a \sin \omega t$, where $a = 14.1$ and $\omega = 120\pi$. So the amplitude of the current is 14.1 and the period T is given by

$$T = \frac{2\pi}{\omega} = \frac{2\pi}{120\pi} = \frac{1}{60}$$

that is, the period is $\frac{1}{60}$ second. The frequency f of the current is given by

$$f = \frac{1}{T} = \frac{1}{\frac{1}{60}} = 60$$

Therefore, the frequency is 60 cycles per second, or 60 hertz (Hz).

Simple harmonic motion occurs when an object is shifted from its rest position and the force exerted to return it to rest is proportional to the shift. The following example describes the motion of a weight moving up and down at the end of a spring.

EXAMPLE 8 Finding the equation of motion of an object suspended from a spring

An object of mass 4 grams suspended from a spring is pulled down 2 centimeters and released at time $t = 0$ so that the object oscillates vertically.

(a) Determine which equation, $y = a \cos \sqrt{k/m}\, t$ or $y = a \sin \sqrt{k/m}\, t$, is better suited to model this situation.
(b) Write an equation of motion if the stiffness coefficient k is 64 and t is the number of elapsed seconds.
(c) Find the period, frequency, and amplitude of the motion.
(d) Give a physical interpretation of the answers in part (c).
(e) Graph three cycles of the displacement y as a function of t.

Solution

(a) First, we compare the two equations available to model the situation:

Equation	$y = a \cos \sqrt{\frac{k}{m}}\, t$	$y = a \sin \sqrt{\frac{k}{m}}\, t$
Value of Equation When $t = 0$	$y = a \cos 0 = a$	$y = a \sin 0 = 0$

For the given situation, the displacement is 2 centimeters below the resting position at the start of the motion; that is, when $t = 0$, $y = -2$. So the equation $y = a \cos \sqrt{k/m}\, t$, with $a = -2$, is a more suitable model.

(b) Using the equation $y = a \cos \sqrt{k/m}\, t$ with $a = -2$, $k = 64$, and $m = 4$, we get the motion equation

$$y = -2 \cos \sqrt{\tfrac{64}{4}}\, t = -2 \cos 4t$$

(c) The period $T = (2\pi)/\omega = (2\pi)/4 = \pi/2$. The frequency $f = 1/T = 2/\pi$ cycles per second, and the amplitude is $|a| = |-2| = 2$ centimeters.

(d) Since the period is $\pi/2$, it takes about 1.57 seconds for one complete oscillation. An amplitude of 2 means the highest and lowest positions are 2 centimeters above and below the resting position. A frequency of $2/\pi$ indicates that it makes about 0.64 oscillation per second.

(e) The graph in Figure 11 shows three cycles of this situation.

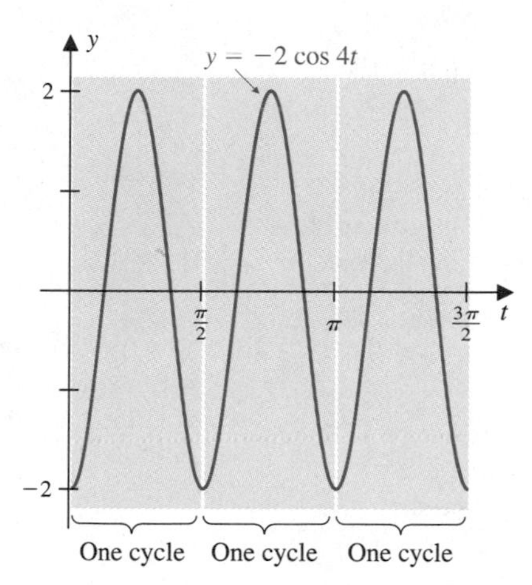

FIGURE 11

PROBLEM SET 5.8

Mastering the Concepts

In problems 1–10, assume that ACB is a right triangle with $\gamma = 90°$ (Figure 12). In each case, solve the triangle.

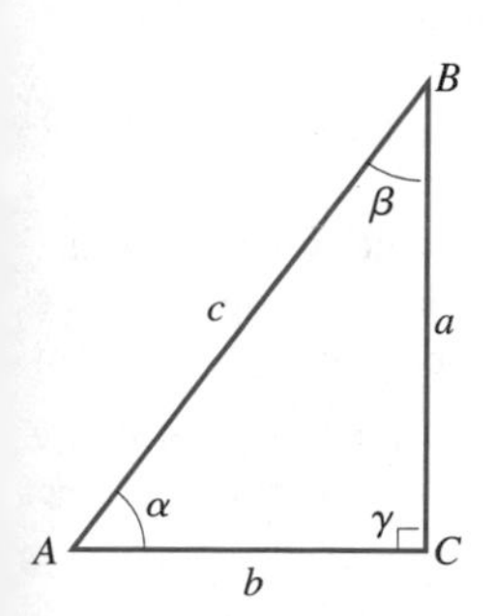

FIGURE 12

Round off angles to the nearest hundredth of a degree and side lengths to two decimal places.

1. $c = 10,\ \beta = 50°$

2. $c = 9,\ \beta = 20°$

3. $b = 4,\ \beta = 43°$

4. $a = 6.5,\ \alpha = 37.2°$

5. $b = 1500,\ \alpha = 31.23°$

6. $b = 567.3,\ \alpha = 67.41°$

7. $a = 13.2,\ b = 4.1$

8. $a = 31,\ b = 4.7$

9. $\alpha = 3\beta,\ a = 1$

10. $\alpha = \frac{2}{3}\beta,\ a = 3$

Applying the Concepts

In problems 11–34, round off the answers to two decimal places.

11. Surveying: The lot for a new home site is uniformly pitched from the horizontal at an angle of 0.51° (Figure 13, at the top of the next page). How many inches will the ground drop if one walks 100 feet down the slope?

FIGURE 13

12. Surveying: A straight sidewalk is inclined to the horizontal at an angle of 5.3°. How far must one walk on the sidewalk to change elevation by 2 meters?

13. Foot of a Ladder: A ladder 24 feet long is leaning against a building. The angle formed by the ladder and the ground is 63°. How far from the building is the foot of the ladder?

14. Length of a Shadow: A monument is 180 meters high. What is the length of the shadow cast by the monument if the angle of elevation of the sun is 58.4°?

15. Utility Pole: A guy wire attached to a vertical utility pole makes an angle of 71.4° with the ground. If the end of the wire attached to the ground is 15.5 feet from the pole, how high up the pole is the other end of the wire attached?

16. Altitude of a Shuttle: A space shuttle rises vertically. A camera 4.3 feet above the ground is located at a point 1000 feet (on the horizontal) away from the base of the launching pad. At a certain instant, the angle of elevation of the camera, focusing on the bottom of the shuttle, is 53° (Figure 14). How high above the ground is the shuttle at that instant?

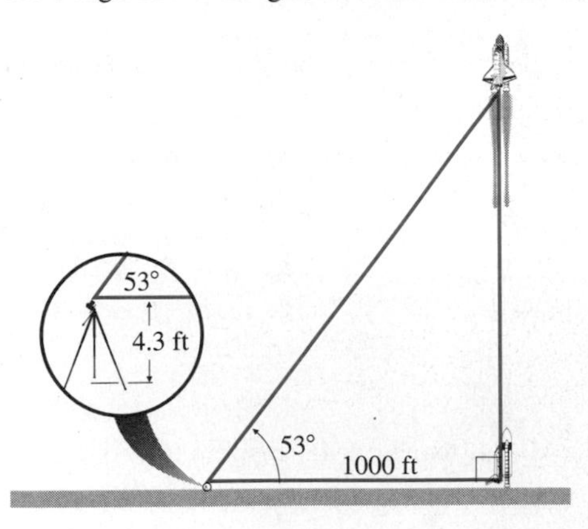

FIGURE 14

17. Highway Patrol: A highway patrol officer monitors traffic on a straight highway from a point A, which is 35 feet from the side of the highway (Figure 15). At a certain time, the patrol officer observed a speeding truck at point B on the highway at an angle of 25.3°. One second later, the truck was observed by the officer at point C at an angle of 10.3° (Figure 15). Find the speed of the truck in miles per hour.

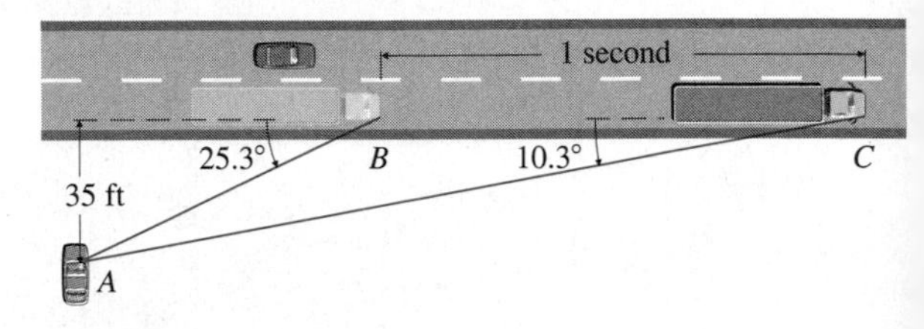

FIGURE 15

18. Forestry: A tree broken by the wind forms a right triangle with the ground. The broken part makes an angle of 38° with the ground, and the top of the tree is now 15 meters from the base, as shown in Figure 16. How tall was the whole tree?

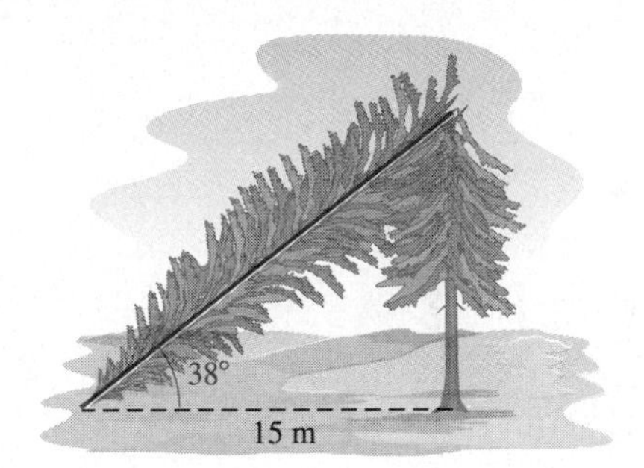

FIGURE 16

19. Escalator: An escalator in a department store is 50 feet long and carries customers through a vertical distance of 20 feet.
    (a) What is the angle that the escalator makes with the floor?
    (b) If it takes 25 seconds to carry a customer from the bottom of the escalator to the top, how fast (in feet per second) is the escalator moving?

20. Surveying: To find the length of an island, a navigator of a small plane flying at an altitude of 1540 meters determines the angles of depression of the extremities of the island to be 26.5° and 79.3° (Figure 17). What is the length of the island in kilometers?

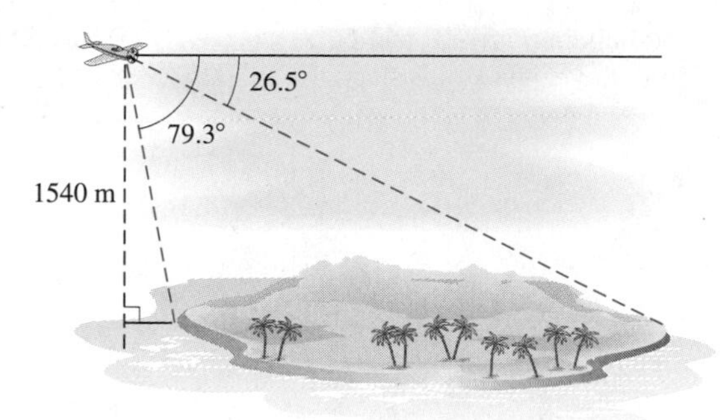

FIGURE 17

21. Computer Monitor: Suppose that a computer screen has a height of 7.5 inches and a width of 10 inches (Figure 18). What is the length of the diagonal d, and what angle does it make with the width?

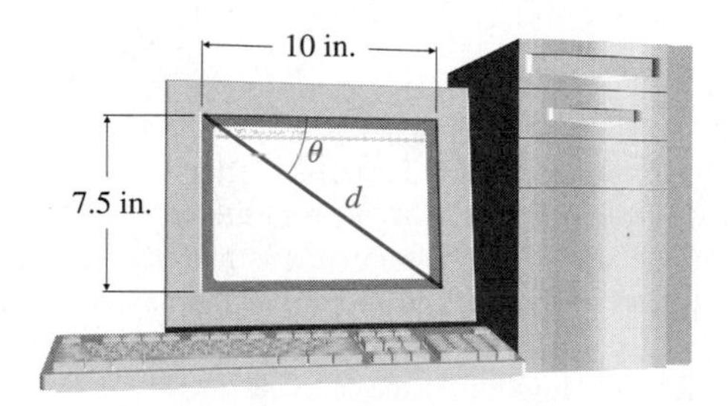

FIGURE 18

22. The Statue of Liberty: The torch of the Statue of Liberty is about 305 feet high. A sighting of the statue is taken from a ship 838 feet away. Find the angle of elevation from the ship to the top of the torch (Figure 19).

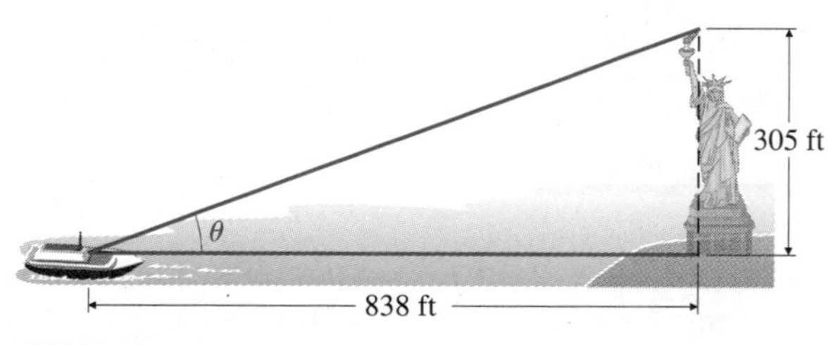

FIGURE 19

23. Angle of Elevation of the Eiffel Tower: The height of the Eiffel Tower (without the television mast added to the top) is approximately 300 meters (Figure 20). If the angle of elevation from a point on the ground to the top of the tower is 62.75°, how far from the center of the base of the tower is the point?

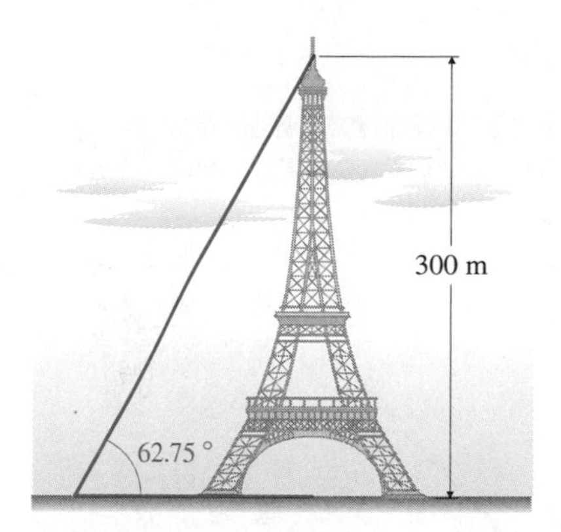

FIGURE 20

24. Light Span of a Lamp: A lamp is suspended 2 feet 10 inches above the center of a circular table. What is the radius of the table if the angle of depression from the lamp to the edge of the table is 56°15′, as shown in Figure 21?

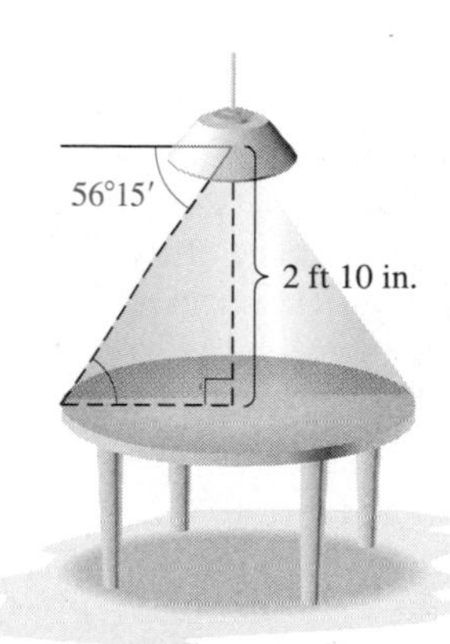

FIGURE 21

25. Tower of Pisa: The Leaning Tower of Pisa was designed to stand 55 meters high when vertical. However, it was built on unstable ground and is now 5.2 meters out of perpendicular (Figure 22). Find the acute angle that the tower makes with the ground.

FIGURE 22

26. Length of a Ladder: A 6 foot fence stands 9 feet from a high wall. The shortest ladder that can reach the wall from outside the fence makes an angle of 41° with the horizontal (Figure 23). Find the length of the ladder.

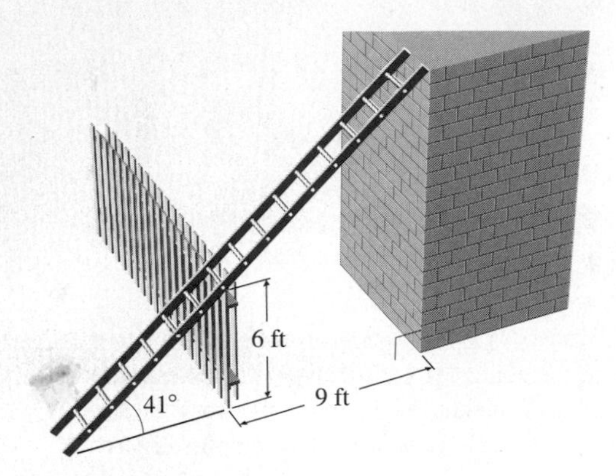

FIGURE 23

27. Height of Al Giza Pyramid: The distance measured along an edge from the original top of the Great Pyramid of Cheops at Al Giza, Egypt, to one of the corners of its square base is 219 meters. The angle of elevation from the corner of the base to the top is 42.06° (Figure 24). Find the original height of the Great Pyramid.

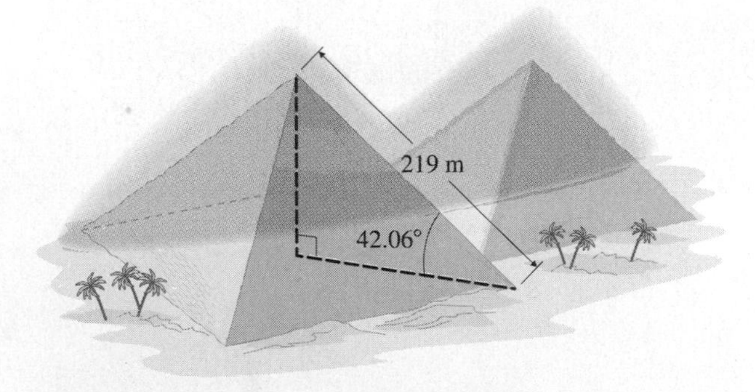

FIGURE 24

28. Height of a Flagpole: From the window of an apartment building 12 meters above ground, an observer determines that the angles of depression of the top and bottom of a flagpole standing on the ground (level with the base of the building) are 43° and 66°, respectively (Figure 25). Find the height of the flagpole.

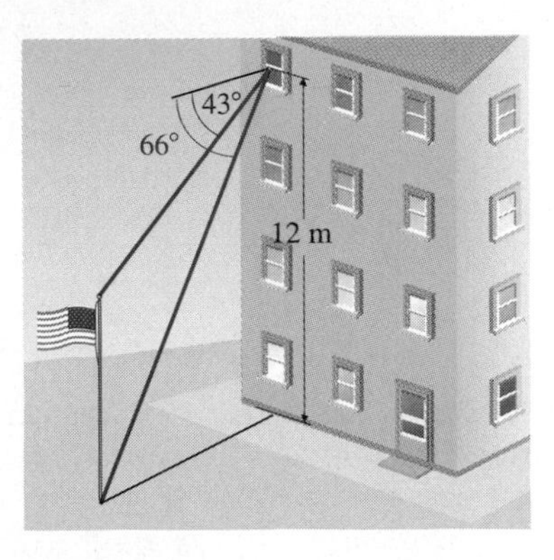

FIGURE 25

29. Length of an Antenna: An antenna is situated at the edge of a flat roof on top of a building that is located on level ground. From a point 100 feet from the base of the building, on the side where the antenna is placed, the angles of elevation of the top and bottom of the antenna measure 24° and 17°, respectively (Figure 26). How high is the building? How tall is the antenna?

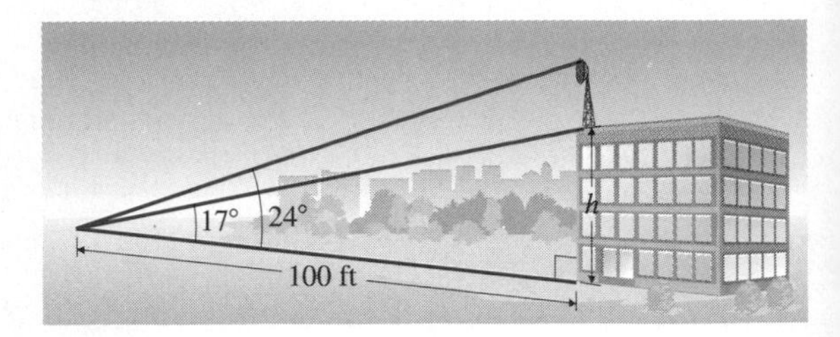

FIGURE 26

30. Surveying: To find the shortest distance d from a baseline l to a point T on an island offshore, a surveyor marks off a distance of 200 meters between the two points B and C on the baseline. She locates the point T on the island and then measures the angles TBC and TCB to be 68.3° and 81.5°, respectively (Figure 27). Find the distance d from the point T on the island to the nearest point on the baseline.

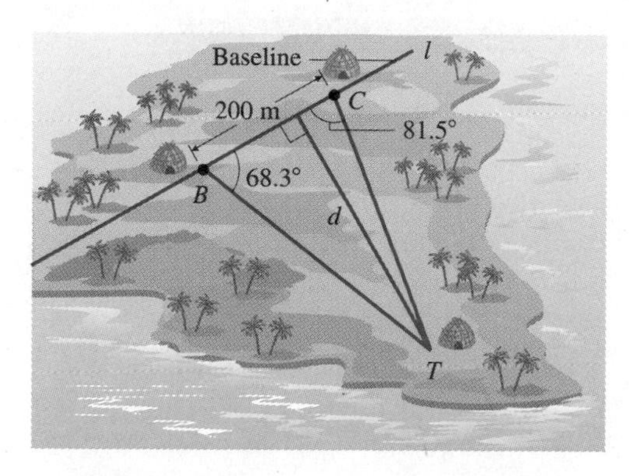

FIGURE 27

31. Height of a Blimp: A blimp is between two spotters who are 522 feet apart. One spotter reports that the angle of elevation of the blimp is 66°, and the other reports that it is 59°. If the blimp is directly over a line from one spotter to the other, how high is the blimp above the ground?

32. Security: A video security monitor is mounted at location A, 10 feet above the floor on a back wall of a drug store. The monitor scans through an angle of 51° in a vertical plane along an aisle of the store. The aisle begins at location B, 5 feet from the foot of the back wall at C and ends at D, the foot of an opposite wall (Figure 28). What is the length of the aisle; that is, what is the distance $|\overline{BD}|$?

FIGURE 28

33. Viewing a Billboard: A billboard is to be built adjacent to a highway so that the top and bottom will be 32 feet and 24 feet, respectively, above the eye level of a passing motorist. Suppose that the motorist passes the billboard at a distance of x feet and that the angle formed by the motorist's eye and the top and bottom of the billboard is θ (Figure 29).

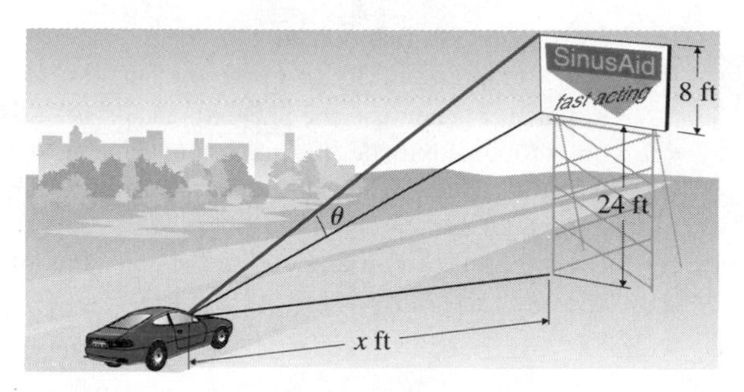

FIGURE 29

(a) Express θ as a function of x.

GU **(b)** Use a graphing utility to graph the function found in part (a). Then use the ZOOM and TRACE features to approximate the maximum value of θ and when it occurs. What is the degree measure for this maximum angle?

(c) Does the maximum angle give the best view of the billboard? Explain.

34. Viewing a Window: A night watchman wants the greatest possible view of a second story window of a building as he makes his rounds. The windows of the building are 4 feet high, with the lower sill 11 feet above the ground. Assume the watchman's eyes are precisely 6 feet above the ground. Suppose that the watchman stands x feet away from the building, and that the angle formed by the watchman's eyes and the top and bottom of the window is θ (Figure 30).

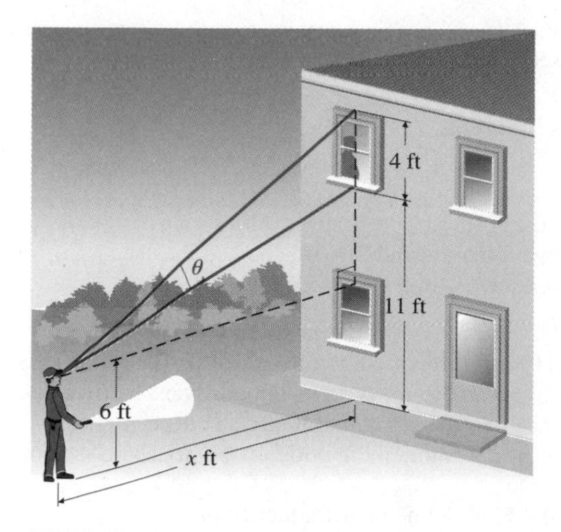

FIGURE 30

(a) Express θ as a function of x.

GU (b) Use a graphing utility to graph the function found in part (a). Then use the ZOOM and TRACE features to approximate the maximum value of θ and when it occurs. What is the degree measure for this maximum angle?

(c) Does the maximum angle give the best view of the window? Explain.

In problems 35–38, a simple harmonic model is described by the given equation, where t is time (in seconds). Determine the amplitude, period, and frequency of each model.

35. $y = \sin 200\pi t$

36. $y = 2 \cos 0.001\pi t$

37. $y = 0.15 \sin 792\pi t$

38. $y = 0.06 \cos 120\pi t$

39. Spring–Weight Model: An object of mass 30 grams suspended from a spring is pulled down 50 centimeters and released at time $t = 0$ so that the mass oscillates vertically.

(a) Write an equation of motion if the stiffness coefficient $k = 8$ and t is the number of elapsed seconds.

(b) Graph two cycles of the displacement y as a function of t.

(c) Find the amplitude, period, and frequency of the motion and interpret them.

40. Spring–Weight Model: A weight suspended from a spring oscillates up and down according to the model

$$y = a \cos \sqrt{\frac{k}{m}}\, t$$

If the stiffness coefficient of the spring $k = 4000$, how large a mass should be attached so that the spring will oscillate 6.4 times per second?

41. Pendulum Swing Model: A pendulum swings uniformly back and forth through an arc length of 0.05 meter, taking 2.2 seconds to move from the position directly above point A to the position directly above point B (Figure 31). Assume the motion of the pendulum is simple harmonic.

(a) Find the angular frequency ω.

(b) Find the equation for y as a function of t, where y denotes the displacement of the mass at t seconds.

(c) Sketch the graph for three cycles.

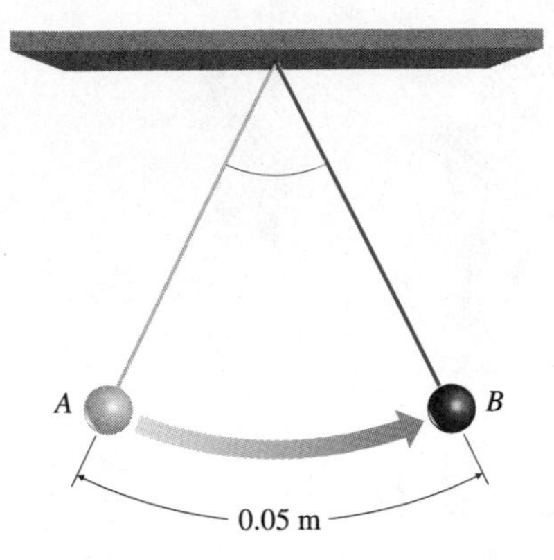

FIGURE 31

42. Grandfather Clock: A grandfather clock has a pendulum of length $L = 0.99$ meter, and it swings through an arc length of 0.25 meter according to the model

$$y = a \sin \omega t$$

(a) Find the period $T = 2\pi\sqrt{L/g}$, where $g = 9.8$ meters per second per second.

(b) Find the angular frequency ω.

(c) Determine the equation for the displacement y as a function of t.

(d) Sketch the graph of the equation for three cycles.

Developing and Extending the Concepts

43. Two tangent lines, each of length T, are drawn from a point P and touch a circle of radius r at points M and N (Figure 32). Assume the angle MPN is θ. Express the length T of each tangent line in terms of θ and r.

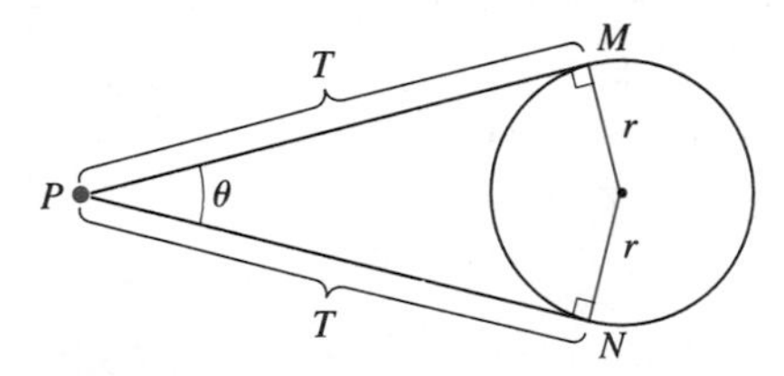

FIGURE 32

44. Figure 33 shows a regular *heptagon* (a seven-sided polygon with sides of equal lengths) incribed in a circle of radius 1. Determine the length of one side of the heptagon.

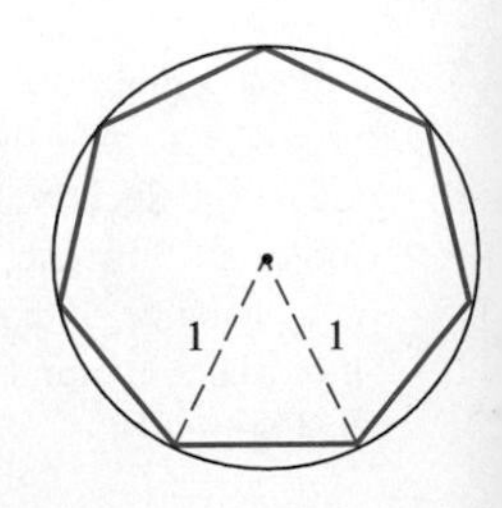

FIGURE 33

45. Explain why it is not possible to have a right triangle with a hypotenuse of 23 inches and an angle of 35° with a side opposite of length:
(a) 11 inches (b) 14 inches

CHAPTER 5 REVIEW PROBLEM SET

1. Express each angle measure in degrees, minutes, and seconds.
(a) 76.25° **(b)** −61.35°
(c) 143.47° **(d)** −14.47°

2. Express each angle as a decimal. Round off the answers to four decimal places.
(a) 4°7′ **(b)** 21°19′13″
(c) −45°35′25″ **(d)** 15″

3. Convert each degree measure to radian measure. Round off the answers to four decimal places.
(a) 15° **(b)** −17.45°
(c) 45°16′51″ **(d)** 36°11′25″

4. Convert each radian measure to degree measure. Round off the answers to four decimal places.
(a) $\frac{17\pi}{4}$ **(b)** $-\frac{4\pi}{7}$
(c) 5.82 **(d)** −7.63

5. Suppose that angle $\theta = 36°$ is in standard position. Find the measure of each of the following angles and sketch the angles.
(a) Complement of θ **(b)** Supplement of θ
(c) Two angles coterminal with θ, one positive and one negative

6. Let s denote the length of the arc intercepted on a circle of radius r by a central angle θ. In each case, find the missing quantity.
(a) $r = 7$ centimeters, $\theta = 75°$, $s = ?$
(b) $r = 2$ inches, $s = 5$ inches, $\theta = ?$
(c) $s = 17$ meters, $\theta = (5\pi)/6$, $r = ?$

7. Find the exact values of the six trigonometric functions of the acute angle θ in Figure 1.

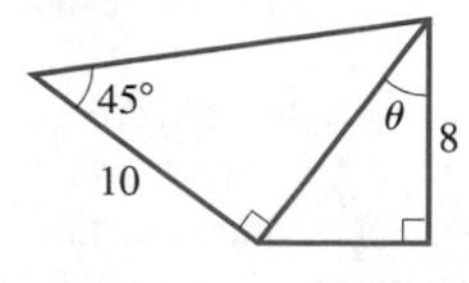

FIGURE 1

8. Find the exact values of the six trigonometric functions of θ if θ is in standard position and the terminal side of θ contains the given point
(a) $(-6, -8)$ **(b)** $(-3, 5)$

9. Sketch an angle θ in standard position and name the quadrant in which θ lies if:
(a) $\theta = -100°$ **(b)** $\sin\theta < 0$ and $\cos\theta < 0$
(c) $\sec\theta > 0$ and $\cot\theta < 0$

10. Use a reference triangle and periodicity to find the exact value of each expression.
(a) $\cos\left(-\frac{5\pi}{3}\right)$ **(b)** $\sec 780°$
(c) $\cot\frac{37\pi}{6}$ **(d)** $\tan(-600°)$

11. Use a calculator to find the approximate value of each expression rounded off to four decimal places.
(a) $\sin 3$ **(b)** $\cos(-208°)$
(c) $\tan\frac{3\pi}{5}$ **(d)** $\sec(-3.92)$

12. Find the exact values of the six trigonometric functions of t.
(a) $P(t) = \left(-\frac{\sqrt{3}}{2}, -\frac{1}{2}\right)$
(b) $P(t) = \left(\frac{4}{\sqrt{17}}, \frac{1}{\sqrt{17}}\right)$

13. Display the location of each point on the unit circle. Then find the coordinates of each point. Round off the answers to four decimal places.
(a) $P\left(\frac{2\pi}{5}\right)$ **(b)** $P(3.35)$ **(c)** $P\left(-\frac{3\pi}{7}\right)$

14. Use the fundamental identities to find the exact values of the other five trigonometric functions under the given conditions.
(a) $\sin\theta = \frac{3}{5}$; θ in quadrant II
(b) $\cos\theta = \frac{3}{8}$; θ in quadrant IV

15. Use the fundamental identities to simplify each expression.
(a) $\csc^2 t \tan^2 t - \tan^2 t$
(b) $\sec t - \sin t \tan t$ **(c)** $\sin(-\theta)\sec(-\theta)$

16. Use the known graphs of the sine and cosine functions to graph one cycle of each function. Indicate the amplitude, period, and phase shift.

(a) $f(x) = 2 \sin\left(\frac{3x}{2} - \frac{\pi}{4}\right)$

(b) $g(x) = -0.2 \cos\left(2x + \frac{\pi}{8}\right)$

17. GU Use a graphing utility to graph each function. Find the period and phase shift.

(a) $f(x) = 0.6 \tan\left(3x - \frac{\pi}{2}\right)$

(b) $g(x) = -1.5 \sec\left(\frac{x}{4} + \pi\right)$

18. The functions for the graphs in Figure 2 have the form $y = a \cos kx$ or $y = a \sin kx$, where $k > 0$. For each graph, specify the period and amplitude, and write a specific equation for the function.

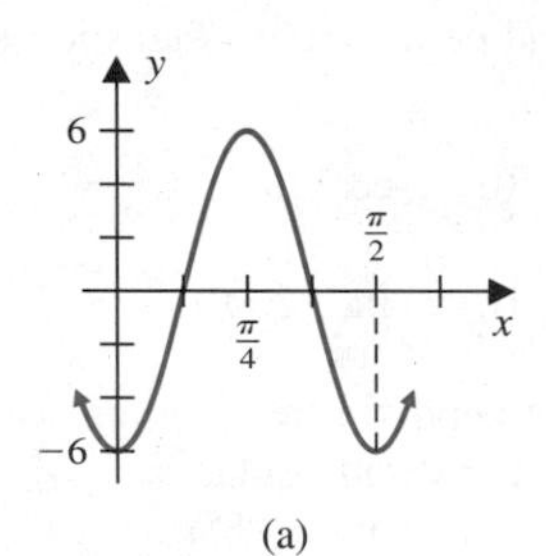

(a)

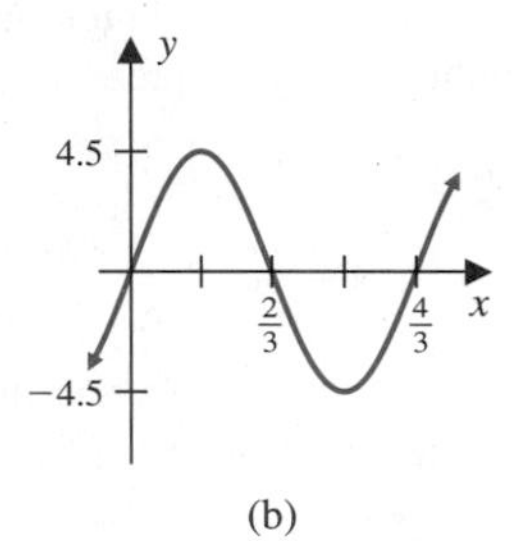

(b)

FIGURE 2

19. GU Use a graphing utility to sketch the graph of each function for $0 \le x \le 4\pi$.

(a) $f(x) = 2 \cos x + 3 \cos \frac{x}{2}$

(b) $g(x) = \frac{\sin x}{x} - x$ **(c)** $h(x) = \frac{1 - \cos x}{x^2}$

20. Find the exact value of each expression.

(a) $\cos^{-1} \frac{1}{2}$ **(b)** $\sin^{-1}\left(-\frac{1}{2}\right)$

(c) $\tan^{-1}\left(\sin \frac{\pi}{2}\right)$ **(d)** $\tan\left(\sin^{-1} \frac{3}{5}\right)$

(e) $\sec\left[\tan^{-1}\left(-\frac{5}{3}\right)\right]$

21. Use a calculator to approximate the value of each expression. Give the answers in radians rounded off to two decimal places.

(a) $\sin^{-1} 0.3741$ **(b)** $\arccos(-0.4901)$

(c) $\sec^{-1} 9.723$ **(d)** $\cot^{-1} 57.29$

22. Sketch the graph of each function by using transformations.

(a) $f(x) = \sin^{-1}(2x - 1)$

(b) $g(x) = 3 \cos^{-1}(-2x)$

23. In each case, solve for x in terms of y. Specify the restrictions on the values of x and y.

(a) $y = \sin^{-1} \frac{x}{2}$ **(b)** $y = \arctan(2x + 3)$

24. Solve the right triangle ABC if $\gamma = 90°$. Round off each answer to two decimal places.

(a) $b = 25$ inches and $\beta = 65°$

(b) $b = 7$ centimeters and $\alpha = 35°$

25. Engineering: A curve on the highway subtends an angle of 31° on a circle of radius 560 meters.

(a) How long is the curve subtended by this angle?

(b) How long will it take a car traveling 64 kilometers per hour to round the curve?

Round off the answers to two decimal places.

26. Geometry: An isosceles triangle ABC with $|\overline{AB}| = |\overline{AC}|$ and angle $\alpha = 48°$ is inscribed in a circle. Find the radius of the circle if the side $\overline{BC}$ intercepts an arc of 6.3 centimeters on the circle. Round off the answer to two decimal places.

27. Space Satellite: A satellite traveling in a circular orbit of radius 6371 kilometers is known to have a linear speed of 30,720 kilometers per hour. Find the angular speed of the satellite in radians per hour. Round off the answer to two decimal places.

28. Simple Harmonic Motion: The position of a particle at time t is given by each equation. Find the amplitude, period, and phase shift of the path of each particle.

(a) $y = 2.7 \sin(7t - 4\pi)$

(b) $y = 4 \cos\left(3t - \frac{\pi}{3}\right)$

29. GU Hours of Daylight: Suppose the number of daylight hours N in a city at a particular time of the year is given by the model

$$N = 12 + 2.5 \sin\left[\frac{2\pi}{365}(t - 81)\right]$$

where t is the number of days, with $t = 1$ corresponding to January 1.

(a) Use a graphing utility to graph the function for $0 \le t \le 365$.

(b) Use the ZOOM and TRACE features to find the maximum and minimum number of hours of daylight and the corresponding values of t rounded to two decimal places.

30. Height of a Blimp: An observer on the ground notices that the angle of elevation of a blimp is 29°. If the observer is 2 miles away from a point directly below the blimp, what is the approximate height of the blimp? Round off the answer to two decimal places.

31. Surveying: A person standing on top of a building that is 120 feet high spots a car parked on top of a parking garage that is 7 feet high. The angle of depression is 57° from the person to the car. What is the approximate distance of the garage from the building? Round off the answer to two decimal places.

32. Balloonist: A balloonist 100 meters above ground level observed a car on the ground. If the horizontal distance from a point on the ground directly under the balloon to the car is 79 meters, find the angle of depression from the balloon to the car (to the nearest tenth of a degree).

CHAPTER 5 TEST

1. **(a)** Convert 520° to radian measure.
(b) Convert $(3\pi)/7$ to degree measure.

2. Suppose that the sides of a central angle intercept an arc of length 15 centimeters on the circumference of a circle of radius 10 centimeters.
(a) Find the radian measure of the central angle.
(b) Express the measure of the central angle in degrees.

3. Suppose that the point $P = (3, -7)$ is on the terminal side of angle θ in standard position.
(a) Draw two angles, one positive and one negative, coterminal with θ.
(b) Find the exact value of each of the six trigonometric functions of θ.

4. Suppose that $P(t) = (a, 2a)$, where $a > 0$.
(a) Determine the value of a.
(b) Use the results from part (a) to find $\sin t$, $\cos t$, and $\tan t$.

5. Find the exact value of each expression.
(a) $\cos(-240°)$ **(b)** $\sin \dfrac{13\pi}{4}$
(c) $\tan \dfrac{17\pi}{3}$ **(d)** $\cos^{-1}\left(-\dfrac{\sqrt{2}}{2}\right)$
(e) $\sin\left(\cos^{-1} \dfrac{3}{5}\right)$

6. Use a calculator to approximate the value of each expression rounded off to two decimal places.
(a) $\sin 23.17°$ **(b)** $\csc 2.85$ **(c)** $\cos^{-1}(-0.8413)$

7. Determine the quadrant that contains the angle θ.
(a) $\sin \theta < 0$ and $\tan \theta > 0$
(b) $\cos \theta > 0$ and $\csc \theta < 0$

8. Suppose that $\cos \theta = -\frac{5}{17}$ and θ is in quadrant II. Find the exact values of the other five trigonometric functions of θ.

9. Find the amplitude, period, and phase shift, and sketch the graph of one cycle of each function.
(a) $f(x) = 3 \sin(4x - \pi)$
(b) $g(x) = -2 \cos\left(\dfrac{\pi x}{2} - \dfrac{\pi}{3}\right)$

10. Simplify the expression $(\sin t + \cos t)^2 - 2 \sin t \cos t$.

11. A straight string of lights is to reach from the top of an 80 meter pole and make an angle of elevation of 61° with the horizontal ground. How far from the pole should the lights be fastened to the ground? Round off the answer to two decimal places.

12. One end of an 80 foot rope is attached to the top of a vertical pole standing in the center of a tent. The other end of the rope is attached to a stake 60 feet away from the base of the pole. Determine the angle of elevation of the rope with the ground.

6

Trigonometric Identities and Equations

CHAPTER CONTENTS

Earlier in Section 5.4, we introduced the fundamental identities. In this section, we'll develop additional identities that relate the trigonometric functions to each other. The identities are important because of their use in solving trigonometric equations and in rewriting certain expressions in calculus.

6.1 Fundamental Identities

OBJECTIVES

1. Verify trigonometric identities.
2. Use trigonometric substitution.

We begin by focusing our attention on the fundamental identities because of their importance in calculus and their use in deriving other identities.

Verifying Trigonometric Identities

The process of manipulating and converting trigonometric expressions can be practiced by verifying trigonometric identities. There is no general rule for proving that a trigonometric equation is an identity; however, we normally start with one side of the equation and try to convert it to the other side by means of a sequence of algebraic manipulations and substitutions utilizing known identities. Often, it is helpful to start with the side containing more terms and try to convert it to the *simpler* side. Some suggestions that may help in carrying out a proof are listed below:

SUGGESTIONS FOR VERIFYING IDENTITIES

1. Combine a sum or difference of fractions into a single fraction.
2. Reduce a fraction.
3. Factor the expression.
4. Combine like terms.
5. Multiply both the numerator and denominator by the same expression.
6. Write all trigonometric expressions in terms of sines and cosines, and then simplify.

There may be more than one way of verifying that an equation is an identity.

EXAMPLE 1 **Verifying a trigonometric identity**
Verify that the equation

$$(\cos^2 t)(1 + \tan^2 t) = 1$$

is an identity by using two different approaches.

Solution We offer the following two approaches to verify that the equation is an identity.

First approach Working with the left side of the equation, we get

$$\begin{aligned}(\cos^2 t)(1 + \tan^2 t) &= \cos^2 t \cdot \sec^2 t && \text{Pythagorean identity}\\ &= (\cos t \cdot \sec t)^2 = \left(\cos t \cdot \frac{1}{\cos t}\right)^2 && \text{Reciprocal identity}\\ &= 1^2 = 1\end{aligned}$$

Since we have transformed the left side into the right side, the equation is an identity.

Second approach This strategy uses the quotient identity to express the left side in terms of sin t and cos t.

$$\begin{aligned}(\cos^2 t)(1 + \tan^2 t) &= \cos^2 t\left(1 + \frac{\sin^2 t}{\cos^2 t}\right) && \text{Quotient identity}\\ &= \cos^2 t + \cancel{\cos^2 t}\left(\frac{\sin^2 t}{\cancel{\cos^2 t}}\right) && \text{Multiply and reduce}\\ &= \cos^2 t + \sin^2 t = 1 && \text{Pythagorean identity}\end{aligned}$$

Graphs can be used to suggest whether an equation is an identity before verifying it analytically.

EXAMPLE 2 **Using multiplication to verify an identity**

(a) Compare the graphs of: $y_1 = \dfrac{\cos t}{1 - \sin t}$ and $y_2 = \dfrac{1 + \sin t}{\cos t}$

(b) Verify the identity: $\dfrac{\cos t}{1 - \sin t} = \dfrac{1 + \sin t}{\cos t}$

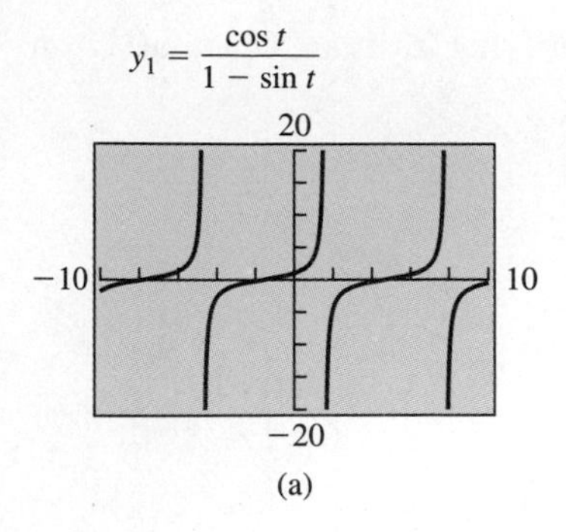

(a)

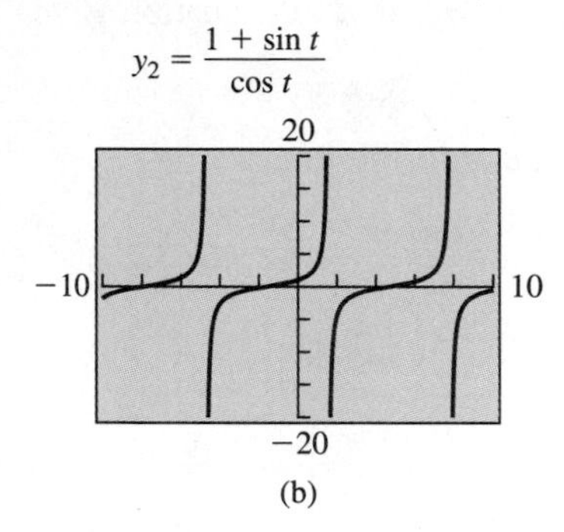

(b)

FIGURE 1

Solution

(a) We use a graphing utility to get two separate graphs for the same window settings:

$$y_1 = \frac{\cos t}{1 - \sin t} \quad \text{(Figure 1a)} \qquad \text{and} \qquad y_2 = \frac{1 + \sin t}{\cos t} \quad \text{(Figure 1b)}$$

We observe that the two graphs appear to be the same. So the two graphs suggest that the equation

$$\frac{\cos t}{1 - \sin t} = \frac{1 + \sin t}{\cos t}$$

is an identity.

(b) To verify this identity analytically, it may be tempting to multiply *both sides* by either $\cos t$ or by $1 - \sin t$. But this step is incorrect because it assumes the equation is an identity before it has been verified. Instead, we work with the left side by multiplying the numerator and denominator by the expression $1 + \sin t$ to obtain

$$\begin{aligned}
\frac{\cos t}{1 - \sin t} &= \frac{\cos t}{1 - \sin t} \cdot \frac{1 + \sin t}{1 + \sin t} \\
&= \frac{(\cos t)(1 + \sin t)}{1 - \sin^2 t} && \text{Multiply} \\
&= \frac{(\cos t)(1 + \sin t)}{\cos^2 t} && \text{Pythagorean identity} \\
&= \frac{1 + \sin t}{\cos t} && \text{Reduce fraction}
\end{aligned}$$

Since we have transformed the left side into the right side, the equation is an identity.

In the next example we verify the identity by converting each side of the equation to the same expression.

EXAMPLE 3 Verifying an identity by simplifying each side

Verify that the equation

$$\tan(-\theta)\sin(-\theta) = \sec\theta - \cos\theta$$

is an identity.

Solution Here we use the even–odd identities to write $\tan(-\theta) = -\tan\theta$ and $\sin(-\theta) = -\sin\theta$, so it is enough to prove that $\tan\theta \sin\theta = \sec\theta - \cos\theta$. In this situation, we verify that the equation is an identity by simplifying each side until the results become identical.

Left Side	Right Side
$\tan\theta \sin\theta = \dfrac{\sin\theta}{\cos\theta} \cdot \sin\theta$	$\sec\theta - \cos\theta = \dfrac{1}{\cos\theta} - \cos\theta$
$= \dfrac{\sin^2\theta}{\cos\theta}$	$= \dfrac{1 - \cos^2\theta}{\cos\theta}$
	$= \dfrac{\sin^2\theta}{\cos\theta}$

They are identical

Therefore, the equation is an identity.

In the next example, we use graphs to show that an equation is not an identity.

EXAMPLE 4 Showing that an equation is not an identity

(a) Use graphs to show that the equation

$$\sin x \cot x = \csc x - \sin x$$

is not an identity.

(b) Verify the conclusion in part (a) numerically.

Solution

(a) Viewing windows of the graphs of $y_1 = \sin x \cot x$ and $y_2 = \csc x - \sin x$ are shown in Figures 2a and 2b, respectively. Clearly, the graphs are not the same, so the two sides of the equation are not the same for all values in the domains of y_1 and y_2. Thus, the equation is not an identity.

(b) In order for the equation to be an identity, the value of each side must turn out to be the same whenever we replace x with any number found in the domains of both sides. This is not the case here. For instance, each side is defined for $x = 2$; however,

A *counterexample* is sufficient to show that the proposed equation is not an identity.

$$\sin 2 \cot 2 = -0.4161 \text{ (approx.)} \qquad \text{and} \qquad \csc 2 - \sin 2 = 0.1905 \text{ (approx.)}$$

are different, so we conclude that the equation is not an identity.

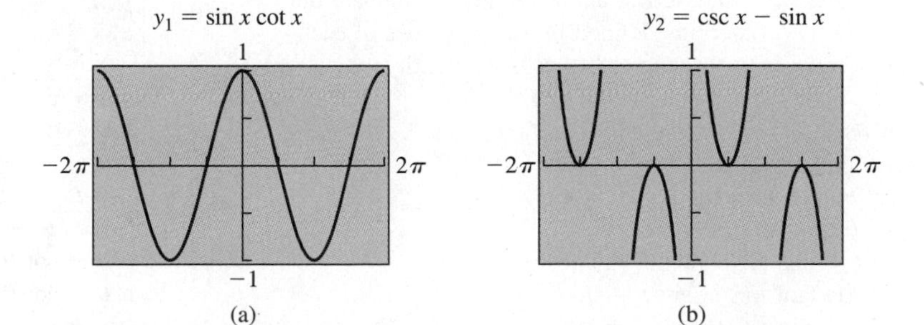

FIGURE 2

Using Trigonometric Substitution

In calculus, algebraic expressions of the form $\sqrt{a^2 - u^2}$, $\sqrt{u^2 + a^2}$, and $\sqrt{u^2 - a^2}$, where a is a positive constant, can be converted to more useful forms by making appropriate substitutions involving trigonometric functions. This technique is called **trigonometric substitution**. The idea is to rewrite radical expressions as trigonometric expressions containing no radical.

EXAMPLE 5 Converting a radical to a trigonometric form

Use the substitution $u = 5 \sin \theta$, where θ is an acute angle, to rewrite $\sqrt{25 - u^2}$ as a trigonometric expression containing no radical by using:

(a) A trigonometic identity (b) Right triangle trigonometry

Solution

(a) We know that $\cos \theta > 0$ for an acute angle θ, so

$$\sqrt{25 - u^2} = \sqrt{25 - (5 \sin \theta)^2} = \sqrt{25 - 25 \sin^2 \theta}$$
$$= \sqrt{25(1 - \sin^2 \theta)} = \sqrt{25 \cos^2 \theta} = 5 \cos \theta$$

(b) Since $u = 5 \sin \theta$, then $\sin \theta = u/5$. First, we construct a right triangle so that θ, u, and 5 satisfy the condition that $\sin \theta = u/5$ (Figure 3). By the Pythagorean theorem, the side adjacent to angle θ is given by $\sqrt{25 - u^2}$. Using right triangle trigonometry, we get

$$\cos \theta = \frac{\text{adj}}{\text{hyp}} = \frac{\sqrt{25 - u^2}}{5}$$

So $\sqrt{25 - u^2} = 5 \cos \theta$.

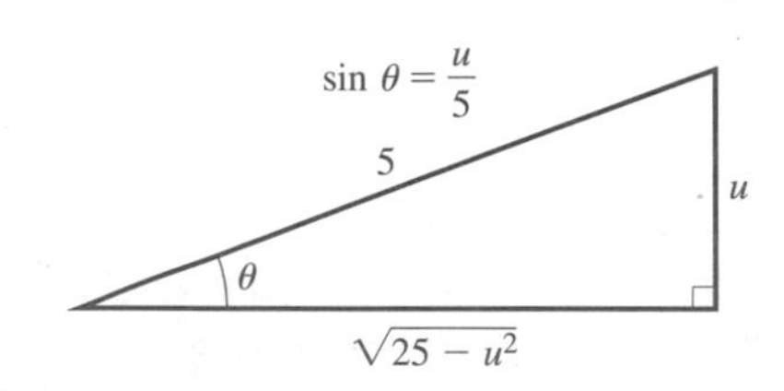

FIGURE 3

PROBLEM SET 6.1

Mastering the Concepts

GU In problems 1–34, use a graphing utility to compare the graphs of the two functions defined by the two sides of each equation. Then verify that the equation is an identity.

1. $\sin \theta \csc \theta = 1$
2. $\tan t \cot t = 1$
3. $\sec^2 \theta (1 - \sin^2 \theta) = 1$
4. $\cot^2 \theta (\sec^2 \theta - 1) = 1$
5. $(\sin t + \cos t)^2 + 2 \sin(-t) \cos(-t) = 1$
6. $\cos(-t) \tan(-t) + \sin t = 0$
7. $(\cos t - \sin t)^2 + (\cos t + \sin t)^2 = 2$
8. $\sec^7 \theta \tan \theta - \tan \theta \sec^5 \theta = \sec^5 \theta \tan^3 \theta$
9. $\frac{\sin \theta}{\tan \theta} + \frac{\cos \theta}{\cot \theta} = \sin \theta + \cos \theta$
10. $\tan^2 \theta - \sin^2 \theta = \frac{\sin^4 \theta}{\cos^2 \theta}$
11. $\frac{\sin \theta}{\cot \theta + \csc \theta} - \frac{\sin \theta}{\cot \theta - \csc \theta} = 2$
12. $\frac{\cos \theta \cot \theta}{\cot \theta - \cos \theta} = \frac{\cot \theta + \cos \theta}{\cos \theta \cot \theta}$
13. $\frac{\sin t}{1 + \cos t} + \frac{1 + \cos t}{\sin t} = 2 \csc t$
14. $\frac{\tan \theta + \cot \theta}{\tan \theta - \cot \theta} = \frac{\sec^2 \theta}{\tan^2 \theta - 1}$
15. $\frac{\cos \theta}{1 - \sin \theta} + \frac{\cos \theta}{1 + \sin \theta} = 2 \sec \theta$

16. $\dfrac{\sin\theta}{\csc\theta - \cot\theta} = 1 + \cos\theta$

17. $\dfrac{1}{1 + \tan\theta} - \dfrac{\cot\theta}{1 + \cot\theta} = 0$

18. $(\sec\theta - \tan\theta)^2 = \dfrac{1 - \sin\theta}{1 + \sin\theta}$

19. $\dfrac{\sec\theta + 1}{\tan\theta} + \dfrac{\tan\theta}{\sec\theta + 1} = 2\csc\theta$

20. $\dfrac{1}{1 + \sin t} + \dfrac{1}{1 - \sin t} = 2\sec^2 t$

21. $\dfrac{1 - \sin\theta}{1 - \sec\theta} + \dfrac{1 + \sin\theta}{1 + \sec\theta} = 2\cot\theta(1 - \cot\theta)$

22. $\dfrac{\cos\theta + \sin^2(-\theta)\sec\theta}{\csc\theta} = \tan\theta$

23. $\dfrac{\sin t}{1 + \cos t} + \dfrac{\sin t}{1 - \cos t} = 2\csc t$

24. $\dfrac{\tan t}{\sin t(1 + \tan^2 t)} = \cos t$

25. $\dfrac{1}{\cos^2\theta} + 1 + \dfrac{\sin^2\theta}{\cos^2\theta} = 2\sec^2\theta$

26. $1 - \dfrac{\cos^2 t}{1 + \sin t} = \sin t$

27. $\dfrac{1 + \tan t}{\sin t} - \sec t = \csc t$

28. $\cot t + \dfrac{\sin t}{1 + \cos t} = \csc t$

29. $\dfrac{\tan t + \cot t}{\tan t \cot t} = \sec t \csc t$

30. $\dfrac{\sec t - \csc t}{\sec t + \csc t} = \dfrac{\tan t - 1}{\tan t + 1}$

31. $\dfrac{\cot t - \tan t}{\cot t + \tan t} = \cos^2 t - \sin^2 t$

32. $\dfrac{\sin t}{1 - \cos t} = \csc t + \cot t$

33. $\dfrac{\sin t + \tan t}{\cot t + \csc t} = \sin t \tan t$

34. $\dfrac{\sin t - \cos t}{\cos^2 t} = \dfrac{\tan^2 t - 1}{\sin t + \cos t}$

In problems 35–42, rewrite each algebraic expression as a trigonometric expression containing no radical by using the given trigonometric substitution. Assume that θ is an acute angle. Illustrate the situation with a right triangle.

35. $\sqrt{4 - u^2};\ u = 2\sin\theta$
36. $\sqrt{u^2 + 4};\ u = 2\tan\theta$
37. $\sqrt{u^2 + 9};\ u = 3\tan\theta$
38. $\sqrt{64 - u^2};\ u = 8\sin\theta$
39. $\sqrt{u^2 - 25};\ u = 5\sec\theta$
40. $\sqrt{9u^2 + 4};\ u = \frac{2}{3}\tan\theta$
41. $\sqrt{4 - 9u^2};\ u = \frac{2}{3}\sin\theta$
42. $\sqrt{u^2 - 81};\ u = 9\sec\theta$

GU In problems 43–46, use a graphing utility to show that the given equation is *not* an identity. Confirm the result by giving a value of t for which the given equation is false.

43. $\sin t\,(1 + \cot t) = 2\sin t + \cos t$
44. $\cos t\,(\tan t + \cot t) = -\csc t$
45. $\sec t - \cos t = 3\tan t \cos t$
46. $\sin t \cot t + \cos t \tan t = \sin t - \cos t$

Developing and Extending the Concepts

GU In problems 47–50, each equation is an identity in certain quadrants associated with t. Use graphs to help determine the quadrants in each case. Then confirm the result analytically.

47. $\dfrac{\sin t}{\sqrt{1 - \sin^2 t}} = -\tan t$

48. $\dfrac{\cos t}{\sqrt{1 - \cos^2 t}} = \cot t$

49. $\dfrac{1}{\cos t} - \dfrac{1}{\cot t} = \sqrt{\dfrac{1 - \sin t}{1 + \sin t}}$

50. $\dfrac{\sec t - 1}{\tan t} = \sqrt{\dfrac{1 - \cos t}{1 + \cos t}}$

51. GU (a) Use a graphing utility to get two separate graphs for $y_1 = \cos^4 t - \sin^4 t$ and $y_2 = \cos^2 t - \sin^2 t$ for the same window settings.
(b) Compare the graphs.
(c) Is there an identity that relates y_1 and y_2? Explain.

52. GU Use a graphing utility to help determine whether or not the equation

$$\ln(1 - \cos t) - 2\ln|\sin t| = -\ln(1 + \cos t)$$

is an identity. If the equation appears to be an identity, then prove it.

In problems 53–56, verify that each equation is an identity.

53. $-2\csc(-\theta) - \dfrac{\sin\theta}{1 + \cos(-\theta)} = \dfrac{1 + \cos(-\theta)}{-\sin(-\theta)}$

54. $\dfrac{\sec(-t)}{\csc(-t)[\tan(-t) + \cot(-t)]} = \sin^2 t$

55. $\ln|\sec t + \tan t| = -\ln|\sec t - \tan t|$

56. $\dfrac{1 - \sin t \cos t}{\cos t(\sec t - \csc t)} \cdot \dfrac{\sin^2 t - \cos^2 t}{\sin^3 t + \cos^3 t} = \sin t$

OBJECTIVES

1. Use the addition and subtraction cosine identities.
2. Use the addition and subtraction sine identities.
3. Simplify trigonometric expressions and verify identities.
4. Solve applied problems.

6.2 Addition and Subtraction Identities

In Section 6.1, we worked with identities that express relationships among trigonometric functions of a *single variable*. In this section, we develop trigonometric identities involving the sum or difference of *two variables*.

Using the Addition and Subtraction Cosine Identities

Suppose we want to express the value of $\cos(u + v)$ in terms of values of trigonometric functions of u and v. We might be tempted to say that $\cos(u + v)$ is the same as $\cos u + \cos v$. To find out whether this is true, we compare the values of the expressions $\cos(30° + 60°)$ and $\cos 30° + \cos 60°$ (Table 1). Clearly, the results are different. Thus, in general, $\cos(u + v) \neq \cos u + \cos v$.

TABLE 1

$\cos(30° + 60°) \neq \cos 30° + \cos 60°$

$\cos(30° + 60°)$	$\cos 30° + \cos 60°$
$= \cos 90°$	$= \dfrac{\sqrt{3}}{2} + \dfrac{1}{2}$
$= 0$	$= \dfrac{\sqrt{3} + 1}{2}$

The next result shows how it is possible to express $\cos(u + v)$ and $\cos(u - v)$ in terms of trigonometric functions of u and v.

ADDITION AND SUBTRACTION IDENTITIES FOR THE COSINE

(i) $\cos(u - v) = \cos u \cos v + \sin u \sin v$
(ii) $\cos(u + v) = \cos u \cos v - \sin u \sin v$

Proof

The results are true for all possible values of u and v.

(i) To prove identity (i), let u and v represent angles in standard position whose terminal sides intersect the unit circle at $P_1 = (\cos u, \sin u)$ and $P_2 = (\cos v, \sin v)$, respectively. Figure 1a illustrates the situation when $0 < v < \pi/2 < u < \pi$. After placing the angle $u - v$ in standard position (Figure 1b), we see that the coordinates of the point of intersection of its terminal side and the unit circle are given by

$$P_3 = (\cos(u - v), \sin(u - v))$$

If $P_0 = (1, 0)$, then the lengths of arcs $\widehat{P_1P_2}$ and $\widehat{P_0P_3}$ are equal. It follows from geometry that the corresponding chord lengths are equal, so $d(P_0, P_3) = d(P_1, P_2)$ or $[d(P_0, P_3)]^2 = [d(P_1, P_2)]^2$. From the distance formula, we get

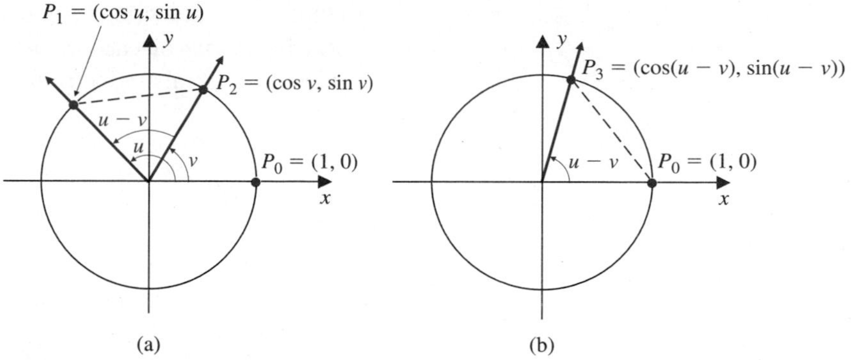

(a) (b)

FIGURE 1

$$\begin{aligned}
[d(P_0, P_3)]^2 &= [\cos(u - v) - 1]^2 + [\sin(u - v) - 0]^2 \\
&= \cos^2(u - v) - 2\cos(u - v) + 1 + \sin^2(u - v) && \text{Multiply} \\
&= 1 + 1 - 2\cos(u - v) && \text{Pythagorean} \\
&= 2 - 2\cos(u - v) && \text{identity}
\end{aligned}$$

Also,

$$\begin{aligned}
[d(P_1, P_2)]^2 &= (\cos u - \cos v)^2 + (\sin u - \sin v)^2 \\
&= \cos^2 u - 2\cos u \cos v + \cos^2 v \\
&\quad + \sin^2 u - 2\sin u \sin v + \sin^2 v && \text{Multiply} \\
&= 1 + 1 - 2\cos u \cos v - 2\sin u \sin v && \text{Pythagorean} \\
&= 2 - 2\cos u \cos v - 2\sin u \sin v && \text{identity}
\end{aligned}$$

By setting $[d(P_0, P_3)]^2 = [d(P_1, P_2)]^2$, we get

$$\begin{aligned}
2 - 2\cos(u - v) &= 2 - 2\cos u \cos v - 2\sin u \sin v \\
-2\cos(u - v) &= -2\cos u \cos v - 2\sin u \sin v && \text{Subtract 2 from each side} \\
\cos(u - v) &= \cos u \cos v + \sin u \sin v && \text{Divide each side by } -2
\end{aligned}$$

What happens to the identity for $\cos(u - v)$ if $u = v$?

(ii) To prove identity (ii), we proceed as follows:

$$\begin{aligned}
\cos(u + v) &= \cos[u - (-v)] \\
&= \cos u \cos(-v) + \sin u \sin(-v) && \text{Identity (i)} \\
&= \cos u \cos v - \sin u \sin v && \text{Cosine is even; sine is odd}
\end{aligned}$$

EXAMPLE 1 **Using the addition and subtraction cosine identities**

Find the exact value of each expression by using the addition and subtraction cosine identities. Compare the result to the calculator value, rounded off to four decimal places.

(a) $\cos 15°$ (b) $\cos \frac{7\pi}{12}$

Solution

(a) Using $15° = 45° - 30°$ and the identity for $\cos(u - v)$, we get

$$\begin{aligned}\cos 15° &= \cos(45° - 30°)\\ &= \cos 45° \cos 30° + \sin 45° \sin 30°\\ &= \frac{\sqrt{2}}{2} \cdot \frac{\sqrt{3}}{2} + \frac{\sqrt{2}}{2} \cdot \frac{1}{2} = \frac{\sqrt{6}}{4} + \frac{\sqrt{2}}{4} = \frac{\sqrt{6} + \sqrt{2}}{4}\end{aligned}$$

Rounding off $(\sqrt{6} + \sqrt{2})/4$ to four decimal places gives 0.9659. Using a calculator to evaluate $\cos 15°$ and rounding off to four decimal places, we again get 0.9659.

(b) Applying the identity for $\cos(u + v)$ and using $(7\pi)/12 = (\pi/3) + (\pi/4)$, we get

$$\begin{aligned}\cos \frac{7\pi}{12} &= \cos\left(\frac{\pi}{3} + \frac{\pi}{4}\right)\\ &= \cos \frac{\pi}{3} \cos \frac{\pi}{4} - \sin \frac{\pi}{3} \sin \frac{\pi}{4}\\ &= \frac{1}{2} \cdot \frac{\sqrt{2}}{2} - \frac{\sqrt{3}}{2} \cdot \frac{\sqrt{2}}{2} = \frac{\sqrt{2}}{4} - \frac{\sqrt{6}}{4} = \frac{\sqrt{2} - \sqrt{6}}{4}\end{aligned}$$

After rounding off $(\sqrt{2} - \sqrt{6})/4$ to four decimal places, we get -0.2588. Using a calculator to evaluate $\cos[(7\pi)/12]$ and rounding off to four decimal places, we get the same approximate value, -0.2588.

We can use the identity for $\cos(u - v)$ to establish some identities that relate the trigonometric functions to their corresponding *cofunctions*.

COFUNCTION IDENTITIES

If v is a real number or the radian measure of an angle, then:

(i) $\cos\left(\frac{\pi}{2} - v\right) = \sin v$ (ii) $\sin\left(\frac{\pi}{2} - v\right) = \cos v$

(iii) $\tan\left(\frac{\pi}{2} - v\right) = \cot v$ (iv) $\cot\left(\frac{\pi}{2} - v\right) = \tan v$

(v) $\sec\left(\frac{\pi}{2} - v\right) = \csc v$ (vi) $\csc\left(\frac{\pi}{2} - v\right) = \sec v$

Proof

(i) Using the subtraction identity for the cosine, we have

$$\cos\left(\frac{\pi}{2} - v\right) = \cos\frac{\pi}{2}\cos v + \sin\frac{\pi}{2}\sin v = 0 + \sin v = \sin v$$

This gives us the first identity.

(ii) If we replace v by $(\pi/2) - v$ in the first identity, we get

$$\cos\left[\frac{\pi}{2} - \left(\frac{\pi}{2} - v\right)\right] = \sin\left(\frac{\pi}{2} - v\right)$$

$$\cos v = \sin\left(\frac{\pi}{2} - v\right)$$

that is,

$$\sin\left(\frac{\pi}{2} - v\right) = \cos v$$

(iii) We use the first two identities to obtain the cofunction identity for the tangent as follows:

$$\tan\left(\frac{\pi}{2} - v\right) = \frac{\sin\left(\frac{\pi}{2} - v\right)}{\cos\left(\frac{\pi}{2} - v\right)} = \frac{\cos v}{\sin v} = \cot v$$

The proofs of the remaining identities are similar (Problem 49). 

Naturally, if we measure an angle θ in degrees instead of radians, the cofunction identities still hold. Thus,

(i) $\cos(90° - \theta) = \sin\theta$ (ii) $\sin(90° - \theta) = \cos\theta$
(iii) $\tan(90° - \theta) = \cot\theta$ (iv) $\cot(90° - \theta) = \tan\theta$
(v) $\sec(90° - \theta) = \csc\theta$ (vi) $\csc(90° - \theta) = \sec\theta$

Using the Addition and Subtraction Sine Identities

Using the cofunction identities and the addition and subtraction identities for the cosine, we can derive identities for the sine.

ADDITION AND SUBTRACTION IDENTITIES FOR THE SINE

(i) $\sin(u - v) = \sin u\cos v - \cos u\sin v$
(ii) $\sin(u + v) = \sin u\cos v + \cos u\sin v$

Proof

(i) To prove the identity for $\sin(u - v)$, we use the cofunction identity

$$\cos\left(\frac{\pi}{2} - v\right) = \sin v$$

Reading from right to left and replacing v with $u - v$, we get

$$\begin{aligned}\sin(u - v) &= \cos\left[\frac{\pi}{2} - (u - v)\right] = \cos\left(\frac{\pi}{2} - u + v\right)\\ &= \cos\left[\left(\frac{\pi}{2} - u\right) + v\right]\end{aligned}$$

So,

$$\begin{aligned}\sin(u - v) &= \cos\left(\frac{\pi}{2} - u\right)\cos v - \sin\left(\frac{\pi}{2} - u\right)\sin v && \text{Addition identity for the cosine}\\ &= \sin u \cos v - \cos u \sin v && \text{Cofunction identities (i) and (ii)}\end{aligned}$$

(ii) After replacing v by $-v$ in part (i), we obtain

$$\begin{aligned}\sin(u + v) &= \sin[u - (-v)]\\ &= \sin u \cos(-v) - \cos u \sin(-v) && \text{Identity (i)}\\ &= \sin u \cos v + \cos u \sin v && \text{Cosine is even; sine is odd}\end{aligned}$$

EXAMPLE 2 Evaluating an expression by recognizing the identity

(a) Find the exact value of the expression

$$\sin 81^\circ \cos 21^\circ - \cos 81^\circ \sin 21^\circ$$

by recognizing the applicable sine identity.

(b) Use a calculator to approximate the value of the given expression to four decimal places. Compare the result to the answer for part (a).

Solution

(a) By recognizing the subtraction sine identity, from right to left, we have

$$\sin 81^\circ \cos 21^\circ - \cos 81^\circ \sin 21^\circ = \sin(81^\circ - 21^\circ) = \sin 60^\circ = \frac{\sqrt{3}}{2}$$

(b) Using a calculator and rounding off to four decimal places, we have

$$\sin 81^\circ \cos 21^\circ - \cos 81^\circ \sin 21^\circ = 0.8660$$

In part (a) we obtained $\sqrt{3}/2$, which is also approximately 0.8660, to four decimal places.

EXAMPLE 3 Finding the exact value of the sine of a difference

Suppose that u and v are angles in standard position, where u is in quadrant I, $\cos u = \frac{4}{5}$, v is in quadrant II, and $\sin v = \frac{3}{5}$. Use identities to determine the exact value of $\sin(u - v)$.

Solution Before we can apply the subtraction identity, we need to determine both $\sin u$ and $\cos v$. Since the terminal side of u is in quadrant I, $\sin u$ is positive, so

$$\sin u = \sqrt{1 - \cos^2 u} = \sqrt{1 - \tfrac{16}{25}} = \tfrac{3}{5}$$

Also, since the terminal side of v is in quadrant II, $\cos v$ is negative, so

$$\cos v = -\sqrt{1 - \sin^2 v} = -\sqrt{1 - \tfrac{9}{25}} = -\tfrac{4}{5}$$

It follows that

$$\sin(u - v) = \sin u \cos v - \cos u \sin v = \left(\tfrac{3}{5}\right)\left(-\tfrac{4}{5}\right) - \left(\tfrac{4}{5}\right)\left(\tfrac{3}{5}\right) = -\tfrac{24}{25}$$

EXAMPLE 4 Finding a value involving inverse functions

Find the exact value of $\sin\left(\cos^{-1} \frac{5}{13} + \sin^{-1} \frac{3}{5}\right)$ by using the addition identity for the sine function and right triangle trigonometry.

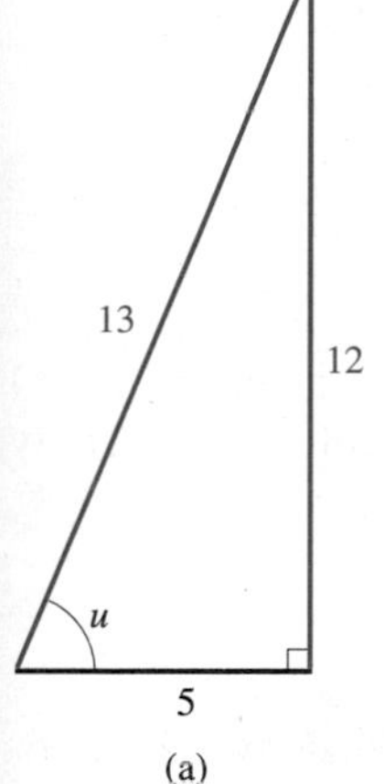

FIGURE 2

Solution First, we let $u = \cos^{-1} \frac{5}{13}$ and $v = \sin^{-1} \frac{3}{5}$, so $\cos u = \frac{5}{13}$, $\sin v = \frac{3}{5}$, and u and v are acute angles. Using right triangles, we see in Figure 2a that $\sin u = \frac{12}{13}$ and in Figure 2b that $\cos v = \frac{4}{5}$. Thus,

$$\begin{aligned}
\sin\left(\cos^{-1} \tfrac{5}{13} + \sin^{-1} \tfrac{3}{5}\right) &= \sin(u + v) \\
&= \sin u \cos v + \cos u \sin v \\
&= \left(\tfrac{12}{13}\right)\left(\tfrac{4}{5}\right) + \left(\tfrac{5}{13}\right)\left(\tfrac{3}{5}\right) \\
&= \tfrac{48}{65} + \tfrac{15}{65} = \tfrac{63}{65}
\end{aligned}$$

Using the addition and subtraction identities for the sine and cosine, we can derive addition and subtraction identities for the tangent function.

ADDITION AND SUBTRACTION IDENTITIES FOR THE TANGENT

(i) $\tan(u - v) = \dfrac{\tan u - \tan v}{1 + \tan u \tan v}$ (ii) $\tan(u + v) = \dfrac{\tan u + \tan v}{1 - \tan u \tan v}$

Proof

(i) We prove the first identity as follows:

$$\tan(u - v) = \frac{\sin(u - v)}{\cos(u - v)}$$

$$= \frac{\sin u \cos v - \cos u \sin v}{\cos u \cos v + \sin u \sin v}$$ Sine and cosine subtraction identities

$$= \frac{\dfrac{\sin u \cos v}{\cos u \cos v} - \dfrac{\cos u \sin v}{\cos u \cos v}}{\dfrac{\cos u \cos v}{\cos u \cos v} + \dfrac{\sin u \sin v}{\cos u \cos v}}$$ Divide numerator and denominator by $\cos u \cos v$

$$= \frac{\tan u - \tan v}{1 + \tan u \tan v}$$ Reduce fractions and use $\tan t = \dfrac{\sin t}{\cos t}$

The proof of part (ii) is left as an exercise (Problem 50).

Care must be taken when applying the trigonometric identities. For example, if we replace u by $\pi/2$ in the subtraction identity for $\tan(u - v)$, we get

$$\tan\left(\frac{\pi}{2} - v\right) = \frac{\tan \dfrac{\pi}{2} - \tan v}{1 + \tan \dfrac{\pi}{2} \tan v}$$

Since $\tan(\pi/2)$ is not defined, we might erroneously conclude that $\tan[(\pi/2) - v]$ is not defined for any value of v. However, we know from the cofunction identity that $\tan[(\pi/2) - v]$ can be written as

$$\tan\left(\frac{\pi}{2} - v\right) = \cot v$$

This illustration emphasizes the important fact that the identities derived in trigonometry are applicable only when the values of the variables are in the domains of all the functions contained in the identities.

Simplifying Trigonometric Expressions and Verifying Identities

We can use the addition and subtraction identities to simplify certain trigonometric expressions and to verify other identities.

EXAMPLE 5 **Simplifying a trigonometric expression**

Write the expression $\sin 5t \cos 2t - \sin 2t \cos 5t$ in terms of the sine function.

Solution We recognize that the expression fits the identity for $\sin(u - v)$, with $u = 5t$ and $v = 2t$. So,

$$\sin 5t \cos 2t - \sin 2t \cos 5t = \sin(5t - 2t)$$
$$= \sin 3t$$

In Section 5.4, we indicated that the tangent function has a period π. In the next example, we confirm this fact analytically.

EXAMPLE 6 **Proving that the tangent function has period π**

Prove that $f(x) = \tan x$ has a period π.

Solution Using the addition identity for the tangent, we get

$$\tan(x + \pi) = \frac{\tan x + \tan \pi}{1 - \tan x \tan \pi}$$

$$= \frac{\tan x + 0}{1 - 0} = \tan x$$

Thus, the tangent function has period π.

EXAMPLE 7 **Verifying an identity**

(a) Use a graphing utility to compare the graphs of

$$y_1 - \sin\left(t + \frac{\pi}{3}\right) \quad \text{and} \quad y_2 = \frac{1}{2}\left(\sin t + \sqrt{3}\cos t\right)$$

(b) Verify that the equation

$$\sin\left(t + \frac{\pi}{3}\right) = \frac{1}{2}\left(\sin t + \sqrt{3}\cos t\right)$$

is an identity.

Solution

(a) Viewing windows for the graphs of the functions $y_1 = \sin[t + (\pi/3)]$ and $y_2 = \frac{1}{2}\left(\sin t + \sqrt{3}\cos t\right)$ are shown in Figure 3a and Figure 3b, respectively. Clearly, the graphs are the same for the selected window.

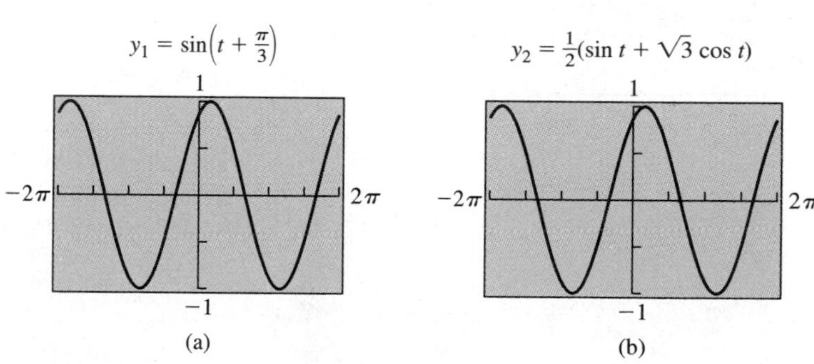

FIGURE 3

(b) The results in part (a) suggest that the equation is an identity. To verify this, we proceed as follows:

$$\begin{aligned}\sin\left(t + \frac{\pi}{3}\right) &= \sin t \cos \frac{\pi}{3} + \cos t \sin \frac{\pi}{3} && \text{Addition identity}\\ &= \sin t \cdot \frac{1}{2} + \frac{\sqrt{3}}{2} \cos t && \text{Evaluate}\\ &= \frac{1}{2}\left(\sin t + \sqrt{3} \cos t\right) && \text{Factor}\end{aligned}$$

We have transformed the left side into the right side, thus verifying the identity.

In some applications it is necessary to transform expressions from the form $P \cos u + Q \sin u$ to the form $A \cos(u - v)$.

EXAMPLE 8 Writing $P \cos u + Q \sin u$ in the form $A \cos(u - v)$

Let P and Q be constants, and let v be an angle in standard position with the point (P, Q) on the terminal side of v.

(a) Show that if $A = \sqrt{P^2 + Q^2}$, then

$$P = A \cos v \qquad \text{and} \qquad Q = A \sin v$$

(b) Use part (a) to show that

$$P \cos u + Q \sin u = A \cos(u - v)$$

for any angle u.

Solution

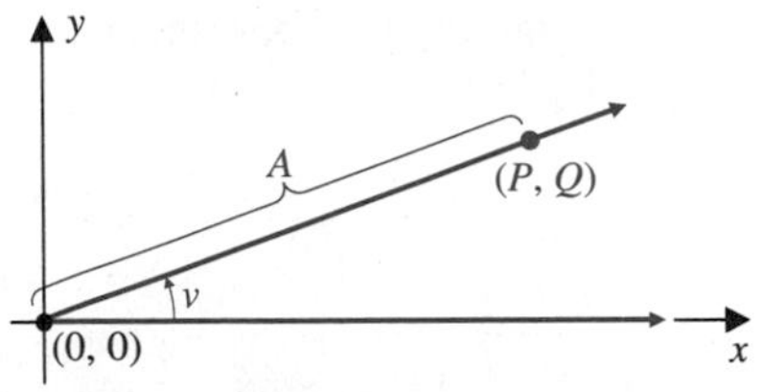

(a) Figure 4 shows angle v in standard position with the point (P, Q) on the terminal side. Notice that the distance A from the point $(0, 0)$ to (P, Q) is given by $A = \sqrt{P^2 + Q^2}$. Using the cosine and sine definitions, we have

$$\cos v = \frac{P}{A} \qquad \text{and} \qquad \sin v = \frac{Q}{A}$$

So

$$P = A \cos v \qquad \text{and} \qquad Q = A \sin v$$

FIGURE 4

(b) Using the results from part (a), we get

$$\begin{aligned}P \cos u + Q \sin u &= (A \cos v)(\cos u) + (A \sin v)(\sin u)\\ &= A(\cos v \cos u + \sin v \sin u)\\ &= A \cos(v - u)\\ &= A \cos[-(u - v)]\\ &= A \cos(u - v)\end{aligned}$$

for any angle u.

Solving Applied Problems

Applied problems from different fields such as physics, engineering, and meteorology are modeled by functions of the form $y = P \cos u + Q \sin u$.

EXAMPLE 9 **Solving a meteorology problem**

The U.S. Weather Bureau conducted a study on the temperature fluctuation for a desert region. Using data collected over several years, it was determined that the temperature T (in degrees Fahrenheit) is approximated by the model

$$T = 80 - 12 \cos \frac{\pi t}{6} + 12\sqrt{3} \sin \frac{\pi t}{6}$$

where t is the number of elapsed months, and $t = 0$ represents the month of March in each year.

(a) Write the equation representing this model in the form

$$T = 80 + A \cos(u - v)$$

GU

(b) Use a graphing utility to approximate (to the nearest integer) the maximum average temperature and when it occurs over a 12 month period starting in March. Confirm the result analytically by finding the exact maximum average temperature and when it occurs.

(c) Determine the exact minimum average temperature and when it occurs.

Solution

(a) We begin by rewriting the given function as

$$T = 80 + y \qquad \text{where } y = -12 \cos \frac{\pi t}{6} + 12\sqrt{3} \sin \frac{\pi t}{6}$$

This expression for y is of the form $P \cos u + Q \sin u$, where $P = -12$, $Q = 12\sqrt{3}$, and $u = (\pi t)/6$. Proceeding as we did in Example 8, we locate the point $(P, Q) = \left(-12, 12\sqrt{3}\right)$ on the terminal side of an angle v in standard position (Figure 5). Since

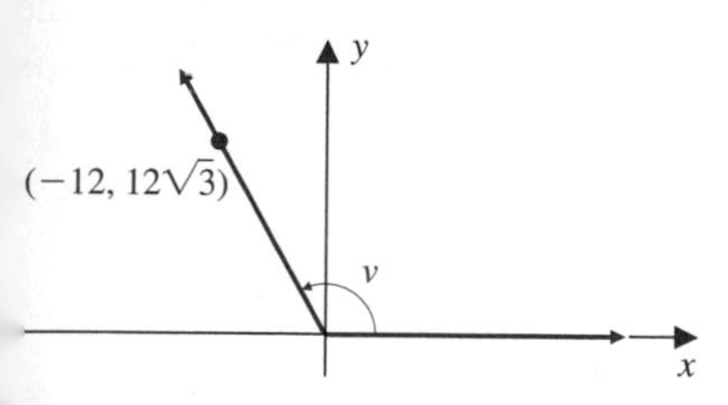

FIGURE 5

$$\tan v = \frac{Q}{P} = \frac{12\sqrt{3}}{-12} = -\sqrt{3}$$

and the point is in quadrant II, it follows that one possibility for angle v is $v = (2\pi)/3$. Also,

$$A = \sqrt{P^2 + Q^2} = \sqrt{(-12)^2 + \left(12\sqrt{3}\right)^2} = 24$$

So

$$y = A \cos(u - v) = 24 \cos\left(\frac{\pi t}{6} - \frac{2\pi}{3}\right)$$

Therefore,

$$T = 80 + y = 80 + 24 \cos\left(\frac{\pi t}{6} - \frac{2\pi}{3}\right)$$

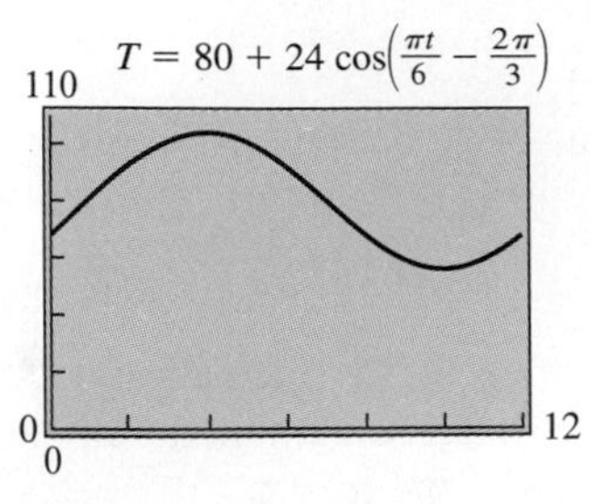

FIGURE 6

(b) Figure 6 shows a viewing window of the graph of

$$T = 80 + 24\cos\left(\frac{\pi t}{6} - \frac{2\pi}{3}\right) \qquad \text{where } 0 \le t \le 12$$

Using the ZOOM and TRACE features, we find that the maximum average temperature is approximately 104°F, and it occurs when $t = 4$, that is, in the month of July. To confirm the result analytically, we observe that the maximum value of T occurs when the cosine value attains its maximum value of 1. So the maximum value of T is given by

$$T = 80 + 24(1) = 104°\text{F}$$

and it occurs when

$$\cos\left(\frac{\pi t}{6} - \frac{2\pi}{3}\right) = 1$$

that is, for the first time, when

$$\frac{\pi t}{6} - \frac{2\pi}{3} = 0 \qquad \text{or} \qquad t = 4$$

(c) The minimum value of T occurs when the cosine value attains its minimum value of -1. This occurs when

$$\cos\left(\frac{\pi t}{6} - \frac{2\pi}{3}\right) = -1$$

that is, for the first time, when

$$\frac{\pi t}{6} - \frac{2\pi}{3} = \pi \qquad \text{or} \qquad t = 10$$

Thus, the minimum value of T is given by

$$T = 80 + 24(-1) = 56°\text{F}$$

and it occurs in the month of January (when $t = 10$).

PROBLEM SET 6.2

Mastering the Concepts

In problems 1 and 2, use an appropriate addition or subtraction identity to find the exact value of each expression. In each case compare this value to the calculator value (rounding off to four decimal places).

1. **(a)** $\sin\left(\frac{\pi}{6} + \frac{3\pi}{4}\right)$ **(b)** $\cos(330° + 45°)$

2. **(a)** $\cos\left(\frac{5\pi}{4} - \frac{\pi}{3}\right)$ **(b)** $\tan(60° - 45°)$

In problems 3 and 4, find the exact value of each expression by using an appropriate addition or subtraction identity and $(5\pi)/12 = (\pi/6) + (\pi/4)$ or $165° = 210° - 45°$. In each case compare this value to the calculator value (rounding off to four decimal places).

3. **(a)** $\sin\frac{5\pi}{12}$ **(b)** $\cos 165°$ **(c)** $\tan\frac{5\pi}{12}$

4. **(a)** $\cos\frac{5\pi}{12}$ **(b)** $\sin 165°$ **(c)** $\tan 165°$

In problems 5 and 6, use an appropriate addition or subtraction identity to find the exact value of each expression. In each case compare this value to the calculator value (rounding off to four decimal places).

5. **(a)** $\cos 105°$ **(b)** $\sin \frac{19\pi}{12}$

6. **(a)** $\tan \frac{13\pi}{12}$ **(b)** $\cos 285°$

In problems 7 and 8, find the exact value of each expression by recognizing the applicable identity. Also, use a calculator to evaluate each expression to four decimal places. Compare the results in both approaches.

7. **(a)** $\sin 33° \cos 27° + \sin 27° \cos 33°$

(b) $\dfrac{\tan \frac{4\pi}{5} - \tan \frac{3\pi}{10}}{1 + \tan \frac{4\pi}{5} \tan \frac{3\pi}{10}}$

8. **(a)** $\cos \frac{5\pi}{7} \sin \frac{2\pi}{7} + \cos \frac{2\pi}{7} \sin \frac{5\pi}{7}$

(b) $\dfrac{\tan 17° + \tan 43°}{1 - \tan 17° \tan 43°}$

In problems 9–12, suppose that u and v are angles in standard position. Use the given information to find the value of each expression. Round off the answers in problems 11 and 12 to three decimal places.

(a) $\sin(u + v)$ **(b)** $\cos(u + v)$ **(c)** $\sin(u - v)$
(d) $\cos(u - v)$ **(e)** $\tan(u + v)$ **(f)** $\tan(u - v)$

9. $\sin u = \frac{12}{13}$, u is in quadrant II, $\cos v = -\frac{4}{5}$, and v is in quadrant II

10. $\sin u = \frac{3}{5}$, u is in quadrant II, $\cos v = \frac{3}{5}$, and v is in quadrant IV

11. $\cos u = 0.47$, u is in quadrant IV, $\sin v = -0.96$, and v is in quadrant III

12. $\tan u = 0.42$, u is in quadrant I, $\tan v = 0.59$, and v is in quadrant III

In problems 13–16, find the exact value of each expression by using the addition and subtraction identities and right triangle trigonometry.

13. $\cos\left(\sin^{-1} \frac{3}{5} - \cos^{-1} \frac{4}{5}\right)$

14. $\sin\left[\sin^{-1}\left(-\frac{3}{5}\right) - \cos^{-1} \frac{4}{5}\right]$

15. $\sin\left[\cos^{-1} \frac{5}{13} + \sin^{-1}\left(-\frac{12}{13}\right)\right]$

16. $\cos\left(\sin^{-1} \frac{8}{17} - \cos^{-1} \frac{15}{17}\right)$

In problems 17–26, rewrite each expression in terms of the sine, cosine, or tangent using the addition and subtraction identities. Simplify each result.

17. **(a)** $\sin\left(\frac{5\pi}{2} - t\right)$ **(b)** $\cos(270° + t)$

18. **(a)** $\cos\left(\frac{\pi}{6} - t\right)$ **(b)** $\sin(\theta - 45°)$

19. **(a)** $\tan\left(t + \frac{\pi}{4}\right)$ **(b)** $\sin(45° - \theta)$

20. **(a)** $\sin(\pi - t)$ **(b)** $\tan(\theta + 60°)$

21. **(a)** $\csc(90° + \theta)$ **(b)** $\sin\left(\frac{3\pi}{2} - t\right)$

22. **(a)** $\cot(10\pi - t)$ **(b)** $\tan(1080° - \theta)$

23. **(a)** $\cos 7t \cos t - \sin 7t \sin t$
(b) $\sin 7t \cos 3t - \cos 7t \sin 3t$

24. **(a)** $\sin(-t) \cos 4t - \cos(-t) \sin 4t$
(b) $\cos 3\theta \cos 2\theta + \sin 3\theta \sin 2\theta$

25. **(a)** $\cos \frac{2x}{3} \cos \frac{x}{3} - \sin \frac{2x}{3} \sin \frac{x}{3}$

(b) $\dfrac{\tan 4t + \tan 3t}{1 - \tan 4t \tan 3t}$

26. **(a)** $\sin(-5t) \cos 2t - \cos(-5t) \sin 2t$

(b) $\dfrac{\tan 5t - \tan 2t}{1 + \tan 5t \tan 2t}$

GU In problems 27–30, express each function in the form

$$y = A \cos(u - v)$$

where v is the smallest positive angle. Use a graphing utility to demonstrate that the function obtained is equal to the given function.

27. $y = \sin t + \cos t$ **28.** $y = \sqrt{3} \cos \pi t + \sin \pi t$

29. $y = \cos 2\pi t - \sqrt{3} \sin 2\pi t$

30. $y = \cos 4t + \sqrt{3} \sin 4t$

GU In problems 31–38, compare the graphs of the two functions defined by the two sides of the given equation. Then verify that the equation is an identity.

31. $\sin(t - 3\pi) = -\sin t$ **32.** $\cos(t - 450°) = \sin t$

33. $\sin(30° + t) = \frac{1}{2} \cos t + \frac{\sqrt{3}}{2} \sin t$

34. $\sin(60° + t) - \cos(30° + t) = \sin t$

35. $\tan(45° + t) = \dfrac{1 + \tan t}{1 - \tan t}$

36. $\sin(\pi - t) - \cot t \sin\left(t - \frac{\pi}{2}\right) = \csc t$

37. $\sin\left(\frac{\pi}{4}+t\right)-\sin\left(\frac{\pi}{4}-t\right)=\sqrt{2}\sin t$

38. $\cos\left(\frac{\pi}{6}+t\right)\cos\left(\frac{\pi}{6}-t\right)-\sin\left(\frac{\pi}{6}+t\right)\sin\left(\frac{\pi}{6}-t\right)=\frac{1}{2}$

In problems 39–48, verify that each equation is an identity.

39. $\sin(t+s)\sin(t-s)=\sin^2 t-\sin^2 s$

40. $\sin t\cos s-\sin\left(t+\frac{\pi}{2}\right)\sin(-s)=\sin(s+t)$

41. $\cos(s+t)\cos(s-t)=\cos^2 s+\cos^2 t-1$

42. $\tan s-\tan t=\dfrac{\sin(s-t)}{\cos s\cos t}$

43. $\cot s-\tan t=\dfrac{\cos(s+t)}{\sin s\cos t}$

44. $\dfrac{1}{\tan s+\tan t}=\dfrac{\csc(s+t)}{\sec s\sec t}$

45. $\dfrac{\cos(u+v)}{\cos(u-v)}=\dfrac{1-\tan u\tan v}{1+\tan u\tan v}$

46. $\sin(\pi-s-t)=\sin s\cos t+\cos s\sin t$

47. $\dfrac{\sin(s+t)}{\sin(s-t)}=\dfrac{\tan s+\tan t}{\tan s-\tan t}$

48. $\cot(u+v)=\dfrac{\cot u\cot v-1}{\cot u+\cot v}$

49. Verify that each equation is an identity.

(a) $\cot\left(\frac{\pi}{2}-v\right)=\tan v$ **(b)** $\sec\left(\frac{\pi}{2}-v\right)=\csc v$

(c) $\csc\left(\frac{\pi}{2}-v\right)=\sec v$

50. Use the identity for $\tan(u-v)$ to derive the identity for $\tan(u+v)$.

Applying the Concepts

51. Height of a Tide: In a certain harbor, the mathematical model

$$h(t)=0.3\cos\frac{\pi t}{6}+0.4\sin\frac{\pi t}{6}$$

is used to predict the height h (in meters) of the tide above mean sea level t hours after midnight, over a 12 hour period.

(a) Write the equation of this model in the form $h(t)=A\cos(u-v)$. Round off v to three decimal places.

GU **(b)** Use a graphing utility to approximate (to one decimal place) the maximum height and when it occurs. Confirm the results analytically by finding the exact maximum height and when it occurs.

(c) Determine the exact minimum height, and indicate when it occurs.

52. Physics: An object attached to a spring vibrates vertically according to the mathematical model

$$d(t)=4\cos 4t+3\sin 4t$$

where $d(t)$ is the distance of the object from its rest position measured in centimeters, t seconds after the start of the motion.

(a) Write the equation of this model in the form $d(t)=A\cos(u-v)$. Round off v to three decimal places.

GU **(b)** Use a graphing utility to graph the function used in this model and interpret it for $0\le t\le 10$.

(c) Determine the amplitude of the vibration of the object.

(d) Determine the period of the vibration of the object.

Developing and Extending the Concepts

53. Let $f(x)=\sin x$.

(a) Show that:

$$\frac{f(t+h)-f(t)}{h}=\sin t\left(\frac{\cos h-1}{h}\right)+\cos t\left(\frac{\sin h}{h}\right)$$

GU **(b)** Use a graphing utility to approximate the value of the expressions

$$\frac{\cos h-1}{h}\quad\text{and}\quad\frac{\sin h}{h}\quad\text{as}\quad h\to 0$$

(c) Explain why

$$\frac{f(t+h)-f(t)}{h}\to\cos t\quad\text{as}\quad h\to 0$$

54. Let $g(x)=\cos x$.

(a) Show that:

$$\frac{g(t+h)-g(t)}{h}=\cos t\left(\frac{\cos h-1}{h}\right)-\sin t\left(\frac{\sin h}{h}\right)$$

(b) Use part (b) of problem 53 to show that

$$\frac{g(t+h)-g(t)}{h}\to -\sin t\quad\text{as}\quad h\to 0$$

55. Consider the triangle ABC in Figure 7.

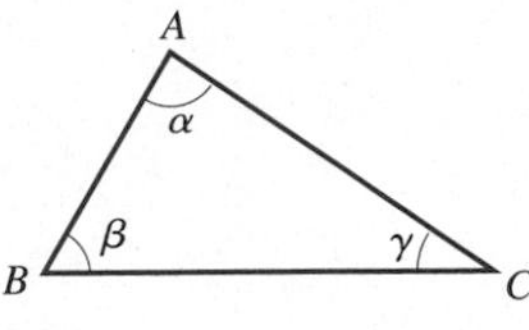

FIGURE 7

(a) Express $\sin(\alpha + \beta)$ in terms of a trigononometric function of γ.
(b) Express $\cos(\alpha + \beta)$ in terms of a trigonometric function of γ.

56. Figure 8 shows intersecting lines L_1 and L_2 with slopes m_1 and m_2, respectively.

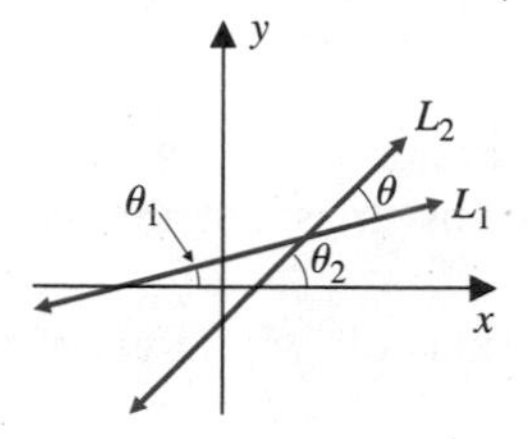

FIGURE 8

(a) Show that: $\tan \theta_1 = m_1$ and $\tan \theta_2 = m_2$
(b) Show that the angle $\theta = \theta_2 - \theta_1$ between L_1 and L_2 is given by

$$\tan \theta = \frac{m_2 - m_1}{1 + m_1 m_2}$$

(c) Use part (b) to show that if L_1 and L_2 are perpendicular, then $m_1 m_2 = -1$.

57. **(a)** Show that: $\cos\left(\frac{\pi}{2} - x\right) = \cos\left(x - \frac{\pi}{2}\right)$
(b) Explain why: $\sin x = \cos\left(x - \frac{\pi}{2}\right)$
(c) In view of the identity in part (b), compare the graphs of

$$f(x) = \sin x \quad \text{and} \quad g(x) = \cos\left(x - \frac{\pi}{2}\right)$$

How can the graph of $y = \cos x$ be used to obtain the graph of $f(x) = \sin x$?

58. **(a)** Find the value of the expression

$$\sin 1° + \sin 2° + \sin 3° + \cdots + \sin 357° + \sin 358° + \sin 359°$$

by using the identity

$$\sin \theta + \sin(360° - \theta) = \sin \theta - \sin \theta = 0$$

(b) Use the identity

$$\cos \theta + \cos(360° - \theta) = 2 \cos \theta$$

to evaluate the expression

$$\cos 1° + \cos 2° + \cos 3° + \cdots + \cos 357° + \cos 358° + \cos 359°$$

OBJECTIVES

1. Use double-angle identities.
2. Use half-angle identities.

6.3 Double-Angle and Half-Angle Identities

In this section we'll study identities that express trigonometric functions of twice an angle and half an angle in terms of trigonometric functions of the angle. These identities follow from the addition identities. They are used frequently in calculus.

Using Double-Angle Identities

Suppose that we want to express the value of $\sin 2t$ in terms of trigonometric functions of t. We might be tempted to say that $\sin 2t = 2 \sin t$. However, upon comparing the graphs of $y_1 = \sin 2t$ and $y_2 = 2 \sin t$ in Figures 1a and 1b, respectively, we see that $\sin 2t$ and $2 \sin t$ are different.

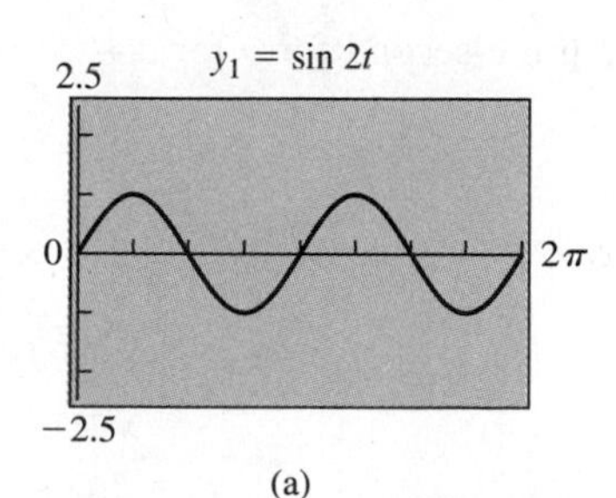

(a)

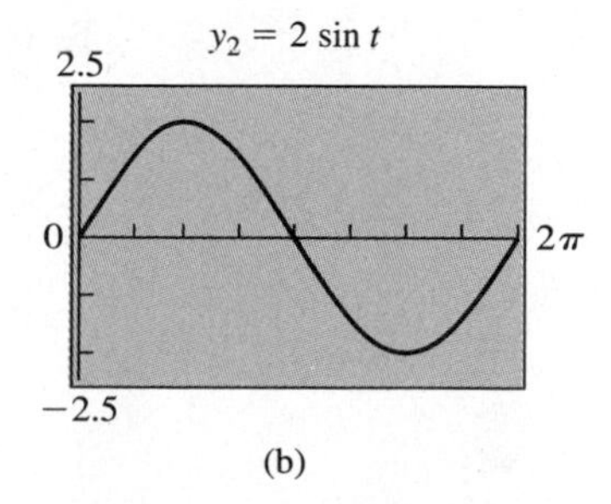

(b)

FIGURE 1

Numerically, if $t = \pi/6$, then

$$\sin 2t = \sin\left[2\left(\frac{\pi}{6}\right)\right] = \sin\frac{\pi}{3} = \frac{\sqrt{3}}{2}$$

whereas

$$2\sin t = 2\sin\frac{\pi}{6} = 2\cdot\frac{1}{2} = 1$$

Again, we see that $\sin 2t$ is not the same as $2\sin t$. Similarly, we can show from both a graphical and numerical point of view that $\cos 2t$ is not the same as $2\cos t$ (Problem 1).

The identities that express the trigonometric functions of $2t$ in terms of functions of t are called **double-angle identities**.

DOUBLE-ANGLE IDENTITIES

(i) $\sin 2t = 2\sin t\cos t$

(ii) $\cos 2t = \cos^2 t - \sin^2 t$
$= 2\cos^2 t - 1$
$= 1 - 2\sin^2 t$

(iii) $\tan 2t = \dfrac{2\tan t}{1 - \tan^2 t}$

Proof

Compare the graphs of $y_1 = \sin 2t$ and $y_2 = 2\sin t\cos t$. What do the graphs suggest regarding the equation $\sin 2t = 2\sin t\cos t$?

(i) To prove the first identity, we start with the addition identity for $\sin(u + v)$ and replace both u and v by t throughout to get

$$\sin(t + t) = \sin t\cos t + \cos t\sin t$$
$$\sin 2t = 2\sin t\cos t$$

(ii) To prove the second identity, we use the identity for $\cos(u + v)$ and replace both u and v by t to obtain

$$\cos(t + t) = \cos t\cos t - \sin t\sin t$$
$$\cos 2t = \cos^2 t - \sin^2 t$$

To obtain the second form for $\cos 2t$, we substitute the Pythagorean identity

$$\sin^2 t = 1 - \cos^2 t$$

into $\cos 2t = \cos^2 t - \sin^2 t$ to get

$$\begin{aligned}\cos 2t &= \cos^2 t - (1 - \cos^2 t)\\ &= \cos^2 t - 1 + \cos^2 t\\ &= 2\cos^2 t - 1\end{aligned}$$

Using the alternate form of the Pythagorean identity, $\cos^2 t = 1 - \sin^2 t$, we have

$$\begin{aligned}\cos 2t &= \cos^2 t - \sin^2 t \\ &= (1 - \sin^2 t) - \sin^2 t \\ &= 1 - 2\sin^2 t\end{aligned}$$

which is the third form of the identity.

(iii) The identity for $\tan 2t$ is obtained by replacing both u and v by t in the addition identity for $\tan(u + v)$ to get

$$\begin{aligned}\tan(t + t) &= \frac{\tan t + \tan t}{1 - \tan t \tan t} \\ \tan 2t &= \frac{2\tan t}{1 - \tan^2 t}\end{aligned}$$

EXAMPLE 1 Using double-angle identities

If $\sin\theta = \frac{12}{13}$ and θ is an angle in quadrant I, find the exact value of each expression.

(a) $\sin 2\theta$ (b) $\cos 2\theta$ (c) $\tan 2\theta$

Solution First, we need to find $\cos\theta$. Because θ is in quadrant I, we know that $\cos\theta$ is positive, so

$$\cos\theta = \sqrt{1 - \sin^2\theta} = \sqrt{1 - \left(\tfrac{12}{13}\right)^2} = \sqrt{\tfrac{25}{169}} = \tfrac{5}{13}$$

Thus, we get the following results:

(a) $\sin 2\theta = 2\sin\theta\cos\theta = 2\left(\frac{12}{13}\right)\left(\frac{5}{13}\right) = \frac{120}{169}$

(b) $\cos 2\theta = \cos^2\theta - \sin^2\theta = \left(\frac{5}{13}\right)^2 - \left(\frac{12}{13}\right)^2 = \frac{25}{169} - \frac{144}{169} = -\frac{119}{169}$

(c) Since

$$\tan\theta = \frac{\sin\theta}{\cos\theta} = \frac{\frac{12}{13}}{\frac{5}{13}} = \frac{12}{5}$$

it follows that

$$\tan 2\theta = \frac{2\tan\theta}{1 - \tan^2\theta} = \frac{2\left(\frac{12}{5}\right)}{1 - \left(\frac{12}{5}\right)^2} = \frac{\frac{24}{5}}{1 - \frac{144}{25}} = \frac{120}{25 - 144} = -\frac{120}{119}$$

A more efficient way of finding $\tan 2\theta$ in this situation is to proceed as follows:

$$\tan 2\theta = \frac{\sin 2\theta}{\cos 2\theta} = \frac{\frac{120}{169}}{-\frac{119}{169}} = -\frac{120}{119}$$

EXAMPLE 2 **Rewriting a trigonometric expression**

GU

Use a double-angle identity to rewrite the expression $\cos^2 3t - \sin^2 3t$ as a single trigonometric function. Demonstrate the validity of the result graphically.

Solution We use the double-angle identity $\cos 2x = \cos^2 x - \sin^2 x$ by replacing x with $3t$ to get

$$\cos^2 3t - \sin^2 3t = \cos[2(3t)] = \cos 6t$$

Figure 2 demonstrates the result, since the graph of $y_1 = \cos^2 3t - \sin^2 3t$ is the same as the graph of $y_2 = \cos 6t$ for the selected viewing window.

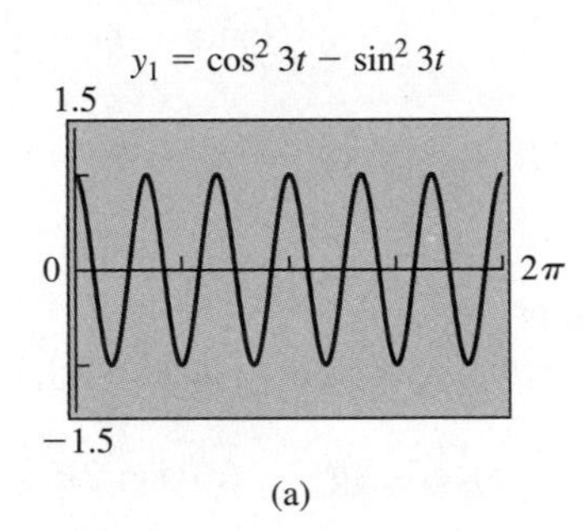

(a)

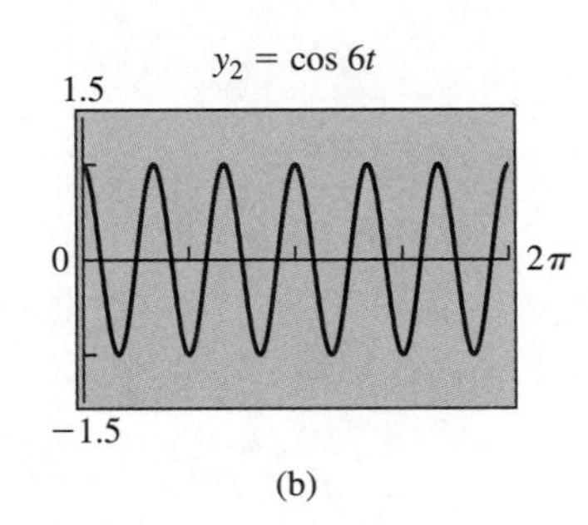

(b)

FIGURE 2

EXAMPLE 3 **Finding a value involving inverse functions**

Find the exact value of $\sin\left(2 \cos^{-1} \frac{8}{17}\right)$. Compare this value to the calculator value, rounded to four decimal places.

Solution First, we let $\theta = \cos^{-1} \frac{8}{17}$, so $\cos \theta = \frac{8}{17}$. Next, we draw a right triangle with an acute angle θ whose cosine is $\frac{8}{17}$ (Figure 3). From right triangle trigonometry, it follows that $\sin \theta = \frac{15}{17}$. Thus,

$$\begin{aligned}\sin\left(2 \cos^{-1} \tfrac{8}{17}\right) &= \sin 2\theta \\ &= 2 \sin \theta \cos \theta \\ &= 2\left(\tfrac{15}{17}\right)\left(\tfrac{8}{17}\right) \\ &= \tfrac{240}{289} \text{ (approx. 0.8304)}\end{aligned}$$

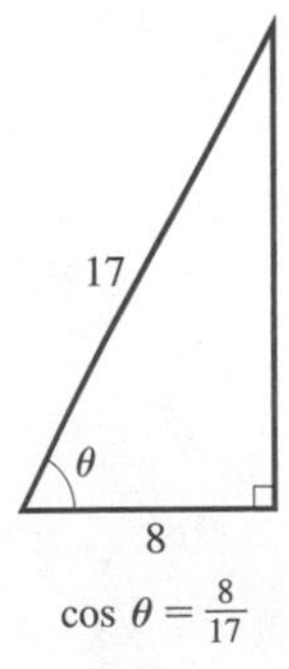

FIGURE 3

The calculator value of the expression $\sin\left(2 \cos^{-1} \frac{8}{17}\right)$ also equals 0.8304, rounded to four decimal places.

EXAMPLE 4 **Verifying an identity**

(a) Use a graphing utility to compare the graphs of

$$y_1 = \frac{2 \cos 2t}{\sin 2t - 2 \sin^2 t} - 1 \qquad \text{and} \qquad y_2 = \cot t$$

(b) Verify the identity: $\dfrac{2 \cos 2t}{\sin 2t - 2 \sin^2 t} - 1 = \cot t$

Solution

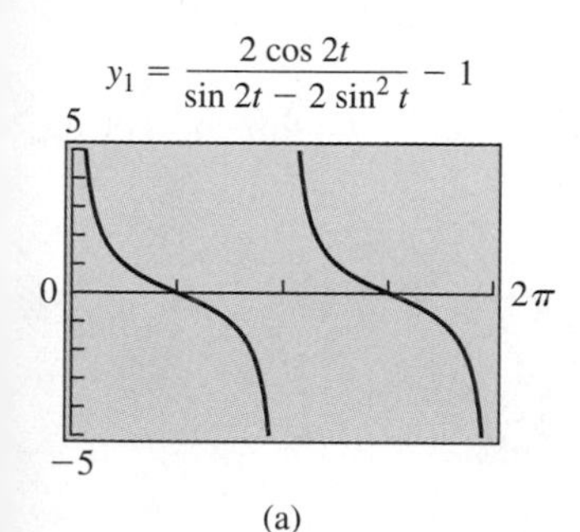

(a)

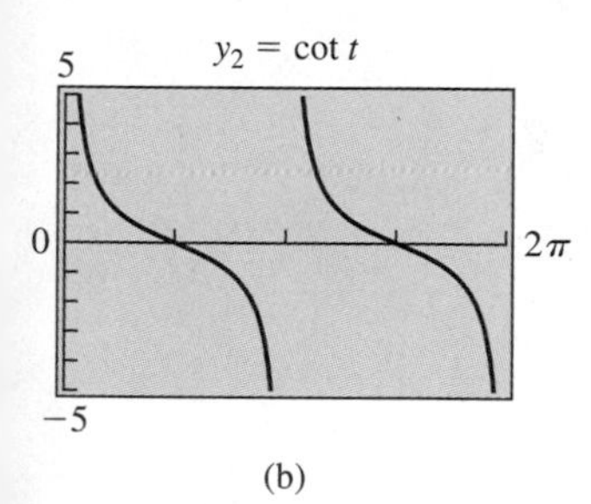

(b)

FIGURE 4

(a) Figure 4a shows a viewing window of the graph of

$$y_1 = \frac{2 \cos 2t}{\sin 2t - 2 \sin^2 t} - 1$$

and Figure 4b shows a viewing window with the same settings for the graph of $y_2 = \cot t$. The two graphs are the same for the selected window, suggesting that the equation in part (b) is an identity.

(b) To verify that the equation is an identity, we proceed as follows:

$$\frac{2 \cos 2t}{\sin 2t - 2 \sin^2 t} - 1 = \frac{2(\cos^2 t - \sin^2 t)}{2 \sin t \cos t - 2 \sin^2 t} - 1 \qquad \cos 2t \text{ and } \sin 2t \text{ identities}$$

$$= \frac{2(\cos t - \sin t)(\cos t + \sin t)}{2(\sin t)(\cos t - \sin t)} - 1 \qquad \text{Factor}$$

$$= \frac{\cos t + \sin t}{\sin t} - 1 \qquad \text{Reduce}$$

$$= \frac{\cos t}{\sin t} + \frac{\sin t}{\sin t} - 1 \qquad \frac{A+B}{C} = \frac{A}{C} + \frac{B}{C}$$

$$= \cot t + 1 - 1 = \cot t$$

Therefore, the equation is an identity.

Using Half-Angle Identities

Consider the double-angle identities

$$\cos 2t = 2 \cos^2 t - 1 \quad \text{and} \quad \cos 2t = 1 - 2 \sin^2 t$$

If we solve the first for $\cos^2 t$ and the second for $\sin^2 t$ in terms of $\cos 2t$, we get the following identities:

SQUARE IDENTITIES

(i) $\cos^2 t = \dfrac{1 + \cos 2t}{2}$ (ii) $\sin^2 t = \dfrac{1 - \cos 2t}{2}$

These identities are commonly used in calculus. For example,

$$\cos^2 3x = \frac{1 + \cos 6x}{2} \quad \text{and} \quad \sin^2 5x = \frac{1 - \cos 10x}{2}$$

By replacing t with $s/2$ in the square identities, we obtain the equivalent forms

$$\cos^2 \frac{s}{2} = \frac{1 + \cos s}{2} \quad \text{and} \quad \sin^2 \frac{s}{2} = \frac{1 - \cos s}{2}$$

Next, we take the square roots of both sides of the latter identities to get the following results:

HALF-ANGLE IDENTITIES

The sign depends on the quadrant in which the terminal side of the angle (in standard position) with measure $s/2$ lies.

$$\cos \frac{s}{2} = \pm\sqrt{\frac{1 + \cos s}{2}} \quad \text{and} \quad \sin \frac{s}{2} = \pm\sqrt{\frac{1 - \cos s}{2}}$$

These two identities can be used to derive the identity

$$\tan \frac{s}{2} = \frac{1 - \cos s}{\sin s}$$

(See Problem 62.)

EXAMPLE 5 Using a half-angle identity

Find the exact value of $\cos(\pi/12)$ by using the half-angle cosine identity. Compare this value to the calculator value, rounded to four decimal places.

Solution Using the half-angle identity for the cosine and the fact that $\cos(\pi/12)$ is positive, we get

$$\cos \frac{\pi}{12} = \cos\left[\frac{1}{2}\left(\frac{\pi}{6}\right)\right] = \sqrt{\frac{1 + \cos(\pi/6)}{2}} = \sqrt{\frac{1 + \left(\sqrt{3}/2\right)}{2}}$$

$$= \sqrt{\frac{2 + \sqrt{3}}{4}} = \frac{\sqrt{2 + \sqrt{3}}}{2} \quad \text{(approx. 0.9659)}$$

The calculator value of $\cos(\pi/12)$ also equals 0.9659, to four decimal places.

EXAMPLE 6 Evaluating functions involving half-angle identities

Suppose that θ is a positive obtuse angle in standard position, measured in degrees, and $\sin \theta = \frac{3}{5}$. Find the exact value of each expression.

(a) $\sin \dfrac{\theta}{2}$ (b) $\cos \dfrac{\theta}{2}$ (c) $\tan \dfrac{\theta}{2}$

Solution Since θ is obtuse, it is in quadrant II, thus, $\cos \theta$ is negative, so

$$\begin{aligned}\cos \theta &= -\sqrt{1 - \sin^2 \theta}\\ &= -\sqrt{1 - \left(\tfrac{3}{5}\right)^2}\\ &= -\sqrt{\tfrac{16}{25}} = -\tfrac{4}{5}\end{aligned}$$

Also, because $90° < \theta < 180°$, we have $45° < \theta/2 < 90°$, so $\theta/2$ is in quadrant I. Thus, $\sin(\theta/2)$ and $\cos(\theta/2)$ are both positive, and we have:

(a) $\sin \dfrac{\theta}{2} = \sqrt{\dfrac{1 - \cos \theta}{2}} = \sqrt{\dfrac{1 - \left(-\frac{4}{5}\right)}{2}} = \sqrt{\dfrac{9}{10}} = \dfrac{3}{\sqrt{10}} = \dfrac{3\sqrt{10}}{10}$

(b) $\cos \dfrac{\theta}{2} = \sqrt{\dfrac{1 + \cos \theta}{2}} = \sqrt{\dfrac{1 + \left(-\frac{4}{5}\right)}{2}} = \sqrt{\dfrac{1}{10}} = \dfrac{1}{\sqrt{10}} = \dfrac{\sqrt{10}}{10}$

(c) $\tan \dfrac{\theta}{2} = \dfrac{\sin(\theta/2)}{\cos(\theta/2)} = \dfrac{\left(3\sqrt{10}\right)/10}{\sqrt{10}/10} = 3$

EXAMPLE 7 **Verifying an identity**

Verify the identity

$$2 \tan \theta \cos^2 \frac{\theta}{2} = \sin \theta + \tan \theta$$

Demonstrate the validity of the result graphically.

Solution

$$\begin{aligned}2 \tan \theta \cos^2 \frac{\theta}{2} &= 2 \tan \theta \left(\frac{1 + \cos \theta}{2}\right) && \text{Square identity}\\ &= \tan \theta + \tan \theta \cos \theta && \text{Multiply}\\ &= \tan \theta + \frac{\sin \theta}{\cos \theta} \cdot \cos \theta && \text{Quotient identity}\\ &= \tan \theta + \sin \theta && \text{Reduce}\\ &= \sin \theta + \tan \theta\end{aligned}$$

We have transformed the left side of the equation into the right side, so the equation is an identity. Figure 5 shows that the graphs of $y_1 = 2 \tan \theta \cos^2(\theta/2)$ and $y_2 = \sin \theta + \tan \theta$ are the same for the same viewing window settings, thus demonstrating the validity of the identity.

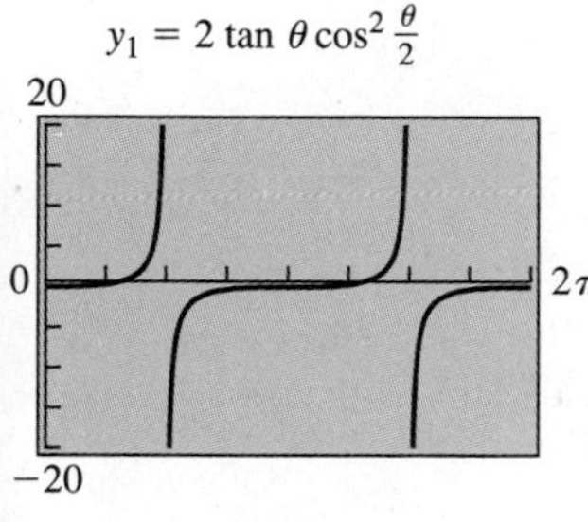
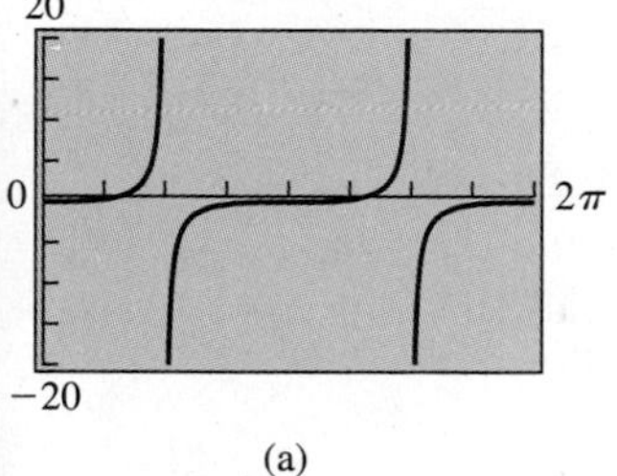

FIGURE 5

PROBLEM SET 6.3

Mastering the Concepts

1. **(a)** Use a graphing utility to graph $y_1 = \cos 2t$ and $y_2 = 2 \cos t$ with the same viewing window settings. What conclusion can be made about $\cos 2t$ and $2 \cos t$?
GU
(b) Evaluate $\cos 2t$ and $2 \cos t$ for $t = \pi/4$. What conclusion can be made about $\cos 2t$ and $2 \cos t$?

2. **(a)** Use a graphing utility to graph $y_1 = \tan 2t$ and $y_2 = 2 \tan t$ with the same viewing window settings. What conclusion can be made about $\tan 2t$ and $2 \tan t$?
GU
(b) Evaluate $\tan 2t$ and $2 \tan t$ for $t = \pi/4$. What conclusion can be made about $\tan 2t$ and $2 \tan t$?

In problems 3–10, use the given information to find the exact value of each expression.
(a) $\sin 2t$ **(b)** $\cos 2t$ **(c)** $\tan 2t$

3. $\sin t = \frac{4}{5}$; t in quadrant I
4. $\cos t = -\frac{12}{13}$; t in quadrant III
5. $\cos t = -\frac{7}{25}$; t in quadrant III
6. $\sin t = -\frac{5}{13}$; t in quadrant IV
7. $\tan t = -\frac{5}{12}$; t in quadrant II
8. $\sec t = -\frac{5}{2}$; t in quadrant III
9. $\cot t = -\frac{7}{25}$; t in quadrant IV
10. $\tan t = \frac{18}{5}$; t in quadrant I

GU In problems 11–18, use the double-angle identities to rewrite each expression in terms of a single trigonometric function. Demonstrate the validity of each result graphically.

11. $2 \sin 3t \cos 3t$
12. $1 - 2 \sin^2 3\pi x$
13. $2 \cos^2 8t - 1$
14. $4 \sin^2 7\theta \cos^2 7\theta$
15. $\cos^2 5t - \sin^2 5t$
16. $2 - 4 \sin^2 \dfrac{t}{2}$
17. $\dfrac{2 \tan 4\theta}{1 - \tan^2 4\theta}$
18. $\dfrac{\tan(t/2)}{\frac{1}{2} - \frac{1}{2}\tan^2(t/2)}$

19. Express $\cos 3t$ in terms of $\cos t$ and simplify. Demonstrate the validity of the result graphically.
GU
20. Express $\cos 4t$ in terms of $\sin t$ and simplify. Demonstrate the validity of the result graphically.
GU
21. Express $\sin 4t$ in terms of $\cos t$, where t is in quadrant I, and simplify. Demonstrate the validity of the result graphically.
GU
22. Express $\sin^4 t \cos^2 t$ in terms of $\cos^2 t$ and simplify. Demonstrate the validity of the result graphically.
GU

In problems 23 and 24, use an identity and right triangle trigonometry to find the exact value of each expression. Compare each value to the calculator value, rounded to four decimal places.

23. **(a)** $\sin\left(2 \sin^{-1} \dfrac{1}{2}\right)$ **(b)** $\sin\left(2 \cos^{-1} \dfrac{\sqrt{3}}{2}\right)$
(c) $\tan\left(2 \tan^{-1} \dfrac{4}{3}\right)$

24 **(a)** $\cos(2 \sin^{-1} 1)$ **(b)** $\cos\left(2 \sin^{-1} \dfrac{\sqrt{2}}{2}\right)$
(c) $\tan\left[2 \cos^{-1}\left(-\dfrac{24}{25}\right)\right]$

In problems 25–30, use the half-angle identities to find the exact value of each expression. Compare this value to the calculator value, rounded to four decimal places.

25. **(a)** $\cos 22.5°$ **(b)** $\sin \dfrac{5\pi}{12}$
26. **(a)** $\sin 75°$ **(b)** $\cos \dfrac{\pi}{8}$
27. **(a)** $\cos \dfrac{5\pi}{12}$ **(b)** $\tan \dfrac{3\pi}{8}$
28. **(a)** $\sin\left(-\dfrac{7\pi}{12}\right)$ **(b)** $\tan \dfrac{5\pi}{8}$
29. **(a)** $\cos 112.5°$ **(b)** $\sin 67.5°$
30. **(a)** $\tan \dfrac{7\pi}{12}$ **(b)** $\cos 67.5°$

In problems 31–38, use the given information to find the exact values of each expression if $0 < t < 2\pi$.

(a) $\sin \dfrac{t}{2}$ **(b)** $\cos \dfrac{t}{2}$ **(c)** $\tan \dfrac{t}{2}$

31. $\sin t = \frac{24}{25}$; t in quadrant I
32. $\cos t = \frac{7}{25}$; t in quadrant IV
33. $\cos t = -\frac{4}{5}$; t in quadrant II
34. $\sin t = -\frac{5}{13}$; t in quadrant III
35. $\cot t = \frac{8}{15}$; t in quadrant III
36. $\sec t = \frac{13}{5}$; t in quadrant I
37. $\tan t = -\frac{8}{15}$; t in quadrant IV
38. $\cot t = \frac{3}{4}$; t in quadrant III

In problems 39 and 40, use the half-angle identities to rewrite each expression as a single trigonometric function.

39. **(a)** $\sqrt{\dfrac{1+\cos 4t}{2}}$ **(b)** $\sqrt{\dfrac{1-\cos 6t}{2}}$

40. **(a)** $\sqrt{\dfrac{1-\cos 2\pi t}{2}}$ **(b)** $-\sqrt{\dfrac{1-\cos 2t}{1+\cos 2t}}$

GU In problems 41–54, compare the graphs of the two functions defined by the two sides of the given equation. Then verify that the equation is an identity.

41. **(a)** $\sin^2 t + \cos 2t = \cos^2 t$
(b) $\csc t \sec t = 2 \csc 2t$

42. **(a)** $\dfrac{\sin 2t}{2 \sin t} = \cos t$
(b) $\cos^2 \theta(1 - \tan^2 \theta) = \cos 2\theta$

43. **(a)** $(\sin 2t + \cos 2t)^2 - \sin 4t = 1$
(b) $(\sin t - \cos t)^2 + 2 \sin 2t = 1 + \sin 2t$

44. **(a)** $\dfrac{1-\cos 2t}{\sin 2t} = \tan t$ **(b)** $\dfrac{\sin 4\theta}{1-\cos 4\theta} = \cot 2\theta$

45. **(a)** $\dfrac{\sin^2 2\theta}{\sin^2 \theta} = 4\cos^2 \theta$
(b) $\sin \theta \tan \dfrac{\theta}{2} \csc^2 \dfrac{\theta}{2} = 2$

46. **(a)** $2 \tan \dfrac{\theta}{2} \csc \theta = \sec^2 \dfrac{\theta}{2}$
(b) $2 \cos \dfrac{\theta}{2} = (1 + \cos \theta) \sec \dfrac{\theta}{2}$

47. $\dfrac{\sin 3\theta}{\sin \theta} - \dfrac{\cos 3\theta}{\cos \theta} = 2$

48. $\dfrac{\sin 5t}{\sin t} - \dfrac{\cos 5t}{\cos t} = 4 \cos 2t$

49. $\dfrac{\cos^3 t - \sin^3 t}{\cos t - \sin t} = \dfrac{2 + \sin 2t}{2}$

50. $\dfrac{\sec^2 t}{2 - \sec^2 t} = \sec 2t$

51. $\sin^2 t \cos^2 t = \frac{1}{8}(1 - \cos 4t)$

52. $\sin^4 \theta = \dfrac{3}{8} - \dfrac{1}{2}\cos 2\theta + \dfrac{\cos 4\theta}{8}$

53. $\dfrac{\sin 5t + \sin t}{\cos t - \cos 5t} = \cot 2t$

54. $\dfrac{\cos 5t + \cos 2t}{\sin 5t + \sin 2t} = \cot \dfrac{7t}{2}$

Applying the Concepts

55. Area: Suppose that triangle PBC is an isosceles triangle where each of the equal sides $\overline{PB}$ and $\overline{PC}$ has a length of 4 centimeters and θ is the angle BPC, as shown in Figure 6.

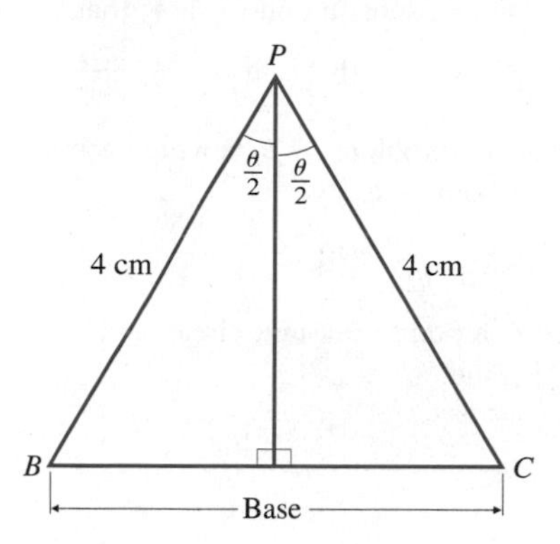

FIGURE 6

(a) Show that the area A of the triangle is given by $A = 8 \sin \theta$.
(b) Find the area A of the triangle when $\theta = 36°$. Round off the answer to two decimal places.
(c) Find the domain of the function in part (a).
GU **(d)** Use a graphing utility to approximate (to two decimal places) the maximum area and the value of θ when it occurs. Confirm the result analytically by finding the exact maximum area and when it occurs.

56. Area: Consider an isosceles trapezoid with the dimensions shown in Figure 7.

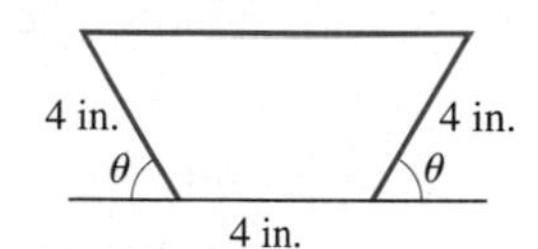

FIGURE 7

(a) Show that the area A of the trapezoid is given by

$$A = 64 \sin \frac{\theta}{2} \cos^3 \frac{\theta}{2}$$

(b) Find the area A of the trapezoid when $\theta = 36°$. Round off the answer to two decimal places.
(c) Find the domain of the function in part (a).
GU **(d)** Use a graphing utility to approximate (to two decimal places) the maximum area and the value of θ when it occurs.

Developing and Extending the Concepts

57. Some techniques in calculus make use of the substitution $z = \tan(\theta/2)$. Use this substitution to show that:

(a) $\cos\theta = \dfrac{1 - z^2}{1 + z^2}$ **(b)** $\sin\theta = \dfrac{2z}{1 + z^2}$

58. Use the results from problem 57 to rewrite each expression in terms of z and simplify.

(a) $\dfrac{3}{2\sin\theta + 3\cos\theta}$ **(b)** $\dfrac{1}{3 + 2\sin\theta}$

59. Suppose that P is a point on a unit circle, and $P(t) = \left(\frac{3}{5}, -\frac{4}{5}\right)$. Find:

(a) $P(2t)$ **(b)** $P(t/2)$

60. **(a)** Suppose that $u + v = \pi/2$. Verify that $\sin(u - v) = -\cos 2u$.
(b) Suppose that $u + v = \pi$. Verify that $\sin(u - v) = -\sin 2u$.

61. **(a)** Express $\sec 2t$ in terms of $\sec t$.
(b) Find $\sec 2t$ if $\tan t = 3$ and $\sec t < 0$.

62. **(a)** Verify that

GU

$$\tan\frac{s}{2} = \frac{1 - \cos s}{\sin s}$$

Demonstrate the validity of the result graphically.
(b) Use part (a) to find the exact value of $\tan(\pi/12)$.

63. Suppose that α and β are acute angles in a right triangle. Show that $\sin 2\alpha = \sin 2\beta$.

OBJECTIVES

1. Use reference angles.
2. Solve trigonometric equations.
3. Solve applied problems.

6.4 Trigonometric Equations

In this section we solve *conditional* trigonometric equations—that is, those equations that are true for some values but false for others. Examples of such equations are

$$\sin t = 1, \quad 2\cos\theta - 1 = 0, \quad \sin 2t = 2\cos t, \quad \text{and} \quad \tan^2\theta + \sec\theta = 1$$

Suppose we want to solve the equation $\sin t = \frac{1}{2}$, where $0 \le t < 2\pi$. We can interpret this solution graphically from two perspectives:

1. The values of t that satisfy the equation are the same as the t coordinates of the points of intersection of the graphs of $y_1 = \sin t$ and $y_2 = \frac{1}{2}$ (Figure 1a).
2. The equation $\sin t = \frac{1}{2}$ is equivalent to $\sin t - \frac{1}{2} = 0$. So the solutions are the same as the t intercepts of the graph of $y = \sin t - \frac{1}{2}$ (Figure 1b).

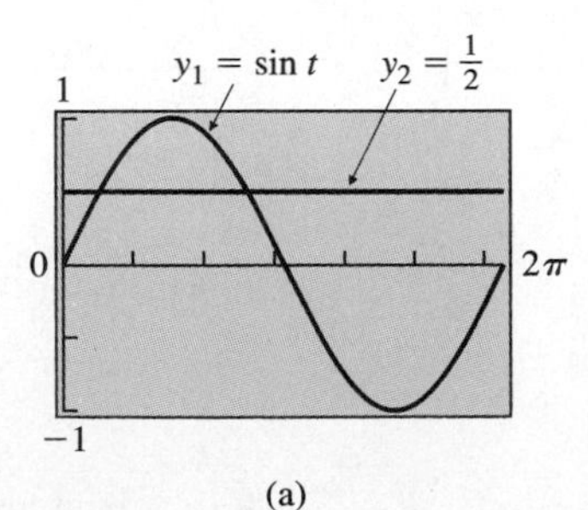

(a)

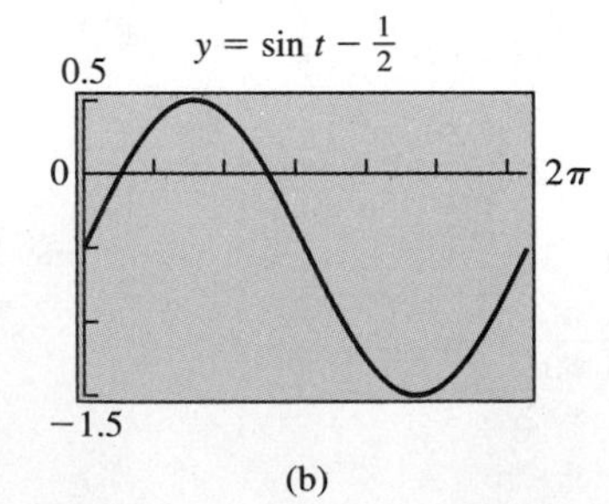

(b)

FIGURE 1

From either view, we see that the equation has exactly two solutions in the interval $[0, 2\pi)$. By using the ZOOM and TRACE features of a graphing utility, we find that the approximate values of the solutions are 0.5236 and 2.6180. The question is, how do we find these two solutions analytically? The answer involves the concept of *reference angles.*

Using Reference Angles

Reference angles are defined as follows:

DEFINITION

REFERENCE ANGLE

If θ is an angle in standard position whose terminal side does not lie on either coordinate axis, then the **reference angle** θ_R for θ is defined to be the *positive acute angle* formed by the terminal side of θ and the x axis.

For instance, the reference angle for $\theta = 115°$ is $\theta_R = 180° - 115° = 65°$ (Figure 2a); the reference angle for $\theta = 3.78$ is given by $\theta_R = 3.78 - \pi$ or approximately 0.64 (Figure 2b).

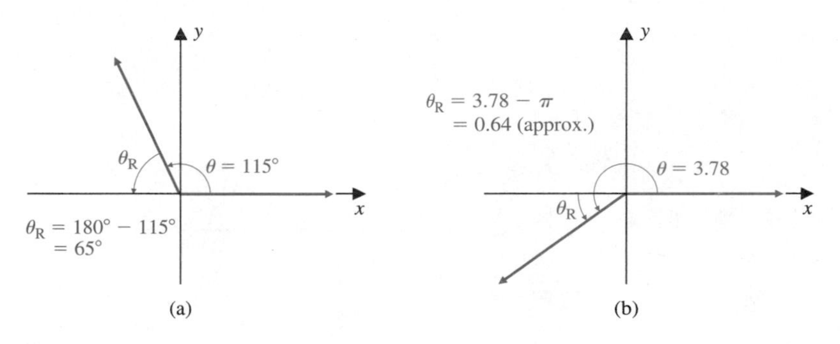

FIGURE 2

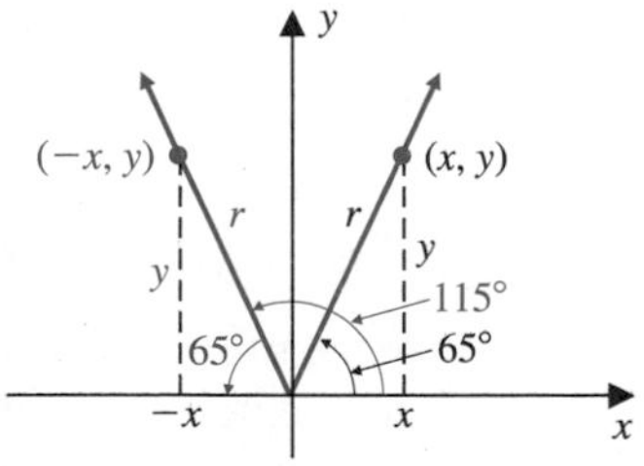

It is possible to express trigonometric function values of an angle in terms of values of its reference angle. For instance, suppose $\theta = 115°$. Its reference angle is $\theta_R = 65°$. As shown in Figure 3, we see that

$$\sin 115° = \frac{y}{r} = \sin 65° \quad \text{and} \quad \cos 115° = \frac{-x}{r} = -\frac{x}{r} = -\cos 65°$$

FIGURE 3

This important observation leads us to the following general result (Problem 63):

If $T(\theta)$ represents any of the six trigonometric functions of θ, and if θ_R is the reference angle of θ, then either

$$T(\theta) = T(\theta_R) \qquad \text{or} \qquad T(\theta) = -T(\theta_R)$$

depending on the function and the quadrant in which the terminal side of θ lies, where θ is in standard position.

EXAMPLE 1 **Relating trigonometric function evaluations to reference angles**

For each expression, draw the angle in standard position, and then determine and display the reference angle. Rewrite the given expression in terms of the reference angle. Use a calculator to confirm the result (round off to four decimal places).

(a) $\cos 163°$ (b) $\tan 5.1416$ (use $\pi = 3.1416$)

Solution

(a) From Figure 4a, we see that the reference angle for $\theta = 163°$ is $\theta_R = 17°$. Since the value of the cosine is negative in quadrant II, it follows that $\cos 163° = -\cos 17°$. By using a calculator, we find that both $\cos 163°$ and $-\cos 17°$ are approximately equal to -0.9563.

(b) Figure 4b shows that the reference angle for $\theta = 5.1416$ is approximately $\theta_R = 1.1416$. Since the tangent is negative in quadrant IV, it follows that $\tan 5.1416 = -\tan 1.1416$. Both $\tan 5.1416$ and $-\tan 1.1416$ are approximately equal to -2.1850.

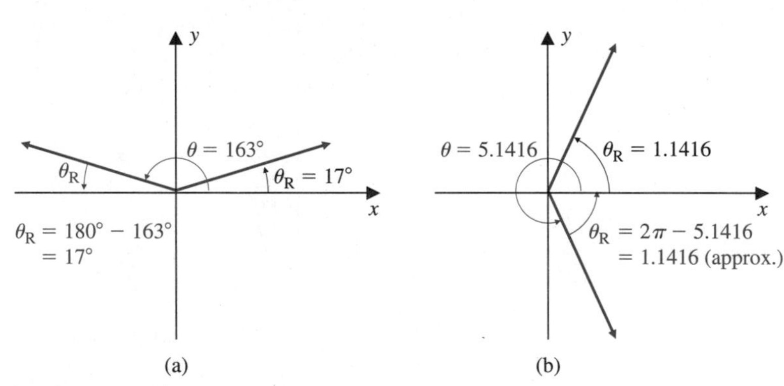

FIGURE 4

Solving Trigonometric Equations

Analytic techniques for solving trigonometric equations involve the use of properties of the trigonometric functions such as periodicity, reference angles, signs of the values, and identities.

EXAMPLE 2 **Solving trigonometric equations analytically**

This is the same equation that was solved graphically on page 390.

(a) Solve: $\sin t = \frac{1}{2}$, for $0 \le t < 2\pi$

(b) Use the result from part (a) to solve: $\sin\left(t - \frac{\pi}{3}\right) = \frac{1}{2}$, for $0 \le t < 2\pi$

Solution

(a) To find the solutions of $\sin t = \frac{1}{2}$, $0 \le t < 2\pi$, analytically, we observe that the values of $\sin t$ are positive for angles in quadrants I and II. Thus, there are two solutions that satisfy the given conditions, one in quadrant I and one in quadrant II. Both solutions have the same reference angle t_R, which is found by calculating

$$t_R = \sin^{-1}\frac{1}{2} = \frac{\pi}{6}$$

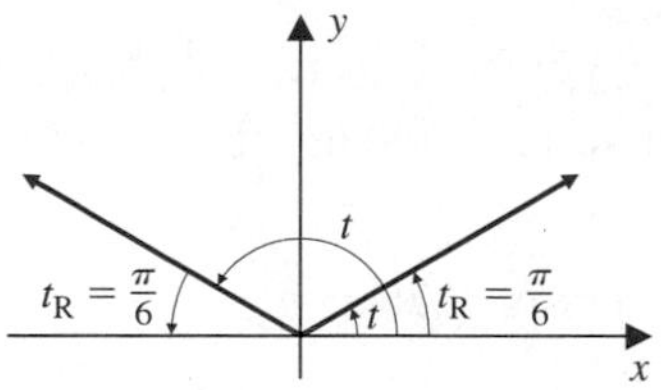

FIGURE 5

(Figure 5). So the solution in quadrant I is $\pi/6$, and the solution in quadrant II is given by

$$\pi - \frac{\pi}{6} = \frac{5\pi}{6}$$

Since $\pi/6$ is approximately 0.5236 and $(5\pi)/6$ is approximately 2.6180, the solutions found analytically here confirm the solution of this equation found graphically. If no restriction is put on t, then, because of periodicity, the equation $\sin t = \frac{1}{2}$ has infinitely many solutions, $(\pi/6) + 2\pi k$ and $[(5\pi)/6] + 2\pi k$, where k is any integer.

(b) Using the results from part (a), the solutions of the equation

$$\sin\left(t - \frac{\pi}{3}\right) = \frac{1}{2}$$

satisfy

$$t - \frac{\pi}{3} = \frac{\pi}{6} \quad \text{or} \quad t - \frac{\pi}{3} = \frac{5\pi}{6}$$

that is,

$$t = \frac{\pi}{3} + \frac{\pi}{6} = \frac{\pi}{2} \quad \text{or} \quad t = \frac{\pi}{3} + \frac{5\pi}{6} = \frac{7\pi}{6}$$

so the solutions are $\pi/2$ and $(7\pi)/6$. ■

EXAMPLE 3 Solving a trigonometric equation

Solve the equation $\tan t = -1.2137$, for $0 \le t < 2\pi$:

(a) Graphically (b) Analytically

Round off the answers to four decimal places.

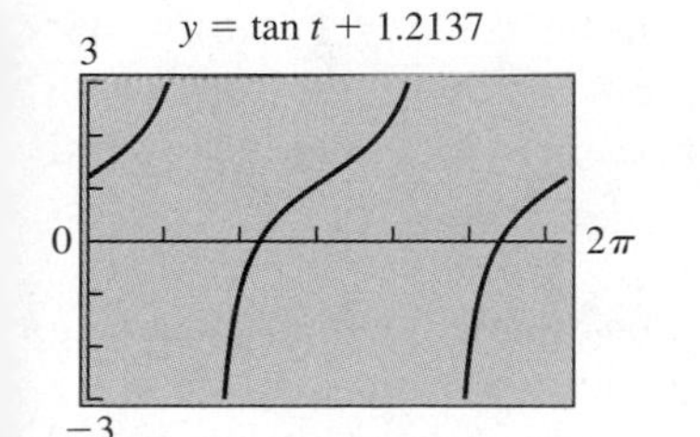

FIGURE 6

Solution

(a) Solving $\tan t = -1.2137$ is equivalent to solving $\tan t + 1.2137 = 0$. The graph of $y = \tan t + 1.2137$ in Figure 6 shows that there are two t intercepts. These intercepts correspond to the solutions of the given equation in the given interval. By using the ZOOM and TRACE features repeatedly, we find the solutions to be approximately 2.2600 and 5.4016.

(b) We solve the equation analytically as follows: The reference angle t_R for the given equation satisfies the condition

$$\tan t = -1.2137 = -\tan t_R$$

that is,

$$\tan t_R = 1.2137$$

A calculator in radian mode gives $t_R = \tan^{-1} 1.2137 = 0.8816$, rounded to four decimal places. Since the tangent is negative in quadrants II and IV, it follows that the solution in quadrant II is

$$t = \pi - t_R = \pi - 0.8816 \quad \text{or approx. } 2.2600$$

and the solution in quadrant IV is given by

$$t = 2\pi - t_R = 2\pi - 0.8816 \quad \text{or approx. } 5.4016$$

In the Cartesian coordinate system, we can graph trigonometric functions with real number domains (corresponding to radian measure). However, graphing utilities also enable us to display the horizontal axis in terms of degrees rather than just real numbers (radians). This feature is used in the next example.

EXAMPLE 4 Solving a trigonometric equation by factoring

Solve the equation $\sin\theta\tan\theta - \sin\theta = 0$, for $0° \le \theta < 360°$:

(a) Graphically (b) Analytically

Solution

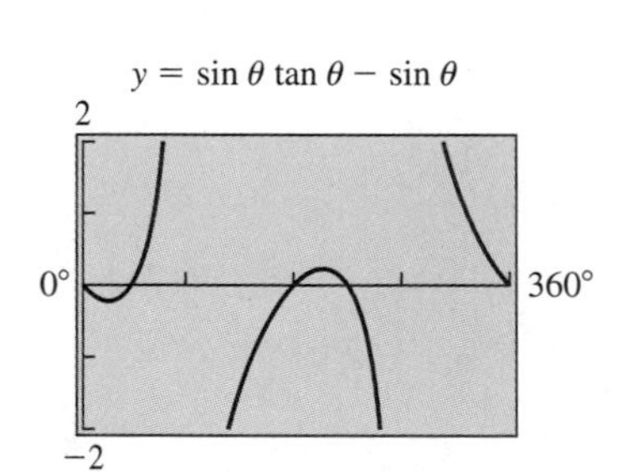

FIGURE 7

(a) First, we use a graphing utility (set in degree mode) to graph the function

$$f(\theta) = \sin\theta\tan\theta - \sin\theta \qquad \text{for } 0° \le \theta < 360°$$

The resulting graph in Figure 7 shows that the function has four intercepts on the horizontal axis, and so the equation $\sin\theta\tan\theta - \sin\theta = 0$ has four solutions. By using the ZOOM and TRACE features, we find the solutions to be 0°, 45°, 180°, and 225°.

(b) First, we factor the given equation $\sin\theta\tan\theta - \sin\theta = 0$ to get $\sin\theta(\tan\theta - 1) = 0$. Setting each factor equal to 0, we have

$$\sin\theta = 0 \qquad \text{or} \qquad \tan\theta - 1 = 0$$

For $\sin\theta = 0$, we get the solutions $\theta = 0°$ or $\theta = 180°$. For $\tan\theta - 1 = 0$, we solve the equivalent equation $\tan\theta = 1$ as follows: Since the value of the tangent is a positive number, it follows that the solutions of $\tan\theta = 1$ are in quadrants I and III. Also, the reference angle for $\tan\theta = 1$ is $\theta_R = 45°$ (Figure 8). So the solution in quadrant I is $\theta = 45°$, and the solution in quadrant III is $\theta = 180° + 45° = 225°$. Therefore, the solutions of the original equation are 0°, 45°, 180°, and 225°.

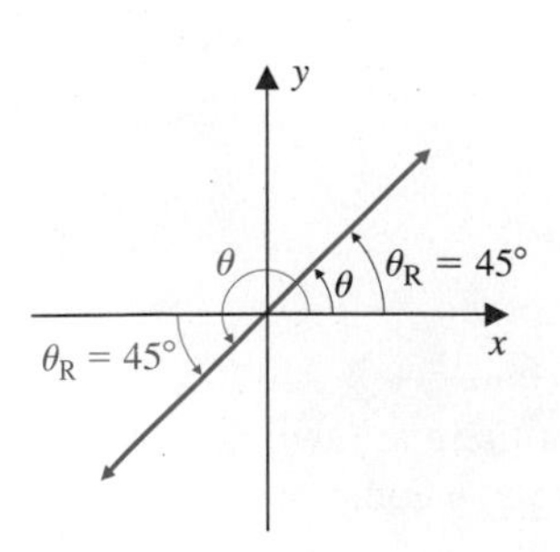

FIGURE 8

Note that we do not divide both sides of an equation by an expression containing a variable. For instance, in Example 4, if we had divided both sides by $\sin\theta$, we would have obtained the equation $\tan\theta - 1 = 0$ and two of the solutions, 0° and 180°, would have been lost.

Trigonometric equations involving multiple angles require special treatment, as illustrated in the next example.

EXAMPLE 5 Solving a trigonometric equation involving a multiple angle

Solve the equation $\cos 3\theta = \sqrt{3}/2$, for $0° \leq \theta < 360°$:

(a) Graphically (b) Analytically

Round off answers to the nearest degree.

Solution

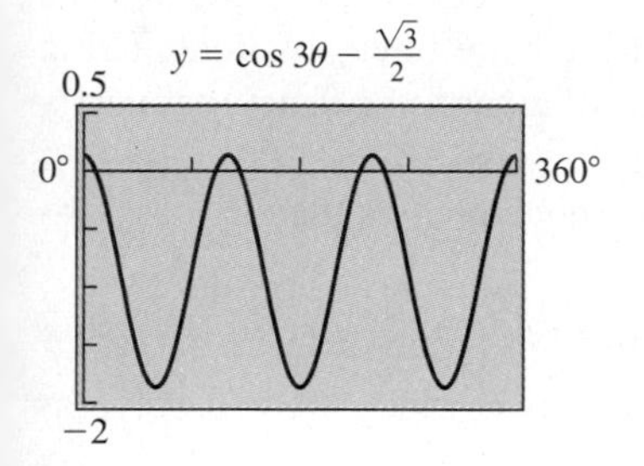

FIGURE 9

(a) The viewing window of the graph of the related function

$$y = \cos 3\theta - \frac{\sqrt{3}}{2}$$

(Figure 9) indicates that there are six solutions for the equation $\cos 3\theta = \sqrt{3}/2$, $0° \leq \theta < 360°$. By repeated use of the ZOOM and TRACE features, we find that the solutions are approximately 10°, 110°, 130°, 230°, 250°, and 350°.

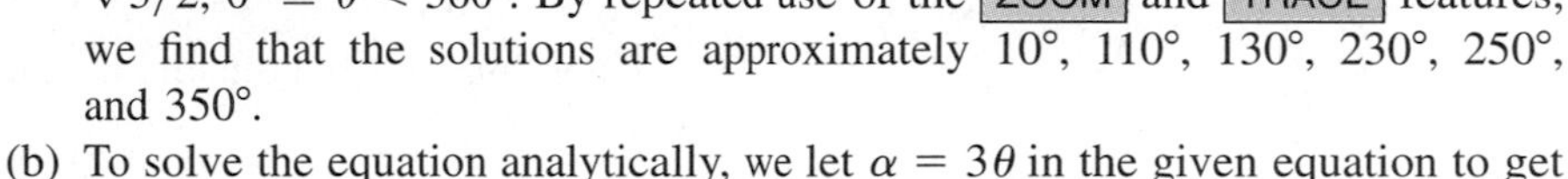

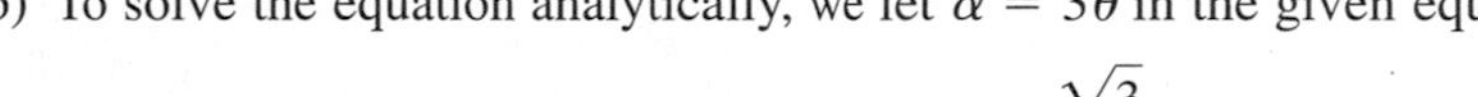

(b) To solve the equation analytically, we let $\alpha = 3\theta$ in the given equation to get

$$\cos 3\theta = \cos \alpha = \frac{\sqrt{3}}{2}$$

Since $0° \leq \theta < 360°$, then $0° \leq 3\theta < 1080°$; that is,

$$\cos \alpha = \frac{\sqrt{3}}{2} \qquad \text{for } 0° \leq \alpha < 1080°$$

The reference angle α_R for the latter equation is given by

$$\alpha_R = \cos^{-1} \frac{\sqrt{3}}{2} = 30°$$

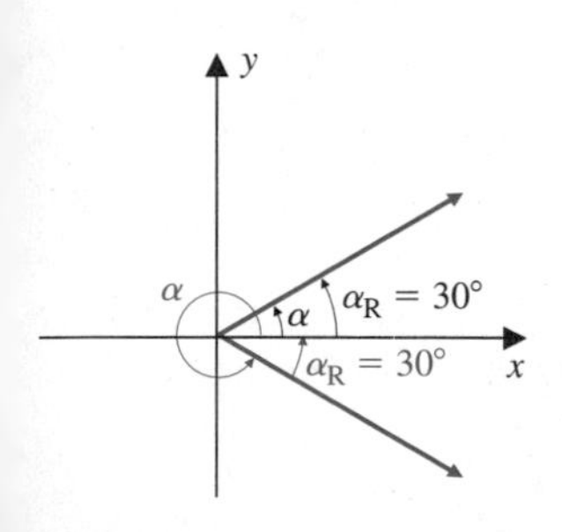

FIGURE 10

Since the cosine is positive in quadrants I and IV, it follows that $\alpha = 30°$ and $\alpha = 330°$ are two solutions (Figure 10). Next, we use periodicity by adding 360° to these two angle measures to get two more solutions:

$$\alpha = 30° + 360° = 390° \qquad \text{and} \qquad \alpha = 330° + 360° = 690°$$

both of which satisfy the condition $0° \leq \alpha < 1080°$. Finally, adding 360° again, we obtain two additional solutions:

$$\alpha = 390° + 360° = 750° \qquad \text{and} \qquad \alpha = 690° + 360° = 1050°$$

Thus, for $0° \leq \alpha < 1080°$, we have

$$\alpha = 3\theta = 30°, 330°, 390°, 690°, 750°, \text{ or } 1050°$$

We stop at 750° and 1050° because adding 360° again would result in angle measures exceeding the 1080° restriction.

The values of θ, the original unknown, are found by dividing each value of $\alpha = 3\theta$ by 3 to obtain the following solutions of the original equation: 10°, 110°, 130°, 230°, 250°, or 350°.

The next example employs an identity to solve a trigonometric equation.

EXAMPLE 6 **Solving an equation by using an identity**

Solve the equation $\cos 2t = \cos t$ for t in the interval $[0, 2\pi)$:

(a) Graphically (b) Analytically

Round off answers to four decimal places.

Solution

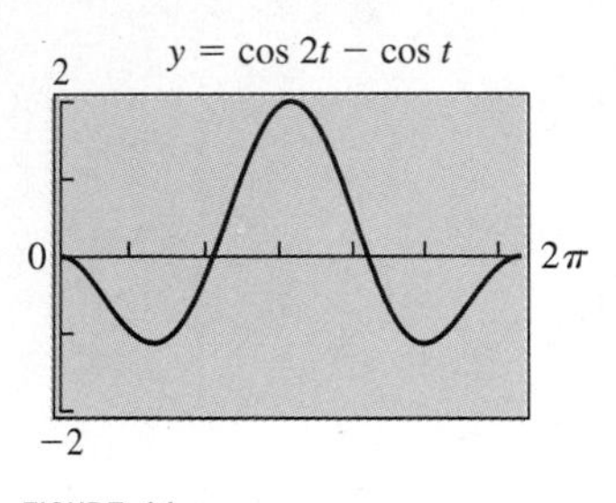

FIGURE 11

(a) We begin by considering the equivalent form $\cos 2t - \cos t = 0$. The graph of $y = \cos 2t - \cos t$ for $0 \le t < 2\pi$ (Figure 11) intersects the t axis at three points in the interval $[0, 2\pi)$, indicating that there are three solutions. Using the ZOOM and TRACE features, we find the solutions to be 0, approximately 2.0944, and approximately 4.1888.

(b) To find these solutions analytically, we use the double-angle identity $\cos 2t = 2\cos^2 t - 1$ to rewrite the equation

$$\cos 2t = \cos t \quad \text{as} \quad 2\cos^2 t - 1 = \cos t$$

The latter equation is a quadratic equation in $\cos t$, which we can solve as follows:

$$2\cos^2 t - 1 = \cos t$$

$$2\cos^2 t - \cos t - 1 = 0 \qquad \text{Subtract } \cos t \text{ from each side}$$

$$(2\cos t + 1)(\cos t - 1) = 0 \qquad \text{Factor}$$

$$\begin{array}{ll|ll} 2\cos t + 1 = 0 & \text{or} & \cos t - 1 = 0 & \\ 2\cos t = -1 & & & \\ \cos t = -\frac{1}{2} & & \cos t = 1 & \\ \text{So} \quad t = \frac{2\pi}{3} \text{ or } \frac{4\pi}{3} & & \text{So} \quad t = 0 & \end{array}$$

Set each factor equal to zero, and solve the resulting equations with the restriction $0 \le t < 2\pi$

Therefore, the solutions are 0, $(2\pi)/3$ (approx. 2.0944) and $(4\pi)/3$ (approx. 4.1888).

The following example is an equation that cannot be solved by the analytic methods studied so far. In this case, we use a graphing utility to approximate the solutions.

EXAMPLE 7 **Solving a trigonometric equation graphically**

(a) Solve the equation $\cos x = x$ graphically. Round off the result to four decimal places.

(b) Use the result from part (a) to solve the inequality $\cos x < x$.

Solution

(a) The equation $\cos x = x$ is equivalent to $\cos x - x = 0$. The viewing window in Figure 12 (page 397) shows that the graph of $y = \cos x - x$ has one x intercept, so there is only one solution. By using the ZOOM and TRACE features, we find that the solution is approximately 0.7391.

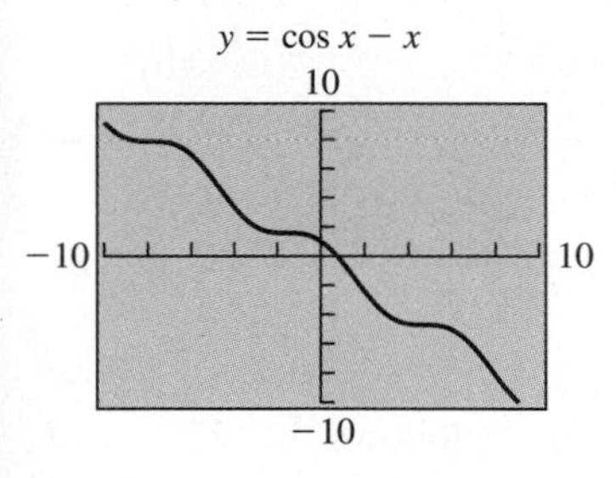

FIGURE 12

(b) The given inequality is equivalent to $\cos x - x < 0$. The solution of this inequality includes all values of x for which the graph of $y = \cos x - x$ is below the x axis. Since the x intercept is approximately 0.7391, it follows that the solution includes all numbers approximately in the interval $(0.7391, \infty)$, that is, $x > 0.7391$.

Solving Applied Problems

At times, mathematical models of real-world situations involve trigonometric equations.

EXAMPLE 8 **Designing a window**

A custom window is to be designed in the shape of a rectangle 2 feet wide, topped by an isosceles triangle whose equal sides make an angle θ with the width. The perimeter of the entire window is to be 12 feet.

(a) Show that the total area A (in square feet) of the window is given by

$$A = 10 - 2 \sec \theta + \tan \theta$$

(b) What are the restrictions on θ?

(c) Use a graphing utility to find the angle that yields a window with area 7.5 square feet. Also find the outside dimensions of this window. Round off the answers to two decimal places.

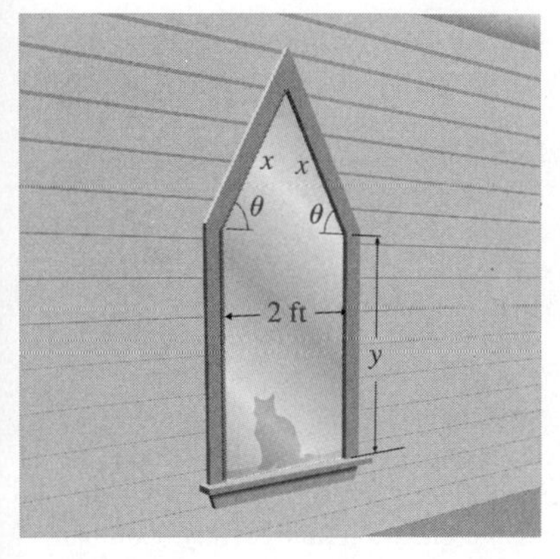

FIGURE 13

Solution

(a) Let y (in feet) be the height of the rectangular part of the window; and let x (in feet) be the length of each of the equal sides of the isosceles triangle (Figure 13). Since the perimeter of the entire window is 12 feet, we have

$$2y + 2x + 2 = 12$$

Figure 14 shows the isosceles triangle that forms the top of the window. If the height of the triangle is h, it follows from right triangle trigonometry that

$$\sec \theta = \frac{x}{1} \qquad \text{and} \qquad \tan \theta = \frac{h}{1}$$

that is,

$$x = \sec \theta \qquad \text{and} \qquad h = \tan \theta$$

Substituting for x into the equation

$$2y + 2x + 2 = 12$$

we get

$$2y + 2 \sec \theta + 2 = 12$$
$$2y + 2 \sec \theta = 10$$

Solving for y, we have

$$y = 5 - \sec \theta$$

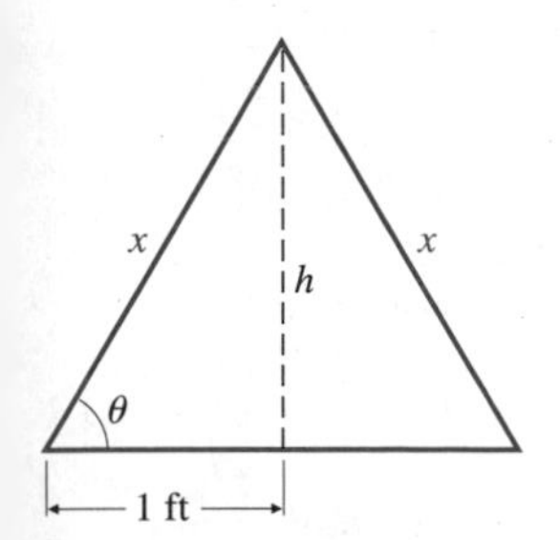

FIGURE 14

So the area of the window is given by

$$\begin{aligned} A &= (\text{Area of rectangle}) + (\text{Area of triangle}) \\ &= 2y \qquad\qquad\qquad + \tfrac{1}{2}(2)h \qquad\qquad \text{Area formulas} \\ &= 2(5 - \sec\theta) \qquad + \tan\theta \end{aligned}$$

Consequently,

$$A = 10 - 2\sec\theta + \tan\theta$$

(b) Since the sum of the angles of a triangle is 180°, it follows that $2\theta < 180°$, that is, $\theta < 90°$. So $0° < \theta < 90°$.

(c) To find the angle when the area $A = 7.5$, we need to solve the equation

$$7.5 = 10 - 2\sec\theta + \tan\theta$$

or, equivalently,

$$0 = 2.5 - 2\sec\theta + \tan\theta$$

Figure 15 shows a viewing window of the graph of

$$f(\theta) = 2.5 - 2\sec\theta + \tan\theta \qquad \text{for } 0° < \theta < 90°$$

(set in degree mode). By using the ZOOM and TRACE features, we find that θ is approximately 63.83°. The length of each of the equal sides of the isosceles triangle is approximately $x = \sec 63.83° = 2.27$ feet, and the height of the rectangle is approximately $y = 5 - \sec 63.83° = 2.73$ feet.

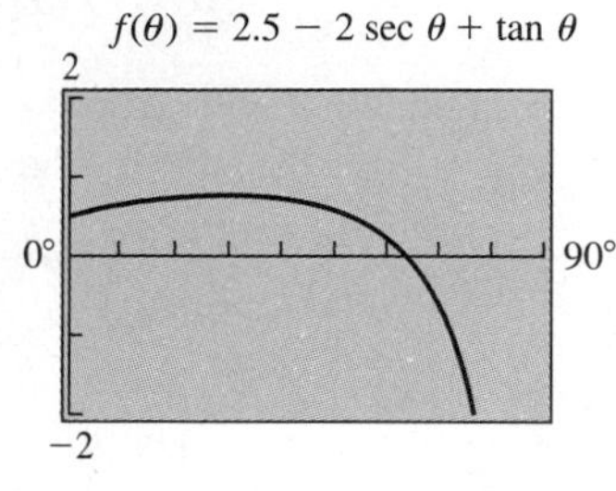

FIGURE 15

PROBLEM SET 6.4

Mastering the Concepts

In problems 1 and 2, verify that the given values are solutions of the indicated equation.

1. **(a)** $2\sin\theta - 1 = 0$; 30°, 150°
(b) $\tan\left(3t - \frac{\pi}{4}\right) = 1$; $-\frac{\pi}{6}, \frac{\pi}{6}$

2. **(a)** $2\sin 2t - \sqrt{3} = 0$; $\frac{7\pi}{6}, \frac{4\pi}{3}$
(b) $2\sin(\theta + 30°) = -1$; $-180°, 300°$

In problems 3–6, draw the angle in standard position. Then determine and display its reference angle. Rewrite the given expression in terms of the reference angle. Use a calculator to confirm the result (rounding off to four decimal places). Use $\pi = 3.1416$ (approx.).

3. **(a)** sin 190° **(b)** tan(−113°) **(c)** sec 372°

4. **(a)** cos 219° **(b)** cot 114° **(c)** cos(−295°)

5. **(a)** csc 3.1 **(b)** cos(−1.9) **(c)** cot 7.2

6. **(a)** $\sin\frac{11\pi}{8}$ **(b)** sec 5.9 **(c)** tan(−4)

GU In problems 7–14, assume that $0 \le t < 2\pi$ and $0° \le \theta < 360°$. Solve each equation graphically. Round off the answers to four decimal places. Then find the exact solutions by solving the same equation analytically. Compare the answers from both approaches.

7. **(a)** $\cos t = -\frac{1}{2}$ **(b)** $\sec t = -\sqrt{2}$

8. **(a)** $\sin t = \sqrt{2}/2$ **(b)** $3\csc\theta = -2\sqrt{3}$

9. **(a)** $\tan\theta = \sqrt{3}$ **(b)** $3\cot\theta = -\sqrt{3}$

10. **(a)** $\csc t = \sqrt{2}$ **(b)** $\cot t = -1$

11. **(a)** $2\sin t - \sqrt{3} = 0$ **(b)** $2\sin\theta + \sqrt{2} = 0$

12. **(a)** $\tan\theta - 1 = 0$ **(b)** $\cot t + \sqrt{3} = 0$

13. **(a)** $\sqrt{2}\cos t = -1$ **(b)** $\sqrt{2}\cos\left(t - \frac{\pi}{4}\right) = -1$

14. **(a)** $\tan\theta = -1$ **(b)** $\tan(\theta + 60°) = -1$

GU In problems 15–18, assume $0 \le t < 2\pi$. Solve each equation both graphically and analytically. Round off the answers to four decimal places and compare the answers from both approaches.

15. **(a)** $\sin t = 0.8134$ **(b)** $\cot t = -6.6173$
16. **(a)** $\cos t = -0.4176$ **(b)** $\tan t = 0.6696$
17. **(a)** $\sec t = -\frac{6}{5}$ **(b)** $\tan t = 10$
18. **(a)** $\csc t = \pi$ **(b)** $\cot t = -10$

GU In problems 19–36, assume that $0 \le t < 2\pi$ and $0° \le \theta < 360°$. Solve each equation both graphically and analytically. Round off the answers to four decimal places and compare the answers from both approaches.

19. $\sin t \cos t - \sin t = 0$
20. $2 \sin \theta \cos \theta - \sqrt{2} \cos \theta = 0$
21. $2 \sin \theta \cos \theta + \sqrt{3} \cos \theta = 0$
22. $\sin t \cos t - \cos t = 0$
23. $\tan^2 t - \sqrt{3} \tan t = 0$
24. $2 \cos^2 \theta - \cos \theta = 0$
25. $\sec^2 \theta + \sec \theta = 0$
26. $\sqrt{3} \csc^2 \theta + 2 \csc \theta = 0$
27. $\tan^2 t - 3 = 0$
28. $\cot^2 t - 1 = 0$
29. $4 \sin^2 \theta - 3 = 0$
30. $\csc^2 \theta - 4 = 0$
31. $3 - 2 \cos^2 t = 3 - \cos t$
32. $1 - 2 \sin^2 t = 1 - \sin t$
33. **(a)** $2 \cos 3\theta = -\sqrt{2}$
(b) $2 \cos(3\theta - 30°) = -\sqrt{2}$
34. **(a)** $2 \sin 2t = 1$ **(b)** $2 \sin\left(2t - \frac{\pi}{4}\right) = 1$
35. **(a)** $\sec \frac{t}{2} = \sqrt{2}$ **(b)** $\sec\left(\frac{t}{2} + \frac{\pi}{5}\right) = \sqrt{2}$
36. **(a)** $\tan \frac{\theta}{2} + \sqrt{3} = 0$
(b) $\tan\left(\frac{\theta}{2} - 40°\right) + \sqrt{3} = 0$

GU In problems 37–52, assume that $0 \le t < 2\pi$ and $0° \le \theta < 360°$. Solve each equation graphically and analytically by using identities. Round off the answers to four decimal places, when applicable, and compare the answers from both approaches.

37. $\sin^2 t = \cos^2 t$
38. $2 \cos^2 t - 1 = 2 \sin^2 t$
39. $\cot^2 \theta = \cot \theta$
40. $2 \tan \theta - \sec^2 \theta = 0$
41. $2 - \sin t = 2 \cos^2 t$
42. $\cos^2 t - \sin^2 t - \sin t = 1$
43. $\cos \theta - \sin \theta = 1$
44. $\tan \theta + \sec \theta = 1$
45. $\sin 2t = \sqrt{2} \cos t$
46. $2 - \cos^2 t = 4 \sin^2 \frac{t}{2}$
47. $\sin \frac{t}{2} + \cos t = 1$
48. $\cos 2\theta = 2 \sin^2 \theta$
49. $\cos 2\theta - 3 \sin \theta = 2$
50. $\cos 2\theta + \sin^2 \theta = 1$
51. $\cos 2t + 2 \cos^2 \frac{t}{2} = 2$
52. $\cos 2\theta - \cos \theta = 0$

GU In problems 53–58:
(a) Solve each equation graphically if $0 \le t < 2\pi$. Round off the answers to three decimal places. Use $\pi = 3.142$.
(b) Use the result from part (a) to solve the corresponding inequality.

53. $15 \sin^2 t - 8 \sin t + 1 = 0$;
$15 \sin^2 t - 8 \sin t + 1 > 0$
54. $\tan^2 t - 3 \tan t + 2 = 0$; $\tan^2 t - 3 \tan t + 2 < 0$
55. $t + \sin t = 0.32$; $t + \sin t \le 0.32$
56. $\sin t + \ln t = 2t$; $\sin t + \ln t > 2t$
57. $e^{-t} - \cos t = 0$; $e^{-t} - \cos t \ge 0$
58. $\tan t = t^2 + 1$; $\tan t \le t^2 + 1$

Applying the Concepts

In problems 59–62, round off each answer to two decimal places.

59. Physics: In physics, Snell's law deals with the change in direction of a ray of light as it passes through a medium, as shown in Figure 16, where the dashed line is perpendicular to the surface of the medium:

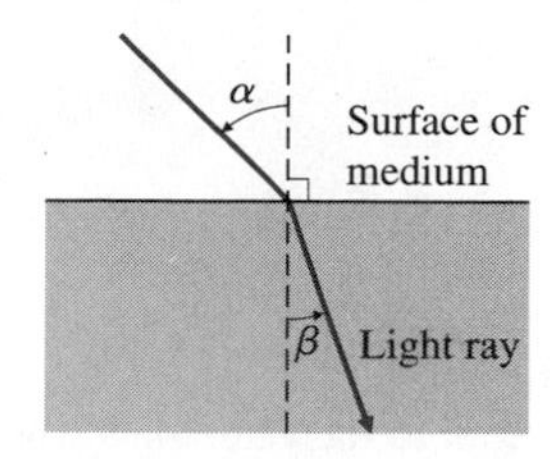

FIGURE 16

The angle α is called the **angle of incidence**, and the angle β is called the **angle of refraction**. **Snell's law** states that for a given medium,

$$\frac{\sin \alpha}{\sin \beta} = C$$

where C is a constant. The constant C is called the **index of refraction** of the medium.

(a) Find the angle of refraction of a light ray traveling through ice if the angle of incidence is 45°, and the index of refraction is 1.31.

(b) Find the angle of incidence of a light ray striking a rock-salt crystal if the angle of refraction of the ray is 35°. The index of refraction of rock-salt is 1.54.

60. Animal Population: Suppose that the number of rodents N in a certain population after t years is given by the mathematical model

$$N(t) = 3200 + 800 \sin \frac{3t}{2} \qquad \text{where } 0 \le t < 5$$

(a) How many rodents are initially in the population?

(b) How many years does it take for the population to drop to 3000 for the first time?

61. Length of a Ladder: A 5 foot fence stands 4 feet away from a high wall. A ladder on the ground leaning against the wall touches the top of the fence and makes an angle θ with the ground (Figure 17).

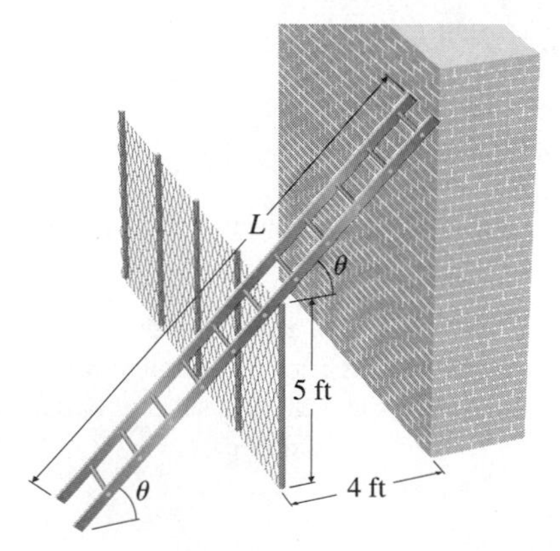

FIGURE 17

(a) Show that the length L of the ladder is given by

$$L = 4 \sec \theta + 5 \csc \theta$$

(b) What are the restrictions on θ?

GU **(c)** Use a graphing utility to find the value of θ when the length of the ladder is 13 feet.

62. Number of Hours of Daylight: Suppose that the number of hours h of daylight for a certain city on day D of the year is given by the model

$$h = 2.3 \sin\left[\frac{2\pi}{365}(D - 80)\right] + 12 \quad \text{where } 1 \le D < 365$$

Approximately on what days D of the year will there be 13 hours of daylight?

Developing and Extending the Concepts

63. **(a)** Let θ be an angle in quadrant II in standard position and let θ_R be its reference angle. Show that the value of any trigonometric function of θ is the same as the value for the function of θ_R, except possibly for a change of algebraic sign.

(b) Repeat part (a) for θ in quadrant III.

(c) Repeat part (a) for θ in quadrant IV.

64. GU Solve each equation graphically. Then explain why the solutions are different.

(a) $\frac{\sin x}{x} = 1, \; -\pi < x < \pi$

(b) $\sin x = x, \; -\pi < x < \pi$

65. Solve the equation $\sin \theta = -\sqrt{2}/2$ if:

(a) $0° \le \theta < 90°$ **(b)** $0° \le \theta < 180°$

(c) $0° \le \theta < 360°$ **(d)** θ has no restriction

66. Solve the equation $\cos t = -\sqrt{2}/2$ if:

(a) $0 \le t < \pi/2$ **(b)** $0 \le t < \pi$

(c) $0 \le t < 2\pi$ **(d)** t has no restriction

GU In problems 67–70, assume $0 \le t < 2\pi$. Solve each equation graphically. Round off the answers to two decimal places. Then use identities to find the exact values of the solutions analytically. Compare the results from both approaches.

67. $\cos\left(t + \frac{\pi}{4}\right) + \cos\left(t - \frac{\pi}{4}\right) = 1$

68. $\sin\left(t + \frac{\pi}{3}\right) + \sin\left(t - \frac{\pi}{3}\right) = \frac{\sqrt{3}}{2}$

69. $\sin 2t \cos t + \cos 2t \sin t = \frac{1}{2}$

70. $2 \sin\left(t + \frac{\pi}{2}\right) = \sec^2 t - \tan^2 t$

GU In problems 71 and 72, assume $0 \le t < 2\pi$. For what values of k does the equation have:

(a) No solution? **(b)** One solution?

(c) Two solutions? **(d)** More than two solutions?

71. $\cos t = k$

72. $\sec t = k$

OBJECTIVES

1. Use product-to-sum identities.
2. Use sum-to-product identities.
3. Use inverse trigonometric identities.

6.5 Other Identities

In calculus it is often necessary to rewrite a product of sine and cosine functions as a sum or difference of such functions, or vice versa. This can be done by using the identities in this section.

Using Product-to-Sum Identities

The addition and subtraction identities can be used to derive identities that express the products of sines and cosines in terms of sums or differences.

PRODUCT-TO-SUM IDENTITIES

(i) $\sin u \cos v = \frac{1}{2}[\sin(u + v) + \sin(u - v)]$
(ii) $\cos u \sin v = \frac{1}{2}[\sin(u + v) - \sin(u - v)]$
(iii) $\cos u \cos v = \frac{1}{2}[\cos(u - v) + \cos(u + v)]$
(iv) $\sin u \sin v = \frac{1}{2}[\cos(u - v) - \cos(u + v)]$

Proof

(i) We use the addition and subtraction identities to prove the first identity:

$$\begin{aligned}\tfrac{1}{2}[\sin(u + v) + \sin(u - v)] &= \tfrac{1}{2}[(\sin u \cos v + \cos u \sin v) \\ &\quad + (\sin u \cos v - \cos u \sin v)] \\ &= \tfrac{1}{2}[2 \sin u \cos v] \\ &= \sin u \cos v\end{aligned}$$

The proofs of parts (ii)–(iv) are left as exercises (Problems 39–41). 

EXAMPLE 1 Using a product-to-sum identity

(a) Find the exact value of

$$\cos \frac{\pi}{12} \sin \frac{7\pi}{12}$$

by using a product-to-sum identity.

(b) Compare the result in part (a) to the calculator value of the product, rounded to four decimal places.

Solution

(a) By using product-to-sum identity (ii), we get

$$\begin{aligned}\cos\frac{\pi}{12}\sin\frac{7\pi}{12} &= \frac{1}{2}\left[\sin\left(\frac{\pi}{12}+\frac{7\pi}{12}\right)-\sin\left(\frac{\pi}{12}-\frac{7\pi}{12}\right)\right]\\ &= \frac{1}{2}\left[\sin\frac{8\pi}{12}-\sin\left(-\frac{6\pi}{12}\right)\right]\\ &= \frac{1}{2}\left[\sin\frac{2\pi}{3}-\sin\left(-\frac{\pi}{2}\right)\right]\\ &= \frac{1}{2}\left[\sin\frac{2\pi}{3}+\sin\frac{\pi}{2}\right] && \text{Sine is an odd function}\\ &= \frac{1}{2}\left[\frac{\sqrt{3}}{2}+1\right]\\ &= \frac{\sqrt{3}}{4}+\frac{1}{2}\end{aligned}$$

(b) Rounding off

$$\frac{\sqrt{3}}{4}+\frac{1}{2}$$

to four decimal places gives 0.9330. Using a calculator to evaluate

$$\cos\frac{\pi}{12}\sin\frac{7\pi}{12}$$

and rounding off to four decimal places, we again get 0.9330.

EXAMPLE 2 Rewriting a product as a sum

Rewrite $\cos 3t \cos 4t$ as a sum. Demonstrate the result graphically.

Solution Using identity (iii), with $u = 3t$ and $v = 4t$, we get

$$\begin{aligned}\cos 3t\cos 4t &= \tfrac{1}{2}[\cos(3t-4t)+\cos(3t+4t)]\\ &= \tfrac{1}{2}[\cos(-t)+\cos 7t]\\ &= \tfrac{1}{2}\cos t+\tfrac{1}{2}\cos 7t && \text{Cosine is an even function}\end{aligned}$$

Figure 1a shows the graph of $y_1 = \cos 3t \cos 4t$, and Figure 1b shows the graph of $y_2 = \frac{1}{2}\cos t + \frac{1}{2}\cos 7t$. The two graphs are the same for the selected viewing window, thus demonstrating the validity of the result.

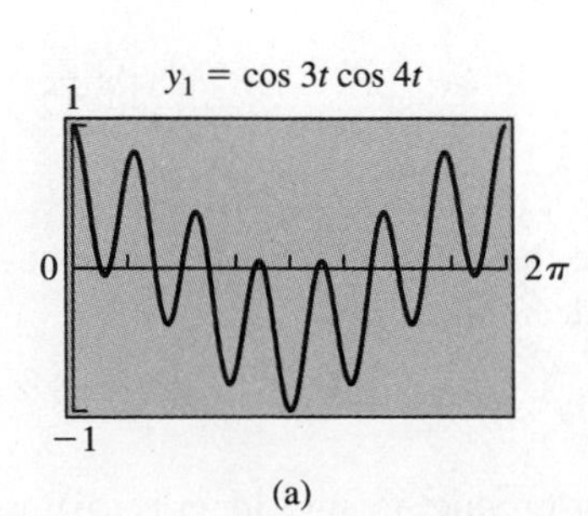

(a)

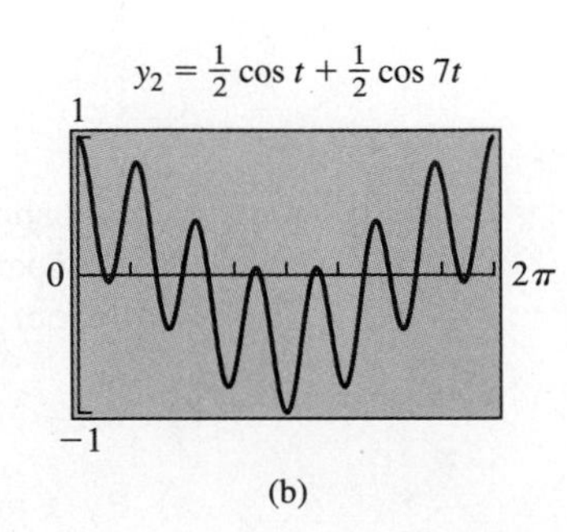

(b)

FIGURE 1

Using Sum-to-Product Identities

The sum-to-product identities enable us to rewrite the sums of sines and cosines as products.

SUM-TO-PRODUCT IDENTITIES

(i) $\sin \omega + \sin t = 2 \sin\left(\dfrac{\omega + t}{2}\right) \cos\left(\dfrac{\omega - t}{2}\right)$

(ii) $\sin \omega - \sin t = 2 \cos\left(\dfrac{\omega + t}{2}\right) \sin\left(\dfrac{\omega - t}{2}\right)$

(iii) $\cos \omega + \cos t = 2 \cos\left(\dfrac{\omega + t}{2}\right) \cos\left(\dfrac{\omega - t}{2}\right)$

(iv) $\cos \omega - \cos t = -2 \sin\left(\dfrac{\omega + t}{2}\right) \sin\left(\dfrac{\omega - t}{2}\right)$

Proof

(i) To prove the first identity, we let $\omega = u + v$ and $t = u - v$. Then

$$\omega + t = (u + v) + (u - v) = 2u$$

So
$$u = \frac{\omega + t}{2}$$

Also,
$$\omega - t = (u + v) - (u - v) = 2v$$

So
$$v = \frac{\omega - t}{2}$$

By substituting $(\omega + t)/2$ for u and $(\omega - t)/2$ for v in product-to-sum identity (i), we have

$$\sin\left(\frac{\omega + t}{2}\right) \cos\left(\frac{\omega - t}{2}\right) = \frac{1}{2}(\sin \omega + \sin t)$$

or
$$\sin \omega + \sin t = 2 \sin\left(\frac{\omega + t}{2}\right) \cos\left(\frac{\omega - t}{2}\right)$$

The proofs of parts (ii)–(iv) are left as exercises (Problems 42–44).

EXAMPLE 3 Using a sum-to-product identity

(a) Use a sum-to-product identity to find the exact value of $\sin 75° + \sin 15°$.
(b) Compare the result from part (a) to the calculator value, rounded to four decimal places.

Solution

(a) We use sum-to-product identity (i) as follows:

$$\begin{aligned}\sin 75^\circ + \sin 15^\circ &= 2\sin\left(\frac{75^\circ + 15^\circ}{2}\right)\cos\left(\frac{75^\circ - 15^\circ}{2}\right)\\ &= 2\sin 45^\circ \cos 30^\circ\\ &= 2\left(\frac{\sqrt{2}}{2}\right)\left(\frac{\sqrt{3}}{2}\right) = \frac{\sqrt{6}}{2}\end{aligned}$$

(b) Rounding $\sqrt{6}/2$ to four decimal places gives 1.2247. The calculator value of $\sin 75^\circ + \sin 15^\circ$, when rounded to four decimal places, also gives 1.2247.

EXAMPLE 4 **Rewriting a difference as a product**

GU

Rewrite $\cos 8\theta - \cos 3\theta$ as a product. Demonstrate the result graphically.

Solution Using sum-to-product identity (iv) with $\omega = 8\theta$ and $t = 3\theta$, we get

$$\begin{aligned}\cos 8\theta - \cos 3\theta &= -2\sin\left(\frac{8\theta + 3\theta}{2}\right)\sin\left(\frac{8\theta - 3\theta}{2}\right)\\ &= -2\sin\frac{11\theta}{2}\sin\frac{5\theta}{2}\end{aligned}$$

The graphs of

$$y_1 = \cos 8\theta - \cos 3\theta \text{ (Figure 2a)} \quad \text{and} \quad y_2 = -2\sin\frac{11\theta}{2}\sin\frac{5\theta}{2} \text{ (Figure 2b)}$$

are identical for the selected viewing window, thus demonstrating the validity of the result.

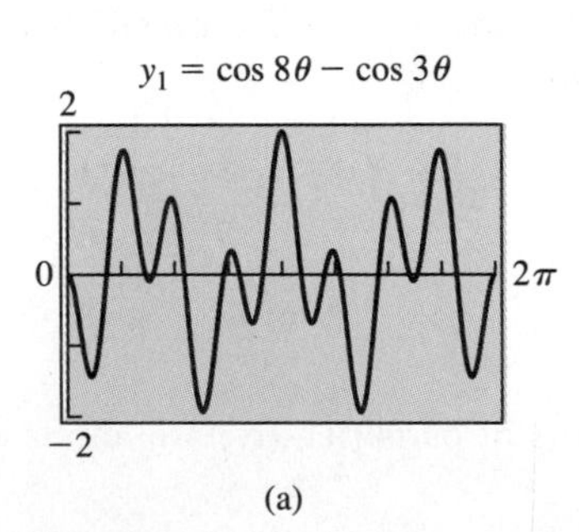

(a)

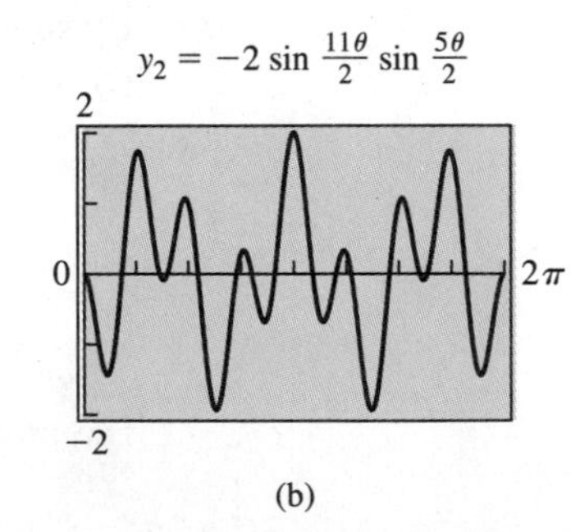

(b)

FIGURE 2

Using Inverse Trigonometric Identities

By applying the restrictions on the inverse functions that were presented in Section 5.7, we get the following identities:

INVERSE SINE IDENTITIES

$$\sin(\sin^{-1} x) = x \qquad \text{if } -1 \le x \le 1$$

$$\sin^{-1}(\sin x) = x \qquad \text{if } -\frac{\pi}{2} \le x \le \frac{\pi}{2}$$

EXAMPLE 5 Using the inverse sine identities

Find the exact value of each expression by using the inverse sine identities.

(a) $\sin^{-1}\left(\sin \frac{\pi}{6}\right)$ (b) $\sin(\sin^{-1} 3)$

Solution

(a) Since $-\pi/2 \le \pi/6 \le \pi/2$, then

$$\sin^{-1}\left(\sin \frac{\pi}{6}\right) = \frac{\pi}{6}$$

As a check, we note that

$$\sin^{-1}\left(\sin \frac{\pi}{6}\right) = \sin^{-1} \frac{1}{2} = \frac{\pi}{6}$$

(b) The number 3 is not in the domain of the inverse sine function. Therefore, the expression $\sin(\sin^{-1} 3)$ is not defined.

INVERSE COSINE IDENTITIES

$$\cos(\cos^{-1} x) = x \qquad \text{if } -1 \le x \le 1$$

$$\cos^{-1}(\cos x) = x \qquad \text{if } 0 \le x \le \pi$$

EXAMPLE 6 Using the inverse cosine identities

Find the exact value of each expression by using the inverse cosine identities.

(a) $\cos\left(\cos^{-1} \frac{1}{2}\right)$ (b) $\cos^{-1}[\cos(-\pi/6)]$

Solution

(a) Since $-1 \le \frac{1}{2} \le 1$, then

$$\cos\left(\cos^{-1} \frac{1}{2}\right) = \frac{1}{2}$$

(b) Since $-\pi/6$ does not satisfy $0 \le x \le \pi$, $\cos^{-1}[\cos(-\pi/6)] \ne -\pi/6$. To evaluate the expression we proceed as follows:

$$\cos^{-1}\left[\cos\left(-\frac{\pi}{6}\right)\right] = \cos^{-1} \frac{\sqrt{3}}{2} = \frac{\pi}{6}$$

INVERSE TANGENT IDENTITIES

$\tan(\tan^{-1} x) = x$	for every real number x
$\tan^{-1}(\tan x) = x$	if $-\frac{\pi}{2} < x < \frac{\pi}{2}$

EXAMPLE 7 **Using the inverse tangent identities**

Find the exact value of each expression by using the inverse tangent identities.

(a) $\tan\left[\tan^{-1}\left(-\frac{2}{3}\right)\right]$ (b) $\tan^{-1}\left(\tan\frac{3\pi}{4}\right)$

Solution

(a) From the first identity above it follows that

$$\tan\left[\tan^{-1}\left(-\frac{2}{3}\right)\right] = -\frac{2}{3}$$

(b) We first observe that

$$\tan^{-1}\left(\tan\frac{3\pi}{4}\right) \neq \frac{3\pi}{4}$$

because $(3\pi)/4$ does not satisfy the restriction $-\pi/2 < x < \pi/2$. Thus, we proceed as follows:

$$\tan^{-1}\left(\tan\frac{3\pi}{4}\right) = \tan^{-1}(-1) = -\frac{\pi}{4}$$

PROBLEM SET 6.5

Mastering the Concepts

In problems 1–8:

(a) Use a product-to-sum identity to find the exact value.

(b) Compare the result from part (a) to the calculator value of the product, rounded to four decimal places.

1. $\sin\frac{5\pi}{12}\cos\frac{7\pi}{12}$ **2.** $\sin\frac{\pi}{12}\cos\frac{5\pi}{12}$

3. $\cos 195° \sin 75°$ **4.** $\cos 112.5° \sin 67.5°$

5. $\cos\frac{17\pi}{12}\cos\frac{7\pi}{12}$ **6.** $\cos\frac{11\pi}{12}\cos\frac{5\pi}{12}$

7. $\sin 67.5° \sin 22.5°$ **8.** $\sin 165° \sin 75°$

GU In problems 9–16, rewrite each product as a sum or difference using the product-to-sum identities. Demonstrate the result graphically.

9. $\sin 4w \sin w$ **10.** $\cos 3\theta \sin 9\theta$

11. $\cos 8v \cos 4v$ **12.** $\cos 7t \cos 3t$

13. $\cos 3t \sin 4t$ **14.** $\cos 5t \cos 2t$

15. $\sin 4t \sin 5t$ **16.** $\cos 3t \sin t$

In problems 17–24:

(a) Use a sum-to-product identity to find the exact value.

(b) Compare the result from part (a) to the calculator value of the expression, rounded to four decimal places.

17. $\sin\frac{7\pi}{18} + \sin\frac{\pi}{18}$ **18.** $\sin\frac{13\pi}{12} + \sin\frac{7\pi}{12}$

19. $\sin 75° - \sin 165°$ **20.** $\sin 105° - \sin 75°$

21. $\cos\frac{7\pi}{12} + \cos\frac{\pi}{12}$ **22.** $\cos\frac{19\pi}{12} + \cos\frac{13\pi}{12}$

23. $\cos 75° - \cos(-15°)$ **24.** $\cos(-75°) - \cos 16.5°$

GU In problems 25–32, rewrite each sum as a product using the sum-to-product identities. Demonstrate graphically.

25. $\sin 3\theta + \sin\theta$
26. $\sin 5x + \sin 4x$
27. $\sin 5t - \sin 7t$
28. $\sin 3v - \sin v$
29. $\cos 2t + \cos 3t$
30. $\cos 3v + \cos v$
31. $\cos 5x - \cos 6x$
32. $\cos 2x - \cos x$

In problems 33–38, use the inverse trigonometric identities, whenever possible, to find the value of each expression. Round off answers to four decimal places when applicable.

33. **(a)** $\sin(\sin^{-1} 0.3)$ **(b)** $\sin^{-1}(\sin 1.3)$
34. **(a)** $\sin\left[\sin^{-1}\left(-\frac{1}{5}\right)\right]$ **(b)** $\sin^{-1}\left(\sin\frac{5\pi}{4}\right)$
35. **(a)** $\cos\left[\cos^{-1}\left(-\frac{\pi}{6}\right)\right]$ **(b)** $\cos^{-1}\left(\cos\frac{7\pi}{6}\right)$
36. **(a)** $\cos(\cos^{-1} 0.8)$ **(b)** $\cos^{-1}\left(\cos\sqrt{2}\right)$
37. **(a)** $\tan(\tan^{-1} 90)$ **(b)** $\tan^{-1}[\tan(-15)]$
38. **(a)** $\tan[\tan^{-1}(-3.2)]$ **(b)** $\tan^{-1}\left(\tan\frac{5\pi}{4}\right)$

Developing and Extending the Concepts

In problems 39–44, prove each identity.

39. $\cos u \sin v = \frac{1}{2}[\sin(u + v) - \sin(u - v)]$
40. $\cos u \cos v = \frac{1}{2}[\cos(u - v) + \cos(u + v)]$
41. $\sin u \sin v = \frac{1}{2}[\cos(u - v) - \cos(u + v)]$
42. $\sin\omega - \sin t = 2\cos\left(\frac{\omega + t}{2}\right)\sin\left(\frac{\omega - t}{2}\right)$
43. $\cos\omega + \cos t = 2\cos\left(\frac{\omega + t}{2}\right)\cos\left(\frac{\omega - t}{2}\right)$
44. $\cos\omega - \cos t = -2\sin\left(\frac{\omega + t}{2}\right)\sin\left(\frac{\omega - t}{2}\right)$

In problems 45 and 46, show that each equation is true.

45. $\sin(-\sin^{-1} x) = -\sin(\sin^{-1} x)$, if $-1 \le x \le 1$
46. $\cos(-\cos^{-1} x) = \cos(\cos^{-1} x)$, if $-1 \le x \le 1$

In problems 47–50, find the required restrictions on x in order for the equation to be true.

47. $\sin[\sin^{-1}(3x + 2)] = 3x + 2$
48. $\cos[\cos^{-1}(5 - 2x)] = 5 - 2x$
49. $\tan^{-1}\left[\tan\left(\frac{x}{3} - 1\right)\right] = \frac{x}{3} - 1$
50. $\cos^{-1}[\cos(x^2 - 4)] = x^2 - 4$

CHAPTER 6 REVIEW PROBLEM SET

GU In problems 1–4, use a graphing utility to graph and compare the functions defined by each side of the equation. Then prove that each equation is an identity.

1. **(a)** $\cos t \sin t \csc t \sec t = 1$
(b) $\cot\theta\cos\theta + \sin\theta = \csc\theta$
2. **(a)** $\sec t - \cos t = \sin t \tan t$
(b) $(\csc t + \cot t)(1 - \cos t) = \sin t$
3. **(a)** $\frac{\sin^2 t \cos t + \cos^3 t}{\cot t} = \sin t$
(b) $\cos^2 t - \sin^2 t = \frac{1 - \tan^2 t}{1 + \tan^2 t}$
4. **(a)** $\frac{\sin^4 t - \cos^4 t}{\sin^2 t - \cos^2 t} = 1$
(b) $\frac{\sec^2 t + 2\tan t}{2 - \sec^2 t} = \frac{1 + \tan t}{1 - \tan t}$

In problems 5–10, simplify each expression by using identities.

5. **(a)** $\sin 7t \cos 2t - \cos 7t \sin 2t$
(b) $\sin(t + s)\cos s - \cos(t + s)\sin s$
6. **(a)** $\cos 51^\circ \cos 39^\circ - \sin 51^\circ \sin 39^\circ$
(b) $\cos 33^\circ \cos 27^\circ - \sin 33^\circ \sin 27^\circ$
7. **(a)** $\sin(3\pi - t)$ **(b)** $2\sin^2 4\theta - 1$
8. **(a)** $\csc(90^\circ - \theta)\cos\theta + \cot^2\theta$
(b) $\cos(270^\circ + \theta)$
9. **(a)** $\cos\left(\frac{\pi}{2} - t\right)\tan\left(\frac{\pi}{2} - t\right)$
(b) $\cos\left(\frac{\pi}{3} - t\right) - \cos\left(\frac{\pi}{3} + t\right)$
10. **(a)** $\frac{\tan 5t + \tan t}{1 - \tan 5t \tan t}$ **(b)** $\frac{\tan 7t - \tan 2t}{1 + \tan 7t \tan 2t}$

In problems 11–16, use an identity to find the exact value of each expression.

11. **(a)** $\cos\left(\frac{\pi}{4} + \frac{\pi}{6}\right)$ **(b)** $\cos\frac{\pi}{4} + \cos\frac{\pi}{6}$

12. **(a)** $\sin\left(\frac{3\pi}{4} - \frac{\pi}{3}\right)$ **(b)** $\sin\frac{3\pi}{4} - \sin\frac{\pi}{3}$

13. **(a)** $\sin 165°$; $165° = 120° + 45°$
(b) $\cos 195°$; $195° = 150° + 45°$

14. **(a)** $\cos 285°$; $285° = 240° + 45°$
(b) $\cos 255°$; $255° = 210° + 45°$

15. **(a)** $\sin 22.5°$; $22.5° = \frac{1}{2}(45°)$
(b) $\cos\frac{11\pi}{12}$; $\frac{11\pi}{12} = \frac{1}{2}\left(\frac{11\pi}{6}\right)$

16. **(a)** $\cos 75°$; $75° = \frac{1}{2}(150°)$
(b) $\sin\frac{7\pi}{12}$; $\frac{7\pi}{12} = \frac{1}{2}\left(\frac{7\pi}{6}\right)$

17. Suppose that $\sin t = \frac{1}{4}$, where t is in quadrant I, and $\cos s = -\frac{5}{17}$, where s is in quadrant II. Find:
(a) $\sin(t - s)$ **(b)** $\cos(t - s)$ **(c)** $\sin 2t$
(d) $\cos(s/2)$ **(e)** $\tan 2t$

18. Suppose that $\tan t = -\frac{4}{3}$, where t is in quadrant II, and $\tan s = -\frac{5}{12}$, where s is in quadrant IV. Find:
(a) $\tan(t + s)$ **(b)** $\sin(t + s)$ **(c)** $\cos(t + s)$
(d) $\cos 2s$ **(e)** $\sin(t/2)$

GU In problems 19–28, use a graphing utility to graph and compare the functions defined by each side of the equation. Then prove that each equation is an identity.

19. $\sin\left(t + \frac{\pi}{6}\right) + \cos\left(t + \frac{\pi}{3}\right) = \cos t$

20. $\sin\left(s - \frac{\pi}{5}\right)\cos\frac{\pi}{5} + \cos\left(s - \frac{\pi}{5}\right)\sin\frac{\pi}{5} = \sin s$

21. $2\csc 2\theta \cot\theta = 1 + \cot^2\theta$

22. $\csc\theta \sin 2\theta = 2\cos\theta$

23. $\tan\left(t + \frac{3\pi}{4}\right) = \frac{\tan t - 1}{\tan t + 1}$

24. $\dfrac{\cos\left(t - \frac{\pi}{4}\right)}{\cos t \sin\frac{\pi}{4}} = \tan t + 1$

25. $\dfrac{1 - \cos 2\theta}{\sin 2\theta} = \tan\theta$

26. $\dfrac{\cos 2\theta}{1 - \sin^2\theta} = 2 - \sec^2\theta$

27. $\sin\frac{t}{2}\cos\frac{t}{2} = \frac{\sin t}{2}$

28. $\dfrac{1 - \tan^2\theta}{1 + \tan^2\theta} = \cos 2\theta$

In problems 29–32, solve each equation graphically and analytically, where $0 \le t < 2\pi$ or $0° \le \theta < 360°$. Round off part (b) of each problem to two decimal places. Compare the solutions in both approaches. Use $\pi = 3.14$.

29. **(a)** $\sin\theta = -1$ **(b)** $\cos\theta = 0.3217$

30. **(a)** $\tan t = \sqrt{3}$ **(b)** $\sec t = 1.4931$

31. **(a)** $5\sin^2 t - 4\sin t - 1 = 0$
(b) $(2\sin t - 1)\left(\sin t - \frac{1}{3}\right) = 0$

32. **(a)** $3\tan^2\theta - \sqrt{3}\tan\theta = 0$
(b) $\tan^2 t - \tan t - 2 = 0$

In problems 33 and 34, find the exact value of each expression without using a calculator.

33. **(a)** $\cos 15° \cos 105°$ **(b)** $\cos\frac{11\pi}{12}\sin\frac{7\pi}{12}$

34. **(a)** $\sin 195° \cos 15°$ **(b)** $\sin\frac{\pi}{12}\cos\frac{7\pi}{12}$

In problems 35 and 36, write each expression as a sum or difference of sine and/or cosine functions.

35. **(a)** $2\sin t\cos 2t$ **(b)** $2\cos 3t\cos 5t$

36. **(a)** $\sin\frac{5t}{2}\cos\frac{t}{2}$ **(b)** $\sin 7t \sin 3t$

In problems 37 and 38, write each expression as a product of sine and/or cosine functions.

37. **(a)** $\sin 4t - \sin t$ **(b)** $\cos 7t + \cos 3t$

38. **(a)** $\sin 10t + \sin 2t$ **(b)** $\cos 6t - \cos 2t$

GU In problems 39 and 40, use the ZOOM and TRACE features on a graphing utility to approximate the solution to each equation (to two decimal places) for $0 \le t < 2\pi$.

39. $\cos t = 3t$ **40.** $2\sin t = 2t - 1$

GU In problems 41–44, write each equation in the form $y = A\cos(u - v)$, where b is a real number and $a > 0$. Use a graphing utility to graph one cycle of the resulting function.

41. $y = \sqrt{3}\sin x + \cos x$ **42.** $y = -\sin x - \sqrt{3}\cos x$

43. $y = -\sin x + \cos x$ **44.** $y = -\sin x - \cos x$

45. Electrical Voltage: Suppose the voltage V (in volts) in an electrical circuit is given by the model
GU

$$V = 80\cos\left(120\pi t - \frac{\pi}{4}\right)$$

where t is in seconds.

(a) Use a graphing utility to graph at least one cycle of the function.
(b) Use the ZOOM and TRACE features to determine the smallest positive value of t when the voltage is 60 volts. Round off to three decimal places.
(c) Solve for t analytically when $V = 60$, and compare the answer to the result from part (b).

46. Business Profit: A company's profit during a 12 month period is modeled by the function

GU

$$p(t) = 100{,}000(t - 2\cos t)$$

where $p(t)$ (in dollars) is the profit after t months.
(a) Use a graphing utility to graph p for $0 \le t < 12$.
(b) Use the graph to approximate the value of t (to three decimal places) when $p(t) = \$650{,}000$.

47. Billiard Ball Motion: Figure 1 shows a billiard ball (at position A) that travels a path to hit a ball at position B. Assume the billiard ball strikes the rail of the table and rebounds so that the angles labeled θ are equal.
(a) Find the distance x denoted in the figure.
(b) Find $\tan\theta$, and then solve the equation for θ. Round off the answer to the nearest degree.

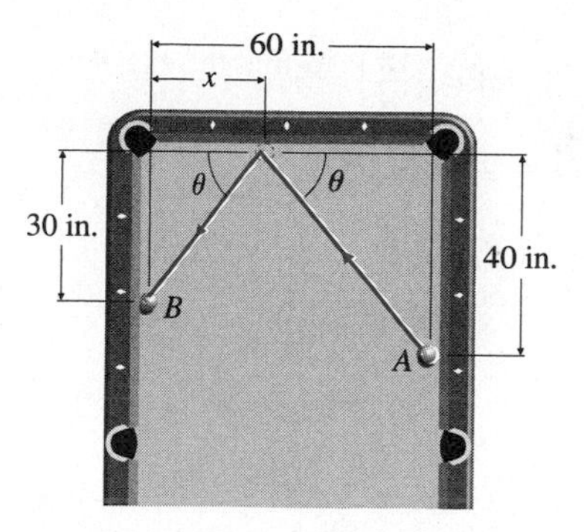

FIGURE 1

48. Hours of Daylight: Suppose the number of hours N of daylight on a given day of the year for a certain area is given by the model

GU

$$N = 12 + 3\sin\left[\frac{2\pi}{365}(t - 85)\right]$$

where t is the day of the year, and $t = 0$ is January 1.
(a) Use a graphing utility to graph the function.
(b) Use the ZOOM and TRACE features to determine which days of the year have 14 hours of daylight.

CHAPTER 6 TEST

1. Use $255° = 225° + 30°$ to find the exact value of each expression.
(a) $\sin 255°$ (b) $\cos 255°$ (c) $\tan 255°$

2. Suppose that $\sin s = -\frac{3}{5}$, where s is in quadrant IV, and $\cos t = \frac{4}{5}$, where t is in quadrant IV. Find the exact value of each expression.
(a) $\sin(s + t)$ (b) $\cos(s + t)$ (c) $\tan(s + t)$
(d) $\sin 2t$ (e) $\cos 2t$ (f) $\sin 4t$
(g) $\sin^2(t/2)$ (h) $\sin(s + 2t)$

3. Use $67.5° = \frac{1}{2}(135°)$ to find the exact value of each expression.
(a) $\sin 67.5°$ (b) $\cos 67.5°$ (c) $\tan 67.5°$

4. Simplify each trigonometric expression by using identities.
(a) $1 - 2\cos^2 47°$
(b) $\sin\frac{5\pi}{7}\cos\frac{5\pi}{7}$
(c) $\sin 25°\cos 35° + \sin 35°\cos 25°$
(d) $\cos\frac{5\pi}{7}\cos\frac{2\pi}{7} - \sin\frac{5\pi}{7}\sin\frac{2\pi}{7}$
(e) $\cos\frac{3t}{4}\cos\frac{t}{4} + \sin\frac{3t}{4}\sin\frac{t}{4}$

5. Solve each equation both graphically and analytically for $0° \le \theta < 360°$ or $0 \le t < 2\pi$. Round off the answers to two decimal places. Compare the results from both approaches.
(a) $4\cos^2 t - 3 = 0$
(b) $\sin 2\theta = \frac{1}{2}$
(c) $(\tan t - 1)(2\tan t + 3) = 0$

6. Prove that each equation is an identity. Use a graphing utility to graph the functions defined by each side of the equation to demonstrate the validity of the identity.
(a) $(\csc^2 t - 1)(1 + \tan^2 t) = \csc^2 t$
(b) $\sin 2t\sec t = 2\sin t$

(c) $\sin\left(t + \frac{\pi}{6}\right) \sin\left(t - \frac{\pi}{6}\right) = \sin^2 t - \frac{1}{4}$

(d) $\frac{1 - \cos 2t}{\sin t} = 2 \sin t$

7. Use a graphing utility to solve the equation $3x + \cos x = 0$. Round off the answer to two decimal places.

8. An observer in a balloon finds that the angle of depression from a balloon to a car is θ (measured in degrees). The balloon is x feet high and the distance horizontal from a point on the ground directly under it to the car is 240 feet.

(a) Express $\cos 2\theta$ as a function of x.

(b) If $x = 296$ feet, what is the measure of θ? Round off the answer to two decimal places.

7

CHAPTER CONTENTS

Additional Applications of Trigonometry

So far, we have used the trigonometric functions to solve right triangles. However, in fields such as surveying, navigation, and engineering, it is often necessary to solve triangles that do not necessarily contain a right angle. In this chapter, we will study ways of solving such triangles. We will also use trigonometry to explore the polar coordinate system, to study vectors, and to extend our understanding of the complex number system.

OBJECTIVES

1. Use the law of cosines to solve triangles.
2. Use the law of sines to solve triangles.
3. Solve the ambiguous case.
4. Solve applied problems.

7.1 The Law of Cosines and the Law of Sines

In this section, we derive two formulas—the *law of cosines* and the *law of sines*—that enable us to solve triangles. When solving a triangle, the given parts will always fall into one of the five categories listed in Table 1.

TABLE 1

Categories of Triangles

Given Parts	Description	Abbreviation
1. Two sides and the angle between them	Side–Angle–Side	SAS
2. Three sides	Side–Side–Side	SSS
3. Two angles and the side between them	Angle–Side–Angle	ASA
4. Two angles and a side (not the included one)	Angle–Angle–Side	AAS
5. Two sides and an angle opposite one of them	Side–Side–Angle	SSA

Using the Law of Cosines to Solve Triangles

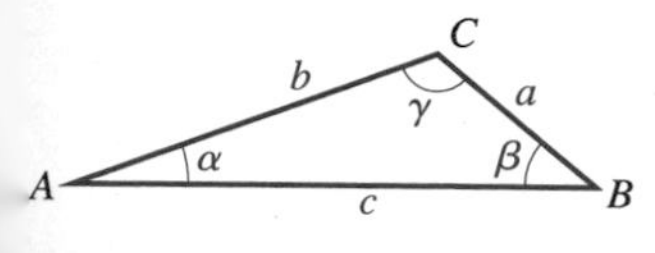

FIGURE 1

The law of cosines is a generalization of the Pythagorean theorem, and it is used to solve triangles for the SAS and SSS cases listed in Table 1.

For convenience, we adopt standard notation to describe the sides and angles of any triangle. As before, in triangle ABC (Figure 1) the angle at vertex A is denoted by α, the angle at B is denoted by β, and the angle at C is denoted by

γ. The length of the side opposite α is a, the length of the side opposite β is b, and the length of the side opposite γ is c.

THE LAW OF COSINES

In any triangle ABC:

$$\text{(i)}\ c^2 = a^2 + b^2 - 2ab\cos\gamma$$
$$\text{(ii)}\ b^2 = a^2 + c^2 - 2ac\cos\beta$$
$$\text{(iii)}\ a^2 = b^2 + c^2 - 2bc\cos\alpha$$

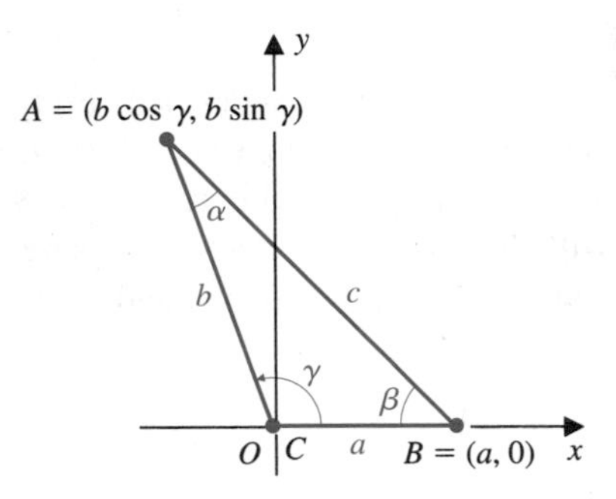

FIGURE 2

Proof of Formula (i) To derive the law of cosines, consider triangle ABC in a Cartesian plane with γ in standard position (Figure 2). By the definitions of the trigonometric functions of angles, point A has coordinates $(b\cos\gamma, b\sin\gamma)$. Point B has coordinates $(a, 0)$.

(i) So we have

$c^2 = (b\cos\gamma - a)^2 + (b\sin\gamma - 0)^2$	Distance formula
$= b^2\cos^2\gamma - 2ab\cos\gamma + a^2 + b^2\sin^2\gamma$	Multiply
$= a^2 + b^2\cos^2\gamma + b^2\sin^2\gamma - 2ab\cos\gamma$	Rearrange terms
$= a^2 + b^2(\cos^2\gamma + \sin^2\gamma) - 2ab\cos\gamma$	Factor
$= a^2 + b^2 - 2ab\cos\gamma$	$\cos^2\gamma + \sin^2\gamma = 1$

Formulas (ii) and (iii) are obtained by placing the other two angles in standard position and repeating the steps in the above proof.

The law of cosines can be stated in words as follows:

The square of the length of any side of a triangle is equal to the sum of the squares of the lengths of the other two sides minus twice the product of these two sides and the cosine of their included angle.

EXAMPLE 1 Using the law of cosines (SAS)

In triangle ABC (Figure 3), find c if $a = 8$, $b = 6$, and $\gamma = 60°$. Round off the answer to two decimal places.

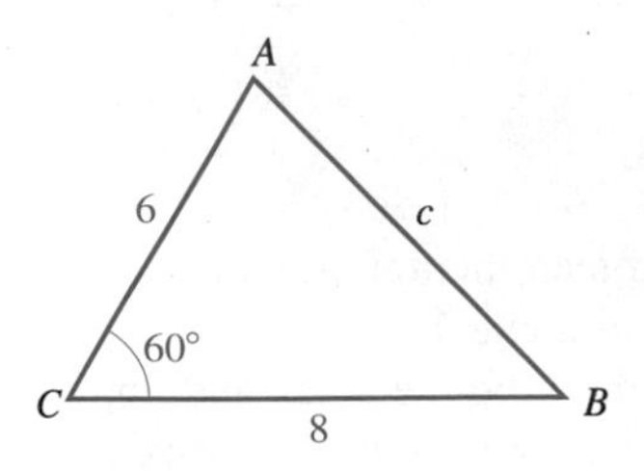

FIGURE 3

Solution We observe in Figure 3 that two sides and the included angle of the triangle are given. This corresponds to the SAS case, so the law of cosines is applicable. To find c, we substitute the given information into formula (i) as follows:

$$c^2 = a^2 + b^2 - 2ab\cos\gamma$$
$$= 64 + 36 - 2(8)(6)\left(\tfrac{1}{2}\right) = 100 - 48 = 52$$

So $c = \sqrt{52}$ or approx. 7.21

The next example shows that when the three sides (SSS) of a triangle are given, we can apply the law of cosines to find any of the three angles.

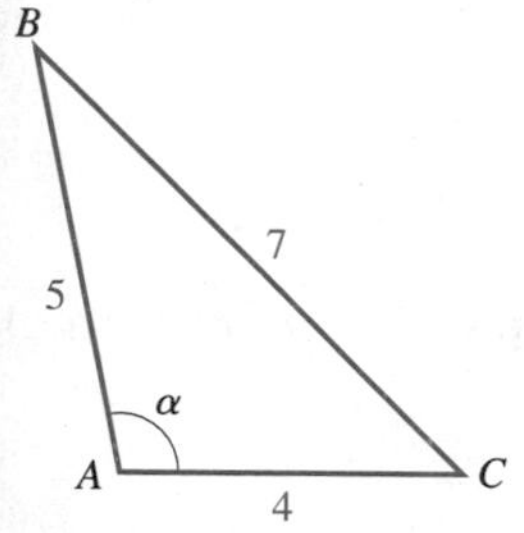

FIGURE 4

EXAMPLE 2 Using the law of cosines (SSS)

In triangle ABC (Figure 4), find α if $a = 7$, $b = 4$, and $c = 5$. Round off the answer to the nearest hundredth of a degree.

Solution Figure 4 shows that the three sides (SSS) of the triangle are given, so the law of cosines is applicable. To find angle α, we substitute the given information into formula (iii) as follows:

$$a^2 = b^2 + c^2 - 2bc \cos \alpha$$

$$7^2 = 4^2 + 5^2 - 2(4)(5) \cos \alpha$$

So

$$\cos \alpha = \frac{4^2 + 5^2 - 7^2}{(2)(4)(5)} = -0.2$$

Thus, $\alpha = \cos^{-1}(-0.2) = 101.54°$ (approx.).

Using the Law of Sines to Solve Triangles

The AAS case can always be converted to the ASA case by first solving for the third angle.

When two angles and any side (ASA or AAS), or two sides and an angle opposite one of them (SSA) are given, then the law of sines is used to solve the triangle.

THE LAW OF SINES

In any triangle ABC:

$$\frac{\sin \alpha}{a} = \frac{\sin \beta}{b} = \frac{\sin \gamma}{c}$$

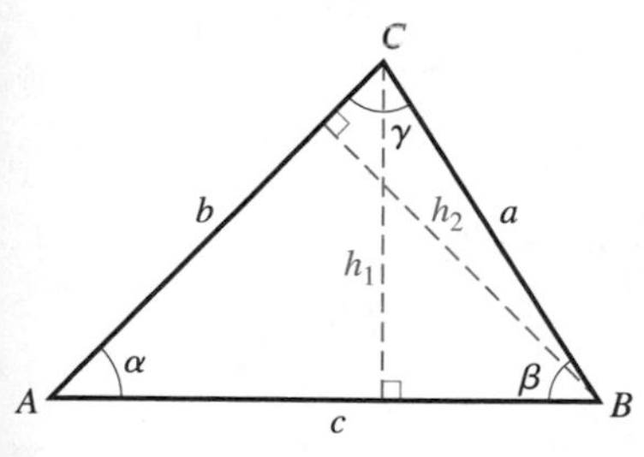

FIGURE 5

Proof To derive the law of sines, let h_1 be the height of triangle ABC from vertex C to side $\overline{AB}$, and let h_2 be the height from vertex B to side $\overline{AC}$ (Figure 5). By using right triangle trigonometry, we have

$$\sin \alpha = \frac{h_1}{b} \quad \text{and} \quad \sin \beta = \frac{h_1}{a}$$

or

$$h_1 = b \sin \alpha \quad \text{and} \quad h_1 = a \sin \beta$$

So $b \sin \alpha = a \sin \beta$; that is,

$$\frac{\sin \alpha}{a} = \frac{\sin \beta}{b}$$

Similarly, we can show that

$$\sin \alpha = \frac{h_2}{c} \quad \text{and} \quad \sin \gamma = \frac{h_2}{a}$$

So $h_2 = c \sin \alpha = a \sin \gamma$, or

$$\frac{\sin \alpha}{a} = \frac{\sin \gamma}{c}$$

Thus,

$$\frac{\sin \alpha}{a} = \frac{\sin \beta}{b} = \frac{\sin \gamma}{c}$$

Note that the law of sines can also be written in the equivalent recriprocal form

$$\boxed{\frac{a}{\sin \alpha} = \frac{b}{\sin \beta} = \frac{c}{\sin \gamma}}$$

EXAMPLE 3 **Using the law of sines (ASA)**
In triangle ABC (Figure 6), $a = 20$, $\gamma = 51°$, and $\beta = 42°$. Solve the triangle for α, b, and c (to two decimal places).

FIGURE 6

Solution Because $\alpha + \beta + \gamma = 180°$,

$$\alpha = 180° - (\beta + \gamma) = 180° - 93° = 87°$$

Since we have an ASA situation, the law of sines is applicable. Thus,

$$\frac{b}{\sin 42°} = \frac{20}{\sin 87°}$$

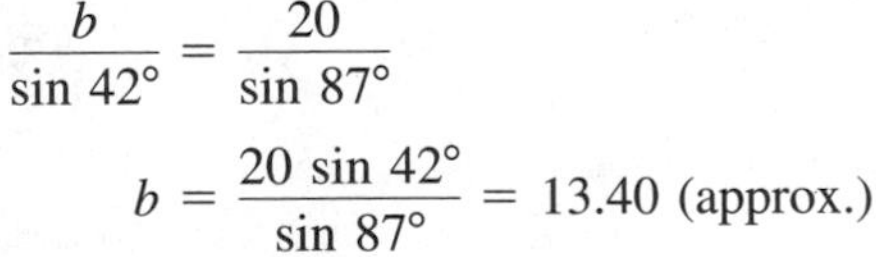

So $$b = \frac{20 \sin 42°}{\sin 87°} = 13.40 \text{ (approx.)}$$

To find c, we again use the law of sines to get

$$\frac{c}{\sin 51°} = \frac{20}{\sin 87°}$$

So $$c = \frac{20 \sin 51°}{\sin 87°} = 15.56 \text{ (approx.)}$$

Solving the Ambiguous Case

When we are given two sides of a triangle and an angle opposite one of them (SSA), a unique triangle is not always determined. In this case, there are three possible outcomes: no triangle; only one triangle; two different triangles. For this reason, we sometimes refer to the SSA situation as the *ambiguous case*.

Memorizing special rules to handle each of these possibilities is unnecessary. Rather, we apply the law of sines and reason our way through the problem, as illustrated in the next example.

EXAMPLE 4 **Using the law of sines (SSA)**

Solve each triangle (if possible). Determine all angles to the nearest hundredth of a degree and all lengths to two decimal places.

(a) $a = 5,\ b = 20,\ \alpha = 30°$ (b) $a = 5,\ b = 9,\ \alpha = 33°$
(c) $a = 20,\ b = 15,\ \alpha = 29°$

Solution

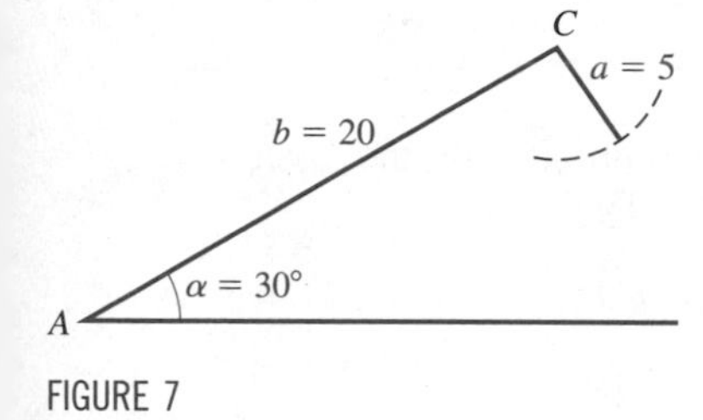

FIGURE 7

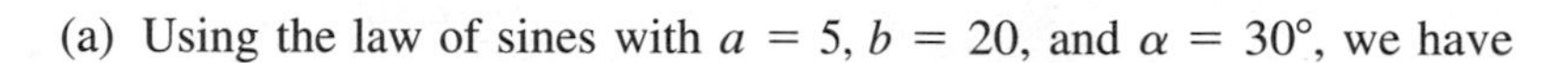

(a) Using the law of sines with $a = 5$, $b = 20$, and $\alpha = 30°$, we have

$$\sin \beta = \frac{b \sin \alpha}{a} = \frac{20 \sin 30°}{5} = 2 > 1$$

Since the sine cannot be greater than 1, there is no possible triangle satisfying the given conditions. Figure 7 illustrates this situation.

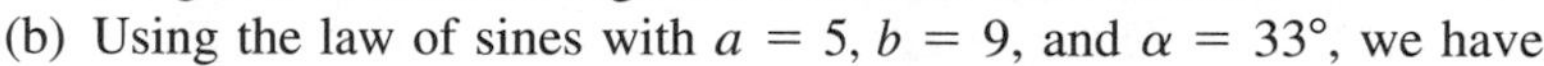

(b) Using the law of sines with $a = 5$, $b = 9$, and $\alpha = 33°$, we have

$$\sin \beta = \frac{b \sin \alpha}{a} = \frac{9 \sin 33°}{5}$$

One possible angle that satisfies this equation is the acute angle β_1 given by

$$\beta_1 = \sin^{-1}\left(\frac{9 \sin 33°}{5}\right) = 78.62° \text{ (approx.)}$$

However, using 78.62° as a reference angle, we find an obtuse angle $\beta_2 = 180° - 78.62° = 101.38°$ (approx.) that also satisfies the equation (Figure 8a). The solution $\beta_1 = 78.62°$ leads to the triangle AB_1C shown in Figure 8b. Here,

$$\gamma_1 = 180° - (33° + 78.62°) = 68.38° \text{ (approx.)}$$

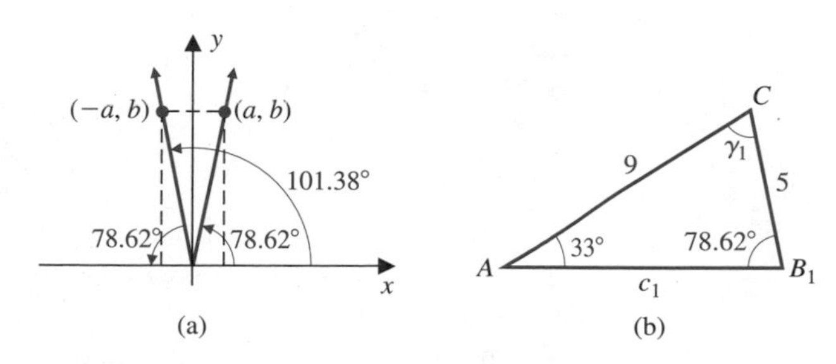

FIGURE 8

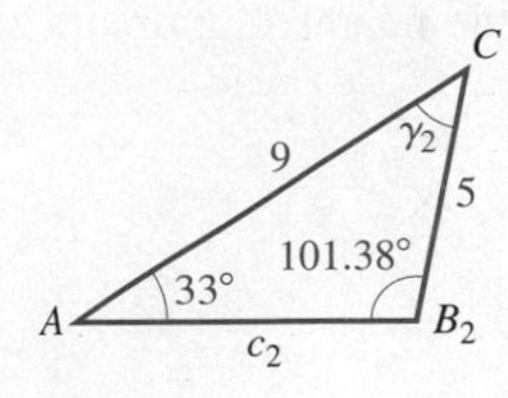

FIGURE 8c

So, by the law of sines,

$$c_1 = \frac{a \sin \gamma_1}{\sin \alpha} = \frac{5 \sin 68.38°}{\sin 33°} = 8.53 \text{ (approx.)}$$

The solution $\beta_2 = 101.38°$ leads to a second possibility, triangle AB_2C shown in Figure 8c. For this triangle, $\gamma_2 = 180° - (33° + 101.38°) = 45.62°$ (approx.). So,

$$c_2 = \frac{a \sin \gamma_2}{\sin \alpha} = \frac{5 \sin 45.62°}{\sin 33°} = 6.56 \text{ (approx.)}$$

Thus, in this situation, we have two possible solutions with the given parts.

(c) Applying the law of sines with $a = 20$, $b = 15$, and $\alpha = 29°$, we have

$$\sin \beta = \frac{b \sin \alpha}{a} = \frac{15 \sin 29°}{20}$$

We find that an acute angle given by

$$\beta_1 = \sin^{-1}\left(\frac{15 \sin 29°}{20}\right) = 21.32° \text{ (approx.)}$$

and an obtuse angle given by

$$\beta_2 = 180° - 21.32° = 158.68° \text{ (approx.)}$$

both give the same value of $\sin \beta$. It is impossible to have angle $\beta_2 = 158.68°$ in a triangle that is known to have an angle of 29°, since $\alpha + \beta_2 = 29° + 158.68° = 187.68° > 180°$. So the β_2 solution must be rejected. There is only one possible triangle (Figure 9) with the given parts, so there is only one solution. The solution $\beta_1 = 21.32°$ leads to

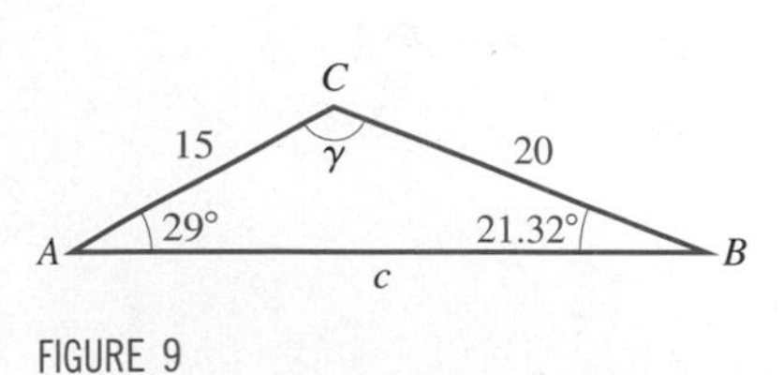

FIGURE 9

$$\gamma = 180° - (29° + 21.32°) = 129.68° \text{ (approx.)}$$

and

$$c = \frac{a \sin \gamma}{\sin \alpha} = \frac{20 \sin 129.68°}{\sin 29°} = 31.75 \text{ (approx.)}$$

The three possible situations in the ambiguous case are summarized in Table 2. We assume that two sides a and b and an acute angle α opposite one of them are given. We also let h be the length of the perpendicular line segment from vertex C to side c, so $h = b \sin \alpha$.

TABLE 2

Ambiguous Case (SSA) Possibilities (Given a, b, and α)

$h = b \sin \alpha$	Number of Possible Triangles	Illustration
$0 < a < h = b \sin \alpha$	0	
$h < a < b$	2: $\triangle AB_1C$ and $\triangle AB_2C$	
$a \geq b$	1: $\triangle ABC$	

Solving Applied Problems

The following examples illustrate applications of the law of cosines and the law of sines.

EXAMPLE 5

Measuring distance indirectly

A straight tunnel with end points A and B is to be blasted through a mountain. The distance between A and B cannot be measured directly. So, from a third point C, a surveyor finds the distance from C to A to be 573 meters and from C to B to be 819 meters (Figure 10). If the measure of angle ACB is 67°, find the length of the tunnel; that is, find $|\overline{AB}|$. Round off the answer to two decimal places.

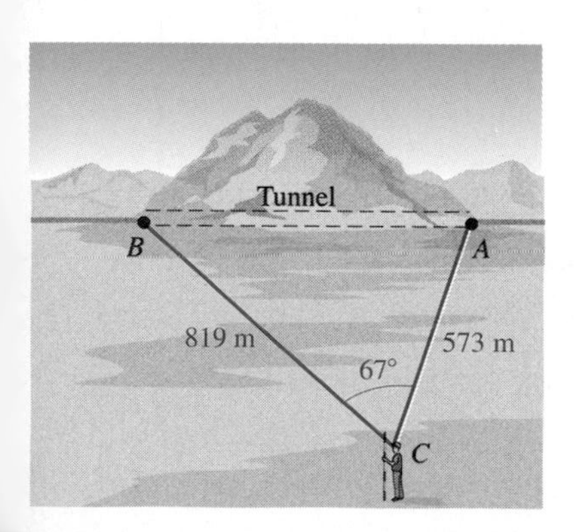

FIGURE 10

Solution Here we have an SAS case, so we apply the law of the cosines to triangle ACB to get

$$|\overline{AB}|^2 = |\overline{CB}|^2 + |\overline{CA}|^2 - 2|\overline{CB}||\overline{CA}| \cos \gamma$$
$$= 819^2 + 573^2 - 2(819)(573) \cos 67°$$

Hence,

$$|\overline{AB}| = \sqrt{819^2 + 573^2 - 2(819)(573) \cos 67°}$$
$$= 795.21 \text{ (approx.)}$$

Therefore, the length of the tunnel is approximately 795.21 meters.

EXAMPLE 6 **Measuring the height of a tower**

The Leaning Tower of Pisa is a marble bell tower that was built on unstable ground. It currently leans at an angle of about 5.4° from the vertical. When the sun makes an angle of 52° with the horizontal, the length of the shadow cast by the tower is 156 feet measured from the center of the base. Find the height of the side of the tower. Round off the answer to two decimal places.

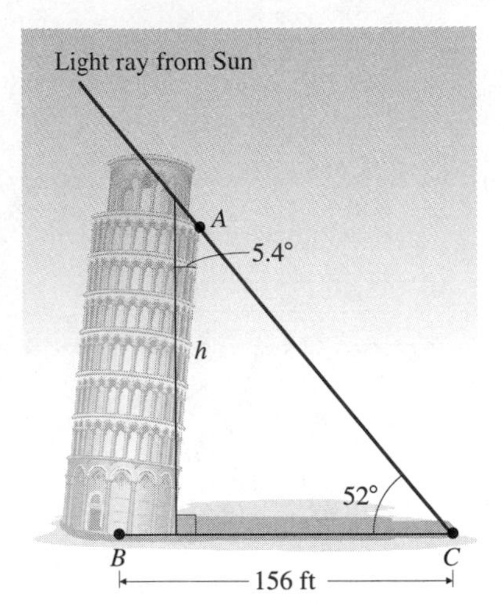

FIGURE 11

Solution The situation is shown in Figure 11, where h represents the height of the side of the tower. Since the tower leans at an angle of 5.4°, it follows that angle CBA is $90° - 5.4° = 84.6°$ and angle CAB is $180° - (84.6° + 52°) = 43.4°$. Thus, we have an ASA situation, so we apply the law of sines to get

$$\frac{h}{\sin 52°} = \frac{156}{\sin 43.4°} \quad \text{or} \quad h = \frac{156 \sin 52°}{\sin 43.4°} = 178.91 \text{ (approx.)}$$

Therefore, the height of the side of tower is about 178.91 feet.

In applications of trigonometry to surveying and navigation, the **direction** or **bearing** of a point Q as viewed from a point P is defined as the positive acute angle θ between the ray from P through Q and the north–south line through P (Figure 12). Bearings are usually measured in degrees east of north, west of north, east of south, or west of south. For instance, the symbol N25°E is read as 25° east of north. Figure 13 shows four bearings.

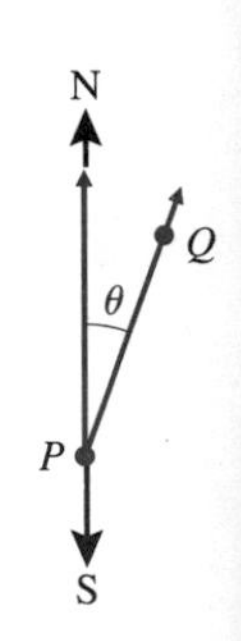

FIGURE 12

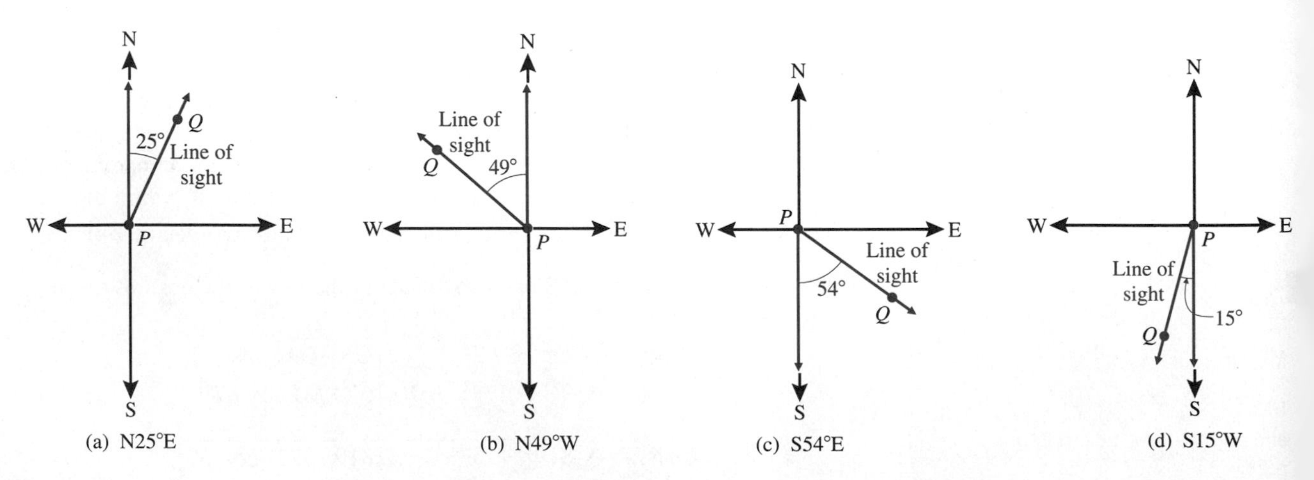

FIGURE 13

EXAMPLE 7 Sighting a fire from observation posts

A forest ranger at observation post A sights a fire at location C in the direction N66.5°E. Another ranger at observation post B, 8 kilometers directly east of A, sights the same fire at N32.2°W. How far is the fire from observation post A? Round off the answer to two decimal places.

Solution Figure 14 illustrates the situation.

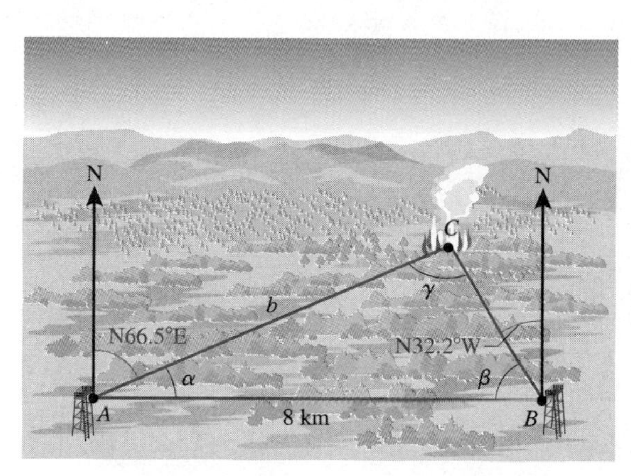

FIGURE 14

Because the interior angles α and β at vertices A and B are complementary to the known angles, we find that they measure 23.5° and 57.8°, respectively. So the angle γ at C is given by

$$\gamma = 180° - (23.5° + 57.8°) = 98.7°$$

This ASA case can now be solved by using the law of sines to get

$$\frac{b}{\sin 57.8°} = \frac{8}{\sin 98.7°}$$

or

$$b = \frac{8 \sin 57.8°}{\sin 98.7°} = 6.85 \text{ (approx.)}$$

Therefore, the fire is about 6.85 kilometers from observation post A.

PROBLEM SET 7.1

Mastering the Concepts

In problems 1–12, indicate whether the case is SAS or SSS, and then use the law of cosines to find the specified unknown part of triangle ABC. Round off angles to the nearest hundredth of a degree and side lengths to two decimal places.

1. a; if $b = 3$, $c = 4$, $\alpha = 30°$

2. c; if $a = 6$, $b = 4$, $\gamma = 59°$

3. b; if $a = 3$, $c = 4$, $\beta = 105°$

4. a; if $c = 2.4$, $b = 3.2$, $\alpha = 117°$

5. c; if $a = 5.8$, $b = 4.8$, $\gamma = 35.25°$

6. b; if $a = 6.1$, $c = 5.2$, $\beta = 5.67°$
7. γ; if $a = 2$, $b = 3$, $c = 4$
8. β; if $a = 14.3$, $b = 10.6$, $c = 8.4$
9. β; if $a = 23.7$, $b = 31.4$, $c = 40.6$
10. α; if $a = 0.6$, $b = 0.3$, $c = 0.5$
11. α; if $a = 81$, $b = 193$, $c = 253$
12. γ; if $a = 6$, $b = 5$, $c = 5.5$

In problems 13–22, indicate whether the case is ASA or AAS, and then use the law of sines to solve each triangle ABC. Round off angles to the nearest hundredth of a degree and side lengths to two decimal places.

13. $a = 20$, $\alpha = 29°$, $\beta = 136°$
14. $c = 32$, $\beta = 57°$, $\alpha = 38°$
15. $b = 30$, $\alpha = 80°$, $\gamma = 41°$
16. $a = 10.5$, $\alpha = 41°$, $\gamma = 77°$
17. $b = 19.7$, $\alpha = 42.17°$, $\gamma = 61.33°$
18. $b = 20$, $\alpha = 30°$, $\gamma = 52°$
19. $c = 5.4$, $\beta = 50.83°$, $\gamma = 70.5°$
20. $a = 13.1$, $\gamma = 100°$, $\alpha = 12.67°$
21. $b = 19$, $\alpha = 40°$, $\gamma = 62°$
22. $b = 38.8$, $\alpha = 103.45°$, $\gamma = 27.19°$

In problems 23–32, if possible, solve the ambiguous case (SSA) by using the law of sines. Round off angles to the nearest hundredth of a degree and side lengths to two decimal places. Be sure to find all possible triangles that satisfy the given conditions.

23. $b = 12$, $c = 11$, $\gamma = 81°$
24. $a = 140$, $c = 115$, $\gamma = 53.54°$
25. $a = 27$, $b = 52$, $\alpha = 70°$
26. $a = 1$, $b = 1.8$, $\alpha = 26°$
27. $a = 12.4$, $b = 8.7$, $\beta = 36.67°$
28. $a = 12.41$, $b = 81.69$, $\beta = 36.67°$
29. $a = 21.3$, $b = 18.9$, $\alpha = 65.18°$
30. $a = 263.6$, $c = 574.3$, $\alpha = 32.32°$
31. $a = 30$, $b = 44.5$, $\alpha = 33.33°$
32. $a = 10.1$, $b = 15.2$, $\alpha = 67.67°$

In problems 33–40, identify the case—SSS, SAS, ASA, AAS, or SSA—and then use the law of sines or the law of cosines to solve triangle ABC. Round off all angles to the nearest hundredth of a degree and side lengths to two decimal places. Be sure to find all possible triangles that satisfy the given conditions.

33. $a = 10$, $b = 50$, $\alpha = 22°$
34. $a = 14$, $\alpha = 21°$, $\beta = 35°$
35. $a = 4$, $b = 4$, $c = 6$
36. $a = 5$, $b = 10$, $c = 10$
37. $a = 14$, $\alpha = 12°$, $\beta = 97°$
38. $a = 10$, $b = 12$, $\gamma = 108°$
39. $b = 3.7$, $\alpha = 100.30°$, $\beta = 5.50°$
40. $b = 98$, $c = 39$, $\alpha = 17°$

Applying the Concepts

In problems 41–58, round off angles to the nearest hundredth of a degree and lengths to two decimal places.

41. Surveying: A vertical tower that is 70 feet high stands on a hill that makes an angle of 14° with the horizontal. A surveyor locates a point C downhill, 110 feet from the base of the tower at A (Figure 15). Find the distance from C to the top of the tower, B; that is, find $|\overline{CB}|$.

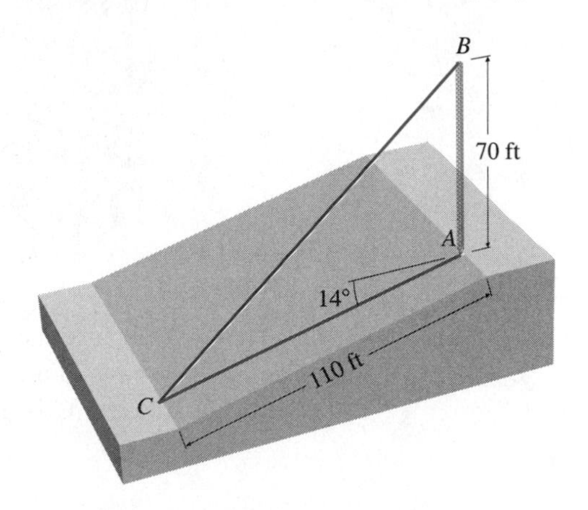

FIGURE 15

42. Distance Between Jets: Two jets leave an air base at the same time and fly at the same speed along straight courses forming an angle of 112.45° with each other (Figure 16). After the jets have each flown 504 kilometers, how far apart are they?

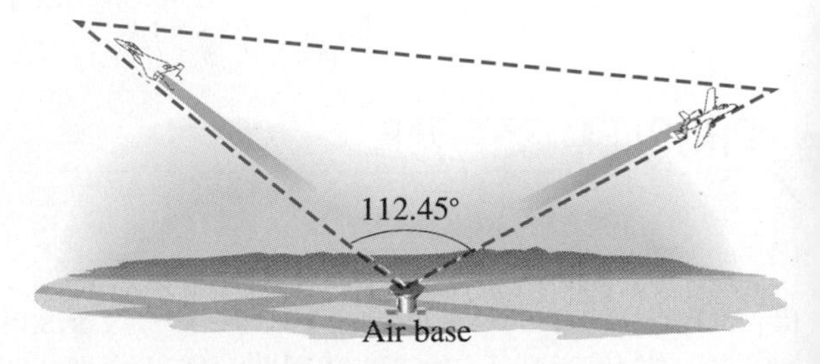

FIGURE 16

43. Surveying: Engineers are planning to construct a straight tunnel from point A on one side of a hill to point B on the other side. Since its length cannot be measured directly, point C is chosen by a surveyor who is 473 meters from A and 367 meters from B (Figure 17). If the measure of angle ACB is 46.2°, find the length of the tunnel.

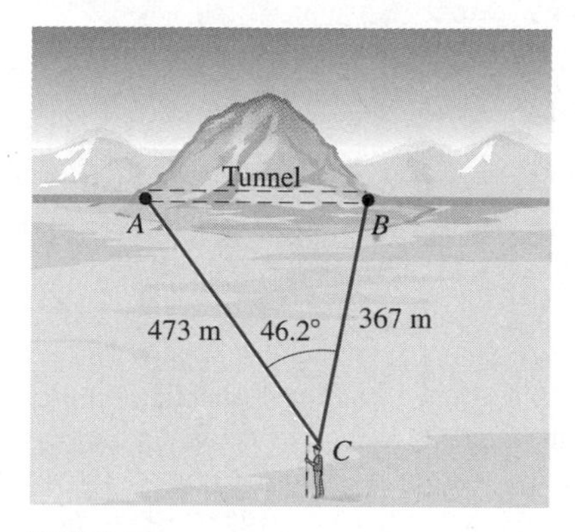

FIGURE 17

44. Surveying: To measure the length of a lake between points A and B, a point C on land is chosen. The distance between C and A is found to be 137 meters, and the distance between C and B is 405 meters (Figure 18). If angle ACB is 42.5°, how long is the lake?

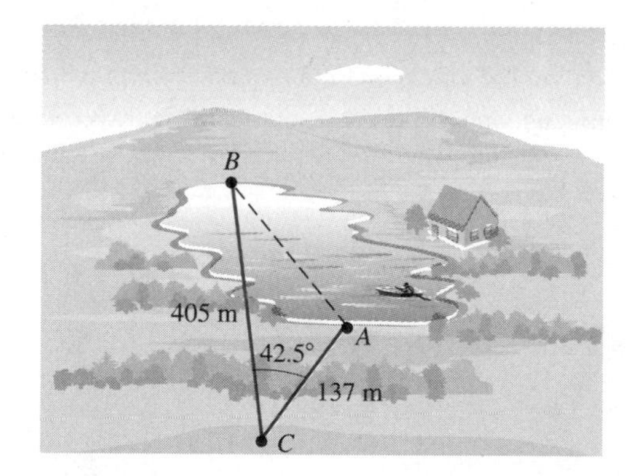

FIGURE 18

45. Joggers' Courses: Two straight roads intersect at 65° (Figure 19). A jogger on one road is 5 miles from the intersection and moving away from it at a constant rate of

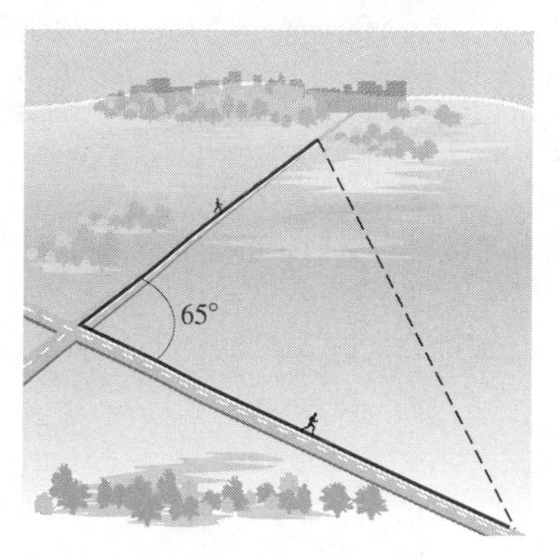

FIGURE 19

7 miles per hour. At the same instant, another jogger on the other road is 4 miles from the intersection and moving away from it at a constant rate of 5 miles per hour. If the two joggers maintain constant speeds, what is the distance between them 30 minutes later?

46. Navigation: A navigator of a ship plots a straight course from point A to point B. Because of an error, the ship proceeds from point A along a slightly different straight-line course. After traveling 3 hours on the wrong course, the error is discovered and the ship is turned through an angle of 22.5° so that it can reach point B (Figure 20). After 5 more hours, the ship reaches point B. Assuming that the ship was proceeding at a constant speed of 10 nautical miles per hour, find the time lost (to the nearest minute) because of the error.

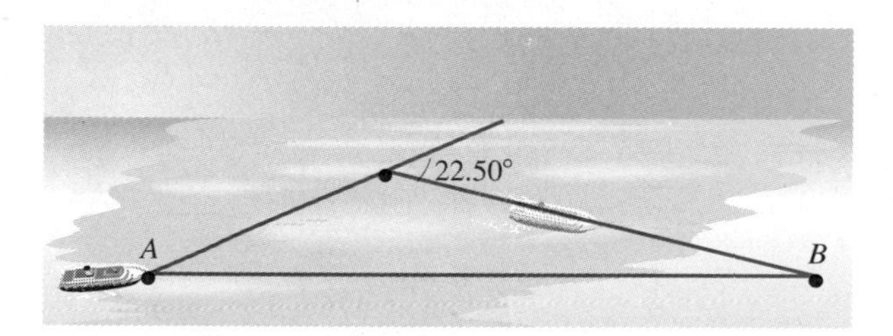

FIGURE 20

47. Lighthouse Lookout: As shown in Figure 21 (page 422), two lighthouses along a straight shoreline of a lake are located 2.43 kilometers apart at points A and B. The keepers of the lighthouses at A and B sight an overturned boat at point C at angles 70.3° and 48.2°, respectively. A Coast Guard rescue boat at D, located 1.2 kilometers downshore from B, is to rescue the boat. How far is the Coast Guard boat from the overturned boat?

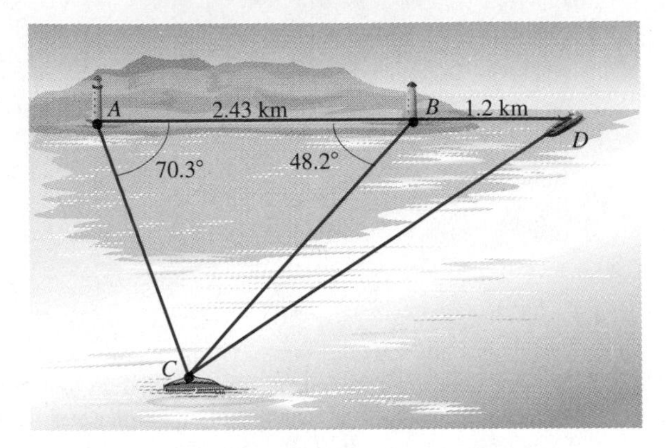

FIGURE 21

48. Baseball: Figure 22 shows a baseball diamond. It is in the form of a square that is 90 feet long on each side. The pitcher's mound at P is 60.5 feet from home plate, H, along the diagonal from home plate to second base. How far is the pitcher's mound P from first base, located at F?

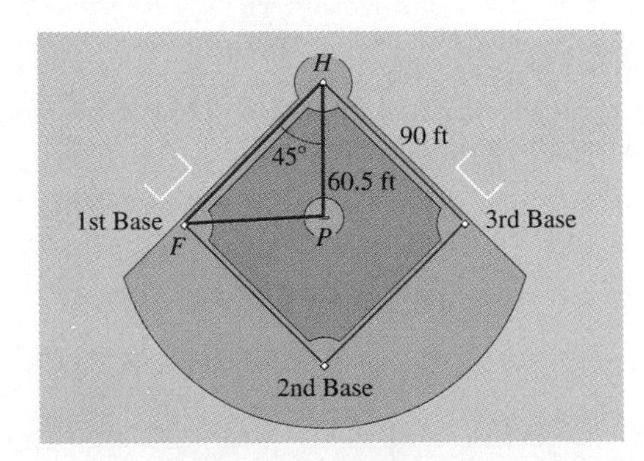

FIGURE 22

49. Fire Lookout: Two forest rangers are 1.2 kilometers apart. One ranger sights a fire at an angle of 41.43° from the line between the two observation points. The other ranger sights the same fire at an angle of 61.4° from the same line between the two observation points (Figure 23). How far is the fire from each observation point?

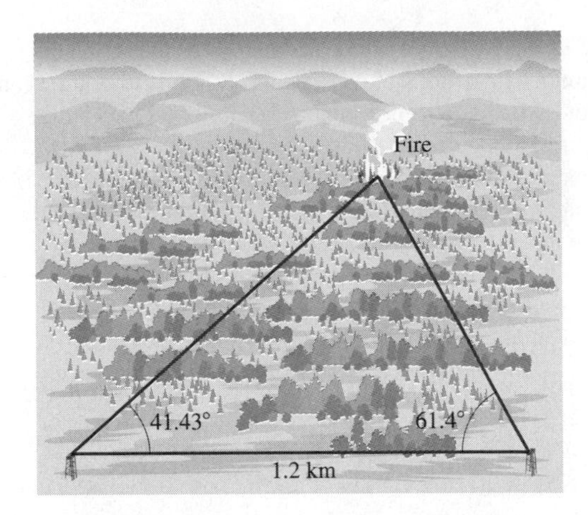

FIGURE 23

50. Distance to Top of a Tree: A biologist standing on the ground on the same horizontal plane as the base of a tree, determines that the angle of elevation from a point A to the top of the tree is 26.2°. After walking 150 feet closer to the tree to a point B, she finds the angle of elevation from point B to the top of the tree to be 37.1°. Find the distance from point B to the top of the tree.

51. Distance from a Parachutist: A sports parachutist is sighted simultaneously by two observers 3 miles apart on opposite sides of the parachutist. The observed angles of elevation are 25.5° and 17.8°. Assuming that the parachutist and the two observers lie in the same vertical plane, find the distance of the parachutist from the farther observer.

52. Engineering: The crankshaft $\overline{OA}$ of an engine is 5.8 centimeters long, and the connecting rod $\overline{AP}$ is 20.3 centimeters long (Figure 24). Find angle AOP at the instant when angle APO is 12°.

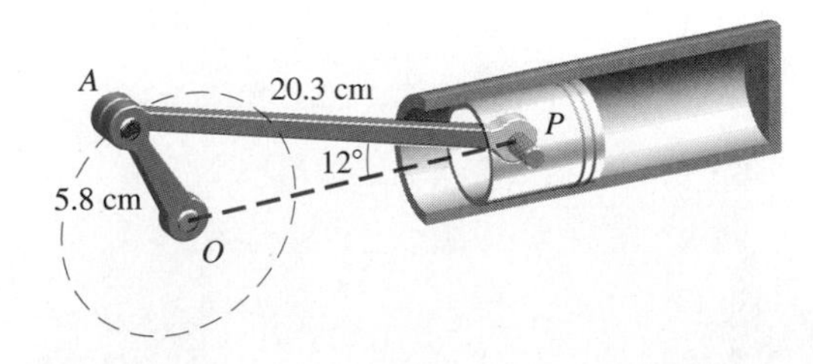

FIGURE 24

53. Streetlight Design: A streetlight is to be mounted on a brick wall. The length of brace $\overline{BC}$ is 1.3 meters, angle BCA

is 24.8°, and angle CAB is 37.8° (Figure 25). Find the length of the supporting brace $\overline{AC}$.

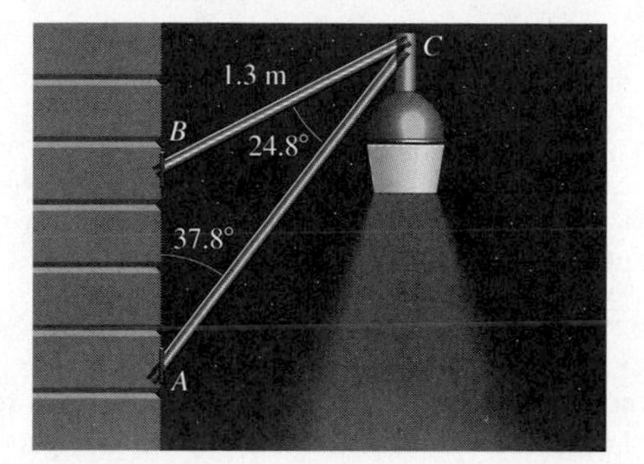

FIGURE 25

54. Communications Satellite: A communications satellite traveling in a circular orbit 1600 kilometers above the surface of the Earth is located at position A by a tracking station B on Earth at a certain time (Figure 26). If the tracking antenna is aimed 35° above the horizon and if the radius of the Earth is 6400 kilometers, what is the distance from the antenna to the satellite?

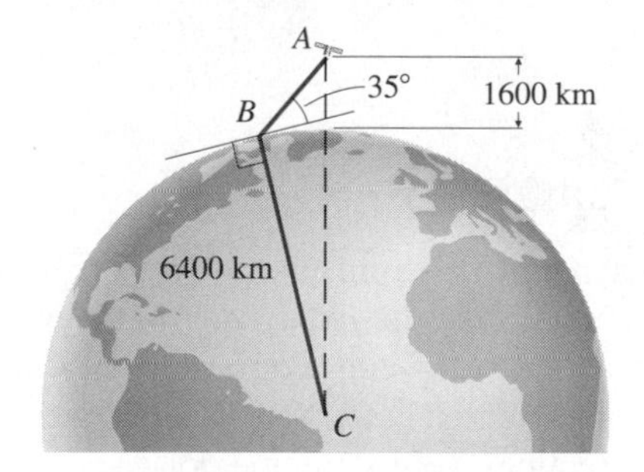

FIGURE 26

55. Navigation: A ship sails 19 nautical miles in the direction S29.3°W, and then turns onto a course S51.7°W and sails 24 nautical miles. How far is the ship from the starting point?

56. Navigation: A ship leaves a harbor at noon and sails S50°W at 18 nautical miles per hour until 2:30 PM. At that time, it changes course and sails N20°W at a reduced speed of 13 nautical miles per hour until 4:00 PM. Find:

(a) The distance of the ship from the harbor at 4:00 PM

(b) The ship's bearing from the harbor at 4:00 PM

57. Distance Model: Two straight roads intersect at an angle of 70°. A bus on one of the roads is 8 kilometers from the intersection and moving away from it at a rate of 90 kilometers per hour. At the same instant, a truck on the other road is 16 kilometers from the intersection and moving away from it at 100 kilometers per hour. Assume the bus and the truck maintain constant speeds.

(a) Express the distance d between them t hours later as a function of t.

GU **(b)** Use a graphing utility to graph the function from part (a) for $0 \leq t \leq 2.5$.

(c) Use the TRACE feature to approximate how long it takes for the distance between the vehicles to exceed 50 kilometers, 100 kilometers, and 150 kilometers. Confirm the results by using the function found in part (a).

58. Navigation–Distance Model: Two ships leave a port at the same time, traveling along straight-line paths at 12 nautical miles per hour and 16 nautical miles per hour, respectively. Assume that the angle between their directions of travel is 65° and both ships maintain constant speeds.

(a) Express the distance d between them t hours later as a function of t.

GU **(b)** Use a graphing utility to graph the function from part (a) for $0 \leq t \leq 1.5$.

(c) Use the TRACE feature to approximate how long it takes for the distance between the ships to exceed 14 nautical miles, 16 nautical miles, and 18 nautical miles. Confirm the results by using the function in part (a).

Developing and Extending the Concepts

59. Geometry: Figure 27 shows three circles with centers at P, Q, and R, and with radii 5, 8, and 8.4 centimeters, respectively. If the circles are tangent to one another, find the three angles of triangle PQR formed by the line segments connecting their centers.

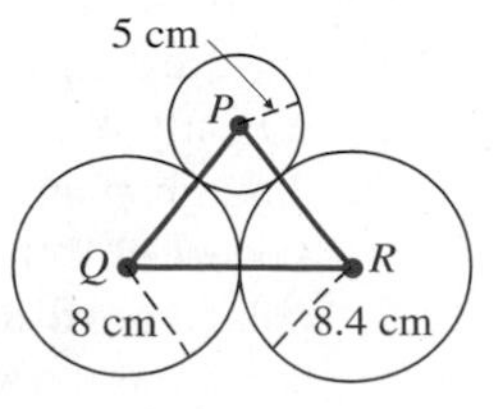

FIGURE 27

60. Geometry: Assume that a diagonal of a parallelogram is 80 inches long and that one end of it forms angles of 35° and 27°, respectively, with the two sides. Find the lengths of the sides of the parallelogram.

61. Find the lengths of the diagonals of the parallelogram $ABCD$ with side $\overline{AB}$ of length 10 inches and side $\overline{BC}$ of length 15 inches, and with angle ABC of 110°.

62. Rewrite formula (i) of the law of cosines if $\gamma = 90°$. Explain the result.

63. Explain why it is not possible to have a triangle with side lengths of 7, 8, and 16. What happens if we attempt to find an angle by using the law of cosines under the SSS case?

64. Assume that triangle ABC has height $h = b \sin \alpha$ (Figure 28).

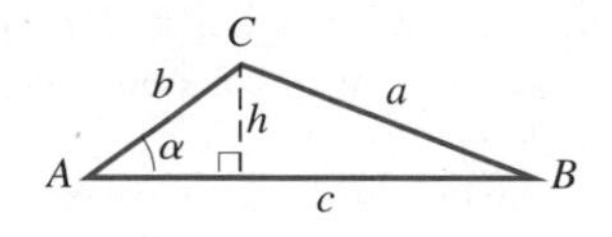

FIGURE 28

Using the law of cosines, we have

$$a^2 = b^2 + c^2 - 2bc \cos \alpha$$

or

$$b^2 + c^2 - a^2 - 2bc \cos \alpha = 0$$

So,

$$c^2 - (2b \cos \alpha)c + (b^2 - a^2) = 0$$

(a) Use the quadratic formula with c as the unknown to show that

$$c = b \cos \alpha \pm \sqrt{a^2 - b^2 \sin^2 \alpha}$$

(b) Under what conditions on a and h are there no real solutions for c? Relate the result to the ambiguous case when no triangles exist for given values of a, b, and α.

(c) If $a^2 - b^2 \sin^2 \alpha > 0$, how many possible triangles exist? Explain.

7.2 Introduction to Polar Coordinates

OBJECTIVES

1. Plot points in the polar coordinate system.
2. Convert points from polar to Cartesian form and vice versa.
3. Convert equations from Cartesian to polar form and vice versa.
4. Graph polar equations.

So far, we have used the Cartesian (rectangular) coordinate system to represent points in the plane. There is another system, called the *polar coordinate system*, where we use angles and radial distances to represent points. For certain procedures in calculus, it is easier to work with equations written in terms of polar coordinates rather than Cartesian coordinates. In this section, we'll study the polar coordinate system and use trigonometry to relate it to the Cartesian coordinate system.

Plotting Points in the Polar Coordinate System

The frame of reference for the polar coordinate system consists of a fixed point O, called the **pole**, and a fixed ray with end point O, called the **polar axis** (Figure 1a). The position of a point P is determined by coordinates r and θ, where θ is any angle having the polar axis as its initial side and the ray from the pole through point P as its terminal side; r is the directed distance along the terminal side of θ from the pole O to P. The position of P is given by (r, θ), called the **polar coordinates** of P (Figure 1b). Figure 1c shows a sample grid for the polar coordinate system. The angle θ can be measured either in degrees or radians. It may be positive or negative, depending on whether it is generated by a counter-clockwise or clockwise rotation.

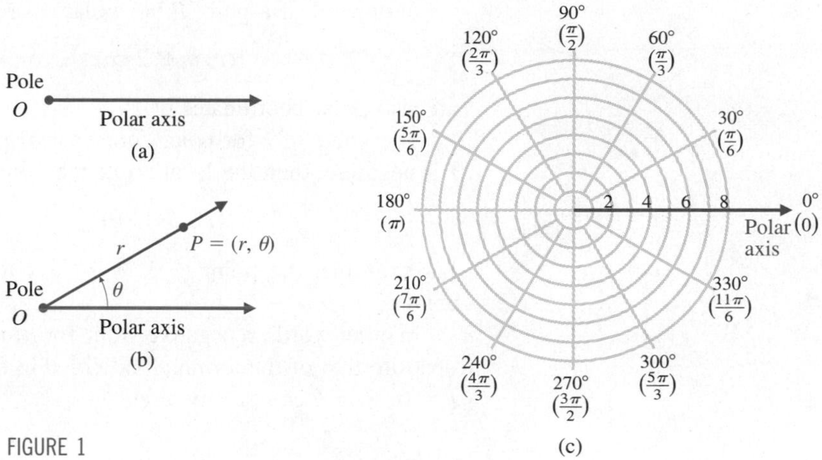

FIGURE 1

EXAMPLE 1 Plotting points in the polar coordinate system

Plot the points with the given polar coordinates.

(a) $(3, 60°)$ (b) $(4, -30°)$ (c) $\left(4, \dfrac{3\pi}{4}\right)$

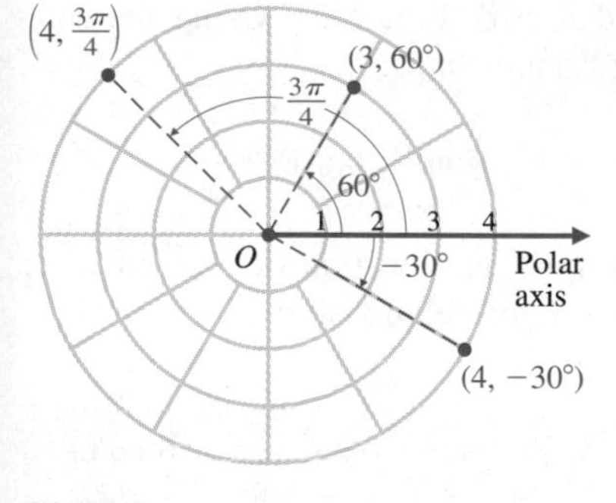

FIGURE 2

Solution

(a) To plot the point $(3, 60°)$, we first form an angle of 60° in such a way that the polar axis is the initial side of the angle. Then we move 3 units from the pole on the terminal side of this angle to locate the point (Figure 2).

(b)–(c) The other points in parts (b) and (c) are plotted in a similar way (Figure 2). ■

Note that in the Cartesian system each point has a unique number pair representation, whereas in the polar system, a point has infinitely many possible number pair representations because of periodicity. For example, the polar coordinates $(2, 315°)$, $(2, -45°)$, and $(2, 675°)$ all represent the same point (Figure 3).

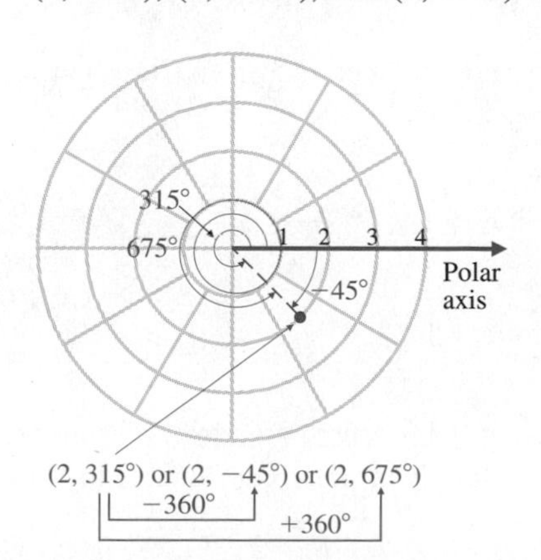

FIGURE 3

In general, if a point P has polar coordinates (r, θ), then for any integer n,

$$(r, \theta + 2\pi n) \qquad \text{or} \qquad (r, \theta + 360°n)$$

are also polar coordinates of P.

The value of r for polar coordinates (r, θ) may be positive, negative, or 0. If r is negative, then the location of the point (r, θ) is the same as the point

$$(|r|, \theta + \pi) \qquad \text{or} \qquad (|r|, \theta + 180°)$$

For example, the point $(-3, \pi/4)$ has the same location as $(3, (5\pi)/4)$ (Figure 4).

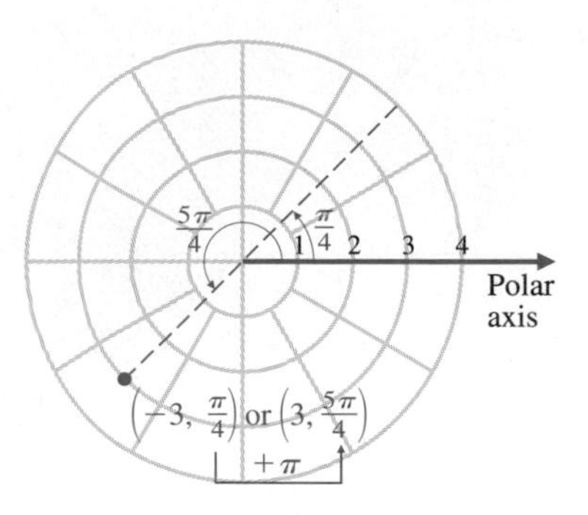

FIGURE 4

In other words, a negative value for r indicates moving $|r|$ units in the direction opposite that of the terminal side of θ in order to locate the point (r, θ). Also, if $r = 0$, then θ can be any angle.

Converting Points from Polar to Cartesian Form and Vice Versa

To establish the relationship between polar coordinates and Cartesian coordinates, we draw the polar axis of a polar coordinate system so that it coincides with the positive x axis of a Cartesian coordinate system. Any point P in the plane can be located either by polar coordinates (r, θ) or by Cartesian coordinates (x, y). For instance, suppose that P is in quadrant I and $r > 0$ (Figure 5). Using the right triangle in Figure 5, we obtain the following relationships:

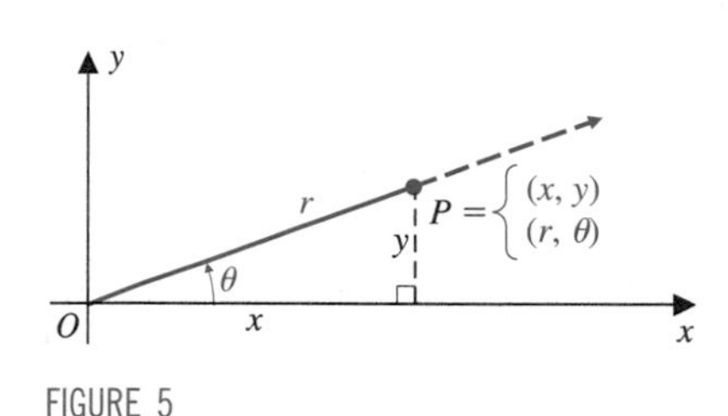

FIGURE 5

$$\cos\theta = \frac{x}{r}, \quad \sin\theta = \frac{y}{r}, \quad x^2 + y^2 = r^2, \quad \text{and} \quad \tan\theta = \frac{y}{x}$$

These relationships are true regardless of the quadrant location of P, even when $r < 0$. In general, we have the following conversion relationships:

CONVERSION OF COORDINATES

1. To convert a point P from polar coordinates (r, θ) to Cartesian coordinates (x, y), use

$$x = r\cos\theta \qquad \text{and} \qquad y = r\sin\theta$$

2. To convert a point P from Cartesian coordinates (x, y) to polar coordinates (r, θ), use

$$r^2 = x^2 + y^2$$

so that if $r > 0$, then $r = \sqrt{x^2 + y^2}$; and also use

$$\tan\theta = \frac{y}{x}$$

EXAMPLE 2 Converting polar coordinates to Cartesian coordinates

Convert the given polar coordinates to Cartesian coordinates.

(a) $(3, 60°)$ (b) $\left(4, -\frac{5\pi}{6}\right)$

Solution To convert these points from polar to Cartesian coordinates, we proceed as follows:

(a) For the point (3, 60°), we have

$$x = r\cos\theta = 3\cos 60° = 3 \cdot \frac{1}{2} = \frac{3}{2}$$

$$y = r\sin\theta = 3\sin 60° = 3 \cdot \left(\frac{\sqrt{3}}{2}\right) = \frac{3\sqrt{3}}{2}$$

So the Cartesian coordinates are $\left(\frac{3}{2}, \frac{3\sqrt{3}}{2}\right)$.

(b) For the point $\left(4, -\frac{5\pi}{6}\right)$, we have

$$x = r\cos\theta = 4\cos\left(-\frac{5\pi}{6}\right) = 4\left(-\frac{\sqrt{3}}{2}\right) = -2\sqrt{3}$$

$$y = r\sin\theta = 4\sin\left(-\frac{5\pi}{6}\right) = 4\left(-\frac{1}{2}\right) = -2$$

So the Cartesian coordinates are $\left(-2\sqrt{3}, -2\right)$. ■

It should be noted that the conversion equation $\tan\theta = y/x$ does not determine θ uniquely in terms of x and y. To determine the polar coordinates of P, we must pay attention to the quadrant in which P lies and the restrictions on θ, as the next example illustrates.

EXAMPLE 3 Converting Cartesian coordinates to polar coordinates

Convert the given Cartesian coordinates to polar coordinates, where $r \geq 0$ and $0° \leq \theta < 360°$.

(a) $(-1, 1)$ (b) $(5.03, -2.76)$ Round off the answers to two decimal places.

Solution Here, we use the conversion equations $r = \sqrt{x^2 + y^2}$ and $\tan\theta = y/x$, keeping in mind that the polar coordinates of a given point are not unique.

(a) For the point $(-1, 1)$, we have

$$r = \sqrt{x^2 + y^2} = \sqrt{(-1)^2 + 1^2} = \sqrt{2} \qquad \text{and} \qquad \tan\theta = -1$$

Because the point $(-1, 1)$ lies in quadrant II, we may use $\theta = 135°$, so polar coordinates of the point are $\left(\sqrt{2}, 135°\right)$.

(b) For the point $(5.03, -2.76)$, we have

$$r = \sqrt{(5.03)^2 + (-2.76)^2} = 5.74 \text{ (approx.)} \qquad \text{and} \qquad \tan\theta = \frac{-2.76}{5.03}$$

Because the point lies in quadrant IV, we have $270° < \theta < 360°$. Since $\tan\theta = -2.76/5.03$, the corresponding reference angle is approximately

$$\theta_R = \tan^{-1}\frac{2.76}{5.03} = 28.75°$$

So, $\theta = 360° - \theta_R = 331.25°$, and polar coordinates are (5.74, 331.25°).

Converting Equations from Cartesian to Polar Form and Vice Versa

We know that an equation in x and y may be represented by a graph in the Cartesian coordinate system. Suppose we superimpose on the xy system a polar system so that the polar axis coincides with the positive x axis. Clearly, the graph maintains its shape. By using the conversion equations we can determine an equation for the graph in terms of polar coordinates r and θ.

EXAMPLE 4 Converting a Cartesian equation to polar form

Sketch the graph of the Cartesian equation

$$(x - 2)^2 + y^2 = 4$$

and then find an equation of the graph in polar form.

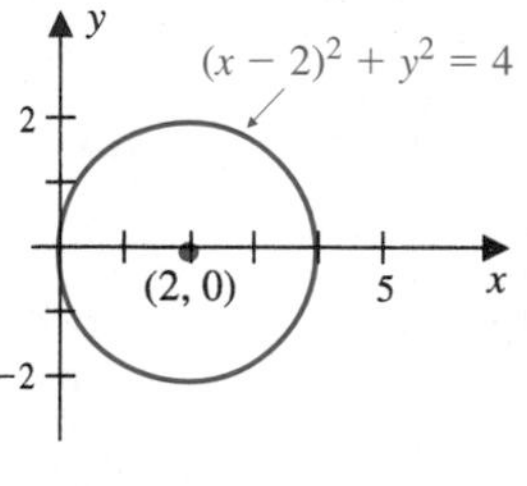

FIGURE 6

Solution We recognize the graph of this equation as a circle with radius 2 and center at (2, 0) (Figure 6). To convert the given equation to polar form, we proceed as follows:

$$\begin{aligned} (x-2)^2 + y^2 &= 4 && \text{Given} \\ x^2 - 4x + 4 + y^2 &= 4 && \text{Multiply} \\ x^2 + y^2 - 4x &= 0 && \text{Simplify and rewrite} \\ r^2 - 4r\cos\theta &= 0 && x^2 + y^2 = r^2 \text{ and } x = r\cos\theta \\ r^2 &= 4r\cos\theta && \text{Add } 4r\cos\theta \text{ to each side} \\ r &= 4\cos\theta && \text{Divide each side by } r \end{aligned}$$

Note that the resulting equation is valid even when $r = 0$.

We can also convert an equation from polar form to Cartesian form. Once again, note that even though the equations are given in different coordinate systems, the graphs are the same.

EXAMPLE 5 Converting a polar equation to Cartesian form

Convert the polar equation $\theta = \pi/3$ to Cartesian form, and then sketch the graph.

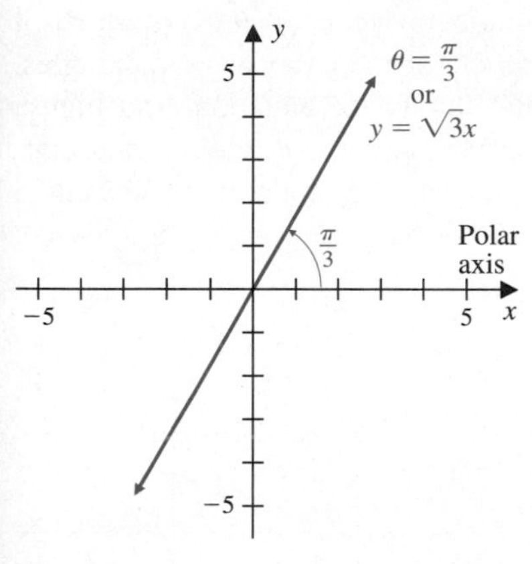

FIGURE 7

Solution Since $\theta = \pi/3$, it follows from the conversion equation $\tan \theta = y/x$ that

$$\tan \frac{\pi}{3} = \frac{y}{x} \qquad \text{or} \qquad \sqrt{3} = \frac{y}{x}$$

$$y = \sqrt{3}\,x$$

This last equation is the equation of a line with slope $\sqrt{3}$ and containing the origin (Figure 7).

Graphing Polar Equations

By the **graph** of a polar equation, we mean the set of all points (r, θ) that satisfy an equation that relates r and θ. In Example 5, we graphed the polar equation by first converting it to Cartesian form, and then we used the Cartesian coordinate system to sketch the graph. Polar equations can also be graphed directly by using the point-plotting method.

EXAMPLE 6 Graphing a polar equation

Use point-plotting to sketch the graph of each equation.

(a) $r = 3$ (b) $r = \theta$, where $\theta \geq 0$ (c) $r = 1 + \sin \theta$

Solution

(a) The equation $r = 3$ indicates that a point is on the graph if and only if it is of the form $(3, \theta)$, where θ is any angle. After plotting a few points of this form and then connecting them with a smooth curve, we see that the graph consists of all points that are 3 units away from the pole. The graph is the circle with center at the pole and radius 3 (Figure 8a).

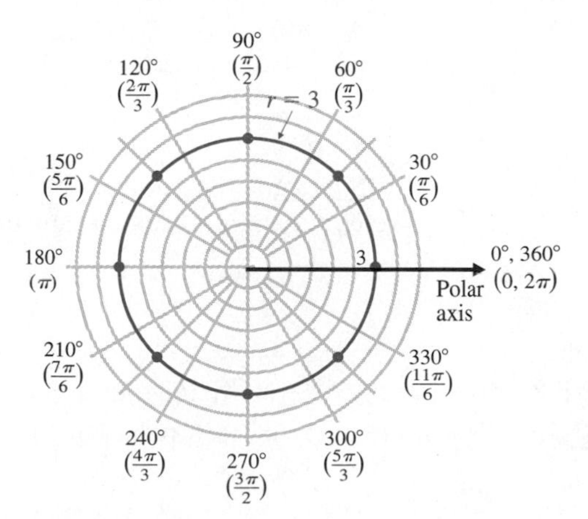

FIGURE 8a

The graph of an equation of the form $r = k\theta$ is called an **Archimedean spiral**.

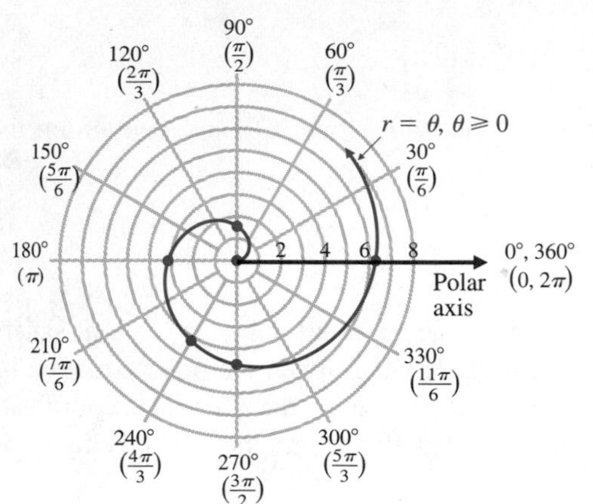

FIGURE 8b

(b) Since r represents distances (real numbers) and $r = \theta$ (for $\theta \geq 0$), it is implied that θ represents real numbers or radian measures of angles. The table lists some coordinates that satisfy the equation, and Figure 8b shows the resulting graph when these points are plotted and connected with a smooth curve. Note that as the angle θ is allowed to increase (rotate counterclockwise), the distances r correspondingly get longer. This results in a spiral-shaped curve.

θ	0	$\frac{\pi}{2}$	π	$\frac{4\pi}{3}$	$\frac{3\pi}{2}$	2π
r	0	1.57 (approx.)	3.14 (approx.)	4.19 (approx.)	4.71 (approx.)	6.28 (approx.)

(c) To graph $r = 1 + \sin\theta$ we first build a table of coordinates that satisfy the equation, then plot the points, and finally connect them with a smooth curve. The results are shown in Figure 8c.

θ	0	$\frac{\pi}{6}$	$\frac{\pi}{3}$	$\frac{\pi}{2}$	$\frac{2\pi}{3}$	$\frac{5\pi}{6}$	π	$\frac{3\pi}{2}$	2π
r	1	$\frac{3}{2} = 1.5$	1.87 (approx.)	2	1.87 (approx.)	$\frac{3}{2} = 1.5$	1	0	1

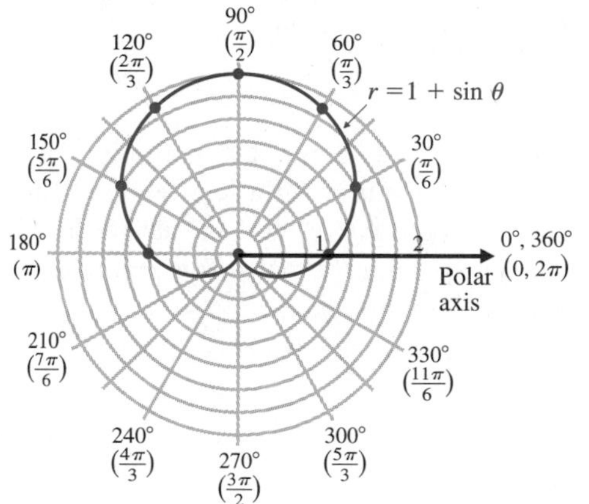

FIGURE 8c

This graph is called a **cardioid** because of its heart shape.

Some graphing utilities have a polar graphing mode that enables us to sketch the graphs of polar equations of the form $r = f(\theta)$.

EXAMPLE 7 **Using a graphing utility to graph a polar equation**

GU

Sketch the graph of each polar equation by using a graphing utility.

(a) $r = 2\cos 2\theta$
(b) $r = 1 + 2\sin\theta$

Solution

(a) A viewing window of the graph of $r = 2 \cos 2\theta$ is shown in Figure 9a.
(b) Figure 9b displays a viewing window of the graph of $r = 1 + 2 \sin \theta$.

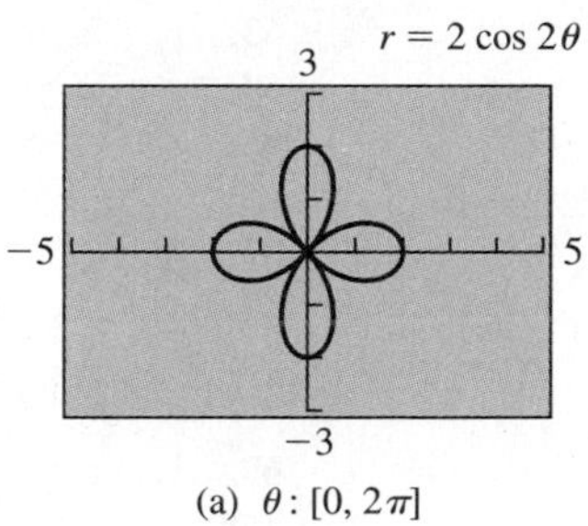

(a) θ: $[0, 2\pi]$

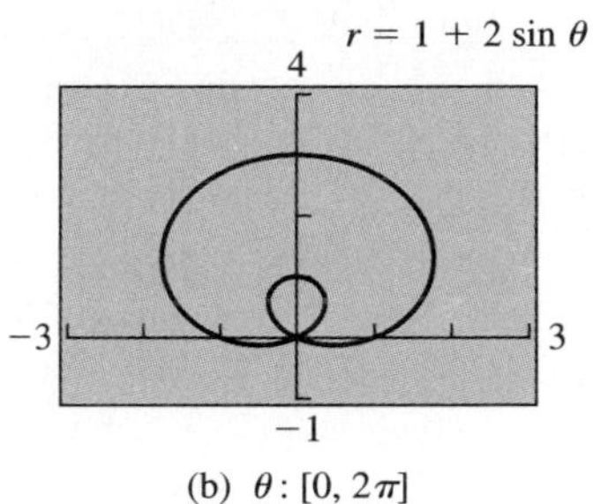

(b) θ: $[0, 2\pi]$

FIGURE 9

PROBLEM SET 7.2

Mastering the Concepts

In problems 1–6, plot points P and Q on the same polar coordinate system.

1. $P = (2, -45°)$; $Q = (2, 315°)$
2. $P = (3, -60°)$; $Q = (3, 300°)$
3. $P = \left(3, \frac{\pi}{6}\right)$; $Q = \left(-3, \frac{\pi}{6}\right)$
4. $P = \left(4, \frac{7\pi}{6}\right)$; $Q = \left(-4, \frac{7\pi}{6}\right)$
5. $P = \left(4, \frac{2\pi}{3}\right)$; $Q = \left(4, -\frac{2\pi}{3}\right)$
6. $P = \left(-5, \frac{\pi}{4}\right)$; $Q = \left(-5, -\frac{\pi}{4}\right)$

In problems 7 and 8, plot the given point P in the polar coordinate system, and then find three additional representations of P such that: **(i)** $r > 0$ **(ii)** $r < 0$

7. **(a)** $P = (3, 100°)$ **(b)** $P = \left(-3, \frac{\pi}{4}\right)$
8. **(a)** $P = (4, 60°)$ **(b)** $P = \left(4, -\frac{\pi}{3}\right)$

In problems 9–14, determine the Cartesian coordinates of the points with the given polar coordinates. Round off the answers to two decimal places in problems 11–14.

9. **(a)** $(6, 30°)$ **(b)** $\left(4, -\frac{\pi}{4}\right)$
10. **(a)** $(-8, 45°)$ **(b)** $\left(4, -\frac{\pi}{6}\right)$
11. **(a)** $\left(3, -\frac{\pi}{7}\right)$ **(b)** $(4, -110°)$
12. **(a)** $\left(-3, \frac{5\pi}{13}\right)$ **(b)** $(-6, 213°)$
13. **(a)** $(2, -3)$ **(b)** $\left(-3, \frac{5\pi}{2}\right)$
14. **(a)** $(-2, 4)$ **(b)** $\left(4, -\frac{\pi}{9}\right)$

In problems 15–18, determine the polar coordinates of the points with the given Cartesian coordinates satisfying the conditions that $r > 0$ and $0 \le \theta < 2\pi$. Round off the answers to two decimal places in problems 17 and 18.

15. **(a)** $\left(-1, \sqrt{3}\right)$ **(b)** $\left(-6, 6\sqrt{3}\right)$
16. **(a)** $(5, 5)$ **(b)** $\left(2\sqrt{3}, -2\right)$
17. **(a)** $(-3, 5.1)$ **(b)** $(-6, -7)$
18. **(a)** $\left(4\sqrt{2}, -\sqrt{3}\right)$ **(b)** $(-4.3, 1.4)$

In problems 19–32, sketch the graph of each Cartesian equation. Then find an equation of the graph in polar form.

19. $x = 2$
20. $y = 3$
21. $y = -3$
22. $x = -2$

23. $y = 3x - 2$
24. $x - y = 5$
25. $x^2 + y^2 = 25$
26. $x^2 + y^2 = 7$
27. $(x + 4)^2 + y^2 = 16$
28. $x^2 + y^2 - 6y = 0$
29. $y = 4x^2$
30. $xy = 4$
31. $y^2 = 8x$
32. $y = \sqrt{x^2 + 4}$

In problems 33–46, find an equation in terms of Cartesian coordinates that corresponds to each polar equation. Then sketch the graph.

33. $r = 3$
34. $r = -2$
35. $r = 2 \csc \theta$
36. $r = -4 \sec \theta$
37. $\theta = \frac{\pi}{4}$
38. $\theta = \frac{5\pi}{3}$
39. $r = -4 \cos \theta$
40. $r = 2 \sin \theta$
41. $r = 4 \tan \theta \sec \theta$
42. $r^2 \sin 2\theta = 4$
43. $4r \cos \theta + 3r \sin \theta = 12$
44. $2r \sin \theta - 3r \cos \theta = 6$
45. $r = 4 \cos \theta + 2 \sin \theta$
46. $r = \cos \theta - \sin \theta$

In problems 47–54, use point-plotting to sketch the graph of each polar equation.

47. $r = 5$
48. $r = 1$
49. $\theta = -\frac{\pi}{3}$
50. $\theta = \frac{\pi}{6}$
51. $r = \cos \theta$
52. $r = -2 \sin \theta$
53. $r = 1 + \cos \theta$
54. $r = 1 - \cos \theta$

GU Problems 55–68 refer to the following: Some graphs of polar equations are symmetric. Table 1 indicates how to recognize certain types of symmetry:

TABLE 1

Testing a Polar Equation for Symmetry
If (r, θ) Lies on the Graph:

Type of Symmetry	Then These Points Lie on the Graph
1. The polar axis	$(-r, \pi - \theta)$ or $(r, -\theta)$
2. The line $\theta = \pi/2$	$(-r, -\theta)$ or $(r, \pi - \theta)$
3. The pole	$(-r, -\theta)$ or $(r, \pi + \theta)$

In problems 55–68, apply the symmetry tests in Table 1 to determine the symmetry (if any) of each graph. Then use a graphing utility to graph each polar equation to demonstrate the symmetry.

55. $r = 1 - \sin \theta$ (cardioid)
56. $r = 1 + 2 \cos \theta$
57. $r = 2 + 3 \cos \theta$ (limaçon)
58. $r = 1 - 2 \sin \theta$ (limaçon)
59. $r = 4 + 3 \sin \theta$ (limaçon)
60. $r = 5 + 3 \cos \theta$ (limaçon)
61. $r = -2 \cos 2\theta$
62. $r = 4 \sin 2\theta$ (four-petaled rose)
63. $r = 2 \sin 3\theta$ (three-petaled rose)
64. $r = 2 \cos 3\theta$ (three-petaled rose)
65. $r^2 = 4 \cos 2\theta$ (lemniscate)
66. $r^2 = 4 \sin 2\theta$ (lemniscate)
67. $r = 1/\theta, \theta > 0$ (spiral)
68. $r = e^\theta, \theta \geq 0$ (spiral)

Developing and Extending the Concepts

69. GU Use a graphing utility to sketch the graphs of $r = 3 \sin \theta$ and $r = 3 + 3 \sin \theta$ in the same viewing window. Can the graph of $r = 3 + 3 \sin \theta$ be obtained by vertically shifting the graph of $r = 3 \sin \theta$ up 3 units? Explain.

70. GU Use a graphing utility to sketch the graphs of the equations $r = 2 \cos 3\theta$ and $r = |2 \cos 3\theta|$ in two viewing windows with the same settings. Compare the two graphs. Is it possible to obtain the graph of $r = |2 \cos 3\theta|$ from the graph of $r = 2 \cos 3\theta$? Explain.

71. Show that the graph of the polar equation

$$r = \frac{c}{a \cos \theta + b \sin \theta}$$

where a, b, and c are constants, and a and b are not both 0, is a straight line. What is the slope of the line?

72. Use the law of cosines to show that the distance d between the polar points $P = (r_1, \theta_1)$ and $Q = (r_2, \theta_2)$ satisfies the equation

$$d^2 = [d(P, Q)]^2 = r_1{}^2 + r_2{}^2 - 2r_1r_2 \cos(\theta_2 - \theta_1)$$

73. Polar Coordinates of a Tile: Find the polar coordinates of the corners of a tile in the shape of a regular hexagon (a polygon with six equal sides) if the center is at the pole, one vertex lies on the polar axis, and the distance from the center to any vertex is a units (Figure 10).

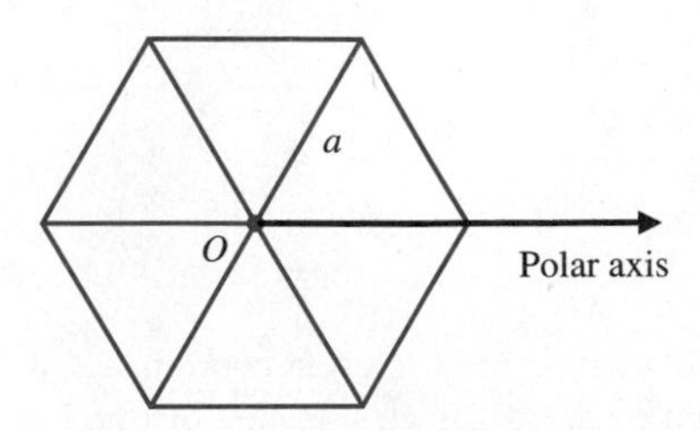

FIGURE 10

OBJECTIVES

1. Describe vector operations geometrically.
2. Perform vector operations analytically.
3. Find unit vectors and direction angles.
4. Solve applied problems.

7.3 Vectors in the Plane

Until now, we have dealt with quantities that can be measured or represented by single real numbers, such as area, volume, time, temperature, and speed. Since real numbers can be represented by points on a number scale, these quantities are often called *scalars*. In this section, we deal with quantities called *vectors*, which cannot be described or represented by a single real number. Vectors are important tools in applications involving force, velocity, acceleration, and displacement. We'll see how trigonometry is used when working with vectors.

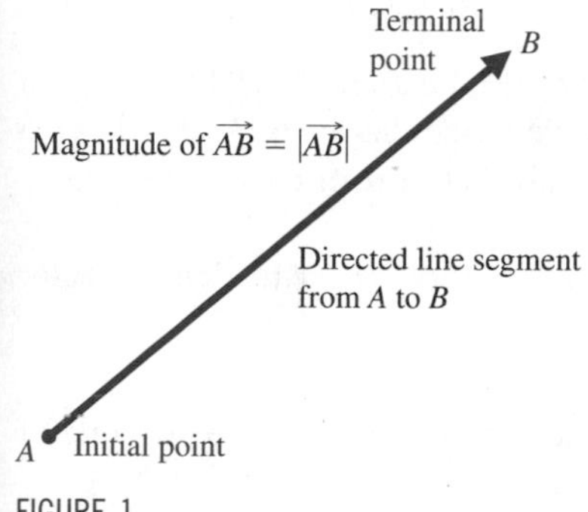

FIGURE 1

Describing Vector Operations Geometrically

Geometrically, a **vector** in the plane is a line segment with a direction usually denoted by an arrowhead at one end of the segment. The end point that contains the arrowhead is called the **terminal point**, and the other end point is called the **initial point** of the vector. The length of a vector is called its **magnitude**. The vector in Figure 1, denoted by $\overrightarrow{AB}$, has initial point A and terminal point B, and its magnitude is denoted by $|\overrightarrow{AB}|$.

The **zero vector**, written **0**, is a vector whose initial and terminal points are the same. Its magnitude is given by $|\mathbf{0}| = 0$. Two vectors are considered to be **equal** if they have the same magnitude and the same direction, such as vectors $\overrightarrow{PQ}$ and $\overrightarrow{RS}$ in Figure 2a. Note that $\overrightarrow{AB}$ and $\overrightarrow{BA}$ in Figure 2b are not equal, even though their lengths are the same, because they have opposite directions. That is, $\overrightarrow{AB} \neq \overrightarrow{BA}$, even though $|\overrightarrow{AB}| = |\overrightarrow{BA}|$.

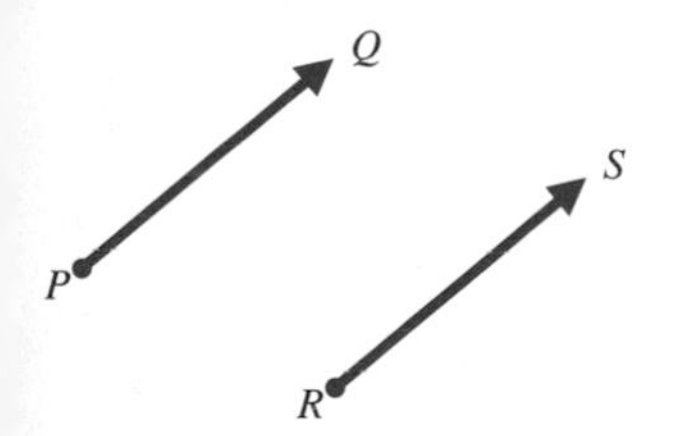

(a) Equal vectors, $\overrightarrow{PQ} = \overrightarrow{RS}$

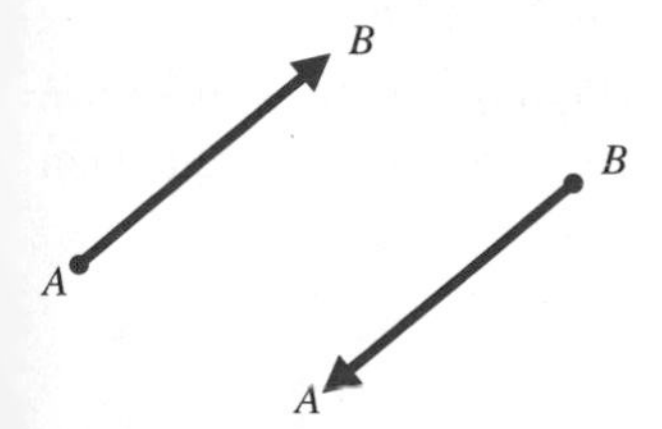

(b) Unequal vectors, $\overrightarrow{AB} \neq \overrightarrow{BA}$

FIGURE 2

It is important to understand that a vector can be shifted from one location to another as long as we do not change its magnitude or its direction. Special notation is used for vectors so that they can be distinguished from real numbers. In this book, we normally use lowercase boldface letters to denote vectors. Hence, we speak of "vector **u**" and just write boldface **u**.

Two vectors **u** and **v** (Figure 3a) may be added to form the *sum* **u** + **v** as follows. First, we shift **v** so that its initial point coincides with the terminal point of **u**. The vector having the same initial point as **u** and the same terminal point as **v** is defined to be the **sum (resultant vector) u + v** of **u** and **v** (Figure 3b). As Figure 3b shows, the resultant vector is a diagonal vector of the parallelogram determined by vectors **u** and **v**. This description of addition is called the *parallelogram law of addition*.

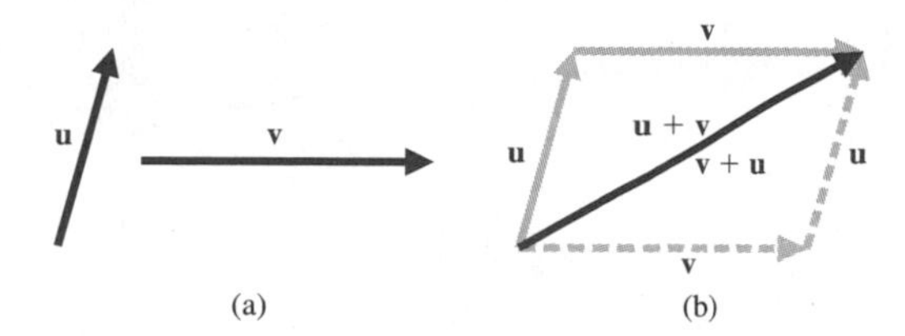

FIGURE 3

Figure 3b displays the fact that **u** + **v** = **v** + **u**; that is, vector addition is commutative.

Vectors can be multiplied by real numbers, called **scalars** in this context, to form other vectors. We refer to this operation as **scalar multiplication**, and we describe it as follows. If $a \neq 0$ is a scalar (real number) and $\mathbf{u}$ is a vector, then the vector $a\mathbf{u}$ is a vector that satisfies the following conditions:

1. **The Magnitude of $a\mathbf{u}$:** $|a\mathbf{u}| = |a||\mathbf{u}|$; that is, the magnitude of $a\mathbf{u}$ is the product of the absolute value of a and the magnitude of $\mathbf{u}$.
2. **The Direction of $a\mathbf{u}$:**
   If $a > 0$, then the direction of $a\mathbf{u}$ is the same as the direction of $\mathbf{u}$.
   If $a < 0$, then the direction of $a\mathbf{u}$ is opposite the direction of $\mathbf{u}$.

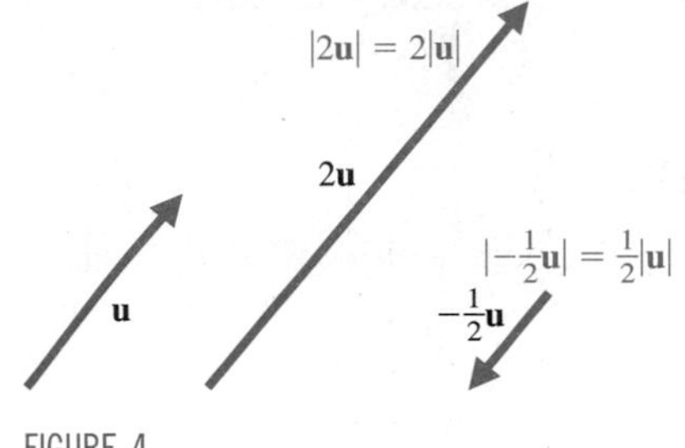
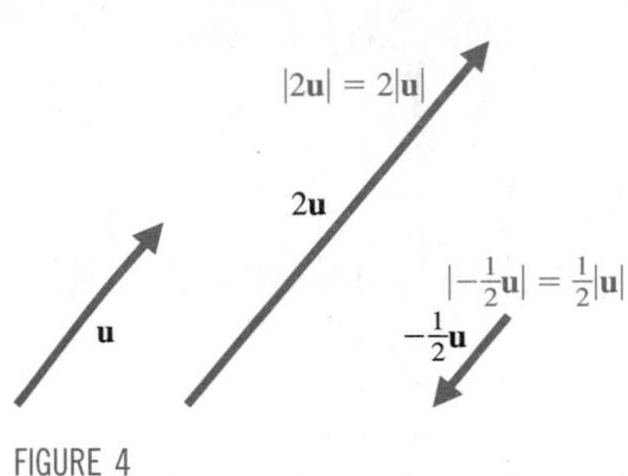

FIGURE 4

Figure 4 displays some illustrations of scalar multiplication. Note that $2\mathbf{u}$ is a vector in the same direction as $\mathbf{u}$, with magnitude twice that of $\mathbf{u}$. And $-\frac{1}{2}\mathbf{u}$ is a vector opposite in direction to $\mathbf{u}$, with a length half that of $\mathbf{u}$. If $a = 0$, then $a\mathbf{u} = 0\mathbf{u} = \mathbf{0}$.

Now we can use vector addition and scalar multiplication to define vector **subtraction** as

$$\mathbf{u} - \mathbf{v} = \mathbf{u} + (-\mathbf{v}) \quad \text{(Figure 5a)}$$

Observe that if we form the parallelogram defined by $\mathbf{u}$ and $\mathbf{v}$ (Figure 5b), one of the diagonal vectors is $\mathbf{u} + \mathbf{v}$ and the other is $\mathbf{u} - \mathbf{v}$.

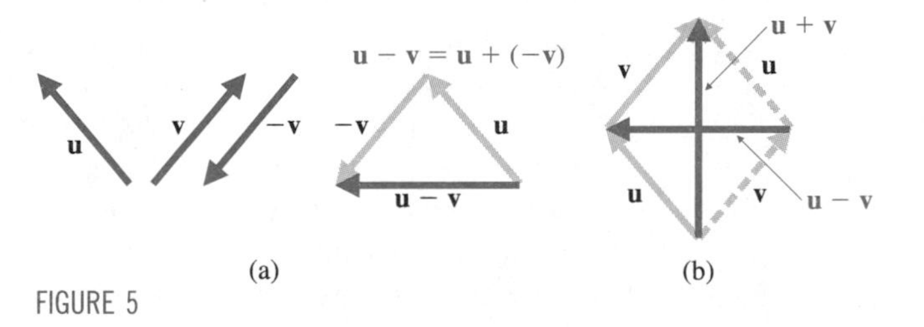

FIGURE 5

EXAMPLE 1 Describing vector operations geometrically

Given vectors $\mathbf{u}$ and $\mathbf{v}$ as shown in Figure 6, display the vector $3\mathbf{u} - 2\mathbf{v}$.

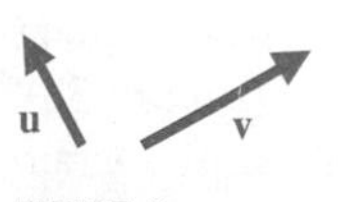

FIGURE 6

Solution To form $3\mathbf{u} - 2\mathbf{v}$, we first form $3\mathbf{u}$ and $2\mathbf{v}$ (Figure 7a). Next, we form the parallelogram defined by $3\mathbf{u}$ and $2\mathbf{v}$. The difference $3\mathbf{u} - 2\mathbf{v}$ is the diagonal vector shown in Figure 7b.

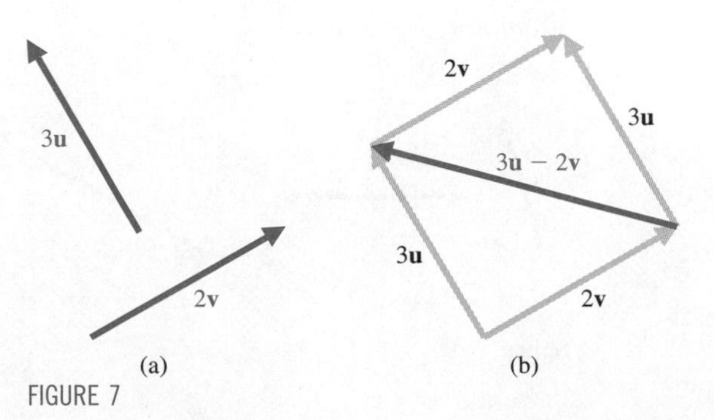

FIGURE 7

Performing Vector Operations Analytically

When a vector is positioned in a Cartesian plane so that its initial point is at the origin, it is called a **radius vector** or **standard position vector**. Such a vector can be uniquely represented by the coordinates of its terminal point. For instance, Figure 8a shows a radius vector $\mathbf{u}$ with terminal point (a, b). We call a the $\boldsymbol{x}$ **component** and b the $\boldsymbol{y}$ **component** of $\mathbf{u}$, and we write $\mathbf{u}$ in the **component form**

$$\mathbf{u} = \langle a, b \rangle$$

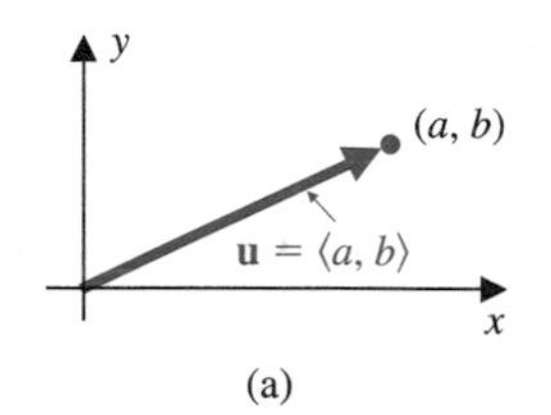

From the Pythagorean theorem, the **magnitude** $|\mathbf{u}|$ of vector $\mathbf{u} = \langle a, b \rangle$ is given by

$$|\mathbf{u}| = \sqrt{a^2 + b^2}$$

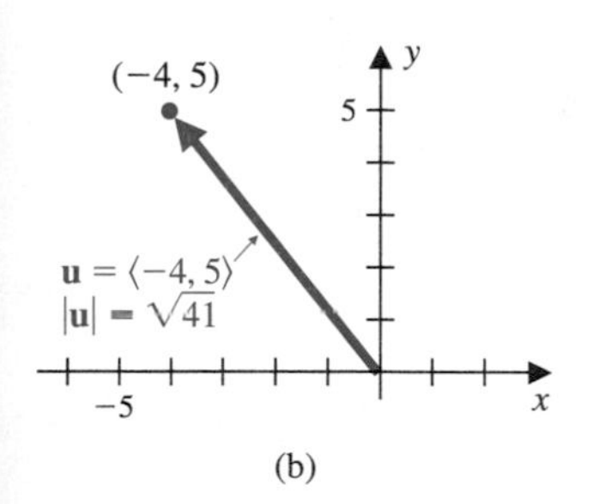

FIGURE 8

For example, the vector $\mathbf{u} = \langle -4, 5 \rangle$ has x component -4, y component 5, and $|\mathbf{u}| = \sqrt{(-4)^2 + 5^2} = \sqrt{41}$ (Figure 8b).

If $\mathbf{u}$ and $\mathbf{v}$ are equal position vectors, they have the same initial point $(0, 0)$, so the components must also be equal; that is:

> If $\mathbf{u} = \langle x_1, y_1 \rangle$ and $\mathbf{v} = \langle x_2, y_2 \rangle$, then $\mathbf{u} = \mathbf{v}$ whenever $x_1 = x_2$ and $y_1 = y_2$.

For instance, if $\mathbf{u} = \mathbf{w}$, where $\mathbf{u} = \langle 1, -3 \rangle$ and $\mathbf{w} = \langle 1, b \rangle$, then $b = -3$.

Suppose that $\mathbf{w}$ is a vector in a Cartesian plane and $\mathbf{w}$ is not a radius vector. Assume that the initial point of $\mathbf{w}$ is (x_1, y_1) and its terminal point is (x_2, y_2), as shown in Figure 9. To find the radius vector representation of $\mathbf{w}$, we shift $\mathbf{w}$ from its given position in such a way that the initial point (x_1, y_1) is positioned at the origin, to obtain the component form

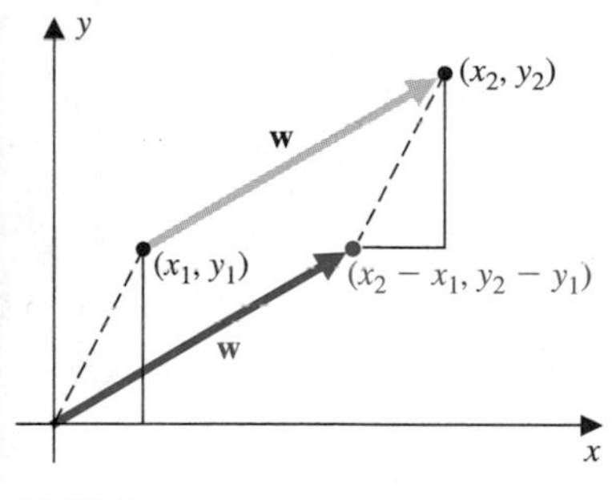

FIGURE 9

$$\mathbf{w} = \langle x_2 - x_1, y_2 - y_1 \rangle$$

EXAMPLE 2 **Finding the component form**

Let $P = (3, -2)$ be the initial point and $Q = (-4, 1)$ be the terminal point of the vector $\mathbf{u} = \overrightarrow{PQ}$. Find its component form, and display $\mathbf{u}$ in both its given position and its standard position.

Solution Let $P = (3, -2) = (x_1, y_1)$ and $Q = (-4, 1) = (x_2, y_2)$. The x and y components of $\mathbf{u} = \overrightarrow{PQ}$ are given, respectively, by

$$x_2 - x_1 = -4 - 3 = -7 \quad \text{and} \quad y_2 - y_1 = 1 - (-2) = 3$$

So $\mathbf{u} = \langle -7, 3 \rangle$. Figure 10 displays the vector in both its given location and its standard position.

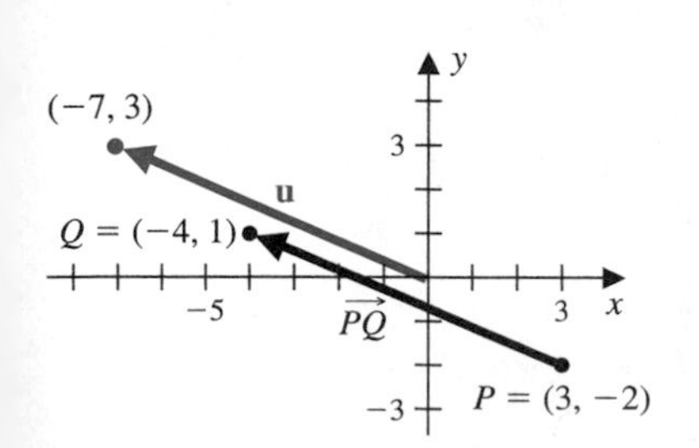

FIGURE 10

When vectors are written in component form, vector addition becomes very simple. Suppose that $\mathbf{u}$ and $\mathbf{v}$ are in standard position with $\mathbf{u} = \langle x_1, y_1 \rangle$ and $\mathbf{v} = \langle x_2, y_2 \rangle$. As displayed in Figure 11, moving the initial point of $\mathbf{v}$ to the terminal point of $\mathbf{u}$ shows that the x component of $\mathbf{u} + \mathbf{v}$ is obtained by adding the x components of $\mathbf{u}$ and $\mathbf{v}$. Similarly, the y component of $\mathbf{u} + \mathbf{v}$ is obtained by adding the y components of $\mathbf{u}$ and $\mathbf{v}$.

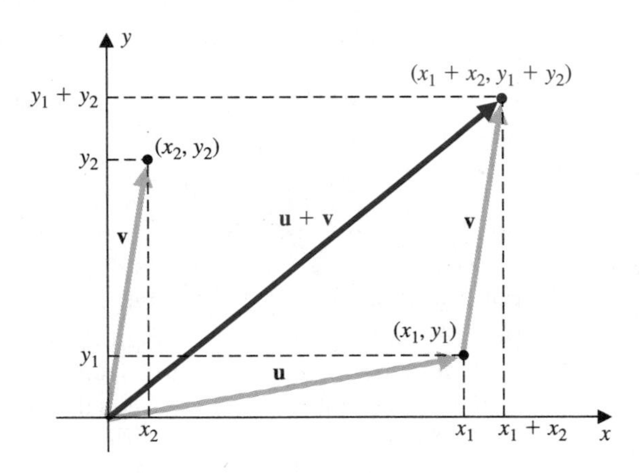

FIGURE 11

Thus, we have the *component rule for adding vectors*:

If $\mathbf{u} = \langle x_1, y_1 \rangle$ and $\mathbf{v} = \langle x_2, y_2 \rangle$, then

$$\mathbf{u} + \mathbf{v} = \langle x_1 + x_2, y_1 + y_2 \rangle$$

As illustrated in Figure 12, the *component rule for scalar multiplication* is given as follows:

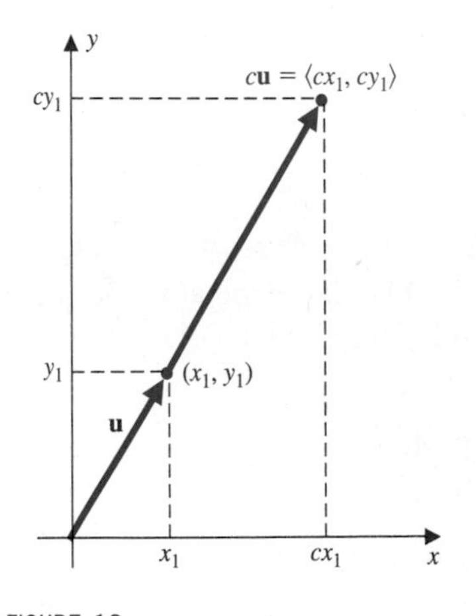

FIGURE 12

If $\mathbf{u} = \langle x_1, y_1 \rangle$ and c is a real number, then

$$c\mathbf{u} = c\langle x_1, y_1 \rangle = \langle cx_1, cy_1 \rangle$$

Finally, we derive the subtraction rule as follows: If $\mathbf{u} = \langle x_1, y_1 \rangle$ and $\mathbf{v} = \langle x_2, y_2 \rangle$, then $-\mathbf{v} = -1 \cdot \mathbf{v} = \langle -x_2, -y_2 \rangle$. So,

$$\begin{aligned} \mathbf{u} - \mathbf{v} &= \mathbf{u} + (-\mathbf{v}) \\ &= \langle x_1, y_1 \rangle + \langle -x_2, -y_2 \rangle \\ &= \langle x_1 - x_2, y_1 - y_2 \rangle \end{aligned}$$

Thus, we are led to the *component rule for subtracting vectors*:

If $\mathbf{u} = \langle x_1, y_1 \rangle$ and $\mathbf{v} = \langle x_2, y_2 \rangle$, then

$$\mathbf{u} - \mathbf{v} = \langle x_1 - x_2, y_1 - y_2 \rangle$$

EXAMPLE 3 Performing vector operations

Let $\mathbf{u} = \langle 3, 4 \rangle$ and $\mathbf{v} = \langle -5, 6 \rangle$. Find:

(a) $\mathbf{u} + \mathbf{v}$ (b) $\mathbf{u} - \mathbf{v}$ (c) $4\mathbf{u} - 3\mathbf{v}$ (d) $|4\mathbf{u} - 3\mathbf{v}|$

Solution

(a) $\mathbf{u} + \mathbf{v} = \langle 3, 4 \rangle + \langle -5, 6 \rangle = \langle 3 - 5, 4 + 6 \rangle = \langle -2, 10 \rangle$
(b) $\mathbf{u} - \mathbf{v} = \langle 3, 4 \rangle - \langle -5, 6 \rangle = \langle 3 - (-5), 4 - 6 \rangle = \langle 8, -2 \rangle$
(c) $4\mathbf{u} - 3\mathbf{v} = 4\langle 3, 4 \rangle - 3\langle -5, 6 \rangle = \langle 12, 16 \rangle + \langle 15, -18 \rangle = \langle 27, -2 \rangle$
(d) Using thc rcsult from part (c),

$$|4\mathbf{u} - 3\mathbf{v}| = \sqrt{(27)^2 + (-2)^2} = \sqrt{729 + 4} = \sqrt{733}$$

Finding Unit Vectors and Direction Angles

A vector of magnitude 1 is called a **unit vector**. There are two special unit vectors, called **basis vectors**, defined by

$$\mathbf{i} = \langle 1, 0 \rangle \quad \text{and} \quad \mathbf{j} = \langle 0, 1 \rangle$$

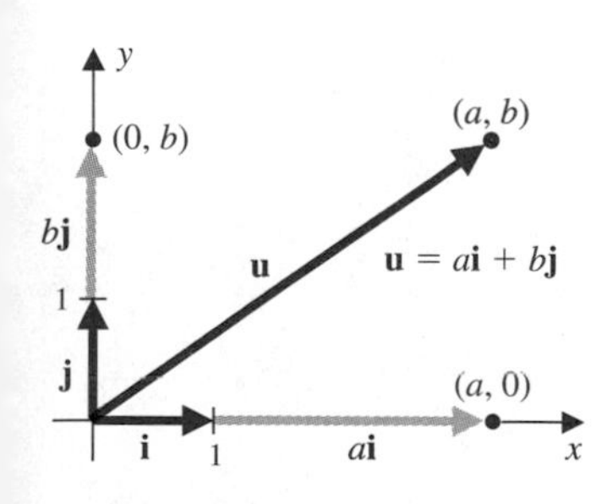

FIGURE 13

Any vector $\mathbf{u} = \langle a, b \rangle$ can be written in terms of $\mathbf{i}$ and $\mathbf{j}$ as follows:

$$\langle a, b \rangle = \langle a, 0 \rangle + \langle 0, b \rangle = a\langle 1, 0 \rangle + b\langle 0, 1 \rangle = a\mathbf{i} + b\mathbf{j} \quad \text{(Figure 13)}$$

Thus, we have

$$\mathbf{u} = \langle a, b \rangle = a\mathbf{i} + b\mathbf{j}$$

The component rules for addition, subtraction, and scalar multiplication apply to vectors written in terms of $\mathbf{i}$ and $\mathbf{j}$. For example,

$$(3\mathbf{i} + 5\mathbf{j}) + (2\mathbf{i} - 7\mathbf{j}) = 5\mathbf{i} - 2\mathbf{j}, \quad -2(3\mathbf{i} + 2\mathbf{j}) = -6\mathbf{i} - 4\mathbf{j},$$
$$(2\mathbf{i} - 3\mathbf{j}) - (\mathbf{i} + 5\mathbf{j}) = \mathbf{i} - 8\mathbf{j}$$

We know that for $a > 0$ and $\mathbf{u} \neq \mathbf{0}$, the vector $a\mathbf{u}$ has the same direction as $\mathbf{u}$, and its magnitude is $|a|$ times the magnitude of $\mathbf{u}$. Thus, the scalar product of $\mathbf{u} \neq \mathbf{0}$ and the reciprocal of its magnitude gives us a vector of magnitude 1 in the same direction as $\mathbf{u}$. That is:

A **unit vector** in the direction of $\mathbf{u} = \langle a, b \rangle$ is given by

$$\frac{1}{|\mathbf{u}|}\mathbf{u} = \left\langle \frac{a}{\sqrt{a^2 + b^2}}, \frac{b}{\sqrt{a^2 + b^2}} \right\rangle$$

The process of forming this vector is referred to as **normalizing the vector**.

EXAMPLE 4 Normalizing a vector

Find a unit vector in the same direction as $\mathbf{u} = 4\mathbf{i} - 3\mathbf{j}$.

Solution The magnitude of $\mathbf{u}$ is given by

$$|\mathbf{u}| = \sqrt{4^2 + (-3)^2} = 5$$

Thus, a unit vector in the same direction as $\mathbf{u}$ is given by

$$\frac{1}{|\mathbf{u}|}\mathbf{u} = \frac{1}{5}(4\mathbf{i} - 3\mathbf{j}) = \frac{4}{5}\mathbf{i} - \frac{3}{5}\mathbf{j}$$

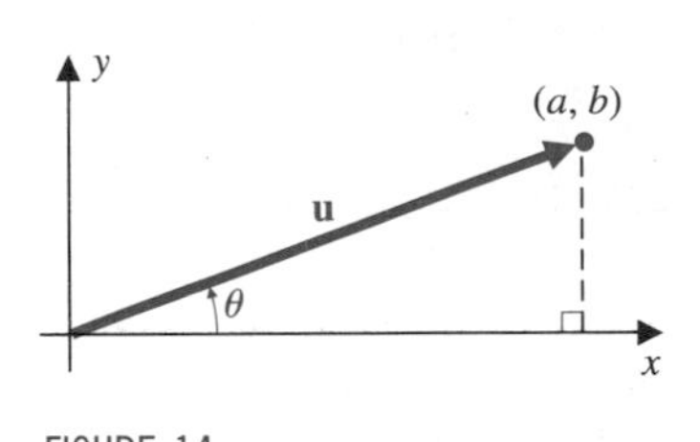

FIGURE 14

An angle θ formed by the vector $\mathbf{u} = \langle a, b \rangle$ and the positive x axis is called the **direction angle** of the vector $\mathbf{u}$ (Figure 14). Using trigonometry, we have

$$\cos \theta = \frac{a}{|\mathbf{u}|} \quad \text{and} \quad \sin \theta = \frac{b}{|\mathbf{u}|}$$

Rewriting these equations, we get

$$a = |\mathbf{u}| \cos \theta \quad \text{and} \quad b = |\mathbf{u}| \sin \theta$$
$$\mathbf{u} = a\mathbf{i} + b\mathbf{j} = |\mathbf{u}|(\cos \theta)\mathbf{i} + |\mathbf{u}|(\sin \theta)\mathbf{j}$$

Also, if $a \neq 0$, it follows that

$$\frac{b}{a} = \frac{|\mathbf{u}| \sin \theta}{|\mathbf{u}| \cos \theta} = \frac{\sin \theta}{\cos \theta} = \tan \theta$$

Thus, a direction angle θ for $\mathbf{u} = a\mathbf{i} + b\mathbf{j}$ can be found by solving

$$\tan \theta = \frac{b}{a}$$

where θ is selected according to the quadrant location of (a, b). For example, a direction angle of $\mathbf{u} = 5\mathbf{i} + 12\mathbf{j}$ satisfies

$$\tan\theta = \frac{b}{a} = \frac{12}{5} = 2.4$$

Since (5, 12) is in quadrant I, it follows that

$$\theta = \tan^{-1} 2.4 \quad \text{or approx. } 67.4° \quad \text{(Figure 15)}$$

is a direction angle.

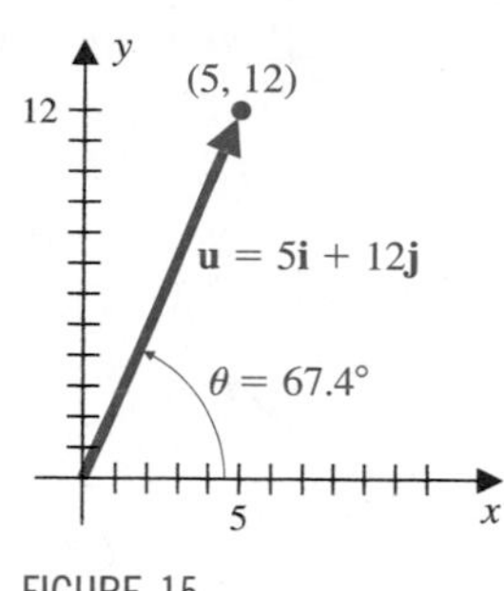

FIGURE 15

EXAMPLE 5 Finding the components of a vector

Find the component form of the vector $\mathbf{u} = a\mathbf{i} + b\mathbf{j}$, if $|\mathbf{u}| = 2$ and a direction angle $\theta = (5\pi)/4$.

Solution The vector $\mathbf{u}$ has magnitude 2 and direction angle $\theta = (5\pi)/4$ (Figure 16). So $\mathbf{u}$ can be written as

$$\begin{aligned}\mathbf{u} = a\mathbf{i} + b\mathbf{j} &= |\mathbf{u}|\left(\cos\frac{5\pi}{4}\right)\mathbf{i} + |\mathbf{u}|\left(\sin\frac{5\pi}{4}\right)\mathbf{j} \\ &= 2\left(-\frac{\sqrt{2}}{2}\right)\mathbf{i} + 2\left(-\frac{\sqrt{2}}{2}\right)\mathbf{j} \\ &= -\sqrt{2}\,\mathbf{i} - \sqrt{2}\,\mathbf{j}\end{aligned}$$

Thus, the x component of $\mathbf{u}$ is $-\sqrt{2}$ and the y component is also $-\sqrt{2}$.

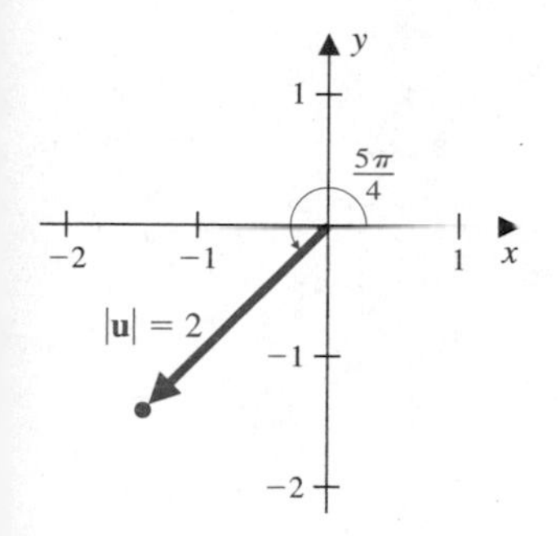

FIGURE 16

Solving Applied Problems

Vectors are used in many applications in science and engineering—for example, force problems.

EXAMPLE 6 Finding the magnitude of a resultant force

A boat is pulled along a canal by two ropes on opposite sides of the canal, as shown in Figure 17. The first rope exerts a force of 350 pounds and makes an angle of 25° with respect to the axis of the boat. The other exerts a force of 200 pounds and makes an angle of 32° with respect to the axis of the boat.

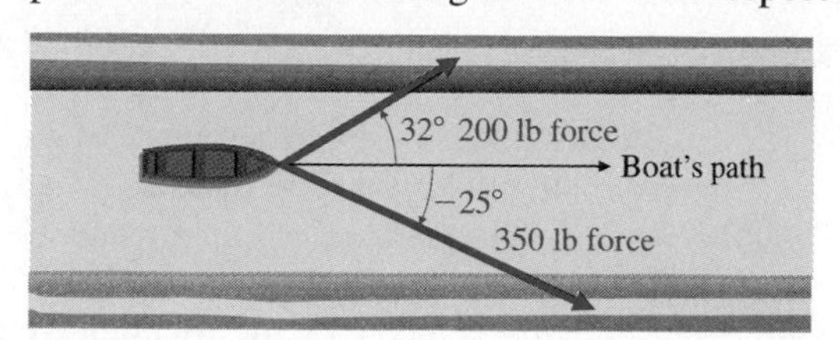

FIGURE 17

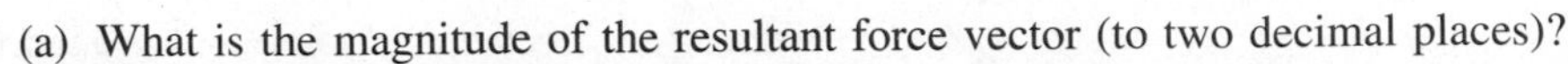

(a) What is the magnitude of the resultant force vector (to two decimal places)?

(b) What is the angle that the resultant vector makes with the axis of the boat (to the nearest hundredth of a degree)?

Solution

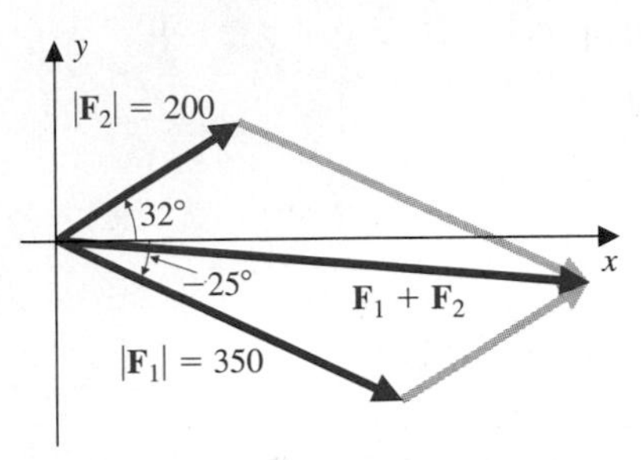

FIGURE 18

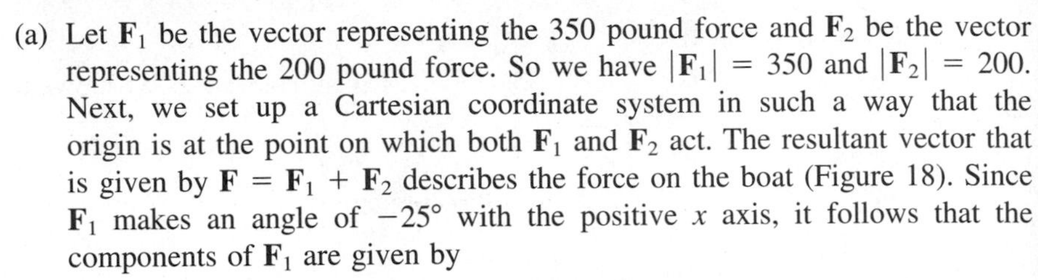

(a) Let $\mathbf{F}_1$ be the vector representing the 350 pound force and $\mathbf{F}_2$ be the vector representing the 200 pound force. So we have $|\mathbf{F}_1| = 350$ and $|\mathbf{F}_2| = 200$. Next, we set up a Cartesian coordinate system in such a way that the origin is at the point on which both $\mathbf{F}_1$ and $\mathbf{F}_2$ act. The resultant vector that is given by $\mathbf{F} = \mathbf{F}_1 + \mathbf{F}_2$ describes the force on the boat (Figure 18). Since $\mathbf{F}_1$ makes an angle of $-25°$ with the positive x axis, it follows that the components of $\mathbf{F}_1$ are given by

$$\begin{aligned}\mathbf{F}_1 &= |\mathbf{F}_1|\cos(-25°)\mathbf{i} + |\mathbf{F}_1|\sin(-25°)\mathbf{j}\\ &= 350(\cos 25°)\mathbf{i} - 350(\sin 25°)\mathbf{j}\\ &= 317.21\mathbf{i} - 147.92\mathbf{j} \text{ (approx.)}\end{aligned}$$

Similarly, $\mathbf{F}_2$ makes an angle of 32° with the positive x axis, so its components are given by

$$\begin{aligned}\mathbf{F}_2 &= |\mathbf{F}_2|(\cos 32°)\mathbf{i} + |\mathbf{F}_2|(\sin 32°)\mathbf{j}\\ &= 200(\cos 32°)\mathbf{i} + 200(\sin 32°)\mathbf{j}\\ &= 169.61\mathbf{i} + 105.98\mathbf{j} \text{ (approx.)}\end{aligned}$$

The resultant vector is given by

$$\begin{aligned}\mathbf{F} = \mathbf{F}_1 + \mathbf{F}_2 &= (317.21\mathbf{i} - 147.92\mathbf{j}) + (169.61\mathbf{i} + 105.98\mathbf{j})\\ &= 486.82\mathbf{i} - 41.94\mathbf{j} \text{ (approx.)}\end{aligned}$$

and its magnitude is

$$\begin{aligned}|\mathbf{F}| &= \sqrt{(486.82)^2 + (-41.94)^2}\\ &= 488.62 \text{ (approx.)}\end{aligned}$$

Thus, the resultant force has a magnitude of about 488.62 pounds.

(b) The angle θ between $\mathbf{F}$ and the axis of the boat (positive x axis) is given by

$$\tan\theta = \frac{-41.94}{486.82}$$

that is,

$$\theta = \tan^{-1}\left(\frac{-41.94}{486.82}\right) = -4.92° \text{ (approx.)}$$

■

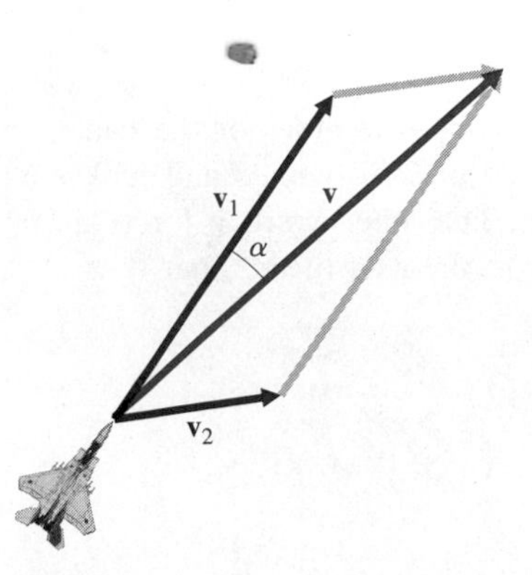

FIGURE 19

Vectors are also used in aeronautical navigation, where the following terminology is commonly used: The **heading** of an airplane is the direction in which it is pointed; its **airspeed** is its speed relative to the air. As shown in Figure 19, the vector $\mathbf{v}_1$, whose magnitude is the airspeed and whose direction is the heading, represents the **velocity of the airplane relative to the air**. The vector $\mathbf{v}_2$ is the

velocity vector for the wind; that is, $|\mathbf{v}_2|$ is the speed of the wind, and the direction angle of $\mathbf{v}_2$ is the direction of the wind. The **course**, or **track**, of the airplane is the direction in which it is actually moving over the ground, and its **ground speed** is its speed relative to the ground. The magnitude of the resultant vector $\mathbf{v} = \mathbf{v}_1 + \mathbf{v}_2$ is the ground speed or velocity of the airplane, and the direction of $\mathbf{v}$ is the course. In this context, vector $\mathbf{v}$ is called the **velocity vector** of the airplane. The angle α between the vectors $\mathbf{v}$ and $\mathbf{v}_1$ is called the **drift angle**.

EXAMPLE 7 Finding the course and speed of an airplane

An airplane is headed N30°E with an airspeed of 500 miles per hour. The wind is blowing S29°E at a speed of 50 miles per hour. Find:

(a) The ground speed (to two decimal places)
(b) The course of the airplane—that is, the direction angle of the velocity vector relative to the ground (to the nearest hundredth of a degree)
(c) The drift angle (to the nearest hundredth of a degree)

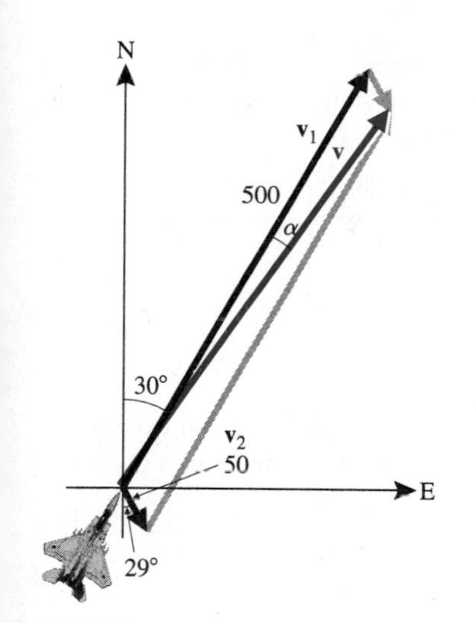

FIGURE 20

Solution Let $\mathbf{v}_1$ represent the velocity and direction of the airplane relative to the air; let $\mathbf{v}_2$ represent the velocity and direction of the wind relative to the ground; and let $\mathbf{v}$ represent the velocity vector of the airplane (Figure 20). Then

$$\begin{aligned}\mathbf{v}_1 &= 500(\cos 60°)\mathbf{i} + 500(\sin 60°)\mathbf{j}\\ &= 250.00\mathbf{i} + 433.01\mathbf{j}\\ \mathbf{v}_2 &= 50[\cos(-61°)]\mathbf{i} + 50[\sin(-61°)]\mathbf{j}\\ &= 24.24\mathbf{i} + (-43.73)\mathbf{j} \text{ (approx.)}\end{aligned}$$

(a) Thus,

$$\begin{aligned}\mathbf{v} &= \mathbf{v}_1 + \mathbf{v}_2\\ &= (250.00\mathbf{i} + 433.01\mathbf{j}) + (24.24\mathbf{i} - 43.73\mathbf{j})\\ &= 274.24\mathbf{i} + 389.28\mathbf{j}\end{aligned}$$

So,

$$\begin{aligned}|\mathbf{v}| &= \sqrt{(274.24)^2 + (389.28)^2}\\ &= 476.18 \text{ (approx.)}\end{aligned}$$

Therefore, the ground speed is approximately 476.18 miles per hour.

(b) A direction angle θ of the velocity vector $\mathbf{v}$ is given by $\tan\theta = 389.28/274.24$. So,

$$\theta = \tan^{-1}\frac{389.28}{274.24} = 54.84° \text{ (approx.)}$$

(c) Upon examining Figure 20 and using the result from part (b), we find that the drift angle α is given by $\alpha = 60° - 54.84° = 5.16°$ (approx.).

PROBLEM SET 7.3

Mastering the Concepts

In problems 1 and 2, let $O = (0, 0)$, $P = (-2, 1)$, $Q = (3, -2)$, $R = (4, 3)$, and $S = (-3, -5)$. Sketch each vector.

1. (a) $\mathbf{u} = \overrightarrow{OP}$ (b) $\mathbf{v} = \overrightarrow{OQ}$
   (c) $2\mathbf{u}$ (d) $3\mathbf{u} - 4\mathbf{v}$
2. (a) $\mathbf{u} = \overrightarrow{OS}$ (b) $\mathbf{v} = \overrightarrow{OR}$
   (c) $-\frac{1}{5}\mathbf{v}$ (d) $-2\mathbf{u} + 5\mathbf{v}$

In problems 3–8, let $\mathbf{u}$ be a vector whose initial point is P and whose terminal point is Q. Draw $\mathbf{u} = \overrightarrow{PQ}$ in standard position. Also, find the component form and the magnitude of $\mathbf{u}$.

3. $P = (8, 6)$, $Q = (3, 4)$
4. $P = (-7, -6)$, $Q = (2, 3)$
5. $P = (-2, 6)$, $Q = (3, -5)$
6. $P = (-3, 2)$, $Q = (1, -3)$
7. $P = (3, 7)$, $Q = (-3, 1)$
8. $P = (1, -3)$, $Q = (5, -1)$

In problems 9–14, use $\mathbf{u} = \langle 3, 4\rangle$, $\mathbf{v} = \langle -2, 4\rangle$, and $\mathbf{w} = \langle 7, 8\rangle$ to find each expression. Display part (a) of each problem graphically.

9. (a) $\mathbf{u} + \mathbf{v}$ (b) $|\mathbf{u} + \mathbf{v}|$
10. (a) $\mathbf{v} - \mathbf{w}$ (b) $|\mathbf{v} - \mathbf{w}|$
11. (a) $3\mathbf{u} - 4\mathbf{w}$ (b) $|3\mathbf{u} - 4\mathbf{w}|$
12. (a) $5\mathbf{w} + 3\mathbf{u}$ (b) $|5\mathbf{w} + 3\mathbf{u}|$
13. (a) $-6\mathbf{v} - \mathbf{u}$ (b) $|-6\mathbf{v} - \mathbf{u}|$
14. (a) $-(2\mathbf{u} + \mathbf{v})$ (b) $|-(2\mathbf{u} + \mathbf{v})|$

In problems 15 and 16, express each vector in terms of $\mathbf{i}$ and $\mathbf{j}$.

15. (a) $\langle -2, 4\rangle$ (b) $\langle 3, 0\rangle$
16. (a) $\langle 0, 2\rangle$ (b) $\langle -1, -5\rangle$
17. Let $\mathbf{w} = \mathbf{u} + \mathbf{v}$. Find $\mathbf{v}$ in terms of $\mathbf{i}$ and $\mathbf{j}$, and sketch the vector $\mathbf{v}$ in each case.
    (a) $\mathbf{u} = 3\mathbf{i} - 2\mathbf{j}$ and $\mathbf{w} = 5\mathbf{i} + 3\mathbf{j}$
    (b) $\mathbf{u} = 2\mathbf{i} - \mathbf{j}$ and $\mathbf{w} = -2\mathbf{i} + 3\mathbf{j}$
18. Let $\mathbf{u} = \langle -2, 2\rangle$ and $\mathbf{v} = \langle -5, 0\rangle$. Suppose $\mathbf{w} = \mathbf{u} - \mathbf{v}$. Express $\mathbf{w}$ in terms of $\mathbf{i}$ and $\mathbf{j}$.

In problems 19 and 20, sketch the resultant vector $\mathbf{w} = \mathbf{u} + \mathbf{v}$. Find the length of $\mathbf{w}$ and the angle that $\mathbf{w}$ makes with $\mathbf{v}$.

19. (−4, 3), (2, 4)

20.

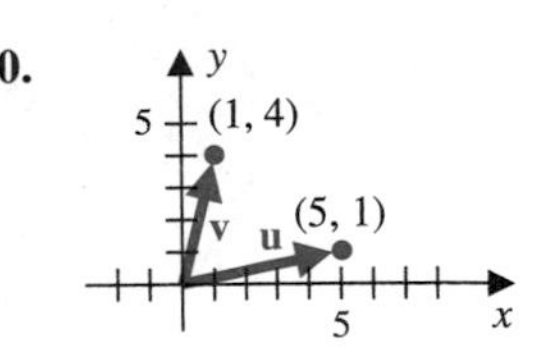

In problems 21–28, normalize each vector. Also, find the direction angle of $\mathbf{u}$ (rounded off to two decimal places).

21. $\mathbf{u} = -\sqrt{3}\,\mathbf{i} + \mathbf{j}$
22. $\mathbf{u} = -2\mathbf{i} - 2\sqrt{3}\,\mathbf{j}$
23. $\mathbf{u} = 5\mathbf{i} + 12\mathbf{j}$
24. $\mathbf{u} = 3\mathbf{i} - 3\mathbf{j}$
25. $\mathbf{u} = \sqrt{2}\,\mathbf{i} - \sqrt{2}\,\mathbf{j}$
26. $\mathbf{u} = 3\mathbf{i} + 7\mathbf{j}$
27. $\mathbf{u} = (3\mathbf{i} + 8\mathbf{j}) - (\mathbf{i} - 2\mathbf{j})$
28. $\mathbf{u} = \langle 3, -2\rangle - 2\langle 4, 1\rangle$

In problems 29–36, find the x and y components of the vector $\mathbf{u} = x\mathbf{i} + y\mathbf{i}$, if θ represents a direction angle of $\mathbf{u}$. Round off the answers to two decimal places.

29. $|\mathbf{u}| = 3$; $\theta = 0°$
30. $|\mathbf{u}| = 2$; $\theta = 180°$
31. $|\mathbf{u}| = 5$; $\theta = \pi/4$
32. $|\mathbf{u}| = 1$; $\theta = (5\pi)/3$
33. $|\mathbf{u}| = 2$; $\theta = -150°$
34. $|\mathbf{u}| = 4$; $\theta = -75°$
35. $|\mathbf{u}| = \frac{1}{2}$; $\theta = 100.2°$
36. $|\mathbf{u}| = 7$; $\theta = 253.3°$

Applying the Concepts

In problems 37–44, compute each scalar value to two decimal places and each angle to the nearest hundredth of a degree.

37. Resultant Force: Two forces $\mathbf{F}_1$ and $\mathbf{F}_2$ act on a point. If $|\mathbf{F}_1| = 40$ pounds, $|\mathbf{F}_2| = 23$ pounds, and the angle between $\mathbf{F}_1$ and $\mathbf{F}_2$ is 58°, find:
    (a) The magnitude of the resultant force $\mathbf{F}$
    (b) The angle between $\mathbf{F}_1$ and $\mathbf{F}$
38. Resultant Force: Two forces acting on the same object produce a resultant force of 17 pounds with an angle of 20° relative to a horizontal axis. Assume that one of the two forces is 20 pounds and has an angle of 50° relative to the horizontal axis. Find the other force and the direction angle relative to the horizontal axis.
39. Pulling a Sled: Two children are pulling a third child across the ice on a sled. The first child pulls a rope with a force of 5 newtons and the second child pulls a rope with a force of 8 newtons. If the angle between the ropes is 28° (Figure 21), find:

(a) The magnitude of the resultant force
(b) The angle the resultant force vector makes with the first child's rope

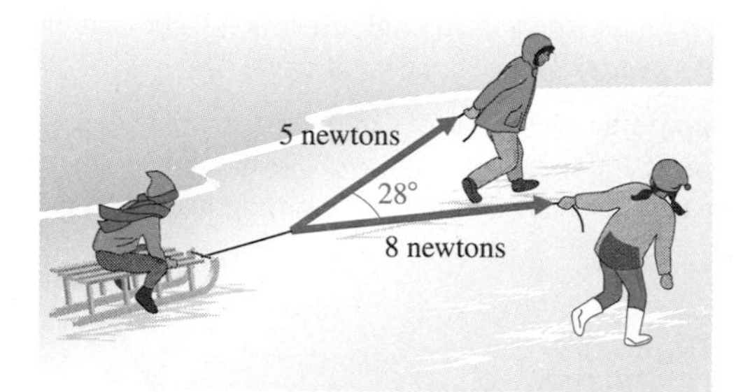

FIGURE 21

40. Helicopter Aviation: A helicopter is flying in the direction N50°W with an airspeed of 165 miles per hour. The wind is blowing at 35 miles per hour in the direction S60°E. Find the course and the ground speed of the helicopter.

41. Airplane Aviation: A commercial airplane with an airspeed of 510 miles per hour is headed N85°E. The wind is blowing N35°E at 45 miles per hour. Find:
(a) The ground speed
(b) The course of the airplane
(c) The drift angle

42. Airplane Aviation: An airplane is flying in the direction N25°E with an airspeed of 450 kilometers per hour. Its ground speed is 500 kilometers per hour, and its course is N40°E. Find:
(a) The speed of the wind
(b) The direction of the wind

43. Boat Navigation: A boat heading S40°E at a still water speed of 40 kilometers per hour is pushed off course by a current of 32 kilometers per hour flowing in the direction S50°W. Find:
(a) The speed of the boat
(b) The course the boat is traveling
(c) The drift angle

44. Motorboat Navigation: Suppose a motorboat that has a speed in still water of 12 miles per hour heads due west across a river that is flowing due south at a speed of 3 miles per hour.
(a) What is the speed of the boat?
(b) What angle does the path of the boat make with the vector representing the current?

Developing and Extending the Concepts

45. Find the terminal point of $\mathbf{u} = \langle 10, -4 \rangle$ if the initial point P is $(3, 5)$.

46. Find the initial point of $\mathbf{u} = \langle -4, 7 \rangle$ if the terminal point Q is $(2, -1)$.

Problems 47 and 48 pertain to the following properties:

Vector Addition and Scalar Multiplication Properties

Let $\mathbf{u}$, $\mathbf{v}$, and $\mathbf{w}$ be vectors, and let t and s be scalars; then the following relationships hold:

1. $\mathbf{u} + \mathbf{v} = \mathbf{v} + \mathbf{u}$
2. $\mathbf{u} + (\mathbf{v} + \mathbf{w}) = (\mathbf{u} + \mathbf{v}) + \mathbf{w}$
3. $\mathbf{u} + \mathbf{0} = \mathbf{u}$
4. $\mathbf{u} + (-\mathbf{u}) = \mathbf{0}$
5. $s(\mathbf{u} + \mathbf{v}) = s\mathbf{u} + s\mathbf{v}$
6. $(s + t)\mathbf{u} = s\mathbf{u} + t\mathbf{u}$
7. $(st)\mathbf{u} = s(t\mathbf{u}) = t(s\mathbf{u})$
8. $1\mathbf{u} = \mathbf{u}$ and $0\mathbf{u} = \mathbf{0}$

47. Illustrate properties 1–4 for $\mathbf{u} = \langle 1, 2 \rangle$, $\mathbf{v} = \langle -3, 2 \rangle$, and $\mathbf{w} = \langle -5, 4 \rangle$.

48. Illustrate properties 5–8 for $\mathbf{u} = \langle -2, 5 \rangle$, $\mathbf{v} = \langle 4, 3 \rangle$, $s = 4$, and $t = -3$.

The **dot product**, or **inner product**, of two vectors $\mathbf{u} = \langle a, b \rangle$ and $\mathbf{v} = \langle c, d \rangle$ is given by

$$\mathbf{u} \cdot \mathbf{v} = ac + bd$$

The angle θ between the vectors $\mathbf{u}$ and $\mathbf{v}$ is given by

$$\cos\theta = \frac{\mathbf{u} \cdot \mathbf{v}}{|\mathbf{u}||\mathbf{v}|} \qquad 0° \le \theta \le 180°$$

In problems 49–54, find:
(a) The dot product of the two vectors $\mathbf{u}$ and $\mathbf{v}$
(b) The angle θ between $\mathbf{u}$ and $\mathbf{v}$ (round off θ to the nearest tenth of a degree)

49. $\mathbf{u} = \langle -3, 2 \rangle$; $\mathbf{v} = \langle 1, -3 \rangle$
50. $\mathbf{u} = \langle 1, -3 \rangle$; $\mathbf{v} = \langle 5, -1 \rangle$
51. $\mathbf{u} = \langle -3, 6 \rangle$; $\mathbf{v} = \langle 2, -1 \rangle$
52. $\mathbf{u} = \langle -2, 6 \rangle$; $\mathbf{v} = \langle 3, -5 \rangle$
53. $\mathbf{u} = 5\mathbf{i} - 2\mathbf{j}$; $\mathbf{v} = 4\mathbf{i} + 3\mathbf{j}$
54. $\mathbf{u} = -4\mathbf{i}$; $\mathbf{v} = 4\mathbf{i} + 5\mathbf{j}$

55. Explain why two nonzero vectors **u** and **v** are **perpendicular** (or **orthogonal**) if and only if

$$\mathbf{u} \cdot \mathbf{v} = 0$$

[*Hint*: Use the formulas given for problems 49–54.]

In problems 56–58, use the result from problem 55.

56. Find the values of the constant k so that **u** and **v** are orthogonal. Sketch **u** and **v** in the same coordinate system.

(a) $\mathbf{u} = \langle k, 3\rangle$; $\mathbf{v} = \langle -4, 2\rangle$
(b) $\mathbf{u} = \langle 3k, -1\rangle$; $\mathbf{v} = \langle k, 4\rangle$

57. Use vectors and the dot product to determine whether the triangle with vertices $A = (6, 1)$, $B = (4, 3)$, and $C = (2, 1)$ is a right triangle. If it is a right triangle, which vertex angle is the right angle?

58. Suppose that vector **u** has initial point (2, 1) and terminal point (5, −4), and $\mathbf{v} = \langle 5, 3\rangle$. Use the dot product to determine whether **u** and **v** are orthogonal.

OBJECTIVES

1. Define the complex plane.
2. Write a complex number in trigonometric form.
3. Find the powers of complex numbers.
4. Find the roots of complex numbers.

7.4 Trigonometric Forms of Complex Numbers

We have already considered complex numbers in Section 8 (page 23), and we have seen how to perform basic operations on them. Now we'll discuss how to represent these numbers in trigonometric form. Such representations make multiplication and division of complex numbers much easier to perform. In this section, we'll also see the usefulness of trigonometric forms in finding powers and roots of complex numbers.

Defining the Complex Plane

We begin by representing complex numbers as points in the plane. Each ordered pair of real numbers (a, b) can be associated with the complex number $z = a + bi$, and each complex number $z = a + bi$ can be associated with the ordered pair of real numbers (a, b). Because of this one-to-one correspondence between the complex numbers and the ordered pairs of real numbers, we can use points in the Cartesian plane to represent the complex numbers. The plane in which the complex numbers are represented is called the **complex plane**; the horizontal axis (x axis) is called the **real axis** and the vertical axis (y axis) is called the **imaginary axis** (Figure 1a). For example, the ordered pairs $(2, -3)$ and $(5, 2)$ are used to represent the complex numbers $z_1 = 2 - 3i$ and $z_2 = 5 + 2i$, respectively (Figure 1b). Complex numbers of the form $z = bi$ are represented by points of the form $(0, b)$, that is, as points on the imaginary axis. For example, the complex number $3i$ is represented by the point $(0, 3)$. Complex numbers of the form $z = a$ are represented by points of the form $(a, 0)$, that is, as points on the real axis. Thus, the number 5 is represented by the point $(5, 0)$ (Figure 1c).

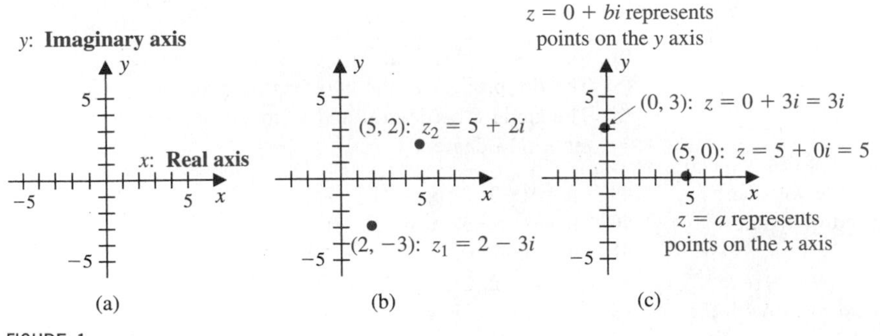

FIGURE 1

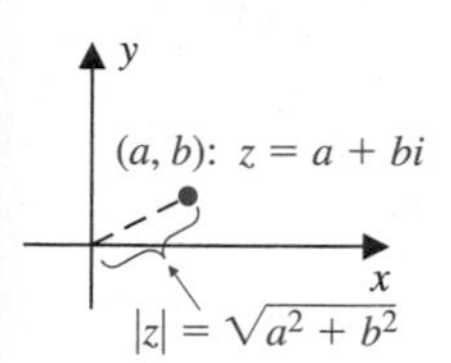

FIGURE 2

Geometrically, we interpret the *absolute value*, or *magnitude*, of a complex number $a + bi$ as the distance between the origin of the complex plane and the point (a, b). We usually denote it by the absolute value $|z| = |a + bi|$ and refer to it as the **modulus** of the complex number $a + bi$. So, by the Pythagorean theorem, if $z = a + bi$, then $|z|$ is given by

$$|z| = \sqrt{a^2 + b^2} \quad \text{(Figure 2)}$$

EXAMPLE 1 Finding absolute values

Let $z_1 = 4 + 3i$ and $z_2 = \sqrt{3} - i$. Find:

(a) $|z_1|$ (b) $|z_2|$ (c) $|z_1 z_2|$

Solution

(a) $|z_1| = \sqrt{4^2 + 3^2} = \sqrt{16 + 9} = \sqrt{25} = 5$

(b) $|z_2| = \sqrt{\left(\sqrt{3}\right)^2 + (-1)^2} = \sqrt{4} = 2$

(c) $|z_1 z_2| = \left|(4 + 3i)\left(\sqrt{3} - i\right)\right| = \left|\left(4\sqrt{3} + 3\right) + \left(3\sqrt{3} - 4\right)i\right|$

$$= \sqrt{\left(4\sqrt{3} + 3\right)^2 + \left(3\sqrt{3} - 4\right)^2} = \sqrt{100} = 10$$

Example 1 leads us to the following results (Problem 60):

(i) $|z_1 z_2| = |z_1||z_2|$ (ii) $\left|\dfrac{z_1}{z_2}\right| = \dfrac{|z_1|}{|z_2|}, \quad z_2 \neq 0$

Writing a Complex Number in Trigonometric Form

It is possible to write complex numbers in trigonometric form. Suppose $z = x + yi$ is a nonzero complex number; then its graphical representation is the point $P = (x, y)$. Let θ be any angle in standard position whose terminal side lies on the line segment $\overline{OP}$ and $r = |z| = \sqrt{x^2 + y^2}$. The angle θ is usually chosen in the interval $[0, 2\pi)$ and is called an **argument** of z (Figure 3).

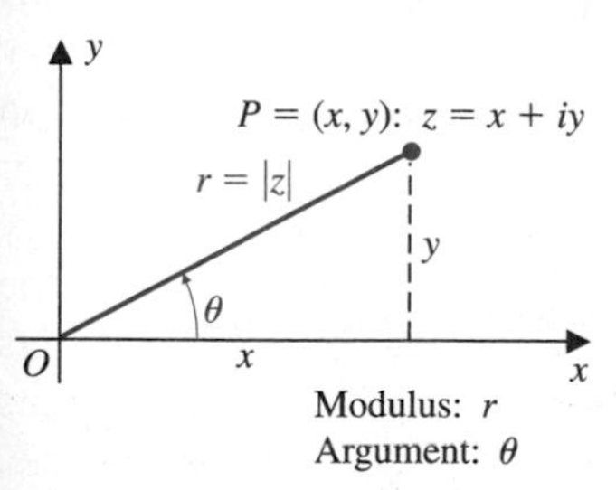

FIGURE 3

Using trigonometry, we obtain

$$\cos\theta = \frac{x}{r} \qquad \text{and} \qquad \sin\theta = \frac{y}{r}$$

or

$$x = r\cos\theta \qquad \text{and} \qquad y = r\sin\theta$$

Thus, we can rewrite any complex number $z = x + yi$ as

$$z = r\cos\theta + (r\sin\theta)i = r(\cos\theta + i\sin\theta)$$

where $r = \sqrt{x^2 + y^2}$ and θ is the argument. This form is called the **trigonometric form**, or **polar form**, of a complex number.

Notice that θ is not unique, since

$$r(\cos \theta + i \sin \theta) = r(\cos \theta_1 + i \sin \theta_1)$$

holds whenever $\theta - \theta_1$ is an integer multiple of 2π. Consequently:

Two complex numbers are equal if and only if their moduli are equal and their arguments differ by an integer multiple of 2π.

Thus, if z has modulus r and argument θ, then

$$\boxed{z = x + iy = r[\cos(\theta + 2\pi k) + i \sin(\theta + 2\pi k)]}$$

where k is any integer. If we use degree measure instead of radians to measure θ, then

$$\boxed{z = r[\cos(\theta + 360°k) + i \sin(\theta + 360°k)]}$$

where k is any integer.

EXAMPLE 2 **Writing a complex number in Cartesian form**

Express the complex number

$$z = 2\left(\cos \frac{\pi}{3} + i \sin \frac{\pi}{3}\right)$$

in the Cartesian form $x + yi$.

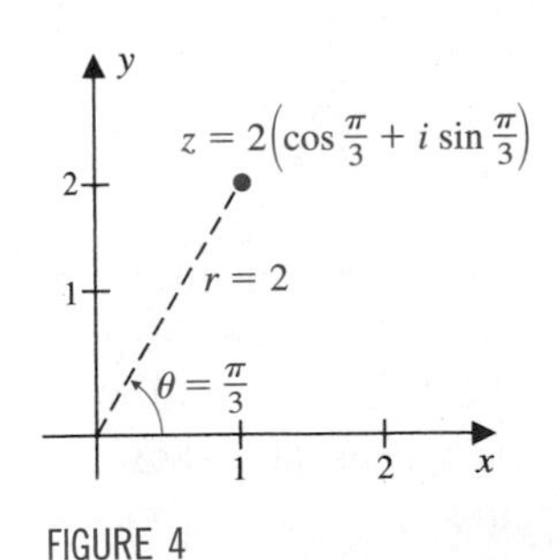

FIGURE 4

Solution The given complex number is in the form $z = r(\cos \theta + i \sin \theta)$, with $r = 2$ and $\theta = \pi/3$. So,

$$z = x + yi = 2 \cos \frac{\pi}{3} + \left(2 \sin \frac{\pi}{3}\right)i$$

$$= 2\left(\frac{1}{2}\right) + 2\left(\frac{\sqrt{3}}{2}\right)i = 1 + \sqrt{3}\, i \quad \text{(Figure 4)}$$

EXAMPLE 3 **Writing a complex number in trigonometric form**

Express $z = -\sqrt{3} - i$ in trigonometric form with $0 \le \theta < 2\pi$.

Solution First, we plot the point $(-\sqrt{3}, -1)$ corresponding to $z = -\sqrt{3} - i$ (Figure 5). Then we find the modulus r and angle θ. From Figure 5, we see that the modulus r is given by

$$r = \sqrt{(-\sqrt{3})^2 + (-1)^2} = 2$$

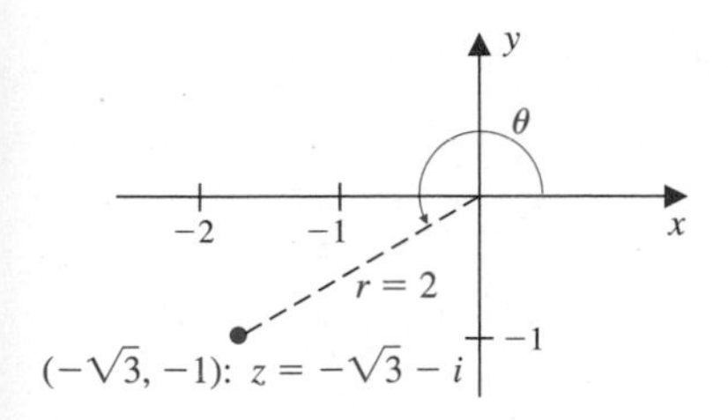

FIGURE 5

The argument θ satisfies the equation $\tan\theta = (-1)/\left(-\sqrt{3}\right) = 1/\sqrt{3}$, and its terminal side is in quadrant III. It follows from trigonometry that $\theta = (7\pi)/6$. Therefore,

$$z = 2\left(\cos\frac{7\pi}{6} + i\sin\frac{7\pi}{6}\right)$$

Finding the Powers of Complex Numbers

Trigonometric forms can be used to multiply and divide complex numbers. Let's examine the product of z_1 and z_2, where

$$z_1 = r_1(\cos\theta_1 + i\sin\theta_1) \quad \text{and} \quad z_2 = r_2(\cos\theta_2 + i\sin\theta_2)$$

We have

$$\begin{aligned} z_1z_2 &= r_1[(\cos\theta_1 + i\sin\theta_1)]\cdot r_2[(\cos\theta_2 + i\sin\theta_2)] \\ &= r_1r_2[(\cos\theta_1\cos\theta_2 - \sin\theta_1\sin\theta_2) + i(\cos\theta_1\sin\theta_2 + \cos\theta_2\sin\theta_1)] \\ &= r_1r_2[\cos(\theta_1+\theta_2) + i\sin(\theta_1+\theta_2)] \quad \text{Sum identities for the cosine and sine} \end{aligned}$$

This result gives us the first of the following two formulas:

MULTIPLICATION AND DIVISION OF COMPLEX NUMBERS

Suppose that trigonometric forms of the complex numbers z_1 and z_2 are given by

$$z_1 = r_1(\cos\theta_1 + i\sin\theta_1) \text{ and } z_2 = r_2(\cos\theta_2 + i\sin\theta_2)$$

Then:

(i) $z_1z_2 = r_1r_2[\cos(\theta_1+\theta_2) + i\sin(\theta_1+\theta_2)]$

(ii) $\dfrac{z_1}{z_2} = \dfrac{r_1}{r_2}[\cos(\theta_1-\theta_2) + i\sin(\theta_1-\theta_2)]$, $z_2 \neq 0$

The derivation of the second formula is left as an exercise (Problem 62).
In words, the first formula states that:

> The product of two complex numbers in trigonometric form is obtained by multiplying their moduli and adding their arguments. Also, the quotient of two complex numbers is obtained by dividing their moduli and subtracting their arguments.

EXAMPLE 4 Multiplying and dividing complex numbers

Let

$$z_1 = \sqrt{2}\left(\cos\frac{3\pi}{4} + i\sin\frac{3\pi}{4}\right) \quad \text{and} \quad z_2 = 4\left(\cos\frac{3\pi}{2} + i\sin\frac{3\pi}{2}\right)$$

Express the following in both trigonometric and Cartesian form:

(a) z_1z_2 (b) $\dfrac{z_1}{z_2}$

Solution

(a) Using the multiplication formula, we have

$$\begin{aligned} z_1z_2 &= 4\sqrt{2}\left[\cos\left(\frac{3\pi}{4}+\frac{3\pi}{2}\right)+i\sin\left(\frac{3\pi}{4}+\frac{3\pi}{2}\right)\right] \\ &= 4\sqrt{2}\left(\cos\frac{9\pi}{4}+i\sin\frac{9\pi}{4}\right) && \text{Trigonometric form} \\ &= 4\sqrt{2}\left(\frac{\sqrt{2}}{2}+i\frac{\sqrt{2}}{2}\right) \\ &= 4+4i && \text{Cartesian form} \end{aligned}$$

(b) Using the division formula, we have

$$\begin{aligned} \frac{z_1}{z_2} &= \frac{\sqrt{2}}{4}\left[\cos\left(\frac{3\pi}{4}-\frac{3\pi}{2}\right)+i\sin\left(\frac{3\pi}{4}-\frac{3\pi}{2}\right)\right] \\ &= \frac{\sqrt{2}}{4}\left[\cos\left(-\frac{3\pi}{4}\right)+i\sin\left(-\frac{3\pi}{4}\right)\right] \\ &= \frac{\sqrt{2}}{4}\left(\cos\frac{3\pi}{4}-i\sin\frac{3\pi}{4}\right) && \text{Trigonometric form} \\ &= \frac{\sqrt{2}}{4}\left(-\frac{\sqrt{2}}{2}-i\frac{\sqrt{2}}{2}\right) \\ &= -\frac{1}{4}-\frac{1}{4}i && \text{Cartesian form} \end{aligned}$$

Repeated use of the multiplication rule for complex numbers in trigonometric form allows us to compute powers of a complex number. For instance, if $z = r(\cos\theta + i\sin\theta)$, then

$$z^2 = z \cdot z = r \cdot r[\cos(\theta+\theta) + i\sin(\theta+\theta)] = r^2(\cos 2\theta + i\sin 2\theta)$$

Since $z^3 = z^2 \cdot z$, then

$$z^3 = r^2 \cdot r[\cos(2\theta+\theta) + i\sin(2\theta+\theta)] = r^3(\cos 3\theta + i\sin 3\theta)$$

If we repeat the process one more time, we get

$$z^4 = z^3 \cdot z = r^4(\cos 4\theta + i\sin 4\theta)$$

This pattern is expressed in the following theorem, which is attributed to Abraham DeMoivre (1667–1754).

DEMOIVRE'S THEOREM

Let $z = r(\cos\theta + i\sin\theta)$. Then, for n any positive integer,

$$z^n = [r(\cos\theta + i\sin\theta)]^n = r^n(\cos n\theta + i\sin n\theta)$$

EXAMPLE 5 Finding powers of complex numbers

Use DeMoivre's theorem to determine each of the given powers. Express the answer in Cartesian form.

(a) $[3(\cos 60° + i\sin 60°)]^4$ (b) $(1 + i)^{20}$

Solution

(a) By DeMoivre's theorem,

$$[3(\cos 60° + i\sin 60°)]^4 = 3^4(\cos 240° + i\sin 240°)$$
$$= 81\left(-\frac{1}{2} - \frac{i\sqrt{3}}{2}\right) = -\frac{81}{2} - \frac{81\sqrt{3}}{2}i$$

(b) In order to use DeMoivre's theorem, we first convert $1 + i$ to trigonometric form:

$$1 + i = \sqrt{2}\left(\cos\frac{\pi}{4} + i\sin\frac{\pi}{4}\right)$$

So the modulus of $1 + i$ is $\sqrt{2}$ and an argument is $\pi/4$. Thus,

$$(1 + i)^{20} = \left[\sqrt{2}\left(\cos\frac{\pi}{4} + i\sin\frac{\pi}{4}\right)\right]^{20} = 2^{10}(\cos 5\pi + i\sin 5\pi)$$
$$= 1024(-1 + 0i) = -1024 + 0i$$

Finding the Roots of Complex Numbers

Recall from Section 3.5 that a polynomial equation of degree n has n roots (or solutions) in the complex number system. Hence, the equation $z^4 = 16$ has four roots. One way to find these roots is to write the equation in the form

$$z^4 - 16 = 0$$

so that

$$(z^2 - 4)(z^2 + 4) = 0$$
$$(z - 2)(z + 2)(z + 2i)(z - 2i) = 0$$

Explain why the graph of the function $f(x) = x^4 - 16$ will not show all four zeros.

It follows that the solutions are 2, -2, $-2i$, and $2i$.

It was easy to find the roots for this equation because of the factorization. In more complicated situations, we can find all the roots of a complex number by using DeMoivre's theorem as follows: Suppose we want to determine all the nth roots of a complex number w. That is, suppose we want to solve the equation

$$z^n = w \qquad \text{where } n \text{ is a positive integer}$$

Assume that trigonometric forms for w and z are given by

$$w = R(\cos \phi + i \sin \phi) \qquad \text{and} \qquad z = r(\cos \theta + i \sin \theta)$$

Then, applying DeMoivre's theorem to $z^n = w$ yields

$$z^n = [r(\cos \theta + i \sin \theta)]^n = r^n(\cos n\theta + i \sin n\theta) = R(\cos \phi + i \sin \phi)$$

So

$$r^n = R \qquad \text{and} \qquad n\theta = \phi + 2k\pi \quad \text{Or } n\theta = \phi + 360°k \text{ if degrees are used}$$

Hence, $z = r(\cos \theta + i \sin \theta)$ is a root of $z^n = w$ whenever

$$r = \sqrt[n]{R} \qquad \text{and} \qquad \theta = \frac{\phi}{n} + \frac{2k\pi}{n} \quad \text{Or } \theta = \frac{\phi}{n} + \frac{360°k}{n} \text{ if degrees are used}$$

where $k = 0, \pm 1, \pm 2, \pm 3, \ldots$.

At first glance, it appears that there are an infinite number of roots for the equation $z^n = w$. However, there are only n *distinct roots*. These n roots can be determined by letting k take on the values $0, 1, 2, 3, 4, \ldots, n - 1$. If we let $k = n$, then

$$\theta = \frac{\phi}{n} + \frac{2n\pi}{n} = \frac{\phi}{n} + 2\pi$$

This angle has the same terminal side as ϕ/n, so we get the same value of z as when $k = 0$. Similarly, the value of θ obtained by letting $k = n + 1$ gives an angle

$$\theta = \frac{\phi}{n} + \frac{2\pi}{n} + 2\pi$$

and the resulting value of z is the same as when $k = 1$; and so on.

*n*TH ROOTS OF A COMPLEX NUMBER

If $w = R(\cos \phi + i \sin \phi)$ is any nonzero complex number, and if n is any positive integer, then the distinct nth roots of w are $z_0, z_1, z_2, \ldots, z_{n-1}$, where

$$z_k = \sqrt[n]{R}\left[\cos\left(\frac{\phi}{n} + \frac{2\pi k}{n}\right) + i \sin\left(\frac{\phi}{n} + \frac{2\pi k}{n}\right)\right] \qquad k = 0, 1, 2, \ldots, n - 1$$

If we use degree measure for ϕ, the nth roots of w are given by

$$\sqrt[n]{R}\left[\cos\left(\frac{\phi}{n} + \frac{360°k}{n}\right) + i \sin\left(\frac{\phi}{n} + \frac{360°k}{n}\right)\right] \qquad k = 0, 1, 2, \ldots, n - 1$$

EXAMPLE 6 **Finding the fourth roots of a complex number**

Find the four fourth roots of $1 + i$; that is, solve $z^4 = 1 + i$. Express the answers in both trigonometric form and Cartesian form (rounded to two decimal places). Represent the roots graphically.

Solution First, we determine $R(\cos \phi + i \sin \phi)$, a trigonometric representation of $1 + i$. Here, $R = \sqrt{1 + 1} = \sqrt{2}$ and $\phi = \pi/4$, so

$$1 + i = \sqrt{2}\left(\cos \frac{\pi}{4} + i \sin \frac{\pi}{4}\right)$$

Since $n = 4$, the fourth roots are given by

$$z_k = \sqrt[4]{\sqrt{2}}\left[\cos\left(\frac{\pi/4}{4} + \frac{2\pi k}{4}\right) + i \sin\left(\frac{\pi/4}{4} + \frac{2\pi k}{4}\right)\right] \qquad \text{where } k = 0, 1, 2, 3$$

Using $\sqrt[4]{\sqrt{2}} = \sqrt[8]{2}$, we substitute each value of k into this expression to obtain

$$z_0 = \sqrt[8]{2}\left(\cos \frac{\pi}{16} + i \sin \frac{\pi}{16}\right) \qquad \text{for } k = 0$$

$$z_1 = \sqrt[8]{2}\left(\cos \frac{9\pi}{16} + i \sin \frac{9\pi}{16}\right) \qquad \text{for } k = 1$$

$$z_2 = \sqrt[8]{2}\left(\cos \frac{17\pi}{16} + i \sin \frac{17\pi}{16}\right) \qquad \text{for } k = 2$$

$$z_3 = \sqrt[8]{2}\left(\cos \frac{25\pi}{16} + i \sin \frac{25\pi}{16}\right) \qquad \text{for } k = 3$$

By using a calculator and rounding off the answers to two decimal places, we get the approximations

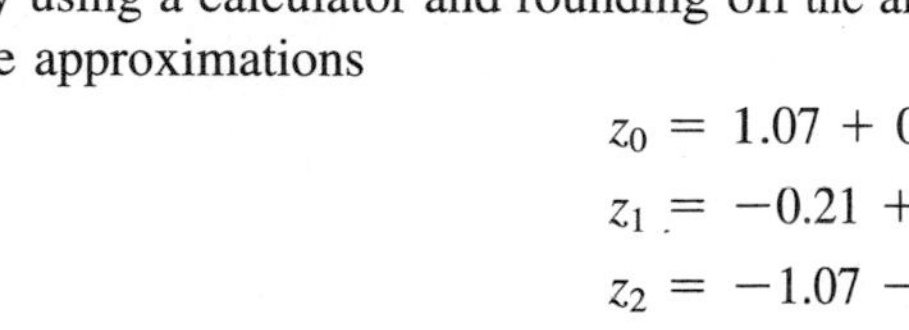

$$z_0 = 1.07 + 0.21i$$
$$z_1 = -0.21 + 1.07i$$
$$z_2 = -1.07 - 0.21i$$
$$z_3 = 0.21 - 1.07i$$

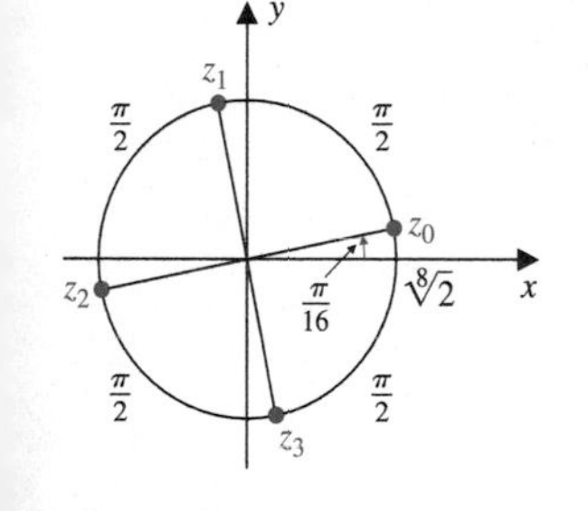

FIGURE 6

These four fourth roots of $1 + i$ are equally spaced on the circle of radius $\sqrt[8]{2}$ centered at the origin; their arguments differ by $\pi/2$ (Figure 6).

The solutions of $z^n = 1$ are called the ***n*th roots of unity**. In the real number system, we find the solution of $x^3 = 1$ by taking the cube root of each side of the equation to get $x = 1$. There is only one real number solution. In the complex number system, we know that the equation $z^3 = 1$ has three possible solutions. We can now find them by using trigonometric forms of complex numbers.

EXAMPLE 7 **Finding the cube roots of unity**

Find the cube roots of unity in the form $x + yi$; that is, find all three roots of the equation $z^3 = 1$.

Solution The number 1 can be written in trigonometric form as $1(\cos 0° + i \sin 0°)$, so $R = 1$, $\phi = 0°$, and $n = 3$. The three roots given by

$$z_k = \sqrt[3]{1}\left[\cos\left(\frac{0° + 360°k}{3}\right) + i \sin\left(\frac{0° + 360°k}{3}\right)\right] \qquad \text{where } k = 0, 1, 2$$

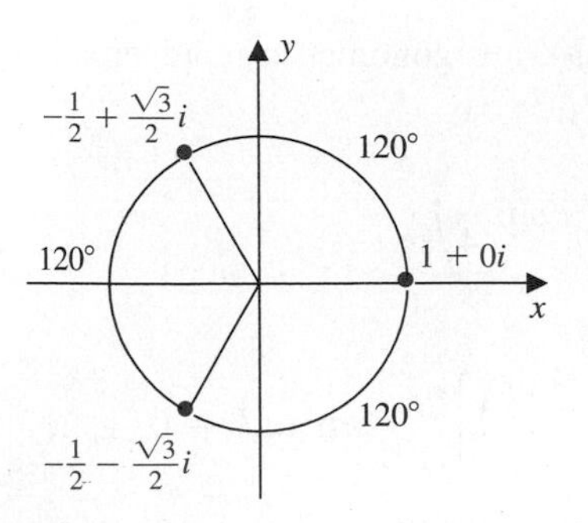

FIGURE 7

Substituting these values for k yields the following three roots:

$$z_0 = 1(\cos 0° + i \sin 0°) = 1 + 0i = 1 \qquad \text{for } k = 0$$
$$z_1 = 1(\cos 120° + i \sin 120°) = -\frac{1}{2} + \frac{\sqrt{3}}{2}i \qquad \text{for } k = 1$$
$$z_2 = 1(\cos 240° + i \sin 240°) = -\frac{1}{2} - \frac{\sqrt{3}}{2}i \qquad \text{for } k = 2$$

Figure 7 illustrates the three cube roots of 1 graphically. They are equally spaced on the circle of radius 1 with center at the origin; their arguments differ by 120°.

PROBLEM SET 7.4

Mastering the Concepts

In problems 1–4, represent each complex number in the complex plane, and find its modulus.

1. **(a)** $z = 2 + 7i$ **(b)** $z = 5 - 2i$
2. **(a)** $z = \sqrt{3} - i$ **(b)** $z = -4 - 4i$
3. **(a)** $z = 3 + 4i$ **(b)** $z = -5 + 3i$
4. **(a)** $z = -4$ **(b)** $z = 2i$

In problems 5–8, find the value of each expression, where $\bar{z} = a - bi$ is the **conjugate** of $z = a + bi$.

(a) $|zw|$ **(b)** $|z\bar{z}|$ **(c)** $|z||w|$ **(d)** $\left|\frac{z}{w}\right|$ **(e)** $\frac{|z|}{|w|}$

5. $z = -1 - i$; $w = 2 - 2i$
6. $z = \sqrt{3}\, i$; $w = -2 - 2i$
7. $z = -4 + 3i$; $w = -2 - 5i$
8. $z = 4 + 2i$; $w = 3 - 5i$

In problems 9–16, express each complex number in the Cartesian form $x + yi$, and graph it. Round off to two decimal places in problems 15 and 16.

9. $z = 2(\cos 30° + i \sin 30°)$
10. $z = 10\left(\cos \frac{3\pi}{4} + i \sin \frac{3\pi}{4}\right)$
11. $z = 2\left(\cos \frac{\pi}{2} + i \sin \frac{\pi}{2}\right)$
12. $z = 7\left[\cos\left(-\frac{3\pi}{2}\right) + i \sin\left(-\frac{3\pi}{2}\right)\right]$
13. $z = 8(\cos 900° + i \sin 900°)$
14. $z = 6(\cos 120° + i \sin 120°)$
15. $z = 2(\cos 10° + i \sin 10°)$
16. $z = 2[\cos(-75°) + i \sin(-75°)]$

In problems 17–22, express each complex number in trigonometric form with $0 \le \theta < 2\pi$. Round off to two decimal places when necessary.

17. **(a)** $z = -1 - i$ **(b)** $z = -\sqrt{3} - i$
18. **(a)** $z = -3 + 4i$ **(b)** $z = 8 + 7i$
19. **(a)** $z = \sqrt{2} + i$ **(b)** $z = -5$
20. **(a)** $z = 3 + 4i$ **(b)** $z = -3i$
21. **(a)** $z = -\frac{1}{2} + \frac{\sqrt{3}}{2}i$ **(b)** $z = \frac{\sqrt{3}}{2} + \frac{1}{2}i$
22. **(a)** $z = \frac{\sqrt{2}}{2} - \frac{\sqrt{2}}{2}i$ **(b)** $z = -\frac{\sqrt{2}}{2} + \frac{\sqrt{2}}{2}i$

In problems 23–30, find:

(a) $z_1 z_2$ **(b)** $\frac{z_1}{z_2}$

Express the answers in both trigonometric form and Cartesian form. In problems 29 and 30, round off answers to two decimal places.

23. $z_1 = 4\left(\cos \frac{5\pi}{6} + i \sin \frac{5\pi}{6}\right)$; $z_2 = 2\left(\cos \frac{\pi}{3} + i \sin \frac{\pi}{3}\right)$
24. $z_1 = 2\left[\cos\left(-\frac{\pi}{6}\right) + i \sin\left(-\frac{\pi}{6}\right)\right]$;
$z_2 = 2\left[\cos\left(-\frac{5\pi}{6}\right) + i \sin\left(-\frac{5\pi}{6}\right)\right]$
25. $z_1 = \cos 30° + i \sin 30°$; $z_2 = \cos 60° + i \sin 60°$
26. $z_1 = 5(\cos 30° + i \sin 30°)$;
$z_2 = 6(\cos 240° + i \sin 240°)$
27. $z_1 = \sqrt{2}(\cos 45° + i \sin 45°)$;
$z_2 = \sqrt{2}(\cos 135° + i \sin 135°)$
28. $z_1 = 4\left(\cos \frac{3\pi}{4} + i \sin \frac{3\pi}{4}\right)$; $z_2 = 2(\cos \pi + i \sin \pi)$

29. $z_1 = 2(\cos 50° + i \sin 50°)$; $z_2 = 3(\cos 40° + i \sin 40°)$

30. $z_1 = 5(\cos 170° + i \sin 170°)$; $z_2 = \cos 55° + i \sin 55°$

In problems 31–44, use DeMoivre's theorem to compute each of the given powers. Express the answer in Cartesian form.

31. $(\cos 30° + i \sin 30°)^7$

32. $(\cos 15° + i \sin 15°)^8$

33. $\left[2\left(\cos \frac{\pi}{6} + i \sin \frac{\pi}{6}\right)\right]^{10}$

34. $\left[3\left(\cos \frac{\pi}{18} + i \sin \frac{\pi}{18}\right)\right]^{6}$

35. $\left[2\left(\cos \frac{5\pi}{4} + i \sin \frac{5\pi}{4}\right)\right]^{8}$

36. $[4(\cos 36° + i \sin 36°)]^5$

37. $(5 + 5i)^6$

38. $\left(1 + \sqrt{3}i\right)^5$

39. $\left(\sqrt{3} - i\right)^4$

40. $\left(-\frac{1}{2} - \frac{\sqrt{3}}{2}i\right)^8$

41. $\left(\sqrt{3} + i\right)^{30}$

42. $(1 + i)^{50}$

43. $\left(\frac{1}{\sqrt{2}} + \frac{1}{\sqrt{2}}i\right)^{100}$

44. $\left(\frac{1}{2} + \frac{\sqrt{3}}{2}i\right)^{30}$

In problems 45–52, find all the indicated roots of each complex number in the form $x + yi$. Round off answers to two decimal places. Represent the roots graphically.

45. Square roots of i

46. Square roots of $3 - 3i$

47. Cube roots of 8

48. Cube roots of i

49. Fifth roots of $32(\cos 315° + i \sin 315°)$

50. Fifth roots of unity

51. Fourth roots of $-8 - 8\sqrt{3}\,i$

52. Fourth roots of -16

In problems 53–58, find all the roots of each equation in both trigonometric and Cartesian form. Round off answers to two decimal places.

53. $z^3 + 8 = 0$

54. $z^3 + 8i = 0$

55. $z^4 + 81 = 0$

56. $z^5 + 1 = 0$

57. $z^6 + 64 = 0$

58. $z^5 - i = 1$

Developing and Extending the Concepts

59. Prove that $|z| = \sqrt{z \cdot \bar{z}}$, where $z = a + bi$ and $\bar{z}$ is the conjugate of z.

60. Prove each of the following statements, where $z_1 = a_1 + b_1 i$ and $z_2 = a_2 + b_2 i$:

(a) $|z_1 z_2| = |z_1||z_2|$ **(b)** $\left|\frac{z_1}{z_2}\right| = \frac{|z_1|}{|z_2|}$, $z_2 \neq 0$

61. Describe geometrically the set of complex numbers $z = x + yi$ that satisfy each equation. Sketch the graph.

(a) $|z| = 2$ **(b)** $|z - i| = 1$

62. Let

$$z_1 = r_1(\cos \theta_1 + i \sin \theta_1) \text{ and } z_2 = r_2(\cos \theta_2 + i \sin \theta_2)$$

Show that

$$\frac{z_1}{z_2} = \frac{r_1}{r_2}[\cos(\theta_1 - \theta_2) + i \sin(\theta_1 - \theta_2)] \qquad z_2 \neq 0$$

63. Let $z = r(\cos \theta + i \sin \theta)$. Show that:

(a) $\bar{z} = r(\cos \theta - i \sin \theta)$
$= r[\cos(-\theta) + i \sin(-\theta)]$

(b) $z^{-1} = r^{-1}(\cos \theta - i \sin \theta)$
$= r^{-1}[\cos(-\theta) + i \sin(-\theta)]$

64. If $z = r(\cos \theta + i \sin \theta)$, show that

$$z^{-n} = r^{-n}(\cos n\theta - i \sin n\theta)$$

where n is a positive integer. [*Hint*: Use the result from problem 63(b).]

65. Use the result from problem 64 to write each expression in the form $x + yi$.

(a) $\left(\frac{\sqrt{3}}{2} + \frac{1}{2}i\right)^{-5}$ **(b)** $(-2 + 2i)^{-3}$

66. Use DeMoivre's theorem to derive formulas for $\cos 3\theta$ in terms of $\cos \theta$, and $\sin 3\theta$ in terms of $\sin \theta$. [*Hint*: Use the identity $\cos 3\theta + i \sin 3\theta = (\cos \theta + i \sin \theta)^3$.]

CHAPTER 7 REVIEW PROBLEM SET

In problems 1 and 2, solve triangle ABC. Round off each angle to the nearest hundredth of a degree and each length to two decimal places.

1. **(a)** $a = 5$, $b = 7$, $\gamma = 30°$
(b) $c = 10$, $\alpha = 45°$, $\beta = 75°$

2. **(a)** $\alpha = 63.3°$, $\gamma = 81.6°$, $c = 20.7$
(b) $a = 36$, $b = 47$, $c = 41$

3. Locate the point with the given polar coordinates. Then find the corresponding Cartesian coordinates of the point.

(a) $\left(5, \frac{\pi}{4}\right)$ **(b)** $\left(2, \frac{\pi}{6}\right)$ **(c)** $\left(\sqrt{2}, -135°\right)$

4. Locate the point with the given Cartesian coordinates. Then find polar coordinates that correspond to the point, where $r > 0$ and $0° \leq \theta < 360°$.
(a) $(-3, 0)$ **(b)** $(-2, 2\sqrt{3})$ **(c)** $(-5\sqrt{3}, -5)$

5. Convert each polar equation to a corresponding equation in Cartesian form. Identify the curve and graph it.
(a) $r + 4 \sin \theta = 0$ **(b)** $r = -3 \cos \theta$

6. Convert each equation to a corresponding equation in polar form. Identify the curve and graph it.
(a) $3x^2 + 3y^2 = 48$ **(b)** $y^2 = 4x$

7. GU Sketch the graph of each polar equation by the point-plotting method. Then use a graphing utility to demonstrate the validity of the result.
(a) $r = 4 \cos \theta$ **(b)** $r = 1 + 3 \cos \theta$

8. GU Use a graphing utility to graph each equation.
(a) $r = 2 \sin 2\theta$ **(b)** $r = -2 \cos 3\theta$

9. Let $\mathbf{u} = \langle 3, 4\rangle$ and $\mathbf{v} = \langle 4, 3\rangle$. Find each of the following vectors and represent them graphically:
(a) $3\mathbf{u}$ **(b)** $\mathbf{v} - \mathbf{u}$
(c) $-2\mathbf{u} + \mathbf{v}$ **(d)** $3\mathbf{u} + 2\mathbf{v}$

10. Determine the components of **u** if θ is a direction angle of **u**.
(a) $|\mathbf{u}| = 5$ and $\theta = 30°$
(b) $|\mathbf{u}| = 6$ and $\theta = 45°$
(c) $|\mathbf{u}| = 8$ and $\theta = \dfrac{5\pi}{6}$

In problems 11 and 12, find the magnitude and direction of each vector. Round off the answers to two decimal places.

11. **(a)** $\mathbf{u} = 12\mathbf{i} + 5\mathbf{j}$ **(b)** $\mathbf{u} = 7\mathbf{i} - 2\mathbf{j}$
12. **(a)** $\mathbf{u} = 6\mathbf{i} + 3\mathbf{j}$ **(b)** $\mathbf{u} = \sqrt{3}\,\mathbf{i} - 2\mathbf{j}$

13. If **u** is a vector whose initial point is the first point given and whose terminal point is the second point given, write **u** in the form $\mathbf{u} = \langle a, b\rangle$ and find $|\mathbf{u}|$.
(a) $(-6, 8)$; $(4, 3)$ **(b)** $(-7, 6)$; $(3, -1)$

14. Suppose that the vector **u** has a magnitude of 10 and a direction of 135°. Find the components of **u**.

In problems 15–18, express each complex number in trigonometric form. Give the exact angle in problems 15 and 16; in problems 17 and 18, round off the angle measure to two decimal places.

15. **(a)** $z = -1 + \sqrt{3}\,i$ **(b)** $z = -1 + i$
16. **(a)** $z = 5 + 5i$ **(b)** $z = 4\sqrt{3} - 12i$
17. **(a)** $z = 5 + 12i$ **(b)** $z = 7 - 6i$
18. **(a)** $z = -12 + 5i$ **(b)** $z = -7 - 8i$

In problems 19–22, find both z_1z_2 and z_1/z_2. Write the answers in both trigonometric and Cartesian form. In problems 21 and 22, round off the answers to two decimal places.

19. $z_1 = 2(\cos \pi + i \sin \pi)$; $z_2 = 3\left(\cos \dfrac{\pi}{2} + i \sin \dfrac{\pi}{2}\right)$
20. $z_1 = \sqrt{2}(\cos 315° + i \sin 315°)$;
$z_2 = 2\sqrt{2}(\cos 135° + i \sin 135°)$
21. $z_1 = 6(\cos 230° + i \sin 230°)$;
$z_2 = 3(\cos 75° + i \sin 75°)$
22. $z_1 = \sqrt{2}\left(\cos \dfrac{\pi}{4} + i \sin \dfrac{\pi}{4}\right)$; $z_2 = 2\left(\cos \dfrac{2\pi}{3} + i \sin \dfrac{2\pi}{3}\right)$

In problems 23–26, use DeMoivre's theorem to write each expression in both trigonometric and Cartesian form. In problems 25 and 26, round off the answers to two decimal places.

23. **(a)** $(\cos 60° + i \sin 60°)^6$
(b) $\left[\sqrt{6}\left(\cos \dfrac{\pi}{5} + i \sin \dfrac{\pi}{5}\right)\right]^5$
24. **(a)** $(1 + i)^{40}$ **(b)** $\left(\dfrac{\sqrt{2}}{2} + \dfrac{\sqrt{2}}{2}i\right)^{100}$
25. **(a)** $[3(\cos 27° + i \sin 27°)]^4$
(b) $\left[\sqrt{5}\left(\cos \dfrac{\pi}{8} + i \sin \dfrac{\pi}{8}\right)\right]^5$
26. **(a)** $[2(\cos 44° + i \sin 44°)]^6$
(b) $\left[\sqrt{6}\left(\cos \dfrac{\pi}{7} + i \sin \dfrac{\pi}{7}\right)\right]^4$

In problems 27 and 28, write all the roots of each equation in trigonometric form.

27. **(a)** $z^3 = 8i$ **(b)** $z^4 = 1 + i$
28. **(a)** $z^4 = 8\sqrt{2} - 8\sqrt{2}\,i$ **(b)** $z^5 = 243i$

In problems 29–34, round off the answers to two decimal places.

29. Surveying: Two points A and B are 50 feet apart on one bank of a straight river. A point C on the bank across the river is located so that angle CAB is 70° and angle ABC is 80°. How wide is the river?

30. Geometry: A diagonal of a parallelogram is 16 inches long and forms angles of 43° and 15°, respectively, with the two sides. How long are the sides of the parallelogram?

31. Engineering: A guy wire attached to the top of a pole is 40 feet long and forms a 50° angle with the ground. How tall is the pole if it is tilted 15° from the vertical directly away from the guy wire?

32. Telephone Pole: A telephone pole AB is 30 feet tall and is located on a 10° slope from a point C on the horizontal plane (Figure 1). If angle ACB is 31°, find the length of the guy wire CB and the distance from C to the base of the pole.

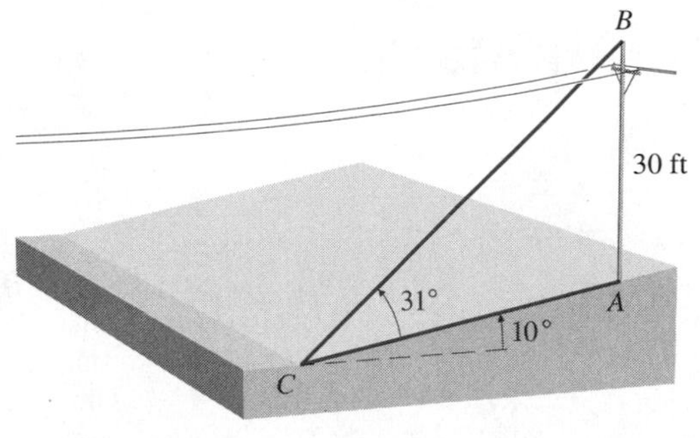

FIGURE 1

33. Engineering: Two forces, one of 63 pounds and one of 45 pounds, yield a resultant force of 75 pounds. Find the angle between the two forces (to the nearest hundredth of a degree).

34. Navigation: An airplane, which has an airspeed of 550 miles per hour, heading N30°W, is affected by a wind blowing at 45 miles per hour from S45°W. Find the velocity vector of the airplane's path. What is the speed of the airplane, and in what direction is it traveling, relative to the direction of the wind (to the nearest hundredth of a degree)?

CHAPTER 7 TEST

1. Use the law of sines to solve triangle ABC, with $a = 70$, $\beta = 36.65°$, and $\alpha = 40°$. Round off the answers to two decimal places.

2. Use the law of cosines to find b in triangle ABC, if $a = 5.47$, $c = 8.94$, and $\beta = 37.2°$. Round off the answer to two decimal places.

3. **(a)** Write the polar point $\left(4, \frac{5\pi}{3}\right)$ in Cartesian form.

(b) Write the Cartesian point $(-3, 3)$ in polar form.

4. Sketch the graph of each polar equation. Convert to rectangular form in each case.

(a) $r = 5$ **(b)** $r = 4 \sin \theta$ **(c)** $\theta = \pi/4$

5. Find the polar form of each equation, and graph the equation.

(a) $3x + 2y = 5$ **(b)** $x^2 + y^2 = 16$

6. Find $4\mathbf{u} - 8\mathbf{v}$ if $\mathbf{u} = 2\mathbf{i} + 3\mathbf{j}$ and $\mathbf{v} = 4\mathbf{i} - \mathbf{j}$.

7. Find a direction angle of the vector $\mathbf{u} = 2\mathbf{i} + 3\mathbf{j}$. Round off the answer to two decimal places.

8. Let $\mathbf{u}$ and $\mathbf{v}$ be vectors such that $|\mathbf{u}| = 6$ and $|\mathbf{v}| = 8$. Assume that angles 110° and 56° are two direction angles of $\mathbf{u}$ and $\mathbf{v}$, respectively. Determine $\mathbf{u} + \mathbf{v}$. Round off to two decimal places.

9. Write $z = 4(\cos 315° + i \sin 315°)$ in Cartesian form.

10. Write $z = -1 - \sqrt{3}\, i$ in trigonometric form.

11. For z_1 and z_2 given below, find the following in both the Cartesian and trigonometric forms:

(a) $z_1 z_2$ **(b)** $\frac{z_1}{z_2}$

$$z_1 = (\cos 139° + i \sin 139°)$$

and

$$z_2 = 2(\cos 71° + i \sin 71°)$$

Round off each answer to four decimal places.

12. Use DeMoivre's theorem to determine $[2(\cos 60° + i \sin 60°)]^{12}$. Express the answer in the form $a + bi$.

13. Find the three roots of $z^3 = i$. Write the answers in Cartesian form.

14. An airplane flies 40 miles due east, then changes direction and flies 60 miles N20°E. How far is the airplane from its starting point? Round off the answer to one decimal place.

8

Systems of Equations and Inequalities

CHAPTER CONTENTS

In science, business, and economics, we often encounter mathematical models that involve more than one equation and contain two or more variables. Such a set of equations is referred to as a *system of equations*. In this chapter, we study several different techniques for solving systems of equations and also *systems of inequalities*.

OBJECTIVES

1. Solve a system by substitution and elimination.
2. Solve a system by Gaussian elimination.
3. Solve a system graphically.
4. Solve applied problems.

8.1 Solutions of Linear Systems by Substitution and Elimination

In this section, we investigate systems that contain only linear equations. Such systems are called *linear systems*. In general, a *solution* of a linear system of equations is a collection of particular values for the variables that satisfy all equations in the system.

For example,

$$\begin{cases} 3x + 4y = 12 \\ 3x - 8y = 0 \end{cases}$$

is a linear system of two equations that contains the two variables x and y. (We use a brace to group together the equations in the system.) If we substitute $\frac{8}{3}$ for x and 1 for y into each equation in this system, we find that these values satisfy both equations:

$$\begin{cases} 3\left(\frac{8}{3}\right) + 4(1) = 8 + 4 = 12 \\ 3\left(\frac{8}{3}\right) - 8(1) = 8 - 8 = 0 \end{cases}$$

We say that the pair of values $\frac{8}{3}$ and 1 for x and y, respectively, is a *solution* of the system. A solution of a system of two equations in two variables is often written as an ordered pair of numbers, so we express the solution above in the form $\left(\frac{8}{3}, 1\right)$.

Solving a System by Substitution and Elimination

The process of finding all solutions of a system is called *solving the system.* To solve a linear system, we can use either substitution or elimination. To solve a linear system of two equations in two variables by *substitution*, we use one of the equations to express one of the variables in terms of the other. Then we substitute this expression for the variable in the other equation to obtain an equation in one variable. This procedure is illustrated in the next example.

EXAMPLE 1 **Solving a system by substitution**

Solve the system by substitution: $\begin{cases} 4x - y = 2 \\ 3x + y = 5 \end{cases}$

Solution We begin by solving the first equation for y in terms of x to get

$$y = 4x - 2$$

Replacing y with $4x - 2$ in the second equation of the system, gives us

$$3x + (4x - 2) = 5$$
$$7x = 7 \quad \text{or} \quad x = 1$$

The value of y that solves the system is obtained by substituting 1 for x in $y = 4x - 2$:

$$y = 4(1) - 2 = 2$$

Thus, the solution is the pair of numbers $x = 1$ and $y = 2$, or $(1, 2)$. Graphically, this means that the two lines with equations $4x - y = 2$ and $3x + y = 5$ intersect at the point $(1, 2)$.

Check To check the solution, replace x by 1 and y by 2 in the original system to get

$$\begin{cases} 4(1) - 2 = 2 \\ 3(1) + 2 = 5 \end{cases}$$

The substitution method is efficient for solving certain systems with two equations. However, it tends to become cumbersome when more than two equations are involved. An alternate method, called the *elimination method*, enables us to solve systems of linear equations in any number of variables more efficiently. The elimination method is based on the idea of *equivalent systems* of equations. Two systems of equations are said to be **equivalent** if they have the same solutions.

For example, the systems

$$\begin{cases} 2x + y = 1 \\ -x + 3y = -4 \end{cases} \quad \text{and} \quad \begin{cases} 2x + y = 1 \\ y = -1 \end{cases}$$

are equivalent, since they have the same solution: $(1, -1)$.

A linear system can be converted to an equivalent form by performing one or more of the following three elementary operations:

ELEMENTARY OPERATIONS YIELDING EQUIVALENT SYSTEMS

1. Interchange the positions of two equations in the system.
2. Replace an equation in the system by a nonzero multiple of itself by multiplying each side of the equation by the same number.
3. Replace an equation in the system by the sum of a nonzero multiple of that equation and another equation in the system.

The basic strategy behind the elimination method is to use these operations to convert a system to an equivalent one that contains fewer variables and is easier to solve. For instance, consider the system

$$\begin{cases} x + y = 6 \\ x - y = 2 \end{cases}$$

If we replace the second equation by the sum of itself and the first equation, we get the equivalent system

$$\begin{cases} x + y = 6 \\ 2x = 8 \end{cases}$$

This form of the system is easier to solve than the original. By inspection, we see that $x = 4$. Substituting this value into the first equation, we get

$$4 + y = 6 \quad \text{or} \quad y = 2$$

So the solution of the system is $(4, 2)$.

EXAMPLE 2 Solving a system by elimination

Solve each system by elimination. Then represent the system graphically; that is, graph both equations in the same coordinate system. Relate the solution to the graph of the system.

$$\text{(a)} \begin{cases} x - y = 1 \\ 3x - y = -1 \end{cases} \quad \text{(b)} \begin{cases} x + 3y = 4 \\ 2x + 6y = -6 \end{cases} \quad \text{(c)} \begin{cases} 3x - y = 1 \\ 6x - 2y = 2 \end{cases}$$

Solution

(a) For

$$\begin{cases} x - y = 1 \\ 3x - y = -1 \end{cases}$$

we'll start by using the third elementary operation to convert to an equivalent system in which the x variable is eliminated in the second equation. This is accomplished by replacing the second equation by the sum of -3 times the first equation and the second to obtain

$$\begin{cases} x - y = 1 \\ 3x - y = -1 \end{cases} \rightarrow \begin{cases} x - y = 1 \\ 2y = -4 \end{cases}$$

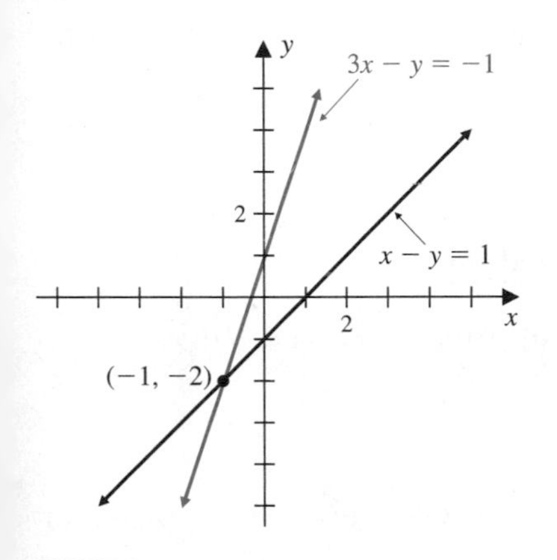

FIGURE 1a

Next, we multiply both sides of the second equation by $\frac{1}{2}$ to obtain an equivalent system that is simpler to solve:

$$\begin{cases} x - y = 1 \\ y = -2 \end{cases}$$

Obviously, $y = -2$. By substituting -2 for y into the first equation, $x - y = 1$, we obtain $x - (-2) = 1$, or $x = -1$. Thus, the solution is $(-1, -2)$. The graph of the system is shown in Figure 1a. Clearly, the graphs intersect at one point. Since the coordinates of this point must satisfy both equations, it follows that the solution of the system and the point are the same: $(-1, -2)$.

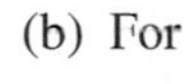

(b) For

$$\begin{cases} x + 3y = 4 \\ 2x + 6y = -6 \end{cases}$$

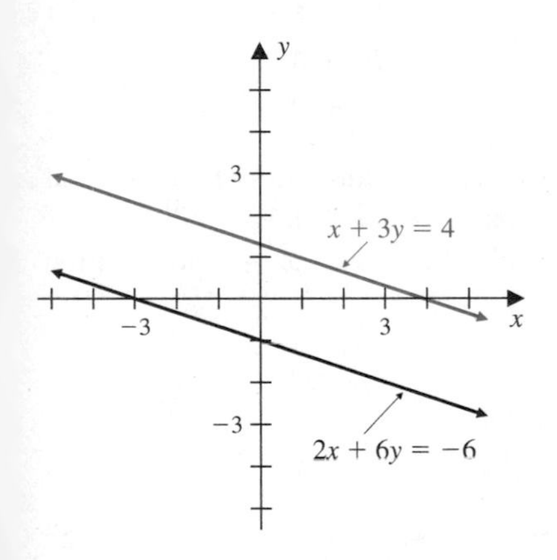

FIGURE 1b

Our strategy is to produce an equivalent system by eliminating x in the second equation. To do this, we replace the second equation by the sum of -2 times the first equation and the second equation to obtain an equivalent system:

$$\begin{cases} x + 3y = 4 \\ 2x + 6y = -6 \end{cases} \rightarrow \begin{cases} x + 3y = 4 \\ 0 = -14 \end{cases}$$

This leads to a contradiction, so the system has no solution. As Figure 1b shows, the two lines are parallel (the lines have the same slope $m = -\frac{1}{3}$), so there is no point common to the two lines. Consequently, no pair of numbers can satisfy both equations simultaneously.

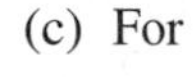

(c) For

$$\begin{cases} 3x - y = 1 \\ 6x - 2y = 2 \end{cases}$$

in order to eliminate the x term in the second equation, we replace the second equation by the sum of -2 times the first equation and the second to get

$$\begin{cases} 3x - y = 1 \\ 6x - 2y = 2 \end{cases} \rightarrow \begin{cases} 3x - y = 1 \\ 0 = 0 \end{cases}$$

We conclude that the solution consists of all pairs of numbers (x, y) that satisfy the equation $3x - y = 1$ or $y = 3x - 1$. It follows that the solution consists of ordered pairs of the form $(x, 3x - 1)$, where x is a real number.

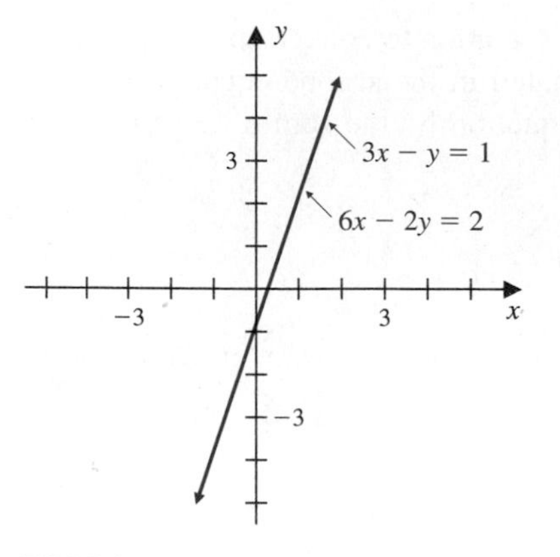

FIGURE 1c

In Figure 1c, we can see that the two equations have the same graph, so there are infinitely many solutions (one for each point on the graph).

Example 2 illustrates that a linear system may have one solution, infinitely many solutions, or no solution. A system that has one solution is said to be **consistent** and **independent**; a system that has infinitely many solutions is said to be **consistent** and **dependent**; and a system that has no solution is said to be **inconsistent**.

Solving a System by Gaussian Elimination

In Example 2a, we solved the given system by using the equivalent form

$$\begin{cases} x - y = 1 \\ y = -2 \end{cases}$$

This latter system is said to be in *triangular form*. Other examples of triangular forms are:

$$\begin{cases} x + 2y = 7 \\ y = 1 \end{cases} \quad \text{and} \quad \begin{cases} x + 2y - 3z = 1 \\ y + 2z = 4 \\ z = 7 \end{cases}$$

Note the staggered pattern of a triangular form. The leading coefficient is 1 for each equation, and each equation after the first one has one less variable than the one above it. When a system is written in triangular form, we can find its solution by using *back-substitution*. For instance, the solution of

$$\begin{cases} x + 2y - 3z = 1 \\ y + 2z = 4 \\ z = 7 \end{cases}$$

is found as follows: Clearly, $z = 7$ by inspection. Substituting this value into the second equation, we get

$$\begin{aligned} y + 2(7) &= 4 \\ y + 14 &= 4 \\ y &= -10 \end{aligned}$$

Finally, we substitute -10 for y and 7 for z into the first equation to get

$$\begin{aligned} x + 2(-10) - 3(7) &= 1 \\ x - 20 - 21 &= 1 \\ x - 41 &= 1 \\ x &= 42 \end{aligned}$$

Thus, the solution is $(42, -10, 7)$.

The process of using elimination to convert a given linear system to an equivalent triangular one and then using back-substitution to find the solution is referred to as the **Gaussian elimination method**.

EXAMPLE 3 Solving a system by Gaussian elimination

Solve system (A) by using the Gaussian elimination method.

Subscripts are often used to denote variables in linear systems.

$$\begin{cases} 2x_1 + x_2 - 2x_3 = 10 \\ 3x_1 + 2x_2 + 2x_3 = 1 \\ 5x_1 + 4x_2 + 3x_3 = 4 \end{cases} \tag{A}$$

Solution First, we eliminate the x_1 term in the second equation by replacing the second equation in system (A) by the sum of -3 times the first equation and 2 times the second equation to obtain the equivalent system (B):

$$\begin{cases} 2x_1 + x_2 - 2x_3 = 10 \\ x_2 + 10x_3 = -28 \\ 5x_1 + 4x_2 + 3x_3 = 4 \end{cases} \tag{B}$$

In the next step, we eliminate the x_1 term in the third equation by replacing the third equation of (B) with the sum of -5 times the first equation and 2 times the third equation to obtain the equivalent system (C):

$$\begin{cases} 2x_1 + x_2 - 2x_3 = 10 \\ x_2 + 10x_3 = -28 \\ 3x_2 + 16x_3 = -42 \end{cases} \tag{C}$$

Now, we eliminate the x_2 term in the third equation by replacing the third equation of (C) by the sum of the third equation and -3 times the second equation to get the equivalent system (D):

$$\begin{cases} 2x_1 + x_2 - 2x_3 = 10 \\ x_2 + 10x_3 = -28 \\ -14x_3 = 42 \end{cases} \tag{D}$$

Next, we multiply the third equation by $-\frac{1}{14}$ to obtain the equivalent system (E), in which the coefficient of x_3 is 1:

$$\begin{cases} 2x_1 + x_2 - 2x_3 = 10 \\ x_2 + 10x_3 = -28 \\ x_3 = -3 \end{cases} \tag{E}$$

Finally, in order to change the coefficient of x_1 in the first equation to 1, we multiply the first equation by $\frac{1}{2}$ to obtain the equivalent triangular system (F):

$$\begin{cases} x_1 + \frac{1}{2}x_2 - x_3 = 5 \\ x_2 + 10x_3 = -28 \\ x_3 = -3 \end{cases} \tag{F}$$

Many graphing utilities are programmed to solve linear systems.

Now we use back-substitution to solve this system. Substituting $x_3 = -3$ into the second equation, we get $x_2 = 2$; then, substituting $x_2 = 2$ and $x_3 = -3$ into the first equation, we get $x_1 = 1$. Thus, the solution is the triple of numbers $x_1 = 1$, $x_2 = 2$, and $x_3 = -3$, or $(1, 2, -3)$.

Solving a System Graphically

Some graphing utilities have a special feature that automatically locates the points of intersection of two graphs.

A graphing utility can be used to approximate the values of the solution of a system of two linear equations in two variables. We first graph the system, and then approximate the coordinates of the point of intersection by using the ZOOM and TRACE features to move the cursor as close as possible to that point. This technique is illustrated in the next example.

EXAMPLE 4

Solving a system graphically

Use a graphing utility to solve the system graphically. Round off the answers to two decimal places. Confirm the result by solving the system algebraically.

$$\begin{cases} 3x - 2y = 3 \\ 2x + 5y = -1 \end{cases}$$

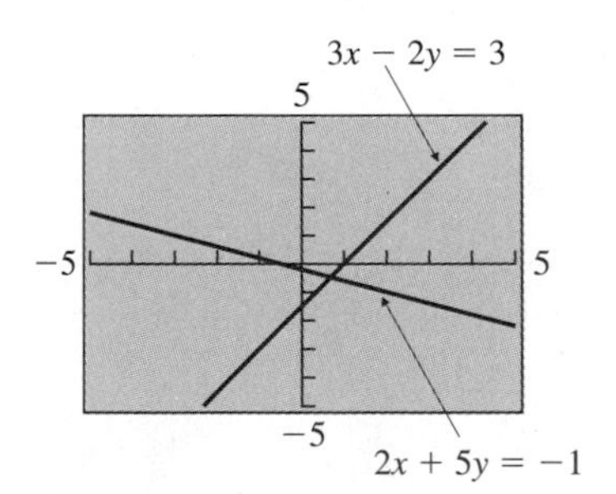

FIGURE 2

Solution The system is graphed in the viewing window of Figure 2. By using the ZOOM and TRACE features repeatedly, we approximate the location of the point of intersection to be $(0.68, -0.47)$. So the solution is approximately $(0.68, -0.47)$.

To solve the system algebraically, we eliminate x in the second equation by replacing the second equation with the sum of $-\frac{2}{3}$ times the first equation and the second equation to get

$$\begin{cases} 3x - 2y = 3 \\ \frac{19}{3}y = -3 \end{cases}$$

Next, multiply the first equation by $\frac{1}{3}$ and the second equation by $\frac{3}{19}$ to obtain the triangular form:

$$\begin{cases} x - \frac{2}{3}y = 1 \\ y = -\frac{9}{19} \end{cases}$$

The value of x may be found by substituting $-\frac{9}{19}$ for y in the first equation to get

$$3x - 2\left(-\tfrac{9}{19}\right) = 3 \quad \text{or} \quad 3x = \tfrac{39}{19}$$

So $x = \frac{13}{19}$, and the solution is $\left(\frac{13}{19}, -\frac{9}{19}\right)$. Since $\frac{13}{19}$ is approximately 0.68 and $-\frac{9}{19}$ is approximately -0.47, we have confirmed the result found graphically.

Although graphs are excellent tools for visualizing solutions of systems of equations in two variables, they do not always give the exact solution, especially if the point of intersection is not located at integer values of x and y. For this reason we rely on algebraic methods for solving these systems.

Solving Applied Problems

Systems of linear equations are often used in modeling real-world situations.

EXAMPLE 5 Using a system to solve a physics model

Suppose that a ball is thrown up in the air from a cliff h_0 feet above ground level with an initial upward speed of v_0 (in feet per second). Its height h (in feet) above ground level t seconds after it was thrown is given by the model

$$h = -16t^2 + v_0t + h_0 \quad \text{(Figure 3)}$$

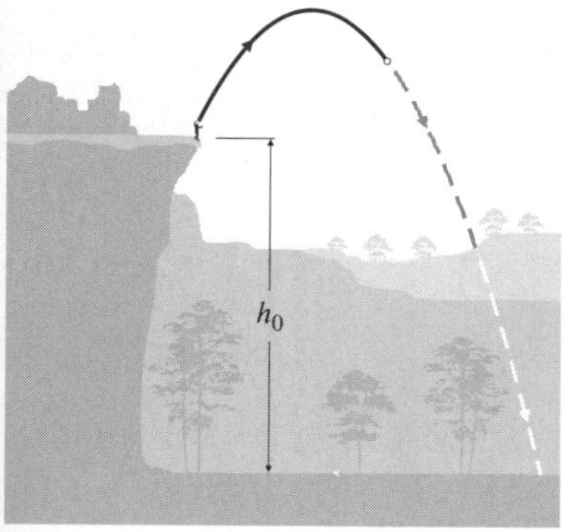

(a) Suppose that the ball is 67 feet above the ground after 1 second and 61 feet above the ground after 1.5 seconds. Use a linear system to determine the initial upward speed v_0 and the height of the cliff h_0.
(b) How high did the ball travel after 2.5 seconds?

FIGURE 3

Solution

(a) Since the ball is 67 feet above the ground after 1 second and 61 feet after 1.5 seconds, we substitute 1 for t when $h = 67$ and 1.5 for t when $h = 61$ in the equation of the model to get the system

$$\begin{cases} 67 = -16(1)^2 + v_0(1) + h_0 \\ 61 = -16(1.5)^2 + v_0(1.5) + h_0 \end{cases} \quad \text{or} \quad \begin{cases} v_0 + h_0 = 83 \\ 1.5v_0 + h_0 = 97 \end{cases}$$

Solving this system, we get $v_0 = 28$ and $h_0 = 55$. Therefore, the initial speed is 28 feet per second and the height of the cliff is 55 feet.

(b) When $t = 2.5$,

$$h = -16(2.5)^2 + 28(2.5) + 55 = 25$$

So the height of the ball is 25 feet after 2.5 seconds.

EXAMPLE 6 Solving a manufacturing problem

A manufacturer of portable radios produces three models of radios, model A, model B, and model C. The manufacturer determines that the total times allocated for production, assembly, and testing are 638 hours, 253 hours, and 159 hours, respectively. Table 1 lists the information for each model. Use a linear system to determine how many of each model can be produced if all the allocated time is used up.

TABLE 1

Model	Production Hours *Per radio*	Assembly Hours *Per radio*	Testing Hours *Per radio*
A	1.4	0.5	0.3
B	1.6	0.6	0.4
C	1.8	0.8	0.5

Solution Let x represent the number of radios of model *A* produced, y the number of model *B*, and z the number of model *C*. Now we analyze the data in Table 1. The production hours add up to 638, the assembly hours add up to 253, and the testing hours add up to 159, so the required system is

$$\begin{cases} 1.4x + 1.6y + 1.8z = 638 \\ 0.5x + 0.6y + 0.8z = 253 \\ 0.3x + 0.4y + 0.5z = 159 \end{cases}$$

Solving this system, we obtain $x = 150$, $y = 110$, and $z = 140$. Therefore, the manufacturer should produce 150 radios of model *A*, 110 of model *B*, and 140 of model *C* in order to use up the allocated times.

PROBLEM SET 8.1

Mastering the Concepts

In problems 1–12, solve each system by both substitution and elimination. Represent the system graphically, and then relate the solution to the graph of the system.

1. $\begin{cases} 3x - 2y = 1 \\ x + y = 5 \end{cases}$

2. $\begin{cases} x - y = 2 \\ 4x - 2y = 6 \end{cases}$

3. $\begin{cases} 2x + 6y = 7 \\ x + 3y = 1 \end{cases}$

4. $\begin{cases} 2x - y = 1 \\ 4x - 2y = 5 \end{cases}$

5. $\begin{cases} x + 2y = 3 \\ 2x - 3y = 1 \end{cases}$

6. $\begin{cases} 3x + 2y = 11 \\ -2x + y = 2 \end{cases}$

7. $\begin{cases} 2x - y = 5 \\ -6x + 3y = 13 \end{cases}$

8. $\begin{cases} 3x + y = 4 \\ 7x - y = 6 \end{cases}$

9. $\begin{cases} y = x - 5 \\ 2y = 2x - 10 \end{cases}$

10. $\begin{cases} 3x + 6y = 1 \\ x - 2y = -6 \end{cases}$

11. $\begin{cases} y = 1 - x \\ 5x - y = 13 \end{cases}$

12. $\begin{cases} 3x - 4y = -5 \\ x = 3 - y \end{cases}$

In problems 13–16, solve each system by elimination.

13. $\begin{cases} x + y = 5 \\ x + z = 1 \\ y + z = 2 \end{cases}$

14. $\begin{cases} x - 3y = -11 \\ 2y - 5z = 26 \\ 7x - 3z = -2 \end{cases}$

15. $\begin{cases} x + y + 2z = 11 \\ x - y + z = 3 \\ 2x + y + 3z = 17 \end{cases}$

16. $\begin{cases} x + 3y + z = 4 \\ 3x - 2y + 4z = 11 \\ 2x + y + 3z = 13 \end{cases}$

In problems 17–34, solve each system by the Gaussian elimination method.

17. $\begin{cases} x + 3y = 9 \\ x - y = 1 \end{cases}$

18. $\begin{cases} x + y = 1 \\ y - 3x = 3 \end{cases}$

19. $\begin{cases} 5x - y = 7 \\ 10x - 4y = 18 \end{cases}$

20. $\begin{cases} 2x + 4y = 3 \\ -x + y = 3 \end{cases}$

21. $\begin{cases} 3y + 2z = 5 \\ 2y = -3z + 1 \end{cases}$

22. $\begin{cases} 4x - 3y = 11 \\ 3x + 4y = 2 \end{cases}$

23. $\begin{cases} 3x + 2y = 4 \\ 5x + 3y = 7 \end{cases}$

24. $\begin{cases} 2x - 3y = 1 \\ 5x + 2y = 12 \end{cases}$

25. $\begin{cases} 2x - 7y = -5 \\ 4x + 3y = 7 \end{cases}$

26. $\begin{cases} x + y + 2z = 4 \\ x + y - 2z = 0 \\ x - y = 0 \end{cases}$

27. $\begin{cases} x + y + z = 2 \\ x + 2y - z = 4 \\ 2x + y + z = 0 \end{cases}$

28. $\begin{cases} x_1 + 2x_2 + 4x_3 = 12 \\ 2x_1 - 3x_2 + x_3 = 10 \\ 3x_1 - x_2 - 2x_3 = 1 \end{cases}$

29. $\begin{cases} x + y + z = 6 \\ x - y + 2z = 12 \\ 2x + y + z = 1 \end{cases}$

30. $\begin{cases} x + y + 2z = 4 \\ x - 5y + z = 5 \\ 3x - 4y + 7z = 24 \end{cases}$

31. $\begin{cases} x_1 + 2x_2 + 5x_3 = 4 \\ 4x_1 + x_2 + 3x_3 = 9 \\ 6x_1 + 9x_2 + x_3 = 21 \end{cases}$

32. $\begin{cases} 2x + y - 3z = 9 \\ x - 2y + 4z = 5 \\ 3x + y - 2z = 15 \end{cases}$

33. $\begin{cases} x + 3y - 2z = -21 \\ 7x - 5y + 4z = 31 \\ 2x + y + 3z = 17 \end{cases}$

34. $\begin{cases} x_1 - 5x_2 + 4x_3 = 8 \\ 3x_1 + x_2 - 2x_3 = 4 \\ 9x_1 - 3x_2 + 6x_3 = 6 \end{cases}$

GU In problems 35–44, solve each system graphically by finding the location of the point of intersection by using a graphing utility. Round off the answers to two decimal places. Confirm the result by solving the system algebraically.

35. $\begin{cases} 5x - y = 13 \\ x + y = 1 \end{cases}$

36. $\begin{cases} \frac{1}{2}x - \frac{3}{4}y = 1 \\ 3x + y = 1 \end{cases}$

37. $\begin{cases} 7x_1 + x_2 = 3 \\ 5x_1 + x_2 = 6 \end{cases}$

38. $\begin{cases} 0.2x - 0.5y = 7 \\ 0.5x + 0.3y = 8 \end{cases}$

39. $\begin{cases} 0.1x - 0.3y = 4 \\ 0.2x + 0.7y = -5 \end{cases}$

40. $\begin{cases} \frac{1}{2}x_1 + \frac{1}{3}x_2 = 13 \\ \frac{1}{5}x_1 + \frac{1}{8}x_2 = 5 \end{cases}$

41. $\begin{cases} -\frac{3}{5}x_1 + x_2 = \frac{2}{3} \\ \frac{4}{7}x_1 + \frac{2}{3}x_2 = 10 \end{cases}$

42. $\begin{cases} \frac{1}{3}x - \frac{1}{4}y = 2 \\ \frac{1}{4}x - \frac{1}{2}y = 7 \end{cases}$

43. $\begin{cases} 0.2x + 0.3y = 0.7 \\ 0.4x - 0.5y = 0.3 \end{cases}$

44. $\begin{cases} 4.01x + 6.07y = 2.33 \\ 7.57x - 3.11y = 13.81 \end{cases}$

Applying the Concepts

In problems 45–60, use a linear system to solve each problem.

45. Grading: Suppose that three-fifths of the male students and two-thirds of the female students have A or B averages. If there are 120 students in the class and 46 have an average of C or lower, how many male and how many female students are in the class?

46. Recreation: A theater group plans to sell 500 tickets for a play. They charge \$45 for each orchestra seat and \$30 for each balcony seat, and they plan to collect \$18,000 for a performance. How many of each type of ticket do they expect to sell?

47. Physics: An arrow is shot straight up in the air from a height of h_0 feet above the ground with an initial upward speed of v_0 (in feet per second). Its height h (in feet) above the ground t seconds after it was shot is given by

$$h = -16t^2 + v_0t + h_0$$

If the arrow is 22 feet above the ground after 1.5 seconds and 10 feet above the ground after 2 seconds, find its initial speed v_0 and its initial height h_0.

48. Physics: A stone is thrown vertically upward from a height of h_0 (in feet) above the ground with an initial upward speed of v_0 (in feet per second). Its height h (in feet) above the ground t seconds after it was thrown is given by

$$h = -16t^2 + v_0t + h_0$$

If the stone is 40 feet above the ground after $\frac{1}{2}$ second and 48 feet above the ground after 1.5 seconds, find its initial speed v_0 and its initial height h_0.

49. Chemistry: A chemist has one solution that is 30% acid and another that is 20% acid. How much of each should be used to make 10 liters of a solution that is 22% acid?

50. Chemistry: A chemist prepares two acid solutions. The first solution is 20% acid and the second solution is 50% acid. How many milliliters of each solution should be mixed to obtain 12 milliliters of a 30% acid solution?

51. Aviation: An airplane flying with the wind takes 3.75 hours to fly the 2500 miles from Los Angeles to New York, but 4.4 hours to fly the same distance from New York to Los Angeles against the wind. Assuming that the airplane travels at a constant airspeed and the wind blows at a constant rate, find the speed of the airplane in still air and the wind speed.

52. Water Speed: It takes a powerboat traveling upstream (against the current) 3 hours to make a 36 mile trip on a river. Returning downstream (with the current), the same trip takes 2 hours. Assuming that the boat travels at a constant speed in still water and the rate of the current is constant, find the speed of the boat in still water and the rate of the current.

53. Shoe Sales: A national sporting gear store sells running shoes and tennis shoes. A pair of tennis shoes sells for \$46 and a pair of running shoes sells for \$79. During a 1 day sale, a total of 260 pairs of tennis and running

shoes were sold. The total receipts were \$14,435. How many pairs of each kind were sold?

54. Carpooling: To encourage carpooling, a city parking lot charges \$7 a day per single driver vehicle or \$4 per day for each carpool vehicle with two or more people. If the city parking lot took in \$649 and 115 vehicles used the lot for 1 day, how many of each kind of car used the lot that day?

55. Investment Portfolio: An investment club invested \$42,000 in three accounts. During one year, the first account earned 5% simple interest, the second earned 7% simple interest, and the third earned 9% simple interest. The total interest from the three investments for the year was \$2600. The interest in the first account was \$200 less than the total interest from the other two accounts. How much was invested in each type of account?

56. Landscape Engineering: A landscape architect wishes to construct a sprinkler system by locating the sprinklers at the centers of three circular regions that are mutually tangent to each other. The center of each circle is labeled A, B, and C in Figure 4. If the distances $|\overline{AB}| = 15$ meters, $|\overline{BC}| = 13$ meters, and $|\overline{CA}| = 16$ meters, find the radii of the circles.

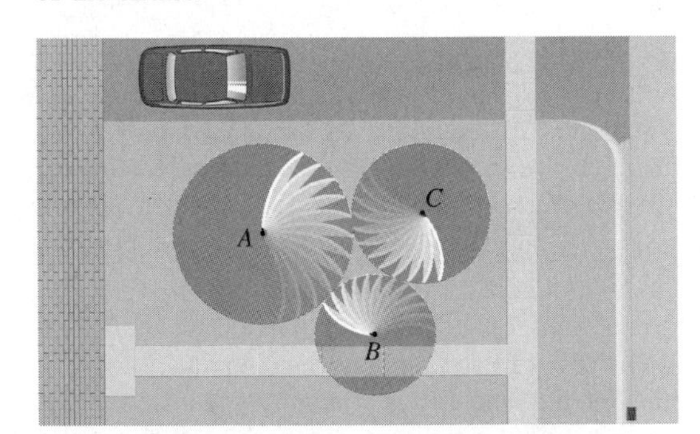

FIGURE 4

57. Airport Shuttle: An airport shuttle service has three sizes of vans. The biggest van holds 12 passengers, the next holds 10 passengers, and the smallest holds 6 passengers. The service manager has 15 vans available to accommodate 152 passengers. However, to reduce fuel costs, the manager wants to use 2 more of the available 12 passenger vans than the 6 passenger vans. How many vans of each kind should be used?

58. Sports: The total number of seats in a basketball sports arena is 18,000. The arena is divided into three sections: courtside, balcony, and end zone. There are three times as many balcony seats as courtside seats. For the conference championship game, ticket prices are \$25 for courtside, \$15 for balcony, and \$10 for end zone. If the arena is sold out and the total receipts are \$270,000, how many seats are courtside?

59. Irrigation: A grower uses three pumps to provide water to irrigate a grove. When pumps A, B, and C are used for 2 hours, they can pump 74,000 gallons. If the grower runs pump A for 4 hours and pump B for 2 hours, then 64,000 gallons of water are pumped. If pump B is used for 5 hours and pump C for 4 hours, then 120,000 gallons are pumped. How many gallons per hour can each pump handle?

60. Manufacturing: An electronics manufacturer places three wires next to each other inside a radio transmitter. The cross section of each wire is circular, and these circles are tangent to each other. Figure 5 shows the centers of these circles labeled as A, B, and C. If $|\overline{AB}| =$ 13 millimeters, $|\overline{BC}| = 22$ millimeters, and $|\overline{DE}| =$ 52 millimeters, find the radius of each wire.

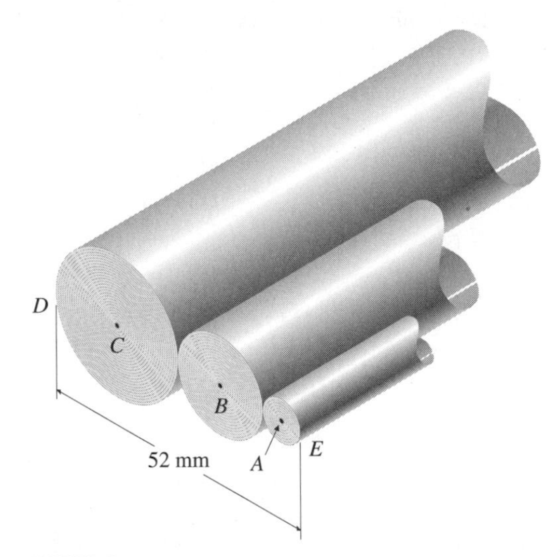

FIGURE 5

Developing and Extending the Concepts

61. Solve each system by using the substitutions $u = 1/x$ and $v = 1/y$.

(a) $\begin{cases} \dfrac{5}{x} - \dfrac{6}{y} = 21 \\ \dfrac{1}{x} - \dfrac{2}{y} = 5 \end{cases}$ (b) $\begin{cases} \dfrac{4}{x} + \dfrac{5}{y} = 6 \\ \dfrac{2}{x} - \dfrac{1}{y} = 10 \end{cases}$

62. Solve the system by using substitutions $u = 1/x$, $v = 1/y$, and $w = 1/z$.

$$\begin{cases} \dfrac{3}{x} + \dfrac{4}{y} - \dfrac{1}{z} = -7 \\ \dfrac{1}{x} - \dfrac{1}{y} + \dfrac{2}{z} = 0 \\ \dfrac{4}{x} + \dfrac{1}{y} - \dfrac{3}{z} = -17 \end{cases}$$

63. Give an example of a linear system that has no solution if the system has:
(a) Two variables
(b) Three variables

64. Give an example of a linear system that has infinitely many solutions if the system has:
(a) Two variables
(b) Three variables

OBJECTIVES

1. Write linear systems in matrix notation.
2. Solve linear systems using row-reduction.
3. Decompose fractions.
4. Solve applied problems.

8.2 Solutions of Linear Systems Using Augmented Matrices

In carrying out the Gaussian elimination method, we wrote the equations in the system in such a way that the same variables were aligned vertically in each step of the process. In this section, we'll develop a procedure based on the Gaussian method in which we solve a system of linear equations by keeping track of the changing coefficients without writing down the variables. This procedure makes use of *matrices*.

A *matrix* is a rectangular array of numbers. The numbers in the matrix are called the *entries* or *elements* of the matrix. Traditionally, we enclose the elements of a matrix in brackets. Matrices are usually denoted by capital letters such as A, B, C, D, X, and Y.

For example,

$$A = \begin{bmatrix} 2 & -1 & 3 \\ 5 & 4 & 7 \end{bmatrix}, \quad B = \begin{bmatrix} 1 & 3 \\ 2 & -1 \end{bmatrix}, \quad \text{and} \quad C = \begin{bmatrix} 4 & 3 & -1 \end{bmatrix}$$

are all matrices. Each horizontal line of numbers is called a *row* of the matrix; each vertical line of numbers is called a *column*. When we refer to a matrix, we usually refer to its *size* or *dimension* by specifying the number of rows and the number of columns in that order. In the above examples, matrix A is a 2 by 3 matrix (2 rows and 3 columns), which is usually written as 2×3. Matrix B is 2×2, and matrix C is 1×3. An entry or element of a matrix can be identified by subscripts to indicate its row and column position. For instance, in matrix A, we write $a_{11} = 2$, $a_{12} = -1$, $a_{13} = 3$, $a_{21} = 5$, $a_{22} = 4$, and $a_{23} = 7$. In general, a_{mn} is the entry in the mth row and nth column of matrix A.

Writing Linear Systems in Matrix Notation

Consider the linear system of equations

$$\begin{cases} 3x - y = 7 \\ 4x + 7y = 15 \end{cases}$$

If we list the coefficients and constants of the system, we obtain the rectangular array of numbers

$$\left[\begin{array}{cc|c} 3 & -1 & 7 \\ 4 & 7 & 15 \end{array}\right]$$

It is customary to draw a vertical line to separate the last column (the column that corresponds to the constants).

This matrix is called the *augmented matrix* form of the system. The matrix of the coefficients of the variables of the system,

$$\begin{bmatrix} 3 & -1 \\ 4 & 7 \end{bmatrix}$$

is called the *coefficient matrix*. The first column of this matrix lists the coefficients of the x variables and the second column lists the coefficients of the y variables. If any variable does not appear in an equation, a 0 is inserted in the appropriate position in the array of coefficients and constants.

EXAMPLE 1 Writing a system in matrix notation

Write the system in matrix notation:

$$\begin{cases} 4x - 3y = 5 \\ x + 7y = 8 \end{cases}$$

Solution The augmented matrix form of the system is

$$\left[\begin{array}{cc|c} 4 & -3 & 5 \\ 1 & 7 & 8 \end{array}\right]$$

EXAMPLE 2 Writing a system corresponding to a matrix

Write the system in terms of x and y corresponding to the following augmented matrix, and solve it.

$$\left[\begin{array}{cc|c} 1 & -2 & 7 \\ 0 & 1 & -3 \end{array}\right]$$

Solution Here, we used a 0 to denote the coefficient of x in the second equation. So the system is

$$\begin{cases} x - 2y = 7 \\ y = -3 \end{cases}$$

Using back-substitution, we see that

$$\begin{aligned} x - 2(-3) &= 7 \\ x + 6 &= 7 \\ x &= 1 \end{aligned}$$

Therefore, the solution is $(1, -3)$.

Solving Linear Systems Using Row-Reduction

In Section 8.1, we used elementary operations to transform a linear system of equations into an equivalent system in triangular form, because the latter form is easier to solve. In this section, we perform the same process, but rather than working with the equations in the system, we work with the rows of the corresponding augmented matrix form. Each row serves as an abbreviation of the resulting equation when an elementary operation is applied. In the following presentation we solve a system of linear equations in two ways. On the left, the elementary operations are used to convert the given system to an equivalent triangular form. On the right, we show the augmented matrix corresponding to each equivalent linear system and describe the elementary operation in terms of a row manipulation. A vertical arrow is used to indicate that each matrix is *row-equivalent* to the one below it in the sense that they both represent equivalent systems.

System of Equations	Augmented Matrix of System	Strategy
(A) $\begin{cases} 3x - y = 15 \\ x + 2y = -2 \end{cases}$	(A) $\left[\begin{array}{cc\|c} 3 & -1 & 15 \\ 1 & 2 & -2 \end{array}\right]$	
$\downarrow$ First, interchange equations	$\downarrow$ First, interchange rows	We want the first equation to have a leading coefficient of 1.
(B) $\begin{cases} x + 2y = -2 \\ 3x - y = 15 \end{cases}$	(B) $\left[\begin{array}{cc\|c} 1 & 2 & -2 \\ 3 & -1 & 15 \end{array}\right]$	
$\downarrow$ Next, replace the second equation with the sum of -3 times the first equation and the second equation	$\downarrow$ Next, replace the second row with the sum of -3 times the first row and the second row	We want to eliminate the x term in the second equation.
(C) $\begin{cases} x + 2y = -2 \\ 0 - 7y = 21 \end{cases}$	(C) $\left[\begin{array}{cc\|c} 1 & 2 & -2 \\ 0 & -7 & 21 \end{array}\right]$	
$\downarrow$ Multiply the second equation by $-\frac{1}{7}$	$\downarrow$ Multiply the second row by $-\frac{1}{7}$	We want the second equation to have a leading coefficient of 1.
(D) $\begin{cases} x + 2y = -2 \\ y = -3 \end{cases}$	(D) $\left[\begin{array}{cc\|c} 1 & 2 & -2 \\ 0 & 1 & -3 \end{array}\right]$	
$\downarrow$ The system is in triangular form	$\downarrow$ The matrix is in triangular form	

Thus, the last matrix (D) is an abbreviated form of the last system form (D). Clearly, $y = -3$. To get x, we substitute -3 for y in the first equation to obtain $x + 2(-3) = -2$, so $x = 4$. So the solution is $(4, -3)$.

The augmented matrix form may be used to input a linear system into a graphing utility that is programmed to solve such systems.

As this illustration shows, the two approaches for solving a linear system are the same. However, we can work more efficiently with augmented matrices than with the actual systems of equations.

In general, the elementary operations performed on the equations of a linear system correspond to analogous elementary row operations performed on the rows of an augmented matrix. These row operations, which allow us to transform a matrix into a row-equivalent matrix, are listed in Table 1. Table 2 gives illustrations of the elementary row operations.

TABLE 1
Elementary Row Operations

Row Operation Performed on a Matrix	Explanation of Operation	Symbolic Representation of Operation
1. Interchange two rows	Interchange row i and row j	$R_i \leftrightarrow R_j$
2. Multiply a row by a nonzero constant	Multiply each entry in row i by a constant k to get new row i, where $k \neq 0$	$kR_i \rightarrow R_i$
3. Add a constant multiple of a row to another row	Multiply each entry of row i by the constant k and add the resulting entries to each entry in row j to get a new row j	$kR_i + R_j \rightarrow R_j$

TABLE 2

Original Matrix	Row Operation	Symbolic Representation	Resulting Row-Equivalent Matrix
1. $\begin{bmatrix} 1 & 2 & -3 \\ 4 & 5 & 6 \end{bmatrix}$	Interchange row 1 and row 2	$R_1 \leftrightarrow R_2$	$\begin{bmatrix} 4 & 5 & 6 \\ 1 & 2 & -3 \end{bmatrix}$
2. $\begin{bmatrix} 1 & 2 & -3 & 5 \\ 4 & 5 & 6 & -1 \end{bmatrix}$	Multiply each entry of row 2 by -3 to get a new row 2	$-3R_2 \rightarrow R_2$	$\begin{bmatrix} 1 & 2 & -3 & 5 \\ -12 & -15 & -18 & 3 \end{bmatrix}$
3. $\begin{bmatrix} 2 & 4 & -2 \\ 3 & 1 & 5 \\ 1 & -6 & -1 \end{bmatrix}$	Multiply each entry in row 1 by 2 and add the resulting entries to the entries in row 3 to get a new row 3	$2R_1 + R_3 \rightarrow R_3$	$\begin{bmatrix} 2 & 4 & -2 \\ 3 & 1 & 5 \\ 5 & 2 & -5 \end{bmatrix}$

EXAMPLE 3 **Solving a system using matrix row operations**
Write the associated augmented matrix of the system and solve it using elementary row operations.

$$\begin{cases} 4x - 3y = 15 \\ x + 2y = 1 \end{cases}$$

Solution We first write the augmented matrix of the system:

$$\left[\begin{array}{cc|c} 4 & -3 & 15 \\ 1 & 2 & 1 \end{array}\right]$$

Next, we convert it into an equivalent matrix in triangular form. The easiest way to get a 1 in the upper left corner is to interchange rows 1 and 2:

$$\left[\begin{array}{cc|c} 4 & -3 & 15 \\ 1 & 2 & 1 \end{array}\right] \xrightarrow{R_1 \leftrightarrow R_2} \left[\begin{array}{cc|c} 1 & 2 & 1 \\ 4 & -3 & 15 \end{array}\right]$$

We want a 0 to be the first entry in row 2, so we multiply row 1 by -4 and then add the result to row 2:

$$\left[\begin{array}{cc|c} 1 & 2 & 1 \\ 4 & -3 & 15 \end{array}\right] \xrightarrow{-4R_1 + R_2 \rightarrow R_2} \left[\begin{array}{cc|c} 1 & 2 & 1 \\ 0 & -11 & 11 \end{array}\right]$$

Finally, to get 1 as the first entry in row 2, we multiply row 2 by $-\frac{1}{11}$:

$$\left[\begin{array}{cc|c} 1 & 2 & 1 \\ 0 & -11 & 11 \end{array}\right] \xrightarrow{-\frac{1}{11}R_2 \rightarrow R_2} \left[\begin{array}{cc|c} 1 & 2 & 1 \\ 0 & 1 & -1 \end{array}\right]$$

The last matrix is in the desired triangular form. By writing the associated system, we have

$$\begin{cases} x + 2y = 1 \\ y = -1 \end{cases}$$

Using back-substitution, we replace y by -1 in the first equation to obtain $x + 2(-1) = 1$ or $x - 2 = 1$, so $x = 3$. Therefore, the solution is $(3, -1)$.

Let us review the strategy used to solve the system in Example 3.

Step 1. Form the augmented matrix of the system.
Step 2. Use the elementary row operations to reduce the augmented matrix to triangular form.
Step 3. Write the corresponding system in triangular form.
Step 4. Use back-substitution to solve the resulting system.

We can avoid the necessity of back-substitution in Example 3 by reducing the last matrix even further. To do this, multiply row 2 by -2 and add the result to row 1 to obtain:

$$\left[\begin{array}{cc|c} 1 & 2 & 1 \\ 0 & 1 & -1 \end{array}\right] \xrightarrow{-2R_2 + R_1 \rightarrow R_1} \left[\begin{array}{cc|c} 1 & 0 & 3 \\ 0 & 1 & -1 \end{array}\right]$$

This final matrix is an especially simple one to interpret because the solution can be read directly from the entries in the last column. This matrix is said to be in *row-reduced echelon form*.

DEFINITION

ROW-REDUCED ECHELON FORM OF A MATRIX

A matrix is said to be in **row-reduced echelon form** if it satisfies the following conditions:

1. The first nonzero entry or leading term in each row is 1.
2. Any rows that consist entirely of 0's are below all rows that do not consist entirely of 0's.
3. The first nonzero entry in each row is to the right of the first nonzero entry in the preceding row.
4. In each column that contains a leading 1 of some row, all other entries are 0.

For example, the matrices

$$\begin{bmatrix} 1 & -2 & 0 \\ 0 & 0 & 1 \end{bmatrix}, \quad \begin{bmatrix} 1 & -1 & 0 & -3 \\ 0 & 0 & 1 & 4 \end{bmatrix}, \quad \text{and} \quad \begin{bmatrix} 1 & -4 & 0 & 3 & 5 \\ 0 & 0 & 1 & -2 & 4 \\ 0 & 0 & 0 & 0 & 0 \end{bmatrix}$$

are in row-reduced echelon form, because each satisfies the conditions of the above definition. However,

$$A = \begin{bmatrix} 0 & 1 \\ 0 & -2 \end{bmatrix}, \quad B = \begin{bmatrix} 0 & 0 & 0 \\ 1 & 0 & 0 \end{bmatrix}, \quad \text{and} \quad C = \begin{bmatrix} 1 & 1 & 0 \\ 0 & 1 & 0 \\ 0 & 0 & 1 \end{bmatrix}$$

are not in row-reduced echelon form. In matrix A, condition 1 is violated, because the leading term in the second row is not 1. In matrix B, condition 2 is violated; and in matrix C, condition 4 is violated, because the entry in row 1, column 2 is 1 rather than 0.

EXAMPLE 4 Interpreting row-reduced echelon matrices

Each of the following augmented matrices is the row-reduced echelon matrix form corresponding to a linear system of equations. Solve each system.

$$\text{(a)} \left[\begin{array}{cc|c} 1 & 0 & 2 \\ 0 & 1 & 3 \end{array}\right] \quad \text{(b)} \left[\begin{array}{cc|c} 1 & 0 & 5 \\ 0 & 0 & -1 \end{array}\right] \quad \text{(c)} \left[\begin{array}{ccc|c} 1 & 0 & 0 & 2 \\ 0 & 1 & 2 & 5 \\ 0 & 0 & 0 & 0 \end{array}\right]$$

Solution

(a) The matrix

$$\left[\begin{array}{cc|c} 1 & 0 & 2 \\ 0 & 1 & 3 \end{array}\right]$$

corresponds to the system

$$\begin{cases} x = 2 \\ y = 3 \end{cases}$$

so the system has one solution, (2, 3).

(b) The matrix

$$\left[\begin{array}{cc|c} 1 & 0 & 5 \\ 0 & 0 & -1 \end{array}\right]$$

corresponds to the system

$$\begin{cases} x = 5 \\ 0 = -1 \end{cases}$$

Because of the contradiction in the second equation, we conclude that the system has no solution.

(c) The matrix

$$\left[\begin{array}{ccc|c} 1 & 0 & 0 & 2 \\ 0 & 1 & 2 & 5 \\ 0 & 0 & 0 & 0 \end{array}\right]$$

corresponds to the system

$$\begin{cases} x = 2 \\ y + 2z = 5 \end{cases}$$

This implies that there are infinitely many solutions (x, y, z), where $x = 2$, $y = 5 - 2z$, and z is any real number. That is, the system has infinitely many solutions of the form $(2, 5 - 2z, z)$, where z is any real number. For instance, if we let $z = 0$, we get the solution $(2, 5, 0)$; if $z = 1$, we get $(2, 3, 1)$, and so forth.

We now can establish the **Gauss–Jordan elimination method** for solving linear systems. This procedure includes the following steps:

Step 1. Represent the linear system with an augmented matrix.

Step 2. Use elementary row operations to transform the original matrix to an equivalent row-reduced echelon form.

Step 3. Change the reduced matrix from step 2 to the equation form of the linear system.

Step 4. Interpret the result in step 3 to get the solution of the system, if there is one.

Many graphing calculators are programmed to transform a matrix into an equivalent matrix in row-reduced echelon form.

EXAMPLE 5 Solving a system by Gauss–Jordan elimination

Solve the system below by using Gauss–Jordan elimination.

$$\begin{cases} x_1 - 2x_2 + 3x_3 = -20 \\ 2x_1 - x_2 + 2x_3 = -14 \\ 3x_1 + x_2 + 2x_3 = -11 \end{cases}$$

Solution First, we write the augmented matrix of the system; then we transform the matrix into an equivalent matrix in row-reduced echelon form as follows:

$$\left[\begin{array}{rrr|r} 1 & -2 & 3 & -20 \\ 2 & -1 & 2 & -14 \\ 3 & 1 & 2 & -11 \end{array}\right] \xrightarrow{-2R_1 + R_2 \to R_2} \left[\begin{array}{rrr|r} 1 & -2 & 3 & -20 \\ 0 & 3 & -4 & 26 \\ 3 & 1 & 2 & -11 \end{array}\right] \xrightarrow{-3R_1 + R_3 \to R_3} \left[\begin{array}{rrr|r} 1 & -2 & 3 & -20 \\ 0 & 3 & -4 & 26 \\ 0 & 7 & -7 & 49 \end{array}\right]$$

These operations are performed in order to obtain a matrix in which there are 0 entries in the column below the leading entry of 1 in row 1.

$$\xrightarrow{\frac{1}{3}R_2 \to R_2} \left[\begin{array}{rrr|r} 1 & -2 & 3 & -20 \\ 0 & 1 & -\frac{4}{3} & \frac{26}{3} \\ 0 & 7 & -7 & 49 \end{array}\right] \xrightarrow{\frac{1}{7}R_3 \to R_3} \left[\begin{array}{rrr|r} 1 & -2 & 3 & -20 \\ 0 & 1 & -\frac{4}{3} & \frac{26}{3} \\ 0 & 1 & -1 & 7 \end{array}\right]$$

These operations produce an equivalent matrix in which the leading entries in rows 2 and 3 are 1's.

$$\xrightarrow{R_2 \leftrightarrow R_3} \left[\begin{array}{rrr|r} 1 & -2 & 3 & -20 \\ 0 & 1 & -1 & 7 \\ 0 & 1 & -\frac{4}{3} & \frac{26}{3} \end{array}\right]$$

This operation puts the row with fractions on the bottom to simplify our work.

(Continued in next column)

$$\xrightarrow{2R_2 + R_1 \to R_1} \left[\begin{array}{rrr|r} 1 & 0 & 1 & -6 \\ 0 & 1 & -1 & 7 \\ 0 & 1 & -\frac{4}{3} & \frac{26}{3} \end{array}\right] \xrightarrow{-R_2 + R_3 \to R_3} \left[\begin{array}{rrr|r} 1 & 0 & 1 & -6 \\ 0 & 1 & -1 & 7 \\ 0 & 0 & -\frac{1}{3} & \frac{5}{3} \end{array}\right]$$

These operations lead to a matrix form in which the entries in column 2 above and below the 1 in row 2 are 0's.

$$\xrightarrow{-3R_3 \to R_3} \left[\begin{array}{rrr|r} 1 & 0 & 1 & -6 \\ 0 & 1 & -1 & 7 \\ 0 & 0 & 1 & -5 \end{array}\right]$$

This operation puts a 1 in the leading entry in row 3.

$$\xrightarrow{R_3 + R_2 \to R_2} \left[\begin{array}{rrr|r} 1 & 0 & 1 & -6 \\ 0 & 1 & 0 & 2 \\ 0 & 0 & 1 & -5 \end{array}\right] \xrightarrow{-R_3 + R_1 \to R_1} \left[\begin{array}{rrr|r} 1 & 0 & 0 & -1 \\ 0 & 1 & 0 & 2 \\ 0 & 0 & 1 & -5 \end{array}\right]$$

These operations produce an equivalent matrix in which the entries in column 3 above the 1 in row 3 are 0's.

This is the row-reduced form.

The last matrix corresponds to the system

$$\begin{cases} x_1 = -1 \\ x_2 = 2 \\ x_3 = -5 \end{cases}$$

so the solution is $(-1, 2, -5)$.

Decomposing Fractions

A rational expression is said to be *proper* if the degree of its numerator is less than the degree of its denominator. Examples of proper fractions are

$$\frac{5x}{(x+1)(x-2)}, \quad \frac{3}{x^2+5}, \quad \text{and} \quad \frac{5x^2+7x+3}{x^3-5x-1}$$

In calculus it is necessary at times to rewrite a rational expression as a sum (or difference) of fractions that contain linear or quadratic factors in the denominators according to the factorization of the denominator of the given rational expression. Such a representation is referred to as a *partial fraction expansion* or *decomposition*. One process of decomposing a proper fraction into partial fractions involves solving linear systems, as illustrated in the next example.

EXAMPLE 6 **Decomposing a fraction**

Given that the fraction

$$\frac{2x-2}{(x+2)(x+5)}$$

has a partial fraction expansion of the form

$$\frac{2x-2}{(x+2)(x+5)} = \frac{A}{x+2} + \frac{B}{x+5}$$

find the constants A and B. Demonstrate the validity of the result graphically.

Solution Given that

$$\frac{2x-2}{(x+2)(x+5)} = \frac{A}{x+2} + \frac{B}{x+5}$$

we proceed to find A and B as follows:

$2x - 2 = A(x+5) + B(x+2)$	Multiply each side by the LCD $(x+2)(x+5)$
$2x - 2 = Ax + 5A + Bx + 2B$	Multiply
$2x - 2 = (A+B)x + (5A+2B)$	Rewrite the right side

Equating the coefficients of the two sides of the latter equation results in

$$\begin{cases} 2 = A + B \\ -2 = 5A + 2B \end{cases}$$

The associated augmented matrix of the system is

$$\left[\begin{array}{cc|c} 1 & 1 & 2 \\ 5 & 2 & -2 \end{array}\right]$$

The equivalent matrix in row-reduced echelon form is

$$\left[\begin{array}{cc|c} 1 & 0 & -2 \\ 0 & 1 & 4 \end{array}\right]$$

so $A = -2$ and $B = 4$. Thus,

$$\frac{2x-2}{(x+2)(x+5)} = \frac{-2}{x+2} + \frac{4}{x+5}$$

The graphs of

$$y_1 = \frac{2x-2}{(x+2)(x+5)} \quad \text{and} \quad y_2 = \frac{-2}{x+2} + \frac{4}{x+5}$$

shown in Figure 1 are the same, thus demonstrating the validity of the result graphically. ■

$y_1 = \frac{2x-2}{(x+2)(x+5)}$

10

−10 5

−5

(a)

$y_2 = \frac{-2}{x+2} + \frac{4}{x+5}$

10

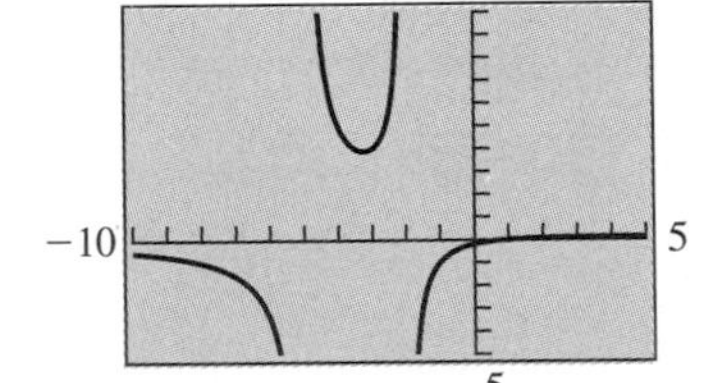

−10 5

−5

(b)

FIGURE 1

Solving Applied Problems

The following example uses row-reduction to solve a linear system of equations that models a real-life situation.

EXAMPLE 7 Solving a diet problem

A dietician prepares a special diet that consists of iron, carbohydrates, and protein. Table 3 lists the information about the content of iron, carbohydrates, and protein in an 8 ounce glass of skim milk, $\frac{1}{4}$ pound of lean red meat, and 2 slices of whole-grain bread. If a person on a special diet must have 10.5 milligrams of iron, 61.0 grams of carbohydrates, and 94.5 grams of protein, how many 8 ounce glasses of skim milk, how many $\frac{1}{4}$ pound servings of lean red meat, and how many 2 slice servings of whole-grain bread will supply this diet?

TABLE 3

Food Source	Iron *Milligrams per serving*	Carbohydrates *Grams per serving*	Protein *Grams per serving*
8 ounce glass of skim milk	0.1	1.0	8.5
$\frac{1}{4}$ pound of lean red meat	3.4	20.0	22.0
2 slices of whole-grain bread	2.2	12.0	10.0

Solution Let x represent the number of 8 ounce glasses of skim milk, let y represent the number of $\frac{1}{4}$ pound servings of lean red meat, and let z represent the number of 2 slice servings of whole-grain bread. The system of linear equations that provides this diet is given as follows:

$$\begin{cases} 0.1x + 3.4y + 2.2z = 10.5 \\ x + 20.0y + 12.0z = 61.0 \\ 8.5x + 22.0y + 10.0z = 94.5 \end{cases}$$

The augmented matrix that corresponds to this system is

$$\left[\begin{array}{ccc|c} 0.1 & 3.4 & 2.2 & 10.5 \\ 1.0 & 20.0 & 12.0 & 61.0 \\ 8.5 & 22.0 & 10.0 & 94.5 \end{array}\right]$$

By using the Gauss–Jordan method, we get the equivalent matrix

$$\left[\begin{array}{ccc|c} 1 & 0 & 0 & 5 \\ 0 & 1 & 0 & 1 \\ 0 & 0 & 1 & 3 \end{array}\right]$$

The associated system for the transformed matrix is

$$\begin{cases} x = 5 \\ y = 1 \\ z = 3 \end{cases}$$

So it takes 5 (8 ounce) glasses of skim milk, 1 serving $\left(\frac{1}{4}\text{ pound}\right)$ of lean red meat, and 3 servings (2 slices each) of whole-grain bread to supply this diet.

PROBLEM SET 8.2

Mastering the Concepts

In problems 1–4, write the augmented matrix of each linear system.

1. $\begin{cases} x + 6y = 7 \\ 3x + 7y = 10 \end{cases}$

2. $\begin{cases} 4x + y = 7 \\ 8x - 3y = 11 \end{cases}$

3. $\begin{cases} 2x + y = 2 \\ x - y = 4 \\ 3x + z = 6 \end{cases}$

4. $\begin{cases} x_1 + 2x_2 - x_3 = 4 \\ -x_1 + 3x_2 - 2x_3 = 5 \\ 2x_1 - x_2 + x_3 = 0 \end{cases}$

In problems 5–10, each matrix is in triangular form and is the augmented matrix form of a corresponding linear system of equations. Solve each system.

5. $\left[\begin{array}{cc|c} 1 & -2 & -5 \\ 0 & 1 & 3 \end{array}\right]$

6. $\left[\begin{array}{cc|c} 1 & 3 & 4 \\ 0 & 1 & 1 \end{array}\right]$

7. $\left[\begin{array}{ccc|c} 1 & 2 & 3 & 6 \\ 0 & 1 & -4 & -3 \\ 0 & 0 & 1 & 1 \end{array}\right]$

8. $\left[\begin{array}{ccc|c} 1 & 8 & 7 & -22 \\ 0 & 1 & 4 & -7 \\ 0 & 0 & 1 & -1 \end{array}\right]$

9. $\left[\begin{array}{ccc|c} 1 & 0 & -1 & 7 \\ 0 & 1 & 3 & -3 \\ 0 & 0 & 1 & -2 \end{array}\right]$ **10.** $\left[\begin{array}{ccc|c} 1 & 1 & 2 & 4 \\ 0 & 1 & 2 & 3 \\ 0 & 0 & 1 & 1 \end{array}\right]$

11. Let

$$A = \begin{bmatrix} 1 & 0 & 3 \\ -3 & 1 & 4 \\ 5 & -1 & 5 \end{bmatrix}$$

Find the matrix obtained by performing the following elementary row operations on A:

(a) Interchange the second and third rows.
(b) Multiply the third row of A by 3.
(c) Add 3 times the first row of A to the third row.

12. Let

$$A = \begin{bmatrix} 2 & 0 & 4 & 2 \\ 3 & -2 & 5 & 6 \\ -1 & 3 & 1 & 1 \end{bmatrix}$$

Find the matrix obtained by performing the following elementary row operations on A:

(a) Interchange the second and third rows.
(b) Multiply the first row of A by $\frac{1}{2}$.
(c) Add 2 times the third row of A to the first row.

In problems 13–16, which of the matrices are in row-reduced echelon form?

13. **(a)** $\begin{bmatrix} 1 & -2 & 0 & 1 \\ 0 & 0 & 1 & -4 \end{bmatrix}$

(b) $\begin{bmatrix} 1 & -1 & 0 & 4 \\ 0 & 0 & 1 & 2 \\ 0 & 0 & 0 & 0 \end{bmatrix}$

14. **(a)** $\begin{bmatrix} 1 & 0 & 0 & -1 \\ 0 & 1 & 0 & -3 \\ 0 & 0 & 1 & 2 \end{bmatrix}$

(b) $\begin{bmatrix} 1 & 0 & 0 \\ 0 & 0 & 1 \\ 0 & 1 & 0 \end{bmatrix}$

15. **(a)** $\begin{bmatrix} 1 & 0 & 1 \\ 0 & 2 & -3 \end{bmatrix}$

(b) $\begin{bmatrix} 1 & 2 & -1 & 0 & 2 \\ 0 & 0 & 0 & 1 & 4 \\ 0 & 0 & 0 & 0 & 0 \end{bmatrix}$

16. **(a)** $\begin{bmatrix} 1 & 2 & 0 & 2 & 5 \\ 0 & 0 & 1 & -3 & 4 \\ 0 & 0 & 0 & 0 & 0 \end{bmatrix}$

(b) $\begin{bmatrix} 1 & 0 & -2 & -2 & 3 \\ 0 & 1 & 4 & 3 & 1 \\ 0 & 0 & 1 & -1 & 5 \end{bmatrix}$

In problems 17–22, each row-reduced echelon matrix is the augmented matrix form of a corresponding linear system of equations. Solve each system.

17. $\left[\begin{array}{cc|c} 1 & 0 & 1 \\ 0 & 1 & 1 \end{array}\right]$ **18.** $\left[\begin{array}{ccc|c} 1 & 0 & 0 & 0 \\ 0 & 1 & 0 & 0 \\ 0 & 0 & 1 & 2 \end{array}\right]$

19. $\left[\begin{array}{ccc|c} 1 & 0 & -1 & 2 \\ 0 & 1 & -2 & 3 \end{array}\right]$ **20.** $\left[\begin{array}{ccc|c} 1 & 0 & 0 & 3 \\ 0 & 1 & 0 & 0 \\ 0 & 0 & 1 & 5 \end{array}\right]$

21. $\left[\begin{array}{ccc|c} 1 & 0 & 0 & -3 \\ 0 & 1 & 0 & 4 \\ 0 & 0 & 0 & 7 \end{array}\right]$ **22.** $\left[\begin{array}{ccc|c} 1 & 0 & 0 & 3 \\ 0 & 1 & 0 & -2 \\ 0 & 0 & 0 & 4 \end{array}\right]$

In problems 23–40, solve each system by the Gauss–Jordan elimination method.

23. $\begin{cases} 3x + y = 14 \\ 2x - y = 1 \end{cases}$ **24.** $\begin{cases} 4x + 3y = 15 \\ 3x + 5y = 14 \end{cases}$

25. $\begin{cases} -2x + 3y = 8 \\ 2x - y = 5 \end{cases}$ **26.** $\begin{cases} x - 2y = 5 \\ 3x - 6y = 4 \end{cases}$

27. $\begin{cases} x + y + z = 6 \\ 3x - y + 2z = 7 \\ 2x + 3y - z = 5 \end{cases}$ **28.** $\begin{cases} 2x + 3y + z = 6 \\ x - 2y + 3z = -3 \\ 3x + y - z = 8 \end{cases}$

29. $\begin{cases} x + y + 2z = 4 \\ x + y - 2z = 0 \\ x - y = 0 \end{cases}$ **30.** $\begin{cases} x + y + z = 4 \\ x - y + 2z = 8 \\ 2x + y - z = 3 \end{cases}$

31. $\begin{cases} 2x + y - z = 7 \\ y - x = 1 \\ z - y = 1 \end{cases}$ **32.** $\begin{cases} x_1 + 2x_2 + 3x_3 = 0 \\ x_1 - 2x_3 = 1 \\ x_2 + x_3 = 1 \end{cases}$

33. $\begin{cases} x - 2y + z = -1 \\ 3x + y - 2z = 4 \\ y - z = 1 \end{cases}$ **34.** $\begin{cases} x + y - 2z = 3 \\ 3x - y + z = 5 \\ 3x + 3y - 6z = 9 \end{cases}$

35. $\begin{cases} 2x + y + z = 1 \\ 4x + 2y + 3z = 1 \\ -2x - y + z = 2 \end{cases}$ **36.** $\begin{cases} x + y + z = 0 \\ 2x - y - 4z = 15 \\ x - 2y - z = 7 \end{cases}$

37. $\begin{cases} 2x - 3y + z = 4 \\ x - 4y - z = 3 \end{cases}$

38. $\begin{cases} 2x + 3y - z = -2 \\ x - y + 2z = 4 \end{cases}$

39. $\begin{cases} 2x_1 - 3x_3 = 5 \\ 4x_2 + 2x_3 = -20 \\ 5x_1 + 2x_2 = -16 \end{cases}$

40. $\begin{cases} x_1 + x_2 = -1 \\ x_1 - 2x_3 = -11 \\ x_2 - 3x_3 = -15 \end{cases}$

GU In problems 41–50, find the partial fraction expansion. Use a graphing utility to demonstrate the validity of the result graphically.

41. $\dfrac{3x - 5}{(x - 1)(x - 2)} = \dfrac{A}{x - 1} + \dfrac{B}{x - 2}$

42. $\dfrac{x + 14}{(x + 4)(x + 12)} = \dfrac{A}{x + 4} + \dfrac{B}{x + 12}$

43. $\dfrac{-9x - 7}{(3x + 1)(x + 1)} = \dfrac{A}{3x + 1} + \dfrac{B}{x + 1}$

44. $\dfrac{13x - 24}{(3x - 5)(x - 2)} = \dfrac{A}{3x - 5} + \dfrac{B}{x - 2}$

45. $\dfrac{-21x + 11}{(x - 1)(x - 2)(x - 3)} = \dfrac{A}{x - 1} + \dfrac{B}{x - 2} + \dfrac{C}{x - 3}$

46. $\dfrac{x^2 - 3x + 4}{(x - 1)(x + 1)(x + 2)} = \dfrac{A}{x - 1} + \dfrac{B}{x + 1} + \dfrac{C}{x + 2}$

47. $\dfrac{5x^2 - 21x + 13}{(x - 3)^2(x + 2)} = \dfrac{A}{x - 3} + \dfrac{B}{(x - 3)^2} + \dfrac{C}{x + 2}$

48. $\dfrac{-x^2 + 13x - 26}{(x + 1)^2(x - 4)} = \dfrac{A}{x + 1} + \dfrac{B}{(x + 1)^2} + \dfrac{C}{x - 4}$

49. $\dfrac{6}{x(x^2 + 1)} = \dfrac{A}{x} + \dfrac{Bx + C}{x^2 + 1}$

50. $\dfrac{30x}{(x - 6)(x^2 + 4)} = \dfrac{A}{x - 6} + \dfrac{Bx + C}{x^2 + 4}$

Applying the Concepts

In problems 51–56, set up an appropriate linear system, and then use the Gauss–Jordan method to solve each system.

51. Nutrition: A nutritionist wants to create a special diet out of three substances *A*, *B*, and *C*. The requirements in the diet are 480 units of calcium, 190 units of iron, and 360 units of vitamins. The calcium, iron, and vitamin contents in substances *A*, *B*, and *C* are listed in Table 4. How many ounces of each substance should be used to prepare this diet?

TABLE 4

Substance	Calcium *Units per ounce*	Iron *Units per ounce*	Vitamin *Units per ounce*
A	30	10	15
B	20	10	10
C	10	5	20

52. Production: Three assembly lines, *A*, *B*, and *C*, can produce 8400 TV dinners per day. Together, lines *A* and *B* can produce 4900 TV dinners, while lines *B* and *C* together can produce 5600 TV dinners. Find the number of TV dinners each assembly line can produce alone.

53. Horticulture: A horticulturist wishes to mix three types of fertilizer which contain 25%, 35%, and 40% nitrogen, respectively, in order to get a mixture of 4000 pounds of $35\frac{5}{8}$% nitrogen. The final mixture should contain three times as much of the 40% type as the 25% type. How much of each type is in the final mixture?

54. Fast-Food Chain: A fast-food chain sells three types of franchises, *A*, *B*, and *C*. Franchise *A* sells for $20,000, franchise *B* for $25,000, and franchise *C* for $30,000. In one year, the company sold 18 franchises for $430,000. If the number sold of franchise *A* is twice the number sold of franchise *C*, how many of each type did the company sell that year?

55. Electrical Engineering: In electronics, the analysis of a circuit leads naturally to a system of linear equations. Suppose an electrical network contains currents I_1, I_2, and I_3 in the circuit, resistors R_1, R_2, and R_3; a 6 volt battery ($E_1 = 6$) and a 12 volt battery ($E_2 = 12$), as shown in Figure 2:

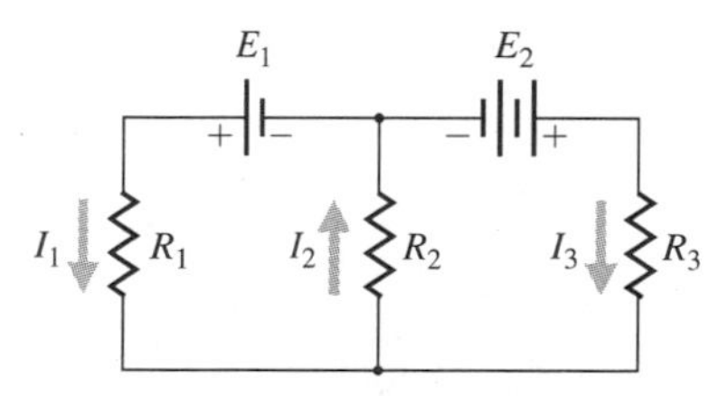

FIGURE 2

Assume the network satisfies the system given by

$$\begin{cases} I_1 - I_2 + I_3 = 0 \\ R_1I_1 + R_2I_2 = E_1 \\ R_2I_2 + R_3I_3 = E_2 \end{cases}$$

If $R_1 = 4$ ohms, $R_2 = 10$ ohms, and $R_3 = 21$ ohms, find the currents I_1, I_2, and I_3 (in amperes).

56. Irrigation Network: Figure 3 is a diagram of a system of irrigation canals that shows the flow of water into the system at junction A and out at junctions B, C, and D. The flows f_1, f_2, f_3, f_4, and f_5 (in gallons per minute) are solutions to the system

$$\begin{cases} f_1 + f_2 + f_3 = 500 & \text{at junction } A \\ f_1 - f_5 = 100 & \text{at junction } B \\ f_2 - f_4 + f_5 = 100 & \text{at junction } C \\ f_3 + f_4 = 300 & \text{at junction } D \end{cases}$$

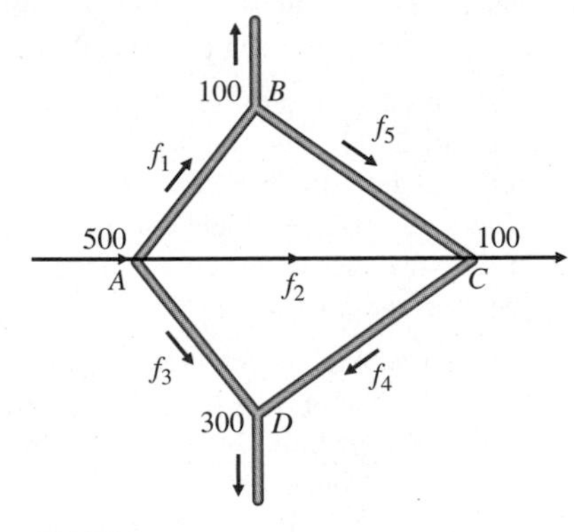

FIGURE 3

(a) Solve the system for f_1, f_2, f_3, f_4, and f_5.
(b) Discuss some choices of flows for this network.

Developing and Extending the Concepts

In problems 57–60, use matrices to solve each system. In each case, write the augmented matrix and reduce it to row-reduced echelon form.

57. $$\begin{cases} x + y + 2z + 3w = 5 \\ x - 2y + z + w = 5 \\ 3x + y + z - w = 2 \\ 2x - y + z + 4w = 8 \end{cases}$$

58. $$\begin{cases} x_1 - x_2 + 2x_3 + x_4 = 8 \\ -x_1 + 2x_2 + 3x_3 - x_4 = 4 \\ x_1 - x_2 + x_3 - 2x_4 = -6 \\ -x_1 + x_2 + x_3 + x_4 = 7 \end{cases}$$

59. $$\begin{cases} 3x + y - 2z + 2w = 7 \\ 2x + z + 4w = 3 \\ y + 3z + 5w = 0 \\ -x + 2y - 3w = 4 \end{cases}$$

60. $$\begin{cases} x_1 + 4x_2 - x_3 + x_4 = 11 \\ 3x_1 + 2x_2 + x_3 + 2x_4 = 9 \\ x_1 - 6x_2 + 3x_4 = 23 \\ x_1 + 14x_2 - 5x_3 + 2x_4 = -13 \end{cases}$$

In problems 61–64, find (if possible) conditions on a, b, and c so that the system has:
(a) No solution **(b)** One solution
(c) Infinite number of solutions

61. $$\begin{cases} 3x_1 + x_2 - x_3 = a \\ x_1 - x_2 + 2x_3 = b \\ 5x_1 + 3x_2 - 4x_3 = c \end{cases}$$

62. $$\begin{cases} x_1 - x_3 = a \\ 2x_1 + x_2 - x_3 = b \\ 2x_2 + 3x_3 = c \end{cases}$$

63. $$\begin{cases} x_1 + x_2 + x_3 = 2 \\ 2x_1 + 3x_2 + 2x_3 = 5 \\ 2x_1 + 3x_2 + (a^2 - 1)x_3 = a \end{cases}$$

64. $$\begin{cases} x_1 + x_2 + x_3 = 2 \\ x_1 + 2x_2 + x_3 = 3 \\ x_1 + x_2 + (c^2 - 5)x_3 = c \end{cases}$$

65. Find a, b, and c, so that $(1, 2, 3)$ is a solution of the system

$$\begin{cases} ax + by + cz = 6 \\ 2bx - cy + 3az = 9 \\ -cx + 2by + 2az = 9 \end{cases}$$

OBJECTIVES

1. Perform matrix arithmetic.
2. Find the inverse of a square matrix.
3. Solve linear systems using matrix inverses.
4. Solve applied problems.

8.3 Solutions of Linear Systems Using Matrix Inverses

In Section 8.2, we used elementary row operations on augmented matrices to solve linear systems of equations. In this section, we examine another method for solving linear systems that involves the algebra of matrices.

We say that two matrices are equal if and only if they are the same size and the corresponding elements in each matrix are equal. For example,

$$\begin{bmatrix} 2 & 3 & -1 \\ 5 & 8 & 3 \end{bmatrix} = \begin{bmatrix} 2 & x & y \\ a & b & c \end{bmatrix}$$

if and only if $x = 3$, $y = -1$, $a = 5$, $b = 8$, and $c = 3$.

A matrix whose elements are all 0's is called a **zero matrix** and is denoted by **0**. For example, the zero matrix of size 2×3 is given by

$$\mathbf{0} = \begin{bmatrix} 0 & 0 & 0 \\ 0 & 0 & 0 \end{bmatrix}$$

Performing Matrix Arithmetic

The addition (or subtraction) of two matrices can be performed *only* if the matrices are the same size. Then, to **add** (or **subtract**) the matrices, we add (or subtract) the corresponding entries.

EXAMPLE 1 Adding and subtracting matrices

Let

$$A = \begin{bmatrix} 3 & 5 & 1 \\ 6 & 2 & -3 \end{bmatrix} \quad \text{and} \quad B - \begin{bmatrix} 1 & 3 & 7 \\ 5 & -2 & 4 \end{bmatrix}$$

Determine:

(a) $A + B$ (b) $A - B$

Solution

(a) $$A + B = \begin{bmatrix} 3 & 5 & 1 \\ 6 & 2 & -3 \end{bmatrix} + \begin{bmatrix} 1 & 3 & 7 \\ 5 & -2 & 4 \end{bmatrix}$$
$$= \begin{bmatrix} 3+1 & 5+3 & 1+7 \\ 6+5 & 2+(-2) & -3+4 \end{bmatrix} = \begin{bmatrix} 4 & 8 & 8 \\ 11 & 0 & 1 \end{bmatrix}$$

(b) $$A - B = \begin{bmatrix} 3 & 5 & 1 \\ 6 & 2 & -3 \end{bmatrix} - \begin{bmatrix} 1 & 3 & 7 \\ 5 & -2 & 4 \end{bmatrix}$$
$$= \begin{bmatrix} 3-1 & 5-3 & 1-7 \\ 6-5 & 2-(-2) & -3-4 \end{bmatrix} = \begin{bmatrix} 2 & 2 & -6 \\ 1 & 4 & -7 \end{bmatrix}$$

If A is a matrix and k is a real number, we define kA to be the matrix obtained by multiplying each entry of A by k. This process is referred to as **multiplication by a scalar**.

EXAMPLE 2 Multiplying a matrix by a real number

Let

$$A = \begin{bmatrix} -5 & 1 \\ 3 & 2 \end{bmatrix}$$

Find the value of each matrix:

(a) $4A$ (b) $-2A$

Solution

(a) $4A = 4\begin{bmatrix} -5 & 1 \\ 3 & 2 \end{bmatrix} = \begin{bmatrix} 4(-5) & 4(1) \\ 4(3) & 4(2) \end{bmatrix} = \begin{bmatrix} -20 & 4 \\ 12 & 8 \end{bmatrix}$

(b) $-2A = -2\begin{bmatrix} -5 & 1 \\ 3 & 2 \end{bmatrix} = \begin{bmatrix} (-2)(-5) & (-2)(1) \\ (-2)(3) & (-2)(2) \end{bmatrix} = \begin{bmatrix} 10 & -2 \\ -6 & -4 \end{bmatrix}$

Because matrices are added by adding their corresponding elements, it follows from the properties of real numbers that the addition of matrices obeys both the *commutative* and *associative properties*. That is, if A, B, and C are matrices of the same size, then we have the following properties:

1. **Commutative Property**: $A + B = B + A$
2. **Associative Property**: $A + (B + C) = (A + B) + C$

A zero matrix acts as the identity for the operation of addition. That is, if A and $\mathbf{0}$ are of the same size, then:

3. **Zero Matrix Property**: $A + \mathbf{0} = \mathbf{0} + A = A$

For example,

$$\begin{bmatrix} 1 & 2 \\ 3 & 8 \end{bmatrix} + \begin{bmatrix} 0 & 0 \\ 0 & 0 \end{bmatrix} = \begin{bmatrix} 1 & 2 \\ 3 & 8 \end{bmatrix}$$

If A is a matrix, the **additive inverse of A**, denoted by $-A$, is the matrix consisting of entries that are opposite in sign to the entries of A. Thus, we have:

The matrix $-A$ is the same as $(-1)A$. Also, $A - B$ can be interpreted as $A - B = A + (-B)$.

4. **Additive Inverse Property**: $A + (-A) = (-A) + A = \mathbf{0}$

Other properties involving multiplication by a scalar and addition can be easily verified.

SCALAR MULTIPLICATION PROPERTIES

If A and B are matrices of the same size, and c and d are real numbers, then:

5. $(cd)A = c(dA)$
6. $(c + d)A = cA + dA$
7. $c(A + B) = cA + cB$

In order to understand how to *multiply two matrices*, we need some background. A matrix consisting of one row is called a **row matrix**, and a matrix consisting of a single column is called a **column matrix**. For example,

$$\begin{bmatrix} 2 & -1 & 4 \end{bmatrix} \quad \text{and} \quad \begin{bmatrix} 1 & 0 & -1 & 5 \end{bmatrix}$$

are row matrices, whereas

$$\begin{bmatrix} 3 \\ 4 \\ -1 \end{bmatrix} \quad \text{and} \quad \begin{bmatrix} 1 \\ 0 \\ -1 \\ 5 \end{bmatrix}$$

are column matrices. Suppose that R is a row matrix, C is a column matrix, and the number of columns in R is the same as the number of rows in C. Then the **product** of R and C is a matrix formed by multiplying the corresponding entries of the two matrices and adding the products. For instance, if

$$R = \begin{bmatrix} 2 & -1 & 3 & 4 \end{bmatrix} \quad \text{and} \quad C = \begin{bmatrix} 5 \\ 2 \\ -3 \\ 1 \end{bmatrix}$$

then

$$RC = \begin{bmatrix} 2 & -1 & 3 & 4 \end{bmatrix} \begin{bmatrix} 5 \\ 2 \\ -3 \\ 1 \end{bmatrix} = 2(5) + (-1)(2) + 3(-3) + 4(1) = [3]$$

Note that, unlike addition and subtraction of matrices, the product of two matrices cannot be determined by multiplying corresponding elements. Also, the two matrices do not have to be of the same size. However, in order to form the product AB of two matrices A and B, the number of columns of A must be the same as the number of rows of B. The product AB will have as many rows as A and as many columns as B. That is:

> If A is an $m \times k$ matrix and B is a $k \times n$ matrix, then the product AB is an $m \times n$ matrix.

The following diagram illustrates these restrictions:

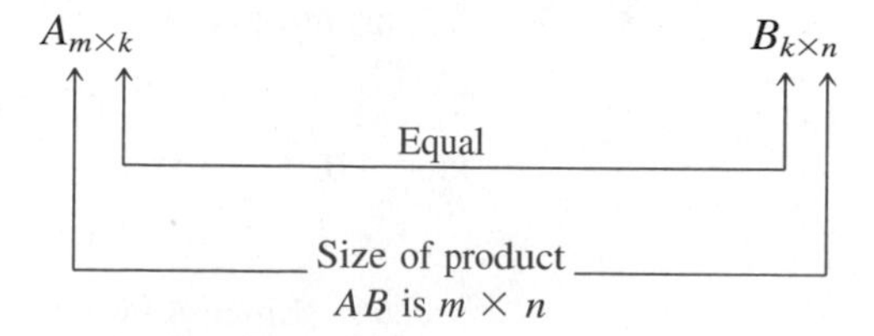

The entry in the ith row and jth column of the product AB is found by multiplying the ith row of a matrix A and the jth column of matrix B, as we did for the row and column matrices. For example, to find the product AB if

$$A = \begin{bmatrix} 1 & 3 & 2 \\ -2 & 4 & 5 \end{bmatrix} \quad \text{and} \quad B = \begin{bmatrix} -4 & 7 \\ 3 & -5 \\ 1 & 6 \end{bmatrix}$$

we first observe that

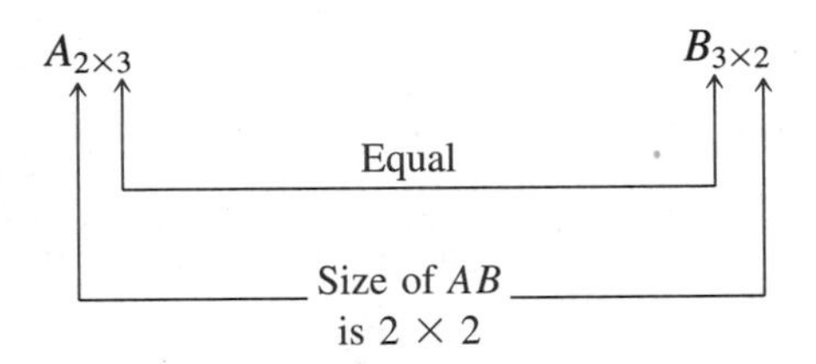

The product AB is determined as follows:

Row 1 in A times column 1 in B — Row 1 in A times column 2 in B

$$AB = \begin{bmatrix} 1 & 3 & 2 \\ -2 & 4 & 5 \end{bmatrix} \begin{bmatrix} -4 & 7 \\ 3 & -5 \\ 1 & 6 \end{bmatrix} = \begin{bmatrix} [1 \quad 3 \quad 2]\begin{bmatrix} -4 \\ 3 \\ 1 \end{bmatrix} & [1 \quad 3 \quad 2]\begin{bmatrix} 7 \\ -5 \\ 6 \end{bmatrix} \\ [-2 \quad 4 \quad 5]\begin{bmatrix} -4 \\ 3 \\ 1 \end{bmatrix} & [-2 \quad 4 \quad 5]\begin{bmatrix} 7 \\ -5 \\ 6 \end{bmatrix} \end{bmatrix}$$

Row 2 in A times column 1 in B — Row 2 in A times column 2 in B

$$= \begin{bmatrix} 1(-4) + 3(3) + 2(1) & 1(7) + 3(-5) + 2(6) \\ -2(-4) + 4(3) + 5(1) & -2(7) + 4(-5) + 5(6) \end{bmatrix}$$

$$= \begin{bmatrix} 7 & 4 \\ 25 & -4 \end{bmatrix}$$

EXAMPLE 3 Multiplying matrices

Find AB if $A = \begin{bmatrix} 3 & -2 \\ 1 & 5 \\ 4 & 6 \end{bmatrix}$ and $B = \begin{bmatrix} 7 & -2 \\ 3 & 0 \end{bmatrix}$

Solution Since A is a 3×2 matrix and B is a 2×2 matrix, then AB is a 3×2 matrix. The product AB is given by

$$AB = \begin{bmatrix} 3 & -2 \\ 1 & 5 \\ 4 & 6 \end{bmatrix} \begin{bmatrix} 7 & -2 \\ 3 & 0 \end{bmatrix}$$

$$= \begin{bmatrix} 3(7) + (-2)(3) & 3(-2) + (-2)(0) \\ 1(7) + 5(3) & 1(-2) + 5(0) \\ 4(7) + 6(3) & 4(-2) + 6(0) \end{bmatrix} = \begin{bmatrix} 15 & -6 \\ 22 & -2 \\ 46 & -8 \end{bmatrix}$$

The product AB may be defined even when BA is not. For instance, the product BA is not defined in Example 3 because the number of rows of A is not the same as the number of columns of B.

Matrix multiplication is not commutative; that is, it is *not* generally true that matrix product AB is the same as matrix product BA. For instance,

$$AB = \begin{bmatrix} 2 & 1 \\ 4 & 2 \end{bmatrix} \begin{bmatrix} -1 & 2 \\ 3 & 0 \end{bmatrix} = \begin{bmatrix} 1 & 4 \\ 2 & 8 \end{bmatrix}$$

whereas

$$BA = \begin{bmatrix} -1 & 2 \\ 3 & 0 \end{bmatrix} \begin{bmatrix} 2 & 1 \\ 4 & 2 \end{bmatrix} = \begin{bmatrix} 6 & 3 \\ 6 & 3 \end{bmatrix}$$

Although matrix multiplication is not commutative, matrix multiplication obeys the associative and distributive properties. That is, if A, B, and C are matrices where multiplication is defined, and if k is a constant real number, then:

1. **Associative Properties:**

   (a) $A(BC) = (AB)C$ (b) $(kA)B = k(AB)$

2. **Distributive Properties:**

   (a) $A(B + C) = AB + AC$ (b) $(B + C)A = BA + CA$

Finding the Inverse of a Square Matrix

The 2×2 matrix

$$\begin{bmatrix} 1 & 0 \\ 0 & 1 \end{bmatrix}$$

and the 3×3 matrix

$$\begin{bmatrix} 1 & 0 & 0 \\ 0 & 1 & 0 \\ 0 & 0 & 1 \end{bmatrix}$$

are referred to as **identity matrices**. If I is an identity matrix and A is an $n \times n$ matrix so that AI and IA are defined, then

$$AI = IA = A$$

For example,

$$\begin{bmatrix} 1 & -3 \\ 2 & 4 \end{bmatrix}\begin{bmatrix} 1 & 0 \\ 0 & 1 \end{bmatrix} = \begin{bmatrix} 1 & -3 \\ 2 & 4 \end{bmatrix} \quad \text{and} \quad \begin{bmatrix} 1 & 0 \\ 0 & 1 \end{bmatrix}\begin{bmatrix} 1 & -3 \\ 2 & 4 \end{bmatrix} = \begin{bmatrix} 1 & -3 \\ 2 & 4 \end{bmatrix}$$

We know from algebra that every nonzero real number a has a multiplicative inverse $a^{-1} = 1/a$ such that

$$aa^{-1} = a^{-1}a = 1$$

The analogous statement for matrix multiplication does not always hold, and some restrictions are needed. An $n \times n$ matrix A is said to be **invertible** if there is an $n \times n$ matrix B such that

$$AB = BA = I$$

The matrix B is called the **inverse matrix** of A and is denoted by A^{-1}; that is,

$$\boxed{AA^{-1} = A^{-1}A = I}$$

EXAMPLE 4 Verifying that one matrix is an inverse of another

Show that the matrix

$$B = \begin{bmatrix} -5 & 2 \\ 3 & -1 \end{bmatrix}$$

is the inverse of the matrix

$$A = \begin{bmatrix} 1 & 2 \\ 3 & 5 \end{bmatrix}$$

Solution Since

$$AB = \begin{bmatrix} 1 & 2 \\ 3 & 5 \end{bmatrix}\begin{bmatrix} -5 & 2 \\ 3 & -1 \end{bmatrix} = \begin{bmatrix} 1 & 0 \\ 0 & 1 \end{bmatrix}$$

and

$$BA = \begin{bmatrix} -5 & 2 \\ 3 & -1 \end{bmatrix}\begin{bmatrix} 1 & 2 \\ 3 & 5 \end{bmatrix} = \begin{bmatrix} 1 & 0 \\ 0 & 1 \end{bmatrix}$$

it follows that $AB = I = BA$, so B is indeed the inverse of A; that is, $B = A^{-1}$. Similarly, $A = B^{-1}$.

Not every square matrix has an inverse. For instance, consider the matrix

$$A = \begin{bmatrix} 1 & 1 \\ 1 & 1 \end{bmatrix}$$

If A has an inverse, say

$$B = \begin{bmatrix} a & b \\ c & d \end{bmatrix}$$

then the product AB would have to equal the 2×2 identity matrix. However,

$$AB = \begin{bmatrix} 1 & 1 \\ 1 & 1 \end{bmatrix} \begin{bmatrix} a & b \\ c & d \end{bmatrix} = \begin{bmatrix} a + c & b + d \\ a + c & b + d \end{bmatrix}$$

Clearly,

$$\begin{bmatrix} a + c & b + d \\ a + c & b + d \end{bmatrix} \quad \text{cannot equal} \quad \begin{bmatrix} 1 & 0 \\ 0 & 1 \end{bmatrix}$$

since it is impossible to have both $a + c = 1$ and $a + c = 0$. Thus, A cannot have an inverse.

Elementary row operations can be used to find the inverse of a square matrix if it exists. For example, to find the inverse of the matrix

$$A = \begin{bmatrix} 3 & 5 \\ 1 & 2 \end{bmatrix}$$

we need to find a matrix

$$B = \begin{bmatrix} x & u \\ y & v \end{bmatrix}$$

with entries x, y, u, and v so that

$$AB = \begin{bmatrix} 3 & 5 \\ 1 & 2 \end{bmatrix} \begin{bmatrix} x & u \\ y & v \end{bmatrix} = \begin{bmatrix} 1 & 0 \\ 0 & 1 \end{bmatrix}$$

Multiplying the left side, we get

$$\begin{bmatrix} 3x + 5y & 3u + 5v \\ x + 2y & u + 2v \end{bmatrix} = \begin{bmatrix} 1 & 0 \\ 0 & 1 \end{bmatrix}$$

Equating the corresponding entries of the two matrices, we have

$$\begin{cases} 3x + 5y = 1 \\ x + 2y = 0 \end{cases} \quad \text{and} \quad \begin{cases} 3u + 5v = 0 \\ u + 2v = 1 \end{cases}$$

Now we can use augmented matrices and row-reduction to solve these two systems. Notice that the coefficient matrices of both systems are the same. Therefore, we can write the matrix so that the first three columns form the augmented matrix of the first system, and the first two columns and last column form the augmented matrix of the second system. That is, we write the matrix

$$\left[\begin{array}{cc|cc} 3 & 5 & 1 & 0 \\ 1 & 2 & 0 & 1 \end{array}\right]$$

Next, we perform elementary row operations to express this matrix in a row-reduced echelon form:

$$\left[\begin{array}{cc|cc} 3 & 5 & 1 & 0 \\ 1 & 2 & 0 & 1 \end{array}\right]$$

$$\xrightarrow{R_1 \leftrightarrow R_2} \left[\begin{array}{cc|cc} 1 & 2 & 0 & 1 \\ 3 & 5 & 1 & 0 \end{array}\right]$$

$$\xrightarrow{-3R_1 + R_2 \to R_2} \left[\begin{array}{cc|cc} 1 & 2 & 0 & 1 \\ 0 & -1 & 1 & -3 \end{array}\right]$$

$$\xrightarrow{(-1)R_2 \to R_2} \left[\begin{array}{cc|cc} 1 & 2 & 0 & 1 \\ 0 & 1 & -1 & 3 \end{array}\right]$$

$$\xrightarrow{-2R_2 + R_1 \to R_1} \left[\begin{array}{cc|cc} 1 & 0 & 2 & -5 \\ 0 & 1 & -1 & 3 \end{array}\right]$$

From the first three columns of the last matrix, we see that $x = 2$ and $y = -1$. Similarly, from the first two columns and the last column, it follows that $u = -5$ and $v = 3$. Therefore,

$$B = A^{-1} = \begin{bmatrix} 2 & -5 \\ -1 & 3 \end{bmatrix}$$

Notice that A^{-1} is the right half of the augmented matrix we found above, and that the left half is the identity matrix I.

This illustration leads us to the following general procedure for finding the inverse of a square matrix A:

PROCEDURE

FINDING THE INVERSE OF A MATRIX A OF SIZE $n \times n$

1. Form the $n \times 2n$ matrix

$$[A \mid I]$$

$n \times n$ matrix (points to A); $n \times n$ identity matrix (points to I)

2. Perform elementary row operations on this matrix to transform it, if possible, to a matrix of the form

$$[I \mid B]$$

$n \times n$ identity matrix (points to I)

3. The matrix B is A^{-1}; that is, $A^{-1} = B$.

Many graphing calculators are programmed to find the inverse of a matrix A.

If the above procedure does not lead to the identity matrix on the left, then the matrix A has no inverse and we say A^{-1} does not exist.

EXAMPLE 5 Finding the inverse of a 3 × 3 matrix

Find A^{-1} where $A = \begin{bmatrix} 1 & 2 & 3 \\ 1 & 1 & 2 \\ 0 & 1 & 2 \end{bmatrix}$

Solution We begin by forming the matrix $[A \mid I]$:

$$\left[\begin{array}{ccc|ccc} 1 & 2 & 3 & 1 & 0 & 0 \\ 1 & 1 & 2 & 0 & 1 & 0 \\ 0 & 1 & 2 & 0 & 0 & 1 \end{array}\right]$$

Next, we perform elementary row operations on the entire matrix until the left half is transformed into the identity matrix:

$$\left[\begin{array}{ccc|ccc} 1 & 2 & 3 & 1 & 0 & 0 \\ 1 & 1 & 2 & 0 & 1 & 0 \\ 0 & 1 & 2 & 0 & 0 & 1 \end{array}\right] \xrightarrow{-R_1 + R_2 \to R_2} \left[\begin{array}{ccc|ccc} 1 & 2 & 3 & 1 & 0 & 0 \\ 0 & -1 & -1 & -1 & 1 & 0 \\ 0 & 1 & 2 & 0 & 0 & 1 \end{array}\right]$$

$$\xrightarrow{R_2 + R_3 \to R_3} \left[\begin{array}{ccc|ccc} 1 & 2 & 3 & 1 & 0 & 0 \\ 0 & -1 & -1 & -1 & 1 & 0 \\ 0 & 0 & 1 & -1 & 1 & 1 \end{array}\right]$$

$$\xrightarrow{2R_2 + R_1 \to R_1} \left[\begin{array}{ccc|ccc} 1 & 0 & 1 & -1 & 2 & 0 \\ 0 & -1 & -1 & -1 & 1 & 0 \\ 0 & 0 & 1 & -1 & 1 & 1 \end{array}\right]$$

$$\xrightarrow{(-1)R_2 \to R_2} \left[\begin{array}{ccc|ccc} 1 & 0 & 1 & -1 & 2 & 0 \\ 0 & 1 & 1 & 1 & -1 & 0 \\ 0 & 0 & 1 & -1 & 1 & 1 \end{array}\right]$$

$$\xrightarrow{(-1)R_3 + R_2 \to R_2} \left[\begin{array}{ccc|ccc} 1 & 0 & 1 & -1 & 2 & 0 \\ 0 & 1 & 0 & 2 & -2 & -1 \\ 0 & 0 & 1 & -1 & 1 & 1 \end{array}\right]$$

$$\xrightarrow{(-1)R_3 + R_1 \to R_1} \left[\begin{array}{ccc|ccc} 1 & 0 & 0 & 0 & 1 & -1 \\ 0 & 1 & 0 & 2 & -2 & -1 \\ 0 & 0 & 1 & -1 & 1 & 1 \end{array}\right]$$

Therefore,

This result can be confirmed by verifying that $AA^{-1} = A^{-1}A = I$.

$$A^{-1} = \begin{bmatrix} 0 & 1 & -1 \\ 2 & -2 & -1 \\ -1 & 1 & 1 \end{bmatrix}$$

Solving Linear Systems Using Matrix Inverses

Inverse matrices can also be used to solve linear systems of equations when the system has the same number of equations as variables—provided that the coefficient matrix has an inverse. Let's consider a linear system consisting of two equations in two variables:

$$\begin{cases} a_1x + a_2y = k_1 \\ b_1x + b_2y = k_2 \end{cases}$$

We can express this system as a product of matrices as follows: Let A be the 2×2 coefficient matrix of the system of equations, let K be the column matrix of the constants on the right side, and let X be the column matrix of unknowns. That is, let

$$A = \begin{bmatrix} a_1 & a_2 \\ b_1 & b_2 \end{bmatrix}, \quad X = \begin{bmatrix} x \\ y \end{bmatrix}, \quad \text{and} \quad K = \begin{bmatrix} k_1 \\ k_2 \end{bmatrix}$$

Then AX is the matrix with 2 rows and 1 column given by

$$AX = \begin{bmatrix} a_1 & a_2 \\ b_1 & b_2 \end{bmatrix} \begin{bmatrix} x \\ y \end{bmatrix} = \begin{bmatrix} a_1x + a_2y \\ b_1x + b_2y \end{bmatrix}$$

Do not change the order of multiplying A^{-1} on each side of the equation $AX = K$, since $A^{-1}(AX)$ and KA^{-1} are not necessarily the same.

Since $a_1x + a_2y = k_1$ and $b_1x + b_2y = k_2$, it follows that $AX = K$. The latter equation is called the **matrix equation form** of the system. To solve this equation, we multiply both sides *on the left* by A^{-1}, assuming it exists, to isolate X on the left side:

$$\begin{aligned} AX &= K \\ A^{-1}(AX) &= A^{-1}K \\ (A^{-1}A)X &= A^{-1}K \\ IX &= A^{-1}K \end{aligned}$$

so

$$\boxed{X = A^{-1}K}$$

The entries in the matrix product $A^{-1}K$ make up the solution of the original system.

EXAMPLE 6 **Solving a 2 × 2 linear system by using the inverse of a matrix**

Solve the system by using the inverse of the coefficient matrix.

$$\begin{cases} 3x + 5y = 11 \\ x + 2y = 4 \end{cases}$$

Solution The given system of equations is equivalent to the matrix equation $AX = K$, where

$$A = \begin{bmatrix} 3 & 5 \\ 1 & 2 \end{bmatrix}, \quad X = \begin{bmatrix} x \\ y \end{bmatrix}, \quad \text{and} \quad K = \begin{bmatrix} 11 \\ 4 \end{bmatrix}$$

From the illustration on page 488, we know that A^{-1} exists and that

$$A^{-1} = \begin{bmatrix} 2 & -5 \\ -1 & 3 \end{bmatrix}$$

It follows that the solution X is given by

$$\begin{aligned} X &= A^{-1}K \\ &= \begin{bmatrix} 2 & -5 \\ -1 & 3 \end{bmatrix} \begin{bmatrix} 11 \\ 4 \end{bmatrix} = \begin{bmatrix} 2 \\ 1 \end{bmatrix} \end{aligned}$$

Therefore, $x = 2$ and $y = 1$, and the solution is $(2, 1)$.

EXAMPLE 7 **Solving a 3 × 3 linear system by using the inverse of a matrix**

Solve the system by using the inverse of the coefficient matrix.

$$\begin{cases} x + 2y + 3z = 4 \\ x + y + 2z = 5 \\ y + 2z = 4 \end{cases}$$

Solution If we let

$$A = \begin{bmatrix} 1 & 2 & 3 \\ 1 & 1 & 2 \\ 0 & 1 & 2 \end{bmatrix}, \quad X = \begin{bmatrix} x \\ y \\ z \end{bmatrix}, \quad \text{and} \quad K = \begin{bmatrix} 4 \\ 5 \\ 4 \end{bmatrix}$$

we can write the system as $AX = K$. From Example 5, we know that

$$A^{-1} = \begin{bmatrix} 0 & 1 & -1 \\ 2 & -2 & -1 \\ -1 & 1 & 1 \end{bmatrix}$$

It follows that the solution X is given by

$$X = A^{-1}K = \begin{bmatrix} 0 & 1 & -1 \\ 2 & -2 & -1 \\ -1 & 1 & 1 \end{bmatrix} \begin{bmatrix} 4 \\ 5 \\ 4 \end{bmatrix} = \begin{bmatrix} 1 \\ -6 \\ 5 \end{bmatrix}$$

Thus, $x = 1$, $y = -6$, and $z = 5$, and the solution is $(1, -6, 5)$.

Solving Applied Problems

The next example illustrates a way in which matrix multiplication is used in the manipulation of data.

EXAMPLE 8 **Solving a business model**

A fruit market packages fruit in three different ways for gift baskets. The economy basket E has 5 apples, 3 grapefruits, and 4 pears. The standard basket S has 4 apples, 4 grapefruits, and 5 pears. The luxury basket L has 6 apples, 5 grapefruits, and 5 pears. The cost is $0.40, $0.70 and $0.50 for each apple, grapefruit, and pear, respectively. What is the cost of preparing each basket of fruit?

Solution We arrange the given data in matrix form as follows:

Cost per fruit: Apple, Grapefruit, Pear

$$A = [\$0.40 \quad \$0.70 \quad \$0.50]$$

Number of items in each basket type (E, S, L); rows: Apples, Grapefruits, Pears

$$B = \begin{bmatrix} 5 & 4 & 6 \\ 3 & 4 & 5 \\ 4 & 5 & 5 \end{bmatrix}$$

To find the cost of each basket, multiply matrix B by matrix A to get

$$AB = [0.40 \quad 0.70 \quad 0.50]\begin{bmatrix} 5 & 4 & 6 \\ 3 & 4 & 5 \\ 4 & 5 & 5 \end{bmatrix}$$

$$= [6.10 \quad 6.90 \quad 8.40]$$

Therefore, the costs of each economy, standard, and luxury basket are $6.10, $6.90, and $8.40, respectively.

PROBLEM SET 8.3

Mastering the Concepts

In problems 1–6, perform each operation on the given matrices A and B.

(a) $A + B$ **(b)** $A - B$ **(c)** $4A$ **(d)** $-5B$ **(e)** $4A - 5B$

1. $A = \begin{bmatrix} 3 & 5 \\ -1 & 4 \end{bmatrix}$; $B = \begin{bmatrix} 5 & 6 \\ -1 & 3 \end{bmatrix}$

2. $A = \begin{bmatrix} -1 & 2 \\ 3 & 0 \end{bmatrix}$; $B = \begin{bmatrix} 2 & -1 \\ 3 & -4 \end{bmatrix}$

3. $A = \begin{bmatrix} 3 & 1 & 2 \\ -1 & 3 & 1 \end{bmatrix}$; $B = \begin{bmatrix} 5 & 11 & 6 \\ 3 & 0 & -1 \end{bmatrix}$

4. $A = \begin{bmatrix} 0 & 1 & 4 & 2 \\ -5 & 6 & 1 & 3 \end{bmatrix}$; $B = \begin{bmatrix} 6 & -1 & 3 & 1 \\ -2 & 0 & 1 & 4 \end{bmatrix}$

5. $A = \begin{bmatrix} -5 & 2 & 1 \\ -2 & 1 & 2 \\ 3 & -1 & 2 \end{bmatrix}$; $B = \begin{bmatrix} -4 & 6 & 2 \\ -1 & 2 & 5 \\ 3 & -1 & 2 \end{bmatrix}$

6. $A = \begin{bmatrix} 1 & 0 & 0 & 1 & -1 \\ 2 & 1 & 3 & 5 & 7 \\ 3 & 6 & 2 & -1 & 4 \end{bmatrix}$; $B = \begin{bmatrix} 6 & 1 & -1 & 3 & 5 \\ 7 & 1 & 2 & 8 & -1 \\ 0 & 1 & 2 & 0 & 1 \end{bmatrix}$

In problems 7–12, if possible, find:
(a) AB **(b)** BA

7. $A = \begin{bmatrix} 1 & 2 \\ -1 & 1 \end{bmatrix}$; $B = \begin{bmatrix} 5 & -1 \\ 7 & 0 \end{bmatrix}$

8. $A = \begin{bmatrix} 1 & -1 \\ 0 & 2 \end{bmatrix}$; $B = \begin{bmatrix} 3 & 1 & -1 \\ -2 & 0 & 1 \end{bmatrix}$

9. $A = \begin{bmatrix} 2 & -3 & 5 \\ -1 & 1 & 3 \end{bmatrix}$; $B = \begin{bmatrix} -3 & 1 \\ 1 & 2 \\ 0 & -5 \end{bmatrix}$

10. $A = \begin{bmatrix} 3 & 2 & 1 \\ 1 & 4 & -1 \\ 2 & 1 & -3 \end{bmatrix}$; $B = \begin{bmatrix} -3 & 0 & 0 \\ 0 & -3 & 0 \\ 0 & 0 & -3 \end{bmatrix}$

11. $A = \begin{bmatrix} 4 & 3 & -1 \\ 8 & -2 & 3 \\ 6 & 5 & 2 \end{bmatrix}$; $B = \begin{bmatrix} 1 & 0 & 0 \\ 0 & 1 & 0 \\ 0 & 0 & 1 \end{bmatrix}$

12. $A = \begin{bmatrix} -1 & 1 & 2 & 0 \\ -2 & 3 & 5 & 1 \\ 2 & -1 & 3 & 2 \end{bmatrix}$; $B = \begin{bmatrix} 1 & -3 & 4 \\ 2 & 1 & 0 \\ -1 & 0 & 2 \\ 0 & -1 & 3 \end{bmatrix}$

In problems 13–18, determine whether A and B are inverses of each other.

13. $A = \begin{bmatrix} 1 & -1 \\ 2 & 1 \end{bmatrix}$; $B = \begin{bmatrix} \frac{1}{3} & \frac{1}{3} \\ -\frac{2}{3} & \frac{1}{3} \end{bmatrix}$

14. $A = \begin{bmatrix} 2 & 3 \\ 5 & 1 \end{bmatrix}$; $B = \begin{bmatrix} -\frac{1}{13} & \frac{3}{13} \\ \frac{5}{13} & -\frac{2}{13} \end{bmatrix}$

15. $A = \begin{bmatrix} 3 & 2 \\ 2 & -3 \end{bmatrix}$; $B = \begin{bmatrix} \frac{3}{13} & \frac{2}{13} \\ \frac{2}{13} & -\frac{3}{13} \end{bmatrix}$

16. $A = \begin{bmatrix} 1 & 1 & 1 \\ 2 & -1 & -1 \\ 1 & -1 & 2 \end{bmatrix}$; $B = \begin{bmatrix} \frac{1}{3} & \frac{1}{3} & 0 \\ \frac{5}{9} & -\frac{1}{9} & -\frac{1}{3} \\ \frac{1}{9} & -\frac{2}{9} & \frac{1}{3} \end{bmatrix}$

17. $A = \begin{bmatrix} 1 & 2 & 4 \\ 2 & -3 & 1 \\ 3 & -1 & -2 \end{bmatrix}$; $B = \begin{bmatrix} \frac{1}{7} & 0 & \frac{2}{7} \\ \frac{1}{7} & -\frac{2}{7} & \frac{1}{7} \\ \frac{1}{7} & \frac{1}{7} & -\frac{1}{7} \end{bmatrix}$

18. $A = \begin{bmatrix} 1 & 1 & 1 \\ 2 & 3 & -1 \\ 3 & 5 & 1 \end{bmatrix}$; $B = \begin{bmatrix} 2 & 1 & -1 \\ -\frac{5}{4} & -\frac{1}{2} & \frac{3}{4} \\ \frac{1}{4} & -\frac{1}{2} & \frac{1}{4} \end{bmatrix}$

In problems 19–26, find the inverse of the given matrix, if it exists.

19. $\begin{bmatrix} 3 & 5 \\ 1 & 2 \end{bmatrix}$

20. $\begin{bmatrix} 2 & -5 \\ -1 & 3 \end{bmatrix}$

21. $\begin{bmatrix} -1 & 1 \\ 1 & 0 \end{bmatrix}$

22. $\begin{bmatrix} 1 & 2 \\ \frac{1}{4} & \frac{3}{4} \end{bmatrix}$

23. $\begin{bmatrix} 1 & 2 & 0 \\ 0 & 2 & 3 \\ 1 & 3 & 1 \end{bmatrix}$

24. $\begin{bmatrix} 2 & 7 & 1 \\ 1 & 4 & -1 \\ 1 & 3 & 0 \end{bmatrix}$

25. $\begin{bmatrix} 7 & 2 & -6 \\ -3 & -1 & 3 \\ 2 & 1 & -2 \end{bmatrix}$

26. $\begin{bmatrix} -1 & 0 & 2 \\ 3 & 1 & 0 \\ 0 & 2 & -3 \end{bmatrix}$

In problems 27–36, use inverse matrices to solve each system.

27. $\begin{cases} x - y = 1 \\ 2x + y = 5 \end{cases}$ (see problem 13)

28. $\begin{cases} 2x_1 + 3x_2 = 7 \\ 5x_1 + x_2 = -2 \end{cases}$ (see problem 14)

29. $\begin{cases} 3x + 2y = 8 \\ 2x - 3y = 14 \end{cases}$ (see problem 15)

30. $\begin{cases} 3x + 5y = 13 \\ x + 2y = 5 \end{cases}$

31. $\begin{cases} 2x - 5y = -3 \\ -x + 3y = 2 \end{cases}$

32. $\begin{cases} 4x + y = 17 \\ 3x + 2y = 17 \end{cases}$

33. $\begin{cases} x + y + z = 2 \\ 2x + 3y - z = 3 \\ 3x + 5y + z = 8 \end{cases}$

34. $\begin{cases} x_1 + 2x_2 + 4x_3 = 12 \\ 2x_1 - 3x_2 + x_3 = 10 \\ 3x_1 - x_2 - 2x_3 = 1 \end{cases}$

35. $\begin{cases} x + y + z = 6 \\ 2x - y - z = 0 \\ x - y + 2z = 7 \end{cases}$

36. $\begin{cases} 2x + 7y + z = 10 \\ x + 4y - z = 4 \\ x + 3y = 4 \end{cases}$

Applying the Concepts

37. Crop Shipments: A nursery raises two kinds of flowers that are shipped in bundles to three different florists. The number of bundles of flower k that are shipped to store m is given by entry a_{km} in the matrix

$$\begin{array}{c} \\ \\ A = \end{array}\begin{array}{c} \text{Store} \quad \text{Store} \quad \text{Store} \\ 1 \qquad\quad 2 \qquad\quad 3 \\ \begin{bmatrix} 200 & 85 & 80 \\ 250 & 170 & 100 \end{bmatrix} \end{array}\begin{array}{l} \\ \\ \text{Flower 1} \\ \text{Flower 2} \end{array}$$

The profit (in dollars) per bundle is represented by the matrix

$$\begin{array}{c} \text{Flower} \quad \text{Flower} \\ 1 \qquad\quad 2 \\ T = [4.50 \quad 5.70] \end{array}$$

(a) Find the product TA.
(b) Indicate what each entry of the product represents.

38. Inventory Levels: A store sells two brands, B_1 and B_2, of microwave ovens. The following matrices give the sales figures and costs for three months. Use matrix multiplication to determine the total dollar sales and the total dollar costs of these items for these three months.

$$\begin{array}{l} \\ \text{Number of brand } B_1 \text{ ovens sold} \\ \text{Number of brand } B_2 \text{ ovens sold} \end{array}\begin{array}{c} \text{May} \quad \text{July} \quad \text{October} \\ \begin{bmatrix} 16 & 10 & 14 \\ 20 & 15 & 16 \end{bmatrix} \end{array}$$

$$\begin{array}{l} \\ \\ \text{Retail price} \\ \text{Dealer cost} \end{array}\begin{array}{c} \text{Brand} \quad \text{Brand} \\ B_1 \qquad B_2 \\ \begin{bmatrix} 210 & 140 \\ 150 & 90 \end{bmatrix} \end{array}$$

39. Total Revenue: An automobile manufacturer produces three different new models of cars, A, B, and C, that are shipped to two different locations for distribution. The following matrices give the number of cars that are shipped as well as the price in dollars per car.

$$\begin{array}{l} \\ \\ \text{Model } A \\ \text{Model } B \\ \text{Model } C \end{array}\begin{array}{c} \text{Number shipped} \\ \text{Site 1} \quad \text{Site 2} \\ \begin{bmatrix} 6{,}000 & 7{,}000 \\ 9{,}000 & 10{,}000 \\ 4{,}000 & 8{,}000 \end{bmatrix} \end{array} = S$$

$$\begin{array}{c} \text{Price} \\ \text{Model } A \quad \text{Model } B \quad \text{Model } C \\ [10{,}000 \quad 12{,}000 \quad 14{,}000] = P \end{array}$$

(a) Find the product PS.
(b) Indicate what each entry of the product represents.

40. Manufacturing: A manufacturer of calculators has two plants, each producing scientific calculators and graphing calculators. The manufacturing time requirements (in hours per calculator) and the assembly and packing costs (in dollars per hour) are given by the following matrices:

$$\begin{array}{c} \\ \\ \\ A = \end{array}\begin{array}{c} \text{Manufacturing time} \\ \text{(hours per unit)} \\ \text{Assembly} \quad \text{Packaging} \\ \begin{bmatrix} 0.2 & 0.1 \\ 0.3 & 0.1 \end{bmatrix} \end{array}\begin{array}{l} \\ \\ \\ \text{Scientific calculator} \\ \text{Graphing calculator} \end{array}$$

$$\begin{array}{c} \\ \\ \\ \\ B = \end{array}\begin{array}{c} \text{Costs} \\ \text{(dollars per hour)} \\ \text{First} \quad \text{Second} \\ \text{plant} \quad \text{plant} \\ \begin{bmatrix} 5 & 6 \\ 4 & 5 \end{bmatrix} \end{array}\begin{array}{l} \\ \\ \\ \\ \text{Assembly} \\ \text{Packaging} \end{array}$$

(a) Compute AB.
(b) Interpret the meaning of each entry in AB.

Developing and Extending the Concepts

41. Let

$$A = \begin{bmatrix} 3 & 2 \\ -1 & 4 \end{bmatrix}, \quad B = \begin{bmatrix} 0 & 1 \\ -1 & 2 \end{bmatrix}, \quad \text{and} \quad C = \begin{bmatrix} 2 & -3 \\ 1 & -4 \end{bmatrix}$$

(a) Demonstrate the commutative property: $A + B = B + A$
(b) Demonstrate the associative property: $A + (B + C) = (A + B) + C$
(c) Find $-A$ and $-C$. Then show that $A + (-A)$ and $C + (-C)$ equal $\mathbf{0}$, the 2×2 zero matrix.
(d) Find matrix E so that $A + E = C$.
(e) Demonstrate that $A(B - C) = AB - AC$.

42. Find the values of x and y if:

(a) $\begin{bmatrix} 3 & 1 \\ 5 & 7 \end{bmatrix} = \begin{bmatrix} 3 & x \\ -y & 7 \end{bmatrix}$

(b) $\begin{bmatrix} 1 & -1 \\ 2 & 6 \end{bmatrix} + \begin{bmatrix} x & 3 \\ y & 2 \end{bmatrix} = \begin{bmatrix} 8 & 2 \\ 0 & 8 \end{bmatrix}$

43. Find the values of x, y, z, and w if:

$$\begin{bmatrix} x & y \\ z & w \end{bmatrix}\begin{bmatrix} 3 & -5 \\ -1 & 2 \end{bmatrix} = \begin{bmatrix} 1 & -1 \\ 2 & 0 \end{bmatrix}$$

44. Let

$$A = \begin{bmatrix} 3 & -1 \\ 0 & -2 \end{bmatrix}$$

Show that

$$A^2 - A - 6I = \begin{bmatrix} 0 & 0 \\ 0 & 0 \end{bmatrix}$$

where

$$I = \begin{bmatrix} 1 & 0 \\ 0 & 1 \end{bmatrix}$$

45. Suppose that A and B are $n \times n$ matrices and $AB = BA$.

(a) Show that: $(A + B)^2 = A^2 + 2AB + B^2$

(b) Show that: $(A - B)(A + B) = A^2 - B^2$

46. If

$$A^{-1} = \begin{bmatrix} 1 & -1 & 3 \\ 2 & 0 & 5 \\ -1 & 1 & 0 \end{bmatrix}$$

find a matrix B such that

$$AB = \begin{bmatrix} 1 & -1 & 2 \\ 0 & 1 & 1 \\ 1 & 0 & 0 \end{bmatrix}$$

47. If

$$A^{-1} = \begin{bmatrix} 1 & -1 & 3 \\ 2 & 1 & 1 \\ 0 & 2 & -2 \end{bmatrix}$$

find A.

48. Let: $A = \begin{bmatrix} 2 & 3 \\ 4 & 5 \end{bmatrix}$ and $B = \begin{bmatrix} 7 & 8 \\ 6 & 7 \end{bmatrix}$

(a) Compute A^{-1}, B^{-1}, and $B^{-1}A^{-1}$.

(b) Compute $(AB)^{-1}$ and compare it to $B^{-1}A^{-1}$.

(c) Can you generalize the result in part (b)? Explain.

8.4 Solutions of Linear Systems Using Determinants

OBJECTIVES

1. Define and evaluate determinants.
2. Use elementary operations to evaluate determinants.
3. Use Cramer's rule to solve linear systems.
4. Solve applied problems.

In this section, we introduce determinants and then use them to solve certain systems of linear equations in two and three variables.

Defining and Evaluating Determinants

We begin by defining a determinant of a 2×2 matrix.

DEFINITION

DETERMINANT OF A 2×2 MATRIX

Let

$$A = \begin{bmatrix} a & b \\ c & d \end{bmatrix}$$

be a 2×2 matrix. The symbol $|A|$ denotes the **determinant** of matrix A. Its value is defined to be the number $ad - cb$, and we write

$$|A| = \begin{vmatrix} a & b \\ c & d \end{vmatrix} = ad - cb$$

The determinant of a square matrix A is also denoted by det A; that is, $|A| = \det A$.

EXAMPLE 1 Evaluating a 2×2 determinant

Evaluate: $\begin{vmatrix} 5 & 3 \\ -3 & -6 \end{vmatrix}$

Solution $\begin{vmatrix} 5 & 3 \\ -3 & -6 \end{vmatrix} = 5(-6) - (-3)(3) = -30 + 9 = -21$ ■

If A is a 3×3 matrix, its determinant value is defined in terms of 2×2 determinants according to the following *expansion formula*:

AN EXPANSION FORMULA FOR THE DETERMINANT OF A 3×3 MATRIX

The value of the determinant of a 3×3 matrix is given by the **expansion formula**:

$$\begin{vmatrix} a_1 & a_2 & a_3 \\ b_1 & b_2 & b_3 \\ c_1 & c_2 & c_3 \end{vmatrix} = a_1\begin{vmatrix} b_2 & b_3 \\ c_2 & c_3 \end{vmatrix} - a_2\begin{vmatrix} b_1 & b_3 \\ c_1 & c_3 \end{vmatrix} + a_3\begin{vmatrix} b_1 & b_2 \\ c_1 & c_2 \end{vmatrix}$$
$$= a_1b_2c_3 - a_1c_2b_3 - a_2b_1c_3 + a_2c_1b_3 + a_3b_1c_2 - a_3c_1b_2$$

Notice that each entry in the first row is multiplied by the 2×2 determinant that remains when the row and column containing the multiplier are crossed out. Thus,

Similar expansion formulas can be used to evaluate $n \times n$ matrices, where $n > 3$; however, most graphing utilities can evaluate determinants more efficiently.

$$a_1 \text{ is multiplied by } \begin{vmatrix} \not{a_1} & \not{a_2} & \not{a_3} \\ \not{b_1} & b_2 & b_3 \\ \not{c_1} & c_2 & c_3 \end{vmatrix} = \begin{vmatrix} b_2 & b_3 \\ c_2 & c_3 \end{vmatrix}$$

$$a_2 \text{ is multiplied by } \begin{vmatrix} \not{a_1} & \not{a_2} & \not{a_3} \\ b_1 & \not{b_2} & b_3 \\ c_1 & \not{c_2} & c_3 \end{vmatrix} = \begin{vmatrix} b_1 & b_3 \\ c_1 & c_3 \end{vmatrix}$$

$$a_3 \text{ is multiplied by } \begin{vmatrix} \not{a_1} & \not{a_2} & \not{a_3} \\ b_1 & b_2 & \not{b_3} \\ c_1 & c_2 & \not{c_3} \end{vmatrix} = \begin{vmatrix} b_1 & b_2 \\ c_1 & c_2 \end{vmatrix}$$

Also, notice the negative sign before a_2 in the formula.

EXAMPLE 2 Evaluating a 3×3 determinant

Evaluate the determinant by using the expansion formula.

$$\begin{vmatrix} 3 & 2 & 7 \\ -1 & 5 & 3 \\ 2 & -3 & -6 \end{vmatrix}$$

Solution

$$\begin{vmatrix} 3 & 2 & 7 \\ -1 & 5 & 3 \\ 2 & -3 & -6 \end{vmatrix} = 3\begin{vmatrix} 5 & 3 \\ -3 & -6 \end{vmatrix} - 2\begin{vmatrix} -1 & 3 \\ 2 & -6 \end{vmatrix} + 7\begin{vmatrix} -1 & 5 \\ 2 & -3 \end{vmatrix}$$
$$= 3(-30 + 9) - 2(6 - 6) + 7(3 - 10)$$
$$= 3(-21) - 2(0) + 7(-7)$$
$$= -63 - 49 = -112$$

■

EXAMPLE 3 Evaluating a 3×3 determinant of a matrix in triangular form

Evaluate the determinant: $\begin{vmatrix} 3 & 4 & 5 \\ 0 & 6 & -1 \\ 0 & 0 & 2 \end{vmatrix}$

Solution
$$\begin{vmatrix} 3 & 4 & 5 \\ 0 & 6 & -1 \\ 0 & 0 & 2 \end{vmatrix} = 3\begin{vmatrix} 6 & -1 \\ 0 & 2 \end{vmatrix} - 4\begin{vmatrix} 0 & -1 \\ 0 & 2 \end{vmatrix} + 5\begin{vmatrix} 0 & 6 \\ 0 & 0 \end{vmatrix}$$
$$= 3(12 - 0) - 4(0 - 0) + 5(0 - 0)$$
$$= 3(12) = 36$$

Notice that the value of the determinant in Example 3 is the product of the entries on its *main diagonal* (the diagonal that runs from upper left to lower right). In general, *the determinant of any square matrix in triangular form is equal to the product of the entries on its main diagonal.*

Using Elementary Operations to Evaluate Determinants

The effect of each elementary row (or column) operation on the value of the determinant is given in Table 1.

The four operations also hold if we replace the word "row" with "column" throughout the table.

TABLE 1

Elementary Row Operations for Determinants

Elementary Row Operation	Example
1. If two rows of a determinant are interchanged, then the algebraic sign of the determinant changes.	$\begin{vmatrix} 3 & 4 \\ 1 & 5 \end{vmatrix} = -\begin{vmatrix} 1 & 5 \\ 3 & 4 \end{vmatrix}$
2. If every entry in one row of a determinant is multiplied by a constant k, the effect is to multiply the value of the determinant by k.	$\begin{vmatrix} 3k & -2k \\ 4 & 7 \end{vmatrix} = k\begin{vmatrix} 3 & -2 \\ 4 & 7 \end{vmatrix}$
3. If one row of a determinant is a multiple of another row, the value of the determinant is 0.	$\begin{vmatrix} 3 & -4 \\ 6 & -8 \end{vmatrix} = 0$
4. If a constant multiple of one row of a determinant is added to any other row, the value of the determinant will not change.	$\begin{vmatrix} -3 & 5 \\ 6 & -4 \end{vmatrix} = \begin{vmatrix} -3 & 5 \\ 6 + 2(-3) & -4 + 2(5) \end{vmatrix} = \begin{vmatrix} -3 & 5 \\ 0 & 6 \end{vmatrix} = -18$

Elementary row operation 2 allows us to factor out a common factor of all elements in a single row of a determinant.

EXAMPLE 4 Evaluating a 4 × 4 determinant

Evaluate the determinant by using elementary row operations.

$$\begin{vmatrix} 1 & 0 & -2 & 0 \\ 0 & 3 & 0 & 2 \\ 1 & 0 & 3 & 1 \\ 0 & 4 & 0 & 5 \end{vmatrix}$$

Solution

$$\begin{vmatrix} 1 & 0 & -2 & 0 \\ 0 & 3 & 0 & 2 \\ 1 & 0 & 3 & 1 \\ 0 & 4 & 0 & 5 \end{vmatrix} = \begin{vmatrix} 1 & 0 & -2 & 0 \\ 0 & 3 & 0 & 2 \\ 0 & 0 & 5 & 1 \\ 0 & 4 & 0 & 5 \end{vmatrix}$$

Replace row 3 by the sum of -1 times row 1 and row 3: $(-1)R_1 + R_3 \rightarrow R_3$ (no change in value, row operation 4)

$$= -\begin{vmatrix} 1 & 0 & -2 & 0 \\ 0 & 3 & 0 & 2 \\ 0 & 4 & 0 & 5 \\ 0 & 0 & 5 & 1 \end{vmatrix}$$

Interchange rows 3 and 4: $R_3 \leftrightarrow R_4$ (change sign, row operation 1)

$$= -3\begin{vmatrix} 1 & 0 & -2 & 0 \\ 0 & 1 & 0 & \frac{2}{3} \\ 0 & 4 & 0 & 5 \\ 0 & 0 & 5 & 1 \end{vmatrix}$$

Factor 3 from row 2: (row operation 2)

$$= -3\begin{vmatrix} 1 & 0 & -2 & 0 \\ 0 & 1 & 0 & \frac{2}{3} \\ 0 & 0 & 0 & \frac{7}{3} \\ 0 & 0 & 5 & 1 \end{vmatrix}$$

Replace row 3 by the sum of -4 times row 2 and row 3: $-4R_2 + R_3 \rightarrow R_3$ (no change, row operation 4)

$$= 3\begin{vmatrix} 1 & 0 & -2 & 0 \\ 0 & 1 & 0 & \frac{2}{3} \\ 0 & 0 & 5 & 1 \\ 0 & 0 & 0 & \frac{7}{3} \end{vmatrix}$$

Interchange rows 3 and 4: $R_3 \leftrightarrow R_4$ (change sign, row operation 1)

$$= 3(1)(1)(5)\left(\tfrac{7}{3}\right) = 35$$

Multiply diagonal entries

Using Cramer's Rule to Solve Linear Systems

Determinants can be used to solve a linear system with n linear equations in n variables by a technique known as **Cramer's rule**. Here, we state two special cases of the rule.

CRAMER'S RULE FOR 2 × 2 SYSTEMS OF EQUATIONS

The solution to the system

$$\begin{cases} ax + by = h \\ cx + dy = k \end{cases}$$

is given by

$$x = \frac{D_x}{D} \quad \text{and} \quad y = \frac{D_y}{D}$$

where

$$D = \begin{vmatrix} a & b \\ c & d \end{vmatrix} \neq 0, \quad D_x = \begin{vmatrix} h & b \\ k & d \end{vmatrix}, \quad \text{and} \quad D_y = \begin{vmatrix} a & h \\ c & k \end{vmatrix}$$

If $D \neq 0$, then the solution of the system is unique.

EXAMPLE 5 Using Cramer's rule to solve a system of two equations

Solve the system by using Cramer's rule.

$$\begin{cases} 5x - 2y = 8 \\ 3x + 4y = 10 \end{cases}$$

Solution The determinant of the coefficient matrix D is given by

$$D = \begin{vmatrix} 5 & -2 \\ 3 & 4 \end{vmatrix} = 20 + 6 = 26$$

Because $D \neq 0$, we can solve the system by applying Cramer's rule as follows:

$$D_x = \begin{vmatrix} 8 & -2 \\ 10 & 4 \end{vmatrix} = 32 + 20 = 52$$

and

$$D_y = \begin{vmatrix} 5 & 8 \\ 3 & 10 \end{vmatrix} = 50 - 24 = 26$$

So,

$$x = \frac{D_x}{D} = \frac{52}{26} = 2 \quad \text{and} \quad y = \frac{D_y}{D} = \frac{26}{26} = 1$$

That is, the solution is (2, 1). ■

Next, we state Cramer's rule for 3×3 systems of linear equations.

CRAMER'S RULE FOR 3×3 SYSTEMS OF EQUATIONS

The solution to the system

$$\begin{cases} a_1x + a_2y + a_3z = k_1 \\ b_1x + b_2y + b_3z = k_2 \\ c_1x + c_2y + c_3z = k_3 \end{cases}$$

is given by

$$x = \frac{D_x}{D}, \quad y = \frac{D_y}{D}, \quad \text{and} \quad z = \frac{D_z}{D}$$

where

$$D = \begin{vmatrix} a_1 & a_2 & a_3 \\ b_1 & b_2 & b_3 \\ c_1 & c_2 & c_3 \end{vmatrix} \neq 0 \qquad D_x = \begin{vmatrix} k_1 & a_2 & a_3 \\ k_2 & b_2 & b_3 \\ k_3 & c_2 & c_3 \end{vmatrix}$$

$$D_y = \begin{vmatrix} a_1 & k_1 & a_3 \\ b_1 & k_2 & b_3 \\ c_1 & k_3 & c_3 \end{vmatrix} \qquad D_z = \begin{vmatrix} a_1 & a_2 & k_1 \\ b_1 & b_2 & k_2 \\ c_1 & c_2 & k_3 \end{vmatrix}$$

EXAMPLE 6 **Using Cramer's rule to solve a system of three equations**

Solve the system by Cramer's rule.

$$\begin{cases} x + y + z = 4 \\ 14x + 13y + 15z = 55 \\ 2x - y = 0 \end{cases}$$

Solution Using Cramer's rule, we have

$$D = \begin{vmatrix} 1 & 1 & 1 \\ 14 & 13 & 15 \\ 2 & -1 & 0 \end{vmatrix} = 5 \qquad D_x = \begin{vmatrix} 4 & 1 & 1 \\ 55 & 13 & 15 \\ 0 & -1 & 0 \end{vmatrix} = 5$$

$$D_y = \begin{vmatrix} 1 & 4 & 1 \\ 14 & 55 & 15 \\ 2 & 0 & 0 \end{vmatrix} = 10 \qquad D_z = \begin{vmatrix} 1 & 1 & 4 \\ 14 & 13 & 55 \\ 2 & -1 & 0 \end{vmatrix} = 5$$

So,

$$x = \frac{D_x}{D} = \frac{5}{5} = 1, \quad y = \frac{D_y}{D} = \frac{10}{5} = 2, \quad \text{and} \quad z = \frac{D_z}{D} = \frac{5}{5} = 1$$

Thus, the solution is (1, 2, 1).

Consider a system of two equations in the variables x and y. If $D = 0$, the system has *no unique solution*. The system is *inconsistent* if $D = 0$ and if $D_x \neq 0$ and $D_y \neq 0$. The system is *dependent* if $D = 0$, $D_x = 0$, and $D_y = 0$.

Solving Applied Problems

Determinants can be used in analytic geometry, as the next example illustrates.

EXAMPLE 7 **Writing an equation of a line in determinant form**

Find an equation of a line containing the points (x_1, y_1) and (x_2, y_2). Then express this formula in determinant form.

Solution The slope m of the line is given by

$$m = \frac{y_2 - y_1}{x_2 - x_1}$$

Using the point–slope form for the equation of a line, we get

$$y - y_1 = \frac{y_2 - y_1}{x_2 - x_1}(x - x_1)$$

To clear the fraction, we multiply each side of this equation by $x_2 - x_1$ to obtain

$$(x_2 - x_1)(y - y_1) = (y_2 - y_1)(x - x_1)$$

Multiplying, we have

$$x_2y - x_2y_1 - x_1y + x_1y_1 = y_2x - y_2x_1 - y_1x + y_1x_1$$

or

$$x_2y - x_2y_1 - x_1y + \cancel{x_1y_1} - y_2x + y_2x_1 + y_1x - \cancel{y_1x_1} = 0$$

That is,

$$x(y_1 - y_2) - y(x_1 - x_2) + (x_1y_2 - x_2y_1) = 0$$

Each of the three parts of this expression defines a 2×2 determinant, which enables us to rewrite this equation as

$$x\begin{vmatrix} y_1 & 1 \\ y_2 & 1 \end{vmatrix} - y\begin{vmatrix} x_1 & 1 \\ x_2 & 1 \end{vmatrix} + \begin{vmatrix} x_1 & y_1 \\ x_2 & y_2 \end{vmatrix} = 0$$

But the left side of this equation is an expansion of a 3×3 determinant, so we rewrite this equation as

$$\begin{vmatrix} x & y & 1 \\ x_1 & y_1 & 1 \\ x_2 & y_2 & 1 \end{vmatrix} = 0$$

For instance, if the points on the line are (2, 3) and (−1, 4), then the determinant form is given by

$$\begin{vmatrix} x & y & 1 \\ 2 & 3 & 1 \\ -1 & 4 & 1 \end{vmatrix} = 0$$

so

$$\begin{vmatrix} x & y & 1 \\ 2 & 3 & 1 \\ -1 & 4 & 1 \end{vmatrix} = x(-1) - y(3) + 1(11) = 0$$

That is, an equation of the line is $-x - 3y + 11 = 0$.

PROBLEM SET 8.4

Mastering the Concepts

In problems 1–10, evaluate each determinant by using an expansion formula.

1. **(a)** $\begin{vmatrix} -1 & 3 \\ -7 & 4 \end{vmatrix}$ **(b)** $\begin{vmatrix} 2 & 3 \\ 9 & 4 \end{vmatrix}$

2. **(a)** $\begin{vmatrix} 2 & -1 \\ 3 & 2 \end{vmatrix}$ **(b)** $\begin{vmatrix} 6 & 9 \\ 8 & 12 \end{vmatrix}$

3. **(a)** $\begin{vmatrix} 3 & -1 \\ 2 & -1 \end{vmatrix}$ **(b)** $\begin{vmatrix} 4 & 3 \\ 3 & 5 \end{vmatrix}$

4. **(a)** $\begin{vmatrix} 14 & -1 \\ 15 & 1 \end{vmatrix}$ **(b)** $\begin{vmatrix} -2 & 5 \\ 7 & 1 \end{vmatrix}$

5. $\begin{vmatrix} 2 & -1 & 3 \\ 9 & -7 & 4 \\ 11 & -6 & 2 \end{vmatrix}$

6. $\begin{vmatrix} 3 & -1 & 2 \\ 0 & 1 & -5 \\ 6 & 7 & 4 \end{vmatrix}$

7. $\begin{vmatrix} 2 & 2 & 2 \\ 3 & 3 & 3 \\ 4 & 4 & 4 \end{vmatrix}$

8. $\begin{vmatrix} \frac{1}{2} & 4 & 7 \\ 1 & -1 & 2 \\ 3 & 2 & 5 \end{vmatrix}$

9. $\begin{vmatrix} 1 & 0 & 2 & 0 \\ 0 & 1 & 0 & 0 \\ 1 & 0 & 3 & 0 \\ 0 & 0 & 0 & 3 \end{vmatrix}$

10. $\begin{vmatrix} -2 & 0 & 1 & 0 \\ -1 & 3 & 0 & 0 \\ 0 & 0 & 2 & 1 \\ 0 & 0 & 0 & 4 \end{vmatrix}$

In problems 11–16, show why each statement is true—not by evaluating each side, but by citing which row operations have been used.

11. $\begin{vmatrix} 4 & 5 \\ 3 & -2 \end{vmatrix} = -\begin{vmatrix} 3 & -2 \\ 4 & 5 \end{vmatrix}$

12. $\begin{vmatrix} 3 & 0 & 1 \\ 1 & 1 & 2 \\ 3 & 0 & 1 \end{vmatrix} = \begin{vmatrix} 0 & 0 & 0 \\ 1 & 1 & 2 \\ 3 & 0 & 1 \end{vmatrix}$

13. $\begin{vmatrix} 3 & -6 & 2 \\ 5 & -3 & 0 \\ 0 & 9 & 18 \end{vmatrix} = 9\begin{vmatrix} 3 & -6 & 2 \\ 5 & -3 & 0 \\ 0 & 1 & 2 \end{vmatrix}$

14. $\begin{vmatrix} 2 & 4 & 12 \\ -1 & 0 & 3 \\ 1 & 0 & 6 \end{vmatrix} = 18\begin{vmatrix} 1 & 2 & 6 \\ -1 & 0 & 3 \\ 0 & 0 & 1 \end{vmatrix}$

15. $\begin{vmatrix} 1 & 1 & 1 \\ 3 & 3 & 3 \\ 2 & 2 & 2 \end{vmatrix} = 6\begin{vmatrix} 0 & 0 & 0 \\ 0 & 0 & 0 \\ 1 & 1 & 1 \end{vmatrix}$

16. $\begin{vmatrix} 7 & -4 & 1 & 2 \\ 21 & 4 & 3 & -1 \\ -35 & 20 & -5 & -10 \\ 14 & 16 & 8 & -2 \end{vmatrix} = 0$

In problems 17–22, use the properties of determinants to evaluate.

17. $\begin{vmatrix} -1 & 0 & 2 \\ 0 & 0 & 0 \\ -1 & 5 & 1 \end{vmatrix}$

18. $\begin{vmatrix} 3 & 1 & 1 \\ -1 & 0 & 3 \\ 2 & 1 & 1 \end{vmatrix}$

19. $\begin{vmatrix} 2 & 1 & 3 \\ 1 & 2 & 1 \\ 4 & 0 & 0 \end{vmatrix}$

20. $\begin{vmatrix} 20 & 12 & 8 \\ 5 & 3 & 2 \\ 5 & 7 & 2 \end{vmatrix}$

21. $\begin{vmatrix} 1 & 0 & 2 & 3 \\ 1 & -3 & 0 & 1 \\ 0 & 3 & 1 & -1 \\ 2 & 1 & 2 & -2 \end{vmatrix}$

22. $\begin{vmatrix} -1 & 0 & 1 & 3 \\ 3 & -2 & 3 & 4 \\ 1 & 4 & -3 & 2 \\ 5 & 2 & -1 & 1 \end{vmatrix}$

In problems 23–36, use Cramer's rule to solve each system.

23. $\begin{cases} 2x - y = 0 \\ x + y = 1 \end{cases}$

24. $\begin{cases} -3x + y = 3 \\ -2x - y = -5 \end{cases}$

25. $\begin{cases} 3x + 2y = 7 \\ -2x + 7y = 12 \end{cases}$

26. $\begin{cases} 4x - y = 7 \\ -2x + 3y = 9 \end{cases}$

27. $\begin{cases} x + 4y = -4 \\ 3x - 2y = -19 \end{cases}$

28. $\begin{cases} 3x + y = 1 \\ -9x + 3y = -4 \end{cases}$

29. $\begin{cases} x + y + 2z = 4 \\ x + y - 2z = 0 \\ x - y = 0 \end{cases}$

30. $\begin{cases} 2x_1 - 3x_2 = 4 \\ x_1 + x_2 - 2x_3 = 1 \\ x_1 - x_2 - x_3 = 5 \end{cases}$

31. $\begin{cases} x + y + z = 4 \\ x - y + 2z = 8 \\ 2x + y - z = 3 \end{cases}$

32. $\begin{cases} 2x + 3y + z = 6 \\ x - 2y + 3z = -3 \\ 3x + y - z = 8 \end{cases}$

33. $\begin{cases} 2x_1 + x_2 + x_3 = 3 \\ -x_1 + 2x_2 - x_3 = 1 \\ 3x_1 + x_2 + 2x_3 = -1 \end{cases}$

34. $\begin{cases} 3x + 2y + 2z = 8 \\ x - 5y + 6z = 8 \\ 6x - 8z = 4 \end{cases}$

35. $\begin{cases} 5x + y - z = 4 \\ 9x + y - z = 1 \\ x - y + 5z = 2 \end{cases}$

36. $\begin{cases} 3x_1 + 4x_2 + 2x_3 = 1 \\ 4x_1 + 6x_2 + 2x_3 = 7 \\ 2x_1 + 3x_2 + x_3 = 11 \end{cases}$

Applying the Concepts

37. Area of a Triangle: Let triangle PQR be located in a Cartesian plane with vertices $P = (x_1, y_1)$, $Q = (x_2, y_2)$, and $R = (x_3, y_3)$. Then the area of the triangle PQR is given by

$$A = \text{Absolute value of } \frac{1}{2}\begin{vmatrix} x_1 & y_1 & 1 \\ x_2 & y_2 & 1 \\ x_3 & y_3 & 1 \end{vmatrix}$$

Sketch triangle PQR, where $P = (1, -5)$, $Q = (-3, -4)$, and $R = (6, 2)$. Then find its area by using the given formula.

38. Area of a Parallelogram: The determinant

$$\begin{vmatrix} a & b \\ c & d \end{vmatrix}$$

can be interpreted as the area A of the parallelogram having the vectors $\mathbf{u} = a\mathbf{i} + b\mathbf{j}$ and $\mathbf{v} = c\mathbf{i} + d\mathbf{j}$ as adjacent sides, provided that the angle between $\mathbf{u}$ and $\mathbf{v}$ is formed by a counterclockwise rotation (Figure 1). Find the area of the parallelogram whose sides are defined by the vectors

$$\mathbf{u} = 5\mathbf{i} - 2\mathbf{j} \quad \text{and} \quad \mathbf{v} = -4\mathbf{i} + 3\mathbf{j}$$

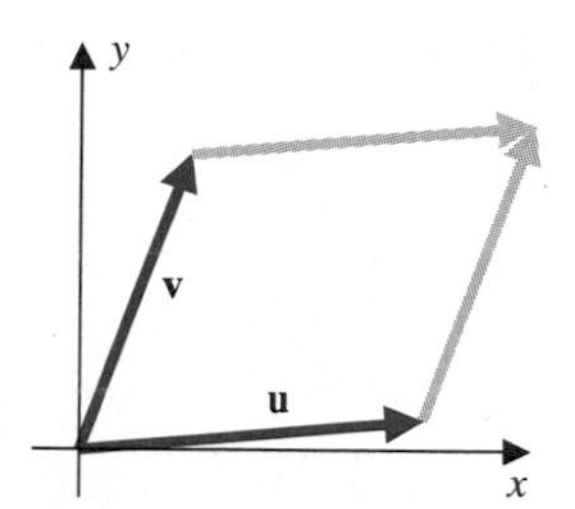

FIGURE 1

Developing and Extending the Concepts

39. Compute each determinant.

(a) $\begin{vmatrix} \cos\theta & -\sin\theta \\ \sin\theta & \cos\theta \end{vmatrix}$ (b) $\begin{vmatrix} \sqrt{6} & -2\sqrt{5} \\ 3\sqrt{5} & 4\sqrt{6} \end{vmatrix}$

(c) $\begin{vmatrix} \sqrt{5} - \sqrt{2} & 2 + \sqrt{3} \\ 2 - \sqrt{3} & \sqrt{5} + \sqrt{2} \end{vmatrix}$

40. Evaluate each determinant by using the properties of determinants.

(a) $\begin{vmatrix} p & q & r \\ p+1 & q+1 & r+1 \\ p-1 & q-1 & r-1 \end{vmatrix}$

(b) $\begin{vmatrix} p & q & r \\ p+q & 2q & r+q \\ 2 & 2 & 2 \end{vmatrix}$

In problems 41–44, solve for x.

41. (a) $\begin{vmatrix} x & -x \\ 5 & 3 \end{vmatrix} = 2$ (b) $\begin{vmatrix} x & 4 & 5 \\ 0 & 1 & x \\ 5 & 2 & 1 \end{vmatrix} = 7$

42. (a) $\begin{vmatrix} x & 0 & 0 \\ 3 & 1 & 2 \\ 0 & 4 & 1 \end{vmatrix} = 5$ (b) $\begin{vmatrix} 5x & 0 & 1 \\ 2x & 1 & 2 \\ 3x & 2 & 3 \end{vmatrix} = 0$

43. $\begin{vmatrix} x-3 & -2 \\ 1 & x \end{vmatrix} = 0$

44. $\begin{vmatrix} x-3 & 2 & 0 \\ 2 & x-3 & 0 \\ 0 & 0 & x-5 \end{vmatrix} = 0$

45. Find a, b, and c so that

$$\begin{vmatrix} 3 & -1 & x \\ 2 & 6 & y \\ -5 & 4 & z \end{vmatrix} = ax + by + cz$$

46. Show that: $\begin{vmatrix} 1 & x & x^2 \\ x^2 & 1 & x \\ x & x^2 & 1 \end{vmatrix} = (x^3 - 1)^2$

47. Let

$$A = \begin{vmatrix} a & b \\ c & d \end{vmatrix}$$

and assume that $|A| \neq 0$. Then show that

$$A^{-1} = \frac{1}{|A|}\begin{bmatrix} d & -b \\ -c & a \end{bmatrix}$$

Use this result to find A^{-1} if:

(a) $A = \begin{bmatrix} 5 & 3 \\ 2 & 4 \end{bmatrix}$ **(b)** $A = \begin{bmatrix} 6 & 7 \\ 7 & 8 \end{bmatrix}$

48. Use Cramer's rule to solve the system

$$\begin{cases} x_1 - 3x_4 = 1 \\ x_1 + 4x_2 + 2x_3 - 2x_4 = -7 \\ 2x_1 + 5x_2 + x_4 = 11 \\ 2x_1 - 2x_2 + 5x_3 + x_4 = 3 \end{cases}$$

Use the properties of determinants to expand each determinant by converting each to triangular form.

49. It is clear that (0, 0, 0) is a solution of the system

$$\begin{cases} x_1 - 4x_2 + 9x_3 = 0 \\ x_1 - x_2 + 3x_3 = 0 \\ 4x_1 - x_2 + 6x_3 = 0 \end{cases}$$

(a) If the determinant of the coefficient matrix does not equal 0, what can you conclude?

(b) If the determinant of the coefficient matrix does equal 0, what can you conclude?

50. Repeat problem 49 for the system

$$\begin{cases} x_1 + x_2 + 2x_3 + x_4 = 0 \\ 2x_1 - x_2 + x_3 - x_4 = 0 \\ 3x_1 + x_2 + 2x_3 + 3x_4 = 0 \\ 2x_1 - x_2 - x_3 + x_4 = 0 \end{cases}$$

51. Given matrix

$$A = \begin{bmatrix} a_1 & a_2 & a_3 \\ b_1 & b_2 & b_3 \\ c_1 & c_2 & c_3 \end{bmatrix}$$

we adjoin the first two columns of A to A to form

$$\begin{bmatrix} a_1 & a_2 & a_3 \\ b_1 & b_2 & b_3 \\ c_1 & c_2 & c_3 \end{bmatrix} \begin{matrix} a_1 & a_2 \\ b_1 & b_2 \\ c_1 & c_2 \end{matrix}$$

Next, we form the sum D of the products of the diagonal entries denoted by the downward arrows to get

$$D = a_1b_2c_3 + a_2b_3c_1 + a_3b_1c_2$$

Similarly, we form the sum V of the products of the diagonal entries denoted by the upward arrows to get

$$V = c_1b_2a_3 + c_2b_3a_1 + c_3b_1a_2$$

(a) Explain why $|A| = D - V$.

(b) Use this technique to find $|A|$ if

$$A = \begin{bmatrix} 1 & 2 & 3 \\ -1 & 4 & -1 \\ 2 & 5 & -7 \end{bmatrix}$$

OBJECTIVES

1. Graph inequalities in two variables.
2. Graph systems of inequalities.
3. Use linear programming.
4. Solve applied problems.

8.5 Solutions of Linear Systems of Inequalities and Linear Programming

In this section, we consider systems of inequalities in two variables. Such systems are represented graphically as regions in the Cartesian plane. We also introduce linear programming, which is used in modeling a variety of situations in areas such as health, transportation, economics, engineering, and science.

Graphing Inequalities in Two Variables

Examples of *linear inequalities* in two variables are

$$y < 8x + 1, \quad 2x + y \leq 3, \quad \text{and} \quad 5x - 3y > 1$$

As with equations in two variables, a **solution** of an inequality involving two variables is an ordered pair of numbers, which, when substituted for x and y, makes the inequality true. For example, (1, 2) is a solution of the inequality $y < 8x + 1$, since $2 < 8(1) + 1$ or $2 < 9$ is true.

The set of all ordered pairs that are solutions of an inequality is called its *solution set*. One way to identify the solution set of an inequality is to show its **graph**, which is the set of all points (x, y) in the xy plane whose coordinates satisfy the inequality. For instance, to graph $y < 8x + 1$, we begin by graphing the linear equation $y = 8x + 1$ (Figure 1a). A dashed line is used to emphasize that points on this *boundary line* do not satisfy $y < 8x + 1$. Next, we select a *test point* in one of the *two half-planes* defined by the line. If the coordinates of the point satisfy the original inequality, then all points in that half-plane also satisfy it. If the coordinates do not satisfy the inequality, then *none* of the others do either. Suppose we select (0, 0) as the test point for $y < 8x + 1$ (Figure 1b). Substituting $x = 0$ and $y = 0$ into the inequality $y < 8x + 1$, we get $0 < 8(0) + 1$ or $0 < 1$, which is true. So all points in the half-plane containing (0, 0) represent the solution set of the inequality. Consequently, the graph of $y < 8x + 1$ is the shaded region in Figure 1c.

If we had selected a point above the boundary line, such as (0, 5), it would not satisfy the inequality ($5 < 8(0) + 1$ or $5 < 1$ is false).

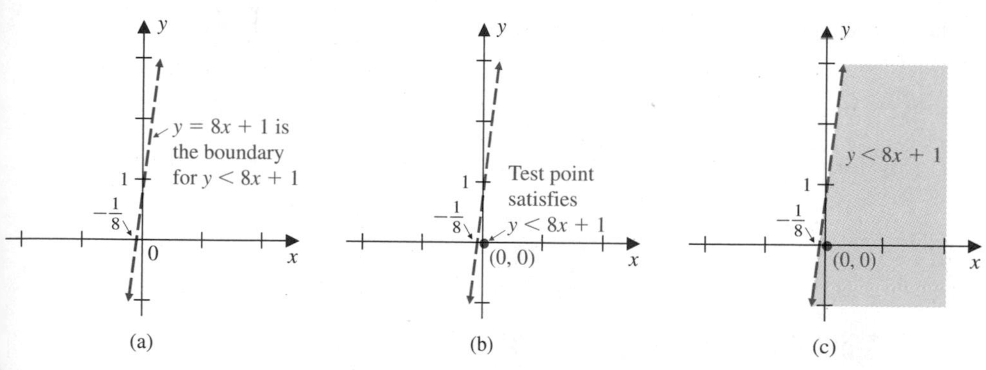

FIGURE 1

The general procedure for graphing a linear inequality is summarized as follows:

PROCEDURE FOR GRAPHING A LINEAR INEQUALITY

Step 1. Graph the associated linear equation. Use a dashed line as a boundary line if it is a *strict* inequality ($<$ or $>$) or use a solid line if it is not a strict inequality ($\leq$ or $\geq$).
Step 2. Pick a test point in either of the half-planes defined by the boundary line. Determine whether the coordinates satisfy the inequality.
Step 3. Shade the half-plane containing the test point if it satisfies the inequality; otherwise, shade the other half-plane.

Inequalities can also be graphed on graphing utilities programmed for this purpose.

EXAMPLE 1 Graphing a linear inequality

Graph the inequality: $2x + y \leq 3$

Solution We'll use the general procedure given above to graph the inequality.

Step 1. We begin by graphing the boundary line $2x + y = 3$. In this case, we use a solid line to indicate that points on the line satisfy the given inequality (Figure 2a).
Step 2. We select a test point, say (4, 1), in the half-plane above the line (Figure 2b). Substituting $x = 4$ and $y = 1$ into the inequality, we get $2(4) + 1 \leq 3$ or $9 \leq 3$, which is false.
Step 3. The graph includes the other half-plane, so we shade the half-plane below the line (Figure 2b).

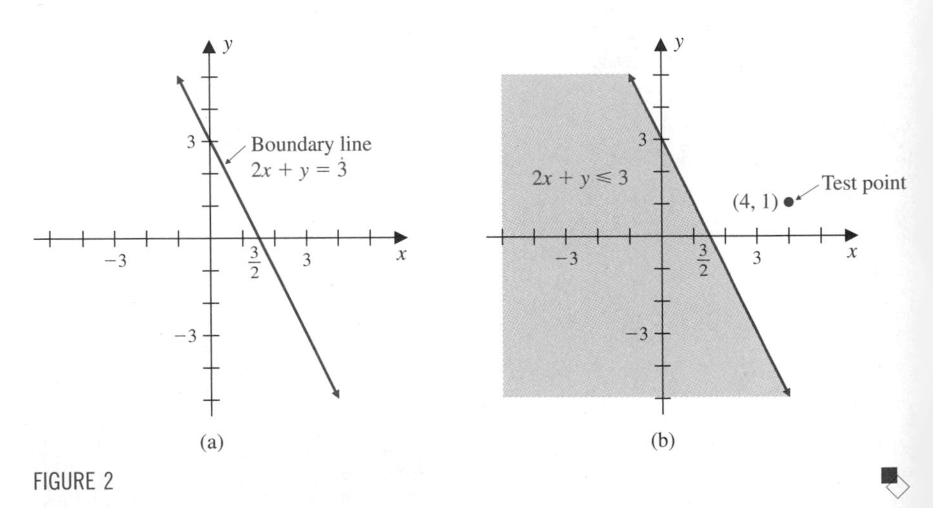

FIGURE 2

Graphing Systems of Inequalities

A *system of inequalities* in two variables is a collection of two or more inequalities in two variables. The **solution** of such a system is the set of all ordered pairs of numbers that satisfy all the inequalities simultaneously. The graph of such a system is obtained by finding the region of points common to the graphs of the individual inequalities in the system.

EXAMPLE 2 **Graphing a linear system of inequalities**

Graph the system of inequalities and identify where the boundary lines intersect.

$$\begin{cases} x + y \leq 5 \\ -x + 2y > 4 \end{cases}$$

Solution Figure 3a shows the graph of the inequality $x + y \leq 5$, while Figure 3b shows the graph of $-x + 2y > 4$. The graph of the system consists of all points common to the two graphs. It is shown as the shaded region in Figure 3c.

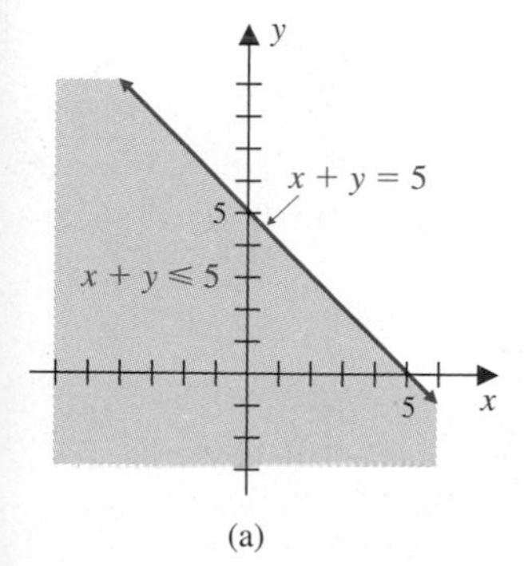

(a)

(b)

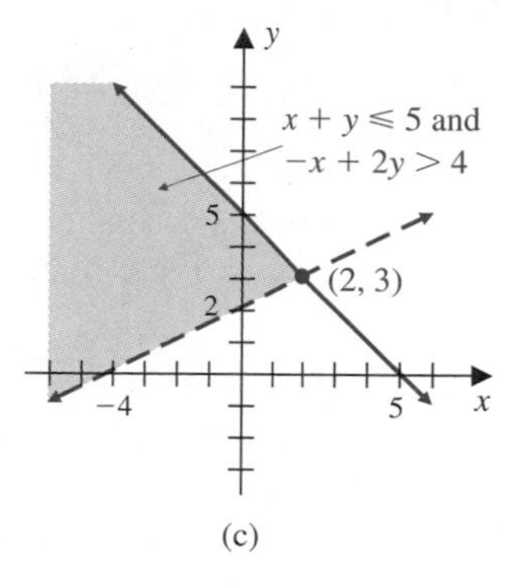

(c)

FIGURE 3

By solving the system of equations

$$\begin{cases} x + y = 5 \\ -x + 2y = 4 \end{cases}$$

Graphing utilities that are programmed to graph inequalitics in two variables can also graph systems of inequalities.

we find that the boundary lines intersect at the point (2, 3). Note that (2, 3) is not in the graph because (2, 3) does not satisfy the given system of inequalities. ■

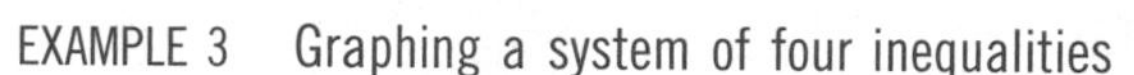

EXAMPLE 3 **Graphing a system of four inequalities**

Graph the system of inequalities and identify the points of intersection of the boundary lines.

$$\begin{cases} -x + 2y \leq 6 \\ x + y \leq 4 \\ x \geq 0 \\ y \geq 0 \end{cases}$$

Solution The graph of the first inequality consists of all points in the plane on or below the line $-x + 2y = 6$. The graph of the second inequality consists of all points in the plane on or below the line $x + y = 4$. The graph of $x \geq 0$ consists of all points on or to the right of the y axis, while the graph of the inequality $y \geq 0$ includes all points in the plane on or above the x axis. The graph of the

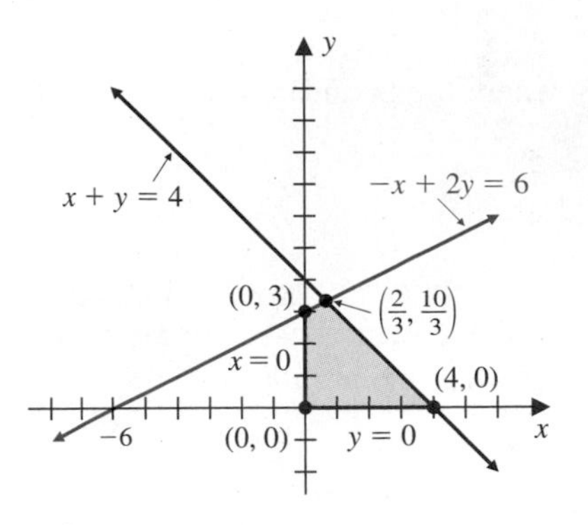

FIGURE 4

system is the shaded plane where all four half-planes intersect (Figure 4). The points of intersection of the lines, which are found by solving the linear systems

$$\begin{cases} x = 0 \\ y = 0 \end{cases} \qquad \begin{cases} x + y = 4 \\ y = 0 \end{cases} \qquad \begin{cases} -x + 2y = 6 \\ x + y = 4 \end{cases} \qquad \begin{cases} -x + 2y = 6 \\ x = 0 \end{cases}$$

are (0, 0), (4, 0), $\left(\frac{2}{3}, \frac{10}{3}\right)$, and (0, 3), respectively. The points of intersection in Figure 4 are called the **vertices** of the region.

Using Linear Programming

Systems of linear inequalities are applied in a field of mathematics called *linear programming*. To introduce this important topic, let's consider an example. Suppose we are given

$$F = 6x + 3y$$

where x and y satisfy the system of four inequalities in Example 3. Now we ask: Is it possible to find the maximum (largest) or the minimum (smallest) value of F under the restrictions imposed on x and y by the given system of inequalities?

In this context, we refer to the inequalities as **constraints**, because they constrain the choice of x and y, and we call the solution of this linear system of inequalities the **feasible region**. Our problem is to choose the point (x, y) of the feasible region so that the expression F has a maximum or minimum value. A problem of this sort is called a **linear programming problem**, and the expression F is called the **objective function**. To determine which ordered pair of the feasible region gives the maximum or minimum value of F, we use the following general procedure:

GENERAL PROCEDURE FOR SOLVING A LINEAR PROGRAMMING PROBLEM

Step 1. Find the feasible region; that is, graph the system of inequalities defined by the constraints.
Step 2. Determine the vertices (corner points) of the feasible region found in step 1.
Step 3. Evaluate the objective function at each vertex of the feasible region.
Step 4. Identify the maximum or minimum value of the objective function at these vertices.

For our illustration, each vertex of the feasible region (see Figure 4), along with the values of F, are listed in Table 1. From the table, we see that the maximum value of F is 24, and it occurs at the point (4, 0). The minimum value is 0, and it occurs at (0, 0).

TABLE 1

Vertex	$F = 6x + 3y$	Value
$(0, 0)$	$F = 6(0) + 3(0)$	0 (Minimum)
$(4, 0)$	$F = 6(4) + 3(0)$	24 (Maximum)
$\left(\frac{2}{3}, \frac{10}{3}\right)$	$F = 6\left(\frac{2}{3}\right) + 3\left(\frac{10}{3}\right)$	14
$(0, 3)$	$F = 6(0) + 3(3)$	9

EXAMPLE 4 **Solving a linear programming problem**

Find the maximum and minimum values of

$$F = 4x + 6y$$

subject to the constraints

$$\begin{cases} x - y \le 4 \\ 2x - y \ge 4 \\ x - 5y \le 2 \\ x \le 5 \end{cases}$$

Solution Figure 5 shows the feasible region. Table 2 lists the systems and corresponding vertices that result from solving pairs of equations. Table 3 shows the values of F at these vertices.

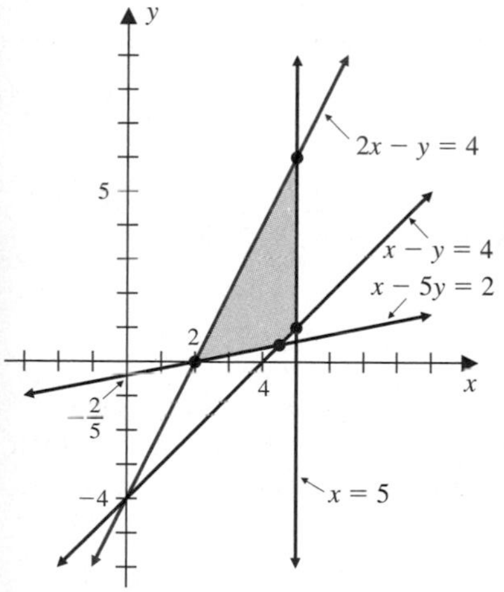

FIGURE 5

TABLE 2

System	Solution
$\begin{cases} x - 5y = 2 \\ 2x - y = 4 \end{cases}$	$(2, 0)$
$\begin{cases} x - 5y = 2 \\ x - y = 4 \end{cases}$	$\left(\frac{9}{2}, \frac{1}{2}\right)$
$\begin{cases} x - y = 4 \\ x = 5 \end{cases}$	$(5, 1)$
$\begin{cases} 2x - y = 4 \\ x = 5 \end{cases}$	$(5, 6)$

TABLE 3

Vertex	$F = 4x + 6y$	Value
$(2, 0)$	$F = 4(2) + 6(0)$	8
$\left(\frac{9}{2}, \frac{1}{2}\right)$	$F = 4\left(\frac{9}{2}\right) + 6\left(\frac{1}{2}\right)$	21
$(5, 1)$	$F = 4(5) + 6(1)$	26
$(5, 6)$	$F = 4(5) + 6(6)$	56

From Table 3, we see that the maximum value of F is 56, and it occurs at $(5, 6)$. The minimum is 8, and it occurs at $(2, 0)$. ■

Solving Applied Problems

Systems of inequalities are used to represent restrictions on the resources available in areas such as business and manufacturing.

EXAMPLE 5 **Modeling a manufacturing problem**

An electronics company manufactures two types of calculators, a scientific calculator and a graphing calculator. In 1 week, the manufacturer can produce no more than 450 calculators in all, and no more than 360 graphing calculators. Suppose that the manufacturer produces at least twice as many graphing calculators as scientific ones.

(a) Find the constraints; that is, form a system of inequalities that represents the restrictions on the number of calculators of each type.
(b) Graph the system and interpret it.

Solution

(a) Let x represent the number of graphing calculators produced in 1 week, and let y represent the number of scientific calculators produced in 1 week. Clearly, $x \geq 0$ and $y \geq 0$. So the system of inequalities that describes the restrictions on x and y is given by

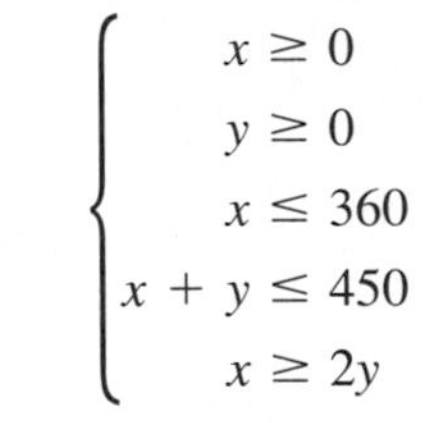

$$\begin{cases} x \geq 0 \\ y \geq 0 \\ x \leq 360 \\ x + y \leq 450 \\ x \geq 2y \end{cases}$$

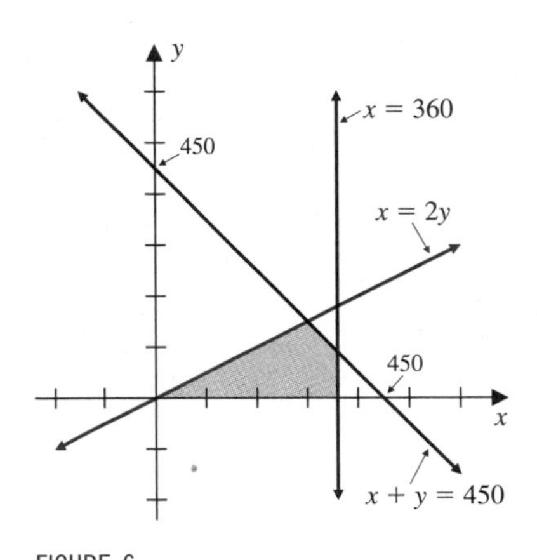

FIGURE 6

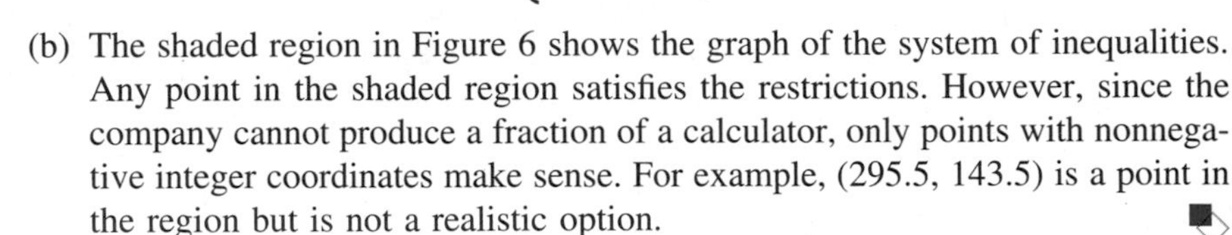

(b) The shaded region in Figure 6 shows the graph of the system of inequalities. Any point in the shaded region satisfies the restrictions. However, since the company cannot produce a fraction of a calculator, only points with nonnegative integer coordinates make sense. For example, (295.5, 143.5) is a point in the region but is not a realistic option.

The technique of linear programming uses systems of linear inequalities as restrictions to model quantities to be maximized or minimized, such as profit or cost.

EXAMPLE 6 Solving a linear programming problem

Suppose that the electronics company in Example 5 sells each graphing calculator for \$85 and each scientific calculator for \$30. Assume it costs the manufacturer \$55 to produce a graphing calculator and \$20 to produce a scientific calculator.

(a) Find the profit function F.
(b) Given the constraints in Example 5, find the number of calculators of each type that would have to be produced and sold in 1 week in order to maximize the profit.

Solution

(a) From Example 5, we are using x to represent the number of graphing calculators produced and sold in 1 week, and y to represent the number of scientific calculators produced and sold. The profit for each graphing calculator is \$85 − \$55 = \$30, and the profit for each scientific calculator is \$30 − \$20 = \$10. The total profit is given by the objective function

$$F = 30x + 10y$$

where x and y are subject to the constraints in Example 5.

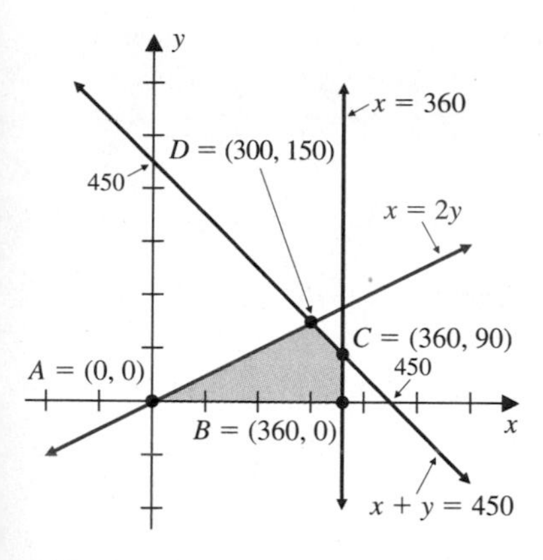

FIGURE 7

(b) Figure 7 shows that the four vertices of the feasible region are $A = (0, 0)$, $B = (360, 0)$, $C = (360, 90)$, and $D = (300, 150)$. Table 4 lists the values of F at each vertex:

TABLE 4

Vertex	$F = 30x + 10y$	Value of F
$A = (0, 0)$	$F = 30(0) + 10(0)$	0
$B = (360, 0)$	$F = 30(360) + 10(0)$	10,800
$C = (360, 90)$	$F = 30(360) + 10(90)$	11,700
$D = (300, 150)$	$F = 30(300) + 10(150)$	10,500

From Table 4, we see that the maximum value of F is 11,700, and it occurs at the point (360, 90). Thus, the maximum profit is \$11,700, and it occurs when 360 graphing calculators and 90 scientific calculators are produced and sold.

PROBLEM SET 8.5

Mastering the Concepts

In problems 1–12, graph each inequality.

1. $y \geq -\frac{1}{2}x$
2. $3y \leq 2x + 1$
3. $2x - 3y < 0$
4. $y > 2 - x$
5. $2x - 3y < 6$
6. $2x + 3y \geq 6$
7. $x - 2y \leq 1$
8. $y + 2x \geq 5$
9. $y \leq -3$ and $x > 4$
10. $y \geq 2$ and $x \leq -1$
11. $2x + 3y < 1$
12. $3x + 4y \geq 1$

In problems 13–16, find the vertices of each shaded region.

13.

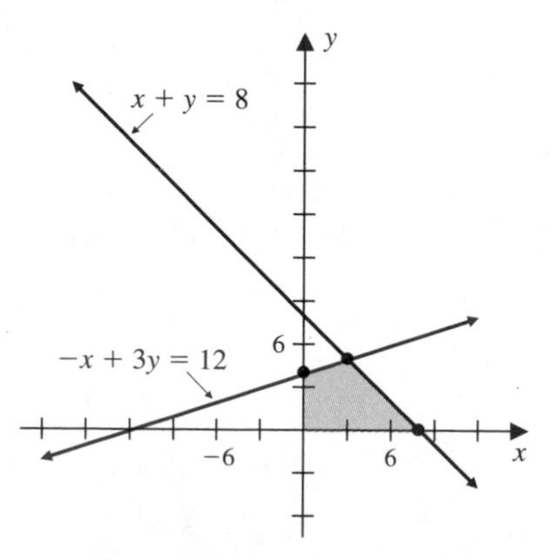

14.

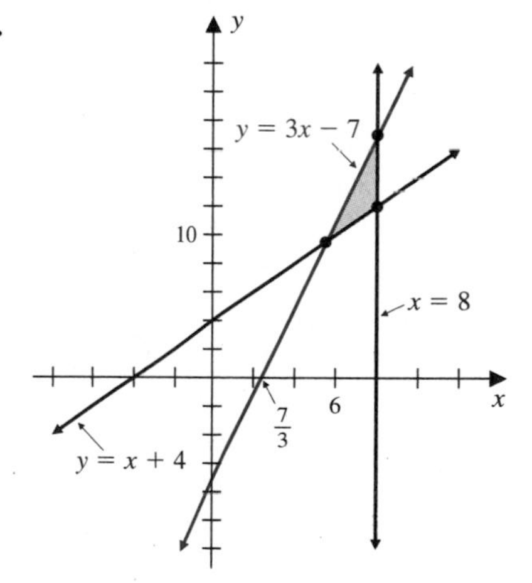

15.

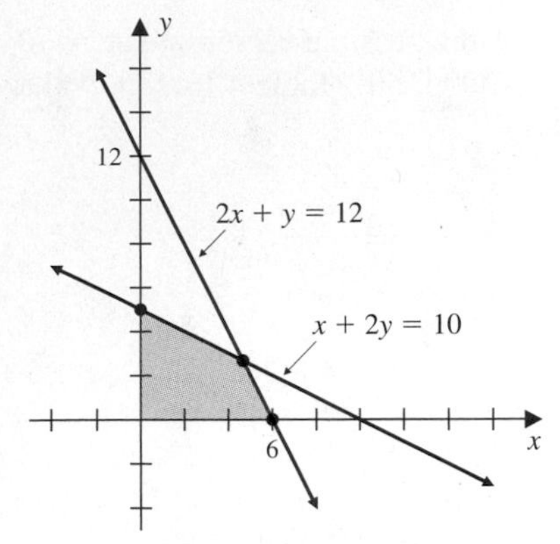

16.

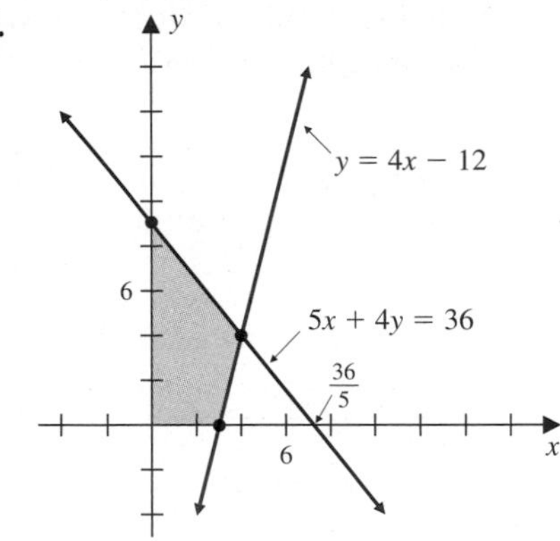

In problems 17–28, graph each system of inequalities, and determine where the boundary lines intersect.

17. $\begin{cases} y \le -5x \\ x \ge 1 \end{cases}$

18. $\begin{cases} y \ge 2x \\ x \le -2 \end{cases}$

19. $\begin{cases} y \ge x \\ y \le 2 \end{cases}$

20. $\begin{cases} y \le x \\ y \ge -1 \end{cases}$

21. $\begin{cases} x + y \le 2 \\ y - 1 > 2x \end{cases}$

22. $\begin{cases} y - x < -1 \\ 3y - x > 4 \end{cases}$

23. $\begin{cases} 5x - 2y < -10 \\ x \ge -4 \end{cases}$

24. $\begin{cases} 2x - y > 0 \\ x - 9y \le 0 \end{cases}$

25. $\begin{cases} y \ge x \\ y \ge -x \end{cases}$

26. $\begin{cases} x + 2y \ge 8 \\ x + 4y \ge 12 \\ x \ge 2 \end{cases}$

27. $\begin{cases} x - y < 4 \\ 2x + y < 12 \\ x \ge 0, y \ge 0 \end{cases}$

28. $\begin{cases} 5x + 2y \ge -10 \\ -2x + 5y \le -10 \\ x \le 1 \end{cases}$

In problems 29–32, find the maximum and minimum value of each objective function subject to the given constraints.

29. $F = 2x + 3y$; a bounded region with vertices (0, 0), (3, 0), (4, 4), (0, 9)

30. $F = 3x + 5y$; a bounded region with vertices (4, 1), (8, 1), (6, 5), (1, 4)

31. $F = 7x + 2y$; the region in Figure 8

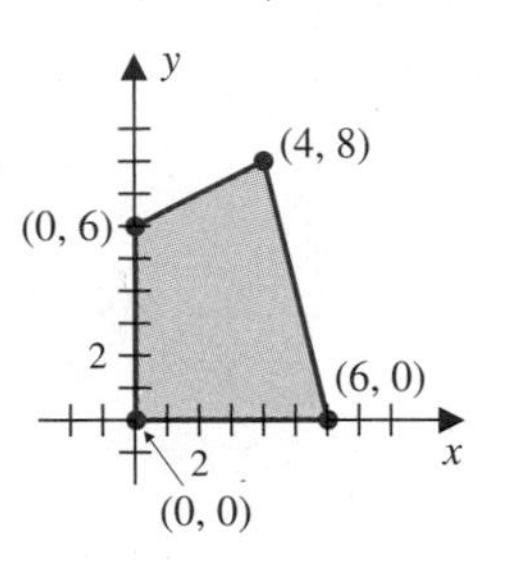

FIGURE 8

32. $F = 4x - 8y$; the region in Figure 9

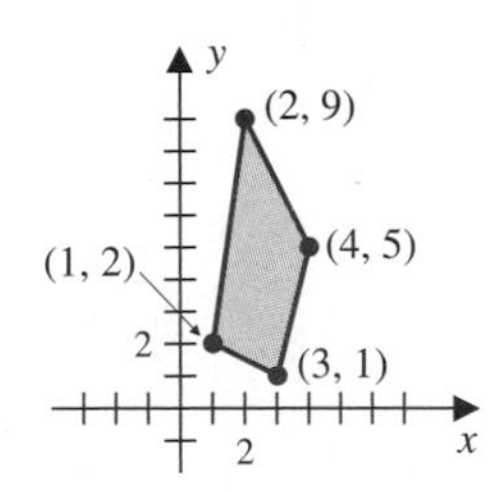

FIGURE 9

In problems 33–40, find the maximum and minimum value of each objective function subject to the given constraints.

33. $F = 3x + 5y$
$\begin{cases} x \ge 0, y \ge 0 \\ 2x + y \le 6 \end{cases}$

34. $F = 2x - y$
$\begin{cases} x + y \ge 3 \\ x \le 3 \\ y \le 3 \end{cases}$

35. $F = 2x + y$
$\begin{cases} x \ge 0, y \ge 0 \\ 4x + y \le 36 \\ 4x + 3y \le 60 \end{cases}$

36. $F = 15x + 25y$
$\begin{cases} x \ge 0, y \ge 0 \\ x + y \le 50 \\ 2x - y \le 40 \\ -3x + y \le 10 \end{cases}$

37. $F = 7x - 3y$

$$\begin{cases} x \geq 0 \\ y \leq 4 \\ x + y \geq 1 \\ x - y \leq 1 \end{cases}$$

38. $F = 5x + 2y$

$$\begin{cases} x \geq 0, y \geq 0 \\ x + 3y \leq 15 \\ 2x + y \leq 10 \end{cases}$$

39. $F = 5x + 4y$

$$\begin{cases} x \geq 0, y \geq 0 \\ x + 2y \geq 3 \\ 2y \leq 5 - x \end{cases}$$

40. $F = x + 2y$

$$\begin{cases} x \geq 0, y \geq 0 \\ 2x - 3y \geq 6 \\ 2x + y \leq 14 \end{cases}$$

Applying the Concepts

41. Investment: A retirement fund uses no more than $200,000 for two investments, one portion at 5% annual simple interest and the rest at 6% annual simple interest. Assume the total interest from both investments for 1 year is at least $10,600.

(a) Write a system of linear inequalities that describes this situation.

(b) Graph the system and interpret it.

42. Retailing: An appliance store stocks washers and dryers. The management discovered that due to demand it is necessary to have at least twice as many washers as dryers. Also, at all times, the store must have at least 10 washers and 5 dryers. Due to limitations in space, the store has room for no more than a total of 30 washers and dryers.

(a) Write a system of inequalities to describe this situation.

(b) Graph the system and interpret it.

43. Nutrition: A nutritionist wishes to determine a formula for the base of an instant breakfast meal. The breakfast must contain at least 12 grams of protein and 8 grams of carbohydrates. A tablespoon of protein powder made from soybeans has 2 grams of protein and 2 grams of carbohydrates. A tablespoon of protein powder made from milk solids has 2 grams of protein and 4 grams of carbohydrates.

(a) Determine the feasible region that describes this situation.

(b) Graph the region and interpret it.

44. Manufacturing: A compact disc manufacturer produces two models of compact disc players, model *A* and model *B*. The company allocates at least 560 hours for production and at least 210 hours for assembly. Suppose that it takes 1.4 hours to produce and 0.4 hour to assemble a unit of model *A*, and it takes 1.3 hours to produce and 0.3 hour to assemble a unit of model *B*.

(a) Determine the feasible region that describes this situation.

(b) Graph the region and interpret it.

45. Maximizing Profit: Because of limited storage capacity, a restaurant owner can order no more than 200 pounds of ground beef per week for making hamburgers and tacos. Each hamburger contains $\frac{1}{3}$ pound of ground beef, while each taco contains $\frac{1}{4}$ pound. The profit is 50¢ on each hamburger and 65¢ on each taco. The labor cost averages 10¢ for each hamburger and 15¢ for each taco. If the owner is willing to pay at most $300 for labor costs, how many tacos and hamburgers must be sold to maximize profit?

46. Forestry: Trees are harvested from a forest by a logging firm that has two kinds of crews. The first crew has 1 driver and 4 loggers. This crew can log 20 trees per day. The second crew has 2 drivers and 6 loggers, and can log 30 trees per day. If the firm employs at most 40 drivers and 150 loggers, how many of each kind of crew would log the maximum number of trees per day?

47. Oil Refinery: A refinery produces a combined maximum of 25,000 barrels of gasoline and diesel oil per day, of which no less than 5000 barrels must be diesel oil. If the profit is $31.50 per barrel of gasoline and $24 per barrel of diesel oil, find the maximum profit and the number of barrels of each product that must be produced to yield this maximum.

48. Purchasing: A manager of a racquetball club must order at least 12 racquets and at most 20 cans of balls, but cannot spend more than $450. At least 34 items must be ordered, but the manager cannot order more cans of balls than racquets. If the racquets cost $18 each and a can of balls costs $4.50 each, how many of each should be ordered if the manager intends to minimize the cost?

49. Manufacturing: A company manufactures two types of electric toothbrushes, one of which is cordless. The cord-type toothbrush requires 2 hours to make, and the cordless model requires 3 hours. The company has only 800 work-hours to use in manufacturing each day, and the packing department can package only 300 toothbrushes per day. If the company sells the cord-type model for $15 and the cordless model for $22, how many of each type should it produce per day to maximize its revenue?

50. Budgeting: A grower has 70 acres of planting fields in which to grow flowers and eucalyptus. It costs the grower \$60 per acre to grow flowers and \$30 per acre to grow eucalyptus. The budget for planting is \$1800. It takes 3 days to plant an acre of flowers and 4 days to plant an acre of eucalyptus. The grower has a maximum of 120 days to plant the flowers. The flowers will bring a profit of \$180 per acre, and the eucalyptus will bring a profit of \$100 per acre. How many acres of each crop should be planted to maximize profit?

Developing and Extending the Concepts

51. Graph the inequality:

$$|x| - |y| \geq 1$$

52. Given a linear programming problem with the objective function

$$F = 5x + 2y$$

and constraints

$$\begin{cases} 4x + 3y \leq 15 \\ x - y \geq -6 \\ x \geq 0,\ y \geq 0 \end{cases}$$

explain which constraints have an effect on the feasible region.

In problems 53 and 54, write a system of inequalities for each solution set shown.

53. **54.**

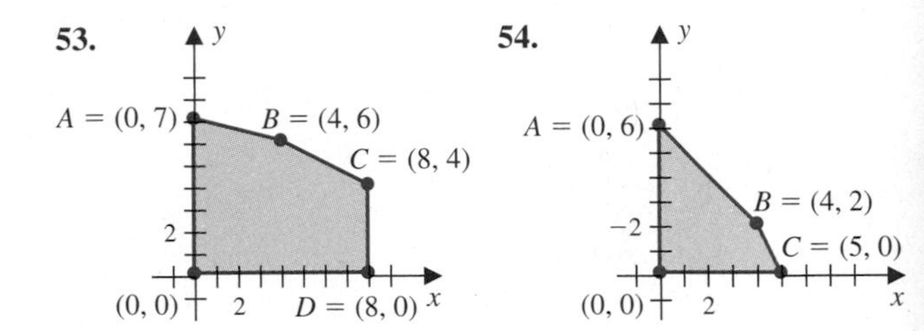

OBJECTIVES

1. Solve nonlinear systems of equations algebraically.
2. Solve nonlinear systems of equations graphically.
3. Solve nonlinear systems of inequalities.
4. Solve applied problems.

8.6 Solutions of Nonlinear Systems of Equations and Inequalities

So far, the systems of equations and inequalities we have solved have been linear. In this section, we consider the solution of systems in which at least one of the equations or inequalities is nonlinear. Some examples are shown below:

$$\begin{cases} x^2 + y = 6 \\ 2x - 5y = -6 \end{cases} \qquad \begin{cases} x + y \leq 16 \\ x^2 + y \geq 4 \end{cases} \qquad \begin{cases} \log(x - 2) + y = 3 \\ \log\left(\dfrac{5x}{6}\right) - y = -1 \end{cases}$$

To solve these systems, we use algebraic methods and graphs.

Solving Nonlinear Systems of Equations Algebraically

At times, we are able to solve nonlinear systems of equations by using various algebraic techniques such as substitution and elimination, as we did when solving linear systems.

EXAMPLE 1 **Solving a nonlinear system by substitution**

Solve the system below by using substitution.

$$\begin{cases} x^2 + y = 6 \\ 2x - 5y = -6 \end{cases}$$

Solution We begin by solving the first equation for y to get $y = 6 - x^2$. Since we are trying to find values for y that satisfy both equations simultaneously, we assume that y is the same in both equations. So we substitute the expression $6 - x^2$ for y in the second equation, $2x - 5y = -6$, to get

$$\left.\begin{aligned} 2x - 5(6 - x^2) &= -6 \\ 2x - 30 + 5x^2 &= -6 \\ 5x^2 + 2x - 24 &= 0 \end{aligned}\right\} \quad \text{Simplify}$$

$$(x - 2)(5x + 12) = 0 \qquad \text{Factor}$$

$$\left.\begin{array}{ccc} x - 2 = 0 & \text{or} & 5x + 12 = 0 \\ x = 2 & | & x = -\frac{12}{5} \end{array}\right\} \quad \text{Solve}$$

If we substitute each of these values of x into the equation

$$2x - 5y = -6 \qquad \text{or} \qquad y = \tfrac{2}{5}x + \tfrac{6}{5}$$

we obtain the two corresponding values for y. When $x = 2$,

$$y = \tfrac{2}{5}(2) + \tfrac{6}{5} = 2$$

When $x = -\frac{12}{5}$,

$$y = \tfrac{2}{5}\left(-\tfrac{12}{5}\right) + \tfrac{6}{5} = \tfrac{6}{25}$$

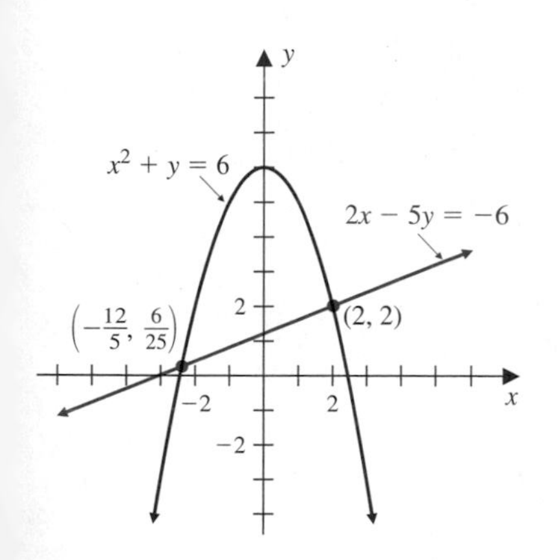

FIGURE 1

Thus, the solutions are (2, 2) and $\left(-\frac{12}{5}, \frac{6}{25}\right)$. Note that the graphs of the two equations, $x^2 + y = 6$ (a parabola) and $2x - 5y = -6$ (a line), show that there are two points of intersection (Figure 1). The coordinates of these two points are (2, 2) and $\left(-\frac{12}{5}, \frac{6}{25}\right)$. ■

When solving a nonlinear system of equations, we may encounter extraneous solutions. This is where graphing the system may help.

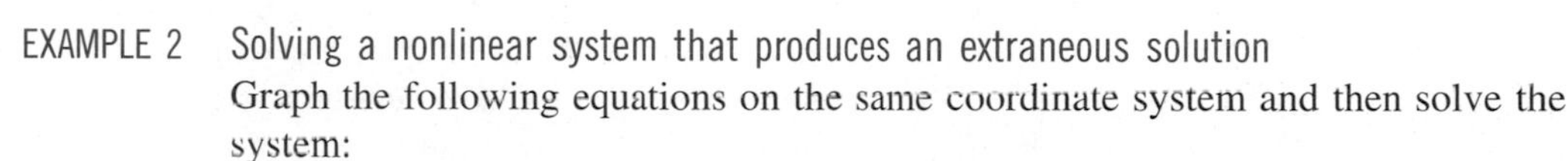

EXAMPLE 2 Solving a nonlinear system that produces an extraneous solution

Graph the following equations on the same coordinate system and then solve the system:

$$\begin{cases} x^2 + y^2 = 13 \\ x + y = 5 \end{cases}$$

Solution Figure 2 shows that the circle $x^2 + y^2 = 13$ and the line $x + y = 5$ intersect at two points, so there are two solutions. To solve this system by substitution, we first solve the second equation, $x + y = 5$, for y in terms of x to get $y = 5 - x$. Then we substitute $5 - x$ for y in the first equation, $x^2 + y^2 = 13$, to get

$$\left.\begin{aligned} x^2 + (5 - x)^2 &= 13 \\ x^2 + 25 - 10x + x^2 &= 13 \\ 2x^2 - 10x + 12 &= 0 \\ x^2 - 5x + 6 &= 0 \end{aligned}\right\} \quad \text{Simplify}$$

$$(x - 2)(x - 3) = 0 \qquad \text{Factor}$$

$$\left.\begin{array}{ccc} x - 2 = 0 & \text{or} & x - 3 = 0 \\ x = 2 & | & x = 3 \end{array}\right\} \quad \text{Solve}$$

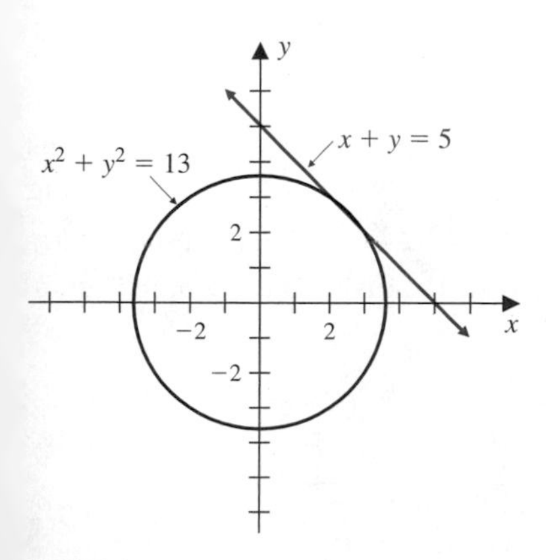

FIGURE 2

If we substitute each of these values of x into the equation $y = 5 - x$, we obtain the corresponding values for y. When $x = 2$,

$$y = 5 - x = 5 - 2 = 3$$

When $x = 3$,

$$y = 5 - x = 5 - 3 = 2$$

Therefore, the solutions of the system are (2, 3) and (3, 2). These ordered pairs satisfy both equations.

Let us see what happens if we substitute each of the x values, $x = 2$ and $x = 3$, into the first equation, $x^2 + y^2 = 13$, instead of the second equation:

When $x = 2$:	When $x = 3$:
$4 + y^2 = 13$	$9 + y^2 = 13$
$y^2 = 9$	$y^2 = 4$
$y = \pm 3$	$y = \pm 2$

This suggests that there are four solutions: (2, 3), (2, −3), (3, 2), (3, −2). If we substitute the two suggested additional solutions, (2, −3) and (3, −2), into the second equation, we have

$x + y = 5$	$x + y = 5$
$2 + (-3) = 5$	$3 - 2 = 5$
$-1 = 5$ False	$1 = 5$ False

Thus, neither (2, −3) nor (3, −2) is a solution of the original system, a fact confirmed by looking at Figure 2. The pairs (2, −3) and (3, −2) are extraneous solutions.

We now solve a nonlinear system of equations by using elimination. A graph of the system is used to help determine how many points of intersection—and thus how many solutions—there are.

EXAMPLE 3 Solving a nonlinear system by elimination

Graph both equations on the same coordinate system and use elimination to solve the system.

$$\begin{cases} x^2 + y^2 = 16 \\ x^2 + y = 4 \end{cases}$$

Solution Figure 3 shows that the two curves $x^2 + y = 4$ and $x^2 + y^2 = 16$ intersect at three points, so there are three solutions. To solve the system, we begin by eliminating the x^2 term as follows:

$$\begin{cases} x^2 + y^2 = 16 \\ x^2 + y = 4 \end{cases} \longrightarrow \begin{cases} x^2 + y^2 = 16 \\ -x^2 - y = -4 \end{cases} \longrightarrow \begin{cases} y^2 - y = 12 \\ -x^2 - y = -4 \end{cases}$$

Multiply the second equation by −1

Add the second equation to the first equation

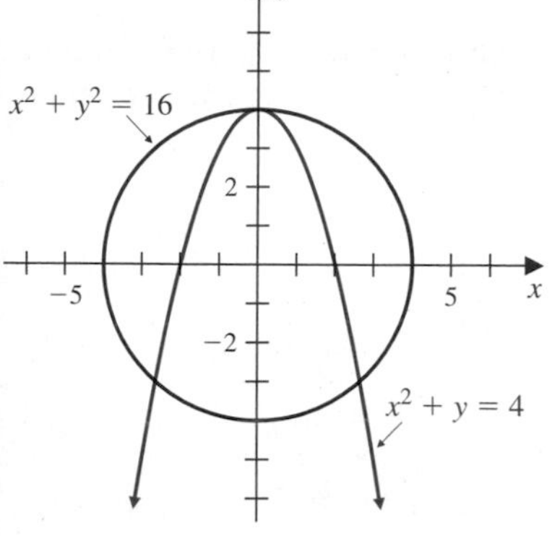

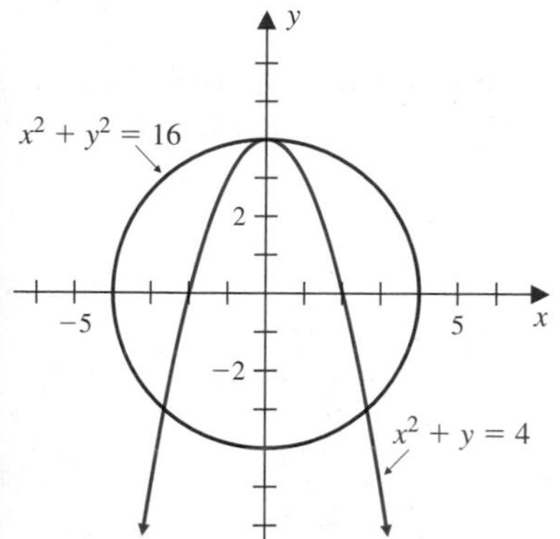

FIGURE 3

Note that we have obtained a total of three solutions, which were anticipated from the graph of the system. Therefore, there is no need to check for extraneous solutions.

Next, we solve the first equation $y^2 - y = 12$:

$$
\begin{aligned}
y^2 - y &= 12 && \text{Given} \\
y^2 - y - 12 &= 0 && \text{Subtract 12 from each side} \\
(y - 4)(y + 3) &= 0 && \text{Factor}
\end{aligned}
$$

$$
\begin{array}{ccc|l}
y - 4 = 0 & \text{or} & y + 3 = 0 & \text{Solve} \\
y = 4 & | & y = -3 &
\end{array}
$$

Now we substitute 4 and -3 for y into the equation $x^2 + y = 4$ to solve for x:

For $y = 4$:	For $y = -3$:
$x^2 + 4 = 4$	$x^2 - 3 = 4$
$x^2 = 0$	$x^2 = 7$
$x = 0$	$x = \pm\sqrt{7}$

Thus, $(0, 4)$, $\left(-\sqrt{7}, -3\right)$. and $\left(\sqrt{7}, -3\right)$ are the three solutions. This is consistent with what was anticipated from the graphs shown in Figure 3. ■

A nonlinear system also may involve logarithmic, exponential, or trigonometric functions. Sometimes these systems can be solved by elimination or substitution.

EXAMPLE 4 GU

Solving a nonlinear system involving exponentials

Graph both equations in the same coordinate system and then solve the system.

$$
\begin{cases}
2^x - y = 0 \\
2^{2x} - y = 2
\end{cases}
$$

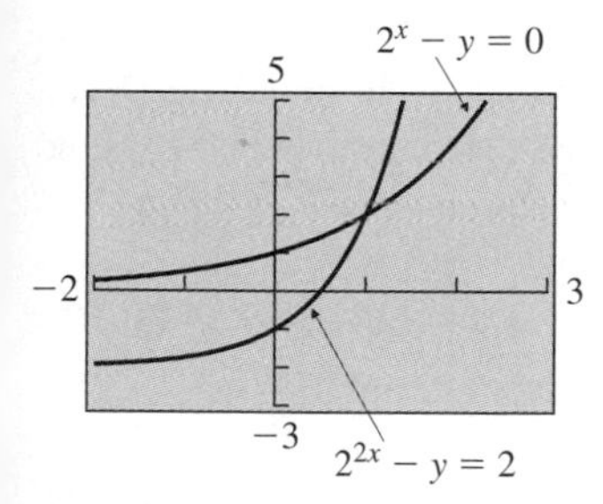

FIGURE 4

Solution For convenience, we use a graphing utility to graph the two equations in the same viewing window (Figure 4). From the graphs, we see that there is only one point of intersection and so there is one solution.

To solve this system algebraically, we could use either substitution or elimination. Here we use substitution. From the first equation, we have $y = 2^x$. Rewriting the second equation in the form

$$y = (2^x)^2 - 2$$

and replacing 2^x with y (from the first equation), we get

$$
\begin{aligned}
y &= y^2 - 2 \\
y^2 - y - 2 &= 0 \\
(y - 2)(y + 1) &= 0
\end{aligned}
$$

$$
\begin{array}{ccc}
y - 2 = 0 & \text{or} & y + 1 = 0 \\
y = 2 & | & y = -1
\end{array}
$$

We reject the solution -1, since $y = 2^x$ cannot equal -1. Also, the graph in Figure 4 clearly shows that the y value of the solution is positive. On the other hand, if $y = 2$, then $y = 2^x = 2$, so $x = 1$. Therefore, $(1, 2)$ is the solution. ■

Solving Nonlinear Systems of Equations Graphically

When solving nonlinear systems of equations, the algebraic approach does not always work. In this case, a graphing utility can help us approximate the solutions.

EXAMPLE 5 **Solving a nonlinear system graphically**

Use a graphing utility to solve the given system.

$$\begin{cases} y = 4 - x^2 \\ y = 3 \sin x \end{cases}$$

Approximate the results to two decimal places.

Solution Figure 5 shows that the graphs of the two equations intersect at two points. Using the ZOOM and TRACE features, we find the solutions to be approximately $(-2.44, -1.94)$ and $(1.13, 2.72)$.

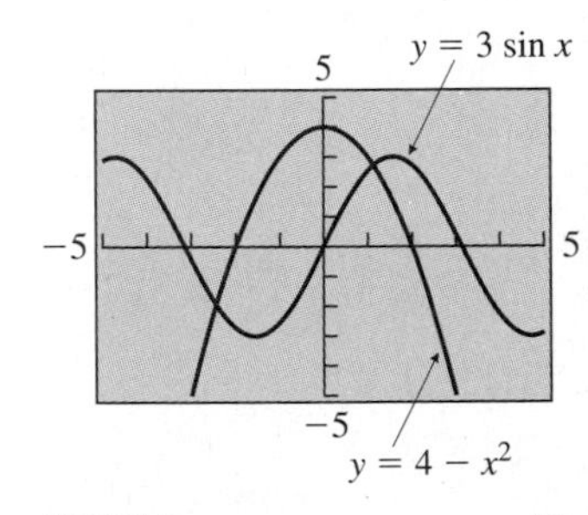

FIGURE 5

Solving Nonlinear Systems of Inequalities

In Section 8.5, we graphed a linear inequality by first using the graph of the associated linear equation as a boundary line and then testing a point in either of the resulting half-planes to determine which one represents the solution. A similar technique can be used to graph nonlinear inequalities. For instance, let us consider the inequality

$$x^2 + y^2 \leq 4$$

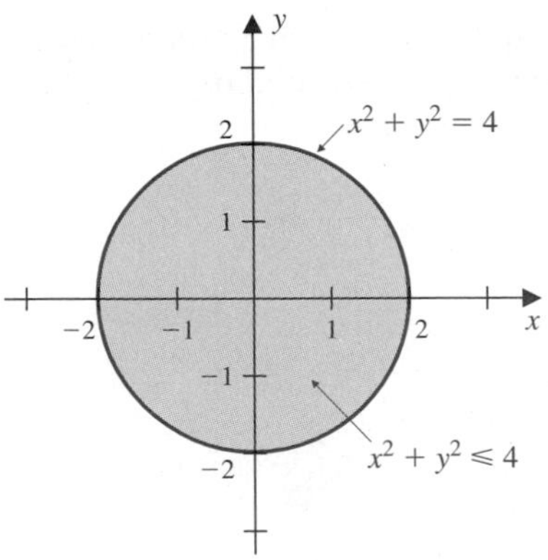

Figure 6 shows the graph of the associated equation $x^2 + y^2 = 4$ as a solid curve, because of the $\leq$ condition. This graph serves as a boundary curve that separates the plane into two regions, one region inside the circle and the other one outside. To test the inside region, we select a point, say (0, 0). Since $0^2 + 0^2 = 0 \leq 4$ is true, this point and all others inside the circle are part of the graph of $x^2 + y^2 \leq 4$. Figure 6 shows the graph, which is the shaded circular region, including the circle.

FIGURE 6

If we had selected (3, 3) as the test point, we would have concluded that the outside region is not part of the solution and so the inside region is the solution.

Also in Section 8.5, we found the solution of a system of linear inequalities with two variables by first graphing each of the inequalities in the same coordinate system. The points common to all graphs define the solution of the system. The same approach is used to graph the solution of nonlinear systems of inequalities with two variables.

EXAMPLE 6 **Graphing a nonlinear system of inequalities**

Graph the system of inequalities: $\begin{cases} y > x^2/4 \\ x + y \leq 4 \end{cases}$

Solution We begin by graphing the solution of each inequality and then determine where those regions overlap. Figure 7a shows the graph of the inequality $y > x^2/4$ as the shaded region above the curve $y = x^2/4$. Figure 7b shows the

As with linear systems, some graphing utilities are programmed to graph nonlinear systems of inequalities.

graph of the inequality $x + y \leq 4$ as the shaded region on and below the line $x + y = 4$. Thus, the graph of the solution set of the system of inequalities consists of points common to both regions, that is, the shaded region in Figure 7c.

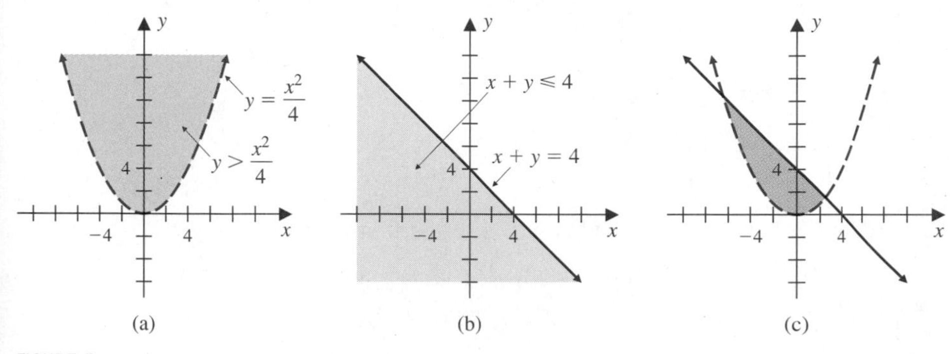

FIGURE 7

Solving Applied Problems

We conclude this section with an applied problem that involves a system of nonlinear equations.

EXAMPLE 7 Determining the center of an earthquake

A seismologist determines that the center of an earthquake is located on a circle 30 miles away from a certain station. From a second station located 40 miles east and 10 miles north of the first, it is determined that the center of the earthquake is on a circle 20 miles away. Find the center of the earthquake, assuming that it is north of both stations. Round off the result to one decimal place.

Solution First we use a coordinate system to model the situation, as shown in Figure 8. Let the first station be located at the origin. Then the second station is located at the point (40, 10). The equation of the circle with center at the origin and radius 30 is $x^2 + y^2 = 30^2$, and the equation of the circle with center at (40, 10) and radius 20 is given by $(x - 40)^2 + (y - 10)^2 = 20^2$.

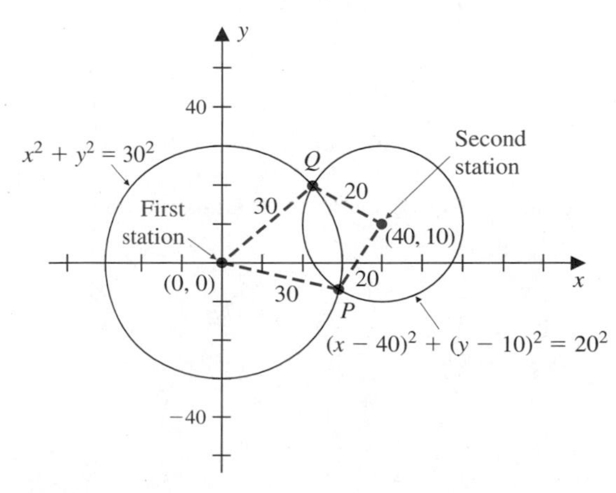

FIGURE 8

To determine where the center of the earthquake is located, we need to find the points of intersection of the two circles. Figure 8 suggests that there are two possible locations of the earthquake's center. We need to solve the system below to find these locations:

$$\begin{cases} (x-40)^2 + (y-10)^2 = 20^2 \\ x^2 + y^2 = 30^2 \end{cases}$$

Rewriting these equations, we have

$$\begin{cases} x^2 - 80x + 1600 + y^2 - 20y + 100 = 400 \\ x^2 + y^2 = 900 \end{cases}$$

Adding -1 times the second equation to the first equation, we get

$$\left.\begin{aligned} -80x + 1600 - 20y + 100 &= -500 \\ -80x + 2200 &= 20y \\ y &= 110 - 4x \end{aligned}\right\} \text{Simplify}$$

Substituting $110 - 4x$ for y into the equation $x^2 + y^2 = 900$, we have

$$\begin{aligned} x^2 + (110 - 4x)^2 &= 900 \\ 17x^2 - 880x + 11{,}200 &= 0 \end{aligned}$$

Using the quadratic formula, we get $x = 29.2$ or $x = 22.6$. For $x = 29.2$, $y = 110 - 4(29.2) = -6.8$. For $x = 22.6$, $y = 110 - 4(22.6) = 19.6$. We reject the solution $(29.2, -6.8)$ because the center of the earthquake is located north of both stations. Thus, the center is located 22.6 miles east and 19.6 miles north of the first station.

PROBLEM SET 8.6

Mastering the Concepts

In problems 1–6, sketch the graphs of both equations in the system in the same coordinate system. Use the result to determine the number of solutions. Then use substitution to solve the system.

1. $\begin{cases} x^2 - 2y = 0 \\ 3x + 2y = 10 \end{cases}$

2. $\begin{cases} x^2 - y = -3 \\ x^2 + y^2 = 9 \end{cases}$

3. $\begin{cases} x - 5y = 0 \\ x^2 + y^2 = 1274 \end{cases}$

4. $\begin{cases} x - 2y = 3 \\ x^2 - 40 = -y^2 \end{cases}$

5. $\begin{cases} 2x + y = 10 \\ xy = 12 \end{cases}$

6. $\begin{cases} y - 2x = 3 \\ x^2 + y^2 = 16 \end{cases}$

In problems 7–12, sketch the graphs of both equations in the system in the same coordinate system. Use the result to determine the number of solutions. Then use the elimination method to solve the system. Round off to two decimal places when necessary.

7. $\begin{cases} x^2 + y = 13 \\ x^2 + y^2 = 25 \end{cases}$

8. $\begin{cases} x^2 + y = 1 \\ y = x^2 - 1 \end{cases}$

9. $\begin{cases} 3x - y^2 = -1 \\ x^2 + y^2 = 5 \end{cases}$

10. $\begin{cases} 2x + y = 1 \\ x^2 + y = 4 \end{cases}$

11. $\begin{cases} 3x + 4y = 12 \\ x^2 - y = -1 \end{cases}$

12. $\begin{cases} x - 2y^2 = 0 \\ x^2 + y^2 = 1 \end{cases}$

In problems 13–20, use the graphs shown for each system to determine the number of real solutions. Then solve each system by an appropriate algebraic method. Round off to two decimal places when necessary.

13. $\begin{cases} x^2 - 25y^2 = 20 \\ 2x^2 + 25y^2 = 88 \end{cases}$

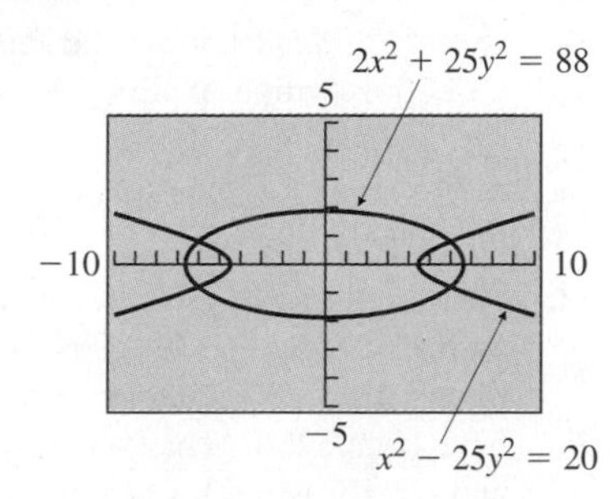

14. $\begin{cases} 3x^2 - 8y^2 = 40 \\ 5x^2 + y^2 = 81 \end{cases}$

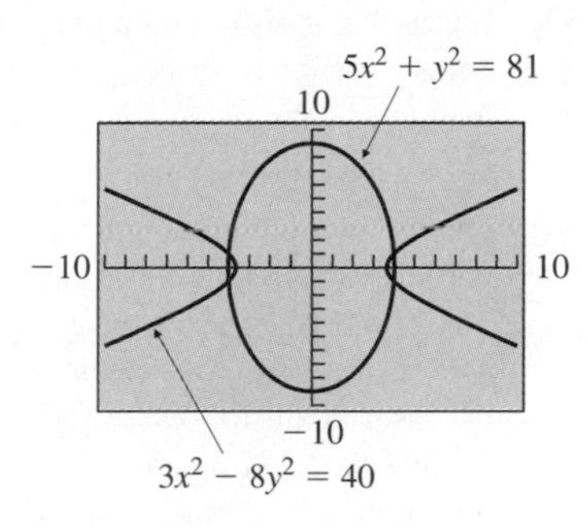

15. $\begin{cases} 2x^2 - 3y^2 = 6 \\ 3x^2 + 2y^2 = 35 \end{cases}$

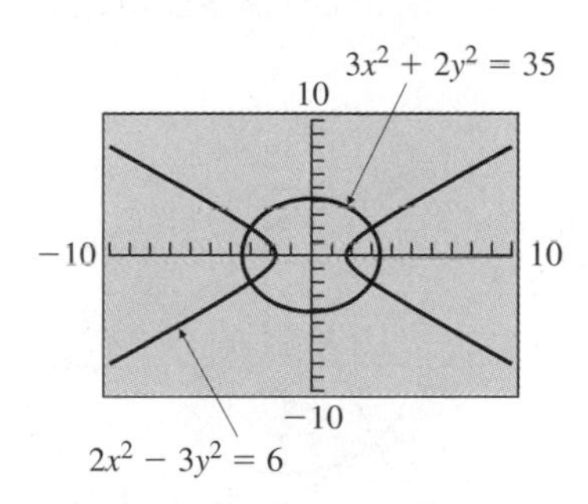

16. $\begin{cases} x^2 - y^2 = 7 \\ x^2 + y^2 = 25 \end{cases}$

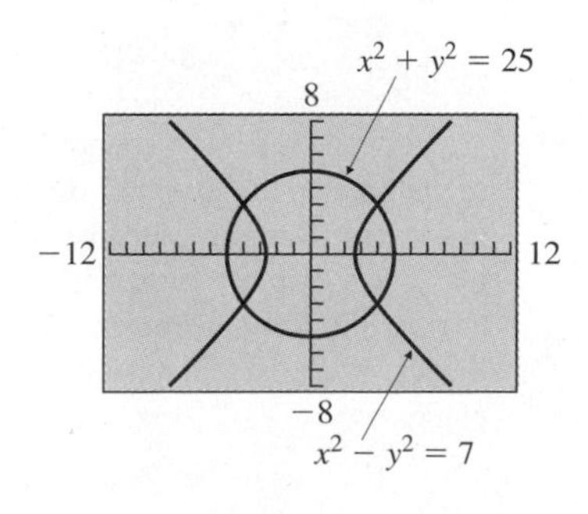

17. $\begin{cases} 6x + 2y = 1 \\ 3x^2 - y^2 = -4 \end{cases}$

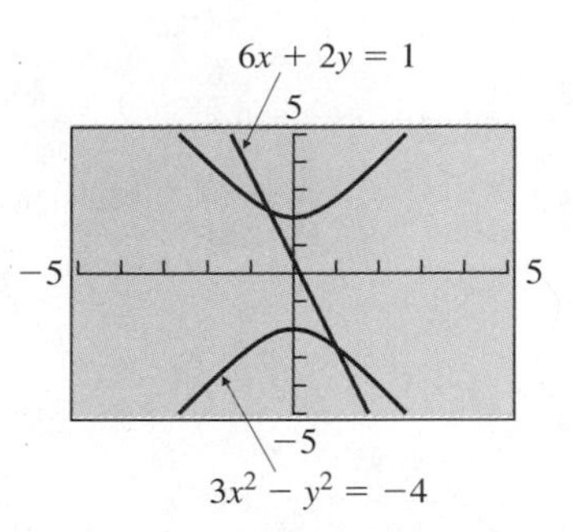

18. $\begin{cases} 5x - 3y = 10 \\ x^2 - y^2 = 4 \end{cases}$

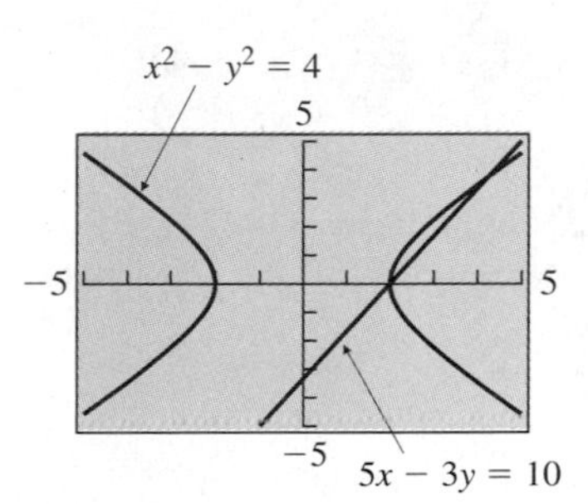

19. $\begin{cases} 2x + 3y = 7 \\ x^2 + y^2 - 4y = 8 \end{cases}$

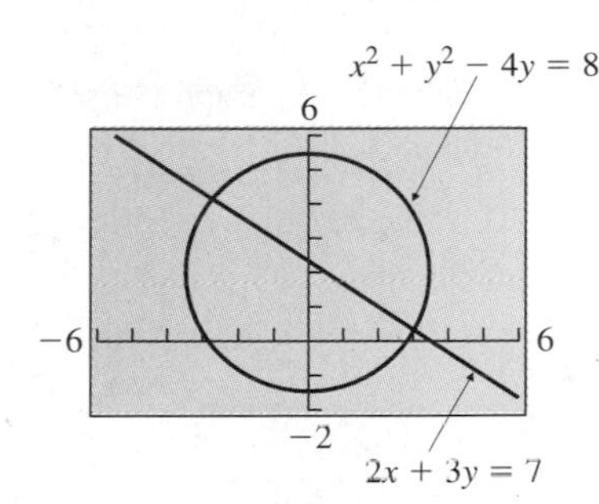

20. $\begin{cases} x - y + 1 = 0 \\ x^2 + 3y^2 = 12 \end{cases}$

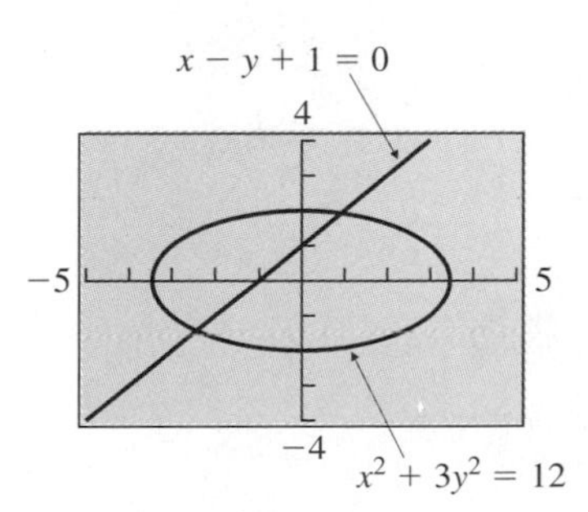

In problems 21–34, solve each system.

21. $\begin{cases} u^2 + 9v^2 = 33 \\ u^2 + v^2 = 25 \end{cases}$

22. $\begin{cases} x^2 + 5y^2 = 70 \\ 3x^2 - 5y^2 = 30 \end{cases}$

23. $\begin{cases} 4x^2 - y^2 = 4 \\ 4x^2 + \frac{5}{3}y^2 = 36 \end{cases}$

24. $\begin{cases} r^2 - 2s^2 = 17 \\ 2r^2 + s^2 = 54 \end{cases}$

25. $\begin{cases} 2x^2 - 3y^2 = 20 \\ x^2 + 2y = 20 \end{cases}$

26. $\begin{cases} 4x^2 + 3y^2 = 43 \\ 3x^2 - y^2 = 3 \end{cases}$

27. $\begin{cases} x^2 - 2y^2 = 1 \\ x^2 + 4y^2 = 25 \end{cases}$

28. $\begin{cases} 2x^2 - 5y + 8 = 0 \\ x^2 - 7y^2 + 4 = 0 \end{cases}$

29. $\begin{cases} x^2 + 4y = 8 \\ x^2 + y^2 = 5 \end{cases}$

30. $\begin{cases} 4x^2 + 7y^2 = 32 \\ -3x^2 + 11y^2 = 41 \end{cases}$

31. $\begin{cases} x^2 + y^2 = 16 \\ x^2 - y^2 = -34 \end{cases}$

32. $\begin{cases} x^2 - 4y^2 = -15 \\ -x^2 + 3y^2 = 11 \end{cases}$

33. $\begin{cases} x^2 + y^2 = 25 \\ (x - 5)^2 + y^2 = 9 \end{cases}$

34. $\begin{cases} x^2 - y = 0 \\ x^2 + (y - 6)^2 = 36 \end{cases}$

GU In problems 35–38, use a graphing utility to solve each system. Round off the answers to two decimal places.

35. $\begin{cases} y = 2 - x^2 \\ y = \cos x \end{cases}$

36. $\begin{cases} y = e^{2x} \\ y = 1 - x \end{cases}$

37. $\begin{cases} 2^x + y = 3 \\ y = \cos x \end{cases}$

38. $\begin{cases} y = x - 4 \\ y = x - 3^x \end{cases}$

In problems 39–48, graph the solution of each system of inequalities.

39. $\begin{cases} -x^2 + y \geq 0 \\ x + y < 1 \end{cases}$

40. $\begin{cases} \sqrt{x} - y > 0 \\ x - 9y \leq 0 \end{cases}$

41. $\begin{cases} x^2 + y \leq 0 \\ x + y > -2 \end{cases}$

42. $\begin{cases} y > x - 3 \\ y \leq \sqrt{x - 1} \\ y \geq 0 \end{cases}$

43. $\begin{cases} x \geq 0,\ y \geq 0 \\ y \leq 4 - x^2 \end{cases}$

44. $\begin{cases} x^2 + y^2 \leq 1 \\ x + y > 1 \end{cases}$

45. $\begin{cases} 4x^2 - y^2 \geq 0 \\ x + y < 9 \end{cases}$

46. $\begin{cases} y \geq (x - 1)^2 + 2 \\ 2x - 3y \leq -9 \end{cases}$

47. $\begin{cases} x^2 + y^2 \leq 9 \\ y - x^2 \leq 0 \end{cases}$

48. $\begin{cases} x^2 + y^2 > 4 \\ x^2 + y^2 < 9 \end{cases}$

Applying the Concepts

In problems 49–56, use a system of equations to solve each problem.

49. Computer Monitor Design: An electronics engineer designs a rectangular computer monitor screen with a 10 inch diagonal and a viewing area of 48 square inches. Find the dimensions of the screen.

50. Template Design: An artist designs a plastic template in the shape of a right triangle with one base length of 60 centimeters and an area of 1500 square centimeters. Find the lengths of the sides of the triangular template.

51. Fencing: A rancher wishes to fence two adjacent rectangular corrals with a total of 100 meters of fence. What are the dimensions of the smallest possible corral, if one corral is a square region and is 144 square meters more than the other?

52. Investment Portfolio: The annual simple interest earned on an investment is \$170. If the interest rate had been 1% higher, the interest earned from this investment would have been \$238. What was the amount of the investment, and what was the lower interest rate?

53. Engineering Design: An engineer designs a rectangular solar collector with a surface area of 750 square feet. If the length of the collector is 30 times its width, find its length and width.

54. Science: In an experiment with a laser beam, the path of a particle orbiting a central object is given by the equation $36x^2 + 49y^2 = 1764$, where x and y are measured (in centimeters) from the center of the object. Suppose that the laser beam follows a path along the line $y = 2x + 6$ (Figure 9). Find the coordinates of the points at which the laser will illuminate the particle. That is, find the points of intersection when the particle will pass through the beam.

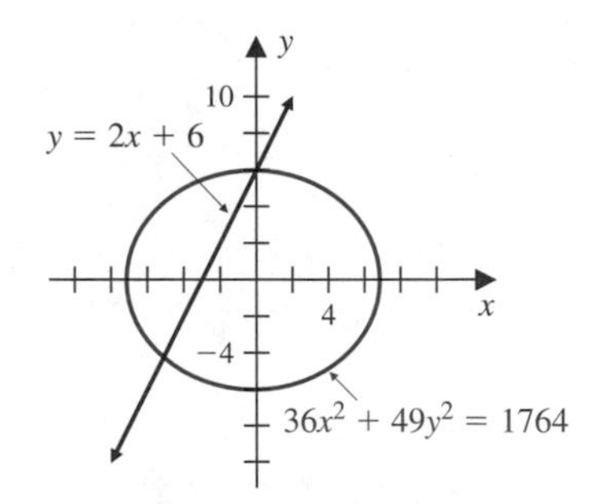

FIGURE 9

55. Construction: A 2 inch square is cut off at each corner of a rectangular piece of cardboard, and an open-topped box is formed by turning up the sides (Figure 10). Suppose that the resulting volume is 448 cubic inches, and the area of the original rectangular cardboard is 360 square inches. Find the dimensions of the original rectangular piece.

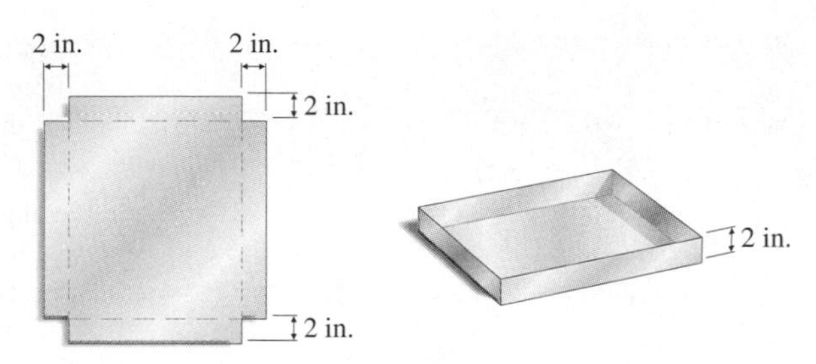

FIGURE 10

56. Angles of Elevation of a Balloon: A hot-air balloon was sighted from two locations, A and B (Figure 11). It was determined that α and β, the two acute angles of elevation of the balloon from the two locations A and B, could be found by solving the system in which the sum of the sines of the two angles is 1 and the sum of the cosines of the two angles is 1.5. Find the degree measures of α and β, rounded off to two decimal places.

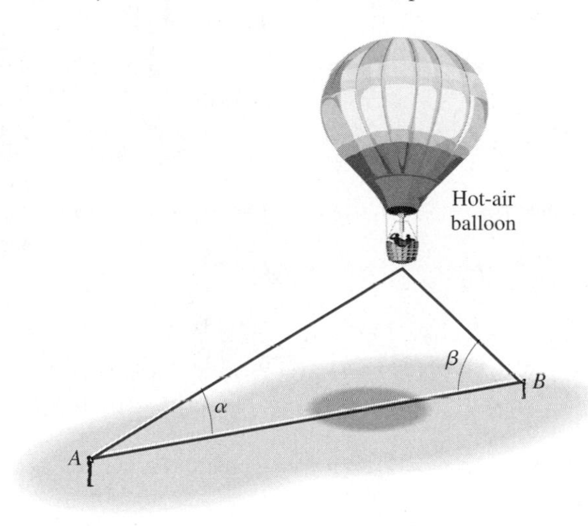

FIGURE 11

Developing and Extending the Concepts

In problems 57–62, solve each system. [*Hint*: In problem 57, let $u = 1/x^2$ and $v = 1/y^2$. Similar substitutions may be helpful in the remaining problems.]

57. $\begin{cases} \dfrac{3}{x^2} + \dfrac{2}{y^2} = 17 \\ \dfrac{4}{x^2} - \dfrac{5}{y^2} = -8 \end{cases}$

58. $\begin{cases} \dfrac{1}{x^3} - \dfrac{1}{y^3} = 63 \\ \dfrac{1}{x} - \dfrac{1}{y} = 3 \end{cases}$

59. $\begin{cases} \sqrt[3]{x^4} + \sqrt[5]{y^2} = 20 \\ \sqrt[3]{x^2} + \sqrt[5]{y} = 6 \end{cases}$

60. $\begin{cases} x + y = 72 \\ \sqrt[3]{x} + \sqrt[3]{y} = 6 \end{cases}$

61. $\begin{cases} 2x + y = 1 \\ y + z = 1 \\ x^2 + z^2 = 5 \end{cases}$

62. $\begin{cases} x^2 - 3y + 4z^2 = 31 \\ 3x + y = 5 \\ 2x - 3z = -7 \end{cases}$

63. Solve the system: $\begin{cases} x_1 - x_2 + x_3 = 2 \\ 2x_2 + x_3 = 1 \\ x_1x_2x_3 = 0 \end{cases}$

64. Solve the system: $\begin{cases} x^2 + 2y^2 - z^2 = 8 \\ x^2 - y^2 + 3z = 6 \\ y^2 - 5z = -6 \end{cases}$

In problems 65–68, solve each system algebraically.

65. $\begin{cases} c + 3(2^d) = 2^{2d} \\ c - 2^{d+1} = -6 \end{cases}$

66. $\begin{cases} \log_3(r^2 + s^2) = 2 \\ r^2 - s = 3 \end{cases}$

67. $\begin{cases} x + |y| = 7 \\ -x + 2|y| = 5 \end{cases}$

68. $\begin{cases} \sin(u + v) = 1 \\ \tan(u - v) = 1 \end{cases}$

GU In problems 69 and 70, use a graphing utility to solve each system. Round off the answers to two decimal places.

69. $\begin{cases} y = e^{-2x} \\ y = x^2 \end{cases}$

70. $\begin{cases} y = \ln(x + 2) \\ y = x^4 \end{cases}$

CHAPTER 8 REVIEW PROBLEM SET

In problems 1–4, solve each system by elimination.

1. $\begin{cases} x - 2y = 3 \\ x + y = -3 \end{cases}$

2. $\begin{cases} 8r - 7s = 28 \\ 5r + 2s = -8 \end{cases}$

3. $\begin{cases} x - y = 3 \\ 2x + y = 3 \end{cases}$

4. $\begin{cases} x - y + 2z = 0 \\ 3x + y + z = 2 \\ 2x - y + 5z = 5 \end{cases}$

In problems 5–8, each row-reduced echelon matrix is the augmented matrix form of a corresponding system of linear equations. Solve each system.

5. $\left[\begin{array}{cc|c} 1 & 0 & -9 \\ 0 & 1 & 5 \end{array}\right]$

6. $\left[\begin{array}{ccc|c} 1 & 0 & 0 & -2 \\ 0 & 1 & 3 & 7 \\ 0 & 0 & 0 & 0 \end{array}\right]$

7. $\left[\begin{array}{ccc|c} 1 & 0 & 3 & -1 \\ 0 & 1 & 1 & 2 \\ 0 & 0 & 0 & 0 \end{array}\right]$

8. $\left[\begin{array}{ccc|c} 1 & 0 & -4 & 1 \\ 0 & 0 & 0 & 2 \end{array}\right]$

In problems 9–12, use row-reduction to solve each system.

9. $\begin{cases} 4x - y = -4 \\ x + 2y = 6 \end{cases}$

10. $\begin{cases} 2x - 3y = 1 \\ -5x + y = 0 \end{cases}$

11. $\begin{cases} 3x - y + 2z = 1 \\ x - 2y + 4z = 2 \\ x - y + z = 0 \end{cases}$

12. $\begin{cases} 2x - y + z = 11 \\ x + y - z = -2 \\ -3x - 2y = -7 \end{cases}$

GU In problems 13 and 14, find the partial fraction expansion. Use a graphing utility to demonstrate the validity of the result graphically.

13. $\dfrac{x+2}{x^2 - 6x - 7} = \dfrac{A}{x-7} + \dfrac{B}{x+1}$

14. $\dfrac{3x+1}{x^2 + x} = \dfrac{A}{x} + \dfrac{B}{x+1}$

In problems 15–18, use the given matrices to perform the indicated matrix operations, if possible.

15. $A = \begin{bmatrix} 2 & 1 \\ 1 & -1 \end{bmatrix}$; $B = \begin{bmatrix} 5 & -2 \\ 3 & 4 \end{bmatrix}$

(a) $A + 2B$ **(b)** $-3A + 4B$

16. $A = \begin{bmatrix} 1 & -4 & -8 \\ 5 & 20 & 3 \end{bmatrix}$; $B = \begin{bmatrix} 7 & -2 & 7 \\ 3 & 5 & 3 \end{bmatrix}$

(a) $3A - 2B$ **(b)** $4A + 2B$

17. $A = \begin{bmatrix} 1 & -2 & 1 \\ 3 & 1 & -2 \\ 0 & 1 & -1 \end{bmatrix}$; $B = \begin{bmatrix} 4 & -1 \\ 3 & 1 \\ 2 & 8 \end{bmatrix}$

(a) AB **(b)** BA

18. $A = \begin{bmatrix} 1 & 3 \\ -1 & 2 \\ 0 & 1 \end{bmatrix}$; $B = \begin{bmatrix} -2 & 3 & 1 \\ 0 & 2 & -3 \\ 1 & -1 & 4 \end{bmatrix}$

(a) AB **(b)** BA

In problems 19 and 20, verify that $AA^{-1} = A^{-1}A = I$.

19. $A = \begin{bmatrix} 3 & -2 \\ 8 & -5 \end{bmatrix}$; $A^{-1} = \begin{bmatrix} -5 & 2 \\ -8 & 3 \end{bmatrix}$

20. $A = \begin{bmatrix} 1 & -1 & 1 \\ 0 & 2 & -1 \\ 2 & 3 & 0 \end{bmatrix}$; $A^{-1} = \begin{bmatrix} 3 & 3 & -1 \\ -2 & -2 & 1 \\ -4 & -5 & 2 \end{bmatrix}$

In problems 21 and 22, use the inverse of the matrix of coefficients to solve each system. (Refer to the inverses given in problems 19 and 20.)

21. $\begin{cases} 3x - 2y = -7 \\ 8x - 5y = -18 \end{cases}$

22. $\begin{cases} x - y + z = 7 \\ 2y - z = -7 \\ 2x + 3y = -1 \end{cases}$

In problems 23–26, evaluate each determinant.

23. $\begin{vmatrix} 1 & -1 \\ 3 & 2 \end{vmatrix}$

24. $\begin{vmatrix} -2 & 1 \\ -1 & 7 \end{vmatrix}$

25. $\begin{vmatrix} 2 & 7 & 4 \\ 3 & -1 & 5 \\ 4 & 14 & 8 \end{vmatrix}$

26. $\begin{vmatrix} 2 & -4 & 7 \\ 0 & 1 & 2 \\ 0 & 0 & 5 \end{vmatrix}$

In problems 27–30, use Cramer's rule to solve each system.

27. $\begin{cases} 3x - y = 7 \\ 2x + 3y = 12 \end{cases}$

28. $\begin{cases} 5x + 3y = 13 \\ 7x - 5y = 18 \end{cases}$

29. $\begin{cases} 2x - 3y + z = 10 \\ 5x + y - z = 6 \\ x - y + 2z = 6 \end{cases}$

30. $\begin{cases} 3x - y + 2z = 5 \\ 2x + 3y + z = 1 \\ 5x + y + 4z = 8 \end{cases}$

In problems 31 and 32, sketch the graph of the solution set of each system of inequalities.

31. $\begin{cases} x + 2y > 5 \\ 3x - y < 9 \end{cases}$

32. $\begin{cases} x^2 + 2y \geq 0 \\ x + 2y \leq 6 \end{cases}$

In problems 33–36, graph the region defined by each constraint system, locate each corner point, and then find the maximum and minimum values of the given linear expression over the region.

33. $F = 7x + 3y$

$\begin{cases} x \geq 0, y \geq 0 \\ 7x + 2y \leq 14 \end{cases}$

34. $F = 2x + y$

$\begin{cases} x \geq 1 \\ y \geq 2 \\ x + 2y \leq 10 \end{cases}$

35. $F = x + 5y$
$$\begin{cases} x \ge 0,\ y \ge 0 \\ x + y \le 4 \\ 4x + y \le 7 \end{cases}$$

36. $F = x + y$
$$\begin{cases} x \ge 0,\ y \ge 0 \\ 3y - 2x \le 6 \\ 3y + 4x \le 24 \end{cases}$$

GU In problems 37–40, sketch the graphs of both equations in the system in the same viewing window. Use the result to determine the number of solutions. Then solve the system.

37. $\begin{cases} 3x - 4y = 25 \\ x^2 + y^2 = 25 \end{cases}$

38. $\begin{cases} 2x - y = 2 \\ x^2 + 2y^2 = 12 \end{cases}$

39. $\begin{cases} x + y^2 = 6 \\ x^2 + y^2 = 36 \end{cases}$

40. $\begin{cases} \log_6 x + \log_6 y = 1 \\ x^2 + y^2 = 13 \end{cases}$

41. Income Tax: An income tax service developed the following relationships among taxes for a corporation:

$$\begin{cases} x + 0.095y = 2{,}584.42 \\ 0.01x + 1.01y + 0.01z = 1{,}529.70 \\ 0.48x + 0.48y + z = 12{,}253.24 \end{cases}$$

where x denotes the state income tax owed, y denotes the city income tax owed, and z denotes the federal income tax owed (all in dollars). Solve the system to determine the taxes owed (to the nearest dollar).

In problems 42–44, use a system of equations to solve each problem.

42. Investment Portfolio: Suppose that \$10,000 is invested in two funds paying 13% and 14% annual simple interest rate. If the annual return from both investments is \$1400, what amount was invested at each rate?

43. Puzzle: The sum of the squares of two numbers is 113. When 5 times the square of one number is added to the square of the other, the sum is 309. What are the numbers?

44. Geometry: The perimeter of a rectangular garden is 46 meters and the area of the garden is 60 square meters. Find the length and the width of the garden.

CHAPTER 8 TEST

1. Solve the system
$$\begin{cases} 4x - y = 3 \\ -2x + 3y = 1 \end{cases}$$
by using:

(a) The elimination method (without matrices)

(b) The elimination method using matrices

2. The row-reduced matrix
$$\left[\begin{array}{ccc|c} 1 & 0 & 0 & 4 \\ 0 & 1 & 0 & -1 \\ 0 & 0 & 1 & 3 \end{array}\right]$$
is the augmented matrix form of a corresponding system of linear equations. Solve the system.

3. Let
$$A = \begin{bmatrix} 3 & -1 \\ 4 & 2 \end{bmatrix} \quad \text{and} \quad B = \begin{bmatrix} 3 & -2 \\ 2 & 7 \end{bmatrix}$$
Perform each matrix operation:

(a) $A + 2B$ **(b)** AB **(c)** $3A - 2B$

4. **(a)** Let $A = \begin{bmatrix} 1 & -2 \\ 1 & 1 \end{bmatrix}$. Verify that $A^{-1} = \begin{bmatrix} \frac{1}{3} & \frac{2}{3} \\ -\frac{1}{3} & \frac{1}{3} \end{bmatrix}$.

(b) Use the inverse of the matrix of coefficients to solve the system
$$\begin{cases} x - 2y = 3 \\ x + y = -3 \end{cases}$$

5. **(a)** Evaluate: $\begin{vmatrix} 5 & 2 \\ 2 & -3 \end{vmatrix}$

(b) Use Cramer's rule to solve the system:
$$\begin{cases} 5x + 2y = 3 \\ 2x - 3y = 5 \end{cases}$$

6. Find the partial fraction decomposition:
$$\frac{x + 4}{x^2 - 3x} = \frac{A}{x} + \frac{B}{x - 3}$$

7. Solve the system: $\begin{cases} x^2 - 2y = 0 \\ x + 2y = 6 \end{cases}$

8. Sketch the region defined by the system

$$\begin{cases} x \geq 0,\ y \geq 0 \\ x + y \leq 4 \\ 2x + y \leq 6 \end{cases}$$

and find the coordinates of the corner points of the boundary of the region.

9. Find the maximum and minimum values of $3x + 4y + 1$ under the constraints

$$\begin{cases} x \geq 0,\ y \geq 0 \\ x + 2y \leq 8 \\ x + \ \ y \leq 5 \end{cases}$$

10. A plumber charges a fixed charge plus an hourly rate for service on a house call. The plumber charged \$70 to repair a water tank that required 2 hours of labor and \$100 to repair a water tank that took 3.5 hours of labor. Find the plumber's fixed charge and hourly rate. (Use a system of equations to solve this problem.)

9

Topics in Analytic Geometry

CHAPTER CONTENTS

Few tools in mathematics have as many varied applications as conic sections. The term *conic sections* (or *conics*, for short) refers to the various figures, or sections, formed by intersecting a right circular cone with a plane. There are two ways to describe the resulting curves of intersection. One way is to keep the cone fixed and position a plane in various locations as shown in Figure 1. When the plane is perpendicular to the axis of the cone, a *circle* is formed (Figure 1a). When the plane is not perpendicular to the axis of the cone, its intersection forms a *parabola* (Figure 1b), an *ellipse* (Figure 1c), or a *hyperbola* (Figure 1d). These plane curves are shown in Figure 2.

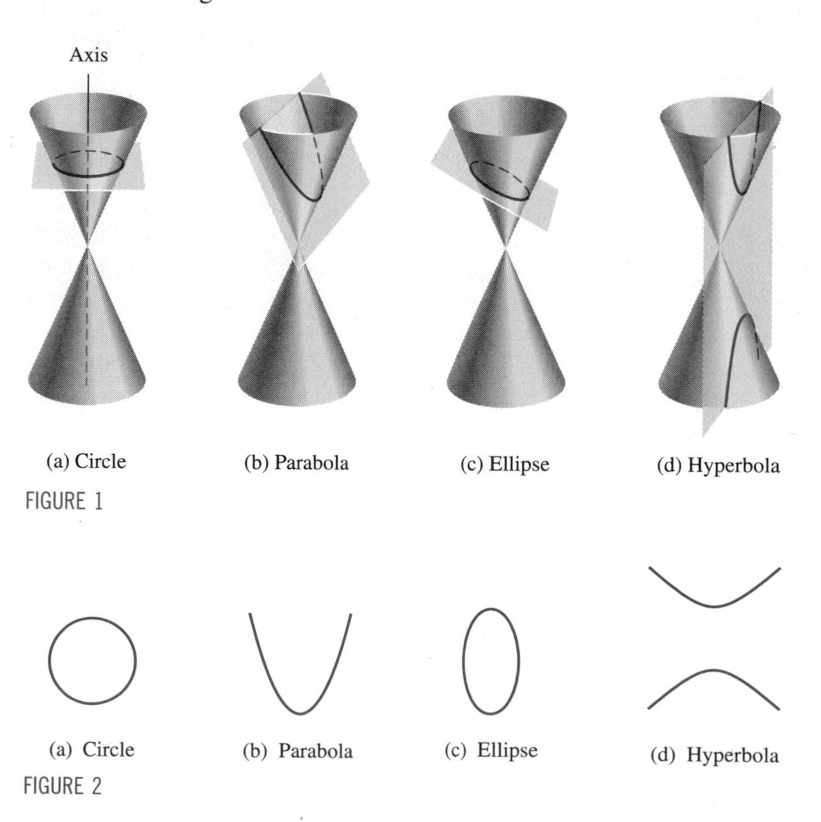

(a) Circle (b) Parabola (c) Ellipse (d) Hyperbola

FIGURE 1

(a) Circle (b) Parabola (c) Ellipse (d) Hyperbola

FIGURE 2

In this chapter, we derive standard equations in Cartesian and polar forms for parabolas, ellipses, and hyperbolas. In addition, we use these curves to illustrate the idea of rotation of axes. Parametric equations are also included in the chapter.

9.1 Parabolas

OBJECTIVES

1. Graph parabolas with vertex at (0, 0).
2. Graph parabolas with vertex at (h, k).
3. Solve applied problems.

In Section 3.1, we indicated that the graphs of quadratic functions are parabolas. In this section we give a proof of this fact based on the following geometric definition:

DEFINITION
PARABOLA

> A **parabola** is the set of all points P in a plane such that the distance from P to a fixed point (the **focus**) is equal to the distance from P to a fixed line (the **directrix**).

FIGURE 1

In Figure 1 the points P_1 and P_2 are on the parabola with focus at point F and vertical directrix d to the left of the focus. The distance from P_1 to d, denoted by d_1, is equal to the distance from P_1 to F, denoted by m_1, that is, $d_1 = m_1$. Also, $d_2 = m_2$. The line containing the focus and perpendicular to the directrix is the **axis of symmetry** of the parabola, and the point V where the parabola intersects its axis of symmetry is called the **vertex**.

Graphing Parabolas with Vertex at (0, 0)

To derive the standard equation of a parabola with vertex at (0, 0), we choose a coordinate system so that the directrix is vertical and the origin is midway between the focus and directrix. Suppose that the distance between the focus F and the origin is c (where $c > 0$); then the distance from the origin to the directrix is also c. Thus, the focus F is located at $(c, 0)$ and the equation of the directrix is $x = -c$ (Figure 2).

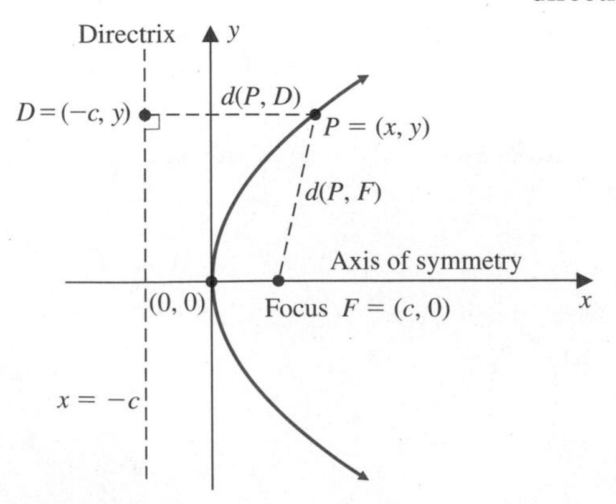

FIGURE 2

Now we pick any point $P = (x, y)$ on the parabola. By definition, the distance from P to F is equal to the distance from P to the directrix; that is,

$$d(P, F) = d(P, D)$$

$$\sqrt{(x - c)^2 + y^2} = |x + c| \qquad \text{Distance formula and absolute value property}$$

$$\left.\begin{aligned}(x - c)^2 + y^2 &= (x + c)^2 \\ x^2 - 2cx + c^2 + y^2 &= x^2 + 2cx + c^2\end{aligned}\right\} \quad \text{Square both sides and multiply}$$

After simplifying the equation, we get

$$\boxed{y^2 = 4cx \qquad c > 0}$$

We refer to this equation (in the box) as the **standard equation** for a parabola with vertex at the origin and a horizontal axis of symmetry. The equation of the directrix is $x = -c$ and the focus is at $(c, 0)$.

Every point (x, y) on the parabola satisfies the equation $y^2 = 4cx$. Conversely, if (x, y) is a point satisfying the equation, then by reversing the steps above, we find that the point (x, y) is on the parabola.

Similar derivations can be given for three other *standard equations* of parabolas with vertices at the origin (Figure 3).

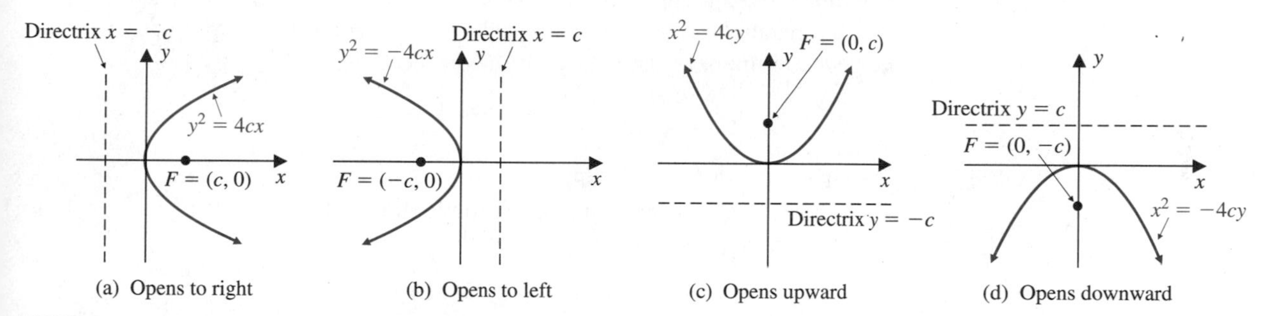

FIGURE 3
Standard equations of parabolas with vertex at the origin, $c > 0$

EXAMPLE 1 **Graphing a parabola**

For the parabola with equation $y^2 = -24x$, find the axis of symmetry, focus, and directrix. Also, sketch the graph.

Solution The standard equation of this parabola has the form $y^2 = -4cx$, with $-4c = -24$ or $c = 6$. Hence, it opens to the left with focus given by

$$F = (-c, 0) = (-6, 0)$$

and its axis of symmetry is the x axis. The vertex is at $(0, 0)$, and the equation of the directrix, which is vertical, is given by $x = c$, or $x = 6$. The graph is shown in Figure 4.

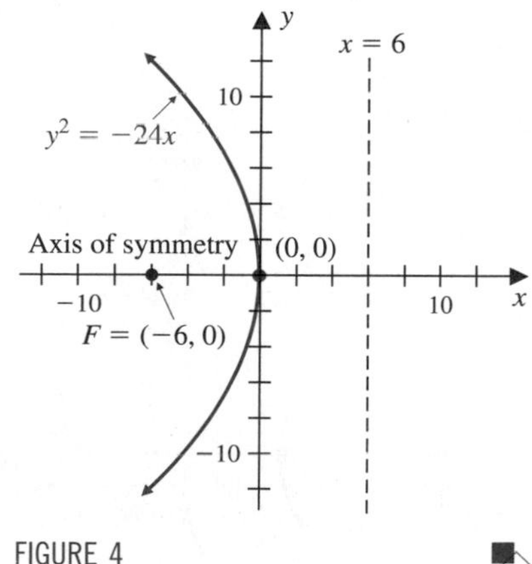

FIGURE 4

EXAMPLE 2 **Finding the standard equation of a parabola**

Find the standard equation of a parabola with vertex at the origin, focus at $(0, 3)$, and axis of symmetry on the y axis. Sketch the graph.

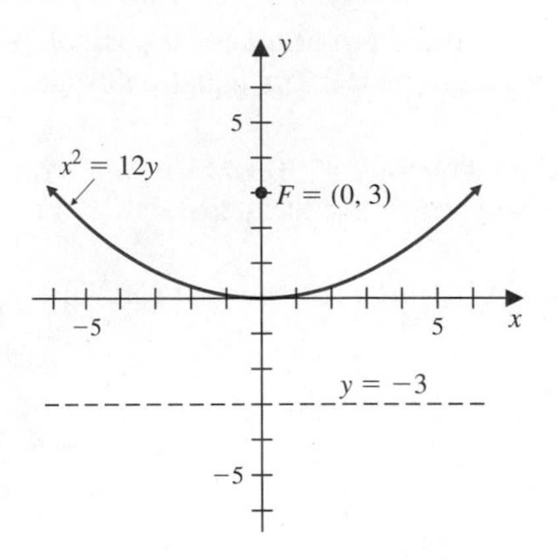

FIGURE 5

Solution The focus (0, 3) is above the vertex (0, 0) and the y axis is the axis of symmetry, so the parabola opens upward and has the standard equation $x^2 = 4cy$. Since c is the distance between the focus and vertex, we have $c = 3$ and $x^2 = 12y$ (Figure 5).

Graphing Parabolas with Vertex at (h, k)

If we apply horizontal and vertical shifting to one of the parabolas in Figure 3, we generate a new parabola whose axis of symmetry is parallel to one of the coordinate axes, and whose vertex is different from the origin. For instance, if we shift the parabola $y^2 = 4cx$ to the right h $(h > 0)$ units and up k $(k > 0)$ units, we generate the new parabola with standard equation

$$(y - k)^2 = 4c(x - h)$$

EXAMPLE 3 **Graphing a parabola by shifting**

Apply horizontal and vertical shifting to the graph of the parabola $y^2 = -24x$ to generate the graph of the equation

$$(y - 3)^2 = -24(x + 9)$$

Locate the vertex, axis of symmetry, focus, and directrix of the new parabola.

Solution Let's examine the equations to determine how to shift the graph of the first one to get the graph of the second, as indicated in the following diagram:

Indicates a vertical shift 3 units up

$$y^2 = -24x \qquad (y - 3)^2 = -24(x + 9)$$

Indicates a horizontal shift 9 units to the left

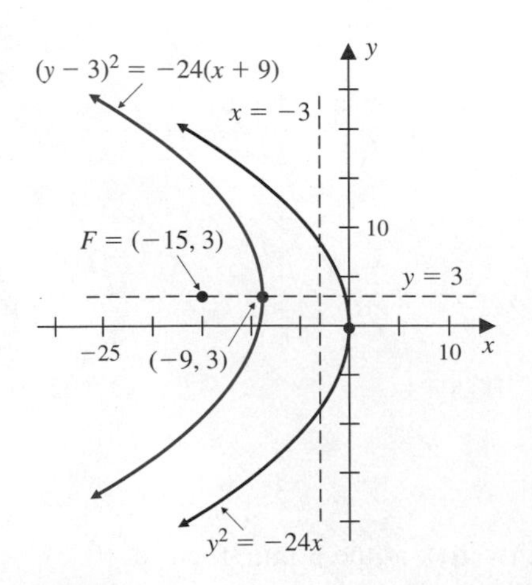

FIGURE 6

The graph of $(y - 3)^2 = -24(x + 9)$ is obtained by shifting the graph of the equation $y^2 = -24x$ to the left 9 units and up 3 units (Figure 6).

The vertex, axis of symmetry, focus, and directrix of both parabolas are listed in Table 1:

TABLE 1

	$y^2 = -24x$	Shift	$(y - 3)^2 = -24(x + 9)$
Vertex	$(0, 0)$	Left 9 units, up 3 units	$(-9, 3)$
Axis of Symmetry	x axis, $y = 0$	Up 3 units	$y = 0 + 3 = 3$
Focus	$(-6, 0)$	Left 9 units, up 3 units	$(-6 - 9, 0 + 3) = (-15, 3)$
Directrix	$x = 6$	Left 9 units	$x = 6 - 9 = -3$

The equation $(y - 3)^2 = -24(x + 9)$ is the standard equation of the parabola graphed in Figure 6. However, if it is expanded algebraically, we get

$$\begin{aligned} (y-3)^2 &= -24(x+9) \\ y^2 - 6y + 9 &= -24x - 216 \\ y^2 + 24x - 6y + 225 &= 0 \end{aligned}$$

This latter equation is referred to as a *general equation* of the parabola. A general equation of a parabola with an axis of symmetry that is vertical or horizontal, respectively, is given by

$$Ax^2 + Dx + Ey + F = 0 \qquad \text{or} \qquad Cy^2 + Dx + Ey + F = 0$$

To convert from a general equation to the standard equation, we complete the square on the squared variable.

EXAMPLE 4 **Writing an equation of a parabola in standard form**

Rewrite the general equation of the parabola

$$y^2 + 2y - 8x - 3 = 0$$

in standard form by completing the square. Graph the parabola. Locate the vertex, axis of symmetry, focus, and directrix.

Solution We rewrite the equation in standard form as follows:

$$\begin{aligned} y^2 + 2y - 8x - 3 &= 0 \\ y^2 + 2y &= 8x + 3 && \text{Isolate the } y \text{ terms on one side} \\ y^2 + 2y + 1 &= 8x + 3 + 1 && \text{Complete the square on the } y \text{ terms} \\ (y+1)^2 &= 8x + 4 && \text{Factor and simplify} \\ (y+1)^2 &= 8\left(x + \tfrac{1}{2}\right) && \text{Factor out 8} \end{aligned}$$

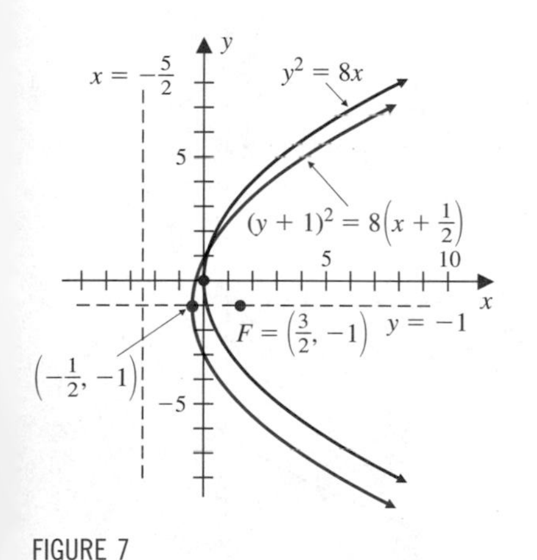

FIGURE 7

We can obtain the desired graph by shifting the graph of $y^2 = 8x$ to the left $\frac{1}{2}$ unit and down 1 unit. The graph of the new parabola is shown in Figure 7 along with the graph of $y^2 = 8x$.

The vertex is $\left(-\frac{1}{2}, -1\right)$ and the axis of symmetry is the line $y = -1$. Since $4c = 8$, it follows that $c = 2$, so the focus shifts from $(c, 0) = (2, 0)$ to $\left(2 - \frac{1}{2}, 0 - 1\right) = \left(\frac{3}{2}, -1\right)$. The directrix, which is vertical, shifts $\frac{1}{2}$ unit to the left of $x = -2$ to become the line $x = -\frac{5}{2}$. ■

Horizontal and vertical shifting of a parabola with vertex at (0, 0) and a horizontal or vertical axis of symmetry, to a parabola with vertex at (h, k) and $c > 0$, results in one of the following situations:

STANDARD EQUATIONS FOR PARABOLAS WITH VERTEX AT (h, k)

(i) $(y - k)^2 = 4c(x - h)$, opens to the right
(ii) $(y - k)^2 = -4c(x - h)$, opens to the left
(iii) $(x - h)^2 = 4c(y - k)$, opens upward
(iv) $(x - h)^2 = -4c(y - k)$, opens downward

It is possible to use certain characteristics of a parabola to find its equations.

EXAMPLE 5 **Finding a standard equation of a parabola**
Find a standard equation of the parabola with vertex at $V = (4, -1)$ and focus at $F = (2, -1)$. Also, sketch the graph.

Solution Let $D = (p, -1)$ be a point on the directrix of the parabola. Figure 8a shows a plot of the points $(4, -1)$, $(2, -1)$, and $(p, -1)$. Since the vertex is the midpoint of the line segment containing these points, it follows that $c = 2$, $p = 6$, and the parabola opens to the left with a horizontal axis of symmetry at $y = -1$. Its standard equation is

$$(y - k)^2 = -4c(x - h)$$

with $h = 4$, $k = -1$, and $c = 2$. Substituting these values into the standard equation, we have

$$[y - (-1)]^2 = -4(2)(x - 4) \qquad \text{or} \qquad (y + 1)^2 = -8(x - 4)$$

The graph is shown in Figure 8b.

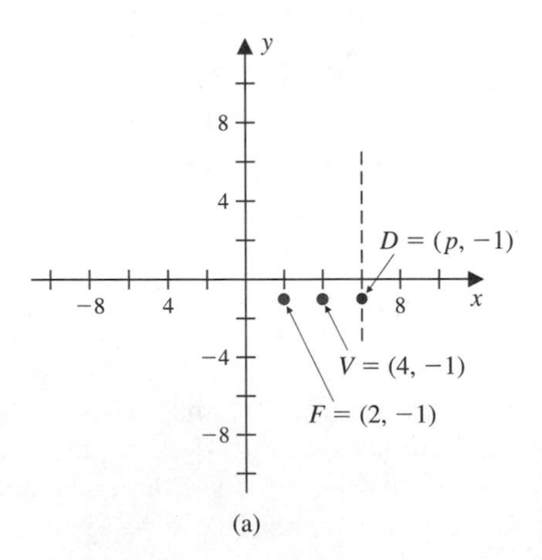

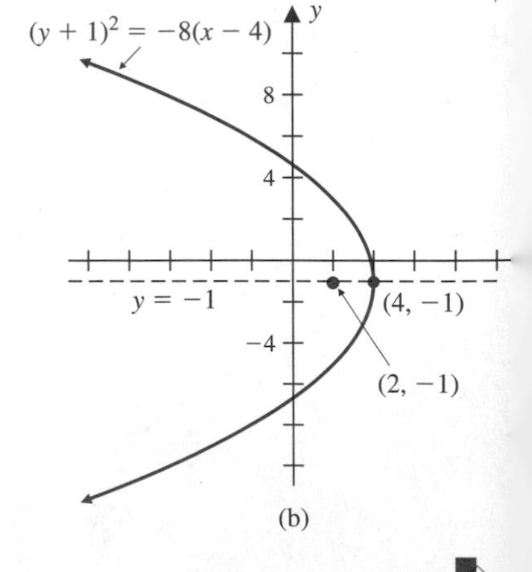

FIGURE 8

Graphs of parabolas can also be obtained by using some graphing utilities especially programmed for drawing conics without having to solve for y in terms of x.

Graphing utilities can be used to graph parabolas. However, it may be necessary to first solve for y in terms of x.

EXAMPLE 6 **Using a graphing utility to graph a parabola**

GU

Use a graphing utility to graph the parabola $(y - 1)^2 = 4x$.

Solution First we solve the equation for y to obtain the two equations

$$y - 1 = -\sqrt{4x} \quad \text{or} \quad y - 1 = \sqrt{4x}$$
$$y - 1 = -2\sqrt{x} \quad \text{or} \quad y - 1 = 2\sqrt{x}$$
$$y = 1 - 2\sqrt{x} \quad \text{or} \quad y = 1 + 2\sqrt{x}$$

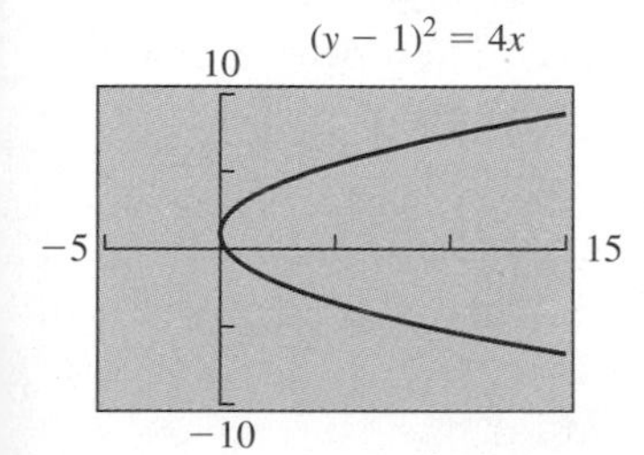

FIGURE 9

Figure 9 shows a viewing window of the graph of $(y - 1)^2 = 4x$, which was obtained by using a graphing utility to graph $y = 1 - 2\sqrt{x}$ and $y = 1 + 2\sqrt{x}$ on the same coordinate axes.

Solving Applied Problems

Parabolas appear often in the real world. A ball thrown up at an angle travels along a parabolic arc (Figure 10a), and the main cables of a suspension bridge form arcs of parabolas (Figure 10b).

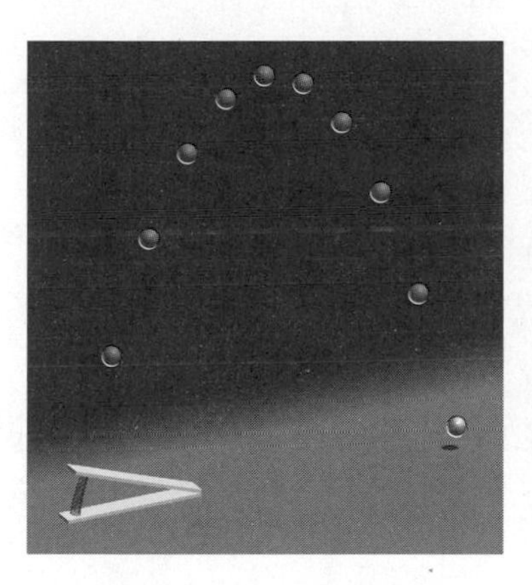

(a) The parabolic path of a ball thrown into the air

(b) Bay Bridge, San Francisco, showing parabolic main cables

FIGURE 10

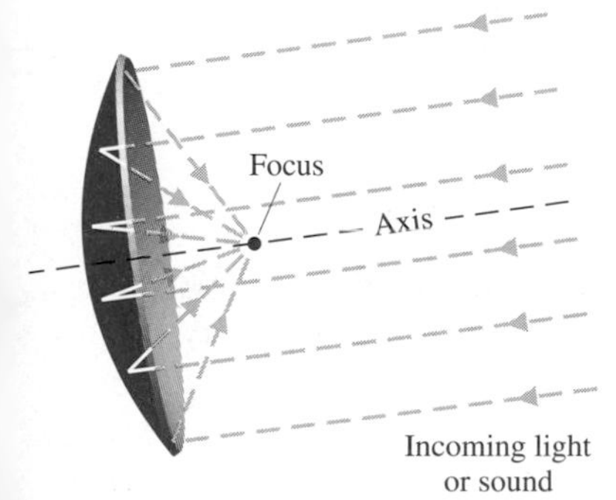

FIGURE 11
Parabolic reflector (paraboloid)

Parabolic-shaped surfaces have a remarkable property called a *reflecting property*. This property is utilized by both automotive and audio engineers. Light and sound emanating from the focus of a parabolic reflector are always reflected in beams parallel to the axis of symmetry. Thus, if an intense source of light such as a carbon arc is placed at the focus of a parabolic mirror, the light is reflected and projected in a beam parallel to the axis of symmetry. The same principle is also used in reverse. That is, rays of light or sound waves from a distant object that are parallel to the axis of symmetry will be reflected off a parabolic reflector and brought together at the focus of the reflector (Figure 11). This is the basis for a reflecting telescope and for the satellite dishes that collect television signals.

EXAMPLE 7 Locating the receiver of a dish antenna

Figure 12a shows a satellite dish antenna with a parabolic cross section. The open end of the dish has a diameter of 8 feet, and the dish is 1.5 feet deep. Because of the reflecting property of parabolas, the receiver is to be placed at the focus of the parabolic cross section.

(a) Sketch a parabolic cross section of the dish in a Cartesian plane so that its vertex is placed at the origin, it opens upward, and the y axis is its axis of symmetry.
(b) Find the standard equation of the parabola in part (a).
(c) How far from the vertex should the receiver be placed?

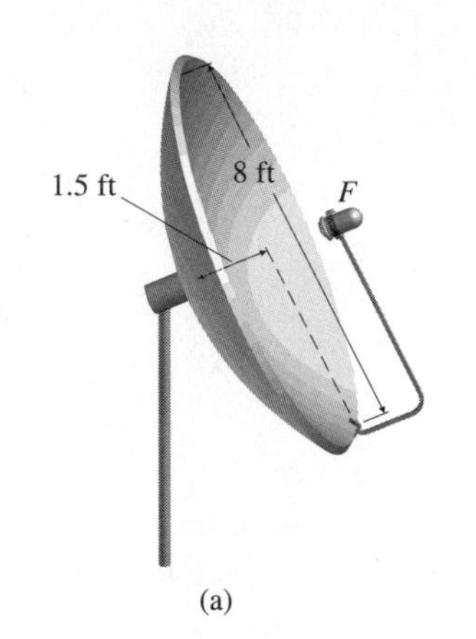

(a)

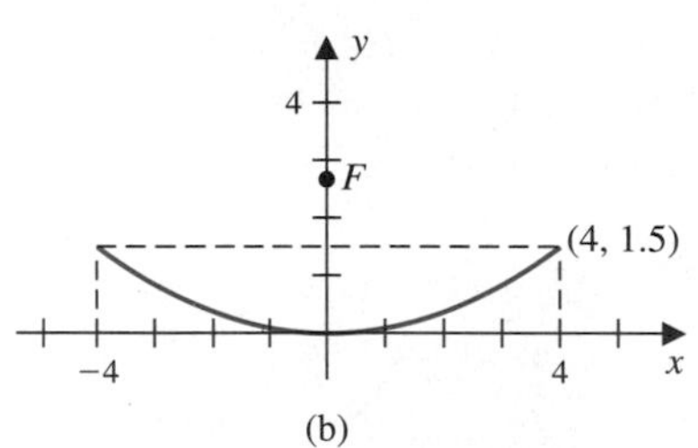

(b)

FIGURE 12

Solution

(a) The cross section is the parabola sketched in the Cartesian plane in Figure 12b. Its vertex is at the origin, it opens upward, and the axis of symmetry is the y axis. The location of the receiver is at the focus F.
(b) The standard equation of the parabola in part (a) is of the form $x^2 = 4cy$, where c is the required distance from the center of the dish to the receiver at the focus. From the given data, (4, 1.5) is a point on the parabola, so its coordinates satisfy the equation $x^2 = 4cy$. Thus,

$$c = \frac{x^2}{4y} = \frac{16}{4(1.5)} = \frac{8}{3}$$

Therefore, the standard equation of the parabola is given by

$$x^2 = 4\left(\tfrac{8}{3}\right)y \qquad \text{or} \qquad x^2 = \tfrac{32}{3}y$$

(c) Since the focus F is at $(0, c) = \left(0, \frac{8}{3}\right)$, it follows that the receiver should be placed 2 feet and 8 inches from the vertex.

PROBLEM SET 9.1

Mastering the Concepts

In problems 1–6, find the axis of symmetry, vertex, focus, and directrix of each parabola. Also, sketch the graph.

1. $y^2 = 8x$
2. $x^2 = 2y$
3. $x^2 + 4y = 0$
4. $y^2 + 4x = 0$
5. $-y^2 - 3x = 0$
6. $2y - 7x^2 = 0$

In problems 7–10, find a standard equation of a parabola with the given information. Identify the axis of symmetry and sketch the graph.

7. Vertex at the origin and focus at (0, 4)
8. Vertex at the origin and focus at (−4, 0)
9. Focus at (0, −2) and directrix $y = 2$
10. Focus at (2, 0) and directrix $x = -2$

In problems 11–16, apply horizontal and vertical shifting to the first equation of a parabola to generate the second. Find the vertex, axis of symmetry, focus, and directrix of the new parabola. Sketch the graphs of both parabolas on the same coordinate system.

11. $y^2 = 8x$; $(y - 2)^2 = 8(x - 3)$
12. $x^2 = 2y$; $(x + 3)^2 = 2(y + 2)$
13. $x^2 = -4y$; $(x + 1)^2 = -4(y + 3)$
14. $y^2 = -4x$; $(y - 3)^2 = -4(x - 1)$
15. $y^2 = -6x$; $(y + 2)^2 = -6(x - 1)$
16. $x^2 = -12y$; $(x - 2)^2 = -12(y + 1)$

In problems 17–22, write each equation in standard form. Sketch the graph of each parabola. Find the vertex, focus, axis of symmetry, and directrix.

17. $y^2 + 2x + 2y + 7 = 0$
18. $2y^2 + 2x + 6y + 9 = 0$
19. $3x^2 + 12x + 6y + 13 = 0$
20. $x^2 + 4x + 5y - 11 = 0$
21. $y^2 - 6x = 4y - 13$
22. $x^2 = \frac{5}{3}y + \frac{2}{3} + 3x$

In problems 23–32, find a standard equation of a parabola with the given characteristics. Identify the axis of symmetry and sketch the graph.

23. Vertex at $(-1, 2)$ and focus at $(3, 2)$
24. Vertex at $(3, -4)$ and focus at $(3, 0)$
25. Vertex at $(3, -5)$ and focus at $(5, -5)$
26. Vertex at $(-2, 3)$ and focus at $(-2, 8)$
27. Vertex at $(1, 3)$, contains point $(5, 7)$, and axis of symmetry parallel to the y axis
28. Vertex at $(3, -4)$, axis of symmetry parallel to the x axis, and x intercept 11
29. Focus at $(-3, 5)$ and directrix $y = -5$
30. Vertex at $(3, -5)$ and directrix $x = 2$
31. Contains the three points $(-2, 0)$, $(0, 3)$, and $(2, 0)$, and opens downward
32. Contains the three points $(0, 0)$, $(-1, 2)$, and $\left(-3, -2\sqrt{3}\right)$, and opens to the left

GU In problems 33–36, solve each equation for y. Then use a graphing utility to graph both of the resulting equations in the same viewing window to obtain a graph of the given equation.

33. **(a)** $(y + 1)^2 = 2(x - 1)$
(b) $(x - 3)^2 = -6(y + 1)$
34. **(a)** $(y - 2)^2 = -4(x + 2)$
(b) $(x + 4)^2 = 16(y - 2)$
35. $y^2 + 2y + 12x - 23 = 0$
36. $y^2 + 10y - x + 21 = 0$

Applying the Concepts

37. Satellite Dish Antenna: Figure 13 shows a satellite dish antenna with a parabolic cross section. The opening of the dish has a diameter of 12 feet, and the dish is 3 feet deep. The receiver is to be placed at the focus.

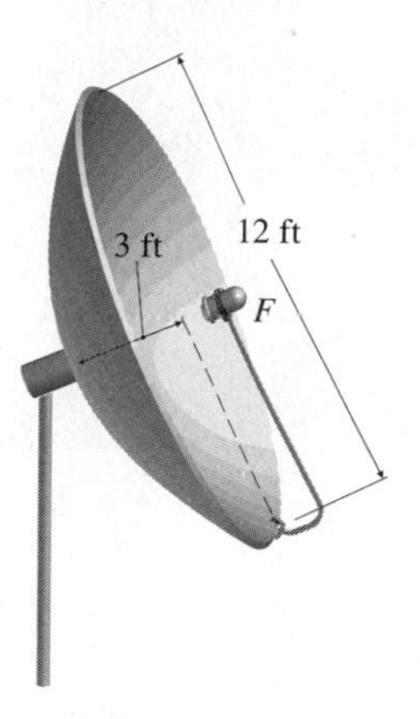

FIGURE 13

(a) Sketch a parabolic cross section of the dish in a Cartesian plane so that its vertex is placed at the origin, it opens to the right, and the x axis is its axis of symmetry.
(b) Find the standard equation of the parabola in part (a).
(c) At what distance from the vertex should the receiver be placed?

38. Parabolic Mirror: Suppose that the diameter of a parabolic mirror is 10 inches, and assume that the mirror is 5 inches deep at the center.
(a) Sketch a parabolic cross section of the mirror in a Cartesian plane so that its vertex is at the origin, it opens to the right, and the x axis is its axis of symmetry.
(b) Find the standard equation of the parabola in part (a).
(c) How far is the focus from the center of the mirror?

39. Reflecting Telescope: One of the world's largest reflecting telescopes is called the Hale telescope in honor of the American astronomer George E. Hale (1868–1938). It is located at the Palomar Mountain Observatory northeast of San Diego, California. It uses a mirror 200 inches in diameter. A cross section of the mirror through a diameter is a parabola. The distance from the focus to the vertex is 666 inches (Figure 14).

FIGURE 14

(a) Sketch the parabola in a Cartesian plane so that its vertex is at the origin and it opens upward.
(b) Find the standard equation of the parabola in part (a).
(c) What is the depth of the parabolic cross section of the mirror?

40. Verrazano Narrows Bridge: The Verrazano Narrows suspension bridge is held up by parabolic main cables. It has twin supporting towers 700 feet above the water level. The distance between these towers is 2627 feet, and the low point on a supporting cable is 225 feet above the water level (Figure 15).

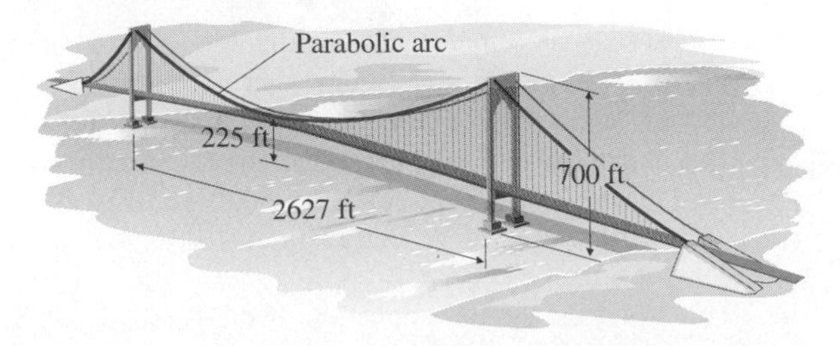

FIGURE 15

(a) Set up a Cartesian system to model the situation so that the x axis coincides with the water level and the vertex of the parabola is on the y axis.
(b) Find the standard equation of the parabolic main cable.
(c) How much higher than the low point is a point on the main cable 900 feet from one supporting tower?

41. Parabolic Stone Bridge: The surface of a roadway over a stone bridge follows a parabolic curve with the vertex in the middle of the bridge. The span of the bridge is 60 feet and the road surface is 1 foot higher in the middle than at the ends (Figure 16). How much higher than the ends is a point on the roadway 15 feet from an end?

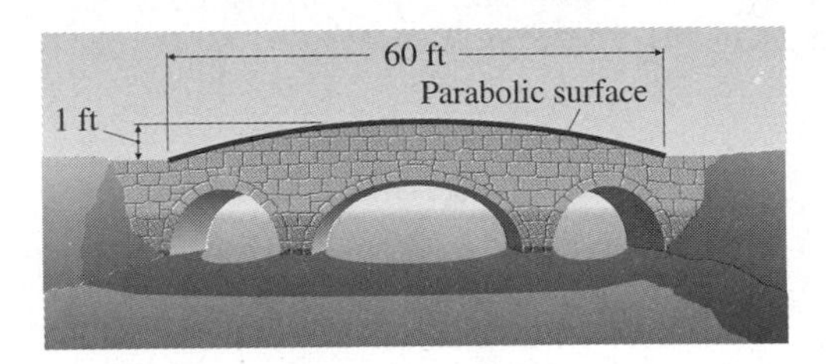

FIGURE 16

42. Trajectory of a Ball: A ball thrown horizontally from the top edge of a building has a parabolic path whose vertex is at the top edge of the building and whose axis of symmetry is along the side of the building. The ball passes through a point 100 feet from the building when it is a vertical distance of 16 feet from the top (Figure 17).

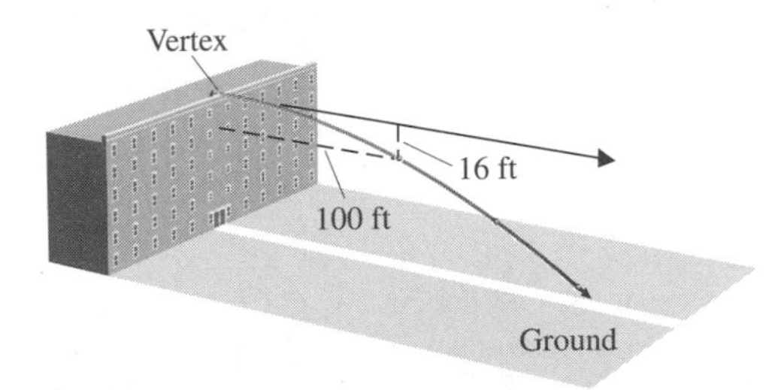

FIGURE 17

(a) Assuming that the building is 64 feet high, sketch the parabola in a Cartesian plane so that the intersection of the ground and the edge of the building is at (0, 0) and the edge of the building lies on the y axis.
(b) Find the standard equation of the parabola in part (a).
(c) How far from the building will the ball land?

Developing and Extending the Concepts

43. The **focal chord** of a parabola, also called the **latus rectum**, is the line segment perpendicular to the axis of symmetry at the focus with end points on the parabola (Figure 18).

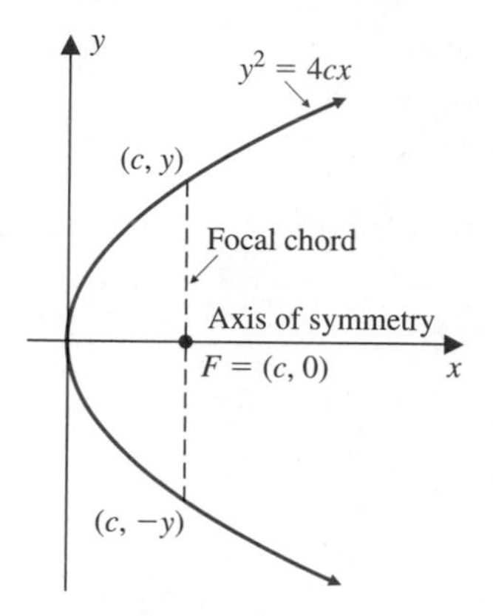

FIGURE 18

(a) Show that if c is the distance between the vertex and the focus of a parabola, then its focal chord has length $4c$.

(b) Find the length of the focal chord for each of the following parabolas:

(i) $y^2 = 8x$ (ii) $x^2 + 4y = 0$
(iii) $y^2 - 4x = 0$ (iv) $2y - 7x^2 = 0$

44. Let A, B, and C be constants with $A > 0$.

(a) Show that $y = Ax^2 + Bx + C$ is an equation of a parabola that has a vertical axis of symmetry and opens upward.

(b) Find the coordinates of the vertex and focus of the parabola in part (a) in terms of A, B, and C.

(c) Find the length of the focal chord.

(d) Find conditions on A, B, and C for the graph to intersect the x axis.

45. GU Use a graphing utility to graph parabolas $x^2 = 4cy$ for $c = 0.25, 0.90, 2$, and 4 in the same viewing window. What happens to the graph as c gets larger and larger?

46. In calculus, it is shown that the slope m of the tangent line to the parabola

$$(y - k)^2 = 4c(x - h)$$

at a point where $y = b$ is given by

$$m = \frac{2c}{b - k}$$

(Figure 19), where $c > 0$. Find an equation of the tangent line to each given parabola at the indicated point.

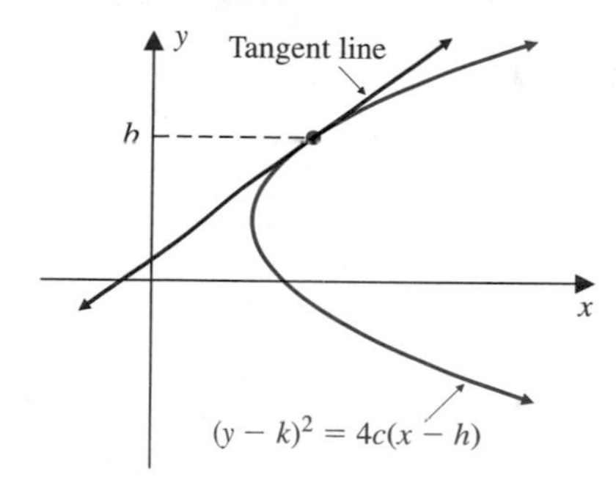

FIGURE 19

(a) $y^2 - 8x$; at $(2, -4)$

(b) $(y + 1)^2 = 4(x - 2)$; at $(3, 1)$

47. Refer to Problems 43 and 46. It can be shown that the slope m of the tangent line to the parabola $x^2 = 4cy$ at the point (a, b) is given by $m = a/(2c)$. Show that the tangent lines at the end points of the focal chord of $x^2 = 8y$ are perpendicular to each other.

OBJECTIVES

1. Graph ellipses with center at (0, 0).
2. Graph ellipses with center at (h, k).
3. Solve applied problems.

9.2 Ellipses

Ellipses are useful in providing mathematical models of a variety of physical phenomena ranging from art to astronomy. The geometric definition of an ellipse follows:

DEFINITION

ELLIPSE

> An **ellipse** is the set of all points P in a plane such that the sum of the distances from P to two fixed points is a constant. The fixed points are called the **foci** (plural of **focus**) of the ellipse.

We can use this definition to draw an ellipse by taking a string of length k and fastening its ends at two fixed points F_1 and F_2. Figure 1a shows two fixed points F_1 and F_2 and a string of length k stretched tightly about them to the point P. Hence, as P moves about, $|\overline{PF_1}| + |\overline{PF_2}|$ always has a constant value k. That is, as Figure 1b shows,

$$d_1 + m_1 = d_2 + m_2 = k \qquad \text{where } k \text{ is a constant}$$

Thus, if a pencil point P is inserted into the string and moved so as to keep the string tight, it traces out an ellipse.

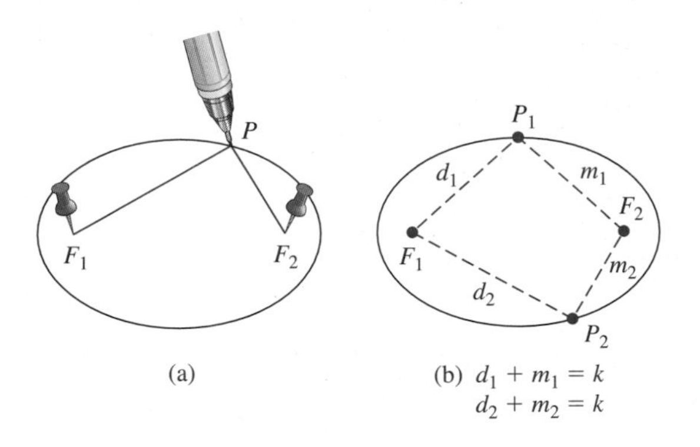

FIGURE 1

Graphing Ellipses with Center at (0, 0)

As with the parabola, we use the definition of the ellipse to derive its standard equation. We choose the x axis as the line containing the foci $F_1 = (-c, 0)$ and $F_2 = (c, 0)$, where $c > 0$. The origin is the midpoint of the line segment $\overline{F_1F_2}$.

The midpoint of the segment $\overline{F_1F_2}$ is called the **center** of the ellipse. Also, we assume that the constant sum of the distances between any point P on the ellipse and the foci is $2a$. (This constant is written in the form $2a$ so that the equation of the ellipse will have a simple form.) In triangle F_1PF_2 (Figure 2) we know from geometry that the sum of the lengths of the two sides of the triangle, $|\overline{PF_1}| + |\overline{PF_2}| = 2a$, is *greater than* the length of the third side, $|\overline{F_1F_2}| = 2c$.

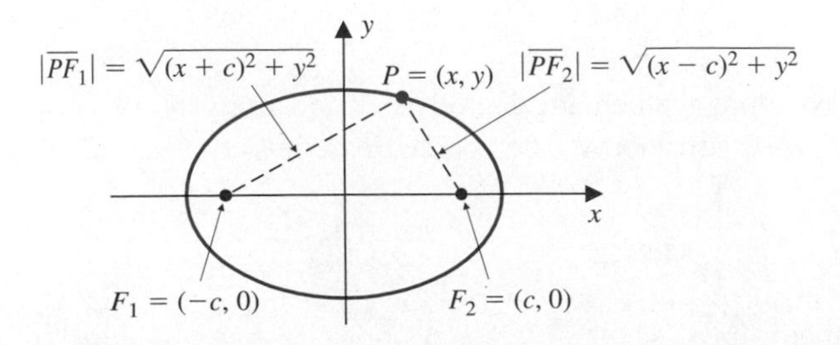

FIGURE 2

Thus, $2a > 2c$ or $a > c$. If we let $P = (x, y)$, we get

$$d(P, F_1) + d(P, F_2) = 2a$$

$$\sqrt{(x+c)^2 + y^2} + \sqrt{(x-c)^2 + y^2} = 2a \qquad \text{Distance formula}$$

$$\sqrt{(x+c)^2 + y^2} = 2a - \sqrt{(x-c)^2 + y^2} \qquad \text{Isolate the first radical}$$

$$x^2 + 2xc + c^2 + y^2 = 4a^2 - 4a\sqrt{(x-c)^2 + y^2} + x^2 - 2cx + c^2 + y^2 \qquad \text{Square each side}$$

$$\left.\begin{aligned} 4cx - 4a^2 &= -4a\sqrt{(x-c)^2 + y^2} \\ cx - a^2 &= -a\sqrt{(x-c)^2 + y^2} \end{aligned}\right\} \quad \text{Isolate radical and simplify}$$

$$\left.\begin{aligned} c^2x^2 - 2a^2cx + a^4 &= a^2(x^2 - 2cx + c^2 + y^2) \\ a^4 - a^2c^2 &= (a^2 - c^2)x^2 + a^2y^2 \end{aligned}\right\} \quad \text{Square each side and rewrite}$$

$$(a^2 - c^2)x^2 + a^2y^2 = a^2(a^2 - c^2) \qquad \text{Factor and rearrange terms}$$

Since $a > c > 0$, $\sqrt{a^2 - c^2}$ is always a *positive* real number and we denote it by b. After substituting $b^2 = a^2 - c^2$ into the latter equation, we obtain

$$b^2x^2 + a^2y^2 = a^2b^2 \qquad a^2 - c^2 = b^2$$

Finally, we divide both sides of this equation by a^2b^2 to get

$$\boxed{\frac{x^2}{a^2} + \frac{y^2}{b^2} = 1}$$

Note that since $b^2 = a^2 - c^2$, it follows that $a^2 = b^2 + c^2$ so that $a^2 > b^2$ and $a > b$ because both a and b are positive numbers.

The boxed equation is called the **standard equation** of an ellipse with center at the origin and foci on the x axis. Every point (x, y) on the ellipse in Figure 3 has coordinates that satisfy the boxed equation; and, conversely, any point with coordinates that satisfy the equation is on the ellipse.

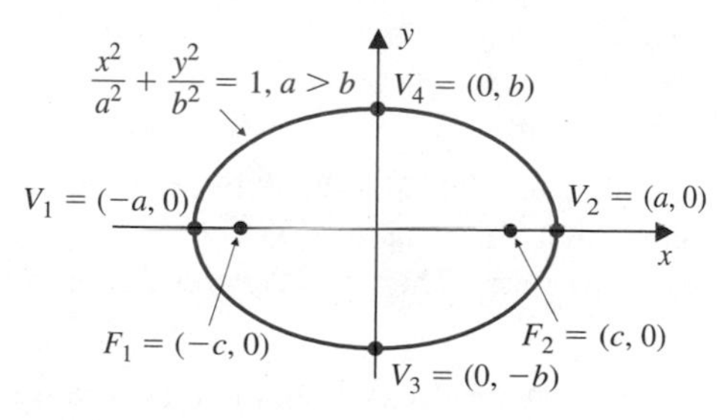

FIGURE 3

From the equation, we can derive the following characteristics of the graph of an ellipse. (Refer to Figure 3.)

1. After substituting $y = 0$ into the equation, we find that the x intercepts of the ellipse are $-a$ and a. The y intercepts are obtained by letting $x = 0$ to get $-b$ and b. The corresponding points $V_1 = (-a, 0)$, $V_2 = (a, 0)$, $V_3 = (0, -b)$, and $V_4 = (0, b)$ on the graph are called the **vertices** of the ellipse.
2. The line segment $\overline{V_1V_2}$ is referred to as the **major axis**, and it has length $2a$; a is called the **semimajor axis**. The segment $\overline{V_3V_4}$ is called the **minor axis** of the ellipse, and it has length $2b$; b is called the **semiminor axis**. The major axis is longer than the minor axis, since $a > b$.
3. The graph is symmetric with respect to both the x axis (major axis) and the y axis (minor axis), as well as the origin. Also, $c^2 = a^2 - b^2$, and the foci lie on the major axis c units from the center at points $(-c, 0)$, and $(c, 0)$.

Using the same analysis (again with $a > b$), we can derive a standard equation of an ellipse with center at the origin and foci on the y axis (Problem 48). The graphs and standard equations that describe the two situations of the ellipse are shown in Figure 4.

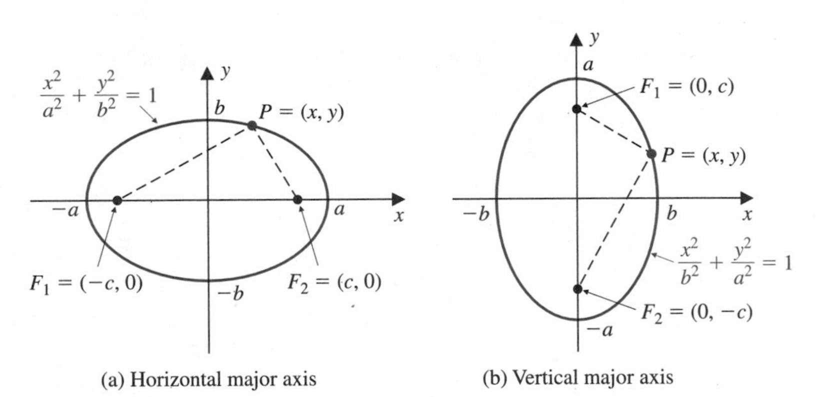

FIGURE 4
Standard equations of ellipses with center at (0, 0), horizontal or vertical major axis, and foci F_1 and F_2; $a > b$, $c^2 = a^2 - b^2$

EXAMPLE 1 Graphing an ellipse with center at (0, 0)

Show that the equation $4x^2 + y^2 = 4$ represents an ellipse by rewriting it in standard form. Then find the vertices and foci, and sketch the graph.

Solution After dividing both sides of the equation by 4, we get the equation

$$\frac{x^2}{1} + \frac{y^2}{4} = 1$$

which is the standard equation of an ellipse with center at the origin and a vertical major axis. Since $b^2 = 1$ and $a^2 = 4$, we have $c^2 = a^2 - b^2 = 4 - 1 = 3$ and $c = \sqrt{3}$, so that the vertices are $V_1 = (0, 2)$, $V_2 = (0, -2)$, $V_3 = (1, 0)$, and $V_4 = (-1, 0)$, and the foci are $F_1 = \left(0, \sqrt{3}\right)$ and $F_2 = \left(0, -\sqrt{3}\right)$. The graph is shown in Figure 5.

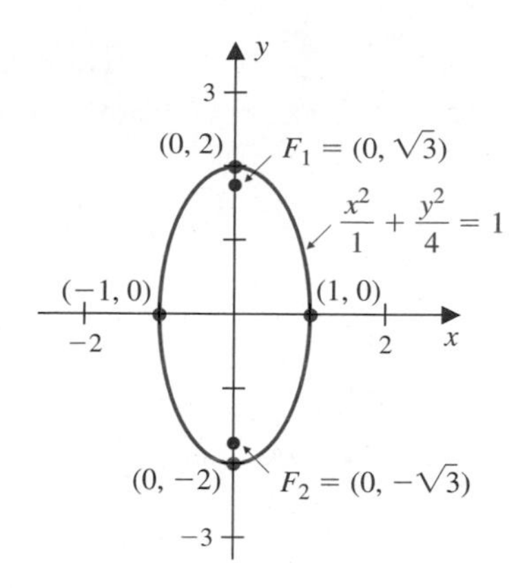

FIGURE 5

The next example shows how to find an equation of an ellipse given certain information.

EXAMPLE 2 **Finding the standard equation of an ellipse**

Find the standard equation of an ellipse with foci (2, 0) and (−2, 0), and vertices (3, 0) and (−3, 0). Also, sketch the graph.

FIGURE 6

Solution Since the vertices and the foci are on the x axis, the major axis is horizontal. Figure 6 shows the graph of the ellipse. Because a is the distance from the center to a vertex on the major axis, $a = 3$. Also, c is the distance from the center to a focus, so $c = 2$. To write the standard equation of the ellipse, we need to find b^2. If we replace a by 3 and c by 2 in the equation $b^2 = a^2 - c^2$, we get

$$b^2 = 9 - 4 = 5$$

Therefore, the standard equation is

$$\frac{x^2}{9} + \frac{y^2}{5} = 1$$

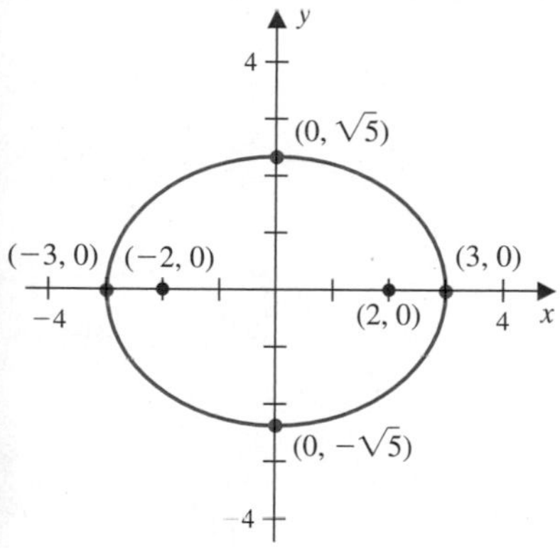

Graphing Ellipses with Center at (h, k)

If we apply horizontal and vertical shifting to the ellipses in Figure 4, we generate new ellipses with center at $(h, k) \neq (0, 0)$. For instance, if we shift the ellipse

$$\frac{x^2}{a^2} + \frac{y^2}{b^2} = 1$$

h $(h > 0)$ units to the right and k $(k > 0)$ units upward, we generate the new ellipse with standard equation

$$\frac{(x - h)^2}{a^2} + \frac{(y - k)^2}{b^2} = 1$$

EXAMPLE 3 Graphing an ellipse with center at (h, k)

Apply horizontal and vertical shifting to the graph of the equation

$$\frac{x^2}{9} + \frac{y^2}{4} = 1$$

to generate the graph of the equation

$$\frac{(x+3)^2}{9} + \frac{(y-5)^2}{4} = 1$$

Find the center, vertices, and foci, and sketch the graph of the new ellipse.

Solution The following diagram shows how to shift the graph of the first equation to get the graph of the second:

Indicates a horizontal shift 3 units to the left

$$\frac{x^2}{9} + \frac{y^2}{4} = 1 \qquad \frac{(x+3)^2}{9} + \frac{(y-5)^2}{4} = 1$$

Indicates a vertical shift 5 units up

The graph of

$$\frac{(x+3)^2}{9} + \frac{(y-5)^2}{4} = 1$$

is obtained by shifting the graph of

$$\frac{x^2}{9} + \frac{y^2}{4} = 1$$

3 units to the left and 5 units up. Figure 7 shows the graphs of both ellipses.

The center, vertices, and foci of both ellipses are listed in Table 1.

TABLE 1

	$\frac{x^2}{9} + \frac{y^2}{4} = 1$	Shift	$\frac{(x+3)^2}{9} + \frac{(y-5)^2}{4} = 1$
Center	$(0, 0)$	Left 3 units, up 5 units	$(-3, 5)$
Vertices	$(-3, 0)$, $(3, 0)$, $(0, 2)$, $(0, -2)$	Left 3 units, up 5 units	$V_1 = (-3-3, 5) = (-6, 5)$ $V_2 = (3-3, 5) = (0, 5)$ $V_3 = (-3, 2+5) = (-3, 7)$ $V_4 = (-3, -2+5) = (-3, 3)$
Foci	$(-\sqrt{5}, 0)$, $(\sqrt{5}, 0)$	Left 3 units, up 5 units	$F_1 = (-3-\sqrt{5}, 5)$ $F_2 = (-3+\sqrt{5}, 5)$

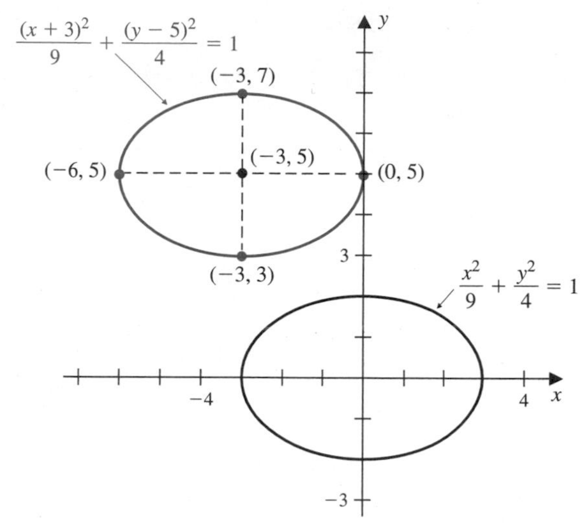

FIGURE 7

Horizontal and vertical shifting of an ellipse centered at the origin with a horizontal or vertical major axis, to an ellipse with center at (h, k) and with $a > 0$, $b > 0$, results in one of the following two situations:

STANDARD EQUATIONS FOR ELLIPSES CENTERED AT (h, k)

(i) $\frac{(x-h)^2}{a^2} + \frac{(y-k)^2}{b^2} = 1$, if $a > b$; the major axis is horizontal

(ii) $\frac{(x-h)^2}{b^2} + \frac{(y-k)^2}{a^2} = 1$, if $b < a$; the major axis is vertical

The *general equation* of an ellipse with axes parallel to the coordinate axes is given by

$$Ax^2 + Cy^2 + Dx + Ey + F = 0$$

where A and C have the same sign. This equation can be rewritten in standard form by completing the squares.

EXAMPLE 4 Writing an equation of an ellipse in standard form

Rewrite the general equation of the ellipse

$$25x^2 + 9y^2 - 100x - 54y = 44$$

in standard form by completing the squares. Find the center, vertices, and foci. Also, sketch the graph.

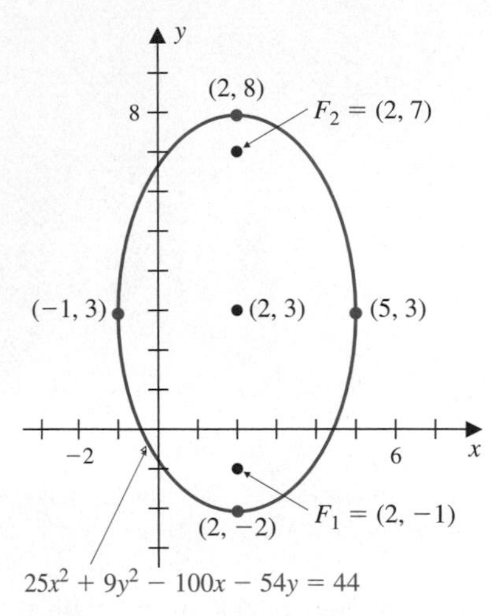

FIGURE 8

Solution Here, we have

$$25(x^2 - 4x \qquad) + 9(y^2 - 6y \qquad) = 44 \quad \text{Factor 25 from the } x \text{ terms and 9 from the } y \text{ terms}$$

$$25(x^2 - 4x + 4) + 9(y^2 - 6y + 9) = 44 + 25(4) + 9(9) \quad \text{Complete the squares}$$

$$25(x - 2)^2 + 9(y - 3)^2 = 225 \quad \text{Factor and simplify}$$

$$\frac{(x - 2)^2}{9} + \frac{(y - 3)^2}{25} = 1 \quad \text{Divide both sides by 225}$$

So the center is at (2, 3), $a = 5$, $b = 3$, and we have a vertical major axis. Also, $c^2 = a^2 - b^2 = 25 - 9 = 16$, so $c = 4$. Thus, the coordinates of the foci are $F_1 = (2, 3 - 4) = (2, -1)$ and $F_2 = (2, 3 + 4) = (2, 7)$. The coordinates of the vertices are

$$V_1 = (2, 3 - 5) = (2, -2) \qquad V_2 = (2, 3 + 5) = (2, 8)$$
$$V_3 = (2 - 3, 3) = (-1, 3) \qquad V_4 = (2 + 3, 3) = (5, 3)$$

The graph is displayed in Figure 8.

To graph an ellipse using a graphing utility, we first solve the equation for y in terms of x and then graph both of the resulting functions in the same viewing window, as we did for parabolas.

EXAMPLE 5 Graphing an ellipse with a graphing utility

Use a graphing utility to sketch the graph of the ellipse whose equation is

$$\frac{(x - 1)^2}{9} + \frac{(y + 2)^2}{4} = 1$$

Solution First, we solve the equation for y in terms of x:

$$\frac{(y + 2)^2}{4} = 1 - \frac{(x - 1)^2}{9}$$

$$\frac{y + 2}{2} = \pm\sqrt{1 - \frac{(x - 1)^2}{9}}$$

$$y + 2 = \pm 2\sqrt{1 - \frac{(x - 1)^2}{9}}$$

$$y = -2 \pm 2\sqrt{\frac{9 - (x - 1)^2}{3}}$$

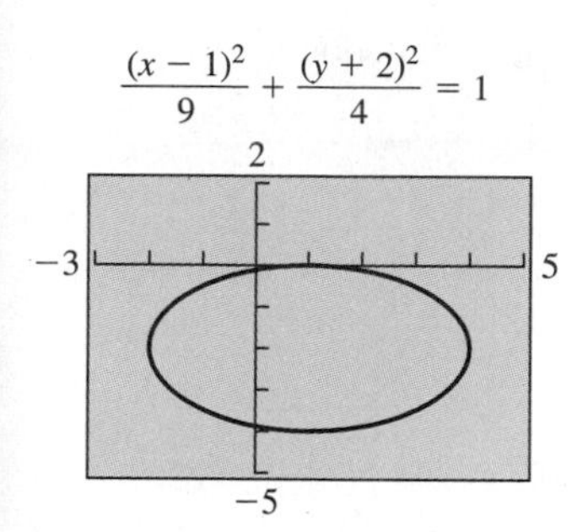

FIGURE 9

Next, we graph both functions,

$$y_1 = -2 + 2\frac{\sqrt{9-(x-1)^2}}{3} \qquad \text{and} \qquad y_2 = -2 - 2\frac{\sqrt{9-(x-1)^2}}{3}$$

in the same viewing window to get the graph of the original equation (Figure 9).

Solving Applied Problems

Elliptically shaped surfaces have a *reflecting property* similar to that of parabolas, in that light and sound produced at one focus of an ellipse will be reflected to the other focus. This property is used in the design of optical apparatus. It also accounts for the "whispering galleries" found in buildings that have ceilings with cross sections that are elliptical in shape. Words whispered at one focus can be heard clearly at the other focus. Some of the most famous whispering galleries are the National Statuary Hall of the old House of Representatives in Washington, DC, the Mormon Tabernacle in Salt Lake City, and St. Paul's Cathedral in London.

EXAMPLE 6 **Solving an overpass problem**

An arch in the shape of the upper half of an ellipse is to support a bridge over a roadway 100 feet wide. The center of the arch is 30 feet above the center of the roadway, and the paved portion of the roadway is within a strip that is 80 feet wide (Figure 10).

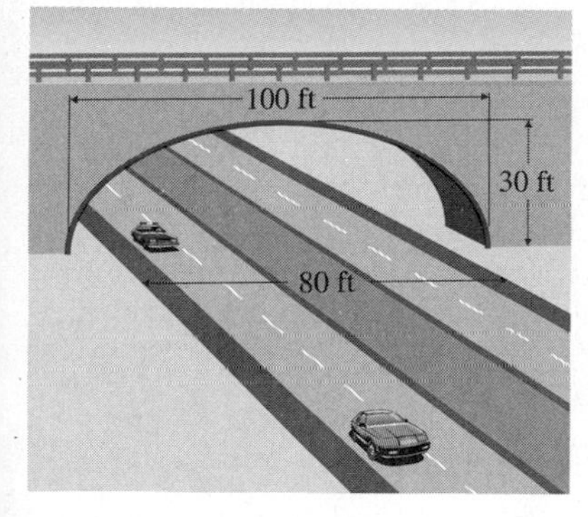

FIGURE 10

(a) Sketch the ellipse in an xy coordinate system, with horizontal major axis along the roadway and center at the origin.
(b) Find the standard equation of the ellipse in part (a).
(c) Confirm that the foci of the ellipse are located at the edges of the paved portion of the roadway.
(d) What is the clearance above the roadway to the overpass at the edge of the paved roadway?

Solution

(a) First, we set up an xy coordinate system so that the center of the ellipse is at the origin, the semimajor axis is horizontal on the x axis and has length 50, and the semiminor axis, which is vertical, has length 30 (Figure 11).

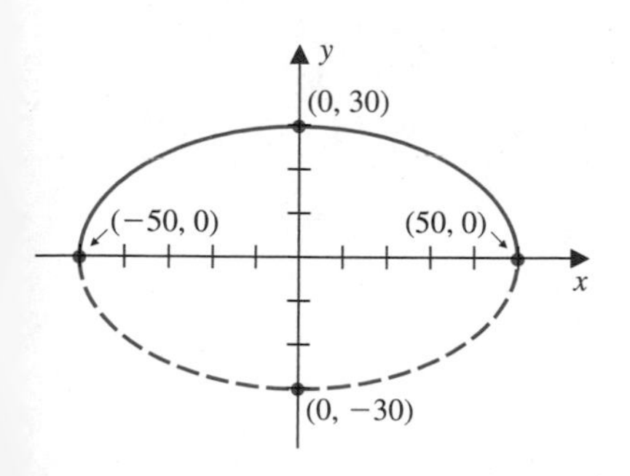

FIGURE 11

(b) The standard equation of the ellipse is

$$\frac{x^2}{2500} + \frac{y^2}{900} = 1$$

(c) To find the coordinates of the foci, we first find c:

$$c = \sqrt{a^2 - b^2} = \sqrt{2500 - 900} = 40$$

So the coordinates of the foci are $(-40, 0)$ and $(40, 0)$. This confirms that the foci of the ellipse are located at the edges of the paved portion of the roadway, which is known to be 80 feet wide.

(d) We want to find the value of y when $x = 40$. Using the equation

$$\frac{x^2}{2500} + \frac{y^2}{900} = 1$$

and replacing x by 40, we have

$$\frac{1600}{2500} + \frac{y^2}{900} = 1$$

$$\frac{y^2}{900} = 1 - \frac{1600}{2500}$$

$$= 1 - \frac{16}{25} = \frac{9}{25}$$

$$y^2 = 324$$

Since y is positive in this context, we get

$$y = \sqrt{324} = 18$$

Therefore, the clearance above the edge of the paved roadway to the overpass is 18 feet.

PROBLEM SET 9.2

Mastering the Concepts

In problems 1–6, find the vertices and foci, and sketch the graph of each ellipse.

1. $\frac{x^2}{16} + \frac{y^2}{9} = 1$
2. $\frac{y^2}{25} + \frac{x^2}{16} = 1$
3. $4y^2 + 16x^2 = 64$
4. $4x^2 + 9y^2 = 36$
5. $0.1x^2 + y^2 - 1 = 0$
6. $0.25x^2 + 0.5y^2 - 1 = 0$

In problems 7–12, find a standard equation of an ellipse with the given characteristics and sketch the graph.

7. Vertices at $(-5, 0)$ and $(5, 0)$; foci at $(-3, 0)$ and $(3, 0)$
8. Vertices at $(0, -5)$, and $(0, 5)$; foci at $(0, -4)$ and $(0, 4)$
9. Foci at $(0, -13)$ and $(0, 13)$; length of semiminor axis = 5
10. Foci at $(0, -12)$ and $(0, 12)$; length of semimajor axis = 13
11. Vertices at $(0, -6)$ and $(0, 6)$; containing the point $(3, 2)$
12. Vertices at $(-5, 0)$ and $(5, 0)$; containing the point $\left(4, \frac{12}{5}\right)$

In problems 13–20, apply horizontal and vertical shifting to the graph of the first equation to generate the graph of the second. Find the center, vertices, and foci of the new ellipse. Sketch the graphs of both equations in the same coordinate system.

13. $\frac{x^2}{16} + \frac{y^2}{9} = 1;\ \frac{(x-2)^2}{16} + \frac{(y+3)^2}{9} = 1$
14. $\frac{x^2}{25} + \frac{y^2}{4} = 1;\ \frac{(x-2)^2}{25} + \frac{(y-1)^2}{4} = 1$
15. $\frac{x^2}{48} + \frac{y^2}{64} = 1;\ \frac{(x-1)^2}{48} + \frac{(y+2)^2}{64} = 1$
16. $\frac{x^2}{10} + \frac{y^2}{25} = 1;\ \frac{(x-3)^2}{10} + \frac{(y+1)^2}{25} = 1$
17. $\frac{x^2}{25} + \frac{y^2}{16} = 1;\ \frac{(x+2)^2}{25} + \frac{(y-1)^2}{16} = 1$
18. $\frac{x^2}{9} + \frac{y^2}{16} = 1;\ \frac{(x-2)^2}{9} + \frac{(y-2)^2}{16} = 1$
19. $\frac{x^2}{3} + \frac{y^2}{15} = 1;\ \frac{(x+1)^2}{3} + \frac{(y-1)^2}{15} = 1$
20. $\frac{x^2}{15} + \frac{y^2}{25} = 1;\ \frac{(x+1)^2}{15} + \frac{(y+1)^2}{25} = 1$

In problems 21–28, rewrite each equation in standard form. Then find the center, vertices, and foci. Also, sketch the graph.

21. $4x^2 + y^2 - 16x - 6y + 21 = 0$
22. $6x^2 + 5y^2 - 60x - 10y + 65 = 0$
23. $x^2 + 4y^2 + 2x - 8y + 1 = 0$
24. $x^2 + 4y^2 - 2x - 16y + 13 = 0$
25. $9x^2 + 4y^2 + 18x - 16y - 11 = 0$
26. $9y^2 + x^2 - 18y + 2x + 9 = 0$
27. $4x^2 + 9y^2 - 32x - 36y + 64 = 0$
28. $4x^2 + 9y^2 - 24x + 36y + 36 = 0$

In problems 29–34, find the standard equation of an ellipse with the given characteristics, and sketch the graph.

29. Vertices at $(1, -2)$, $(5, -2)$, $(3, -7)$, $(3, 3)$
30. Vertices at $(0, -1)$, $(12, -1)$, $(6, -4)$, $(6, 2)$
31. Center at $(-3, 1)$; major axis parallel to the y axis and 10 units long; minor axis 2 units long
32. Center at $(0, 0)$; axes parallel to the coordinate axes; containing the points $\left(\frac{3\sqrt{3}}{2}, 1\right)$ and $\left(2, \frac{2\sqrt{5}}{3}\right)$
33. Foci at $(1, 4)$ and $(3, 4)$; major axis 4 units long
34. Foci at $(-2, 3)$ and $(-2, 7)$; minor axis 1 unit long

GU In problems 35–40, use a graphing utility to graph each given equation.

35. $16x^2 + 25y^2 = 400$
36. $16x^2 + 15y^2 = 240$
37. $0.8(x + 0.1)^2 + 1.6y^2 = 1.28$
38. $3.2(x - 0.2)^2 + 1.6y^2 = 5.12$
39. $4x^2 + y^2 - 2y = 0$
40. $4x^2 + 9y^2 - 32x = 0$

Applying the Concepts

41. Dimensions of an Arch: An arch in the shape of the upper half of an ellipse with a horizontal major axis supports a railroad bridge over a river. The base of the arch is 26 meters across, and the highest part of the arch is 5 meters above the water level of the river (Figure 12).
(a) Sketch the ellipse in an xy coordinate system with the horizontal major axis on the x axis at the level of the river, and center at the origin.
(b) Find the standard equation of the ellipse in part (a).
(c) What is the clearance between the river and the arch at a point where one of the foci is located?
(d) Find the height of the arch 3 meters from the center of the base.

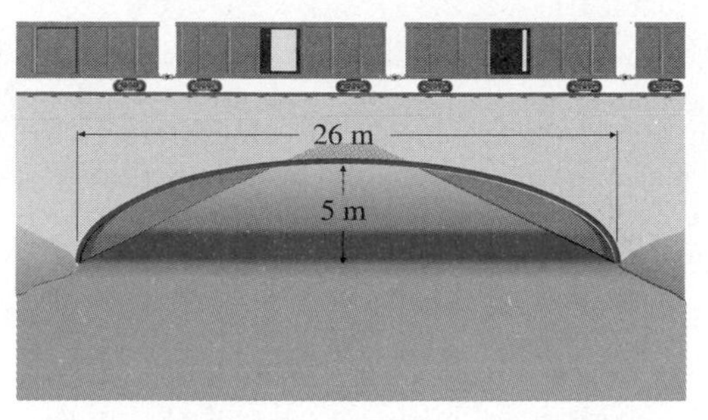

FIGURE 12

42. Dimensions of a Park: The Ellipse, a park located near the Washington Monument, is bounded by an elliptical path with a semimajor axis of 744 feet and a semiminor axis of 634 feet.
(a) Sketch the ellipse in an xy coordinate system with the horizontal major axis on the x axis and center at the origin.
(b) Find the standard equation of the ellipse in part (a).
(c) What is the distance between the two foci of this ellipse?

43. Earth's Orbit: The Earth moves in an elliptical orbit with the Sun at one focus. In early July, the Earth is farthest from the Sun, at a distance of 94,448,000 miles. In early January, it is closest to the Sun, at a distance of 91,341,000 miles.
(a) Sketch the orbit in an xy coordinate system with the horizontal major axis on the x axis and center at the origin.
(b) Find the standard equation of the ellipse in part (a).
(c) What is the distance between the Sun and the other focus on the Earth's orbit?

44. Communications Satellite: A communications satellite travels in an elliptical orbit around the Earth. The radius of the Earth is about 6400 kilometers, and its center is located at one focus of the orbit. Suppose that the satellite moves in such a way that when it is closest to the center of the Earth, it is 2000 kilometers from the surface, and when it is farthest from the center, it is 10,000 kilometers from the surface.
(a) Sketch the orbit in an xy coordinate system with major axis lying on the x axis and with center at the origin.
(b) Determine the standard equation of the ellipse in part (a).
(c) What is the height of the satellite above the surface of the Earth at a point corresponding to the focus of the elliptical orbit?

45. Whispering Gallery: The Mormon Tabernacle in Salt Lake City, Utah, has a whispering gallery. A vertical cross section forms the upper half of an ellipse with a horizontal major axis 250 feet long and a semiminor axis measured vertically from floor to ceiling of 75 feet.

(a) Sketch the ellipse in an xy coordinate system with the horizontal major axis on the x axis and center at the origin.
(b) Find the standard equation of the ellipse in part (a).
(c) What is the distance between the foci of the ellipse?

46. Pool Table: Suppose that a pool table is constructed so that its shape is elliptical, with a major axis of 6 feet and a minor axis of 4 feet. Assume that a single pocket is located at one of the foci.

(a) Sketch the ellipse in an xy coordinate system with the horizontal major axis on the x axis and center at the origin.
(b) Find the standard equation of the ellipse in part (a).
(c) Where should one of the balls be located so that when it is hit, it bounces off the cushion directly into the pocket? Explain.

Developing and Extending the Concepts

47. Find the standard equation of an ellipse so that if $P = (x, y)$ is a point on the ellipse, then the sum of the distances from P to $F_1 = (0, -12)$ and to $F_2 = (0, 12)$ is 26 units.

48. Suppose that the foci of an ellipse are $F_1 = (0, -c)$ and $F_2 = (0, c)$, where $c > 0$, and $2a$ is the constant referred to in the definition of an ellipse. If we let $P = (x, y)$ be any point on the ellipse (Figure 13), then in triangle F_1PF_2, we have $d(P, F_1) + d(P, F_2) = 2a$.

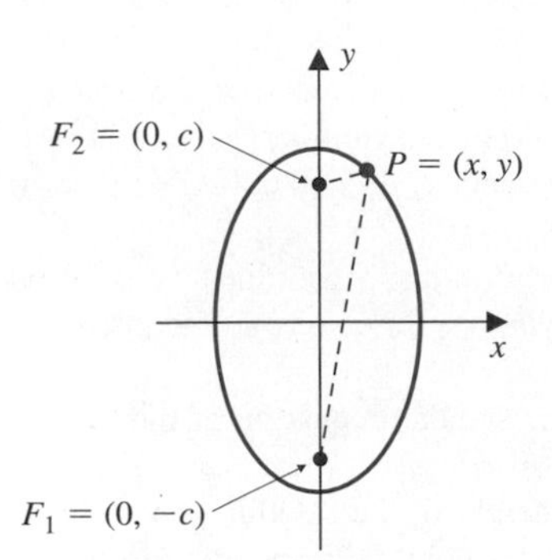

FIGURE 13

(a) Use the distance formula to derive an equation that relates x and y.
(b) Simplify the equation found in part (a) and use the substitution $b^2 = a^2 - c^2$ to show that it is equivalent to the equation $(x^2/b^2) + (y^2/a^2) = 1$.

49. Given the equation: $\dfrac{x^2}{a^2} + \dfrac{y^2}{b^2} = 1$

(a) What is the graph of this equation if $a = b$?
(b) Is this the equation of an ellipse? If so, where are the foci? Explain.

50. **(a)** A line segment that contains a focus, is perpendicular to the major axis, and has end points on the ellipse is called a **focal chord**, or **latus rectum**, of the ellipse (Figure 14). Show that the length of a focal chord of the ellipse with the equation $(x^2/a^2) + (y^2/b^2) = 1$ is $(2b^2)/a$.

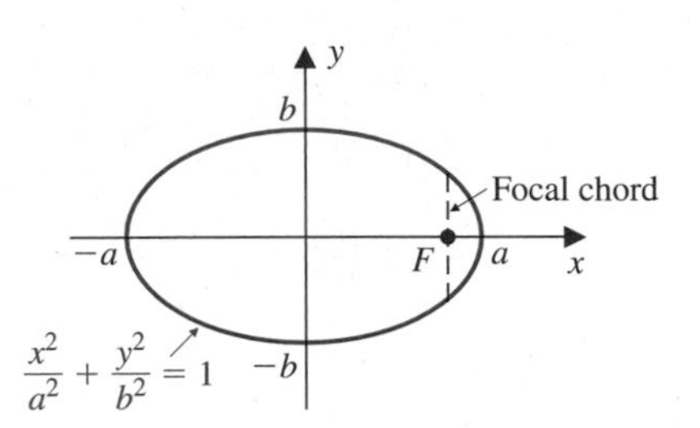

FIGURE 14

(b) Find the length of a focal chord of an ellipse whose equation is given by $9x^2 + 16y^2 = 144$.

51. Figure 15 shows a rectangle inscribed in the region above the x axis and under the half-ellipse with equation $9x^2 + 16y^2 = 144$, where $y \geq 0$.

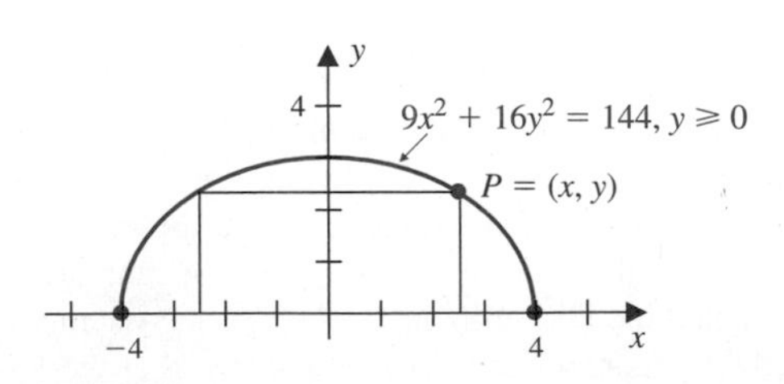

FIGURE 15

(a) Find the area A of the rectangle in terms of the coordinates of the point $P = (x, y)$ where the upper right corner of the rectangle is on the graph of the ellipse.

(b) Rewrite the equation in part (a) so that the area A is expressed as a function of x.

GU (c) Use a graphing utility to graph the function in part (b).

(d) From the graph in part (c), what value of x gives the maximum area? What is the maximum area? Round off the answers to two decimal places.

52. (a) Find the domain of the function

$$y = \sqrt{b^2\left(1 - \frac{x^2}{a^2}\right)}$$

(b) What is the relationship between the graph of this function and the graph of the ellipse with equation $b^2x^2 + a^2y^2 = a^2b^2$?

OBJECTIVES

1. Graph hyperbolas with center at (0, 0).
2. Graph hyperbolas with center at (h, k).
3. Solve applied problems.

9.3 Hyperbolas

Hyperbolas are of practical importance in fields ranging from engineering to navigation. For instance, a spacecraft moving with more than enough kinetic energy to escape the Sun's gravitational pull traces out one branch of a hyperbola. The geometric definition of a hyperbola follows:

DEFINITION
HYPERBOLA

A **hyperbola** is the set of all points P in the plane such that the absolute value of the difference of the distances from P to two points F_1 and F_2 is a constant positive number. Here, F_1 and F_2 are called the **focal points**, or **foci**, of the hyperbola. The midpoint C of the line segment $\overline{F_1F_2}$ is called the **center** of the hyperbola.

Figure 1 shows a geometric description of a hyperbola with foci F_1 and F_2 and with center C. The points P_1 and P_2 are on the hyperbola, and

$$|d_1 - m_1| = |d_2 - m_2| = k$$

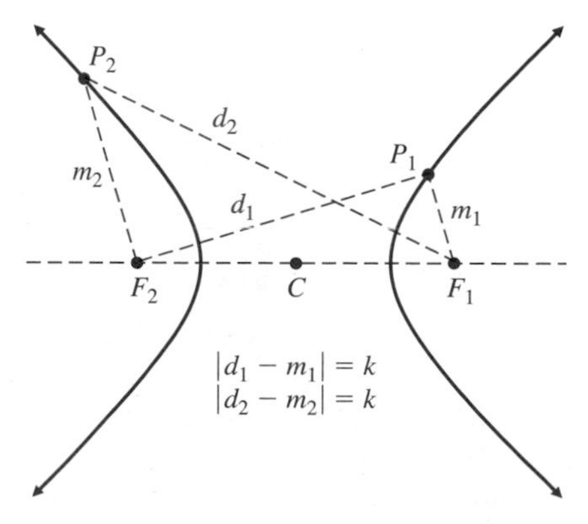

FIGURE 1

Graphing Hyperbolas with Center at (0, 0)

Suppose that a hyperbola is positioned in a Cartesian coordinate system in such a way that the center is at the origin and two foci $F_1 = (-c, 0)$ and $F_2 = (c, 0)$ lie on the x axis, with $c > 0$. To derive the standard equation of the hyperbola analytically, we let $P = (x, y)$ be a point on the hyperbola, and assume that the constant difference k equals $2a$ (Figure 2).

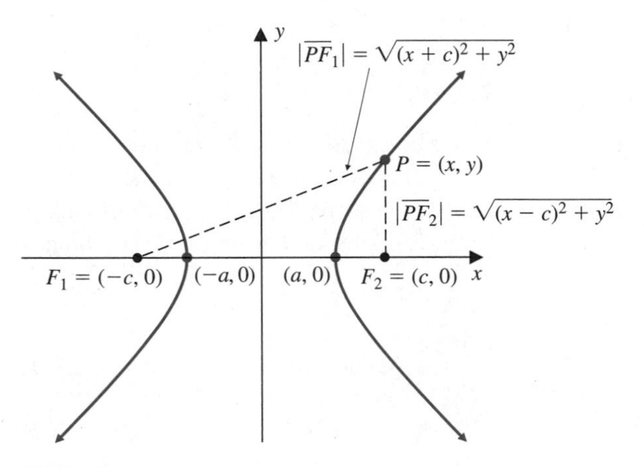

FIGURE 2

Since $P = (x, y)$, we have

$$|d(P, F_1) - d(P, F_2)| = 2a$$

That is,

$$\left|\sqrt{(x + c)^2 + y^2} - \sqrt{(x - c)^2 + y^2}\right| = 2a$$

If we manipulate this equation in the same way we did with the ellipse, we get

$$\frac{x^2}{a^2} - \frac{y^2}{c^2 - a^2} = 1$$

Note that $c > a$. If we let $b^2 = c^2 - a^2$, then we have

$$\boxed{\frac{x^2}{a^2} - \frac{y^2}{b^2} = 1}$$

This equation is called the **standard equation** of a hyperbola. In contrast to an equation of an ellipse (where a was *always* greater than or equal to b), a may be greater than, equal to, or less than b in an equation of a hyperbola. The coordinates of every point (x, y) on the hyperbola in Figure 2 satisfy the boxed equation; and conversely, if the coordinates (x, y) satisfy the boxed equation, then by reversing the preceding steps, we can show that the point (x, y) is on the hyperbola.

Using the boxed equation, we can derive the following characteristics of the graph of a hyperbola (see Figure 2):

1. We determine the x intercepts by setting $y = 0$ to obtain $-a$ and a. The corresponding points on the graph, $(-a, 0)$ and $(a, 0)$, are referred to as the **vertices** of the hyperbola.
2. The **transverse axis** is the line segment that has the vertices as its end points. Note that the length of the transverse axis is $2a$, and a is the distance between the center of the hyperbola and each of its two vertices. The line segment of length $2b$ that contains the center of the hyperbola as its midpoint and is perpendicular to the transverse axis is called the **conjugate axis**. The hyperbola does not intersect the conjugate axis.
3. Because of the y^2 and x^2 terms, the graph is symmetric with respect to both the x axis (transverse axis) and the y axis (conjugate axis), as well as the origin. The line determined by the transverse axis is called the **axis of symmetry** of the hyperbola. In this case, the axis of symmetry is the x axis. Also, the foci lie on the axis of symmetry c units from the center at points $(-c, 0)$ and $(c, 0)$, where $c^2 = a^2 + b^2$.
4. A distinguishing feature of the hyperbola is that it has two **oblique asymptotes**. To determine the equations of these asymptotes, we rewrite

$$\frac{x^2}{a^2} - \frac{y^2}{b^2} = 1 \quad \text{as} \quad y^2 = b^2\left(\frac{x^2}{a^2} - 1\right)$$

and solve for y in terms of x as follows:

$$y^2 = b^2\left(\frac{x^2}{a^2} - 1\right) = \left(\frac{b^2x^2}{a^2}\right)\left(1 - \frac{a^2}{x^2}\right)$$

$$y = \pm\left(\frac{bx}{a}\right)\sqrt{1 - \frac{a^2}{x^2}}$$

provided that $x \geq a$. Note that, as $x \to \infty$, the value $a^2/x^2 \to 0$, and the value of the expression under the radical will come closer and closer to 1. In other words, for large values of $|x|$, points on the hyperbola come very close to the lines

$$y = \frac{b}{a}x \quad \text{and} \quad y = -\frac{b}{a}x$$

Thus, these lines are asymptotes of the hyperbola.

For example, as Figure 3 illustrates, the graph of the equation

$$\frac{x^2}{16} - \frac{y^2}{9} = 1$$

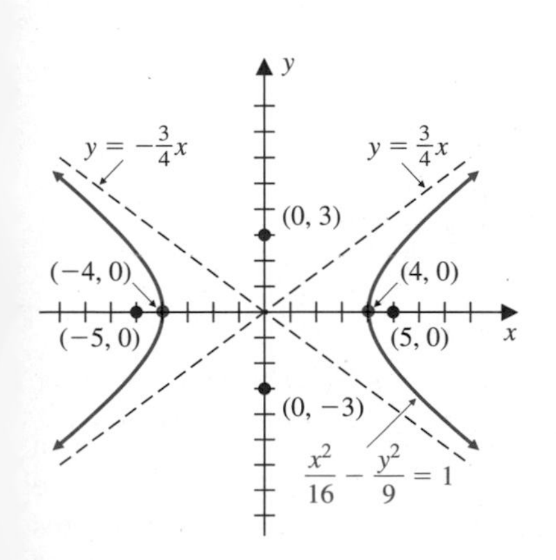

FIGURE 3

is a hyperbola with $a^2 = 16$ and $b^2 = 9$, so that $a = 4$ and $b = 3$. Using $c^2 = a^2 + b^2$, we have $c = \sqrt{16 + 9} = 5$. Therefore, the vertices of this hyperbola

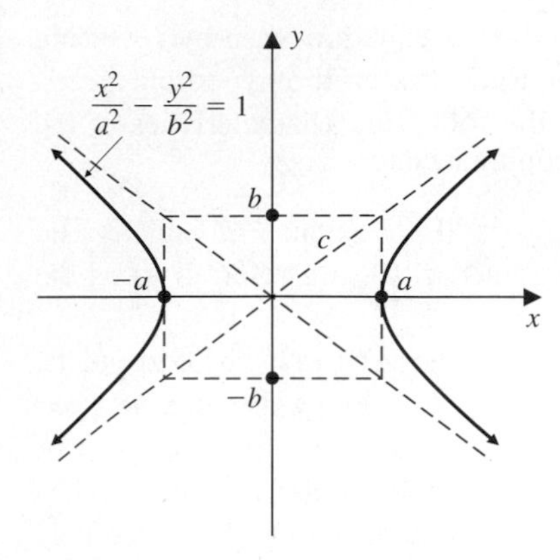

FIGURE 4

are (4, 0) and (−4, 0), and the foci are (5, 0) and (−5, 0). The equations of the asymptotes are

$$y = \frac{b}{a}x = \frac{3}{4}x \quad \text{and} \quad y = -\frac{b}{a}x = -\frac{3}{4}x$$

Although the asymptotes of the hyperbola are not part of the hyperbola itself, they are helpful guides in sketching its graph. For instance, to sketch the graph of the equation $(x^2/a^2) - (y^2/b^2) = 1$, begin by sketching the rectangle with height $2b$ and horizontal base $2a$ whose center is at the origin (Figure 4). The asymptotes are then drawn through the two diagonals of this rectangle. If we keep in mind that the vertices of the hyperbola are located at the midpoints of the left and right sides of the rectangle, and that the hyperbola approaches the asymptotes as it moves out away from the vertices, then it is easy to sketch the graph (Figure 4).

To find an equation of a hyperbola that has a vertical transverse axis, center at the origin, vertices $V_1 = (0, -a)$ and $V_2 = (0, a)$, and foci $F_1 = (0, -c)$ and $F_2 = (0, c)$, we interchange x with y in the above standard equation. Figure 5 shows the standard equations and graphs of hyperbolas with centers at the origin and horizontal or vertical transverse axes.

There is no requirement that $a \geq b$ in the hyperbola equations.

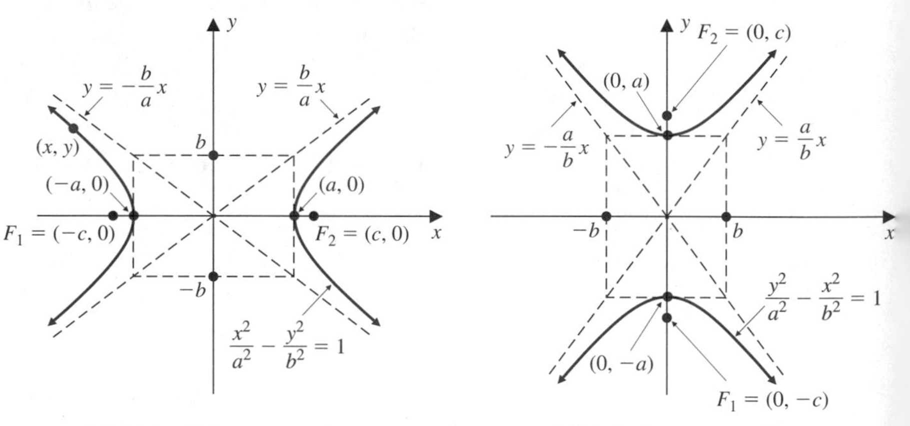

FIGURE 5
Standard equations of hyperbolas with center at (0, 0), horizontal or vertical transverse axis, and foci F_1 and F_2; $c^2 = a^2 + b^2$

EXAMPLE 1 **Graphing a hyperbola with center at (0, 0)**

Show that the equation $25x^2 - 16y^2 = 400$ represents a hyperbola by rewriting it in standard form. Then find the vertices, foci, and equations of the asymptotes. Also, sketch the graph.

Solution After dividing both sides of the equation by 400, we obtain the equation

$$\frac{x^2}{16} - \frac{y^2}{25} = 1$$

which is the standard equation of a hyperbola with center at the origin, horizontal transverse axis, and vertical conjugate axis. Since $a = 4$ and $b = 5$, it follows that the transverse axis has length $2a = 8$, and the conjugate axis has length $2b = 10$. The end points of the transverse and conjugate axes determine a rectangle whose extended diagonals give us the asymptotes (Figure 6). The vertices are $V_1 = (-4, 0)$ and $V_2 = (4, 0)$. To find the foci, we calculate

$$c^2 = a^2 + b^2 = 16 + 25 = 41$$

so $c = \sqrt{41}$, and the foci are $F_1 = \left(-\sqrt{41}, 0\right)$ and $F_2 = \left(\sqrt{41}, 0\right)$. The equations of the asymptotes are given by

$$y = \frac{b}{a}x = \frac{5}{4}x \qquad \text{and} \qquad y = -\frac{b}{a}x = -\frac{5}{4}x$$

The graph of the hyperbola is displayed in Figure 6.

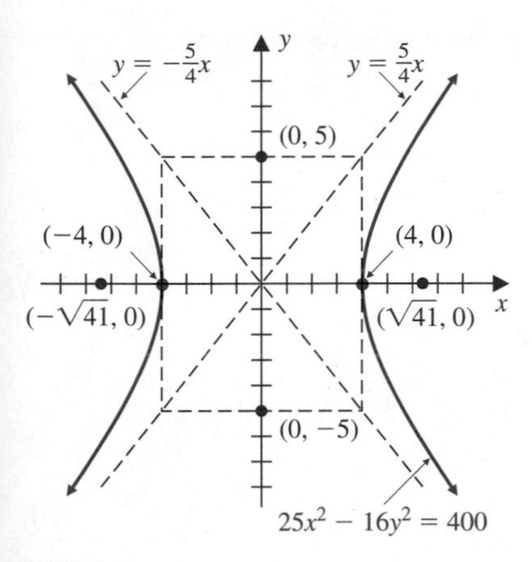

FIGURE 6

EXAMPLE 2 **Finding an equation of a hyperbola**

Find a standard equation of a hyperbola with vertices at (0, 3) and (0, −3) and foci at (0, 5) and (0, −5). Sketch the graph and find its asymptotes.

Solution Since the vertices and foci are on the y axis, the transverse axis is on the y axis. We know that $a = 3$ and $c = 5$, because the distance between the center (0, 0) and each vertex is 3, and the distance between the center (0, 0) and each focus is 5. So, $b^2 = c^2 - a^2 = 25 - 9 = 16$, or $b = 4$. We use the standard equation

$$\frac{y^2}{a^2} - \frac{x^2}{b^2} = 1$$

to get

$$\frac{y^2}{9} - \frac{x^2}{16} = 1$$

Because $b = 4$ and $a = 3$, the equations of the asymptotes are

$$y = \frac{a}{b}x = \frac{3}{4}x \qquad \text{and} \qquad y = -\frac{a}{b}x = -\frac{3}{4}x \quad \text{(Figure 7)}$$

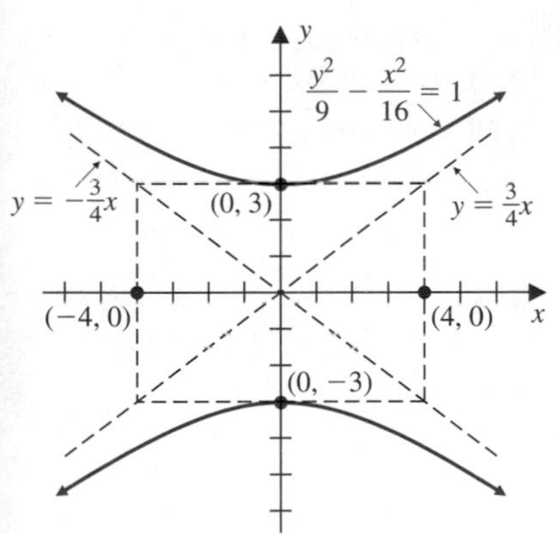

FIGURE 7

Graphing Hyperbolas with Center at (h, k)

If we apply horizontal and vertical shifting to the hyperbolas in Figure 5, we generate new hyperbolas with centers at $(h, k) \neq (0, 0)$. For instance, if we shift the hyperbola $(x^2/a^2) - (y^2/b^2) = 1$, h $(h > 0)$ units to the right and k $(k > 0)$ units upward, we generate the new hyperbola with standard equation

$$\frac{(x-h)^2}{a^2} - \frac{(y-k)^2}{b^2} = 1$$

EXAMPLE 3 Graphing a hyperbola with center at (h, k)

Apply horizontal and vertical shifting to the graph of the equation

$$\frac{x^2}{16} - \frac{y^2}{9} = 1$$

to generate the graph of the equation

$$\frac{(x-2)^2}{16} - \frac{(y-3)^2}{9} = 1$$

Find the center, vertices, foci, and equations of the asymptotes of the new hyperbola.

Solution The diagram indicates the shifts:

Indicates a horizontal shift 2 units to the right

$$\frac{x^2}{16} - \frac{y^2}{9} = 1 \qquad \frac{(x-2)^2}{16} - \frac{(y-3)^2}{9} = 1$$

Indicates a vertical shift 3 units up

The new hyperbola is obtained by shifting the first hyperbola 2 units to the right and 3 units up. Figure 8 shows the graphs of both hyperbolas.

The center, vertices, foci, and asymptotes of both hyperbolas are listed in Table 1:

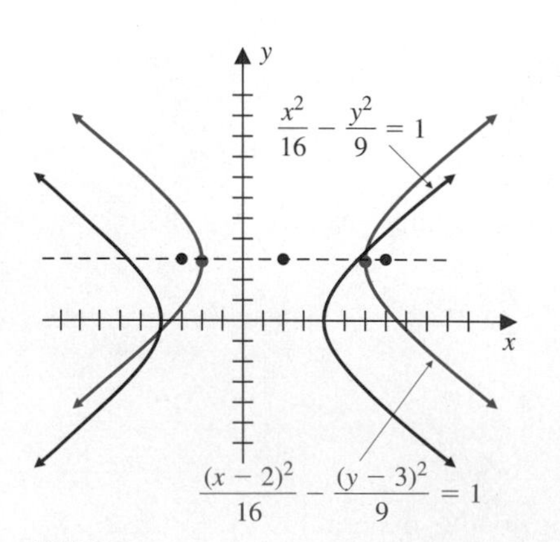

FIGURE 8

TABLE 1

	$\frac{x^2}{16} - \frac{y^2}{9} = 1$	Shift	$\frac{(x-2)^2}{16} - \frac{(y-3)^2}{9} = 1$
Center	(0, 0)	Right 2 units, up 3 units	(2, 3)
Vertices	(−4, 0) and (4, 0)	Right 2 units, up 3 units	$V_1 = (-4+2, 0+3) = (-2, 3)$ $V_2 = (4+2, 0+3) = (6, 3)$
Foci	(−5, 0), (5, 0)	Right 2 units, up 3 units	$F_1 = (-5+2, 0+3) = (-3, 3)$ $F_2 = (5+2, 0+3) = (7, 3)$
Asymptotes	$y = \frac{3}{4}x$, $y = -\frac{3}{4}x$	Right 2 units, up 3 units	$y - 3 = \frac{3}{4}(x-2)$ or $y = \frac{3}{4}x + \frac{3}{2}$ $y - 3 = -\frac{3}{4}(x-2)$ or $y = -\frac{3}{4}x + \frac{9}{2}$

Horizontal and vertical shifting of a hyperbola centered at (0, 0) with horizontal or vertical transverse axis to a hyperbola with center (h, k) results in one of the following two situations:

STANDARD EQUATIONS FOR HYPERBOLAS CENTERED AT (h, k)

(i) $\dfrac{(x-h)^2}{a^2} - \dfrac{(y-k)^2}{b^2} = 1$ The transverse axis is horizontal

(ii) $\dfrac{(y-k)^2}{a^2} - \dfrac{(x-h)^2}{b^2} = 1$ The transverse axis is vertical

The *general equation* of a hyperbola with horizontal or vertical transverse axis is

$$Ax^2 + Cy^2 + Dx + Ey + F = 0$$

with A and C having opposite signs. This equation can be written in standard form by completing the squares.

EXAMPLE 4 Writing an equation of a hyperbola in standard form

Rewrite the general equation of the hyperbola

$$4x^2 - y^2 - 8x + 2y + 7 = 0$$

in standard form by completing the squares. Find the center, vertices, foci, and equations of the asymptotes. Also, sketch the graph.

Solution First, we rewrite the equation in standard form as follows:

$$\begin{aligned}
(4x^2 - 8x \qquad) - (y^2 - 2y \qquad) &= -7 && \text{Isolate variable terms}\\
4(x^2 - 2x + 1) - (y^2 - 2y + 1) &= -7 + 4 - 1 && \text{Complete the squares}\\
4(x-1)^2 - (y-1)^2 &= -4 && \text{Factor and simplify}\\
\frac{(y-1)^2}{4} - \frac{(x-1)^2}{1} &= 1 && \text{Divide each side by } -4
\end{aligned}$$

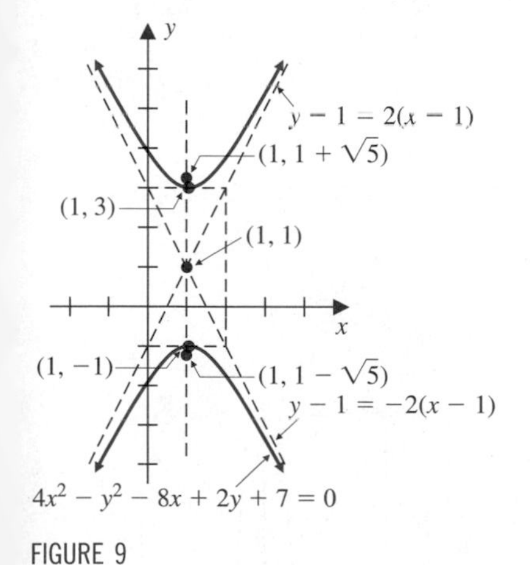

FIGURE 9

This is the standard equation of a hyperbola with center $(h, k) = (1, 1)$. In this situation, $a = 2$ and $b = 1$, so $c^2 = a^2 + b^2 = 4 + 1 = 5$, or $c = \sqrt{5}$. Since the transverse axis is vertical, the vertices are $V_1 = (1, 1 + 2) = (1, 3)$ and $V_2 = (1, 1 - 2) = (1, -1)$, and the foci are $F_1 = \left(1, 1 + \sqrt{5}\right)$ and $F_2 = \left(1, 1 - \sqrt{5}\right)$. The equations of the asymptotes are

$$y - 1 = \tfrac{2}{1}(x - 1) \quad \text{and} \quad y - 1 = -\tfrac{2}{1}(x - 1)$$

or

$$y = 2x - 1 \quad \text{and} \quad y = -2x + 3$$

Figure 9 shows the graph of the hyperbola, obtained by drawing a rectangle with center at the point (1, 1) and dimensions equal to the lengths of the transverse and conjugate axes, which are 4 and 2, respectively. The asymptotes are extensions of the diagonals of the rectangle. ■

To graph a hyperbola using a graphing utility, we first solve the equation for y in terms of x and then graph both of the resulting functions in the same viewing window.

EXAMPLE 5 Graphing a hyperbola by using a graphing utility

GU

Use a graphing utility to sketch the graph of

$$-5x^2 + 3y^2 + 20x + 6y - 32 = 0$$

Solution First, we rewrite the equation as a standard quadratic equation in y:

$$3y^2 + 6y + (-5x^2 + 20x - 32) = 0$$

Next, we use the quadratic formula to solve for y in terms of x:

$$y = \frac{-6 \pm \sqrt{36 - 4(3)(-5x^2 + 20x - 32)}}{6}$$

$$= \frac{-6 \pm \sqrt{36 + 60x^2 - 240x + 384}}{6}$$

$$= \frac{-6 \pm \sqrt{60x^2 - 240x + 420}}{6}$$

to obtain

$$y_1 = \frac{-6 + \sqrt{60x^2 - 240x + 420}}{6} \quad \text{and} \quad y_2 = \frac{-6 - \sqrt{60x^2 - 240x + 420}}{6}$$

By graphing these last two equations in the same viewing window, we get the graph of the hyperbola (Figure 10).

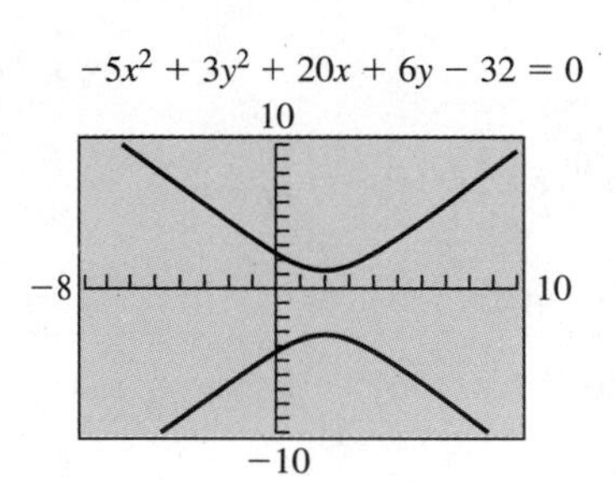

FIGURE 10

Solving Applied Problems

Like parabolas and ellipses, hyperbolas have a reflecting property. This property is used in modern navigational devices. The *long-range navigation system (LORAN)* enables a ship or an aircraft to determine its exact location. The reflecting property of hyperbolas is also used in the design of camera and telescope lenses. Cooling towers at large atomic power or steam stations also have hyperbolic cross sections (Figure 11).

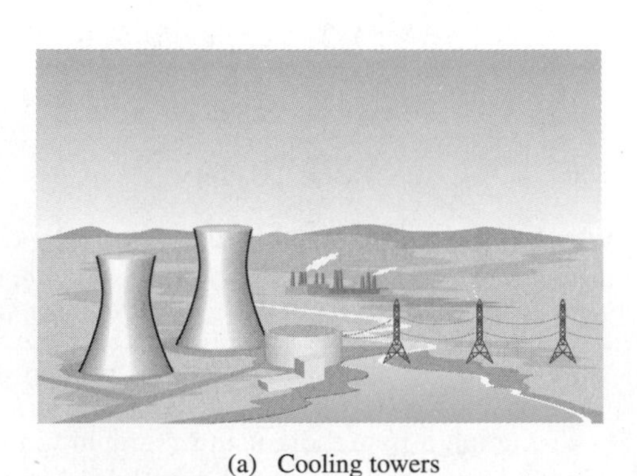

(a) Cooling towers

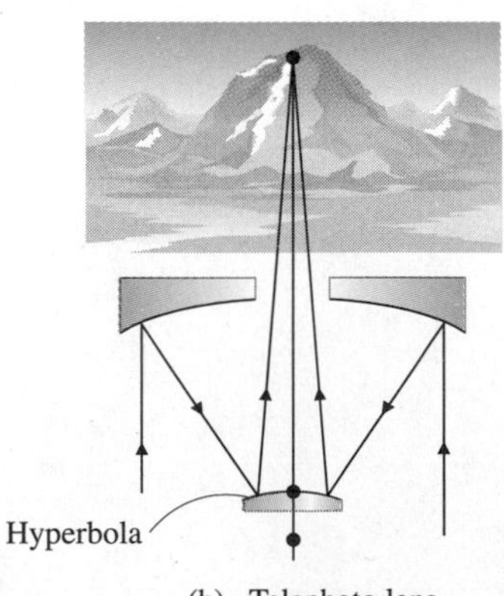

(b) Telephoto lens

FIGURE 11

EXAMPLE 6 Locating a ship using LORAN

Two Coast Guard stations, S_1 and S_2, are located 150 miles apart on an east–west line (Figure 12a). A distress signal from a ship is received at slightly different times by the two stations. It is determined that the difference in the distances from the ship to each station is consistently 100 miles.

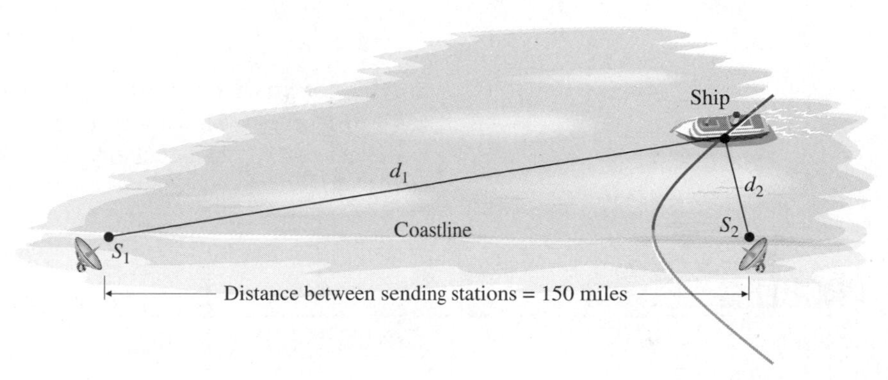

FIGURE 12a

(a) Sketch the position of the ship in a Cartesian plane so that the two stations lie on the x axis and the midpoint between them coincides with the origin. Describe the path of the ship.
(b) Find the standard equation of the curve in part (a).
(c) If, at a certain time, the ship is east of the line containing the conjugate axis and 40 miles north of the line between S_1 and S_2, locate the position of the ship. Round off the answer to the nearest mile.

Solution

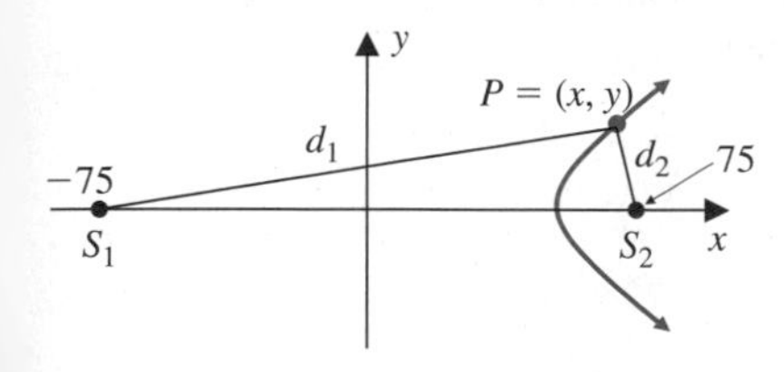

FIGURE 12b

(a) First, we sketch an xy coordinate system, with the origin midway between two stations S_1 and S_2 on the x axis. The ship is located at some point $P = (x, y)$ in the plane, where $|d(P, S_1) - d(P, S_2)| = |d_1 - d_2| = 100$ (Figure 12b).
(b) By definition, the path of the ship is hyperbolic. Because of the orientation, the standard equation of the hyperbolic path has the form

$$\frac{x^2}{a^2} - \frac{y^2}{b^2} = 1$$

Because the stations are 150 miles apart, they are located at the foci 75 miles from each side of the origin. Therefore, $c = 75$. Since $|d_1 - d_2| = 2a$, then $2a = 100$, and $a = 50$. Also, $c^2 = a^2 + b^2$, so

$$b^2 = c^2 - a^2 = (75)^2 - (50)^2 = 3125$$

Thus, the standard equation is

$$\frac{x^2}{2500} - \frac{y^2}{3125} = 1$$

(c) To locate the position of the ship, we replace the y coordinate of P by 40 to get

$$\frac{x^2}{2500} - \frac{1600}{3125} = 1$$

Solving for x and rounding off the answer to the nearest mile, we find that $x = -61$ or $x = 61$. However, since the ship is east of the conjugate axis, we select $x = 61$. Thus, relative to the selected coordinate, the ship is located at (61, 40).

PROBLEM SET 9.3

Mastering the Concepts

In problems 1–8, find the coordinates of the vertices and foci, and find the equations of the asymptotes of each hyperbola. Also, sketch the graph.

1. $\frac{x^2}{9} - y^2 = 1$
2. $x^2 - \frac{y^2}{9} = 1$
3. $y^2 - x^2 = 9$
4. $16y^2 - 4x^2 = 48$
5. $36x^2 - 9y^2 = 1$
6. $36x^2 - 49y^2 = 1764$
7. $4y^2 + 20 - 5x^2 = 0$
8. $3y^2 = x^2 + 1$

In problems 9–14, find the standard equation for the hyperbola that satisfies the given conditions and sketch the graph.

9. Vertices at $(-16, 0)$ and $(16, 0)$; asymptotes $y = \pm\frac{5}{4}x$
10. Vertices at $(-3, 0)$ and $(3, 0)$; asymptotes $y = \pm 2x$
11. Transverse axis of length 8 units; foci at $(0, 5)$ and $(0, -5)$
12. Center at the origin; a vertex at $(3, 0)$; a focus at $(4, 0)$
13. Center at $(0, 0)$; a horizontal transverse axis; contains the points $(2, 5)$ and $(3, -10)$
14. Center at $(0, 0)$; a horizontal transverse axis; contains the points $(4, 3)$ and $(-7, 6)$

In problems 15–20, apply horizontal and vertical shifting to the graph of the first equation to generate the graph of the second. Also, find the center, vertices, foci, and asymptotes of the new graph.

15. $\frac{x^2}{25} - \frac{y^2}{4} = 1$; $\frac{(x-5)^2}{25} - \frac{(y-4)^2}{4} = 1$
16. $\frac{y^2}{16} - \frac{x^2}{25} = 1$; $\frac{(y-3)^2}{16} - \frac{(x+3)^2}{25} = 1$
17. $\frac{y^2}{16} - \frac{x^2}{16} = 1$; $\frac{(y+3)^2}{16} - \frac{(x+2)^2}{16} = 1$
18. $\frac{x^2}{4} - \frac{y^2}{20} = 1$; $\frac{(x+1)^2}{4} - \frac{(y+2)^2}{20} = 1$
19. $\frac{x^2}{4} - \frac{y^2}{9} = 1$; $\frac{(x-1)^2}{4} - \frac{(y+2)^2}{9} = 1$
20. $\frac{x^2}{9} - \frac{y^2}{\frac{9}{4}} = 1$; $\frac{(x+3)^2}{9} - \frac{(y+2)^2}{\frac{9}{4}} = 1$

In problems 21–28, rewrite each equation in standard form by completing the squares. Then find the coordinates of the center, foci, and vertices. Also, find the equations of the asymptotes and sketch the graph.

21. $4x^2 - 3y^2 - 32x + 6y + 73 = 0$
22. $4x^2 - 9y^2 - 32x + 36y + 27 = 0$
23. $9y^2 - 25x^2 + 72y - 100x + 269 = 0$
24. $9x^2 - 16y^2 - 90x - 256y = 223$
25. $4x^2 - y^2 + 8x - 2y + 6 = 0$
26. $4y^2 - x^2 + 40y - 4x + 60 = 0$
27. $4x^2 - 9y^2 - 8x + 36y - 68 = 0$
28. $9x^2 - 16y^2 - 36x - 64y + 116 = 0$

In problems 29–34, find the standard equation for the hyperbola that satisfies the given conditions and sketch the graph.

29. Center at $(3, -4)$; horizontal transverse axis of length 6; conjugate axis of length 4

30. Center at $(-1, 2)$; vertical transverse axis of length 10; conjugate axis of length 5

31. Vertices at $(-1, 4)$ and $(-1, 6)$; foci at $(-1, 3)$ and $(-1, 7)$

32. x intercepts at -5 and 5; asymptotes $y = \pm 3x$

33. Vertices at $(6, 3)$ and $(2, 3)$; foci at $(7, 3)$ and $(1, 3)$

34. Foci at $(-1, -2)$ and $(9, -2)$; horizontal transverse axis of length 6

GU In problems 35–40, use a graphing utility to graph each hyperbola by solving for y in terms of x.

35. $\frac{x^2}{4} - \frac{y^2}{9} = 1$

36. $\frac{y^2}{1} - \frac{x^2}{10} = 1$

37. $\frac{(y-1)^2}{9} - \frac{(x+1)^2}{4} = 1$

38. $\frac{(x+2)^2}{25} - \frac{(y-1)^2}{4} = 1$

39. $x^2 - 2y^2 - 2x - 12y - 35 = 0$

40. $-3x^2 + 2y^2 + 4x + 6y - 49 = 0$

Applying the Concepts

41. LORAN Navigation: Two LORAN stations are 180 miles apart on an east–west line. A ship receiving signals from these stations determines that the difference in the distances from the ship to each station is consistently 70 miles.

(a) Sketch the path of the ship in an xy coordinate system. Locate the position of the ship so that the origin is midway between the stations, which lie on the x axis.

(b) Find the standard equation of the curve in part (a).

(c) If, at a certain time, the ship is east of the conjugate axis and 50 miles north of the line between the two stations, locate the position of the ship. Round off the answer to the nearest mile.

42. Locating Explosions: Two microphones are located at two detection stations, A and B, 1150 meters apart on an east–west line (Figure 13). An explosion occurs at an unknown location P. The sound of the explosion is detected by the microphone at station A exactly 2 seconds before it is detected by the microphone at station B. Assume that the sound travels at a uniform speed of 330 meters per second.

(a) Sketch the curve that describes the possible location of P in an xy coordinate system so that the origin is at the middle of the horizontal axis between A and B.

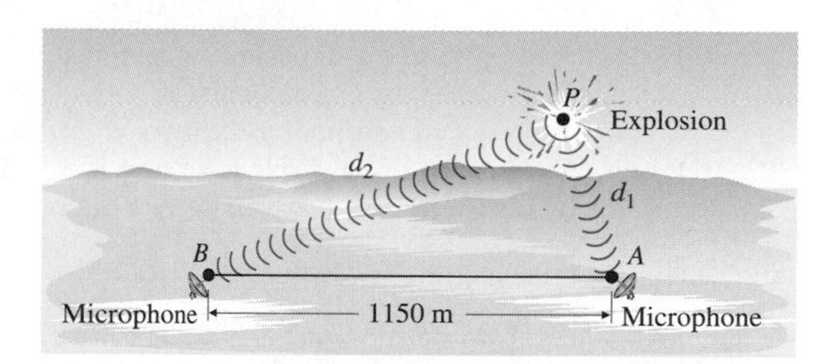

FIGURE 13

(b) Find the standard equation of the curve in part (a).

(c) Can the explosion at P be pinpointed at an exact location? Explain.

43. Comet Orbit: Figure 14 shows a hyperbolic orbit in a Cartesian plane of a comet passing through the solar system with the Sun at one focus. Suppose that the comet is moving away from the Earth (located at the origin) on a path that is asymptotic to the line $y = 2x$.

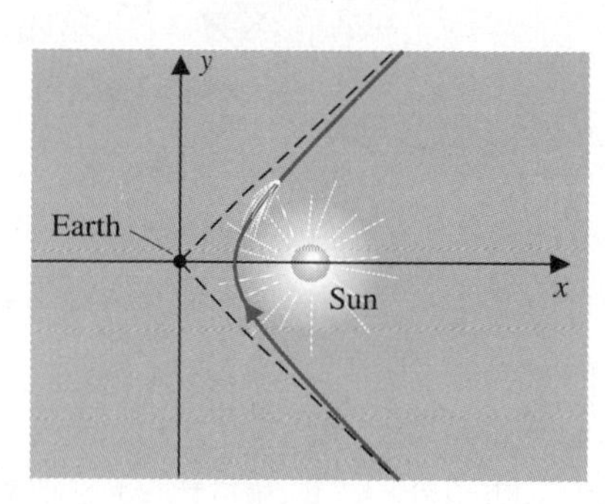

FIGURE 14

(a) Find an equation of the hyperbola if the closest distance between the comet and the Earth is 41.6 million miles.

(b) Approximate the distance between the Sun and the Earth to the nearest million miles.

44. Airplane Flight: An airplane starts at a point directly north of the origin and flies northeasterly in a hyperbolic path. Transmitters located at the origin and 150 miles directly north of the origin send out synchronized radio signals. Instruments in the airplane measure the difference between the arrival times of the signals. By knowing the speed of the radio signals, it is determined that the distance from the plane to the transmitter at the origin is 50 miles farther than the distance to the other transmitter.

(a) Sketch the path of the airplane in a coordinate system that satisfies the given conditions.
(b) Find the standard equation of the curve in part (a).
(c) Find the location of the airplane when it is 40 miles directly east of the line between the two transmitters.

Developing and Extending the Concepts

45. Find the standard equation of a hyperbola such that the difference of the distances from a point $P = (x, y)$ on the hyperbola to the two points $F_1 = (-5, 0)$ and $F_2 = (5, 0)$ is 8.

46. **(a)** The **focal chord** of a hyperbola is a line segment with end points on the hyperbola, containing a focus, and perpendicular to the transverse axis (Figure 15). Show that the length of the focal chord of the hyperbola $(x^2/a^2) - (y^2/b^2) = 1$ is $(2b^2)/a$.

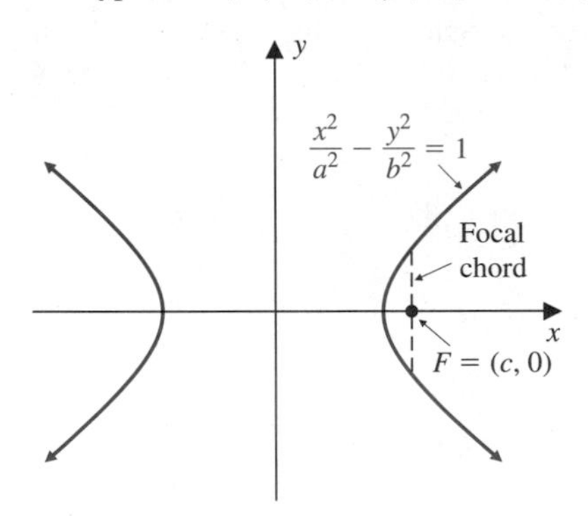

FIGURE 15

(b) Find the length of the focal chord for each of the following hyperbolas:

(i) $\frac{x^2}{9} - \frac{y^2}{4} = 1$ **(ii)** $\frac{y^2}{25} - \frac{x^2}{10} = 1$

47. GU Use a graphing utility to display the points of intersection of the ellipse with equation $4x^2 + 7y^2 = 32$ and the hyperbola with equation $11y^2 - 3x^2 = 41$.
(a) Approximate the locations of the points of intersection by using the ZOOM and TRACE features. Round off to the nearest integer.
(b) Approximate the locations of the points of intersection algebraically.
(c) Compare the results in part (a) to the results in part (b).

48. **(a)** Graph the *unit hyperbola* $x^2 - y^2 = 1$ and determine equations for the asymptotes.
(b) Verify that any point of the form

$$\left(\frac{e^u + e^{-u}}{2}, \frac{e^u - e^{-u}}{2}\right) \quad \text{where } u \text{ represents any real number}$$

lies on the graph of $x^2 - y^2 = 1$.

49. The standard equations of hyperbolas centered at the origin can be written in one of the forms

$$\frac{x^2}{a^2} - \frac{y^2}{b^2} = 1 \quad \text{or} \quad \frac{y^2}{a^2} - \frac{x^2}{b^2} = 1$$

(a) Sketch the graphs of both hyperbolas in the same coordinate plane for $a = 4$ and $b = 3$.
(b) Discuss the relationship between the two graphs.

50. Given the hyperbola with standard equation

$$\frac{x^2}{a^2} - \frac{y^2}{b^2} = 1 \quad \text{where } c^2 = a^2 + b^2$$

(a) Draw the rectangle with center at the origin and side lengths determined by the transverse and conjugate axes.
(b) Graph the circle with equation $x^2 + y^2 = c^2$ in the same coordinate system used in part (a).
(c) Locate the points of intersection of the rectangle and circle drawn in parts (a) and (b).
(d) Explain why the circle in part (b) contains both foci of the hyperbola.

51. Determine the type of curve represented by the equation

$$\frac{x^2}{h} + \frac{y^2}{h - 9} = 1$$

in each of the following cases:
(a) $h > 9$ **(b)** $0 < h < 9$ **(c)** $h < 0$

OBJECTIVES

1. Describe a conic using eccentricity.
2. Identify polar equations of conics.
3. Solve applied problems.

9.4 Eccentricity and Polar Equation Forms

So far, we have derived equations in Cartesian form for the parabola, ellipse, and hyperbola by using three different geometric definitions—one for each type of curve. By using the idea of *eccentricity*, it is possible to give a unified geometric definition for these curves.

It is traditional to use the letter e to denote the eccentricity. This e should not be confused with the natural logarithm base.

Describing a Conic Using Eccentricity

The basic idea here is to describe a conic in terms of a fixed straight line l called the **directrix**, a fixed point F not on l called the **focus**, and a positive constant e called the **eccentricity**.

A point P in the plane belongs to a conic if and only if

$$\frac{|\overline{PF}|}{|\overline{PD}|} = e$$

where D is the point on l and on the perpendicular line segment from P to l (Figure 1).

The value of e determines the particular conic section:

1. For a *parabola*, $e = 1$ (since $|\overline{PF}| = |\overline{PD}|$).
2. For an *ellipse*, $0 < e < 1$.
3. For a *hyperbola*, $e > 1$.

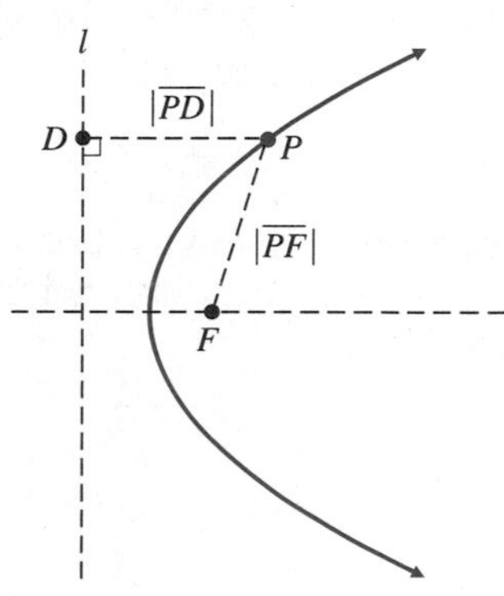

FIGURE 1

EXAMPLE 1 Using the eccentricity to find an equation of a conic

Find the standard equation of the conic with focus (3, 0), directrix $x = 0$, and eccentricity $e = 2$. Sketch the graph.

Solution Assume that $P = (x, y)$ is a point on the graph of the conic as shown in Figure 2. Then

$$\frac{|\overline{PF}|}{|\overline{PD}|} = 2$$

So,

$$|\overline{PF}|^2 = 4|\overline{PD}|^2$$

$$(x-3)^2 + y^2 = 4x^2 \qquad \text{Distance formula}$$

$$3x^2 + 6x - y^2 = 9 \qquad \text{Multiply and isolate variable terms}$$

$$3(x^2 + 2x \qquad) - y^2 = 9 \qquad \text{Factor out 3 from the } x \text{ terms}$$

$$3(x^2 + 2x + 1) - y^2 = 9 + 3 \qquad \text{Complete the square}$$

$$3(x+1)^2 - y^2 = 12 \qquad \text{Factor and simplify}$$

$$\frac{(x+1)^2}{4} - \frac{y^2}{12} = 1 \qquad \text{Divide by 12}$$

This latter equation is the standard equation for a hyperbola. Its graph is shown in Figure 2.

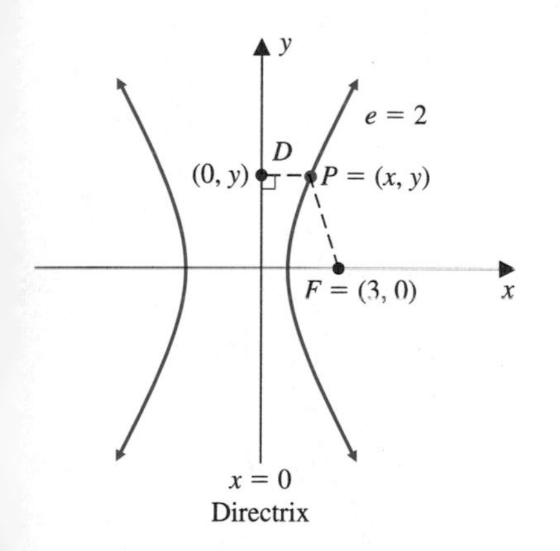

FIGURE 2

In Example 1, notice that $a = 2$, $b = 2\sqrt{3}$, and $c = 4$, so

$$\frac{\text{Distance between foci}}{\text{Distance between vertices}} = \frac{2c}{2a} = \frac{8}{4} = 2 = e$$

It follows that the eccentricity e for the hyperbola is $e = c/a$. In fact, an ellipse or a hyperbola is a conic with eccentricity

$$e = \frac{c}{a}$$

and a parabola is a conic with eccentricity 1. Figure 3 shows graphs of an ellipse and hyperbola, illustrating their eccentricities.

The results in Figure 3 also hold for an ellipse or a hyperbola whose foci lie on a vertical line.

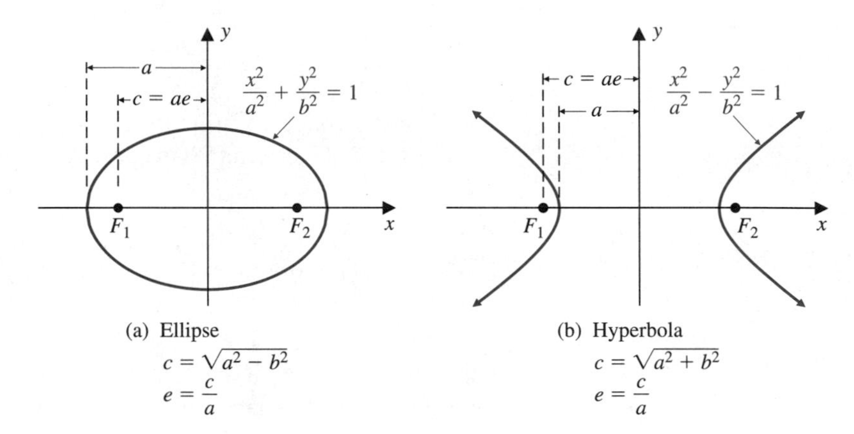

FIGURE 3
Eccentricities of conics

If the center of an ellipse or hyperbola is at a point (h, k) different from the origin, the eccentricity is still given by $e = c/a$.

EXAMPLE 2 Finding the eccentricity of a conic
Find the eccentricity of the conic

$$\frac{(x-2)^2}{9} + \frac{(y+1)^2}{4} = 1$$

Solution The equation describes an ellipse in standard form, with $a = 3$ and $b = 2$, so

$$c = \sqrt{a^2 - b^2} = \sqrt{9 - 4} = \sqrt{5}$$

Therefore, the eccentricity is

$$e = \frac{c}{a} = \frac{\sqrt{5}}{3} = 0.75 \text{ (approx.)}$$

Identifying Polar Equations of Conics

Conics also have standard equations in the polar coordinate system. To derive such a polar equation, we first assume that the conic has one focus located at the

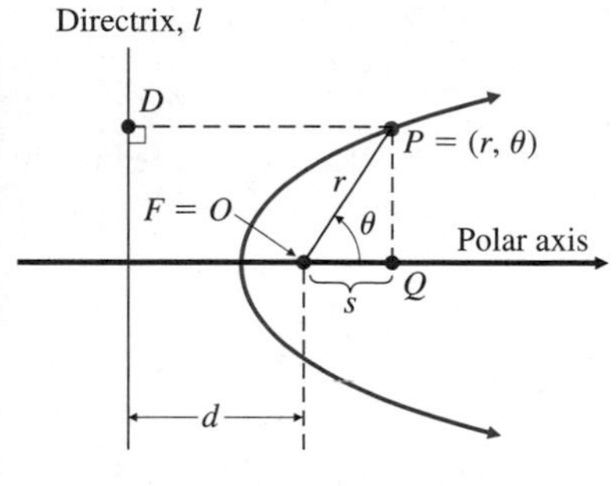

FIGURE 4

pole. That is, let $P = (r, \theta)$ be a point on the graph of a conic with a focus F at the pole O, eccentricity e, and a directrix l that is a vertical line d units to the left of the focus, as displayed in Figure 4.

If P satisfies the condition

$$\frac{|\overline{PF}|}{|\overline{PD}|} = e$$

then

$$(\text{Distance from } P \text{ to } F) = e \cdot (\text{Distance from } P \text{ to } l)$$

That is,

$$\begin{aligned} r &= e(d + s) \\ r &= e(d + r\cos\theta) \\ r &= ed + er\cos\theta \\ r - er\cos\theta &= ed \\ r(1 - e\cos\theta) &= ed \end{aligned}$$

Solving for r yields the *standard polar equation* for the conic:

$$r = \frac{ed}{1 - e\cos\theta}$$

Similar derivations lead to the standard polar equations listed in Table 1. The conics have eccentricity e, vertical or horizontal directrix d units from the focus, and a focus at the pole.

TABLE 1

Standard Polar Equations for Conics (Focus at the Pole and Eccentricity e)

Standard Polar Equation	Description of Directrix	Equation of Directrix
1. $r = \dfrac{ed}{1 + e\cos\theta}$	Vertical; that is, perpendicular to the polar axis at a distance d units to the right of the pole	$x = d$ or $r\cos\theta = d$
2. $r = \dfrac{ed}{1 - e\cos\theta}$	Vertical; that is, perpendicular to the polar axis at a distance d units to the left of the pole	$x = -d$ or $r\cos\theta = -d$
3. $r = \dfrac{ed}{1 + e\sin\theta}$	Horizontal; that is, parallel to the polar axis at a distance d units above the pole	$y = d$ or $r\sin\theta = d$
4. $r = \dfrac{ed}{1 - e\sin\theta}$	Horizontal; that is, parallel to the polar axis at a distance d units below the pole	$y = -d$ or $r\sin\theta = -d$

Note that when the equation of the conic involves $\cos\theta$, the polar axis is an axis of symmetry. When the equation involves $\sin\theta$, the line $\theta = \pi/2$ is an axis of symmetry.

In the next example, we use a graphing utility to study variations of eccentricity for conics.

EXAMPLE 3 Examining eccentricities

Use a graphing utility to sketch the graph of the conic

$$r = \frac{3e}{1 - e\cos\theta}$$

for the given eccentricity. Describe the conics as e gets closer to 0 and as e gets larger and larger.

(a) $e = 0.1$ (b) $e = 0.9$ (c) $e = 1$
(d) $e = 1.1$ (e) $e = 1.5$ (f) $e = 4$

Solution Figure 5 displays viewing windows for the graphs of the given conics.

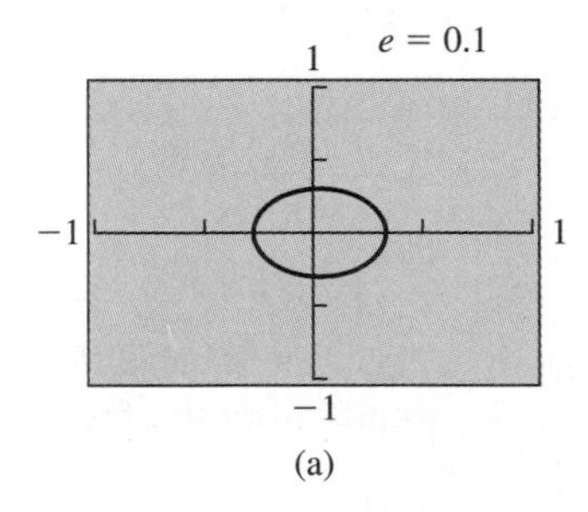

(a)

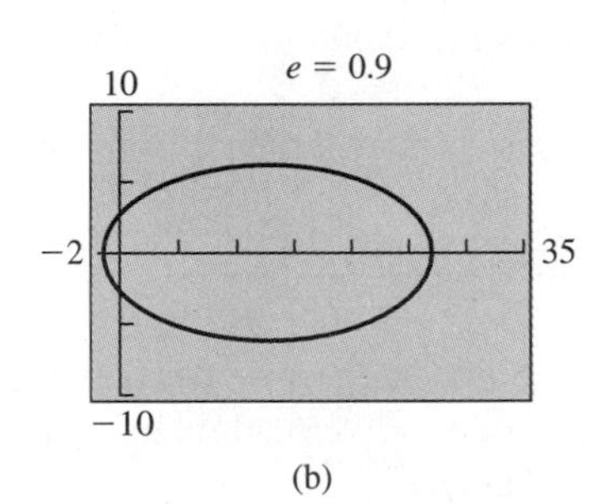

(b)

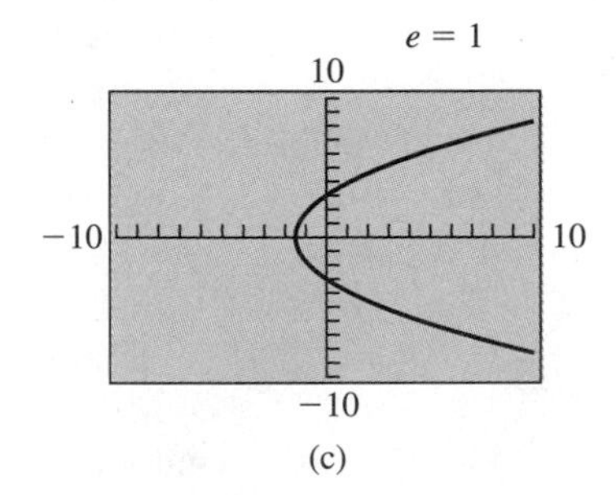

(c)

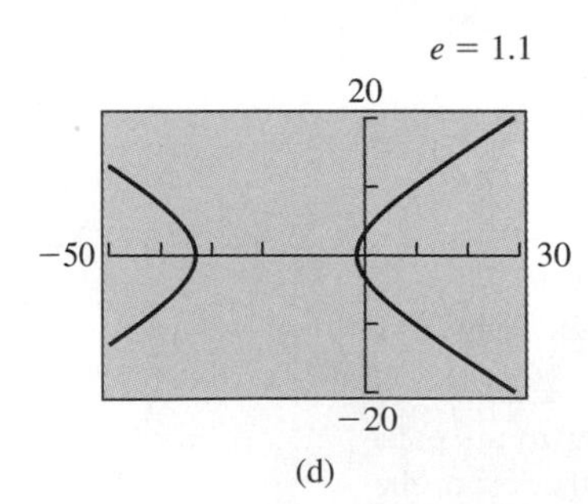

(d)

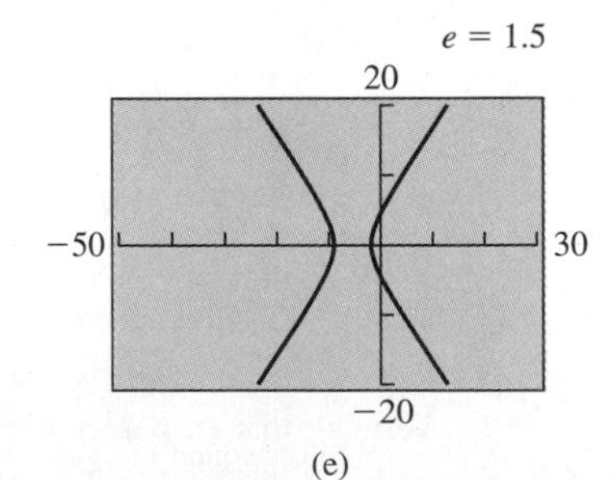

(e)

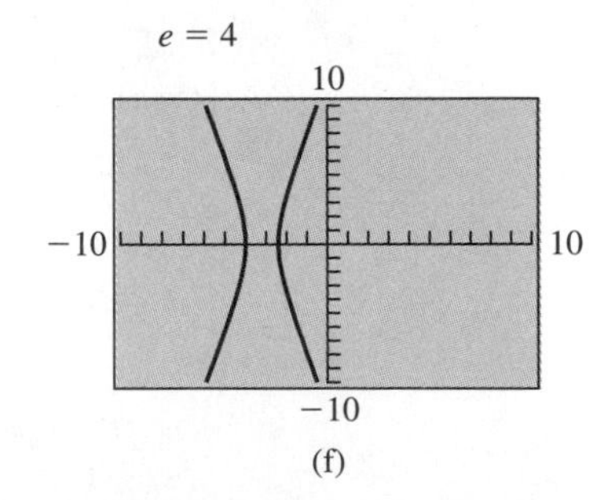

(f)

FIGURE 5

Let us examine how the values of e are related to the shapes of these conics. In a sense, the eccentricity of an ellipse measures its "roundness," and the eccentricity of a hyperbola measures how "flat" its branches are.

It is evident from Figure 5a that when e is close to 0, the ellipse is close to being a circle, whereas it becomes more flattened as e gets closer to 1 (Figure

5b). When $e = 1$, of course, the conic is a parabola (Figure 5c). Figures 5d–f show that as $e > 1$ gets larger and larger, the hyperbola begins to look like a pair of parallel lines.

EXAMPLE 4 Identifying and graphing a conic

A conic is given by the polar equation

$$r = \frac{10}{3 - 2\cos\theta}$$

Find the eccentricity, identify the conic, and use a graphing utility to sketch the graph. Also, use the equation to locate the center and vertices.

Solution Dividing the numerator and denominator by 3, we obtain the second standard equation in Table 1:

$$r = \frac{\frac{10}{3}}{1 - \frac{2}{3}\cos\theta} = \frac{\frac{2}{3}(5)}{1 - \frac{2}{3}\cos\theta}$$

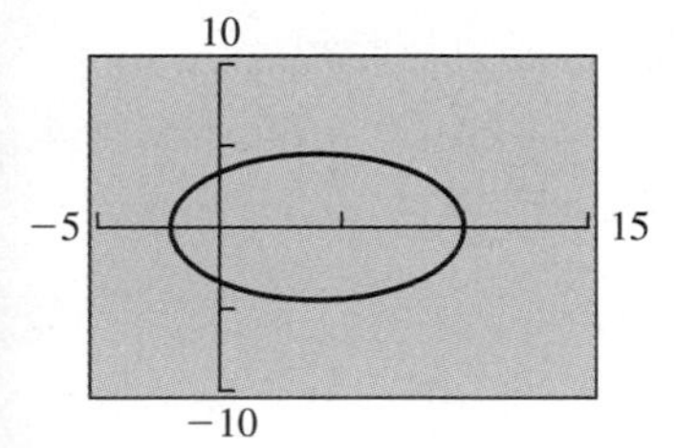

FIGURE 6

Thus, $e = \frac{2}{3}$ and the conic is an ellipse with a focus at the pole and major axis along the polar axis. A viewing window of its graph is shown in Figure 6.

The two vertices along the polar axis can be found by substituting $\theta = 0$ and $\theta = \pi$ into the original equation to obtain $r = 10$ and $r = 2$, respectively. Thus, the vertices are the polar points $(10, 0)$ and $(2, \pi)$. Hence, $2a = 12$, or $a = 6$.

The center of the ellipse is located at the midpoint of the line segment determined by the vertices—that is, $(4, 0)$. Using $e = c/a$, we obtain $c = ae = 6\left(\frac{2}{3}\right) = 4$, so $b^2 = a^2 - c^2 = 36 - 16 = 20$. It follows that $b = \sqrt{20} = 2\sqrt{5}$ and the other two vertices are located $2\sqrt{5}$ units above and below the center (Figure 6).

Solving Applied Problems

Mathematical models of the solar system have been revised several times to accommodate new discoveries. In the sixteenth century, the German astronomer Johannes Kepler (1571–1630) found that planets revolve around the Sun in elliptical orbits, with the Sun at one focus (Table 2).

Because these eccentricities are so close to 0, the orbits are almost circular.

TABLE 2

Eccentricities of Some Planetary Orbits Around the Sun

Planet	Venus	Earth	Mars	Mercury	Pluto
Eccentricity	0.01	0.02	0.09	0.21	0.25

In the next example, we use *astronomical units* to describe the elliptical orbit of Halley's Comet. By definition, 1 **astronomical unit** (or **AU**), which is the length of the semimajor axis of the Earth's elliptical orbit around the Sun, is about 9.26×10^7 miles.

EXAMPLE 5 **Approximating distance in an elliptical path**

The orbit of Halley's Comet is an ellipse that has a major axis 36.2 AU long and a minor axis 9.1 AU wide. The center of the Sun is located at one focus of the elliptical orbit.

(a) Illustrate the comet's orbit by sketching an xy coordinate system with the center of the orbit at the origin and the major axis on the x axis.
(b) Find the eccentricity of this orbit (to two decimal places).
(c) Find the minimum and maximum distances between the center of the Sun and the center of the comet.

Solution

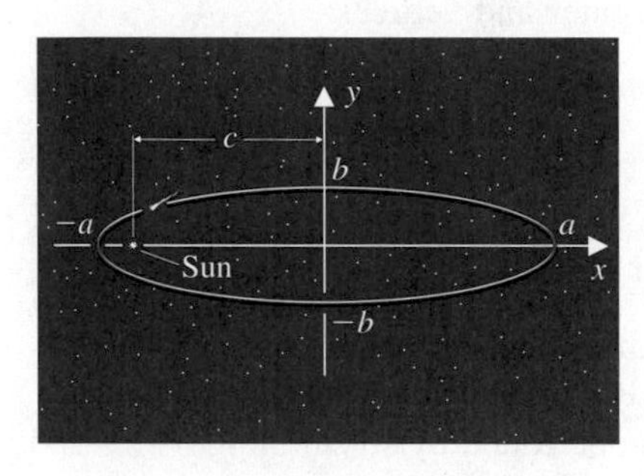

FIGURE 7

(a) The comet's orbit is sketched in Figure 7, where the center of the Sun is located at one focus. The length of the major axis is $2a$ and the length of the minor axis is $2b$. The distance between the center of the ellipse and the Sun is c.

(b) Here, $2a = 36.2$ and $2b = 9.1$, so $a = 18.1$ and $b = 4.55$. Thus, c is given by

$$c = \sqrt{a^2 - b^2} = \sqrt{(18.1)^2 - (4.55)^2} = 17.52 \text{ (approx.)}$$

so

$$e = \frac{c}{a} = \frac{17.52}{18.10} = 0.97 \text{ (approx.)}$$

(c) The minimum distance between the center of the Sun and the center of the comet is given by

$$a - c = 18.10 - 17.52 = 0.58 \text{ AU (approx.)}$$

or about 53,708,000 miles. The maximum distance between the center of the Sun and the center of the comet is given by

$$a + c = 18.10 + 17.52 = 35.62 \text{ AU (approx.)}$$

or about 3,298,412,000 miles.

PROBLEM SET 9.4

Mastering the Concepts

In problems 1–10, determine the standard equation for the conic that satisfies the given conditions and sketch the graph.

1. Focus at (3, 0); directrix $y = \frac{5}{3}$; eccentricity $e = 3/\sqrt{5}$

2. Focus at (5, 0); directrix $x = \frac{16}{5}$; eccentricity $e = \frac{5}{4}$

3. Focus at (3, 0); directrix $x = \frac{25}{3}$; eccentricity $e = \frac{3}{5}$

4. Focus at (1, 2); directrix $y = -2$; eccentricity $e = \frac{1}{2}$

5. Focus at (2, 0); directrix $x = 4$; eccentricity $e = 1$

6. Directrix has the polar equation $r = 2 \csc \theta$ that corresponds to the focus at the pole; eccentricity $e = 3$

7. Center at (0, 0); eccentricity $e = \frac{1}{2}$; foci at (−3, 0) and (3, 0)

8. Center at (2, −1); eccentricity $e = 2$; foci at (2, 1) and (2, −3)

9. Center at (0, 0); eccentricity $e = 3$; containing the point (5, 0)
10. Eccentricity $e = 1$; containing points (2, 3) and (3, 4); symmetric with respect to the line $x = 1$

In problems 11–22, find the eccentricity of the given conic.

11. $25x^2 + 16y^2 = 400$
12. $x^2 + 2y^2 = 1$
13. $x^2 + 3y^2 = 4$
14. $y^2 - 8y + 3x + 5 = 0$
15. $9x^2 - 16y^2 + 144 = 0$
16. $4x^2 - 4y^2 + 1 = 0$
17. $16(x - 1)^2 - 9(y + 2)^2 = 144$
18. $4(x + 2)^2 - 25(y + 3)^2 = 100$
19. $3x^2 - 5x + y^2 + 22y = 1$
20. $x^2 + 16x - y + 7 = 0$
21. $x^2 + 2x - y^2 + 8y = 16$
22. $2x^2 + 12x + y^2 - 8y + 32 = 0$

In problems 23–28, find the standard polar equation of the conic with a focus at the pole, eccentricity e as given, and a directrix with the given equation.

23. $e = 1$; $r = -4 \csc \theta$
24. $e = \frac{1}{3}$; $r = 4 \sec \theta$
25. $e = 2$; $r \cos \theta = -2$
26. $e = 1$; $r \sin \theta = 4$
27. $e = \frac{2}{5}$; $r = 2 \csc \theta$
28. $e = \frac{4}{3}$; $r = -2 \csc \theta$

29. GU Use a graphing utility to sketch the graph of the conic

$$r = \frac{2e}{1 - e \cos \theta}$$

for the given eccentricity. What happens to the shape of the conic as e gets close to 0 and as e gets larger and larger?
(a) $e = 0.25$ (b) $e = 0.5$ (c) $e = 0.90$
(d) $e = 1$ (e) $e = 1.5$ (f) $e = 10$

30. GU Use a graphing utility to sketch the graph of the conic

$$r = \frac{2e}{1 + e \sin \theta}$$

for the given eccentricity. What happens to the shape of the conic as e gets close to 0 and as e gets larger and larger?
(a) $e = 0.1$ (b) $e = 0.7$ (c) $e = 0.8$
(d) $e = 1$ (e) $e = 1.4$ (f) $e = 6$

GU In problems 31–44, find the eccentricity, identify the conic, and use a graphing utility to sketch the graph. Also, use the equation to locate the center (if applicable) and vertices in Cartesian form.

31. $r = \dfrac{5}{1 + \sin \theta}$
32. $r = \dfrac{2}{-1 + 3 \sin \theta}$
33. $r = \dfrac{2}{2 - \cos \theta}$
34. $r = \dfrac{3}{4 + 2 \sin \theta}$
35. $r = \dfrac{5}{3 + 9 \cos \theta}$
36. $r = \dfrac{10}{2 - 5 \cos \theta}$
37. $r = \dfrac{2}{2 + \cos \theta}$
38. $r = \dfrac{2}{1 + \sin\left(\theta + \frac{\pi}{2}\right)}$
39. $r = \dfrac{6 \csc \theta}{3 \csc \theta + 2}$
40. $r = \dfrac{8 \sec \theta}{2 \sec \theta - 1}$
41. $r = \dfrac{2}{\cos \theta - 2}$
42. $r = \dfrac{3}{2 - \sin \theta}$
43. $r = \dfrac{1}{2 - \cos(\theta - \pi)}$
44. $r = \dfrac{1}{2 - 2 \cos\left(\theta - \frac{\pi}{3}\right)}$

Applying the Concepts

45. Earth's Orbit: The orbit of the Earth is an ellipse with the Sun at one focus. The minimum and maximum distances from the center of the Earth to the center of the Sun have a ratio of about $\frac{29}{30}$.
(a) Find the eccentricity of this elliptical orbit.
(b) The closest distance between the center of the Earth and the center of the Sun is approximately 91.3 million miles. What is the farthest distance between the Earth and the Sun?

46. Kahoutek's Comet: Kahoutek's Comet has an elliptical orbit with eccentricity $e = 0.999925$, with the Sun at a focus. The minimum distance from the center of the comet to the center of the Sun is about 0.13 AU.
(a) What is the maximum distance from the center of the comet to the center of the Sun (in miles)?
(b) Find the standard equation for the orbit of the comet, with the Sun at the origin and the major axis along the x axis.
(c) How far from the center of the Sun is the center of the comet when a line drawn to the Sun is perpendicular to the major axis?
(d) Where is the directrix of the ellipse that is nearest the Sun?

47. Window Design: A designer of stained glass windows wishes to make a window in the shape of an ellipse with eccentricity $e = 0.5$. As shown in Figure 8, a rectangular pane of clear glass measuring 1 foot by 2 feet is to be positioned in the center of the elliptical window so that its four corners touch the outer frame. The remaining space of the window is to be filled with a design in stained glass.

FIGURE 8

(a) Sketch a coordinate system with the center of the ellipse at the origin and major axis on the x axis.
(b) Find the standard Cartesian equation of the ellipse sketched in part (a).
(c) What are the lengths of the minor and major axes of the ellipse?
(d) The designer is preparing a worktable by nailing down a rectangular wood enclosure that will allow a minimum distance of 6 inches from the edge of the window. What should be the minimum dimensions of this rectangle?

48. Mercury's Orbit: The planet Mercury travels in an elliptical orbit with eccentricity $e = 0.21$, with the Sun at one focus. Suppose that the length of the major axis is about 11.6×10^7 kilometers. Use the polar equation of an ellipse,

$$r = \frac{a(1 - e^2)}{1 - e \cos \theta}$$

to find the polar equation for the elliptical orbit of Mercury.

Developing and Extending the Concepts

49. What is the relationship between the eccentricities of the following conics?

$$\frac{x^2}{a^2} + \frac{y^2}{b^2} = 1 \qquad \frac{x^2}{a^2} + \frac{y^2}{b^2} = 5$$

50. What is the relationship between the eccentricities of the following conics?

$$\frac{x^2}{a^2} - \frac{y^2}{b^2} = 2 \qquad \frac{x^2}{a^2} - \frac{y^2}{b^2} = 8$$

51. How many directrixes does each type of conic have?
(a) Ellipse **(b)** Hyperbola **(c)** Parabola
Give an example to illustrate each case.

52. How many possible ellipses are there with eccentricity $e = \frac{1}{2}$ and foci at $(0, -2)$ and $(0, 2)$? Explain and illustrate.

53. Consider the equation of the hyperbola, $x^2 - y^2 = 1$. Express this equation in polar equation form and sketch its graph.

54. What is the smallest angle between the asymptotes of the following hyperbola?

$$r = \frac{1}{1 - 2 \cos \theta}$$

55. GU Identify the conic with equation

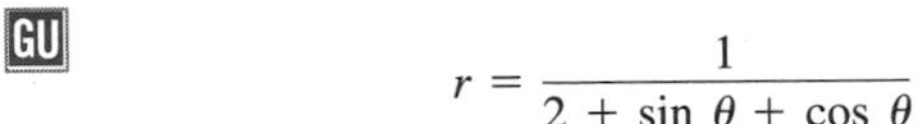

$$r = \frac{1}{2 + \sin \theta + \cos \theta}$$

and use a graphing utility to sketch its graph.

56. Find the two foci and the equations of the directrixes of the ellipse with polar equation

$$r = \frac{1}{2 - \cos \theta}$$

57. GU What are the points of intersection of the following conics? Round off the results to two decimal places.

$$r = \frac{2}{2 + \cos \theta} \qquad r = \frac{2}{2 + \sin \theta}$$

Use a graphing utility to display the points of intersection.

58. GU What point on the conic

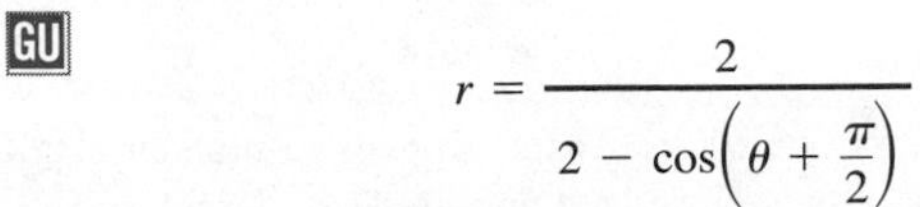

$$r = \frac{2}{2 - \cos\left(\theta + \frac{\pi}{2}\right)}$$

is closest to the origin? Use a graphing utility to demonstrate your result.

OBJECTIVES

1. Use formulas for rotation of axes.
2. Eliminate the xy term.
3. Identify conics given their general equations.

Some choices of A, B, C, D, E, and F may result in a *degenerate conic*. That is, the graph of the equation may be the whole plane, a pair of parallel or intersecting lines, a straight line, a point, or nothing (Problems 41–44).

9.5 Rotation of Axes

So far, we have learned how to graph a conic whose general equation is given by

$$Ax^2 + Cy^2 + Dx + Ey + F = 0$$

where the axes of symmetry of the conics are parallel to the x and y axes.

In this section, we consider conics with second-degree equations of the general form

$$Ax^2 + Bxy + Cy^2 + Dx + Ey + F = 0 \qquad B \neq 0$$

Because of the *cross term xy* in such equations, we will see that the graphs turn out to be *rotated* conics, in the sense that the axes of symmetry are neither horizontal nor vertical.

Using Formulas for Rotation of Axes

Suppose the axes of an xy coordinate system are rotated through an angle θ to form a new $\bar{x}\bar{y}$ coordinate system. Assume P is a point that has coordinates (x, y) in the original system and coordinates $(\bar{x}, \bar{y})$ in the new rotated system (Figure 1a). How are the coordinates (x, y) related to the coordinates $(\bar{x}, \bar{y})$? The easiest way to answer this question is to use polar coordinates. Suppose that the polar coordinates of P relative to the $\bar{x}\bar{y}$ coordinate system are $(\bar{r}, \alpha)$, where $\bar{r} = r = |\overline{OP}|$ (Figure 1b). Then the Cartesian coordinates of P relative to the $\bar{x}\bar{y}$ coordinate system are given by

$$\bar{x} = |\overline{OP}| \cos \alpha \quad \text{and} \quad \bar{y} = |\overline{OP}| \sin \alpha$$

or

$$\bar{x} = \bar{r} \cos \alpha \quad \text{and} \quad \bar{y} = \bar{r} \sin \alpha$$

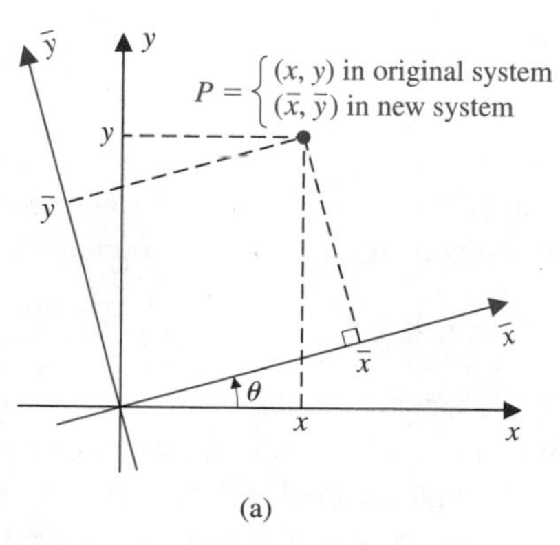

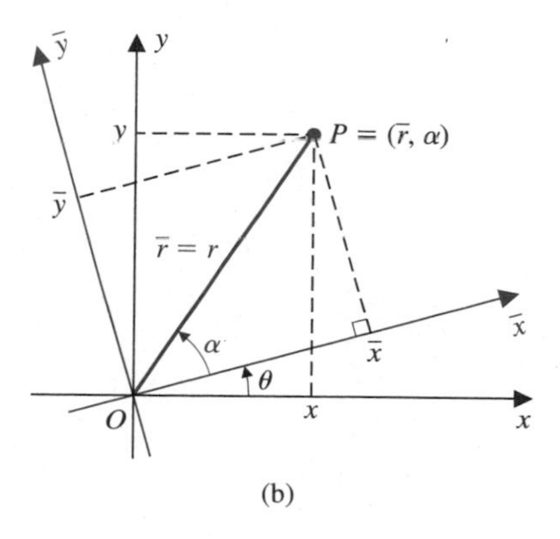

FIGURE 1

Now we consider the same point P relative to the xy coordinate system. To find its polar coordinates, we observe that one possible angle is $\theta + \alpha$. Thus, Cartesian coordinates of P relative to the original unrotated xy axes are

$$x = |\overline{OP}| \cos(\alpha + \theta) \quad \text{and} \quad y = |\overline{OP}| \sin(\alpha + \theta)$$
$$x = \bar{r} \cos(\alpha + \theta) \quad \text{and} \quad y = \bar{r} \sin(\alpha + \theta)$$

That is,

$$\begin{aligned} x = \bar{r}\cos(\alpha + \theta) &= \bar{r}\cos\alpha\cos\theta - \bar{r}\sin\alpha\sin\theta && \text{Cosine of a sum identity} \\ &= \bar{x}\cos\theta - \bar{y}\sin\theta && \text{Substitute } \bar{x} \text{ and } \bar{y} \end{aligned}$$

and

$$\begin{aligned} y = \bar{r}\sin(\alpha + \theta) &= \bar{r}\sin\alpha\cos\theta + \bar{r}\cos\alpha\sin\theta && \text{Sine of a sum identity} \\ &= \bar{y}\cos\theta + \bar{x}\sin\theta && \text{Substitute } \bar{y} \text{ and } \bar{x} \end{aligned}$$

Thus, we obtain the following rotation formulas:

$$\boxed{\begin{aligned} x &= \bar{x}\cos\theta - \bar{y}\sin\theta \\ y &= \bar{x}\sin\theta + \bar{y}\cos\theta \end{aligned}}$$

These formulas express the original coordinates (x, y) in terms of the new coordinates $(\bar{x}, \bar{y})$.

The rotation formulas given above can be solved for $\bar{x}$ and $\bar{y}$ (Problem 45) to obtain

$$\boxed{\begin{aligned} \bar{x} &= x\cos\theta + y\sin\theta \\ \bar{y} &= -x\sin\theta + y\cos\theta \end{aligned}}$$

These latter formulas express the new coordinates $(\bar{x}, \bar{y})$ relative to the rotated axes in terms of the original coordinates (x, y).

EXAMPLE 1 Finding the coordinates of a point relative to rotated axes

Assume that a new $\bar{x}\bar{y}$ coordinate system is formed by rotating the xy coordinate system through an angle of 60°. Find the coordinates of P with respect to the $\bar{x}\bar{y}$ coordinate system if $P = (-1, 7)$ with respect to the xy system. Display the point relative to both pairs of axes.

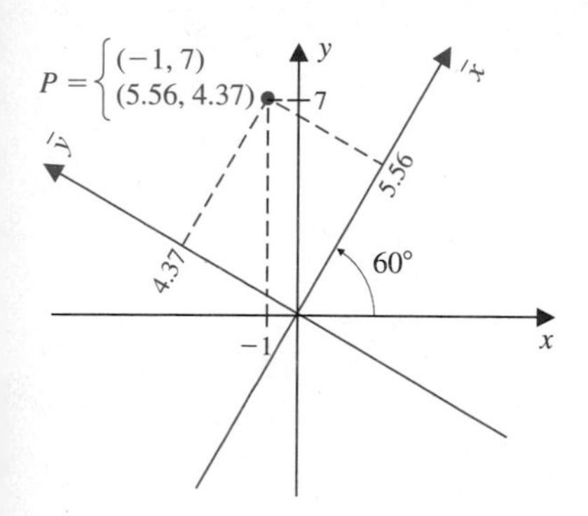

FIGURE 2

Solution Here, we can use the second set of rotation formulas with $x = -1$, $y = 7$, and $\theta = 60°$ to get

$$\begin{cases} \bar{x} = -1 \cos 60° + 7 \sin 60° = -\dfrac{1}{2} + \dfrac{7\sqrt{3}}{2} = \dfrac{-1 + 7\sqrt{3}}{2} \\ \bar{y} = -(-1) \sin 60° + 7 \cos 60° = \dfrac{\sqrt{3}}{2} + \dfrac{7}{2} = \dfrac{\sqrt{3} + 7}{2} \end{cases}$$

Thus

$$P = \left(\frac{-1 + 7\sqrt{3}}{2}, \frac{\sqrt{3} + 7}{2}\right) = (5.56, 4.37) \text{ approx.}$$

in the $\bar{x}\bar{y}$ system. The point P is shown in Figure 2 relative to both coordinate systems.

EXAMPLE 2 Eliminating the xy term from an equation

Assume that an xy coordinate system is rotated through a 45° angle to form an $\bar{x}\bar{y}$ coordinate system. Rewrite the equation $xy = 2$ in terms of the $\bar{x}\bar{y}$ coordinate system and sketch the graph.

Solution For $\theta = 45°$, the rotation formulas for x and y yield

$$\begin{cases} x = \bar{x} \cos 45° - \bar{y} \sin 45° = \bar{x}\left(\dfrac{\sqrt{2}}{2}\right) - \bar{y}\left(\dfrac{\sqrt{2}}{2}\right) = \dfrac{\sqrt{2}}{2}(\bar{x} - \bar{y}) \\ y = \bar{x} \sin 45° + \bar{y} \cos 45° = \bar{x}\left(\dfrac{\sqrt{2}}{2}\right) + \bar{y}\left(\dfrac{\sqrt{2}}{2}\right) = \dfrac{\sqrt{2}}{2}(\bar{x} + \bar{y}) \end{cases}$$

A rotation of axes does not alter the shape or geometric characteristics of the conic.

Next, we substitute the expressions for x and y in the equation $xy = 2$ to get

$$\left[\frac{\sqrt{2}}{2}(\bar{x} - \bar{y})\right] \cdot \left[\frac{\sqrt{2}}{2}(\bar{x} + \bar{y})\right] = 2$$

$$\frac{1}{2}(\bar{x}^2 - \bar{y}^2) = 2 \quad \text{Multiply}$$

$$\frac{\bar{x}^2}{4} - \frac{\bar{y}^2}{4} = 1 \quad \text{Divide by 2}$$

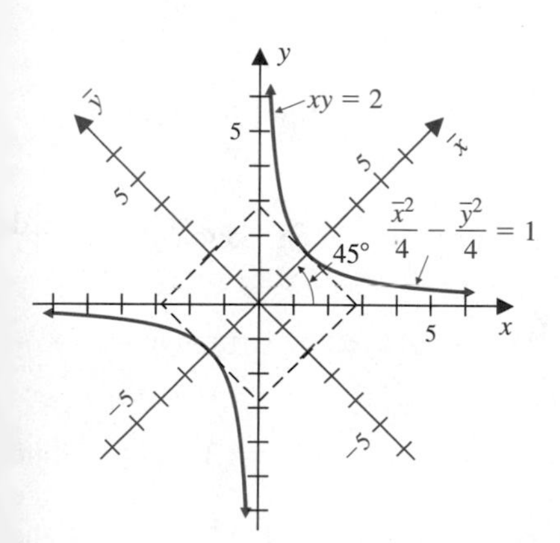

FIGURE 3

This is the standard equation of a hyperbola, with $a = 2$, $b = 2$, and $c = \sqrt{a^2 + b^2} = \sqrt{4 + 4} = \sqrt{8} = 2\sqrt{2}$. To graph the hyperbola in the $\bar{x}\bar{y}$ coordinate system, we start at the origin and find the vertices by moving 2 units in each direction along the $\bar{x}$ axis. We note that the extremities of the conjugate axis are found by moving 2 units in each direction from the origin along the $\bar{y}$ axis. Using this information, we sketch the graph (Figure 3). Notice that the asymptotes for this graph are the original xy coordinate axes.

Eliminating the *xy* Term

Suppose that a conic has the general equation

$$Ax^2 + Bxy + Cy^2 + Dx + Ey + F = 0 \qquad B \neq 0$$

By using the rotation formulas

$$\begin{cases} x = \bar{x}\cos\theta - \bar{y}\sin\theta \\ y = \bar{x}\sin\theta + \bar{y}\cos\theta \end{cases}$$

with an appropriate selection for θ, we can transform the given equation to an equation of the form

$$\bar{A}\bar{x}^2 + \bar{C}\bar{y}^2 + \bar{D}\bar{x} + \bar{E}\bar{y} + \bar{F} = 0$$

with no $\bar{x}\bar{y}$ term ($\bar{B} = 0$). We refer to this process as *eliminating the xy term*, and we can accomplish this if we select an angle θ that satisfies

$$\boxed{\cot 2\theta = \frac{A - C}{B}}$$

(Problem 46).

For instance, in Example 2, the equation of the conic is given by $xy = 2$. Thus, its general equation is

$$xy - 2 = 0$$

that is, $A = 0$, $B = 1$, and $C = 0$. Consequently, we choose θ to satisfy

$$\cot 2\theta = \frac{0 - 0}{1} = 0$$

so $2\theta = 90°$, or $\theta = 45°$, will work.

EXAMPLE 3 Finding an angle of rotation to eliminate the *xy* term

Find an acute angle of rotation that eliminates the xy term from the equation

$$21x^2 - 10\sqrt{3}\,xy + 31y^2 = 144$$

Rewrite the equation in terms of the rotated axes $\bar{x}\bar{y}$. Then sketch the graph, showing both sets of coordinate axes.

Solution The equation $21x^2 - 10\sqrt{3}\,xy + 31y^2 = 144$ can be written in the form

$$Ax^2 + Bxy + Cy^2 + Dx + Ey + F = 0$$
$$\text{with } A = 21,\ B = -10\sqrt{3},\ C = 31,\ F = -144$$

So,

$$\cot 2\theta = \frac{A - C}{B} = \frac{21 - 31}{-10\sqrt{3}} = \frac{-10}{-10\sqrt{3}} = \frac{1}{\sqrt{3}}$$

Thus, $2\theta = 60°$, or $\theta = 30°$. The rotation formulas that will eliminate the xy term are given by

$$\begin{cases} x = \bar{x} \cos 30° - \bar{y} \sin 30° = \dfrac{\sqrt{3}}{2}\bar{x} - \dfrac{1}{2}\bar{y} \\ y = \bar{x} \sin 30° + \bar{y} \cos 30° = \dfrac{1}{2}\bar{x} + \dfrac{\sqrt{3}}{2}\bar{y} \end{cases}$$

To rewrite the equation in terms of $\bar{x}$ and $\bar{y}$, we substitute these expressions for x and y in the given equation to get

$$21\left(\frac{\sqrt{3}}{2}\bar{x} - \frac{1}{2}\bar{y}\right)^2 - 10\sqrt{3}\left(\frac{\sqrt{3}}{2}\bar{x} - \frac{1}{2}\bar{y}\right)\left(\frac{1}{2}\bar{x} + \frac{\sqrt{3}}{2}\bar{y}\right) + 31\left(\frac{1}{2}\bar{x} + \frac{\sqrt{3}}{2}\bar{y}\right)^2 = 144$$

After simplifying the left side of the equation, we obtain

$$16\bar{x}^2 + 36\bar{y}^2 = 144 \qquad \text{or} \qquad \frac{\bar{x}^2}{9} + \frac{\bar{y}^2}{4} = 1$$

This latter equation is the standard equation of an ellipse. After drawing the $\bar{x}\bar{y}$ coordinate axes by rotating the xy coordinate axes 30°, we sketch the ellipse in the $\bar{x}\bar{y}$ system by using $a = 3$ and $b = 2$, where the major axis lies on the $\bar{x}$ axis (Figure 4).

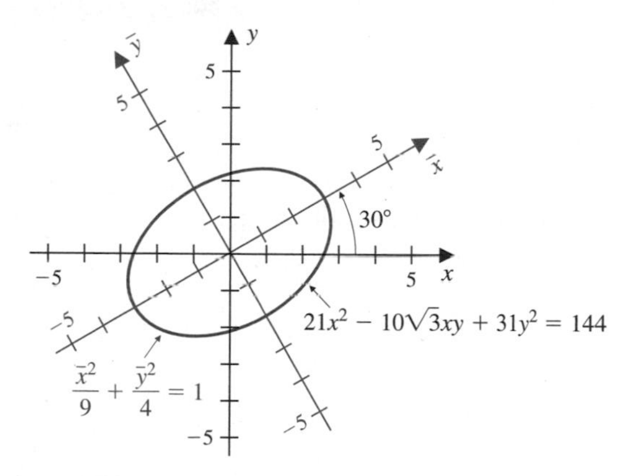

FIGURE 4

EXAMPLE 4 Eliminating the xy term by using trigonometric identities

Find the values of $\sin\theta$ and $\cos\theta$ for the rotation formulas that eliminate the xy term in the equation

$$9x^2 + 12xy + 4y^2 + 2x - 3y = 0$$

Rewrite the equation in terms of the rotated $\bar{x}\bar{y}$ axes. Approximate the acute angle of rotation (to two decimal places) and sketch the graph, displaying both sets of coordinate axes.

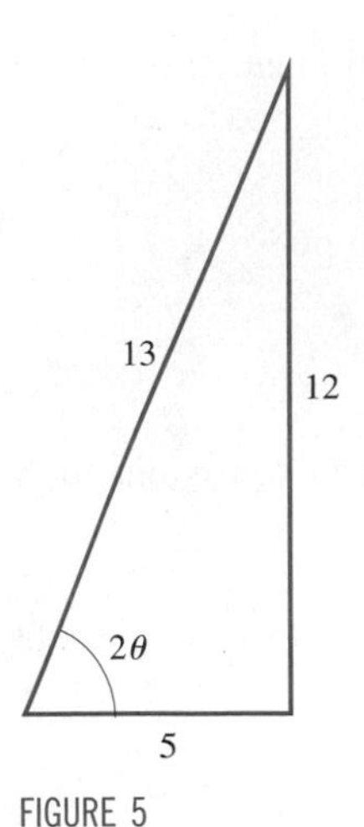

FIGURE 5

Solution In this case, $A = 9$, $B = 12$, and $C = 4$, so

$$\cot 2\theta = \frac{A - C}{B} = \frac{9 - 4}{12} = \frac{5}{12}$$

We'll use right triangle trigonometry and trigonometric identities to determine the appropriate rotation formulas as follows: First, we construct a right triangle that satisfies

$$\cot 2\theta = \frac{5}{12} \quad \text{(Figure 5)}$$

Upon examining the triangle, we see that $\cos 2\theta = \frac{5}{13}$. Next, we use the half-angle trigonometric formulas to get

$$\sin \theta = \sqrt{\frac{1 - \cos 2\theta}{2}} = \sqrt{\frac{1 - \frac{5}{13}}{2}} = \sqrt{\frac{\frac{8}{13}}{2}} = \sqrt{\frac{4}{13}} = \frac{2}{\sqrt{13}}$$

and

$$\cos \theta = \sqrt{\frac{1 + \cos 2\theta}{2}} = \sqrt{\frac{1 + \frac{5}{13}}{2}} = \sqrt{\frac{\frac{18}{13}}{2}} = \sqrt{\frac{9}{13}} = \frac{3}{\sqrt{13}}$$

So the rotation formulas become

$$\begin{cases} x = \bar{x} \cos \theta - \bar{y} \sin \theta = \dfrac{3}{\sqrt{13}}\bar{x} - \dfrac{2}{\sqrt{13}}\bar{y} \\ y = \bar{x} \sin \theta + \bar{y} \cos \theta = \dfrac{2}{\sqrt{13}}\bar{x} + \dfrac{3}{\sqrt{13}}\bar{y} \end{cases}$$

To rewrite the given equation, we substitute these expressions into the original equation and simply to get

$$\bar{y} = \sqrt{13}\,\bar{x}^2$$

This is the equation of a parabola. Since $\cos \theta = 3/\sqrt{13}$, then

$$\theta = \cos^{-1} \frac{3}{\sqrt{13}} = 33.69° \text{ (approx.)}$$

After rotating the xy system, we graph the parabola in the $\bar{x}\bar{y}$ system (Figure 6).

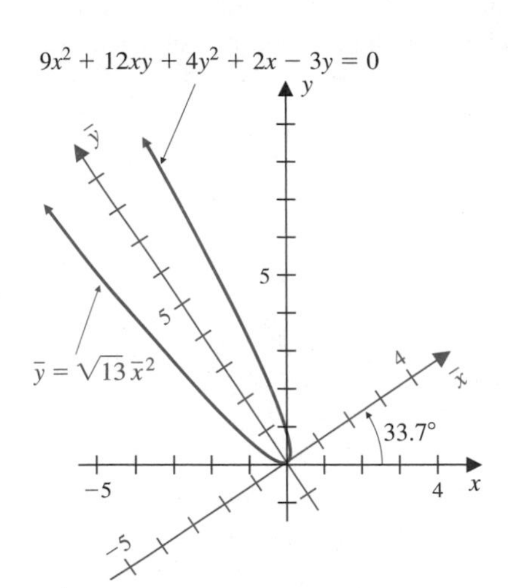

FIGURE 6

We can also use a graphing utility to graph equations of the form

$$Ax^2 + Bxy + Cy^2 + Dx + Ey + F = 0 \qquad B \neq 0$$

EXAMPLE 5 **Using a graphing utility to graph a conic**

Solve the equation $5x^2 - 8xy + 5y^2 = 10$ for y in terms of x. Then graph the given equation by using a graphing utility to graph the resulting equations in the same viewing window.

Solution First, we rewrite the equation as a standard quadratic equation by treating y as the variable and the other terms as constants:

$$5y^2 + (-8x)y + (5x^2 - 10) = 0$$

Next, we use the quadratic formula to solve for y. By letting $a = 5$, $b = -8x$, and $c = 5x^2 - 10$, we get

$$y = \frac{-b \pm \sqrt{b^2 - 4ac}}{2a}$$

$$= \frac{8x \pm \sqrt{64x^2 - 4(5)(5x^2 - 10)}}{10}$$

So

$$y_1 = \frac{8x - \sqrt{200 - 36x^2}}{10} \quad \text{or} \quad y_2 = \frac{8x + \sqrt{200 - 36x^2}}{10}$$

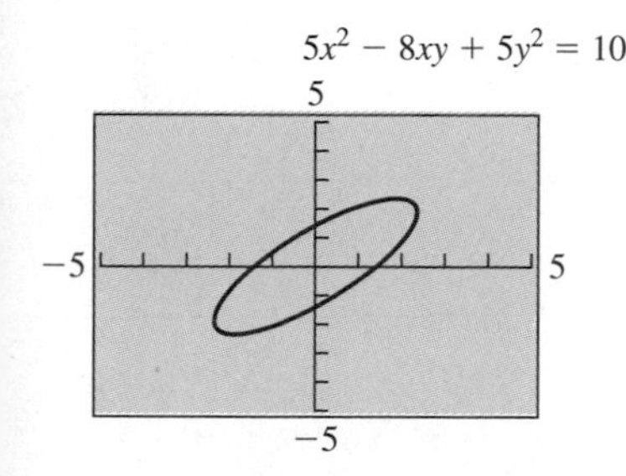

FIGURE 7

Figure 7 shows a viewing window containing the graphs of both functions in the same coordinate system. Together, they give the graph of the original equation.

Identifying Conics Given Their General Equations

Assume that the general equation of a conic is given by

$$Ax^2 + Bxy + Cy^2 + Dx + Ey + F = 0$$

It is possible to determine whether the conic is an ellipse, a parabola, or a hyperbola before graphing it by using the following property.

PROPERTY

DISCRIMINANT OF A CONIC

The graph of the equation

$$Ax^2 + Bxy + Cy^2 + Dx + Ey + F = 0$$

is a conic or one of its degenerate forms. We can identify the type of conic by using the **discriminant** $B^2 - 4AC$. The conic is:

1. An *ellipse* if $B^2 - 4AC < 0$
2. A *parabola* if $B^2 - 4AC = 0$
3. A *hyperbola* if $B^2 - 4AC > 0$

EXAMPLE 6 Identifying a conic

Use the discriminant to identify the conic with general equation

$$2x^2 + 4\sqrt{3}\,xy - 2y^2 - 4 = 0$$

Display the result by using a graphing utility to graph the equation.

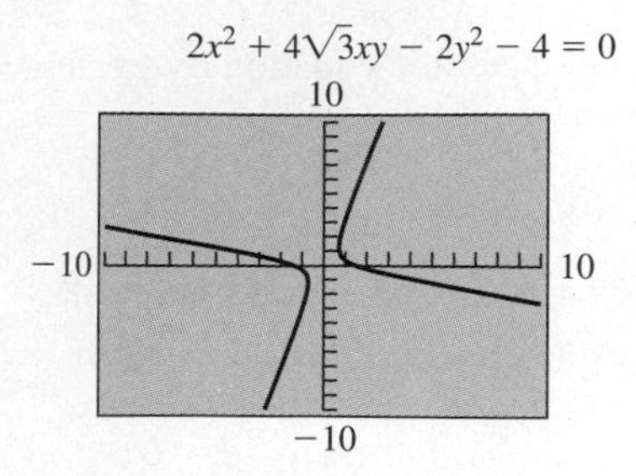

FIGURE 8

Solution Using the discriminant $B^2 - 4AC$, with $A = 2$, $B = 4\sqrt{3}$, and $C = -2$, we have

$$B^2 - 4AC = 48 - 4(2)(-2)$$
$$= 48 + 16 = 64 > 0$$

So the conic is a hyperbola. Figure 8 shows a viewing window of the graph of the given equation.

PROBLEM SET 9.5

Mastering the Concepts

In problems 1–6, assume that new coordinate axes $\bar{x}\bar{y}$ are obtained by rotating the old coordinate axes xy through angle θ. Let P be a point with coordinates (x, y) in the old system and coordinates $(\bar{x}, \bar{y})$ in the new system. Display the point relative to both pairs of axes. In problems 5 and 6, round off each answer to two decimal places.

1. If $(x, y) = (-2, -5)$ and $\theta = 30°$, find $(\bar{x}, \bar{y})$.
2. If $(\bar{x}, \bar{y}) = (0, 3\sqrt{2})$ and $\theta = 60°$, find (x, y).
3. If $(\bar{x}, \bar{y}) = (1, -10)$ and $\theta = 45°$, find (x, y).
4. If $(x, y) = (-2, 5)$ and $\theta = 30°$, find $(\bar{x}, \bar{y})$.
5. If $(x, y) = (3.52, 5.73)$ and $\theta = 55°$, find $(\bar{x}, \bar{y})$.
6. If $(\bar{x}, \bar{y}) = (-4.71, 2.13)$ and $\theta = 22°$, find (x, y).

In problems 7–14, assume an xy coordinate system is rotated through angle θ to form an $\bar{x}\bar{y}$ coordinate system. Rewrite the given equation in terms of the $\bar{x}\bar{y}$ system and sketch the graph. Display both sets of coordinate axes.

7. $xy = 1$; $\theta = 45°$
8. $x^2 - 4xy + y^2 - 6 = 0$; $\theta = \pi/4$
9. $x^2 - \sqrt{3}\,xy + 2y^2 = 4$; $\theta = \pi/6$
10. $5x^2 - 2xy + 5y^2 = 12$; $\theta = 45°$
11. $7x^2 - 6\sqrt{3}\,xy + 13y^2 = 16$; $\theta = 60°$
12. $7x^2 - 2\sqrt{3}\,xy + 5y^2 = 16$; $\theta = 30°$
13. $31x^2 + 10\sqrt{3}\,xy + 21y^2 = 144$; $\theta = 30°$
14. $2x + 3y = 6$; $\theta = \pi/6$

In problems 15–20, determine an angle θ for a rotation of axes that will eliminate the xy term from the given equation. Rewrite the equation in terms of the rotated axes $\bar{x}\bar{y}$ and graph it, showing both sets of coordinate axes.

15. $5x^2 - 4xy + 5y^2 = 9$
16. $2x^2 + 4xy + 2y^2 = 14$
17. $11\bar{x}^2 + 10\sqrt{3}\,xy + y^2 = 64$
18. $xy = 8$
19. $2x^2 + 4\sqrt{3}\,xy - 2y^2 = 4$
20. $3x^2 + 6\sqrt{3}\,xy + 9y^2 = 32$

In problems 21–26, find the values of $\sin\theta$ and $\cos\theta$ for rotation formulas that eliminate the xy term in each equation. Write the equation in terms of the rotated $\bar{x}\bar{y}$ system. Determine the angle of rotation (to one decimal place) and sketch the graph, displaying both sets of coordinate axes.

21. $8x^2 - 4xy + 5y^2 = 144$
22. $x^2 - 3xy + 5y^2 - 22 = 0$
23. $5x^2 - 4xy + 8y^2 = 144$
24. $16x^2 - 24xy + 9y^2 = 80$
25. $4x^2 + 4xy + y^2 + 20x - 10y = \sqrt{5}$
26. $41x^2 - 24xy + 34y^2 = 29$

GU In problems 27–32, solve each equation for y in terms of x. Then use a graphing utility to graph the resulting equations in the same viewing window to obtain a graph of the given equation.

27. $x^2 - xy + y^2 = 4$
28. $5x^2 - 6xy + y^2 = 8$
29. $10xy - y^2 = 32$
30. $4x^2 + 4xy - y^2 = 10$
31. $2x^2 + 3xy + y^2 = 18$
32. $31x^2 + 10xy - y^2 = 14$

GU In problems 33–38, use the discriminant to determine whether the graph of the given equation is an ellipse, a parabola, or a hyperbola. Use a graphing utility to display the result.

33. $x^2 - xy + y^2 = 5$
34. $x^2 + 2xy + y^2 = 16$
35. $x^2 + 24xy - 6y^2 = 30$
36. $17x^2 - 12xy + 8y^2 = 60$
37. $17x^2 - 12xy + 8y^2 = 12$
38. $24xy - 7y^2 + 36 = 0$

Developing and Extending the Concepts

39. Find an angle θ (if it exists) for which the rotation formulas give the following equations:
(a) $\bar{x} = y$ and $\bar{y} = -x$ **(b)** $\bar{x} = -y$ and $\bar{y} = -x$

40. Matrix Form of a Rotation of Axes:
(a) Multiply

$$\begin{bmatrix} \cos\theta & -\sin\theta \\ \sin\theta & \cos\theta \end{bmatrix}\begin{bmatrix} \bar{x} \\ \bar{y} \end{bmatrix}$$

and compare the result to the first set of rotation formulas on page 570.
(b) Multiply

$$\begin{bmatrix} \cos\theta & \sin\theta \\ -\sin\theta & \cos\theta \end{bmatrix}\begin{bmatrix} x \\ y \end{bmatrix}$$

and compare the result to the second set of rotation formulas on page 570.
(c) Use the matrix in part (b) to determine a matrix that will yield the $\bar{x}\bar{y}$ coordinates of a given point (x, y) if the coordinate axes are rotated 45°.

In problems 41–44, each equation has a graph that is a degenerate conic. Verify that the graph is as described.

41. $x^2 + y^2 + 1 = 0$; the empty set (that is, no points)
42. $x^2 + y^2 - 6x + 4y + 13 = 0$; a single point
43. $2x^2 - 4xy + 2y^2 = 0$; a straight line
44. $0x^2 + 0xy + 0y^2 + 0x + 0y + 0 = 0$; the whole x,y plane

45. Given the rotation formulas

$$\begin{cases} x = \bar{x}\cos\theta - \bar{y}\sin\theta \\ y = \bar{x}\sin\theta + \bar{y}\cos\theta \end{cases}$$

use the elimination method to solve for $\bar{x}$ and $\bar{y}$ in terms of x, y, $\cos\theta$, and $\sin\theta$, thus obtaining the rotation formulas

$$\begin{cases} \bar{x} = x\cos\theta + y\sin\theta \\ \bar{y} = -x\sin\theta + y\cos\theta \end{cases}$$

46. **(a)** Substitute the rotation formulas that express x and y (old coordinates) in terms of $\bar{x}$ and $\bar{y}$ (new coordinates) into the general quadratic form

$$Ax^2 + Bxy + Cy^2 + Dx + Ey + F = 0$$

to determine that $\bar{B}$ in the transformed equation

$$\bar{A}\bar{x}^2 + \bar{B}\bar{x}\bar{y} + \bar{C}\bar{y}^2 + \bar{D}\bar{x} + \bar{E}\bar{y} + \bar{F} = 0$$

is given by

$$\bar{B} = 2(C - A)\sin\theta\cos\theta + B(\cos^2\theta - \sin^2\theta)$$

(b) Set $\bar{B} = 0$ and verify that

$$\cot 2\theta = \frac{A - C}{B}$$

GU In problems 47–50, use the graph of the given equation to determine the region that contains the solution of the associated inequality.

47. $xy = 2$; $xy < 2$ (see Example 2)
48. $2x^2 + \sqrt{3}\,xy + y^2 = 10$; $2x^2 + \sqrt{3}\,xy + y^2 \le 10$
49. $2y^2 - \sqrt{3}\,xy + x^2 - 4 = 0$; $2y^2 - \sqrt{3}\,xy + x^2 - 4 \ge 0$
50. $x^2 - 4xy + y^2 - 25 = 0$; $x^2 - 4xy + y^2 - 25 > 0$

OBJECTIVES

1. Graph parametric equations by point-plotting.
2. Eliminate the parameter.
3. Represent Cartesian equations parametrically.
4. Solve applied problems.

9.6 Parametric Equations

So far, we have dealt with curves as graphs of equations relating x and y or r and θ. There is another way of dealing with curves, which is especially useful in studying motion. For these curves, it is useful to express both x and y as functions of a third variable, say t, which represents time in many applications. For example, suppose that a particle moves in the xy plane, and suppose that its x and y coordinates are functions of time t, say

$$\begin{cases} x = f(t) \\ y = g(t) \end{cases} \quad \text{for } t \text{ in } I$$

Then these functions taken together describe the path that the particle traces out in the xy plane as time t elapses.

More formally, we have the following definition:

DEFINITION
PARAMETRIC EQUATIONS IN THE PLANE

Suppose that the points (x, y) on a curve in the plane are defined by the functions

$$\begin{cases} x = f(t) \\ y = g(t) \end{cases}$$

where t is a real number in a given interval. The functions are called **parametric equations** for the curve, and t is referred to as the **parameter**.

Some examples of parametric equations for curves in the plane are

$$\begin{cases} x = t + 1 \\ y = t^2 \end{cases} \qquad \begin{cases} x = 2\cos t \\ y = 3\sin t \end{cases} \quad 0 \le t \le 2\pi \qquad \begin{cases} x = t^2 \\ y = t^2 + 1 \end{cases} \quad 0 \le t \le 1$$

Graphing Parametric Equations by Point-Plotting

We can use point-plotting to graph a curve represented by parametric equations. This is done by evaluating x and y for some values of t, plotting the corresponding points, and then connecting them with a smooth curve, if appropriate. An advantage of studying curves parametrically is that as t varies, we can see how the curve is traced out, not just its shape.

EXAMPLE 1 Graphing parametric equations by point-plotting

Sketch the curve represented by the parametric equations by using point-plotting.

$$\begin{cases} x = t + 1 \\ y = t^2 \end{cases} \qquad 0 \le t \le 5$$

Solution We begin by making a table of values of t and the corresponding values of x and y (Table 1). By plotting the points in Table 1 and then connecting them with a smooth curve, we get a sketch of the graph (Figure 1).

TABLE 1

t	$x = t + 1$	$y = t^2$	(x, y)
0	1	0	(1, 0)
1	2	1	(2, 1)
2	3	4	(3, 4)
3	4	9	(4, 9)
4	5	16	(5, 16)
5	6	25	(6, 25)

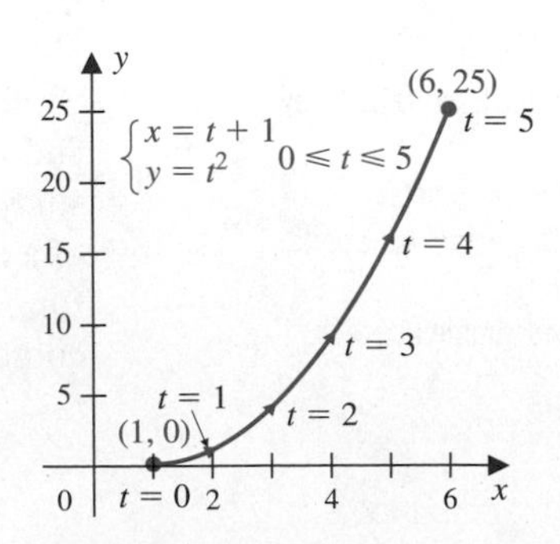

FIGURE 1

Note that as t varies from 0 to 5, the corresponding points on the graph trace out the curve starting at the point (1, 0) and ending at the point (6, 25). This *orientation* is denoted by the arrowheads in Figure 1.

Eliminating the Parameter

We can often derive an equation in terms of x and y for the curve defined by parametric equations. This process is referred to as *eliminating the parameter*.

For instance, in Example 1 where

$$\begin{cases} x = t + 1 \\ y = t^2 \end{cases} \qquad 0 \leq t \leq 5$$

Eliminating the parameter often helps us to recognize the shape of the curve. However, it is important to examine the parametric form when determining the orientation of the curve.

we eliminate the parameter algebraically as follows: First, solve the equation $x = t + 1$ for t to get $t = x - 1$. Then substitute $x - 1$ for t in the second equation, $y = t^2$, to obtain

$$y = (x - 1)^2$$

The condition $0 \leq t \leq 5$ requires that $0 \leq x - 1 \leq 5$, or $1 \leq x \leq 6$. So the graph in Figure 1 is actually a portion of the parabola $y = (x - 1)^2$, where $1 \leq x \leq 6$.

The process of eliminating the parameter sometimes involves trigonometric identities.

EXAMPLE 2 Eliminating the parameter trigonometrically

Eliminate the parameter and then sketch the curve of the parametric equations

In many applications, t represents time, but it could denote other quantities such as an angle measure.

$$\begin{cases} x = 2 \cos t \\ y = 3 \sin t \end{cases} \qquad 0 \leq t \leq 2\pi$$

Describe the orientation of a point on the curve as t varies from 0 to 2π.

Solution First, we solve the equation $x = 2 \cos t$ for $\cos t$ and the second equation for $\sin t$ to obtain

$$\cos t = \frac{x}{2} \qquad \text{and} \qquad \sin t = \frac{y}{3}$$

Squaring both sides of each equation produces

$$\cos^2 t = \frac{x^2}{4} \qquad \text{and} \qquad \sin^2 t = \frac{y^2}{9}$$

Adding these two equations, we have

$$\cos^2 t + \sin^2 t = \frac{x^2}{4} + \frac{y^2}{9}$$

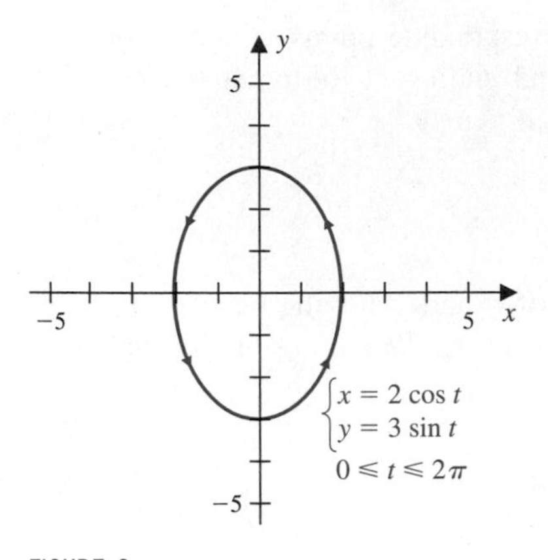

FIGURE 2

Since $\cos^2 t + \sin^2 t = 1$, we get

$$\frac{x^2}{4} + \frac{y^2}{9} = 1$$

which is an equation of an ellipse with center (0, 0), semimajor axis of length 3 on the y axis, and semiminor axis of length 2 on the x axis (Figure 2). The restriction $0 \le t \le 2\pi$ suggests that the orientation of a point moving along the elliptical curve, as t increases from 0 to 2π, starts at point $(2 \cos 0, 3 \sin 0) = (2, 0)$, passes through points $(2 \cos(\pi/2), 3 \sin(\pi/2)) = (0, 3)$ and $(2 \cos \pi, 3 \sin \pi) = (-2, 0)$, and continues around the ellipse to end at $(2 \cos 2\pi, 2 \sin 2\pi) = (2, 0)$. In other words, a point moves along the ellipse in a counterclockwise direction and traverses the ellipse once as t increases from 0 to 2π (Figure 2).

Representing Cartesian Equations Parametrically

We now consider the reverse problem—that is, finding a parametric representation for the graph of an equation in x and y.

If a curve is described by the function $y = f(x)$, we can let $x = t$ to obtain the following parametric equations for this curve:

$$\begin{cases} x = t \\ y = f(t) \end{cases} \qquad \text{where } t \text{ is in the domain of } f$$

For example, the curve defined by $y = 3x^2 + 7$, where $0 \le x \le 10$, has the following parametric equations:

$$\begin{cases} x = t \\ y = 3(t)^2 + 7 \end{cases} \qquad 0 \le t \le 10$$

Parametric representations for curves are not unique. For instance, if we let $x = 2t$, then another set of parametric equations for $y = 3x^2 + 7$, where $0 \le x \le 10$, is given by

$$\begin{cases} x = 2t \\ y = 3(2t)^2 + 7 \end{cases} \quad \text{or} \quad \begin{cases} x = 2t \\ y = 12t^2 + 7 \end{cases}$$

Since $x = 2t$ and $0 \le x \le 10$, then $0 \le 2t \le 10$ or $0 \le t \le 5$.

In other cases, different techniques may be needed. For instance, at times, trigonometric functions are used in parametric representations of curves.

EXAMPLE 3 **Representing an equation parametrically**

Represent the equation of the circle

$$x^2 + y^2 = 4$$

parametrically if the orientation of points on this curve starts at (2, 0) and traverses the circle counterclockwise once.

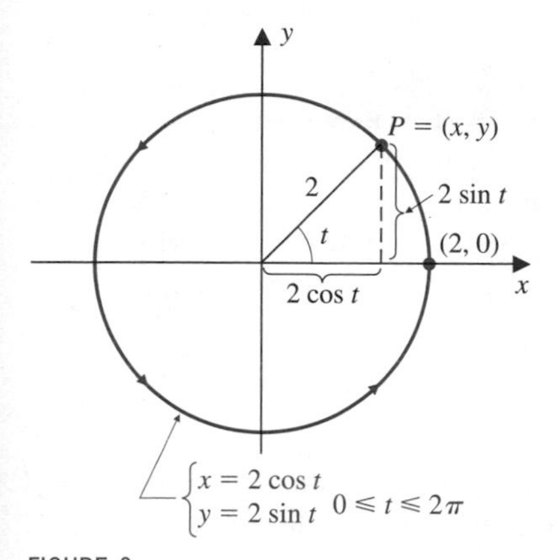

FIGURE 3

Solution Using Figure 3 and right angle trigonometry, we see that the equations

$$\begin{cases} x = 2\cos t \\ y = 2\sin t \end{cases} \qquad 0 \le t \le 2\pi$$

give a parametric representation for every point P on the circle. We verify this result by observing that

$$\begin{aligned} x^2 + y^2 &= 4\cos^2 t + 4\sin^2 t \\ &= 4(\cos^2 t + \sin^2 t) = 4 \end{aligned}$$

The condition $0 \le t \le 2\pi$ suggests that a point moves along the circle starting at the point $(2, 0)$, when $t = 0$, and traverses the circle in a counterclockwise direction as t increases from 0 to 2π, ending at $(2, 0)$, when $t = 2\pi$.

Parametric equations can provide representations of curves generated by physical motion in cases where it may be difficult to find a Cartesian equation. For instance, suppose a ball at point $(0, 0)$ is allowed to drop and then slide down a curve to a point B not directly under $0 = (0, 0)$ (Figure 4). It can be shown that the ball takes the fastest path along a curve whose parametric equations are

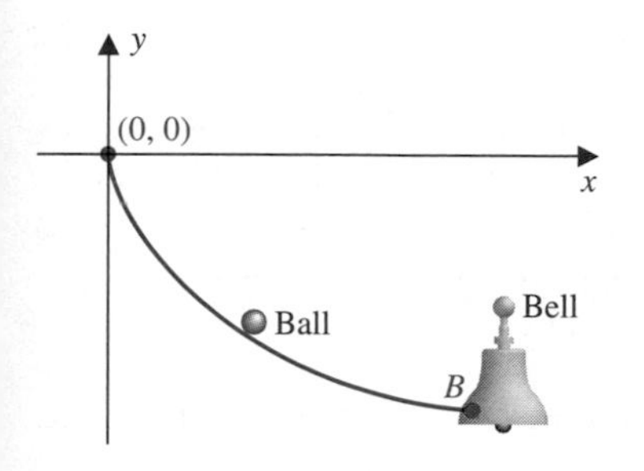

FIGURE 4

$$\begin{cases} x = a(t - \sin t) \\ y = a(1 - \cos t) \end{cases}$$

where a is a constant and t represents time. The curve represented by these equations is called a *cycloid*. Point-plotting is impractical to graph these equations; however, it is easy to use a graphing utility to do this.

EXAMPLE 4  

Graphing parametric equations using a graphing utility

Use a graphing utility to graph the cycloid represented by the parametric equations

$$\begin{cases} x = 2(t - \sin t) \\ y = 2(1 - \cos t) \end{cases} \qquad 0 \le t \le 4\pi$$

When using a graphing utility to graph parametric equations, it is important to take into account the given restrictions on the parameter.

Solution We set the graphing utility in parametric mode, making certain that the restrictions on t are properly entered. Figure 5 shows a viewing window of the graph of the curve. The curve is periodic with period 4π.

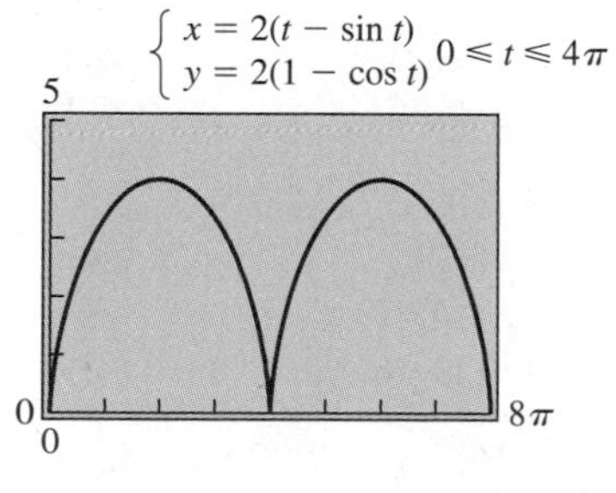

FIGURE 5

Solving Applied Problems

Parametric equations are useful in modeling real-world situations, especially those in which more than one variable is dependent upon time. For example, if an object is launched from a height of s_0 feet into the air at an angle θ with the ground and with an initial speed of v_0 feet per second, then physicists have determined the position of the object after t seconds is given by the parametric equations

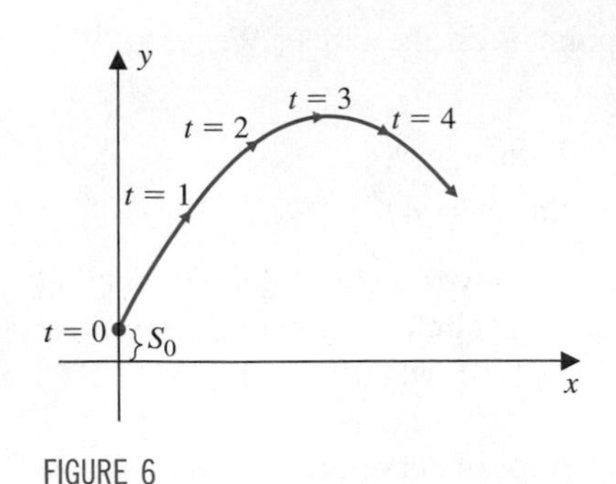

FIGURE 6

$$\begin{cases} x = (v_0 \cos \theta)t \\ y = -16t^2 + (v_0 \sin \theta)t + s_0 \end{cases} \qquad t \geq 0$$

In this model, $v_0 \cos \theta$ is the horizontal component of the initial speed and $v_0 \sin \theta$ is the vertical component of the initial speed. The constant s_0 is the vertical distance (in feet) between the ground and the point from which the object is propelled. Figure 6 shows a general graph that describes the path of the object.

EXAMPLE 5 **Solving a space shuttle problem**

A space shuttle rises from its launch pad and reaches a height of 30,000 feet. Because of a computer malfunction, the booster rockets are turned off prematurely, at a moment when the shuttle is moving at a speed of 32,000 feet per second at an angle of 42° with an imaginary reference plane tangent to the Earth directly below.

(a) Use the model given above to determine the specific parametric equations that describe the shuttle's flight.

(b) Use a graphing utility to graph the curve and determine how long it takes the space shuttle to reach the maximum height above the Earth. Also, find the maximum height of the shuttle. Round off the time to the nearest 10 seconds, and the height to the nearest 1000 feet.

Solution

(a) We use the parametric equations from above:

$$\begin{cases} x = (v_0 \cos \theta)t \\ y = -16t^2 + (v_0 \sin \theta)t + s_0 \end{cases}$$

where $v_0 = 32{,}000$, $\theta = 42°$, and $s_0 = 30{,}000$. Then the horizontal position of the space shuttle is represented by

$$\begin{aligned} x &= (v_0 \cos \theta)t \\ &= (32{,}000 \cos 42°)t = 24{,}000t \quad \text{Rounded off to the nearest thousand} \end{aligned}$$

The vertical position is represented by

$$\begin{aligned} y &= -16t^2 + (32{,}000 \sin 42°)t + 30{,}000 \\ y &= -16t^2 + 21{,}000t + 30{,}000 \quad \text{Rounded off to the nearest thousand} \end{aligned}$$

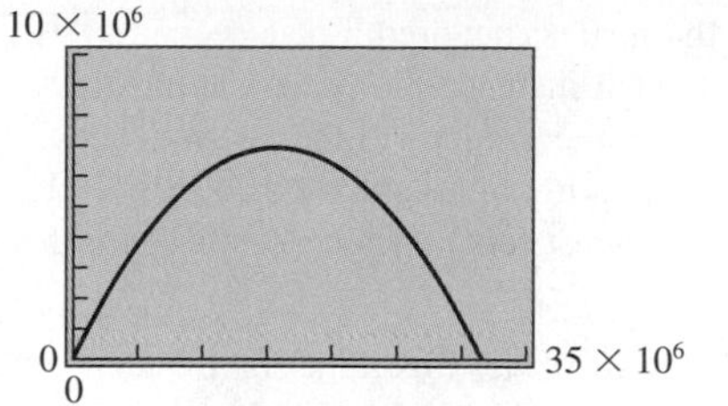

FIGURE 7

(b) Figure 7 shows a viewing window of the graph of the parametric equations from part (a). Using the ZOOM and TRACE features, we find that the approximate maximum point on the graph is $(x, y) = (15{,}750{,}000, 6{,}921{,}000)$. Therefore, the maximum height of the shuttle is 6,921,000 feet, or approximately 1310.8 miles. To find the time when the shuttle is at the maximum height, we solve the equation $15{,}750{,}000 = 24{,}000t$ for t; so $t = 656.25$, or approximately 660 seconds (about 11 minutes).

PROBLEM SET 9.6

Mastering the Concepts

In problems 1–4, sketch the curve represented by the parametric equations by using point-plotting. Describe the orientation of a point on the curve as t varies over its values.

1. $\begin{cases} x = 2t + 1 \\ y = t - 1 \end{cases} \quad 0 \le t \le 4$

2. $\begin{cases} x = \frac{1}{2}t + 1 \\ y = 3t - 4 \end{cases} \quad 0 \le t \le 5$

3. $\begin{cases} x = \cos t \\ y = 4 \sin t \end{cases} \quad 0 \le t \le 2\pi$

4. $\begin{cases} x = 2 \cos t \\ y = 5 \sin t \end{cases} \quad 0 \le t \le 2\pi$

In problems 5–20, eliminate the parameter and then sketch the curve for the given parametric equations. Describe the orientation of a point on the curve as t varies over its values.

5. $\begin{cases} x = 1 - 3t \\ y = 2 + \frac{5}{2}t \end{cases} \quad -3 \le t \le 3$

6. $\begin{cases} x = -2 - 4t \\ y = -5 + 3t \end{cases} \quad 1 \le t \le 8$

7. $\begin{cases} x = 1 + 2t \\ y = -1 + 4t^2 \end{cases} \quad -2 \le t \le 2$

8. $\begin{cases} x = 3t^2 \\ y = 2t \end{cases} \quad -2 \le t \le 2$

9. $\begin{cases} x = t \\ y = \sqrt{1 - t^2} \end{cases} \quad -1 \le t \le 1$

10. $\begin{cases} x = -3t \\ y = \sqrt{4 + 2t} \end{cases} \quad -2 \le t \le 3$

11. $\begin{cases} x = 3 \cos t \\ y = 3 \sin t \end{cases} \quad 0 \le t \le \pi$

12. $\begin{cases} x = 3 \cos t \\ y = -4 \sin t \end{cases} \quad 0 \le t \le 2\pi$

13. $\begin{cases} x = -3 + \cos t \\ y = 4 - \sin t \end{cases} \quad 0 \le t \le 2\pi$

14. $\begin{cases} x = 1 + \sin t \\ y = -1 + 4 \cos t \end{cases} \quad 0 \le t \le 2\pi$

15. $\begin{cases} x = \csc t \\ y = \cot t \end{cases} \quad -\frac{\pi}{2} < t < 0 \text{ or } 0 < t < \frac{\pi}{2}$

16. $\begin{cases} x = \sin t \\ y = \csc t \end{cases} \quad 0 < t < \pi$

17. $\begin{cases} x = e^t \\ y = e^{-2t} \end{cases} \quad t \ge 0$

18. $\begin{cases} x = t^2 \\ y = 3 \ln t \end{cases} \quad t > 0$

19. $\begin{cases} x = 2 \sec t \\ y = 3 \tan t \end{cases} \quad -\frac{\pi}{2} < t < \frac{\pi}{2}$

20. $\begin{cases} x = 2 \cos^3 t \\ y = 2 \sin^3 t \end{cases} \quad 0 \le t \le 2\pi$

In problems 21–30, find parametric equations that represent the given oriented curve.

21. $y = 3x^2 + 5$; from $(0, 5)$ to $(2, 17)$
22. $y = x^3$; from $(-1, -1)$ to $(2, 8)$
23. $y = x^2 - 4x + 3$; from $(-1, 8)$ to $(1, 0)$
24. $y^2 = x$; from $(1, -1)$ to $(4, 2)$
25. Along the line segment from $(1, 2)$ to $(3, 5)$
26. Along the line segment from $(0, 2)$ to $(2, 0)$

27. Around the circle $x^2 + y^2 = 9$ once, counterclockwise starting at (3, 0)

28. Around the ellipse $(x^2/9) + (y^2/4) = 1$ once, counterclockwise starting at (0, 2)

29. Around the circle $x^2 + (y - 1)^2 = 1$ once, counterclockwise starting at (1, 1)

30. Along the hyperbola $x^2 - 4y^2 = 4$, where $x > 0$

GU In problems 31–36, use a graphing utility to graph each curve represented by the parametric equations.

31. $\begin{cases} x = 3t - 2 \\ y = 4t^2 - 3 \end{cases} \quad 0 \le t \le 4$

32. $\begin{cases} x = t^2 - 2t \\ y = t^3 - 3t \end{cases} \quad 0 \le t \le 4$

33. $\begin{cases} x = 2 - \dfrac{1}{t} \\ y = 2t + \dfrac{1}{t} \end{cases} \quad 1 < t \le 10$

34. $\begin{cases} x = \dfrac{1}{t^2 - 1} \\ y = -5t + 3 \end{cases} \quad 2 \le t \le 10$

35. $\begin{cases} x = 3t - 3 \sin t \\ y = 3 - 3 \cos t \end{cases} \quad 0 \le t \le 10\pi$

36. $\begin{cases} x = 2 \cos^2 t \\ y = 4 \sin^2 t \end{cases} \quad 0 \le t \le 2\pi$

Applying the Concepts

37. Path of a Projectile: A projectile is launched from a height of $s_0 = 10$ feet at an angle $\theta = 50°$ with the horizontal and with an initial velocity $v_0 = 200$ feet per second.
(a) Use the model on page 582 to determine the parametric equations that describe the projectile's flight.
GU **(b)** Use a graphing utility to graph the curve.
(c) Determine how long it takes the projectile to return to ground level (rounded off to the nearest tenth of a second). Also, find the horizontal distance the projectile has traveled (to the nearest foot) when it hits the ground.
(d) Determine the maximum height reached by the projectile (to the nearest foot). When does it attain that height?

38. Path of a Projectile: A projectile is launched from ground level with an initial velocity of 64 feet per second and at an angle of 45° with the horizontal.
(a) Determine the parametric equations that model the projectile's path.
GU **(b)** Use a graphing utility to graph the curve.
(c) Find the time it takes for the projectile to hit the ground (to the nearest hundredth of a second). Also, find the horizontal distance the projectile has traveled when it hits the ground (to the nearest foot).
(d) Determine the maximum height reached by the projectile (to the nearest foot). When does it attain this height?

39. Baseball Path: A baseball player hits a baseball 4 feet above home plate at an angle of 25° with the horizontal and with an initial velocity of 125 feet per second.
(a) Find parametric equations that model this path.
GU **(b)** Use a graphing utility to graph its path.
(c) The outfield fence is 380 feet from home plate and the fence is 8 feet high. Determine how high the ball is when it is 380 feet from home plate. Will the ball clear the fence for a home run if it is not caught?

40. Sports: The parametric equations

GU

$$\begin{cases} x = 2(t - \sin t) \\ y = -2(1 - \cos t) \end{cases}$$

describe the shape of a skateboard ramp.
(a) Use a graphing utility to graph these equations, where $0 \le t \le 2\pi$.
(b) Locate the low point of the ramp.

Developing and Extending the Concepts

41. Find parametric equations for the curve represented by

$$2x^2 - 3xy + 2y^2 = 6$$

42. **(a)** Show that

$$\begin{cases} x = a \cos t + h \\ y = b \sin t + k \end{cases} \quad 0 \le t \le 2\pi$$

where $a > 0$, $b > 0$, and $a \ne b$, are parametric equations of an ellipse with center at (h, k), horizontal axis of length $2a$, and vertical axis of length $2b$.
(b) What is the curve if $a = b$?

43. Given parametric equations: $\begin{cases} x = 3 + 5t \\ y = 4 - 2t \end{cases}$
(a) Graph the parametric equations to discover that the graph is a straight line.
(b) Eliminate the parameter to find the equation of the line in terms of x and y.

(c) How does the slope of the line found in part (b) relate to the parametric equations in part (a)?

(d) Can you generalize the above results to relate the parametric equations of a line,

$$\begin{cases} x = a + bt \\ y = c + dt \end{cases}$$

to the slope of the line?

44. Show that

$$\begin{cases} x = h + a \sec t \\ y = k + b \tan t \end{cases} \qquad -\frac{\pi}{2} < t < \frac{3\pi}{2},\ t \neq \frac{\pi}{2}$$

where $a > 0$ and $b > 0$, are parametric equations of a hyperbola with center at (h, k), transverse axis of length $2a$, and conjugate axis of length $2b$.

45. Given parametric equations: $\begin{cases} x = t^2 \\ y = t^2 \end{cases}$

(a) Eliminate the parameter for the given system.

(b) Explain why the graph of the given parametric equations is different from the graph of $y = x$.

46. Do the following sets of parametric equations represent the same curve? Explain.

$$\begin{cases} x = \cos t \\ y = \sin t \end{cases} \qquad 0 \le t \le 2\pi$$

$$\begin{cases} x = \sin t \\ y = \cos t \end{cases} \qquad 0 \le t \le 2\pi$$

47. Give three different sets of parametric equations that represent the curve given by $y = x^2 + 1$, where $0 \le x \le 5$ and the orientation starts at $(0, 1)$ and ends at $(5, 26)$.

CHAPTER 9 REVIEW PROBLEM SET

In problems 1 and 2, sketch the graph of each parabola, and identify the focus and directrix.

1. **(a)** $x^2 = -8y$ **(b)** $y^2 = 36x$

2. **(a)** $y^2 = -5x$ **(b)** $x^2 = 7y$

In problems 3–6, write the standard equation of the parabola using the given information.

3. Focus at $(-4, 0)$; directrix $y = 3$

4. Focus at $(0, 3)$; directrix $x = \frac{5}{2}$

5. Vertex at $(-2, -3)$; focus at $(-5, -3)$

6. Vertex at $(1, -1)$; directrix $y = 2$

In problems 7 and 8, sketch the graph of each ellipse and identify the vertices and foci.

7. **(a)** $\frac{x^2}{9} + \frac{y^2}{36} = 1$ **(b)** $\frac{x^2}{6} + \frac{y^2}{4} = 1$

8. **(a)** $\frac{x^2}{16} + \frac{y^2}{12} = 1$ **(b)** $\frac{x^2}{2} + \frac{y^2}{12} = 1$

In problems 9–12, sketch the graph of each hyperbola and identify the vertices and foci. Also, find the equations of the asymptotes.

9. $\frac{x^2}{4} - \frac{y^2}{9} = 1$

10. $\frac{y^2}{9} - \frac{x^2}{4} = 1$

11. $16x^2 - 4y^2 = 64$

12. $16y^2 - 4x^2 = 64$

In problems 13–18, write the standard equation of each conic using the given information.

13. An ellipse with foci at $(-2, 0)$ and $(2, 0)$, and vertices at $(-3, 0)$ and $(3, 0)$

14. An ellipse with foci at $(3, 0)$ and $(1, 0)$, and a vertex at $(0, 0)$

15. An ellipse with foci at $(3, 3)$ and $(3, -1)$, and containing the point $(4, 0)$

16. A hyperbola with foci at $(-1, 4)$ and $(5, 4)$, and a vertex at $(0, 4)$

17. A hyperbola with foci at $(-5, 0)$ and $(5, 0)$, and vertices at $(-3, 0)$ and $(3, 0)$

18. A hyperbola with vertices at $(-2, 3)$ and $(2, 3)$, and containing the point $(8, 8)$

In problems 19–24, apply horizontal and vertical shifting to the graph of the first equation to generate the graph of the second equation. Find the center (when applicable), vertices (or vertex), and foci (or focus) of the new conic. Also, sketch the graphs of both equations in the same coordinate system.

19. $y = 2x^2$; $y - 2 = 2(x + 1)^2$

20. $9y^2 = -16x$; $9(y + 1)^2 = -16(x - 2)$

21. $\frac{x^2}{9} + \frac{y^2}{16} = 1$; $\frac{(x - 1)^2}{9} + \frac{(y + 2)^2}{16} = 1$

22. $\frac{x^2}{4} + \frac{y^2}{9} = 1$; $\frac{(x - 4)^2}{4} + \frac{(y + 2)^2}{9} = 1$

23. $\frac{y^2}{4} - \frac{x^2}{9} = 1$; $\frac{(y-1)^2}{4} - \frac{(x-2)^2}{9} = 1$

24. $\frac{x^2}{9} - \frac{y^2}{16} = 1$; $\frac{(x+2)^2}{9} - \frac{(y-1)^2}{16} = 1$

In problems 25–32, identify and sketch the graph of the conic section described. Find the vertices (or vertex), foci (or focus), and equations of the asymptotes when applicable.

25. $y^2 + 2y - 2x + 7 = 0$
26. $x^2 - 2x - 6y - 7 = 0$
27. $9x^2 + 4y^2 - 90x - 16y + 205 = 0$
28. $9x^2 + 4y^2 - 90x - 16y - 83 = 0$
29. $4x^2 - 9y^2 - 16x - 90y + 16 = 0$
30. $4x^2 - 9y^2 - 16x - 90y - 210 = 0$
31. $x^2 - 2x + y^2 + 6y - 37 = 0$
32. $9y^2 - 16x^2 - 90y - 256x = 233$

In problems 33–38, determine the standard equation for the conic that satisfies the given conditions.

33. Focus at (2, 0); directrix $x = \frac{9}{2}$; eccentricity $e = \frac{2}{3}$
34. Focus at (0, 0); directrix $y = -2$; eccentricity $e = 3$
35. Foci at (−10, 0) and (10, 0); eccentricity $e = \frac{5}{4}$
36. Focus at (6, −10); directrix $x = 2$; eccentricity $e = 1$
37. Vertex at (−5, −3); directrix $y = -5$; eccentricity $e = 1$
38. Foci at $(-\sqrt{17}, 0)$ and $(\sqrt{17}, 0)$; major axis on the x axis; eccentricity $e = \sqrt{17}/5$

In problems 39–42, find the eccentricity of each conic.

39. **(a)** $8x^2 + 3y^2 = 12$ **(b)** $12(x+3)^2 + 8y^2 = 1$
40. **(a)** $18x^2 - 9y^2 = 18$ **(b)** $24(x-3)^2 - y^2 = 1$
41. **(a)** $y^2 - 6y - 2x + 1 = 0$
(b) $x^2 + 4y^2 + 4x + 40y + 100 = 0$
42. **(a)** $x^2 - 4y - 6x + 8y - 11 = 0$
(b) $4x^2 + 4y^2 - 8x - 16y + 4 = 0$

In problems 43 and 44, find the standard polar equations of the conic that satisfies the given conditions. (Assume the focus is at the pole.)

43. **(a)** $e = 1$; directrix $r = -\frac{1}{2}\csc\theta$
(b) $e = 2$; directrix $r = 4\sec\theta$
44. **(a)** $e = 3$; directrix $r = 2\csc\theta$
(b) $e = \frac{1}{3}$; directrix $r = \sec\theta$

GU In problems 45 and 46, find the eccentricity, identify the conic, and use a graphing utility to sketch the graph. Also, locate the center and vertices of the conic.

45. **(a)** $r = \frac{3}{3 - 2\cos\theta}$ **(b)** $r = \frac{5}{2 - 5\sin\theta}$
46. **(a)** $r = \frac{2}{4 + \cos\theta}$ **(b)** $r = \frac{2}{3 - 3\sin\theta}$

In problems 47–52, use the discriminant to determine whether the graph of the given equation is an ellipse, a parabola, or a hyperbola. Perform a rotation to eliminate the xy term and sketch the graph.

47. $x^2 - xy + y^2 = 2$
48. $14x^2 + 24xy + 7y^2 = 1$
49. $17x^2 - 6xy + 9y^2 = 5$
50. $4x^2 - 4xy + 7y^2 = 24$
51. $x^2 - \sqrt{3}\,xy = 1$
52. $6x^2 - 6xy + 14y^2 = 45$

GU In problems 53–56, use a graphing utility to graph each equation.

53. $2y^2 + 8y + 3x - 4 = 0$
54. $4x^2 + 9y^2 - 8x + 36y + 4 = 0$
55. $x^2 - y^2 - 6x + 8y - 3 = 0$
56. $x^2 - 4xy - 2y^2 = 4$

In problems 57–60, eliminate the parameter and sketch the curve for the given parametric equations. Describe the orientation.

57. $\begin{cases} x = 2t \\ y = 2t^2 - 3t \end{cases}$ $0 \le t \le 4$
58. $\begin{cases} x = 1 - \sin t \\ y = 1 + \cos t \end{cases}$ $0 \le t \le 2\pi$
59. $\begin{cases} x = t^2 \\ y = 2\ln t \end{cases}$ $t \ge 1$
60. $\begin{cases} x = 5\sin 2t \\ y = 5\cos t \end{cases}$ $t \ge 0$

61. Engineering: A bridge has a parabolic arch that is 17 meters high in the center and 50 meters wide at the bottom. Find the height of the arch 10 meters from the center.

62. Space Science: A spacecraft, in one of its orbits about Earth, had a minimum altitude of 200 miles and a maximum altitude of 1000 miles. The path of the spacecraft is elliptical, with the center of the Earth at one focus. Find the standard equation of the path if the radius of the Earth is assumed to be about 4000 miles and the center of the Earth is positioned at the origin.

63. Navigation: Two Coast Guard stations are located 600 kilometers apart at points $A = (0, 0)$ and $B = (0, 600)$. A distress signal from a ship is received at slightly different times by the two stations. It is determined that the ship is 200 kilometers farther from station A than it is from station B. Determine the standard equation of a hyperbola that contains the location of the ship.

CHAPTER 9 TEST

1. Find the vertices (or vertex) and the foci (or focus) of the given conics. If the conic is a hyperbola, find the equations of its asymptotes. Sketch the graph.

(a) $y^2 = -16x$

(b) $9x^2 + y^2 = 9$

(c) $x^2 - 4y^2 = 4$

(d) $4x^2 + y^2 + 16x + 7 = 0$

(e) $y^2 + 6y - x + 21 = 0$

2. Find the standard equation of the conic that satisfies the given conditions.

(a) An ellipse with vertices at $(-2, 0)$ and $(2, 0)$; foci at $\left(-\sqrt{3}, 0\right)$ and $\left(\sqrt{3}, 0\right)$

(b) A parabola with vertex at $(-3, -2)$; containing the origin; axis of symmetry parallel to the x axis

(c) A parabola with vertex at $(2, -3)$; directrix $x = -8$

(d) A hyperbola with vertices at $(-4, 0)$ and $(4, 0)$; foci at $(-6, 0)$ and $(6, 0)$

3. Determine an acute angle θ for a rotation of axes that will eliminate the xy term from $x^2 - 4xy + y^2 = 6$. Express the equation in terms of $\overline{x}$ and $\overline{y}$, and graph the rotated equation.

4. Eliminate the parameter and sketch the curve given by the parametric equations.

$$\begin{cases} x = 2 + \sin t \\ y = 1 - \cos t \end{cases} \qquad 0 \le t \le 2\pi$$

5. Find the eccentricity of the ellipse whose equation is

$$2y^2 + 9x^2 = 18$$

6. Consider the conic given in polar coordinates by the equation

$$r = \frac{4}{2 + \sin\theta}$$

(a) Identify the conic.

(b) Find the eccentricity of the conic.

(c) Find the foci of the conic.

(d) Sketch the conic.

7. Identify the conic given by the equation

$$3x^2 + 4xy + y^2 + y - 10 = 0$$

Use a graphing utility to graph the conic.

8. A parabolic communications antenna has a focus 2 feet from the vertex of the antenna. Find the width of the antenna 3 feet from the vertex in the direction of the focus.

10

Topics in Discrete Mathematics

CHAPTER CONTENTS

In this chapter we discuss *sequences* and *series*. These topics help us to identify important types of patterns in mathematics that have applications in areas such as business and biology, as well as in the study of calculus. We also introduce the principle of *mathematical induction*, a method of proof. In addition, we discover efficient ways of expanding a positive integral power of a binomial.

10.1 Sequences and Summation Notation

OBJECTIVES

1. Describe a sequence.
2. Use recursive formulas.
3. Graph sequences.
4. Use summation notation.
5. Solve applied problems.

Many applications of mathematics involve ordered lists of numbers called *sequences*.

Describing a Sequence

Sequences are described in various ways. For example, the pattern given by the following positive odd integers describes a sequence:

$$1, 3, 5, 7, \ldots$$

The list of paired numbers

$$(1, 3), (2, 6), (3, 9), (4, 12), (5, 15), \ldots$$

describes a sequence in a way that lends itself to graphing. Tables, such as Table 1, are also used to describe a sequence.

TABLE 1

Year	Interest
1	\$131.20
2	\$137.10
3	\$141.30
4	\$118.40
5	\$145.70

It should be noted that in each situation the sequence establishes a correspondence between positive integers, which indicate order or position, and other numbers. These illustrations lead us to the notion of a sequence.

DEFINITION
SEQUENCE

A function whose domain consists of consecutive positive integers and whose range is a set of numbers is a **sequence**.

At times, only the range of a sequence is given. For instance, the sequence of positive odd integers is simply listed as 1, 3, 5, 7, ... This is an example of an *infinite sequence*, since the domain is the set of *all* positive integers. On the other hand, Table 1 illustrates a *finite sequence* whose domain consists of only 1, 2, 3, 4, and 5.

To denote the particular numbers in a sequence we often give a rule defined by a formula. For instance, if we let $f(n) = 3n$, where n is a positive integer, then this sequence is given by

$$f(1), f(2), f(3), f(4), \ldots, f(n), \ldots$$

or

$$3, \quad 6, \quad 9, \quad 12, \ldots, \quad 3n, \ldots$$

In this case, the three dots at the end are used to indicate that the sequence continues indefinitely, following the pattern defined by $f(n) = 3n$.

The numbers in the range of a sequence are called the **terms** of the sequence. The symbolism

$$a_1, a_2, a_3, \ldots, a_n, \ldots$$

is often used to denote a sequence, where a_1 is the **first term**, a_2 is the **second term**, and a_n is the ***n*th**, or **general**, **term** of the sequence. The more compact notation $\{a_n\}$, in which the general term is enclosed in braces, is also used to denote the sequence. For example, $\{3n\}$ conveys that the sequence can be generated by successively substituting the integer values $n = 1, 2, 3, \ldots$ into the formula $3n$. Thus, for $\{a_n\} = \{3n\}$, we have

$$\begin{aligned} a_1 &= 3(1) = 3 \\ a_2 &= 3(2) = 6 \\ a_3 &= 3(3) = 9 \\ a_4 &= 3(4) = 12 \\ &\vdots \end{aligned}$$

EXAMPLE 1 **Determining terms of a sequence**

Determine the first five terms of the sequence with the given nth term.

(a) $a_n = (-1)^n$ (b) $b_n = 3 - \dfrac{1}{n}$

Solution To find the first five terms of each sequence, we substitute the positive integers 1, 2, 3, 4, and 5, in turn, for n in the formula for the general term.

(a)
$$a_1 = (-1)^1 = -1,\ a_2 = (-1)^2 = 1,\ a_3 = (-1)^3 = -1,$$
$$a_4 = (-1)^4 = 1, \text{ and } a_5 = (-1)^5 = -1$$

So the first five terms of this sequence are $-1, 1, -1, 1$, and -1.

(b) For $b_n = 3 - \dfrac{1}{n}$, we have

$$b_1 = 3 - \tfrac{1}{1} = 2,\ b_2 = 3 - \tfrac{1}{2} = \tfrac{5}{2},\ b_3 = 3 - \tfrac{1}{3} = \tfrac{8}{3},$$
$$b_4 = 3 - \tfrac{1}{4} = \tfrac{11}{4}, \text{ and } b_5 = 3 - \tfrac{1}{5} = \tfrac{14}{5}$$

Hence, the first five terms of this sequence are 2, $\frac{5}{2}$, $\frac{8}{3}$, $\frac{11}{4}$, and $\frac{14}{5}$.

In Example 1, we used the formulas for the nth term to determine the first few terms of a sequence. It may be possible to reverse the process—that is, use the first few terms of a sequence to find its general nth term.

EXAMPLE 2 Finding the nth term of a sequence

Find a formula for the general term of the sequence: 5, 10, 15, 20, 25, . . .

Solution By examining the given terms and looking for a pattern, we can see by inspection that the first five terms are multiples of 5, so we can write

$$a_1 = 5 = 5 \cdot 1,\ a_2 = 10 = 5 \cdot 2,\ a_3 = 15 = 5 \cdot 3,$$
$$a_4 = 20 = 5 \cdot 4, \text{ and } a_5 = 25 = 5 \cdot 5$$

Not every sequence can be expressed in terms of a formula for its nth general term.

The pattern of these five terms suggests that one formula for the general term is given by $a_n = 5n$.

Using Recursive Formulas

So far we have considered sequences whose nth terms are given by *explicit formulas*. Another way that is used to specify the terms of a sequence is by a formula that relates the general nth term a_n to terms that precede it. Such a formula is called a **recursive formula**. This type of formula gives one or more initial terms of a sequence and then defines a_n in terms of the preceding terms.

EXAMPLE 3 Finding terms of a sequence by using a recursive formula

The sequence $\{a_n\}$ is defined by $a_1 = 1$, $a_2 = 1$, and $a_n = a_{n-1} + a_{n-2}$, where $n \geq 3$. Find the first six terms of this sequence.

Solution The first six terms of this sequence are found as follows:

$$\begin{aligned}
a_1 &= 1 \\
a_2 &= 1 \\
a_3 &= a_{3-1} + a_{3-2} = a_2 + a_1 = 1 + 1 = 2 \\
a_4 &= a_{4-1} + a_{4-2} = a_3 + a_2 = 2 + 1 = 3 \\
a_5 &= a_{5-1} + a_{5-2} = a_4 + a_3 = 3 + 2 = 5 \\
a_6 &= a_{6-1} + a_{6-2} = a_5 + a_4 = 5 + 3 = 8
\end{aligned}$$

So the first six terms of the sequence are 1, 1, 2, 3, 5, and 8.

The sequence found in Example 3 is a **Fibonacci sequence**, and the numbers that appear in it are called **Fibonacci numbers**. This sequence has interesting mathematical properties and it is used in a wide variety of applications in fields such as physics and ecology.

Graphing Sequences

Since a sequence is a function, the graph of the sequence $\{a_n\}$ is the graph of the function $y = f(n)$, where $f(n) = a_n$ and n is a positive integer. As usual, we use the horizontal axis for the inputs n and the vertical axis for the outputs $a_n = f(n)$.

For example, the graph of the sequence $\{1/n\}$ is the graph of the function

$$f(n) = \frac{1}{n} \qquad \text{where } n = 1, 2, 3, \ldots$$

The graph consists of a succession of isolated points, since the right side of the equation is defined only for positive integer values of n (Figure 1a). This is in marked contrast to the graph of

$$f(x) = \frac{1}{x} \qquad \text{where } x \geq 1$$

which is a continuous (unbroken) curve (Figure 1b).

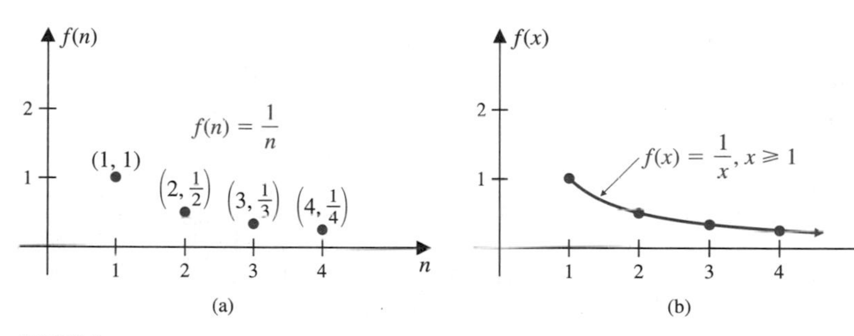

FIGURE 1

EXAMPLE 4 **Graphing a sequence**

(a) Graph the first five terms of the sequence whose general term is given by

$$a_n = \frac{2}{n + 1}$$

for $n = 1, 2, 3, 4, 5$.

(b) Graph the function

$$f(x) = \frac{2}{x + 1}$$

for $x \geq 1$, and describe the limit behavior as $x \to \infty$.

Solution

(a) To find the first five terms of the sequence, we substitute $n = 1, 2, 3, 4$, and 5 into the general term $a_n = 2/(n + 1)$ to get

$$a_1 = \frac{2}{1+1} = 1,\ a_2 = \frac{2}{2+1} = \frac{2}{3},\ a_3 = \frac{2}{3+1} = \frac{2}{4} = \frac{1}{2},$$
$$a_4 = \frac{2}{4+1} = \frac{2}{5},\ \text{and } a_5 = \frac{2}{5+1} = \frac{2}{6} = \frac{1}{3}$$

Because the sequence is defined only for positive integer values of n, the graph of this sequence consists of the following discrete points (Figure 2a):

$$(1, 1),\ \left(2, \tfrac{2}{3}\right),\ \left(3, \tfrac{1}{2}\right),\ \left(4, \tfrac{2}{5}\right),\ \text{and } \left(5, \tfrac{1}{3}\right)$$

(b) The graph of $f(x) = 2/(x + 1)$, for $x \geq 1$, is shown in Figure 2b. Observe that the limit behavior is given by $f(x) \to 0$ as $x \to \infty$.

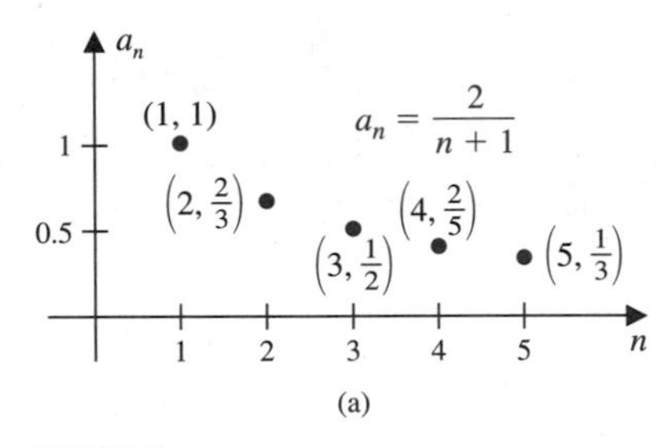
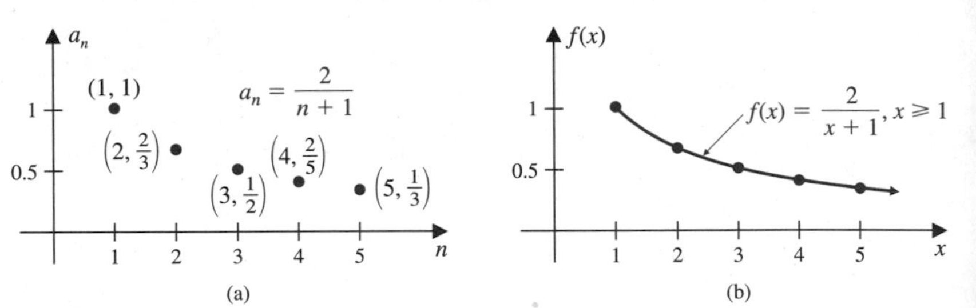

FIGURE 2

Note that the limit behavior of

$$a_n = \frac{2}{n+1} \quad \text{as} \quad n \to +\infty$$

is the same as the limit behavior of

$$f(x) = \frac{2}{x+1} \quad \text{as} \quad x \to +\infty$$

The limit behavior of a sequence $\{a_n\}$ as $n \to \infty$ is called the **limit of the sequence**, which is written in symbols as

$$\lim_{n \to \infty} a_n$$

Thus,

$$\lim_{n \to +\infty} \frac{2}{n+1} = \lim_{x \to +\infty} \frac{2}{x+1} = 0$$

Using Summation Notation

The sum of the first n terms of an infinite sequence $\{a_n\}$ is given by

$$a_1 + a_2 + a_3 + \cdots + a_n$$

Frequently, such a sum is written more compactly by using the following **sigma**, or **summation, notation**:

The symbol Σ is a stylized version of the Greek capital letter sigma.

$$\sum_{k=1}^{n} a_k = a_1 + a_2 + a_3 + \cdots + a_n$$

Here, Σ indicates a sum and the letter k is called the **index of summation**. The symbols n and 1 above and below Σ are called the **upper** and **lower limits** of the summation, respectively. In this case, they indicate that k takes on integer values from 1 to n, inclusive. Any letter can be used for the index; however, i, j, and k are the most commonly used. For instance,

$$\sum_{k=1}^{n} 5^k = \sum_{i=1}^{n} 5^i = \sum_{j=1}^{n} 5^j = 5^1 + 5^2 + 5^3 + \cdots + 5^n$$

Note that the lower limit of a summation may also be 0 or any positive integer. It does not have to be 1.

EXAMPLE 5 **Using sigma notation**

Evaluate each sum.

(a) $\displaystyle\sum_{k=1}^{3} (4k^2 - 3k)$ (b) $\displaystyle\sum_{k=2}^{5} \frac{k-1}{k+1}$

Solution

(a) Here, we have $a_k = 4k^2 - 3k$. To find the sum, we substitute the integers 1, 2, and 3 for k in succession, and then add the resulting numbers:

$$\begin{aligned}\sum_{k=1}^{3} (4k^2 - 3k) &= [4(1^2) - 3(1)] + [4(2^2) - 3(2)] + [4(3^2) - 3(3)]\\ &= 1 + 10 + 27 = 38\end{aligned}$$

(b) Here, $a_k = \dfrac{k-1}{k+1}$, and the index starts at $k = 2$. Thus,

$$\begin{aligned}\sum_{k=2}^{5} \frac{k-1}{k+1} &= \frac{2-1}{2+1} + \frac{3-1}{3+1} + \frac{4-1}{4+1} + \frac{5-1}{5+1}\\ &= \frac{1}{3} + \frac{2}{4} + \frac{3}{5} + \frac{4}{6} = \frac{21}{10}\end{aligned}$$

In the box at the top of page 594, we state some basic properties of summation that are used in areas such as calculus, probability, and statistics.

BASIC PROPERTIES OF SUMMATION

If $\{a_n\}$ and $\{b_n\}$ are given sequences and if c represents a constant, then we have the following results:

1. **Constant Property**: $\sum_{k=1}^{n} c = nc$
2. **Homogeneous Property**: $\sum_{k=1}^{n} ca_k = c \sum_{k=1}^{n} a_k$
3. **Additive Property**: $\sum_{k=1}^{n} (a_k + b_k) = \sum_{k=1}^{n} a_k + \sum_{k=1}^{n} b_k$
4. **Sum of Successive Integers**:
$$\sum_{k=1}^{n} k = 1 + 2 + 3 + \cdots + n = \frac{n(n+1)}{2}$$
5. **Sum of Successive Squares**:
$$\sum_{k=1}^{n} k^2 = 1^2 + 2^2 + 3^2 + \cdots + n^2 = \frac{n(n+1)(2n+1)}{6}$$

The first three properties can be verified by first expanding the summation and then using basic properties of algebra. For instance, Property 2 can be proved as follows:

$$\begin{aligned}\sum_{k=1}^{n} ca_k &= ca_1 + ca_2 + \cdots + ca_n \\ &= c(a_1 + a_2 + \cdots + a_n) \\ &= c \sum_{k=1}^{n} a_k\end{aligned}$$

Properties 4 and 5 will be proved in Section 10.4.

EXAMPLE 6 Using the basic properties to find a sum

Use the basic summation properties to evaluate: $\sum_{k=1}^{20} (2k^2 - 3k + 4)$

Solution We evaluate the given sum by applying the basic summation properties as follows:

$$\begin{aligned}\sum_{k=1}^{20} (2k^2 - 3k + 4) &= \sum_{k=1}^{20} 2k^2 + \sum_{k=1}^{20} (-3k) + \sum_{k=1}^{20} 4 && \text{Property 3} \\ &= 2\sum_{k=1}^{20} k^2 + (-3)\sum_{k=1}^{20} k + 20 \cdot 4 && \text{Properties 1 and 2} \\ &= 2\left[\frac{(20)(21)(41)}{6}\right] - 3\left[\frac{(20)(21)}{2}\right] + 80 && \text{Properties 4 and 5} \\ &= 5190\end{aligned}$$

Solving Applied Problems

Sequences are used to model data involving ordered events.

EXAMPLE 7 Modeling a savings pattern

A newspaper carrier saved \$1 the first week of delivering newspapers, \$2 the second week, \$4 the third week, and \$7 the fourth week. For the fifth week, the amount saved is obtained by adding \$4 to the amount for the preceding week; for the sixth week, the amount saved is obtained by adding \$5 to the amount for the preceding week; and so on.

(a) Find the sequence that describes the savings for each week.
(b) Find a recursive formula for this sequence.
(c) If this pattern continues, predict the newspaper carrier's savings for the seventh week.
(d) Find the sum saved after 8 weeks.

Solution

(a) By examining the sequence, we observe that the second number is obtained by adding 1 to the first, the third is obtained by adding 2 to the second, the fourth by adding 3 to the third, and so on. Therefore, the sequence that describes the savings is given by

$$\underbrace{1, \quad 2}_{+1}, \; \underbrace{\quad 4}_{+2}, \; \underbrace{\quad 7}_{+3}, \; \underbrace{\quad 11}_{+4}, \; \underbrace{\quad 16}_{+5}, \ldots$$

(b) If we denote the sequence by $\{a_n\}$, then we have

$$\begin{aligned}
a_1 &= 1 \\
a_2 &= 2 = 1 + 1 = a_1 + 1 \\
a_3 &= 4 = 2 + 2 = a_2 + 2 \\
a_4 &= 7 = 4 + 3 = a_3 + 3 \\
a_5 &= 11 = 7 + 4 = a_4 + 4 \\
a_6 &= 16 = 11 + 5 = a_5 + 5 \\
&\vdots \\
a_n &= a_{n-1} + (n - 1)
\end{aligned}$$

So the recursive formula is given by $a_n = a_{n-1} + (n - 1)$, where $n > 1$ and $a_1 = 1$.

(c) If the pattern continues, then

$$\begin{aligned}
a_7 &= a_6 + (7 - 1) \\
&= a_6 + 6 \\
&= 16 + 6 = 22
\end{aligned}$$

So the seventh week's savings will be \$22.

(d) The total of the first 8 terms of the sequence is given by the sum

$$\sum_{k=1}^{8} a_k = 1 + 2 + 4 + 7 + 11 + 16 + 22 + 29 = 92$$

So the total amount saved after 8 weeks is \$92.

PROBLEM SET 10.1

Mastering the Concepts

In problems 1–6, indicate whether the sequence is finite or infinite. Then write the general term of the sequence and express it in both subscript and function notation. Indicate the domain and range of each sequence.

1. 1, 3, 5, 7, 9
2. 2, 4, 6, 8
3. 1, 4, 9, 16, 25, . . .
4. 1, 8, 27, 64, . . .
5. 1, −1, 1, −1, 1, . . .
6. −2, 2, −2, 2, −2, . . .

In problems 7–12, find the first five terms of the sequence whose general term is a_n.

7. $a_n = 2n + 3$
8. $a_n = \dfrac{1}{n^2}$
9. $a_n = \dfrac{n}{n+2}$
10. $a_n = \dfrac{2^n}{1 + 2^{-n}}$
11. $a_n = e^{-n}$
12. $a_n = (-1)^n + 1$

In problems 13–18, find the first five terms of each infinite, recursively defined sequence.

13. $a_1 = 1;\ a_n = \frac{1}{2}a_{n-1},\ n \geq 2$
14. $a_1 = 2;\ a_n = 2a_{n-1} - 1,\ n \geq 2$
15. $a_1 = -2;\ a_n = 3a_{n-1},\ n \geq 2$
16. $a_1 = 1,\ a_2 = 2;\ a_n = a_{n-1} + a_{n-2},\ n \geq 3$
17. $a_1 = 1,\ a_2 = 3;\ a_n = a_{n-2} + a_{n-1},\ n > 2$
18. $a_1 = 2,\ a_2 = -1;\ a_n = a_{n-2} - a_{n-1},\ n > 2$

In problems 19–24, find a formula for the general term of each sequence.

19. 3, 5, 7, . . .
20. 2, 5, 8, 11, . . .
21. $\frac{1}{3}, \frac{1}{5}, \frac{1}{7}, \frac{1}{9}, \ldots$
22. $\frac{1}{2}, \frac{3}{4}, \frac{5}{6}, \frac{7}{8}, \ldots$
23. 1, −4, 9, −16, . . .
24. 1, −8, 27, −64, . . .

In problems 25–34:

(a) Find the first five terms of each sequence. Then plot the corresponding pairs to show the pattern of the graph. Round off to two decimal places (when necessary).

GU **(b)** Graph the associated continuous function of each sequence. Use a graphing utility for problems 31–34.

(c) Use the graph from part (b) to find

$$\lim_{n \to \infty} a_n$$

25. $a_n = 3n + 2$
26. $a_n = 1 - 4n$
27. $a_n = 1 + \dfrac{n}{2}$
28. $a_n = \dfrac{2n + 1}{n}$
29. $a_n = 2^n$
30. $a_n = 2^{-n}$
31. $a_n = \dfrac{\sin n}{n}$
32. $a_n = \dfrac{\ln(n + 1)}{n + 1}$
33. $a_n = \dfrac{1 - \cos n}{n}$
34. $a_n = \sqrt[n]{n}$

In problems 35–42, expand and then evaluate each sum.

35. $\sum_{k=1}^{10} 5k$
36. $\sum_{k=0}^{4} \dfrac{2^k}{k + 1}$
37. $\sum_{k=0}^{4} 3^{2k}$
38. $\sum_{k=2}^{5} 2^{k-2}$
39. $\sum_{k=1}^{3} (2k + 1)^2$
40. $\sum_{i=2}^{6} \dfrac{1}{i(i + 1)}$
41. $\sum_{k=1}^{4} \dfrac{(-1)^k + 1}{k}$
42. $\sum_{k=1}^{5} \left(-\dfrac{1}{3}\right)^{k-1}$

In problems 43–46, use the basic properties of summation to evaluate each expression.

43. $\sum_{k=1}^{20} (4k + 3)$
44. $\sum_{k=1}^{30} (7 - 3k)$
45. $\sum_{k=1}^{100} (5k^2 - 3)$
46. $\sum_{k=1}^{10} (3k^2 - 5k + 1)$

Applying the Concepts

47. Mixing Paint: A house painter wishes to mix paint into a 1 gallon bucket from some cans of colored paint. The first can contains $\frac{1}{2}$ gallon of paint, the second can contains $\frac{1}{4}$ gallon of paint, the third can contains $\frac{1}{8}$ gallon of paint, and so on, so that each successive can holds half the amount of the preceding one.
- **(a)** Find the first five terms of the sequence for the amount of colored paint in each can according to the specified order.
- **(b)** Find a formula for this sequence.
- **(c)** If the first five such cans were poured into the 1 gallon bucket, use summation notation to represent the total amount of paint in the bucket. Then evaluate this summation.
- **(d)** Is the bucket full after the contents of the first five cans are poured into it? If this process continues, will the can ever be filled?

48. Telemarketing: An advertising agency begins a telemarketing campaign by making 50 phone calls on the first day. Each day after, the agency will make 10 more calls than were made the day before.
- **(a)** Find the first five terms of this sequence.
- **(b)** Let a_n be the total number of calls made on the nth day. Find a formula that defines a_n.
- **(c)** Find the total number of phone calls made during the first 10 days.

49. Botany: A botanist measured the height of a plant once a month and charted its growth. Suppose that the initial height is recorded as 4 inches and the height increases 10% per month for a year.
- **(a)** Find the first five terms of the sequence.
- **(b)** Find a formula a_n for the height of the plant in the nth month.
- **(c)** What is the total height of the plant after 9 months? Round off the answer to two decimal places.

50. Computer Fund: To save for a computer, a teacher made an initial deposit of \$40 into a savings account. Then \$15 was deposited into the account after 1 week, \$20 after 2 weeks, \$25 after 3 weeks, \$30 after 4 weeks, and so on, until enough money was saved for the computer.
- **(a)** Find the first eight terms of the sequence that shows the amount saved each week.
- **(b)** Write a formula a_n for the sequence.
- **(c)** What is the total amount saved after 8 weeks, including the initial deposit?

Developing and Extending the Concepts

51. Find a formula for the sequence defined by the recursive formula:

$$a_1 = 1 \quad \text{and} \quad a_n = a_{n-1} + 8, \quad n > 1$$

52. Find the first five terms of the sequence defined by the recursive formula:

$$a_1 = 10 \quad \text{and} \quad a_n = \sqrt{a_{n-1}}, \quad n > 1$$

Round off each term to four decimal places. Graph the sequence.

53. Solve each equation for t.

(a) $\sum_{k=1}^{10} kt = 110$ **(b)** $\sum_{k=1}^{20} k(k - t) = 1820$

54. Consider the sequence defined recursively by

$$a_1 = 2 \quad \text{and} \quad a_n = 5 - a_{n-1}, \quad n \geq 2$$

- **(a)** Find the first eight terms of the sequence $\{a_n\}$ and graph it.
- **(b)** Find the first eight terms of the sequence

$$\left\{\frac{a_n}{a_n - 1}\right\}$$

and graph it.

In problems 55–62, suppose that

$$s_n = \sum_{k=1}^{n} a_k$$

Then a recursive formula for a_n is given by

$$a_n = s_n - s_{n-1} \quad \text{where } n \geq 2 \text{ and } a_1 = s_1$$

Use this result to find an explicit formula for the general term a_n of each sequence.

55. $s_n = 3n + 4$

56. $s_n = 4 - 3n$

57. $s_n = n(n + 2)$

58. $s_n = 3n(n + 1)$

59. $s_n = \frac{n}{n + 2}$

60. $s_n = \frac{2n}{3n + 1}$

61. $s_n = 2^n$

62. $s_n = 2^{n+1} - 3$

In problems 63–68, find s_3 and s_6.

63. $s_n = \sum_{k=1}^{n} \frac{k(k + 1)}{2}$

64. $s_n = \sum_{k=1}^{n} \frac{k(k + 1)(2k + 1)}{6}$

65. $s_n = \sum_{k=1}^{n} \frac{k^2(k+1)^2}{4}$

66. $s_n = \sum_{k=1}^{n} (2^{k+1} - 8)$

67. $s_n = \sum_{k=1}^{n} (-1)^k(k+2)$

68. $s_n = \sum_{k=1}^{n} \frac{(-1)^{k+1}}{k+2}$

In problems 69 and 70, use

$$\begin{aligned} s_n &= \sum_{k=1}^{n} (a_k - a_{k+1}) \\ &= (a_1 - a_2) + (a_2 - a_3) + (a_3 - a_4) + \cdots + (a_n - a_{n+1}) \\ &= a_1 - a_{n+1} \end{aligned}$$

to find an explicit formula for s_n in terms of n.

69. **(a)** $s_n = \sum_{k=1}^{n} \left(\frac{1}{k} - \frac{1}{k+1}\right)$

(b) $s_n = \sum_{k=1}^{n} \left(\frac{1}{k+1} - \frac{1}{k+2}\right)$

70. **(a)** $s_n = \sum_{k=1}^{n} \left(\frac{1}{3k-1} - \frac{1}{3k+2}\right)$

(b) $s_n = \sum_{k=1}^{n} \left(\frac{1}{k+2} - \frac{1}{k+3}\right)$

10.2 Arithmetic Sequences

OBJECTIVES

1. Identify an arithmetic sequence and find its general term.
2. Find the sum of an arithmetic sequence.
3. Solve applied problems.

In this section we will investigate a particular type of sequence called an *arithmetic sequence*.

Identifying an Arithmetic Sequence and Finding Its General Term

Consider the sequence

$$2, 6, 10, 14, 18, \ldots$$

Note that the difference between successive terms is always 4. This sequence is an example of an *arithmetic sequence*.

DEFINITION
ARITHMETIC SEQUENCE

A sequence $a_1, a_2, a_3, \ldots, a_n, \ldots$ is called an **arithmetic sequence** if each term (after the first term) differs from the preceding term by a constant number. That is, there is a constant d for which $a_n - a_{n-1} = d$ for all integers $n > 1$. The constant number d is called the **common difference** for the sequence.

For the arithmetic sequence given above the common difference is 4, so each term of the sequence can be obtained by adding 4 to its predecessor. That is,

$$\begin{aligned} a_1 &= 2 \\ a_2 &= 6 = 2 + 4 \\ a_3 &= 10 = 2 + 2 \cdot 4 \\ a_4 &= 14 = 2 + 3 \cdot 4 \\ &\vdots \\ a_n &= 2 + (n-1) \cdot 4 \end{aligned}$$

In general, any term a_n of an arithmetic sequence is the sum of the first term a_1 and a multiple of the common difference d. We can find a general formula for the nth term of an arithmetic sequence as follows:

$$\begin{aligned} a_1 &= a_1 \\ a_2 &= a_1 + d \\ a_3 &= a_2 + d = (a_1 + d) \quad + d = a_1 + 2d \\ a_4 &= a_3 + d = (a_1 + 2d) + d = a_1 + 3d \\ &\vdots \qquad\qquad\qquad\qquad\qquad\qquad \vdots \end{aligned}$$

By repeating this pattern, we discover that the *general term* a_n of an arithmetic sequence with a first term a_1 and a common difference d is given by

$$\boxed{a_n = a_1 + (n - 1)d}$$

Table 1 shows a few examples of arithmetic sequences.

TABLE 1
Examples of Arithmetic Sequences

Sequence	First Term, a_1	Common Difference, d	General Term, a_n
3, 7, 11, 15, . . .	3	$7 - 3 = 4$	$3 + (n - 1)4$
4, 9, 14, 19, . . .	4	$9 - 4 = 5$	$4 + (n - 1)5$
$1, -4, -9, -14, \ldots$	1	$-4 - 1 = -5$	$1 + (n - 1)(-5)$
$\frac{10}{7}, \frac{1}{7}, -\frac{8}{7}, -\frac{17}{7}, \ldots$	$\frac{10}{7}$	$\frac{1}{7} - \frac{10}{7} = -\frac{9}{7}$	$\frac{10}{7} + (n - 1)\left(-\frac{9}{7}\right)$

EXAMPLE 1 **Finding the general term of an arithmetic sequence**

Find the general term of each arithmetic sequence.

(a) $2, -1, -4, -7, \ldots$
(b) The third term is 7 and the seventh term is 15.

Solution

(a) For $2, -1, -4, -7, \ldots$, the common difference d is given by

$$d = (-1) - (2) = -3$$

Using the formula $a_n = a_1 + (n - 1)d$, with $a_1 = 2$ and $d = -3$, we get the general term

$$\begin{aligned} a_n &= 2 + (n - 1)(-3) \\ &= 2 - 3n + 3 \\ &= -3n + 5 \end{aligned}$$

(b) Using the formula $a_n = a_1 + (n - 1)d$ twice, first with $n = 3$ and $a_3 = 7$, and then with $n = 7$ and $a_7 = 15$, we obtain the following system of linear equations in the unknowns a_1 and d:

$$\begin{cases} 7 = a_1 + (3 - 1)d \\ 15 = a_1 + (7 - 1)d \end{cases} \quad \text{or} \quad \begin{cases} a_1 + 2d = 7 \\ a_1 + 6d = 15 \end{cases}$$

Solving for d and a_1 gives us $a_1 = 3$ and $d = 2$. Therefore, the general term a_n of this sequence is given by

$$\begin{aligned} a_n &= 3 + (n - 1)2 \\ &= 3 + 2n - 2 \\ &= 2n + 1 \end{aligned}$$

EXAMPLE 2 **Finding n when a_n is given**

Which term of the arithmetic sequence 7, 3, −1, −5, −9, . . . is −389?

Solution Here we have $a_n = -389$, and we need to find n. First, we note that $a_1 = 7$ and $d = 3 - 7 = -4$. Substituting these numbers into the formula $a_n = a_1 + (n - 1)d$, and solving for n, we get

$$\begin{aligned} a_n &= a_1 + (n - 1)d \\ -389 &= 7 + (n - 1)(-4) \\ -389 &= 7 - 4n + 4 \\ -400 &= -4n \\ 100 &= n \end{aligned}$$

Therefore, the 100th term is −389; that is, $a_{100} = -389$.

The graph of an arithmetic sequence consists of discrete points that define a linear pattern (Problem 51).

EXAMPLE 3 **Graphing an arithmetic sequence**

(a) Graph the arithmetic sequence

$$1, \tfrac{3}{2}, 2, \tfrac{5}{2}, 3, \ldots$$

(b) Compare the graph in part (a) to the continuous graph of the line

$$y = \tfrac{1}{2}x + \tfrac{1}{2} \qquad \text{where } x \geq 1$$

Solution

(a) The graph of the given sequence is displayed in Figure 1a.

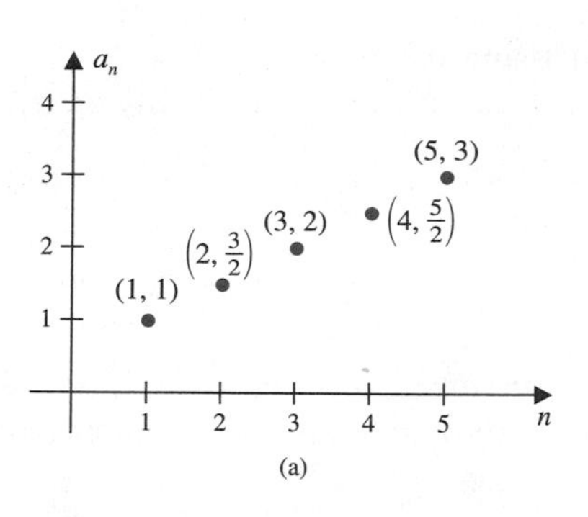

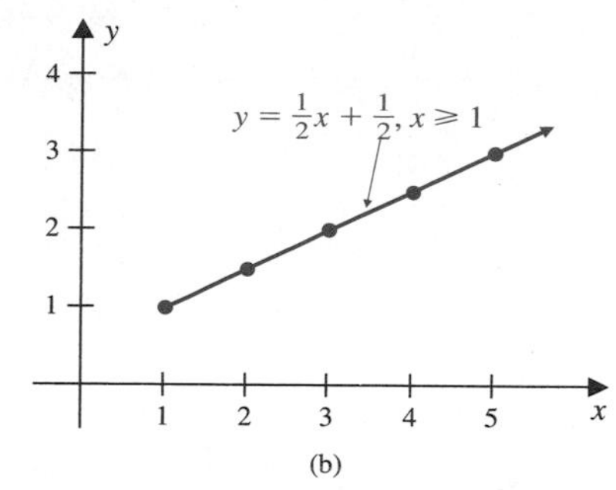

FIGURE 1

(b) The continuous graph of the line $y = \frac{1}{2}x + \frac{1}{2}$, where $x \geq 1$, is shown in Figure 1b. For the given sequence, $a_1 = 1$ and $d = \frac{3}{2} - 1 = \frac{1}{2}$. Thus, the general term of the sequence is given by

$$\begin{aligned} a_n &= a_1 + (n - 1)d \\ &= 1 + (n - 1)\left(\tfrac{1}{2}\right) \\ &= 1 + \tfrac{1}{2}n - \tfrac{1}{2} \\ &= \tfrac{1}{2}n + \tfrac{1}{2} \end{aligned}$$

Upon comparing $a_n = \frac{1}{2}n + \frac{1}{2}$ to $y = \frac{1}{2}x + \frac{1}{2}$, where $x \geq 1$, we conclude that the graph of the arithmetic sequence consists of discrete points that lie on the graph of the line $y = \frac{1}{2}x + \frac{1}{2}$.

Finding the Sum of an Arithmetic Sequence

Consider the arithmetic sequence given by

$$3, 8, 13, 18, \ldots, 5n - 2, \ldots$$

By adding successive terms of this sequence, we create a sequence $\{S_n\}$, called the *sequence of partial sums*, whose terms are given by

$$\begin{aligned} S_1 &= 3 \\ S_2 &= 3 + 8 = 11 \\ S_3 &= 3 + 8 + 13 = 24 \\ S_4 &= 3 + 8 + 13 + 18 = 42 \\ &\vdots \\ S_n &= 3 + 8 + 13 + 18 + \cdots + (5n - 2) \end{aligned}$$

In general, if

$$a_1, a_2, a_3, \ldots, a_n, \ldots$$

is an arithmetic sequence, then the partial sum S_n for the sequence is given by

$$S_n = a_1 + a_2 + \cdots + a_n = \sum_{k=1}^{n} a_k$$

We can derive a formula for S_n as follows: If d is the common difference, by using the basic properties of summation given in Section 10.1, we get

$$\begin{aligned}
S_n &= \sum_{k=1}^{n} a_k \\
&= \sum_{k=1}^{n} [a_1 + (k-1)d] && a_k = a_1 + (k-1)d \\
&= \sum_{k=1}^{n} [a_1 + kd - d] && \text{Multiply} \\
&= \sum_{k=1}^{n} a_1 + \sum_{k=1}^{n} kd - \sum_{k=1}^{n} d && \text{Property 3} \\
&= na_1 + d\sum_{k=1}^{n} k - nd && \text{Properties 1 and 2} \\
&= na_1 + d\left[\frac{n(n+1)}{2}\right] - nd && \text{Property 4} \\
&= \frac{n}{2}[2a_1 + d(n+1) - 2d] && \text{Factor out } \frac{n}{2} \\
&= \frac{n}{2}[2a_1 + (n-1)d] && \text{Simplify}
\end{aligned}$$

Thus, the formula for the partial sum S_n of an arithmetic sequence $\{a_1 + (n-1)d\}$ is

$$S_n = \frac{n}{2}[2a_1 + (n-1)d]$$

Using the fact that $a_n = a_1 + (n-1)d$, we can rewrite this formula as follows:

$$S_n = \frac{n}{2}[2a_1 + (n-1)d] = \frac{n}{2}[a_1 + a_1 + (n-1)d] = \frac{n}{2}(a_1 + a_n)$$

Therefore,

$$S_n = \frac{n}{2}(a_1 + a_n)$$

EXAMPLE 4 Finding the partial sum of an arithmetic sequence

Find the common difference and the sum of the first twenty terms of the arithmetic sequence

$$2, 6, 10, 14, \dots$$

Solution Here, the common difference $d = 6 - 2 = 4$. So, using the formula

$$S_n = \left(\frac{n}{2}\right)[2a_1 + (n - 1)d]$$

with $a_1 = 2$, $d = 4$, and $n = 20$, we have

$$\begin{aligned} S_{20} &= \tfrac{20}{2}[2(2) + (20 - 1)4] \\ &= 10(4 + 76) = 800 \end{aligned}$$

EXAMPLE 5 Finding the number of terms of a finite arithmetic sequence

How many terms are there in the arithmetic sequence $\{a_n\}$ if $a_1 = 3$, the common difference $d = 5$, and the partial sum $S_n = 255$?

Solution Using the formula

$$S_n = \left(\frac{n}{2}\right)[2a_1 + (n - 1)d]$$

with $a_1 = 3$, $d = 5$, and $S_n = 255$, we have

$$\begin{aligned} 255 &= \left(\frac{n}{2}\right)[6 + (n - 1)5] \\ 510 &= n(6 + 5n - 5) && \text{Multiply each side by 2} \\ 510 &= n(1 + 5n) && \text{Simplify} \\ 510 &= n + 5n^2 && \text{Multiply} \\ 5n^2 + n - 510 &= 0 && \text{Subtract 510 from each side} \\ (5n + 51)(n - 10) &= 0 && \text{Factor} \end{aligned}$$

so

$$n = 10 \qquad \text{or} \qquad n = -\tfrac{51}{5}$$

Since n must be a positive integer, $n = 10$ and the sequence has ten terms.

Solving Applied Problems

Sequences are used to record and analyze discrete data, as illustrated in the next example.

EXAMPLE 6 Finding the seating capacity of an auditorium

A school auditorium used for lectures and concerts has 25 rows of seats. The first row contains 30 seats, the second contains 32 seats, the third contains 34 seats, and so on.

(a) Assuming this pattern continues, find a sequence that models the number of seats in each row of the auditorium.
(b) Find the number of seats in the twentieth row.
(c) How many seats are there in the auditorium?

Solution

(a) If the pattern given by 30, 32, 34, . . . continues, the sequence $\{a_n\}$ that models the situation is arithmetic. Specifically, if we let a_n represent the number of seats in the nth row, then the first term is $a_1 = 30$ and the common difference is $d = 2$, so

$$a_1 = 30,\ a_2 = 32,\ a_3 = 34, \ldots,\ a_n = 30 + (n - 1)(2), \ldots,$$

where $1 \leq n \leq 25$.

(b) The number of seats in the twentieth row is given by

$$\begin{aligned} a_{20} &= 30 + (20 - 1)(2) \\ &= 30 + 38 = 68 \end{aligned}$$

(c) Since the sequence $\{a_n\}$ found in part (a) specifies the number of seats in each of the 25 rows, we need to find S_{25}, the sum of the first 25 terms of the sequence. Using the formula

$$S_n = \left(\frac{n}{2}\right)[2a_1 + (n - 1)d]$$

with $n = 25$, $a_1 = 30$, and $d = 2$, we get

$$S_{25} = \tfrac{25}{2}[2(30) + (25 - 1)2] = 1350$$

Therefore, the auditorium has 1350 seats.

PROBLEM SET 10.2

Mastering the Concepts

In problems 1–10, find the general term of each arithmetic sequence and graph it.

1. 2, 5, 8, 11, . . .
2. $-9, -5, -1, 3, \ldots$
3. 12, 9, 6, 3, . . .
4. $2, \frac{8}{3}, \frac{10}{3}, 4, \ldots$
5. $-13, -6, 1, 8, \ldots$
6. 19, 17, 15, 13, . . .
7. The third term is 6 and the eighth term is 16.
8. The fourth term is 6 and the tenth term is -12.
9. The tenth term is -11 and the fortieth term is -71.
10. The fifteenth term is 16 and the eighteenth term is -12.

In problems 11–14, which term of the arithmetic sequence is the given term?

11. $a_n = 85$ for 5, 13, 21, 29, . . .
12. $a_n = 143$ for $-7, -2, 3, 8, \ldots$
13. $a_n = -185$ for $3, -1, -5, -9, \ldots$
14. $a_n = -20$ for $0, -\frac{1}{2}, -1, -\frac{3}{2}, \ldots$

In problems 15–20, find the common difference and the partial sum S_n of each arithmetic sequence for the given value of n.

15. 1, 4, 7, 10, . . . ; $n = 10$
16. 5, 9, 13, 17, . . . ; $n = 15$

17. $4, 7, 10, 13, \ldots; n = 20$
18. $50, 45, 40, 35, \ldots; n = 12$
19. $-5, -\frac{32}{7}, -\frac{29}{7}, -\frac{26}{7}, \ldots; n = 8$
20. $\frac{1}{2}, 1, \frac{3}{2}, 2, \ldots; n = 10$

In problems 21–26, find the sum of the first forty terms of the arithmetic sequence that has the given general term.

21. $a_n = 5n - 4$ **22.** $a_n = -5n - 3$
23. $a_n = 4n + 4$ **24.** $a_n = -4n - 1$
25. $a_n = -3n - 2$ **26.** $a_n = 4n - 7$

In problems 27–30, certain information is given about an arithmetic sequence. Find the indicated unknown(s).

27. n, if $a_1 = 3$, $d = 2$, and $S_n = 143$
28. S_n, if $a_1 = 38$, $d = -2$, and $n = 25$
29. d and a_{18}, if $a_1 = 17$ and $S_{18} = 2310$
30. n and d, if $a_1 = 27$, $a_n = 48$, and $S_n = 1200$

In problems 31–36, find the sum of the first sixty terms of each arithmetic sequence.

31. $11, 13, 15, 17, \ldots$ **32.** $7, 10, 13, 16, \ldots$
33. $2, 6, 10, 14, \ldots$ **34.** $2, 7, 12, 17, \ldots$
35. $\frac{3}{2}, 2, \frac{5}{2}, 3, \ldots$ **36.** $-\frac{1}{3}, \frac{1}{3}, 1, \frac{5}{3}, \ldots$

Applying the Concepts

37. Marching Band: A formation of a marching band has 7 marchers in the front row, 10 in the second row, 13 in the third row, and so on, for 10 rows.
(a) What kind of sequence does this pattern define? Find the general term of a sequence to model the number of marchers in each row.
(b) Find the number of marchers in the seventh row.
(c) How many marchers are in the band?

38. Flower Planting: A farmer has a triangular piece of land to grow plants for floral purposes. There are 100 plants in the first row, 105 in the second, 110 in the third, and so on, for 90 rows.
(a) What kind of sequence does this pattern create? Find the general term of a sequence to model the number of plants in each row.
(b) Find the number of plants in the twelfth row.
(c) How many plants are there altogether?

39. Sales: A small company had sales of \$300,000 during its first year of operation, and the sales increased by \$40,000 per year during each successive year.
(a) List the first five terms of the sequence of increasing sales. What kind of sequence does this pattern create? Find the general term of a sequence to model the sales of the company for each year.
(b) What were the sales during the sixth year?
(c) What were the total sales of the company during its first 10 years?

40. Advertising: During a sales promotion, a salesperson receives a guaranteed salary plus a bonus of \$60 for selling one TV, \$90 for selling two TV's, \$120 for selling three TV's, and so on, in an arithmetic sequence.
(a) Find the sequence that represents the potential bonuses.
(b) How much bonus money will the salesperson earn for selling thirty TV's in 1 week?
(c) How many TV's must be sold to earn a bonus of \$510?
(d) Use the graph of this sequence to display and discuss the trend of the bonus incentive.

41. Physics: An object dropped from a certain height will have fallen vertically 12 feet after the first second, 44 feet after the second second, 76 feet after the third, and so on.
(a) How many feet will the object fall during the ninth second?
(b) What is the total number of feet the object has fallen at the end of 9 seconds?

42. Geometry: The sum of the interior angles of a triangle is 180°, the sum of the interior angles of a quadrilateral is 360°, the sum is 540° for a pentagon, and so on. Assuming this pattern continues, find the sum of the interior angles for a ten-sided polygon.

Developing and Extending the Concepts

43. The first three terms of an arithmetic sequence have the form x, $2x + 3$, $5x - 2$ for some real number x.
(a) Find x.
(b) Find the numerical value of the tenth term.

44. Assume 16, x, and $\frac{9}{4}$ are the first three terms of an arithmetic sequence.
(a) Find x. **(b)** Find the tenth term of the sequence.

45. Find the sixth and ninth terms of the arithmetic sequence

$$7a^2 - 4b, 2a^2 + 7b, -3a^2 + 18b, \ldots$$

46. Find the fifth and tenth terms of the arithmetic sequence

$$t + 1, 3t - 1, 3t + 3, \ldots$$

47. Suppose that the nth term of an infinite sequence is given by $a_n = n^2$. If we are given another infinite sequence whose nth term is given by $b_n = a_{n+1} - a_n$, show that $\{b_n\}$ is an arithmetic sequence.

48. Find the sum of all even numbers between 20 and 484, inclusive.

In problems 49 and 50, find the indicated partial sum for each arithmetic sequence.

49. S_7, for $6, 3b + 1, 6b - 4, \ldots$

50. S_{10}, for $x + 2y, 3y, -x + 4y, \ldots$

51. Explain why the graph of an arithmetic sequence with general term $a_n = a_1 + (n - 1)d$ consists of discrete points that lie on the graph of the continuous line $y = dx + b$, where $b = a_1 - d$ and $x \geq 1$. Give two specific examples to illustrate this result.

10.3 Geometric Sequences and Series

OBJECTIVES

1. Identify a geometric sequence and find its general term.
2. Find the partial sum of a geometric sequence.
3. Find the sum of an infinite geometric series.
4. Solve applied problems.

Another type of sequence is called a *geometric sequence.*

Identifying a Geometric Sequence and Finding Its General Term

Each term of a geometric sequence is obtained from the preceding term by multiplying by a constant value.

DEFINITION

GEOMETRIC SEQUENCE

A sequence $a_1, a_2, a_3, \ldots, a_n, \ldots$ is called a **geometric sequence** if each term (after the first) is obtained by multiplying the preceding term by a constant number. That is, there is a constant r for which $a_n/a_{n-1} = r$ for all integers $n > 1$. The constant number r is called the **common ratio** of the geometric sequence.

For example, the sequence

$$3, 6, 12, 24, \ldots, 3(2^{n-1}), \ldots$$

is a geometric sequence with common ratio 2, since the ratio of any two successive terms is 2. That is,

$$\frac{a_2}{a_1} = \frac{6}{3} = 2, \frac{a_3}{a_2} = \frac{12}{6} = 2, \ldots, \text{ and } \frac{3(2^{n-1})}{3(2^{n-2})} = 2$$

Consider a geometric sequence with common ratio r and with first term a_1. Then the nth term a_n may be expressed in terms of r and a_1 as follows:

$$\begin{aligned}
a_1 &= a_1 \\
a_2 &= a_1 r \\
a_3 &= a_2 r = (a_1 r)r = a_1 r^2 \\
a_4 &= a_3 r = (a_1 r^2)r = a_1 r^3 \\
a_5 &= a_4 r = (a_1 r^3)r = a_1 r^4 \\
&\vdots
\end{aligned}$$

In general,

$$a_n = a_1 r^{n-1}$$

Table 1 shows a few examples of geometric sequences.

TABLE 1
Examples of Geometric Sequences

Sequence	First Term, a_1	Common Ratio, r	General Term, a_n
$2, -4, 8, -16, \ldots$	2	$-\frac{4}{2} = -2$	$2(-2)^{n-1}$
$24, -12, 6, -3, \ldots$	24	$-\frac{12}{24} = -\frac{1}{2}$	$24\left(-\frac{1}{2}\right)^{n-1}$
$3, 0.3, 0.03, 0.003, \ldots$	3	$\frac{0.3}{3} = \frac{1}{10}$	$3\left(\frac{1}{10}\right)^{n-1}$

EXAMPLE 1 Finding the general term of a geometric sequence

Find the general term of the given geometric sequence.

(a) $\frac{1}{3}, \frac{1}{6}, \frac{1}{12}, \frac{1}{24}, \ldots$
(b) The third term is 8, the fifth term is 32, and all terms are positive.

Solution

(a) For $\frac{1}{3}, \frac{1}{6}, \frac{1}{12}, \frac{1}{24}, \ldots$, the common ratio r is given by

$$r = \frac{\frac{1}{6}}{\frac{1}{3}} = \frac{1}{2}$$

Using the formula $a_n = a_1 r^{n-1}$ with $a_1 = \frac{1}{3}$ and $r = \frac{1}{2}$, we get the general term

$$a_n = \left(\frac{1}{3}\right)\left(\frac{1}{2}\right)^{n-1} = \left(\frac{1}{3}\right)(2^{-1})^{n-1}$$
$$= \left(\frac{1}{3}\right)(2^{-n+1}) = \left(\frac{1}{3}\right)(2^{-n} \cdot 2)$$

which is equivalent to

$$a_n = \left(\frac{2}{3}\right)2^{-n}$$

(b) Here, we are given $a_3 = 8$ and $a_5 = 32$. Using the formula $a_n = a_1 r^{n-1}$ twice, we get

$$\begin{cases} a_3 = a_1 r^{3-1} \\ a_5 = a_1 r^{5-1} \end{cases} \quad \text{or} \quad \begin{cases} 8 = a_1 r^2 \\ 32 = a_1 r^4 \end{cases}$$

By dividing the corresponding terms of the two latter equations, we get

$$\frac{1}{4} = \frac{1}{r^2} \quad \text{or} \quad r^2 = 4$$

so $r = 2$. Notice that we used only the positive value for r, because all the terms of the sequence are positive (as stated in the problem). Substituting $r = 2$ into the first equation in the system, $8 = a_1 r^2$, we obtain $8 = a_1 \cdot 2^2$, or $a_1 = 2$. So the general term of the sequence is

$$a_n = 2 \cdot 2^{n-1} \qquad \text{or} \qquad a_n = 2^n$$

EXAMPLE 2 **Graphing a geometric sequence**

(a) Graph the geometric sequence

$$4, \tfrac{12}{5}, \tfrac{36}{25}, \tfrac{108}{125}, \ldots$$

(b) Compare the graph in part (a) to the graph of

$$y = 4\left(\tfrac{3}{5}\right)^{x-1} \qquad \text{where } x \geq 1$$

Solution

(a) Figure 1a shows the graph of the given sequence.
(b) The graph of $y = 4\left(\frac{3}{5}\right)^{x-1}$, where $x \geq 1$, is displayed in Figure 1b.

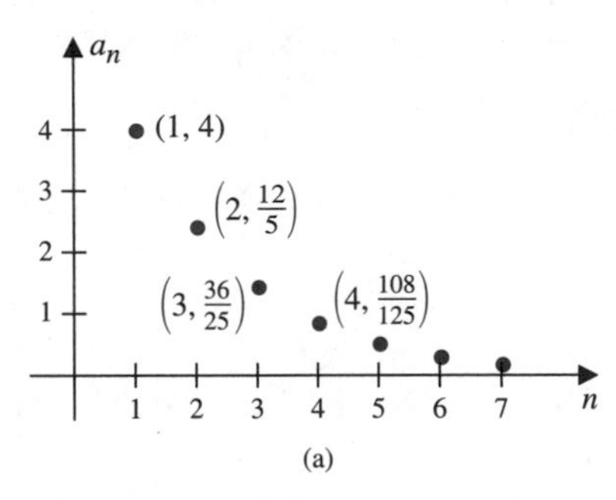

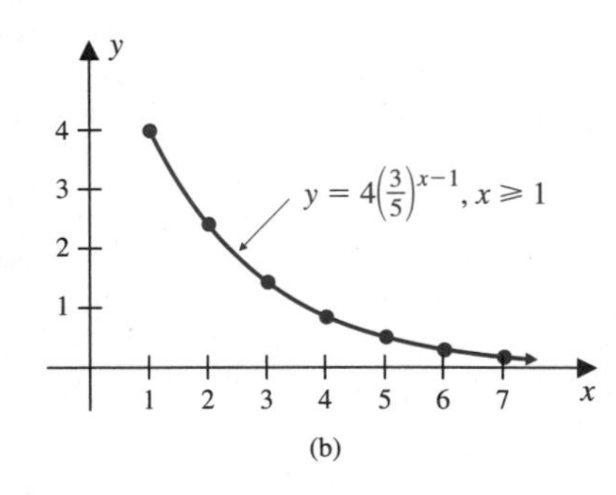

FIGURE 1

For the given sequence,

$$a_1 = 4 \qquad \text{and} \qquad r = \frac{\frac{12}{5}}{4} = \frac{3}{5}$$

So the general term is given by

$$a_n = 4\left(\tfrac{3}{5}\right)^{n-1}$$

Upon comparing $a_n = 4\left(\frac{3}{5}\right)^{n-1}$ to $y = 4\left(\frac{3}{5}\right)^{x-1}$, where $x \geq 1$, we conclude that the graph of the geometric sequence consists of discrete points that lie on the graph of the exponential function $y = 4\left(\frac{3}{5}\right)^{x-1}$.

Finding the Partial Sum of a Geometric Sequence

Consider the geometric sequence

$$1, 3, 9, 27, \ldots, 3^{n-1}, \ldots$$

By adding successive terms of this sequence, we create the sequence of partial sums $\{S_n\}$ as follows:

$$\begin{aligned}
S_1 &= 1 \\
S_2 &= 1 + 3 = 4 \\
S_3 &= 1 + 3 + 9 = 13 \\
S_4 &= 1 + 3 + 9 + 27 = 40 \\
&\vdots \\
S_n &= 1 + 3 + 9 + 27 + \cdots + 3^{n-1}
\end{aligned}$$

Using sigma notation, the latter expression can be rewritten as

$$S_n = \sum_{k=1}^{n} 3^{k-1}$$

To find a formula for the partial sum S_n of a geometric sequence with first term a_1 and common ratio r we proceed as follows:

$$\begin{aligned}
S_n &= a_1 + a_1 r + a_1 r^2 + \cdots + a_1 r^{n-1} \\
rS_n &= a_1 r + a_1 r^2 + a_1 r^3 + \cdots + a_1 r^n && \text{Multiply each side by } r \\
S_n - rS_n &= a_1 - a_1 r^n && \text{Subtract the two equations}
\end{aligned}$$

By factoring each side of this latter equation, we have

$$(1 - r)S_n = a_1(1 - r^n) \qquad \text{or} \qquad S_n = \frac{a_1(1 - r^n)}{1 - r} \qquad \text{if } r \neq 1$$

Thus, the general formula for the partial sum S_n of the geometric sequence $\{a_1 r^{n-1}\}$ is given by

$$S_n = \sum_{k=1}^{n} a_1 r^{k-1} = \frac{a_1(1 - r^n)}{1 - r} \qquad r \neq 1$$

If $r = 1$, then the partial sum S_n is given by

$$S_n = a_1 + a_1 + \cdots + a_1 = na_1$$

For example, if $a_1 = 3$ and $r = 3$, then

$$\begin{aligned}
\sum_{k=1}^{n} 3(3^{k-1}) &= \frac{3(1 - 3^n)}{1 - 3} \\
&= \left(-\frac{3}{2}\right)(1 - 3^n)
\end{aligned}$$

EXAMPLE 3 **Finding a partial sum of a geometric sequence**

Find the common ratio and the sum of the first ten terms of the geometric sequence

$$\tfrac{1}{2}, 1, 2, 4, 8, \ldots$$

Solution Here,

$$a_1 = \frac{1}{2} \quad \text{and} \quad r = \frac{1}{\frac{1}{2}} = 2$$

So, if we let $n = 10$ in the formula

$$S_n = \frac{a_1(1 - r^n)}{1 - r}$$

we get the sum of the first ten terms

$$S_{10} = \frac{\frac{1}{2}(1 - 2^{10})}{1 - 2} = \frac{\frac{1}{2}(-1023)}{-1} = \frac{1023}{2} = 511.5$$

EXAMPLE 4 **Finding terms of a geometric sequence**

The sum of the first five terms of a geometric sequence is $\frac{61}{27}$ and the common ratio is $-\frac{1}{3}$. Find the first four terms of the sequence.

Solution Using the formula for S_n with $r = -\frac{1}{3}$ and $S_5 = \frac{61}{27}$, we have

$$\frac{61}{27} = \frac{a_1\left[1 - \left(-\frac{1}{3}\right)^5\right]}{1 - \left(-\frac{1}{3}\right)}$$

Thus,

$$\frac{61}{27} = \left[\frac{\frac{244}{243}}{\frac{4}{3}}\right]a_1 \quad \text{or} \quad \frac{61}{27} = \frac{61}{81}a_1$$

so

$$a_1 = \frac{81}{61} \cdot \frac{61}{27} = 3$$

Hence, the first four terms of the sequence are 3, $3\left(-\frac{1}{3}\right)$, $3\left(-\frac{1}{3}\right)^2$, and $3\left(-\frac{1}{3}\right)^3$ or $3, -1, \frac{1}{3}$, and $-\frac{1}{9}$.

Finding the Sum of an Infinite Geometric Series

Suppose we have a sequence $\{a_n\}$ and we use it to form the sequence of partial sums $\{S_n\}$:

$$\begin{aligned}
S_1 &= a_1 \\
S_2 &= a_1 + a_2 \\
S_3 &= a_1 + a_2 + a_3 \\
&\vdots \\
S_n &= a_1 + a_2 + a_3 + \cdots + a_n
\end{aligned}$$

Continuing this process indefinitely, we are led to an expression of the form

$$a_1 + a_2 + a_3 + \cdots + a_n + \cdots$$

This latter expression is written more compactly using sigma notation as

$$\sum_{k=1}^{\infty} a_k$$

Such an expression is called an **infinite series**, or simply a **series**.

The general study of series is taken up in calculus. For now, we'll only investigate series that are formed from geometric sequences. Such series are called **geometric series**. Examples of geometric series are

$$\sum_{k=1}^{\infty} \left(\frac{1}{2}\right)^{k-1} \quad \text{and} \quad \sum_{k=1}^{\infty} \frac{4}{5}\left(\frac{1}{5}\right)^{k-1}$$

Let's examine how infinite geometric series are interpreted by considering the series

$$\sum_{k=1}^{\infty} \frac{4}{5}\left(\frac{1}{5}\right)^{k-1} = \frac{4}{5}\left(1 + \frac{1}{5} + \frac{1}{5^2} + \frac{1}{5^3} + \cdots + \frac{1}{5^n} + \cdots\right)$$

This series was formed from the geometric sequence $\{a_1 r^{n-1}\}$, where $a_1 = \frac{4}{5}$ and $r = \frac{1}{5}$. By using the formula for the partial sum,

$$S_n = \sum_{k=1}^{n} a_1 r^{k-1} = \frac{a_1(1 - r^n)}{1 - r}$$

with $a_1 = \frac{4}{5}$ and $r = \frac{1}{5}$, we get

$$S_n = \frac{4}{5}\left[\frac{1 - \left(\frac{1}{5}\right)^n}{1 - \frac{1}{5}}\right] = \frac{4}{5} \cdot \frac{5}{4}\left(1 - \frac{1}{5^n}\right) = 1 - \left(\frac{1}{5}\right)^n$$

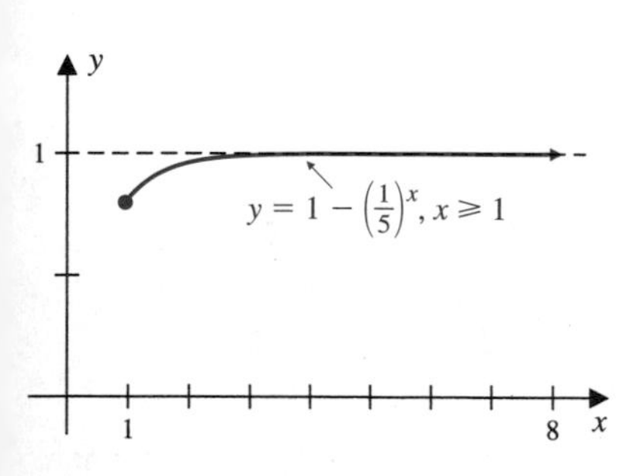

FIGURE 2

Figure 2 shows the graph of the continuous function $f(x) = 1 - \left(\frac{1}{5}\right)^x$, where $x \geq 1$. The limit behavior of the graph indicates that as $x \to +\infty$, $f(x) \to 1$.

Since the graph of the sequence $S_n = 1 - \left(\frac{1}{5}\right)^n$ consists of discrete points on the graph of $f(x) = 1 - \left(\frac{1}{5}\right)^x$, where $x \geq 1$, it follows that the limit behavior of S_n is the same as the limit behavior of $f(x)$. That is, as $n \to +\infty$, $S_n \to 1$. But as $n \to +\infty$,

$$S_n \to a_1 + a_2 + a_3 + \cdots a_n + \cdots = \sum_{k=1}^{\infty} a_k$$

So we assign the value 1 to the infinite series, and write

$$\sum_{k=1}^{\infty} \frac{4}{5}\left(\frac{1}{5}\right)^{k-1} = \frac{4}{5}\left(1 + \frac{1}{5} + \frac{1}{5^2} + \cdots \frac{1}{5^n} + \cdots\right) = \frac{4}{5} \cdot \frac{5}{4}(1) = 1$$

The use of the term *sum* and the infinity symbol, ∞, is misleading, because we cannot actually add an infinite number of terms.

We refer to 1 as the *sum of the series.*

In general, given a geometric series $\sum_{k=1}^{\infty} a_1 r^{k-1}$, the partial sum

$$S_n = a_1\left(\frac{1 - r^n}{1 - r}\right) = \frac{a_1 - a_1 r^n}{1 - r}$$

can be written as

$$S_n = \frac{a_1}{1 - r} - \frac{a_1 r^n}{1 - r}$$

Under the condition that $|r| < 1$, it follows that as $n \to +\infty$, $r^n \to 0$, so

$$S_n \to \frac{a_1}{1 - r} - 0 = \frac{a_1}{1 - r}$$

This result is summarized as follows:

SUM OF AN INFINITE GEOMETRIC SERIES

If $|r| \geq 1$, the series has no sum.

If $|r| < 1$, then

$$\sum_{k=1}^{\infty} a_1 r^{k-1} = a_1 + a_1 r + a_1 r^2 + \cdots + a_1 r^{n-1} + \cdots = \frac{a_1}{1 - r}$$

EXAMPLE 5 Finding the sum of an infinite geometric series

Find the sum of each infinite geometric series.

(a) $\sum_{k=1}^{\infty} 5\left(-\frac{1}{3}\right)^{k-1}$ (b) $\sum_{k=1}^{\infty} \frac{3}{2^{k-1}}$

Solution

(a) Using the formula $a_1/(1 - r)$ for the sum of the series with $a_1 = 5$ and $r = -\frac{1}{3}$, we have

$$\sum_{k=1}^{\infty} 5\left(-\frac{1}{3}\right)^{k-1} = \frac{5}{1 - \left(-\frac{1}{3}\right)} = \frac{5}{\frac{4}{3}} = \frac{15}{4}$$

(b) The series

$$\sum_{k=1}^{\infty} \frac{3}{2^{k-1}}$$

can be rewritten as

$$\sum_{k=1}^{\infty} \frac{3(1)^{k-1}}{2^{k-1}} \quad \text{or} \quad \sum_{k=1}^{\infty} 3\left(\frac{1}{2}\right)^{k-1}$$

So in this case, we have a geometric series of the form $\sum_{k=1}^{\infty} a_1 r^{k-1}$, where $a_1 = 3$ and $r = \frac{1}{2}$. Thus,

$$\sum_{k=1}^{\infty} \frac{3}{2^{k-1}} = \sum_{k=1}^{\infty} 3\left(\frac{1}{2}\right)^{k-1} = \frac{3}{1 - \frac{1}{2}} = \frac{3}{\frac{1}{2}} = 6$$

Solving Applied Problems

The following example illustrates how geometric sequences are used in modeling real-world phenomena.

EXAMPLE 6 **Finding the distance traveled by a bob of a pendulum**

Suppose that the distance traveled by a bob of a pendulum in each swing is 83% of the previous swing (Figure 3).

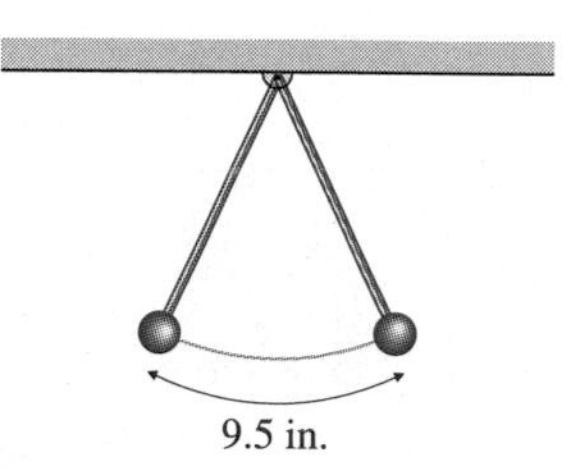

FIGURE 3

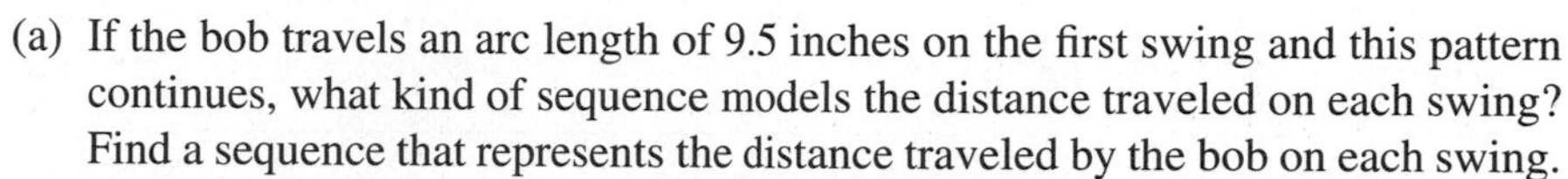

(a) If the bob travels an arc length of 9.5 inches on the first swing and this pattern continues, what kind of sequence models the distance traveled on each swing? Find a sequence that represents the distance traveled by the bob on each swing.

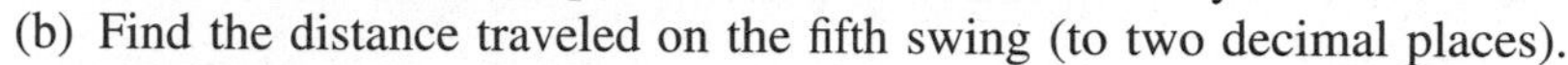

(b) Find the distance traveled on the fifth swing (to two decimal places).

(c) Determine the total distance traveled by the bob after eight swings.

(d) What is the total distance the bob travels as $n \to +\infty$?

Solution

(a) The distances traveled by the bob on each swing follow the pattern

$$9.5,\ 9.5(0.83),\ [9.5(0.83)](0.83) = 9.5(0.83)^2,\ \ldots$$

If this pattern continues, the sequence that models the situation is geometric; its first term is 9.5 and each subsequent term is 0.83 times the prior term. Specifically, if we let a_n represent the distance traveled during the nth swing, then $a_1 = 9.5$ and the common ratio is $r = 0.83$ so

$$a_1 = 9.5,\ a_2 = 9.5(0.83),\ a_3 = 9.5(0.83)^2,\ \ldots,\ a_n = 9.5(0.83)^{n-1},\ \ldots$$

(b) The distance traveled on the fifth swing is given by

$$\begin{aligned} a_5 &= 9.5(0.83)^4 \\ &= 4.51 \text{ inches (approx.)} \end{aligned}$$

(c) Since each term in the sequence found in part (a) specifies the distance traveled on the nth swing, we need to find S_8, the sum of the first eight terms of the geometric sequence. Using the formula

$$S_n = \frac{a_1(1 - r^n)}{1 - r}$$

with $n = 8$, $a_1 = 9.5$, and $r = 0.83$, we get

$$S_8 = \frac{9.5(1 - 0.83^8)}{1 - 0.83} = 43.30 \text{ (approx.)}$$

So the bob has traveled a total of about 43.30 inches by the end of the eighth swing.

(d) The total distance the bob travels as $n \to +\infty$ is represented by the infinite geometric series

$$\sum_{k=1}^{\infty} 9.5(0.83)^{k-1} = \frac{9.5}{1 - 0.83} = 55.88 \text{ (approx.)}$$

So the bob travels a total distance of about 55.88 inches.

PROBLEM SET 10.3

Mastering the Concepts

In problems 1–14, find the general term of each geometric sequence.

1. $2, 6, 18, 54, \ldots$
2. $1, \frac{1}{5}, \frac{1}{25}, \frac{1}{125}, \ldots$
3. $6, 12, 24, 48, \ldots$
4. $10, 10^2, 10^3, 10^4, \ldots$
5. $32, 16, 8, 4, \ldots$
6. $\frac{1}{8}, \frac{1}{4}, \frac{1}{2}, 1, \ldots$
7. $1, 1.03, (1.03)^2, (1.03)^3, \ldots$
8. $10^{-5}, 10^{-7}, 10^{-9}, 10^{-11}, \ldots$
9. $\frac{1}{18}, -\frac{1}{6}, \frac{1}{2}, -\frac{3}{2}, \ldots$
10. $\frac{2}{3}, -\frac{4}{3}, \frac{8}{3}, -\frac{16}{3}, \ldots$
11. The second term is 4 and the twelfth term is 4096.
12. The third term is 3 and the sixth term is $\frac{25}{81}$.
13. The third term is $\frac{9}{2}$ and the sixth term is $-\frac{243}{16}$.
14. The third term is -4 and the seventh term is -64.

In problems 15–24, determine whether the given sequence is arithmetic or geometric. Then find a formula for the general term a_n for each geometric sequence.

15. $1, 8, 15, 22, \ldots$
16. $1, 5, 25, 125, \ldots$
17. $3, 6, 12, 24, \ldots$
18. $3, 8, 13, 18, \ldots$
19. $1, -2, 4, -8, \ldots$
20. $5, 2, -1, -4, \ldots$
21. $1, 0.3, 0.09, 0.027, \ldots$
22. $-3, 9, -27, 81, \ldots$
23. $\frac{1}{4}, \frac{1}{8}, \frac{1}{16}, \frac{1}{32}, \ldots$
24. $6, 2, \frac{2}{3}, \frac{2}{9}, \ldots$

In problems 25–32, graph the sequence. Find the common ratio and the sum of the first n terms of each geometric sequence for the given value of n.

25. $5, 20, 80, 320, \ldots; n = 8$
26. $54, 36, 24, 16, \ldots; n = 7$
27. $-3, 15, -75, 375, \ldots; n = 6$
28. $-4, 8, -16, 32, \ldots; n = 12$
29. $-\frac{1}{3}, -\frac{1}{9}, -\frac{1}{27}, -\frac{1}{81}, \ldots; n = 10$
30. $5, -\frac{5}{2}, \frac{5}{4}, -\frac{5}{8}, \ldots; n = 8$
31. $0.3, 0.03, 0.003, 0.0003, \ldots; n = 6$
32. $1, 1.04, (1.04)^2, (1.04)^3, \ldots; n = 6$

In problems 33–36, find the sum of each infinite geometric series.

33. $\sum_{k=1}^{\infty} \left(\frac{1}{3}\right)^k$
34. $\sum_{k=1}^{\infty} 9\left(\frac{1}{8}\right)^k$
35. (a) $\sum_{k=1}^{\infty} \frac{4}{2^k}$ (b) $\sum_{k=1}^{\infty} \frac{2(-1)^k}{3^{k-1}}$
36. (a) $\sum_{k=1}^{\infty} (0.7)^k$ (b) $\sum_{k=1}^{\infty} (-1)^k\left(\frac{2}{3}\right)^{k-1}$

GU In problems 37 and 38, use the graph and limit behavior of the function f defined by

$$f(x) = \frac{a_1(1 - r^x)}{1 - r} \quad \text{where } x \geq 1$$

to validate each given sum of the infinite geometric series.

37. (a) $\sum_{k=1}^{\infty} 2\left(\frac{3}{5}\right)^{k-1} = 5$ (b) $\sum_{k=1}^{\infty} (-1)^k\left(\frac{3}{4}\right)^{k-1} = -\frac{4}{7}$
38. (a) $\sum_{k=1}^{\infty} \left(\frac{5}{6}\right)^k = 5$ (b) $\sum_{k=1}^{\infty} \left(\frac{4}{7}\right)^k = \frac{4}{3}$

Applying the Concepts

39. Rebounding Ball: Suppose that a ball dropped from the top of a 100 foot tower rebounds $\frac{3}{5}$ of the distance it falls each time.
    (a) Find a geometric sequence that represents the distance traveled on each bounce for this pattern.
    (b) How far (up and down) will the ball have traveled by the time it hits the ground for the eighth time?
    (c) What is the total distance (up and down) this bouncing ball has traveled by the time it hits the ground for the nth time?
    (d) What is the total distance in part (c) as $n \to +\infty$?
40. Baseball Bonus: A baseball pitcher agreed to a contract that pays him a guaranteed salary of \$750,000 plus a bonus

incentive of \$10,000 for winning one game, \$13,500 for winning two games, \$18,225 for winning three games, and so on, up to and including 20 games.

(a) Represent the bonus incentive by a geometric sequence.
(b) How much money would he be paid if he won sixteen games?
(c) How many games must he win to earn a total of about \$3 million?
(d) Use a graph to display and discuss the trend of the bonus incentive.

41. Football Bonus: A running back agreed to a contract that pays him a guaranteed salary of \$1.8 million plus a bonus incentive of \$1000 for scoring one touchdown, \$1800 for scoring two touchdowns, \$3240 for scoring three touchdowns, and so on, up to and including 12 touchdowns.

(a) Represent the bonus incentive as a geometric sequence.
(b) How much money will he be paid if he scores nine touchdowns?
(c) How many touchdowns must he score to earn a total of at least \$2 million?
(d) Use a graph to display and discuss the trend of the bonus incentive.

42. Depreciation: An automobile worth \$18,000 depreciates in value each year such that its value at the end of the year is $\frac{5}{6}$ of its value at the beginning of the year.

(a) Represent the depreciation as a sequence.
(b) How much is the automobile worth after 5 years?
(c) Assuming the automobile is maintained properly, after how many years will it be worth approximately \$6000?
(d) Use the graph of this sequence to display and discuss the depreciation pattern.

43. College Enrollment: In 1993 the enrollment at a college was 25,000. Each successive year the enrollment has dropped to 95% of the preceding year. Suppose this trend continues over a period of 10 years.

(a) Represent the enrollment as a sequence.
(b) Given this pattern, predict the enrollment in 1999.
(c) If this pattern continues, what will the enrollment be in 2003?
(d) Use the graph of this sequence to discuss the pattern of enrollment.

44. Drug Dosage: A patient receives 10 milligrams of a drug. Each hour 50% of the amount of the drug present in the body is eliminated.

(a) Model the amount of drug eliminated as a sequence.
(b) How much of the drug is present after 6 hours?
(c) If this pattern continues, what will be the amount of drug present after 10 hours?
(d) Use the graph of this sequence to describe the pattern.
(e) Will the drug ever be totally eliminated from the patient's body according to the model found in part (a)? Explain.

Developing and Extending the Concepts

In problems 45 and 46, express each sum using sigma notation; then evaluate using the partial sum formula.

45. $\frac{1}{2} + \frac{1}{4} + \frac{1}{8} + \frac{1}{16} + \frac{1}{32}$

46. $\frac{3}{5} + \frac{9}{25} + \frac{27}{125} + \frac{81}{625}$

In problems 47 and 48, find the indicated sum of each geometric sequence.

47. S_6, for $\frac{a}{b}, -1, \frac{b}{a}, \ldots$

48. S_8, for $\frac{k}{b}, \frac{k}{b^2}, \frac{k}{b^3}, \ldots$

In problems 49 and 50, rewrite each repeating decimal as a geometric series. Then find its sum. Convert the resulting sum to decimal form to confirm the result.

49. **(a)** $0.\overline{4}$ **(b)** $0.\overline{32}$ **(c)** $0.\overline{561}$

50. **(a)** $0.\overline{7}$ **(b)** $0.0\overline{49}$ **(c)** $0.0\overline{72}$

51. Suppose that a_n and b_n are the general terms of two geometric sequences. Is $c_n = a_n b_n$ a general term of a geometric sequence? Explain and give an example to support your assertion.

52. Given the geometric sequence

$$1, \frac{1}{10}, \frac{1}{100}, \frac{1}{1000}, \ldots, a_k, \ldots$$

find a formula for a_k. Then determine each given sum.

(a) $\sum_{k=1}^{8} a_k$ **(b)** $\sum_{k=1}^{8} 3a_k$ **(c)** $\sum_{k=1}^{\infty} a_k$

In problems 53 and 54, use the result

$$\begin{aligned} S_n &= \sum_{k=1}^{n} (a_k - a_{k+1}) \\ &= (a_1 - a_2) + (a_2 - a_3) + \cdots + (a_n - a_{n+1}) \\ &= a_1 - a_{n+1} \end{aligned}$$

to find a formula for each partial sum.

53. **(a)** $S_n = \sum_{k=1}^{n} (2^k - 2^{k+1})$

(b) $S_n = \sum_{k=1}^{n} \left[\left(\frac{1}{2}\right)^k - \left(\frac{1}{2}\right)^{k+1}\right]$

54. **(a)** $S_n = \sum_{k=1}^{n} (3^k - 3^{k+1})$

(b) $S_n = \sum_{k=1}^{n} \left[\left(\frac{1}{3}\right)^k - \left(\frac{1}{3}\right)^{k+1}\right]$

55. Explain why the graph of the geometric sequence with general term a_1r^{n-1} consists of discrete points that lie on the graph of $y = a_1r^{x-1}$, where $x \geq 1$. Give two specific examples to illustrate this result.

OBJECTIVES

1. Formulate mathematical induction.
2. Apply mathematical induction.

10.4 Mathematical Induction

Many important results in mathematics and other sciences have been discovered by first observing patterns. To illustrate how we might arrive at these patterns, let's examine the pattern shown in the table for adding the first n odd positive integers for specific values of n.

n	Sum		Total	n^2
1	1	$=$	1	1
2	$1 + 3$	$=$	4	2^2
3	$1 + 3 + 5$	$=$	9	3^2
4	$1 + 3 + 5 + 7$	$=$	16	4^2
5	$1 + 3 + 5 + 7 + 9$	$=$	25	5^2

We might conjecture from this pattern that in general

$$1 + 3 + 5 + \cdots + (2n - 1) = n^2$$

However, we can see that it is impossible to test all positive integers in order to prove the validity of this formula. For this kind of situation, we use a method of proof called *mathematical induction.*

Formulating Mathematical Induction

The method of **mathematical induction** is based on the following principle:

PRINCIPLE OF MATHEMATICAL INDUCTION

Suppose that $S_1, S_2, S_3, \ldots$ is a sequence of statements; that is, suppose that for each positive integer n we have a corresponding statement S_n. Assume that the following two conditions hold:

(i) S_1 is true.
(ii) For each fixed positive integer k, the truth of the statement S_k implies the truth of the following statement S_{k+1}.

Then it follows that every statement $S_1, S_2, S_3, \ldots$ is true; that is, S_n is true for all positive integers.

We may understand the principle of mathematical induction more clearly by making the following analogy. Visualize a communications network set up in such a way that each person in a group is put on a numbered list. When a message is to be communicated (for instance, a bad weather report), each person is to contact the next person whose name is on the list. We can reach a conclusion about this model in the following way:

(i) Show that the first person is given a message. If we denote this by S_1, then we need to show that S_1 is true.

(ii) (a) Assume that some person on the list, say the kth person, knows the message, denoted by S_k.

(b) Show that when the kth person knows the message, S_k, then the next $(k + 1)$th person, denoted by S_{k+1}, also knows the message. That is, show that if S_k is true, then S_{k+1} is also true.

Notice that statement (ii) of the definition is given two parts: First assume that S_k is true, and then show that S_{k+1} is true. The fact that the first person knows the message and that the $(k + 1)$th person is informed when the kth person is informed, is enough to support the claim that all persons are informed.

If we suppose that S_1 indicates that the first person knows the message, then S_2 indicates that the first person informs the second. But when the second person is informed, then the next one, S_3, is also informed, and so on. Continuing this approach, we may see more clearly the underlying logic of the induction principle.

Applying Mathematical Induction

In order to prove a statement using mathematical induction, we may need to rephrase the statement in a form that is more suitable for proof.

EXAMPLE 1 **Using mathematical induction**

Use mathematical induction to prove Property 4 in Section 10.1; that is, prove that

$$1 + 2 + 3 + \cdots + n = \frac{n(n + 1)}{2}$$

for any positive integer n.

Solution Using the principle of mathematical induction, we can prove the statement by verifying conditions (i) and (ii). Let S_n be the statement

$$1 + 2 + 3 + \cdots + n = \frac{n(n + 1)}{2}$$

Condition (i) can be verified by direct computation, since

$$S_1: \quad 1 = \frac{1 \cdot (1 + 1)}{2} = \frac{2}{2} = 1$$

is clearly true.

To prove condition (ii), we must show that S_k implies S_{k+1}; that is, we must show that if S_k is assumed to be true, then S_{k+1} must be true. To this end, assume that S_k is true; that is, assume that the assertion

$$S_k\text{:} \quad 1 + 2 + 3 + \cdots + k = \frac{k(k+1)}{2}$$

is true. Since S_k is a true statement, we can add $k + 1$ to both sides of this equation, to get

$$\begin{aligned} 1 + 2 + 3 + \cdots + k + (k+1) &= \frac{k(k+1)}{2} + (k+1) \\ &= (k+1)\left(\frac{k}{2} + 1\right) \\ &= (k+1)\left(\frac{k+2}{2}\right) \\ &= \frac{(k+1)(k+2)}{2} \end{aligned}$$

But the latter statement is precisely S_{k+1}. Hence, we have proved condition (ii). So by the principle of mathematical induction, we conclude that S_n is true for any positive integer n. That is,

$$1 + 2 + 3 + \cdots + n = \frac{n(n+1)}{2}$$

for any positive integer n.

EXAMPLE 2 **Using mathematical induction**

Use mathematical induction to prove Property 5 in Section 10.1; that is, prove that

$$1^2 + 2^2 + \cdots + n^2 = \frac{n(n+1)(2n+1)}{6}$$

for any positive integer n.

Solution

(i) In this situation, the statement S_1 is given by

$$S_1\text{:} \quad 1^2 = \tfrac{1}{6}(1)(1+1)(2 \cdot 1 + 1) = \tfrac{1}{6}(1)(2)(3) = 1$$

which is true.

(ii) Assume that S_k is true; that is, assume that for any fixed positive integer k,

$$S_k\text{:} \quad 1^2 + 2^2 + 3^2 + \cdots + k^2 = \frac{k(k+1)(2k+1)}{6}$$

We must now prove that S_{k+1} is true, where S_{k+1} is the statement

$$1^2 + 2^2 + 3^2 + \cdots + k^2 + (k+1)^2 = \frac{(k+1)(k+2)(2k+3)}{6}$$

After adding $(k + 1)^2$ to both sides of the equation for S_k, we have

$$\begin{aligned}
1^2 + 2^2 + 3^2 + \cdots + k^2 + (k + 1)^2 &= \frac{k(k + 1)(2k + 1)}{6} + (k + 1)^2 \\
&= (k + 1)\left[\frac{k(2k + 1)}{6} + (k + 1)\right] \\
&= (k + 1)\left(\frac{2k^2 + k + 6k + 6}{6}\right) \\
&= (k + 1)\left(\frac{2k^2 + 7k + 6}{6}\right) \\
&= \frac{(k + 1)(k + 2)(2k + 3)}{6}
\end{aligned}$$

Hence, S_{k+1} is true, and we have proved condition (ii). Thus, by the principle of mathematical induction, we conclude that S_n is true for any positive integer n. That is, for any positive integer n,

$$1^2 + 2^2 + 3^2 + \cdots + n^2 = \frac{n(n + 1)(2n + 1)}{6}$$

The principle of mathematical induction can also be used to prove certain inequalities.

EXAMPLE 3 **Using induction to prove an inequality**
Use mathematical induction to prove that $2^n > n$ for any positive integer n.

Solution

(i) The statement

$$S_1: \quad 2^1 > 1$$

is a true inequality.

(ii) Assume that S_k is true; that is, assume that for any fixed positive integer k,

$$2^k > k$$

We must prove that S_{k+1} is true, where S_{k+1} is the statement

$$2^{k+1} > k + 1$$

Multiplying each side of the inequality $2^k > k$ by 2, we have

$$2^k \cdot 2 > 2k \qquad \text{or} \qquad 2^{k+1} > k + k$$

But $k + k > k + 1$ for $k > 1$, so

$$2^{k+1} > k + 1$$

Thus, for any positive integer n, $2^n > n$.

PROBLEM SET 10.4

Mastering the Concepts

In problems 1–12, use mathematical induction to prove each statement for all positive integers n.

1. $1 + 3 + 5 + \cdots + (2n - 1) = n^2$

2. $1^3 + 2^3 + 3^3 + \cdots + n^3 = \dfrac{n^2(n + 1)^2}{4}$

3. $2 + 4 + 6 + \cdots + 2n = n^2 + n$

4. $\dfrac{1}{1 \cdot 2} + \dfrac{1}{2 \cdot 3} + \dfrac{1}{3 \cdot 4} + \cdots + \dfrac{1}{n(n + 1)} = \dfrac{n}{n + 1}$

5. $4 + 4^2 + 4^3 + \cdots + 4^n = \dfrac{4}{3}(4^n - 1)$

6. $1 + 5 + 5^2 + \cdots + 5^{n-1} = \dfrac{1}{4}(5^n - 1)$

7. $1 + 2 \cdot 2 + 3 \cdot 2^2 + \cdots + n \cdot 2^{n-1} = 1 + (n - 1) \cdot 2^n$

8. $(-1)^1 + (-1)^2 + (-1)^3 + \cdots + (-1)^n = \dfrac{(-1)^n - 1}{2}$

9. $1^2 + 3^2 + 5^2 + \cdots + (2n - 1)^2 = \dfrac{n(2n - 1)(2n + 1)}{3}$

10. $1 \cdot 2 + 2 \cdot 3 + 3 \cdot 4 + \cdots + n(n + 1) = \dfrac{n(n + 1)(n + 2)}{3}$

11. $x^0 + x^1 + x^2 + \cdots + x^n = \dfrac{1 - x^{n+1}}{1 - x}$, for $x \neq 1$

12. $\cos n\pi = (-1)^n$

In problems 13–18, use mathematical induction to prove that each inequality is true.

13. $3^n > n^2$ for all positive integers n

14. $(n + 2)^2 < n^3$ for all positive integers $n \geq 3$

15. $n^2 \leq 2^n$ for integers $n \geq 4$

16. $n^2 > n + 1$ for integers $n \geq 2$

17. $1 + 2n \leq 3^n$ for all positive integers n

18. $4^n \geq 4n$ for all positive integers n

Developing and Extending the Concepts

In problems 19–36, use mathematical induction to prove that the given statement is true for all positive integers n unless otherwise stated.

19. 3 is a factor of $4^n - 1$.

20. 4 is a factor of $5^n - 1$.

21. 8 is a factor of $9^n + 7$.

22. 5 is a factor of $6^n - 1$.

23. 6 is a factor of $n(n^2 - 1)$, $n \geq 2$.

24. 3 is a factor of $2^{2n-1} + 1$.

25. $11^n - 4^n$ is divisible by 7.

26. $n^3 - 4n + 6$ is divisible by 3.

27. $\sin(t + n\pi) = (-1)^n \sin t$

28. $\cos(t + n\pi) = (-1)^n \cos t$

29. $\left(1 + \dfrac{1}{1}\right)\left(1 + \dfrac{1}{2}\right)\left(1 + \dfrac{1}{3}\right) \cdot \cdots \cdot \left(1 + \dfrac{1}{n}\right) = n + 1$

30. $\dfrac{1}{n + 1} + \dfrac{1}{n + 2} + \dfrac{1}{n + 3} + \cdots + \dfrac{1}{2n} \geq \dfrac{1}{2}$

31. $\dfrac{1}{\sqrt{1}} + \dfrac{1}{\sqrt{2}} + \dfrac{1}{\sqrt{3}} + \cdots + \dfrac{1}{\sqrt{n}} \geq \sqrt{n}$

32. $\left(1 - \dfrac{1}{2^2}\right)\left(1 - \dfrac{1}{3^2}\right)\left(1 - \dfrac{1}{4^2}\right) \cdot \cdots \cdot \left(1 - \dfrac{1}{n^2}\right) = \dfrac{n + 1}{2n}$, $n \geq 2$

33. $\displaystyle\sum_{k=1}^{n} ar^{k-1} = \dfrac{a(1 - r^n)}{1 - r}$, $r \neq 1$

34. $\displaystyle\sum_{k=1}^{n} kx^{k-1} = \dfrac{1 - x^n}{(1 - x)^2} - \dfrac{nx^n}{1 - x}$, $x \neq 1$

35. $(\cos\theta + i \sin\theta)^n = \cos n\theta + i \sin n\theta$

36. $\ln(x_1 \cdot x_2 \cdot x_3 \cdot \cdots \cdot x_n) = \ln x_1 + \ln x_2 + \ln x_3 + \cdots + \ln x_n$

OBJECTIVES

1. Examine the pattern in a binomial expansion.
2. Use factorial notation.
3. Use the binomial theorem.

10.5 The Binomial Theorem

In this section, we examine expansions of binomial expressions. First we observe patterns that are common to all binomial expansions. Then we introduce *factorial* notation and present the *binomial theorem*, which provides us with a general formula for the expansion of a binomial.

Examining the Pattern in a Binomial Expansion

Our objective is to construct a formula that generates the expansion of $(a + b)^n$ for a specified positive integer n. Before presenting this formula, we explore the pattern of coefficients that exists in the following binomial expansions obtained by direct calculations:

		$(a + b)^n$
$n = 1$	$(a + b)^1$	$a + b$
$n = 2$	$(a + b)^2$	$a^2 + 2ab + b^2$
$n = 3$	$(a + b)^3$	$a^3 + 3a^2b + 3ab^2 + b^3$
$n = 4$	$(a + b)^4$	$a^4 + 4a^3b + 6a^2b^2 + 4ab^3 + b^4$
$n = 5$	$(a + b)^5$	$a^5 + 5a^4b + 10a^3b^2 + 10a^2b^3 + 5ab^4 + b^5$

The above expansion leads us to discover the following patterns of the products of the powers of a and b in the expansion of $(a + b)^n$, $n \geq 0$:

1. Each expansion has $n + 1$ terms. The first term is a^n, and the last term is b^n.
2. The power of a decreases by 1 for each term, and the power of b increases by 1 for each term, so that the sum of the exponents of a and b in each term is n.
3. A pattern for the coefficients of the terms of an expanded binomial can be found by writing the coefficients in a triangular array of numbers known as *Pascal's triangle.*

Coefficients of Expanded Form of $(a + b)^n$

$n = 0$ 1

$n = 1$ 1 1

$n = 2$ 1 2 1

$n = 3$ 1 3 3 1

$n = 4$ 1 4 6 4 1

$n = 5$ 1 5 10 10 5 1

The values in Pascal's triangle are obtained recursively and obey the following patterns:

(i) The first and last coefficient for each expansion is 1.
(ii) Each of the other coefficients can be found by adding the coefficients that are diagonally above it, as indicated by the arrows in the diagram above.

EXAMPLE 1 **Using Pascal's triangle**

Use the patterns described above to find the expansion of $(x + 2y)^3$.

Solution We know that the expansion has $n + 1 = 3 + 1 = 4$ terms. The first term is $a^3 = x^3$, and the last term is $b^3 = (2y)^3$. By letting $x = a$ and $2y = b$, we can use the description of the products of the powers of a and b given earlier to get the following pattern:

$$\begin{array}{cccccccc} \_\_\ a^3 & + & \_\_\ a^2b & + & \_\_\ ab^2 & + & \_\_\ b^3 \\ \updownarrow & & \updownarrow\nwarrow\searrow & & \updownarrow\nwarrow\searrow & & \searrow\nwarrow \\ \_\_\ x^3 & + & \_\_\ x^2(2y) & + & \_\_\ x(2y)^2 & + & \_\_\ (2y)^3 \end{array}$$

Finally, we use Pascal's triangle to obtain the numerical coefficients for $(a + b)^3$. For $n = 3$:

$$1 \quad 3 \quad 3 \quad 1$$

By combining the patterns, we get the expansion:

$$\begin{aligned} (x + 2y)^3 &= \underline{\ 1\ }\ x^3 + \underline{\ 3\ }\ x^2(2y) + \underline{\ 3\ }\ x(2y)^2 + \underline{\ 1\ }\ (2y)^3 \\ &= x^3 + 6x^2y + 12xy^2 + 8y^3 \end{aligned}$$

As with many sequences that are defined recursively, it is possible to find an explicit formula for the coefficients that Pascal's triangle provides. Before developing the formula, we introduce special notation that simplifies the appearance of the formula.

Using Factorial Notation

Quite often in mathematics it is useful to calculate the product of the first n positive integers. The symbol $n!$, which is read as *n factorial*, is used to denote this product.

DEFINITION

FACTORIAL NOTATION

The symbol $n!$, or **n factorial**, is defined for all nonnegative integers as

$$0! = 1 \qquad \text{and} \qquad n! = n(n - 1)(n - 2) \cdot \cdots \cdot 2 \cdot 1$$

if n is a positive integer.

For instance,

$$5! = 5 \cdot 4 \cdot 3 \cdot 2 \cdot 1 = 120$$

Notice that $5!$ can also be written as $5! = 5 \cdot 4!$. In general, $n!$ can be defined recursively as

$$\boxed{n! = n(n - 1)!}$$

for $n \geq 1$.

In calculus, it is often necessary to simplify expressions involving factorial notation.

EXAMPLE 2 **Simplifying expressions involving factorials**

Simplify the expression: a_{n+1}/a_n if $a_k = 3^k/k!$.

Solution Since

$$a_n = \frac{3^n}{n!} \quad \text{and} \quad a_{n+1} = \frac{3^{n+1}}{(n+1)!}$$

it follows that

$$\frac{a_{n+1}}{a_n} = \frac{\frac{3^{n+1}}{(n+1)!}}{\frac{3^n}{n!}} = \frac{3^{n+1}}{(n+1)!} \cdot \frac{n!}{3^n}$$

$$= \frac{3^{n+1}}{3^n} \cdot \frac{n!}{(n+1)n!} = \frac{3}{n+1}$$

Factorial notation is also used to define the symbol $\binom{n}{k}$.

DEFINITION $\binom{n}{k}$

Assume k and n are integers such that $0 \le k \le n$. Then

$$\binom{n}{k} = \frac{n!}{k!(n-k)!}$$

It should be noted that

$$\binom{n}{0} = \binom{n}{n} = 1$$

EXAMPLE 3 **Calculating $\binom{n}{k}$**

Calculate each expression.

(a) $\binom{5}{3}$ (b) $\binom{17}{14}$ (c) $\binom{n}{n-1}$

Solution

(a) $\binom{5}{3} = \frac{5!}{3!(5-3)!} = \frac{5!}{3!2!} = \frac{5 \cdot 4 \cdot 3!}{3!2!} = 10$

(b) $\binom{17}{14} = \frac{17!}{14!(17-14)!} = \frac{17!}{14!3!} = \frac{17 \cdot 16 \cdot 15 \cdot 14!}{14!3!} = 680$

(c) $\binom{n}{n-1} = \frac{n!}{(n-1)![n-(n-1)]!} = \frac{n!}{(n-1)!1!} = \frac{n(n-1)!}{(n-1)!} = n$

Using the Binomial Theorem

Now that we have convenient notation, we return to the problem of finding a formula that produces all the terms of the expansion of $(a + b)^n$.

THE BINOMIAL THEOREM

Let a and b be real numbers, and let n be a positive integer. Then

$$(a + b)^n = \binom{n}{0}a^n + \binom{n}{1}a^{n-1}b + \cdots + \binom{n}{k}a^{n-k}b^k + \cdots + \binom{n}{n}b^n$$

The binomial theorem can be proved by using mathematical induction. Using summation notation, the binomial theorem can be written as

$$(a + b)^n = \sum_{k=0}^{n} \binom{n}{k} a^{n-k}b^k$$

for n a positive integer.

EXAMPLE 4 Applying the binomial theorem

Use the binomial theorem to determine the expansion of each binomial.

(a) $(x + y)^5$ (b) $\left(3x^2 - \frac{1}{2}\sqrt{y}\right)^4$

Solution

(a) By the binomial theorem,

$$\begin{aligned}(x + y)^5 &= \sum_{k=0}^{5} \binom{5}{k} x^{5-k}y^k \\ &= \binom{5}{0}x^5 + \binom{5}{1}x^4y + \binom{5}{2}x^3y^2 + \binom{5}{3}x^2y^3 + \binom{5}{4}xy^4 + \binom{5}{5}y^5 \\ &= x^5 + \frac{5!}{1!4!}x^4y + \frac{5!}{2!3!}x^3y^2 + \frac{5!}{3!2!}x^2y^3 + \frac{5!}{4!1!}xy^4 + y^5 \\ &= x^5 + 5x^4y + 10x^3y^2 + 10x^2y^3 + 5xy^4 + y^5\end{aligned}$$

(b) By the binomial theorem,

$$\begin{aligned}\left(3x^2 - \frac{1}{2}\sqrt{y}\right)^4 &= \sum_{k=0}^{4} \binom{4}{k}(3x^2)^{4-k}\left(-\frac{1}{2}\sqrt{y}\right)^k \\ &= \binom{4}{0}(3x^2)^4 + \binom{4}{1}(3x^2)^3\left(-\frac{1}{2}\sqrt{y}\right) + \binom{4}{2}(3x^2)^2\left(-\frac{1}{2}\sqrt{y}\right)^2 \\ &\quad + \binom{4}{3}(3x^2)\left(-\frac{1}{2}\sqrt{y}\right)^3 + \binom{4}{4}\left(-\frac{1}{2}\sqrt{y}\right)^4\end{aligned}$$

$$= (3x^2)^4 - 4(3x^2)^3\left(\frac{1}{2}\sqrt{y}\right) + 6(3x^2)^2\left(\frac{1}{2}\sqrt{y}\right)^2$$
$$- 4(3x^2)\left(\frac{1}{2}\sqrt{y}\right)^3 + \left(\frac{1}{2}\sqrt{y}\right)^4$$
$$= 81x^8 - 54x^6\sqrt{y} + \frac{27}{2}x^4y - \frac{3}{2}x^2y\sqrt{y} + \frac{1}{16}y^2$$

EXAMPLE 5 **Finding a term in a binomial expansion**

Use the binomial theorem to find the sixth term of the expansion of $(2x - y^2)^8$.

Solution The binomial theorem indicates that the sixth term is of the form

$$\binom{8}{5}a^{8-5}b^5 = \binom{8}{5}a^3b^5$$

For the given expression, we observe that

$$(2x - y^2)^8 = [(2x) + (-y^2)]^8$$

so $a = 2x$ and $b = -y^2$. Hence, the sixth term is

$$\binom{8}{5}(2x)^3(-y^2)^5 = -\binom{8}{5}8x^3y^{10}$$
$$= -\frac{8 \cdot 7 \cdot 6}{3 \cdot 2 \cdot 1}(8x^3y^{10}) = -448x^3y^{10}$$

PROBLEM SET 10.5

Mastering the Concepts

In problems 1–6, use the patterns in Pascal's triangle to find the expansion of each binomial.

1. $(x + y)^4$ **2.** $(x - y)^6$ **3.** $(2x + 3y)^4$
4. $(2x - 5y)^3$ **5.** $(5x - 3y)^5$ **6.** $(3x + 4y)^4$

In problems 7 and 8, the given numbers correspond to a specific row in Pascal's triangle. Write the next row of numbers in the triangle and indicate the exponent n of the expression $(a + b)^n$ corresponding to this new row.

7. 1, 4, 6, 4, 1 **8.** 1, 5, 10, 10, 5, 1

In problems 9–12, evaluate each binomial coefficient.

9. **(a)** $\binom{15}{10}$ **(b)** $\binom{6}{2}$

10. **(a)** $\binom{15}{5}$ **(b)** $\binom{n}{3}$

11. **(a)** $\binom{52}{5}$ **(b)** $\binom{52}{50}$

12. **(a)** $\binom{52}{52}$ **(b)** $\binom{n}{2}$

In problems 13–16, simplify the expression

$$\left|\frac{a_{n+1}}{a_n}\right|$$

for each given sequence.

13. $a_k = \dfrac{5^k}{(k-1)!}$ **14.** $a_k = \dfrac{(-1)^k(k-1)!}{e^k}$

15. $a_k = \dfrac{3^k}{k!}$ **16.** $a_k = \dfrac{(-1)^k x^{k+1}}{(k+1)!}$

In problems 17–22, find the first four terms of the expansion of the given binomial.

17. $(x + y)^{28}$ **18.** $(x - y)^{49}$
19. $(x - 2y)^{35}$ **20.** $(x + 3y)^{57}$
21. $(2x + 3y)^{78}$ **22.** $(3x - y)^{103}$

In problems 23–30, use the binomial theorem to expand each expression as specified.

23. $(x + 3)^5$; all terms **24.** $(2z + x)^4$; all terms
25. $(x - 2)^4$; all terms **26.** $\left(\frac{1}{a} + \frac{x}{2}\right)^3$; all terms
27. $(x + y)^{12}$; the x^5y^7 term **28.** $(x - 3y)^7$; the x^4y^3 term
29. $(a^{3/2} - 2x^2)^8$; the a^3x^{12} term
30. $\left(x + \frac{1}{2}\right)^{10}$; the x^4 term

In problems 31–38, use the binomial theorem to find the indicated term.

31. $\left(\frac{x^2}{2} + a\right)^{15}$; fourth term **32.** $(y^2 - 2z)^{10}$; sixth term
33. $\left(2x^2 - \frac{a^2}{3}\right)^9$; seventh term
34. $\left(x + \sqrt{a}\right)^{12}$; middle term
35. $\left(a + \frac{x^2}{3}\right)^9$; term containing x^{12}
36. $\left(2\sqrt{y} - \frac{x}{2}\right)^{10}$; term containing y^4
37. $(x - x^{-1})^7$; fourth term **38.** $(2y - y^{-2})^9$; sixth term

Developing and Extending the Concepts

39. Verify that:
(a) $\binom{n}{0} = \binom{n+1}{0}$ **(b)** $\binom{n}{n} = \binom{n+1}{n+1}$
(c) $\binom{n}{k} = \binom{n}{n-k}$

40. Verify that: $\binom{n}{k-1} + \binom{n}{k} = \binom{n+1}{k}$

41. Prove or disprove each equation.
(a) $\frac{2n!}{(2n)!} = 1$
(b) $\frac{(n+2)!}{(n+4)!} = \frac{1}{n^2 + 7n + 12}$
(c) $(n!)(n!) = (2n)!$

42. Rewrite the binomial expansion of $(a + b)^n$ if $a = b = 1$.

In problems 43 and 44, use the summation

$$(a + b)^n = \sum_{k=0}^{n} \binom{n}{k} a^k b^{n-k}$$

to write each binomial expansion using summation notation.

43. $(x + y)^5$ **44.** $(x - 2y)^8$

CHAPTER 10 REVIEW PROBLEM SET

In problems 1 and 2, find the first five terms of each sequence whose general term is a_n.

1. **(a)** $a_n = n^2 - 1$ **(b)** $a_n = (-1)^n(3n - 2)$
2. **(a)** $a_n = 4^{n/2}$ **(b)** $a_n = \frac{1}{n^2 + 3n}$

In problems 3 and 4, find the first five terms of the sequence defined by each recursive formula.

3. **(a)** $a_1 = 4,\ a_n = a_{n-1} + 4,\ n \geq 2$
(b) $a_1 = 8,\ a_n = a_{n-1} + 5,\ n \geq 2$
4. **(a)** $a_1 = 1,\ a_2 = 5,\ a_n = 2a_{n-1} + 3a_{n-2},\ n \geq 3$
(b) $a_0 = 1,\ a_1 = 1,\ a_n = na_{n-1},\ n \geq 2$

In problems 5 and 6, find a formula for the general term of each sequence.

5. **(a)** $2, 4, 6, 8, 10, \ldots$ **(b)** $-1, 1, -1, 1, -1, 1, \ldots$
6. **(a)** $1, 4, 9, 16, 25, \ldots$ **(b)** $\frac{1}{2}, \frac{2}{3}, \frac{3}{4}, \frac{4}{5}, \ldots$

In problems 7 and 8, graph each sequence for $n = 1, 2, 3, 4, 5$. Then graph the associated continuous curve, and find

$$\lim_{n \to \infty} a_n$$

7. **(a)** $a_n = 3n + 7$ **(b)** $a_n = 2^n - 5$
8. **(a)** $a_n = \frac{1}{n^3}$ **(b)** $a_n = (-1)^n + 3n$

In problems 9 and 10, evaluate each sum.

9. **(a)** $\sum_{k=1}^{5} k(2k - 1)$ **(b)** $\sum_{k=1}^{4} 2k^2(k - 3)$

10. **(a)** $\sum_{k=0}^{4} \frac{2}{3^k}$ **(b)** $\sum_{k=2}^{6} (k + 1)(k + 2)$

11. Use the properties of summation to evaluate the expression

$$\sum_{k=1}^{20} (3k^2 - 4k + 7)$$

12. Express the sum $1^2 + 2^2 + 3^2 + 4^2 + 5^2$ in sigma notation.

In problems 13 and 14, find:
(i) The nth term, and graph the sequence.
(ii) The sum of the first n terms of each arithmetic sequence for the given value of n.

13. **(a)** $2, 6, 10, 14, \ldots$; for $n = 15$
(b) $8, 16, 24, 32, \ldots$; for $n = 10$

14. **(a)** $5, 13, 21, 29, \ldots$; for $n = 12$
(b) $20, 12, 4, -4, -12, \ldots$; for $n = 20$

In problems 15 and 16, find the nth partial sum S_n of the arithmetic sequence $\{a_n\}$ with common difference d.

15. **(a)** $a_1 = 2, d = 5, n = 6$
(b) $a_1 = 15, d = -6, n = 8$

16. **(a)** $a_1 = -4, d = 8, n = 7$
(b) $a_1 = 7, d = 5, n = 10$

In problems 17 and 18, find the sum of the first n terms S_n of each arithmetic sequence for the given value of n.

17. **(a)** $\frac{3}{4}, \frac{1}{4}, -\frac{1}{4}, -\frac{3}{4}, \ldots$; $n = 12$
(b) $2a + 3b, 3a + 2b, 4a + b, \ldots$; $n = 6$

18. **(a)** $-8, -3, 2, 7, \ldots$; $n = 30$
(b) $a, a + 2, a + 4, a + 6, \ldots$; $n = 20$

In problems 19 and 20, find the general term a_n of each geometric sequence.

19. **(a)** $5, 15, 45, 135, \ldots$ **(b)** $4, 8, 16, 32, \ldots$

20. **(a)** $\frac{1}{4}, \frac{1}{8}, \frac{1}{16}, \frac{1}{32}, \ldots$ **(b)** $-6, 5, -\frac{25}{6}, \ldots$

In problems 21 and 22, graph the sequence. Also, find the common ratio and the sum of the first n terms of each geometric sequence for the given value of n.

21. **(a)** $48, 96, 192, \ldots$; $n = 8$
(b) $-81, -27, -9, \ldots$; $n = 12$

22. **(a)** $\frac{3}{4}, 3, 12, \ldots$; $n = 10$
(b) $0.2, 0.002, 0.00002, \ldots$; $n = 15$

In problems 23 and 24, find the sum of each infinite geometric series.

23. **(a)** $\sum_{k=1}^{\infty} 3\left(\frac{2}{3}\right)^{k-1}$ **(b)** $\sum_{k=1}^{\infty} 3\left(-\frac{2}{3}\right)^{k-1}$

24. **(a)** $\sum_{k=1}^{\infty} \left(\sqrt{\frac{5}{7}}\right)^{k-1}$ **(b)** $\sum_{k=0}^{\infty} (-1)^k\left(\frac{3}{4}\right)^k$

25. Use the graph and the limit behavior of

$$f(x) = \frac{a_1(1 - r^x)}{1 - r} \qquad x \geq 1, r \neq 1$$

to justify the sum of each geometric series.

(a) $\sum_{k=1}^{\infty} 2\left(\frac{1}{3}\right)^{k-1} = 3$ **(b)** $\sum_{k=1}^{\infty} (-0.4)^{k-1} = \frac{5}{7}$

26. Write each rational number as the quotient of two integers in reduced form.
(a) $0.\overline{45}$ **(b)** $0.\overline{422}$

In problems 27–30, use mathematical induction to prove that the assertion is true for all positive integers n.

27. $n^2 + 3n$ is an even integer

28. 3 is an exact integer divisor of $n^3 + 6n^2 + 11n$

29. $1 \cdot 2 \cdot 3 + 2 \cdot 3 \cdot 4 + 3 \cdot 4 \cdot 5 + \cdots + n(n + 1)(n + 2) = \frac{1}{4}n(n + 1)(n + 2)(n + 3)$

30. $2 \cdot 4 + 4 \cdot 6 + 6 \cdot 8 + \cdots + 2n(2n + 2) = \frac{n}{3}(2n + 2)(2n + 4)$

In problems 31–34, simplify each expression.

31. **(a)** $\binom{12}{5}$ **(b)** $\binom{34}{30}$

32. **(a)** $\binom{n + 7}{n + 5}$ **(b)** $\binom{n + 3}{n + 1}$

33. $\frac{a_{n+1}}{a_n}$ if $a_k = \frac{(k + 1)!}{7}$

34. $\frac{a_{n+1}}{a_n}$ if $a_k = \frac{3^k}{k!}$

In problems 35 and 36, expand each expression.

35. **(a)** $(2 + x)^5$ **(b)** $(3x - 4y)^4$

36. **(a)** $(3x^2 - 2y^2)^7$ **(b)** $(2a + 3b)^6$

In problems 37 and 38, find and simplify the specified term in the binomial expansion of the expression.

37. **(a)** The fourth term of $(2x + y)^9$
(b) The term containing x^5 in $(3x + y)^{10}$

38. **(a)** The fifth term of $(x - 2y)^7$
(b) The term containing x^{10} in $(2x^2 - 3)^{11}$

39. Physics: Assume an object falls vertically 16 feet during the first second, 48 feet during the next second, 80 feet during the third, and so on.
(a) How far will the object fall during the tenth second?
(b) What is the total number of feet the object has fallen at the end of 10 seconds?

40. Biology: Suppose that a colony of bacteria reproduces to become 300 bacteria after the first day. After 2 days, they reproduce to become 700 bacteria. After 3 days, they reproduce to become 1100 bacteria, and so on.
(a) Find a recursive formula for the number of bacteria a_n after n days.
(b) How many bacteria are there after 7 days?

41. Real Estate: A home purchased for \$130,000 appreciates at a rate of 3% per year.
(a) If a_n represents the value of the house after n years, write an explicit formula for a_n.
(b) Find a sequence representing the value of the house after 7 years. Round off to two decimal places.

42. Sports: A runner training for a track meet begins on March 1 by running 1 mile, and increases the length of the run by 0.1 mile a day. Find the length of the run on March 25.

43. Economics: A firm offers a job candidate a starting salary of \$30,000 with a guaranteed minimum annual raise of \$1200. What would be the minimum salary this offer represents for the candidate at the end of 7 years with the firm?

44. Business: An investor received \$230,000 from an oil field after the first year, and two-thirds as much as in the immediately preceding year each year thereafter. How much did the investor realize after 5 years?

CHAPTER 10 TEST

1. Write the first five terms of each sequence with the given general term.
(a) $a_n = 3n - 4$ **(b)** $a_n = \dfrac{n + 1}{n^2}$
(c) $a_1 = 3;\ a_n = -2a_{n-1},\ n \geq 2$

2. Find the general term a_n of each sequence.
(a) 6, 10, 14, 18, . . . **(b)** $-3, 9, -27, 81, \ldots$

3. Evaluate each sum.
(a) $\sum_{k=1}^{5} (3k + 2)$ **(b)** $\sum_{k=2}^{6} (k^2 + 2k)$

4. Find the indicated unknown for each arithmetic sequence.
(a) a_n; if $a_1 = 1$, $d = 2$, $n = 6$
(b) n; if $a_1 = 13$, $d = -4$, $a_n = -7$
(c) S_n; if $a_1 = 2$, $a_n = 14$, $n = 7$
(d) d; if $S_n = 15$, $a_1 = 10$, $n = 6$

5. Find the indicated unknown for each geometric sequence.
(a) a_{10}; if $a_1 = 3$, $a_4 = -24$
(b) r; if $a_5 = 81$, $S_5 = 61$
(c) a_1; if $r = -1$, $a_n = -1$, $n = 7$

6. Evaluate the sum of each infinite geometric series.
(a) $\sum_{k=1}^{\infty} 5\left(\frac{4}{5}\right)^{k-1}$ **(b)** $\sum_{k=1}^{\infty} \frac{1}{8 \cdot 3^k}$

7. Use mathematical induction to prove each formula.
(a) $2 + 2^2 + 2^3 + \cdots + 2^n = 2(2^n - 1)$
(b) $1 \cdot 3 + 2 \cdot 4 + 3 \cdot 5 + \cdots + n(n + 2) = \frac{1}{6}n(n + 1)(2n + 7)$

8. Expand each expression.
(a) $(x - 3)^4$ **(b)** $(2x + 1)^5$

9. Find the seventh term of the expansion of $(x + x^2)^{12}$.

10. Give an example to show that the statement

$$(n^2)! = (n!)^2$$

is not true.

11. The tip of a pendulum travels 15 centimeters on its first swing. On each subsequent swing it travels 86% of the previous distance. How far has it traveled after 8 swings?

APPENDIX I

Plane Geometry Review

Angles

When two lines ℓ and p intersect so that the angles formed are right angles, we say that the lines are **perpendicular lines** (Figure 1a). A common notation for this is the symbol $p \perp \ell$. The line ℓ that is perpendicular to $\overline{AB}$ and intersects $\overline{AB}$ at its midpoint M is called the **perpendicular bisector** of $\overline{AB}$ (Figure 1b).

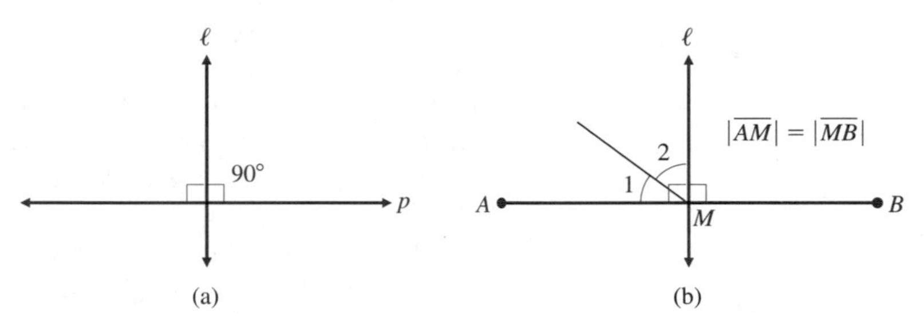

FIGURE 1

Two intersecting lines form four nonstraight angles in the plane. In Figure 2, the four angles are $\angle 1$, $\angle 2$, $\angle 3$, and $\angle 4$. Other angles are classified as follows:

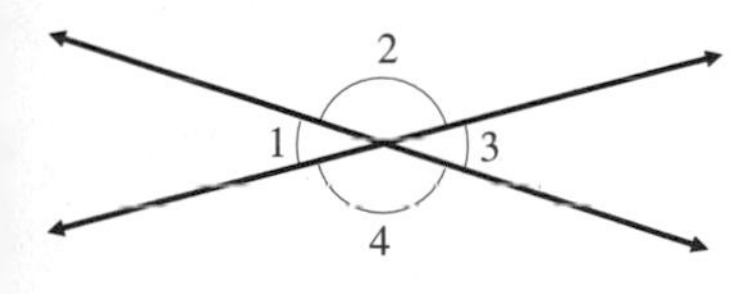

FIGURE 2

1. Angles 1 and 2 (Figure 2) are called **adjacent angles**. Adjacent angles have a common vertex and a common side between them.
2. Nonadjacent angles formed by two intersecting lines, such as angles 1 and 3 or angles 2 and 4, are called **vertical angles** or **opposite angles** (Figure 2). Vertical angles formed by intersecting lines are *equal*.
3. Two angles are called **supplementary angles** if the sum of their measures is 180°. In Figure 2, angles 1 and 2, 2 and 3, 3 and 4, and 4 and 1 are supplementary angles.
4. Two angles are called **complementary angles** if the sum of their measures is 90°. Angles 1 and 2 in Figure 1b are complementary angles.
5. Positive angles less than 90° are **acute angles**, such as angles 1 and 3 in Figure 2. Angles between 90° and 180° are **obtuse angles**, such as angles 2 and 4 in Figure 2.

Any line that intersects two or more lines is called a **transversal** of these lines. The angles made by a transversal that cuts two lines are shown (and numbered) in Figure 3. These angles are named as follows:

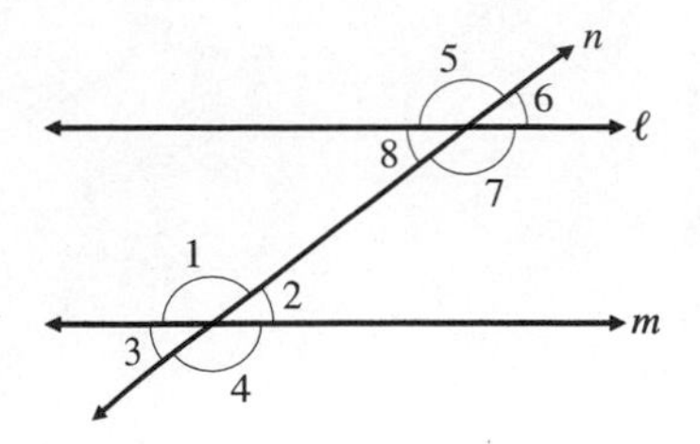

FIGURE 3

6. **Corresponding angles** determined by a transversal are pairs of angles on the "same side" of the transversal and on the "same relative sides" of the given lines. In Figure 3, the pairs of corresponding angles are 6 and 2, 7 and 4, 5 and 1, and 8 and 3.
7. **Interior angles** are those angles between the lines cut by the transversal. Angles 1, 2, 7, and 8 are interior angles in Figure 3.
8. **Alternate interior angles** are certain pairs of angles on opposite sides of the transversal. In Figure 3, the pairs of alternate interior angles are 1 and 7, and 2 and 8.

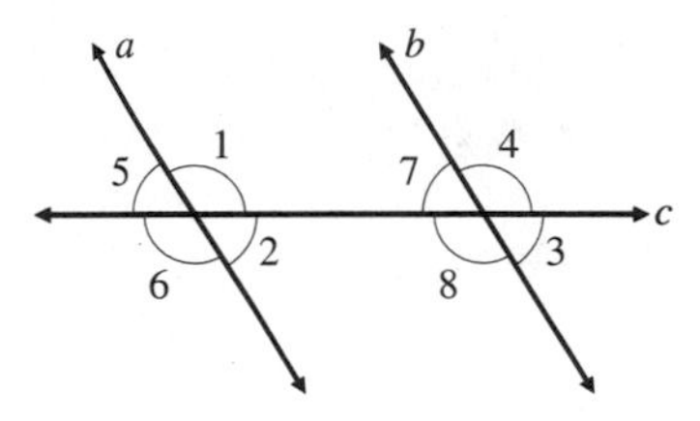

FIGURE 4

In Figure 4, line a is parallel to line b, and c is a transversal; angles 2 and 7 and angles 1 and 8 are alternate interior angles; and angles 5 and 7, angles 6 and 8, angles 2 and 3, and angles 1 and 4 are corresponding angles.

The following properties are noteworthy:

1. **Corresponding angles** (determined by a transversal) of parallel lines are *equal* (Figure 4), and conversely, if two lines are cut by a transversal so that the corresponding angles are equal, then the lines are parallel.
2. **Alternate interior angles** (determined by a transversal) of parallel lines are *equal* (Figure 4), and conversely, if two lines are cut by a transversal so that alternate interior angles are equal, then the lines are parallel.

EXAMPLE 1 **Using angle terminology**

In Figure 5, C, E, and F lie on a straight line, and D, E, and A lie on a different straight line. The angle BEF is a right angle.

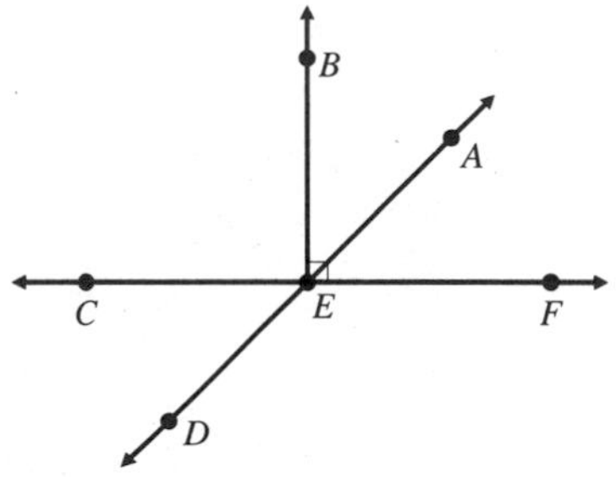

(a) Name a pair of vertical angles.
(b) Name an obtuse angle.
(c) Name an acute angle.
(d) Which angle is the complement of $\angle FEA$?
(e) Which angle is the supplement of $\angle FEA$?

FIGURE 5

Solution

(a) Angles AEF and CED are a pair of vertical angles.
(b) Angle BED is an obtuse angle.
(c) Angle AEF is an acute angle.
(d) The complement of $\angle FEA$ is $\angle AEB$.
(e) The supplement of $\angle FEA$ is $\angle AEC$.

Triangles

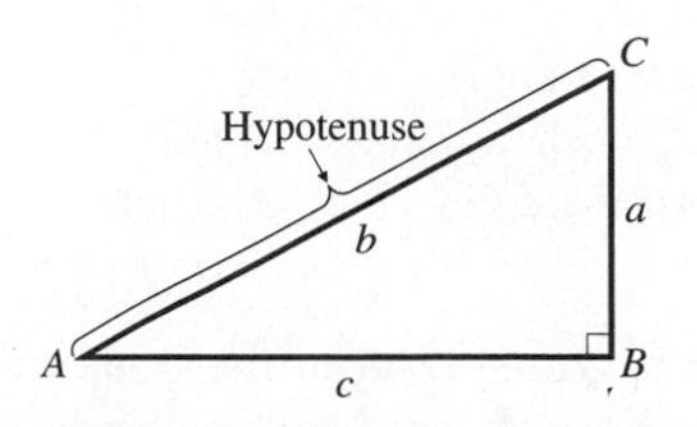

FIGURE 6

A triangle is a **right triangle** if one of its angles is a right angle—that is, a 90° angle. In Figure 6, ABC is a right triangle and $\angle B$ is a right angle. In a right triangle, the side opposite the right angle is called the **hypotenuse** (Figure 6). The **Pythagorean theorem** states that *the square of the length of the hypotenuse is* *equal to the sum of the squares of the lengths of the other two sides*. Thus, in Figure 6, by the Pythagorean theorem,

$$b^2 = a^2 + c^2$$

EXAMPLE 2 **Using the Pythagorean theorem**

In right triangle ABC (Figure 6), suppose $a = 5$ and $b = 13$. Find c.

Solution By the Pythagorean theorem,

$$c^2 = b^2 - a^2 = 13^2 - 5^2 = 169 - 25 = 144$$

so

$$c = \sqrt{144} = 12$$

■

Triangles are classified as follows:

1. A triangle is an **isosceles triangle** if two of its sides are equal in length (Figure 7a). It can be shown that the angles opposite the equal sides of an isosceles triangle are equal. In Figure 7a, $\angle B = \angle C$.
2. A triangle is an **equilateral triangle** if all three of its sides are equal in length (Figure 7b). All three angles are equal.
3. A triangle is an **oblique triangle** if none of the angles of the triangle is a right angle (Figure 7c).

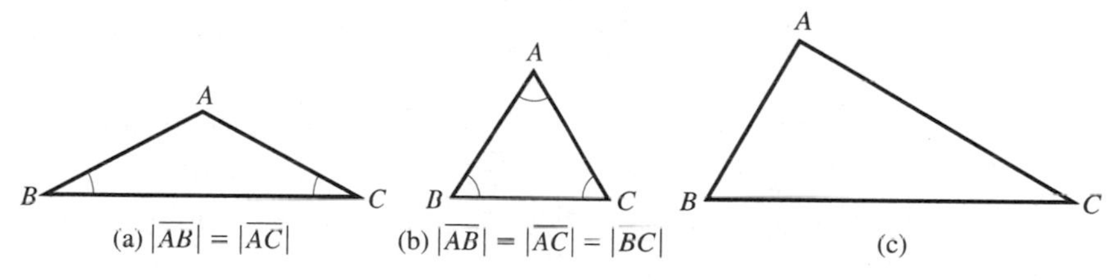

FIGURE 7

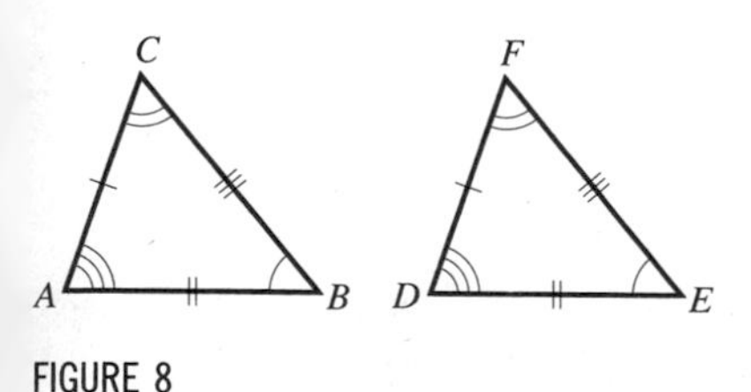

FIGURE 8

Two triangles, say $\triangle ABC$ and $\triangle DEF$, are **congruent**, which is written in symbols as $\triangle ABC \cong \triangle DEF$, if they can be made to coincide. In other words, congruent triangles have the same size and shape (Figure 8). That is, $\triangle ABC \cong \triangle DEF$ means that each of the following holds:

$$\angle A = \angle D \qquad |\overline{CB}| = |\overline{FE}|$$
$$\angle B = \angle E \qquad |\overline{AC}| = |\overline{DF}|$$
$$\angle C = \angle F \qquad |\overline{AB}| = |\overline{DE}|$$

Note that *equal sides are opposite equal angles*. (We use a shorthand notation in Figure 8 to indicate equal sides and equal angles. The single lines drawn through $\overline{AC}$ and $\overline{DF}$, for example, indicate that these sides are equal, as do the double lines through $\overline{AB}$ and $\overline{DE}$, and the triple lines through $\overline{CB}$ and $\overline{FE}$.)

The concept of congruence does not give us any practical way of deciding when two triangles are congruent. The following properties are useful:

1. If three sides of one triangle equal three sides of another triangle (SSS), then the triangles are *congruent.*
2. If two sides and the included angle of one triangle equal two sides and the included angle of another triangle (SAS), then the triangles are *congruent.*
3. If two angles and the included side of one triangle equal two angles and the included side of another triangle (ASA), then the triangles are *congruent.*

EXAMPLE 3 **Determining congruent triangles**

In Figure 9, let $\angle 1 = \angle 2$, and $\angle EAB = \angle CBA$. Show that $\triangle AEB \cong \triangle BCA$. What can be concluded about the remaining corresponding parts?

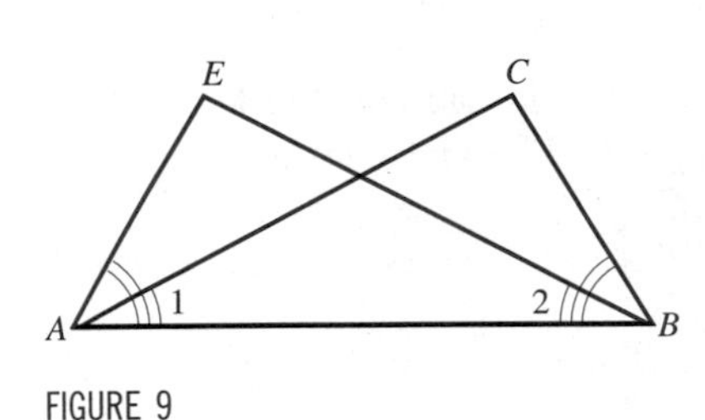

FIGURE 9

Solution In the figure, we have

$$|\overline{AB}| = |\overline{BA}|$$
$$\angle 1 = \angle 2$$
$$\angle EAB = \angle CBA$$

Therefore

$$\triangle AEB \cong \triangle BCA \quad \text{(ASA)}$$

Thus,

$$|\overline{AE}| = |\overline{BC}|, \quad |\overline{AC}| = |\overline{BE}|, \quad \text{and} \quad \angle E = \angle C$$

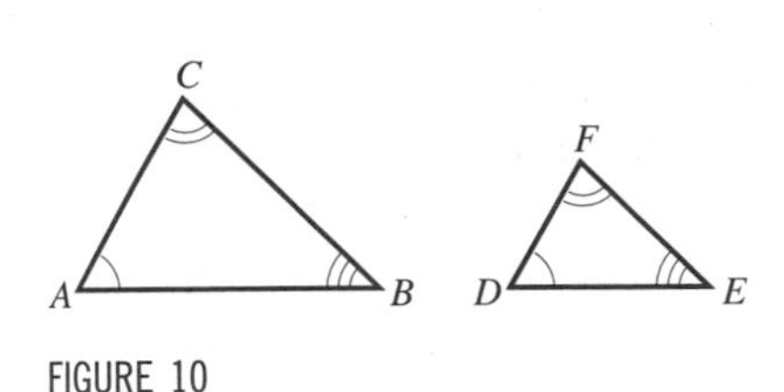

FIGURE 10

Two triangles are **similar** if they have the same shape. Hence, $\triangle ABC$ is similar to $\triangle DEF$, which is denoted by $\triangle ABC \sim \triangle DEF$, if the triangles have equal angles and the corresponding sides are proportional (Figure 10). Since $\triangle ABC \sim \triangle DEF$, it follows that:

$$\angle A = \angle D, \quad \angle B = \angle E, \quad \text{and} \quad \angle C = \angle F$$

Also,

$$\frac{|\overline{AB}|}{|\overline{DE}|} = \frac{|\overline{AC}|}{|\overline{DF}|} = \frac{|\overline{BC}|}{|\overline{EF}|}$$

Note that corresponding sides are opposite equal angles.

In order to show that *two triangles are similar, it is enough to show that two angles of one triangle equal two angles of the other triangle.*

EXAMPLE 4 **Determining similar triangles**

Suppose that one triangle is inside another triangle and the sides of the triangles are parallel. The smaller triangle has sides 5, 8, and 10 inches in length, and the larger triangle has a side of 15 inches parallel to the 10 inch side (Figure 11). Show that the triangles are similar, and find the lengths of the other two sides of the larger triangle.

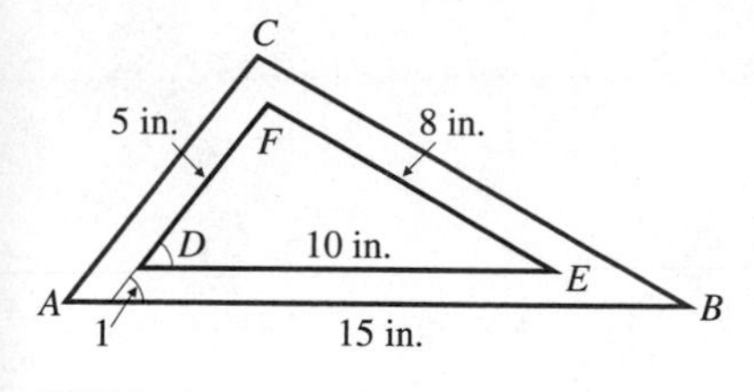

FIGURE 11

Solution In Figure 11, $\angle A = \angle 1$ and $\angle D = \angle 1$, because corresponding angles of parallel lines cut by a transversal are equal. Hence, $\angle D = \angle A$. Similarly, $\angle E = \angle B$, so that $\triangle DEF \sim \triangle ABC$. Therefore,

$$\frac{|\overline{DE}|}{|\overline{AB}|} = \frac{|\overline{FE}|}{|\overline{CB}|} = \frac{|\overline{DF}|}{|\overline{AC}|} \quad \text{or} \quad \frac{10}{15} = \frac{8}{|\overline{CB}|} = \frac{5}{|\overline{AC}|}$$

so that $|\overline{AC}| = 7.5$ inches and $|\overline{CB}| = 12$ inches.

Answers to Selected Problems

REVIEW OF BASIC ALGEBRA: PROBLEM SET 1 (ON PAGE 3)

1. **(a)** 0.75 **(b)** -1.75 **3.** **(a)** 5.3125 **(b)** 0.3 **5.** **(a)** $\frac{7}{10}$ **(b)** $\frac{11}{20}$ **7.** **(a)** $\frac{2}{5}$ **(b)** $\frac{3}{4}$

9. **(a)** $-2 \quad -1 \quad 0 \quad 1 \quad 2 \quad 3 \quad 4 \quad 5 \quad 6$ **(b)** $-5 \quad -4 \quad -3 \quad -2 \quad -1 \quad 0 \quad 1 \quad 2$

(c) $-\frac{1}{2} \quad \frac{2}{3} \quad 2.3 \quad 3\frac{5}{8} \quad 4.7$; $3 \quad 2 \quad 1 \quad 0 \quad 1 \quad 2 \quad 3 \quad 4 \quad 5 \quad 6$ **(d)** $\sqrt{3} \quad \sqrt[3]{17} \quad \pi$; $1 \quad 0 \quad 1 \quad 2 \quad 3 \quad 4$

11. **(a)** 0.6199 **(b)** 3282.8064 **13.** **(a)** 0.5176 **(b)** 0.9344 **15.** 4.19 in.^{3}; 82.45 in.^{3}; 5575.28 in.^{3} **17.** 147.40 cm^2

PROBLEM SET 2 (ON PAGE 8)

1. **(a)** Commutative **(b)** Distributive **(c)** Additive inverse **(d)** Multiplicative inverse **3.** 14.77
5. **(a)** Polynomial; degree = 2; coefficients: 4, -1; binomial **(b)** Not a polynomial **7.** **(a)** $3xy^2 - 3x^2y^2$ **(b)** $3x^3 - 7x^2 + 2$
9. $4x^4 - x^3 + 3x^2 + 2x - 5$ **11.** **(a)** $-14x^9$ **(b)** $6x^3y^8$ **13.** **(a)** $-10x^2 - 19x - 6$ **(b)** $y^3 + 10y^2 + 22y + 7$
15. **(a)** $27x^3 - 64$ **(b)** $6x^4 + x^3 + 12x^2 + 4x + 5$

PROBLEM SET 3 (ON PAGE 10)

1. **(a)** $y^3(5y + 2)$ **(b)** $4x^2(2 + 3x^2)$ **3.** **(a)** $17x^2y(xy - 2)$ **(b)** $3x^2y(27x + 4y^5)$
5. **(a)** $(3y - 5)(3y + 5)$ **(b)** $(2x - 7y)(2x + 7y)$ **7.** **(a)** $(x - 2)(x^2 + 2x + 4)$ **(b)** $(2x - y)(4x^2 + 2xy + y^2)$
9. **(a)** $(x + 2)^2$ **(b)** $(x - 3)^2$ **11.** **(a)** $(x - 7)(x - 9)$ **(b)** $(x - 2y)(x - 5y)$
13. **(a)** $-x(2x - 3)(x + 4)$ **(b)** $2x(3x + 5)(4x - 1)$ **15.** **(a)** $y(y^2 - 3y - 25)$ **(b)** $2t(t^2 + t - 25)$

PROBLEM SET 4 (ON PAGE 14)

1. **(a)** -7 **(b)** $-1, 3$ **3.** **(a)** $\dfrac{y^5}{5xa^2}$ **(b)** $-\dfrac{1}{x + 2}$ **5.** **(a)** $\dfrac{x}{x + 2}$ **(b)** $\dfrac{1}{7x + 5y}$ **7.** $\dfrac{3}{5(x + 3)}$; $x \neq -1, -2, -3$

9. $\dfrac{4x(x - 1)}{(3x - 5)(x + 2)}$; $x \neq -2, -\frac{5}{3}, 0, \frac{5}{3}, 10$ **11.** $x - 3$; $x \neq -3$ **13.** $\dfrac{2x - 5}{(x - 5)(x + 5)}$; $x \neq 5, -5$

15. $\dfrac{-4x + 3}{(x - 2)(x - 3)(x + 3)}$; $x \neq 2, 3, -3$ **17.** **(a)** $\dfrac{3}{28x}$ **(b)** x **(c)** $\dfrac{-2x - 4}{5x + 3}$

PROBLEM SET 5 (ON PAGE 17)

1. **(a)** 1 **(b)** $\dfrac{1}{25}$ **(c)** $\dfrac{x^4}{y^7}$ **(d)** $\dfrac{1}{3125}$ **(e)** x^{2m} **3.** **(a)** $\dfrac{3}{32}$ **(b)** $\dfrac{1}{320}$ **(c)** x^5 **5.** **(a)** 4096 **(b)** 25 **7.** $\dfrac{x^{20}}{81y^{16}z^{12}}$

9. $\dfrac{y^8}{x^2}$ **11.** x **13.** **(a)** -6 **(b)** $-\dfrac{144}{7}$ **15.** **(a)** $-\dfrac{1}{ab}$ **(b)** $\dfrac{b^2 + a^2}{ab}$ **17.** $\dfrac{-2x - h}{x^2(x + h)^2}$

PROBLEM SET 6 (ON PAGE 20)

1. **(a)** -4 **(b)** 2 **(c)** $-\frac{5}{12}$ **(d)** Not a real number **3.** **(a)** x^2 **(b)** x^2 **(c)** $(x + 1)^2$ **(d)** x^2 **5.** **(a)** 4 **(b)** 16
7. **(a)** 7^2 **(b)** x^3 **9.** **(a)** x **(b)** 8^4 **11.** **(a)** $-2\sqrt[5]{2}$ **(b)** $-10\sqrt[3]{2}$ **13.** **(a)** $2x^2y^3\sqrt[4]{x^3y}$ **(b)** $2x\sqrt[3]{4x^2}$
15. $\frac{2v^2}{u}$ **17.** $\frac{t\sqrt[3]{9}}{2}$

PROBLEM SET 7 (ON PAGE 22)

1. **(a)** $\sqrt{3}$ **(b)** $7\sqrt{5}$ **3.** **(a)** $\sqrt{2x}$ **(b)** $6\sqrt{5x}$ **5.** $-11\sqrt{2}$ **7.** $16x\sqrt{3}$ **9.** $-9x\sqrt[3]{x}$ **11.** $16x - 81y$ **13.** $15 + 10\sqrt{2}$
15. $2\sqrt{3} - 4\sqrt{7} + 2$ **17.** **(a)** $\sqrt{2}$ **(b)** $-\frac{3\sqrt{14}}{7}$ **19.** **(a)** $\frac{1 - 2\sqrt{x} + x}{1 - x}$ **(b)** $\frac{x + 2\sqrt{xy} + y}{x - y}$ **21.** $\frac{1}{\sqrt{x + h} + \sqrt{x}}$

PROBLEM SET 8 (ON PAGE 25)

1. **(a)** $0 + 4i$ **(b)** $3 + 2\sqrt{2}\,i$ **(c)** $7 - 9i$ **(d)** $\frac{1}{3} - \frac{1}{3}i$ **(e)** $0 + xi$ **(f)** $4 + 0i$ **3.** **(a)** $10 - 5i$ **(b)** $-3 + 2i$
5. **(a)** $12 - 9i$ **(b)** $11 - 16i$ **7.** **(a)** $3 + 4i$ **(b)** $-7 - 24i$ **9.** **(a)** $1 - i$ **(b)** $-\frac{23}{29} - \frac{44}{29}i$

CUMULATIVE PROBLEM SET (ON PAGE 26)

1. **(a)** $-1.\overline{6}$ **(b)** $0.1\overline{518}$ **(c)** -1.8 **(d)** $113.\overline{513}$ **2.** **(a)** $\frac{11}{100}$ **(b)** $\frac{11}{50}$ **(c)** $\frac{161}{20}$ **(d)** $-\frac{17}{8}$

3. 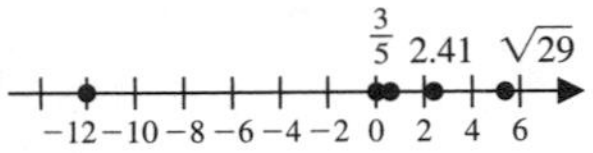 (number line) $\frac{3}{5}$ 2.41 $\sqrt{29}$; -12 -10 -8 -6 -4 -2 0 2 4 6 **4.** **(a)** Commutative **(b)** Distributive **(c)** Additive inverse **(d)** Associative

5. **(a)** $V = 1788.44$ in.^{3}; $S = 712.52$ in.^{2} **(b)** $V = 0.02$ in.^{3}; $S = 0.36$ in.^{2}
6. **(a)** Binomial; degree $= 1$; $8, -5$ **(b)** Monomial; degree $= 2$; -7
(c) Not a monomial, binomial or trinomial; degree $= 3$; $-2, -4, 7, 3$
7. 103.3
8. **(a)** $-3x^2 - 1$ **(b)** $x^2 + 10x + 1$ **(c)** $-16x^9$ **(d)** y^{12} **(e)** $8x^5 + 12x^4 - 4x^3$ **(f)** $2x^2 + 11x - 21$
(g) $4x^2 - 4x + 1$ **(h)** $8x^3 - y^3$
9. **(a)** $3x^2(y^2 + 2z^2 - 3)$ **(b)** $(x + 9)(x - 2)$ **(c)** $(x - 2y)(x + 2y)(x^2 + 4y^2)$ **(d)** $(3x + y)(9x^2 - 3xy + y^2)$ **10.** $\frac{3x + 1}{2x - 1}$
11. **(a)** $\frac{c + 1}{c^2}$ **(b)** $\frac{(x + 3)(x - 2)}{(x + 2)^2}$ **(c)** $\frac{11x - 20}{x(x - 4)(x + 2)}$ **(d)** $\frac{10}{x - 3}$ **(e)** $\frac{xy - 1}{y - x}$
12. **(a)** $\frac{1}{27x^3}$ **(b)** 1 **(c)** $\frac{16}{225}$ **(d)** $\frac{36}{13}$ **(e)** $\frac{1}{625}$ **(f)** $\frac{1}{4x^2}$ **(g)** $64x^2y^9$ **13.** **(a)** $x^2 + 7$ **(b)** $-x^2$ **(c)** $x + 1$
14. **(a)** $3\sqrt{3}$ **(b)** $-3\sqrt[3]{2}$ **(c)** $-\frac{\sqrt[7]{2}}{x^2}$ **(d)** $2x^2\sqrt{x}$, $x \geq 0$
15. **(a)** $6\sqrt{2}$ **(b)** $10\sqrt[3]{2}$ **(c)** $50 + 8\sqrt{6}$ **(d)** $4 - \sqrt{6}$ **(e)** 15
16. **(a)** $\frac{5\sqrt{6}}{6}$ **(b)** $\frac{5\sqrt{3} + 5}{2}$ **17.** **(a)** $6i$ **(b)** $11 - 5i$ **(c)** $4 - 19i$ **(d)** $-\frac{9}{25} + \frac{12}{25}i$

CHAPTER 1: PROBLEM SET 1.1 (ON PAGE 34)

1. **(a)** $-2 \leq 3$ **(b)** $-5 > -7$ **3.** **(a)** $6 < x < a$ **(b)** $x \geq 0$
5. **(a)** $0 < x < 6$ (number line: 0, 6) **(b)** $-1 \leq x < 4$ (number line: -1, 4) **(c)** $x > 1$ (number line: 1) **(d)** $x \leq 5$ (number line: 5)

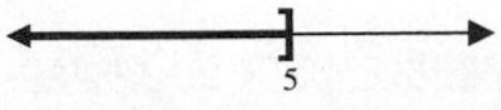

7. **(a)** $x \leq -2$, $(-\infty, -2]$ (number line: -2) **(b)** $x \geq -2$, $[-2, \infty)$ (number line: -2) **9.** **(a)** $x < 3$, $(-\infty, 3)$ (number line: 3) **(b)** $x \geq \frac{9}{4}$, $\left[\frac{9}{4}, \infty\right)$ (number line: $\frac{9}{4}$)

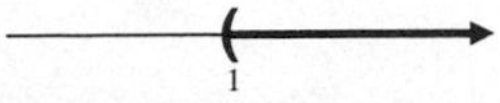

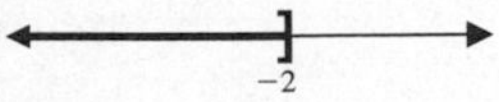

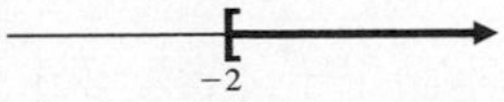

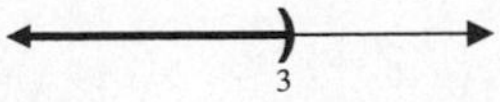

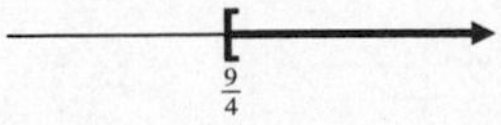

11. **(a)** $w \geq 3$, $[3, \infty)$

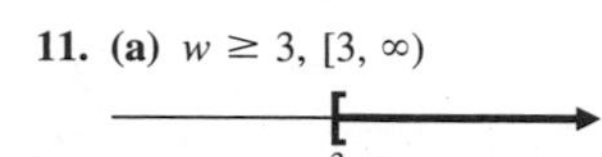

(b) $t \geq 2$, $[2, \infty)$

13. $-\frac{7}{2} < x < -\frac{3}{2}$, $\left(-\frac{7}{2}, -\frac{3}{2}\right)$ **15.** $x \leq -\frac{24}{25}$, $\left(-\infty, -\frac{24}{25}\right]$

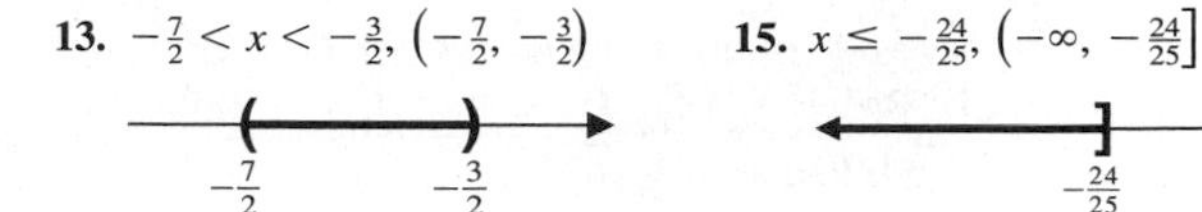

17. **(a)** $5 - \pi$ **(b)** $\sqrt{7} - 2$ **19.** **(a)** $\begin{cases} x - 5 & \text{if } x \geq 5 \\ 5 - x & \text{if } x < 5 \end{cases}$ **(b)** $\begin{cases} x + 3 & \text{if } x \geq -3 \\ -x - 3 & \text{if } x < -3 \end{cases}$ **21.** $\begin{cases} 2x + 1 & \text{if } x \geq -\frac{1}{2} \\ -2x - 1 & \text{if } x < -\frac{1}{2} \end{cases}$ **23.** 8

25. 8 **27.** $10|t|$

29. **(a)** $-3 \leq x \leq 3$, $[-3, 3]$ **(b)** $x > 2$ or $x < -2$, $(-\infty, -2)$ or $(2, \infty)$

31. **(a)** $-1 < x < 5$, $(-1, 5)$ **(b)** $x \geq 5$ or $x \leq -1$, $(-\infty, -1]$ or $[5, \infty)$

33. **(a)** $x \geq 3$ or $x \leq -7$, $(-\infty, -7]$ or $[3, \infty)$ **(b)** $-7 < x < 3$, $(-7, 3)$

35. $-2 < x < -1$, $(-2, -1)$

37. $x \geq 1$ or $x \leq \frac{1}{5}$, $\left(-\infty, \frac{1}{5}\right]$ or $[1, \infty)$

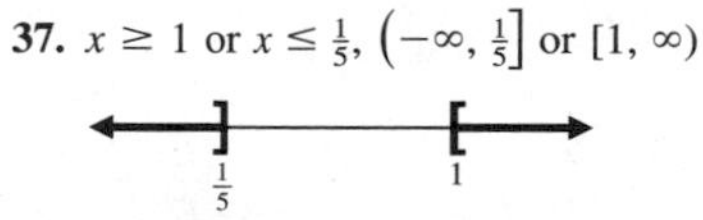

39. **(a)** $72° < T \leq 98°$ **(b)** $65° \leq T \leq 91°$ **41.** At least 3 hr **43.** $16\frac{2}{3}$ weeks

45. $x \leq 17$ or $x \geq 19.50$; alert is set if price drops to \$17 or less or rises to \$19.50 or more.

47. **(a)** $0 \leq x < 2$

(b) $-2 < x \leq 8$

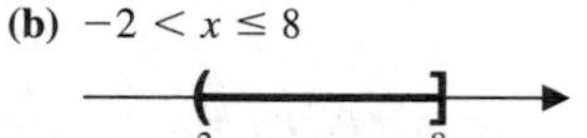

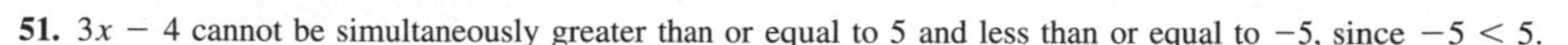

51. $3x - 4$ cannot be simultaneously greater than or equal to 5 and less than or equal to -5, since $-5 < 5$.

PROBLEM SET 1.2 (ON PAGE 44)

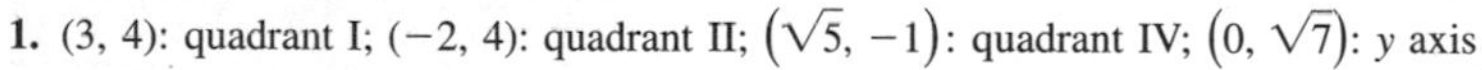

1. $(3, 4)$: quadrant I; $(-2, 4)$: quadrant II; $\left(\sqrt{5}, -1\right)$: quadrant IV; $\left(0, \sqrt{7}\right)$: y axis

3. **(a)** (i) $d = \sqrt{17} = 4.12$; (ii) $(-1, 1.5)$ **(b)** (i) $d = \sqrt{61} = 7.81$; (ii) $(0.5, 2)$

5. **(a)** (i) $d = \sqrt{73} = 8.54$; (ii) $(-1, -5.5)$ **(b)** (i) $d = \sqrt{0.52} = 0.72$; (ii) $\left(-\frac{1}{10}, \frac{2}{5}\right)$

7. **(a)** (i) $d = \sqrt{2} = 1.41$; (ii) $\left(\frac{2t + 1}{2}, \frac{2u + 1}{2}\right)$ **(b)** (i) $d = \sqrt{78.6344} = 8.87$; (ii) $(0.54, 6.77)$

9. **(a)** $(x - 3)^2 + (y + 2)^2 = 25$

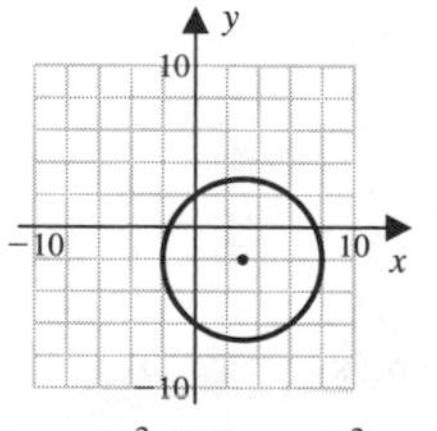

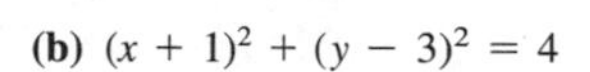

(b) $(x + 1)^2 + (y - 3)^2 = 4$

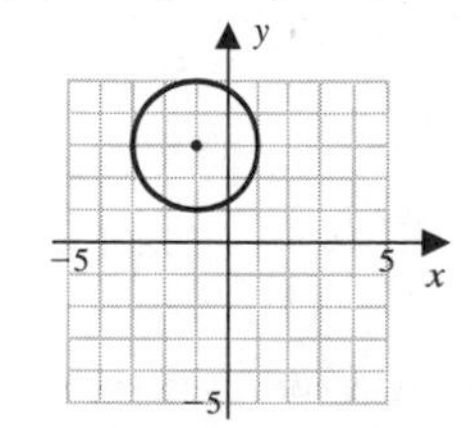

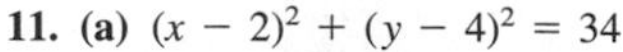

11. **(a)** $(x - 2)^2 + (y - 4)^2 = 34$

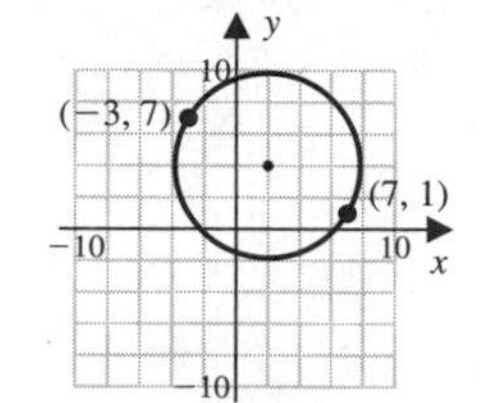

(b) $(x + 1)^2 + (y - 2)^2 = 34$

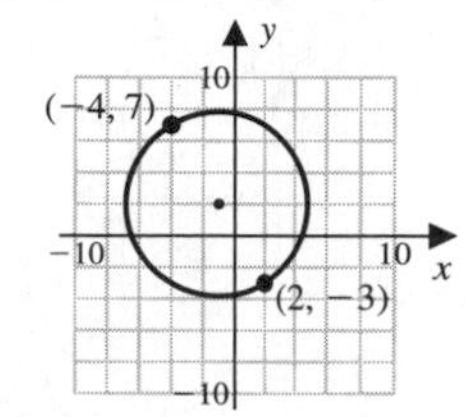

13. **(a)** $(x-1)^2+(y-0)^2=1$ **(b)** $(x+1)^2+(y+1)^2=1$ **15.** Center at $(1,-2)$; Radius $=5$
17. Center at $(-2,1)$; Radius $=4$ **19.** Center at $\left(\frac{3}{2},-2\right)$; Radius $=\frac{3}{2}$ **21.** Center at $(2,3)$; Radius $=\sqrt{13}$
23. Center at $\left(-\frac{25}{2},-5\right)$; Radius $=\frac{1}{2}\sqrt{677}$

25. **(a)**

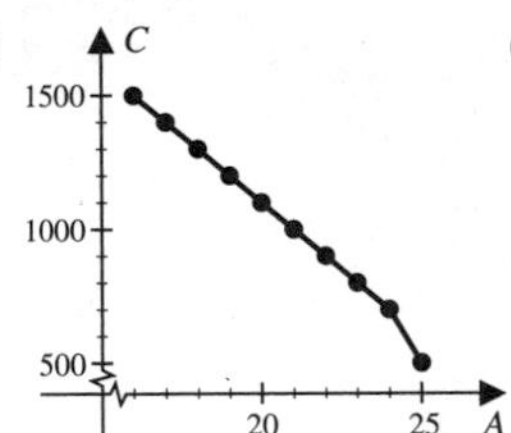

(b) As the temperature increases, so does the humidity.

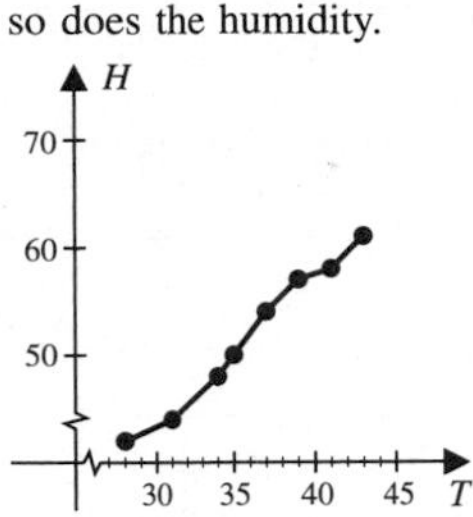

27. **(a)**

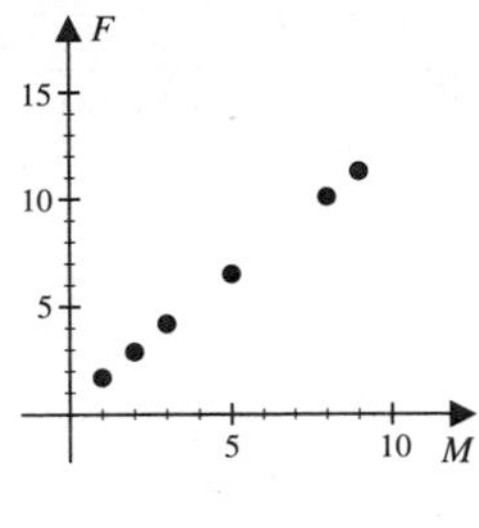

(b) As the miles increase, so does the fare.

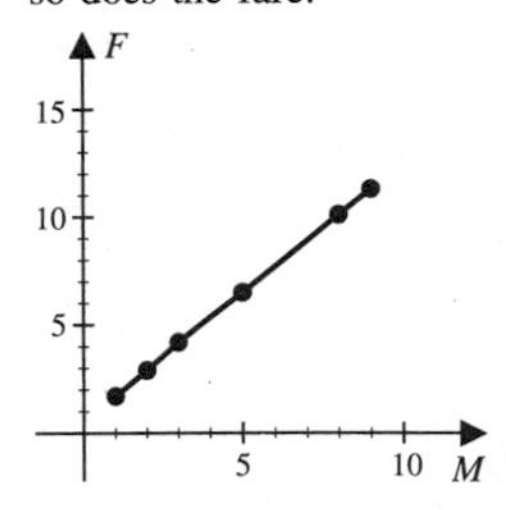

29. 32 **31.** **(a)** $y=x^2$ **(b)** y is multiplied by a factor of 9. **(c)** y is multiplied by a factor of 16.

33. **(a)**

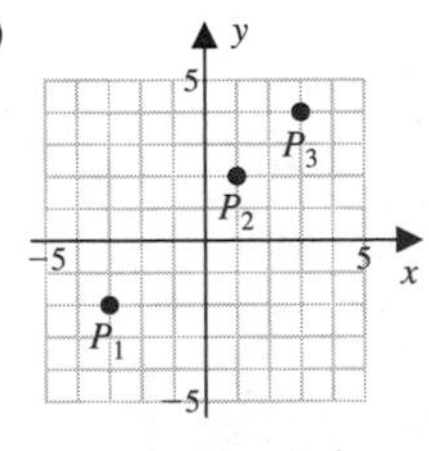

(b) As age increases, cost decreases.

35. **(a)** $V=\frac{20}{\sqrt{22}}\sqrt{d}$ **(b)** $V=56.08$ mph, exceeds speed limit.
37. **(a)** $A=0.7N$ **(b)** 420,000 tons
39. **(a)** IV **(b)** I **(c)** III **(d)** II
41. **(a)** $P=(1,5)$ **(b)** Yes

43. **(a)**

(b) Since $d(P_1,P_2)+d(P_2,P_3)=d(P_1,P_3)$, points are collinear.

45. **(a)** Origin **(b)** No graph **(c)** Circle of radius r centered at origin
47. $(x-6)^2+(y-1)^2=64$ **(a)** Outside **(b)** Inside **(c)** On circle
49. **(a)** (i) $y=\sqrt{16-x^2}$ (ii) $y=-\sqrt{16-x^2}$ (iii) $x=\sqrt{16-y^2}$ (iv) $x=-\sqrt{16-y^2}$
(b) (i) $y=\sqrt{5-x^2+4x}$ (ii) $y=-\sqrt{5-x^2+4x}$ (iii) $x=2+\sqrt{9-y^2}$ (iv) $x=2-\sqrt{9-y^2}$

PROBLEM SET 1.3 (ON PAGE 57)

1. $m=\frac{5}{3}$; rises to the right **3.** $m=0$; horizontal **5.** $m=-\frac{4}{3}$; falls to the right **7.** $m=\frac{1}{3}$; rises to the right
9. $m=-2$; falls to the right **11.** $m=0$; horizontal **13.** $m=$ undefined; vertical

15.

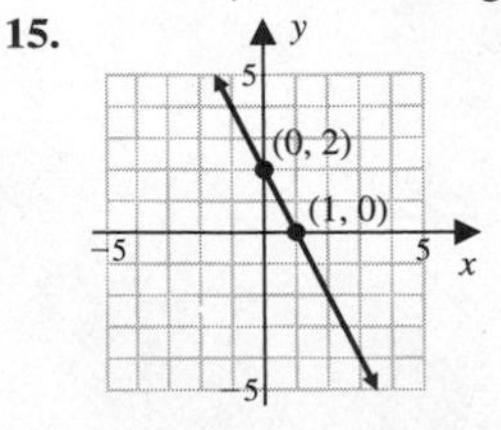

17.

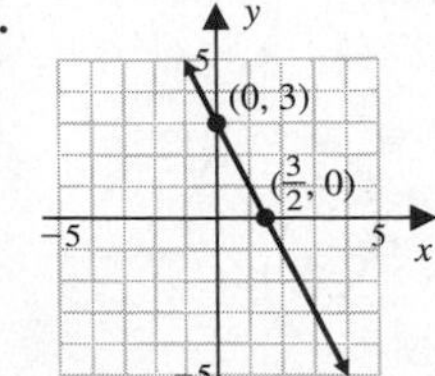

19.

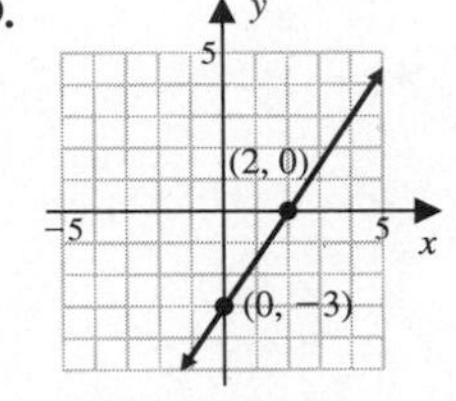

21.

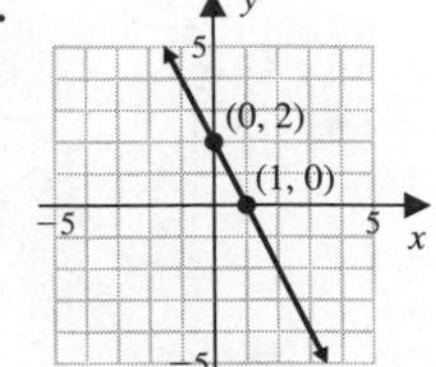

23.

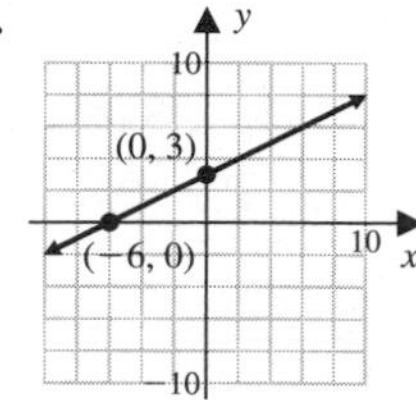

25. No y intercept

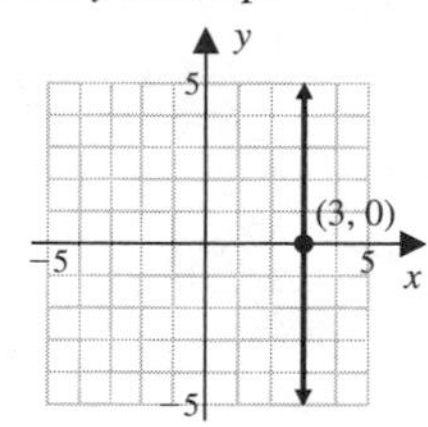

27. No x intercept

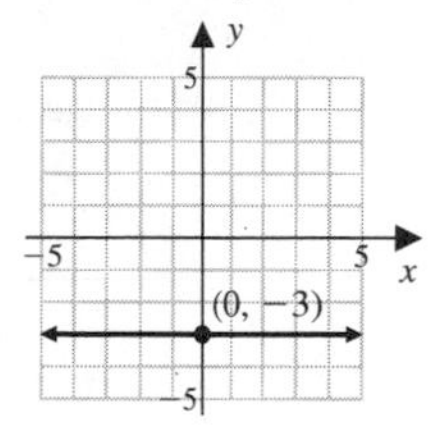

29.

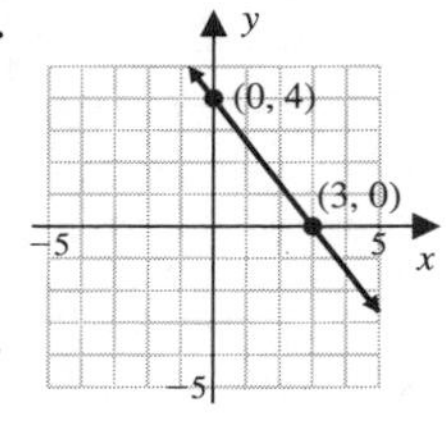

31. **(a)** $y + 3 = 3(x - 4)$

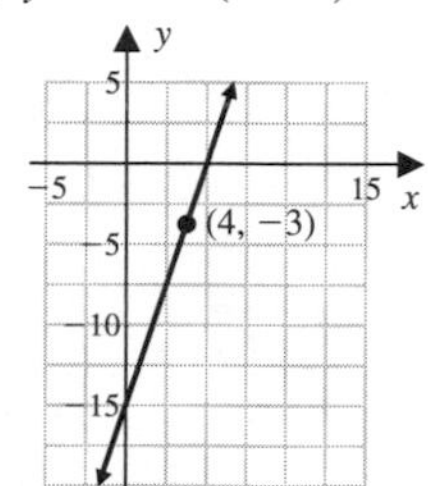

(b) $y - 3 = 0(x + 2)$

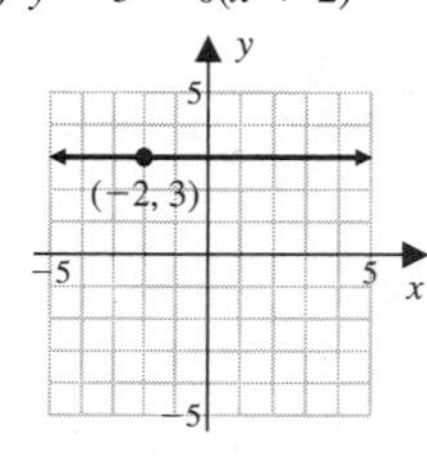

33. **(a)** $y = -\frac{2}{3}x$ **(b)** $y = -2x + 2$

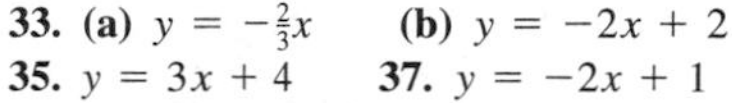

35. $y = 3x + 4$ **37.** $y = -2x + 1$

39. $y = \frac{1}{3}x + \frac{1}{3}$ **41.** $y = -\frac{1}{2}x + \frac{1}{2}$

43. Vertical line, $x = 3$ **45.** 2.32%

47. **(a)** Linear equation **(b)** 0°F corresponds to $-\frac{160}{9}$°C; 0°C corresponds to 32°F. **(c)** $m = \frac{5}{9}$; as F increases by 1, C increases by $\frac{5}{9}$.
(d) Yes; -40

49. **(a)** $C = 80n + 5000$ **(b)** Fixed cost is \$5000 and each printer adds \$80 to the cost.
(c) Rate of increase is large. Each printer adds \$80 to the cost. **(d)** \$21,000 **(e)** \$80 per item; each printer adds \$80 to the cost.

51. **(a)** $N = \frac{5}{3}V + 55$ **(b)** As the speed V increases by 1 ft/sec, N increases by $\frac{5}{3}$ beats/min.
(c) $m = \frac{5}{3}$; heart rate increases $\frac{5}{3}$ times as fast as speed increases. **(d)** 18 ft/sec

53. **(a)** $R = -n + 565$ **(b)** For each dollar increase in rent, one less apartment is rented. **(c)** \$490/month **(d)** 40

55. **(a)** $V = -\frac{2950}{3}t + 6400$ **(b)** $0 \le t \le 6$ **(c)** $m = -\frac{2950}{3}$; value drops about \$983.33 each year. **(d)** \$983.33

57. **(a)** $C = \frac{31}{30}t + \frac{383}{30}$ **(b)** Slope $= \frac{31}{30}$; pollution grows at a rate of 1.03 ppm/yr. **(c)** 24.13 ppm

59. Parallel **61.** $y = -\frac{3}{4}x + \frac{25}{4}$ **63.** Both lines rise from left to right, or both fall from left to right.

65. **(b)** $\dfrac{|P_2'P_3'|}{|P_1'P_3'|}$ **(c)** Slope is independent of any two points on the line.

PROBLEM SET 1.4 (ON PAGE 69)

1.

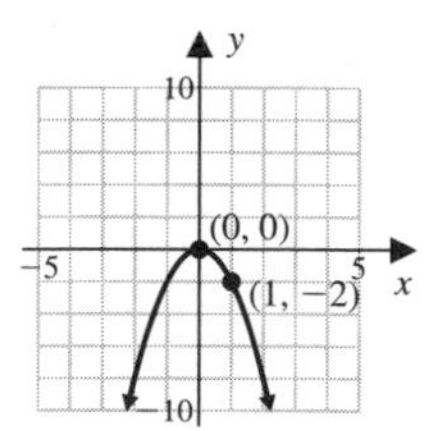

3.

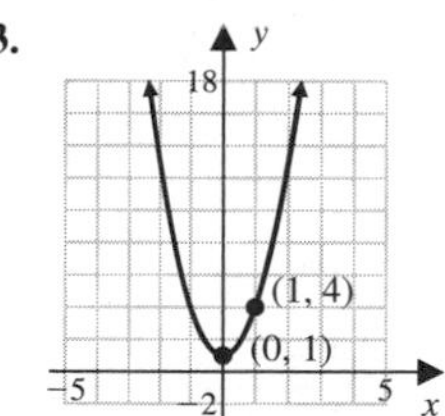

5.

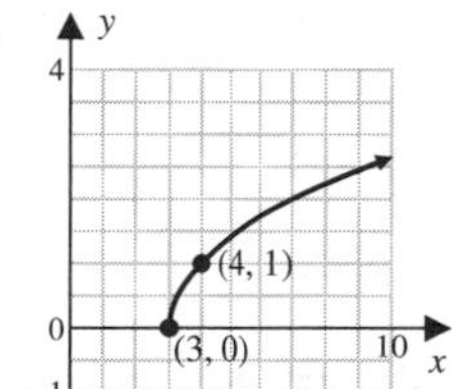

7.

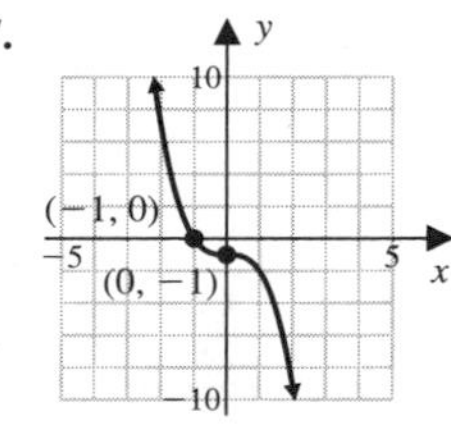

9. $y = \frac{7}{2}x - \frac{5}{2}$

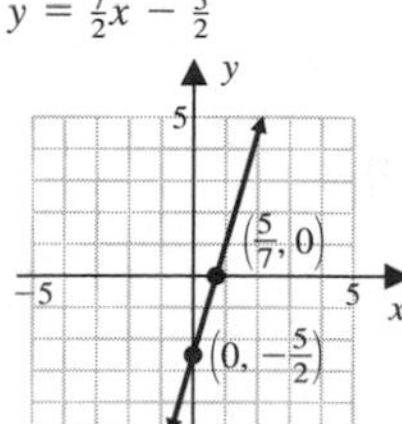

11. $y = \pm\sqrt{\frac{1}{5}(x + 6)}$

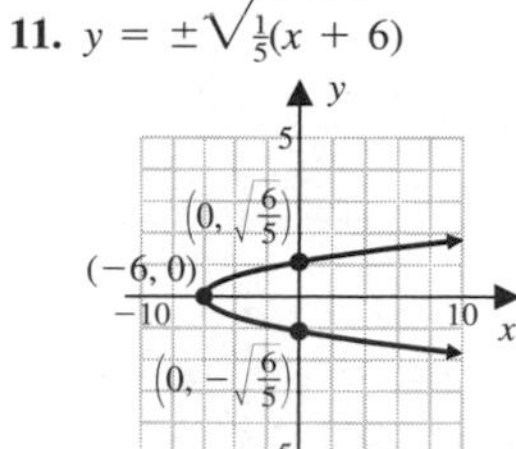

13. **(a)** $Q = (1, -4)$ **(b)** $R = (-1, 4)$ **(c)** $S = (-1, -4)$

15. **(a)** $Q = (-5, -2)$ **(b)** $R = (5, 2)$ **(c)** $S = (5, -2)$

17. **(a)**

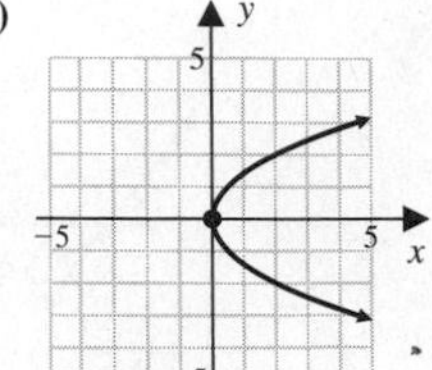

(b)

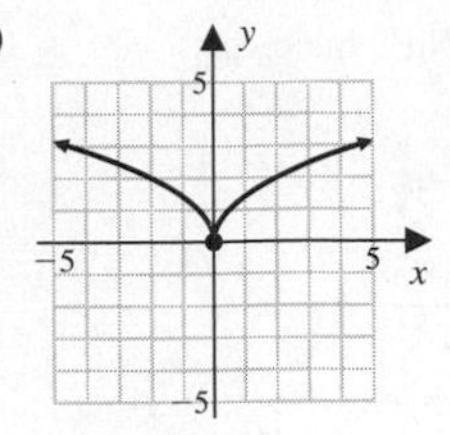

(c)

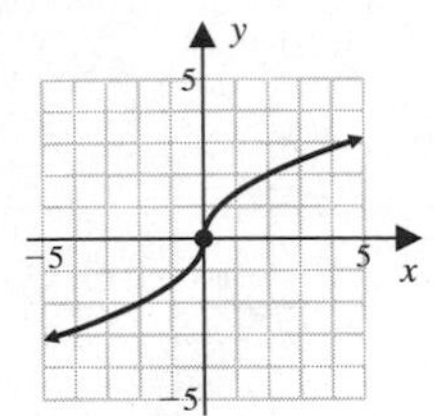

19. **(a)**

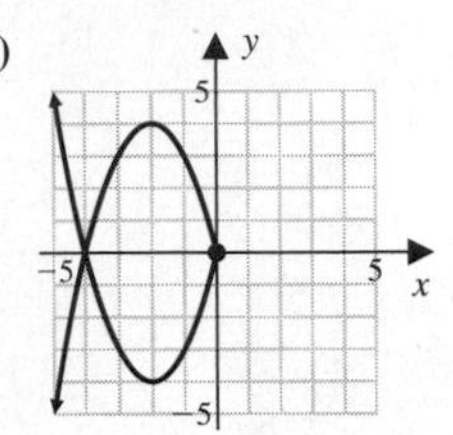

(b)

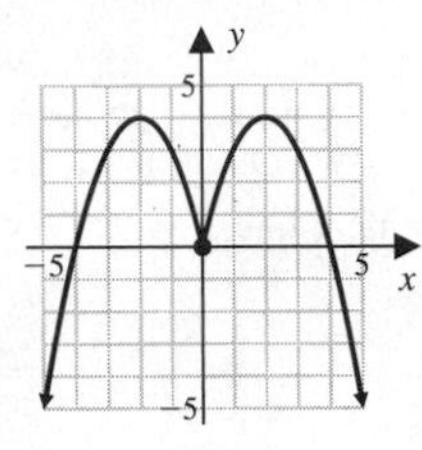

(c)

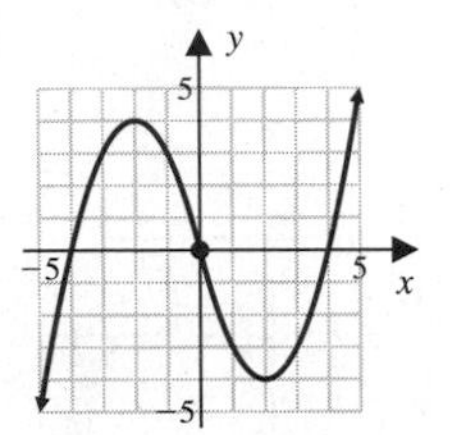

21. **(a)**

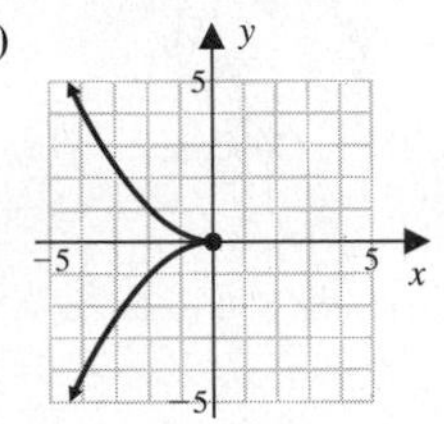

(b)

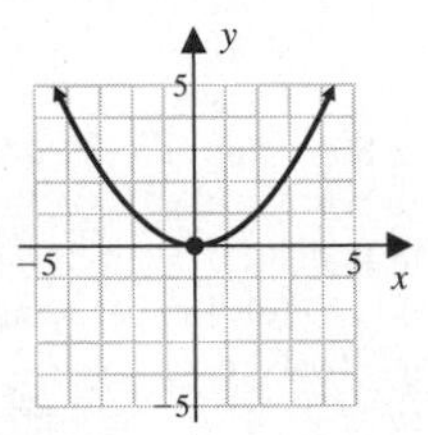

(c)

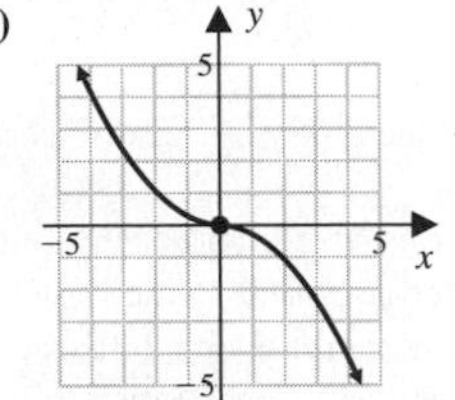

23. Symmetric with respect to the y axis

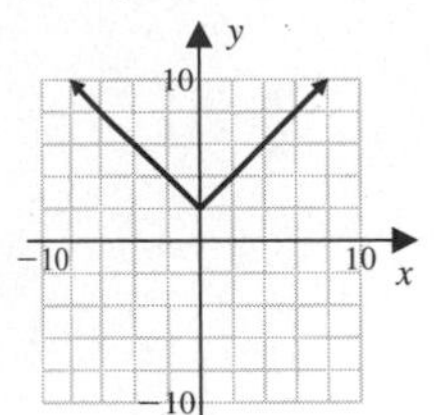

25. Symmetric with respect to the x axis

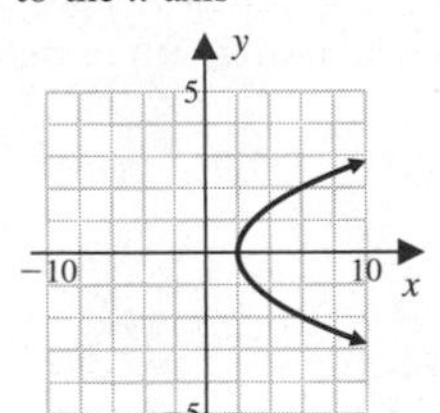

27. Symmetric with respect to the y axis

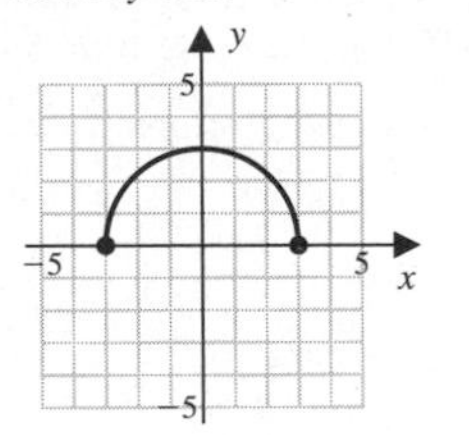

29. Symmetric with respect to the origin

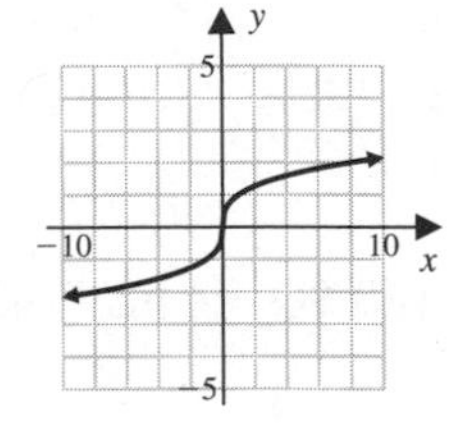

31. E **33.** D **35.** B

37. **(a)**

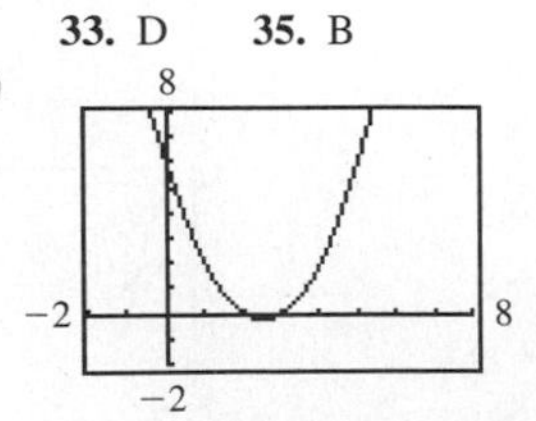

(b)

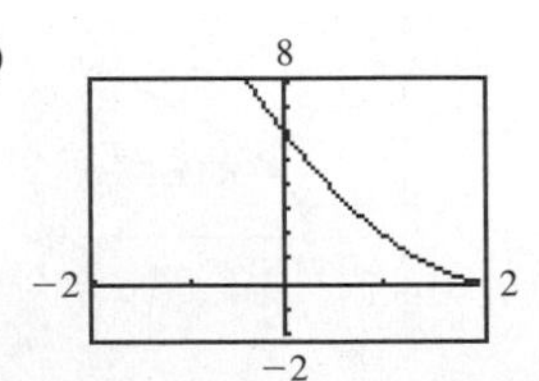

39. **(a)**

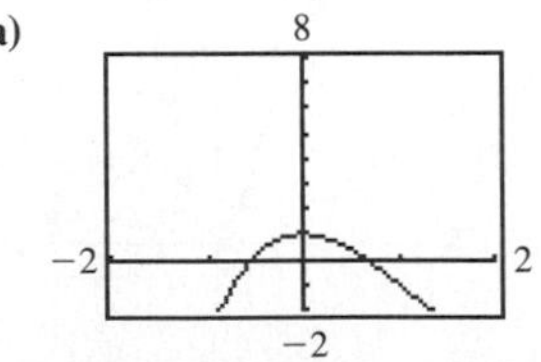

(b)

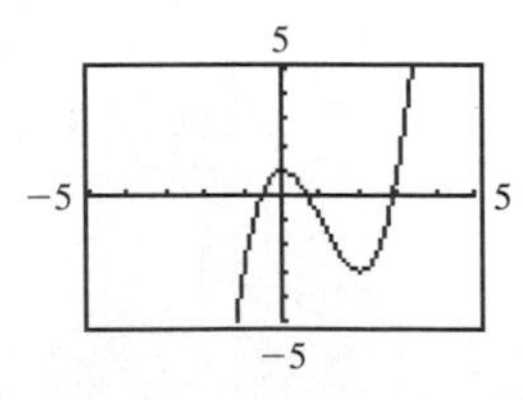

41. Symmetric with respect to the origin

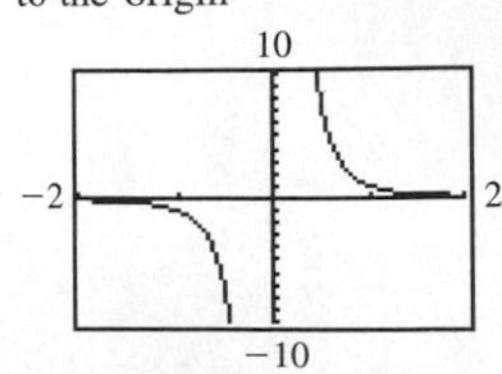

43. No symmetry

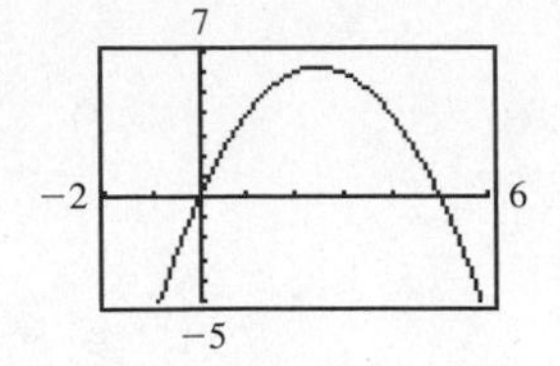

45. Symmetric with respect to the y axis

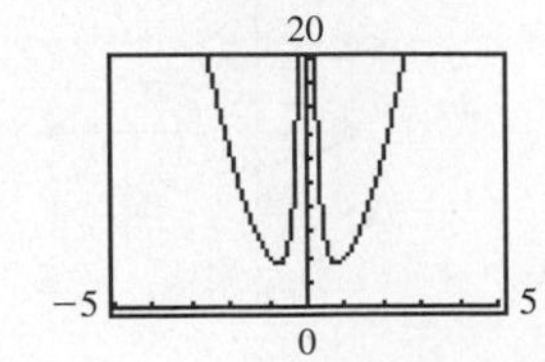

47. $y = -\frac{5}{8}x + \frac{7}{8}$

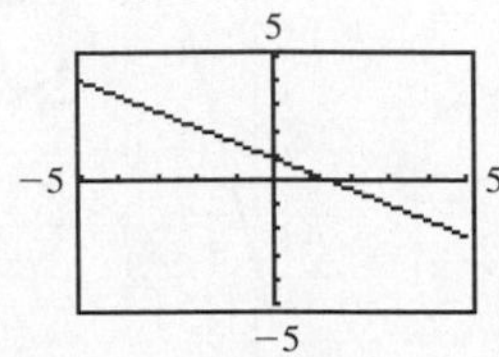

49. $y = x^2 - 4x$

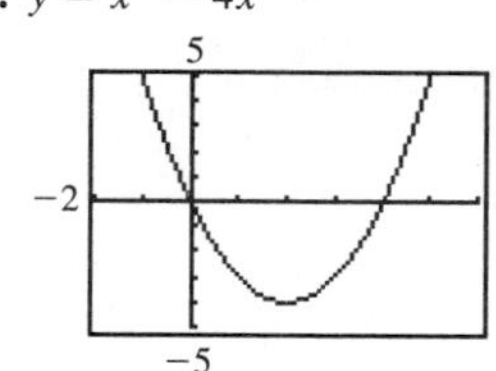

51. $y = \pm\sqrt{\dfrac{x-4}{4}}$

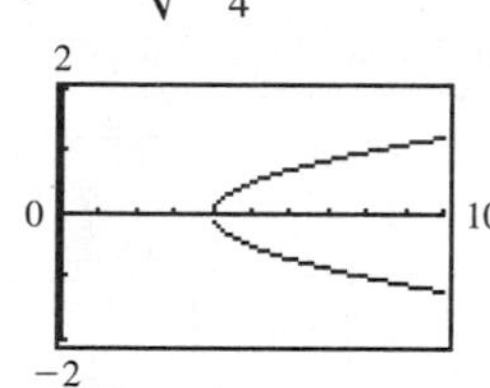

53. $y = \pm 5\sqrt{1 - 4x^2}$

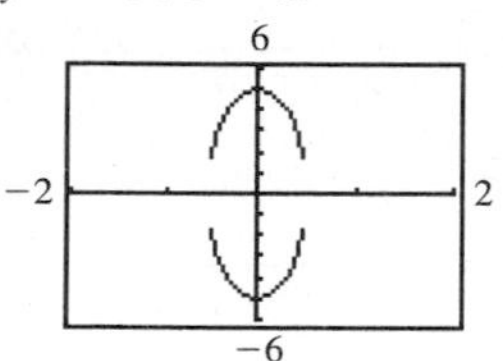

55. **(a)** $w = 6/t$ **(b)** w is halved. **(c)** w is divided by 4.

57. **(a)** $R > 0$ **(b)** $I = 110/R$ **(c)** 13.75 A; 11 A; $9\frac{1}{6}$ A **(d)** Current is higher when resistance is lower.

59. **(a)** $V = \dfrac{2\pi}{3}r^3$ **(b)** $r > 0$ **(c)** 2.09 in.^{3}; 7.07 in.^{3}; 16.76 in.^{3} **(e)** As the radius r increases, the volume V increases.

61. y_1 and y_2 are symmetric with respect to the y axis; y_3 and y_4 are symmetric with respect to the y axis; y_2 and y_4 are symmetric with respect to the x axis; y_1 and y_3 are symmetric with respect to the x axis; y_1 and y_4 are symmetric with respect to the origin; y_2 and y_3 are symmetric with respect to the origin.

63. **(a)** All symmetries **(b)**

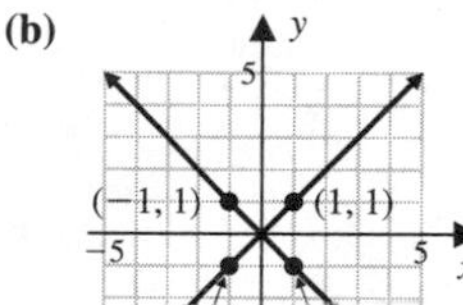

(c) Graph $y = x$ and $y = -x$ in the same viewing window.

PROBLEM SET 1.5 (ON PAGE 82)

1. **(a)** 1 **(b)** −1 **3.** **(a)** 1 **(b)** 1 **5.** 2 **7.** **(a)** 2 **(b)** $\frac{3}{10}$ **9.** −12 **11.** **(a)** −2, 6 **(b)** −12, 6
13. **(a)** −6, 9 **(b)** −10, 25 **15.** **(a)** 1, 7 **(b)** −12, −8 **17.** **(a)** −3, 4 **(b)** −4, 5 **19.** **(a)** −4, 2 **(b)** −3, 2
21. **(a)** −0.73, 2.73 **(b)** −9.69, −0.31 **23.** **(a)** 1, 2.33 **(b)** −1.21, 2.81 **25.** **(a)** −1, 0.5 **(b)** 0.23, 1.43 **27.** −0.52, 5.52
29. **(a)** 9 **(b)** 5 **31.** **(a)** 14 **(b)** 2 **33.** 2, 18 **35.** 6 **37.** 1 **39.** No solution **41.** −9, 3
43. \$4200 at 7%; \$5800 at 6.25% **45.** \$4000 at 5.5%; \$2200 at 5.8% **47.** 30 cc **49.** 9 mph **51.** 3 min **53.** 2 cm
55. **(a)** 0.106 sec; 5.89 sec **(b)** 6 sec **(c)** The rocket reaches its maximum height halfway into its flight.

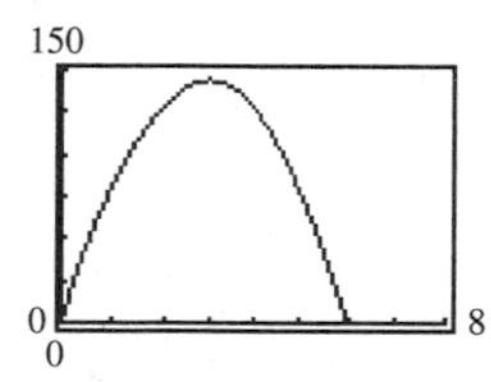

57. **(a)** $x^2 + x$ **(b)** 5 in. by 6 in.

59. **(a)** Two distinct real roots **(b)** Two equal real roots **(c)** No real roots

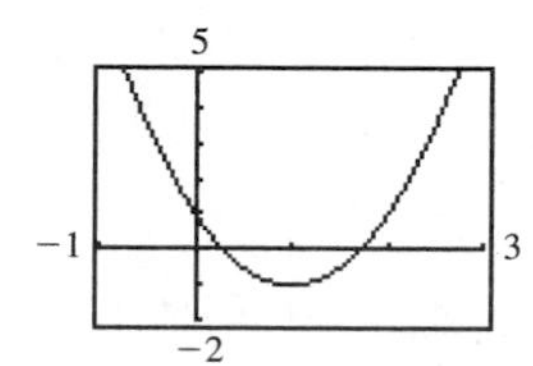

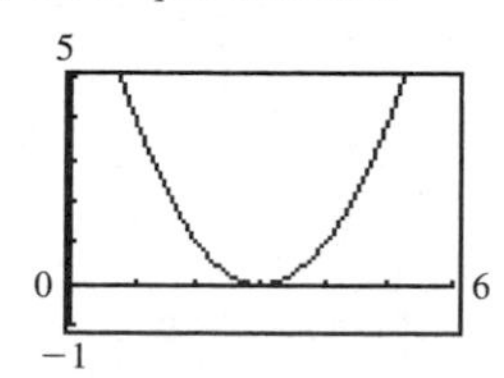

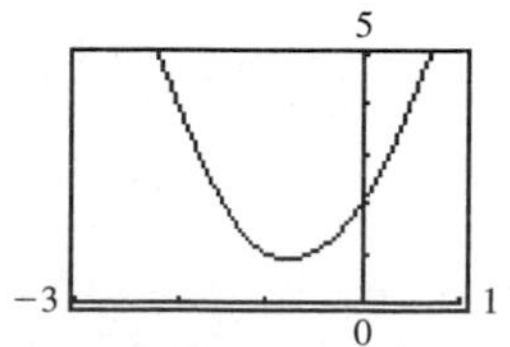

61. **(a)** No solution if $a = 0$ and $c \neq 0$; 0 if $a \neq 0$ and $c = 0$; all real and complex numbers if $a = 0$ and $c = 0$. **(b)** Two real solutions

63. **(a)** $-\frac{1}{2}, 1$ **(b)** $1, 16$ **65.** $-2, 3, \dfrac{1 \pm \sqrt{7}\,i}{2}$

67. **(a)** Graph of $y = \sqrt{x+3} + 5$ never intersects the x axis.

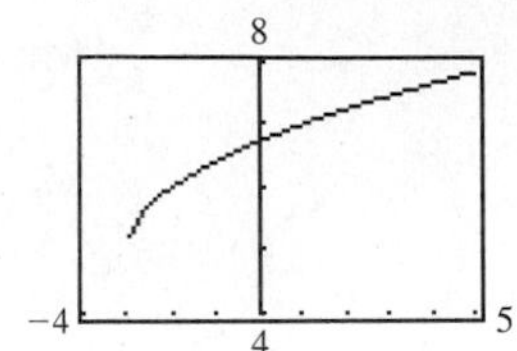

(b) Graph of $y = |x^2 - 4| + 3$ never intersects the x axis.

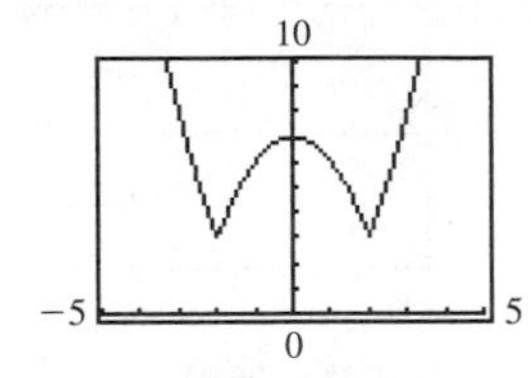

CHAPTER 1 REVIEW PROBLEM SET (ON PAGE 85)

1. **(a)** $>$ **(b)** $<$ **(c)** $>$ **(d)** $=$ **3.** **(a)** $4 \le x \le 5$ **(b)** $-\infty < x \le -2$ **(c)** $1 < x < 3$

5. **(a)** $x > 4$; $(4, \infty)$ **(b)** $x < -2$; $(-\infty, -2)$ **(c)** $x < 0$; $(-\infty, 0)$

7. **(a)** $3 - \sqrt{7}$ **(b)** $\begin{cases} 5x - 3 & \text{if } x \ge \frac{3}{5} \\ 3 - 5x & \text{if } x < \frac{3}{5} \end{cases}$ **(c)** $\begin{cases} 1 - 7x & \text{if } x \le \frac{1}{7} \\ 7x - 1 & \text{if } x > \frac{1}{7} \end{cases}$

9. **(a)** $2 < t < \frac{8}{3}$; $\left(2, \frac{8}{3}\right)$ **(b)** $x \ge 2$ or $x \le -\frac{10}{3}$; $\left(-\infty, -\frac{10}{3}\right] \cup [2, \infty)$ **(c)** $x \le -\frac{3}{2}$; $\left(-\infty, -\frac{3}{2}\right]$

11. **(a)** (i) $d = 2\sqrt{10}$ (ii) $(3, -2)$ **(b)** (i) $d = 3\sqrt{10}$ (ii) $\left(\frac{3}{2}, \frac{1}{2}\right)$

13. **(a)** Center at $(-1, -3)$; Radius $= 4$ **(b)** Center at $(3, -1)$; Radius $= 6$ **15.** **(a)** 6 **(b)** 48 **17.** **(a)** $\frac{3}{2}$ **(b)** $-\frac{2}{5}$

19. **(a)** x intercept $= -3$; y intercept $= \frac{3}{2}$; $m = \frac{1}{2}$ **(b)** x intercept $= \frac{5}{2}$; y intercept $= \frac{5}{3}$; $m = -\frac{2}{3}$ **21.** **(a)** 21 **(b)** -2

23. **(a)** $y = 3x - 7$ **(b)** $y = 4x - 14$ **25.** $-\frac{1}{6}$

27. **(a)** x intercept $= -5$;
y intercept $= 5$;
no symmetry

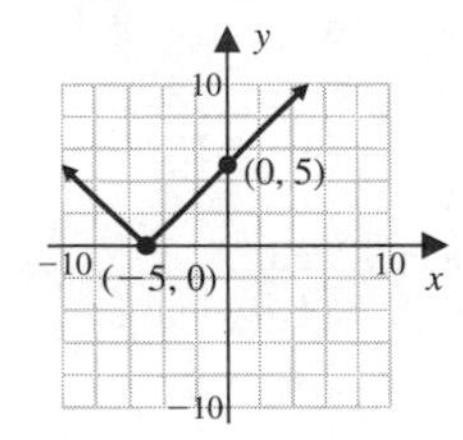

(b) x intercepts $= -1, 0, 1$;
y intercept $= 0$;
symmetric with respect to the y axis

29. **(a)** $y = \pm\sqrt{1 - \dfrac{x}{4}}$

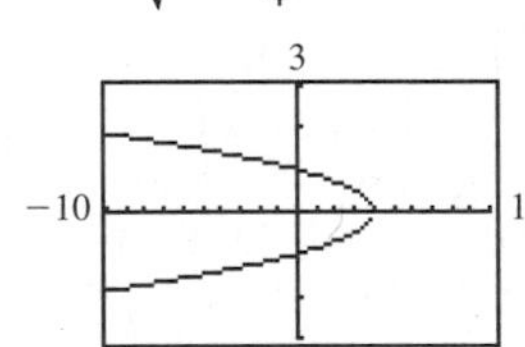

(b) $y = \pm\sqrt{1 - x - x^2}$

31. **(a)** 5 **(b)** $\frac{3}{2}$ **(c)** $-\frac{2}{3}, 2$ **(d)** $-\frac{1}{2}, 1$ **33.** **(a)** $-1, \frac{7}{3}$ **(b)** 0 **(c)** $-3, 5$ **(d)** $-2.83, 0.83$

35. **(a)**

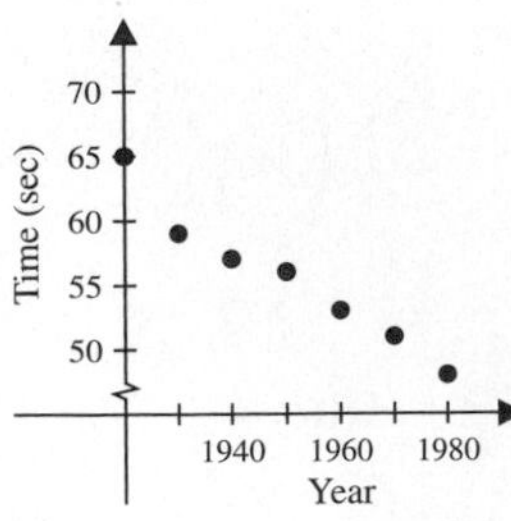

(b) As years go by, record times have been dropping.

37. **(a)** $F = \frac{9}{5}C + 32$ **(b)** 125.6°F **(d)** For each increase of 1°C, the Fahrenheit temperature increases 1.8°F. **39.** \$14,000

41. 75 boys **43.** **(a)** $0 \le t \le 7$ **(c)** 192 ft **(d)** At 7 sec **(e)** 1.70 sec

CHAPTER 1 TEST (ON PAGE 88)

1. (a) (i) $-2 < x \leq 5$

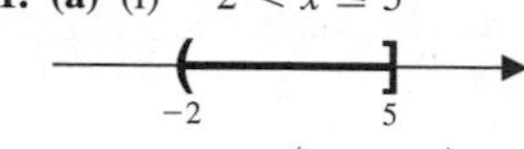

(ii) $x \leq 4$

4

(b) (i) $[-2, 4)$

−2 4

(ii) $[2, \infty)$

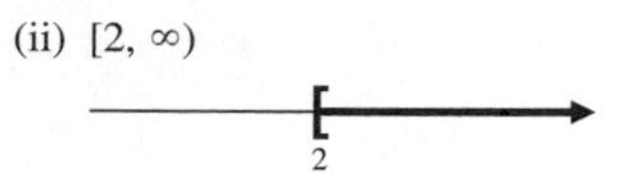

2. (a) $x < -\frac{6}{5}$; $\left(-\infty, -\frac{6}{5}\right)$

$-\frac{6}{5}$

(b) $-12 \leq x \leq 2$; $[-12, 2]$

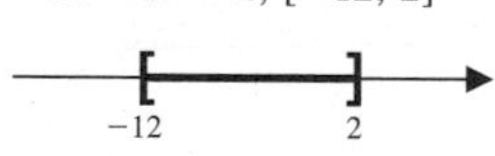

(c) $x > 2$ or $x < -\frac{2}{3}$; $\left(-\infty, -\frac{2}{3}\right) \cup (2, \infty)$

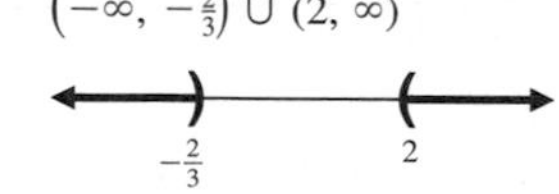

3. Center: $(-3, 1)$; Radius: 5

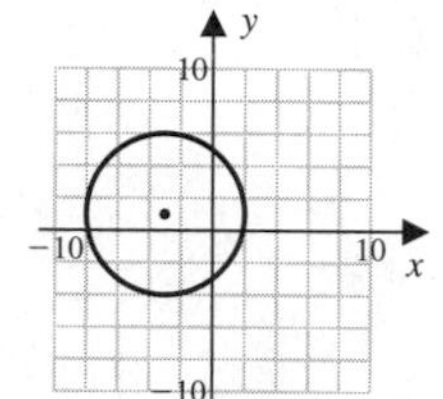

4. (a) $d = \sqrt{13}$
(b) $\left(-4, \frac{5}{2}\right)$
(c) $\frac{3}{2}$
(d) $y = \frac{3}{2}x + \frac{17}{2}$
(e)

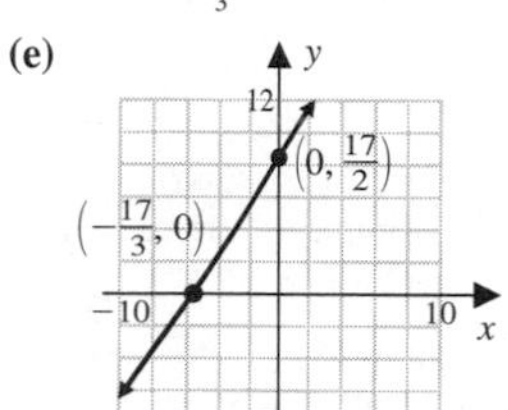

(f) (i) $y - 4 = -3(x + 3)$
(ii) $y - 4 = \frac{1}{3}(x + 3)$

5. (a) No symmetry with respect to the x axis, y axis, or origin; x intercept $= 2$; y intercept $= 2$

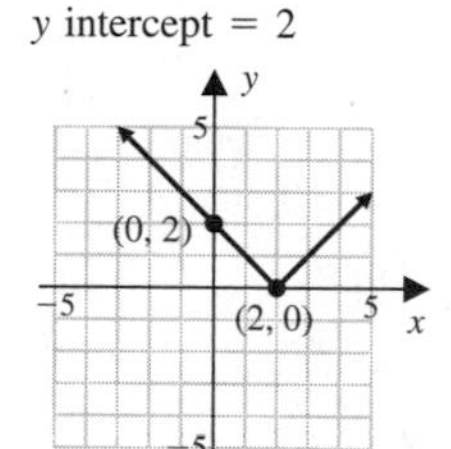

(b) Symmetric with respect to the y axis; x intercept $= 0$; y intercept $= 0$

(0, 0) (−1, −3) (1, −3)

(c) Symmetric with respect to the origin; x intercept $= 0$; y intercept $= 0$

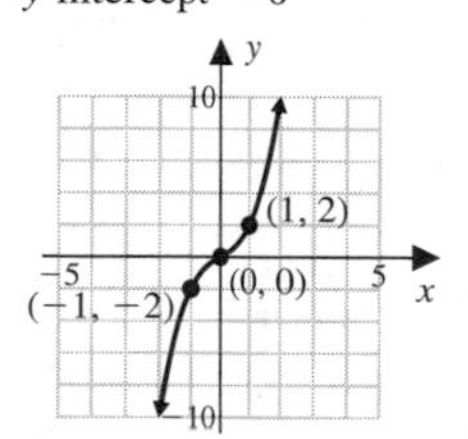

6. (a) $\frac{5}{2}$

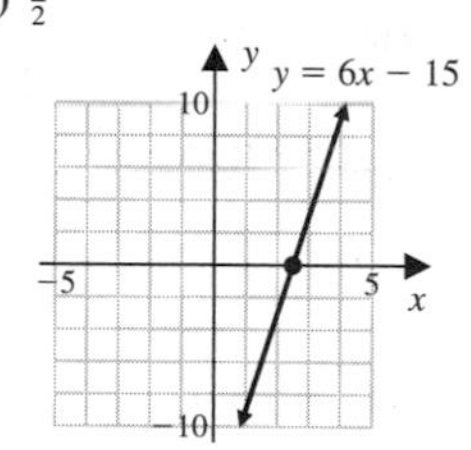

(b) $-5, 1$

$y = (x + 6)(x - 2) + 7$

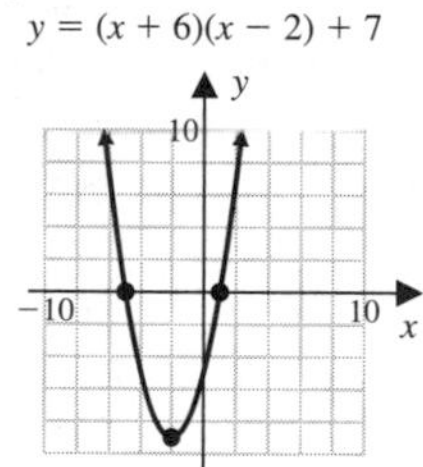

(c) $-\frac{7}{3}, 1$

$y = |3x + 2| - 5$

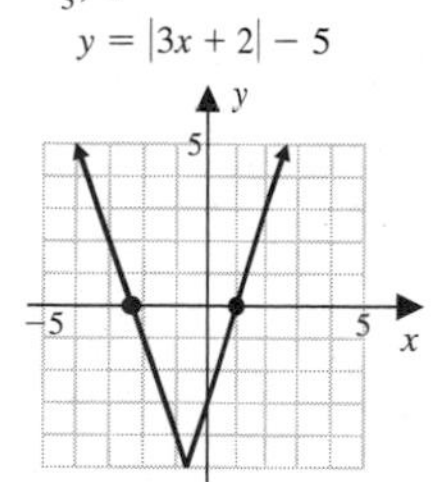

(d) 6

$y = \sqrt{x - 2} - (x - 4)$

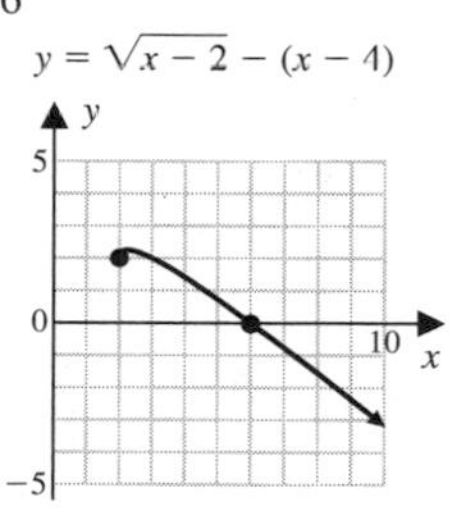

7. \$14,250 **8.** 6 ft^3 **9.** \$2500 at 4%; \$5500 at 7%

CHAPTER 2: PROBLEM SET 2.1 (ON PAGE 99)

1. (a) A function **(b)** Not a function **3. (a)** Yes **(b)** No **5. (a)** -7 **(b)** 13 **(c)** $\frac{3}{2}$ **(d)** $\frac{9}{2}$
7. (a) 23 **(b)** -1 **(c)** $-\frac{2}{9}$ **(d)** $\frac{46}{9}$ **9. (a)** -3 **(b)** $\frac{1}{3}$ **(c)** $-\frac{4}{3}$ **(d)** 0 **11. (a)** 4 **(b)** 5 **(c)** 4.502 **(d)** 4.133
13. (a) 1 **(b)** 5 **(c)** 2 **(d)** 8 **15. (a)** $-2a - 11$ **(b)** $-2a - 16$ **(c)** $\frac{-5a - 23}{a + 3}$ **(d)** $\frac{13}{2a + 11}$
17. (a) $2a^2 - a - 2$ **(b)** $2a^2 - 5a - 1$ **(c)** $-4a^2 + 1$ **(d)** $2a^4 - 5a^2 + 1$
19. (a) $3 - 14|a|$ **(b)** $21 - 14|a|$ **(c)** $3 - 2a^2$ **(d)** $3 - 2a^2$ **21. (a)** $\frac{y + 3}{3y + 10}$ **(b)** $\frac{y + 6}{4}$ **(c)** $\frac{6y + 18}{y + 9}$ **(d)** $\frac{3y + 9}{2y + 18}$

23. **(a)** 4 **(b)** $2t + h$ **25.** **(a)** $\dfrac{-1}{t^2 + ht}$ **(b)** $\dfrac{-3}{(t + h - 1)(t - 1)}$

27. **(a)** Domain: $-2, -1, 0, 1, 2, 3$; Range: $-3, -1, 1, 3, 5, 7$

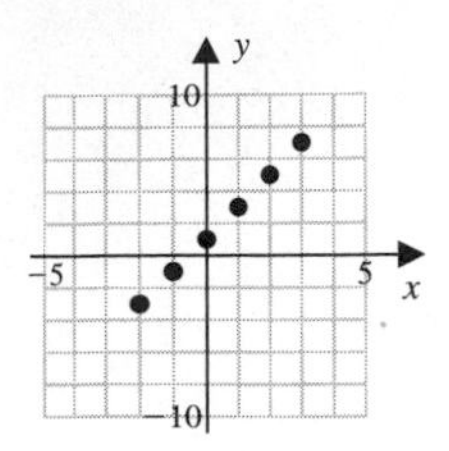

(b) Domain: $-2, -1, 0, 1, 2, 3$; Range: $-8, -3, 0, 1$

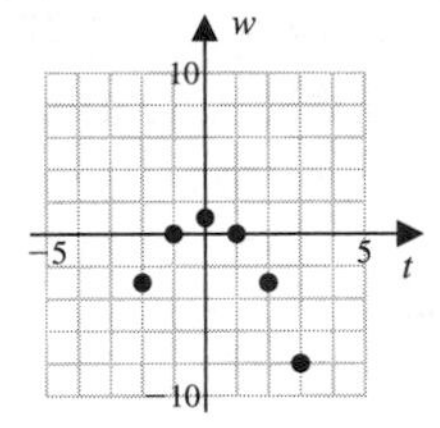

29. **(a)** Straight line; slope $-\frac{1}{2}$; y intercept 3; Domain and range: $\mathbb{R}$
(b) Semicircle; radius = 7; center = (0, 0); Domain: $[-7, 7]$; Range: $[0, 7]$

31. Domain and range: $\mathbb{R}$

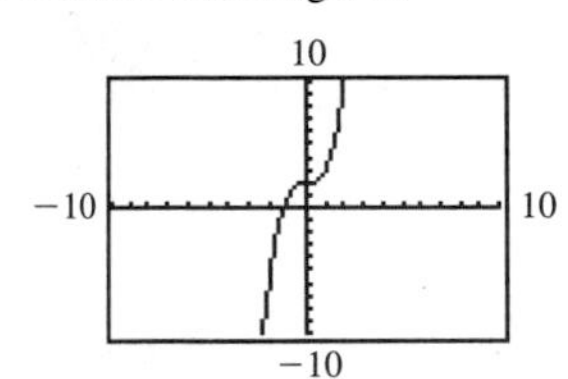

33. Domain: $[5, \infty)$; Range: $[0, \infty)$

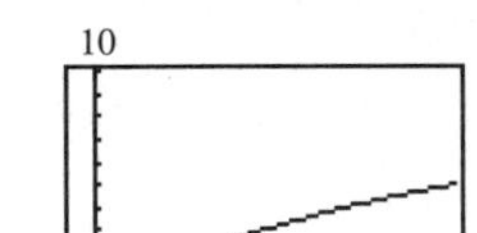
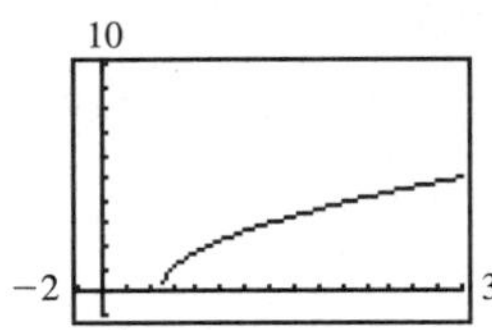

35. Domain: $\mathbb{R}$; Range: $[-5, \infty)$

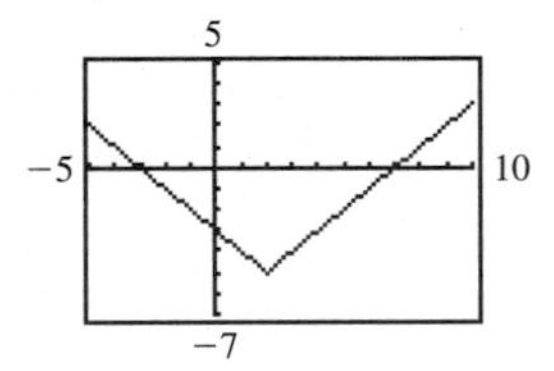

37. **(a)** Yes **(b)** No **39.** **(a)** Not a function of x. **(b)** Function **41.** **(a)** Function **(b)** Function **43.** Domain: $\mathbb{R}$

45. Domain: $x \neq -4$ **47.** Domain: $x \neq \pm 3$ **49.** Domain: $\left[\frac{2}{3}, \infty\right)$ **51.** Domain: $x \neq 0, 1$

53. **(a)** Domain: $\mathbb{R}$; Range: $(-\infty, 5]$ **(b)** Domain:$(-2, 1) \cup (1, 4)$; Range: $(-\infty, -3] \cup [3, \infty)$

55. **(a)** For every value of V there is one and only one value of F. **(b)** Domain: [40, 80]; Range: [27.93, 46.33]
(c) 46.33 mpg; 36.70 mpg; 29.93 mpg **(d)** 50 mph; 63.56 mph; 74.83 mph

57. **(a)** For every value of t there is one and only one value of T.
(b)

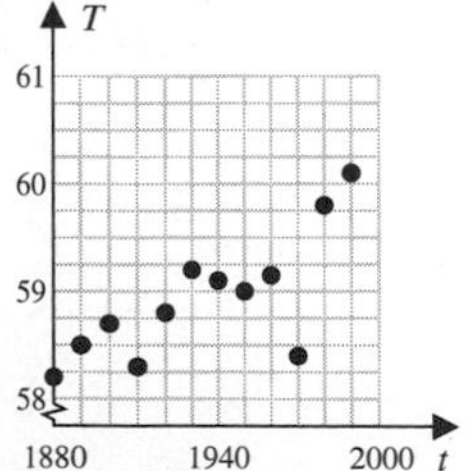

(c) Slowly increasing as time goes on.

59. **(a)** $\ell = 40 - 2w$ **(b)** $A = 40w - 2w^2$ **(c)** $0 \le w \le 20$
(d) As the width increases, the area increases to a highest value (200 m^2), then decreases.

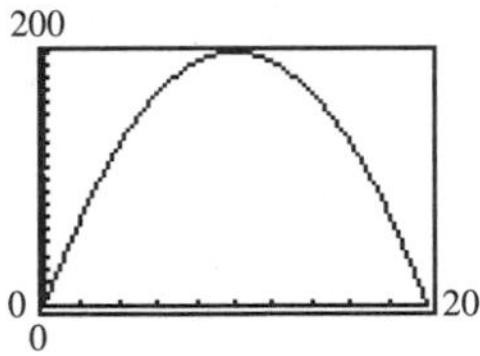

61. **(a)** $C = (\$90{,}000)\sqrt{169 + x^2} + (\$60{,}000)(10 - x)$
(b) $0 \le x \le 10$
(c)

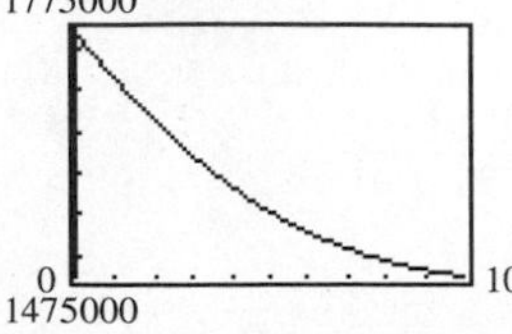

(d) \$1,770,000
(e) \$1,476,109.75
(f) Cost decreases as x increases, cost is the lowest when $x = 10$ km.

63. **(a)** A function **(b)** Not a function **65.** **(a)** Yes **(b)** No **67.** **(a)** -12 **(b)** $327\frac{11}{32}$

69. **(a)** For each calendar date, there may be more than one person. **(b)** Function **71.** Slope is the same as the difference quotient.

73. **(a)–(b)**

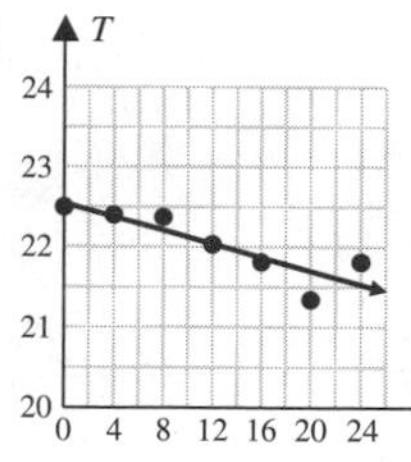

(c) Fairly close
(d) 21.36 sec; 21.19 sec

PROBLEM SET 2.2 (ON PAGE 112)

1. Decreasing: $(-\infty, -2]$ and $[0, 2]$; increasing: $[-2, 0]$ and $[2, \infty)$

3. Decreasing: $[-2, 1]$ and $\left[\frac{5}{2}, \infty\right)$; increasing: $[-4, -2]$ and $\left[1, \frac{5}{2}\right]$; constant: $(-\infty, -4]$

5. Increasing: $[-\pi, -\pi/2)$, $(-\pi/2, \pi/2)$, $(\pi/2, \pi]$

7. **(a)** Even
(b)

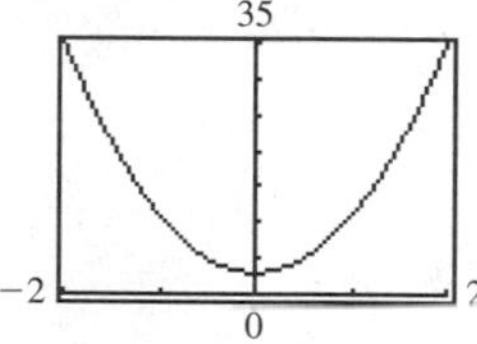

9. **(a)** Even
(b)

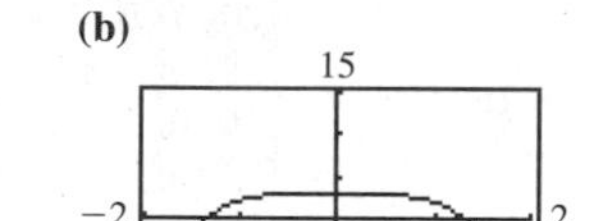

11. **(a)** Odd
(b)

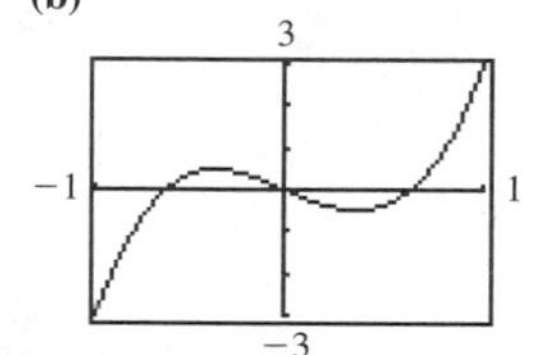

13. **(a)** Even
(b)

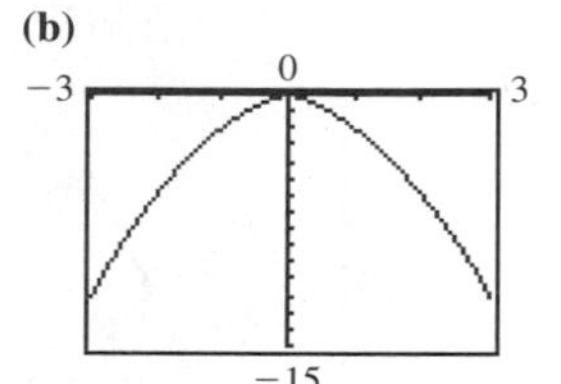

15. **(a)** Neither even nor odd
(b)

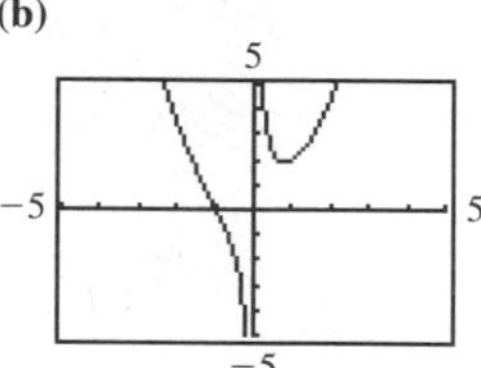

17. Odd; increasing on $\mathbb{R}$

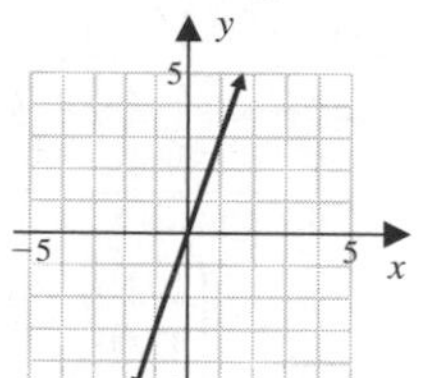

19. Neither; decreasing on $\mathbb{R}$

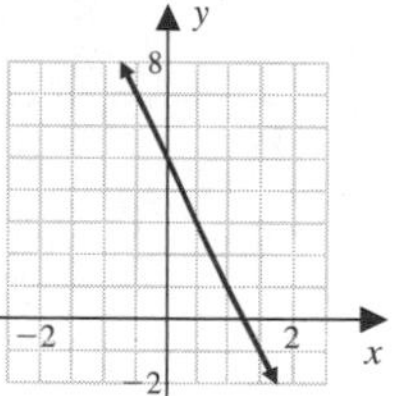

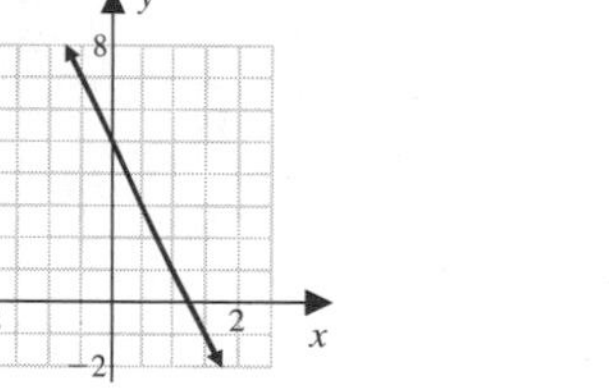

21. Even; decreasing: $[0, \infty)$; increasing $(-\infty, 0]$

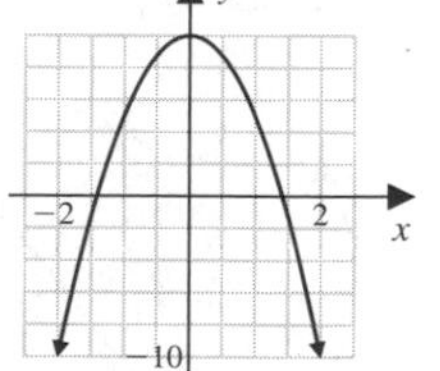

23. Even; decreasing: $[0, 2]$; increasing $[-2, 0]$

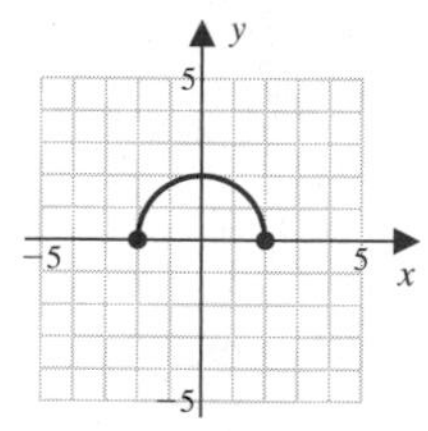

25. Even; decreasing: $[0, \infty)$; increasing $(-\infty, 0]$

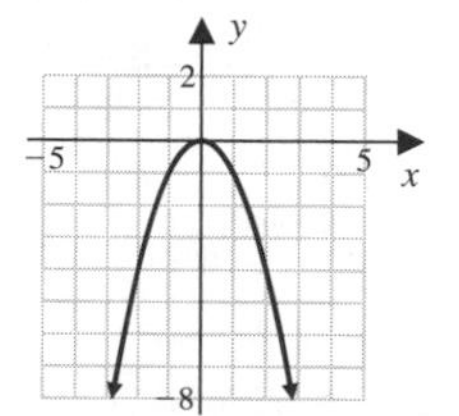

27. $g(x)$ is above $f(x)$ on $(0, 1)$ and below on $(1, \infty)$.

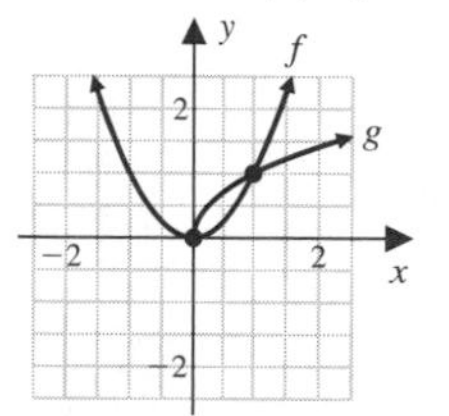

29. $g(x)$ is above $f(x)$ on $(1, \infty)$ and below on $(0, 1)$.

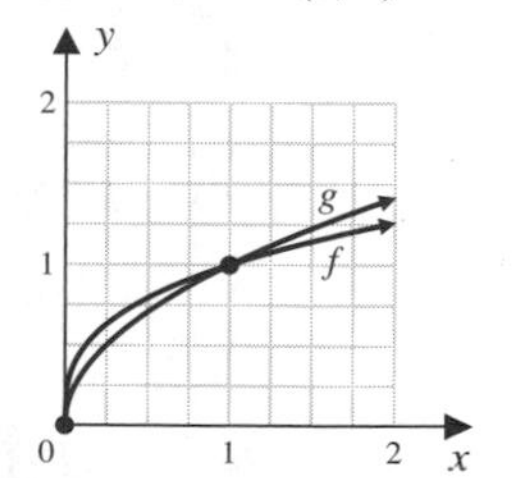

31.

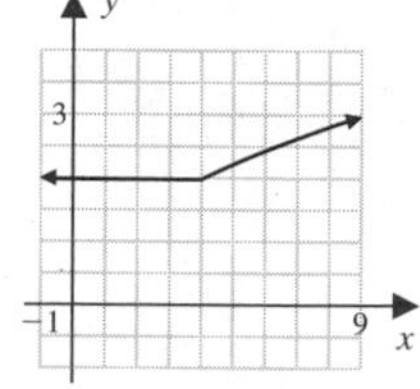

33.

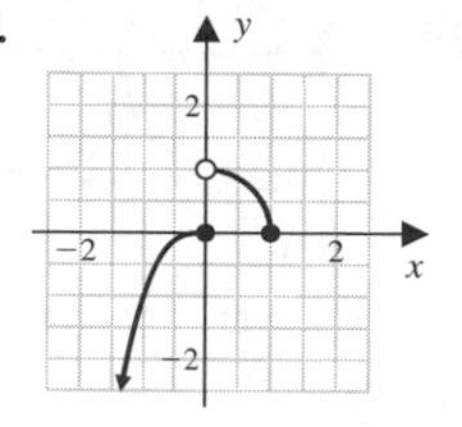

35. Domain: $\mathbb{R}$;
Range: $\mathbb{R}$;
no discontinuity

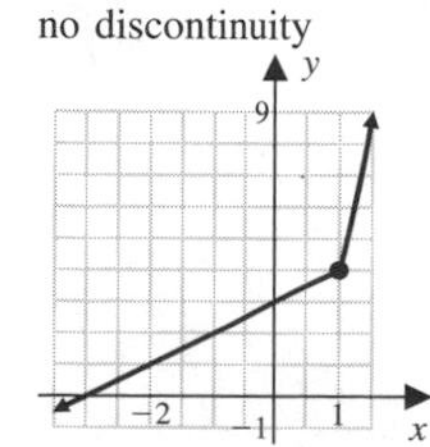

37. Domain: $\mathbb{R}$;
Range: $(-\infty, -1)$ or 0;
discontinuities at −2, 2

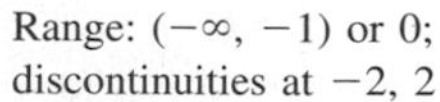

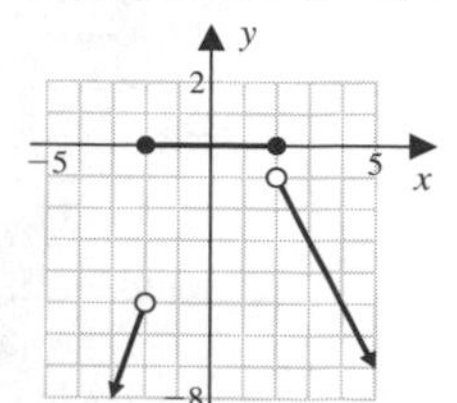

39. Domain: $\mathbb{R}$;
Range: $(-\infty, -7]$
or -2 or $[3, 9)$;
discontinuities at 1, 3

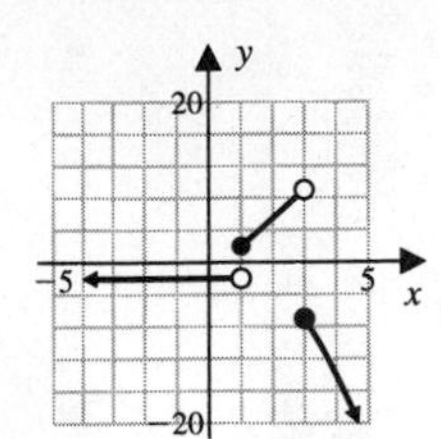

41. Range: all integers

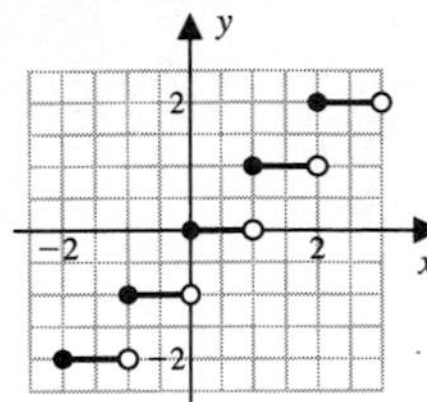

43. Range: all integers

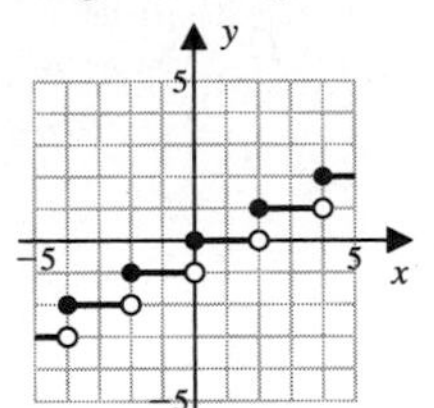

45. **(a)** \$180,000; \$200,000; \$170,000; \$150,000; \$100,000 **(b)** 850 items **(c)** Increasing: [0, 4]; Decreasing: [4, 10]
(d) Lowest: \$100,000 when $x = 10$ (1000 items); Highest: \$200,000 when $x = 4$ (400 items)

47. **(a)**

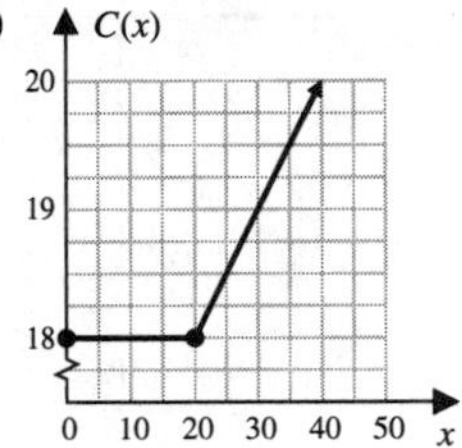

(b) \$18

(c) $K = \begin{cases} 12 & \text{if } 0 \le x \le 15 \\ 12 + 0.25(x - 15) & \text{if } x > 15 \end{cases}$

(d)

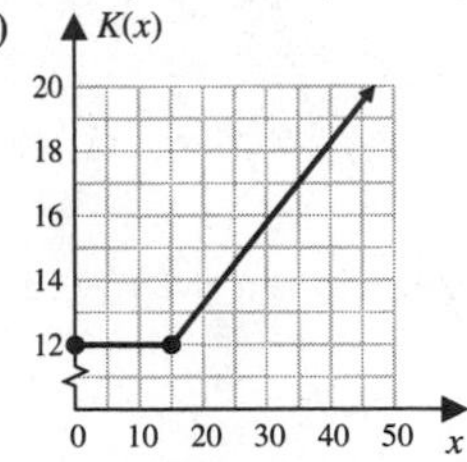

(e) $K < C$ for $x < 51\frac{2}{3}$ hundred gallons;
$C < K$ for $x > 51\frac{2}{3}$ hundred gallons.

49. **(a)** $C(N) = \begin{cases} 20 & \text{if } 0 < N \le 1 \\ 38 & \text{if } 1 < N \le 2 \\ 56 & \text{if } 2 < N \le 3 \\ 74 & \text{if } 3 < N \le 4 \end{cases}$

(b) $0 < N \le 4$

(c)

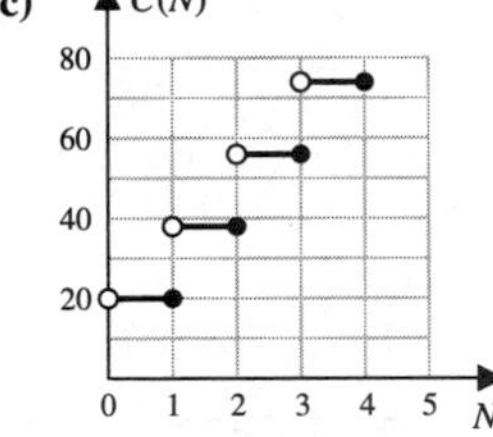

(d) $K(N) = \begin{cases} 19.50 & \text{if } 0 < N \le 1 \\ 39.00 & \text{if } 1 < N \le 2 \\ 58.50 & \text{if } 2 < N \le 3 \\ 78.00 & \text{if } 3 < N \le 4 \end{cases}$

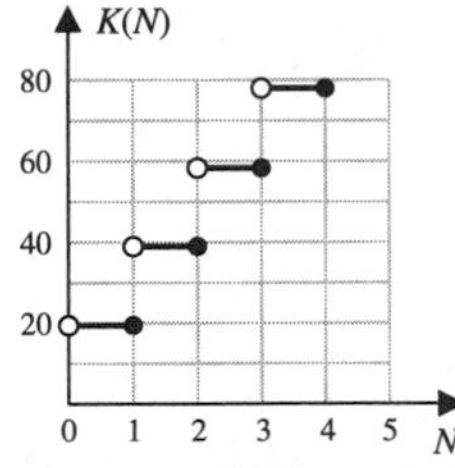

(e) $C < K$ for $N > 1$.

51. **(a)** $R(w) = 0.32 - 0.23[\![1 - w]\!]$
(b) $0 < w \le 6$
(c)

(d) \$1.70; \$2.39; \$2.39

53. **(a)** No; only if $x = 0$ is in the domain.
(b)

(c) Yes

55. (b) No

57. (a)

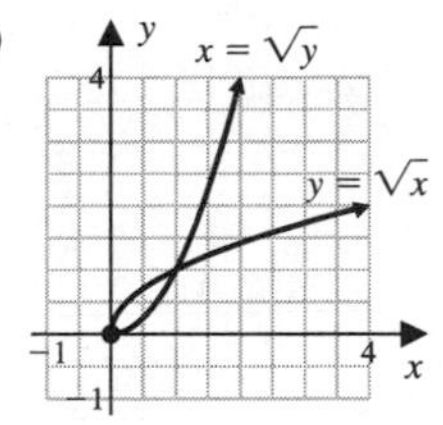

(b) Yes; Domain and Range for both functions: $[0, \infty)$

59. (a)

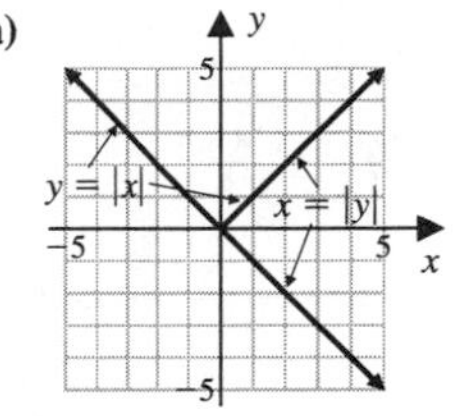

(b) No

61. (a) No, because y has 2 values for $1 < x \leq 5$. **(b)** Yes **(c)** $k \geq 5$

63. (a) No, because there would be 2 values for some x's. (Exception: $y = 0$) **(b)** Yes; consider $y = 0$.

65. (a)

10
−6 2
0

(b)

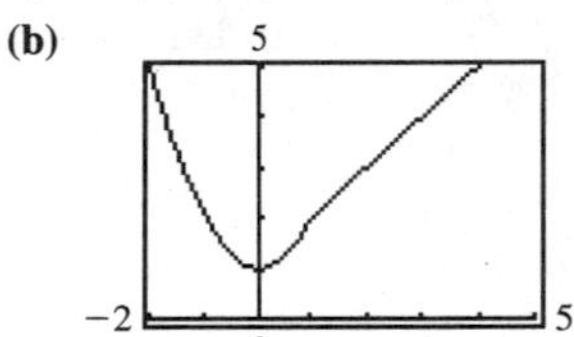

PROBLEM SET 2.3 (ON PAGE 125)

1.

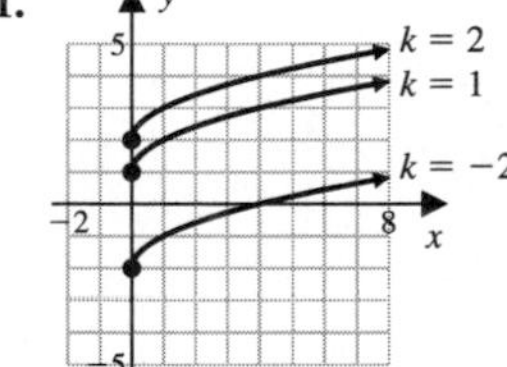

3.

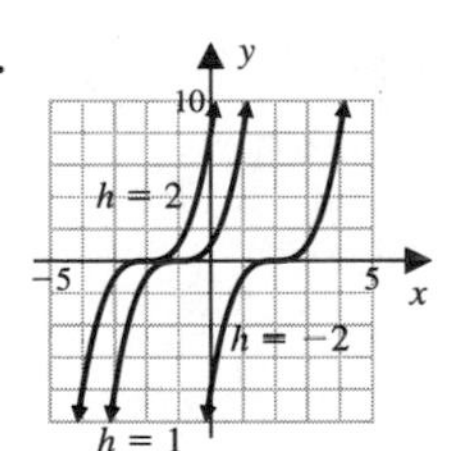

5.

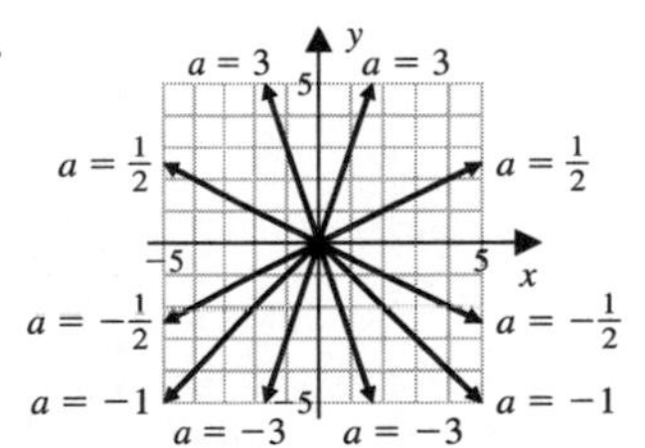

7. Vertically stretch y by 2, then shift up 3 units. **9.** Vertically stretch y by 3, shift to right 1 unit, and shift down 2 units.

11. Vertically stretch y by 2, reflect about the x axis, shift to right 1 unit, and shift down 3 units.

13. (a)

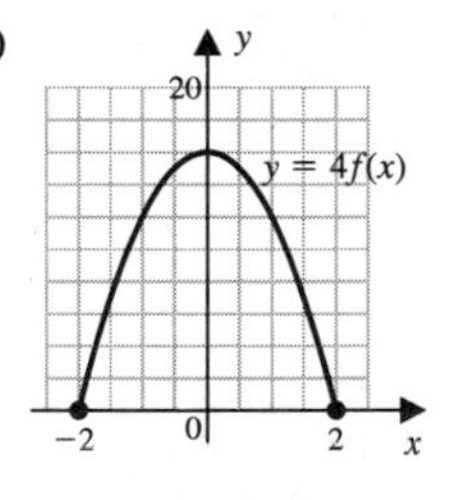

(b)

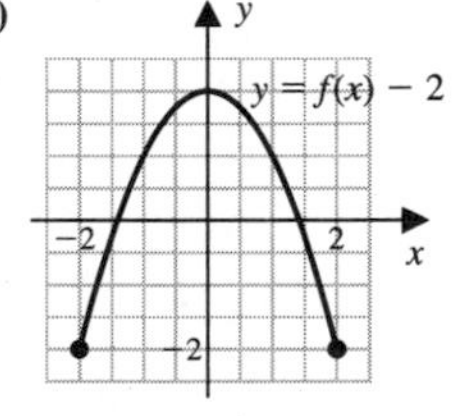

(c)

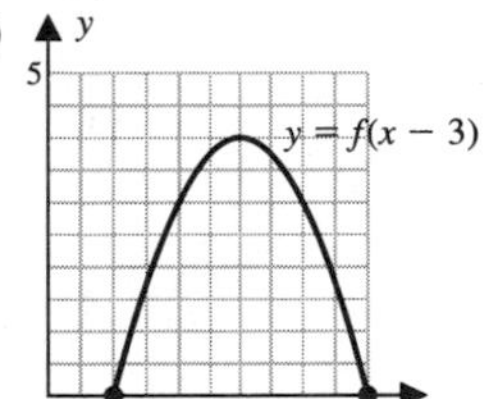

(d)

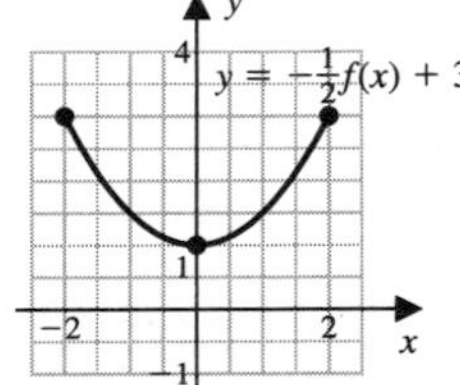

(e)

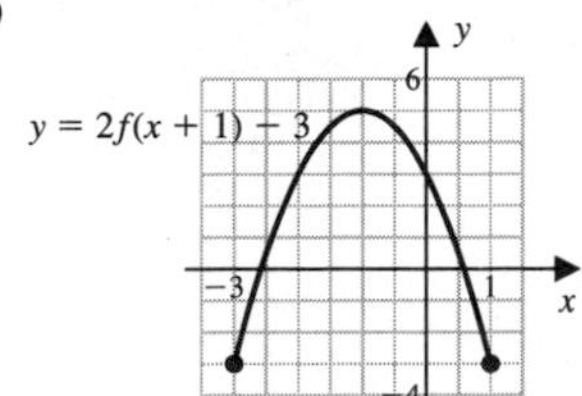

15. $G(x) = |x - 2|$; $T(x) = -|x + 1|$; $S(x) = |x - 2| + 2$

17. $G(x) = x + 3$

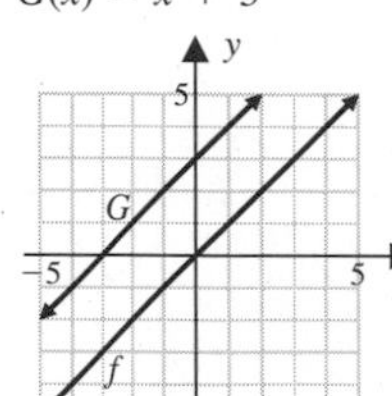

19. $G(x) = (x - 2)^2$

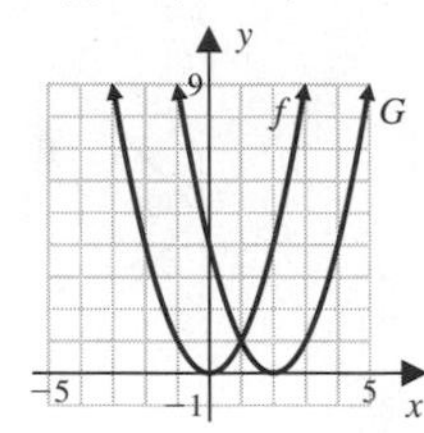

21. $G(x) = -\sqrt{x - 5}$

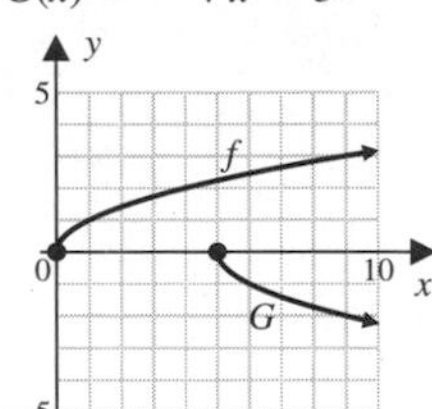

23. $G(x) = \frac{1}{3}[(x + 2)^2 + 1]$

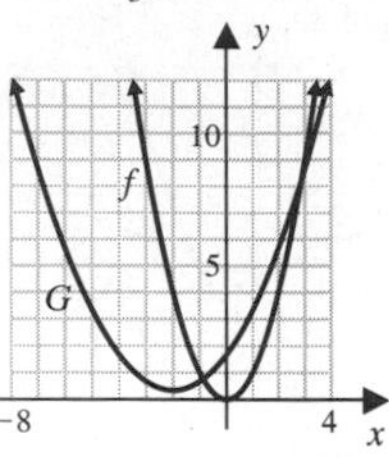

25. Shift the graph of $y = x$ up 2 units; Domain and Range: $\mathbb{R}$ **27.** Reflect $y = x^3$ about the x axis, and shift up 1 unit; Domain and Range: $\mathbb{R}$
29. Shift the graph of $y = \sqrt{x}$ to left 4 units; Domain: $[-4, \infty)$; Range: $[0, \infty)$
31. Vertically stretch y by 2, and shift to right 1 unit; Domain: $\mathbb{R}$; Range: $[0, \infty)$
33. Vertically stretch y by 5, reflect about the x axis, shift to left 1 unit, and shift up 1 unit; Domain: $[-1, \infty)$; Range: $(-\infty, 1]$
35. Vertically compress y by $\frac{1}{3}$, reflect about the x axis, shift to left 1 unit; Domain and Range: $\mathbb{R}$
37. Vertically stretch y by 4, and shift down 1 unit; Domain: $\mathbb{R}$; Range: $[-1, \infty)$
39. Vertically stretch y by 2, shift to right 3 units, and shift up 4 units; Domain: $\mathbb{R}$; Range: $[4, \infty)$
41. Vertically stretch y by 3, shift to right $\frac{1}{2}$ unit, and shift down 1 unit; Domain and Range: $\mathbb{R}$
43. Vertically compress y by $\frac{1}{2}$; Domain: $[-1, 1]$; Range: $\left[0, \frac{1}{2}\right]$
45.

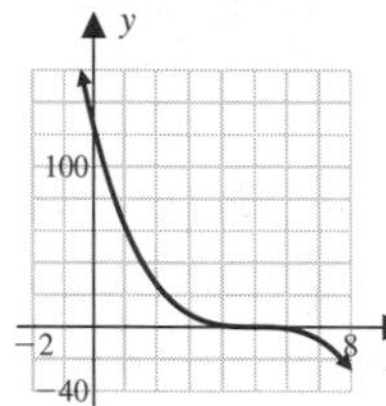

47. Vertically stretch y by 2, reflect about the x axis, shift to left 1 unit, and shift up 3 units.
49. Vertically stretch y by 3, shift to right 2 units, and shift down 4 units. **51.** **(a)** $(a + 2, b)$ **(b)** $(a - 4, b)$ **(c)** $(-a, -b + 1)$
53. **(a)** Yes **(b)** Yes **(c)** Yes if $a > 0$, no if $a < 0$ **55.** Functions (a) and (c) are even; (b) is even if $h \neq 0$.

57. **(a)** $f(x) = \sqrt{-x - 2}$

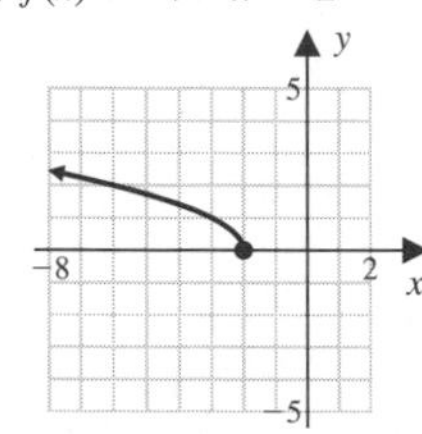

(b) The order of operations is different.
(c) Do transformation in the correct order.

59. **(a)** $G(x) = \sqrt[3]{x - 1}(x + 1)$
(b)

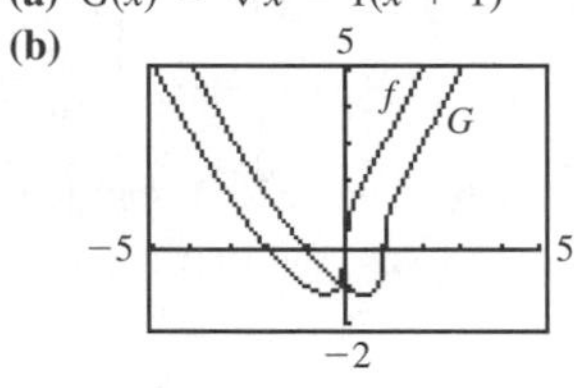

(c) Shift the graph of f to right 1 unit.

61. **(a)** $G(x) = \dfrac{x^2 + 1}{x}$
(b)

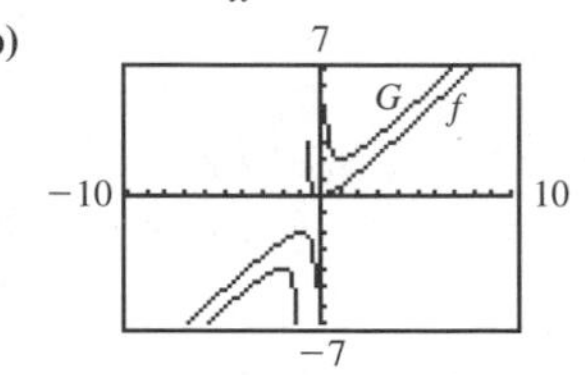

(c) Shift the graph of f to right 1 unit and up 2 units.

PROBLEM SET 2.4 (ON PAGE 134)

1. **(a)** $x - 6$ **(b)** $5x + 8$ **(c)** $-6x^2 - 23x - 7$ **3.** **(a)** $5x^2 - 3x + 3$ **(b)** $-3x^2 - 3x + 1$ **(c)** $4x^4 - 12x^3 + 9x^2 - 3x + 2$
5. **(a)** $3x^3 + 2x + 4$ **(b)** $-x^3 + 8x - 4$ **(c)** $2x^6 + 7x^4 + 4x^3 - 15x^2 + 20x$
7. **(a)** $\dfrac{x^2 - 13x + 6}{-4x^2 + 23x - 15}$ **(b)** $\dfrac{-x^2 - 3x + 6}{-4x^2 + 23x - 15}$ **(c)** $\dfrac{2x}{-4x^2 + 23x - 15}$ **9.** **(a)** $\dfrac{3x + 1}{-2x - 7}, x \neq -\frac{7}{2}$ **(b)** $\dfrac{-2x - 7}{3x + 1}, x \neq -\frac{1}{3}$
11. **(a)** $\dfrac{x + 3}{x^2 - x + 1}, x \neq -1$ **(b)** $\dfrac{x^2 - x + 1}{x + 3}, x \neq -1, -3$
13. **(a)** $\dfrac{(x - 1)\sqrt{2 - x}}{x}, x \leq 2, x \neq 0, 1$ **(b)** $\dfrac{x}{(x - 1)\sqrt{2 - x}}, x < 2, x \neq 1$
15. **(a)** 518 **(b)** 100 **(c)** 398 **(d)** 114 **(e)** 268 **(f)** 1064 **(g)** 2744 **(h)** 394 **(i)** 134 **(j)** −33
17. **(a)** 2 **(b)** −5 **(c)** 3 **(d)** 2 **(e)** 9 **(f)** −3 **19.** **(a)** 15 **(b)** 3 **(c)** −1 **(d)** −15

21. **(a)** $10x - 6$; Domain: $\mathbb{R}$ **(b)** $10x - 3$; Domain: $\mathbb{R}$ **(c)** $f \circ g \neq g \circ f$
23. **(a)** $98x + 5$; Domain: $[0, \infty)$ **(b)** $7\sqrt{2x^2 + 5}$; Domain: $\mathbb{R}$ **(c)** $f \circ g \neq g \circ f$
25. **(a)** x; Domain: $\mathbb{R}$ **(b)** x; Domain: $\mathbb{R}$ **(c)** $f \circ g = g \circ f$
27. **(a)** $\sqrt{x + 4}$; Domain: $[-4, \infty)$ **(b)** $\sqrt{x - 1} + 5$; Domain: $[1, \infty)$ **(c)** $f \circ g \neq g \circ f$
29. **(a)** $\dfrac{2x - 5}{4x - 1}$; Domain: $x \neq \frac{1}{4}, \frac{5}{2}$ **(b)** $\dfrac{9x + 6}{-15x - 8}$; Domain: $x \neq -\frac{2}{3}, -\frac{8}{15}$ **(c)** $f \circ g \neq g \circ f$
31. One possible answer is $g(x) = 5x - 2$; $f(x) = x^3$ **33.** One possible answer is $g(t) = t^2 - 2$; $f(t) = t^{-2}$
35. One possible answer is $g(x) = x + \dfrac{1}{x}$; $f(x) = \sqrt[3]{x}$ **37.** **(a)** $-15x + 7$ **(b)** $x = \frac{1}{3}$
39. **(c)** $[0, \infty)$ **(d)** Use only the first quadrant and the origin. **41.** **(c)** $x \neq 0$ **(d)** Delete $(0, 0)$ from graph.
43. **(a)** $f[g(x)] = \sqrt{2 + x}$ **(c)** $[-2, 2]$ **(d)** Use only part of graph from $x = -2$ to $x = 2$.
45. **(a)** $D(p) = 1.5445p$ (pounds to U.S. dollars); $F(d) = 4.9453d$ (U.S. dollars to French francs); $F[D(p)] = 7.6380p$ (pounds to French francs)
(b) $D(s) = 0.8604s$ (Swiss francs to U.S. dollars); $C(d) = 1.3627d$ (U.S. dollars to Canadian dollars); $C[D(s)] = 1.1725s$ (Swiss francs to Canadian dollars)
47. **(a)** $V = f[g(t)] = 26.52t^3$ in.^{3}/min (approx.); volume increases in proportion to (time)^3.
(b) $S = h[g(t)] = 43t^2$ in.^{2}/min (approx.); surface area increases in proportion to (time)^2.
49. **(a)** $x = g(t) = 55t$; $y = f(x) = \sqrt{90^2 + x^2}$
(b) $f[g(t)] = \sqrt{8100 + 3025t^2}$ is the distance from the ball to first base as a function of time $t \geq 0$.
51. **(a)** $R(x) = 90x$; revenue exceeds cost for $10 < x \leq 45$ units.

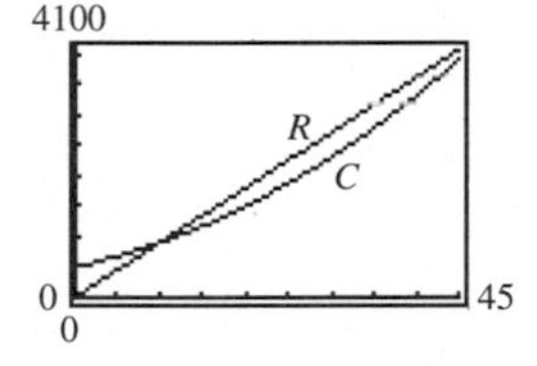

(b) $P(x) = -x^2 + 60x - 500$
A profit occurs for $P > 0$, that is, when x is more than 10 units; a loss occurs when fewer than 10 units are sold.

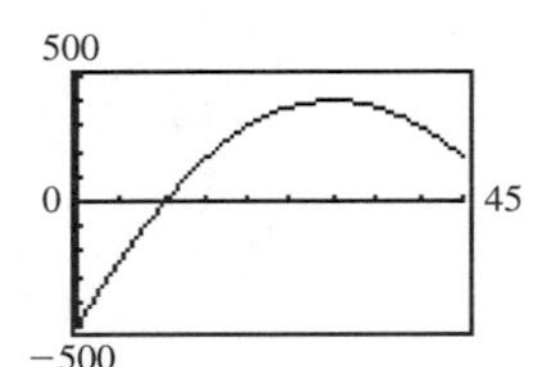

53. **(a)** Yes; as x increases, f and g both increase, so the total $f + g$ increases. **(b)** Let $f(x) = 2x$, $g(x) = 3x$; then $(f - g)(x) = -x$.
55. Yes; all are even. **57.** **(a)** Yes **(b)** Yes **59.** **(a)** $x^{3/5} - 3$ **(b)** $36x^2 - 156x + 173$

PROBLEM SET 2.5 (ON PAGE 146)

1.

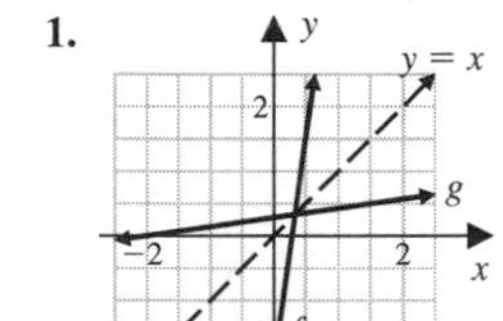

3.

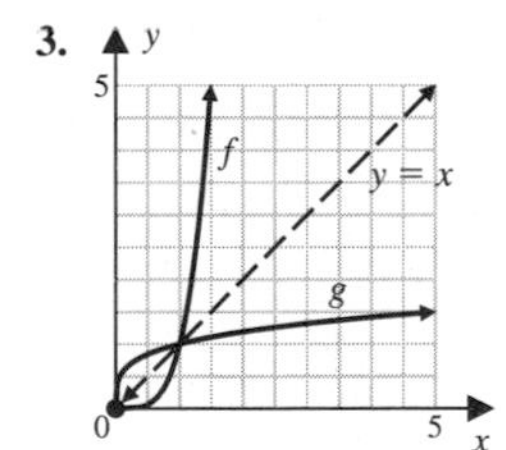

5.

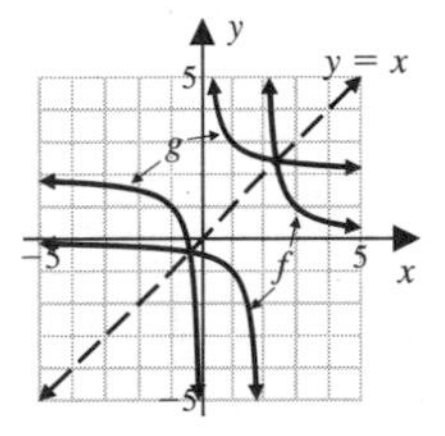

7. **(a)** Yes
(b) No

9. **(a)** Domain of f and Range of f^{-1}: $\mathbb{R}$;
Domain of f^{-1} and Range of f: $(0, \infty)$

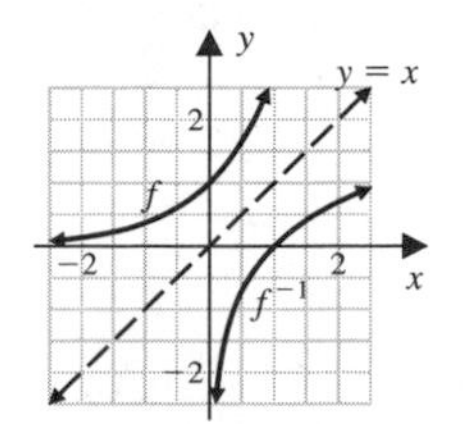

(b) Domain of f and Range of f^{-1}: $[-2, \infty)$;
Domain of f^{-1} and Range of f: $[0, \infty)$

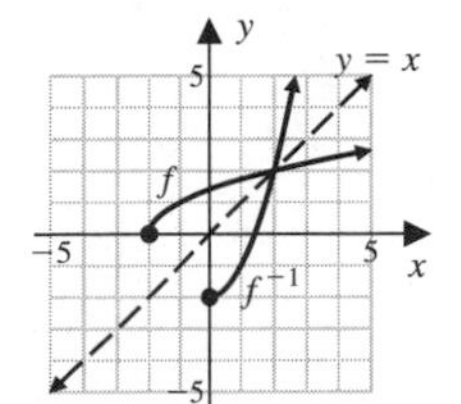

11.

x	2	-3	3	5	7	100	-4	1
$f^{-1}(x)$	1	2	3	4	5	6	7	8

Domain of f and Range of f^{-1}: 1, 2, 3, 4, 5, 6, 7, 8; Domain of f^{-1} and Range of f: $-4, -3, 1, 2, 3, 5, 7, 100$

13. One-to-one **15.** Not one-to-one **17.** Not one-to-one

19. $\dfrac{x-5}{7}$

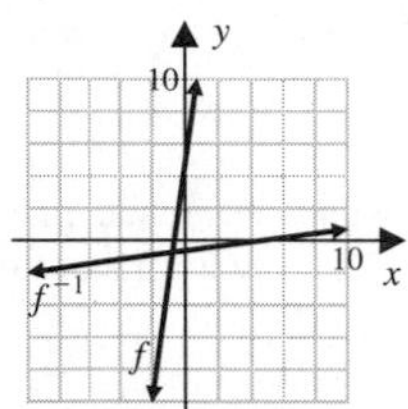

21. $\dfrac{3+x}{x}$

23. $x^2 + 3,\ x \geq 0$

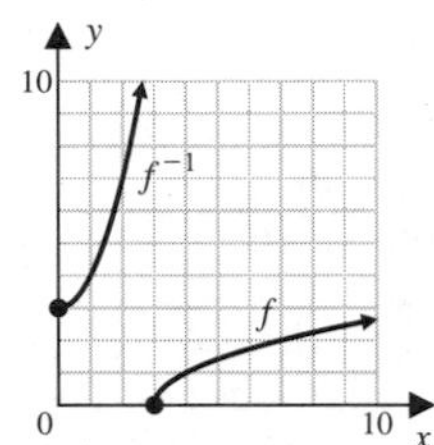

25. $\sqrt[3]{x+8}$

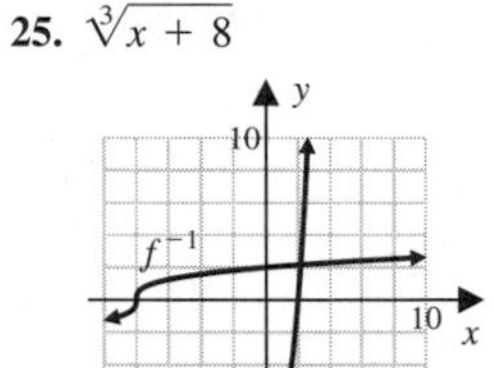

27. f^{-1} does not exist—f fails the horizontal-line test.

29. $(2-x)^3$

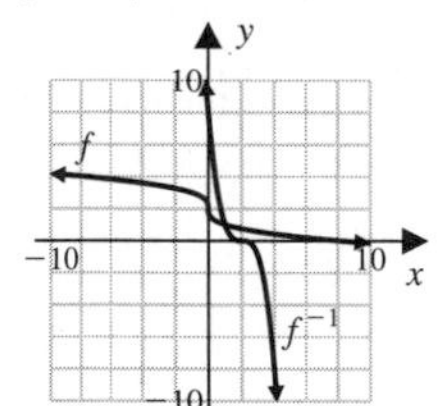

31. $\sqrt{-x},\ x \leq 0$

33. **(a)** f does not have an inverse; g has an inverse

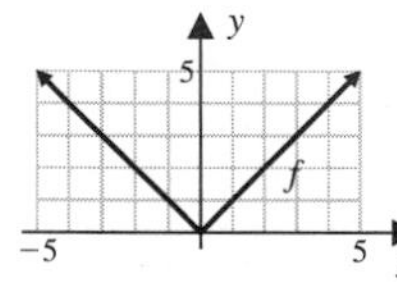

(b) $g^{-1}(x) = -x$, $x \geq 0$

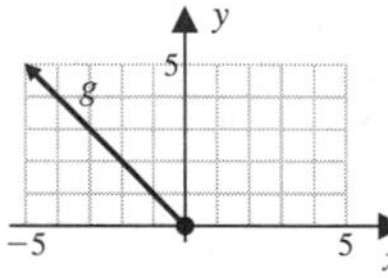

35. **(a)** f does not have an inverse; g has an inverse

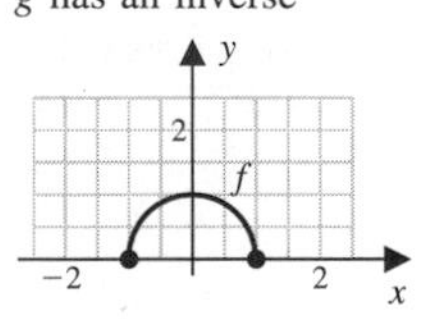

(b) $g^{-1}(x) = \sqrt{1-x^2}$, $0 \leq x \leq 1$

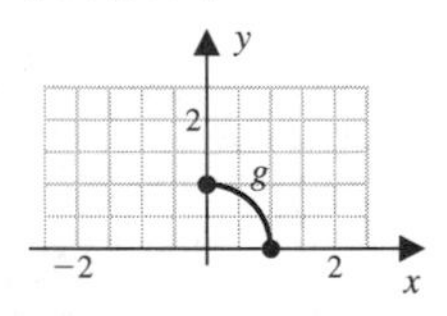

37. No; fails the horizontal-line test. **39.** No; fails the horizontal-line test. **41.** **(a)** B **(b)** A **(c)** C **(d)** D

43. **(a)** **(b)** $y = f^{-1}(x)$ **(c)**

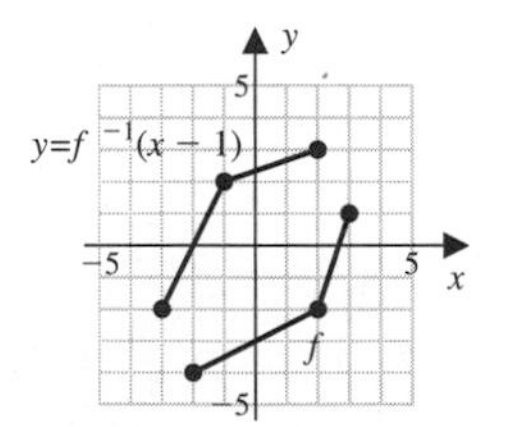

(d)

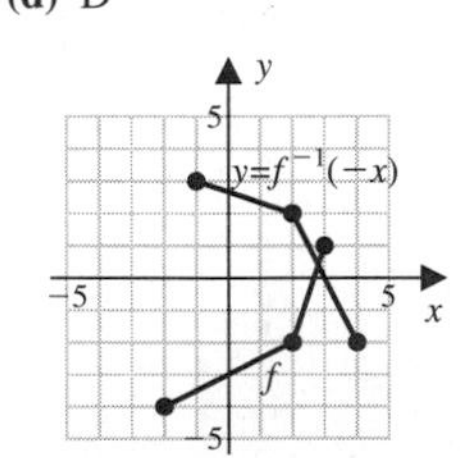

45. **(a)** $W = f(P) = 0.85P$ **(b)** $P = \frac{20}{17}W$

47. **(a)** $C = f^{-1}(F) = \frac{5}{9}(F - 32)$

(b)

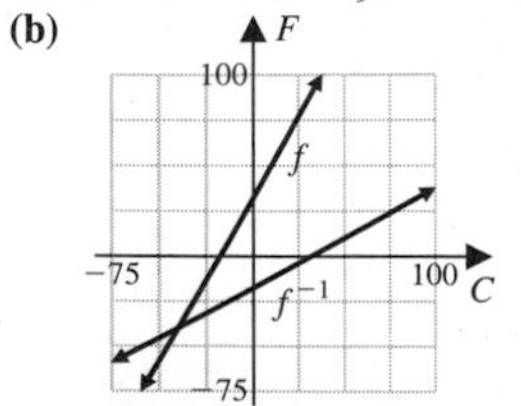

(c) $29\frac{4}{9}$°C; 45°C

49. **(a)** $y = \$25{,}000 + 0.40x,\ x \geq 0$

(b) $f^{-1}(x) = 2.5x - 62{,}500$

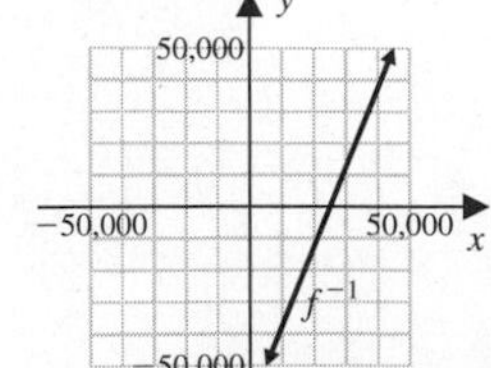

(c) f^{-1} represents sales in terms of the salesperson's income.

51. **(a)** $f^{-1}(x) = \begin{cases} 2x + 8 & \text{if } x < -4 \\ x + 4 & \text{if } x \geq -4 \end{cases}$ Domain and range of both f and f^{-1}: $\mathbb{R}$ **(b)** f^{-1} does not exist (fails horizontal-line test).

53. Reflect $y = f(x)$ about $y = x$, then reflect the result about $y = x$ to get $f(x)$.

55. **(a)** $f^{-1}(x) = \frac{1}{2}x - 2$; $g^{-1}(x) = \dfrac{x+1}{3}$; $(f \circ g)^{-1}(x) = \dfrac{x-2}{6}$ **(d)** $(g \circ f)^{-1} = f^{-1} \circ g^{-1}$; this is the same result as for $f^{-1} \circ g^{-1}$ in part (b).

57. **(a)** $f^{-1}(x) = \dfrac{1-x}{7}$ **(b)** $f^{-1}(x) = x^3 - 1$

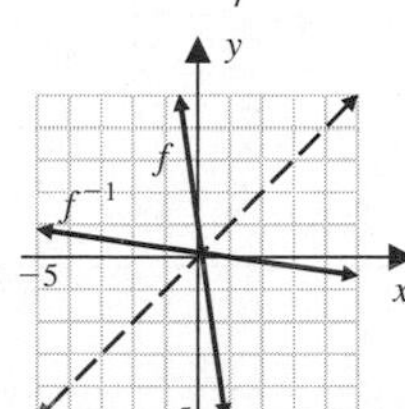

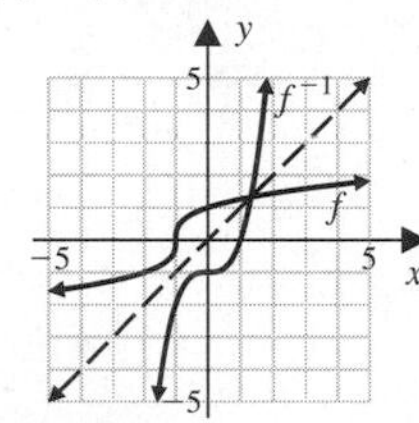

CHAPTER 2 REVIEW PROBLEM SET (ON PAGE 149)

1. **(a)** Function **(b)** Not a function **3.** **(a)** -1 **(b)** 1 **(c)** 50 **(d)** 2 **(e)** $2 - 5\sqrt{2}$ **(f)** $2b$ **(g)** 2

5. **(a)** -8; 0; 1; 2
(b) Domain and Range: $\mathbb{R}$
(c)

7. **(a)** Function
(b) Not a function
(c) Not a function
(d) Function

9. **(a)** Not a function **(b)** Not a function

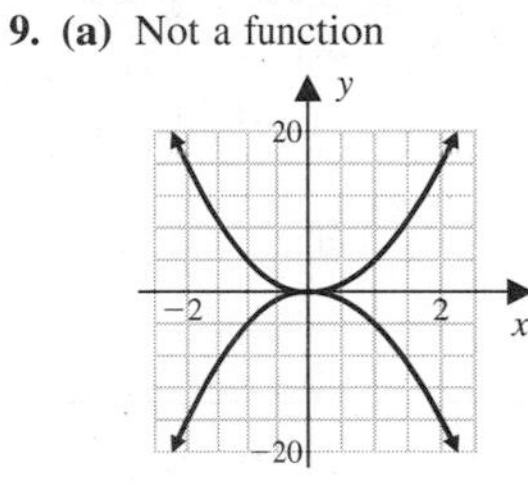

11. **(a)** Shift graph 2 units to left and then shift graph up 3 units. **(b)** Shift graph 1 unit to left and then shift graph 3 units up.
(c) Reflect the graph across the x axis. **(d)** Vertically stretch y by 2, reflect about x axis, shift to left 4 units, and shift up 2 units.

13. **(a)** $g(x) = -(x-3)^2 - 4$ **(b)** $g(x) = 2\sqrt[3]{x+5} + 3$ **(c)** $g(x) = -\sqrt{-x-4}$

15. **(a)** (i) $2x + 1$ (ii) 3 (iii) $x^2 + x - 2$ (iv) $x + 1$ (v) $x + 1$ (vi) $(x+2)/(x-1)$
(b) (i) $x^3 - x$ (ii) $x^3 + x$ (iii) $-x^4$ (iv) $-x^3$ (v) $-x^3$ (vi) $x^3/(-x) = -x^2$
(c) (i) $x^2 + 2x + 1$ (ii) $x^2 - 2x - 1$ (iii) $2x^3 + x^2$ (iv) $4x^2 + 4x + 1$ (v) $2x^2 + 1$ (vi) $x^2/(2x+1)$
(d) (i) $4x^2$ (ii) -2 (iii) $4x^4 - 1$ (iv) $8x^4 + 8x^2 + 1$ (v) $8x^4 - 8x^2 + 3$ (vi) $(2x^2 - 1)/(2x^2 + 1)$

17. **(a)** $f(x) = x^5$; $g(x) = 7x + 2$ **(b)** $f(t) = \sqrt{t}$; $g(t) = t^2 + 17$

21. **(a)** $f^{-1}(x) = \dfrac{7-x}{13}$ **(b)** $f^{-1}(x) = x^5$

23. **(a)** $f^{-1}(x) = x^2 - 2,\ x \geq 0$

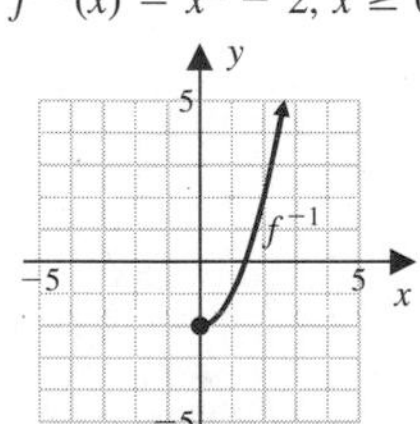

(b) $f^{-1}(x) = \begin{cases} -\sqrt{x+1} & \text{if } x \geq -1 \\ -2 - 2x & \text{if } x < -1 \end{cases}$

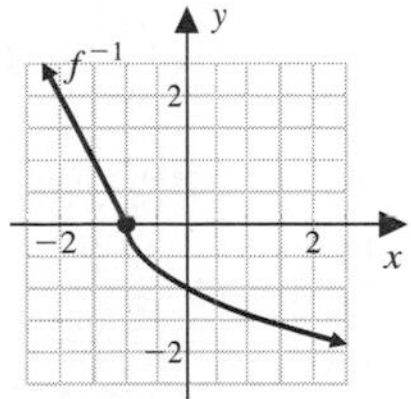

25. **(a)** $P = \dfrac{-x}{4000} + 6.5$ **(b)**

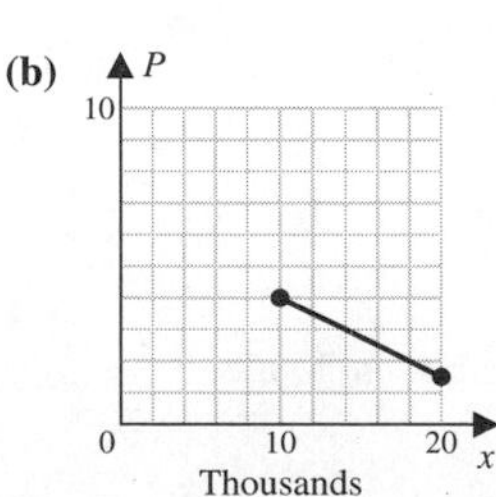

(c) The larger the demand, the lower the price; \$2.50 per bag. **(d)** 18,000 bags

27. **(a)** Vertical-line test is applicable. **(b)** Domain: [0, 45]; Range: [−60, 250] **(c)** −\$50,000; \$0; \$250,000; \$0 **(d)** \$25,000
(e) Increasing profit occurs for $10 \le A \le 25$; decreasing profit occurs for $25 \le A \le 40$; there is a loss if more than \$40,000 is spent.

29. **(a)** $R = 10x$
(b) $P = 10x - 10{,}000\left(5 + \sqrt[3]{x+1}\right)$
(c) Initially there is a loss ($P < 0$), then break even ($P = 0$), and finally a profit ($P > 0$) when $x > 38{,}877$.

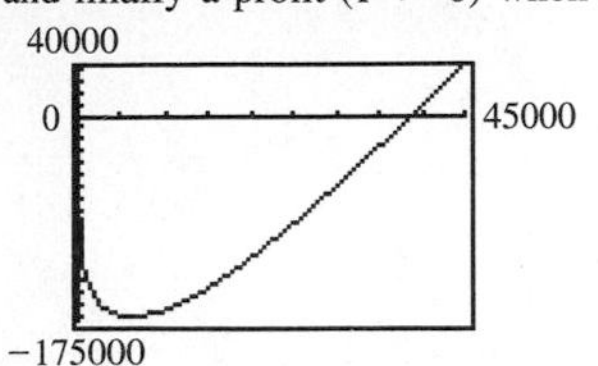

31. **(a)** $C^{-1}(x) = 100/(x - 205)$
(b) C^{-1} is used to find how many people would result in a given cost.

CHAPTER 2 TEST (ON PAGE 152)

1. **(a)** Yes **(b)** No

2. **(a)**

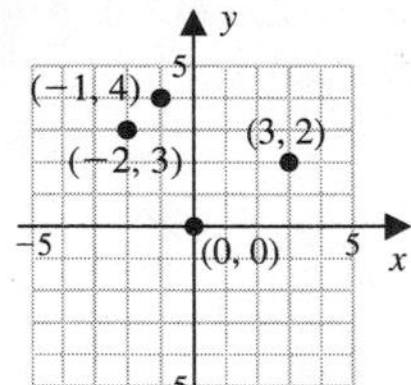

(b)

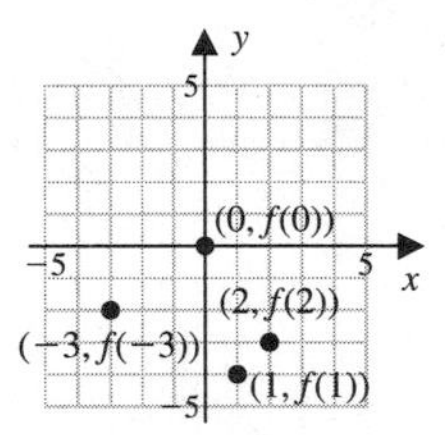

3.

x	−2	−1	0	1	2	$-a$	a	$3t$
$f(x)$	19	7	3	7	19	$4a^2 + 3$	$4a^2 + 3$	$36t^2 + 3$

4. 2; 18, $\frac{47}{64}$

5. **(a)** Domain: $(-\infty, 4]$

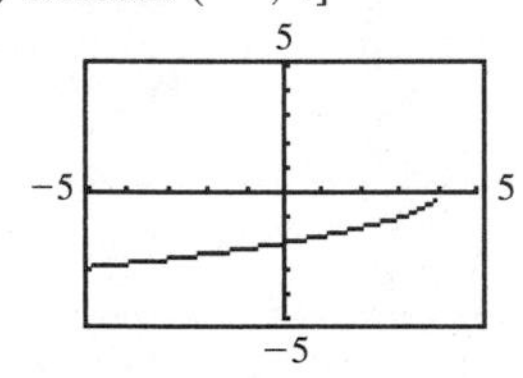

(b) Domain: $[1, \infty)$

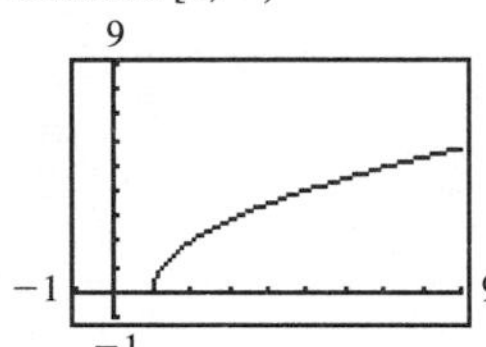

6. Domain: $\mathbb{R}$
Range: $[0, \infty)$

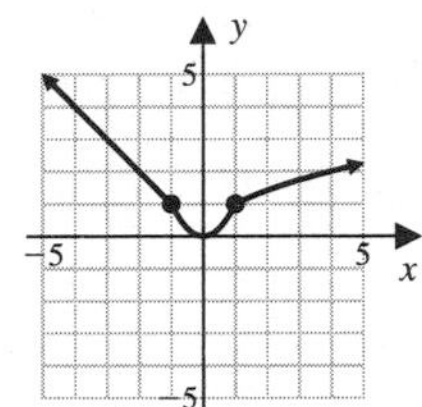

7. **(a)** $2x - 7 - x^2$ **(b)** $-2x^3 + 8x^2 + 2x - 8$ **(c)** $\dfrac{2x - 8}{1 - x^2}$ **(d)** $-6 - 2x^2$ **(e)** −3 **(f)** 2

8. **(a)** 0; −2; 3; 3; 0 **(b)** [−4, −1] **(c)** $(-\infty, -4]$ and $[2, \infty)$ **(d)** [−1, 2] **(e)** $\mathbb{R}$ **(f)** $\mathbb{R}$

9.

x	1	2	3	4
$f[g(x)]$	1	3	2	4
x	1	2	3	4
$g[f(x)]$	3	2	1	4

10. **(a)** $f^{-1}(x) = \dfrac{x+1}{4}$

(c)

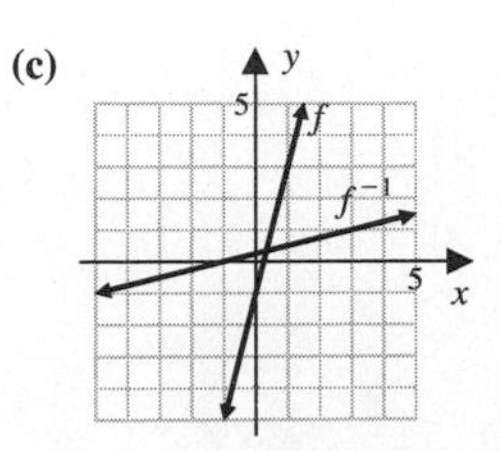

11. Vertically stretch y by 2, reflect about the x axis, shift to left 1 unit, and then shift up 3 units.

12. **(a)** $c = \sqrt{4 + a^2}$ **(b)** $a = \sqrt{c^2 - 4}$

13. **(a)** $[0, 14]$ **(b)** $t = 1, 12$ **(c)** $t = 6$ **(d)** $A = 50$ mg/dL when $t = 6$.

14. **(a)** $P = 5S - 60{,}000$ **(b)** \$140,000 **(c)**

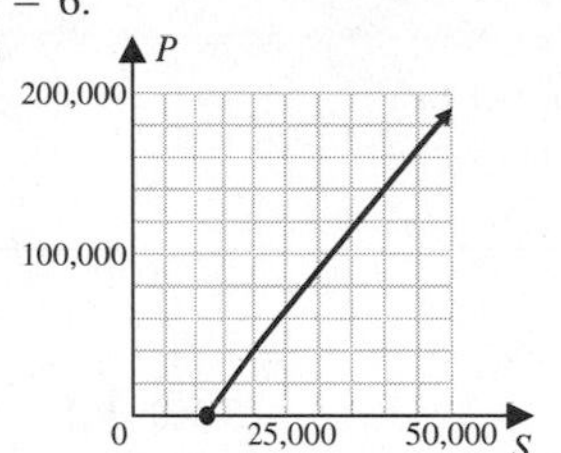

(d) 12,000

CHAPTER 3: PROBLEM SET 3.1 (ON PAGE 163)

1. **(a)** B **(b)** C **(c)** D **(d)** A

3. Domain: $\mathbb{R}$;
Range: $[2, \infty)$;
Vertex: $(3, 2)$;
Axis of symmetry: $x = 3$

5. Domain: $\mathbb{R}$;
Range: $(-\infty, -3]$;
Vertex: $(-2, -3)$;
Axis of symmetry: $x = -2$

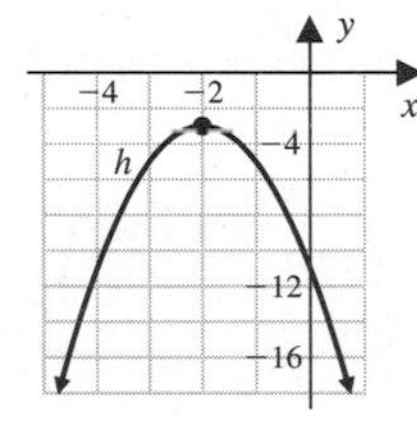

7. Domain: $\mathbb{R}$;
Range: $[1, \infty)$;
Vertex: $(-3, 1)$;
Axis of symmetry: $x = -3$

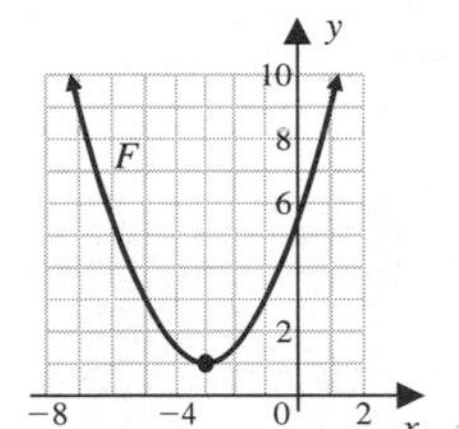

9. $f(x) = (x - 2)^2 - 9$;
Range: $[-9, \infty)$;
Vertex: $(2, -9)$;
Axis of symmetry: $x = 2$

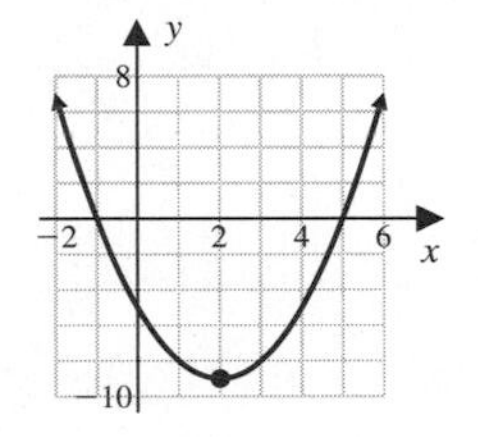

11. $f(x) = -(x + 3)^2 + 1$;
Range: $(-\infty, 1]$;
Vertex: $(-3, 1)$;
Axis of symmetry: $x = -3$

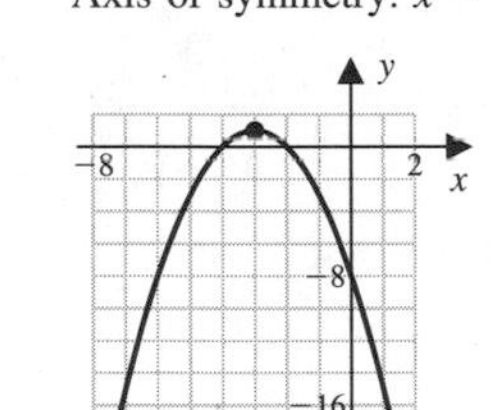

13. $f(x) = 3\left(x - \frac{5}{6}\right)^2 - \frac{277}{12}$;
Range: $\left[-\frac{277}{12}, \infty\right)$;
Vertex: $\left(\frac{5}{6}, -\frac{277}{12}\right)$;
Axis of symmetry: $x = \frac{5}{6}$

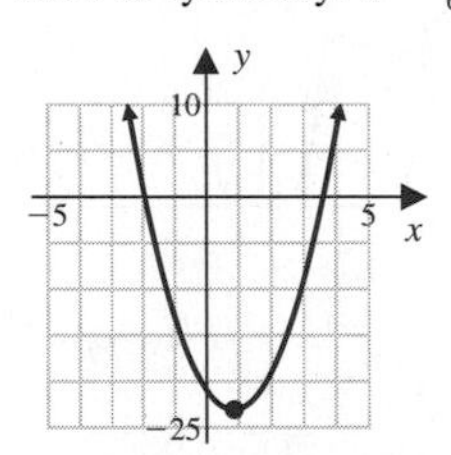

15. $f(x) = -5(x - 0.3)^2 + 4.45$;
Range: $(-\infty, 4.45]$;
Vertex: $(0.3, 4.45)$;
Axis of symmetry: $x = 0.3$

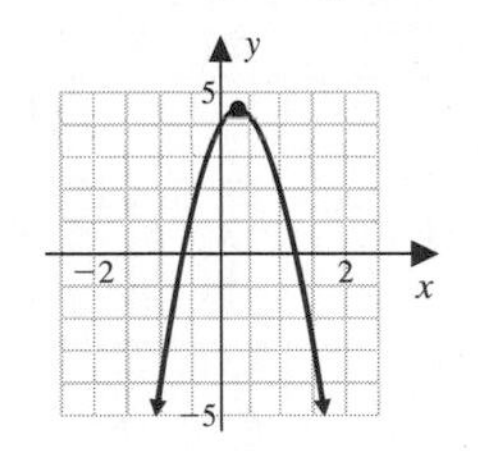

17. $f(x) = -\frac{1}{2}(x + 2)^2 + 3$;
Range: $(-\infty, 3]$;
Vertex: $(-2, 3)$;
Axis of symmetry: $x = -2$

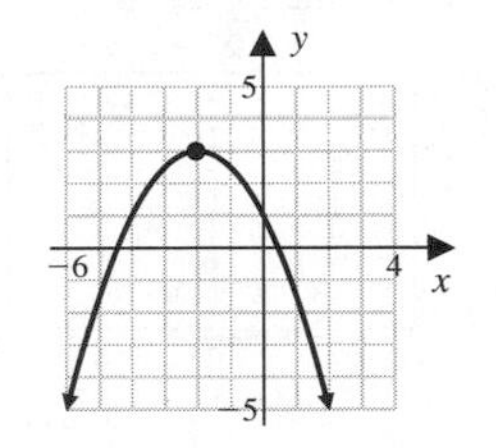

19. **(a)** $y = -x^2 + 4x$
(b) $y = \frac{2}{9}x^2 - 2$

21. $f(x) = \frac{1}{2}(x - 2)^2 - 2$

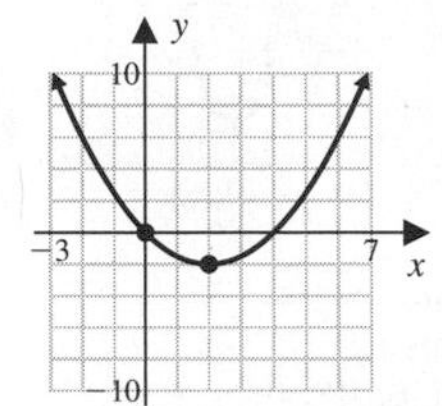

23. $f(x) = -x^2 + 1$

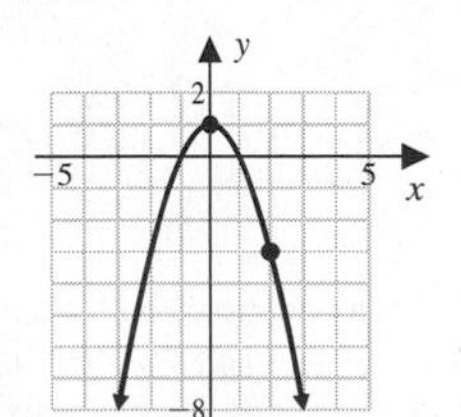

25. $(-\infty, -7)$ or $(2, \infty)$

27. $(-\infty, 1]$ or $\left[\frac{3}{2}, \infty\right)$ **29.** $\left(-\infty, -\frac{5}{2}\right]$ or $\left[\frac{4}{3}, \infty\right)$ **31.** Minimum $= -4$ **33.** Maximum $= -1$ **35.** Minimum $= -\frac{9}{8}$

37. **(a)** Minimum point = (3.79, −12.9141)

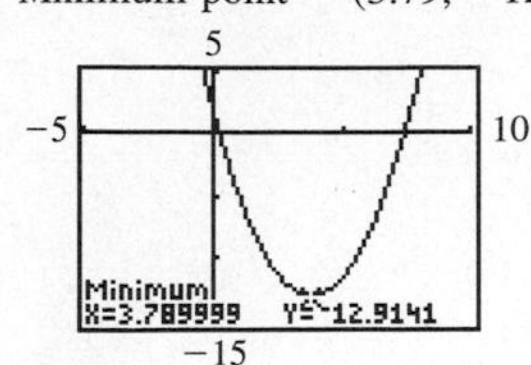

(b) Maximum point = (0.7111, −3.0898)

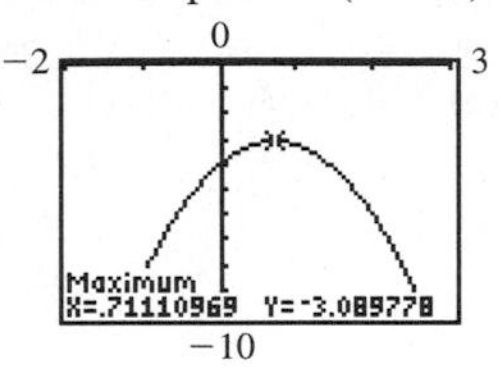

39. **(a)** $P = 6y + y^2$ **(b)** −3 and 3; −9

41. **(a)** $P(5)$ = \$750 thousand; $P(10)$ = \$1437.50 thousand; $P(15)$ = \$2062.50 thousand
(b) At 6250 employees, P = \$4882.81 (in thousands) **(c)** As x increases, P increases to a maximum at \$4882.81 thousand, then decreases.

43. **(a)** $h(50) = 43.50$ ft; $h(100) = 63.50$ ft; $h(200) = 43.50$ ft **(b)** Maximum height 66 ft (approx.); travels 125 ft (approx.) horizontally
(d) 253.45 ft (approx.)

45. **(a)** 124.67 ft; 126.68 ft; 116.7 ft **(b)** Peak at approx. $x = 450.45$, $y = 126.76$ **(d)** 23.24 ft (approx.)

47. **(a)** $A = 500x - x^2$ **(b)** $0 \le x \le 500$ **(c)** 250 ft by 250 ft; 62,500 ft^2 **49.** 112.5 ft by 225 ft; maximum area is 25,312.5 ft^2

51. $y = 2$ ft, $x = 2$ ft **53.** **(a)** Two x intercepts **(b)** One x intercept **(c)** No x intercepts

55. $a < 0$, opens to left; $a > 0$, opens to right

57. **(a)** **(b)** The function fits the data closely. **(c)** **(d)** $T = 25$

PROBLEM SET 3.2 (ON PAGE 175)

1. **(a)** Symmetric with respect to the y axis; resembles $y = x^2$

(b) Symmetric with respect to the origin; resembles $y = x^3$

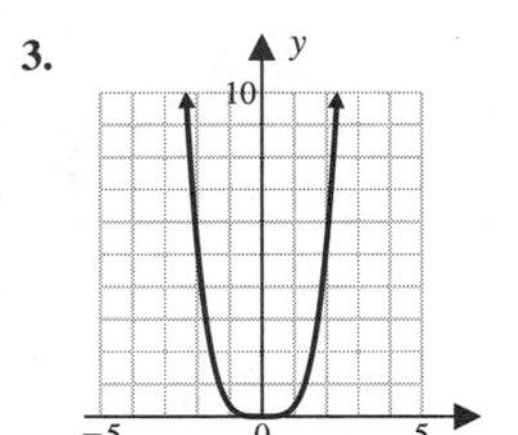

3.

5.

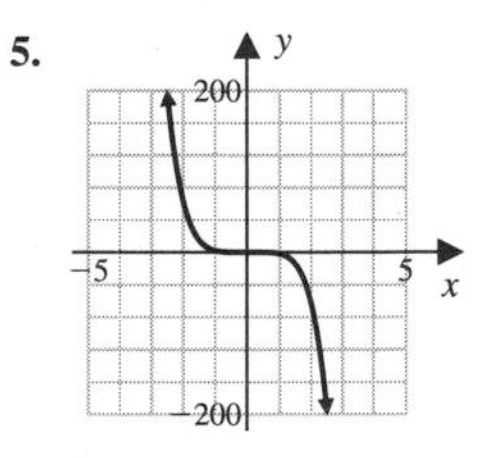

7.

9.

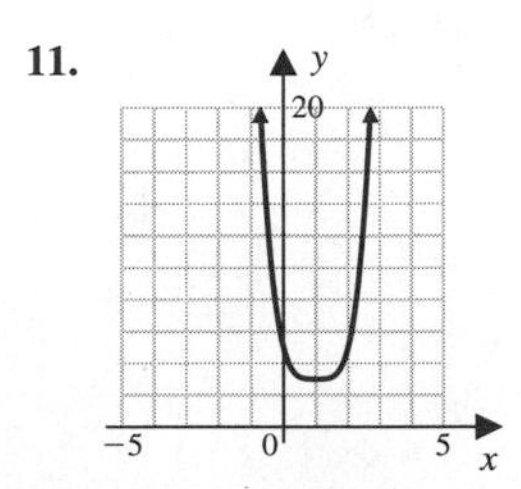

11.

13.

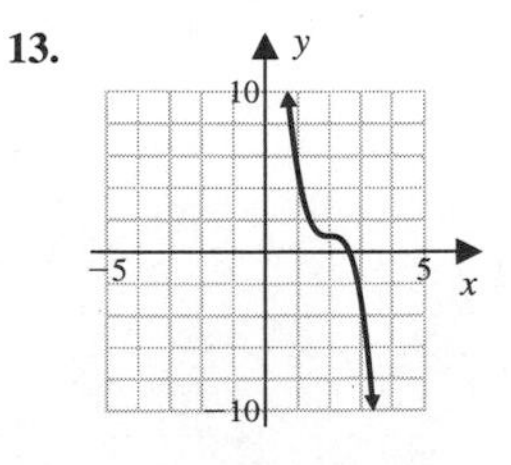

15. **(a)** $f(x) \to -\infty$ as $x \to +\infty$ **(b)** $f(x) \to -\infty$ as $x \to -\infty$ One turning point; a peak

17. **(a)** $f(x) \to +\infty$ as $x \to +\infty$ **(b)** $f(x) \to -\infty$ as $x \to -\infty$ No turning points

19. **(a)** $f(x) \to -\infty$ as $x \to +\infty$ **(b)** $f(x) \to +\infty$ as $x \to -\infty$ Two turning points; 1 valley, 1 peak

21. **(a)** Model: $y = -x^3$; (A) **(b)** Model: $y = x^4$; (B)

23. **(a)** Does not show limit behavior as $x \to +\infty$. **(b)** Graph should be continuous.

25. Model: $y = x^4$
(a) $f(x) \to +\infty$ as $x \to +\infty$
(b) $f(x) \to +\infty$ as $x \to -\infty$
Three turning points;
1 peak, 2 valleys

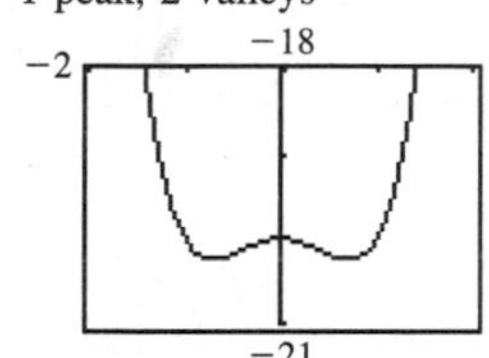

27. Model: $y = x^3$
(a) $f(x) \to +\infty$ as $x \to +\infty$
(b) $f(x) \to -\infty$ as $x \to -\infty$
Two turning points;
1 peak, 1 valley

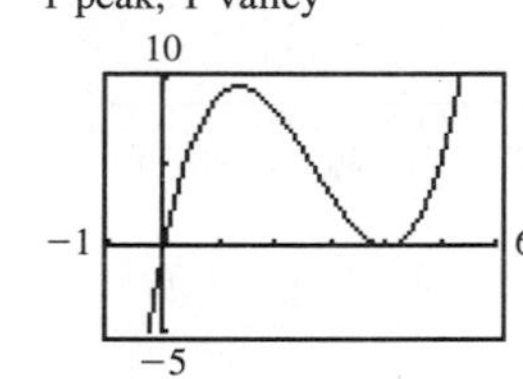

29. Model: $y = -x^4$
(a) $f(x) \to -\infty$ as $x \to +\infty$
(b) $f(x) \to -\infty$ as $x \to -\infty$
One turning point; peak

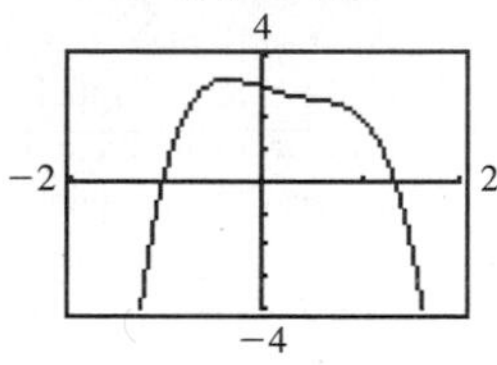

31. Model: $y = -x^3$
(a) $f(x) \to -\infty$ as $x \to +\infty$
(b) $f(x) \to +\infty$ as $x \to -\infty$
Two turning points;
1 peak, 1 valley

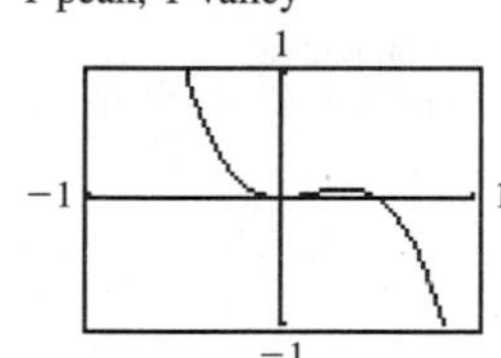

33. Model: $y = x^3$
(a) $f(x) \to +\infty$ as $x \to +\infty$
(b) $f(x) \to -\infty$ as $x \to -\infty$
Two turning points; 1 peak,
1 valley

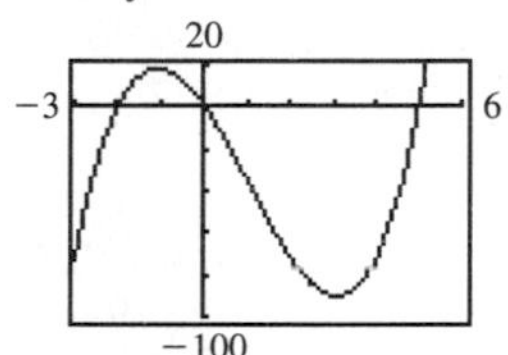

35. **(a)** $\{0\}$ or $[1, \infty)$ **(b)** $(-\infty, 0)$ or $(0, 1)$ **37.** **(a)** $(-\infty, -1]$ or $\{0\}$ or $[1, \infty)$ **(b)** $(-1, 0)$ or $(0, 1)$
39. $-2, 0, 1$; $(-2, 0)$ or $(1, \infty)$ **41.** $-3, 0, 4$; $(-\infty, -3)$ or $(0, 4)$
43. $1, \frac{5}{2}$; $\{1\}$ or $\left[\frac{5}{2}, \infty\right)$ **45.** **(a)** $0 \le t \le 48$ **(c)** 3.82% at $t = 40$ **(d)** 6.8% at $t = 9$
47. **(a)** $V = x(20 - 2x)^2$ **(b)** $0 < x < 10$
(d) Maximum volume is 592.59 in.^{3} (approx.) when $x = 3.33$ in. (approx.)

49. **(a)** For $a > 0$ and $b > 0$: $f(x), g(x) \to +\infty$ as $x \to +\infty$; $f(x), g(x) \to +\infty$ as $x \to -\infty$.
For $a < 0$ and $b < 0$: $f(x), g(x) \to -\infty$ as $x \to +\infty$; $f(x), g(x) \to -\infty$ as $x \to -\infty$.
(b) For $a > 0$ and $b < 0$: $f(x) \to +\infty$ as $x \to +\infty$; $g(x) \to -\infty$ as $x \to +\infty$; $f(x) \to +\infty$ as $x \to -\infty$; $g(x) \to -\infty$ as $x \to -\infty$.
For $a < 0$ and $b > 0$: $f(x) \to -\infty$ as $x \to +\infty$; $g(x) \to +\infty$ as $x \to +\infty$; $f(x) \to -\infty$ as $x \to -\infty$; $g(x) \to +\infty$ as $x \to -\infty$.

51. **(a)** Concave down on $(-\infty, 0)$; concave up on $(0, \infty)$; point of inflection at $(0, 1)$ **(b)** Concave down on $(4, \infty)$; concave up on $(-\infty, 4)$; point of inflection at $(4, 1)$

53. $n > 1$

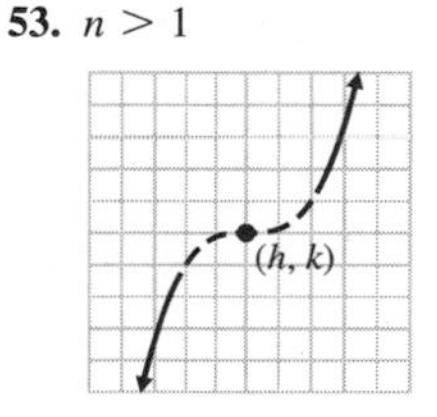

55.

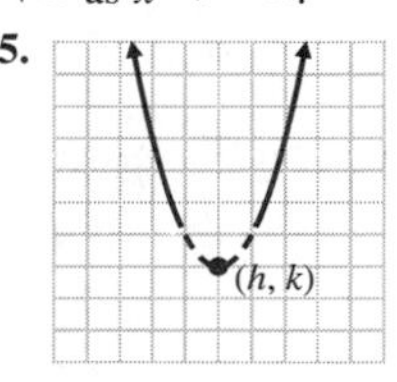

57. $f(x) \to g(x) = x^3$ as $x \to -\infty$ and as $x \to +\infty$ **59.** Degree is at least 4.

PROBLEM SET 3.3 (ON PAGE 186)

1. **(a)** $f(x) = (x + 1)(4x^2 - 6x + 13) - 14$ **3.** **(a)** $f(x) = (x - 3)(3x^3 + 7x^2 + 25x + 78) + 241$

5. **(a)** $f(x) = (2x + 3)\left(3x^3 - \frac{9}{2}x^2 + \frac{47}{4}x - \frac{141}{8}\right) + \frac{479}{8}$ **7.** **(a)** $f(x) = (2x^2 + 1)\left(2x^2 - x - \frac{21}{2}\right) + \left(8x + \frac{23}{2}\right)$
9. **(a)** -14 **(b)** 63 **(c)** 241 **11.** $2x^2 + 23x + 58$; 153 **13.** $4x^3 + x^2 + 3x + 3$; 8 **15.** $-5x^3 + 10x^2 - 20x + 40$; -80
17. $-5x^4 + 10x^3 - 20x^2 + 41x - 82$; 167 **19.** 35 **21.** 12.5
23. **(a)** Yes; $(x + 2)(x + 4)(x - 6)$ **(b)** Yes; $(x + 4)(x - 6)(x + 2)$ **(c)** No **(d)** Yes; $(x - 6)(x + 4)(x + 2)$
25. **(a)** $f(x) = (x - 1)(x - 2)(x - 3)$ **27.** **(a)** $f(x) = (x + 2)(2x^3 - 4x^2 + 8x - 13)$
33. **(a)** 1 is a zero of multiplicity 3; 2 is a zero of multiplicity 2.
(b) 1 is a zero of multiplicity 3; 2 is a zero of multiplicity 1; 5 is a zero of multiplicity 1.
35. **(a)** 0 is a zero of multiplicity 1; 3 is a zero of multiplicity 3; -3 is a zero of multiplicity 1.
(b) 4 is a zero of multiplicity 2; -2 is a zero of multiplicity 2.
37. -10 **39.** 6 **41.** **(a)** $f(1) = 0$ for $f(x) = x^n - 1$ **(b)** n must be even **(c)** Yes; No
43. **(a)** $f(x) = x^2 + 1$ **(b)** $f(x) = x^2$ **(c)** $f(x) = x^2 - 1$ **(d)** x intercepts are the same as the zeros. **45.** No; if $D(x) = 0$
47. At least 3, but may be greater depending on the multiplicities of the zeros.

PROBLEM SET 3.4 (ON PAGE 198)

1. **(a)** 3 positive real zeros; 1 negative real zero; 0 imaginary zeros **(b)** 3 positive real zeros; 1 negative real zero; 2 imaginary zeros
3. 1 positive real zero; 1 negative real zero; 2 imaginary zeros **5.** 1 negative real zero; 2 positive real zeros; 0 imaginary zeros
7. 1 negative real zero; 1 positive real zero; 2 imaginary zeros **9.** 1 positive real zero; 0 or 2 negative real zeros; 0 or 2 imaginary zeros
11. 0 or 2 positive real zeros; 1 negative real zero; 0 or 2 imaginary zeros
13. 3 or 1 positive real zeros; 1 negative real zero; 0 or 2 imaginary zeros
15. 1 positive real zero; 0 or 2 negative real zeros; 2 or 4 imaginary zeros **17.** $-2, 1, 2$; $f(x) = (x - 1)(x - 2)(x + 2)$
19. 1; $h(x) = (x - 1)(5x^2 - 7x + 10)$ **21.** $-1, 1, 2$; $f(x) = (x + 1)(x - 1)^2(x - 2)^2$
23. $-2, -\frac{1}{2}, \frac{1}{2}, 3$; $f(x) = (x + 2)\left(x + \frac{1}{2}\right)\left(x - \frac{1}{2}\right)(x - 3)$ **25.** $-1, \frac{2}{5}, \frac{1}{2}$; $g(x) = \frac{2}{3}(x + 1)\left(x - \frac{2}{5}\right)\left(x - \frac{1}{2}\right)$
27. $-1, -\frac{1}{2}, 1, \frac{3}{2}$; $h(x) = 4(x + 1)\left(x + \frac{1}{2}\right)(x - 1)\left(x - \frac{3}{2}\right)$ **29.** $-1, 4$; $f(x) = (x + 1)(x - 4)(x^2 + 4)$
31. $-\frac{1}{2}$; $f(x) = \left(x + \frac{1}{2}\right)(x^2 - 4x + 5)$ **33.** $-2, \frac{1}{4}, 3$; $g(x) = 2\left(x - \frac{1}{4}\right)(x + 2)(x - 3)$ **35.** -2 is a rational zero; zeros are $-2, -\sqrt{3}, \sqrt{3}$
37. 1 is a rational zero; zeros are $1, -1 - \sqrt{2}, -1 + \sqrt{2}$ **39.** $-\frac{1}{2}$ is a rational zero; zeros are $-\frac{1}{2}, -2\sqrt{3}, 2\sqrt{3}$
41. **(a)** $V = 21\pi x^2 - \dfrac{\pi}{3}x^3$ **(b)** $0 < x \le 40$ **(c)** Maximum volume is $\dfrac{36{,}800\pi}{3}$ ft^3. **(d)** $x = 36$ ft
43. **(a)** $V = 30\pi r^2 - \frac{1}{3}\pi r^3$ **(b)** $0 < r \le 30$ **(c)** Maximum volume is $18{,}000\pi$ ft^3. **(d)** 6 ft **45.** False **47.** True **49.** False

PROBLEM SET 3.5 (ON PAGE 208)

1. **(a)** $f(1) = -3, f(3) = 7$ **3.** **(a)** $f(-3) = -42, f(-2) = 5$
5. **(a)** $f(0.9) = 7.7$ (approx.), $f(2.1) = -7.6$ (approx.); $f(2.1) < 0 < f(0.9)$ **7.** **(a)** Upper bound: 2; lower bound: -2
9. **(a)** Upper bound: 6; lower bound: -2 **11.** **(a)** Upper bound: 5; lower bound: -5 **13.** $c = -1, 2$ **15.** $c = -1.4142, -1, 1, 1.4142$
17. $c = 2.1320$ **19.** $(2, \infty)$ **21.** Approx. $[-1.41, -1]$ or $[1, 1.41]$ **23.** $[2.13, \infty)$ **25.** Zeros of multiplicity 1: $-3i, -2i, 2i, 3i$
27. Zeros of multiplicity 2: $-i, i$ **29.** Zeros of multiplicity 1: $-1 - i, -\sqrt{2}, \sqrt{2}, -1 + i$ **31.** $f(x) = x^3 - 5x^2 + 5x + 3$
33. $f(x) = x^2 - 2x + 2$ **35.** $f(x) = x^4 - 4x^3 + 24x^2 - 40x + 100$ **37.** **(a)** No real factors **(b)** $\left(x - \dfrac{1}{2} - \dfrac{i}{2}\sqrt{3}\right)\left(x - \dfrac{1}{2} + \dfrac{i}{2}\sqrt{3}\right)$
39. **(a)** $f(x) = 2x(x^2 + 2)$ **(b)** $f(x) = 2x\left(x + i\sqrt{2}\right)\left(x - i\sqrt{2}\right)$
41. **(a)** $f(x) = (x + 3)(x - 3)(x^2 + 9)$ **(b)** $f(x) = (x + 3)(x - 3)(x + 3i)(x - 3i)$
43. **(a)** $f(x) = 3\left(x - \frac{1}{3}\right)(x^2 + 1)$ **(b)** $f(x) = 3\left(x - \frac{1}{3}\right)(x + i)(x - i)$ **45.** 1.83%; 4.13%: 10%
47. **(a)**
4200 ft
525 ft
(c) $0 < x \le 525$
(d) Approx. 8.88 ft
49. $f(x) = x^2 - 1, f(-2) > 0, f(2) > 0$, but $f(1) = 0$

51. **(a)** i is a zero, but $-i$ is not a zero. **(b)** No, coefficients must be real numbers.
53. $f(x) = x^3 - 6x^2 + 11x - 6$; $g(x) = 2x^3 - 12x^2 + 22x - 12$ **55.** **(a)** 4 **(b)** $(4, \infty)$ **57.** $c = 0.3473$ **59.** $c = -1.1545$
61. $c = 4.7958$

PROBLEM SET 3.6 (ON PAGE 220)

1. (a) Vertical asymptote: $x = 0$; Horizontal asymptote: $y = 0$

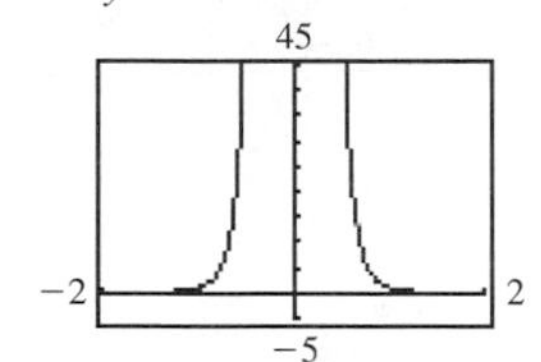

(b) Vertical asymptote: $x = 0$; Horizontal asymptote: $y = 0$

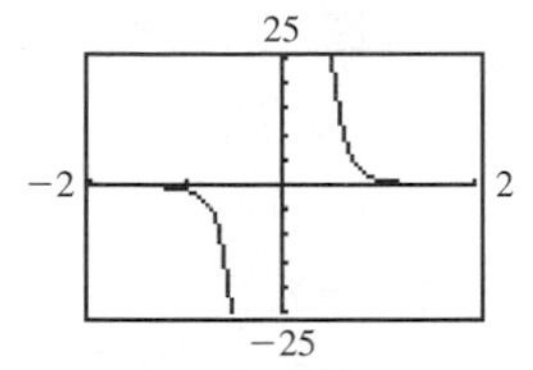

3. Vertical asymptote: $x = 0$; Horizontal asymptote: $y = -4$. As $x \to 0$, $y \to +\infty$. As $x \to +\infty$ and as $x \to -\infty$, $y \to -4$.

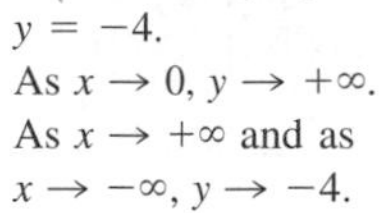

5. Vertical asymptotes: $x = -2$ and $x = 2$; Horizontal asymptote: $y = 0$. As $x \to -2^-$, $y \to +\infty$; as $x \to -2^+$, $y \to -\infty$. As $x \to 2^-$, $y \to +\infty$; as $x \to 2^+$, $y \to -\infty$. As $x \to \infty$, $y \to 0$; as $x \to -\infty$, $y \to 0$.

7. (a) Vertically stretch $1/x$ by 4, shift to the left 3 units. Vertical asymptote: $x = -3$; Horizontal asymptote: $y = 0$

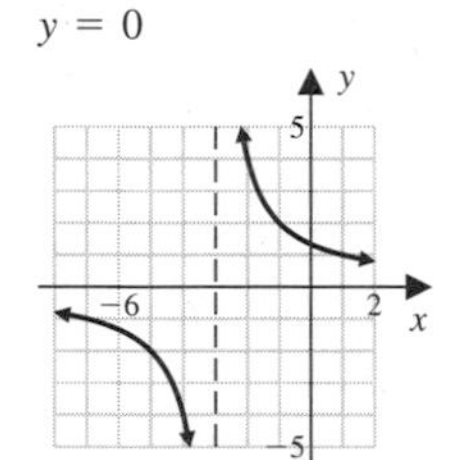

(b) Shift $1/x$ up 3 units. Vertical asymptote: $x = 0$; Horizontal asymptote: $y = 3$

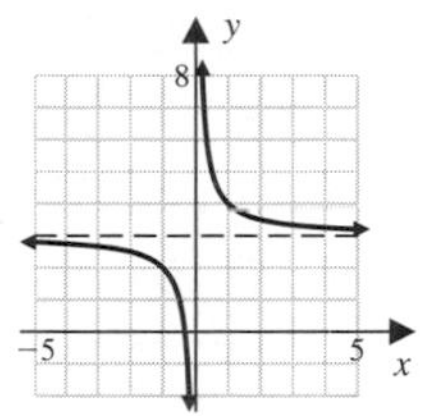

9. (a) Vertically stretch $1/x^2$ by 3, reflect about the x axis, and shift to left 1 unit. Vertical asymptote: $x = -1$; Horizontal asymptote: $y = 0$

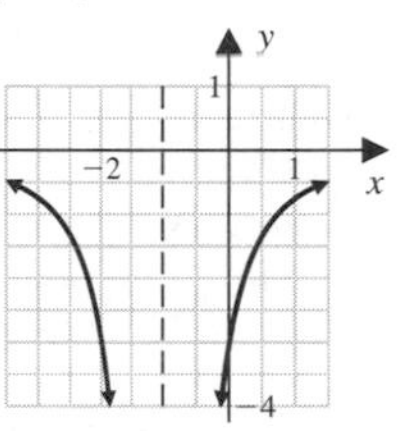

(b) Vertically stretch $1/x$ by 2, shift to the right 1 unit and up 3 units. Vertical asymptote: $x = 1$; Horizontal asymptote: $y = 3$

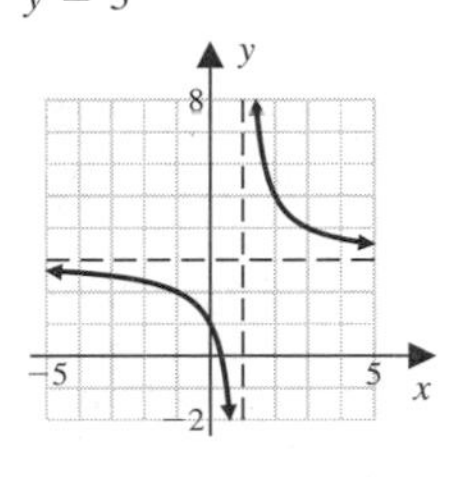

11. (a) Vertically stretch $1/x^2$ by 2, shift to the left 3 units, then down 1 unit. Vertical asymptote: $x = -3$; Horizontal asymptote: $y = -1$

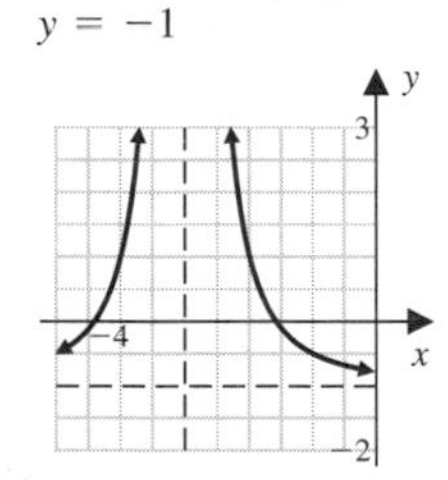

(b) Shift $1/x^2$ to the left 4 units, then up 2 units. Vertical asymptote: $x = -4$; Horizontal asymptote: $y = 2$

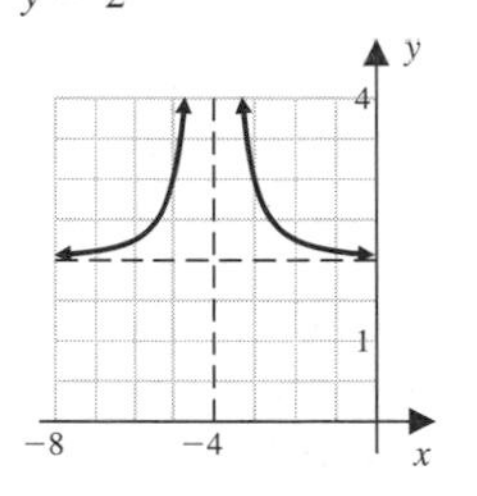

13. (a) No vertical asymptotes; hole at (1, 2)

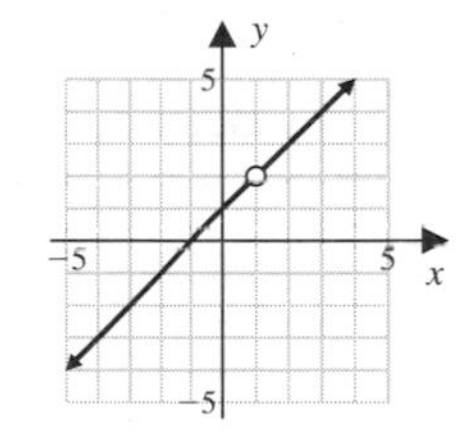

(b) No vertical asymptotes; hole at (3, 4)

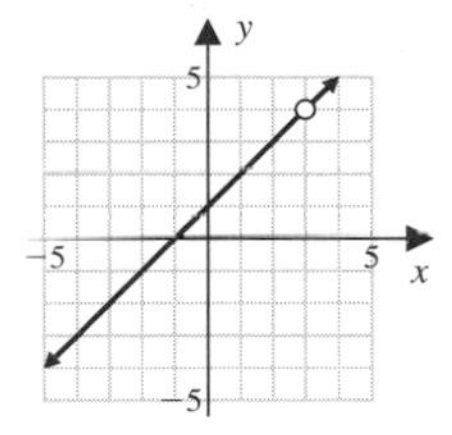

15. (a) Horizontal asymptote: $y = -2$

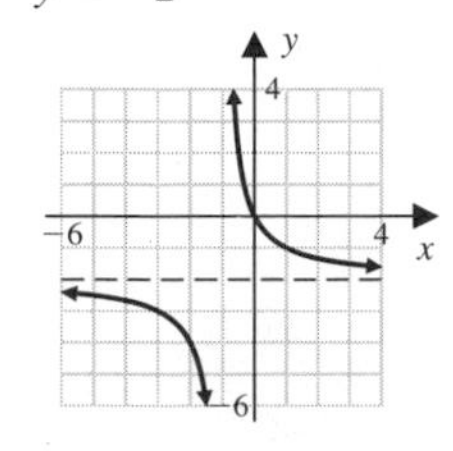

(b) Horizontal asymptote: $y = \frac{3}{2}$

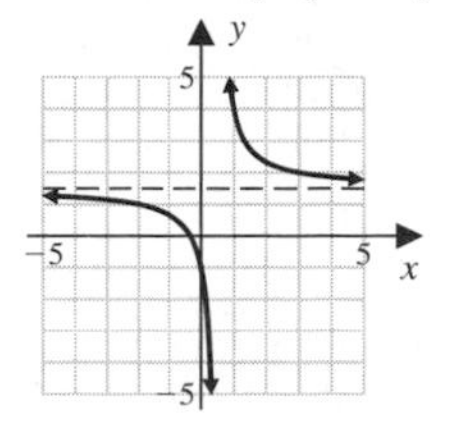

17. (a) Oblique asymptote $y = -4x + 8$

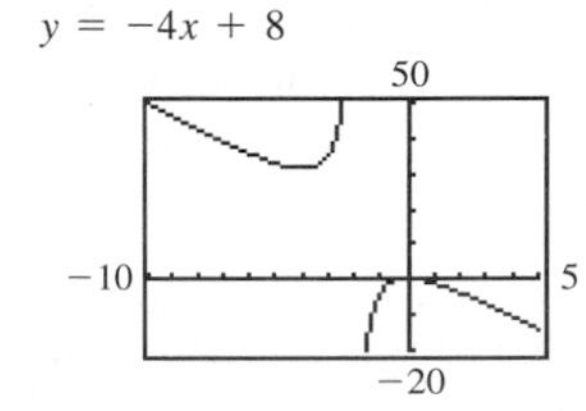

(b) Oblique asymptote $y = x - 3$

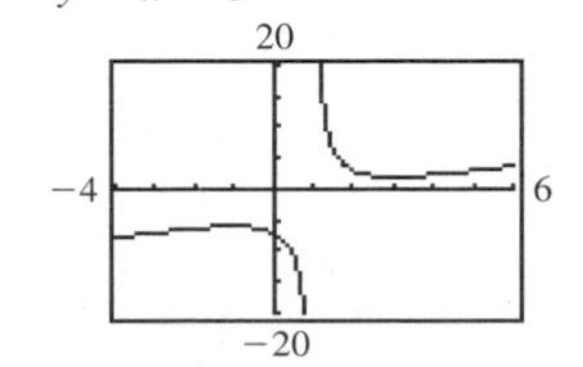

19. **(a)** C **(b)** Vertical asymptotes: $x = -1$ and $x = 0$; Horizontal asymptote: $y = 0$

21. **(a)** D **(b)** Vertical asymptote: $x = -1$; Horizontal asymptote: $y = 2$ **23.** Vertical asymptote: $x = -1$; Oblique asymptote: $y = \frac{1}{2}x - 1$

25. **(a)**

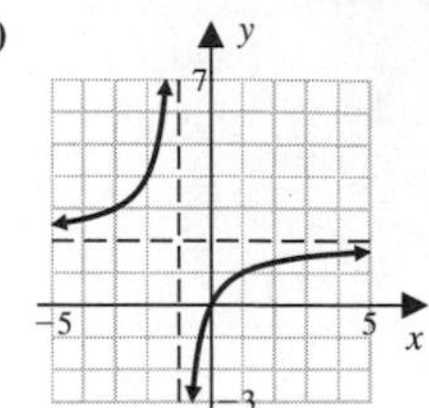

27. **(a)**

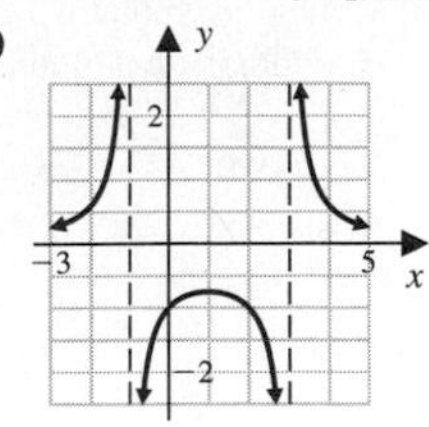

29. **(a)**

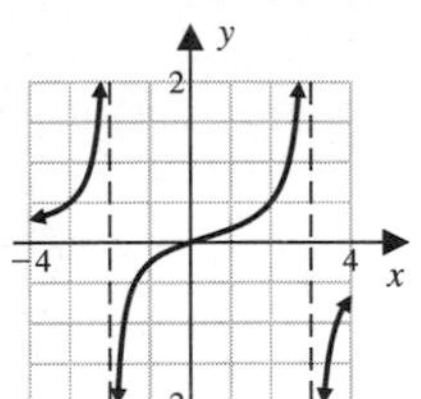

31. **(a)**

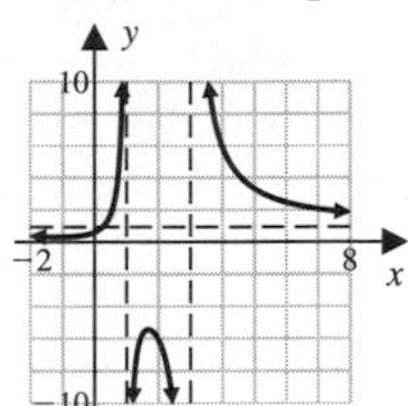

33. **(a)**

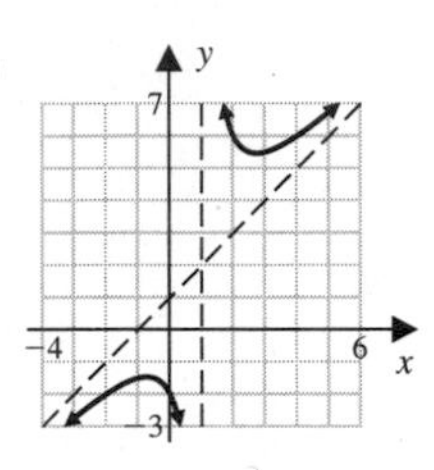

35. **(a)** $x \geq 0$ **(b)** Horizontal asymptote: $S(x) = 5000$ **(c)** As expenditures increase, sales approach 5000 units.

37. **(b)** As $x \to 60\%$, the cost approaches ∞. **39.** $-\frac{1}{2x}, x \neq 2$

41. **(a)** x intercepts: $-\frac{1}{3}, \frac{1}{3}$; no y intercept; Vertical asymptote: $x = 0$; Horizontal asymptote: $y = -3$

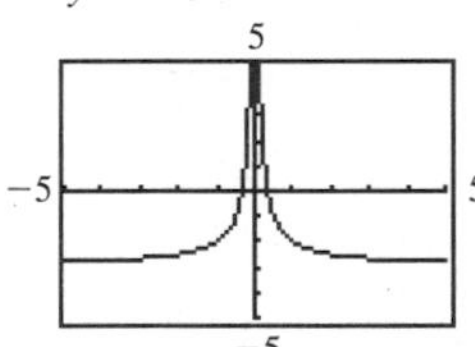

(b) x intercept $\frac{1}{3}$; no y intercept; Vertical asymptote: $x = 0$; Horizontal asymptote: $y = 3$

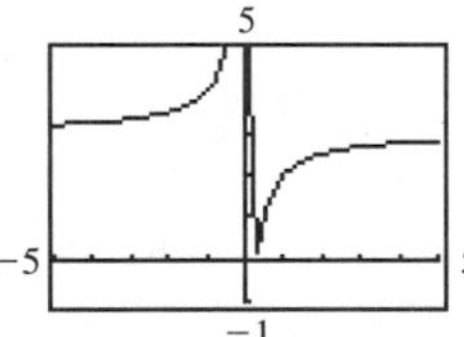

43. **(a)** x intercept $\frac{15}{4}$

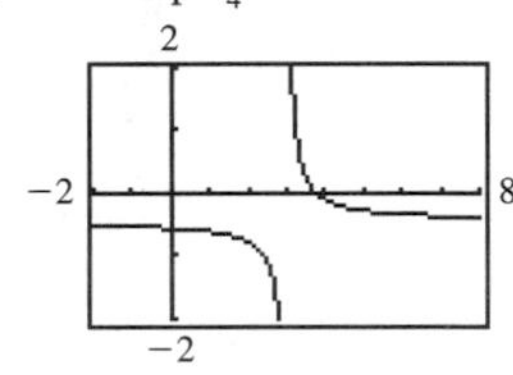

(b) $\frac{15}{4}$; $(-\infty, 3)$ or $\left(\frac{15}{4}, \infty\right)$.

45. **(a)** If k is in the domain of $f(x)$, then $f(k)$ is defined, but $f(x) \to -\infty$ or $f(x) \to +\infty$ as $x \to k^-$ or $x \to k^+$.
(b) A graph can cross its horizontal asymptote; thus, k can be in the range of $f(x)$.

CHAPTER 3 REVIEW PROBLEM SET (ON PAGE 223)

1. Vertex: $(-5, 3)$; axis of symmetry: $x = -5$; Domain: $\mathbb{R}$; Range: $(-\infty, 3]$

3. $f(x) = (x + 2)^2 - 11$; vertex: $(-2, -11)$; axis of symmetry: $x = -2$ **5.** $f(x) = (x + 1)^2 - 9$ **7.** $\left(-\frac{1}{2}, 1\right)$

9.

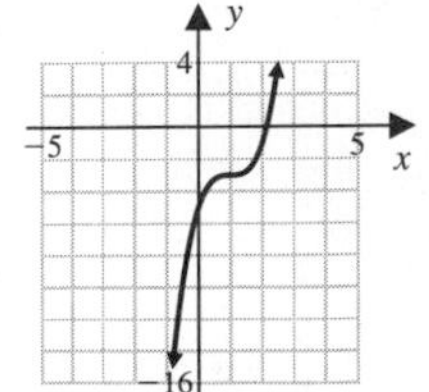

11. **(a)** $f(x) \to -\infty$
(b) $f(x) \to +\infty$

13. x intercepts: $-1, 0, 1$; $h(x) \to +\infty$ as $x \to +\infty$; $h(x) \to -\infty$ as $x \to -\infty$; $(-\infty, -1)$ or $(0, 1)$

15. $Q(x) = \frac{3x^3}{2} + \frac{13x^2}{4} - \frac{15x}{8} - \frac{23}{16}$; $R(x) = -\frac{7}{16}$ **17.** -14 **19.** **(a)** -122 **(b)** 3262

21. $f(x) = (x - 2)\left(x + 4 + \sqrt{11}\right)\left(x + 4 - \sqrt{11}\right)$

23. 3 positive zeros; 1 negative zero; upper bound: 3; lower bound: -1; $-1 \leq c < 0$, $0 \leq c < 1$, $1 \leq c < 2$, $2 \leq c < 3$

25. Rational zeros: $-1, 2, 3$ **27.** $f(0) < 0 < f(2)$ **29.** Zeros: $-2.6458, -1.4142, 1.4142, 2.6458$ (approx.)

31. -1 and 1, each with multiplicity 1; -4 with multiplicity 2; $i, -i$, each with multiplicity 1 **33.** $5 - i, i\sqrt{11}, -i\sqrt{11}$

35. $f(x) = x^3 - 6x^2 + 13x - 10$ **37.** 1 positive zero; 2 or 0 negative zeros; 0 or 2 imaginary zeros.

39. **(a)** $f(x) = (x - 1)(x + 1)(x^2 - 8x + 17)$ **(b)** $f(x) = (x - 1)(x + 1)(x - 4 - i)(x - 4 + i)$ **41.** All real numbers except 0, −3, 3

43. Vertical asymptote: $x = 1$; horizontal asymptote: $y = -5$; x intercepts: $1 + \sqrt{\frac{3}{5}}$, $1 - \sqrt{\frac{3}{5}}$; y intercept: −2

45. Vertical asymptote: $x = 7$; horizontal asymptote: $y = 4$; x intercept: $-\frac{1}{2}$; y intercept: $-\frac{2}{7}$

47. **(a)** $h(0) = 5$, $h(1) = 53$ **(b)** $0 \le t \le 4.08$ (approx.)
(c) The height increases during the first 2 sec, then decreases until the ball hits the ground. **(d)** 2 sec after being hit **(e)** 69 ft

49. **(a)** $N(0) = 20$, $N(10) = 620$, $N(25) = 957$ **(b)** 14.8°C and 35.2°C (approx.) **(c)** Maximum population at 25°C
(d) Maximum population is 957 insects (approx.)

51. **(a)** $0 < x < 100$ **(b)** $p = 2x + \dfrac{200}{x}$ **(c)** $C = 13.2x + \dfrac{1320}{x}$ **(d)** Cost is minimum when $x = 10$ ft; minimum cost is \$264.

CHAPTER 3 TEST (ON PAGE 226)

1. **(a)** C **(b)** A **(c)** B

2. **(a)** $f(x) = -2\left(x + \frac{5}{4}\right)^2 + \frac{49}{8}$ **(b)** Axis of symmetry: $x = -\frac{5}{4}$; vertex: $\left(-\frac{5}{4}, \frac{49}{8}\right)$ **(c)** Opens downward; maximum
(d) Domain: $\mathbb{R}$; Range: $\left(-\infty, \frac{49}{8}\right]$ **(e)** $(-\infty, -3]$ or $\left[\frac{1}{2}, \infty\right)$

3. Vertically stretch x^4 by 2, reflect about the x axis, shift to right 1 unit, and shift up 1 unit.
(a)

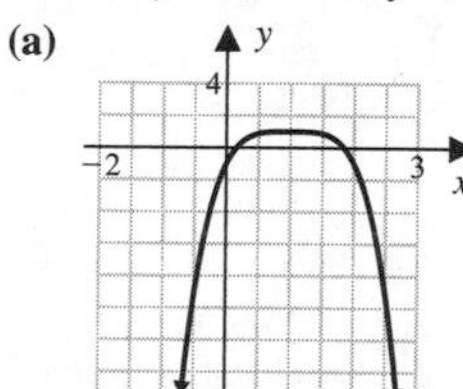

(b) As $x \to +\infty$, $f(x) \to -\infty$; as $x \to -\infty$, $f(x) \to -\infty$.

4. 0 (multiplicity 4), 1 (multiplicity 1), −1 (multiplicity 2)

5. **(a)** $Q(x) = x^4 - 8x^3 + 27x^2 - 56x + 100$; $R = -192$ **(b)** −192 **(c)** $x + 1$ is a factor.

6. **(a)**

(b) 2 negative real zeros; 1 positive real zero
(c) Rational zeros: $-1, -\frac{1}{3}, \frac{2}{3}$
(d) $f(x) = 9(x + 1)\left(x - \frac{2}{3}\right)\left(x + \frac{1}{3}\right)$
(e) $(-\infty, -1)$ or $\left(-\frac{1}{3}, \frac{2}{3}\right)$

7. $f(x) = x^4 - 7x^3 + 18x^2 - 22x + 12$

8. **(a)** If $x < 0$, $f(x) < 0$. **(b)** $f(0)$ is undefined. **(c)** As $x \to +\infty$, $f(x) \to 0$. **(d)** If $k < 0$, $f(x)$ lies in quadrants II and IV.

9. **(a)** Vertical asymptote: $x = 2$; horizontal asymptote: $y = 4$
(b) x intercept: $\frac{5}{4}$; y intercept: $\frac{5}{2}$
(c)

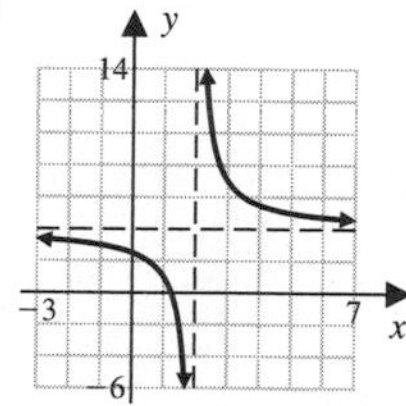

(d) Domain: All real numbers except 2; Range: All real numbers except 4

10. **(a)** $h(0) = 206$, $h(2) = 1078$, $h(4) = 1822$
(b)

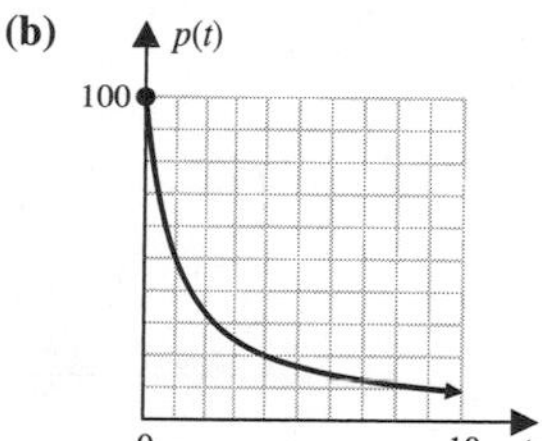

(c) $t = 14\frac{5}{8}$ sec
(d) 3628.25 ft

11. **(a)** $t \ge 0$
(b)

(c) As $t \to +\infty$, $p(t) \to 0$. As time goes on, the amount of information that is remembered decreases.

CHAPTER 4: PROBLEM SET 4.1 (ON PAGE 235)

1. **(a)** 8.8250 **(b)** 0.3752 **(c)** 2.5894 **(d)** 6.3697 **(e)** 85.3490
3. **(a)** 2 **(b)** 4 **(c)** 0.25 **(d)** 0.81 **(e)** 2.30 **(f)** 0.30 **5.** $(0, \infty)$ **7.** $(-\infty, 0)$ **9.** **(a)** B **(b)** C **(c)** D **(d)** A

11. Shift the graph of $y = 5^x$ left 1 unit.
Domain: $\mathbb{R}$;
Range: $(0, \infty)$;
increasing;
horizontal asymptote: x axis

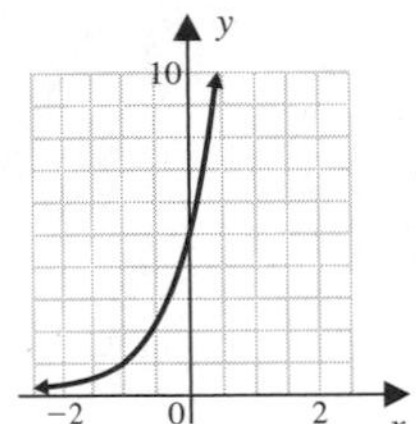

13. Shift the graph of $y = 2^x$ up 3 units.
Domain: $\mathbb{R}$;
Range: $(3, \infty)$;
increasing;
horizontal asymptote: $y = 3$

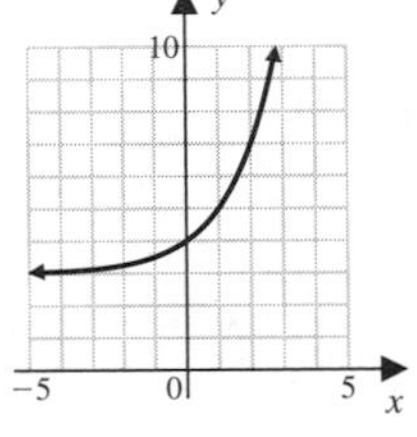
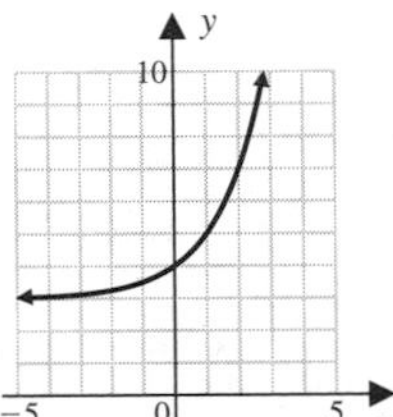

15. Stretch the graph of $y = 5^x$ vertically by a factor of 4.
Domain: $\mathbb{R}$;
Range: $(0, \infty)$;
increasing;
horizontal asymptote: x axis

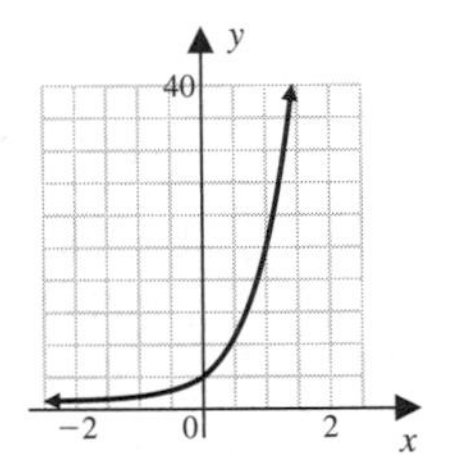

17. Reflect the graph of $y = 2^x$ across the y axis and shift down 3 units.
Domain: $\mathbb{R}$;
Range: $(-3, \infty)$;
decreasing;
horizontal asymptote:
$y = -3$

19. Vertically stretch the graph of $y = 3^x$ by 2, reflect about the x axis, shift left 1 unit, and shift down 1 unit.
Domain: $\mathbb{R}$;
Range: $(-\infty, -1)$;
decreasing;
horizontal asymptote:
$y = -1$

21. **(a)** $b = 3$
(b) $b = \frac{1}{2}$

23. Reflect the graph of f about the y axis.

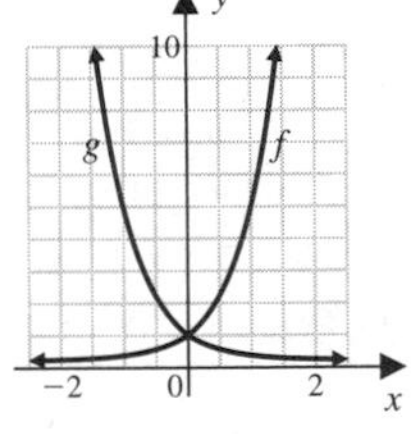
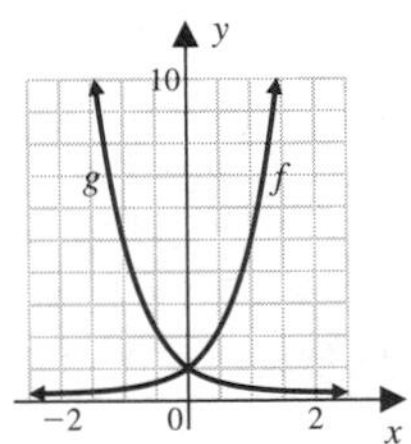

25. Domain: $\mathbb{R}$;
Range: approx. $[-0.2826, \infty)$;
no horizontal asymptote

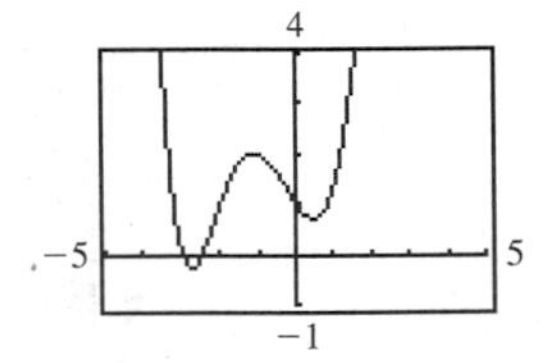

27. Domain: $\mathbb{R}$;
Range: $(0, 1]$;
horizontal asymptote: x axis

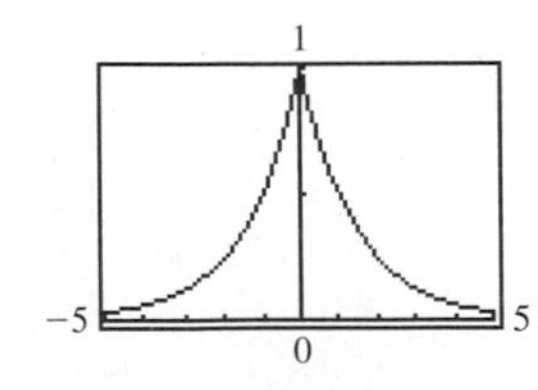

29. Domain: $\mathbb{R}$;
Range: $[0.01314, \infty)$ (approx.);
no horizontal asymptote

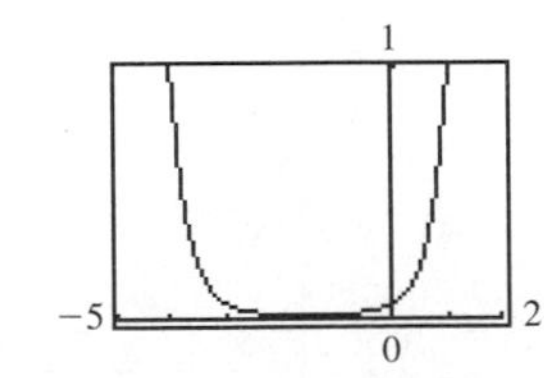

31. **(a)** $P(t) = 866(1.019)^t$
(b) 1026 million (approx.)
(c) 1999

33. **(a)** \$6638.48
(b) \$6651.82
(c) \$6660.88
(d) \$6664.39

35. **(a)** $S(t) = 10{,}450(1.0035417)^{12t}$ **(b)** 39 months; 102 months; 184 months
(c) 4.25% compounded monthly: \$11,868.37; 4.015% compounded daily: \$11,787.57
37. \$7169 **39.** **(a)** $V = \$19{,}545(0.80)^t$ **(b)** \$15,636; \$10,007; \$6405 **(c)** 3.0 yr
41. **(a)** 1.63 million; 0.69 million; 0.29 million **(b)** 2.1 yr; 3.7 yr; 5.3 yr **43.** $b^0 = 1$; $b^1 = b$ **45.** **(a)** $-\frac{1}{2}$, 4
47. **(b)** Horizontal line, $y = 1$ **(d)** The graph shows that $g(x) = 1$ for all x, as found in part (c).
49. **(a)** $2^{|x|}$ **(b)** $|2^x|$ **(c)** No

PROBLEM SET 4.2 (ON PAGE 246)

1. **(a)** 29.9641 **(b)** 0.0432 **(c)** 4.1133 **(d)** 0.1510 **(e)** −1.3544 **3.** $(-\infty, 0)$ **5.** $(0, \infty)$ **7.** $\mathbb{R}$

9. Vertically stretch by 2, reflect about the x axis; horizontal asymptote: x axis

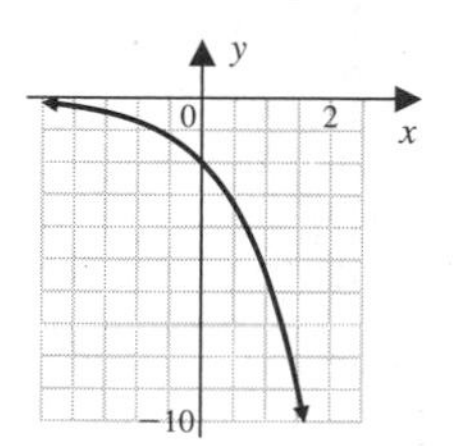

11. Shift up 1 unit; horizontal asymptote: $y = 1$

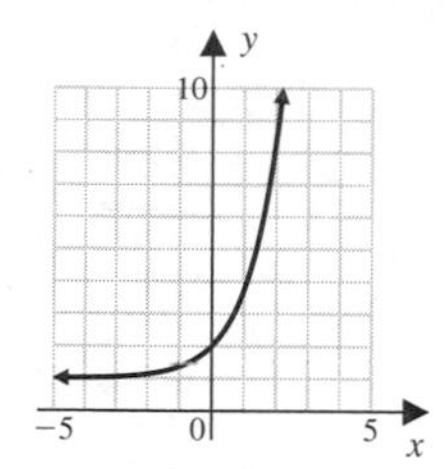

13. Reflect about the y axis, shift down 2 units; horizontal asymptote: $y = -2$

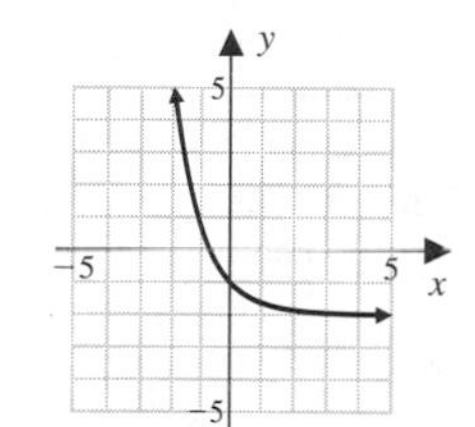

15. Vertically stretch by 3, reflect about the y axis, reflect about the x axis, shift up 2 units; horizontal asymptote: $y = 2$

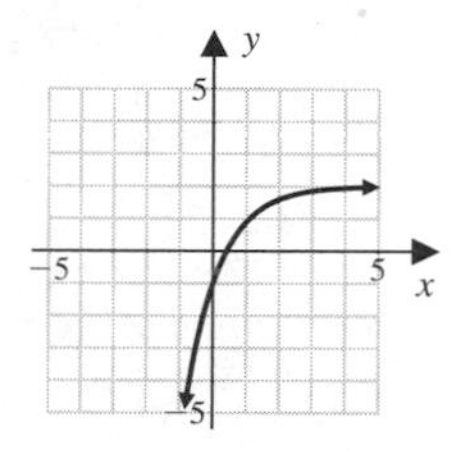

17. **(a)** B **(b)** C **(c)** D **(d)** A

19. Horizontal asymptote: x axis

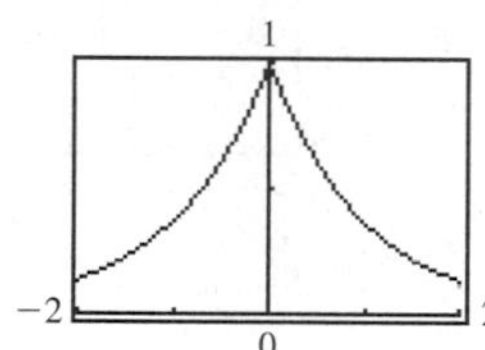

21. No horizontal asymptote

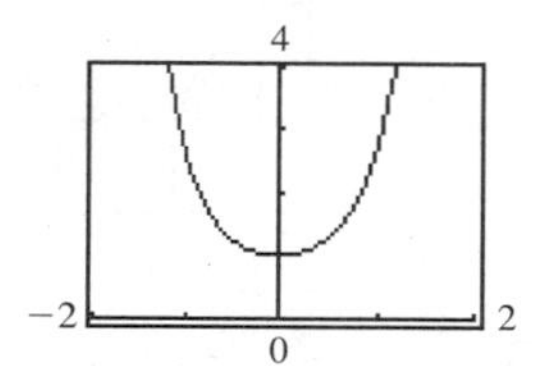

23. Horizontal asymptote: x axis

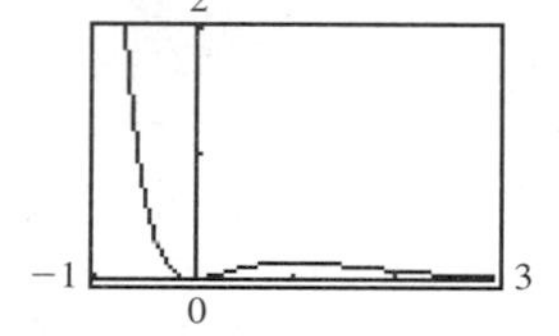

25. **(a)** \$9846.86 **(b)** \$12,303.74 **27.** **(a)** $S = 5000e^{0.0575t}$ **(b)** S increases exponentially as t increases. **(c)** \$6293.00 **(d)** Less
29. \$7225.27 **31.** **(a)** $P = 5.4e^{0.017t}$ **(b)** 6.29 billion **(c)** 2002; 2006; 2027 **33.** **(a)** 10% per day **(b)** 10.99 days
35. **(a)** 32g; 25.60 g; 13.11 g **(b)** A decreases exponentially as time elapses **(c)** 36.06 days **(d)** No; it approaches 0 as time elapses.
37. **(a)** 83.86%; 76.80% **(b)** P decreases exponentially as the depth increases. Range: $(0, 1]$ **(c)** 15.75 ft
39. **(b)** When $x = 0$, $y = 630$ ft. **(c)** 630 ft **41.** **(a)** 114; 3113; 3274 **(b)** 6 hr
43. For values of x near 0, $f(x)$ appears to be approximately equal to $g(x)$. **47.** $4(e^x + e^{-x})^{-2}$

PROBLEM SET 4.3 (ON PAGE 258)

1. **(a)** $\log_5 125 = 3$ **(b)** $\log_{32} 2 = \frac{1}{5}$ **(c)** $\log_e 17 = t$ or $\ln 17 = t$ **(d)** $\log_b(13z + 1) = x$
3. **(a)** $9^2 = 81$ **(b)** $10^{-4} = 0.0001$ **(c)** $c^w = 9$ **(d)** $e^{-1-3x} = \frac{1}{2}$
5. **(a)** 1 **(b)** $-\frac{1}{3}$ **(c)** $-\frac{4}{99}$ **(d)** $-1, 0$ **7.** **(a)** -3 **(b)** 4 **(c)** $\frac{1}{3}$ **(d)** $\frac{2}{3}$ **9.** **(a)** 36 **(b)** 16
11. **(a)** 3 **(b)** 8 **13.** **(a)** 10^{3-y} **(b)** $\frac{1}{2}(e^{y-8} - 1)$
15. **(a)** 1.5933; $10^{1.5933} = 39.2$ **(b)** 6.8680; $e^{6.8680} = 961$ **(c)** -0.1249; $10^{-0.1249} = \frac{3}{4}$ **(d)** -1.4697; $e^{-1.4697} = 0.23$
(e) 2.3026; $e^{2.3026} = 10$
17. **(a)** 10.6974 **(b)** 2.5924 **(c)** 0.0821 **(d)** 0.5112
19. **(a)** $\left(-\frac{1}{5}, \infty\right)$ **(b)** $(-\infty, 0)$ or $(1, \infty)$ **(c)** $(-\infty, -1)$ or $(1, \infty)$ **(d)** $(-\infty, -1)$ or $(1, \infty)$

21. $f^{-1}(x) = \log_5 x$;
Domain: $(0, \infty)$;
Range: $\mathbb{R}$

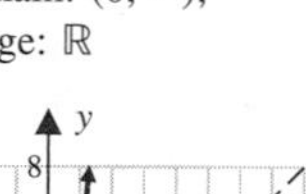

23. $h^{-1}(x) = \left(\frac{1}{10}\right)^x$;
Domain: $\mathbb{R}$;
Range: $(0, \infty)$

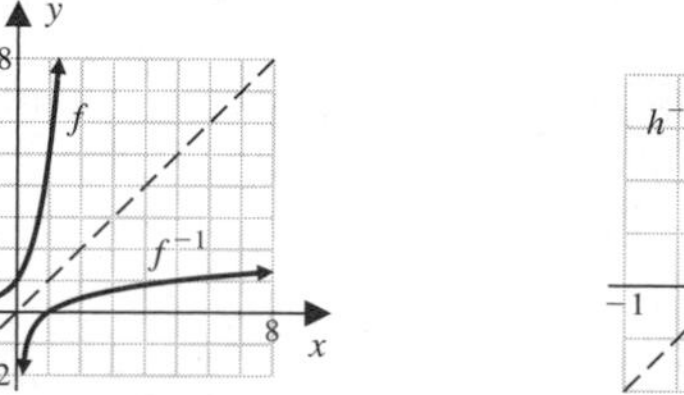

25. **(a)** $b = 4$ **(b)** $b = \frac{1}{16}$ **(c)** $b = e$

27. $f(x)$ is steeper than $g(x)$; $g(x)$ is steeper than $h(x)$; $f(x) \to +\infty$, $g(x) \to +\infty$, and $h(x) \to +\infty$ as $x \to +\infty$.

29. Domain: $(0, \infty)$; vertical asymptote: y axis

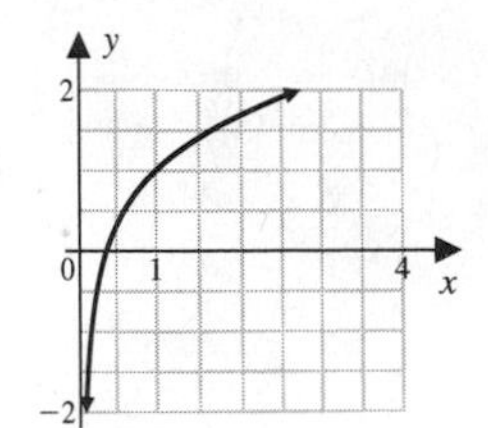

31. Domain: $(0, \infty)$; vertical asymptote: y axis

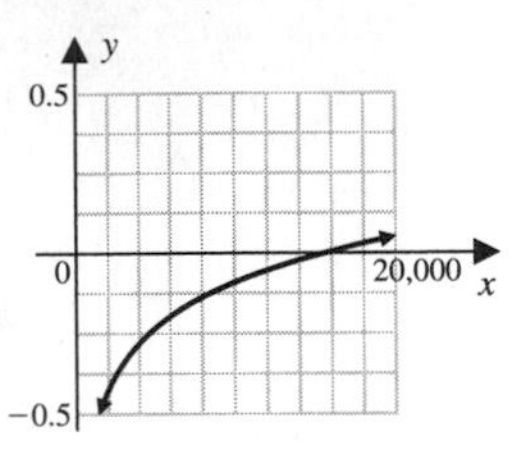

33. Domain: $(-3, \infty)$; vertical asymptote: $x = -3$

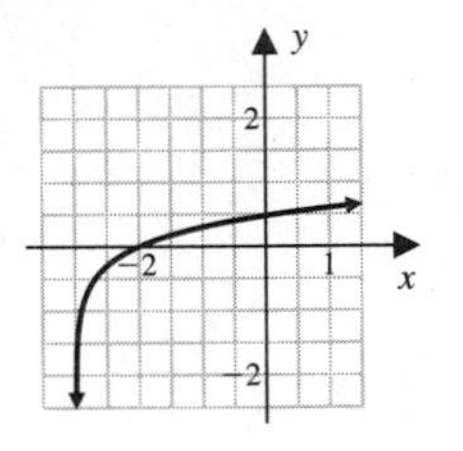

35. Domain: (e, ∞); vertical asymptote: $x = e$

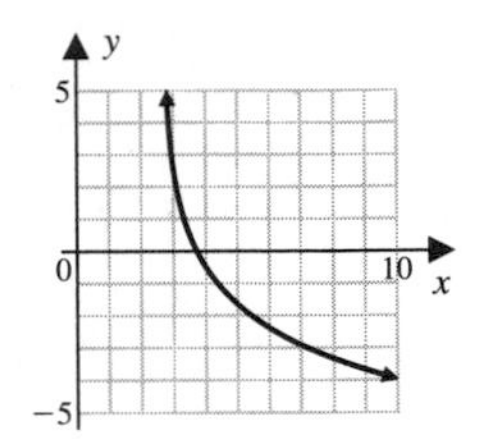

37. Domain: $(2, \infty)$; vertical asymptote: $x = 2$

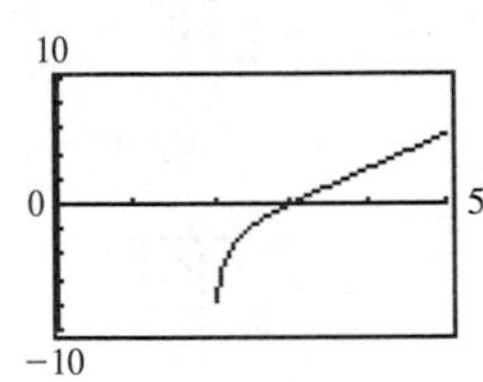

39. Domain: $(-\infty, 1)$ or $(4, \infty)$; vertical asymptotes: $x = 1$, $x = 4$

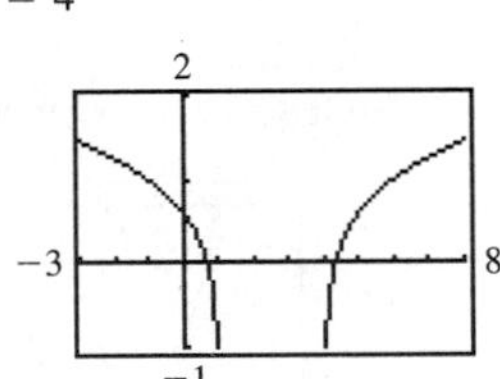

41. Domain: $\mathbb{R}$; no vertical asymptotes

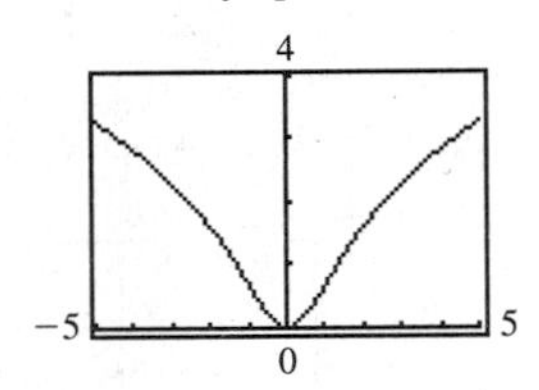

43. 7.7

45. **(a)** 2.5 **(b)** 7.8 **(c)** 6.4 **47.** **(a)** 90 db **(b)** 149.19 db
49. **(a)** \$2,352,000; \$3,481,000; \$4,042,000 **(b)** \$5000; \$27,000; \$95,000 **51.** $10a \log a < 10b \log b$ for $a > 0.3679$ **53.** P_1
55. **(a)** 1 **(b)** $(0, 1)$ **(c)** $(1, \infty)$

PROBLEM SET 4.4 (ON PAGE 267)

1. $\log_3 x + \log_3(x + 1)$ **3.** $9 \log_b x - 7 \log_b y$ **5.** $4 \log_b(x + 3)$ **7.** $\frac{5}{6} \log x - \frac{1}{3} \log y$ **9.** $\ln y + \frac{2}{3} \ln(3x + 1)$
11. $-\log_5(x + 2)$ **13.** $\log 5 + \log x + 2 \log(x^2 + 1) - \log(x + 1) - \frac{1}{2} \log(7x + 3)$ **15.** $\log_5 \frac{8}{7}$ **17.** $\log \frac{a}{7}$ **19.** $\ln\left(\frac{x + 3}{x - 3}\right)$

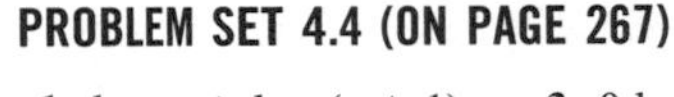

21. $\log\left(\frac{x + 2}{2x}\right)$ **23.** **(a)** 5 **(b)** $\frac{1}{9}$ **(c)** e^3x^4 **(d)** $1/x$ **(e)** $e^{-7}x^{-1}$ **(f)** $x^2 - 4$ **(g)** $e^{-1}x^2$ **(h)** xy^{-3} **25.** 3
27. 4 **29.** 7 **31.** $\frac{1}{3}$ **33.** 22 **35.** **(a)** 2.322 **(b)** 0.545 **(c)** −1.469 **(d)** 0.910

37. Domain: $(0, \infty)$; vertical asymptote: y axis

39. Domain: $\left(-\frac{1}{5}, \infty\right)$; vertical asymptote: $x = -\frac{1}{5}$

41. Domain: $\left(-\infty, -\frac{1}{2}\right)$ or $\left(\frac{1}{2}, \infty\right)$; vertical asymptotes: $x = -\frac{1}{2}$, $x = \frac{1}{2}$

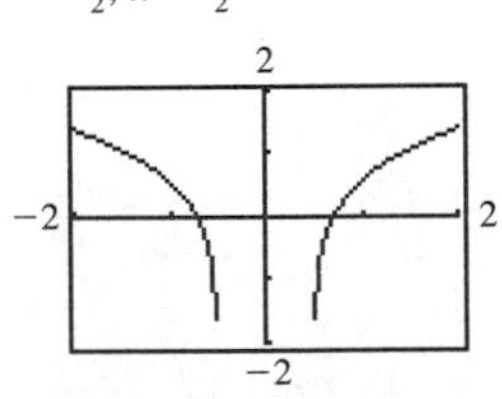

43. Domain: $(-\infty, -1)$ or $(1, \infty)$; vertical asymptotes: $x = -1$, $x = 1$

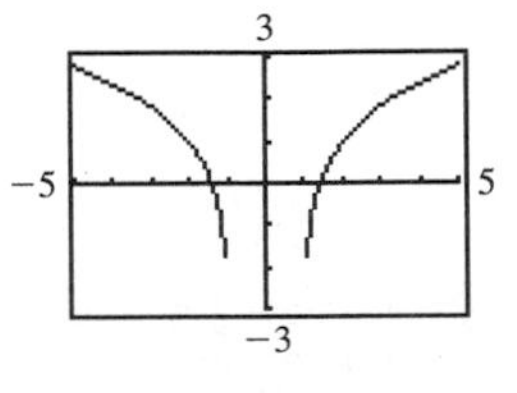

45. **(a)** $A = \dfrac{Pr(1 + r)^n}{(1 + r)^n - 1}$ **(b)** $A = 18{,}000\dfrac{\frac{0.1}{12}\left(1 + \frac{0.1}{12}\right)^n}{\left(1 + \frac{0.1}{12}\right)^n - 1}$; \$580.81; \$456.53; \$382.45

47. **(a)** $N = \dfrac{230}{1 + 6.931e^{-0.1702t}}$ **(b)** 29; 102; 158; 221 **(c)** 16 days; 23 days; never **51.** For instance, $a = b = 2$

55. f and g have different domains.

PROBLEM SET 4.5 (ON PAGE 272)

1. 1.3869 **3.** −0.3892 **5.** 0.6090 **7.** 1.3026 **9.** 2.5604 **11.** −3.3049 **13.** 0.6129 **15.** 0, 3.538 **17.** 0
19. −0.907, 1.500 **21.** 1.314 **23.** 1.310 **25.** $(-\infty, -0.69)$ **27.** $(1.70, \infty)$ **29.** $[0.12, \infty)$ **31.** $(1, \infty)$ **33.** 0.88

35. **(a)** 14.77 yr **(b)** 14.68 yr **(c)** 14.62 yr **(d)** 14.59 yr **37.** **(a)** $t = \dfrac{\log k}{4 \log 1.012125}$ **(b)** 14.38 yr; 22.79 yr; 28.76 yr
39. **(a)** 40 **41.** **(a)** 300 **(b)** 6.39 yr **43.** **(a)** $k = 0.15$ **(b)** 12.64 min **(c)** No; the graph of T never reaches $T = 32$.
45. $x = \ln 3$ **47.** $x = 1$ **49.** No; 1.27, 14.77

CHAPTER 4 REVIEW PROBLEM SET (ON PAGE 274)

1. **(a)** 2.5 **(b)** 0.16 **(c)** 0.760 **(d)** 4.889 **(e)** 0.211
3. **(a)** Shift the graph of $y = 2^x$ up 1 unit. Domain: $\mathbb{R}$; Range: $(1, \infty)$; increasing; horizontal asymptote: $y = 1$
(b) Reflect the graph of $y = 2^x$ about the x axis. Domain: $\mathbb{R}$; Range: $(-\infty, 0)$; decreasing; horizontal asymptote: x axis
(c) Vertically stretch the graph of $y = 2^x$ by 3. Domain: $\mathbb{R}$; Range: $(0, \infty)$; increasing; horizontal asymptote: x axis
(d) Shift the graph of $y = 2^x$ down 3 units. Domain: $\mathbb{R}$; Range: $(-3, \infty)$; increasing; horizontal asymptote: $y = -3$
(e) Shift the graph of $y = 2^x$ left 1 unit and shift down 3 units. Domain: $\mathbb{R}$; Range: $(-3, \infty)$; increasing; horizontal asymptote: $y = -3$
(f) Vertically stretch the graph of $y = 2^x$ by 3, reflect about the x axis, shift left 1 unit, and shift up 4 units. Domain: $\mathbb{R}$; Range: $(-\infty, 4)$; decreasing; horizontal asymptote: $y = 4$
5. **(a)** 24.533; 23.141; 0.243; 2.309 **(b)** $\dfrac{e^2(e^h - 1)}{h}$
(d) Vertically stretch the graph of f by 3, reflect about the x axis, and shift up 2 units.
7. **(a)** $\ln a = 3.1$ **(b)** $\ln(a + b) = x^2$ **(c)** $\log_5 b = -2.7$ **(d)** $\ln c = a + b$ **(e)** $\log t = a - b$
9. **(a)** $-\frac{11}{4}$ **(b)** $-4, 4$ **(c)** 2 **(d)** $\sqrt{\frac{3}{2}}$ **(e)** $8/(e - 1)$ **(f)** 0 **(g)** $e^{(b-t)/3}$ **(h)** $\frac{1}{2}(u - y)$
11. $\frac{1}{2}\ln\left(\dfrac{u + 1}{u - 1}\right)$ **13.** **(a)** False **(b)** False **(c)** False **(d)** False
15. **(a)** Domain: $(-\infty, 0)$; vertical asymptote: y axis **(b)** Domain: $(-\infty, -3)$ or $(4, \infty)$; vertical asymptotes: $x = -3, 4$
(c) Domain: $(-\infty, -2)$; vertical asymptote: $x = -2$ **(d)** Domain: $(-\infty, 0)$ or $(1, \infty)$; vertical asymptotes: y axis, $x = 1$
17. **(a)** $(-\infty, 0.46)$ **(b)** $(0, 0.16)$, $(1.14, \infty)$ **(c)** $[-8.13, \infty)$
19. **(a)** At the end of 3, 7, and 10 yr, the accumulated balance will be \$11,576.25, \$14,071.00, and \$16,288.95, respectively.
(b) 25.68 yr (approx.)
21. **(a)** After 2 and 3 yr, the value of the car will be \$8437.50 and \$6328.13, respectively. **(b)** 1.24 yr (approx.)
23. **(a)** 95.12 g **(b)** 1204 yr (approx.) **25.** **(a)** 15.5 yr **(b)** 15.4 yr (approx.) **27.** **(a)** 2.80 **(b)** 10.50
29. **(a)** 298; 333 **(b)** No

CHAPTER 4 TEST (ON PAGE 276)

1. **(a)** 9 **(b)** 9 **(c)** 0.794 **(d)** 20.832
2. **(a)**

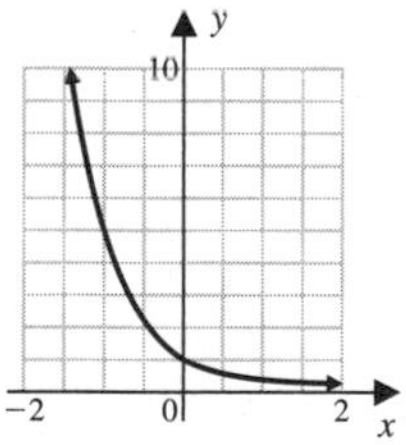

(b) Domain: $\mathbb{R}$; Range: $(0, \infty)$
(c) Decreasing
(d) Horizontal asymptote: x axis
(e) Vertically stretch the graph of f by 2 and shift down 3 units. Horizontal asymptote: $y = -3$

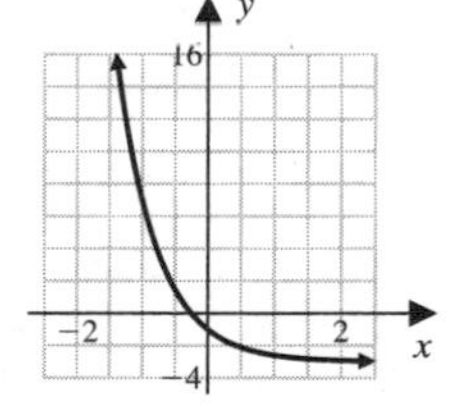

3. **(a)** $\log_6 7 = u$ **(b)** $3^b = x$ **4.** **(a)** -3 **(b)** $\frac{1}{7}$ **(c)** $\sqrt{x + 3}$ **(d)** x^2
5. **(a)** -1; 0; 0.4771; 0.8451
(b) Domain: $(0, \infty)$; Range: $\mathbb{R}$

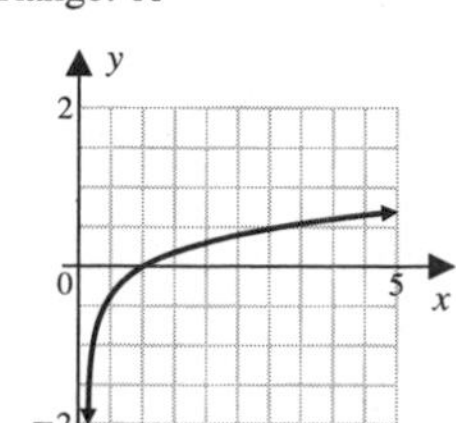

(c) $f^{-1}(x) = 10^x$

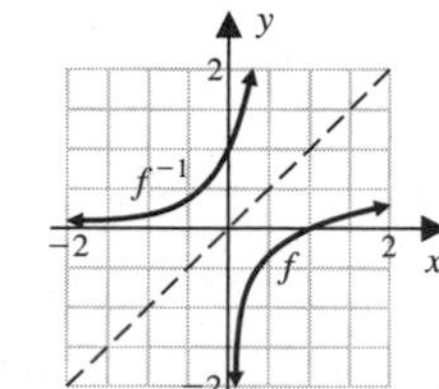

(d) Reflect the graph of f about the x axis and shift left 7 units. Vertical asymptote: $x = -7$

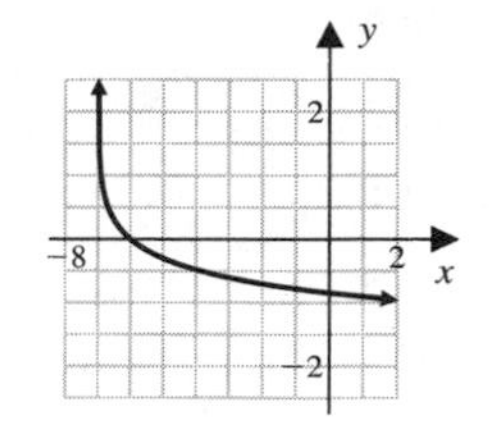

6. $A = 1, B = 5$ **7. (a)** True **(b)** False **(c)** True **(d)** False **8. (a)** 5 **(b)** 3 **(c)** 1 **(d)** -0.23

9. $\frac{3}{2}$ **10. (a)** \$4420.65 **(b)** 22 yr (approx.) **11. (a)** 16,970 **(b)** N increases exponentially over time.

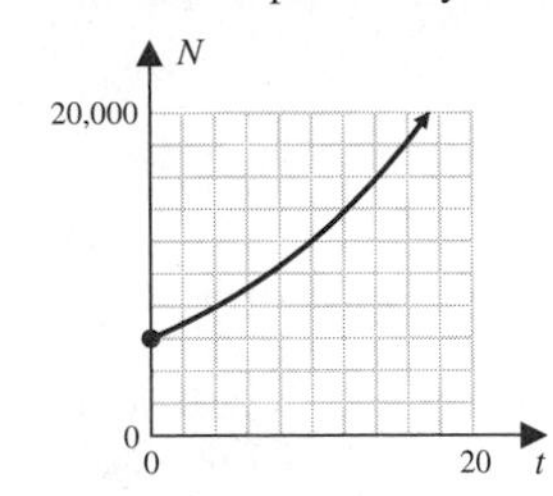

(c) 12.22 hr (approx.)

CHAPTER 5: PROBLEM SET 5.1 (ON PAGE 285)

1. (a) 16°18′36″ **(b)** −87°48′36″ **3. (a)** −156°37′48″ **(b)** 89°44′24″ **5. (a)** 38.3° **(b)** 65.19°

7. (a) −141.47° **(b)** 244.77° **9. (a)** $\frac{5\pi}{12} = 1.31$ **(b)** $-\frac{3\pi}{4} = -2.36$ **11. (a)** $-\frac{19\pi}{36} = -1.66$ **(b)** $\frac{37\pi}{15} = 7.75$

13. (a) $\frac{3\pi}{8}$, 1.18 **(b)** $\frac{339\pi}{2000}$, 0.53 **15. (a)** 120° **(b)** 1290° **17. (a)** 105° **(b)** −80°

19. (a) 286.48° **(b)** −131.78° **21. (a)** (i) 54° (ii) 144° **(b)** (i) 12° (ii) 102°

23. (a) (i) $\frac{5\pi}{14}$ (ii) $\frac{6\pi}{7}$ **(b)** (i) $\frac{\pi}{18}$ (ii) $\frac{5\pi}{9}$ **25.** 4.71 in. **27.** 6.82 m **29.** 0.95 yd

31. (a) Quadrant II; coterminal: $\frac{11\pi}{4}, -\frac{5\pi}{4}$ **(b)** Quadrant IV; coterminal: $\frac{11\pi}{6}, -\frac{13\pi}{6}$

33. (a) Quadrant IV; coterminal: 315°, −405° **(b)** Quadrant II; coterminal: 480°, −240°

35. (a) Quadrant IV; coterminal: $\frac{5\pi}{3}, -\frac{\pi}{3}$ **(b)** Quadrant I; coterminal: 20°, −340° **37. (a)** 1.15 mi **(b)** 6082.12 ft

39. 20.94 ft **41.** 10.05 cm/sec **43. (a)** 5280 rad/min **(b)** 840.34 rotations/min **45. (a)** 863.94 in./min **(b)** 30.56 rotations/min

47. $\omega = 0.63$ rad/sec; $v = 5.03$ ft/sec **49.** Radian measure depends on s as well as r. **51. (a)** 16.49 in.^{2} **(b)** 22.62 m^2

53. $\theta_1 \neq \theta_2$; in general, for example, $\theta_1 = \frac{\pi}{3}$ and $\theta_2 = \frac{7\pi}{3}$

PROBLEM SET 5.2 (ON PAGE 296)

1. $\sin\theta = \frac{3}{5} = \frac{12}{20}$; $\cos\theta = \frac{4}{5} = \frac{16}{20}$; $\tan\theta = \frac{3}{4} = \frac{12}{16}$; $\csc\theta = \frac{5}{3} = \frac{20}{12}$; $\sec\theta = \frac{5}{4} = \frac{20}{16}$; $\cot\theta = \frac{4}{3} = \frac{16}{12}$

3. $\sin\theta = 0.625$; $\cos\theta = 0.781$; $\tan\theta = 0.801$; $\csc\theta = 1.600$; $\sec\theta = 1.281$; $\cot\theta = 1.249$

5. $\sin\theta = 0.324$; $\cos\theta = 0.946$; $\tan\theta = 0.342$; $\csc\theta = 3.086$; $\sec\theta = 1.057$; $\cot\theta = 2.920$

7. $\sin\theta = 0.882$; $\cos\theta = 0.471$; $\tan\theta = 1.875$; $\csc\theta = 1.133$; $\sec\theta = 2.125$; $\cot\theta = 0.533$

9. $\cos\theta = \frac{3}{5}$; $\tan\theta = \frac{4}{3}$; $\csc\theta = \frac{5}{4}$; $\sec\theta = \frac{5}{3}$; $\cot\theta = \frac{3}{4}$ **11.** 1 **13.** $\sqrt{2}/2$ **15.** 1

17. $\sin\theta = \frac{3}{5}$; $\cos\theta = \frac{4}{5}$; $\tan\theta = \frac{3}{4}$; $\csc\theta = \frac{5}{3}$; $\sec\theta = \frac{5}{4}$; $\cot\theta = \frac{4}{3}$

19. $\sin\theta = -\frac{4}{5}$; $\cos\theta = -\frac{3}{5}$; $\tan\theta = \frac{4}{3}$; $\csc\theta = -\frac{5}{4}$; $\sec\theta = -\frac{5}{3}$; $\cot\theta = \frac{3}{4}$

21. $\sin\theta = -0.781$; $\cos\theta = 0.625$; $\tan\theta = -1.250$; $\csc\theta = -1.281$; $\sec\theta = 1.601$; $\cot\theta = -0.800$

23. $\sin\theta = -0.898$; $\cos\theta = 0.439$; $\tan\theta = -2.044$; $\csc\theta = -1.113$; $\sec\theta = 2.276$; $\cot\theta = -0.489$

25. $\sin\theta = -\frac{15}{17}$; $\cos\theta = \frac{8}{17}$; $\tan\theta = -\frac{15}{8}$; $\csc\theta = -\frac{17}{15}$; $\sec\theta = \frac{17}{8}$; $\cot\theta = -\frac{8}{15}$

27. $\sin\theta = -\sqrt{2}/2$; $\cos\theta = -\sqrt{2}/2$; $\tan\theta = 1$; $\csc\theta = -\sqrt{2}$; $\sec\theta = -\sqrt{2}$; $\cot\theta = 1$

29. $\sin\theta = \frac{\sqrt{1+\pi^4}}{1+\pi^4}$; $\cos\theta = \frac{\pi^2\sqrt{1+\pi^4}}{1+\pi^4}$; $\tan\theta = \frac{1}{\pi^2}$; $\csc\theta = \sqrt{1+\pi^4}$; $\sec\theta = \frac{\sqrt{1+\pi^4}}{\pi^2}$; $\cot\theta = \pi^2$

31. (a) $\sin 90° = 1$; $\cos 90° = 0$; $\tan 90°$ is undefined; $\csc 90° = 1$; $\sec 90°$ is undefined; $\cot 90° = 0$
(b) $\sin(-\pi/2) = -1$; $\cos(-\pi/2) = 0$; $\tan(-\pi/2)$ is undefined; $\csc(-\pi/2) = -1$; $\sec(-\pi/2)$ is undefined; $\cot(-\pi/2) = 0$

33. (a) IV **(b)** IV **35. (a)** II **(b)** IV **37. (a)** $\frac{1}{2}$ **(b)** $-\sqrt{2}/2$ **39. (a)** $\sqrt{3}$ **(b)** $-\sqrt{3}$

41. **(a)** $\sqrt{2}$ **(b)** $-(2\sqrt{3})/3$ **43.** **(a)** -1 **(b)** $\sqrt{3}$ **45.** **(a)** 0.6820 **(b)** -1.1371 **47.** **(a)** 0.0230 **(b)** -0.9738
49. **(a)** 0.7265 **(b)** 0.3127 **51.** **(a)** -0.4612 **(b)** 1.2892 **53.** Because $|\sin\theta| \le 1$ and $|\cos\theta| \le 1$
55. For example, let $\alpha = \pi/6$ and $\beta = \pi/3$.

PROBLEM SET 5.3 (ON PAGE 305)

1. **(a)** $b = \frac{15}{17}$ **(b)** $a = \frac{8}{17}$ **3.** **(a)** $a = \sqrt{5}/5$ **(b)** $b = \frac{1}{5}$ **5.** $a = -0.4534$ **7.** 10 arcs of $\pi/5$
9. 3 arcs of 2 and 1 arc of 0.28 **11.** 12 arcs of 0.5 and 1 arc of 0.28 **13.** **(a)** I **(b)** III
15. **(a)** II **(b)** II **17.** **(a)** IV **(b)** I **19.** **(a)** I **(b)** III
21. $\sin t = \frac{1}{2}$; $\cos t = -\frac{\sqrt{3}}{2}$; $\tan t = -\frac{\sqrt{3}}{3}$; $\csc t = 2$; $\sec t = -\frac{2\sqrt{3}}{3}$; $\cot t = -\sqrt{3}$
23. $\sin t = -\frac{12}{13}$; $\cos t = \frac{5}{13}$; $\tan t = -\frac{12}{5}$; $\csc t = -\frac{13}{12}$; $\sec t = \frac{13}{5}$; $\cot t = -\frac{5}{12}$
25. $\sin t = \frac{2\sqrt{13}}{13}$; $\cos t = -\frac{3\sqrt{13}}{13}$; $\tan t = -\frac{2}{3}$; $\csc t = \frac{\sqrt{13}}{2}$; $\sec t = -\frac{\sqrt{13}}{3}$; $\cot t = -\frac{3}{2}$
27. **(a)** $(0, -1)$; $\sin t = -1$; $\cos t = 0$; $\tan t$ is undefined; $\csc t = -1$; $\sec t$ is undefined; $\cot t = 0$ **(b)** Same as part (a)
29. **(a)** $(1, 0)$; $\sin t = 0$; $\cos t = 1$; $\tan t = 0$; $\csc t$ is undefined; $\sec t = 1$; $\cot t$ is undefined
(b) $(0, -1)$; $\sin t = -1$; $\cos t = 0$; $\tan t$ is undefined; $\csc t = -1$; $\sec t$ is undefined; $\cot t = 0$
31. **(a)** $\sqrt{2}/2$ **(b)** $-\frac{1}{2}$ **33.** **(a)** $-\frac{1}{2}$ **(b)** $-\sqrt{3}/2$ **35.** **(a)** $-\sqrt{2}/2$ **(b)** $-\sqrt{2}/2$ **37.** **(a)** $\frac{1}{2}$ **(b)** $-\sqrt{2}/2$
39. **(a)** 0.6765 **(b)** -0.6737 **41.** **(a)** 0.6438 **(b)** 1.6797 **43.** **(a)** 1.4965 **(b)** -1.1261
45. **(a)** $(-0.9041, 0.4274)$ **(b)** $(-0.5076, 0.8616)$ **47.** **(a)** $(0.9081, -0.4187)$ **(b)** $(0.4047, -0.9145)$ **49.** Tangent and secant
51. $\sin t = -(3\sqrt{10})/10$; $\cos t = -\sqrt{10}/10$; $\tan t = 3$; $\csc t = -\sqrt{10}/3$; $\sec t = -\sqrt{10}$; $\cot t = \frac{1}{3}$
53. **(a)** 0.9983 **(b)** 1.0000 **(c)** 1.0000 **(d)** 1.0000 **(e)** $f(t) \to 1$ as $t \to 0^+$

55.

t	$\cos t$	$f(t)$
0.1	0.9950	0.9950
0.2	0.9801	0.9801
−0.3	0.9553	0.9553
1	0.5403	0.5417

PROBLEM SET 5.4 (ON PAGE 314)

1. **(a)** $-\sqrt{3}/2$ **(b)** $\sqrt{3}/2$ **3.** **(a)** $-\sqrt{2}/2$ **(b)** 1 **5.** **(a)** $-\sqrt{2}$ **(b)** $-\sqrt{3}$ **7.** **(a)** -1 **(b)** $\sqrt{3}/2$
9. **(a)** -1 **(b)** $\sqrt{3}/3$ **11.** **(a)** $(2\pi)/3$ **(b)** $\pi/2$ **13.** **(a)** 6π **(b)** 3 **15.** **(a)** 8 **(b)** $\pi/3$
17. $\cos t = \frac{8}{17}$; $\tan t = \frac{15}{8}$; $\csc t = \frac{17}{15}$; $\sec t = \frac{17}{8}$; $\cot t = \frac{8}{15}$ **19.** $\sin t = \frac{12}{13}$; $\tan t = \frac{12}{5}$; $\csc t = \frac{13}{12}$; $\sec t = \frac{13}{5}$; $\cot t = \frac{5}{12}$
21. $\sin\theta = \frac{8}{17}$; $\cos\theta = -\frac{15}{17}$; $\tan\theta = -\frac{8}{15}$; $\sec\theta = -\frac{17}{15}$; $\cot\theta = -\frac{15}{8}$
23. $\sin t = \frac{-3\sqrt{10}}{10}$; $\cos t = \frac{\sqrt{10}}{10}$; $\csc t = -\frac{\sqrt{10}}{3}$; $\sec t = \sqrt{10}$; $\cot t = -\frac{1}{3}$
25. $\cos t = -0.82$; $\tan t = -0.69$; $\csc t = 1.75$; $\sec t = -1.22$; $\cot t = -1.44$
27. $\sin t = 0.84$; $\cos t = -0.55$; $\csc t = 1.19$; $\sec t = -1.83$; $\cot t = -0.65$ **29.** **(a)** $\sin t$ **(b)** $8 - 5\sin^2 t$ **(c)** $\sin^2 t$
31. **(a)** $\cos t$ **(b)** $\cos t + 1$ **(c)** $2\cos^2 t - 1$ **33.** **(a)** $2\cos^4\theta$ **(b)** $-\csc\theta$ **35.** **(a)** $\tan t$ **(b)** $\cos\theta$
37. **(a)** $\csc t$ **(b)** $\sec\theta$ **39.** **(a)** $\csc\theta$ **(b)** 4
41. **(a)** (i) $P(3)$ − \$700,000 = $P(15)$ (ii) $P(6)$ = \$1,400,000 = $P(18)$ (iii) $P(9)$ − \$700,000 = $P(21)$ (iv) $P(0)$ = \$0 = $P(12)$
(b) Yes, the period is 12 months. **(c)** Sales increase to a maximum in the middle of the summer; then decrease steadily.
43. **(a)** 72.38 V; −52.54 V; −148.98 V; 169.99 V **(b)** The period is $(2\pi)/377$. The voltage repeats every $(2\pi)/377$ sec. **(c)** 60
49. **(a)** $\sin(-3.752) = 0.5732 = -\sin 3.752$ **(b)** $\cos(-5.384) = 0.6222 = \cos 5.384$

PROBLEM SET 5.5 (ON PAGE 325)

1. Minimum points: $(\pi + k \cdot 2\pi, -1)$, where k is any integer; Maximum points: $(k \cdot 2\pi, 1)$, where k is any integer;
x intercepts: $x = (\pi/2) + k \cdot \pi$, where k is any integer.

3. **(a)** Range: $[-3, 3]$;
Amplitude $= 3$;
Period $= 2\pi$

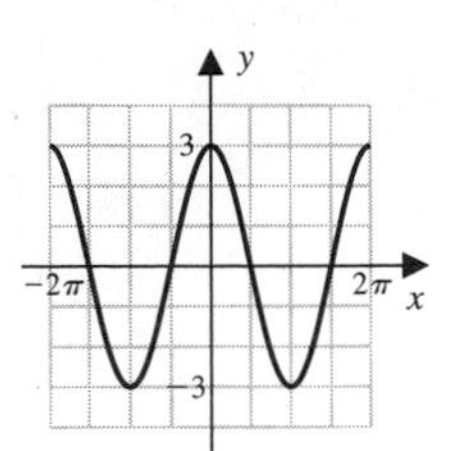

(b) Range: $[-3, 3]$;
Amplitude $= 3$;
Period $= 2\pi$

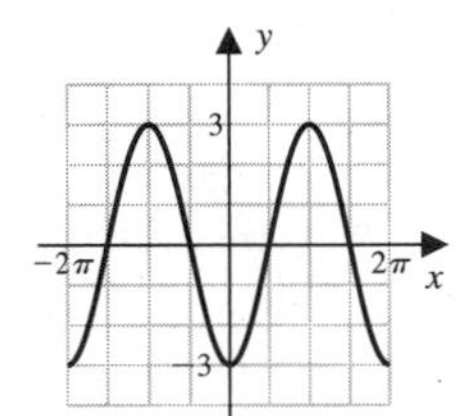

5. **(a)** Range: $\left[-\frac{2}{3}, \frac{2}{3}\right]$;
Amplitude $= \frac{2}{3}$;
Period $= 2\pi$

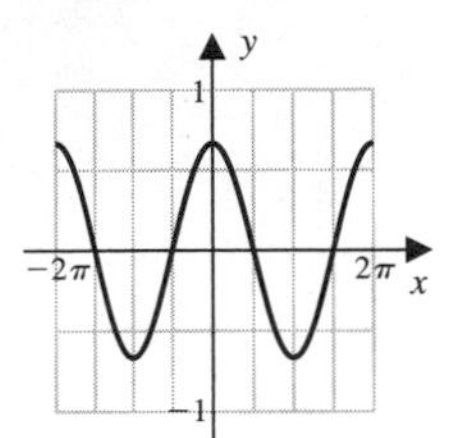

(b) Range: $\left[-\frac{1}{2}, \frac{1}{2}\right]$;
Amplitude $= \frac{1}{2}$;
Period $= 2\pi$

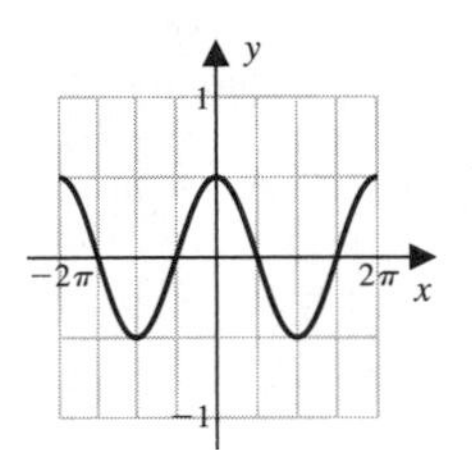

7. **(a)** Range: $[-1, 1]$;
Amplitude $= 1$;
Period $= \pi$

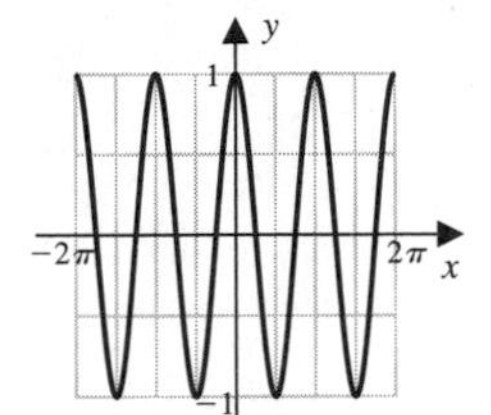

(b) Range:$[-1, 1]$;
Amplitude $= 1$;
Period $= 4\pi$

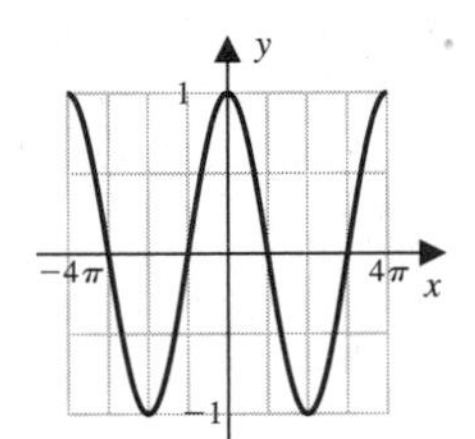

9. **(a)** Range: $[-1, 1]$;
Amplitude $= 1$;
Period $= 6$

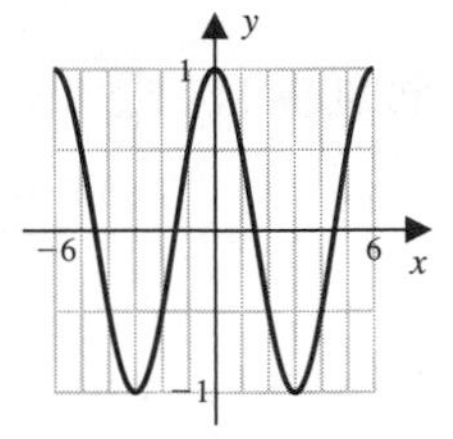

(b) Range: $[-1, 1]$;
Amplitude $= 1$;
Period $= 1$

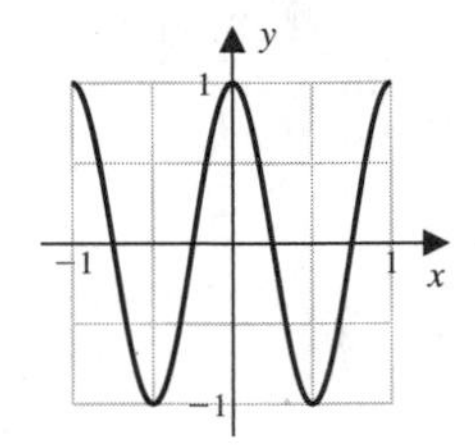

11. **(a)** Range: $\left[-\frac{5}{3}, \frac{5}{3}\right]$;
Amplitude $= \frac{5}{3}$;
Period $= \pi/3$

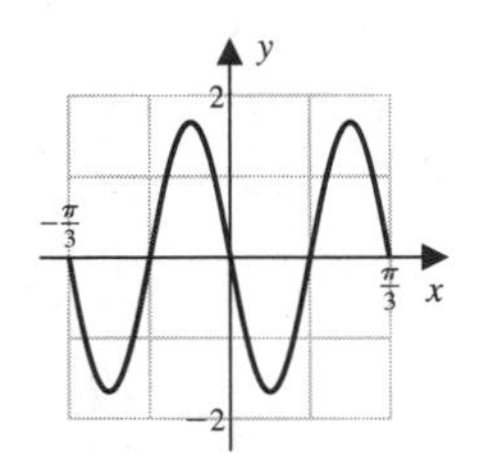

(b) Range: $[-5, 5]$;
Amplitude $= 5$;
Period: $= \pi/5$

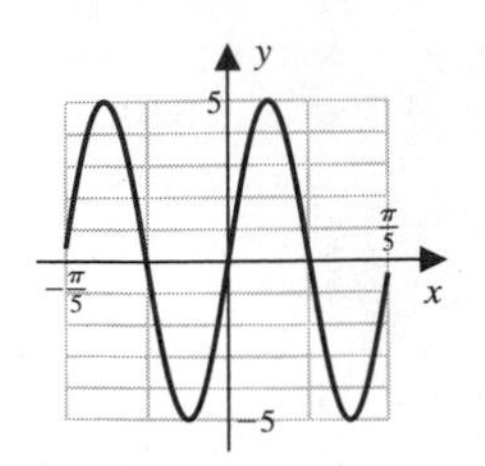

13. **(a)** Range: $[-3, 3]$;
Amplitude $= 3$;
Period $= 4$

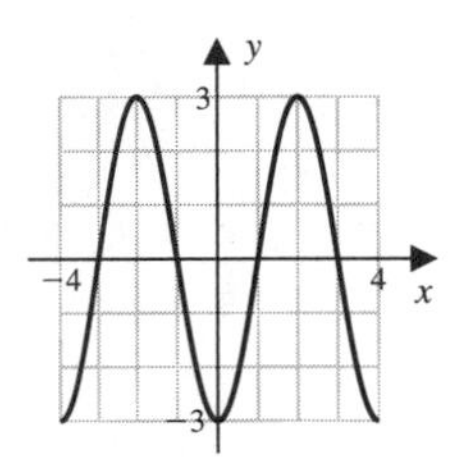

(b) Range: $[-\pi, \pi]$;
Amplitude $= \pi$;
Period $= (4\pi)/5$

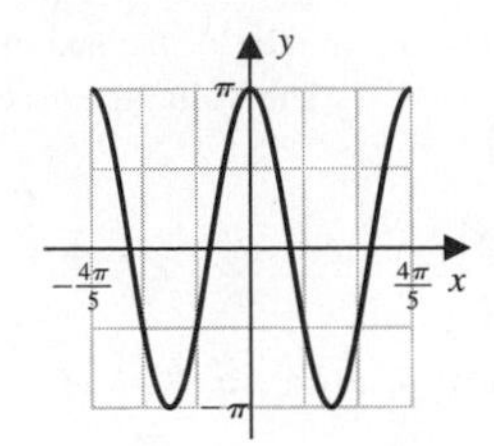

15. **(a)** Range: $[0, 2]$;
Amplitude $= 1$;
Period $= 2\pi$

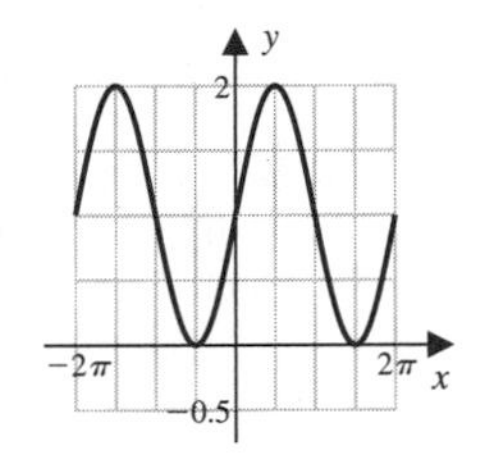

(b) Range: $[2, 4]$;
Amplitude $= 1$;
Period $= 2\pi$

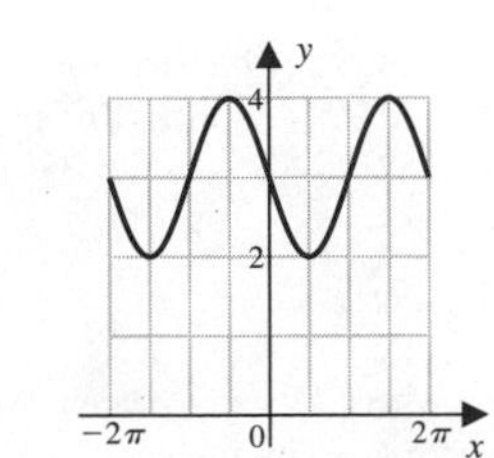

17. **(a)** Range: $[-1, 5]$;
Amplitude $= 3$;
Period $= 8$

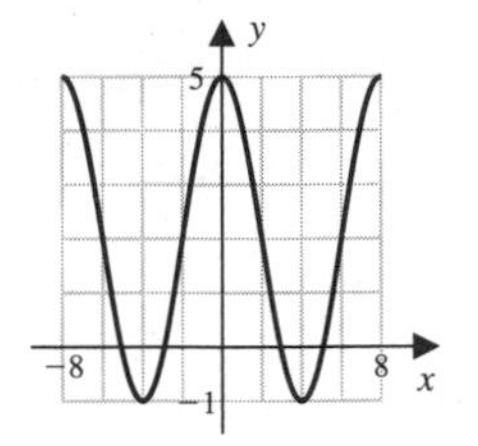

(b) Range: $[-5, 1]$;
Amplitude $= 3$;
Period $= 1$

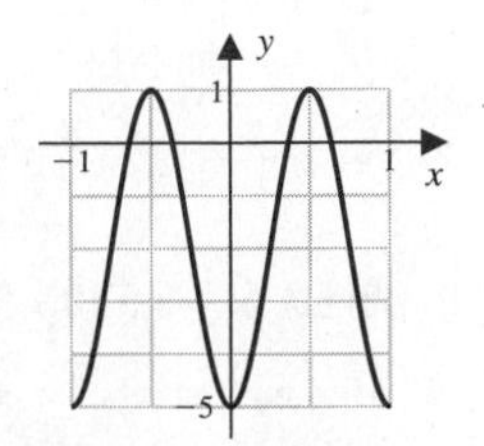

19. C; B; A

21. Amplitude = 2;
Period = 2π;
Phase shift = $\pi/4$

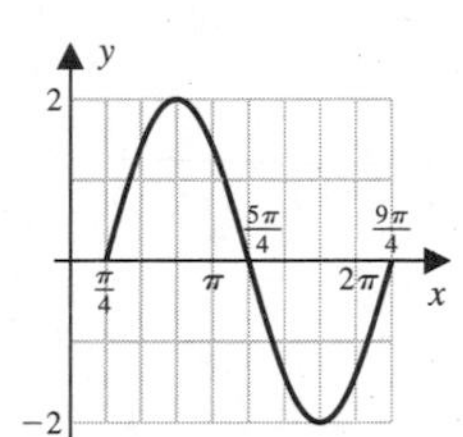

23. Amplitude = 4;
Period = 2π;
Phase shift = $-\pi/6$

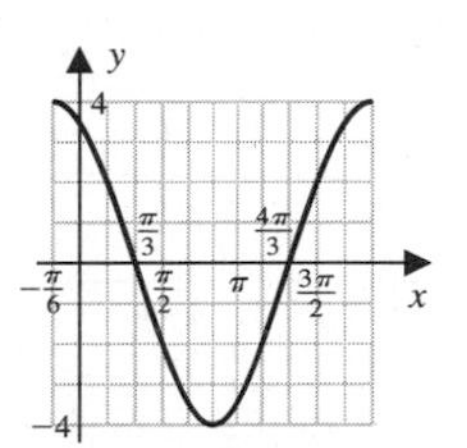

25. Amplitude = 2;
Period = π;
Phase shift = $\pi/2$

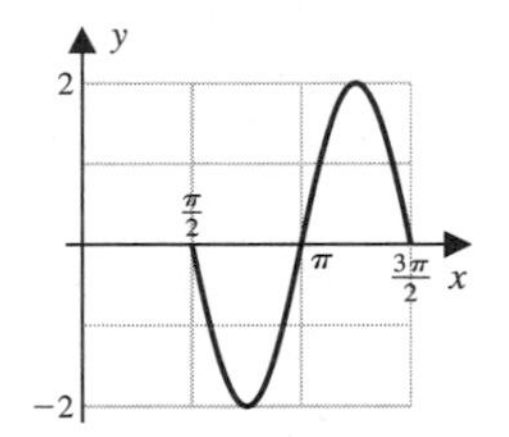

27. Amplitude = 2;
Period = 2;
Phase shift = 1

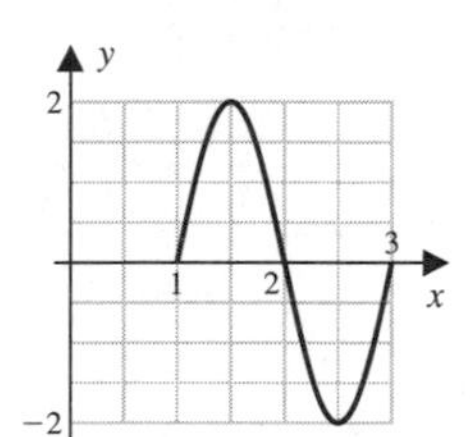

29. Amplitude = 3;
Period = $(2\pi)/3$;
Phase shift = $-\frac{2}{3}$

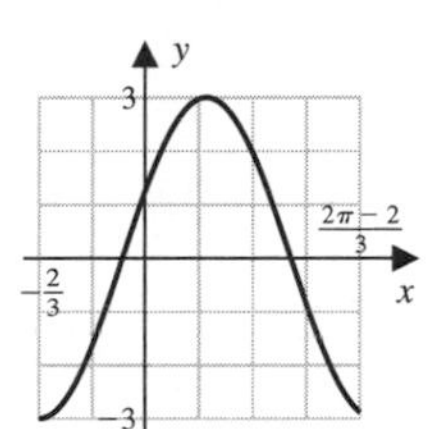

31. Amplitude = 2;
Period = 4;
Phase shift = $1/(2\pi)$

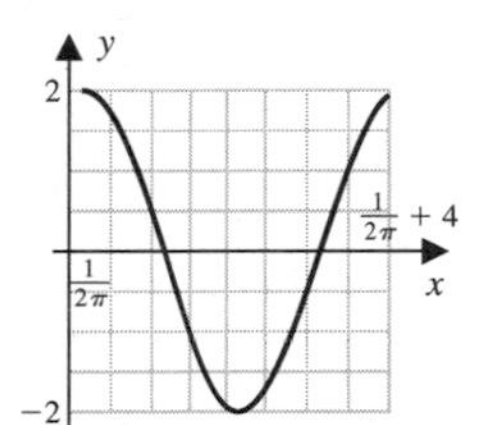

33. Amplitude = 2;
Period = $(2\pi)/3$;
Phase shift = $\frac{5}{3}$

35. Amplitude = 3;
Period = π;
Phase shift = π

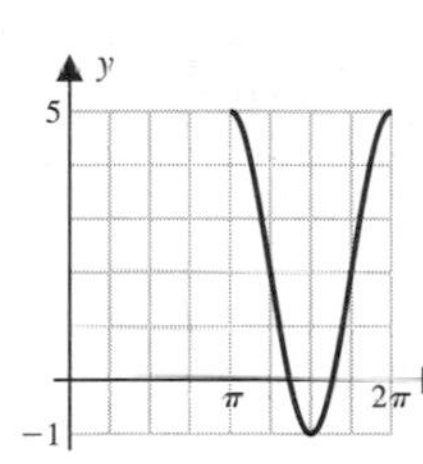

37. **(a)** $y = 2\sin\left(2x - \frac{\pi}{3}\right)$;
Amplitude = 2;
Period = π;
Phase shift = $\pi/6$
(b) $y = 5\sin(4x - \pi)$;
Amplitude = 5;
Period = $\pi/2$
Phase shift = $\pi/4$

39. $y = 3\sin 6x$

41. The line $y = 1$

43. **(a)** No
(b) $f(x) \to 1$ as $x \to 0$

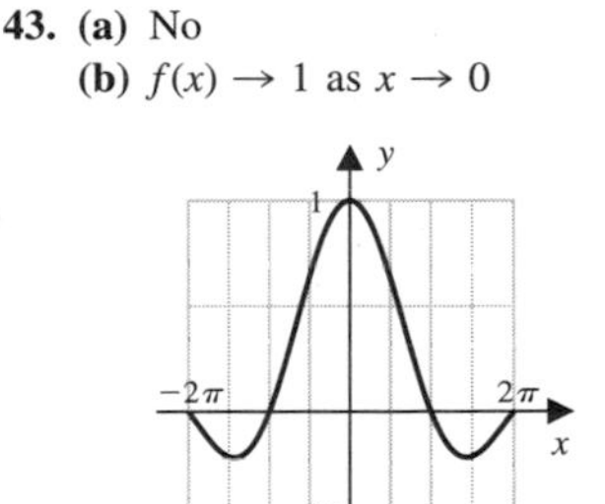

45. **(b)** $\frac{5}{6}$ **(c)** Maximum: 140 mm Hg; Minimum: 80 mm Hg

47. **(a)** 12.24 ft; −0.37 ft; 14.37 ft **(b)** 10 sec **(d)** Minimum: −1 ft; Maximum: 15 ft

49. **(b)** Amplitude: 700; Period: 12 months **(c)** \$900,000; \$900,000; \$200,000; \$1,600,000; \$200,000
(d) \$1,600,000 on July 1, 1994 and July 1, 1995

51. **(b)** Amplitude: 2.3; Period: 365 days **(c)** Maximum: 14.30 hr on June 20 ($t = 171$); Minimum: 9.70 hr on December 20 ($t = 354$)

53. **(a)** The absolute value reflects the negative values of the graph of $y = -3\cos 2x$ about the x axis.
(b) The absolute value combined with the negative sign reflects the positive values of the graph of $y = 2\sin\frac{\pi x}{2}$ about the x axis.

55. They are approximately the same.

57. Period = 2π

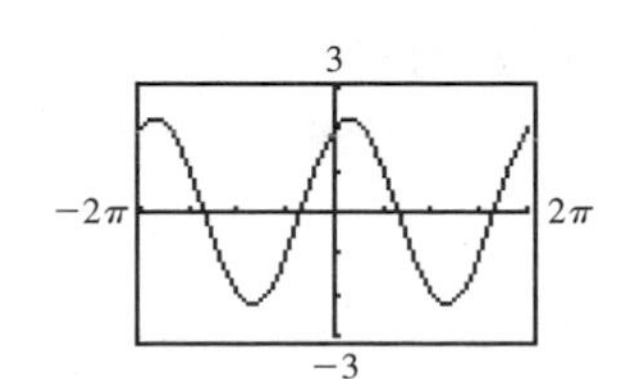

59. Period = 2π

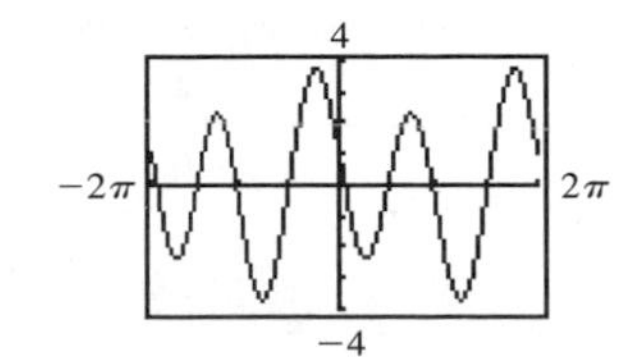

PROBLEM SET 5.6 (ON PAGE 336)

1. Period $= \pi$;
Phase shift $= 0$

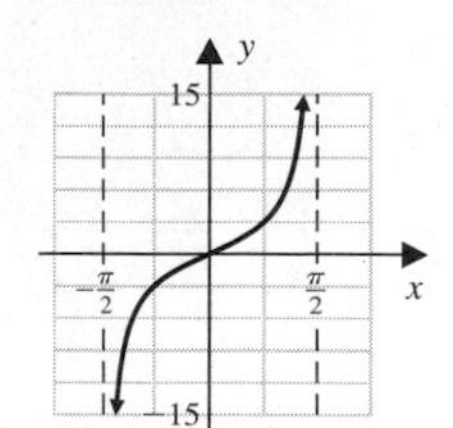

3. Period $= \pi$;
Phase shift $= 0$

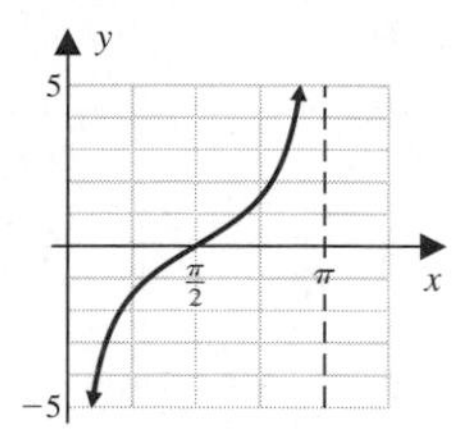

5. Period $= 2\pi$;
Phase shift $= 0$

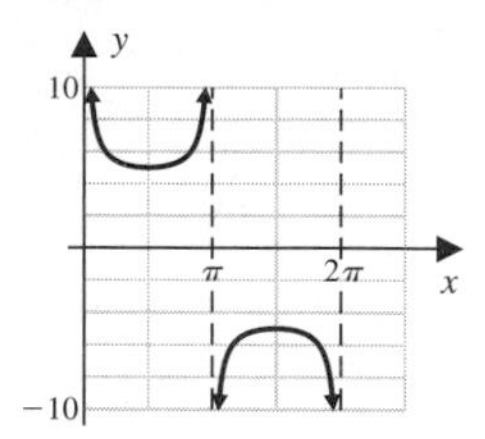

7. Period $= 2\pi$;
Phase shift $= 0$

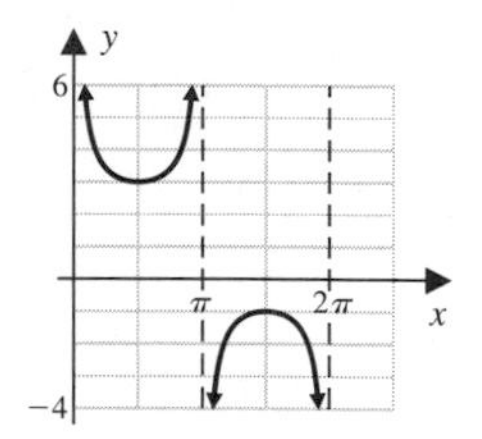

9. Period $= \pi/4$;
Phase shift $= 0$

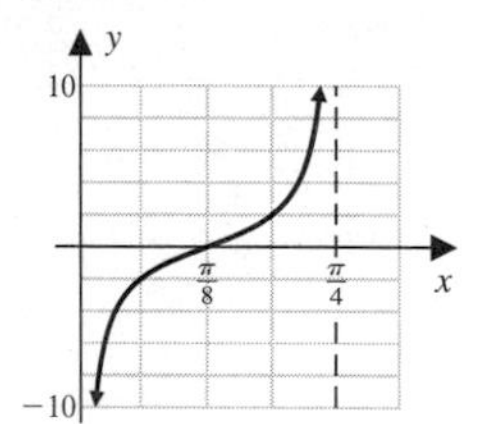

11. Period $= 3\pi$;
Phase shift $= 0$

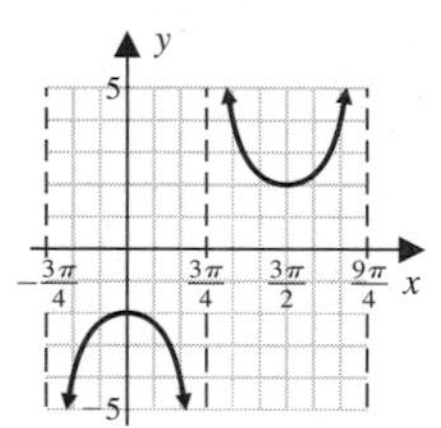

13. Period $= \pi/2$;
Phase shift $= 0$

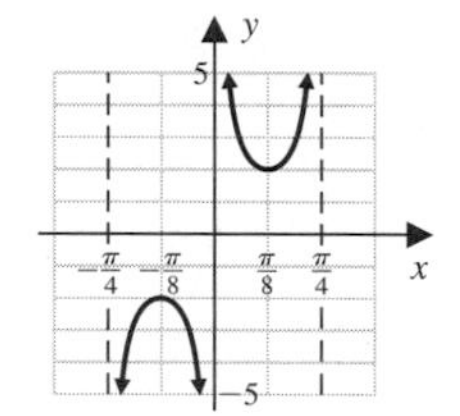

15. Period $= 2\pi$;
Phase shift $= \pi/6$

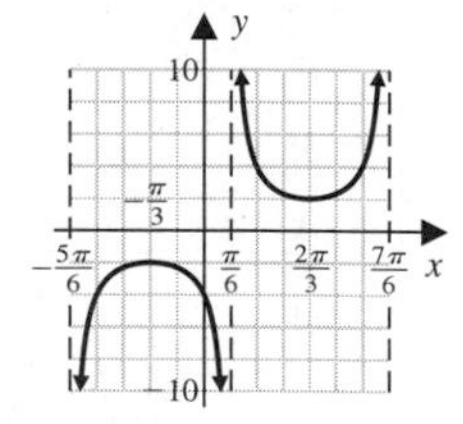

17. Period $= \pi$;
Phase shift $= \pi/4$

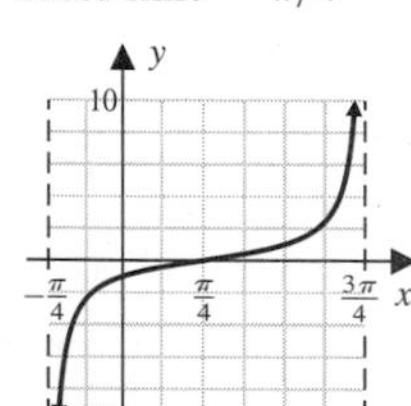

19. Period $= \pi$;
Phase shift $= \pi/2$

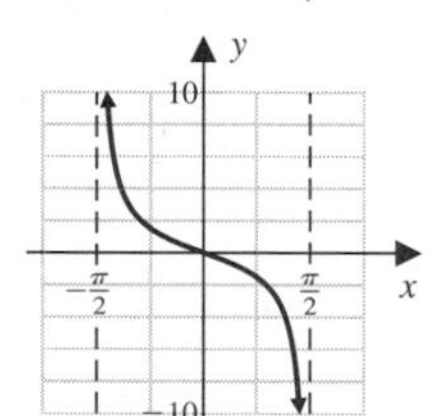

21. Period $= 8\pi$;
Phase shift $= -\pi$

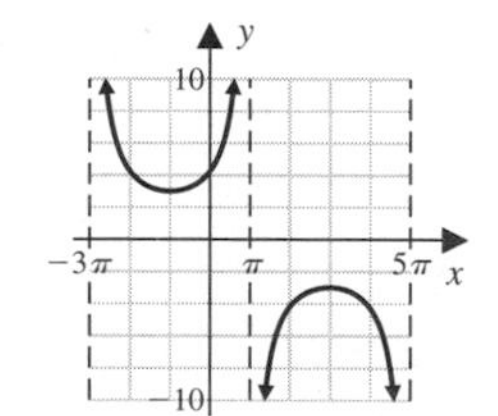

23. Period $= 4$;
Phase shift $= 1$

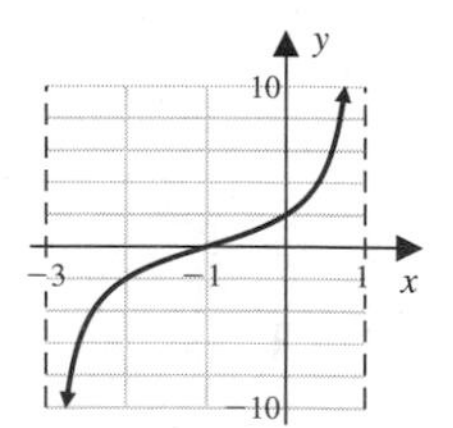

25.

	As x increases from: 0 to $\pi/2$	$\pi/2$ to π	π to $(3\pi)/2$	$(3\pi)/2$ to 2π
(a)	Increasing	Increasing	Increasing	Increasing
(b)	Decreasing	Decreasing	Decreasing	Decreasing
(c)	Increasing	Increasing	Decreasing	Decreasing
(d)	Decreasing	Increasing	Increasing	Decreasing

27. (a) 1 **(b)** $-\infty$ **29. (a)** ∞ **(b)** 1 **31.** $y = \tan(x/2)$ **33.** $y = 2\csc(2x + \pi)$
35. The graph appears to be $y = 1$, however, it is not in the domain of the function.
37. The graph appears to be the same; functions are not equal (different domains).
39. $f(x) \neq g(x)$; $f(x) = -g(x)$ because the cosecant function is odd. **41.** The graph of f is the same as the graph of g.

43.

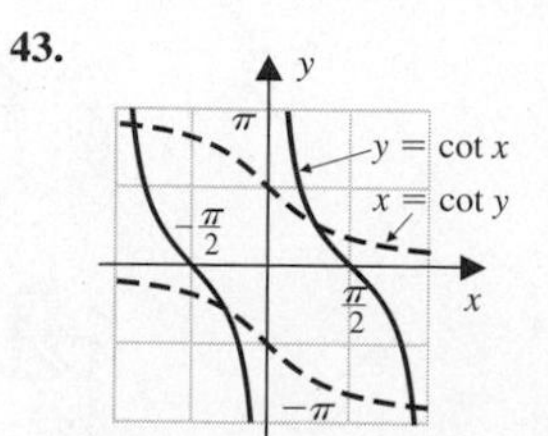

45. (a) $y = \tan \dfrac{\pi x}{2}$ **(b)** $y = \csc \dfrac{4\pi x}{3}$

PROBLEM SET 5.7 (ON PAGE 344)

1. **(a)** $\pi/4$ **(b)** $-\pi/6$ **3.** **(a)** $\pi/6$ **(b)** $-\pi/6$ **5.** **(a)** $-\pi/3$ **(b)** $(5\pi)/6$ **7.** **(a)** $\pi/6$ **(b)** $\pi/4$
9. **(a)** $-\pi/4$ **(b)** -1 **11.** **(a)** 0.22 **(b)** -0.89 **13.** **(a)** 0.82 **(b)** -1.55 **15.** **(a)** 2.36 **(b)** 1.56
17. $x = \sin y - 1$; $-\pi/2 \le y \le \pi/2$, $-2 \le x \le 0$ **19.** $x = \frac{1}{2}\cos y + 4$; $0 \le y \le \pi$, $\frac{7}{2} \le x \le \frac{9}{2}$
21. $x = \frac{1}{3}[2 - \tan(y/3)]$; $-(3\pi)/2 < y < (3\pi)/2$, x is real **23.** $x = \frac{1}{2}\cos 3y + 2$; $0 \le y \le \pi/3$, $\frac{3}{2} \le x \le \frac{5}{2}$ **25.** **(a)** $\frac{3}{5}$ **(b)** $\frac{12}{13}$
27. **(a)** 2 **(b)** $(-2\sqrt{5})/5$ **29.** **(a)** $(4\sqrt{15})/15$ **(b)** $\pi/3$ **31.** **(a)** $\sqrt{1-x^2}$ **(b)** $\sqrt{1-x^2}/x$ **(c)** $1/x$
33. **(a)** $\pi/4$ **(b)** $-\pi/4$ **35.** **(a)** $\pi/4$ **(b)** $-\pi/3$ **37.** **(a)** 0.80 **(b)** 1.27 **39.** **(a)** 2.00 **(b)** -0.11
41. For $0 \le x \le 1$, $\sin^{-1} x + \cos^{-1} x = \pi/2$.
43.

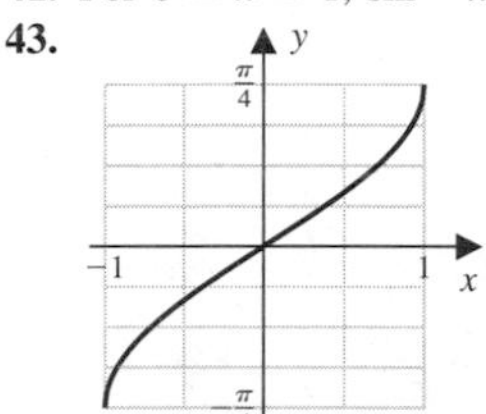

45.

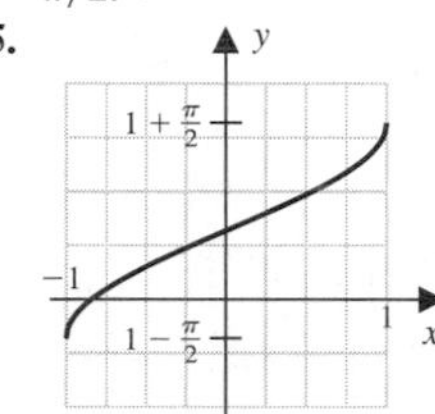

47.

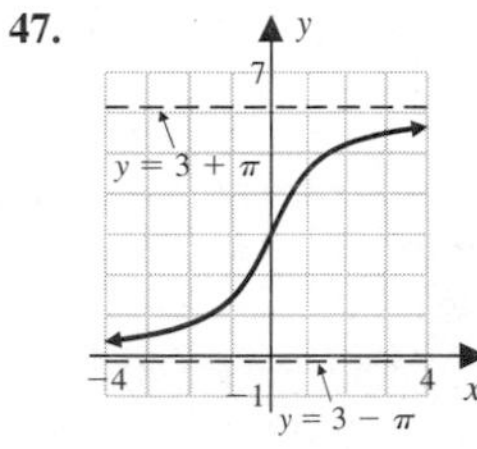

49.

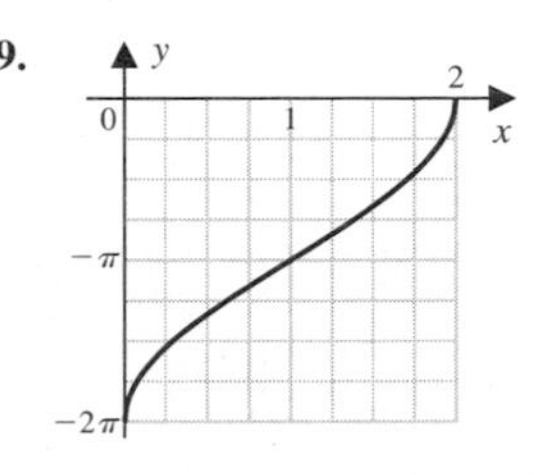

53. No **55.** They are approximately equal. **57.** $y = \sin(\sin^{-1} x) = x$ for $-1 \le x \le 1$; Domain: $[-1, 1]$

PROBLEM SET 5.8 (ON PAGE 353)

1. $\alpha = 40°$, $a = 6.43$, $b = 7.66$ **3.** $\alpha = 47°$, $a = 4.29$, $c = 5.87$ **5.** $\beta = 58.77°$, $a = 909.51$, $c = 1754.20$
7. $\alpha = 72.74°$, $\beta = 17.26°$, $c = 13.82$ **9.** $\alpha = 67.5°$, $\beta = 22.5°$, $b = 2.41$, $c = 2.61$ **11.** 10.68 in. **13.** 10.90 ft
15. 46.06 ft **17.** 80.83 mph **19.** **(a)** 23.58° **(b)** 2 ft/sec **21.** $d = 12.5$ in., $\theta = 36.87°$ **23.** 154.51 m
25. 84°36′ **27.** 146.71 m **29.** Building: 30.57 ft; antenna: 13.95 ft **31.** 499.00 ft
33. **(a)** $\theta = \tan^{-1}(32/x) - \tan^{-1}(24/x)$ **(b)** $\theta = 8.21°$ when $x = 27.71$ ft **(c)** Yes
35. Amplitude = 1; Period = $\frac{1}{100}$ sec; Frequency = 100 **37.** Amplitude = 0.15; Period = $\frac{1}{396}$ sec; Frequency = 396
39. **(a)** $y = -50\cos\sqrt{\frac{8}{30}}\,t$ **(c)** Amplitude = 50; Period = 12.17 (approx.); Frequency = 0.08 (approx.)
41. **(a)** $\dfrac{\pi}{2.2}$ **(b)** $y = 0.025\sin\left(\dfrac{\pi}{2.2}t\right)$ **43.** $T = r\cot(\theta/2)$
45. **(a)–(b)** Side opposite would have length 23 sin 35°, or approx. 13.19 in.

CHAPTER 5 REVIEW PROBLEM SET (ON PAGE 359)

1. **(a)** 76°15′ **(b)** −61°21′ **(c)** 143°28′12″ **(d)** −14°28′12″ **3.** **(a)** 0.2618 **(b)** −0.3046 **(c)** 0.7903 **(d)** 0.6316
5. **(a)** 54° **(b)** 144° **(c)** 396°, −324° **7.** $\sin\theta = \frac{3}{5}$; $\cos\theta = \frac{4}{5}$; $\tan\theta = \frac{3}{4}$; $\csc\theta = \frac{5}{3}$; $\sec\theta = \frac{5}{4}$; $\cot\theta = \frac{4}{3}$
9. **(a)** III **(b)** III **(c)** IV **11.** **(a)** 0.1411 **(b)** −0.8829 **(c)** −3.0777 **(d)** −1.4044
13. **(a)** (0.3090, 0.9511) **(b)** (−0.9784, −0.2069) **(c)** (0.2225, −0.9749) **15.** **(a)** 1 **(b)** $\cos t$ **(c)** $-\tan\theta$
17. **(a)** Period = $\pi/3$; Phase shift = $\pi/6$ **(b)** Period = 8π; Phase shift = -4π
19. **(a)**

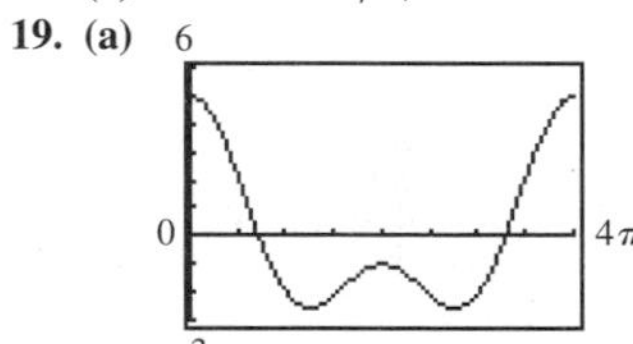

(b)

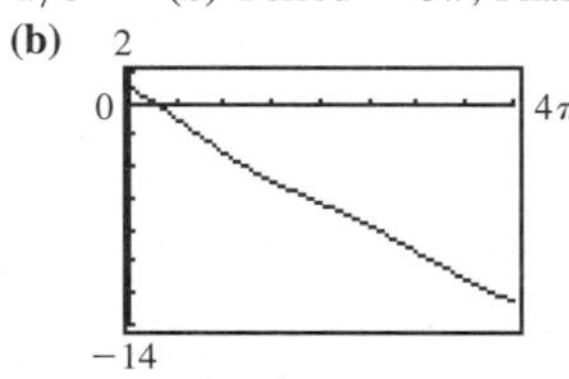

(c)

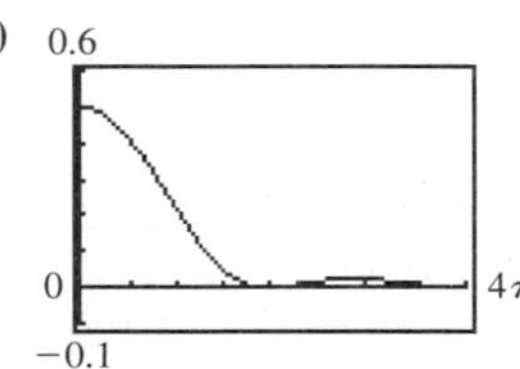

21. **(a)** 0.38 **(b)** 2.08 **(c)** 1.47 **(d)** 0.02
23. **(a)** $x = 2\sin y$; $-2 \le x \le 2$, $-\pi/2 \le y \le \pi/2$ **(b)** $x = \frac{1}{2}\tan y - \frac{3}{2}$; x is real; $-\pi/2 < y < \pi/2$
25. **(a)** 302.99 m **(b)** 17.04 sec **27.** 4.82 rad/hr
29. **(a)**

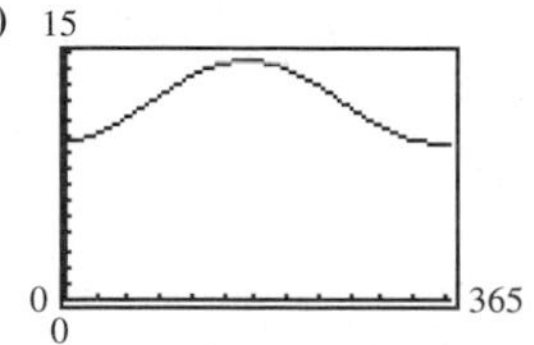

(b) Maximum: 14.5 hr for $t = 172.25$; Minimum: 9.5 hr for $t = 354.75$

31. 73.38 ft

CHAPTER 5 TEST (ON PAGE 361)

1. (a) $(26\pi)/9$ or 9.08 (approx.) **(b)** $\frac{540}{7}$ or 77.14° (approx.) **2. (a)** 1.5 **(b)** $\frac{270}{\pi}$ or 85.94° (approx.)

3. (a)

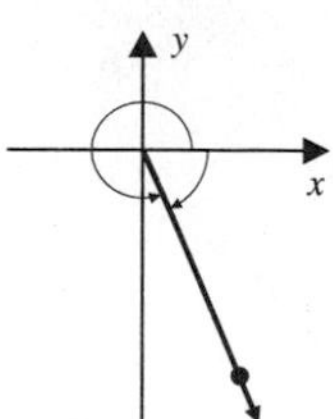

(b) $\sin\theta = -\frac{7\sqrt{58}}{58}$; $\cos\theta = \frac{3\sqrt{58}}{58}$; $\tan\theta = -\frac{7}{3}$; $\csc\theta = -\frac{\sqrt{58}}{7}$; $\sec\theta = \frac{\sqrt{58}}{3}$; $\cot\theta = -\frac{3}{7}$

4. (a) $\frac{\sqrt{5}}{5}$ **(b)** $\sin t = \frac{2\sqrt{5}}{5}$; $\cos t = \frac{\sqrt{5}}{5}$; $\tan t = 2$ **5. (a)** $-\frac{1}{2}$ **(b)** $-\frac{\sqrt{2}}{2}$ **(c)** $-\sqrt{3}$ **(d)** $\frac{3\pi}{4}$ **(e)** $\frac{4}{5}$

6. (a) 0.39 **(b)** 3.48 **(c)** 2.57 **7. (a)** III **(b)** IV

8. $\sin\theta = \frac{2\sqrt{66}}{17}$; $\tan\theta = -\frac{2\sqrt{66}}{5}$; $\csc\theta = \frac{17\sqrt{66}}{132}$; $\sec\theta = -\frac{17}{5}$; $\cot\theta = -\frac{5\sqrt{66}}{132}$

9. (a) Amplitude = 3; Period = $\pi/2$; Phase shift = $\pi/4$ **(b)** Amplitude = 2; Period = 4; Phase shift = $\frac{2}{3}$ **10.** 1 **11.** 44.34 m **12.** 41.41°

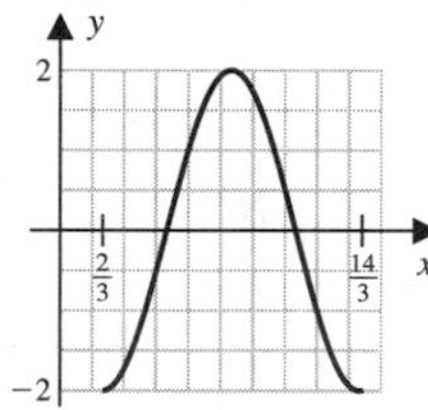

CHAPTER 6: PROBLEM SET 6.1 (ON PAGE 366)

35. $2\cos\theta$ **37.** $3\sec\theta$ **39.** $5\tan\theta$ **41.** $2\cos\theta$ **43.** False for $t = \pi/2$ **45.** False for $t = \pi/4$ **47.** Quadrants II and III **49.** Quadrants I and IV **51. (b)** Graphs are the same. **(c)** Yes

PROBLEM SET 6.2 (ON PAGE 378)

1. (a) $\frac{\sqrt{6}-\sqrt{2}}{4} = 0.2588$ **(b)** $\frac{\sqrt{6}+\sqrt{2}}{4} = 0.9659$

3. (a) $\frac{\sqrt{6}+\sqrt{2}}{4} = 0.9659$ **(b)** $\frac{-\sqrt{6}-\sqrt{2}}{4} = -0.9659$ **(c)** $2+\sqrt{3} = 3.7321$

5. (a) $\frac{\sqrt{2}-\sqrt{6}}{4} = -0.2588$ **(b)** $\frac{-\sqrt{6}-\sqrt{2}}{4} = -0.9659$ **7. (a)** $\frac{\sqrt{3}}{2} = 0.8660$ **(b)** $\tan\frac{\pi}{2}$; undefined

9. (a) $-\frac{63}{65}$ **(b)** $-\frac{16}{65}$ **(c)** $-\frac{33}{65}$ **(d)** $\frac{56}{65}$ **(e)** $\frac{63}{16}$ **(f)** $-\frac{33}{56}$

11. (a) −0.204 **(b)** −0.979 **(c)** 0.698 **(d)** 0.716 **(e)** 0.208 **(f)** 0.976 **13.** 1 **15.** 0 **17. (a)** $\cos t$ **(b)** $\sin t$

19. (a) $\frac{\tan t + 1}{1 - \tan t}$ **(b)** $\frac{\sqrt{2}}{2}(\cos\theta - \sin\theta)$ **21. (a)** $\frac{1}{\cos\theta}$ **(b)** $-\cos t$ **23. (a)** $\cos 8t$ **(b)** $\sin 4t$

25. (a) $\cos x$ **(b)** $\tan 7t$ **27.** $y = \sqrt{2}\cos\left(t - \frac{\pi}{4}\right)$ **29.** $y = 2\cos\left(2\pi t - \frac{5\pi}{3}\right)$

51. (a) $h(t) = 0.5\cos\left(\frac{\pi t}{6} - 0.927\right)$ **(b)** Maximum height is 0.5 m when t is about 1.8.
(c) Minimum height is −0.5 m when t is about 7.8.

53. (b) 0 and 1 **55. (a)** $\sin\gamma$ **(b)** $-\cos\gamma$ **57. (c)** Same; shift the graph of $y = \cos x$ to the right $\pi/2$ units.

PROBLEM SET 6.3 (ON PAGE 388)

1. **(a)** and **(b)** $\cos 2t \neq 2\cos t$ **3.** **(a)** $\frac{24}{25}$ **(b)** $-\frac{7}{25}$ **(c)** $-\frac{24}{7}$ **5.** **(a)** $\frac{336}{625}$ **(b)** $-\frac{527}{625}$ **(c)** $-\frac{336}{527}$
7. **(a)** $-\frac{120}{169}$ **(b)** $\frac{119}{169}$ **(c)** $-\frac{120}{119}$ **9.** **(a)** $-\frac{175}{337}$ **(b)** $-\frac{288}{337}$ **(c)** $\frac{175}{288}$ **11.** $\sin 6t$ **13.** $\cos 16t$ **15.** $\cos 10t$ **17.** $\tan 8\theta$
19. $4\cos^3 t - 3\cos t$ **21.** $(8\cos^3 t - 4\cos t)\sqrt{1-\cos^2 t}$; t is in quadrant I
23. **(a)** $\sqrt{3}/2 = 0.8660$ **(b)** $\sqrt{3}/2 = 0.8660$ **(c)** $-\frac{24}{7} = -3.4286$ **25.** **(a)** $\dfrac{\sqrt{2+\sqrt{2}}}{2} = 0.9239$ **(b)** $\dfrac{\sqrt{2+\sqrt{3}}}{2} = 0.9659$
27. **(a)** $\dfrac{\sqrt{2-\sqrt{3}}}{2} = 0.2588$ **(b)** $\sqrt{3+2\sqrt{2}} = 2.4142$ **29.** **(a)** $-\dfrac{\sqrt{2-\sqrt{2}}}{2} = -0.3827$ **(b)** $\dfrac{\sqrt{2+\sqrt{2}}}{2} = 0.9239$
31. **(a)** $t/2$ in quadrant I: $\frac{3}{5}$ **(b)** $t/2$ in quadrant I: $\frac{4}{5}$ **(c)** $t/2$ in quadrant I: $\frac{3}{4}$
33. **(a)** $t/2$ in quadrant I: $3/\sqrt{10}$ **(b)** $t/2$ in quadrant I: $1/\sqrt{10}$ **(c)** $t/2$ in quadrant I: 3
35. **(a)** $t/2$ in quadrant II: $5/\sqrt{34}$ **(b)** $t/2$ in quadrant II: $-3/\sqrt{34}$ **(c)** $t/2$ in quadrant II: $-\frac{5}{3}$
37. **(a)** $t/2$ in quadrant II: $1/\sqrt{17}$ **(b)** $t/2$ in quadrant II: $-4/\sqrt{17}$ **(c)** $t/2$ in quadrant II: $-\frac{1}{4}$
39. **(a)** $\cos 2t$ **(b)** $\sin 3t$ **55.** **(b)** 4.70 cm² **(c)** $0 < \theta < 180°$ **(d)** Maximum area is 8 cm² when $\theta = 90°$.
59. **(a)** $\left(-\frac{7}{25}, -\frac{24}{25}\right)$ **(b)** $\left(-2/\sqrt{5}, 1/\sqrt{5}\right)$ if t is positive **61.** **(a)** $\dfrac{\sec^2 t}{2-\sec^2 t}$ **(b)** $-\frac{5}{4}$

PROBLEM SET 6.4 (ON PAGE 398)

3. **(a)** $\sin 190° = -\sin 10° = -0.1736$ **(b)** $\tan(-113°) = \tan 67° = 2.3559$ **(c)** $\sec 372° = \sec 12° = 1.0223$
5. **(a)** $\csc 3.1 = 24.0496$; $\csc 0.0416 = 24.0454$ **(b)** $\cos(-1.9) = -0.3233$; $-\cos 1.2416 = -0.3233$
(c) $\cot 7.2 = 0.7665$; $\cot 0.9168 = 0.7665$
7. **(a)** 2.0944 or $\frac{2\pi}{3}$; 4.1888 or $\frac{4\pi}{3}$ **(b)** 2.3562 or $\frac{3\pi}{4}$; 3.9270 or $\frac{5\pi}{4}$ **9.** **(a)** 60° or 240° **(b)** 120° or 300°
11. **(a)** 1.0472 or $\frac{\pi}{3}$; 2.0944 or $\frac{2\pi}{3}$ **(b)** 225° or 315° **13.** **(a)** 2.3562 or $\frac{3\pi}{4}$; 3.9270 or $\frac{5\pi}{4}$ **(b)** 3.1416 or π; 4.7124 or $\frac{3\pi}{2}$
15. **(a)** 0.9500 or 2.1916 **(b)** 2.9916 or 6.1332 **17.** **(a)** 2.5559 or 3.7273 **(b)** 1.4711 or 4.6127 **19.** 0, 3.1416 or π
21. 90°, 240°, 270°, 300° **23.** 0, 1.0472 or $\pi/3$, 3.1416 or π, 4.1888 or $(4\pi)/3$ **25.** 180°
27. 1.0472 or $\pi/3$, 2.0944 or $(2\pi)/3$, 4.1888 or $(4\pi)/3$; 5.2360 or $(5\pi)/3$ **29.** 60°, 120°, 240°, 300°
31. 1.0472 or $\pi/3$, 1.5708 or $\pi/2$, 4.7124 or $(3\pi)/2$, 5.2360 or $(5\pi)/3$
33. **(a)** 45°, 75°, 165°, 195°, 285°, 315° **(b)** 55°, 85°, 175°, 205°, 295°, 325° **35.** **(a)** 1.5708 or $\pi/2$ **(b)** 0.3142 or $\pi/10$
37. 0.7854 or $\pi/4$, 2.3562 or $(3\pi)/4$, 3.9270 or $(5\pi)/4$, 5.4978 or $(7\pi)/4$ **39.** 45°, 90°, 225°, 270°
41. 0, 0.5234 or $\pi/6$, 2.6180 or $(5\pi)/6$, 3.1416 or π **43.** 0°, 270°
45. 0.7854 or $\pi/4$, 1.5708 or $\pi/2$, 2.3562 or $(3\pi)/4$, 4.7124 or $(3\pi)/2$ **47.** 0, 1.0472 or $\pi/3$, 5.2360 or $(5\pi)/3$ **49.** 210°, 270°, 330°
51. 0.6749 or 5.6083 **53.** **(a)** 0.201, 0.340, 2.802, 2.940 **(b)** (0, 0.201) or (0.340, 2.802) or (2.940, 6.283)
55. **(a)** 0.160 **(b)** [0, 0.160] **57.** **(a)** 0, 1.293, 4.721; $t = 0$ **(b)** [1.293, 4.721]; $t = 0$ **59.** **(a)** 32.67° **(b)** 62.04°
61. **(b)** $0° < \theta < 90°$ **(c)** 40° or 54.17° (approx.)
65. **(a)** No solution **(b)** No solution **(c)** 225°, 315° **(d)** $225° + (360°)k$ or $315° + (360°)k$, where k is an integer
67. 0.79 or $\pi/4$, 5.5 or $(7\pi)/4$
69. 0.17 or $\pi/18$, 0.87 or $(5\pi)/18$, 2.27 or $(13\pi)/18$, 2.97 or $(17\pi)/18$, 4.36 or $(25\pi)/18$, 5.06 or $(29\pi)/18$
71. **(a)** $k < -1$ or $k > 1$ **(b)** $k = -1$ or $k = 1$ **(c)** $-1 < k < 1$ **(d)** No values of k

PROBLEM SET 6.5 (ON PAGE 406)

1. **(a)** $-\frac{1}{4}$ **(b)** -0.2500 **3.** **(a)** $\left(-2-\sqrt{3}\right)/4$ **(b)** -0.9330 **5.** **(a)** $\left(2-\sqrt{3}\right)/4$ **(b)** 0.0670 **7.** **(a)** $\sqrt{2}/4$ **(b)** 0.3536
9. $\frac{1}{2}(\cos 3w - \cos 5w)$ **11.** $\frac{1}{2}(\cos 4v + \cos 12v)$ **13.** $\frac{1}{2}(\sin 7t + \sin t)$ **15.** $\frac{1}{2}(\cos t - \cos 9t)$ **17.** **(a)** $\sqrt{3}\sin\frac{2\pi}{9}$ **(b)** 1.1133
19. **(a)** $\sqrt{2}/2$ **(b)** 0.7071 **21.** **(a)** $\sqrt{2}/2$ **(b)** 0.7071 **23.** **(a)** $-\sqrt{2}/2$ **(b)** -0.7071 **25.** $2\sin 2\theta\cos\theta$
27. $-2\cos 6t\sin t$ **29.** $2\cos\frac{5t}{2}\cos\frac{t}{2}$ **31.** $2\sin\frac{11x}{2}\sin\frac{x}{2}$ **33.** **(a)** 0.3 **(b)** 1.3 **35.** **(a)** $-\pi/6$ **(b)** $(5\pi)/6$
37. **(a)** 90 **(b)** 0.7080 **47.** $-1 \le x \le -\frac{1}{3}$ **49.** $3 - \frac{3\pi}{2} < x < 3 + \frac{3\pi}{2}$

CHAPTER 6 REVIEW PROBLEM SET (ON PAGE 407)

5. **(a)** $\sin 5t$ **(b)** $\sin t$ **7.** **(a)** $\sin t$ **(b)** $-\cos 8\theta$ **9.** **(a)** $\cos t$ **(b)** $\sqrt{3}\sin t$

11. **(a)** $\frac{\sqrt{6}-\sqrt{2}}{4}$ **(b)** $\frac{\sqrt{2}+\sqrt{3}}{2}$ **13.** **(a)** $\frac{\sqrt{6}-\sqrt{2}}{4}$ **(b)** $\frac{-\sqrt{6}-\sqrt{2}}{4}$ **15.** **(a)** $\frac{\sqrt{2-\sqrt{2}}}{2}$ **(b)** $\frac{-\sqrt{2+\sqrt{3}}}{2}$

17. **(a)** $\frac{-5-6\sqrt{110}}{68}$ **(b)** $\frac{2\sqrt{66}-5\sqrt{15}}{68}$ **(c)** $\frac{\sqrt{15}}{8}$ **(d)** $\frac{\sqrt{102}}{17}$ **(e)** $\frac{\sqrt{15}}{7}$ **29.** **(a)** 270° **(b)** 71.23°, 288.77°

31. **(a)** $\pi/2$, 3.34, 6.08 **(b)** $\pi/6$, $\sin^{-1}\frac{1}{3}$, $\pi - \sin^{-1}\frac{1}{3}$, $(5\pi)/6$ or $t = 0.34, 0.52, 2.62, 2.80$ **33.** **(a)** $-\frac{1}{4}$ **(b)** $(-2-\sqrt{3})/4$

35. **(a)** $\sin 3t - \sin t$ **(b)** $\cos 2t + \cos 8t$ **37.** **(a)** $2\cos\frac{5t}{2}\sin\frac{3t}{2}$ **(b)** $2\cos 5t\cos 2t$ **39.** 0.32

41. $y = 2\cos\left(x - \frac{\pi}{3}\right)$ **43.** $y = \cos\left(x - \frac{7\pi}{4}\right)$ or $y = \cos\left(x + \frac{\pi}{4}\right)$ **45.** **(b)** 0.004 sec **(c)** 0.004 sec

47. **(a)** 25.71 in. **(b)** $\tan\theta = \frac{7}{6}$; $\theta = 49°$

CHAPTER 6 TEST (ON PAGE 409)

1. **(a)** $(-\sqrt{6}-\sqrt{2})/4$ **(b)** $(\sqrt{2}-\sqrt{6})/4$ **(c)** $2+\sqrt{3}$

2. **(a)** $-\frac{24}{25}$ **(b)** $\frac{7}{25}$ **(c)** $-\frac{24}{7}$ **(d)** $-\frac{24}{25}$ **(e)** $\frac{7}{25}$ **(f)** $-\frac{336}{625}$ **(g)** $\frac{1}{10}$ **(h)** $-\frac{117}{125}$

3. **(a)** $(\sqrt{2+\sqrt{2}})/2$ **(b)** $(\sqrt{2-\sqrt{2}})/2$ **(c)** $\sqrt{2}+1$ **4.** **(a)** $-\cos 94°$ **(b)** $\frac{1}{2}\sin\frac{10\pi}{7}$ **(c)** $\sin 60°$
(d) $\cos\pi$ **(e)** $\cos\frac{t}{2}$

5. **(a)** 0.52 or $\pi/6$, 2.62 or $(5\pi)/6$, 3.67 or $(7\pi)/6$, 5.76 or $(11\pi)/6$ **(b)** 15°, 75°, 195°, 255°
(c) $\pi/4$, $(5\pi)/4$, 2.16, 5.30; or 0.79, 2.16, 3.93, 5.30

7. -0.32 **8.** **(a)** $\frac{57{,}600 - x^2}{57{,}600 + x^2}$ **(b)** 50.96°

CHAPTER 7: PROBLEM SET 7.1 (ON PAGE 419)

1. SAS; $a = 2.05$ **3.** SAS; $b = 5.59$ **5.** SAS; $c = 3.35$ **7.** SSS; $\gamma = 104.48°$ **9.** SSS; $\beta = 50.50°$ **11.** SSS; $\alpha = 14.15°$
13. AAS; $\gamma = 15°$, $b = 28.66$, $c = 10.68$ **15.** ASA; $\beta = 59°$, $a = 34.47$, $c = 22.96$ **17.** ASA; $\beta = 76.5°$, $a = 13.60$, $c = 17.78$
19. AAS; $\alpha = 58.67°$, $a = 4.89$, $b = 4.44$ **21.** ASA; $\beta = 78°$, $a = 12.49$, $c = 17.15$ **23.** No possible triangle
25. No possible triangle **27.** Case 1: $\alpha = 58.34°$, $\gamma = 84.99°$, $c = 14.51$; Case 2: $\alpha = 121.66°$, $\gamma = 21.67°$, $c = 5.38$
29. $\beta = 53.65°$, $\gamma = 61.17°$, $c = 20.56$ **31.** Case 1: $\beta = 54.59°$, $\gamma = 92.08°$, $c = 54.56$; Case 2: $\beta = 125.41°$, $\gamma = 21.26°$, $c = 19.80$
33. SSA; no possible triangle **35.** SSS; $\alpha = 41.41°$, $\beta = 41.41°$, $\gamma = 97.18°$ **37.** AAS; $\gamma = 71°$, $b = 66.83$, $c = 63.67$
39. AAS; $\gamma = 74.2°$, $a = 37.98$, $c = 37.15$ **41.** 143.96 ft **43.** 343.68 m **45.** 8.23 mi **47.** 3.52 km **49.** 1.08 km; 0.81 km
51. 1.88 mi **53.** 1.88 m **55.** 42.19 nautical mi **57.** **(a)** $d = \sqrt{11943.64t^2 + 3107.75t + 232.44}$ **(c)** 0.32 hr; 0.78 hr; 1.24 hr
59. Angle PQR: 52.70°; angle PRQ: 50.51°; angle QPR: 76.79° **61.** 20.68 in.; 14.91 in. **63.** Because of the triangle inequality; $|\cos\theta| > 1$

PROBLEM SET 7.2 (ON PAGE 431)

7. Some points are: **(a)** **(i)** (3, −260°), (3, 460°), (3, 820°) **(ii)** (−3, −80°), (−3, 280°), (−3, 640°)
(b) **(i)** $\left(3, -\frac{3\pi}{4}\right), \left(3, \frac{5\pi}{4}\right), \left(3, \frac{13\pi}{4}\right)$ **(ii)** $\left(-3, -\frac{7\pi}{4}\right), \left(-3, \frac{9\pi}{4}\right), \left(-3, \frac{17\pi}{4}\right)$

9. **(a)** $(3\sqrt{3}, 3)$ **(b)** $(2\sqrt{2}, -2\sqrt{2})$ **11.** **(a)** (2.70, −1.30) **(b)** (−1.37, −3.76) **13.** **(a)** (−1.98, −0.28) **(b)** (0, −3)

15. **(a)** $\left(2, \frac{2\pi}{3}\right)$ **(b)** $\left(12, \frac{2\pi}{3}\right)$ **17.** **(a)** (5.92, 2.10) **(b)** (9.22, 4.00) **19.** $r = 2\sec\theta$ **21.** $r = -3\csc\theta$

23. $r = \frac{2}{3\cos\theta - \sin\theta}$ **25.** $r = 5$ **27.** $r = -8\cos\theta$ **29.** $r = \frac{1}{4}\tan\theta\sec\theta$ **31.** $r = 8\cot\theta\csc\theta$ **33.** $x^2 + y^2 = 9$

35. $y = 2$ **37.** $y = x$ **39.** $(x+2)^2 + y^2 = 4$ **41.** $y = \frac{1}{4}x^2$ **43.** $4x + 3y = 12$ **45.** $(x-2)^2 + (y-1)^2 = 5$

47. Circle

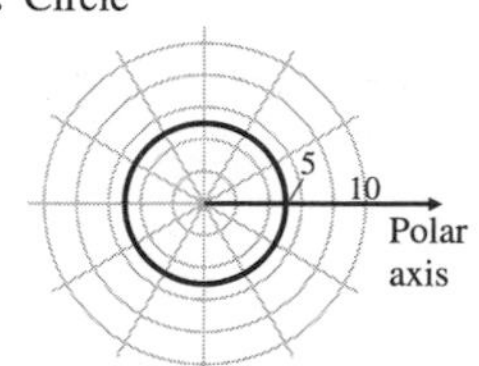

49. Line

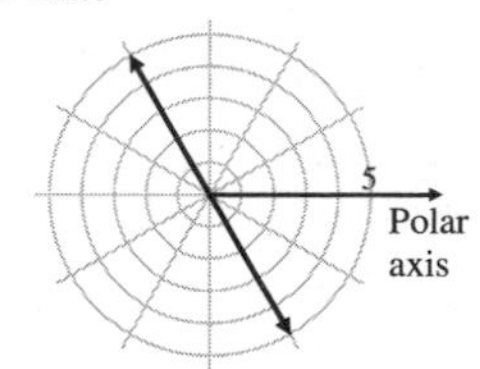

51. Circle

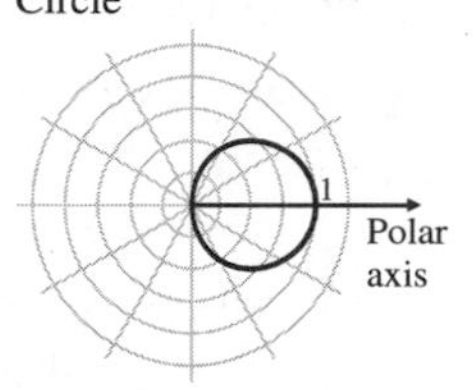

53. Cardioid

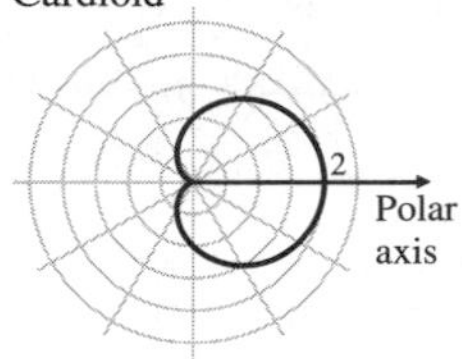

55. Symmetry: line $\theta = \pi/2$

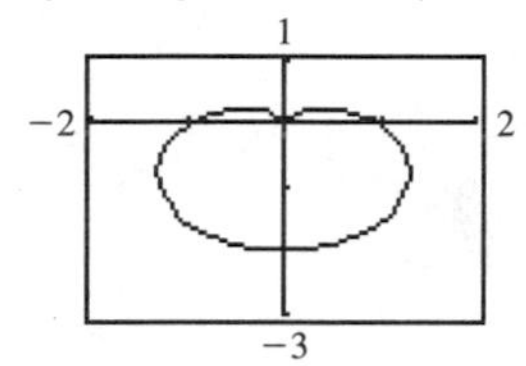

57. Symmetry: polar axis

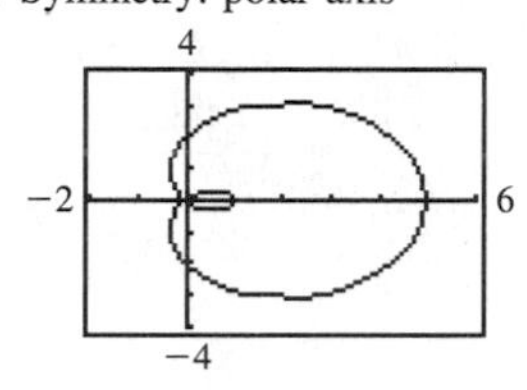

59. Symmetry: line $\theta = \pi/2$

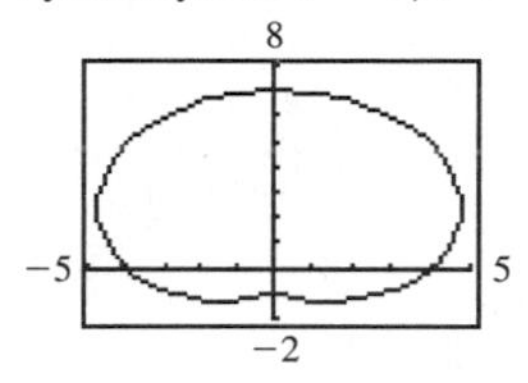

61. Symmetry: pole (and both axes)

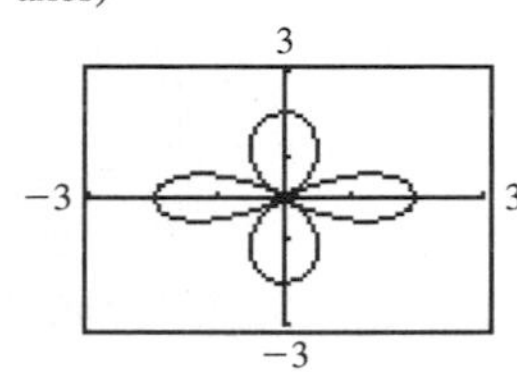

63. Symmetry: line $\theta = \pi/2$

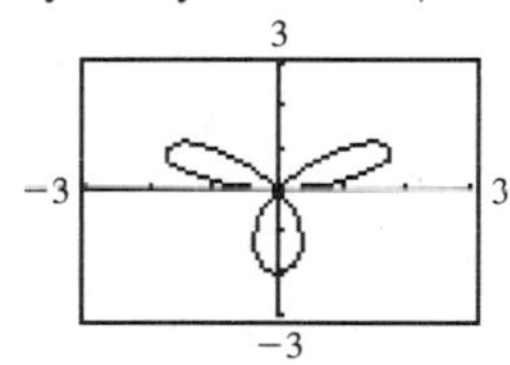

65. Symmetry: polar axis, line $\theta = \pi/2$, and the pole

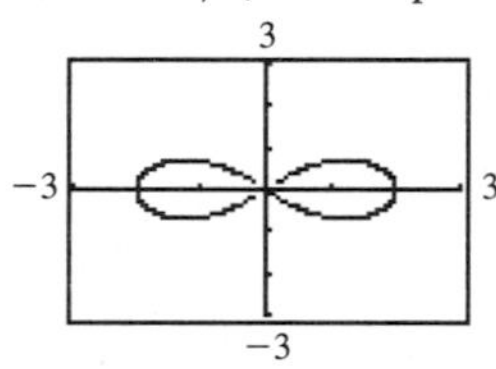

67. No symmetry

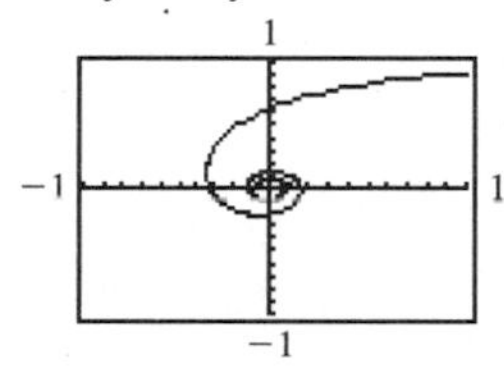

69. No

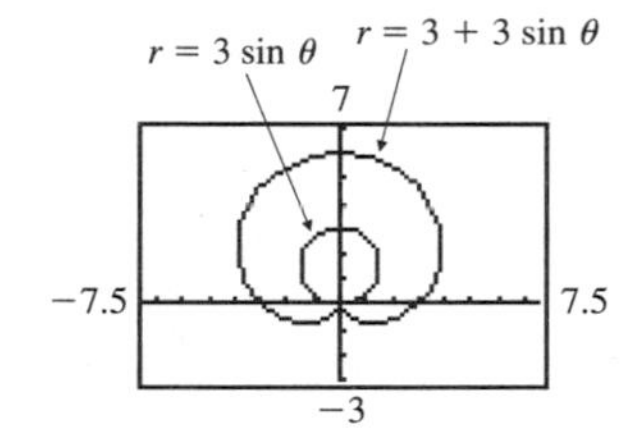

73. $(a, 0), \left(a, \frac{\pi}{3}\right), \left(a, \frac{2\pi}{3}\right), (a, \pi), \left(a, \frac{4\pi}{3}\right), \left(a, \frac{5\pi}{3}\right)$

PROBLEM SET 7.3 (ON PAGE 442)

3. $\mathbf{u} = \langle -5, -2 \rangle$; $|\mathbf{u}| = \sqrt{29}$ **5.** $\mathbf{u} = \langle 5, -11 \rangle$; $|\mathbf{u}| = \sqrt{146}$ **7.** $\mathbf{u} = \langle -6, -6 \rangle$; $|\mathbf{u}| = 6\sqrt{2}$ **9. (a)** $\langle 1, 8 \rangle$ **(b)** $\sqrt{65}$
11. (a) $\langle -19, -20 \rangle$ **(b)** $\sqrt{761}$ **13. (a)** $\langle 9, -28 \rangle$ **(b)** $\sqrt{865}$ **15. (a)** $-2\mathbf{i} + 4\mathbf{j}$ **(b)** $3\mathbf{i}$ **17. (a)** $2\mathbf{i} + 5\mathbf{j}$ **(b)** $-4\mathbf{i} + 4\mathbf{j}$
19. $|\mathbf{w}| = \sqrt{53}$; approx. 37.18° **21.** $-\frac{\sqrt{3}}{2}\mathbf{i} + \frac{1}{2}\mathbf{j}$; 150° **23.** $\frac{5}{13}\mathbf{i} + \frac{12}{13}\mathbf{j}$; 67.38° **25.** $\frac{1}{\sqrt{2}}\mathbf{i} - \frac{1}{\sqrt{2}}\mathbf{j}$; −45°
27. $\frac{1}{\sqrt{26}}\mathbf{i} + \frac{5}{\sqrt{26}}\mathbf{j}$; 78.69° **29.** $x = 3, y = 0$ **31.** $x = 3.54, y = 3.54$ **33.** $x = -1.73, y = -1$ **35.** $x = -0.09, y = 0.49$
37. (a) 55.71 lb **(b)** 20.49° **39. (a)** 12.63 N **(b)** 17.29° **41. (a)** 540.03 mph **(b)** N81.34°E **(c)** 3.66°
43. (a) 51.22 kph **(b)** S1.34°E **(c)** 38.66° **45.** (13, 1) **49. (a)** −9 **(b)** 142.1° **51. (a)** −12 **(b)** 143.1°
53. (a) 14 **(b)** 58.7° **57.** Yes; vertex angle *B* is the right angle.

PROBLEM SET 7.4 (ON PAGE 452)

1. (a) $|z| = \sqrt{53}$ **(b)** $|z| = \sqrt{29}$ **3. (a)** $|z| = 5$ **(b)** $|z| = \sqrt{34}$ **5. (a)** 4 **(b)** 2 **(c)** 4 **(d)** $\frac{1}{2}$ **(e)** $\frac{1}{2}$
7. (a) $5\sqrt{29}$ **(b)** 25 **(c)** $5\sqrt{29}$ **(d)** $5/\sqrt{29}$ **(e)** $5/\sqrt{29}$ **9.** $\sqrt{3} + i$ **11.** $2i$ **13.** −8 **15.** $1.97 + 0.35i$
17. (a) $\sqrt{2}\left(\cos\frac{5\pi}{4} + i\sin\frac{5\pi}{4}\right)$ **(b)** $2\left(\cos\frac{7\pi}{6} + i\sin\frac{7\pi}{6}\right)$
19. (a) $\sqrt{3}(\cos 0.62 + i\sin 0.62)$ **(b)** $5(\cos\pi + i\sin\pi)$ **21. (a)** $1\left(\cos\frac{2\pi}{3} + i\sin\frac{2\pi}{3}\right)$ **(b)** $1\left(\cos\frac{\pi}{6} + i\sin\frac{\pi}{6}\right)$
23. (a) $8\left(\cos\frac{7\pi}{6} + i\sin\frac{7\pi}{6}\right) = -4\sqrt{3} - 4i$ **(b)** $2\left(\cos\frac{\pi}{2} + i\sin\frac{\pi}{2}\right) = 2i$
25. (a) $1(\cos 90° + i\sin 90°) = i$ **(b)** $1[\cos(-30°) + i\sin(-30°)] = \frac{\sqrt{3}}{2} - \frac{1}{2}i$

27. **(a)** $2(\cos 180° + i \sin 180°) = -2$ **(b)** $1[\cos(-90°) + i \sin(-90°)] = -i$

29. **(a)** $6(\cos 90° + i \sin 90°) = 6i$ **(b)** $\frac{2}{3}(\cos 10° + i \sin 10°) = 0.66 + 0.12i$ **31.** $-\frac{\sqrt{3}}{2} - \frac{1}{2}i$ **33.** $512 - 512\sqrt{3}\,i$

35. 256 **37.** $-125{,}000i$ **39.** $-8 - 8\sqrt{3}\,i$ **41.** $-2^{30} = -1{,}073{,}741{,}824$ **43.** -1 **45.** $0.71 + 0.71i, -0.71 - 0.71i$

47. $2, -1 + 1.73i, -1 - 1.73i$ **49.** $0.91 + 1.78i, -1.41 + 1.41i, -1.78 - 0.91i, 0.31 - 1.98i, 1.98 - 0.31i$

51. $1 + 1.73i, -1.73 + i, -1 - 1.73i, 1.73 - i$

53. $2(\cos 60° + i \sin 60°) = 1 + 1.73i$; $2(\cos 180° + i \sin 180°) = -2$; $2(\cos 300° + i \sin 300°) = 1 - 1.73i$

55. $3(\cos 45° + i \sin 45°) = 2.12 + 2.12i$; $3(\cos 135° + i \sin 135°) = -2.12 + 2.12i$; $3(\cos 225° + i \sin 225°) = -2.12 - 2.12i$; $3(\cos 315° + i \sin 315°) = 2.12 - 2.12i$

57. $2(\cos 30° + i \sin 30°) = 1.73 + i$; $2(\cos 90° + i \sin 90°) = 2i$; $2(\cos 150° + i \sin 150°) = -1.73 + i$; $2(\cos 210° + i \sin 210°) = -1.73 - i$; $2(\cos 270° + i \sin 270°) = -2i$; $2(\cos 330° + i \sin 330°) = 1.73 - i$

61. **(a)** On the circle $x^2 + y^2 = 4$ **(b)** On the circle $x^2 + (y - 1)^2 = 1$ **65.** **(a)** $-\frac{\sqrt{3}}{2} - \frac{1}{2}i$ **(b)** $\frac{1}{32} - \frac{1}{32}i$

CHAPTER 7 REVIEW PROBLEM SET (ON PAGE 453)

1. **(a)** $c = 3.66, \alpha = 43.12°, \beta = 106.88°$ **(b)** $\gamma = 60°, a = 8.16, b = 11.15$ **3.** **(a)** $\left(\frac{5\sqrt{2}}{2}, \frac{5\sqrt{2}}{2}\right)$ **(b)** $\left(\sqrt{3}, 1\right)$ **(c)** $(-1, -1)$

5. **(a)** $x^2 + (y + 2)^2 = 4$; circle **(b)** $\left(x + \frac{3}{2}\right)^2 + y^2 = \frac{9}{4}$; circle

7. **(a)**

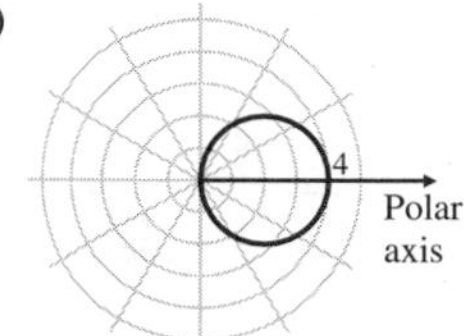

(b)

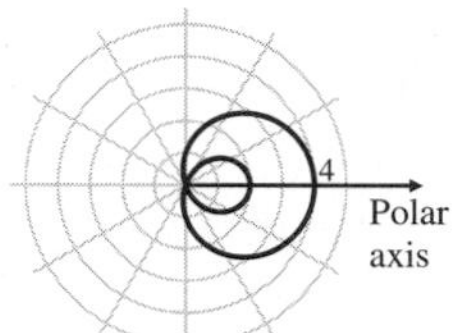

9. **(a)** $\langle 9, 12\rangle$ **(b)** $\langle 1, -1\rangle$ **(c)** $\langle -2, -5\rangle$ **(d)** $\langle 17, 18\rangle$

11. **(a)** $|\mathbf{u}| = 13$; $\theta = 22.62°$ **(b)** $|\mathbf{u}| = 7.28$; $\theta = 344.05°$

13. **(a)** $\mathbf{u} = \langle 10, -5\rangle$; $|\mathbf{u}| = 5\sqrt{5}$ **(b)** $\mathbf{u} = \langle 10, -7\rangle$; $|\mathbf{u}| = \sqrt{149}$

15. **(a)** $2\left(\cos \frac{2\pi}{3} + i \sin \frac{2\pi}{3}\right)$ **(b)** $\sqrt{2}\left(\cos \frac{3\pi}{4} + i \sin \frac{3\pi}{4}\right)$ **17.** **(a)** $13(\cos 1.18 + i \sin 1.18)$ **(b)** $\sqrt{85}(\cos 319.40° + i \sin 319.40°)$

19. $z_1z_2 = 6\left(\cos \frac{3\pi}{2} + i \sin \frac{3\pi}{2}\right) = -6i$; $\frac{z_1}{z_2} = \frac{2}{3}\left(\cos \frac{\pi}{2} + i \sin \frac{\pi}{2}\right) = \frac{2}{3}i$

21. $z_1z_2 = 18(\cos 305° + i \sin 305°) = 10.32 - 14.74i$; $\frac{z_1}{z_2} = 2(\cos 155° + i \sin 155°) = -1.81 + 0.85i$

23. **(a)** $1(\cos 360° + i \sin 360°) = 1$ **(b)** $36\sqrt{6}(\cos \pi + i \sin \pi) = -36\sqrt{6}$

25. **(a)** $81(\cos 108° + i \sin 108°) = -25.03 + 77.04i$ **(b)** $25\sqrt{5}\left(\cos \frac{5\pi}{8} + i \sin \frac{5\pi}{8}\right) = -21.39 + 51.65i$

27. **(a)** $2\left(\cos \frac{\pi}{6} + i \sin \frac{\pi}{6}\right)$; $2\left(\cos \frac{5\pi}{6} + i \sin \frac{5\pi}{6}\right)$; $2\left(\cos \frac{3\pi}{2} + i \sin \frac{3\pi}{2}\right)$

(b) $\sqrt[8]{2}\left(\cos \frac{\pi}{16} + i \sin \frac{\pi}{16}\right)$; $\sqrt[8]{2}\left(\cos \frac{9\pi}{16} + i \sin \frac{9\pi}{16}\right)$; $\sqrt[8]{2}\left(\cos \frac{17\pi}{16} + i \sin \frac{17\pi}{16}\right)$; $\sqrt[8]{2}\left(\cos \frac{25\pi}{16} + i \sin \frac{25\pi}{16}\right)$

29. 92.54 ft **31.** 31.72 ft **33.** 93.73°

CHAPTER 7 TEST (ON PAGE 455)

1. $b = 65.01°, c = 105.96°, \gamma = 103.35°$ **2.** $b = 5.65$ **3.** **(a)** $\left(2, -2\sqrt{3}\right)$ **(b)** $\left(3\sqrt{2}, \frac{3\pi}{4}\right)$

4. **(a)** $x^2 + y^2 = 25$

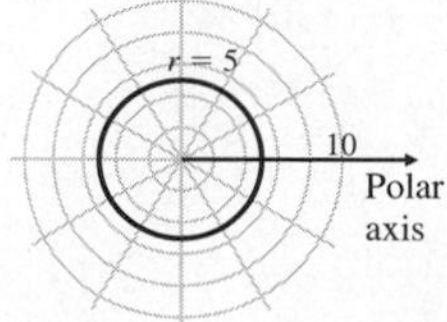

(b) $x^2 + (y - 2)^2 = 4$

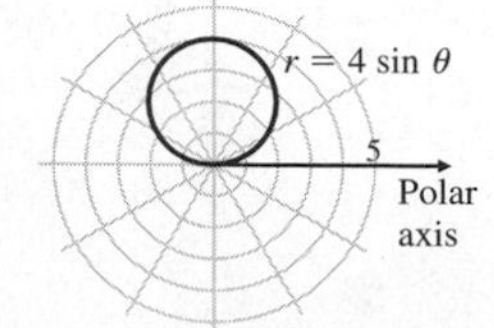

(c) $y = x$

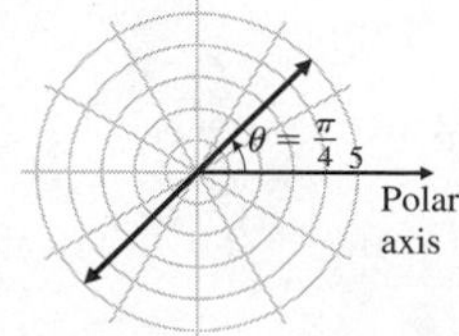

5. (a) $r = \dfrac{5}{3\cos\theta + 2\sin\theta}$ 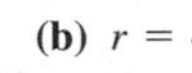 **(b)** $r = 4$ **6.** $-24\mathbf{i} + 20\mathbf{j}$ **7.** $56.31°$ **8.** $2.42\mathbf{i} + 12.27\mathbf{j}$

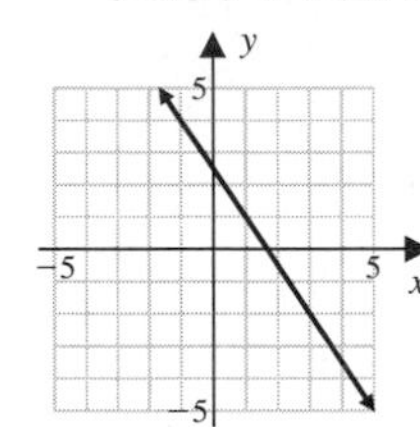

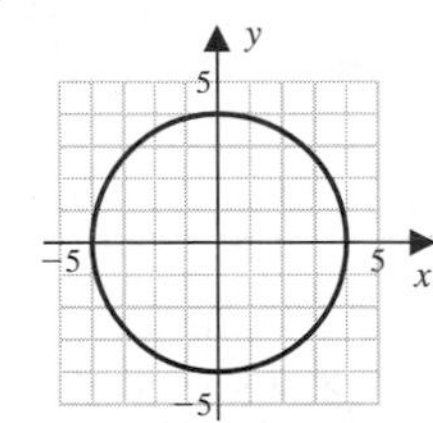

9. $2\sqrt{2} - 2\sqrt{2}\,i$ **10.** $2\left(\cos\dfrac{4\pi}{3} + i\sin\dfrac{4\pi}{3}\right)$ **11. (a)** $-1.7321 - i$ **(b)** $0.1873 + 0.4636i$
12. $2^{12} + 0i$ **13.** $\dfrac{\sqrt{3}}{2} + \dfrac{1}{2}i,\ -i,\ -\dfrac{\sqrt{3}}{2} + \dfrac{1}{2}i$ **14.** 82.7 mi

CHAPTER 8: PROBLEM SET 8.1 (ON PAGE 464)

1. $\left(\frac{11}{5}, \frac{14}{5}\right)$ **3.** No solution **5.** $\left(\frac{11}{7}, \frac{5}{7}\right)$ **7.** No solution **9.** $(a, a - 5)$, where a is a real number **11.** $\left(\frac{7}{3}, -\frac{4}{3}\right)$ **13.** $(2, 3, -1)$
15. $(4, 3, 2)$ **17.** $(3, 2)$ **19.** $(1, -2)$ **21.** $(y, z) = \left(\frac{13}{5}, -\frac{7}{5}\right)$ **23.** $(2, -1)$ **25.** $(1, 1)$ **27.** $\left(-2, \frac{10}{3}, \frac{2}{3}\right)$ **29.** $\left(-5, \frac{5}{3}, \frac{28}{3}\right)$
31. $(2, 1, 0)$ **33.** $(-1, -2, 7)$ **35.** $(2.33, -1.33)$ **37.** $(-1.50, 13.50)$ **39.** $(10, -10)$ **41.** $(9.84, 6.57)$ **43.** $(2, 1)$
45. 30 females, 90 males **47.** $v_0 = 32$ ft/sec, $h_0 = 10$ ft **49.** 2 L 30% solution, 8 L 20% solution
51. Airplane: approx. 617 mph; wind: approx. 49 mph **53.** 185 pairs tennis shoes; 75 pairs running shoes
55. \$24,000 at 5% interest; \$11,000 at 7% interest; \$7000 at 9% interest
57. 3 of the 12 passenger vans, 11 of the 10 passenger vans, 1 of the 6 passenger vans
59. A pumps 10,000 gal/hr; B pumps 12,000 gal/hr; C pumps 15,000 gal/hr **61. (a)** $x = \frac{1}{3}, y = -1$ **(b)** $x = \frac{1}{4}, y = -\frac{1}{2}$

PROBLEM SET 8.2 (ON PAGE 477)

1. $\left[\begin{array}{cc|c} 1 & 6 & 7 \\ 3 & 7 & 10 \end{array}\right]$ **3.** $\left[\begin{array}{ccc|c} 2 & 1 & 0 & 2 \\ 1 & -1 & 0 & 4 \\ 3 & 0 & 1 & 6 \end{array}\right]$ **5.** $(1, 3)$ **7.** $(1, 1, 1)$ **9.** $(5, 3, -2)$

11. (a) $\begin{bmatrix} 1 & 0 & 3 \\ 5 & -1 & 5 \\ -3 & 1 & 4 \end{bmatrix}$ **(b)** $\begin{bmatrix} 1 & 0 & 3 \\ -3 & 1 & 4 \\ 15 & -3 & 15 \end{bmatrix}$ **(c)** $\begin{bmatrix} 1 & 0 & 3 \\ -3 & 1 & 4 \\ 8 & -1 & 14 \end{bmatrix}$ **13.** Both (a) and (b) **15.** (b) **17.** $(1, 1)$
19. $(2 + t, 3 + 2t, t)$, where t is a real number **21.** No solution **23.** $(3, 5)$ **25.** $\left(\frac{23}{4}, \frac{13}{2}\right)$ **27.** $(1, 2, 3)$ **29.** $(1, 1, 1)$
31. $(4, 5, 6)$ **33.** $(1, 1, 0)$ **35.** No solution **37.** $\left(\frac{7}{5} - \frac{7}{5}t, -\frac{2}{5} - \frac{3}{5}t, t\right)$, where t is a real number **39.** $\left(-\frac{23}{13}, -\frac{93}{26}, -\frac{37}{13}\right)$
41. $A = 2, B = 1$ **43.** $A = -1, B = -6$ **45.** $A = -5, B = 31, C = -26$ **47.** $A = 2, B = -1, C = 3$
49. $A = 6, B = -6, C = 0$ **51.** 10 oz A, 5 oz B, 8 oz C **53.** 500 lb 25%, 2000 lb 35%, 1500 lb 40%
55. $I_1 = \frac{33}{167}$ A, $I_2 = \frac{87}{167}$ A, $I_3 = \frac{54}{167}$ A **57.** $(1, -1, 1, 1)$ **59.** $(3, 2, 1, -1)$
61. (a) $2a - b - c \neq 0$ **(b)** Not possible **(c)** $2a - b - c = 0$ **63. (a)** $a = -\sqrt{3}, \sqrt{3}$ **(b)** $a \neq -\sqrt{3}, \sqrt{3}$ **(c)** Not possible
65. $a = b = c = 1$

PROBLEM SET 8.3 (ON PAGE 492)

1. (a) $\begin{bmatrix} 8 & 11 \\ -2 & 7 \end{bmatrix}$ **(b)** $\begin{bmatrix} -2 & -1 \\ 0 & 1 \end{bmatrix}$ **(c)** $\begin{bmatrix} 12 & 20 \\ -4 & 16 \end{bmatrix}$ **(d)** $\begin{bmatrix} -25 & -30 \\ 5 & -15 \end{bmatrix}$ **(e)** $\begin{bmatrix} -13 & -10 \\ 1 & 1 \end{bmatrix}$

3. (a) $\begin{bmatrix} 8 & 12 & 8 \\ 2 & 3 & 0 \end{bmatrix}$ **(b)** $\begin{bmatrix} -2 & -10 & -4 \\ -4 & 3 & 2 \end{bmatrix}$ **(c)** $\begin{bmatrix} 12 & 4 & 8 \\ -4 & 12 & 4 \end{bmatrix}$ **(d)** $\begin{bmatrix} -25 & -55 & -30 \\ -15 & 0 & 5 \end{bmatrix}$ **(e)** $\begin{bmatrix} -13 & -51 & -22 \\ -19 & 12 & 9 \end{bmatrix}$

5. (a) $\begin{bmatrix} -9 & 8 & 3 \\ -3 & 3 & 7 \\ 6 & -2 & 4 \end{bmatrix}$ **(b)** $\begin{bmatrix} -1 & -4 & -1 \\ -1 & -1 & -3 \\ 0 & 0 & 0 \end{bmatrix}$ **(c)** $\begin{bmatrix} -20 & 8 & 4 \\ -8 & 4 & 8 \\ 12 & -4 & 8 \end{bmatrix}$ **(d)** $\begin{bmatrix} 20 & -30 & -10 \\ 5 & -10 & -25 \\ -15 & 5 & -10 \end{bmatrix}$ **(e)** $\begin{bmatrix} 0 & -22 & -6 \\ -3 & -6 & -17 \\ -3 & 1 & -2 \end{bmatrix}$

7. **(a)** $\begin{bmatrix} 19 & -1 \\ 2 & 1 \end{bmatrix}$ **(b)** $\begin{bmatrix} 6 & 9 \\ 7 & 14 \end{bmatrix}$ **9.** **(a)** $\begin{bmatrix} -9 & -29 \\ 4 & -14 \end{bmatrix}$ **(b)** $\begin{bmatrix} -7 & 10 & -12 \\ 0 & -1 & 11 \\ 5 & -5 & -15 \end{bmatrix}$ **11.** **(a)** $\begin{bmatrix} 4 & 3 & -1 \\ 8 & -2 & 3 \\ 6 & 5 & 2 \end{bmatrix}$ **(b)** $\begin{bmatrix} 4 & 3 & -1 \\ 8 & -2 & 3 \\ 6 & 5 & 2 \end{bmatrix}$

13. Inverses **15.** Inverses **17.** Inverses **19.** $\begin{bmatrix} 2 & -5 \\ -1 & 3 \end{bmatrix}$ **21.** $\begin{bmatrix} 0 & 1 \\ 1 & 1 \end{bmatrix}$ **23.** $\begin{bmatrix} 7 & 2 & -6 \\ -3 & -1 & 3 \\ 2 & 1 & -2 \end{bmatrix}$ **25.** $\begin{bmatrix} 1 & 2 & 0 \\ 0 & 2 & 3 \\ 1 & 3 & 1 \end{bmatrix}$

27. (2, 1) **29.** (4, −2) **31.** (1, 1) **33.** (−1, 2, 1) **35.** (2, 1, 3)

37. **(a)** [2325 1351.5 930] **(b)** Profit from store 1 is $2325; profit from store 2 is $1351.50; profit from store 3 is $930.

39. **(a)** [224,000,000 302,000,000] **(b)** Value of cars shipped to site 1 is $224,000,000; value of cars shipped to site 2 is $302,000,000.

43. $x = 1, y = 2, z = 4, w = 10$ **47.** $\begin{bmatrix} -1 & 1 & -1 \\ 1 & -\frac{1}{2} & \frac{5}{4} \\ 1 & -\frac{1}{2} & \frac{3}{4} \end{bmatrix}$

PROBLEM SET 8.4 (ON PAGE 502)

1. **(a)** 17 **(b)** −19 **3.** **(a)** −1 **(b)** 11 **5.** 63 **7.** 0 **9.** 3 **17.** 0 **19.** −20 **21.** −2 **23.** $\left(\frac{1}{3}, \frac{2}{3}\right)$ **25.** (1, 2)
27. $\left(-6, \frac{1}{2}\right)$ **29.** (1, 1, 1) **31.** (3, −1, 2) **33.** $\left(\frac{17}{2}, -\frac{3}{2}, -\frac{25}{2}\right)$ **35.** $\left(-\frac{3}{4}, \frac{83}{8}, \frac{21}{8}\right)$ **37.** $A = \frac{33}{2}$ **39.** **(a)** 1 **(b)** 54 **(c)** 2
41. **(a)** $\frac{1}{4}$ **(b)** $\frac{21 \pm \sqrt{185}}{4}$ **43.** 1, 2 **45.** $a = 38, b = -7, c = 20$
49. **(a)** (0, 0, 0) is the only solution. **(b)** There are infinitely many solutions. **51.** **(b)** −80

PROBLEM SET 8.5 (ON PAGE 511)

13. (0, 0), (0, 4), (3, 5), (8, 0) **15.** (0, 0), (0, 5), $\left(\frac{14}{3}, \frac{8}{3}\right)$, (6, 0)

17.

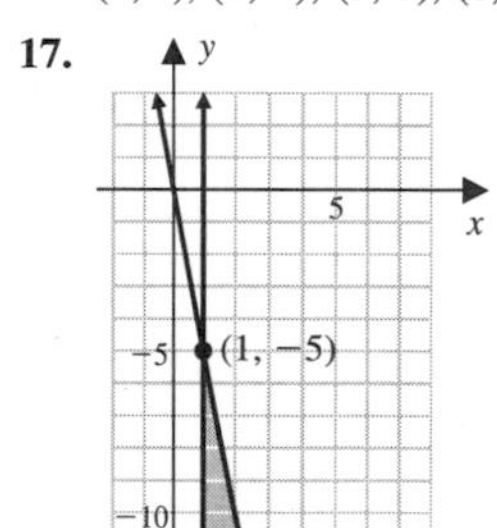

19.

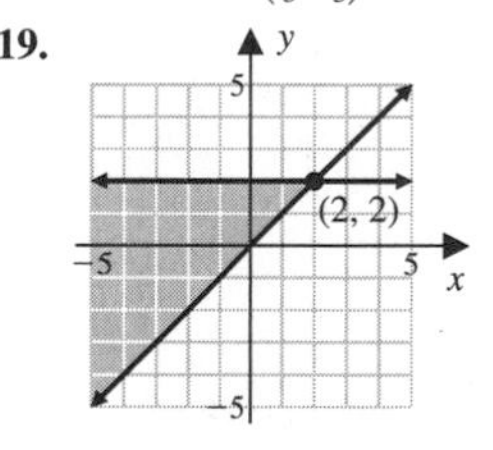

21.

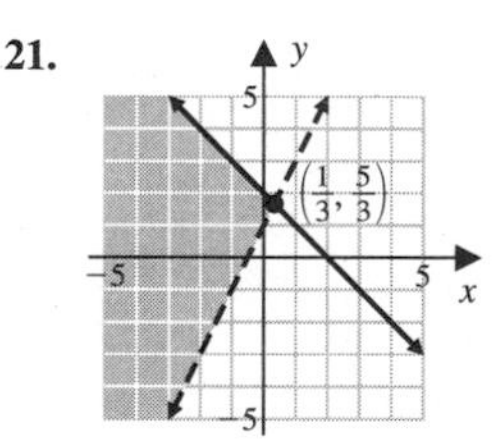

23.

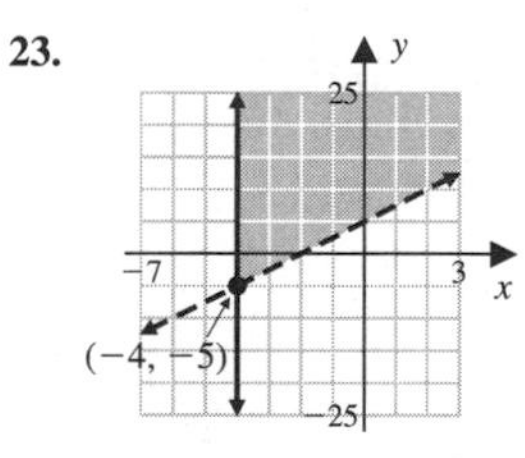

25.

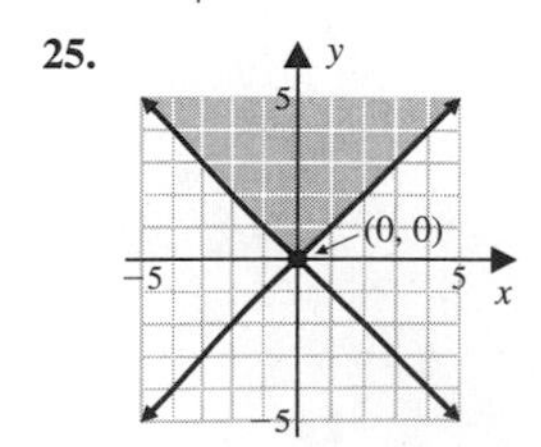

27.

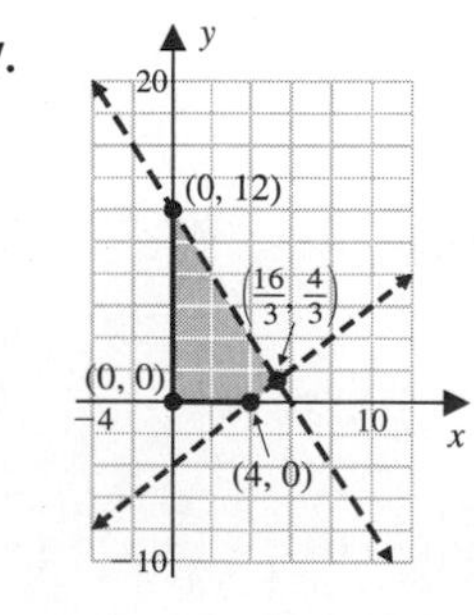

29. Maximum: 27; Minimum: 0 **31.** Maximum: 44; Minimum: 0
33. Maximum: 30; Minimum: 0 **35.** Maximum: 24; Minimum: 0
37. Maximum: 23; Minimum: −12 **39.** Maximum: 25; Minimum: 6

41. **(a)** $\begin{cases} x + y \leq 200{,}000 \\ 0.05x + 0.06y \geq 10{,}600 \\ x \geq 0, y \geq 0 \end{cases}$ **(b)** Maximum revenue of $12,000 when $200,000 is invested at 6% interest

43. **(a)** Let S and M be the amounts of protein powder made from soybeans and milk, respectively (in grams); $\begin{cases} 2S + 2M \geq 12 \\ 2S + 4M \geq 8 \\ S \geq 0, M \geq 0 \end{cases}$
(b) Any combination that supplies adequate protein also supplies adequate carbohydrates.
45. 800 tacos and 0 hamburgers **47.** \$750,000; 5000 barrels of diesel oil and 20,000 barrels of gasoline
49. 100 cord-type and 200 cordless toothbrushes
51.

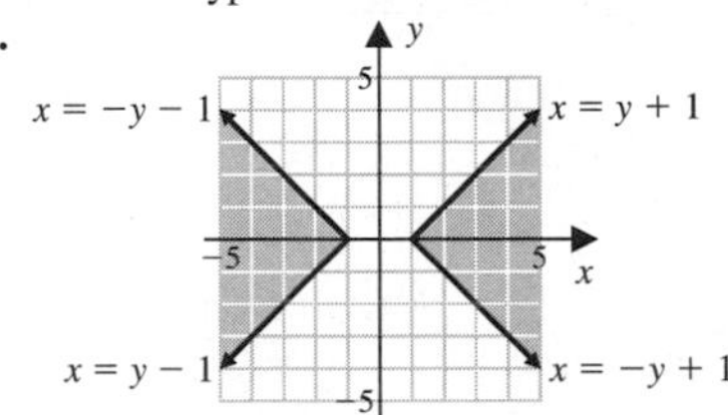

53. $\begin{cases} x + 4y \leq 28 \\ x + 2y \leq 16 \\ x \leq 8 \\ x \geq 0, y \geq 0 \end{cases}$

PROBLEM SET 8.6 (ON PAGE 520)

1. Two solutions: $(2, 2), \left(-5, \frac{25}{2}\right)$ **3.** Two solutions: $(35, 7), (-35, -7)$ **5.** Two solutions: $(2, 6), (3, 4)$
7. Four solutions: $(3, 4), (-3, 4), (4, -3), (-4, -3)$ **9.** Two solutions: $(1, 2), (1, -2)$
11. Two solutions: $(1.09, 2.18), (-1.84, 4.38)$ **13.** Four real solutions: $\left(6, \frac{4}{5}\right), \left(6, -\frac{4}{5}\right), \left(-6, \frac{4}{5}\right), \left(-6, -\frac{4}{5}\right)$
15. Four real solutions: $(3, 2), (3, -2), (-3, 2), (-3, -2)$ **17.** Two real solutions: $(1.08, -2.74), (-0.58, 2.24)$
19. Two real solutions: $(-2.72, 4.15), (3.03, 0.32)$ **21.** $\left(2\sqrt{6}, 1\right), \left(2\sqrt{6}, -1\right), \left(-2\sqrt{6}, 1\right), \left(-2\sqrt{6}, -1\right)$
23. $\left(2, 2\sqrt{3}\right), \left(2, -2\sqrt{3}\right), \left(-2, 2\sqrt{3}\right), \left(-2, -2\sqrt{3}\right)$ **25.** $(4, 2), (-4, 2)\ \left(4\sqrt{\frac{5}{3}}, -\frac{10}{3}\right), \left(-4\sqrt{\frac{5}{3}}, -\frac{10}{3}\right)$
27. $(3, 2), (3, -2), (-3, 2), (-3, -2)$ **29.** $(2, 1), (-2, 1)$ **31.** No real solutions **33.** $\left(\frac{41}{10}, \frac{3\sqrt{91}}{10}\right), \left(\frac{41}{10}, -\frac{3\sqrt{91}}{10}\right)$
35. $(-1.33, 0.24), (1.33, 0.24)$ **37.** $(1.60, -0.03)$
39.

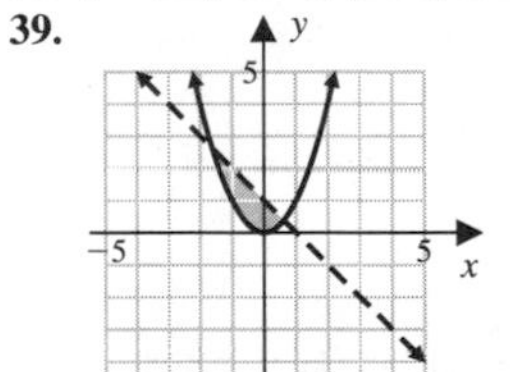

41.

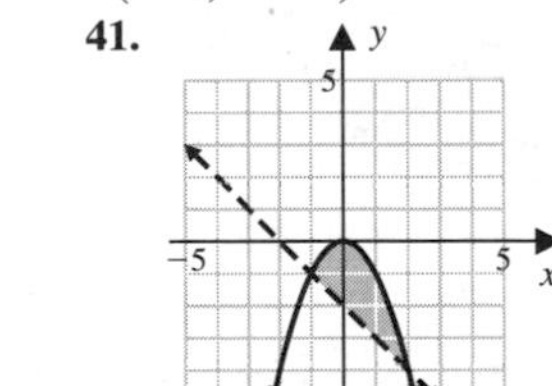

43.

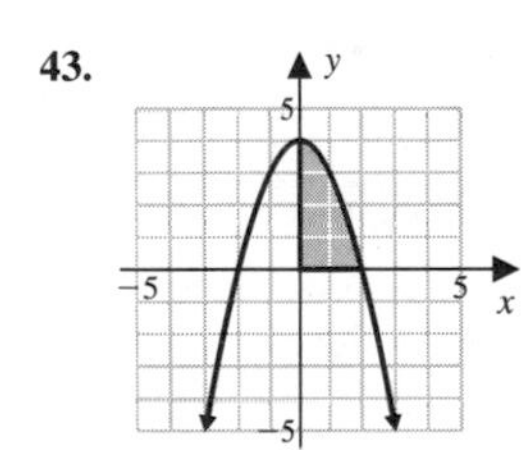

45.

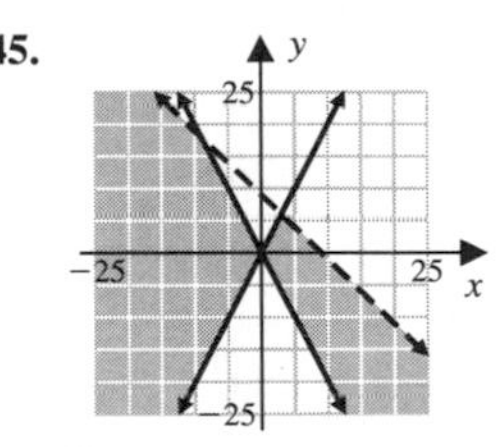

47.

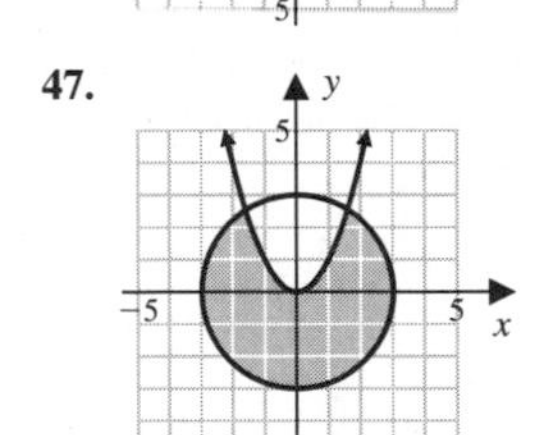

49. 6 in. by 8 in. **51.** 40 m by 4 m and 4 m by 4 m **53.** 150 ft, 5 ft **55.** 18 in. by 20 in.
57. $\left(1/\sqrt{3}, \frac{1}{2}\right), \left(1/\sqrt{3}, -\frac{1}{2}\right), \left(-1/\sqrt{3}, \frac{1}{2}\right), \left(-1/\sqrt{3}, -\frac{1}{2}\right)$
59. $(8, 32), \left(\sqrt{8}, 1024\right), (-8, 32), \left(-\sqrt{8}, 1024\right)$ **61.** $(1, -1, 2), (-1, 3, -2)$
63. $\left(0, -\frac{1}{3}, \frac{5}{3}\right), (1, 0, 1), \left(\frac{5}{2}, \frac{1}{2}, 0\right)$ **65.** $(-2, 1), \left(0, \frac{\ln 3}{\ln 2}\right)$ **67.** $(3, 4), (3, -4)$ **69.** $(0.57, 0.32)$

CHAPTER 8 REVIEW PROBLEM SET (ON PAGE 523)

1. $(-1, -2)$ **3.** $(2, -1)$ **5.** $(-9, 5)$ **7.** $(-1 - 3t, 2 - t, t)$, where t is a real number **9.** $\left(-\frac{2}{9}, \frac{28}{9}\right)$ **11.** $(0, 1, 1)$

13. $A = \frac{9}{8}, B = -\frac{1}{8}$ **15.** **(a)** $\begin{bmatrix} 12 & -3 \\ 7 & 7 \end{bmatrix}$ **(b)** $\begin{bmatrix} 14 & -11 \\ 9 & 19 \end{bmatrix}$ **17.** **(a)** $\begin{bmatrix} 0 & 5 \\ 11 & -18 \\ 1 & -7 \end{bmatrix}$ **(b)** Not possible **21.** $(-1, 2)$

23. 5 **25.** 0 **27.** $(3, 2)$ **29.** $\left(\frac{52}{29}, -\frac{50}{29}, \frac{36}{29}\right)$

31.

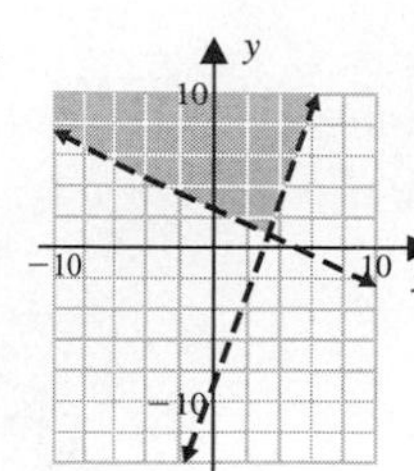

33. Maximum: 21; Minimum: 0 **35.** Maximum: 20; Minimum: 0 **37.** One solution: $(3, -4)$
39. Three solutions: $\left(-5, \sqrt{11}\right)$, $(6, 0)$, $\left(-5, -\sqrt{11}\right)$
41. \$2453 in state income tax, \$1387 in city income tax, \$10,410 in federal income tax
43. 7 and 8, 7 and -8, -7 and 8, or -7 and -8

CHAPTER 8 TEST (ON PAGE 525)

1. (a)–(b) $(1, 1)$ **2.** $(4, -1, 3)$ **3. (a)** $\begin{bmatrix} 9 & -5 \\ 8 & 16 \end{bmatrix}$ **(b)** $\begin{bmatrix} 7 & -13 \\ 16 & 6 \end{bmatrix}$ **(c)** $\begin{bmatrix} 3 & 1 \\ 8 & -8 \end{bmatrix}$ **4. (b)** $(-1, -2)$
5. (a) -19 **(b)** $(1, -1)$ **6.** $A = -\frac{4}{3}, B = \frac{7}{3}$ **7.** $(2, 2), \left(-3, \frac{9}{2}\right)$ **8.** $(0, 4), (2, 2), (3, 0), (0, 0)$ **9.** Maximum: 19; Minimum: 1
10. Fixed charge = \$30; hourly rate = \$20

CHAPTER 9: PROBLEM SET 9.1 (ON PAGE 534)

1. $y = 0$; $V = (0, 0)$; $F = (2, 0)$; d: $x = -2$ **3.** $x = 0$; $V = (0, 0)$; $F = (0, -1)$; d: $y = 1$
5. $y = 0$; $V = (0, 0)$; $F = \left(-\frac{3}{4}, 0\right)$; d: $x = \frac{3}{4}$ **7.** $x^2 = 16y$; $x = 0$ **9.** $x^2 = -8y$; $x = 0$

11. $V = (3, 2)$; $y = 2$; $F = (5, 2)$; d: $x = 1$

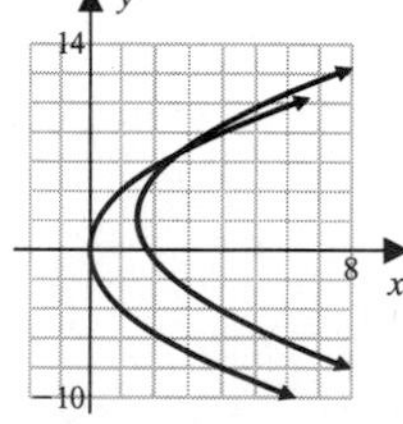

13. $V = (-1, -3)$; $x = -1$; $F = (-1, -4)$; d: $y = -2$

15. $V = (1, -2)$; $y = -2$; $F = \left(-\frac{1}{2}, -2\right)$; d: $x = \frac{5}{2}$

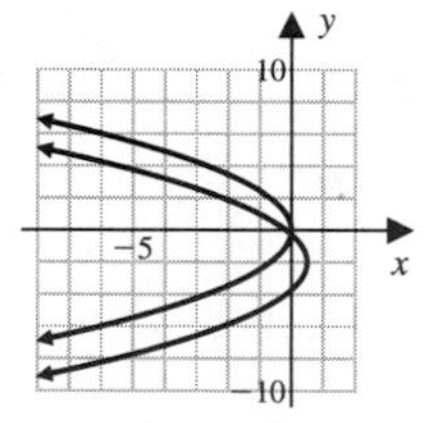

17. $(y + 1)^2 = -2(x + 3)$; $V = (-3, -1)$; $F = \left(-\frac{7}{2}, -1\right)$; $y = -1$; d: $x = -\frac{5}{2}$

19. $(x + 2)^2 = -2\left(y + \frac{1}{6}\right)$; $V = \left(-2, -\frac{1}{6}\right)$; $F = \left(-2, -\frac{2}{3}\right)$; $x = -2$; d: $y = \frac{1}{3}$
21. $(y - 2)^2 = 6\left(x - \frac{3}{2}\right)$; $V = \left(\frac{3}{2}, 2\right)$; $F = (3, 2)$; $y = 2$; d: $x = 0$ **23.** $(y - 2)^2 = 16(x + 1)$; $y = 2$ **25.** $(y + 5)^2 = 8(x - 3)$; $y = -5$
27. $(x - 1)^2 = 4(y - 3)$; $x = 1$ **29.** $(x + 3)^2 = 20y$; $x = -3$ **31.** $x^2 = -\frac{4}{3}(y - 3)$; $x = 0$

33. (a) $y = -1 - \sqrt{2x - 2}$ or $y = -1 + \sqrt{2x - 2}$

(b) $y = -\dfrac{(x - 3)^2}{6} - 1$

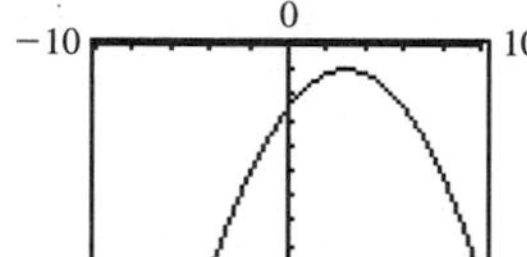

35. $y = -1 - 2\sqrt{6 - 3x}$ or $y = -1 + 2\sqrt{6 - 3x}$

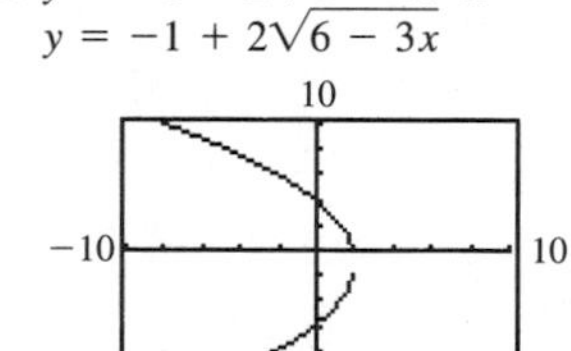

37. (b) $y^2 = 12x$ **(c)** 3 ft **39. (b)** $x^2 = 2664y$ **(c)** Approx. 3.75 in. **41.** 0.75 ft **43. (b)** (i) 8 (ii) 4 (iii) 4 (iv) $\frac{2}{7}$
45. As c gets larger, the graphs get closer to the x axis.

PROBLEM SET 9.2 (ON PAGE 546)

1. Vertices: $(-4, 0), (4, 0), (0, -3), (0, 3)$; Foci: $\left(-\sqrt{7}, 0\right), \left(\sqrt{7}, 0\right)$
3. Vertices: $(-2, 0), (2, 0), (0, -4), (0, 4)$; Foci: $\left(0, -2\sqrt{3}\right), \left(0, 2\sqrt{3}\right)$
5. Vertices: $\left(-\sqrt{10}, 0\right), \left(\sqrt{10}, 0\right), (0, -1), (0, 1)$; Foci: $(-3, 0), (3, 0)$ **7.** $\dfrac{x^2}{25} + \dfrac{y^2}{16} = 1$ **9.** $\dfrac{x^2}{25} + \dfrac{y^2}{194} = 1$ **11.** $\dfrac{x^2}{\frac{81}{8}} + \dfrac{y^2}{36} = 1$

13. Center: $(2, -3)$; Vertices: $(-2, -3)$, $(6, -3)$, $(2, -6)$, $(2, 0)$; Foci: $(2 - \sqrt{7}, -3)$, $(2 + \sqrt{7}, -3)$

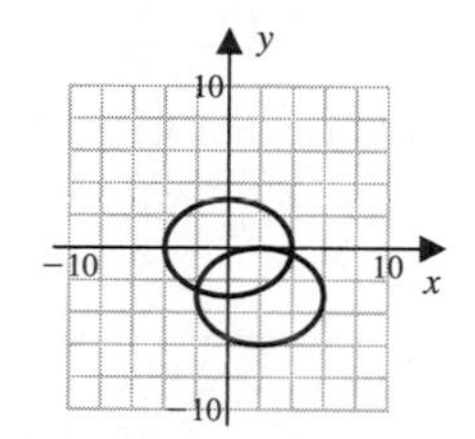

15. Center: $(1, -2)$; Vertices: $(1 - 4\sqrt{3}, -2)$, $(1 + 4\sqrt{3}, -2)$, $(1, -10)$, $(1, 6)$; Foci: $(1, -6)$; $(1, 2)$

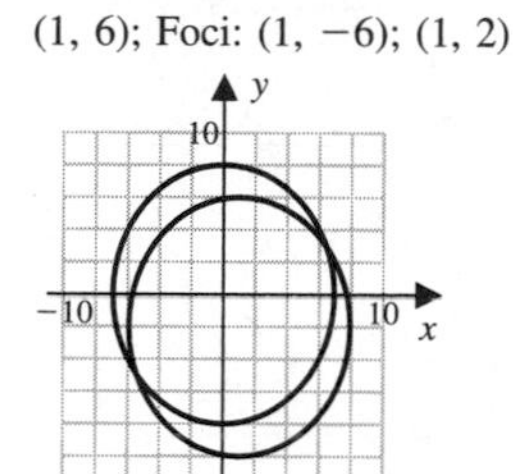

17. Center: $(-2, 1)$;

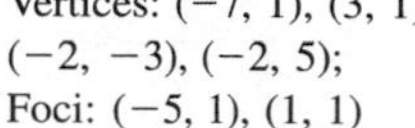

Vertices: $(-7, 1)$, $(3, 1)$, $(-2, -3)$, $(-2, 5)$; Foci: $(-5, 1)$, $(1, 1)$

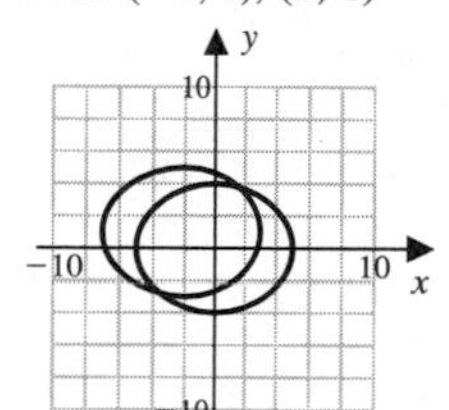

19. Center: $(-1, 1)$; Vertices: $(-1 - \sqrt{3}, 1)$, $(-1 + \sqrt{3}, 1)$, $(-1, 1 - \sqrt{15})$, $(-1, 1 + \sqrt{15})$; Foci: $(-1, 1 - 2\sqrt{3})$, $(-1, 1 + 2\sqrt{3})$

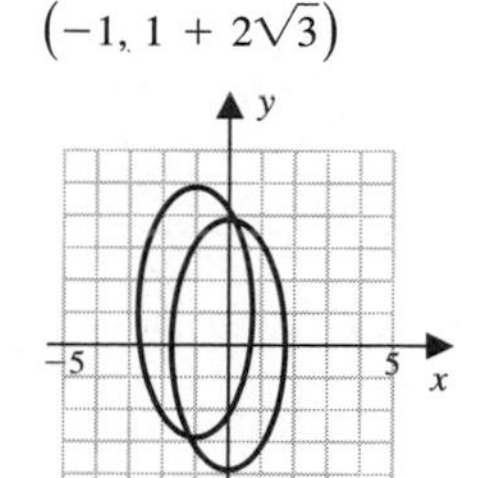

21. $\frac{(x-2)^2}{1} + \frac{(y-3)^2}{4} = 1$; Center: $(2, 3)$; Vertices: $(1, 3)$, $(3, 3)$, $(2, 1)$, $(2, 5)$; Foci: $(2, 3 - \sqrt{3})$, $(2, 3 + \sqrt{3})$

23. $\frac{(x+1)^2}{4} + \frac{(y-1)^2}{1} = 1$; Center: $(-1, 1)$; Vertices: $(-3, 1)$, $(1, 1)$, $(-1, 0)$, $(-1, 2)$; Foci: $(-1 - \sqrt{3}, 1)$, $(-1 + \sqrt{3}, 1)$

25. $\frac{(x+1)^2}{4} + \frac{(y-2)^2}{9} = 1$; Center: $(-1, 2)$; Vertices: $(-3, 2)$, $(1, 2)$, $(-1, -1)$, $(-1, 5)$; Foci: $(-1, 2 - \sqrt{5})$, $(-1, 2 + \sqrt{5})$

27. $\frac{(x-4)^2}{9} + \frac{(y-2)^2}{4} = 1$; Center: $(4, 2)$; Vertices: $(1, 2)$, $(7, 2)$, $(4, 0)$, $(4, 4)$; Foci: $(4 - \sqrt{5}, 2)$, $(4 + \sqrt{5}, 2)$

29. $\frac{(x-3)^2}{4} + \frac{(y+2)^2}{25} = 1$ **31.** $\frac{(x+3)^2}{1} + \frac{(y-1)^2}{25} = 1$ **33.** $\frac{(x-2)^2}{4} + \frac{(y-4)^2}{3} = 1$

35.

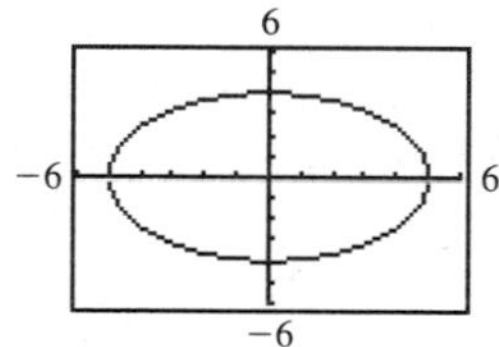

37.

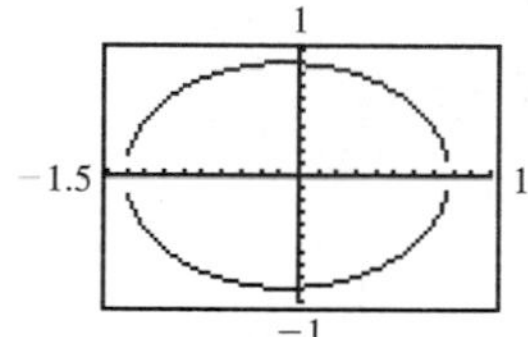

39.

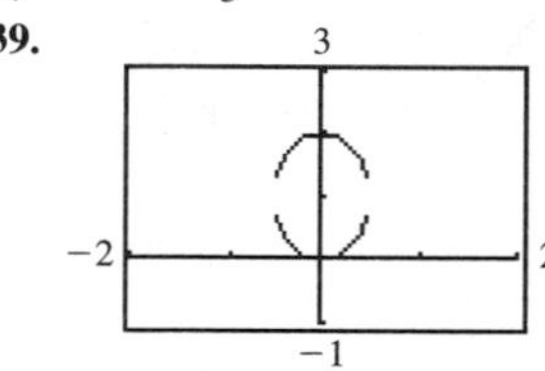

41. **(b)** $\frac{x^2}{169} + \frac{y^2}{25} = 1$ **(c)** $\frac{25}{13}$ m **(d)** Approx. 4.87 m

43. **(b)** $\frac{x^2}{8{,}629{,}388{,}130{,}250{,}000} + \frac{y^2}{8{,}626{,}974{,}768{,}000{,}000} = 1$ **(c)** 3,107,000 mi **45.** **(b)** $\frac{x^2}{15{,}625} + \frac{y^2}{5625} = 1$ **(c)** 200 ft

47. $\frac{x^2}{25} + \frac{y^2}{169} = 1$ **49.** **(a)** Circle of radius a and center at the origin **(b)** Yes; Foci = Center

51. **(a)** $A = 2xy$
(b) $A = \frac{3}{2}x\sqrt{16 - x^2}$
(c)

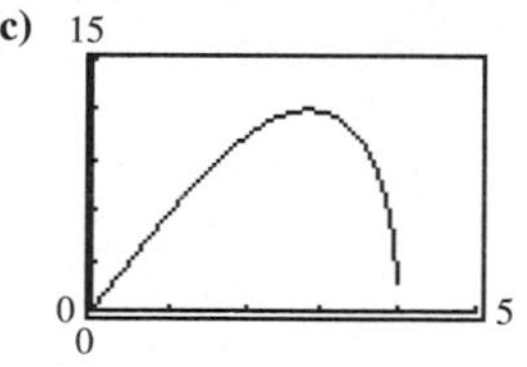

(d) $x = 2.83$; $A = 12$

PROBLEM SET 9.3 (ON PAGE 558)

1. Vertices: $(-3, 0)$, $(3, 0)$; Foci: $(-\sqrt{10}, 0)$, $(\sqrt{10}, 0)$; Asymptotes: $y = -\frac{1}{3}x$ and $y = \frac{1}{3}x$

3. Vertices: $(0, -3)$, $(0, 3)$; Foci: $(0, -3\sqrt{2})$, $(0, 3\sqrt{2})$; Asymptotes: $y = -x$ and $y = x$

5. Vertices: $(-\frac{1}{6}, 0)$, $(\frac{1}{6}, 0)$; Foci: $(-\sqrt{5}/6, 0)$, $(\sqrt{5}/6, 0)$; Asymptotes: $y = -2x$ and $y = 2x$

7. Vertices: $(-2, 0)$, $(2, 0)$; Foci: $(-3, 0)$, $(3, 0)$; Asymptotes: $y = -(\sqrt{5}/2)x$ and $y = (\sqrt{5}/2)x$

9. $\frac{x^2}{256} - \frac{y^2}{400} = 1$ **11.** $\frac{y^2}{16} - \frac{x^2}{9} = 1$ **13.** $\frac{x^2}{\frac{7}{3}} - \frac{y^2}{35} = 1$

15. Center: $(5, 4)$;
Vertices: $(0, 4)$, $(10, 4)$;
Foci: $(5 - \sqrt{29}, 4)$, $(5 + \sqrt{29}, 4)$;
Asymptotes:
$y - 4 = -\frac{2}{5}(x - 5)$
and $y - 4 = \frac{2}{5}(x - 5)$

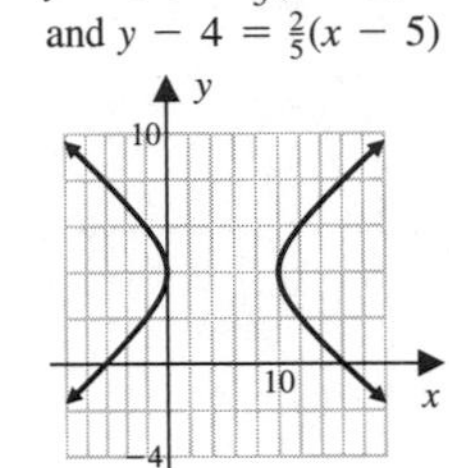

17. Center: $(-2, -3)$;
Vertices: $(-2, -7)$, $(-2, 1)$;
Foci: $(-2, -3 - 4\sqrt{2})$; $(-2, -3 + 4\sqrt{2})$;
Asymptotes:
$y + 3 = -(x + 2)$
and $y + 3 = x + 2$

19. Center: $(1, -2)$;
Vertices: $(-1, -2)$, $(3, -2)$;
Foci: $(1 - \sqrt{13}, -2)$, $(1 + \sqrt{13}, -2)$;
Asymptotes:
$y + 2 = -\frac{3}{2}(x - 1)$
and $y + 2 = \frac{3}{2}(x - 1)$

21. $\frac{(y - 1)^2}{4} - \frac{(x - 4)^2}{3} = 1$;
Center: $(4, 1)$;
Vertices: $(4, -1)$, $(4, 3)$;
Foci: $(4, 1 - \sqrt{7})$, $(4, 1 + \sqrt{7})$;
Asymptotes:
$y - 1 = -\frac{2}{\sqrt{3}}(x - 4)$
and $y - 1 = \frac{2}{\sqrt{3}}(x - 4)$

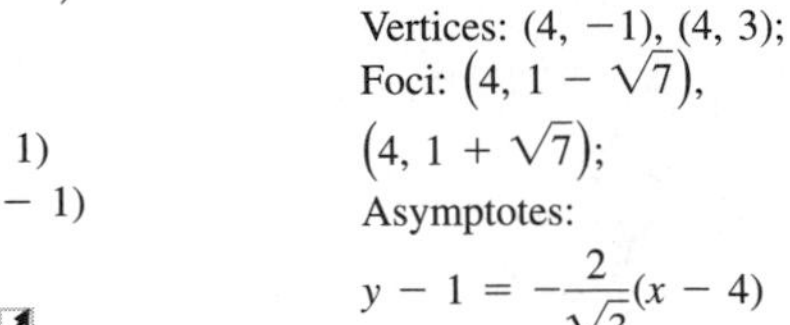

23. $\frac{(x + 2)^2}{9} - \frac{(y + 4)^2}{25} = 1$; Center: $(-2, -4)$; Vertices: $(-5, -4)$, $(1, -4)$; Foci: $(-2 - \sqrt{34}, -4)$, $(-2 + \sqrt{34}, -4)$;
Asymptotes: $y + 4 = -\frac{5}{3}(x + 2)$ and $y + 4 = \frac{5}{3}(x + 2)$

25. $\frac{(y + 1)^2}{3} - \frac{(x + 1)^2}{\frac{3}{4}} = 1$; Center: $(-1, -1)$; Vertices: $(-1, -1 - \sqrt{3})$, $(-1, -1 + \sqrt{3})$; Foci: $\left(-1, -1 - \frac{\sqrt{15}}{2}\right)$, $\left(-1, -1 + \frac{\sqrt{15}}{2}\right)$;
Asymptotes: $y + 1 = -2(x + 1)$ and $y + 1 = 2(x + 1)$

27. $\frac{(x - 1)^2}{9} - \frac{(y - 2)^2}{4} = 1$; Center: $(1, 2)$; Vertices: $(-2, 2)$, $(4, 2)$; Foci: $(1 - \sqrt{13}, 2)$, $(1 + \sqrt{13}, 2)$;
Asymptotes: $y - 2 = -\frac{2}{3}(x - 1)$ and $y - 2 = \frac{2}{3}(x - 1)$

29. $\frac{(x - 3)^2}{9} - \frac{(y + 4)^2}{4} = 1$ **31.** $\frac{(y - 5)^2}{1} - \frac{(x + 1)^2}{3} = 1$ **33.** $\frac{(x - 4)^2}{4} - \frac{(y - 3)^2}{5} = 1$

35.

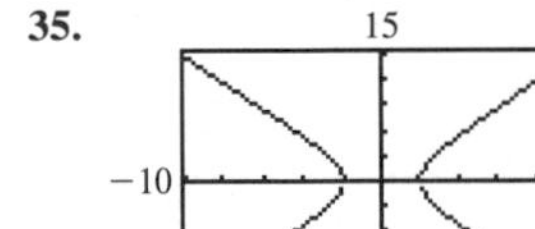

37.

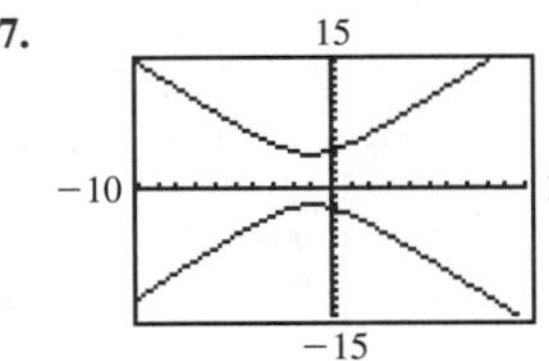

39.

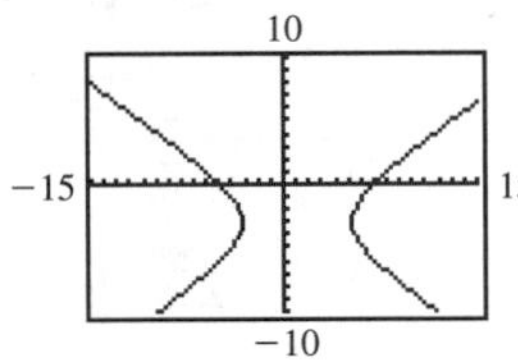

41. (b) $\frac{x^2}{1225} - \frac{y^2}{6875} = 1$ **(c)** $(41, 50)$ **43. (a)** $\frac{x^2}{1730.56} - \frac{y^2}{6922.24} = 1$ **(b)** 93 million mi **45.** $\frac{x^2}{16} - \frac{y^2}{9} = 1$

47. (a) $(-1, -2)$, $(1, -2)$, $(-1, 2)$, $(1, 2)$ **(b)** $(-1, -2)$, $(1, -2)$, $(-1, 2)$, $(1, 2)$ **(c)** The results are the same.

49. (b) The graphs are reflections of each other about the line $y = x$ and have the same asymptotes.

51. (a) Ellipse **(b)** Hyperbola **(c)** Not possible

PROBLEM SET 9.4 (ON PAGE 566)

1. $\frac{\left(y - \frac{15}{4}\right)^2}{\frac{125}{16}} - \frac{(x - 3)^2}{\frac{25}{4}} = 1$ **3.** $\frac{x^2}{25} + \frac{y^2}{16} = 1$ **5.** $y^2 = -4(x - 3)$ **7.** $\frac{x^2}{36} + \frac{y^2}{27} = 1$ **9.** $\frac{x^2}{25} - \frac{y^2}{200} = 1$ **11.** $\frac{3}{5}$ **13.** $\sqrt{\frac{2}{3}}$

15. $\frac{5}{3}$ **17.** $\frac{5}{3}$ **19.** $\sqrt{\frac{2}{3}}$ **21.** $\sqrt{2}$ **23.** $r = \frac{4}{1 - \sin\theta}$ **25.** $r = \frac{4}{1 - 2\cos\theta}$ **27.** $r = \frac{4}{5 + 2\sin\theta}$

29. As e gets close to 0, the conic appears closer to a circle. As e gets larger and larger, the conic appears closer to two parallel vertical lines.

31. $e = 1$; parabola; $V = \left(0, \frac{5}{2}\right)$

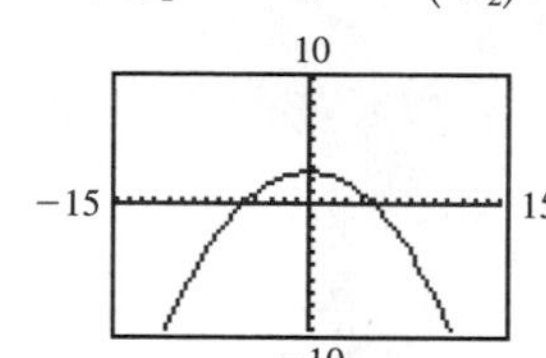

33. $e = \frac{1}{2}$; ellipse;
Center $\left(\frac{2}{3}, 0\right)$;
Vertices: $\left(-\frac{2}{3}, 0\right)$, $(2, 0)$, $\left(\frac{2}{3}, -2/\sqrt{3}\right)$, $\left(\frac{2}{3}, 2/\sqrt{3}\right)$

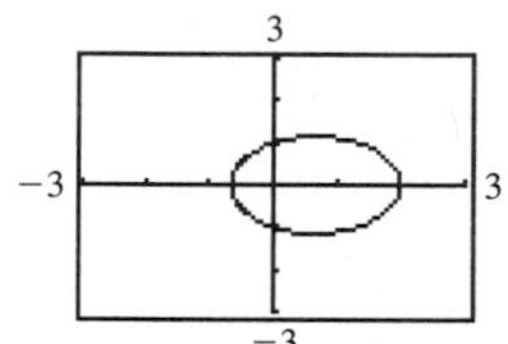

35. $e = 3$; hyperbola;
Center: $\left(\frac{5}{8}, 0\right)$;
Vertices: $\left(\frac{5}{12}, 0\right)$, $\left(\frac{5}{6}, 0\right)$

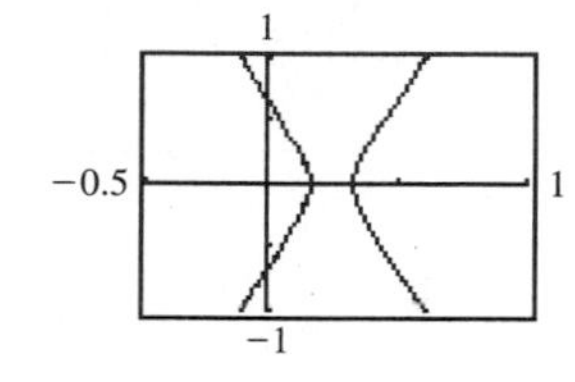

37. $e = \frac{1}{2}$; ellipse;
Center: $\left(-\frac{2}{3}, 0\right)$;
Vertices: $(-2, 0)$, $\left(\frac{2}{3}, 0\right)$, $\left(-\frac{2}{3}, -2/\sqrt{3}\right)$, $\left(-\frac{2}{3}, 2/\sqrt{3}\right)$

39. $e = \frac{2}{3}$; ellipse;
Center: $\left(0, -\frac{12}{5}\right)$;
Vertices: $(0, -6)$, $\left(0, \frac{6}{5}\right)$, $\left(-6/\sqrt{5}, -\frac{12}{5}\right)$, $\left(6/\sqrt{5}, -\frac{12}{5}\right)$

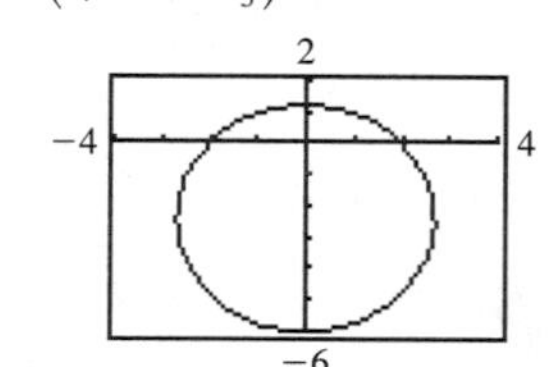

41. $e = \frac{1}{2}$; ellipse;
Center: $\left(-\frac{2}{3}, 0\right)$;
Vertices: $(-2, 0)$, $\left(\frac{2}{3}, 0\right)$, $\left(-\frac{2}{3}, -2/\sqrt{3}\right)$, $\left(-\frac{2}{3}, 2/\sqrt{3}\right)$

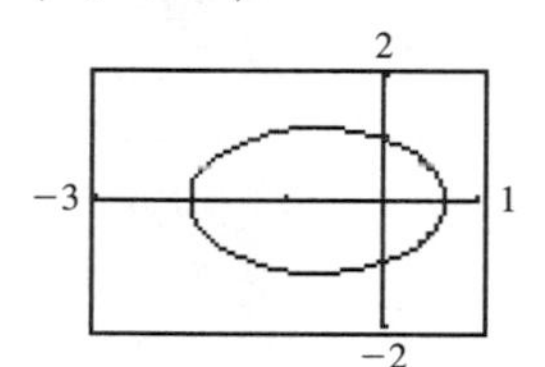

43. $e = \frac{1}{2}$; ellipse;
Center: $\left(-\frac{1}{3}, 0\right)$;
Vertices: $(-1, 0)$, $\left(\frac{1}{3}, 0\right)$, $\left(-\frac{1}{3}, -1/\sqrt{3}\right)$, $\left(-\frac{1}{3}, 1/\sqrt{3}\right)$

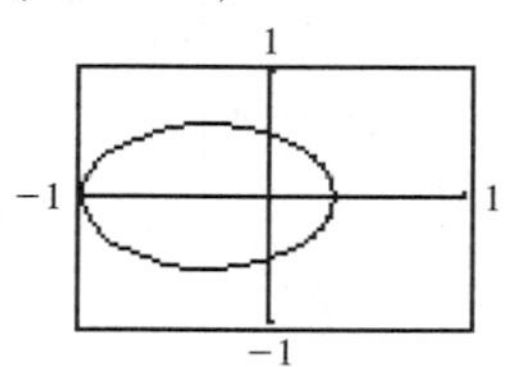

45. **(a)** $e = \frac{1}{59}$ **(b)** Approx. 94.4 million mi **47.** **(b)** $\frac{x^2}{\frac{4}{3}} + \frac{y^2}{1} = 1$ **(c)** 2 ft; $4/\sqrt{3}$ ft **(d)** 3 ft by 3.31 ft (approx.)

49. The eccentricities are the same. **51.** **(a)** Two **(b)** Two **(c)** One **53.** $r^2 = \sec 2\theta$

55. Ellipse

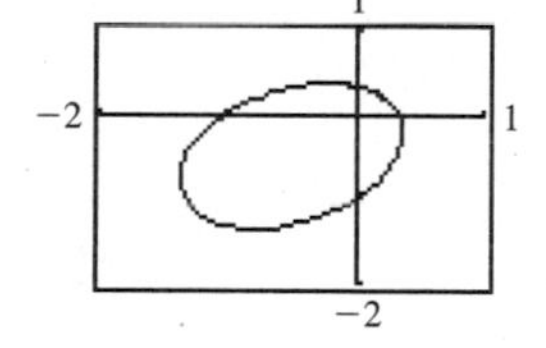

57. In polar coordinates: (0.74, 0.79), (1.55, 3.93);
in Cartesian coordinates: (0.52, 0.52), (−1.09, −1.09)

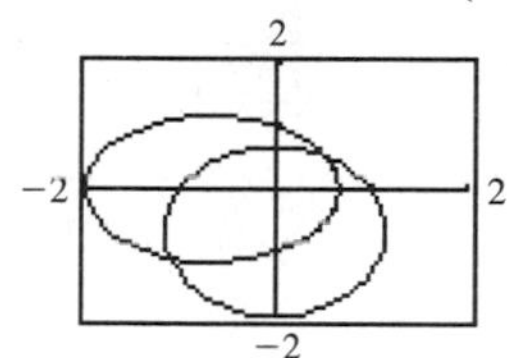

PROBLEM SET 9.5 (ON PAGE 576)

1. $\left(-\sqrt{3} - \frac{5}{2}, 1 - \frac{5\sqrt{3}}{2}\right)$ **3.** $\left(\frac{11}{\sqrt{2}}, -\frac{9}{\sqrt{2}}\right)$ **5.** (6.71, 0.40)

7. $\frac{\bar{x}^2}{2} - \frac{\bar{y}^2}{2} = 1$

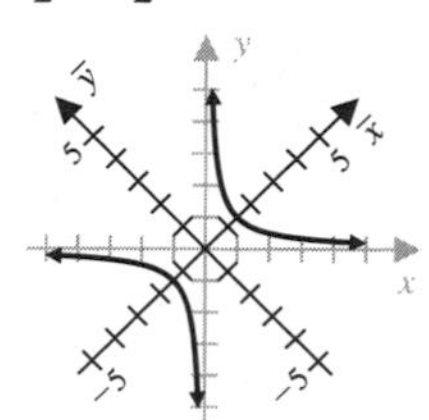

9. $\frac{\bar{x}^2}{8} + \frac{\bar{y}^2}{\frac{8}{5}} = 1$

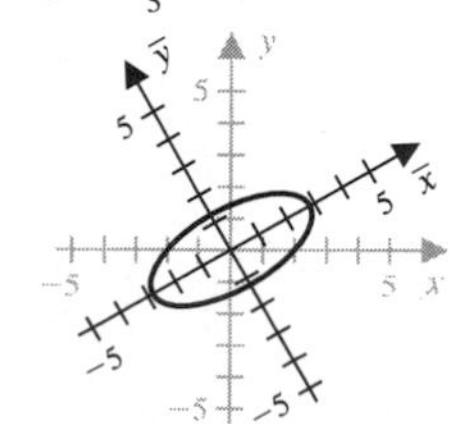

11. $\frac{\bar{x}^2}{4} + \frac{\bar{y}^2}{1} = 1$

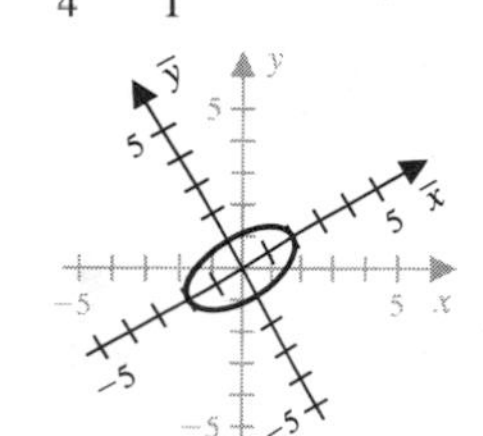

13. $\frac{\bar{x}^2}{4} + \frac{\bar{y}^2}{2} = 1$

15. $45°; \dfrac{\bar{x}^2}{3} + \dfrac{\bar{y}^2}{\frac{9}{7}} = 1$

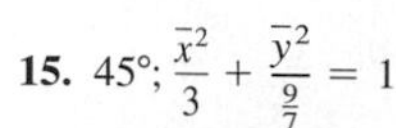

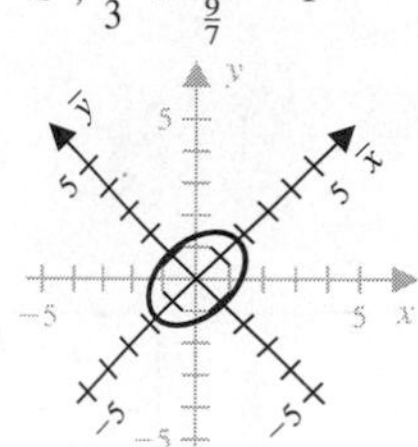

17. $30°; \dfrac{\bar{x}^2}{4} - \dfrac{\bar{y}^2}{16} = 1$

19. $30°; \bar{x}^2 - \bar{y}^2 = 1$

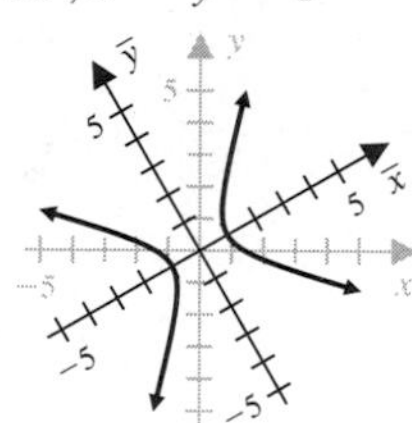

21. $\sin\theta = \dfrac{2}{\sqrt{5}}$; $\cos\theta = \dfrac{1}{\sqrt{5}}$; $\theta = 63.4°$; $\dfrac{\bar{x}^2}{36} + \dfrac{\bar{y}^2}{16} = 1$

23. $\sin\theta = \dfrac{1}{\sqrt{5}}$; $\cos\theta = \dfrac{2}{\sqrt{5}}$; $\theta = 26.6°$; $\dfrac{\bar{x}^2}{36} + \dfrac{\bar{y}^2}{16} = 1$

25. $\sin\theta = \dfrac{1}{\sqrt{5}}$; $\cos\theta = \dfrac{2}{\sqrt{5}}$; $\theta = 26.6°$; $\left(\bar{x} + \dfrac{3}{\sqrt{5}}\right)^2 = \dfrac{8}{\sqrt{5}}\left(\bar{y} + \dfrac{9}{8\sqrt{5}} + \dfrac{1}{8}\right)$

27. $y = \dfrac{x - \sqrt{16 - 3x^2}}{2}$ or $y = \dfrac{x + \sqrt{16 - 3x^2}}{2}$

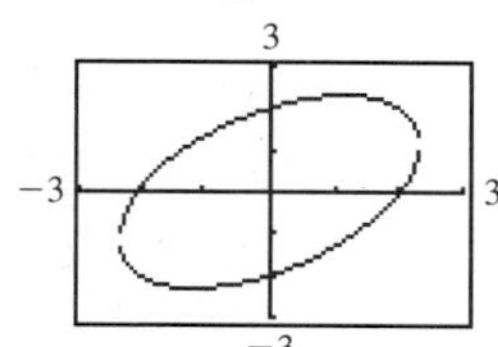

29. $y = 5x - \sqrt{25x^2 - 32}$ or $y = 5x + \sqrt{25x^2 - 32}$

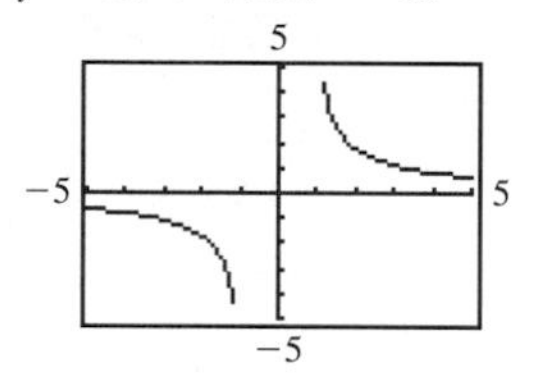

31. $y = \dfrac{-3x - \sqrt{x^2 + 72}}{2}$ or $y = \dfrac{-3x + \sqrt{x^2 + 72}}{2}$

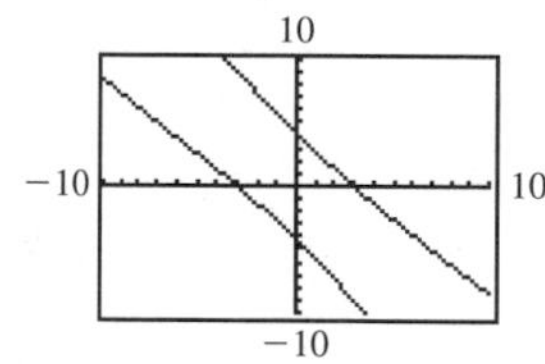

33. Ellipse
35. Hyperbola
37. Ellipse
39. **(a)** $\theta = 90°$
(b) Does not exist.

47.

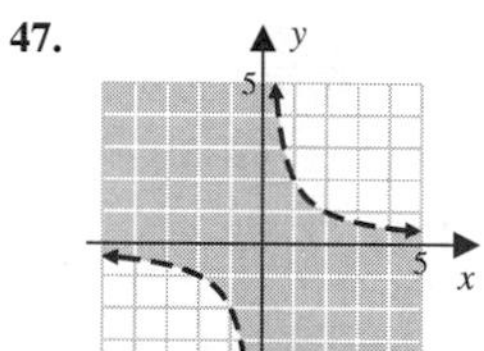

49.

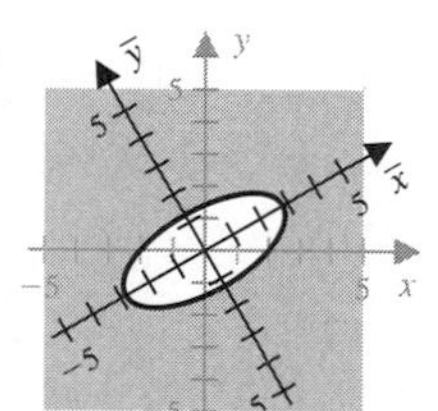

PROBLEM SET 9.6 (ON PAGE 583)

1.

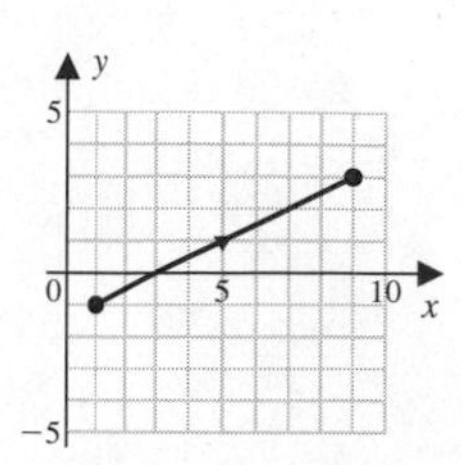

3.

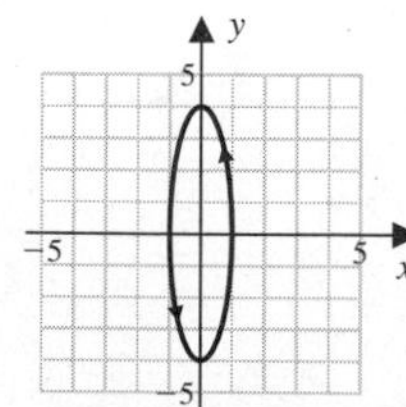

5. $y = \frac{17}{6} - \frac{5}{6}x$ **7.** $y = x^2 - 2x$ **9.** $y = \sqrt{1 - x^2}$

11. $x^2 + y^2 = 9, y \geq 0$ **13.** $(x + 3)^2 + (y - 4)^2 = 1$

15. $x^2 - y^2 = 1, x \geq 1$ or $x \leq -1$ **17.** $y = \dfrac{1}{x^2}, x \geq 1$

19. $\dfrac{x^2}{4} - \dfrac{y^2}{9} = 1, x \geq 2$

21. One possibility: $\begin{cases} x = t \\ y = 3t^2 + 5 \end{cases}$ $0 \leq t \leq 2$

23. One possibility: $\begin{cases} x = t \\ y = t^2 - 4t + 3 \end{cases}$ $-1 \leq t \leq 1$

25. One possibility: $\begin{cases} x = t \\ y = \frac{3}{2}t + \frac{1}{2} \end{cases}$ $1 \leq t \leq 3$

27. One possibility: $\begin{cases} x = 3\cos t \\ y = 3\sin t \end{cases}$ $0 \leq t \leq 2\pi$

29. One possibility: $\begin{cases} x = \cos t \\ y = 1 + \sin t \end{cases}$ $0 \leq t \leq 2\pi$

31.

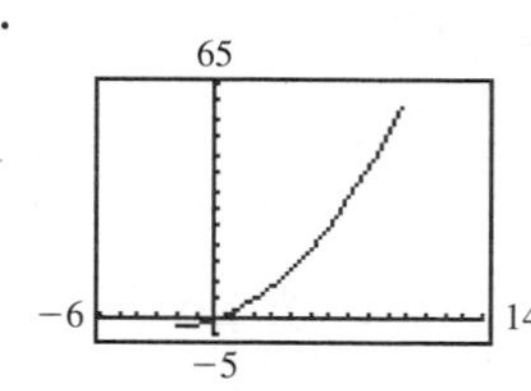

33.

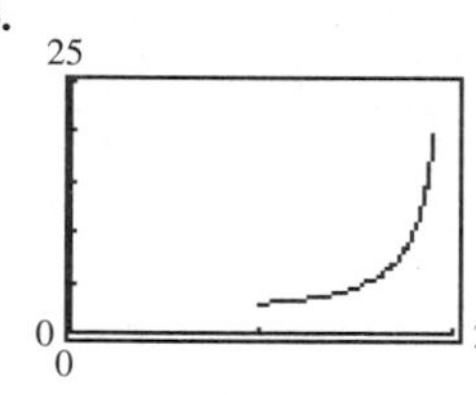

35.

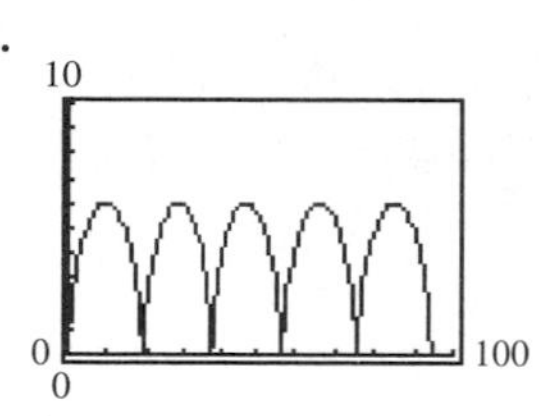

37. **(a)** $\begin{cases} x = (200 \cos 50°)t \\ y = -16t^2 + (200 \sin 50°)t + 10 \end{cases}$ **(c)** 9.6 sec; 1239 ft **(d)** 377 ft; 4.8 sec

39. **(a)** $\begin{cases} x = (125 \cos 25°)t \\ y = -16t^2 + (125 \sin 25°)t + 4 \end{cases}$ **(c)** Approx. 1.2 ft; no

41. One possibility: $\begin{cases} x = \sqrt{6} \cos t - \sqrt{\frac{6}{7}} \sin t \\ y = \sqrt{6} \cos t + \sqrt{\frac{6}{7}} \sin t \end{cases}$ $0 \le t \le 2\pi$

43. **(b)** $y = -\frac{2}{5}x + \frac{26}{5}$ **(c)** Slope $= \dfrac{\text{Coefficient of } t \text{ values in } y}{\text{Coefficient of } t \text{ values in } x}$ **(d)** The slope is d/b.

45. **(a)** $y = x$ **(b)** Since t^2 cannot be negative, $x \ge 0$ and $y \ge 0$.

47. (i) $\begin{cases} x = t \\ y = t^2 + 1 \end{cases}$ $0 \le t \le 5$; (ii) $\begin{cases} x = 5t \\ y = 25t^2 + 1 \end{cases}$ $0 \le t \le 1$; (iii) $\begin{cases} x = \tan t \\ y = \sec^2 t \end{cases}$ $0 \le t \le \tan^{-1} 5$

CHAPTER 9 REVIEW PROBLEM SET (ON PAGE 585)

1. **(a)** $F = (0, -2)$; D: $y = 2$ **(b)** $F = (9, 0)$; D: $x = -9$ **3.** $(x + 4)^2 = -6\left(y - \frac{3}{2}\right)$ **5.** $(y + 3)^2 = -12(x + 2)$

7. **(a)** Vertices: $(-3, 0)$, $(3, 0)$, $(0, -6)$, $(0, 6)$; Foci: $\left(0, -3\sqrt{3}\right)$, $\left(0, 3\sqrt{3}\right)$

(b) Vertices: $\left(-\sqrt{6}, 0\right)$, $\left(\sqrt{6}, 0\right)$, $(0, -2)$, $(0, 2)$; Foci: $\left(-\sqrt{2}, 0\right)$, $\left(\sqrt{2}, 0\right)$

9. Vertices: $(-2, 0)$, $(2, 0)$; Foci: $\left(-\sqrt{13}, 0\right)$, $\left(\sqrt{13}, 0\right)$; Asymptotes: $y = -\frac{3}{2}x$ and $y = \frac{3}{2}x$

11. Vertices: $(-2, 0)$, $(2, 0)$; Foci: $\left(-2\sqrt{5}, 0\right)$, $\left(2\sqrt{5}, 0\right)$; Asymptotes: $y = -2x$ and $y = 2x$ **13.** $\dfrac{x^2}{9} + \dfrac{y^2}{5} = 1$

15. $\dfrac{(x - 3)^2}{\sqrt{5} - 1} + \dfrac{(y - 1)^2}{3 + \sqrt{5}} = 1$ **17.** $\dfrac{x^2}{9} - \dfrac{y^2}{16} = 1$ **19.** $V = (-1, 2)$; $F = \left(-1, \frac{5}{2}\right)$

21. Center: $(1, -2)$; Vertices: $(-2, -2)$, $(4, -2)$, $(1, -6)$, $(1, 2)$; Foci: $\left(1, -2 - \sqrt{7}\right)$, $\left(1, -2 + \sqrt{7}\right)$

23. Center: $(2, 1)$; Vertices: $(2, -1)$, $(2, 3)$; Foci: $\left(2, 1 - \sqrt{13}\right)$, $\left(2, 1 + \sqrt{13}\right)$ **25.** Parabola; $V = (3, -1)$; $F = \left(\frac{7}{2}, -1\right)$

27. Ellipse; Vertices: $(3, 2)$, $(7, 2)$, $(5, -1)$, $(5, 5)$; Foci: $\left(5, 2 - \sqrt{5}\right)$, $\left(5, 2 + \sqrt{5}\right)$

29. Hyperbola; Vertices: $(2, -10)$, $(2, 0)$; Foci: $\left(2, \dfrac{5\sqrt{13}}{2} - 5\right)$, $\left(2, -\dfrac{5\sqrt{13}}{2} - 5\right)$ **31.** Circle **33.** $\dfrac{x^2}{9} + \dfrac{y^2}{5} = 1$ **35.** $\dfrac{x^2}{64} - \dfrac{y^2}{36} = 1$

37. $(x + 5)^2 = 8(y + 3)$ **39.** **(a)** $\dfrac{\sqrt{10}}{4}$ **(b)** $1/\sqrt{3}$ **41.** **(a)** 1 **(b)** $\sqrt{3}/2$ **43.** **(a)** $r = \dfrac{\frac{1}{2}}{1 - \sin\theta}$ **(b)** $r = \dfrac{8}{1 + 2\cos\theta}$

45. **(a)** $\frac{2}{3}$; Ellipse; Center: $\left(\frac{6}{5}, 0\right)$; Vertices in Cartesian form: $\left(-\frac{3}{5}, 0\right)$, $(3, 0)$, $\left(\frac{6}{5}, -3/\sqrt{5}\right)$, $\left(\frac{6}{5}, 3/\sqrt{5}\right)$

(b) $\frac{5}{2}$; Hyperbola; Center: $\left(0, -\frac{25}{21}\right)$; Vertices: $\left(0, -\frac{5}{3}\right)$, $\left(0, -\frac{5}{7}\right)$

47. Ellipse **49.** Ellipse **51.** Hyperbola

57. $y = \frac{1}{2}x^2 - \frac{3}{2}x$ **59.** $y = \ln x$ **61.** 14.28 m

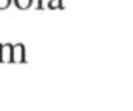

63. $\dfrac{(y - 300)^2}{10{,}000} - \dfrac{x^2}{80{,}000} = 1$

53.

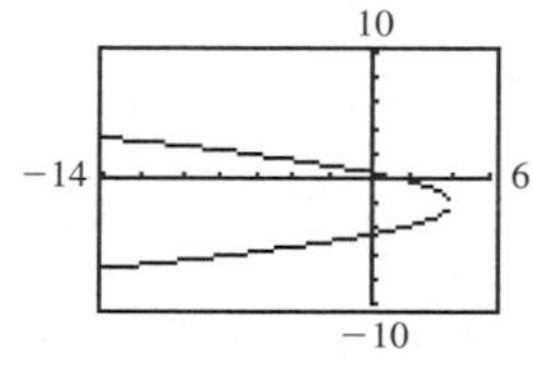

55.

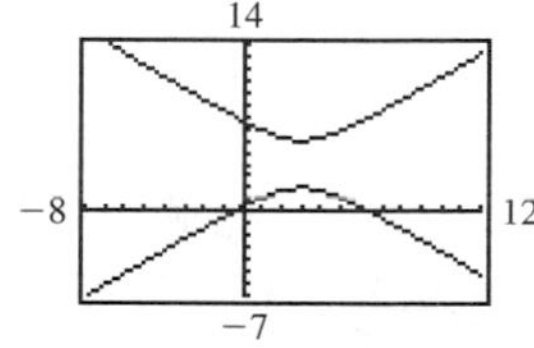

CHAPTER 9 TEST (ON PAGE 587)

1. **(a)** $V = (0, 0)$; $F = (-4, 0)$ **(b)** Vertices: $(-1, 0)$, $(1, 0)$, $(0, -3)$, $(0, 3)$; Foci: $\left(0, -\sqrt{8}\right)$, $\left(0, \sqrt{8}\right)$

(c) Vertices: $(-2, 0)$, $(2, 0)$; Foci: $\left(-\sqrt{5}, 0\right)$, $\left(\sqrt{5}, 0\right)$; Asymptotes: $y = -\frac{1}{2}x$ and $y = \frac{1}{2}x$

(d) Vertices: $\left(-\frac{7}{2}, 0\right)$, $\left(-\frac{1}{2}, 0\right)$, $(-2, -3)$, $(-2, 3)$; Foci: $\left(-2, -(3\sqrt{3})/2\right)$, $\left(-2, (3\sqrt{3})/2\right)$ **(e)** $V = (12, -3)$; $F = \left(\frac{49}{4}, -3\right)$

2. **(a)** $\frac{x^2}{4} + \frac{y^2}{1} = 1$ **(b)** $(y + 2)^2 = \frac{4}{3}(x + 3)$ **(c)** $(y + 3)^2 = 40(x - 2)$ **(d)** $\frac{x^2}{16} - \frac{y^2}{20} = 1$

3. $\theta = 45°$; $-\bar{x}^2 + 3\bar{y}^2 = 6$ or $3\bar{x}^2 - \bar{y}^2 = 6$ **4.** $(x - 2)^2 + (y - 1)^2 = 1$ **5.** $\sqrt{7}/3$

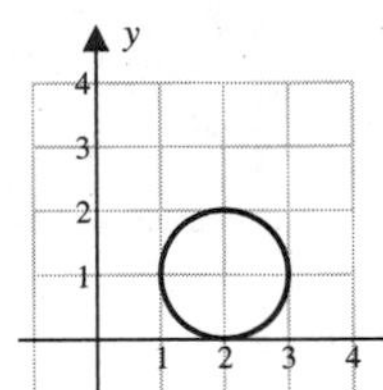
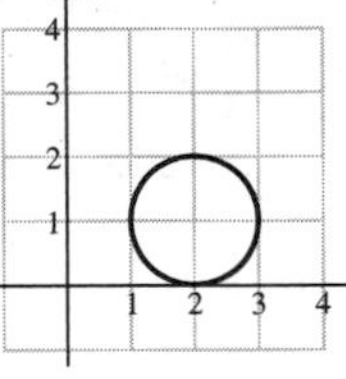

6. **(a)** Ellipse
(b) $\frac{1}{2}$
(c) Foci: $(0, 0)$, $\left(0, -\frac{8}{3}\right)$
(d)

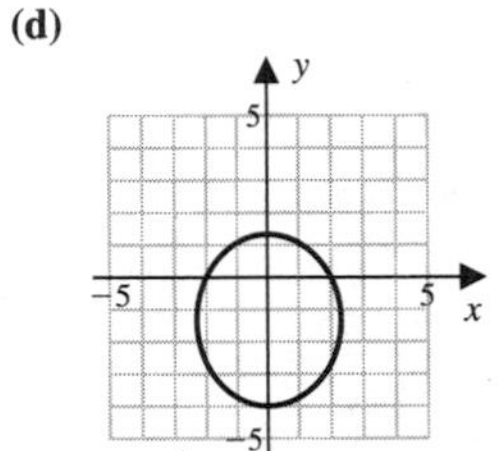

7. Hyperbola

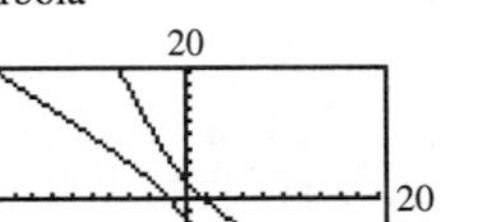

8. $4\sqrt{6}$ ft

CHAPTER 10: PROBLEM SET 10.1 (ON PAGE 596)

1. Finite; $a_n = 2n - 1$; $f(n) = 2n - 1$; Domain: {1, 2, 3, 4, 5}; Range: {1, 3, 5, 7, 9}
3. Infinite; $a_n = n^2$; $f(n) = n^2$; Domain: {1, 2, 3, 4, 5, . . .}; Range: {1, 4, 9, 16, 25, . . .}
5. Infinite; $a_n = (-1)^{n+1}$; $f(n) = (-1)^{n+1}$; Domain: {1, 2, 3, 4, 5, . . .}; Range: {−1, 1} **7.** 5, 7, 9, 11, 13 **9.** $\frac{1}{3}, \frac{1}{2}, \frac{3}{5}, \frac{2}{3}, \frac{5}{7}$
11. $1/e, 1/e^2, 1/e^3, 1/e^4, 1/e^5$ **13.** $1, \frac{1}{2}, \frac{1}{4}, \frac{1}{8}, \frac{1}{16}$ **15.** −2, −6, −18, −54, −162 **17.** 1, 3, 4, 7, 11 **19.** $a_n = 2n + 1$
21. $a_n = 1/(2n + 1)$ **23.** $a_n = (-1)^{n+1} n^2$ **25.** **(a)** 5, 8, 11, 14, 17 **(c)** ∞
27. **(a)** $\frac{3}{2}, 2, \frac{5}{2}, 3, \frac{7}{2}$ **(c)** ∞ **29.** **(a)** 2, 4, 8, 16, 32 **(c)** ∞
31. **(a)** 0.84, 0.45, 0.05, −0.19, −0.19 **(b)** **(c)** 1

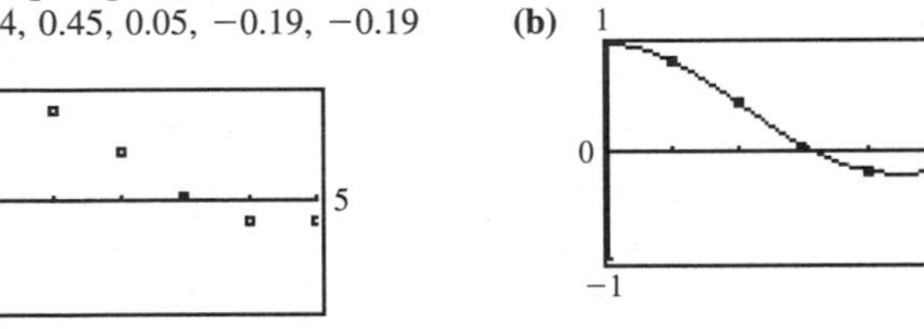

33. **(b)** **(c)** 0

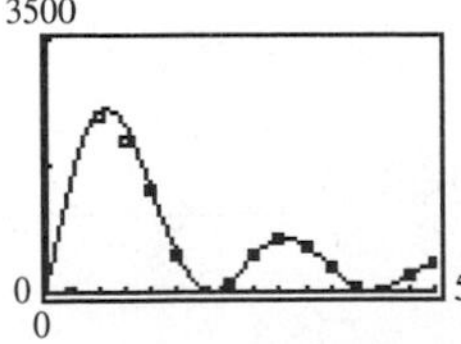

35. 275 **37.** 7381 **39.** 83 **41.** $\frac{3}{2}$ **43.** 900 **45.** 1,691,450

47. **(a)** $\frac{1}{2}, \frac{1}{4}, \frac{1}{8}, \frac{1}{16}, \frac{1}{32}$ **(b)** $a_n = \frac{1}{2^n}$ **(c)** $\sum_{k=1}^{5} \frac{1}{2^k} = \frac{31}{32}$ gal
(d) The bucket will never be full. Each successive can contains only half as much paint as is needed to fill the bucket.
49. **(a)** 4, 4.4, 4.84, 5.324, 5.8564 **(b)** $a_n = 4(1.1)^{n-1}$ **(c)** Approx. 8.57 in. **51.** $a_n = 8n - 7$ **53.** **(a)** 2 **(b)** 5
55. $a_n = 3, n \geq 2$ **57.** $a_n = 1 + 2n, n \geq 2$ **59.** $a_n = \frac{2}{(n + 1)(n + 2)}, n \geq 2$ **61.** $a_n = 2^{n-1}, n \geq 2$ **63.** $s_3 = 10, s_6 = 56$
65. $s_3 = 46, s_6 = 812$ **67.** $s_3 = -4, s_6 = 3$ **69.** **(a)** $s_n = \frac{n}{n + 1}$ **(b)** $s_n = \frac{n}{2n + 4}$

PROBLEM SET 10.2 (ON PAGE 604)

1. $a_n = 3n - 1$ **3.** $a_n = 15 - 3n$ **5.** $a_n = 7n - 20$ **7.** $a_n = 2n$ **9.** $a_n = 9 - 2n$ **11.** 11 **13.** 48
15. $d = 3$; $S_{10} = 145$ **17.** $d = 3$, $S_{20} = 650$ **19.** $d = \frac{3}{7}$; $S_8 = -28$ **21.** 3940 **23.** 3440 **25.** −2540 **27.** 11
29. $d = \frac{668}{51}$; $a_{18} = \frac{719}{3}$ **31.** 4200 **33.** 7200 **35.** 975
37. **(a)** Arithmetic sequence; $a_n = 4 + 3n$, $1 \le n \le 10$ **(b)** 25 **(c)** 205
39. **(a)** 300,000, 340,000, 380,000, 420,000, 460,000; arithmetic sequence; $a_n = 260{,}000 + 40{,}000n$ **(b)** \$500,000 **(c)** \$4,800,000
41. **(a)** 268 ft **(b)** 1260 ft **43.** **(a)** 4 **(b)** 67 **45.** $a_6 = -18a^2 + 51b$; $a_9 = -33a^2 + 84b$ **49.** $63b - 63$

PROBLEM SET 10.3 (ON PAGE 614)

1. $a_n = 2(3)^{n-1}$ **3.** $a_n = 6(2)^{n-1}$ **5.** $a_n = 2^{6-n}$ **7.** $a_n = (1.03)^{n-1}$ **9.** $a_n = \frac{1}{18}(-3)^{n-1}$ **11.** $a_n = 2^n$ **13.** $a_n = 2\left(-\frac{3}{2}\right)^{n-1}$
15. Arithmetic; $a_n = 7n - 6$ **17.** Geometric; $a_n = 3(2)^{n-1}$ **19.** Geometric; $a_n = (-2)^{n-1}$ **21.** Geometric; $a_n = (0.3)^{n-1}$
23. Geometric; $a_n = 2^{-n-1}$ **25.** $r = 4$; $S_8 = 109{,}225$ **27.** $r = -5$, $S_6 = 7812$ **29.** $r = \frac{1}{3}$; $S_{10} = -\frac{29{,}524}{59{,}049}$
31. $r = 0.1$; $S_6 = 0.333333$ **33.** $\frac{1}{2}$ **35.** **(a)** 4 **(b)** $-\frac{3}{2}$
39. **(a)** $a_n = 100$ **(b)** 391.6 ft (approx.) **(c)** $100 + 300(1 - 0.6^{n-1})$ ft **(d)** 400 ft
41. **(a)** $a_n = 1000(1.8)^{n-1}$ **(b)** \$1,910,200 **(c)** 10
43. **(a)** $a_n = 25{,}000(0.95)^{n-1}$, $1 \le n \le 10$ **(b)** $a_6 = 18{,}377$ **(c)** $a_{10} = 14{,}968$ **45.** $\sum_{k=1}^{5} \left(\frac{1}{2}\right)^k = \frac{31}{32}$ **47.** $\frac{a^6 - b^6}{a^5b + a^4b^2}$
49. **(a)** $0.4 + 0.04 + 0.004 + 0.0004 + \cdots = \frac{4}{9} = 0.\overline{4}$ **(b)** $0.32 + 0.0032 + 0.000032 + \cdots = \frac{32}{99} = 0.\overline{32}$
(c) $0.561 + 0.000561 + 0.000000561 + \cdots = \frac{187}{333} = 0.\overline{561}$
51. Yes **53.** **(a)** $S_n - 2(1 - 2^n)$ **(b)** $S_n - \frac{2^n - 1}{2^{n+1}}$

PROBLEM SET 10.4 (ON PAGE 620)

We give answers for problems 1, 19, and 27. The remaining solutions are included in the solutions manual.
1. S_1: $1 = 1^2$; S_k: $1 + 3 + 5 + \cdots + (2k - 1) = k^2$; S_{k+1}: $1 + 3 + 5 + \cdots + (2k - 1) + (2k + 1) = k^2 + (2k + 1) = (k + 1)^2$
19. S_1: If $n = 1$, then $4^n - 1 = 4^1 - 1 = 3$; so 3 is a factor.
S_k: Assume 3 is a factor of $4^k - 1$; that is, assume that $4^k - 1 = 3p$ for some integer p.
S_{k+1}: Show that 3 is a factor of $4^{k+1} - 1$: $4^{k+1} - 1 = 4^k \cdot 4 - 1 = 4(3p + 1) - 1 = 3 \cdot 4p + 3$, and 3 is a factor of $4^{k+1} - 1$.
27. S_1: $\sin(t + \pi) = \sin t \cos \pi + \cos t \sin \pi = (-1)^1 \sin t$; S_k: $\sin(t + k\pi) = (-1)^k \sin t$;
S_{k+1}: $\sin[t + (k + 1)\pi] = \sin[(t + k\pi) + \pi] = \sin(t + k\pi) \cos \pi + \cos(t + k\pi) \sin \pi = (-1)^k \sin t(-1) = (-1)^{k+1} \sin t$

PROBLEM SET 10.5 (ON PAGE 625)

1. $x^4 + 4x^3y + 6x^2y^2 + 4xy^3 + y^4$ **3.** $16x^4 + 96x^3y + 216x^2y^2 + 216xy^3 + 81y^4$
5. $3125x^5 - 9375x^4y + 11{,}250x^3y^2 - 6750x^2y^3 + 2025xy^4 - 243y^5$ **7.** 1, 5, 10, 10, 5, 1; $n = 5$ **9.** **(a)** 3003 **(b)** 15
11. **(a)** 2,598,960 **(b)** 1326 **13.** $5/k$ **15.** $3/(k + 1)$ **17.** $x^{28} + 28x^{27}y + 378x^{26}y^2 + 3276x^{25}y^3$
19. $x^{35} + (-70x^{34}y) + 2380x^{33}y^2 + (-52{,}360x^{32}y^3)$ **21.** $2^{78}x^{78} + 234 \cdot 2^{77}x^{77}y + 27{,}027 \cdot 2^{76}x^{76}y^2 + 2{,}054{,}052 \cdot 2^{75}x^{75}y^3$
23. $x^5 + 15x^4 + 90x^3 + 270x^2 + 405x + 243$ **25.** $x^4 - 8x^3 + 24x^2 - 32x + 16$ **27.** $792x^5y^7$ **29.** $1792a^3x^{12}$
31. $\frac{455}{4096}x^{24}a^3$ **33.** $\frac{224}{243}x^6a^{12}$ **35.** $\frac{28}{243}a^3x^{12}$ **37.** $-35x$ **41.** **(a)** and **(c)** are false **43.** $\sum_{k=0}^{5} \binom{5}{k} x^{5-k}y^k$

CHAPTER 10 REVIEW PROBLEM SET (ON PAGE 626)

1. **(a)** 0, 3, 8, 15, 24 **(b)** −1, 4, −7, 10, −13 **3.** **(a)** 4, 8, 12, 16, 20 **(b)** 8, 13, 18, 23, 28 **5.** **(a)** $a_n = 2n$ **(b)** $a_n = (-1)^n$
7. **(a)** ∞ **(b)** ∞ **9.** **(a)** 95 **(b)** 20 **11.** 7910 **13.** **(a)** (i) $a_{15} = 58$ (ii) $S_{15} = 450$ **(b)** (i) $a_{10} = 80$ (ii) $S_{10} = 440$
15. **(a)** 87 **(b)** −48 **17.** **(a)** −24 **(b)** $27a + 3b$ **19.** **(a)** $a_n = 5(3)^{n-1}$ **(b)** $a_n = 4(2)^{n-1} = 2^{n+1}$
21. **(a)** $r = 2$; $S_8 = 12{,}240$ **(b)** $r = \frac{1}{3}$; $S_{12} = -\frac{265{,}720}{2187}$ **23.** **(a)** 9 **(b)** $\frac{9}{5}$ **31.** **(a)** 792 **(b)** 46,376 **33.** $n + 2$
35. **(a)** $32 + 80x + 80x^2 + 40x^3 + 10x^4 + x^5$ **(b)** $81x^4 - 432x^3y + 864x^2y^2 - 768xy^3 + 256y^4$
37. **(a)** $5376x^6y^3$ **(b)** $61{,}236x^5y^5$ **39.** **(a)** 304 ft **(b)** 1600 ft
41. **(a)** $a_n = 130{,}000(1.03)^{n-1}$ **(b)** 130,000, 133,900, 137,917, 142,054.51, 146,316.15, 150,705.63, 155,226.80 **43.** \$37,200

CHAPTER 10 TEST (ON PAGE 628)

1. (a) $-1, 2, 5, 8, 11$ **(b)** $2, \frac{3}{4}, \frac{4}{9}, \frac{5}{16}, \frac{6}{25}$ **(c)** $3, -6, 12, -24, 48$ **2. (a)** $a_n = 2 + 4n$ **(b)** $a_n = (-3)^n$
3. (a) 55 **(b)** 130 **4. (a)** $a_6 = 11$ **(b)** $n = 6$ **(c)** $S_7 = 56$ **(d)** $d = -3$
5. (a) $a_{10} = -1536$ **(b)** $r = -3$ **(c)** $a_1 = -1$ **6. (a)** 25 **(b)** $\frac{1}{16}$
7. (a) S_1: $2^1 = 2(2^1 - 1) = 2 \cdot 1$; S_k: $2 + 2^2 + 2^3 + \cdots + 2^k = 2(2^k - 1)$;
S_{k+1}: $2 + 2^2 + 2^3 + \cdots + 2^k + 2^{k+1} = 2(2^k - 1) + 2^{k+1} = 2^{k+1} - 2 + 2^{k+1} = 2(2^{k+1} - 1)$
(b) S_1: $1 \cdot 3 = \frac{1}{6} \cdot 1 \cdot (1 + 1)(2 + 7) = \frac{1}{6} \cdot 2 \cdot 9 = 3$; S_k: $1 \cdot 3 + 2 \cdot 4 + 3 \cdot 5 + \cdots + k(k + 2) = \frac{1}{6}k(k + 1)(2k + 7)$;
S_{k+1}: $1 \cdot 3 + 2 \cdot 4 + 3 \cdot 5 + \cdots + k(k + 2) + (k + 1)(k + 3) = \frac{1}{6}k(k + 1)(2k + 7) + (k + 1)(k + 3)$
$= \frac{1}{6}(k + 1)[k(2k + 7) + 6(k + 3)] = \frac{1}{6}(k + 1)(k + 2)[2(k + 1) + 7]$
8. (a) $x^4 - 12x^3 + 54x^2 - 108x + 81$ **(b)** $32x^5 + 80x^4 + 80x^3 + 40x^2 + 10x + 1$ **9.** $924x^{18}$
10. $n = 2$: $(2^2)! = 4! = 24$; $(2!)^2 = (2)^2 = 4$ **11.** Approx. 75.08 cm

Index

Trigonometric Functions

ACUTE ANGLES

$$\sin\theta = \frac{\text{opp}}{\text{hyp}} \qquad \csc\theta = \frac{\text{hyp}}{\text{opp}}$$

$$\cos\theta = \frac{\text{adj}}{\text{hyp}} \qquad \sec\theta = \frac{\text{hyp}}{\text{adj}}$$

$$\tan\theta = \frac{\text{opp}}{\text{adj}} \qquad \cot\theta = \frac{\text{adj}}{\text{opp}}$$

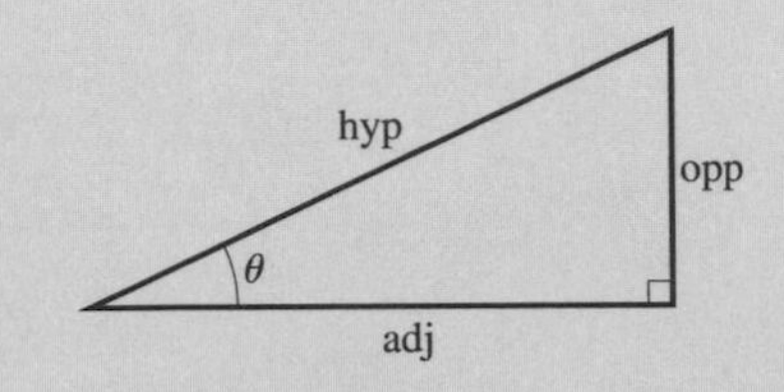

GENERAL ANGLES

$$\sin\theta = \frac{y}{r} \qquad \csc\theta = \frac{r}{y}$$

$$\cos\theta = \frac{x}{r} \qquad \sec\theta = \frac{r}{x}$$

$$\tan\theta = \frac{y}{x} \qquad \cot\theta = \frac{x}{y}$$

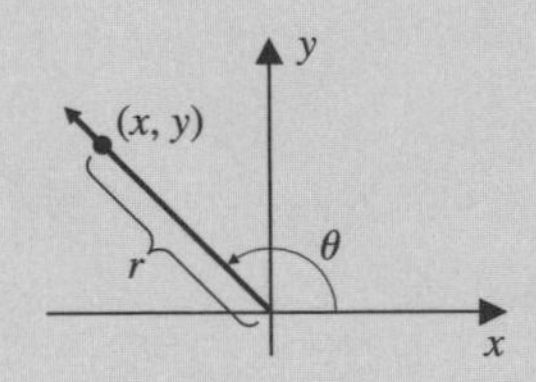

Trigonometric Identities

FUNDAMENTAL IDENTITIES

1. $\csc\theta = \dfrac{1}{\sin\theta}$
2. $\sec\theta = \dfrac{1}{\cos\theta}$
3. $\cot\theta = \dfrac{1}{\tan\theta}$
4. $\tan\theta = \dfrac{\sin\theta}{\cos\theta}$
5. $\cot\theta = \dfrac{\cos\theta}{\sin\theta}$
6. $\cos^2\theta + \sin^2\theta = 1$
7. $1 + \tan^2\theta = \sec^2\theta$
8. $1 + \cot^2\theta = \csc^2\theta$

EVEN–ODD IDENTITIES

1. $\sin(-\theta) = -\sin\theta$
2. $\cos(-\theta) = \cos\theta$
3. $\tan(-\theta) = -\tan\theta$
4. $\cot(-\theta) = -\cot\theta$
5. $\sec(-\theta) = \sec\theta$
6. $\csc(-\theta) = -\csc\theta$

ADDITION FORMULAS

1. $\sin(\alpha + \beta) = \sin\alpha\cos\beta + \sin\beta\cos\alpha$
2. $\cos(\alpha + \beta) = \cos\alpha\cos\beta - \sin\alpha\sin\beta$
3. $\tan(\alpha + \beta) = \dfrac{\tan\alpha + \tan\beta}{1 - \tan\alpha\tan\beta}$

SUBTRACTION FORMULAS

1. $\sin(\alpha - \beta) = \sin\alpha\cos\beta - \sin\beta\cos\alpha$
2. $\cos(\alpha - \beta) = \cos\alpha\cos\beta + \sin\alpha\sin\beta$
3. $\tan(\alpha - \beta) = \dfrac{\tan\alpha - \tan\beta}{1 + \tan\alpha\tan\beta}$